ARSENIC RESEARCH AND GLOBAL SUSTAINABILITY

Arsenic in the Environment – Proceedings

Series Editors

Jochen Bundschuh

Deputy Vice-Chancellor's Office (Research and Innovation) & Faculty of Health, Engineering and Sciences, University of Southern Queensland, Toowoomba, Queensland, Australia
KTH-International Groundwater Arsenic Research Group, Department of Sustainable Development, Environmental Sciences and Engineering, KTH Royal Institute of Technology, Stockholm, Sweden

Prosun Bhattacharya

KTH-International Groundwater Arsenic Research Group, Department of Sustainable Development, Environmental Science and Engineering, KTH Royal Institute of Technology, Stockholm, Sweden
International Center for Applied Climate Science, University of Southern Queensland, Toowoomba, Queensland, Australia

ISSN: 2154-6568

PROCEEDINGS OF THE 6TH INTERNATIONAL CONGRESS ON ARSENIC IN THE ENVIRONMENT, STOCKHOLM, SWEDEN, 19–23 JUNE 2016

Arsenic Research and Global Sustainability

As 2016

Editors

Prosun Bhattacharya
KTH-International Groundwater Arsenic Research Group, Department of Sustainable Development, Environmental Sciences and Engineering, KTH Royal Institute of Technology, Stockholm, Sweden
International Center for Applied Climate Science, University of Southern Queensland, Toowoomba, Queensland, Australia

Marie Vahter
Institute of Environmental Medicine, Karolinska Institutet, Stockholm, Sweden

Jerker Jarsjö
Department of Physical Geography and Quaternary Geology, Stockholm University, Stockholm, Sweden
Bolin Centre for Climate Research, Stockholm University, Stockholm, Sweden

Jurate Kumpiene
Waste Science & Technology, Department of Civil, Environmental and Natural Resources Engineering, Luleå University of Technology, Luleå, Sweden

Arslan Ahmad
KWR Watercycle Research Institute, Nieuwegein, The Netherlands

Charlotte Sparrenbom
Department Geology, Quaternary Sciences, Lund University, Lund, Sweden

Gunnar Jacks
KTH-International Groundwater Arsenic Research Group, Department of Sustainable Development, Environmental Sciences and Engineering, KTH Royal Institute of Technology, Stockholm, Sweden

Marinus Eric Donselaar
Department of Geoscience and Engineering, Delft Univ. of Technology, Delft, The Netherlands

Jochen Bundschuh
Faculty of Health, Engineering and Sciences, The University of Southern Queensland, Toowoomba, Australia
KTH-International Groundwater Arsenic Research Group, Department of Sustainable Development, Environmental Sciences and Engineering, KTH Royal Institute of Technology, Stockholm, Sweden

Ravi Naidu
Global Centre for Environmental Remediation (GCER), Faculty of Science & Information Technology, The University of Newcastle, Callaghan, NSW, Australia
Research Centre for Contamination Assessment and Remediation of the Environment (CRC CARE), University of Newcastle, Newcastle, New South Wales, Australia

CRC Press
Taylor & Francis Group
Boca Raton London New York

CRC Press is an imprint of the
Taylor & Francis Group, an **informa** business

A BALKEMA BOOK

ISGSD

*International Society of
Groundwater for
Sustainable Development*

Cover photo

The cover photo is from the oxidized tailings deposit at Adak mine in Västerbotten county in Northern Sweden. During the period 1940 to 1977 6.3 M tons of ore were mined containing 2 % copper, 0.6 g/ton gold and 6 g/ton silver. The main sulphides were pyrite, chalcopyrite, arsenopyrite and pyrrhotite. While a common practice has been to deposit tailings in lakes and keep them under water, the tailings at Adak mine has been deposited on land allowing the oxidation and formation of acid mine drainage. The drainage water seen in the cover photo had a pH of 2–3. Arsenic is a major pollutant in connection to the tailings and is recorded at high levels in downstream surface water. Up to 2900 g/L As has been recorded in water and 900 mg/kg in soil. As(III) makes up the larger portion of the total arsenic content in water. A specific arsenic reducing actinomycete has been found in the mine tailings area. Plants show elevated trace element concentrations. Salix species have up to 700 mg/kg of zinc and the silica rich Equisetum species have up to 200 mg/kg of arsenic. Ferric precipitates constitute the most efficient trap for arsenic when the pH gradually increases downstreams.

Published by:
CRC Press/Balkema
P.O. Box 447, 2300 AK Leiden, The Netherlands
e-mail: Pub.NL@taylorandfrancis.com
www.crcpress.com – www.taylorandfrancis.com

First issued in paperback 2020

© Prosun Bhattacharya, 2001
© 2016 by Taylor & Francis Group, LLC
CRC Press/Balkema is an imprint of the Taylor & Francis Group, an informa business

No claim to original U.S. Government works

Typeset by V Publishing Solutions Pvt Ltd., Chennai, India

ISBN 13: 978-0-367-73705-4 (pbk)
ISBN 13: 978-1-138-02941-5 (hbk)

**Visit the Taylor & Francis Web site at
http://www.taylorandfrancis.com**

**and the CRC Press Web site at
http://www.crcpress.com**

Arsenic Research and Global Sustainability – Bhattacharya, Vahter, Jarsjö, Kumpiene, Ahmad, Sparrenbom, Jacks, Donselaar, Bundschuh & Naidu (Eds)
© 2016 Taylor & Francis Group, London, ISBN 978-1-138-02941-5

Table of contents

Section 2: Arsenic in the food chain

2.1   Arsenic in rice

Section 3: Arsenic and health

3.1   Epidemiology

3.2   Biomarkers of exposure and metabolism

Section 5: Societal and policy implications, mitigation and management

5.1   Societal and policy implications of long term exposure

5.2   Risk assessment and remediation of contaminated land and water environments—case studies

Arsenic Research and Global Sustainability – Bhattacharya, Vahter, Jarsjö, Kumpiene,
Ahmad, Sparrenbom, Jacks, Donselaar, Bundschuh & Naidu (Eds)
© 2016 Taylor & Francis Group, London, ISBN 978-1-138-02941-5

About the book series

Although arsenic has been known as a 'silent toxin' since ancient times, and the contamination of drinking water resources by geogenic arsenic was described in different locations around the world long ago—e.g. in Argentina in 1914—it was only two decades ago that it received overwhelming worldwide public attention. As a consequence of the biggest arsenic calamity in the world, which was detected more than twenty years back in West Bengal, India and other parts of Southeast Asia, there has been an exponential rise in scientific interest that has triggered high quality research. Since then, arsenic contamination (predominantly of geogenic origin) of drinking water resources, soils, plants and air, the propagation of arsenic in the food chain, the chronic affects of arsenic ingestion by humans, and their toxicological and related public health consequences, have been described in many parts of the world, and every year, even more new countries or regions are discovered to have elevated levels of arsenic in environmental matrices.

Arsenic is found as a drinking water contaminant, in many regions all around the world, in both developing as well as industrialized countries. However, addressing the problem requires different approaches which take into account, the differential economic and social conditions in both country groups. It has been estimated that 200 million people worldwide are at risk from drinking water containing high concentrations of As, a number which is expected to further increase due to the recent lowering of the limits of arsenic concentration in drinking water to 10 µg/L, which has already been adopted by many countries, and some authorities are even considering decreasing this value further.

The book series "Arsenic in the Environment—Proceedings" is an inter- and multidisciplinary source of information, making an effort to link the occurrence of geogenic arsenic in different environments and the potential contamination of ground- and surface water, soil and air and their effect on the human society. The series fulfills the growing interest in the worldwide arsenic issue, which is being accompanied by stronger regulations on the permissible Maximum Contaminant Levels (MCL) of arsenic in drinking water and food, which are being adopted not only by the industrialized countries, but increasingly by developing countries.

Consequently, we see the book series *Arsenic in the Environment-Proceedings* with the outcomes of the international congress series Arsenic in the Environment, which we organize biannually in different parts of the world, as a regular update on the latest developments of arsenic research. It is further a platform to present the results from other from international or regional congresses or other scientific events. This Proceedings series acts as an ideal complement to the books of the series *Arsenic in the Environment*, which includes authored or edited books from world-leading scientists on their specific field of arsenic research, giving a comprehensive information base. Supported by a strong multi-disciplinary editorial board, book proposals and manuscripts are peer reviewed and evaluated. Both of the two series will be open for any person, scientific association, society or scientific network, for the submission of new book projects.

We have an ambition to establish an international, multi- and interdisciplinary source of knowledge and a platform for arsenic research oriented to the direct solution of problems with considerable social impact and relevance rather than simply focusing on cutting edge and breakthrough research in physical, chemical, toxicological and medical sciences. It shall form a consolidated source of information on the worldwide occurrences of arsenic, which otherwise is dispersed and often hard to access. It will also have role in increasing the awareness and knowledge of the arsenic problem among administrators, policy makers and company executives and improving international and bilateral cooperation on arsenic contamination and its effects.

Both of the book series cover all fields of research concerning arsenic in the environment and aims to present an integrated approach from its occurrence in rocks and mobilization into the ground- and surface water, soil and air, its transport therein, and the pathways of arsenic introduction into the food chain including uptake by humans. Human arsenic exposure, arsenic bioavailability, metabolism and toxicology are treated together with related public health effects and risk assessments in order to better manage the contaminated land and aquatic environments and to reduce human arsenic exposure. Arsenic removal technologies and other methodologies to mitigate the arsenic problem are addressed not only from the technological perspective, but also from an economic and social point of view. Only such inter- and multidisciplinary approaches will allow a case-specific selection of optimal mitigation measures for each specific arsenic problem and provide the local population with arsenic-safe drinking water, food, and air.

Jochen Bundschuh
Prosun Bhattacharya
(Series Editors)

Arsenic Research and Global Sustainability – Bhattacharya, Vahter, Jarsjö, Kumpiene,
Ahmad, Sparrenbom, Jacks, Donselaar, Bundschuh & Naidu (Eds)
© 2016 Taylor & Francis Group, London, ISBN 978-1-138-02941-5

Dedication

Ramiro Rodríguez Castillo, D.Sc.
Formerly Senior Research Scientist, Hydrogeology,
Geophysics Institute. Universidad Nacional Autónoma de México
* 1953 † 2015

We dedicate this Volume of Proceedings of the International Congress of Arsenic in the Environment (As 2016) to our colleague and friend Dr. Ramiro Rodríguez Castillo, who had been an active researcher and scientist working in the field of arsenic in groundwater for more than 23 years. Born in Córdoba, Veracruz, México in the year 1953, he obtained a bachelor degree in Physics and a Master on Sciences from the Universidad Nacional Autónoma de México (UNAM). In 1983, he received a PhD in Geology from the University of Bucarest.

After his return to Mexico, he joined the Geophysics Institute (UNAM) as a researcher and leader in the Natural Resources Department. He coordinated numerous hydrogeological and contamination studies on groundwater availability, subsidence-induced infrastructure damage, risk evaluation from water pollution, and aquifer vulnerability. In these research areas, he published more than 75 papers in international journals, books, and technical reports to Mexican water authorities, receiving more than 500 citations. He contributed to the development of hydrogeology in Mexico through professional's formation, organization of national and international congress and more than 200 conferences.

He had an outstanding participation on the stablishment of the As in the Environment Conference series as an organizer of the first one held in Mexico City, in 2006. He also had an enthousiastic participation in the Iberoarsen network aimed to contribute to solve As problems in Latin America and the Iberian Peninsula. As a professor, besides numerous courses in Earth Science programs, he supervised 27 Bachelor, M.Sc. and PhD thesis of students from diverse national and international universities, seven of them receiving awards.

His studies on groundwater throughout the country oriented decision-makers to solve water availability and contamination problems in México.

The arsenic community will always remember his important contribution to the solution of As-related problems and will miss his supportive, hard-worker and optimistic company.

Arsenic Research and Global Sustainability – Bhattacharya, Vahter, Jarsjö, Kumpiene,
Ahmad, Sparrenbom, Jacks, Donselaar, Bundschuh & Naidu (Eds)
© 2016 Taylor & Francis Group, London, ISBN 978-1-138-02941-5

Organizers

ORGANIZERS OF BIANNUAL CONGRESS SERIES: ARSENIC IN THE ENVIRONMENT

Jochen Bundschuh
University of Southern Queensland (USQ), Toowomba, QLD,
Australia International Society of Groundwater for Sustainable
Development (ISGSD), Stockholm, Sweden

Prosun Bhattacharya
KTH-International Groundwater Arsenic Research Group
Department of Sustainable Development, Environmental Sci-
ences and Engineering, KTH Royal Institute of Technology,
Stockholm, Sweden
International Center for Applied Climate Science, University
of Southern Queensland (USQ), Toowomba, QLD, Australia
International Society of Groundwater for Sustainable Develop-
ment (ISGSD), Stockholm, Sweden

Local Organizing Committee

Prosun Bhattacharya
KTH-International Groundwater Arsenic Research Group
Department of Sustainable Development, Environmental Sci-
ences and Engineering, KTH Royal Institute of Technology,
Stockholm, Sweden
International Center for Applied Climate Science, University
of Southern Queensland (USQ), Toowomba, QLD, Australia
International Society of Groundwater for Sustainable Develop-
ment (ISGSD), Stockholm, Sweden

Marie Vahter
Institute of Environmental Medicine, Karolinska Institutet,
Stockholm, Sweden

Jerker Jarsjö
Department of Physical Geography, Stockholm University,
Stockholm, Sweden

Jurate Kumpiene
Waste Science & Technology, Department of Civil, Environ-
mental and Natural Resources Engineering, Luleå University
of Technology, Luleå, Sweden

Charlotte Sparrenbom
Department Geology, Quaternary Sciences, Lund University, Lund, Sweden

 Gunnar Jacks
KTH-International Groundwater Arsenic Research Group, Department of Sustainable Development, Environmental Sciences and Engineering, KTH Royal Institute of Technology, Stockholm, Sweden

 Arslan Ahmad
KWR Watercycle Research Institute, Nieuwegein, The Netherlands

Mattias Bäckström
Man-Technology-Environment Research Centre, Örebro University, Örebro, Sweden

*Arsenic Research and Global Sustainability – Bhattacharya, Vahter, Jarsjö, Kumpiene,
Ahmad, Sparrenbom, Jacks, Donselaar, Bundschuh & Naidu (Eds)*
© 2016 Taylor & Francis Group, London, ISBN 978-1-138-02941-5

Sponsors and contributors

Arsenic Research and Global Sustainability – Bhattacharya, Vahter, Jarsjö, Kumpiene,
Ahmad, Sparrenbom, Jacks, Donselaar, Bundschuh & Naidu (Eds)
© 2016 Taylor & Francis Group, London, ISBN 978-1-138-02941-5

Scientific committee

K. Matin Ahmed, *Department of Geology, University of Dhaka, Dhaka, Bangladesh*
A. Ahmad, *KWR Watercycle Research Institute, Nieuwegein, The Netherlands*
Ma.Teresa Alarcón-Herrera, *Centro de Investigación en Materiales Avanzados (CIMAV), Chihuahua, Chih., Mexico*
M. Alauddin, *Department of Chemistry, Wagner College, Staten Island, NY, USA*
S. Anac, *Ege University, Izmir, Turkey*
M.A. Armienta, *National Autonomous University of Mexico, Mexico D.F., Mexico*
M. Auge, *Buenos Aires University, Argentina*
A. Baba, *Geothermal Energy Research and Application Center, Izmir Institute of Technology, Izmir, Turkey*
M. Berg, *Eawag, Swiss Federal Institute of Aquatic Science and Technology, Duebendorf, Switzerland*
P. Bhattacharya, *KTH-International Groundwater Arsenic Research Group, Department of Sustainable Development, Environmental Sciences and Engineering, KTH Royal Institute of Technology, Stockholm, Sweden; International Center for Applied Climate Science, University of Southern Queensland, Toowoomba, Queensland, Australia*
M. Biagini, *Salta, Argentina*
P. Birkle, *Gerencia de Geotermia, Instituto de Investigaciones Eléctricas (IIE), Cuernavaca, Mor., Mexico*
M. del Carmen Blanco, *National University of the South, Bahía Blanca, Argentina*
M. Blarasin, *Río Cuarto National University, Río Cuarto, Argentina*
A. Boischio, *Pan American Health Organization, USA*
K. Broberg, *Karolinska Institutet, Solna, Sweden*
J. Bundschuh, *University of Southern Queensland (USQ), Toowomba, Queensland, Australia; International Society of Groundwater for Sustainable Development (ISGSD), Stockholm, Sweden*
M. Bäckström, *Man-Technology-Environment Research Centre, Örebro University, Örebro, Sweden*
Y. Cai, *Florida International University, Miami, USA*
A.A. Carbonell Barrachina, *Miguel Hernández University, Orihuela, Alicante, Spain*
M.L. Castro de Esparza, *CEPIS, Lima, Peru*
J.A. Centeno, *Joint Pathology Center, Malcolm Grow Medical Clinic, Joint Base Andrews Air Naval Facility, Washington DC, USA*
D. Chandrasekharam, *Department of Earth Sciences, Indian Institute of Technology-Bombay, Mumbai, India*
D. Chatterjee, *Department of Chemistry, Universityof Kalyani, Kalyani, India*
C.-J. Chen, *Academia Sinica, Taipei City, Taiwan*
L. Charlet, *Institut des Sciences de La Terre (ISTerre), Universite Grenoble Alpes and CNRS, France*
V. Ciminelli, *Department of Metallurgical and Materials Engineering, Universidade Federal de Minas Gerais, Belo Horizonte, Minas Gerais, Brazil*
L. Cornejo, *University of Tarapacá, Arica, Chile*
L.H. Cumbal, *Escuela Politécnica del Ejército, Sangolquí, Ecuador*
A. F. Danil de Namor, *University of Surrey, UK*
S. Datta, *Kansas State University, Manhattan, Kansas, USA*
J. W.V. De Mello, *Federal University of Vicosa, Vicosa, Minas Gerais, Brazil*
D. De Pietri, *Ministry of Health, Buenos Aires, Argentina*
L.M. Del Razo, *Cinvestav-IPN, México D.F., Mexico*
E. de Titto, *Health Ministry, Buenos Aires, Argentina*
V. Devesa, *IATA-CSIC, Valencia, Spain*
B. Dousova, *ICT, Prague, Czech Republic*
Ö. Ekengren, *IVL, Swedish Environmental Reseasrch Institute, Stockholm, Sweden*
M. Ersoz, *Department of Chemistry, Selcuk University, Konya Turkey*
M.L. Esparza, *CEPIS, Lima, Peru*

S. Farías, *National Atomic Energy Commission, Buenos Aires, Argentina*
J. Feldman, *University of Aberdeen, Aberdeen, Scotland, UK*
R. Fernández, *National University of Rosario, Rosario, Argentina*
A. Fernández Cirelli, *University of Buenos Aires, Buenos Aires, Argentina*
A. Figoli, *Institute on Membrane Technology, ITM-CNR c/o University of Calabria, Rende (CS), Italy*
B. Figueiredo, *UNICAMP, Campinas, SP, Brazil*
R.B. Finkelman, *US Geological Survey, Menlo Park, CA, USA*
A. Fiúza, *University of Porto, Porto, Portugal*
A.E. Fryar, *University of Kentucky, Lexington, KY, USA*
S.E. Garrido Hoyos, *Mexican Institute of Water Technology, Jiutepec, Mor., Mexico*
M. Gasparon, *The University of Queensland, Australia*
A.K. Ghosh, *A.N. College, Patna, Bihar, India*
A.K. Giri, *CSIR-Indian Institute of Chemical Biology, Kolkata, India*
W. Goessler, *University of Graz, Austria*
D.N. Guha Mazumder, *DNGM Research Foundation, Kolkata, India*
L.R. Guimaraes Guilherme, *Federal University of Lavras, Lavras, M.G., Brazil*
X. Guo, *Peking University, Peking, PR China*
J.P. Gustafsson, *Swedish University of Agricultural Sciences, Uppsala, Sweden*
S. Hahn-Tomer, *UFZ, Leipzig, Germany*
B. Hendry, *Cape Peninsula University of Technology, Cape Town, South Africa*
M. Hernández, *National University of La Plata, La Plata, Argentina*
J. Hoinkis, *Karlsruhe University of Applied Sciences, Karlsruhe, Germany*
C. Hopenhayn, *University of Kentucky, Lexington, KY, USA.*
M.F. Hughes, *Environmental Protection Agency, Research Triangle Park, NC, USA*
A.M. Ingallinella, *Centro de Ingeniería Sanitaria (CIS), Facultad de Ciencias Exactas, Ingeniería y Agri-mensura, Universidad Nacional de Rosario, Rosario, Prov. de Santa Fe, Argentina*
G. Jacks, *Department of Sustainable Development, Environmental Sciences and Engineering, KTH Royal Institute of Technology, Stockholm, Sweden*
J. Jarsjö, *Stockholm University, Stockholm, Sweden*
J.-S. Jean, *National Cheng Kung University, Tainan, Taiwan*
B. Johnson, *Dundee Precious Metals, Toronto, ON, Canada*
R. Johnston, *World Health Organization, Geneva, Switzerland*
N. Kabay, *Chemical Engineering Department, Engineering Faculty, Ege University, Izmir, Turkey*
I.B. Karadjova, *Faculty of Chemistry, University of Sofia, Sofia, Bulgaria*
A. Karczewska, *Institute of Soil Sciences and Environmental Protection, Wroclaw University of Environ-mental and Life Sciences, Poland*
G. Kassenga, *Ardhi University, Dar es Salaam, Tanzania*
D.B. Kent, *US Geological Survey, Menlo Park, CA, USA*
N.I. Khan, *The Australian National University, Canberra, Australia*
K.-W. Kim, *Department of Environmental Science and Engineering, Gwangju Institute of Science and Tech-nology, Gwangju, South Korea*
W. Klimecki, *Department of Pharmacology and Toxicology, University of Arizona, Tucson, Arizona, USA*
J. Kumpiene, *Department of Civil, Environmental and Natural Resources Engineering, Luleå University of Technology, Luleå, Sweden*
M.I. Litter, *Comisión Nacional de Energía Atómica, and Universidad de Gral. San Martín, San Martín, Argentina*
D.L. López, *Ohio University, Athens, Ohio, USA*
L.Q. Ma, *Universidad de Florida, Florida, USA*
M. Mallavarapu, *Faculty of Science and Information Technology, The University of Newcastle Callaghan, NSW, Australia*
N. Mañay, *De la República University, Montevideo, Uruguay*
A. Martin, *KTH Royal Instituite of Technology, Stockholm, Sweden*
R.R. Mato, *Ardhi University, Dar es Salaam, Tanzania*
J. Matschullat, *Interdisciplinary Environmental Research Centre (IÖZ), TU Bergakademie Freiberg, Freib-erg, Germany*
F. Mtalo, *Department of Water Resources Engineering, University of Dar es Salaam, Dar es Salaam, Tanzania*
M. Mörth, *Stockholm University, Sweden*

A. Meharg, *Institute of Biological and Environmental Sciences, University of Aberdeen, Aberdeen, UK*

A. Mukherjee, *Department of Geology and Geophysics, Indian Institute of Technology (IIT) – Kharagpur, India*

R. Naidu, *CRC Care, University of Newcastle, NSW, Australia*

J. Ng, *National Research Centre for Environmental Toxicology, The University of Queensland, Brisbane, Australia*

H.B. Nicolli, *Instituto de Geoquímica (INGEOQUI), San Miguel, Prov. de Buenos Aires, Argentina and Consejo Nacional de Investigaciones Científicas y Técnicas (CONICET), Argentina*

B. Noller, *The University of Queensland, Australia*

D.K. Nordstrom, *U. S. Geological Survey, Menlo Park, CA, USA*

G. Owens, *Mawson Institute, University of South Australia, Australia*

P. Pastén González, *Pontificia Universidad Católica de Chile, Chile*

C.A. Pérez, *LNLS -Brazilian Synchrotron Light Source Laboratory, Campinas, SP, Brazil*

A.L. Pérez Carrera, *University of Buenos Aires, Buenos Aires, Argentina*

C. Pérez Coll, *National San Martín University, San Martín, Argentina*

B. Petrusevski, *UNESCO-IHE, Institute for Water Education,Delft, The Netherlands*

B. Planer-Friedrich, *University Bayreuth, Bayreuth, Germany*

D.A. Polya, *The University of Manchester, UK*

T. Pradeep, *Indian Institute of Technology Madras, Chennai, India*

L. Pflüger, *Secretary of Environment, Buenos Aires, Argentina*

G.E. Pizarro Puccio, *Pontificia Universidad Católica de Chile, Chile*

B. Planer-Friedrich, *University Bayreuth, Bayreuth, Germany*

T. Pradeep, *Indian Institute of Technology Madras, Chennai, India*

I. Queralt, *Institute of Earth Sciences Jaume Almera – CSIC, Spain*

J. Quintanilla, *Institute of Chemical Research, Universidad Mayor de San Andrés, La Paz, Bolivia*

M.M. Rahman, *The University of Newcastle Callaghan, NSW, Australia*

AL. Ramanathan, *School of Environmental Science, Jawaharlal Nehru University, New Delhi, India*

B. Rosen, *Florida International University, Miami, USA*

J. Routh, *Linköping University, Linköping, Sweden*

D. Saha, *Central Ground Water Board, New Delhi, India*

A.M. Sancha, *Department of Civil Engineering, University of Chile, Santiago de Chile, Chile*

M. Schreiber, *Virginia Polytechnic Institute and State University, VA, USA*

C. Schulz, *La Pampa National University, Argentina*

O. Selinus, *Linneaus University, Kalmar, Sweden*

A. SenGupta, *Lehigh University, Bethlehem, PA, USA*

V.K. Sharma, *Florida Institute of Technology, Florida, USA*

A. Shraim, *The University of Queensland, Brisbane, Australia*

M. Sillanpää, *Lappeenranta University of Technology, Finland*

P.L. Smedley, *British Geological Survey, Keyworth, UK*

A.H. Smith, *University of California, Berkeley, CA, USA*

M. Styblo, *University of North Carolina, Chapel Hill, USA*

C. Tsakiroglou, *Foundation for Research and Technology, Hellas, Greece*

M. Vahter, *Karolinska Institutet Stockholm, Sweden*

D. van Halem, *Delft University of Technology, Delft, The Netherlands*

D. Velez, *Institute of Agrochemistry and Food Technology, Valencia, Spain*

M. Vithanage, *Institute of Fundamental Studies, Hantana Road, Kandy, Sri Lanka*

Y. Zheng, *Lamont-Doherty Earth Observatory, Columbia University, New York, NY, USA*

Zhu, Y.-G.: *State Key Lab of Urban and Regional Ecology, Research Center for Eco-environmental Sciences, Chinese Academy of Sciences, P.R. China; Key Lab of Urban Environment and Health, Institute of Urban Environment, Chinese Academy of Sciences, Xiamen, P.R. China; Key Lab of Urban Environment and Health, Institute of Urban Environment, Chinese Academy of Sciences, Xiamen, P.R. China*

Arsenic Research and Global Sustainability – Bhattacharya, Vahter, Jarsjö, Kumpiene,
Ahmad, Sparrenbom, Jacks, Donselaar, Bundschuh & Naidu (Eds)
© *2016 Taylor & Francis Group, London, ISBN 978-1-138-02941-5*

Foreword (President, KTH)

**ROYAL INSTITUTE
OF TECHNOLOGY**

PRESIDENT

Arsenic is a natural or anthropogenic contaminant in many areas around the globe, where human subsistence is at risk. It is considered as a class 1 carcinogen, and its presence in groundwater has emerged as a major environmental calamity in several parts of the world. It has been estimated that nearly 137 million people drink water contaminated with arsenic globally. In Sweden, approximately 250 000 people rely on drinking water from private wells with arsenic concentrations above the drinking water guideline value of 10 µg/L. In several regions of the world especially in different countries of Asia such as Bangladesh, Cambodia, China, India, Nepal, Pakistan, Taiwan, Thailand and Vietnam, the situation of arsenic toxicity is alarming and severe health problems are reported amongst the inhabitants relying on groundwater as drinking water. However recent investigations have shown that the arsenic problem in many Latin American countries is of the same order of significance. The use of arsenic contaminated groundwater for irrigation and its bioavailability to food crops and ingestion by humans and livestock through the food chain has presented additional pathways for arsenic exposure. The widespread discovery of arsenic in Asia has paved the way to the discovery of the presence of this element in different environmental compartments as a "silent" toxin globally. New areas with elevated arsenic occurrences are reported in groundwater exceeding the maximal contamination levels set by the WHO and other national and international regulatory organizations are identified each year. It therefore requires innovative solutions to ensure access to clean drinking water. The Netherlands is now focusing on reducing their arsenic levels to below 1 µg/L, as there is a healthy arsenic content, and therefore assumes that the requirements will be tightened.

Since 2000, we have witnessed a remarkable rise in interest on research in the field of arsenic. Many research councils and international donor organizations have provided significant support to local and international research teams to develop strategies to address the problem with an aim to minimize the risk of arsenic exposure among the population. As a consequence, there has been a radical increase in the number of scientific publications that give a holistic overview on the nature and fate of arsenic in natural environment, and its impact on human health.

The WHO/FAO Joint Expert Committee (JEC) review document on Food Additives, resulted in withdrawal of the provisional tolerable weekly intake (PTWI) since 2010. The other important gaps identified by the JEC is particularly related to the need for accurate quantification of arsenic in dietary and other exposure routes as well as the speciation of arsenic and bioavailability that account for the total daily intake. Long-term exposure to arsenic is related to non-specific pathological irreversible effects and has significant social and economic impacts. The presence of arsenic in rice and rice products available in the markets has raised a critical concern – and this includes rice cakes, breakfast cereals as well as plain rice. Daily intake of inorganic arsenic in small quantities in rice and all rice products leads to high levels of arsenic exposure—and especially to the group of population with rice as the staple diet. Children are vulnerable to arsenic exposure, where the risk of arsenic exposure is exceptionally high due to the consumption of rice cakes especially in the pre-schools. Thus, arsenic in environment is clearly a concern that needs an inter- and multi-disciplinary and cross-disciplinary platform of research including hydrogeology and hydrogeochemistry, environmental sciences, food and nutrition, toxicology, health and medical sciences, remediation technologies and social sciences.

The biennial International Congress Series on Arsenic in the Environment is providing a common platform for sharing knowledge and experience on multidisciplinary issues on arsenic occurrences in groundwater and other environmental compartments on a worldwide scale to identify, assess, develop and promote approaches for management of arsenic in the environment and health effects. Since the first International Congress on "Arsenic in the Environment" at the UNAM, Mexico City in 2006, there has been an overwhelming response from the scientific community engaged with multidisciplinary facets of arsenic research to participate and present their research findings on this platform. The conference has been taken a form of biennial congress series with rotating venues at different continents. The following three events namely the 2nd International Congress (As 2008), with the theme "Arsenic from Nature to Humans" (Valencia, Spain) and the 3rd International Congress (As 2010) with the theme "Arsenic in Geosphere and Human Diseases" (Tainan, Taiwan), the 4th International Congress on Arsenic in the Environment (As 2012) with a theme "Understanding the Geological and Medical Interface" (Cairns, Australia) and the 5th International Congress on Arsenic in the Environment (As 2014) with a theme "One Century of the Discovery of Arsenicosis in Latin America (1914–2014)" have been successfully organized and participated by a leading scientific community around the globe. The upcoming 6th International Congress on Arsenic in the Environment (As 2016) is envisioned with a theme "Arsenic Research and Global Sustainability" to be organized in Stockholm, Sweden between 19th and 23rd June, 2016, with an aim to provide another international, multi- and interdisciplinary discussion platform for the presentation of cutting edge scientific research involving arsenic in natural systems, food chain, health impacts, clean water technology and other related social issues linked with environmental arsenic by bringing together scientific, medical, engineering and regulatory professionals.

I feel proud to write this foreword to this Volume of Arsenic in the Environment-Proceedings Series that contains the extended abstracts of the presentations to be made during the forthcoming 6th International Congress on Arsenic in the Environment-As 2016. The present volume "Arsenic Research and Global Sustainability" being published as a new volume of the book series "Arsenic in the Environment-Proceedings under the auspices of the International Society of Groundwater for Sustainable Development (ISGSD), will be an important updated contribution, comprising a large number of over 230 extended abstracts submitted by various researchers, health workers, technologists, students, legislators, and decision makers around the world that would be discussed during the conference. Apart from exchanging ideas, and discovering common interests, the scientific community involved in this this specialized field needs to carry out researches, which not only address academic interests but also contribute to the societal needs through prevention or reduction of exposure to arsenic and its toxic effects in millions of exposed people throughout the world.

I deeply appreciate the efforts of the International Organizers and Local Organizers from KTH International Groundwater Arsenic Research Group, Department of Sustainable Development, Environmental Science and Engineering, School of Architecture and Built Environment KTH Royal Institute of Technology and the University of Southern Queensland, Toowoomba, Australia, the Institute of Environmental Medicine, Karolinska Institutet, Stockholm University and Bolin Center for Climate Research, Luleå University of Technology, Lund University, and the Örebro University together with The Technical University of Delft and the KWR Watercycle Research Institute (KWR) and the entire editorial team for their efforts to bring together for their untiring work with this volume. I hope that the book will reflect the update on the current state-of-the-art of our knowledge on the interdisciplinary facets of arsenic in the environment required for the management of arsenic in the environment for protecting human health.

Professor Dr. Peter Gudmundson
KTH Royal Institute of Technology
Stockholm, Sweden
April, 2016

Arsenic Research and Global Sustainability – Bhattacharya, Vahter, Jarsjö, Kumpiene,
Ahmad, Sparrenbom, Jacks, Donselaar, Bundschuh & Naidu (Eds)
© 2016 Taylor & Francis Group, London, ISBN 978-1-138-02941-5

Foreword (Vice-Chancellor and President, USQ)

The University of Southern Queensland (USQ) has great pleasure in co-organising the 6th International Congress on Arsenic in the Environment (As 2016) themed 'Arsenic Research and Global Sustainability', in June 2016.

Worldwide more than 200 million people are currently suffering from arsenic contamination of water resources or other geoenvironments. Evidence suggests the incidence of arsenic contamination of drinking and irrigation water has doubled in the last ten years, with reports of contamination from over 75 countries. Hence, arsenic is an increasing global problem requiring global solutions. Research into the occurrence, mobility and bioavailability of arsenic in different environments, including aquifers, soils, sediments as well as the food chain, and arsenic mitigation, including treatment, of drinking water resources will be increasingly important.

The University of Southern Queensland provides education and research services to the local, regional and global communities. A particular focus is the conduct of applied research to support sustainable development and fulfilling lives. USQ has an emerging interdisciplinary research capability in Environmental Science and Management, which corresponds to code 0502 of the Excellence in Research for Australia (ERA) where USQ has a 5 star ranking (i.e., well above world standard). This includes capabilities for (hydro)geochemical and groundwater research and sustainable renewable energy powered and low-energy demanding water and wastewater treatment research, supported by existing internationally recognised expertise across the Faculty of Health, Engineering and Sciences and seven key research centres. The National Centre for Engineering in Agriculture and the International Centre for Applied Climate Science are both actively involved in research to utilise waste streams, mitigate the impacts of mining, rehabilitate contaminated and degraded landscapes, performs research on the water-energy-food nexus and is world-leader in climate variability forecasting while the Centre for Health Sciences Research conducts health related research in rural and remote communities. Much of this research is conducted in collaboration with international partners and is directed towards safeguarding environments while contributing to the sustainable utilisation of our mineral, energy and water resources to optimise the economic, social and ecological benefits for society.

I congratulate the organisers for their success in bringing this congress, to Sweden and acknowledge the collaborative and cooperative efforts of the Royal Institute of Technology (KTH) – with which we reciprocally share several adjunct professors – in co-organising this international congress and look forward to expand and accelerate the USQ-KTH cooperation. I also hope that these proceedings will serve as a lasting record of our improving knowledge base to better manage groundwater and geoenvironments as well as protect communities.

Jan Thomas
Vice-Chancellor and President
The University of Southern Queensland
Toowoomba, Australia
January, 2016

Foreword (Institute of Environmental Medicine, KI)

The Institute of Environmental Medicine, Karolinska Institutet, has great pleasure in co-organizing the 6th International Congress on Arsenic in the Environment (As2016) with the theme "Arsenic Research and Global Sustainability" in Stockholm, June 2016.

Exposure to inorganic arsenic through drinking water and certain food is an increasing global health problem. Arsenic is a potent toxicant and carcinogen. It gives rice to a multitude of health effects, even at fairly low exposure levels, causing enormous suffering. Recent research indicates that fetuses and young children are particularly at risk and that early-life exposure also increases the disease risk, including cancer, much later in life. Research concerning different chemical forms of arsenic in water and food, bioavailability, metabolism, modes and mechanisms of action, susceptibility factors, dose-response relationships and the increasing medical use of arsenic will be progressively important.

The Institute of Environmental Medicine (IMM) is a department at Karolinska Institutet, and also a national and international resource in environmental health risk assessment. It is an interdisciplinary research organization with internationally competitive research in the fields of epidemiology, toxicology, physiology and occupational and environmental medicine. A main focus is applied research to support sustainable development and improved environmental health in a global sense. IMM has a long history of research on the health effects of arsenic—since the late 1970ies when the focus was on occupational and environmental health effects of the emissions from a copper smelter in the north of the country and many small factories employing preservation of wood with the Boliden salt. We have also repeatedly been engaged in health risk assessment of arsenic for our national agencies as well as the World Health Organization, the European Food Safety Authority, and the U.S. National Research Council, the National Academies of Sciences. To note, Karolinska Institutet also sponsored the first International Conference on Environmental Arsenic, which took place almost exactly 40 years ago, October 1976, in Fort Lauderdale, Florida, USA.

I congratulate the organizers for their success in bringing the congress to Sweden and acknowledge the hard work of the international organizers, the strong coordinating function of the Royal Institute of Technology (KTH), and the effort by many colleagues. I'm convinced that the conference will shed important new light on the puzzling world of arsenic and move our understanding of occurrence, biochemistry, toxicity and health effects of arsenic quite a bit forward. That is certainly needed for improving the present situation with the exposure of many millions of people world-wide. Possibly, it will also raise new important questions about arsenic. I also hope that these proceedings will be a useful lasting record of that important gain in knowledge.

Professor Ulla Stenius
Institutet for Environmental Medicine
Karolinska Institutet
Stockholm, Sweden
April 2016

Arsenic Research and Global Sustainability – Bhattacharya, Vahter, Jarsjö, Kumpiene,
Ahmad, Sparrenbom, Jacks, Donselaar, Bundschuh & Naidu (Eds)
© 2016 Taylor & Francis Group, London, ISBN 978-1-138-02941-5

Foreword (Vice Chancellor, LTU)

Arsenic in the environment and related health risks is a challenge affecting millions of people worldwide. Even though the extent of the problem in Sweden is substantially smaller than in other regions of the world (Bangladesh, India, Argentina, China, etc.), arsenic is among the priority contaminants and we still need to take measures to reduce risks related with contaminated soil and groundwater. Due to the industrial activities, such as glass works and wood impregnation, multiple point sources of arsenic contamination are scattered over the country with arsenic concentrations in soil reaching alarming levels (e.g. >10 000 mg/kg As). In addition, naturally elevated arsenic concentration in the bedrock (e.g. Skellefteå region) is the reason that several percent of individual groundwater wells used for drinking water contain arsenic concentrations above the World Health Organisation's limit of 10 µg/L.

Negative effects on human health caused by this toxic element are widely acknowledged, but efficient and comprehensive risk mitigation measures are not fully realised. Although technical solutions for arsenic removal from drinking water have highly advanced over the years, the long-lasting and environmentally sound management of solid materials enriched in arsenic are still to be developed. The separated toxic element has to be handled further by placing it in a final sink (e.g. landfill) in order to prevent its return to the circulation. Arsenic contaminated solid medium (soil, used sorbents, industrial waste products, etc.) might be a secondary source of environmental contamination and should be treated to prevent arsenic leaching to groundwater. This means that water treatment and soil remediation does not finish with arsenic separation. Arsenic management solutions that would render arsenic immobile are required even if the material is destined for final disposal. To make the arsenic management process complete, we need solutions covering entire process chain: removal, concentration, stabilisation and safe long-lasting storage (disposal).

It is my pleasure to write this foreword to the Proceedings Series on Arsenic in the Environment containing the extended abstracts of the presentations of the 6th International Congress on Arsenic in the Environment-As 2016. This is an important and relevant event that would further strengthen our research in this area as well as bring the research outcomes closer to industry and society at large.

Professor Johan Sterte
Vice Chancellor
Luleå University of Technology
Luleå, Sweden
April, 2016

*Arsenic Research and Global Sustainability – Bhattacharya, Vahter, Jarsjö, Kumpiene,
Ahmad, Sparrenbom, Jacks, Donselaar, Bundschuh & Naidu (Eds)
© 2016 Taylor & Francis Group, London, ISBN 978-1-138-02941-5*

Foreword (Director, KWR Watercycle Research Institute)

KWR Watercycle Research Institute

It is with great pleasure and expectations that I write this Foreword to the Proceedings of the 6th International Congress on Arsenic in the Environment (As 2016), themed 'Arsenic Research and Global Sustainability' held in Stockholm, Sweden, June 19–23, 2016.

The International Congress on Arsenic in the Environment has been previously held five times: Mexico 2006, Spain 2008, Taiwan 2010 Australia, 2012 and Argentina 2014. The Congress series has evolved into a highly reputable platform for sharing and assessing global knowledge on various aspects of arsenic research. Arsenic in drinking water is a global problem affecting populations on all five continents. Despite historical recognition of arsenic toxicity, more than 200 million people around the world are still exposed to above acceptable arsenic levels. This situation is alarming. Arsenic contamination of drinking water can be caused both by natural and anthropogenic processes. For example, in Poland and Brazil, arsenic contamination of groundwater due to anthropogenic mining activities have been reported. On the other hand, in some parts of Turkey elevated arsenic in groundwater is attributed to natural geothermal factors, and in Bangladesh geogenic processes are the major cause of large scale arsenic contamination. Whatever the origin may be, once detected in drinking water sources, suitable arsenic remediation measures should be taken to ensure supply of safe drinking water—as this is the fundamental right of every human being.

In the Netherlands, drinking water companies have recently updated their policy on arsenic and they will present their rationale at As 2016. KWR Watercycle Research Institute is collaborating with the water companies in various fundamental and applied research projects to support the realization of this policy. Recognizing the global significance of arsenic for safe water supply, KWR has gladly invested in realizing As2016 via participation in the organizing committee and the scientific board of As 2016, by our research scientist, Mr. Arslan Ahmad, from our Knowledge Group Water Systems and Technology.

I congratulate all the authors, reviewers and editors for providing excellent content and structure to this book. I hope that these proceedings will serve as a deep-rooted record of the state-of-the-arsenic-related-science in the year 2016 and serve as a reference base for future research and support water suppliers and policy makers all over the world in addressing the arsenic problem efficiently and effectively.

Prof. Dr. Wim van Vierssen
Director
KWR Watercycle Research Institute
Nieuwegein, The Netherlands
March, 2016

Arsenic Research and Global Sustainability – Bhattacharya, Vahter, Jarsjö, Kumpiene,
Ahmad, Sparrenbom, Jacks, Donselaar, Bundschuh & Naidu (Eds)
© 2016 Taylor & Francis Group, London, ISBN 978-1-138-02941-5

Editors' foreword

Occurrence of elevated arsenic concentrations in ground water used for drinking purpose, and associated health risks, were reported at first international conference on environmental arsenic, which was held in Fort Lauderdale, USA, almost exactly 40 year ago; October, 1976. Over the past 2 to 3 decades arsenic in drinking water, and more recently, in plant based foods, especially rice, has been recognized as a major public health concern in many parts of the world. Latest surveys estimated that currently more than 200 million people around the world are exposed to unacceptably high arsenic levels. The geological, geomorphological and geochemical reasons for high arsenic concentrations in groundwater vary from place to place and require different mitigation policies and practices. Although, the high income countries may invest in research and development of suitable remediation techniques, arsenic in private water sources is not always tested. On the other hand, low to lower-middle income countries, such as many areas in South-East Asia, Africa and South America, where millions of people still use arsenic-contaminated drinking water, are still coping with stagnated mitigation efforts and slow progress towards safe drinking water. It is disturbing to enter almost any village of the Bengal basin today and find that groundwater drawn from untested shallow wells continues to be used routinely for drinking and cooking, given that the arsenic problem was already recognized in the mid-1980s in West Bengal and the mid-1990s in Bangladesh. Equally problematic is the fact that hundreds of millions of wells world-wide are not yet tested for arsenic. Moreover, many low and lower-middle income countries have yet not been able to revise their standards for arsenic in drinking water to 10 µg/L, the guideline value of the World Health Organization. We sincerely believe that sharing knowledge and experience on arsenic related science and practices on a world-wide scale and across varied disciplines can serve as an effective strategy to support global arsenic management and mitigation efforts.

The biannual International Congress Series on Arsenic in the Environment aims at providing a common platform for sharing knowledge and experience on multidisciplinary issues on arsenic occurrences in groundwater and other environmental compartments on a worldwide scale for identifying and promoting optimal approaches for the assessment and management of arsenic in the environment. The International Congress on Arsenic in the Environment has previously been held five times; Mexico 2006, Spain 2008, Taiwan 2010, Australia, 2012 and Argentina 2014. The sixth International Congress on Arsenic in the Environment (As2016) is being organized in Stockholm, the Capital of Sweden, between 19–23 June, 2016 and is themed as "Arsenic Research and Global Sustainability", owing to the fact that in 2016 the 17 Sustainable Development Goals (SDGs) of the 2030 Agenda for Sustainable Development, adopted by world leaders in September 2015 at a historic UN Summit, officially came into force. The increased emphasis on holistic management of drinking water services and monitoring of drinking water quality in the 2030 Agenda for Sustainable Development will further raise the global profile of arsenic in order to achieve universal and equitable access to safe and affordable drinking water for all. We envision(ed) As2016 as a global level interdisciplinary information exchange event to improve our understanding of the occurrence, mobility, bioavailability, toxicity and relationships between the dose and various health effects of arsenic, and to brainstorm with the colleagues from around the world about our role towards meeting the targets of the 2030 Agenda(for sustainable development).

We have received a large number of (over 300) extended abstracts which were submitted mainly from researchers, but also health workers, technologists, students, legislators, government officials. The topics to be covered in the Congress As 2016 have been grouped under the five general thematic areas:

Theme 1: Arsenic in environmental matrices (air, water and soil)
Theme 2: Arsenic in food
Theme 3: Arsenic and health
Theme 4: Clean water technologies for control of arsenic
Theme 5: Societal and Policy Implications, Mitigation and Management

We thank the international scientific committee members, for their efforts on reviewing the extended abstracts. Further, we thank the sponsors of the Congress from around the world: KTH Royal Institute of Technology (Sweden), especially KTH Sustainability and the School of Architecture and Built Environment, Institute of Environmental Medicine-Karolinska Institute (Sweden), Bolin Center for Climate Research, Stockholm University, Luleå University of Technology (LTU), University of Southern Queensland (Australia), KWR Watercycle Research Institute (The Netherlands), the Swedish Environmental Research Institute (IVL) Stockholm, Lavaris Technologies GmbH (Germany), GEH Wasserchemie GmbH & Co. KG (Germany) and Carus Europe S.L. We are extremely grateful to the CRC-CARE, At the University of South Australia and OPCW for their generous support—Thank you all sponsors for your support that contributed to the success of the congress As2016.

The International Organizers Prosun Bhattacharya and Jochen Bundschuh would like thank Instutute of Environmental Medicine, Karolinska Institute, Bolin Center for Climate Research, Stockholm University, Luleå University of Technology, University of Lund and KWR Watercycle Research Institute, The Netherlands, the Technical University of Delft, The Netherlands and the University of Southern Queensland, Australia for their support to organize the sixth International Congress, Arsenic in the Environment (As2016). We thank the KTH Royal Institute of Technology, especially the KTH Sustainability, the School of Architecture and Built Environment for supporting the KTH-International Groundwater Arsenic Research Group at the Department of Sustainable Development, Environmental Sciences and Engineering, Stockholm to organize this Congress. Lastly, the editors thank Lukas Goosen of the CRC Press/Taylor and Francis (A.A. Balkema) Publishers, The Netherlands for their patience and skill for the final production of this volume.

Prosun Bhattacharya
Marie Vahter
Jerker Jarsjö
Jurate Kumpiene
Arslan Ahmad
Charlotte Sparrenbom
Gunnar Jacks
Marinus Eric Donselaar
Jochen Bundschuh
Ravi Naidu
(Editors)

List of contributors

Abhinav, S.: *L.N. Mithila University, Darbhanga, Bihar, India*

Abramsson, L.: *National Food Agency, Uppsala, Sweden*

Adak, S.: *Department of Zoology, Seth Anandaram Jaipuria College, Kolkata, India*

Afzal, B.: *Environmental Geochemistry Laboratory, Department of Environmental Sciences, Faculty of Biological Sciences, Quaid-i-Azam University, Islamabad, Pakistan*

Aguayo, A.: *Instituto de Geofísica, Universidad Nacional Autónoma de México, Ciudad de México, D.F., Mexico*

Ahmad, A.: *KWR Watercycle Research Institute. Nieuwegein, The Netherlands.*

Ahmad, H.B.: *Institute of Chemical Sciences, Bahauddin Zakariya University Multan, Pakistan*

Ahmed, A.: *Columbia University Arsenic Research Project, Dhaka, Bangladesh*

Ahmed, K.M.: *Department of Geology, Faculty of Earth and Environmental Sciences, University of Dhaka, Dhaka, Bangladesh*

Ahmed, S.: *Institute of Environmental Medicine, Karolinska Institutet, Stockholm, Sweden; International Centre for Diarrhoeal Disease Research, Bangladesh (ICDDR, B), Dhaka, Bangladesh*

Ahmed, T.: *Functional Materials Division, Materials and Nano Physics Department, School of ICT, KTH Royal Institute of Technology, Kista, Stockholm, Sweden*

Ahsan, H.: *Departments of Health Studies, Medicine and Human Genetics and Cancer Research Center, The University of Chicago, Chicago, IL, USA*

Aitken, J.B.: *School of Chemistry, The University of Sydney, Sydney, NSW, Australia*

Akter, F.: *Research and Evaluation Division, BRAC, Mohakhali, Dhaka, Bangladesh*

Akter, T.: *Research and Evaluation Division, BRAC, Mohakhali, Dhaka, Bangladesh*

Al Bualy, A.A.N.: *School of Earth, Atmospheric and Environmental Sciences and Williamson Research Centre for Molecular Environmental Science, University of Manchester, Manchester, UK*

Alakangas, L.: *Department of Civil, Environmental and Natural Resources Engineering, Division of Geosciences and Environmental Engineering, Luleå University of Technology, Luleå, Sweden*

Alam, N.: *Department of Science and Technology, Government of India, New Delhi, India*

Alam, O.: *Department of Civil and Environmental Engineering, Birla Institute of Technology, Mesra, Ranchi, Jharkhand, India*

Alarcón-Herrera, M.T.: *Advanced Materials Research Center (CIMAV), Durango, Mexico*

Alauddin, S.: *Department of Chemistry, Wagner College, Staten Island, NY, USA*

Alauddin, M.: *Department of Chemistry, Wagner College, Staten Island, NY, USA*

Ali, W.: *Environmental Geochemistry Laboratory, Faculty of Biological Sciences, Department of Environmental Sciences, Quaid-i-Azam University, Islamabad, Pakistan*

Ali, M.: *Mahavir Cancer Institute and Research Centre, Patna, Bihar, India*

Ali, S.: *Department of Environmental Sciences and Engineering, Government College University, Faisalabad, Pakistan*

Allegretta, I.: *Department of Soil, Plant and Food Sciences, University of Bari "Aldo Moro", Bari, Italy*

Alpaslan, M.N.: *Dokuz Eylül University, Faculty of Eng., Dept. of. Env.Eng. Buca, İzmir, Turkey.*

Alvarez, C.: *Toxicology. DEC, Faculty of Chemistry. Universidad de la República (UdelaR), Montevideo, Uruguay*

Alvarez, J.H.: *INTI Textiles—UT Comercialización y Diseño - Laboratorio químico textil, Buenos Aires, Argentina*

Álvarez,, C.: *Toxicology. DEC, Faculty of Chemistry. Universidad de la República (UdelaR), Montevideo, Uruguay*

Alvarez- Gonçalvez, C.V.: *Instituto de Investigaciones en Producción Animal (INPA) UBA—CONICET, Centro de Estudios Transdisciplinarios del Agua (CETA – UBA), Facultad de Ciencias Veterinarias, Universidad de Buenos Aires, Argentina*

Ameer, S.S.: *Division of Occupational and Environmental Medicine, Lund University, Lund, Sweden*

Amini, M.: *Eawag, Swiss Federal Institute of Aquatic Science and Technology, Dübendorf, Switzerland*

Anantharaman, G.: *Department of Chemistry IIT Kanpur, Kanpur, India*

Annaduzzaman, M.: *KTH-International Groundwater Arsenic Research Group, Department of Sustainable Development, Environmental Science and Engineering, KTH Royal Institute of Technology, Stockholm, Sweden*

Araki, H.: *JDC Corporation/ Saga University, Japan*

Arancibia, V.: *Chemistry Faculty, Pontificia Universidad Católica de Chile, Santiago, Chile*

Arancibia-Miranda, N.: *Department of Materials Chemistry, Faculty of Chemistry and Biology, University of Santiago, Chile*

Arao, T.: *Central Region Agricultural Research Center, NARO, Tsukuba, Ibaraki, Japan*

Arellano, F.E.: *Instituto de Investigaciones en Producción Animal (UBA-CONICET), Universidad de Buenos Aires, Buenos Aires, Argentina; Centro de Estudios Transdisciplinarios del Agua, Facultad de Ciencias Veterinarias, Universidad de Buenos Aires, Buenos Aires, Argentina*

Arıkan, S.: *Dokuz Eylül University, Faculty of Eng., Dept. of. Env.Eng. Buca, İzmir, Turkey*

Armienta, M.A.: *Instituto de Geofísica, Universidad Nacional Autónoma de México, Ciudad de México, D.F., Mexico*

Arroyo-Herrera, I.: *Departamento de Microbiología, Escuela Nacional de Ciencias Biológicas, Instituto Politécnico Nacional, México D. F., México*

Arshad, M.: *Institute of Soil and Environmental Sciences, University of Agriculture Faisalabad, Faisalabad, Pakistan*

Baba, A.: *İzmir Institute of Technology, Izmir, Turkey*

Baba, K.: *Institute for Agro-Environmental Sciences, NARO, Ibaraki, Japan*

Backman, B.: *Geological Survey of Finland, Espoo, Finland*

Bagade, A.V.: *Department of Chemistry, Savitribai Phule Pune University, Pune, India*

Baisch, P.: *Geological Oceanography Laboratory, Institute of Oceanography, Federal University of Rio Grande, Rio Grande RS, Brazil*

Bajpai, S.: *Department of Science and Technology, Government of India, New Delhi, India*

Baken, K.: *KWR Watercycle Research Institute, Nieuwegein, The Netherlands*

Baker, J.: *SELOR, Amsterdam, The NetherlandsA. Fiúza, A. Futuro & M. Guimarães*

Balakrishnan, P.: *Johns Hopkins University, Baltimore, Maryland, USA*

Balint, R.: *Department of Agricultural, Forest and Food Sciences, University of Torino, Grugliasco, Torino, Italy*

Ballentine, C.J.: *Department of Earth Sciences, University of Oxford, Oxford, UK*

Ballinas-Casarrubias, M.L.: *Facultad de Ciencias Químicas, Universidad Autónoma de Chihuahua, Chihuahua, México*

Balomajumder, C.: *Department of Chemical Engineering, Indian Institute of Technology, Roorkee, Roorkee, India*

Bandyopadhyay, A.K.: *Molecular Genetics Division, CSIR-Indian Institute of Chemical Biology, Kolkata, India*

Bano, S.: *Environmental Geochemistry Laboratory, Department of Environmental Sciences, Faculty of Biological Sciences, Quaid-i-Azam University, Islamabad, Pakistan*

Bansiwal, A.: *Environmental Materials Division, CSIR-NEERI, Nehru Marg, Nagpur, Maharashtra, India*

Barats, A.: *Université Nice Sophia Antipolis, CNRS, IRD, Observatoire de la Côte d'Azur, Géoazur, Campus Azur CNRS, Valbonne, France*

Barberis, E.: *Department of Agricultural, Forest and Food Sciences, University of Torino, Grugliasco, Torino, Italy*

Bardelli, F.: *Institute of Earth Science, University of Grenoble, Grenoble, France*

Barla, A.: *IISER Kolkata, Mohanpur, Nadia, West Bengal, India*

Barman, S.: *Department of Chemistry, University of Kalyani, Kalyani, West Bengal, India*

Battaglia-Brunet, F.: *BRGM, Water, Environment and Ecotechnology Division, Environmental Biogeochemistry and Water Quality Unit, 3, Orléans, France*

Bea, S.A.: *Instituto de Hidrología de Llanuras "Dr. Eduardo J. Usunoff" (IHLLA), Azul, Argentina*

Beattie, D.A.: *Future Industries Institute, University of South Australia, Mawson Lakes, South Australia*

Belluzzi Muiños, M.: *Analytical Chemistry. DEC, Faculty of Chemistry. Universidad de la República (UdelaR), Montevideo, Uruguay*

Berg, M.: *Eawag, Swiss Federal Institute of Aquatic Science and Technology, Dübendorf, Switzerland*

Bergmann, M.: *Institute of Chemistry, Analytical Chemistry, University of Graz, Graz, Austria*

Berner, Z.: *Institute of Mineralogy and Geochemistry, Karlsruhe Institute of Technology, KIT, Karlsruhe, Germany*

Bernier-Latmani, R.: *Ecole Polytechnique Federale de Lausanne, Lausanne, Switzerland*

Berube, M.: *Department of Geology, Kansas State University, Manhattan, KS, USA*

Best, L.: *Missouri Breaks Industries Research, Inc, Timber Lake, South Dakota, USA*

Bewsher, A.: *School of Earth, Atmospheric and Environmental Sciences and Williamson Research Centre for Molecular Environmental Science, University of Manchester, Manchester, UK*

Bhadury, P.: *Department of Biological Sciences, Indian Institute of Science Education and Research Kolkata, Mohanpur, India*

Bhatia, S.: *Disaster Management, Tata Institute of Social Sciences, Mumbai, India*

Bhatnagar, A.: *Department of Environmental Science, University of Eastern Finland, Kuopio, Finland*

Bhattacharya, P.: *KTH-International Groundwater Arsenic Research Group, Department of Sustainable Development, Environmental Science and Engineering, KTH Royal Institute of Technology, Stockholm, Sweden, International Center for Applied Climate Science, University of Southern Queensland, Toowoomba, Queensland, Australia*

Bhattacharya, S.: *School of Environmental Studies, Faculty of Interdisciplinary Studies (Law & Management), Jadavpur University, Kolkata, West Bengal, India*

Bhattacharya, A.: *Public Health Engineering Department, Government of West Bengal, Kolkata, India*

Bhowmick, S.: *Department of Chemistry, University of Kalyani, Kalyani, West Bengal, India*

Bia, G.L.: *Centro de Investigaciones en Ciencias de la Tierra (CICTERRA), CONICET and Universidad Nacional de Córdoba, Argentina.*

Bibi, I.: *Institute of Soil and Environmental Sciences, University of Agriculture Faisalabad, Faisalabad, Pakistan*

Biswas, U.: *Department of Chemistry, University of Kalyani, Kalyani, West Bengal, India*

Biswas, A.: *AKVO, New Delhi, India*

Biswas, A.: *Department of Chemistry, University of Kalyani, Kalyani, West Bengal, India; KTH-International Groundwater Arsenic Research Group, Department of Sustainable Development, Environmental Science and Engineering, KTH Royal Institute of Technology, Stockholm, Sweden*

Bjerselius, R.: *National Food Agency, Uppsala, Sweden*

Blanes, P.S.: *Área Química General, FCBioyF, IQUIR-CONICET, Universidad Nacional de Rosario, Argentina*

Blokker, E.J.M.: *KWR Watercycle Research Institute, Nieuwegein, The Netherlands*

Blondes, M.S.: *U.S. Geological Survey, Reston, VA, USA*

Boemo, A.: *Facultad de Ciencias Exactas & Consejo de Investigación, Universidad Nacional de Salta, Salta, Argentina*

Bohan, M.T.: *Department of Mining and Materials Engineering, McGill University, Montreal, Canada*

Borgnino, L.: *Centro de Investigaciones en Ciencias de la Tierra (CICTERRA). CONICET, and FCEFyN Universidad Nacional de Córdoba, Córdoba, Argentina*

Bosch, J.: *Brabant Water's-Hertogenbosch, The Netherlands*

Bose, N.: *Department of Environment and Water Management, A.N. College, Patna, India*

Bose, S.: *IISER Kolkata, Mohanpur, Nadia, West Bengal, India*

Bostick, B.: *Lamont-Doherty Earth Observatory, Columbia University, New York, New York, USA*

Boyce, A.: *Scottish Universities Environmental Research Centre, East Kilbride, UK*

Braeuer, S.: *Institute of Chemistry – Analytical Chemistry, University of Graz, Graz, Austria*

Brandão, P.F.B.: *Laboratorio de Microbiología Ambiental y Aplicada, Departamento de Química, Universidad Nacional de Colombia, Bogotá, Colombia*

Bretzler, A.: *Eawag, Swiss Federal Institute of Aquatic Science and Technology, Dübendorf, Switzerland*

Broberg, K.: *Institute of Environmental Medicine, Karolinska Institutet, Stockholm, Sweden*

Brown, C.: *Ryan Institute, Environmental, Marine and Energy Research, National University of Ireland, Galway, Ireland*

Buachidze, Z.E.: *Georgian Technical University, Institute for Problems of Engineering Physics, Tbilisi, Georgia*

Buccheri, S.: *Institute of Applied Research, Karlsruhe University of Applied Sciences, Karlsruhe, Germany*

Bundschuh, J.: *Deputy Vice-Chancellor's Office (Research and Innovation), University of Southern Queensland, Toowoomba, Queensland, Australia; Faculty of Health, Engineering and Sciences, University of Southern Queensland, Toowoomba, Queensland, Australia*

Butler, A.P.: *Department of Civil and Environmental Engineering, Imperial College, London, UK*

Buzek, F.: *Republic Czech Geological Survey, Prague, Czech Republic*

Bühl, V.: *Analytical Chemistry, DEC, Faculty of Chemistry. Universidad de la República (UdelaR), Montevideo, Uruguay*

Bäckström, M.: *Man-Technology-Environment Research Centre, Örebro University, Örebro, Sweden*
Cabrera, F.E.: *Departamento de Ciencias Básicas y Aplicadas, Universidad Nacional del Chaco Austral, Chaco, Argentina*
Cacciabue, L.: *Instituto de Hidrología de Llanuras "Dr. Eduardo J. Usunoff" (IHLLA), Azul, Argentina*
Calatayud, M.: *Laboratory of Microbial Ecology and Technology (LabMET), Ghent University, Ghent, Belgium*
Calderón, E.: *Centro de Estudios Transdisciplinarios del Agua (Universidad de Buenos Aires), Centro de Estudios Transdisciplinarios del Agua (CETA- UBA), Buenos Aires, Argentina*
Cao, Y.: *State Key Laboratory of Biogeology and Environmental Geology, China University of Geosciences, Beijing, P.R. China; School of Water Resources and Environment, China University of Geosciences, Beijing, P.R. China*
Cao, W.: *Institute of Hydrogeology and Environment al Geology, Chinese Academy of Geological Sciences, Zhengding, Hebei, P.R. China*
Cappuyns, V.: *Centre for Economics and Corporate Sustainability, KU Leuven, Belgium*
Cardenas, M.B.: *Jackson School of Geosciences, University of Texas at Austin, Austin, TX, USA*
Casanueva-Marenco, M.J.: *School of Earth, Atmospheric and Environmental Sciences and Williamson Research Centre for Molecular Environmental Science, University of Manchester, Manchester, UK*
Casimiro, E.: *Hospital Dr. Nicolás Cayetano Pagano, San Antonio de los Cobres, Salta, Argentina*
Casiot, C.: *HydroSciences Montpellier, Université de Montpellier, Montpellier, France*
Castillo, E.: *Laboratorio de Química Ambiental, Grupo de Estudios para la Remediación y Mitigación de Impactos Negativos al Ambiente (GERMINA), Departamento de Química, Universidad Nacional de Colombia, Bogotá, Colombia*
Cebrián, M.E.: *Departamento de Toxicología, Centro de Investigación y Estudios Avanzados del Instituto Politécnico Nacional, D.F., México*
Cejkova, B.: *Republic Czech Geological Survey, Prague, Czech Republic*
Cekovic, R.: *Department of Chemistry, Wagner College, Staten Island, NY, USA*
Celi, L.: *Department of Agricultural, Forest and Food Sciences, University of Torino, Grugliasco, Torino, Italy*
Ceniceros, N.: *Instituto de Geofísica, Universidad Nacional Autónoma de México, Ciudad de México, D.F., Mexico*
Cesio, M.V.: *Pharmacognosy and Natural Products. DQO, Faculty of Chemistry. Universidad de la República (UdelaR), Montevideo, Uruguay*
Chakraborty, S.: *Department of Civil and Environmental Engineering, Birla Institute of Technology, Mesra, Ranchi, Jharkhand, India*
Chakraborty, Sudipta.: *Department of Chemistry, Kanchrapara College, Kanchrapara, West Bengal, India*
Chalov, S.: *Lomonosov Moscow State University, Faculty of Geography, Moscow, Russia*
Chana, G.: *School of Earth, Atmospheric and Environmental Sciences and Williamson Research Centre for Molecular Environmental Science, University of Manchester, Manchester, UK*
Chandrasekharam, D.: *Department of Earth Sciences, IIT Bombay, Mumbai, India*
Chandrashekhar, A.K.: *Department of Earth Sciences, IIT Bombay, Mumbai, India*
Charlet, L.: *Institut des Sciences de La Terre (ISTerre), Université Grenoble Alpes and CNRS, France*
Chatterjee, D.: *Department of Chemistry, University of Kalyani, Kalyani, West Bengal, India*
Chatterjee, D.: *Heritage Institute of Technology, Chowbaga Road, Anandpur, East Kolkata Township, Kolkata, India*
Chatterjee, P.K.: *Society for Technology with a Human Face (NGO), Kolkata, India*
Chaudhary, S.: *Tilka Manjhi Bhagalpur University, Bhagalpur, India*
Chavez, M.: *Laboratorio Químico Toxicológico-Centro Nacional de Salud Ocupacional y Protección del Ambiente para la salud (CENSOPAS)-Instituto Nacional de Salud, Lima, Perú*
Chen, C.-Y.: *Department of Earth and Environmental Sciences, National Chung Cheng University, Chiayi, Taiwan*
Chen, C.-J.: *Genomics Research Center, Academia Sinica, Taipei, Taiwan*
Chen, Y.: *Departments of Environmental Medicine, New York University School of Medicine, New York, NY, USA*
Chen, B.D.: *State Key Laboratory of Urban and Regional Ecology, Research Center for Eco-Environmental Sciences, Chinese Academy of Sciences, Beijing, P.R. China*
Chi, Q.Q.: *Key Lab of Urban Environment and Health, Institute of Urban Environment, Chinese Academy of Sciences, Xiamen, P. R. China*
Chiou, H.-Y.: *School of Public Health, Taipei Medical University, Taipei, Taiwan*

Chirakadze, A.A.: *Georgian Technical University, Institute for Problems of Engineering Physics, Tbilisi, Georgia*
Cho, N.: *Chemical Safety Division, National Institute of Agricultural Science, Wanju, Republic of Korea*
Choudhury, R.: *Department of Civil Engineering, Indian Institute of Technology Guwahati, India*
Choudhury, I.: *Department of Geology, University of Dhaka, Dhaka, Bangladesh*
Chowdhury, T.R.: *Research and Evaluation Division, BRAC, Mohakhali, Dhaka, Bangladesh*
Chowdhury, M.T.A.: *Institute of Biological and Environmental Sciences, University of Aberdeen, Aberdeen, UK*
Chung, J.-H.: *Research Institute of Pharmaceutical Sciences, Seoul National University, Seoul, Korea*
Chunxiang, T.: *China Conservation and Research Centre for the Giant Panda, China*
Ciminelli, V.S.T.: *Department of Metallurgical and Materials Engineering, Universidade Federal de Minas Gerais-UFMG, Belo Horizonte, Brazil; National Institute of Science and Technology on Mineral Resources, Water and Biodiversity, INCT-Acqua, Brazil*
Cole, S.: *Texas Biomedical Research Institute, San Antonio, Texas, USA*
Concha, G.: *National Food Agency, Uppsala, Sweden*
Cornejo, D.: *Universidad de Antofagasta, Instituto Antofagasta, Biotechnology Department, Campus Coloso, Antofagasta, Chile*
Cornelis, G.: *Department of Soil and Environment, Swedish University of Agricultural Sciences, Uppsala, Sweden*
Corroto, C.E.: *Agua y Saneamientos Argentinos S.A. (AySA S.A.), Buenos Aires, Argentina*
Cortada, U.: *Department of Geology, Campus Científico Tecnológico de Linares, University of Jaen, Spain*
Cote, I.: *U.S. Environmental Protection Agency, National Center for Environmental Assessment, Research Triangle Park, USA*
Couture, R.M.: *Ecohydrology Research Group, Department of Earth and Environmental Sciences, University of Waterloo, Waterloo, Canada*
Cozzarelli, I. M.: *U.S. Geological Survey, Reston, VA, USA*
Cruz, O.: *Instituto de Geofísica, Universidad Nacional Autónoma de México, Ciudad de México, D.F., Mexico*
Cui, S.: *Neo Environmental Business Company, Korea*
Das, A.: *Department of Environmental Sciences, Tezpur Central University, Assam, India*
Das, C.: *Hijli Inspiration, Kolkata, India*
Das, N.: *Department of Environmental Sciences, Tezpur Central University, Assam, India*
Das, S.: *Department of Earth Sciences, National Cheng Kung University, Tainan, Taiwan*
Datta, S.: *Department of Geology, Kansas State University, Manhattan, KS, USA*
Dave, S.N.: *Unicef, Kolkata, India*
de la Torre, M.J.: *Department of Geology, Campus Científico Tecnológico de Linares, University of Jaen, Spain*
de Lillo, E.: *Department of Soil, Plant and Food Sciences, University of Bari "Aldo Moro", Bari, Italy*
De Mello, J.W.V.: *Department of Soils, Universidade Federal de Viçosa, Viçosa, Brazil, National Institute of Science and Technology on Mineral Resources, Water and Biodiversity, INCT-Acqua, Brazil*
de Meyer, C.: *Eawag, Swiss Federal Institute of Aquatic Science and Technology, Dübendorf, Switzerland*
Deacon, C.: *Institute of Biological and Environmental Sciences, University of Aberdeen, Aberdeen, UK*
Del Razo, L.M.: *Departamento de Toxicología, Centro de Investigación y de Estudios Avanzados del IPN, México City DF, Mexico*
Delbem, I.: *Universidade Federal de Minas Gerais, Center of Microscopy, Department of Chemistry, Belo Horizonte, Brazil*
Demopoulos, G.P.: *Department of Mining and Materials Engineering, McGill University, Montreal, Canada*
Dennis De Nora, R.S.: *Water Technologies, Inc., Rodenbach, Germany*
Deowan, S.A.: *Mechatronics Engineering Department, Dhaka University, Dhaka, Bangladesh*
Desheng, L.: *China Conservation and Research Centre for the Giant Panda, China*
Dhali, P.: *School of Environmental Studies, Faculty of Interdisciplinary Studies (Law & Management), Jadavpur University, Kolkata, West Bengal, India*
Dhanachandra, W.: *Department of Earth Sciences, IIT Bombay, Mumbai, India*
Dhotre, D.: *Microbial Culture Collection, National Centre for Cell Science, Pune, India*
Dietrich, S.: *Instituto de Hidrología de Llanuras "Dr. Eduardo J. Usunoff" (IHLLA), Azul, Buenos Aires, Argentina*

Dimova, N.: *Department of Geological Sciences, University of Alabama, Tuscaloosa, AL, USA*
Dinis, M.L.: *Centro de Recursos Naturais e Ambiente (CERENA), Polo do Porto, Universidade do Porto, Faculdade de Engenharia*
Dold, B.: *SUMIRCO (Sustainable Mining Research & Consult EIRL), San Pedro de la Paz, Chile*
Dong, H.: *Department of Cellular Biology and Pharmacology, Herbert Wertheim College of Medicine, Florida International University, Miami, Florida, USA*
Dong, L.J.: *Hawaii Institute of Interdisciplinary Research, USA*
Dong, Q.: *Institute of Hydrogeology and Environment al Geology, Chinese Academy of Geological Sciences, Zhengding, Hebei, P.R. China*
Donselaar, M.E.: *Department of Applied Geoscience and Engineering, Delft University of Technology, Delft, The Netherlands*
Dorador, C.: *Universidad de Antofagasta, Instituto Antofagasta, Biotechnology Department, Campus Coloso, Antofagasta, Chile*
Douglas, A.: *University of Aberdeen, The Institute of Biological and Environmental Sciences, Aberdeen, Scotland*
Douillet, C.: *University of North Carolina at Chapel Hill, Chapel Hill, North Carolina, USA*
Dousova, B.: *University of Chemistry and Technology Prague, Prague, Czech Republic*
Drobná, Z.: *University of North Carolina at Chapel Hill, Chapel Hill, North Carolina, USA*
Du Laing, G.: *Laboratory of Analytical Chemistry and Applied Ecochemistry (ECOCHEM), Ghent University, Ghent, Belgium*
Duchkova, E.: *University of Chemistry and Technology Prague, Prague, Czech Republic*
Dungl, E.: *Vienna Zoo, Austria*
Dupraz, S.: *BRGM, ISTO, UMR 7327, 45060 Orléans, France*
Durrieu, G.: *Université de Toulon, PROTEE, La Garde, France*
Dutta, D.: *All India Coordinated Research Project on Water Management, Bidhan Chandra Krishi Viswavidyalaya, Gayeshpur, Nadia, West Bengal, India*
Dutta, J.: *Functional Materials Division, Materials and Nano Physics Department, School of ICT, KTH Royal Institute of Technology, Kista, Stockholm, Sweden*
Duxbury, J.: *School of Integrative Plant Science, Cornell University, Ithaca, New York, USA*
Dölgen, D.: *Dokuz Eylül University, Faculty of Engineering, Department of Environmental Engineering, Buca, İzmir, Turkey*
Elçi, A.: *Dokuz Eylul University, Izmir, Turkey*
Elmes, M.: *The University of Queensland, School of Earth Sciences, Brisbane, Australia*
Engström, K.: *Institute of Environmental Medicine, Karolinska Institutet, Stockholm, Sweden*
Eqrar, M.N.: *Geosciences Faculty, Kabul University, Kabul, Afghanistan*
Eriksson, A.K.: *Department of Soil and Environment, Swedish University of Agricultural Sciences, Uppsala, Sweden*
Esteller, M.V.: *Universidad Autonoma del Estado de Mexico, Centro Interamericano de Recursos del Agua, Cerro Coatepec S/N Ciudad Universitaria, Toluca, México*
Ettler, V.: *Institute of Geochemistry, Mineralogy and Mineral Resources, Faculty of Science, Charles University in Prague, Praha, Czech Republic*
Eunus, M.: *Columbia University Arsenic Research Project, Dhaka, Bangladesh*
Even, E.: *Department of Geosciences, Osaka City University, Osaka, Japan*
Fanaian, S.: *SaciWATERs, Hyderabad, India*
Farid, L.M.: *Department of Environmental Sciences and Engineering, Government College University, Faisalabad, Pakistan*
Farooq, S.H.: *School of Earth, Ocean and Climate Sciences, IIT Bhubaneswar, Bhubaneswar, India*
Farooqi, A.: *Environmental Geochemistry Laboratory, Department of Environmental Sciences, Faculty of Biological Sciences, Quaid-i-Azam University, Islamabad, Pakistan*
Fasoli, H.J.: *Facultad de CFM e Ingeniería (UCA) y Facultad de Ingeniería del Ejercito, Escuela Superior Técnica, M.N. Savio (IUE), Argentina*
Feldmann, J.: *TESLA (Trace Element Speciation Laboratory), Department of Chemistry, University of Aberdeen, Aberdeen, UK*
Fernádez Cirelli, A.: *Centro de Estudios Transdisciplinarios del Agua (Universidad de Buenos Aires) Centro de Estudios Transdisciplinarios del Agua (CETA- UBA), Buenos Aires, Argentina*
Fernández, R.G.: *Sanitary Engineering of Center, Faculty of Exact Sciences, Engineering and Surveying, National University of Rosario, Argentina*
Fernández, A.E.: *SEGEMAR - Centro de Procesamiento de Minarales, Buenos Aires, Argentina*

Ferreccio, C.: *School of Medicine, Pontificia Universidad Católica de Chile, Santiago, Chile; School of Public Health, University of California, Berkeley, Berkeley, CA, USA*

Figoli, A.: *Institute on Membrane Technology ITM-CNR, Rende (CS), Italy*

Finger, A.: *Institute of Earth Surface Dynamics, University of Lausanne, Lausanne, Switzerland*

Fiuza, A.: *Centro de Recursos Naturais e Ambiente (CERENA), Polo do Porto, Universidade do Porto, Faculdade de Engenharia*

Fletcher, T.: *Centre for Radiation, Chemicals and Environmental Hazards (CRCE), Public Health England, Chilton, Didcot, Oxfordshire, UK*

Fletcher, T.: *London School of Hygiene and Tropical Medicine, London, UK*

Flores Cabrera, S.A.: *Departamento de Ciencias Básicas y Aplicadas, Universidad Nacional del Chaco Austral, Chaco, Argentina*

Folens, K.: *Laboratory of Analytical Chemistry and Applied Ecochemistry (ECOCHEM), Ghent University, Ghent, Belgium*

Franceschini, N.: *University of North Carolina, Chapel Hill, North Carolina, USA*

Francesconi, K.: *University of Graz, Graz, Austria*

Frank, A.: *Swedish University of Agricultural Sciences, Uppsala, Sweden*

Frape, S.K.: *Department of Earth and Environmental Sciences, University of Waterloo, Waterloo, Canada*

Freitas, E.: *National Institute of Science and Technology on Mineral Resources, Water and Biodiversity, INCT-Acqua, Brazil*

Frutschi, M.: *Ecole Polytechnique Federale de Lausanne, Lausanne, Switzerland*

Fry, R.C.: *University of North Carolina at Chapel Hill, Chapel Hill, North Carolina, USA*

Fuchida, S.: *National Institute of Environmental Science, Tsukuba, Japan*

Futuro, A.: *Centro de Recursos Naturais e Ambiente (CERENA), Polo do Porto, Universidade do Porto, Faculdade de Engenharia*

Gahlot, V.: *Mahavir Cancer Institute & Research Centre, Patna, Bihar, India*

Gajdosechova, Z.: *TESLA (Trace Element Speciation Laboratory), Department of Chemistry, University of Aberdeen, Aberdeen, UK*

Galdámez, A.: *Department of Chemistry, Faculty of Sciences, University of Chile, Chile*

Gamboa-Loira, B.: *Instituto Nacional de Salud Pública, Cuernavaca, México*

García, K.: *Instituto Mexicano de Tecnología del Agua, Jiutepec, México*

García, M.G.: *Centro de Investigaciones en Ciencias de la Tierra (CICTERRA), CONICET and Universidad Nacional de Córdoba, Argentina*

García-Martínez, A.: *Instituto Nacional de Salud Pública, Cuernavaca, México*

García-Vargas, G.G.: *Universidad Juárez del Estado de Durango, Gómez Palacio, Durango, México*

Gardon, J.: *IRD, Hydrosciences, Montpellier, HSM, France*

Garrido Hoyos, S.E.: *Instituto Mexicano de Tecnologia del Agua, Paseo Cuauhnahuac, Jiutepec, México*

Gasparon, M.: *The University of Queensland, School of Earth Sciences, Queensland, Australia*

Gautret, P.: *Université d'Orléans, CNRS, BRGM, ISTO, UMR 7327, Orléans, France*

Gayen, A.: *Rajiv Gandhi National Ground Water Training and Research Institute (RGNGWTRI), Raipur, Chhattisgarh, India*

Geboy, N.: *U.S. Geological Survey, Reston, VA, USA*

Geirnaert, A.: *Laboratory of Microbial Ecology and Technology (LabMET), Ghent University, Ghent, Belgium*

German, M.: *Department of Civil and Environmental Engineering, Lehigh University, Pennsylvania, USA*

Ghosh, A.K.: *A.N. College, Patna, Bihar, India*

Ghosh, A.K.: *Department of Environment and Water Management, A.N. College, Patna, India; Mahavir Cancer Institute & Research Centre, Patna, Bihar, India*

Ghosh, D.: *Department of Biological Sciences, Indian Institute of Science Education and Research Kolkata, Mohanpur, India*

Ghosh, D.: *Department of Thematic Studies- Environmental Change, Linköping University, Linköping, Sweden*

Ghosh, U.C.: *Department of Chemistry, Presidency University, Kolkata, India*

Gift, J.S.: *U.S. Environmental Protection Agency, National Center for Environmental Assessment, Research Triangle Park, USA*

Giménez , M.C.: *Departamento de Ciencias Básicas y Aplicadas, Universidad Nacional del Chaco Austral, Chaco, Argentina*

Giri, A.K.: *Molecular Genetics Division, CSIR-Indian Institute of Chemical Biology, Kolkata, India*

Giri, A.P.: *Division of Biochemical Sciences, CSIR-National Chemical Laboratory, Pune, India*
Glasauer, S.: *School of Environmental Sciences, University of Guelph, Guelph, ON, Canada*
Gnagnarella, P.: *European Oncology Institute, Milan, Italy*
Goessler, W.: *Institute of Chemistry, Analytical Chemistry, University of Graz, Graz, Austria*
González, A.: *Sanitary Engineering of Center, Faculty of Exact Sciences, Engineering and Surveying, National University of Rosario, Argentina*
González-Horta, C.: *Facultad de Ciencias Químicas, Universidad Autónoma de Chihuahua, Chihuahua, México*
González-Rodríguez, B.: *Universidad Autónoma de San Luis Potosí, San Luis Potosí, México*
Gorra, R.: *Department of Agricultural, Forest and Food Sciences, University of Torino, Grugliasco, Torino, Italy*
Goswami, R.: *Department of Environmental Science, Tezpur University, Tezpur, Assam, India*
Goswamy, S.: *School of Environmental Studies, Faculty of Interdisciplinary Studies (Law & Management), Jadavpur University, Kolkata, West Bengal, India*
Goudour, J.P.: *Université Nice Sophia Antipolis, CNRS, IRD, Observatoire de la Côte d'Azur, Géoazur, Campus Azur CNRS, Valbonne, France*
Graziano, J.H.: *Department of Environmental Health Sciences, Mailman School of Public Health, Columbia University, New York City, NY, USA*
Greger, M.: *Department of Ecology, Environment and Plant Sciences, Stockholm University, Stockholm, Sweden*
Gribble, M.: *Emory University, Atlanta, Georgia, USA*
Groenendijk, M.: *Evides Waterbedrijf N.V., Rotterdam, The Netherlands*
Grootaert, C.: *Laboratory of Food Chemistry and Human Nutrition, Ghent University, Ghent, Belgium*
Gräfe, M.: *Departamento del Manejo Suelos y Aguas, Instituto Nacional de Investigaciones Agropecuarias, Quito, Ecuador*
Guadalupe Peñaloza, L.: *Facultad de Ciencias Exactas & Consejo de Investigación, Universidad Nacional de Salta, Salta, Argentina*
Gude, J.C.J.: *Delft University of Technology, Delft, The Netherlands*
Guerrero Aguilar, A.: *University of Guanajuato, Engineering Division, Environmental Engineering Department, Guanajuato, Mexico*
Guha Mazumder, D.N.: *DNGM Research Foundation, Kolkata, India*
Guo, H.: *State Key Laboratory of Biogeology and Environmental Geology, China University of Geosciences, Beijing, P.R. China; School of Water Resources and Environment, China University of Geosciences, Beijing, P.R. China*
Guo, Y.-Q.: *Key Lab of Urban Environment and Health, Institute of Urban Environment, Chinese Academy of Sciences, Xiamen, P.R. China*
Gupta, K.: *Department of Chemistry, Presidency University, Kolkata, India*
Gupta, V.: *National Institute of Industrial Engineering (NITIE), Mumbai, India*
Gustafsson, J.P.: *Department of Soil and Environment, Swedish University of Agricultural Sciences, Uppsala, Sweden*
Gustafsson, K.: *National Food Agency, Uppsala, Sweden*
Gündüz, O.: *Dokuz Eylul University, Izmir, Turkey*
Gürleyük, H.: *Brooks Land Labs, Seattle, WA, USA*
Gvakharia, V.G.: *St. Andrew the First Called Georgian University, Tbilisi, Georgia*
Haack, K.: *Texas Biomedical Research Institute, San Antonio, Texas, USA*
Hahn-Tomer, S.: *Aquacheck GmbH, Hof, Germany; Helmholtz Centre for Environmental Research–UFZ, Leipzig, Germany*
Haimi, H.: *Department of Built Environment, School of Engineering, Aalto University, Aalto, Finland*
Halldin Ankarberg, E.: *National Food Agency, Uppsala, Sweden*
Hamadani, J.: *International Center for Diarrhoeal Disease Research in Bangladesh, Dhaka, Bangladesh*
Hamberg, R.: *Department of Civil, Environmental and Natural Resources Engineering, Division of Geosciences and Environmental Engineering, Luleå University of Technology, Luleå, Sweden*
Handziuk, E.: *Ramboll Environ, Seattle, WA, USA*
Harari, F.: *Institute of Environmental Medicine, Karolinska Institutet, Stockholm, Sweden*
Harris, H.H.: *Department of Chemistry, Adelaide University, Adelaide, South Australia, Australia*
Harvey, C.F.: *Department of Civil and Environmental Engineering, MIT, Cambridge, Massachusetts, USA*
Hasan, M.A.: *Department of Geology, Dhaka University, Dhaka, Bangladesh*
Hasan, R.: *Columbia University Arsenic Research Project, Dhaka, Bangladesh*
Hass, A.E.: *Department of Geosciences and Natural Resource Management, University of Copenhagen, Copenhagen, Denmark*

Herath, I.: *Faculty of Health, Engineering and Sciences, University of Southern Queensland, Toowoomba, Queensland, Australia; Chemical and Environmental Systems Modeling Research Group, National Institute of Fundamental Studies, Hantana Road, Kandy, Sri Lanka*

Herbert Jr., R.B.: *Department of Earth Sciences, Uppsala University, Uppsala, Sweden*

Hermosillo-Muñoz , M.C.: *Autonomous University of Chihuahua, Faculty of Agrotechnological Science, Chihuahua, Mexico*

Hernández-Mendoza, H.: *Laboratorio Nacional de Investigaciones en Forense Nuclear, Instituto Nacional de Investigaciones Nucleares, La Marquesa, Ocoyoacac, Edo. de México, México*

Hernández-Ramírez, R.U.: *Instituto Nacional de Salud Pública, Cuernavaca, México*

Heuser, S.: *Helmholtz Centre for Environmental Research–UFZ, Leipzig, Germany*

Hidalgo, M.: *Department of Chemistry, University of Girona, Campus Montilivi, Girona, Spain*

Hidalgo, M.C.: *Department of Geology, Campus Científico Tecnológico de Linares, University of Jaen, Spain*

Hidetoshi, Y.: *Graduate School of Economics, Hitotsubashi University, Kunitachi, Tokyo, Japan*

Himeno, S.: *Laboratory of Molecular Nutrition and Toxicology, Faculty of Pharmaceutical Sciences, Tokushima Bunri University, Tokushima, Japan*

Hoan, H.V.: *Hanoi University of Mining and Geology, Hanoi, Vietnam*

Hoffmann, A.H.: *Department of Geosciences and Natural Resource Management, University of Copenhagen, Copenhagen, Denmark*

Hoffmann, W.: *Vienna Zoo, Austria*

Hofman-Caris, R.: *KWR Watercycle Research Institute. Nieuwegein, The Netherlands.*

Hoinkis, J.: *Institute of Applied Research, Karlsruhe University of Applied Sciences, Karlsruhe, Germany*

Hong, C.-H.: *Department of Dermatology, Kaohsiung Veterans General Hospital, Kaohsiung, Taiwan; Department of Dermatology and Faculty of Medicine, National Yang-Ming University, Taipei, Taiwan*

Honma, T.: *Niigata Agricultural Research Institute, Nagaoka, Japan*

Hoque, B.A.: *Environment and Population Research Centre (EPRC), Dhaka, Bangladesh*

Hoque, M.A.: *Department of Civil and Environmental Engineering, Imperial College, London, UK*

Hosain, A.: *Department of Geology, University of Dhaka, Dhaka, Bangladesh*

Hosomi, M.: *Department of Chemical Engineering, Tokyo University of Agriculture and Technology, Koganei, Tokyo, Japan*

Hossain, A.: *Department of Geology, University of Dhaka, Dhaka, Bangladesh*

Hossain, K.: *Department of Biochemistry and Molecular Biology, University of Rajshahi, Rajshahi, Bangladesh*

Hossain, M.: *KTH-International Groundwater Arsenic Research Group, Department of Sustainable Development Environmental Science and Technology, KTH Royal Institute of Technology, Stockholm, Sweden*

Hossain, S.: *Department of Geology, University of Dhaka, Dhaka, Bangladesh*

Hossain, S.: *Department of Biochemistry and Molecular Biology, University of Rajshahi, Rajshahi, Bangladesh*

Hsu, K.-H.: *Department of Health Care Management, Chang-Gung University, Taoyuan, Taiwan*

Hsu, L.-I.: *Genomics Research Center, Academia Sinica, Taipei, Taiwan*

Hsueh, Y.-M.: *School of Public Health, College of Public Health and Nutrition, Taipei Medical University, Taipei, Taiwan*

Htway, S.M.: *Department of Physiology, University of Medicine, Magway, Myanmar*

Huamaní, C.: *Laboratorio Químico Toxicológico-Centro Nacional de Salud Ocupacional y Protección del Ambiente para la salud (CENSOPAS)-Instituto Nacional de Salud, Lima, Perú*

Huamaní, J.: *Laboratorio Químico Toxicológico-Centro Nacional de Salud Ocupacional y Protección del Ambiente para la salud (CENSOPAS)-Instituto Nacional de Salud, Lima, Perú*

Huang, M.C.: *University of North Carolina at Chapel Hill, Chapel Hill, North Carolina, USA*

Hube, D.: *BRGM, ISTO, UMR 7327, Orléans, France*

Hug, S.J.: *Eawag, Swiss Federal Institute of Aquatic Science and Technology, Dübendorf, Switzerland*

Huhmann, B.L.: *Department of Civil and Environmental Engineering, MIT, Cambridge, Massachusetts, USA*

Huque, S.: *Environment and Population Research Centre (EPRC), Dhaka, Bangladesh*

Hussain, I.: *Environmental Geochemistry Laboratory, Department of Environmental Sciences, Faculty of Biological Sci-ences, Quaid-i-Azam University, Islamabad, Pakistan*

Iglesias, M.: *Department of Chemistry, University of Girona, Campus de Montilivi, Girona, Spain*

Ijumulana, J.: *KTH-International Groundwater Arsenic Research Group, Department of Sustainable Development Environmental Science and Technology, KTH Royal Institute of Technology, Stockholm, Sweden; Department of Water Resource Engineering, University of Dar es Salaam, Dar es Salaam, Tanzania*

Ingallinella, A.M.: *Sanitary Engineering of Center, Faculty of Exact Sciences, Engineering and Surveying, National University of Rosario, Argentina*

Iriel, A.: *Centro de Estudios Transdisciplinarios del Agua (Universidad de Buenos Aires) Centro de Estudios Transdisciplinarios del Agua (CETA- UBA), Buenos Aires, Argentina*

Irunde, R.: *Department of Chemistry, College of Natural and Applied Sciences, University of Dar es Salaam, Dar es Salaam, Tanzania*

Ishii, H.: *University of Hawaii, Hawai, USA*

Ishikawa, S.: *Institute for Agro-Environmental Sciences, NARO, Ibaraki, Japan*

Islam, M.R.: *Department of Soil Science, Bangladesh Agricultural University, Mymensingh, Bangladesh*

Islam, M.S.: *Department of Applied Nutrition and Food Technology, Islamic University, Kushtia, Bangladesh*

Islam, S.: *Global Centre for Environmental Remediation (GCER), Faculty of Science and Information Technology, The University of Newcastle, Callaghan NSW, Australia; Cooperative Research Centre for Contamination Assessment and Remediation of the Environment (CRC CARE), The University of Newcastle, Callaghan, Australia; Department of Soil Science, Bangladesh Agricultural University, Mymensingh, Bangladesh*

Islam, T.: *Columbia University Arsenic Research Project, Dhaka, Bangladesh*

Ito, T.: *Kokusai Kogyo Co., Ltd. and Mekong Group, Tokyo, Japan*

Jacks, G.: *KTH-International Groundwater Arsenic Research Group, Department of Sustainable Development Environmental Science and Technology, KTH Royal Institute of Technology, Stockholm, Sweden*

Jagals, P.: *The University of Queensland, School of Population Health, Brisbane, Queensland, Australia*

Jakobsen, R.: *Department of Geochemistry, Geological Survey of Denmark and Greenland (GEUS), Copenhagen, Denmark*

Janiszewski, M.: *Department of Civil Engineering, School of Engineering, Aalto University, Aalto, Finland*

Jarošíková, A.: *Institute of Geochemistry, Mineralogy and Mineral Resources, Faculty of Science, Charles University in Prague, Praha, Czech Republic*

Jarsjö, J.: *Department of Physical Geography and Quaternary Geology and the Bolin Centre for Climate Research, Stockholm University, Stockholm, Sweden*

Javed, A.: *Environmental Geochemistry Laboratory, Department of Environmental Sciences, Faculty of Biological Sciences, Quaid-i-Azam University, Islamabad, Pakistan, Department of Earth and Environmental Sciences, Bahria University, Islamabad, Pakistan*

Jean, J.-S.: *Department of Earth Sciences, National Cheng Kung University, Tainan, Taiwan*

Jentner, J.: *Institute of Applied Research, Karlsruhe University of Applied Sciences, Karlsruhe, Germany*

Jewell, K.: *Department of Geology and Geophysics, Texas A&M University, College Station, TX, USA*

Jhohura, F.T.: *Research and Evaluation Division, BRAC, Mohakhali, Dhaka, Bangladesh*

Jia, X.: *Key Lab of Urban Environment and Health, Institute of Urban Environment, Chinese Academy of Sciences, Xiamen, P.R. China*

Jia, Y.: *State Key Laboratory of Biogeology and Environmental Geology, China University of Geosciences, Beijing, P.R. China; School of Water Resources and Environment, China University of Geosciences, Beijing, P.R. China*

Jiang, Y.: *State Key Laboratory of Biogeology and Environmental Geology, China University of Geosciences, Beijing, P.R. China; School of Water Resources and Environment, China University of Geosciences, Beijing, P.R. China*

Jiang, Z.: *China University of Geosciences, Wuhan, P.R. China*

Jing, C.: *Research Center for Eco-Environmental Sciences, Chinese Academy of Sciences, Beijing China*

Johannesson, K.: *Department of Earth and Environmental Sciences, Tulane University, New Orleans, USA*

John, V.: *WSSO, Public Health Engineering Department, Government of West Bengal, Kolkata, India*

Johnson, B.D.: *Vice-President: Environment, Dundee Precious Metals, Toronto, Canada*

Johnston, R.B.: *World Health Organization, Geneva, Switzerland*

Joulian, C.: *BRGM, Water, Environment and Ecotechnology Division, Environmental Biogeochemistry and Water Quality Unit, 3, Orléans, France*

Jovanović, D.D.: *Institute of Public Health of Serbia "Dr Milan Jovanović Batut", Belgrade, Serbia*

Kader, M.: *Global Centre for Environmental Remediation (GCER), The University of Newcastle, Newcastle, New South Wales, Australia; CRC for Contamination Assessment and Remediation of the Environment (CRC CARE), University of Newcastle, Newcastle, New South Wales, Australia*

Karim, M.R.: *Department of Applied Nutrition and Food Technology, Islamic University, Kushtia, Bangladesh*

Karthe, D.K.: *Department of Aquatic Ecosystem Analysis and Management, Helmholtz Centre for Environmental Research Magdeburg, Germany*

Katou, H.: *National Institute for Agro-Environmental Sciences, Tsukuba, Japan*

Kawasaki, A.: *Institute for Agro-Environmental Sciences, NARO, Ibaraki, Japan*

Kay, P.: *water@leeds, School of Geography, University of Leeds, UK*

Kazmierczak, J.: *Department of Geochemistry, Geological Survey of Denmark and Greenland (GEUS), Copenhagen, Denmark*

Kempton, P.: *Department of Geology, Kansas State University, Manhattan, Kansas, USA*

Kent, D.B.: *U.S. Geological Survey, Menlo Park, CA, USA*

Kent, Jr, J.: *Texas Biomedical Research Institute, San Antonio, Texas, USA*

Kew, G.P.: *EOH Workplace Health and Wellness, Cape Town, South Africa*

Khan, E.U.: *Department of Energy Technology, KTH Royal Institute of Technology, Stockholm, Sweden*

Khanam, S.: *Environment and Population Research Centre (EPRC), Dhaka, Bangladesh*

Khattak, J.A.: *Environmental Geochemistry Laboratory, Department of Environmental Sciences, Faculty of Biological Sciences, Quaid-i-Azam University, Islamabad, Pakistan*

Kim, C.K.: *Neo Environmental Business Company, Korea; Korea Research Institute of Standards and Science, Daejeon, Korea*

Kim, J.-Y.: *Hazardous Substance Analysis Division, Gwangju Regional FDA, Gwangju, 61012, Republic of Korea*

Kim, K.S.: *University of North Carolina at Chapel Hill, Chapel Hill, North Carolina, USA*

Kim, W.-I.: *Chemical Safety Division, National Institute of Agricultural Science, Wanju, Republic of Korea*

Kippler, M.: *Institute of Environmental Medicine, Karolinska Institutet, Stockholm, Sweden*

Kirschbaum, A.M.: *IBIGEO-CONICET, Rosario de Lerma, Salta, Argentina*

Kleja, D.B.: *Department of Soil and Environment, Swedish University of Agricultural Sciences, Uppsala, Sweden*

Knappett, P.S.K.: *Department of Geology and Geophysics, Texas A&M University, College Station, TX, USA*

Kodam, K.: *Department of Chemistry, Savitribai Phule Pune University, Pune, India*

Koen, H.: *PIDPA Department of Process Technology and Water Quality, Antwerp, Belgium*

Koen, J.: *PIDPA Department of Process Technology and Water Quality, Antwerp, Belgium*

Kokovkin, V.V.: *Nikolaev Institute of Inorganic Chemistry, Siberian Branch of Russian Academy of Sciences, Novosibirsk State University, Novosibirsk, Russia*

Kollander, B.: *National Food Agency, Uppsala, Sweden*

Kopp, J.F.: *TESLA (Trace Element Speciation Laboratory), Department of Chemistry, University of Aberdeen, Aberdeen, UK*

Kosugi, T.: *Department of Chemical Engineering, Tokyo University of Agriculture and Technology, Koganei, Tokyo, Japan*

Krohn, R.M.: *Faculty of Veterinary Medicine, University of Calgary, Calgary, Alberta, Canada*

Kruis, F.: *UNESCO-IHE Institute for Water Education, Delft, The Netherlands*

Krupp, E.: *TESLA (Trace Element Speciation Laboratory), Department of Chemistry, University of Aberdeen, Aberdeen, UK*

Kuehnelt, D.: *Institute of Chemistry, Analytical Chemistry, NAWI Graz, University of Graz, Graz, Austria*

Kumar, A.: *Mahavir Cancer Institute and Research Centre, Patna, Bihar, India*

Kumar, M.: *Department of Environmental Science, Tezpur University, Tezpur, Assam, India*

Kumar, Manoj: *School of Environmental Sciences, Jawaharlal Nehru University, New Delhi, India*

Kumar, N.: *Department of Geological Sciences, Stanford University, Stanford, USA*

Kumar, R.: *Mahavir Cancer Institute and Research Centre, Patna, Bihar, India*

Kumar, V.: *Water Action, West Champaran, India*

Kumpiene, J.: *Waste Science and Technology, Luleå University of Technology, Luleå, Sweden*

Kundu, A.K.: *Department of Chemistry, University of Kalyani, Kalyani, West Bengal, India*

Kunhikrishnan, A.: *Chemical Safety Division, National Institute of Agricultural Science, Wanju, Republic of Korea*

Kwon, S.-W.: *Department of Chemistry, Mokpo National University Muan-Gun, Chonnam, South Korea*

Königsberger, L.C.: *Chemical and Metallurgical Engineering and Chemistry, Murdoch University, Murdoch, WA, Australia*

Königsberger, E.: *Chemical and Metallurgical Engineering and Chemistry, Murdoch University, Murdoch, WA, Australia*

Labastida, I.: *Universidad Autónoma Metropolitana Unidad Azcapotzalco, Reynosa Tamaulipas, Ciudad de México, D.F., Mexico*

Lal, V.: *National Research Centre for Environmental Toxicology (Entox), The University of Queensland, Brisbane, Queensland, Australia; CRC for Contamination Assessment and Remediation of the Environment (CRC CARE), University of Newcastle, Newcastle, New South Wales, Australia*

Lamb, D.: *Global Centre for Environmental Remediation (GCER), The University of Newcastle, Newcastle, New South Wales, Australia; CRC for Contamination Assessment and Remediation of the Environment (CRC CARE), University of Newcastle, Newcastle, New South Wales, Australia*

Lan, V.M.: *Research Centre for Environmental Technology and Sustainable Development (CETASD), Hanoi University of Science, Hanoi, Vietnam*

Landberg, T.: *Department of Ecology, Environment and Plant Sciences, Stockholm University, Stockholm, Sweden*

Lara, R.H.: *Universidad Juárez del Estado de Durango, Durango, Mexico*

Larsbo, M.: *Department of Soil and Environment, Swedish University of Agricultural Sciences, Uppsala, Sweden*

Larsen, F.: *Department of Geochemistry, Geological Survey of Denmark and Greenland (GEUS), Copenhagen, Denmark*

Laston, S.: *Texas Biomedical Research Institute, San Antonio, Texas, USA*

Laxman, K.: *Water Research Center, Sultan Qaboos University, Muscat, Oman*

Le Forestier, L.: *Université d'Orléans, CNRS, BRGM, ISTO, UMR 7327, Orléans, France*

LeBlanc, D. R.: *U.S. Geological Survey, Northborough, MA, USA*

Lee, C.-H.: *Department of Dermatology, Kaohsiung Chang Gung Memorial Hospital and Chang Gung University College of Medicine, Kaohsiung, Taiwan*

Lee, E.: *University of Oklahoma Health Sciences Center, Oklahoma City, Oklahoma, USA*

Lee, J.H.: *Korea Research Institute of Standards and Science, Daejeon. Korea*

Lee, J.-F.: *Beamline 17C1, National Synchrotron Radiation Research Centre, Hsinchu, Taiwan*

Lee, J.S.: *U.S. Environmental Protection Agency, National Center for Environmental Assessment, Research Triangle Park, USA*

Lee, K.-W.: *Department of Biotechnology, Korea University, Seoul, South Korea*

Lee, S.G.: *Department of Biotechnology, Korea University, Seoul, South Korea*

Leiva, E.D.: *Departament of Inorganic Chemistry, Faculty of Chemistry, Pontificia Universidad Católica de Chile, Santiago, Chile; Department of Hydraulic and Environmental Engineering, Pontificia Universidad Católica de Chile, Santiago, Chile and CEDEUS, Centro de Desarrollo Urbano Sustentable, Chile*

Lemaire, M.: *Faculty of Veterinary Medicine, University of Calgary, Calgary, Alberta, Canada*

Leonardi, G.S.: *London School of Hygiene and Tropical Medicine, London, UK*

Lesafi, F.J.: *Department of Chemistry, College of Natural and Applied Sciences, University of Dar es Salaam, Dar es Salaam, Tanzania*

Leus, K.: *Center for Ordered Materials, Organometallics and Catalysis (COMOC), Ghent University, Ghent, Belgium*

Leyva-Ramos, R.: *Centro de Investigación y Estudios de Posgrado, Facultad de Ciencias Químicas, Universidad Autónoma de San Luis Potosí, San Luis Potosí, México*

Lhotka, M.: *University of Chemistry and Technology Prague, Prague, Czech Republic*

Li, H.: *Division of Occupational and Environmental Medicine, Lund University, Lund, Sweden*

Li, J.J.: *Department of Cellular Biology and Pharmacology, Herbert Wertheim College of Medicine, Florida International University, Miami, Florida, USA*

Li, J.N.: *Department of Chemical Engineering, Tokyo University of Agriculture and Technology, Koganei, Tokyo, Japan*

Li, P.: *China University of Geosciences, Wuhan, P.R. China*

Li, S.: *State Key Laboratory of Biogeology and Environmental Geology, China University of Geosciences, Beijing, P.R. China; School of Water Resources and Environment, China University of Geosciences, Beijing, P.R. China*

Li, X.: *Michigan Technological University, Houghton, MI, USA*

Li, Y.: *Institute of Hydrogeology and Environmental Geology, Chinese Academy of Geological Sciences, Shijiazhuang, P. R. China*

Li, Y.H.: *School of Medicine, Hangzhou Normal University, Hangzhou, Zhejiang, P.R. China*

Ligate, F.J.: *KTH-International Groundwater Arsenic Research Group, Department of Sustainable Development Environmental Science and Technology, KTH Royal Institute of Technology, Stockholm, Sweden*

Lillo, E.: *Department of Soil, Plant and Food Sciences, University of Bari "Aldo Moro", Bari, Italy*

Lim, K.-M.: *College of Pharmacy, Ewha Womans University, Seoul, Korea*

Lin, W.-X.: *Fujian Provincial Key Laboratory of Agroecological Processing and Safety Monitoring, College of Life Sciences, Fujian Agriculture and Forestry University, Fuzhou, Fujian, P.R. China*

Lipsi, M.: *Department of Geology, University of Dhaka, Dhaka, Bangladesh*

Liu, S.: *Research Center for Eco-Environmental Sciences, Chinese Academy of Sciences, Beijing China*

Liu, Y.: *Institute of Hydrogeology and Environmental Geology, Chinese Academy of Geological Sciences, Shijiazhuang, P.R. China*

Lloyd, J.R.: *School of Earth, Atmospheric and Environmental Sciences, The University of Manchester, Manchester, UK*

López-Carrillo, L.: *Instituto Nacional de Salud Pública, Cuernavaca, México*

Löv, Å.: *Department of Soil and Environment, Swedish University of Agricultural Sciences, Uppsala, Sweden*

Lu, H.: *State Key Laboratory of Pollution Control and Resource Reuse, Tongji University, Shanghai, P.R. China*

Lu, Y.: *Institute of Environmental Medicine, Karolinska Institutet, Stockholm, Sweden*

Luoma, S.: *Geological Survey of Finland, Helsinki, Finland*

Lv, Z.-M.: *Fujian Provincial Key Laboratory of Agroecological Processing and Safety Monitoring, College of Life Sciences, Fujian Agriculture and Forestry University, Fuzhou, Fujian, P.R. China*

Lythgoe, P.R.: *School of Earth, Atmospheric and Environmental Sciences and Williamson Research Centre for Molecular Environmental Science, University of Manchester, Manchester, UK*

Ma, T.: *School of Environmental Studies & State Key Laboratory of Biogeology and Environmental Geology, China University of Geosciences, Wuhan, P.R. China*

MacCluer, J.: *Texas Biomedical Research Institute, San Antonio, Texas, USA*

Machado, I.: *Analytical Chemistry. DEC, Faculty of Chemistry. Universidad de la República (UdelaR), Montevideo, Uruguay*

Machovic, V.: *University of Chemistry and Technology Prague, Prague, Czech Republic*

Madegowda, M.: *Department of Cellular Biology and Pharmacology, Herbert Wertheim College of Medicine, Florida International University, Miami, Florida, USA.*

Magnone, D.: *School of Earth, Atmospheric and Environmental Sciences and Williamson Research Centre for Molecular Environmental Science, University of Manchester, Manchester, UK*

Mahagaonkar, A.: *Center for the Environment, Indian Institute of Technology Guwahati, India*

Mahanta, C.: *Department of Civil Engineering, Indian Institute of Technology Guwahati, India*

Mahoney, J.J.: *Mahoney Geochemical Consulting LLC, Lakewood, Colorado, USA*

Maity, J.P.: *International Center for Applied Climate Studies, University of Southern Queensland, Toowomba, QLD, Australia; Department of Earth and Environmental Sciences, National Chung Cheng University, Chiayi, Taiwan*

Majumdar, A.: *IISER Kolkata, Mohanpur, Nadia, West Bengal, India*

Majumder, K.K.: *Department of Community Medicine, KPC Medical College & Hospital, Jadavpur, Kolkata, India*

Majumder, S.: *Department of Environmental Management, International Centre for Ecological Engineering, University of Kalyani, Kalyani, West Bengal, India, Department of Chemistry, University of Girona, Campus Montilivi, Girona, Spain*

Makavipour, F.: *Department of Chemistry, School of Physical, Environmental and Mathematical Science, University of New South Wales, Canberra, Australia*

Makino, T.: *Institute for Agro-Environmental Sciences, NARO, Ibaraki, Japan*

Mallavarapu, M.: *Global Centre for Environmental Remediation (GCER), The University of Newcastle, Newcastle, New South Wales, Australia; CRC for Contamination Assessment and Remediation of the Environment (CRC CARE), University of Newcastle, Newcastle, New South Wales, Australia*

Mañay, N.: *Toxicology. DEC, Faculty of Chemistry. Universidad de la República (UdelaR), Montevideo, Uruguay*

Mandal, N.: *Department of Environmental Science, Bidhan Chandra Krishi Viswavidyalaya, Mohanpur, Nadia, West Bengal, India*

Mann, K.K.: *Lady Davis Institute for Medical Research, McGill University, Montréal, Canada*

Manojlović, D.D.: *Institute of Chemistry, Technology and Metallurgy, Center of Chemistry, Belgrade, Serbia*

Marinich, L.G.: *Departamento de Ciencias Básicas y Aplicadas, Universidad Nacional del Chaco Austral, Chaco, Argentina*

Markelova, E.: *Institut des Sciences de La Terre (ISTerre), Université Grenoble Alpes and CNRS, France*

Martin, A.R.: *Department of Energy Technology, KTH Royal Institute of Technology, Stockholm, Sweden*

Martin, E.: *University of North Carolina at Chapel Hill, Chapel Hill, North Carolina, USA*

Martin, H.S.: *Universidad Nacional Autónoma de México, Instituto de Ciencias Físicas, Cuernavaca, México*

Martin, M.: *Department of Agricultural, Forest and Food Sciences, University of Torino, Grugliasco, Torino, Italy*

Martínez, J.: *Department of Mechanical and Mining Engineering, Campus Científico Tecnológico de Linares, University of Jaen, Spain*

Martinez, L.C.: *INTI Textiles - UT Comercialización y Diseño - Laboratorio químico textil, Buenos Aires, Argentina*

Masuda, H.: *Department of Geosciences, Osaka City University, Osaka, Japan*

Matoušek, T.: *Institute of Analytical Chemistry of the ASCR, v. v. i., Brno, Czech Republic*

Matsumoto, S.: *Shimane University, Shimane, Japan*

Maurice, C.: *Department of Civil, Environmental and Natural Resources Engineering, Division of Geosciences and Environmental Engineering, Luleå University of Technology, Luleå, Sweden*

McCleskey, R.B.: *United States Geological Survey, Boulder, CO, USA*

McGrory, E.: *Earth and Ocean Sciences, School of Natural Sciences, National University of Ireland, Galway, Ireland*

Meharg, A.A.: *Institute for Global Food Security, Queen's University, Belfast, Belfast, Ireland*

Meher, A.K.: *Environmental Materials Division, CSIR-NEERI, Nehru Marg, Nagpur, Maharashtra, India*

Mendez, M.A.: *University of North Carolina at Chapel Hill, Chapel Hill, North Carolina, USA*

Mendoza-Barron, J.: *Centro de Investigación y Estudios de Posgrado, Facultad de Ciencias Químicas, Universidad Autónoma de San Luis Potosí, San Luis Potosí, México*

Menzies, A.: *Universidad Católica del Norte (UCN), Geology Department, Antofagasta, Chile*

Middleton, D.R.S.: *School of Earth, Atmospheric and Environmental Sciences and Williamson Research Centre for Molecular Environmental Science, University of Manchester, Manchester, UK; Inorganic Geochemistry, Centre for Environmental Geochemistry, British Geological Survey, Nicker Hill, Keyworth, Nottinghamshire, UK*

Mihaljevič, M.: *Institute of Geochemistry, Mineralogy and Mineral Resources, Faculty of Science, Charles University in Prague, Praha, Czech Republic*

Mirlean, N.: *Geological Oceanography Laboratory, Institute of Oceanography, Federal University of Rio Grande, Rio Grande RS, Brazil*

Mishima, Y.: *JDC Corporation/ Saga University, Japan*

Mistry, S.K.: *Research and Evaluation Division, BRAC, Mohakhali, Dhaka, Bangladesh*

Miyataka, H.: *Laboratory of Molecular Nutrition and Toxicology, Faculty of Pharmaceutical Sciences, Tokushima Bunri University, Tokushima, Japan*

Mohan, D.: *School of Environmental Sciences, Jawaharlal Nehru University, New Delhi, India*

Mojumdar, S.: *WASH Specialist, UNICEF, ICO, New Delhi, India*

Mondal, P.: *Ceramic Membrane Division, CSIR-Central Glass and Ceramic Research Institute, Kolkata, India*

Montoro, L.A.: *Department of Chemistry, Universidade Federal de Minas Gerais-UFMG, Belo Horizonte, Brazil*

Morales, I.: *Instituto de Geofísica, Universidad Nacional Autónoma de México, Ciudad de México, D.F., Mexico*

Morrison, L.: *Earth and Ocean Sciences, School of Natural Sciences, National University of Ireland, Galway, Ireland*

Mtalo, F.: *Department of Water Resource Engineering, University of Dar es Salaam, Dar es Salaam, Tanzania*

Mukherjee, A.: *Department of Geology and Geophysics, Indian Institute of Technology (IIT) – Kharagpur, Kharagpur, India*

Mukherjee, Arunangshu: *Central Ground Water Board, Bhujal Bhawan, Faridabad, India*

Muñoz, C.: *Chemistry Faculty, Pontificia Universidad Católica de Chile, Santiago, Chile*
Muñoz-Lira, D.: *Department of Materials Chemistry, Faculty of Chemistry and Biology, University of Santiago, Chile*
Murathan, A.: *State Hydraulic Works, 2nd Regional Directorate, Izmir, Turkey*
Murray, J.: *IBIGEO-CONICET, Rosario de Lerma, Salta, Argentina*
Mushtaq, N.: *Environmental Geochemistry Laboratory, Faculty of Biological Sciences, Department of Environmental Sciences, Quaid-i-Azam University, Islamabad, Pakistan*
Muthusamy, S.: *The University of Queensland, National Research Centre for Environmental Toxicology (Entox), Brisbane, Queensland, Australia; CRC for Contamination Assessment and Remediation of the Environment (CRC CARE), University of Newcastle, Newcastle, New South Wales, Australia*
Myers, K.: *Department of Geology and Geophysics, Texas A&M University, College Station, TX, USA*
Myers, J.E.: *Consultant, Dundee Precious Metals, Cape Town, South Africa*
Naidu, R.: *Global Centre for Environmental Remediation (GCER), University of Newcastle, Callaghan Campus, Callaghan, NSW, Australia; Cooperative Research Centre for Contamination Assessment and Remediation of the Environment (CRC CARE), Australia*
Nakamura, K.: *National Institute for Agro-Environmental Sciences, Tsukuba, Japan*
Nam, S.-H.: *Department of Chemistry, Mokpo National University Muan-Gun, Chonnam, South Korea*
Nambiar, O.G.B.: *Former Scientist, National Chemical Laboratory, Pune, India*
Nan, X.: *School of Medicine, Hangzhou Normal University, Hangzhou, Zhejiang, P.R. China*
Nath, B.: *Lamont-Doherty Earth Observatory of Columbia University, Palisades, NY, USA*
Navas-Acien, A.: *Johns Hopkins University, Baltimore, Maryland, USA; Department of Environmental Health Sciences, Johns Hopkins University, Baltimore, USA and Department of Environmental Health Sciences, Columbia University, New York, USA*
Navin, S.: *L.N. Mithila University, Darbhanga, Bihar, India*
Neidhardt, H.: *Institute of Mineralogy and Geochemistry, Karlsruhe Institute of Technology, KIT, Karlsruhe, Germany*
Ng, G-H.: *Department of Earth Sciences, University of Minnesota, Minneapolis, MN, USA*
Ng, J.C.: *The University of Queensland, National Research Centre for Environmental Toxicology (Entox), Brisbane, Queensland, Australia; CRC for Contamination Assessment and Remediation of the Environment (CRC CARE), University of Newcastle, Newcastle, New South Wales, Australia*
Nhan, P.Q.: *Hanoi University of Mining and Geology, Hanoi, Vietnam*
Niazi, N.K.: *Southern Cross GeoScience, Southern Cross University, Lismore, NSW, Australia*
Nicome, N.R.: *Laboratory of Analytical Chemistry and Applied Ecochemistry (ECOCHEM), Ghent University, Ghent, Belgium*
Niero, L.: *Waste Science and Technology, Luleå University of Technology, Luleå, Sweden*
Nieva, N. E.: *Centro de Investigaciones en Ciencias de la Tierra (CICTERRA). CONICET, and FCEFyN Universidad Nacional de Córdoba, Córdoba, Argentina*
Nilsson, B.: *Department of Sustainable Development, Environmental Science and Engineering, KTH Royal Institute of Technology, Stockholm, Sweden*
Nóbrega, J.A.: *Group of Applied Instrumental Analysis, Department of Chemistry. Federal University of São Carlos. São Carlos, SP, Brazil*
Noller, B.N.: *The University of Queensland, Centre For Mined Land Rehabilitation, Brisbane, Queensland, Australia*
Nordstrom, D.K.: *United States Geological Survey, Boulder, CO, USA*
Norra, S.: *Institute of Mineralogy and Geochemistry, KIT, Karlsruhe, Germany*
North, K.: *University of North Carolina, Chapel Hill, North Carolina, USA*
Norton, G.J.: *Institute of Biological and Environmental Sciences, University of Aberdeen, Aberdeen, UK*
Nuñez, C.: *Chemistry Faculty, Pontificia Universidad Católica de Chile, Santiago, Chile*
Nuruzzaman, M.: *Global Centre for Environmental Remediation (GCER), Faculty of Science and Information Technology, The University of Newcastle, Callaghan NSW, Australia; CRC for Contamination Assessment and Remediation of the Environment (CRC CARE), University of Newcastle, Newcastle, New South Wales, Australia*
Ohba, H.: *Niigata Agricultural Research Institute, Nagaoka, Japan*
Ohnmar: *Department of Physiology, University of Medicine, Yangon, Myanmar*
Ohno, M.: *JDC Corporation/ Saga University, Japan*
Olopade, C.: *Departments of Health Studies, Medicine and Human Genetics and Cancer Research Center, The University of Chicago, Chicago, IL, USA*

Orani, A.M.: *Université Nice Sophia Antipolis, CNRS, IRD, Observatoire de la Côte d'Azur, Géoazur, Campus Azur CNRS, Valbonne, France; International Atomic Energy Agency, Environment Laboratories, Monaco*

Öhrvik, V.: *National Food Agency, Uppsala, Sweden*

Önnby, L.: *Department of Biotechnology, Lund University, Lund, Sweden*

Pacini, V.A.: *Center of Sanitary Engineering, Faculty of Exact Sciences, Engineering and Surveying, National University of Rosario, Argentina*

Packianathan, C.: *Department of Cellular Biology and Pharmacology, Herbert Wertheim College of Medicine, Florida International University, Miami, Florida, USA.*

Paik, M.-K.: *Chemical Safety Division, National Institute of Agricultural Science, Wanju, Republic of Korea*

Paknikar, K.: *Agharkar Research Institute (an autonomous body under the Department of Science and Technology, Government of India), Pune, India*

Panzarino, O.: *Department of Soil, Plant and Food Sciences, University of Bari "Aldo Moro", Bari, Italy*

Parada, L.A.: *Institute of Experimental Pathology, UNSa - CONICET, Argentina*

Park, J.-D.: *College of Medicine, Chung-Ang University, Seoul, Korea*

Park, M.-Y.: *Department of Chemistry, Mokpo National University Muan-Gun, Chonnam, South Korea*

Parvez, F.: *Department of Environmental Health Sciences, Mailman School of Public Health, Columbia University, New York City, NY, USA*

Pashley, R.M.: *Department of Chemistry, School of Physical, Environmental and Mathematical Science, University of New South Wales, Canberra, Australia*

Pasten, P.A.: *Department of Hydraulic and Environmental Engineering, Pontificia Universidad Católica de Chile, Santiago, Chile; CEDEUS, Centro de Desarrollo Urbano Sustentable, Chile*

Patki, S.: *Department of Chemistry, Savitribai Phule Pune University, Pune, India*

Paul, D.: *Microbial Culture Collection, National Centre for Cell Science, Pune, India*

Paunović, K.: *Institute of Hygiene and Medical Ecology, Faculty of Medicine, University of Belgrade, Belgrade, Serbia*

Pawar, S.: *Microbial Culture Collection, National Centre for Cell Science, Pune, India*

Pellizzari, E.E.: *Departamento de Ciencias Básicas y Aplicadas, Universidad Nacional del Chaco Austral, Chaco, Argentina*

Peña, J.: *Institute of Earth Surface Dynamics, University of Lausanne, Lausanne, Switzerland*

Peng, C.: *National Research Centre for Environmental Toxicology (Entox), The University of Queensland, Brisbane, Queensland, Australia; CRC for Contamination Assessment and Remediation of the Environment (CRC CARE), University of Newcastle, Callaghan, New South Wales, Australia*

Penížek, V.: *Department of Soil Science and Soil Protection, Faculty of Environmental Sciences, Czech University of Life Sciences Prague, Praha, Suchdol, Czech Republic*

Penke, Y.K.: *Materials Science Programme, IIT Kanpur, Kanpur, India*

Pérez Carrera, A.L.: *Instituto de Investigaciones en Producción Animal (INPA) UBA – CONICET, Centro de Estudios Transdisciplinarios del Agua (CETA – UBA), Facultad de Ciencias Veterinarias, Universidad de Buenos Aires, Buenos Aires, Argentina*

Persson, I.: *Department of Chemistry and Biotechnology, Swedish University of Agricultural Sciences, Uppsala, Sweden*

Peterson, J.: *Department of Geology and Geophysics, Texas A&M University, College Station, TX, USA*

Petrusevski, B.: *UNESCO-IHE Institute for Water Education, Delft, The Netherlands*

Phan, T.H.V.: *Institute of Earth Science, University of Grenoble, Grenoble, France*

Pi, K.: *School of Environmental Studies & State Key Laboratory of Biogeology and Environmental Geology, China University of Geosciences, Wuhan, P.R. China*

Pillewan, P.: *Environmental Materials Division, CSIR-NEERI, Nehru Marg, Nagpur, Maharashtra, India*

Pinheiro, F.C.: *Group of Applied Instrumental Analysis, Department of Chemistry. Federal University of São Carlos. São Carlos, SP, Brazil*

Pistón, M.: *Analytical Chemistry. DEC, Faculty of Chemistry. Universidad de la República (UdelaR), Montevideo, Uruguay*

Planer-Friedrich, B.: *Environmental Geochemistry Group, University of Bayreuth, Bayreuth, Germany*

Podder, M.S.: *Department of Chemical Engineering, Indian Institute of Technology, Roorkee. Roorkee, India*

Podgorski, J.: *Eawag, Swiss Federal Institute of Aquatic Science and Technology, Dübendorf, Switzerland*

Polya, D.A.: *School of Earth, Atmospheric and Environmental Sciences and Williamson Research Centre for Molecular Environmental Science, University of Manchester, Manchester, UK*

Porfido, C.: *Department of Soil, Plant and Food Sciences, University of Bari "Aldo Moro", Bari, Italy*

Postma, D.: *Department of Geochemistry, Geological Survey of Denmark and Greenland (GEUS), Copenhagen, Denmark*

Prasad, E.: *Megh Pyne Abhiyan, Patna, India*

Price, A.H.: *Institute of Biological and Environmental Sciences, University of Aberdeen, Aberdeen, UK*

Qiu, Y.: *Key Laboratory of Yangtze River Water Environment, Ministry of Education, Tongji University, Shanghai, P.R. China*

Qiu, Z.-Q.: *Fujian Provincial Key Laboratory of Agroecological Processing and Safety Monitoring, College of Life Sciences, Fujian Agriculture and Forestry University, Fuzhou, Fujian, P.R. China*

Quevedo, H.: *Sanitary Engineering of Center, Faculty of Exact Sciences, Engineering and Surveying, National University of Rosario, Argentina.*

Quiroga Flores, R.: *Department of Biotechnology, Lund University, Lund, Sweden*

Raab, A.: *TESLA (Trace Element Speciation Laboratory), Department of Chemistry, University of Aberdeen, Aberdeen, UK*

Rahman, A.: *Environment and Population Research Centre (EPRC), Dhaka, Bangladesh*

Rahman, M.: *Research and Evaluation Division, BRAC, Mohakhali, Dhaka, Bangladesh*

Rahman, M.: *Department of Biochemistry and Molecular Biology, University of Rajshahi, Rajshahi, Bangladesh*

Rahman, M.M.: *Global Centre for Environmental Remediation (GCER), University of Newcastle, Callaghan Campus, Callaghan, NSW, Australia; Cooperative Research Centre for Contamination Assessment and Remediation of the Environment (CRC CARE), University of Newcastle, Callaghan, Australia*

Rahman Sarker, M.M.: *Department of Agricultural Statistics, Sher-e-Bangla Agricultural University, Dhaka, Bangladesh*

Rahman, S.M.: *Institute of Environmental Medicine, Karolinska Institutet, Stockholm, Sweden*

Raju, N.J.: *School of Environmental Sciences, Jawaharlal Nehru University, New Delhi, India*

Ramanathan, A.L.: *School of Environmental Sciences, Jawaharlal Nehru University, New Delhi, India*

Ramkumar, J.: *Department of Mechanical Engineering, IIT Kanpur*

Ramos Arroyo, Y.R.: *University of Guanajuato, Engineering Division, Hydraulic Engineering Department, Guanajuato, Mexio*

Raputa, V.F.: *Institute of Computational Mathematics and Mathematical Geophysics, Siberian Branch of Russian Academy of Sciences, Novosibirsk State University, Novosibirsk, Russia*

Raqib, R.: *International Centre for Diarrhoeal Disease Research, Bangladesh (ICDDR,B), Dhaka, Bangladesh*

Rasheed, H.: *water@leeds, School of Geography, University of Leeds, UK*

Rasic-Milutinović, Z.: *Departments of Endocrinology, University Hospital Zemun, Belgrade, Serbia*

Rathnayake, S.: *Texas A&M University, College Station, TX, USA*

Rayalu, S.: *Environmental Materials Division, CSIR-NEERI, Nehru Marg, Nagpur, Maharashtra, India*

Raza, A.: *Environmental Geochemistry Laboratory, Department of Environmental Sciences, Faculty of Biological Sciences, Quaid-i-Azam University, Islamabad, Pakistan.*

Ren, B.H.: *State Key Laboratory of Urban and Regional Ecology, Research Center for Eco-Environmental Sciences, Chinese Academy of Sciences, Beijing, P.R. China*

Renac, C.: *Université Nice Sophia Antipolis, CNRS, IRD, Observatoire de la Côte d'Azur, Géoazur, Campus Azur CNRS, Valbonne, France*

Repert, D.: *U.S. Geological Survey, Boulder, CO USA*

Rey, J.: *Department of Geology, Campus Científico Tecnológico de Linares, University of Jaen, Spain*

Rhodes, K.: *Department of Geology and Geophysics, Texas A&M University, College Station, TX, USA*

Richards, L.A.: *School of Earth, Atmospheric and Environmental Sciences and Williamson Research Centre for Molecular Environmental Science, University of Manchester, Manchester, UK*

Rietveld, L.C.: *Sanitary Engineering Section, Department of Water Management, Delft University of Technology, Delft, The Netherlands*

Rikame, T.: *Department of Chemistry, Savitribai Phule Pune University, Pune, India*

Rios, R.: *Centro de Recursos Naturais e Ambiente (CERENA), Universidade do Porto, Faculdade de Engenharia, Polo do Porto, Portugal*

Ríos-Lugo, M.J.: *Universidad Autónoma de San Luis Potosí, San Luis Potosí, México*

Riya, S.: *Department of Chemical Engineering, Tokyo University of Agriculture and Technology, Koganei, Tokyo, Japan*

Rizwan, M.: *Department of Environmental Sciences and Engineering, Government College University, Faisalabad, Pakistan*

Rodríguez, C.M.: *Laboratorio de Microbiología Ambiental y Aplicada, Departamento de Química, Universidad Nacional de Colombia, Bogotá, Colombia*

Rodríguez Arce, E.: *Analytical Chemistry. DEC, Faculty of Chemistry. Universidad de la República (UdelaR), Montevideo, Uruguay.*

Rodriguez†, R.: *Instituto de Geofísica, Universidad Nacional Autónoma de México, Ciudad de México, D.F., Mexico † Deceased after submission*

Rodríguez-Díaz, R.: *Departamento de Microbiología, Escuela Nacional de Ciencias Biológicas, Instituto Politécnico Nacional, México D. F., México*

Rodriguez-Lado, L.: *Eawag, Swiss Federal Institute of Aquatic Science and Technology, Dübendorf, Switzerland*

Rojas, D.: *Department of Geology, Campus Científico Tecnológico de Linares, University of Jaen, Spain*

Román Ponce, B.: *Departamento de Microbiología, Escuela Nacional de Ciencias Biológicas, Instituto Politécnico Nacional, México D. F., México*

Romero Guzmán, E.T.: *Laboratorio Nacional de Investigaciones en Forense Nuclear, Instituto Nacional de Investigaciones Nucleares, La Marquesa, Ocoyoacac, Edo. de México, México*

Rosado, C.B.: *Universidade Federal de Viçosa, Department of Soils, Viçosa-MG, Brazil*

Rosen, B.P.: *Department of Cellular Biology and Pharmacology, Herbert Wertheim College of Medicine, Florida International University, Miami, Florida, USA*

Routh, J.: *Department of Thematic Studies- Environmental Change, Linköping University, Linköping, Sweden*

Roychowdhury, N.: *School of Environmental Studies, Faculty of Interdisciplinary Studies (Law & Management), Jadavpur University, Kolkata, West Bengal, India*

Roychowdhury, T.: *School of Environmental Studies, Faculty of Interdisciplinary Studies (Law & Management), Jadavpur University, Kolkata, West Bengal, India*

Rubio, M.A.: *Department of Environment, Faculty of Chemistry and Biology Sciences, University of Santiago, Chile*

Rubio-Andrade, M.: *Universidad Juárez del Estado de Durango, Gómez Palacio, Durango, México*

Saffi, M.H.: *DACCAR, Wazirabad, Kabul, Afghanistan*

Saha, A.: *Regional Occupational Health Center (Eastern), (ICMR), Kolkata, India*

Saha, D.: *Central Ground Water Board, Government of India, New Delhi, India*

Saha, I.: *Department of Chemistry, Sripat Singh College, Jiaganj, Murshidabad, West Bengal, India*

Sahu, S.: *Central Ground Water Board, MOWR, RD & GR, SER, Bhujal Bhawan, Khandagiri, Bhubaneswar, Odisha, India*

Salgado, V.A.: *Delft University of Technology, Sanitary Engineering Section, Delft, Netherlands*

Samanta, J.: *School of Environmental Studies, Faculty of Interdisciplinary Studies (Law & Management), Jadavpur University, Kolkata, West Bengal, India*

Sánchez-Ramírez, B.: *Facultad de Ciencias Químicas, Universidad Autónoma de Chihuahua, Chihuahua, México*

Sand, S.: *National Food Agency, Uppsala, Sweden*

Sanguinetti, G.S.: *Sanitary Engineering of Center, Faculty of Exact Sciences, Engineering and Surveying, National University of Rosario, Argentina*

Santdasani, N.T.: *WASH Officer, UNICEF, Patna, India*

Sareewan, J.: *Graduate School of Science and Engineering, Saga University, Japan*

Sarma, K.P.: *Department of Environmental Sciences, Tezpur Central University, Assam, India*

Sarwar, G.: *Columbia University Arsenic Research Project, Dhaka, Bangladesh*

Sathe, S.S.: *Department of Civil Engineering, Indian Institute of Technology Guwahati, India*

Sato, H.: *Faculty of Symbiotic Systems Science, Fukushima University and Mekong Group, Fukushima, Japan*

Schoof, R.A.: *Ramboll Environ, Seattle, WA, USA*

Schriks, M.: *KWR Watercycle Research Institute, Nieuwegein, The Netherlands*

Schwab, A.P.: *Texas A&M University, College Station, TX, USA*

SenGupta, A.K.: *Department of Civil and Environmental Engineering, Lehigh University, Pennsylvania, USA*

Sepúlveda, P.: *Department of Materials Chemistry, Faculty of Chemistry and Biology, University of Santiago, Chile*

Shaha, S.: *Exonics Technology Center, Dhaka, Bangladesh*

Shahid, M.: *Department of Environmental Sciences, COMSATS Institute of Information Technology, Vehari, Pakistan*

Shaikh, W.A.: *Department of Civil and Environmental Engineering, Birla Institute of Technology, Mesra, Ranchi, Jharkhand, India*

Shakoor, M.B.: *Institute of Soil and Environmental Sciences, University of Agriculture Faisalabad, Faisalabad, Pakistan*

Shakya, S.K.: *Environmental Specialist, School of Environmental Science and Management, Kathmandu, Nepal*

Sheshan, K.: *Arghyam, Bengaluru, India*

Shibasaki, N.: *Faculty of Symbiotic Systems Science, Fukushima University and Mekong Group, Fukushima, Japan*

Shrivas, K.: *Department of Chemistry, Guru Ghasidas Vishwavidyalaya, Bilaspur, Chhattisgarh, India*

Shrivastava, A.: *IISER Kolkata, Mohanpur, Nadia, West Bengal, India*

Shuai, P.: *Department of Geology and Geophysics, Texas A&M University, College Station, TX, USA*

Shuai, P.-Y.: *Fujian Provincial Key Laboratory of Agroecological Processing and Safety Monitoring, College of Life Sciences, Fujian Agriculture and Forestry University, Fuzhou, Fujian, P.R. China*

Shuvaeva, O.V.: *Nikolaev Institute of Inorganic Chemistry, Siberian Branch of Russian Academy of Sciences, Novosibirsk State University, Novosibirsk, Russia*

Siddik, M.A.: *Environment and Population Research Centre (EPRC), Dhaka, Bangladesh*

Siegers, W.: *KWR Watercycle Research Institute. Nieuwegein, The Netherlands*

Siegfried, K.: *Helmholtz Centre for Environmental Research–UFZ, Leipzig, Germany; Aquacheck GmbH, Hof, Germany*

Sierra, L.: *Instituto de Hidrología de Llanuras "Dr. Eduardo J. Usunoff" (IHLLA), Azul, Argentina*

Şimşek, C.: *Dokuz Eylul University, Izmir, Turkey*

Singh, A.: *Department of Research and Planning, Xavier Institute of Social Service, Ranchi, Jharkhand, India*

Singh, P.: *School of Environmental Sciences, Jawaharlal Nehru University, New Delhi, India*

Singh, S.: *IISER Kolkata, Mohanpur, Nadia, West Bengal, India*

Singh, S.: *School of Environmental Sciences, Jawaharlal Nehru University, New Delhi, India*

Singh, S.K.: *Department of Environment and Water Management, A.N. College, Patna, Magadh University, Bodh-Gaya, India*

Sjöstedt, C.: *Department of Soil and Environment, Swedish University of Agricultural Sciences, Uppsala, Sweden*

Skröder Löveborn, H.: *Institute of Environmental Medicine, Karolinska Institutet, Stockholm, Sweden*

Slack, R.: *water@leeds, School of Geography, University of Leeds, UK*

Slavkovich, V.: *Department of Environmental Health Sciences, Mailman School of Public Health, Columbia University, New York City, NY, USA*

Slokar, Y.M.: *UNESCO-IHE Institute for Water Education, Delft, The Netherlands*

Smeester, L.: *University of North Carolina at Chapel Hill, Chapel Hill, North Carolina, USA*

Smith, A.H.: *School of Public Health, University of California, Berkeley, Berkeley, CA, USA*

Smith, R.L.: *U.S. Geological Survey, Boulder, CO, USA*

Smits, J.E.G.: *Faculty of Veterinary Medicine, University of Calgary, Calgary, Alberta, Canada*

Sohel, N.: *Clinical Epidemiology and Biostatistics Department, McMaster University, Hamilton, Canada*

Son, S.-H.: *Department of Chemistry, Mokpo National University Muan-Gun, Chonnam, South Korea*

Sosa, N.N.: *Centro de Investigaciones Geológicas (CONICET-UNLP), La Plata, Argentina*

Sovann, C.: *Department of Environmental Sciences, Royal University of Phnom Penh, Cambodia*

Sozbilir, H.: *Dokuz Eylul University, Izmir, Turkey*

Sø, H.U.: *Department of Geochemistry, Geological Survey of Denmark and Greenland (GEUS), Copenhagen, Denmark*

Spagnuolo, M.: *Department of Soil, Plant and Food Sciences, University of Bari "Aldo Moro", Bari, Italy*

Speak, A.: *School of Earth, Atmospheric and Environmental Sciences and Williamson Research Centre for Molecular Environmental Science, University of Manchester, Manchester, UK*

Srivastava, P.K.: *Environmental Sciences, CSIR – National Botanical Research Institute, Lucknow, India*

Steinmaus, C.: *School of Public Health, University of California, Berkeley, Berkeley, CA, USA*

Sthiannopkao, S.: *Department of Environmental Engineering, Dong A University, Busan, Korea*

Stosnach, H.: *Bruker Nano GmbH, Berlin, Germany*

Stüben, D.: *Institute of Mineralogy and Geochemistry, KIT, Karlsruhe, Germany*

Stýblo, M.: *University of North Carolina at Chapel Hill, Chapel Hill, North Carolina, USA*

Su, C.: *School of Environmental Studies & State Key Laboratory of Biogeology and Environmental Geology, China University of Geosciences, Wuhan, P.R. China*

Suazo, J.: *Department of Materials Chemistry, Faculty of Chemistry and Biology, University of Santiago, Chile*

Suhara, H.: *JDC Corporation/ Saga University, Japan*

Sumi, D.: *Laboratory of Molecular Nutrition and Toxicology, Faculty of Pharmaceutical Sciences, Tokushima Bunri University, Tokushima, Japan*

Sumon, M.H.: *Department of Soil Science, Bangladesh Agricultural University, Mymensingh, Bangladesh*

Sun, G.: *Department of Occupational and Environmental Health, China Medical University, Shenyang, P.R. China*

Sundström, B.: *National Food Agency, Uppsala, Sweden*

Suresh Kumar, P.: *Department of Biotechnology, Lund University, Lund, Sweden*

Sültenfuß, J.: *Institute of Environmental Physics, University of Bremen, Bremen, Germany*

Szlachta, M.: *Faculty of Environmental Engineering, Wrocław University of Technology, Wrocław, Poland*

Taga, R.B.: *The University of Queensland, National Research Centre for Environmental Toxicology (Entox), Brisbane, Queensland, Australia*

Tallving, P.: *Division of Occupational and Environmental Medicine, Lund University, Lund, Sweden*

Tan, X.H.: *School of Medicine, Hangzhou Normal University, Hangzhou, Zhejiang, P.R. China*

Tapase, S.: *Department of Chemistry, Savitribai Phule Pune University, Pune, India*

Tapia, J.: *Universidad de Antofagasta (UANTOF), Mines Engineering Department, Campus Coloso, Antofagasta, Chile*

Terada, A.: *Department of Chemical Engineering, Tokyo University of Agriculture and Technology, Koganei, Tokyo, Japan*

Terzano, R.: *Department of Soil, Plant and Food Sciences, University of Bari "Aldo Moro", Bari, Italy*

Thakur, B.K.: *National Institute of Industrial Engineering (NITIE), Mumbai, India*

Theis, K.: *School of Earth, Atmospheric and Environmental Sciences and Williamson Research Centre for Molecular Environmental Science, University of Manchester, Manchester, UK*

Thorslund, J.: *Department of Physical Geography and Quaternary Geology and the Bolin Centre for Climate Research, Stockholm University, Stockholm, Sweden*

Thouin, H.: *Université d'Orléans, CNRS, BRGM, ISTO, UMR 7327, Orléans, France*

Tiberg, C.: *Department of Soil and Environment, Swedish University of Agricultural Sciences, Uppsala, Sweden*

Tirado, N.: *Institute of Genetic, School of Medicine, Universidad Mayor de San Andrés, La Paz, Bolivia*

Tirkey, P.: *Department of Civil and Environmental Engineering, Birla Institute of Technology, Mesra, Ranchi, Jharkhand, India*

Tisserand, D.: *Institute of Earth Science, University of Grenoble, Grenoble, France*

Tofail, F.: *International Center for Diarrhoeal Disease Research in Bangladesh, Dhaka, Bangladesh*

Tohmori, C.: *Laboratory of Molecular Nutrition and Toxicology, Faculty of Pharmaceutical Sciences, Tokushima Bunri University, Tokushima, Japan*

Torre, M.H.: *Inorganic Chemistry. DEC, Faculty of Chemistry. Universidad de la República (UdelaR), Montevideo, Uruguay*

Tran, A.T.K.: *Faculty of Chemical and Food Technology, HCM University of Technical Education, Vietnam*

Trang, P.T.K.: *Research Centre for Environmental Technology and Sustainable Development (CETASD), Hanoi University of Science, Hanoi, Vietnam*

Travis, T.J.: *University of Aberdeen, The Institute of Biological and Environmental Sciences, Aberdeen, Scotland*

Trung, D.: *Hanoi University of Mining and Geology, Hanoi, Vietnam*

Tsuchida, T.: *Niigata Agricultural Research Institute, Nagaoka, Japan*

Uddin, A.: *Department of Geology, University of Dhaka, Dhaka, Bangladesh*

Uheida, A.S.: *Functional Materials Division, Materials and Nano Physics Department, School of ICT, KTH Royal Institute of Technology, Kista, Stockholm, Sweden*

Ullrich, M.K.: *Environmental Geochemistry Group, University of Bayreuth, Bayreuth, Germany*

Umans, J.: *Georgetown University, Washington, DC, USA*

Vahala, R.: *Department of Built Environment, School of Engineering, Aalto University, Aalto, Finland*

Vahter, M.: *Institute of Environmental Medicine, Karolinska Institutet, Stockholm, Sweden*

Vaidya, D.: *Johns Hopkins University, Baltimore, Maryland, USA*

Valderrama , M.: *Chemistry Faculty, Pontificia Universidad Católica de Chile, Santiago, Chile*

Valles-Aragón , M.C.: *Autonomous University of Chihuahua, Faculty of Agrotechnological Science, Chihuahua, Mexico*

Van Daele, J.: *Department of Earth and Environmental Sciences, KU Leuven, Belgium*

van de Wetering, S.: *Brabant Water N.V., Hertogenbosch, The Netherlands*

Van de Wiele, T.: *Laboratory of Microbial Ecology and Technology (LabMET), Ghent University, Ghent, Belgium*

Van der Bruggen, B.: *Department of Chemical Engineering, Process Engineering for Sustainable Systems (ProCESS), KU Leuven, Leuven, Belgium*

van der Haar, M.: *Brabant Water's-Hertogenbosch, The Netherlands*

van der Meer, J.: *Department of Fundamental Microbiology, University of Lausanne, Lausanne, Switzerland*

van der Wens, P.: *Brabant Water, Magistratenlaan 200, Den Bosch, The Netherlands*

Van Der Voort, P.: *Center for Ordered Materials, Organometallics and Catalysis (COMOC), Ghent University, Ghent, Belgium*

van Dijk, T.: *Brabant Water N.V., Hertogenbosch, The Netherlands*

van Dongen, B.E.: *School of Earth, Atmospheric and Environmental Sciences and Williamson Research Centre for Molecular Environmental Science, University of Manchester, Manchester, UK*

van Geen, A.: *Lamont-Doherty Earth Observatory of Columbia University, Palisades, New York, USA*

van Genuchten, C.M.: *Institute of Earth Surface Dynamics, University of Lausanne, Lausanne, Switzerland*

van Halem, D.: *Faculty of Civil Engineering and Geosciences, Water Management Department, Delft University of Technology, Delft, The Netherlands*

van Hoorn, W.A.C.: *Sanitary Engineering Section, Department of Water Management, Delft University of Technology, Delft, The Netherlands*

van Vossen, J.: *KWR Watercycle Research Institute, Nieuwegein, The Netherlands.*

Varela, N.M.: *Departamento de. Química Analítica, Universidad Nacional del Chaco Austral, Argentina*

Vargas, I.T.: *CEDEUS, Centro de Desarrollo Urbano Sustentable, Chile*

Vásquez-Murrieta, M.S.: *Departamento de Microbiología, Escuela Nacional de Ciencias Biológicas, Instituto Politécnico Nacional, México D. F., México*

Vega, M.: *Department of Geology, Kansas State University, Manhattan, Kansas, USA*

Veloso, R.W.: *Universidade Federal de Viçosa, Department of Soils, Viçosa-MG, Brazil*

Verma, S.: *Department of Geology and Geophysics, Indian Institute of Technology (IIT) – Kharagpur, India*

Viet, P.H.: *Research Centre for Environmental Technology and Sustainable Development (CETASD), Hanoi University of Science, Hanoi, Vietnam*

Vila, M.C.: *Centro de Recursos Naturais e Ambiente (CERENA), Universidade do Porto, Faculdade de Engenharia, Polo do Porto, Portugal*

Villa, G.: *Laboratorio Químico Toxicológico-Centro Nacional de Salud Ocupacional y Protección del Ambiente para la salud (CENSOPAS)-Instituto Nacional de Salud, Lima, Perú*

Villela-Martinez, D.E.: *Centro de Investigación y Estudios de Posgrado, Facultad de Ciencias Químicas, Universidad Autónoma de San Luis Potosí, San Luis Potosí, México*

Violante, A.: *Department of Agriculture, University of Naples Federico II, Portici (Napoli), Italy*

Vithanage, M.: *Chemical and Environmental Systems Modeling Research Group, National Institute of Fundamental Studies, Kandy, Sri Lanka; International Center for Applied Climate Science, University of Southern Queensland, Toowoomba, Queensland, Australia*

von Brömssen, M.: *Department of Water Resources, Ramböll Water, Ramböll Sweden AB, Stockholm, Sweden*

Voruganti, V.S.: *University of North Carolina, Chapel Hill, North Carolina, USA*

Wai, K.M.: *Michigan Technological University, Houghton, MI, USA*

Wang, D.: *School of Public Health, China Medical University, Shenyang, P.R. China*

Wang, E.T.: *Departamento de Microbiología, Escuela Nacional de Ciencias Biológicas, Instituto Politécnico Nacional, México D. F., México*

Wang, J.: *Key Lab of Urban Environment and Health, Institute of Urban Environment, Chinese Academy of Sciences, Xiamen, P. R. China*

Wang, Y.: *School of Environmental Studies & State Key Laboratory of Biogeology and Environmental Geology, China University of Geosciences, Wuhan, P.R. China*

Wang, Y.H.: *Graduate Institute of Clinical Medicine, College of Medicine, Taipei Medical University, Taipei, Taiwan*

Watts, M.J.: *Inorganic Geochemistry, Centre for Environmental Geochemistry, British Geological Survey, Nicker Hill, Keyworth, Nottinghamshire, UK*

Weinzettel, P.A.: *Instituto de Hidrología de Llanuras "Dr. Eduardo J. Usunoff" (IHLLA), Azul, Argentina*

Williams, P.N.: *Institute for Global Food Security, Queen's University, Belfast, Belfast, Ireland*

Winke, L.H.E.: *Eawag, Swiss Federal Institute of Aquatic Science and Technology, Dübendorf, Switzerland*

Wójtowicz, P.: *Faculty of Environmental Engineering, Wrocław University of Technology, Wrocław, Poland*

Wu, C.-C.: *Department of Urology, School of Medicine, College of Medicine, Taipei Medical University, Taipei, Taiwan*

Wu, F.: *Departments of Environmental Medicine, New York University School of Medicine, New York, NY, USA*

Wu, S.: *Michigan Technological University, Houghton, MI, USA*

Xia, Q.: *National Research Centre for Environmental Toxicology (Entox), The University of Queensland, Brisbane, Queensland, Australia; CRC for Contamination Assessment and Remediation of the Environment (CRC CARE), University of Newcastle, Newcastle, New South Wales, Australia*

Xie, X.: *School of Environmental Studies & State Key Laboratory of Biogeology and Environmental Geology, China University of Geosciences, Wuhan, P.R. China*

Xiu, W.: *MOE Key Laboratory of Groundwater Circulation & Environment Evolution and School of Water Resources and Environment, China University of Geosciences, Beijing, P.R. China*

Xu, X.L.: *School of Medicine, Hangzhou Normal University, Hangzhou, Zhejiang, P.R. China*

Xu, Y.Y.: *Division of Occupational and Environmental Medicine, Lund University, Lund, Sweden*

Xue, X.-M.: *Key Lab of Urban Environment and Health, Institute of Urban Environment, Chinese Academy of Sciences, Xiamen, P.R. China*

Yan, Y.: *Key Lab of Urban Environment and Health, Institute of Urban Environment, Chinese Academy of Sciences, Xiamen, P.R. China*

Yang, G.-D.: *Fujian Provincial Key Laboratory of Agroecological Processing and Safety Monitoring, College of Life Sciences, Fujian Agriculture and Forestry University, Fuzhou, Fujian, P.R. China*

Yang, L.: *School of Medicine, Hangzhou Normal University, Hangzhou, Zhejiang, P.R. China*

Yazdani, M.: *Department of Built Environment, School of Engineering, Aalto University, Aalto, Finland*

Ye, H.P.: *College of Resources and Environmental Science, South-central University for Nationalities, Wuhan, P.R. China*

Ye, J.: *Key Lab of Urban Environment and Health, Institute of Urban Environment, Chinese Academy of Sciences, Xiamen, P.R. China*

Yonni, F.: *Facultad de CFM e Ingeniería (UCA) y Facultad de Ingeniería del Ejercito, Escuela Superior Técnica, M.N. Savio (IUE), Argentina*

Yoo, J.-H.: *Chemical Safety Division, National Institute of Agricultural Science, Wanju, Republic of Korea*

Yoshinishi, H.: *Department of Geosciences, Osaka City University, Osaka, Japan*

Younus, A.: *Environmental Geochemistry Laboratory, Faculty of Biological Sciences, Department of Environmental Sciences, Quaid-i-Azam University, Islamabad, Pakistan*

Yracheta, J.: *Missouri Breaks Industries Research, Inc, Timber Lake, South Dakota, USA*

Yu, H.-S.: *National Environmental Toxicology Center, National Health Research Institutes, Zhunan, Taiwan*

Yu, Q.: *College of Resources and Environmental Science, South-central University for Nationalities, Wuhan, P.R. China*

Yuan, R.: *State Key Laboratory of Biogeology and Environmental Geology, China University of Geosciences, Beijing, P.R. China*

Yuan, R.: *School of Water Resources and Environment, China University of Geosciences, Beijing, P.R. China*

Yuan, Z.: *School of Medicine, Hangzhou Normal University, Hangzhou, Zhejiang, P.R. China*

Yunus, M.: *International Centre for Diarrhoeal Disease Research, Dhaka, Bangladesh*

Zahid, M.A.: *Environment and Population Research Centre (EPRC), Dhaka, Bangladesh*

Zahir, S.: *Environmental Geochemistry Laboratory, Faculty of Biological Sciences, Department of Environmental Sciences, Quaid-i-Azam University, Islamabad, Pakistan*

Zaib, A.: *Institute of Chemical Sciences, Bahauddin Zakariya University Multan, Pakistan*

Zaman, A.: *Department of Agronomy, Bidhan Chandra Krishi Viswavidyalaya (BCKV), Mohanpur, West Bengal, India*

Zarate, M.: *INCITAP (CONICET-UNLPam), Santa Rosa, La Pampa, Argentina*

Zhang, D.: *State Key Laboratory of Biogeology and Environmental Geology, China University of Geosciences, Beijing, P.R. China; School of Water Resources and Environment, China University of Geosciences, Beijing, P.R. China*

Zhang, H.: *State Key Laboratory of Pollution Control and Resource Reuse, Tongji University, Shanghai, P.R. China*

Zhang, J.X.: *School of Medicine, Hangzhou Normal University, Hangzhou, Zhejiang, P.R. China*

Zhang, Q.: *Department of Occupational and Environmental Health, China Medical University, Shenyang, P.R. China*

Zhang, X.: *State Key Laboratory of Urban and Regional Ecology, Research Center for Eco-Environmental Sciences, Chinese Academy of Sciences, Beijing, P.R. China; Key Lab of Urban Environment and Health, Institute of Urban Environment, Chinese Academy of Sciences, Xiamen, P. R. China*

Zhang, Y.: *Institute of Hydrogeology and Environment al Geology, Chinese Academy of Geological Sciences, Zhengding, Hebei, P.R. China*

Zhang, Z.: *Institute of Hydrogeology and Environmental Geology, Chinese Academy of Geological Sciences, Shijiazhuang, P.R. China*

Zheng, Q.: *Department of Occupational and Environmental Health, China Medical University, Shenyang, P.R. China*

Zhou, Q.: *State Key Laboratory of Pollution Control and Resource Reuse, Tongji University, Shanghai, P.R. China*

Zhu, X.: *School of Earth Sciences and Engineering, Nanjing University, Nanjing, P.R. China*

Zhu, Y.-G.: *State Key Lab of Urban and Regional Ecology, Research Center for Eco-environmental Sciences, Chinese Academy of Sciences, P.R. China; Key Lab of Urban Environment and Health, Institute of Urban Environment, Chinese Academy of Sciences, Xiamen, P.R. China; Key Lab of Urban Environment and Health, Institute of Urban Environment, Chinese Academy of Sciences, Xiamen, P.R. China*

Zhu, Z.: *State Key Laboratory of Pollution Control and Resource Reuse, Tongji University, Shanghai, P.R. China*

Zou, F.: *University of North Carolina at Chapel Hill, Chapel Hill, North Carolina, USA*

Plenary presentations

Arsenic Research and Global Sustainability – Bhattacharya, Vahter, Jarsjö, Kumpiene,
Ahmad, Sparrenbom, Jacks, Donselaar, Bundschuh & Naidu (Eds)
© 2016 Taylor & Francis Group, London, ISBN 978-1-138-02941-5

On the spatial variation of arsenic in groundwater of the Red River floodplain, Vietnam

D. Postma[1], F. Larsen[1], R. Jakobsen[1], H.U. Sø[1], J. Kazmierczak[1], P.T.K. Trang[2], V.M. Lan[2], P.H. Viet[2], H.V. Hoan[3], D.T. Trung[3] & P.Q. Nhan[3]

[1]*Department of Geochemistry, Geological Survey of Denmark and Greenland (GEUS), Copenhagen, Denmark*
[2]*Research Centre for Environmental Technology and Sustainable Development (CETASD), Hanoi University of Science, Hanoi, Vietnam*
[3]*Hanoi University of Mining and Geology, Hanoi, Vietnam*

ABSTRACT: We have explored the origin of spatial variations in the arsenic concentration within Holocene aquifers of the Red River floodplain. On the scale of the entire floodplain the groundwater arsenic concentration is related to the sedimentary environment of the aquifer. On a smaller local scale within the floodplain, the arsenic concentration is related to the burial age of the aquifer sediment. A 1D reactive transport model was constructed to predict groundwater arsenic contents over time. Results suggest the meandering history of the Red River determines the arsenic content.

1 INTRODUCTION

The biggest unsolved problem concerning the contamination of groundwater in the floodplains of SE Asia with arsenic concerns the high spatial variability of the arsenic content. Wells with a high arsenic concentration in the groundwater may be located within a few hundreds of meters from wells containing very little arsenic without any apparent explanation. In order to obtain effective management strategies for the arsenic contamination problem, we clearly need a proper understanding of the mechanisms controlling the spatial arsenic variability. Here we use the Red River flood plain in Vietnam as the example to address this issue.

2 RESULTS AND DISCUSSION

The spatial variability of arsenic must be addressed at several scales. At the scale of the entire flood plain we can demonstrate how the arsenic distribution is connected to the sedimentary environment. Maps on the distribution of the sedimentary environment, based on satellite images (Mathers & Zalasiewicz, 1999) overlain on maps of the arsenic distribution in the groundwater (Winkel *et al.*, 2011) clearly show that marine influenced deposits, consisting of beach ridge deposits and tidal deposits are low in arsenic, while the area covered by alluvial deposits contains all the localities where a high groundwater arsenic concentration was found. The reason why marine influenced groundwater has a very low arsenic content is apparently the high sulfate concentration of seawater which under anoxic groundwater conditions becomes reduced to sulfide which reacts with sedimentary Fe-oxides and subsequently precipitates as iron sulfides containing arsenic. Within the area containing the alluvial deposits, the highest arsenic content is found where meander belts were identified on satellite images. The relation between high arsenic groundwater and meander belt deposits could either indicate that specific fluvial deposits are able to generate high arsenic groundwater or the presence of young deposits where the sediment structure can still be easily identified on satellite images and which because of their young age can generate arsenic.

Satellite images at a much smaller scale show that the surface geology of these floodplain sediments is very complex. Abandoned channels, oxbow lakes, point bar progressions are intersecting into a complex mosaic. Previous studies have shown that particularly abandoned channels may be associated with high arsenic groundwater which indicates that the sedimentary architecture may have an important effect on the groundwater arsenic content (Sahu & Saha, 2015). At a small scale we have studied localities with high and low arsenic groundwater with the objective of resolving the controlling parameters for the groundwater arsenic content (Postma *et al.*, 2007, 2012). By dating the burial age of the aquifer sediment, using optical stimulated luminescence, we could show that young sediments (400–600 years) contain groundwater with a much higher arsenic content than older (3,500–6,000 years) sediments. From this observation it follows that the geochemical properties of these very young aquifer sediments change over time with respect to their ability to mobilize arsenic from the sediment and into the groundwater. An important aspect of the system is that groundwater dating shows that most of the groundwater is generally very young (< 40 years), reflecting limited local flow systems discharging predominantly into the nearest surface water body rather than larger regional flow systems.

The changes in the geochemical properties of the aquifer sediment over time are very similar to those occurring during early diagenesis of anoxic marine sediment. Thus over the period of 6,000 years the reactivity of sedimentary organic matter was measured to decrease from 3 mM C/year to 0.2 mM C/year, and ultimately it is the reactivity of organic matter and its oxidation, coupled to the reduction of iron oxides, that controls the release of arsenic to the groundwater. Initially arsenic is present in iron oxide fraction of the sediment with an approximate As/Fe ratio of 1.2 mmol/mol. However, both the amount and the reactivity of these As-containing iron oxides will vary over time as reductive dissolution proceeds, while releasing arsenic and iron to the groundwater. Progressively the most reactive iron oxides will react first, leaving a more recalcitrant remnant. Both the rates of sedimentary organic matter oxidation and of iron oxide reduction constitute important kinetic controls on the arsenic mobilization. Furthermore, arsenic mobilized into the groundwater may adsorb to the surfaces of iron oxides as well as other sediment particles. Additional processes are the precipitation and dissolution of secondary Fe(II) phases as well as the dissolution of sedimentary calcium carbonate. All together, these chemical processes constitute a complex network of chemical reactions. We therefore constructed a 1D reactive transport model, using the code PHREE-QC-3 (Parkhurst & Appelo, 2013) that can quantify the interacting processes and is able to calculate the evolution in water chemistry, including arsenic content, over the last 6,000 years. The model is based on present day flow conditions with a vertical infiltration of 0.5 m/yr and calibrated against three sites with the dated sediment burial age.

The results show that the groundwater arsenic content increases over the first 1,200 years as the result of iron oxide reduction becoming relatively more important than methanogenesis as a pathway for organic matter oxidation. The filling of arsenic adsorption sites on the sediment surface results in an additional delay in the built-up of arsenic in the groundwater. Most of the released Fe(II) will precipitate as siderite but little arsenic is incorporated in secondary precipitates. After 1,200 years, the groundwater arsenic content will gradually decrease from a maximum of about 600 µg/L As due to diminishing reactivity of organic matter and the rate of iron oxide reduction. A delay in the decrease of the groundwater arsenic content is caused by desorption of As from the sediment. Even after 6,000 years, the arsenic concentration is approximately 20 µg/L or twice the WHO recommended maximum concentration in drinking water.

3 CONCLUSIONS

Understanding how the groundwater arsenic content depends on the burial age of the aquifer sediment, provides a powerful tool for spatially predicting the groundwater arsenic content. Intuitively one can perceive how a meandering river erodes into older deposits and thereby locates deposits of very different age next to each other and accordingly results in a large variability in the groundwater arsenic contents. Presently we use analyses of satellite data in combination with geophysical data and borehole data to derive the historical development of the river deposits and we will try to relate it to the groundwater arsenic content.

ACKNOWLEDGEMENTS

This research has been funded by the European Research Council under the ERC Advanced Grant ERG-2013-ADG. Grant Agreement Number 338972.

REFERENCES

Mathers, S., Zalasiewicz, J. 1999. Holocene sedimentary architecture of the Red River Delta, Vietnam. *J. Coastal Res.* 15: 314–325.

Parkhurst D.L., Appelo C.A.J. 2013. Description of input and examples for PHREEQC version 3—A computer program for speciation, batch-reaction, one-dimensional transport, and inverse geochemical calculations: *U.S. Geological Survey Techniques and Methods*, Book 6, Chap. A43, 497 p., available at http://pubs.usgs.gov/tm/06/a43.

Postma, D. Larsen, F., Hue, N.T.M., Duc, M.T., Viet, P.H., Nhan, P.Q., Jessen, S. 2007. Arsenic in groundwater of the Red River floodplain, Vietnam: controlling geochemical processes and reactive transport modeling. *Geochim. Cosmochim. Acta* 71: 5054–5071.

Postma, D., Larsen, F., Thai, N.T., Trang, P.T.K., Jacobsen, R., Nhan, P.Q., Long, T.V., Hung, P., Murray, A.S. 2012. Groundwater arsenic concentrations in Vietnam controlled by sediment age. *Nature Geosci.* 5: 656–661.

Sahu, S., Saha, D. 2015. Role of shallow alluvial stratigraphy and Holocene geomorphology on groundwater arsenic contamination in the Middle Ganga Plain, India. *Environ. Earth Sci.* 673: 3523–35.

Winkel, L., Kim, P.T., Lan, V.M., Stengel, C., Amini, M., Ha, N.T., Viet, P.H., Berg, M. 2011. Arsenic pollution of groundwater in Vietnam exacerbated by deep aquifer exploitation for more than a century. *P. Natl. Acad. Sci. USA (PNAS)* 108: 1246–1251.

Arsenic Research and Global Sustainability – Bhattacharya, Vahter, Jarsjö, Kumpiene,
Ahmad, Sparrenbom, Jacks, Donselaar, Bundschuh & Naidu (Eds)
© 2016 Taylor & Francis Group, London, ISBN 978-1-138-02941-5

Arsenic biotransformation: From genes to biogeochemical cycling

Y.-G. Zhu[1,2], J. Ye[1] & X.-M. Xue[1]
[1]*Key Laboratory of Urban Environment and Health, Institute of Urban Environment, Chinese Academy of
Sciences, P.R. China*
[2]*State Key Laboratory of Urban and Regional Ecology, Research Center for Eco-environmental Sciences,
Chinese Academy of Sciences, P.R. China*

ABSTRACT   Arsenic is a ubiquitous toxin in the environment and causes health problems. Part of the
arsenic biogeochemical cycle involves the biotransformations between arsenic species. These biotransformation processes are driven by enzymes encoded by various genes. Species and fate of arsenic are determined by
the combined effect of all these genes from diverse organisms under environmental conditions. In this presentation, we will provide a review on the status of knowledge on gene discoveries and their expression in relation
to arsenic biotransformation. Furthermore, arsenic biogeochemical cycling is coupled with other elements. It
is clear that the biogeochemistry of arsenic in the environment is determined by a network of genes directly
involved in arsenic biotransformation and genes involved in the cycling of other elements. Although individual
gene studies are essential to our understanding of arsenic biotransformations, more systematic approaches are
needed to elucidate behaviors of arsenicals in real environments and their underpinning mechanisms.

1 INTRODUCTION

Arsenic is a ubiquitously present environmental toxin
and enters the environment from either natural or
anthropogenic sources. It causes various health problems, such as cardiovascular disease, diabetes, neurological disorders, and lung, bladder, and skin cancers.

The toxicity, mobility, and fate of arsenic are determined by its speciation. Arsenic speciation, to some
extent, is driven by biological processes. Ultimately,
these biological processes are controlled by genes
encoding proteins and the regulation of these genes.
In addition, the bioavailability of arsenic is affected
by physicochemical properties of compounds/min-
erals present. Here we investigate some of the genes
involved in arsenic biotransformation and the effect
of other elements on arsenic fate. Although mechanisms of many individual genes have been demonstrated, we propose a more systematic approach,
including genomics, meta-genomics, transcriptomics,
and proteomics, to understand the speciation and
fate of arsenic in the environment.

2 ARSENIC BIOTRANSFORMATION
AND GENES

The majority of arsenic in terrestrial and oceanic
environments is present in inorganic forms, mostly
arsenate (As(V)) and arsenite (As(III)), and released
from the lithosphere via natural or anthropogenic
activities. Although inorganic arsenic species in the
environment are highly toxic and not essential to
living organisms, they enter living organisms by
uptake pathways for analogous essential or beneficial

nutrients. Therefore numerous studies have investigated how living organisms detoxify and transform
inorganic arsenic through oxidation, reduction, and
methylation. These biotransformations are driven by
genes. Some of these genes are reviewed below.

Microbes have been observed to oxidize arsenite.
The first identified arsenite oxidase contains two
heterologous subunits, AioA and AioB, encoded by
aioA and *aioB*, respectively (Anderson *et al.*, 1992).
An enzyme Arx has also been shown to have arsenite
oxidase activity (Zargar *et al.*, 2010).

There are various types of arsenate reductases. Arsenate respiratory reductase complex, encoded by *arrAB*
(Krafft & Macy, 1998), catalyzes energy-generating
respiratory reduction. Bacteria also use reductases to
detoxify inorganic arsenate (Zhu *et al.*, 2014). One type
of the reductases is the ArsC enzyme family represented
by the ArsC encoded in plasmid R773. Another type
of the arsenate reductases is represented by the ArsC
from *Staphylococcus aureus*. There is another family
of arsenate reductases found in eukaryotes and related
to the CDC25 phosphatases, represented by the *Sac-
charomyces cerevisiae* Acr2. Recently, an organoarsenical oxidase, ArsH, was found to confer resistance to
trivalent forms of monosodium methylarsenate and
roxarsone (Chen *et al.*, 2015).

Arsenic methylation is catalyzed by the As(III)
S-adenosylmethionine (SAM) methyltransferases
designated ArsM in microbes and AS3MT in higher
eukaryotes. The first microbial *arsM* gene was identified and characterized from *Rhodopseudomonas
palustris* (Qin *et al.*, 2006). Since then, *arsM* genes
have been characterized from various organisms.

The enzyme catalyzes demethylation of methylarsenicals was only identified and characterized very

recently. The *arsI* gene, which encodes a C-As lyase, was identified from an environmental MAs(III)-demethylating isolate, *Bacillus* sp. MD1 (Yoshinaga & Rosen, 2014).

3 ENVIRONMENTAL ARSENIC AND OMICS

Arsenic is subject to various physicochemical processes and interactions in the environment. The bioavailability of arsenic depends on soil Fe, P, organic matter, Si, Al, pH, texture, and other properties.

Arsenate is a phosphate analog and the phosphorous concentration in soil solution affects arsenic uptake and accumulation by plants. Our previous study showed that low phosphorous rice plants cause the oxidation of Fe(II) to form iron plaque on the root surface by releasing more O_2 and oxidants (Liu *et al.*, 2004). Iron oxides or hydroxides have very strong binding affinity for arsenate, and reduce arsenic uptake by rice.

Organic matter affects the arsenic fate by changing microbial communities and arsenic occurrences in soil. Our study (Huang *et al.*, 2012) revealed that organic matter addition in paddy soil significantly increased arsenic methylation and volatilization, as well as the physiologically active bacteria capable of arsenic transformation and mediated the arsenic speciation in soil and soil solution.

Microbial mediated methylation is probably the main source of methylated arsenic in plants, which likely occurs in soils prior to plant uptake. We designed primers to survey the prokaryotic *arsM* gene in 14 tested soils with wide range of arsenic concentrations (Jia *et al.*, 2013). Our results showed that microbes containing *arsM* genes were phylogenetically diverse and branched into various phyla.

In another study, we targeted genes beyond *arsM* to include *aioA*, *arrA*, and *arsC* (Zhang *et al.*, 2015). We found that these genes are mainly from rice rhizosphere bacteria and a significant correlation of gene abundance between *arsC* and *arsM*. In our study based on high-throughput sequencing and metagenomic analysis, we characterizes arsenic metabolism genes in five paddy soils with low-As contents (Xiao *et al.*, 2016). The results showed that arsenate reduction genes (*ars*) dominated in all soil samples, and significant correlation existed between the abundance of *arr*, *aio*, and *arsM* genes. These studies just began to show their potential in determining speciation and fate of arsenic in the environment. With so many genes involved in arsenic transformation, and more to be discovered, a thorough understanding of the behavior of arsenic in the environment requires quantification of all related genes. This can only be achieved by more systematic approaches, such as genomics, meta-genomics, transcriptomics, and proteomics. These "omics" approaches should not only include genes directly involved in arsenic metabolism, but also consider genes involved in metabolism of other elements that interact with arsenic.

4 CONCLUSIONS

Many genes involved in arsenic biotransformation have been well studied. These studies have greatly strengthened our understanding of the arsenic biogeochemical cycle. However, the complex nature of speciation and fate of arsenic in the environment calls for more "omics" approaches.

ACKNOWLEDGEMENTS

This work is supported by the National Natural Science foundation of China (21507125 and 31270161).

REFERENCES

Anderson, G.L., Williams, J., Hille, R. 1992. The purification and characterization of arsenite oxidase from Alcaligenes faecalis, a molybdenum-containing hydroxylase. *J. Biol. Chem.* 267: 23674–23682

Chen, J., Bhattacharjee, H., Rosen, B.P. 2015. ArsH is an organoarsenical oxidase that confers resistance to trivalent forms of the herbicide monosodium methylarsenate and the poultry growth promoter roxarsone. *Mol Microbiol.* 96: 1042–1052.

Huang, H., Jia, Y., Sun, G.X., Zhu, Y.G. 2012. Arsenic speciation and volatilization from flooded paddy soils amended with different organic matters. *Environ. Sci. Technol.* 46: 2163–2168.

Jia, Y., Huang, H., Zhong, M., Wang, F.H., Zhang, L.M., Zhu, Y.G. 2013. Microbial arsenic methylation in soil and rice rhizosphere. *Environ. Sci. Technol.* 47: 3141–3148.

Krafft, T., Macy, J.M. 1998. Purification and characterization of the respiratory arsenate reductase of Chrysiogenes arsenatis. *Eur. J. Biochem.* 255: 647–653.

Liu, W.J., Zhu, Y.G., Smith, F.A., Smith, S.E. 2004. Do phosphorus nutrition and iron plaque alter arsenate (As) uptake by rice seedlings in hydroponic culture? *New Phytol.* 162: 481–488.

Qin, J., Rosen, B.P., Zhang, Y., Wang, G., Franke, S., Rensing, C. 2006. Arsenic detoxification and evolution of trimethylarsine gas by microbial arsenite S-adenosylmethionine methyltransferase. *Proc. Natl. Acad. Sci. U.S.A.* 103: 2075–2080.

Xiao, K.-Q., Li, L.-G., Ma, L.-P., Zhang, S.-Y., Bao, P., Zhang, T, Zhu, Y.G. 2016. Metagenomic analysis revealed highly diverse microbial arsenic metabolism genes in paddy soils with low-arsenic contents. *Environ. Pollut.* 211: 1–8.

Yoshinaga, M., Rosen, B.P. 2014. A C*As lyase for degradation of environmental organoarsenical herbicides and animal husbandry growth promoters. *Proc. Natl. Acad. Sci. U.S.A.* 111: 7701–7706.

Zargar, K., Hoeft, S., Oremland, R., Saltikov, C.W. 2010. Identification of a novel arsenite oxidase gene, arxA, in the haloalkaliphilic, arsenite-oxidizing bacterium Alkalilimni-cola ehrlichii strain MLHE-1. *J. Bacteriol.* 192: 3755–3762.

Zhang, S.Y., Zhao, F.J., Sun, G.X., Su, J.Q., Yang, X.R., et al. 2015. Diversity and abundance of arsenic biotransformation genes in paddy soils from southern China. *Environ. Sci. Technol.* 49: 4138–4146.

Zhu, Y.-G., Yoshinaga, M., Zhao, F.-J., Rosen, B.P. 2014. Earth Abides Arsenic Biotransformations. *Annu. Rev. Earth Pl. Sc.* 42: 443–467.

*Arsenic Research and Global Sustainability – Bhattacharya, Vahter, Jarsjö, Kumpiene,
Ahmad, Sparrenbom, Jacks, Donselaar, Bundschuh & Naidu (Eds)*
© 2016 Taylor & Francis Group, London, ISBN 978-1-138-02941-5

Arsenic and rice: The problems and the potential solutions

A.A. Meharg

Institute for Global Food Security, Queen's University Belfast, Northern Ireland

ABSTRACT: Over the last decade it has become increasingly clear that the dominant route of inorganic
arsenic to the global population is through rice consumption, and these exposure rates are associated a
range of negative health effects according to the most up-to-date risk assessments. Rice is, in the main,
elevated in inorganic arsenic due to being grown under flooded conditions that mobilize arsenic from soil
stores. Additional pollution can further elevate inorganic arsenic in rice such as irrigation with arsenic
contaminated waters or growing on soils impacted by base and precious mining. This talk will consider
how inorganic arsenic is elevated in rice and will consider potion solutions.

1 INTRODUCTION

Rice is circa. 10-fold elevated in arsenic compared
to most other foods, and importantly, compared to
all other dietary staples (Meharg & Zhao, 2012).
As rice is the dominant dietary staple globally this
makes rice the dominant vector for human exposure to inorganic arsenic. In regions where arsenic
in drinking water is high drinking water may be the
highest exposure route, but in the main, these sub-
populations also tend to be high rice consumers.

Identifying why rice is high in inorganic arsenic
through integration of soil biogeochemical cycling
and plant physiology and genetics is essential to
develop the best mitigation strategies.

2 METHODS/EXPERIMENTAL

The presentation will give an overview of why
inorganic arsenic is problematic in rice and what
potential mitigation procedures can be developed,
and where these are appropriate. New data on the
global extent of the arsenic in rice problem will be
presented as well as new findings on human arsenic
exposure. Particular emphasis will be placed on the
most affected sub-populations, with young children being to the fore. Mitigation strategies will be
considered. These vary from paddy management,
cultivar breeding/selection, geographic sourcing
through to processing and cooking.

3 RESULTS AND DISCUSSION

The sub-populations most at risk from inorganic
arsenic exposure from rice are young children,
those on restricted diets that rely heavily on rice
and those whose diets traditionally have a large
reliance on rice.

The largest populations exposed to inorganic
arsenic through rice are those that traditionally
consume large concentrations of rice. These populations are of most concern as mitigating strategies
have to be found that work on a large-scale and
that are highly effective. Yong children from such
populations have the highest exposure to inorganic
from rice on a body weight basis.

Children under 4.5 years old consume circa.
3-fold more food on a body weight basis than adults
and thus for any given food item their arsenic exposure will be three-fold higher than an adult. The
range of rice products targeted at young children is
ever growing as rice is seen as an allergen (i.e. gluten) free food. Because of the dominance of rice
foods in the toddler market and because toddlers
consume 3-times more food per body weight than
adults specific legislation has been issued by the
EU to account for this and from the 1st of January 2016 rice based foods for young children will
have to have half the inorganic arsenic compared
to adult foods, 0.1 mg/kg versus 0.2 mg/kg. What
counts as a baby food is open to some debate and is
a loophole in these new laws. Our studies show that
40% of rice based baby foods currently on the market will be illegal from 2016 unless they are refor-
mulated. The gluten free nature of rice also means
that rice based products are now widely promoted
to those with gluten intolerances and these populations are of concern.

While plant breeding and field management
offer the greatest hope in the long-term to produce
low inorganic arsenic rice crops we are still a long
way off from achieving this. The genetics, and the
environmental interaction of these genes, the variability of soils, climate and cultivation practice all

make producing low inorganic arsenic rice highly challenging. Upland rice has low inorganic arsenic due to being grown under non-flooded conditions, but, unfortunately, cadmium export to grain is efficient under upland conditions.

Processing (milling, parboiling, cooking) offer the greatest potential for removing inorganic arsenic from rice, with up to 50% inorganic arsenic removed by milling, and a further ~70% through optimized cooking.

Sourcing low inorganic arsenic rice from within or between rice growing regions can be highly effective in removing inorganic arsenic from the diet. Globally rice varies by two-orders of magnitude in its inorganic arsenic content. For specialized baby or gluten intolerance foods this low inorganic arsenic rice is an obvious choice to produce low inorganic rice based foods.

Finally, the simplest way to decrease inorganic arsenic from the diets is to eat less rice. This is easiest in cultures that traditionally do not rely on rice, and is obviously much more challenging for rice based cultures.

4 CONCLUSIONS

To remove inorganic arsenic from rice multiple strategies must be taken. These range from simple to complex, and are dependent on regional differences in dietary culture. A combination of: alteration of farming practices, breeding for low arsenic, sourcing of rice grain, alteration of milling practice and modifying cooking practice all can be used to lower grain arsenic.

REFERENCE

Meharg, A.A., Zhao, F.J. 2012. Arsenic & Rice. Springer, 155p.

Arsenic Research and Global Sustainability – Bhattacharya, Vahter, Jarsjö, Kumpiene,
Ahmad, Sparrenbom, Jacks, Donselaar, Bundschuh & Naidu (Eds)
© 2016 Taylor & Francis Group, London, ISBN 978-1-138-02941-5

Tolerance and susceptibility to arsenic—the role of genetics

K. Broberg
Institute of Environmental Medicine, Karolinska Institutet, Stockholm, Sweden

ABSTRACT: Organisms have adapted to tolerate toxic substances in their environments, often by evolving metabolic enzymes that can efficiently detoxify the toxic compound. Inorganic arsenic is one of the most toxic substances in the environment, but many organisms, including humans, can metabolise inorganic arsenic to less toxic metabolites. This multistep process produces mono-, di-, and trimethylated arsenic metabolites, which the organism then can excrete. In this study, we performed a genome-wide association study using well-characterized arsenic metabolism phenotypes to demonstrate that *AS3MT* is likely to be the leading gene for arsenic methylation in humans. Further we identified that certain populations with long-term exposure to arsenic shows dramatic signs of selection, indicating adaptation to arsenic-rich environments. Finally, we present data from phylogenetic and epidemiologic analyses that *AS3MT* correlates with the arsenic methylation capacity and susceptibility to arsenic toxicity. Adaptation to the very toxic element arsenic is a model for how organisms have evolved in a harsh environment.

1 INTRODUCTION

Organisms have adapted to tolerate toxic chemicals in their environments, however, we know little about adaptation to toxic chemicals in humans. Exposure to inorganic arsenic is associated with multiple adverse health effects, including increased morbidity and mortality in early life (Rahman et al., 2010, 2011), cancer, cardiovascular disease and liver toxicity, and likely diabetes (IARC, 2012, Maull et al., 2012, Moon et al., 2012). In a few areas in the world, such as the Andean highlands, human populations have lived for thousands of years with drinking water contaminated with arsenic. This raises the question as to whether such populations could have adapted over time to their toxic environment.

The efficiency of arsenic metabolism strongly affects susceptibility to arsenic toxicity (Antonelli et al., 2014). In the body, cellular enzymes methylate inorganic arsenic to monomethylarsonic acid (MMA) and then dimethylarsinic acid (DMA). The fraction of arsenic present as MMA shows a positive association with arsenic toxicity, indicating that MMA is more toxic than DMA. By contrast, DMA is more readily excreted in urine and expelled from the body (Gardner et al., 2010). The fractions of arsenic metabolites in human urine vary in different populations (MMA: 2–30%) (Vahter, 2002). Indigenous populations in Andes, including the Argentinean village of San Antonio de los Cobres (SAC), show uniquely low urinary excretion of MMA.

The enzyme arsenite methyltransferse [formerly arsenic (+3 oxidation state) methyltransferase], AS3MT, plays a key role in methylation of arsenic (Engström et al., 2011). Single Nucleotide Polymorphisms (SNP) in *AS3MT* are associated with the arsenic methylation in humans as shown in several study groups, e.g. in Bangladesh, Argentina (Engstrom et al., 2011, Pierce et al., 2012). In particular, the *AS3MT* alleles associated with efficient arsenic methylation vary markedly in frequency (Schlebusch et al., 2013). Individuals from SAC and surrounding villages have higher frequencies of inferred protective *AS3MT* haplotypes that other Native American and Asian populations. This observation led to the hypothesis that natural selection has favored *AS3MT* haplotypes that associate with more efficient arsenic metabolism in populations that have lived with arsenic exposure for many generations. In this study, we performed a Genome-Wide Association Study (GWAS) using dense, genome-wide markers and well-characterized arsenic metabolism phenotypes to demonstrate that *AS3MT* is likely to be the leading gene for arsenic methylation in humans. In strong support of our hypothesis we found that in the people from SAC, the genomic region around *AS3MT* shows dramatic signs of selection, indicating adaptation to arsenic-rich environments. Further we explore by phylogenetic analysis the evolutionary origin of *AS3MT*, and relationship between *AS3MT* genotype and production of methylated arsenic metabolites in other species than humans.

2 METHODS

The study site SAC is situated at 3800 meters above sea level in the Puna area of the Andes Mountains in Argentina. Sampling of women in SAC was performed during 2004–5 and in 2008. Speciation of arsenic metabolites in urine was performed using HPLC hyphenated with hydride generation and inductively coupled plasma mass spectrometry at Institute of Environmental Medicine, Karolinska Institutet. For the GWAS study we selected non-first-degree-related women with a wide range of percentage of MMA in urine and genotyped them using Illumina chip. We compared the genetic data from SAC with genetic data from individuals from the 1000 Genomes Project in Peru and Colombia, both from populations with a presumably lower historical exposure to arsenic than in SAC.

3 RESULTS AND DISCUSSION

We used the Illumina 5M Omni chip to genotype 124 arsenic-exposed women from SAC who had a wide range of percentages of MMA in their urine. We performed GWAS on 1,258,737 filtered SNPs for all arsenic metabolism phenotypes. In a GWA scan unadjusted as well as adjusted for further covariates and population structure, we found clear associations for chromosome (chr.) 10 (MMA FDR corrected q = 0.00072; DMA q = 0.0035, unadj. analyses) and chr. 21 (MMA q = 0.019; DMA q = 0.040) in all scans involving %DMA and %MMA. Zoomed views of the peaks show many SNPs that were elevated, in particular upstream of *AS3MT* on chr. 10, where active regulatory elements are found and fewer SNPs for the peak on chr. 21.

To detect regions of the genome that may have been targets of selection, we used a statistic that captures greater than typical levels of differentiation in one population compared with two other populations: the Locus Specific Branch Length (LSBL) statistic. The LSBL values in SAC, using Peruvian and Colombian individuals from the 1000 Genomes Project data as comparative populations, revealed a strong peak on chromosome 10 in the region of *AS3MT*. This peak was absent from the LSBL scan focusing on the Peruvian population (compared with Colombian and SAC populations), despite the strong genetic similarities between the Peruvian and SAC populations (genome-wide average FST = 0.014). The SNP in the peak with the greatest LSBL value for SAC had a genome-wide percentile value of 99.99% (only 34 SNPs out of the 1,456,054 SNPs that were polymorphic in all three populations have a greater LSBL value). Among the 100 SNPs with greatest genome-wide LSBL values, 13 occurred within the particular peak near *AS3MT*. Additional evidence for positive selection for arsenic tolerance in the SAC population was revealed by the haplotype homozygosity statistic iHS. We found elevated iHS values in the *AS3MT* region for the SAC population but not in the other populations.

Further aspects of how genetics contribute to tolerance to arsenic will be discussed during the presentation, i.e.: 1) how genetic variation in *AS3MT* correlates with the tolerance to arsenic toxicity in humans; as well as the role of *AS3MT* for arsenic tolerance in other species than in humans by 2) recent analysis of the evolutionary origin of *AS3MT*.

4 CONCLUSIONS

Adaptation to an arsenic-rich habitat through the *AS3MT* gene is the first case of human selection for tolerance to a toxic chemical. Further phylogenetic and epidemiological studies support the strong role of *AS3MT* for tolerance to arsenic in nature.

ACKNOWLEDGEMENTS

This research was supported by the Swedish Research Council FORMAS, Swedish Research Council, Karolinska Institutet and the Eric Philip Sörensens stiftelse.

REFERENCES

Antonelli, R., Shao, K., Thomas, D.J., Sams, R. 2nd, Cowden, J. 2014. *AS3MT, GSTO,* and *PNP* polymorphism:impact on arsenic methylation and implications for disease susceptibility. *Environ. Res.* 132. 156–167.

Engström, K., Vahter, M., Mlakar, S.J., Concha, G., Nermell, B., Raqib, R. 2011. Polymorphisms in arsenic(+III)methyltransferase (AS3MT) strongly influence gene expression of AS3MT as well as arsenic metabolism. *Environ. Health Perspect.* 119: 182–188.

Gardner, R.M., Nermell, B., Kippler, M., Grander, M., Li, L., Ekstrom, E.C. et al. 2010. Arsenic methylation efficiency increases during the first trimester of pregnancy independent of folate status. *Reprod. Toxicol.* 31: 210–218.

IARC. 2012. *Arsenic, metals, fibres, and dusts.* Volume 100C. International Agency for Research on Cancer (World Health Organization, Lyon, France).

Maull, E.A. et al. 2012. Evaluation of the association between arsenic and diabetes: a National Toxicology Program workshop review. Environ Health Perspect. 120:1658–1670.

Moon, K. Guallar, E., Navas-Acien, A. 2012. Arsenic exposure and cardiovascular disease: an updated review. *Curr. Atheroscler Rep.* 14: 542–555.

Pierce, B.L. Kibriya, M.., Tong, L., Jasmine, F., Argos, M. et al. 2012. Genome-wide association study identifies chromosome 10q24.32 variants associated with arsenic metabolism and toxicity phenotypes in Bangladesh. *PLoS Genet.* 8(2): e1002522.

Rahman, A., Persson, L.Å., Nermell, B., El Arifeen, S., et al. 2010. Arsenic exposure and risk of spontaneous abortion, stillbirth, and infant mortality. *Epidemiology* 21: 797–804.

Rahman, A., Vahter, M., Ekström, E.C., Persson, L.Å. 2011. Arsenic exposure in pregnancy increases the risk of lower respiratory tract infection and diarrhea during infancy in Bangladesh. *Environ. Health Persp.* 119(5): 719–724.

Schlebusch, C.M., Lewis Jr., C.M., Vahter, M., Engström, K., Tito, R.Y., Obregón-Tito, A.J., Huerta, D., Polo, S.I., Medina, A.C., Brutsaert, T.D., Concha, G., Jakobsson, M., Broberg, K. 2013. Possible positive selection for an arsenic-protective haplotype in humans. *Environ. Health Perspect.* 121: 53–58.

Schlebusch, C.M., Gattepaille, L.M., Engström, K., Vahter, M., Jakobsson, M., Broberg, K. 2015. Human adaptation to arsenic rich environments. *Mol. Biol. Evol.* 32:1544–1555.

Vahter, M. 2002. Mechanisms of arsenic biotransformation. *Toxicology* 181–182: 211–217.

11

Arsenic Research and Global Sustainability – Bhattacharya, Vahter, Jarsjö, Kumpiene,
Ahmad, Sparrenbom, Jacks, Donselaar, Bundschuh & Naidu (Eds)
© 2016 Taylor & Francis Group, London, ISBN 978-1-138-02941-5

Arsenic and the 2030 Agenda for Sustainable Development

R.B. Johnston
World Health Organization, Geneva, Switzerland

ABSTRACT: The 2030 Agenda for Sustainable Development agreed by UN Member States in 2015 will drive global and national policies for the coming fifteen years. Arsenic will feature more prominently in this development agenda, both in terms of direct influence on the target for universal access to safe drinking water, and indirect impacts on the realization of other targets. For states to successfully track progress towards the new targets, surveillance and monitoring systems will need to be strengthened at national and sub-national levels. Development and strengthening of such systems will take many years, and will certainly still be in progress by the end of the Sustainable Development Goals in 2030.

1 INTRODUCTION

Global approaches to development are at a crossroads, as the Millennium Development Goals (MDGs), established by Member States of the United Nations, concluded in 2015, and a new agenda for sustainable development is being developed. An ambitious set of global goals and targets has been adopted by the United Nation General Assembly in September 2015 (UNGA, 2015), but the operationalization of this framework at the national level, and the means of monitoring progress, are still being developed. Arsenic was largely absent in the MDGs, but will take on more importance in the Sustainable Development Goals (SDGs), due to the adoption of more ambitious indicators for drinking-water services and the recognition of arsenic's links with other development targets.

2 A NEW GLOBAL FRAMEWORK

2.1 *The water goal*

Water has a higher prominence in the 2030 Agenda for Sustainable Development than it did under the MDGs, where one target under the "Environmental Sustainability" goal addressed drinking water. In the 2030 Agenda, a dedicated goal for Water and Sanitation aims to "ensure availability and sustainable management of water and sanitation for all", and addresses water in a much more holistic way, including resource protection, drinking water, wastewater management, environmental water quality, water use efficiency, and ecosystem health. Six technical targets and two 'means of implementation' targets cover the full spectrum of water in Sustainable Development Goal 6 of the 2030 Agenda (UNGA, 2015).

The target text in the MDGs and the SDGs addressing drinking water is similar, with both making reference to 'safe drinking water'. However, a key difference is in the indicators corresponding to the targets: in the MDGs safety was measured only indirectly through the proxy of 'improved sources', whereas in the SDGs water quality will also be included in the indicator.

Use of improved sources of drinking water is a valid proxy for exposure to microbiologically safe water. Although improved sources are often faecally contaminated, unimproved sources are more likely to be contaminated, and more likely to have high levels of indicator bacteria (Bain *et al.*, 2014).

Table 1. Comparison of drinking water in MDGs and SDGs.

MDGs	SDGs
Target 7c: "to halve, by 2015, the proportion of the population without sustainable access to safe drinking water and basic sanitation"	Target 6.1: "by 2030, achieve universal and equitable access to safe and affordable drinking water for all"
Indicator 7.8: proportion of population using improved drinking-water sources	Indicator 6.1.1: proportion of population using safely-managed drinking-water services (defined as an improved source of drinking water which is located on premises, available when needed and free of faecal and priority chemical contamination).

Table 2. Arsenic links with other SDG targets.

SDG target	Link with arsenic
2.2 End all forms of malnutrition	Arsenic exposure has been linked to malnutrition
2.3 Double agricultural productivity	Arsenic in irrigation water increases arsenic in soils, which lead to reductions in crop yields
3.2 Reduce under-five mortality rate and neo-natal mortality rate	Arsenic exposure has been linked to increases in child and neonatal mortality
3.4 Reduce mortality from non-communicable diseases	Arsenic causes a range of non-communicable diseases, with a heavy mortality burden due to internal cancers and cardiovascular disease
3.9 Reduce the number of deaths and illnesses from hazardous chemicals and air, water and soil pollution and contamination	Arsenic exposure in drinking water and foods is a significant cause of morality and illness

However, this proxy relationship breaks down for chemical contamination. Indeed, arsenic is more likely to be found in protected groundwater than in unimproved sources such as surface water.

The SDG indicator of 'safely managed drinking-water services' will provide a more accurate presentation of drinking-water quality than the proxy indicator used in the MDGs. The 'priority chemical contaminants' considered in the definition, at the global level, are fluoride and arsenic, because these are the two chemical which occur widely in drinking water, resulting in a substantial burden of disease globally.

Monitoring of the drinking-water target will therefore require information not only about microbial water quality, but also about arsenic and fluoride levels in drinking water.

2.2 *Arsenic in other targets*

Monitoring of arsenic will be required for tracking target 6.1, but reducing exposure to arsenic will also be essential for progress towards other SDG targets, especially within Goal 3 ("Ensure healthy lives and promote well-being for all at all ages").

3 NATIONAL FRAMEWORKS

While the global agenda of goals and targets has been agreed by the United Nations General Assembly in September 2015 (UNGA, 2015), the monitoring framework is still under development, and the way in which countries will interpret the global agenda remains to be seen.

3.1 *Global and national targets*

The UNGA agreement applies at a global level, but specifies that countries are to set their own corresponding targets (UNGA, 2015):

"Targets are defined as aspirational and global, with each Government setting its own national targets guided by the global level of ambition but taking into account national circumstances. Each Government will also decide how these aspirational and global targets should be incorporated into national planning processes, policies and strategies."

While many governments set targets for universal access to drinking water, these targets do not always include water quality. The global agenda will provide an incentive for states to include water quality in future targets.

3.2 *Monitoring*

Even when targets include measures of drinking-water quality, surveillance systems are weak, especially in low—and middle-income countries. Arsenic in drinking water has been measured in nationally-representative household surveys in Bangladesh and Ghana, but in other countries data from national authorities are sparser. China and India, in particular, have large populations potentially exposed to arsenic in drinking water (Amini et al., 2008, Rodriguez-Lado et al., 2013), and will have large impacts on progress towards Target 6.1.

A complication is that drinking-water quality standards vary from state to state. The WHO issues guidelines for drinking-water quality (WHO, 2011), but stresses that member states should set their own standards, taking into account local needs and constraints. Thus, many countries have adopted standards for arsenic in drinking water that are either less restrictive or more restrictive than the WHO provisional guideline value of 10 µg/L.

The 2030 Agenda envisages a clear shift in responsibility to the national level, where both target-setting and monitoring are the responsibility of Member States. Monitoring of all targets will be country-led and country-driven. The Agenda 2030 recognizes that monitoring systems in developing countries are currently unable to meet the demands of tracking the SDGs. Target 17.18 therefore calls for capacity-building support to developing countries to increase significantly the availability of high-quality, timely and reliable data.

The emergence of drinking-water regulatory authorities in many countries holds promise for more systematic monitoring of the quality of services accessed by populations, including the question of chemical quality of drinking water. Future global reports will increasingly draw on data collected and compiled by national regulatory authorities.

4 CONCLUSIONS

The increasing emphasis on monitoring drinking-water quality under SDG target 6.1 will raise the global profile of arsenic (and fluoride) and will require investments in national monitoring systems in order to accurately monitor populations exposed to unsafe drinking water. Improving monitoring and regulation of drinking-water services is not only a mean towards tracking SDGs, but an end in itself, as independent and robust surveillance and regulation result in improved services. Progress towards achieving Target 6.1 will contribute towards achieving other SDG targets with links to arsenic, particularly the health-related targets in Goal 3.

REFERENCES

Amini, M., Abbaspour, K.C., Berg, M., Winkel, L., Hug, S.J., Hoehn, E., Yang, H., Johnson, C.A. 2008. Statistical modeling of global geogenic arsenic contamination in groundwater. *Environ. Sci. Technol.* 42: 3669–3675.

Bain, R., Cronk, R., Wright, J., Yang, H., Slaymaker, T., Bartram, J. 2014. Fecal contamination of drinking water in low—and middle-income countries: a systematic review and meta-analysis. *PLoS Medicine.* 11: e1001644.

Rodriguez-Lado, L., Sun, G., Berg, M., Zhang, Q., Xue, H., Zheng, Q., Johnson, C.A. 2013. Groundwater arsenic contamination throughout China. *Science* 341: 866–868.

UNGA. 2015. Transforming our world: the 2030 Agenda for Sustainable Development. In: Resolution 70/1. United Nations General Assembly, New York.

WHO. 2011. Guidelines for drinking water quality, 4th edition. World Health Organization, Geneva.

Arsenic Research and Global Sustainability – Bhattacharya, Vahter, Jarsjö, Kumpiene,
Ahmad, Sparrenbom, Jacks, Donselaar, Bundschuh & Naidu (Eds)
© 2016 Taylor & Francis Group, London, ISBN 978-1-138-02941-5

Arsenic and cardiovascular disease: 100 years advancing epidemiologic research

A. Navas-Acien[1,2]
[1]*Department of Environmental Health Sciences, Johns Hopkins University, Baltimore, USA*
[2]*Department of Environmental Health Sciences, Columbia University, New York, USA*

ABSTRACT: Evidence indicates that arsenic may be a risk factor for cardiovascular disease. The cardiovascular effects of arsenic were first mentioned in reports of cases occurring in Argentina and Taiwan around 1917–1920. Ecological studies in Northern Chile advanced our understanding of the impact of adding and removing arsenic in drinking water for the rise and fall of an epidemic of cardiovascular disease in the population, and the importance of early life exposure. Prospective cohort studies from Taiwan and Bangladesh showed a dose-response relationship of chronic moderate-high arsenic exposure in drinking water with the development of coronary heart disease and maybe stroke. At low-moderate arsenic exposure levels, prospective cohort studies in the US have shown an association with coronary heart disease at levels above 10 µg/L. Evidence is needed to assess cardiovascular risk at levels below 10 µg/L, the shape of the dose-response, and the role of arsenic sources of exposure beyond groundwater.

1 INTRODUCTION

Chronic exposure to arsenic, an established poison and carcinogen, has been implicated as a risk factor for cardiovascular disease. Several biological mechanisms support the possibility that arsenic causes cardiovascular disease, including increased reactive oxygen species, endothelial dysfunction, up-regulation of inflammatory signals, enhanced thrombosis and platelet aggregation, and epigenetic modifications.

Traditionally, risk assessment agencies have considered cardiovascular outcomes only qualitatively in risk analyses to establish arsenic standards in drinking water and food. In 2013, however, the National Research Council in the US recommended the US Environmental Protection Agency (EPA) to include coronary heart disease, a common form of clinical cardiovascular disease characterized by underlying atherosclerosis, in its ongoing quantitative risk assessment of the health effects or arsenic.

The objective of this communication is to review the epidemiologic evidence evaluating the relationship between arsenic exposure and cardiovascular outcomes in populations around the world.

2 METHODS/EXPERIMENTAL

2.1 Study selection

We reviewed the epidemiologic evidence on arsenic and cardiovascular disease using previously published systematic reviews supplemented with manuscripts identified through hand search and review of the investigator's files. To provide a historical perspective, from the first reports of cardiovascular outcomes to the most advanced epidemiologic studies, all study designs were considered, including case reports, case series, ecological studies, case-control studies, cross-sectional studies, prospective cohort studies, and nested designs.

2.2 Statistical analyses

Measures of association (odds ratios, prevalence ratios, relative risks, relative hazards, comparisons of means) and their standard errors were abstracted or derived using data reported in the publications. Pooled estimates published in previous systematic reviews are also presented, while a dose-response evaluation was reported descriptively.

3 RESULTS AND DISCUSSION

3.1 Case reports, case series and ecological studies

The first report of the connection between arsenic and cardiovascular disease was possibly reported in 1917/1918 in northern Argentina, an area known to be affected by arsenic in drinking water since 1913 near Cordoba city (Jean *et al.*, 2010).

In Southwestern Taiwan, arsenic is believed to be the cause of blackfoot disease, a severe form of peripheral arterial disease characterized by thromboangeitis obliterans, severe arteriosclerosis, and high arsenic levels in the arterial wall. Blackfoot disease was first described in the medical literature in the late 1950s, however, report of cases date back to shortly after the installation of artesian wells in Southwestern Taiwan around 1920. A similar form of peripheral arterial disease had also been described in the first part of the 20th century among German vintners exposed to inorganic arsenic pesticides.

In Antofagasta, Chile, histopathologic studies of children and young adults exposed to very high levels of arsenic in drinking water (>600 µg/L) conducted in the 1970s showed fibrous intimal thickening of small and medium-sized arteries. The ecologic evaluation of myocardial infarction mortality over several decades comparing region II to V of Chile showed that myocardial infarction mortality increased shortly after the change to a water source in Antofagasta that contained extremely high arsenic levels from the late 1950s to 1970. Myocardial infarction mortality in Antofagasta also decreased shortly after the installation of an arsenic removal plant in both men and women.

Ecological studies in the US and Spain, and recent administrative cohort studies in Italy and Bangladesh have also identified higher cardiovascular mortality in communities with higher arsenic levels in drinking water, supporting the possible role of arsenic in cardiovascular disease at lower levels of exposure than those typically found in Argentina, Taiwan, Chile, and German vintners.

3.2 Prospective cohort studies

The first prospective cohort studies evaluating the association of arsenic exposure with incident cardiovascular disease were conducted in Taiwan. Two cohorts were established, one in Southwestern Taiwan and another in Northeastern Taiwan. These cohorts controlled for relevant confounders. A major limitation was the use of average drinking water arsenic. The lack of individual measures of arsenic make it possible to systematically underestimate the association with cardiovascular disease. The association of arsenic with clinical cardiovascular in those cohorts from Taiwan was most evident at levels above 500 µg/L.

A major advance in the epidemiologic investigation of arsenic related cardiovascular disease was the conduction and reporting of the Health Effect of Arsenic Longitudinal Study (HEALS), an ongoing prospective cohort study in rural Bangladesh. HEALS included arsenic exposure assessment at the individual level measured both in drinking water and urine, together with long-follow-up and detailed outcome assessment. This study found a clear increased risk of coronary heart disease and maybe stroke at arsenic levels in drinking water

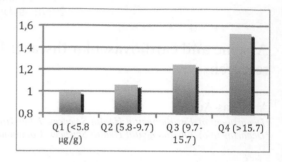

Figure 1. Hazard ratio of coronary heart disease mortality by baseline urine arsenic concentrations categorized in quartiles adjusted for demographic and cardiovascular disease risk factors in the Strong Heart Study.

above 100 µg/L. A positive dose-response relationship was also observed at levels below 100 µg/L, although power was insufficient to precisely estimate the dose-response relationship at low-moderate arsenic exposure levels.

In the United States, two prospective studies have evaluated the association of arsenic exposure with incident cardiovascular disease. The Strong Heart Study is a prospective cohort study of adult men and women from 13 American Indian communities in Arizona, Oklahoma and North and South Dakota. Arsenic levels in drinking water ranges from <10 µg/L to more than 50 µg/L. In the Strong Heart Study, baseline arsenic exposure, as measured in urine, was associated with incident coronary heart disease and stroke, with a stronger association for mortality (Fig. 1) as compared to fatal and non-fatal events combined.

The second prospective study in the US is a case-cohort study nested within the San Luis Valley Diabetes Study, conducted in rural Colorado. This study also found a positive association between arsenic levels in drinking water and coronary heart disease after adjustment for sociodemographic and cardiovascular risk factors, although the reference category included individual exposed to arsenic levels in drinking water of 20 µg/L and lower.

4 CONCLUSIONS

Epidemiologic evidence largely supports the cardiovascular effects of arsenic. Remaining questions include the characterization of the dose-response relationship at low levels of exposure, in particular <10 µg/L in drinking water and food.

ACKNOWLEDGEMENTS

Funded by R01ES021367 and R01ES025216.

REFERENCES

Chen, Y., Graziano, J.H., Parvez, F., Liu, M., Slavkovich, V., Kalra, T., Argos, M., Islam, T., Ahmed, A., Rakibuz-Zaman, M., Hasan, R., Sarwar, G., Levy, D., van Geen, A., Ahsan, H. 2011. Arsenic exposure from drinking water and mortality from cardiovascular disease in Bangladesh: prospective cohort study. *BMJ* 342: d2431.

Jean, J.S., Bundschuh, J., Chen, C.J., Guo, H.R., Liu, C.W., Lin, T.F., Chen, Y.H. 2010. The Taiwan Crisis: a showcase of the global arsenic problem. *Arsenic in the Environment.* Volume 3. CRC Press/Balkema, Leiden, 234p.

Navas-Acien, A., Sharrett, A.R., Silbergeld, E.K., Schwartz, B.S., Nachman, K.E., Burke, T.A., Guallar, E. 2005. Arsenic exposure and cardiovascular disease: a systematic review of the epidemiologic evidence. *Am. J. Epidemiol.* 162: 1037–1049.

Yuan, Y., Marshall, G., Ferreccio, C., Steinmaus, C., Liaw, J., Bates, M. et al. 2007. Acute myocardial infarction mortality in comparison with lung and bladder cancer mortality in arsenic-exposed region II of Chile from 1950 to 2000. *Am. J. Epidemiol.* 166: 1381–1391.

Section 1: Arsenic in environmental matrices and interactions (air, water, soil and biological matrices)

1.1   Origin, distribution of arsenic in groundwater systems

Arsenic Research and Global Sustainability – Bhattacharya, Vahter, Jarsjö, Kumpiene, Ahmad, Sparrenbom, Jacks, Donselaar, Bundschuh & Naidu (Eds)
© 2016 Taylor & Francis Group, London, ISBN 978-1-138-02941-5

Regional to sub-continental prediction modeling of groundwater arsenic contamination

M. Berg[1], L.H.E. Winkel[1], M. Amini[1], L. Rodriguez-Lado[1], S.J. Hug[1],
J. Podgorski[1], A. Bretzler[1], C. de Meyer[1], P.T.K. Trang[2], V.M. Lan[2], P.H. Viet[2],
G. Sun[3], Q. Zhang[3] & Q. Zheng[3]

[1]*Eawag, Swiss Federal Institute of Aquatic Science and Technology, Dübendorf, Switzerland*
[2]*Center for Environmental Technology and Sustainable Development, Hanoi University of Science, Vietnam*
[3]*Department of Occupational and Environmental Health, China Medical University, Shenyang, P.R. China*

ABSTRACT: We developed statistical risk models at regional to sub-continental scales that classify safe and unsafe areas with respect to geogenic arsenic (As) contamination for the threshold of 10 micrograms per liter. The advantage of this approach lies in the fact that predictions can even be made for regions where groundwater has not yet been tested. The resulting prediction maps are valuable resource-saving tools that guide both scientists and policy makers in initiating early mitigation measures to protect people from As-related health problems as well as to efficiently guide water resources management.

1 INTRODUCTION

Roughly one third of the world's population depends on groundwater for drinking with an estimated 10% of wells being contaminated with the natural geogenic contaminants arsenic (As) and/or fluoride. This causes severe health effects for the exposed people. The first step in any mitigation strategy is to determine the extent of the problem. Although many arsenic-affected areas have been identified and investigated in the past decades, there are still many regions around the globe where groundwater quality has not been tested.

With the aim to identify unrecognized areas at risk of groundwater As contamination, we developed probability maps by combining geological and surface environmental parameters in geostatistical models calibrated with available groundwater As concentrations from existing surveys in Bangladesh, Cambodia, Vietnam and China (Amini *et al.* 2008; Lado *et al.* 2008; Winkel *et al.* 2008a and 2011; Rodriguez-Lado *et al.* 2013). The concept of prediction modeling is illustrated in Figure 1.

2 METHODS

The first step made in developing prediction maps was an in-depth assessment of depositional environments in Southeast Asia. We then applied logistic regression analyses to evaluate and quantify relationships between surface proxies (sedimentary depositional environments, soil data and other

environmental information from digital maps) and measured groundwater As concentrations.

Geostatistical prediction modeling required three steps: i) aggregation of measured As concentrations to reduce spatial variability with the dependent variables, and binary-coding using the WHO threshold value of 10 µg/L; ii) logistic regression to obtain the weighting coefficients for the environmental proxies; and iii) calculation of the probability of As exceeding the WHO threshold. The geospatial datasets used as independent variables in e.g. the Southeast Asia model (Winkel *et al.* 2008a) were sedimentary depositional environments as a proxy for aquifer conditions and soil variables as a proxy for the drainage and chemical maturity of sediments.

3 RESULTS

3.1 *Predictions based on surface parameters (2D models)*

The predicted areas at risk of As contamination agree well with known spatial contamination patterns in Bangladesh, Cambodia, Vietnam and China with 70–85% correctly classified As concentrations and deviations of 7–15% between expected and modeled probabilities of As being ≤10 or >10 µg/L. This is an excellent result considering that neither well depths nor aquifer hydrological data were part of the modeling.

In addition, the probability maps highlight risk areas that were previously unreported or even

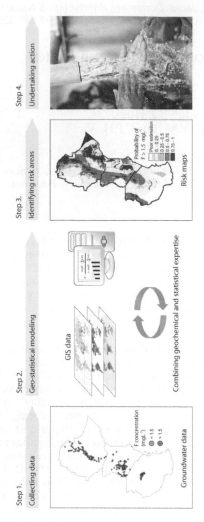

Figure 1. Geospatial modeling concept for predicting and mitigating the consumption of contaminated groundwater. Illustration using the example of fluoride contamination in the East African Rift Valley.

entirely unknown, namely in Myanmar, Sumatra and in parts of China. Ground-truthing surveys conducted thereafter in Sumatra (Winkel *et al.* 2008b), Myanmar (van Geen *et al.* 2014) and China confirmed the presence of elevated As levels.

3.2 *Three-dimensional model based on geology at depth (3D)*

All of the above is based on two-dimensional data (e.g. surface information). We have hence explored avenues of three-dimensional predictions (where depth is the 3rd dimension) in the severely arsenic-affected Red River Delta (Vietnam) where the probability of groundwater As being >10 µg/L was modeled with 3D geological information.

3.3 *Arsenic risk maps for Pakistan and Burkina Faso*

We are currently modeling probabilities of groundwater As contamination in Pakistan and Burkina Faso, two countries with very different geological and environmental conditions. Latest results of these ongoing studies will be presented.

4 OUTLOOK

Prediction modeling of geogenic As contamination is increasingly recognized as a valuable and resource-saving tool that can serve both scientists and policy makers to initiate early mitigation measures in order to protect people from As-related health problems as well as to support water resources management.

This geospatial modeling approach provides a blueprint for further modeling and prediction of As-tainted aquifers around the globe. We plan to expand these modeling capabilities for both arsenic and fluoride and create global risk maps.

REFERENCES

Amini, M., Abbaspour, K.C., Berg, M., Winkel, L., Hug, S.J., Hoehn, E., Yang, H., Johnson, C.A. 2008. Statistical Modeling of Global Geogenic Arsenic Contamination in Groundwater. *Environ. Sci. Technol.* 42: 3669–3675.

Lado, L.R., Polya, D., Winkel, L., Berg, M., Hegan, A. 2008. Modelling arsenic hazard in Cambodia—A geostatistical approach using ancillary data. *Appl. Geochem.* 23: 3010–3018.

Rodriguez-Lado, L., Sun, G., Berg, M., Zhang, Q., Xue, H., Zheng, Q, Johnson, C.A. 2013. Groundwater Arsenic Contamination Throughout China. *Science* 341: 866–868.

van Geen, A., Win, K.H., Zaw, T., Naing, W., Mey, J.L., Mailloux, B. 2014. Confirmation of elevated arsenic levels in groundwater of Myanmar. *Sci. Total Environ.* 478: 21–24.

Winkel, L., Berg, M., Amini, M., Hug, S.J., Johnson, C.A. 2008a. Predicting groundwater arsenic contamination in Southeast Asia from surface parameters. *Nature Geosci.* 1: 536–542.

Winkel, L., Berg, M., Stengel, C., Rosenberg, T. 2008b. Hydrogeological survey assessing arsenic and other groundwater contaminants in the lowlands of Sumatra, Indonesia. *Appl. Geochem.* 23: 3019–3028.

Winkel, L.H.E., Trang, P.T.K., Lan, V.M., Stengel, C., Amini, M., Ha, N.T., Viet, P.H., Berg, M. 2011. Arsenic pollution of groundwater in Vietnam exacerbated by deep aquifer exploitation for more than a century. *P. Natl. Acad. Sci. USA (PNAS)* 108: 1246–1251.

Arsenic Research and Global Sustainability – Bhattacharya, Vahter, Jarsjö, Kumpiene,
Ahmad, Sparrenbom, Jacks, Donselaar, Bundschuh & Naidu (Eds)
© *2016 Taylor & Francis Group, London, ISBN 978-1-138-02941-5*

Geochemical processes for mobilization of arsenic in groundwater

I. Herath[1], J. Bundschuh[1], M. Vithanage[2,4] & P. Bhattacharya[3,4]

[1]*Faculty of Health, Engineering and Sciences, University of Southern Queensland, Toowoomba, Queensland, Australia*
[2]*Chemical and Environmental Systems Modeling Research Group, National Institute of Fundamental Studies, Kandy, Sri Lanka*
[3]*KTH-International Groundwater Arsenic Research Group, Department of Sustainable Development, Environmental Science and Engineering, KTH Royal Institute of Technology, Stockholm, Sweden*
[4]*International Center for Applied Climate Science, University of Southern Queensland, Toowoomba, Queensland, Australia*

ABSTRACT: This paper presents an overview of natural geochemical processes that can mobilize arsenic (As) from aquifer sediments into groundwater. The primary source of As in groundwater is predominantly natural (geogenic), and can be mobilized via complicated biogeochemical interactions within various aquifer solids and water. Oxidation of sulfide minerals such as arsenopyrite, As-substituted pyrite and other sulfide minerals are susceptible to oxidation in near surface environment, and release significant quantities of As from sediments. Reductive dissolution of As-bearing Fe(III) oxides/hydroxides 2and sulfide oxidation are the most common and significant geochemical triggers that tend to release As from aquifer sediments into groundwater. Hence, strengthening direct action and implementation of the best practices to target the safe groundwater sources for installing safe drinking water wells are an urgent necessity for sustainable As mitigation on a global scale.

1 INTRODUCTION

Existence of arsenic (As) in groundwater at elevated concentrations has recently received much attention due to its serious threat for public health as well as agricultural crop productivity. Over 97% of global freshwater is groundwater, which is currently used as the main source of drinking water supply in many parts of the world (Maity *et al.*, 2012). Naturally occurring As due to geochemical processes in sedimentary aquifers may result in significant quantities of As in groundwater.

Natural geochemical processes play a vital role in the mobilization of As into the groundwater. The release of As from solid or semi-solid phases to groundwater is triggered by several geochemical processes such as adsorption/desorption, dissolution, and oxidation/reduction reactions. Iron oxide deposited in aquifer sediments is considered to be the most common source of As, which may lead to As rich groundwater (Bhattacharya *et al.*, 2002). Sulfide containing minerals also can act as a source of groundwater As. Moreover, the mobility and transformation of As in groundwater are predominately controlled by the adsorption onto metal oxyhydroxides and clay minerals, depending on the surface properties of the solid phase, pH, concentration and speciation of As, as well as the presence of competing ions and As speciation (Bundschuh & Maity, 2015). The contamination of groundwater by geogenic As is of particular concern, and hence, the understanding of major geochemical pathways that involve release of As from aquifer

sediments are an urgent necessity. Hence, this paper provides a concise description of important naturally occurring geochemical processes that involve in the mobilization of As in groundwater systems.

2 DISCUSSION

2.1 Geochemical processes to release arsenic

Natural geochemical processes tend to create the risk of drinking water contamination with elevated concentrations of As. Four major naturally occurring geochemical processes including reductive dissolution, alkali desorption, sulfide oxidation and geothermal can initiate the release of As from aquifer materials into the groundwater (Fig. 1).

The reductive dissolution of As bearing Fe(III) oxides and sulfide oxidation in aquifer sediments have

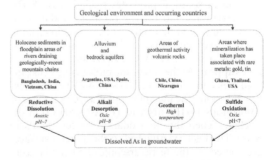

Figure 1. Naturally occurring geochemical processes that involve in releasing As into groundwater.

been recognized as the most significant geochemical triggers in order to release As into groundwater.

2.2 Reductive dissolution of iron oxide

Iron hydroxides is the most prominent in the aquifer sediments. Biologically mediated reactions in the presence of organic carbon generated from sedimentary organic matter and anthropogenic organic compounds are capable of releasing As from iron oxide. This process is referred to as dissimilatory iron reduction (Welch et al., 2000). Dissimilatory iron-reducing bacterium, Shewanella alga promotes the release of As(V) from a crystalline mineral scorodite ($FeAsO_4 \cdot 2H_2O$) via dissimilatory reduction of Fe(III) to Fe(II). The primary mechanism for the release of As from soil and sediments is the competition between As and other anions present in sorption sites. In aqueous medium, the reaction of iron oxides can be stimulated depending on the pH and at high pH values desorption processes become stronger and thereby resulting high levels of As in the groundwater. Furthermore, sulfides, particularly H_2S, accumulated in sediments can also reductively dissolve As bearing iron oxide minerals releasing As into groundwater at large extent.

2.3 Oxidation of sulfide minerals

The primary source of As in the environment is the oxidation of As-bearing sulfides, arsenopyrite (FeAsS) and pyrite (FeS_2). The FeAsS is the most common As bearing mineral, and FeS_2 is commonly found associated with sulfidic ore deposits which have severe implication on acid mine drainage.

The FeAsS can be oxidized by Fe(III) and this process is catalyzed in the presence of microorganisms including T. ferrooxidans. The first step of this process involves a strong adhesion of bacteria onto mineral surfaces, the appearance of a surface phase of elemental sulfur, the weak solubilization of Fe(II), As(III), and As(V) and the presence of the first corrosion patterns, which follow the fragility zones and the crystallographic orientation of mineral grains (Marcos et al., 1995). Growth of the unattached bacteria in solution promotes oxidation of Fe(II) to Fe(III) ions, and thereby promote oxidation of sulfides regenerating Fe(II). At the same time, a bioleaching cycle takes place and a coarse surface phase of ferric arsenate ($FeAsO_4.2H_2O$) and deep ovoid pores appear. At the end of the bioleaching cycle, the high concentration of Fe(III) and As(V) in the solution promotes the precipitation of a second phase of amorphous Fe(III) arsenate ($FeAsO_4.4H_2O$) in the leachate. Then the bio-oxidation process tends to cease. The bacteria attached to the mineral surfaces are coated by the ferric arsenates and the concentration of Fe(III) on the leachate tends to decrease strongly (Marcos et al., 1995).

Oxidation of FeS_2 takes place via several reaction pathways and the first step involves the chemical oxidation of pyrite in the presence of dissolved oxygen (reaction 1).

$$2FeS_2 + 7O_2 + 2H_2O \rightarrow 2Fe^{2+} + 4SO_4^{2-} + 4H^+ \quad (1)$$

The second possible pathway is the oxidation of FeS_2 by aqueous nitrate at pH>5. In the groundwater, nitrates can act as an oxidant for the oxidation of sulfide minerals deposited in aquifer sediments (reaction 2).

$$5FeS_2 + 14NO_3^- + 4H^+ \rightarrow 5Fe^{2+} + 10SO_4^{2-} + 7N_2 + 2H_2O \quad (2)$$

The oxidation of Fe(II) to Fe(III) and As adsorption on iron hydroxides depends on the Dissolved Organic Carbon (DOC) and pH in water aquifers. Waters having neutral pH, the majority of Fe(III) can exist in the solid or colloidal phase which leads the release and transport of As in drinking water aquifers. At near-neutral pH, dissolved Fe(III) would not be able to present at high concentrations as long as it is not interacted with DOC and would consequently be a less important oxidant than dissolved oxygen.

3 CONCLUSIONS

The primary source of As is natural, which is predominately derived from the interactions between groundwater and aquifer sediments of minerals including pyrite, arsenopyrite, and other sulfide minerals. The geochemical behavior of As is generally governed by its multiple oxidation status, speciation and redox transformations. Mobilization of As from aquifer sediments into the groundwater is mainly governed by reductive dissolution of iron oxide and oxidation of arsenopyrite and pyrite minerals.

REFERENCES

Bhattacharya, P., Frisbie, S.H., Smith, E., Naidu, R., Jacks, G., Sarkar B. 2002. Arsenic in the Environment: A Global Perspective. In: B.Sarkar (Ed.) Handbook of Heavy Metals in the Environment,. Marcell Dekker, NY, 147–215p.

Bundschuh, J., Maity, J.P. 2015. Geothermal arsenic: Occurrence, mobility and environmental implications. Renew. Sust. Energ. Rev. 42: 1214–1222.

Marcos, G.M.F. Mustin, C., Donato, P., Barres, O., Marion, P., Berthelin, J. 1995. Occurrences at mineral–bacteria interface during oxidation of arsenopyrite by Thiobacillus ferrooxidans. Biotechnol. Bioeng. 46(1): 13–21.

Maity, J.P. Nath, B. Kar, S. Chen, C.Y. Banerjee, S. Jean, J.S. Liu, M.Y. Centeno, J.A. Bhattacharya. P., Chang, C.L. 2012. Arsenic-induced health crisis in peri-urban Moyna and Ardebok villages, West Bengal, India: an exposure assessment study. Environ. Geochem. Hlth. 34(5): 563–574.

Welch, A.H. Westjohn, D. Helsel D.R. & Wanty R.B. 2000. Arsenic in ground water of the United States: occurrence and geochemistry. Groundwater 38(4): 589–604.

Arsenic Research and Global Sustainability – Bhattacharya, Vahter, Jarsjö, Kumpiene,
Ahmad, Sparrenbom, Jacks, Donselaar, Bundschuh & Naidu (Eds)
© *2016 Taylor & Francis Group, London, ISBN 978-1-138-02941-5*

Sewage disposal, petroleum spills, eutrophic lakes, and wastewater from oil and gas production: Potential drivers of arsenic mobilization in the sub-surface

D.B. Kent[1], M.S. Blondes[2], I.M. Cozzarelli[2], N. Geboy[2], D.R. LeBlanc[3], G-H. Ng[4],
D. Repert[5] & R.L. Smith[5]
[1] *U.S. Geological Survey, Menlo Park, CA, USA*
[2] *U.S. Geological Survey, Reston, VA, USA*
[3] *U.S. Geological Survey, Northborough, MA, USA*
[4] *Department of Earth Sciences, University of Minnesota, Minneapolis, MN, USA*
[5] *U.S. Geological Survey, Boulder, CO, USA*

ABSTRACT: Organic carbon from treated-sewage disposal, hyper-eutrophic lake sediments, and petroleum hydrocarbon spills can mobilize naturally occurring arsenic (As) by stimulating microbial reductive dissolution of Fe(III) oxyhydroxides These observations are stimulating investigations of the potential for past and present intentional or accidental releases of wastewater from oil and gas production to mobilize As in the subsurface.

1 INTRODUCTION

Arsenic (As) can be mobilized in the sub-surface when Fe(III) oxyhydroxides with sorbed As(V) dissolves under anaerobic, Fe-reducing conditions caused by input of bioavailable organic carbon. We illustrate similarities between sites where As mobilization results from inputs of organic carbon derived from sewage disposal, organic-rich sediments in a hyper-eutrophic lake, and petroleum hydrocarbon spills and explore the potential for sub-surface As contamination and mobilization by wastewater from oil and gas production.

2 METHODS

2.1 *Groundwater sampling and analysis*

Groundwater samples from the USGS Cape Cod research site were collected, processed, and preserved following the detailed protocols in Savoie *et al.* (2004). Field parameters were determined using standard methods (Savoie *et al.*, 2004). Concentrations of inorganic elements, including As, were determined by inductively coupled atomic emission spectroscopy (ICP-AES) and quadrupole inductively coupled atomic mass spectrometry ICP-MS after eliminating chloride (Kent & Fox, 2004).

2.2 *Produced water*

Arsenic and other elemental concentrations in water associated with natural gas and petroleum production (produced water) were obtained from an extensive database of produced water compositions in the United States (Blondes *et al.*, 2015).

3 RESULTS AND DISCUSSION

3.1 *Arsenic mobilization from domestic wastewater disposal*

Synoptic sampling conducted in 2000 and 2002 of a contaminant plume caused by land-disposal of secondary domestic wastewater effluent to a sand glacial outwash aquifer on Cape Cod, Massachusetts, USA (MA) showed As concentrations exceeding the WHO drinking water limit of 0.13 μM, (~10 μg/L). Arsenic concentrations up to 0.2 μM were observed along with 100–500 μM dissolved Fe(II) in the anaerobic zone of the plume (Kent and Fox, 2004). Background sediments had total As concentrations in the range 10–30 nmol/g sediment (~0.75–2.3 μg As/g), with up to 50% of the naturally occurring sediment-bound As present as labile As(V) sorbed on mineral surfaces (Kent & Fox, 2004). Approximately half of the dissolved As in contaminated groundwater was present as As(III), indicating the importance of reduction of As(V) to As(III) in addition to reductive dissolution of Fe(III) oxyhydroxides in As mobilization. Elevated concentrations of dissolved As(V) in the plume resulted from competitive sorption with wastewater-derived phosphate (Kent & Fox, 2004).

Subsequent sampling of the wastewater plume has shown the persistence of As at concentrations of 0.1 to 0.2 μM in the Fe-reducing zone despite the end of disposal in 1995. Dissolved As(III) and As(V) have continued to be present at about the same concentration. The end of disposal has also resulted in decreases in nitrate concentrations at the upper boundary of the Fe-reducing zone and concomitant increases in dissolved As, Fe(II), and phosphate concentrations.

Santuit Pond is a 70-hectare, hyper-eutrophic, groundwater flow-through lake on Cape Cod, MA. Hyper-eutrophic conditions likely result from nutrients released from septic systems serving private residences on the upgradient end of the lake and result in the accumulation of organic-rich sediments. Lake water recharges the aquifer through these organic rich sediments at the south end of the lake. Groundwater sampled approximately 20 meters downgradient from the lake in September 2015 was anoxic, with up to 170 μM Fe(II) (Fig. 1). Arsenic concentrations up to 0.21 μM, along with 2–5 μM phosphate (not shown), were detected where dissolved Fe(II) concentrations exceeded 50 μM (Fig. 1). These results suggest that sediment-bound As was mobilized under Fe-reducing conditions caused by lake-derived organic carbon.

3.2 *Arsenic from oil and gas development*

Dissolved As concentrations up to 3.1 μM were detected in a contaminant plume in a sandy glacial outwash aquifer near Bemidji, Minnesota, USA, resulting from a crude oil spill (Cozzarelli et al., 2015). The crude oil had an As concentration of 0.47 μM, insufficient to have caused the observed As concentrations. The range of As concentrations on sediments was similar to the range on Cape Cod sediments. Arsenic contamination likely resulted from mobilization of sediment-bound As by reductive dissolution of Fe(III) oxyhydroxides, coupled to reduction of As(V) to As(III), driven by oxidation of benzene and other organic compounds leaching into groundwater from the oil body. Association of As mobilization with a petroleum hydrocarbon spill prompts us to consider potential risks of sub-surface As contamination from wastewater generated during oil and natural gas production.

Large volumes of wastewater are generated during oil and gas production, can enter groundwater through land application or disposal (e.g., Engle et al., 2014) and accidental leaks or spills. Wastewater can contain organic chemicals derived from the formation and injected during hydraulic fracturing. Organics from guar breakdown and other chemicals used for hydraulic fracturing can mobilize naturally occurring As (Savoie et al., 2004). The produced water itself may also contain elevated concentrations of As. Analyses extracted from a database of produced water compositions from the United States show As concentrations in numerous produced waters range from 3–10 μM; dissolved salt concentrations can exceed 300 g/L (Blondes et al., 2014).

Improving the capability to determine concentrations of As and other trace elements in produced water with high dissolved salt concentrations is an area of active research. Samples with high dissolved salt concentrations must be diluted up to 1000-fold prior to analysis in order to diminish the concentrations of salts admitted to the plasma. Even once diluted, potential inter-element interferences, such as the interference between As, which has a mass of 75, and $^{40}Ar^{35}Cl^+$, must be addressed.

4 CONCLUSIONS

Arsenic-contaminated groundwater often results from microbially driven reductive dissolution of naturally occurring Fe(III)-oxyhydroxides and As(V) to As(III) coupled to organic carbon degradation. The organic carbon that fuels these processes can be derived from many sources, including domestic wastewater, petroleum hydrocarbons, and organic-rich sediments in eutrophic lakes. Better understanding of the coupling between hydrologic and biogeochemical controls on As-plume dynamics will help improve our ability to manage groundwater resources to minimize As concentrations in water supplies. The potential impact of releases of produced water from oil and gas development on As contamination in the subsurface poses significant challenges.

REFERENCES

Blondes, M.S., Gans, K.D., Thordsen, J.J., et al. 2015. U.S. Geological Survey National Produced Waters Geochemical Database v2.1 (PROVISIONAL). energy.usgs.gov.
Cozzarelli, I. M, Schreiber M.E., Erickson M. et al. 2015. Arsenic cycling in hydrocarbon plumes: secondary effects of natural attenuation, *Ground Water* in press.
Engle, M.A., Cozzarelli, I.M., Smith, B.D. 2014. USGS Investigations of water produced during hydrocarbon reservoir development. *USGS Fact Sheet* 3104, 3 pp.
Kent, D.B., Fox, P.M. 2004. The influence of groundwater chemistry on arsenic concentrations and speciation in a sand and gravel aquifer. *Geochem. Trans.* 5: 1–12.
Savoie, J.G., Kent, D.B., Smith, R L., LeBlanc, D.R., Hubble, D.W. 2004. Changes in ground-water quality near two granular-iron permeable reactive barriers, Cape Cod, Massachusetts, 1997–2000. *USGS WRIR 03–4309*, 84 p.

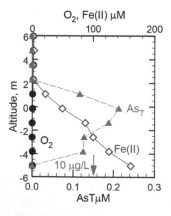

Figure 1. Vertical profiles of dissolved oxygen, Fe(II), and AsT in groundwater downgradient from Santuit Pond, MA, USA.

Arsenic Research and Global Sustainability – Bhattacharya, Vahter, Jarsjö, Kumpiene,
Ahmad, Sparrenbom, Jacks, Donselaar, Bundschuh & Naidu (Eds)
© *2016 Taylor & Francis Group, London, ISBN 978-1-138-02941-5*

On the geological conditioning of arsenic pollution in shallow aquifers

M.E. Donselaar
Department of Applied Geoscience and Engineering, Delft University of Technology, Delft, The Netherlands

ABSTRACT: Strong spatial variability of arsenic (As) concentrations over short distances and depth intervals characterizes shallow aquifers in fluvial and deltaic deposits in As-affected sedimentary basins. A literature review of key publications on arsenic concentration levels shows that the dominant geomorphological setting of the As-prone areas consists of meandering river deposits with sand-prone fluvial point-bar deposits surrounded by clay-filled (*clay plug*) former meander bends (*oxbow lakes*). Analysis of the lithofacies succession and organic matter in two wells in such geomorphology along the Ganges River reveals that the clay-plug deposits have a high TOC of 1.5%. Present-day, partly in-filled oxbow lakes have a rich plant and animal life which will be preserved in the anoxic environment at the poorly oxygenized lower part of the lake (*hypolimnion*). The clay-plug sediment is identified as the locus of microbial respiration that triggers the reductive dissolution of FeOOH and associated desorption of As.

1 INTRODUCTION

As concentration levels in shallow aquifers of Holocene fluvial and deltaic floodbasins are characterized by large lateral variability over distances of 100 s of meters and a general strong vertical decrease when the wells penetrate deeper Pleistocene strata (e.g. McArthur *et al.*, 2004, Zheng *et al.*, 2004, 2005, Shah, 2008). Various authors relate the spatial variability in As concentrations to the geomorphological setting of the shallow aquifers. Ahmed *et al.* (2004) and McArthur *et al.* (2004) related the release of sorbed As from FeOOH by reductive dissolution in the shallow aquifers to the occurrence of organic matter-rich peat layers in the shallow subsurface of the meandering Hugli and Sunti River morphology. Hoque *et al.* (2014) concluded that palaeosols act as shields to prevent As to move to shallow palaeo-interfluvial aquifers in the Ganges River floodplain of the Bengal Basin. Sahu & Saha (2015) correlated the variability in As concentrations with the different sediment types in the floodplain of the meandering Ganges River. The present paper aims to highlight the importance of the meandering river geomorphology in conditioning the aquifer flux and the release and entrapment of As.

2 METHODS

The study areas of key publications on the spatial variability of As pollution were mapped out in Google Earth-Pro, and from this an inventory was made for the depositional setting and specific geomorphological conditions. Two fully-cored, 50-m-deep wells were drilled in the fluvial geomorphology along the Ganges River, Bhojpur District in Bihar with the aim to analyze the lithofacies distribution in the shallow aquifer domain. Clay samples from the boreholes were analyzed for organic carbon content with the Walkley-Black procedure.

3 RESULTS AND DISCUSSION

3.1 *Depositional process and geomorphology*

The As-polluted areas in a fluvial floodplain setting are characterized by multiple abandoned river bends. Abandonment occurred after expansion of the meander bends and associated increase in sinuosity of the active river, and subsequent neck cut-off of the meander loop. The following main geomorphological elements are distinguished:

1. The abandoned river bend (*oxbow lake*) became part of the floodplain and gradually filled up with clay and organic material from plant and animal remains. The filled-in oxbow lake (*clay plug*) forms a continuous, non-permeable rim along the point bar;
2. The sandy point bar on the inner bend of the river. The specific river flow processes in the bend resulted in a fining-upward sequence with a coarse sandy lag at the base of the point bar, fining up-slope to sand, silt and floodplain clay. Inclined lateral accretion surfaces and a distinct wedge shape define the two dimensional architecture of point-bar. The vertical succession and lateral accretion surfaces reflect the expansion in space and time of the inner bend of a meandering river over its channel floor deposits.

3.2 *Core analysis*

The stratigraphic succession in both wells consists of two sequences with a sharp break at ~ 28 m depth (Fig. 1). The lower sequence consists of thin—to thick-bedded gravel layers and coarse-grained gravelly sand formed in a braided river environment. Permeability is very high, to the point that the drilling mud (bentonite) has completely invaded the core. The upper sequence consists of medium—to fine-grained, laminated sand, silt and clay, organized in three 5–12 m-thick, fining-upward units,

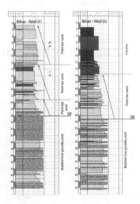

Figure 1. Lithofacies description and interpretation of the two cored wells. Well spacing is 2.28 km. SB: Sequence Boundary.

formed by vertical stacking of successive generations of Ganges River point bar sediment. The top of the sequence in Well 02 consists of a 12-m-thick succession of silt and black clay, which is the clay-plug fill of the oxbow-lake adjacent to the point bar sand in Well 1. The sharp break between both sequences is interpreted as a sequence boundary which marks the southward shift of the Ganges River belt to this area, with truncation of the upper part of the underlying braided river deposits.

3.3 Organic carbon content in the clay plug

The silt and black clay of the clay plug in the upper part of Well 2 are rich in organic carbon; total organic carbon is 1.5%. Analysis of time-lapse Google Earth-Pro imagery from 2003 to 2014, and corroborated with a site visit, shows that a close-by present-day oxbow lake (Suhiya, Bhojpur District, Bihar) is rapidly infilled at the upstream and downstream end by small streams that end in deltas. Moreover, water plants grow from all edges to the lake center and form platforms for new vegetation. Plant and animal life and fecal contamination from the hamlets bordering the oxbow lake all contribute to the high organic matter content in the clay plug sediment.

3.4 Discussion

Many authors reported that microbial respiration in an anoxic environment triggers the reductive dissolution of FeOOH and the desorption of As (e.g. McArthur et al., 2004). Ghosh et al. (2015) reported TOC of 0.7% in clay deposits at shallow depths (6–12 m) in Haringhata (Nadia district, West Bengal, India), which has a clear meandering river geomorphology. The TOC peak can be interpreted as a clay plug sediment, similar to the one found in the present study. From the high TOC levels in the clay plugs and the concentration of organic matter in present-day, partly in-filled oxbow lakes it is interpreted that clay plugs may be the locus of microbial respiration in an anoxic environment. Lack of water circulation in the lower part of the oxbow lake (the *hypolimnion*)

contributes to the development of anoxic conditions. Compaction of accumulating clay in the up to 12-m-thick clay plug succession triggers the expulsion of pore fluids to the adjacent permeable point-bar sands, and the diffusion process of reductive dissolution of FeOOH. Arsenic concentration levels in the aquifers are characterized by large lateral and vertical variability, which is here interpreted as directly related to the lithofacies heterogeneity.

4 CONCLUSIONS

Arsenic (As) pollution is widespread in shallow aquifers of Holocene fluvial and deltaic flood basins. An inventory of studies dealing with the spatial variability of As pollution in different flood basins shows that in all cases the geomorphological setting consists of meandering river deposits with sand-prone fluvial point-bar deposits surrounded by clay-filled former meander bends.

ACKNOWLEDGEMENTS

The author thanks his colleagues at A.N College, Patna, India (A.G. Bhatt, N. Bose and A.K. Ghosh) and TU Delft (J. Bruining, D. van Halem) for discussions on the geology of As pollution, and for help with data acquisition and analysis. Financial support for this study came from the EU Erasmus Mundus EURINDIA—Lot 13 program.

REFERENCES

Ahmed, K.M., Bhattacharya, P., Hasan, M.A., et al. 2004. Arsenic contamination in groundwater of alluvial aquifers in Bangladesh: an overview. *Appl. Geochem.* 19: 181–200.
Ghosh, D., Routh, J., Dario, M., Bhadury, P. 2015. Elemental and biomarker characteristics in a Pleistocene aquifer vulnerable to arsenic contamination in the Bengal Delta Plain, India. *Appl. Geochem.* 61: 87–98.
Hoque, M.A., McArthur, J.M., Sikdar, P.K. 2014. Sources of low-arsenic groundwater in the Bengal Basin: investigating the influence of the last glacial maximum palaeosol using a 115-km traverse across Bangladesh. *Hydrogeol. J.* 22: 1535–1547.
McArthur, J.M., Banerjee, D.M., Hudson-Edwards, K.A., Mishra, R., et al. 2004. Natural organic matter in sedimentary basins and its relation to arsenic in anoxic groundwater: the example of West Bengal and its worldwide implications. *Appl. Geochem.* 19: 1255–1293.
Sahu, S., Saha, D. 2015. Role of shallow alluvial stratigraphy and Holocene geomorphology on groundwater arsenic contamination in the Middle Ganga Plain, India. *Environ. Earth Sci.* 73: 3523–3536.
Shah, B.A. 2008. Role of Quaternary stratigraphy on arsenic-contaminated groundwater from parts of Middle Ganges Plain, UP–Bihar, India. *Environ. Geol.* 53: 1553–1561.
Zheng, Y., Stute, M., van Geen, A., Gavrieli, I., et al. 2004. Redox control of arsenic mobilization in Bangladesh groundwater. *Appl. Geochem.* 19: 201–214.
Zheng Y., van Geen, A., Stute, M., Dhar, R.K., et al. 2005. Geochemical and hydrogeological contrasts between shallow and deeper aquifers in the two villages of Araihazar, Bangladesh: implications for deeper aquifers as drinking water sources. *Geochim. Cosmochim. Acta* 69: 5203–5218.

Arsenic Research and Global Sustainability – Bhattacharya, Vahter, Jarsjö, Kumpiene, Ahmad, Sparrenbom, Jacks, Donselaar, Bundschuh & Naidu (Eds)
© *2016 Taylor & Francis Group, London, ISBN 978-1-138-02941-5*

Natural arsenic and its distribution in global geothermal systems

J. Bundschuh[1,2], I. Herath[2,3] & M. Vithanage[3,4]

[1]*Deputy Vice-Chancellor's Office (Research and Innovation), University of Southern Queensland, Toowoomba, Queensland, Australia*
[2]*Faculty of Health, Engineering and Sciences, University of Southern Queensland, Toowoomba, Queensland, Australia*
[3]*Chemical and Environmental Systems Modeling Research Group, National Institute of Fundamental Studies, Kandy, Sri Lanka*
[4]*International Center for Applied Climate Science, University of Southern Queensland, Toowoomba, Queensland, Australia*

ABSTRACT: This paper highlights the occurrence and distribution of geothermal As worldwide and its mobilization in geothermal fluids focusing different geological settings and processes. The mobilization of As from rocks and mineral phases into geothermal fluids depends on existing As sources, geochemical conditions and microbiological activity. In deep geothermal reservoirs, As mobilization is predominantly occurred from As-bearing pyrite at low temperatures (150–250°C), whereas at higher temperatures (>250°C), from arsenopyrite. The highest As concentrations were found in the volcanic geothermal systems of Los Humeros (Mexico) with up to 162,000 µg/L. Hence, updated and novel information on the present topic would be very much beneficial in order to understand the role of geothermal As in the genesis of As-rich ground- and surface water resources which are used for human consumption worldwide.

1 INTRODUCTION

Occurrence of arsenic (As) in geothermal fluids and springs has become a major environmental problem in many parts of the world over the recent decades. The mobilization of As from surrounding rocks into geothermal fluids takes place prominently along the active tectonic plate boundaries (Chandrasekharam & Bundschuh, 2008). The existence of As in geothermal fluids and respective surface manifestations such as hot springs, fumaroles, solfataras may contaminate cold aquifers, vadose zone, and surface and ground waters. Moreover, terrestrial geothermal activity may pose the pollution of groundwater systems at significant levels by natural As and such naturally occurring As in groundwaters can be seen in many areas of the world including Alaska, western USA, Mexico, Central America, northern Chile, Kamchatka, Japan, Taiwan, Philippines, Indonesia, Papua New Guinea, New Zealand, Iceland and France (Nordstrom & Archer, 2003).

The contamination of ground and surface waters with naturally occurring As by geothermal activities may lead to environmental impacts of geothermal systems as well as serious consequences on water resources used for drinking or irrigation purposes. Therefore, the understanding of occurrence, mobility and distribution of As species in geothermal systems would be beneficial to setup sustainable As mitigation strategies on a global scale. Hence, the present paper provides available data on As concentrations in terrestrial geothermal systems, and the mechanisms for immobilization or release of As into freshwater bodies.

2 DISCUSSION

2.1 Occurrence and distribution of geothermal arsenic

The fluids of geothermal reservoirs in volcanic rocks along with active plate boundaries show the

Table 1. Major geothermal areas and As concentrations (Bundschuh & Maity, 2015; Chandrasekharam & Bundschuh, 2008).

Geothermal field/area	As concentration (µg/L or µg/kg)
Chile: El Tatio, Chile (VOL; H)	30000–40000
Costa Rica: Miravalles (VOL; H)	11900–29100
Rincón de la Vieja (VOL; H)	6000–13000
Mexico: Cerro Prieto (SED; H)	250–1500
Los Azufres (VOL; H)	5100–49600
Los Humeros (VOL; H)	500–162000
New Zealand: Broadlands (VOL; H)	5700–8900
Kawerau (VOL; H)	539–4860
Orakei Korako (VOL; H)	599–802
Waiotapu (VOL; H)	2900–3100
Wairakei (VOL; H)	4100–4800
Philippines: Tongonan (VOL; H)	20000–34000
USA: Lassen Nat Park (VOL; H)	2000–19000
Russia: Kamchatka (VOL; H)	2000–30000
Ebeko volcano, Kuril Is. (VOL)	190–28000
Tibet: Yangbajing GTP (CC)	5700
Japan: Hachoubaru GTP, Oita	3230

VOL: volcanic rocks; CC: Continental collision zone; L: low temperature reservoir, T < 150°C; H: high temperature reservoir, T > 150°C; mix: mixed with shallow groundwater or surface water: GTP: Geothermal power plant; Is.: Island.

highest As concentrations. The concentrations of As in geothermal wells in different areas worldwide are summarized in Table 1.

Table 2 summarizes the As concentrations in fluids of major geothermal springs. Geothermal springs, having NaCl water type have the highest As concentrations and these waters correspond to original reservoir waters which were not significantly altered during its ascent. These NaCl waters can significantly contaminate freshwater aquifers and surface environments when discharging as geothermal springs.

2.2 Mobilization of geothermal arsenic

Arsenic is released from the host rocks of the geothermal reservoir. In deep geothermal reservoirs, As is mainly released from pyrite at 150–250°C, whereas at higher temperatures (>250°C), from arsenopyrite.

Table 2. The As concentrations in major geothermal springs in different parts of the world (Bundschuh & Maity, 2015, Chandrasekharam & Bundschuh, 2008).

Geothermal springs	As concentrations (µg/L)
El Tatio, Chile (VOL)	47000
Tambo river area (VOL)	1090–7850
North-central Andean – water	2–969
– Sediments	1600–717600
Costa Rica	
Miravalles (VOL; H)	5–4650
Rincón de la Vieja (VOL; H)	5–10900
Mexico—Los Azufres (VOL; H)	<3900
USA Salton Sea (granite intrusion; H)	30–12000
Soda Dam/Valles Caldera, (VOL; H)	1700
Hot Creek, Eastern Sierra Nevada	157–15000
Yellowstone, WY (VOL; H)	≈1500
Bath Spring, Y NP.	1500
Hot Spring Ojo Caliente, YNP	2500
New Zealand	996
Broadlands (VOL; H)	712–6470
Waiotapu (VOL; H)	710–6500
Orakei Korako (VOL; H)	3740–5110
Wairakei (VOL; H)	230–3000
Philippines–Mt. Apo (VOL; H)	3100–6200
Japan	
Tamagawa (VOL)	2300–2600
Beppu hot spring (VOL)	210–1360
Russia–Kamchatka (VOL, H)	2000–3600
Geothermal Spring Valley	1070–4210
Italy–Phlegraean Fields (VOL)	12–12600
Turkey	
Balcova	163.5–1419.8
Heybeli	1249
Balcova-Narlidera	1400
Hamambogazi	6936

VOL: volcanic rocks; CC: Continental collosion zone; L: low temperature reservoir, T < 150°C; H: high temperature reservoir, T > 150°C; mix: mixed with shallow groundwater or surface water; YNP: Yellowstone National Park.

minerals. Under reducing conditions, As mobility is governed by dissolution of Fe- and Mn-oxyhydroxides in the aquifer sediments due to microbial mediated biogeochemical interactions. Arsenic is metabolized by bacterial assimilation, methylation, detoxification, and anaerobic respiration. Along the detoxification/metabolism pathway of microorganism, As(III) is transformed into As(V), thereby reducing sulfur compounds or serving As(III) as electron donor. The thermophilic bacterial community can change the oxidation state of As(III) to As (V) in geothermal systems. On the other hand, bacterial isolates such as *Clostridium sulfidigenes* and *Desulfovibrio psychrotolerans* are capable of reducing both sulfate and arsenate, thereby releasing of As(III) into the hot spring fluids under reducing conditions (Maity et al., 2011). Because of arsenite-oxidation in sulfate-chloride springs, arsenate concentration in geothermal fluids tends to increase with the increase of the O_2 concentration, while decreasing dissolved sulfide at the same time. As a result active deposition of As(V)-Fe(III)-oxides occurred in the hot spring (Maity et al., 2011). The immobilization of As(V) species is controlled by the pH and availability of adequate sorption sites of minerals such as oxide minerals amorphous Al-, Mn- and Fe oxides and hydroxides. Furthermore, sulfate reducing bacteria play an important role in the transformation of As(V) to As(III) in geothermal systems with temperatures below 70°C.

3 CONCLUSIONS

In geothermal systems, As occurs predominantly as As-bearing pyrite at temperatures of 150–250°C and as arsenopyrite at >250°C. The As can mainly be released from the host rocks of the deep-seated geothermal reservoirs in active geothermal systems. The highest As concentrations in geothermal reservoir fluids occur through the active plate boundaries associated with volcanic rocks where the typical As concentrations are 500–162,000 µg/L. The microbial community of hot spring environments plays a crucial role in mobilizing As via oxidation and reduction process in geothermal systems.

REFERENCES

Bundschuh, J., Maity, J.P. 2015. Geothermal arsenic: Occurrence, mobility and environmental implications. *Renew. Sust. Energ. Rev.* 42: 1214–1222.

Chandrasekharam, D., Bundschuh, J. 2008. *Low-enthalpy geothermal resources for power generation* (Vol. 172): CRC Press Leiden.

Maity, J.P., Liu, C.C., Nath, B., Bundschuh, J., Kar, S., Jean, J.S. Chen, C.Y. 2011. Biogeochemical characteristics of Kuan-Tzu-Ling, Chung-Lun and Bao-Lai hot springs in southern Taiwan. *J. Environ. Sci. Heal. A* 46(11): 1207–1217.

Nordstrom, D.K., Archer, D.G. 2003. Arsenic thermodynamic data and environmental geochemistry *Arsenic in ground water*. 1–25: Springer.

*Arsenic Research and Global Sustainability – Bhattacharya, Vahter, Jarsjö, Kumpiene,
Ahmad, Sparrenbom, Jacks, Donselaar, Bundschuh & Naidu (Eds)*
© *2016 Taylor & Francis Group, London, ISBN 978-1-138-02941-5*

Arsenic in bedrock groundwater in Tampere region, South Finland

B. Backman & S. Luoma
Geological Survey of Finland, Espoo, Finland

ABSTRACT: Altogether, 1272 groundwater samples were collected from domestic wells drilled into fractured crystalline Precambrian bedrock in the Tampere Region, southern Finland. The Arsenic (As) concentration in the drilled bedrock wells varied from <0.05 to 2230 µg/L with median of 2.5 µg/L. 22.5% of total samples exceeds the safe drinking water limit of 10 µg/L and is much higher than the median for bedrock groundwater from all of Finland (0.16 µg/L, n = 263). High As groundwater was found in the felsic and mafic metavolcanic bedrock areas. Piper diagram showed that the majority of groundwater samples are Ca–HCO₃ type. Principal Component Analysis (PCA) and Hierarchical Cluster Analysis (HCA), showed that As is related to Mo, F⁻, B, Br⁻, Na, HCO₃⁻, Ca, Na, U, Sr, pH, EC, and well depth. This may reflect their origin host rocks and groundwater evolution mechanism in both oxidizing and reducing conditions.

1 INTRODUCTION

The groundwater arsenic was possible to determine at Geological Survey of Finland (GTK) since 1994. Arsenic determination has subsequently been carried out on a routine basis from all water samples. However, the history of arsenic research in Finland began in connection with ore exploration. During an intensive gold exploration programme in South Finland in the 1980s, high arsenic concentrations were observed in association with the gold deposits. Gold-bearing veins often also contained arsenopyrite. Elevated arsenic concentrations were additionally observed in till fines in certain regions of southern Finland during nationwide geochemical mapping in the 1980s. As a consequence of these findings, the arsenic problem in Finland was recognized, and in 1994 the Ministry of Social Affairs and Health contracted GTK to conduct a more detailed investigation of arsenic concentrations in groundwater in southern Finland. This revealed arsenic values up to 1040 µg/L in bedrock groundwater from the Tampere Region (Backman *et al.*, 2006). These values were extremely high compared to the maximum allowable concentration in drinking water of 10 µg/L set by WHO 2011. After that several groundwater sampling campaigns were made in the Tampere region and during 2004–2007 within the framework of the wide-ranging EU Life Environment project RAMAS (http://newprojects.gtk.fi/ramasfi/) and generated enormous volume of new data as discussed in Parviainen *et al.* (2016).

2 MATERIALS AND METHODS

2.1 Sampling

Groundwater samplings for arsenic studies in the Tampere Region were carried out many times during different projects since 1994. The largest sampling set was carried out during the RAMAS project. All the bedrock groundwater samples were taken through the household tap. Before sampling the water was allowed to run until the readings of water quality parameters e.g. Temperature (T), pH and Electrical Conductivity (EC) were constant. The wells were mainly sampled once but in 35 wells the water was monitored. Total number of drilled well water samples available in this study is 1272.

2.2 Laboratory analysis and data processing

The analyses were carried out in the Geolaboratory of GTK (Labtium Ltd since 2007). In the laboratory pH, conductivity and total alkalinity (as mmol HCO₃⁻/L) were determined using an automatic titrator. The main anions F⁻, Cl⁻, Br⁻, NO₃⁻, and SO₄²⁻ were analysed using the ion chromatographic technique and PO₄³⁻ by spectrophotometric method. Chemical oxygen demand of groundwater samples was determined using the titrimetric method with KMnO₄. All cations and other elements were analysed by ICP-MS/AES technique. The chemical characterization of arsenic in bedrock groundwater was identified by utilising the integrations of multivariate statistical approaches; PCA and HCA, in conjunction Piper diagrams and bi-variant correlation analyses as well as the hydrogeological and geological background.

3 RESULTS AND DISCUSSION

The dissolved arsenic concentrations in the bedrock groundwater vary between <0.05–2230 µg/L, with median of 2.5 µg/L and 22.5% of total 1272 samples exceeds the safe drinking water limit of 10 µg/L. Piper diagram showed that the majority of groundwater samples are generally Ca–HCO₃ type but Na-Cl and Na-HCO3 types were also found. Arsenic had poor correlations with other water quality parameters, but the data reveal a strong geological control of groundwater arsenic on a regional scale. The Tampere Region was divided to three different geological subdivisions and high arsenic contents were found in TB and PB zones (Fig. 1). High arsenic concentrations are clearly focused in the bedrock areas composed of felsic and mafic metavolcanic rocks, mica schist and mica gneiss. The highest arsenic concentrations are found in the gabbros and its intrusive area (Fig. 2). The median value of arsenic content in gabbros in Tampere Region were 167 mg/kg, and maximum of 1090 mg/kg, n = 17 (Parviainen *et al.* 2015). PCA and HCA showed that As is related to Mo, F⁻, B, Br⁻, Na, HCO_3^-, Ca, Na, U, Sr, pH, EC, and well depth. According to the results it was possible to find own factors for high arsenic in both oxidizing conditions with short delay and reducing conditions with longer delay.

Arsenic concentrations are elevated in wide area in Tampere Region in different bedrock type areas and according to the geochemistry of the well waters arsenic occurs in different form and dissolve in either oxidizing or reducing conditions. The research concerning As sources and the hydrogeological conditions favourable for dissolve of As in crystalline fractured bedrock will still continue at GTK.

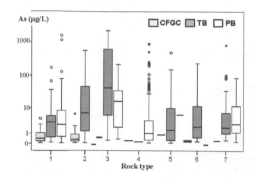

Figure 2. Box plot of arsenic concentrations (μg/L) in drilled well water (excluded monitoring data) from seven different bedrock areas in three different geological subdivisions: CFGC, TB and PB (Fig.1). Numbers of samples in bedrock type are 1 = Granodiorite, tonalite and quatzdiorite (255), 2 = Granite (91), 3 = Gabbro and diorite (45), 4 = Mica schist and mica gneiss (342), 5 = Mica schist (101), 6 = Intermediate and felsic metavolcanic rocks (26), 7 = Mafic metavolcanic rock (105).

4 CONCLUSIONS

Arsenic concentrations in bedrock groundwater in Tampere Region are exceptional high in comparison to the rest of Finland. High arsenic concentrations in bedrock groundwater are found in the areas of felsic and mafic metavolcanic rock types, especially gabbros, mica schist and mica gneiss. Elevated arsenic concentrations were found in waters both from oxidize and reduced hydrogeological conditions.

ACKNOWLEDGEMENTS

Research Program at Geological Survey of Finland (GTK) and most of the data were received from the RAMAS Project (Risk assessment and Risk Management Procedure for Arsenic in the Tampere Region) which was part-financed by the EU LIFE Environment Programme 2004–2007.

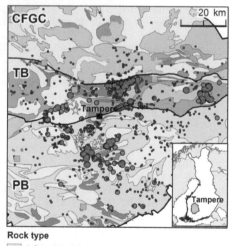

Rock type

1. Granodiorite, tonalite, quartzdiorite
2. Granite
3. Gabbro and diorite
4. Mica schist and mica gneiss
5. Mica schist
6. Intermediate and felsic metavolcanic rocks
7. Mafic metavolcanic rock

Arsenic (μg/L) in drilled bedrock wells

- < 10
- 10 - 50
- > 50
- Mines
- Gold prospects

Figure 1. Bedrock type and arsenic concentrations in drilled bedrock wells in the study area. Three geological subdivisions in Tampere Region consists of CFGC = Central Finland Granitoid Complex; TB = Tampere Belt; PB = Pirkanmaa Belt.

REFERENCES

Backman, B., Luoma, S., Ruskeeniemi, T., Karttunen, V., Talikka, M. and Kaija, J. 2006. *Natural Occurrence of Arsenic in the Pirkanmaa region in Finland.* Geological Survey of Finland. Espoo. Finland. 80 p. Available on the internet: http://projects.gtk.fi/ramasfi/index.html
Parviainen, A., Loukola-Ruskeeniemi, K., Tarvainen, T., Hatakka, T., Härmä, P., Backman, B., Ketola, T., Kuula, P., Lehtinen, H., Sorvari, J., Pyy, O., Ruskeeniemi, T. and Kaija, J. 2015: Arsenic in bedrock, soil and groundwater – The first guidelines for aggregate production established in Finland. *Earth-Sci. Rev.* 150: 709–723.

Arsenic Research and Global Sustainability – Bhattacharya, Vahter, Jarsjö, Kumpiene, Ahmad, Sparrenbom, Jacks, Donselaar, Bundschuh & Naidu (Eds)
© 2016 Taylor & Francis Group, London, ISBN 978-1-138-02941-5

National assessment of arsenic within groundwater: A case study with Ireland

L. Morrison[1,2], E. McGrory[1,2] & C. Brown[2]

[1]*Earth and Ocean Sciences, School of Natural Sciences, National University of Ireland, Galway, Ireland*
[2]*Ryan Institute, Environmental, Marine and Energy Research, National University of Ireland, Galway, Ireland*

ABSTRACT: The presence of arsenic in groundwater affects the safe distribution of potable water. Groundwater arsenic data from Ireland was analysed using various geostatistical techniques (including spatial interpolation methods). In some locations around Ireland, elevated arsenic is potentially linked to certain rock groups. In addition interpolation mapping has aided in identifying potential regional locations of arsenic contamination. This present study has, for the first time, mapped arsenic within groundwater in an EU nation to assess contamination and spatial distribution. It is envisioned that the methods and approach utilised in this study could be used by other nations for similar evaluations of arsenic in groundwater.

1 INTRODUCTION

Globally the presence of arsenic (As) within the groundwater system has caused both environmental and health related issues mainly in South East Asia and America. Within these regions several studies have focused on national assessments of As (Welch *et al.*, 2000, He & Charlet, 2013), however the presence of a similar study within an EU state does not exist. Recently elevated As concentrations have been reported as a result of routine testing of drinking water supplies as part of WFD obligations by local government around Ireland (elevated As concentrations defined as ≥ 7.5 µg/L, the Groundwater Threshold Value (GTV) as per EC (2010)). The primary aim of the present study was to explore if statistically significant differences existed between As levels across different lithologies, aquifer and vulnerability classifications in Ireland. This research will strive to facilitate in guiding future research needs and opportunities for monitoring As in drinking water together with the associated groundwater management recommendations as part of the WFD. Finally this methodology can be used as a framework for other nations for national assessments of arsenic distribution within groundwater.

2 METHODS

2.1 Data overview

A national collaboration was established between government agencies and the Environmental Protection Agency (EPA) in order to collate reported As concentrations in groundwater samples collected across Ireland. EPA data collected (1993–2012) was from a spatially distributed monitoring network that was established to represent different groundwater usage patterns and hydrogeological settings. This monitoring network was expanded in 2007 for the criteria of the WFD, in order to assess water quality in regions with Poorly Productive Aquifers (PPA) (which underline approximately 70% of Ireland). This resulted in the installation of 60 piezometers in seven hydrogeologically different catchments. However, because of different usage patterns and abstractions this PPA data was separated into a data base containing PPA data (EPA-PPA – 8339 measurements) and a data base without PPA data (EPA-NPPA – 7334 measurements).

2.2 Statistical analysis

The EPA database was used as a tool to monitor the assessment of As at a national level. In order to determine if statistically significant differences existed between the distribution functions of the rock type, the NADA package (Lee, 2013) in the open source statistical program R (version 3.0.1) was used. Additionally hot-spot analysis was performed to potentially address trends, clusters or outliners. Local Indicator of Spatial Association (LISA) maps were created in GeoDa™ 1.4.6 (Anselin *et al.*, 2006). Additionally a surface interpolation using the technique of Inverse Distance Weighting (IDW) was generated using ArcGIS® 10.2 (ESRI®, Colorado). Interval mapping, IDW and LISA were performed on both the EPA-PPA and EPA-NPPA data separately.

3 RESULTS AND DISCUSSION

3.1 Statistics

The EPA-PPA and EPA-NPPA datasets comprise of 294 and 222 unique monitoring locations. A geochemical classification using five intervals was completed, (0,1], (1,7.5], (7.5,10], (10,100] and (100, 234]

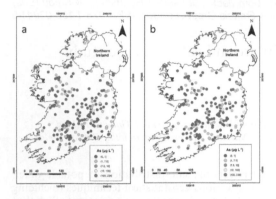

Figure 1. Spatial distribution of EPA arsenic data with five geochemical classifications of: a) EPA-PPA data, and b) EPA-NPPA data.

Table 1. Breakdown of sites with arsenic values for EPA-PPA and EPA-NPPA datasets.

EPA-PPA		EPA-NPPA	
Interval	N	Interval	N
(0,1]	173	(0,1]	150
(1,7.5]	95	(1,7.5]	63
(7.5,10]	4	(7.5,10]	3
(10,100]	19	(10,100]	5
(100, 234]	3	(100, 234]	1

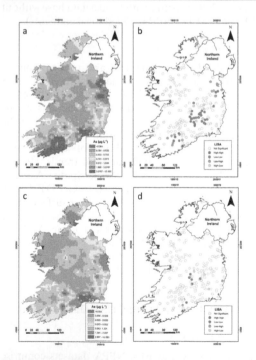

Figure 2. Geostatistical analysis of arsenic data using: a) IDW interpolation (EPA-PPA), b) LISA hot-spot analysis (EPA-PPA), c) IDW interpolation (EPA-NPPA), d) LISA hot-spot analysis (EPA-NPPA).

which is illustrated in Figs. 1a and 1b with the proportion of sites in each interval classification shown in Table 1. While the majority of sites fall into the (0,1] and (1,7.5] intervals, some do ex ceed these. Removal of the PPA data reduced the number of sites exceeding the As GTV value from 26 to 9.

Significant differences were observed among different aquifer-hosting lithologies, indicating that rhyolite, sandstone, and shale and impure limestone presented greater risk of elevated As in groundwaters (data not shown).

3.2 Interpolation methods

In addition to previous statistical analysis, surface interpolation and cluster analysis was performed to determine potential local As hot-spots on both the EPA-PPA and EPA-NPPA datasets (Fig. 2a-d). From the removal of the PPA data there is a reduction in potential hot-spots with some PPA locations showing elevated levels of As. IDW analysis revealed potential As contamination to the east and south of Ireland.

4 CONCLUSION

This present study, while not showing grossly elevated As concentrations, shows persistent low-to-medium levels of As around certain locations in Ireland. This study also shows that the presence of As within groundwater presented at a national level can highlight potential areas of contamination for future monitoring. The framework of this study can be implemented within other EU states to help achieve objectives of the WFD.

ACKNOWLEDGEMENTS

Funding based on research grant-aided by the Department of Communications, Energy and Natural Resources under the National Geoscience Programme 2007–2013 (Griffiths Award). The authors would like to thank the EPA, GSI & HSE for assistance in data collection. © Ordnance Survey Ireland. All rights reserved. License number NUIG220212.

REFERENCES

Anselin, L., Syabri, I. and Kho, Y. 2006. GeoDa: an introduction to spatial data analysis. *Geogr. Anal.* 38: 5–22.

European Communities (EC). 2010. European Communities Environmental Objective (Groundwater) Regulations, S.I. No. 9 of 2010, p. 41.

He, J. and Charlet, L., 2013. A review of arsenic presence in China drinking water. *J. Hydrol.* 492: 79–88.

Lee, L. 2013. NADA: Nondetects And Data Analysis for environmental data. R package version 1.5–6. https://cran.r-project.org/web/packages/NADA/index.html

Welch, A.H., Westjohn, D.B., Helsel, D.R. and Wanty, R.B., 2000. Arsenic in ground water of the United States: occurrence and geochemistry. *Ground Water* 38: 589–604.

Arsenic Research and Global Sustainability – Bhattacharya, Vahter, Jarsjö, Kumpiene,
Ahmad, Sparrenbom, Jacks, Donselaar, Bundschuh & Naidu (Eds)
© 2016 Taylor & Francis Group, London, ISBN 978-1-138-02941-5

High arsenic levels in groundwater resources of Gediz Graben, Western Turkey

A. Baba[1], O. Gündüz[2], C. Şimşek[2], A. Elçi[2], A. Murathan[3] & H. Sozbilir[2]

[1]İzmir Institute of Technology, Izmir, Turkey
[2]Dokuz Eylul University, Izmir, Turkey
[3]State Hydraulic Works, 2nd Regional Directorate, Izmir, Turkey

ABSTRACT: Gediz Graben situated in western Turkey is an area containing extensional structures with active tectonics and geothermal systems that serves as a suitable environment for the presence of high levels of arsenic (As) in groundwater. The results of a comprehensive monitoring program in the basin revealed that the maximum As concentration detected during the monitoring program was 3086 ppb and the average value was calculated to be 23.63 ppb. In addition, As levels in 28% of the water samples were above the 10 ppb limit value making them unsuitable for drinking water supply. These high As concentrations in the basin was mostly associated with long detention times of water in altered rocks located along detachment faults, strong water–rock interaction processes in the alteration zones and anthropogenic influences made within the geothermal fluid reservoir.

1 INTRODUCTION

Occurrence of arsenic (As) in groundwater has been a major problem worldwide since the last century. Arsenic related problems have been encountered recently in several countries including Bangladesh, India, Indonesia, Nepal, Myanmar, Mexico Pakistan, Vietnam, Cambodia, China (Mukherjee et al., 2006) and Turkey (Baba & Gündüz, 2010). It is estimated that nearly 125–150 million people are exposed to the potential risk of As toxicity (Bhattacharya et al., 2002). Recently, As pollution is becoming a major topic in the agenda of Turkey as well. Particularly, high As levels have been detected in water resources flowing through volcanic rocks in western part of Turkey such as Afyon, Çanakkale, Isparta, Kütahya, Manisa, Muğla and Izmir (Gündüz et al., 2010). In many of these locations, As is naturally found in the subsurface strata within volcanic, metamorphic and sedimentary formations as well as in areas of geothermal systems related to tectonic activity. Based on this background, this study focuses on high As levels in groundwater resources of the Gediz Graben located in western Turkey, which is one such area of complex geology with active tectonics and high geothermal potential.

2 GEOLOGY OF STUDY AREA

The Gediz Graben is 140 km long and 3–40 km wide and has a WNW–ESE trending structure bounded by active normal faults (Fig. 1). The Gediz detachment fault is one of several crustal-scale detachment faults

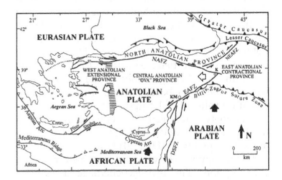

Figure 1. Simplified tectonic map of Turkey showing major neotectonic structures and neotectonic provinces (from Barka, 1992; Bozkurt, 2001).

that forms the southern margin of the Gediz Graben (Sözbilir, 2001, 2002). Many geothermal fields (i.e., Turgutlu (Urganlı), Salihli and Alaşehir) are present along the Gediz Graben with measured reservoir temperatures reaching as high as 287°C.

Based on the tectonic characteristics and the geological structure of Gediz Graben, wells drilled in As containing geological formations and boreholes that are close to geothermal resources are likely to contain high As levels. Majority of these wells cut hydrothermally altered and fractured metamorphic and sedimentary rocks due to the influence of active faults. These rocks are characterized by phyllic, argillic, and silicic-hematitic alteration zones in Gediz Graben (Baba, 2010; Gündüz et al., 2010; Baba & Sözbilir, 2012). Especially along the detachment fault, there are wide areas of alteration with mineral assemblages that range from advanced argillic type through silicification to deeper level propylitization. Such areas are also enriched by As.

3 METHODS

A comprehensive water quality monitoring program with three sampling campaigns and a total of 750 groundwater quality monitoring stations was conducted in the Gediz River Basin during the wet and dry seasons of 2013–2014. About 60% of these stations were situated within the Gediz Graben area. Groundwater samples were collected into 50 mL plastic bottles and were later acidified with 0.5 N HNO$_3$ to prevent the complex formation. In the laboratory, samples were filtered through 0.45 µm filter paper and measured with ICP-MS.

4 SOURCE OF ARSENIC IN GEDİZ GRABEN

Groundwater As was found to be above 10 µg/L in all parts of the basin. The levels of concentrations were particularly high in deep wells located in the central portions of the graben area (Fig. 2). These locations are located around the geothermal fields and within alteration zones. The maximum As concentration detected during the monitoring program was 3086 µg/L whereas the average value was calculated to be 23.6 µg/L with a standard deviation value of 131.5 µg/L.

The analysis revealed that As concentrations in the basin exceeded 10 µg/L limit value for drinking water quality (WHO, 2011) and the majority of water supply wells were not suitable for human consumption. The hydrogeochemical assessment in the basin also revealed that the main reason for obtaining high As concentrations in the basin was associated with the long detention times of water in altered rocks located along the detachment faults and strong water–rock interaction processes in these alteration zones as well as the strong anthropogenic influences within the geothermal fluid reservoir.

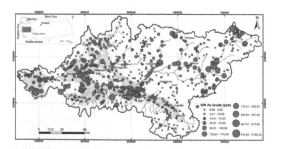

Figure 2. Distribution of groundwater arsenic in Gediz River Basin.

5 CONCLUSIONS

Due to its active extensional tectonics and presence of rich geothermal resources, Gediz Graben contains various altered rock that contain elevated levels of As. Dissolution from these rocks are mostly responsible for the elevated As concentrations in groundwater that exceed the maximum allowable limits for drinking water. Thus, it is important to continuously monitor As in the basin and to introduce As removal practices for all residential areas in Gediz Graben that depend on local groundwater.

ACKNOWLEDGEMENTS

The authors acknowledge the support of State Hydraulic Works (DSİ) on this study.

REFERENCES

Baba, A. 2010. High arsenic levels in water resources resulting from alteration zones: a case study from Biga Peninsula, Turkey. *Proceedings of As2010: The Third International Congress on Arsenic in the Environment*, 17–21 May, 2010, Taiwan.
Baba, A. and Gündüz, O. 2010. Effect of alteration zones on water quality: A Case Study from Biga Peninsula, Turkey. *Arch. Environ. Cont. Toxicol.* 58(3): 499–513.
Baba, A. and Sözbilir, H. 2012. source of arsenic based on geological and hydrogeochemical properties of geothermal systems in Western Turkey, *Chem. Geol.* 334: 364–377.
Barka A.A. 1992. The North Anatolian Fault zone. *Annales Tecton*. 6: 164–195.
Bhattacharya, P., Frisbie, S.H. and Smith, E., et al. 2002. Arsenic in the Environment: A Global Perspective. In: Sarkar B (ed) *Handbook of Heavy Metals in the Environment:* Marcell Dekker Inc., New York. pp. 145–215.
Bozkurt, E. 2001. Neotectonics of Turkey-a synthesis. *Geodin. Acta* 14:3–30.
Gündüz, O., Baba, A. and Elpit, H. 2010. Arsenic in groundwater in Western Anatolia, Turkey: a review. *XXVIII IAH Congress, Groundwater Quality Sustainability*, 12–17 September 2010, Krakow, Poland, pp. 183–191
Mukherjee, A, Sengupta, M.K., Hossain, M.A., Ahamed, S., Das, B., Nayak, B., Lodh, D., Rahman, M.M. and Chakraborti, D. 2006. Arsenic contamination in groundwater: a global perspective with emphasis on the Asian scenario, *J Health Popul Nutr,* 24(2):142–163.
Sözbilir, H. 2001. Extensional tectonics and the geometry of related macroscopic structures: field evidence from the Gediz detachment, western Turkey. *Turk. J. .Earth Sci.* 10: 51–67.
Sözbilir, H. 2002. Geometry and origin of folding in the Neogene sediments of the Gediz Graben, western Anatolia, Turkey. *Geodin. Acta* 15:277–288.
WHO. 2011. *Guidelines for Drinking-Water Quality*. 4th ed. World Health Organization, Geneva, 541p.

Arsenic Research and Global Sustainability – Bhattacharya, Vahter, Jarsjö, Kumpiene,
Ahmad, Sparrenbom, Jacks, Donselaar, Bundschuh & Naidu (Eds)
© *2016 Taylor & Francis Group, London, ISBN 978-1-138-02941-5*

Tracing the relative distribution of arsenic species in groundwater and its association with soil arsenic levels in the Simav Graben area, Turkey

O. Gündüz[1], A. Baba[2], C. Şimşek[1], Alper Elçi[1] & H. Gürleyük[3]
[1]*Dokuz Eylul University, Izmir, Turkey*
[2]*İzmir Institute of Technology, Izmir, Turkey*
[3]*Brooks Land Labs, Seattle, WA, USA*

ABSTRACT: A comprehensive hydrogeochemical assessment was conducted in the Simav Plain, Turkey where major health concerns were reported and elevated arsenic (As) levels were previously found. Boreholes drilled in the plain were used to take core samples from the alluvial aquifer from which most groundwater was extracted. The results of core analyses revealed several orders of magnitude higher values than global average. Groundwater samples were later collected from these boreholes and analyzed for arsenic and its species as well as other related hydrochemical parameters. The results of groundwater quality assessment revealed that the groundwater in the plain was enriched with arsenic that exceed 1 mg/L level and was mostly under reducing conditions. The dominant arsenic specie in groundwater was arsenite in more than 70% of all samples.

1 INTRODUCTION

Arsenic (As) is a proven carcinogen that causes internal organ and skin cancers when it enters the body through oral or dermal route. Throughout the world, many countries including Bangladesh, Taiwan, India, Chile, Argentina and USA are experiencing geogenic As problems in groundwaters (Bhattacharya *et al.*, 2012). Similar As-related problems have emerged in Turkey in recent years and difficulties in drinking water supply were experienced in some residential areas (Gündüz *et al.*, 2010a, Baba & Sözbilir, 2012). The Simav Plain is one of those areas with above standard As concentrations in surface and subsurface waters (Gündüz *et al.*, 2010b). This study presents the results of the hydrogeochemical assessments of subsurface waters in the region as well as the sources of As and its dominant species.

2 THE STUDY AREA

The Simav Graben area is located in west-central Anatolia and is a semi-closed basin that contained the Simav Lake (Fig. 1) which was drained in the 1960s and was converted to agricultural land. It is a E–W trending Pliocene to Quaternary asymmetric depression that was developed on the older NE–SW trending Miocene basins in Western Anatolia (Seyitoğlu, 1997). The graben is bounded from the south by an active oblique-slip normal Simav fault. The graben fill is composed of semi-lithified boulder conglomerate and sandstone.

The Simav geothermal field that is situated to the north of the Simav district center has three geothermal outflows (Fig. 1) with reservoir temperatures reaching as high as 170°C. (Gündüz *et al*, 2010b).

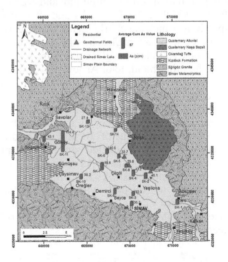

Figure 1. Geology of the study area and distribution of core As.

3 METHODS

Research boreholes were drilled in 21 locations to a total depth of 846 m and 157 core samples were taken to determine the depth-integrated geochemical status of the plain. Elemental analysis of these core samples were conducted by acid extraction followed by detection using ICP-MS. The boreholes were then converted to groundwater monitoring wells by locating filters at elevations that correspond to high core As levels. Samples were later collected from these 21 subsurface water monitoring boreholes as well as 7 other wells that were previously drilled in the area. In addition, samples were also taken from 3 geothermal waters in three different sampling periods during the study. All water samples were then analyzed for total As, arsenite, arsenate and organic arsenic species (DMA, MMA) with ICP-MS and LC-ICP-MS tandem techniques.

4 RESULTS AND DISCUSSIONS

Based on previous studies, the metamorphic rocks of the region has strong iron and sulfide oxidation due to hydrothermal alteration and arsenic is typically observed in sulfide oxide sediments transported from these rocks that create the Quaternary alluvium sediments (Gündüz et al., 2010b). A comprehensive study was conducted on the sediments and groundwaters of the plain to determine the vertical layering of arsenic containing sediments. Geochemical analysis conducted in 21 boreholes revealed that the alluvial sediments of the plain had As levels that are 2 to 3 orders of magnitude higher than world average value. The 157 core samples taken from 5–18 different depths had maximum, minimum and average As values of 833.9 mg/kg, 7.1 mg/kg and 48.99 mg/kg, respectively, which are clearly higher than the world average value of 1.5 mg/kg. Spatial distribution of core sample As values are given in Figure 1.

On the other hand, total As levels in groundwater samples ranged from 0.48 to 1000 µg/L with an average value of 227.8 µg/L (Fig. 2). Thus, the majority of the wells sampled had total As concentrations that are 2–3 orders of magnitude higher than the standard value of 10 µg/L. In addition, speciation studies conducted on these samples also showed that dominant As specie in 20 of 28 wells was the more toxic As(III) with values ranging from 0.3 to 909 µg/L with average value of 183.4 µg/L. As(V) was the second most dominant specie, and DMA and MMA species were below detection limits (Fig. 3).

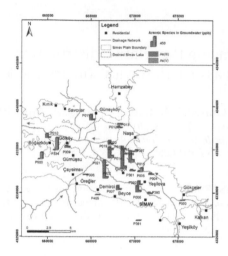

Figure 3.   Dominance of As species in Simav Plain.

5 CONCLUSIONS

Total As values were found to be several orders of magnitude higher than the guideline value of 10 µg/L in Simav Plain. This result is strongly supported by the presence of elevated values of As in core samples. As groundwater in the plain was mostly under reducing conditions arsenite was the dominant specie and reached to about 1 mg/L, indicating significant As pollution in the region. These results indicate an eminent risk for direct human consumption of groundwater and further necessitate the use of As removal technologies for all residential areas supplying their drinking water from the plain.

REFERENCES

Baba, A. and Sözbilir, H. 2012. Source of arsenic based on geological and hydrogeochemical properties of geothermal systems in western Turkey. Chem. Geol. 334: 364–377.
Bhattacharya, P., Frisbie, S.H., Smith, E., Naidu, R., Jacks, G. and Sarkar, B. 2002. Arsenic in the Environment: A Global Perspective. In: B. Sarkar (Ed.) Handbook of Heavy Metals in the Environment: Marcell Dekker Inc., New York, pp. 145–215.
Gündüz, O., Baba, A. and Elpit, H. 2010a. Arsenic in groundwater in Western Anatolia, Turkey: a review. XXVIII IAH Congress, Groundwater Quality Sustainability, 12–17 September 2010, Krakow, Poland, pp. 183–191.
Gündüz, O., Şimşek, C. and Hasözbek, A. 2010b. Arsenic pollution in the groundwater of Simav Plain, Turkey: Its impact on water quality and human health. Water Air Soil Poll. 205: 43–62.
Seyitoğlu, G. 1997. Late Cenozoic tectono-sedimentary development of Selendi and Uşak-Güre basins: a contribution to the discussion on the development of east–west and north-trending basins in western Turkey. Geol. Mag. 134: 163–175.

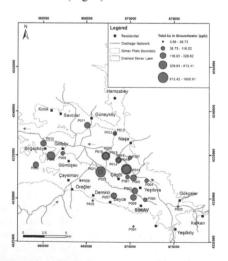

Figure 2.   Total As in groundwater in Simav Plain.

Arsenic Research and Global Sustainability – Bhattacharya, Vahter, Jarsjö, Kumpiene,
Ahmad, Sparrenbom, Jacks, Donselaar, Bundschuh & Naidu (Eds)
© 2016 Taylor & Francis Group, London, ISBN 978-1-138-02941-5

Arsenic exposure in drinking water a growing health threat: Well testing in outskirts of Lahore to identify wells low in arsenic to mitigate the As crisis in Pakistan

A. Farooqi[1], W. Ali[1], N. Mushtaq[1], S. Zahir[1], A. Younus[1] & A. van Geen[2]

[1]Environmental Geochemistry Laboratory, Faculty of Biological Sciences, Department of Environmental
Sciences, Quaid-i-Azam University, Islamabad, Pakistan
[2]Lamont Doherty Earth Observatory, Columbia University, Palisades, New York, USA

ABSTRACT: For the first time blanket testing was done in 1818 wells from 10 villages along a transect line selected on the basis of an ongoing study in India. As was tested by Arsenic Econo-Quick (™) (EQ) kit with 10% samples collected for quality control. 100% of the wells installed were untested. As concentrations in 82% of the samples exceeded the WHO guidelines (>10 µg/L) for drinking water. 42% of wells were save when compared with National Environmental Quality Standards (NEQS) which is 50 µg/L. The mechanism of arsenic enrichment seems different form that of the other side of border on the basis of the basic parameters done in the study. The research is ongoing and approximately 40,000 wells will be tested along the two selected transect lines.

1 INTRODUCTION

An estimated 200 million people worldwide are exposed to arsenic (As) concentrations in drinking water that exceed the recommended limit of 10 µg/L as set out in the guidelines of the World Health Organization (WHO) (UNICEF/BBS, 2009). Concerns about elevated As concentrations in Pakistan groundwater were first raised in 2002. Initial screening was done by Pakistan Council of Water Resources (PCRWR) in 2003–2006. It has been more than nine years since screening to test tube wells for As was conducted by PCRWR. It is well-known that distribution of As in groundwater is highly variable spatially, and temporal variability is limited. The problem is that most households do not know the status of their well because the vast majority of hand pumps have never been tested. It has been demonstrated in As affected areas that distinguishing individual hand-pumps that are safe from those that are not using field-kits as an effective way of lowering exposure of the rural population in the short- to medium term (van Geen et al., 2002, 2003). The lack of testing of hand pumps in Pakistan has led people to drink their groundwater without knowing that whether it is safe or unsafe. There is an urgent need for Pakistan to identify the As contaminated Tube-wells (TWs) in order to assess the health risks and initiate appropriate mitigation measures. The long term objective of this research is to provide a sound scientific basis for predicting where in the Indus plain in Pakistan the composition of groundwater is likely to be hazardous to human health and how to target safe aquifers for the installation of safe community wells. This is a cross boundary study, the second objective is to understand the geochemistry for enrichment of arsenic on Pakistan side of the border.

2 METHODS/EXPERIMENTAL

2.1 Study area

Well testing was done in ten villages. These villages were selected using Google Earth imagery along the Pakistan part of the transect, parallel to Ravi river along the low lying floodplain area 31°.22′42″.66 N and 74°.0612.62 (Fig. 1). The aquifer system under is underlain by unconsolidated alluvial sediments deposited by Indus river tributaries during Pleistocene-Recent age. The alluvial complex has an average thickness of more than 400 meters. The sediments consist of brownish-grey, fine to medium sand, silt and clay in varying proportions with quartz, muscovite, biotite, and chlorite making up the chief mineral constituents of the sediments. The shifting course of tributaries have imparted heterogeneous nature to alluvial complex resulting in little to no vertical or lateral continuity. But still the sediments on regional scale behave as homogeneous aquifer (Farooqi, 2009).

2.2 Sampling and methodology

We tested 1818 wells from 10 villages were tested for As by Arsenic Econo-Quick (™) (EQ) kit. pH, ORP, and EC was measured for each well on site. For quality control and quality assurance, 10% of groundwater samples were randomly collected from wells in 20 mL scintillation vials for laboratory analysis at Lamont-Doherty Earth Observatory (LDEO). The samples were acidified with 1% high-purity Optima HCl at LDEO at least 48 hours prior to analysis using an ICP-MS at the LDEO.

Figure 1. Red color for As > 50, green 10–50 and blue <10 ug/L. On left side is the close up for Pakistan Punjab first blanket testing in one of the 10 villages. Whereas a considerable number of households own a well that is elevated in As, most live within walking distance of a well that is low in As. Green lines show the one of the proposed transects for Pakistan where future testing is underway.

3 RESULTS AND DISCUSSION

3.1 Arsenic distribution

The household survey showed that 100% of the wells installed were untested and no households knew the status of their wells. As concentrations in 77% of the samples exceeded the WHO guidelines (>10 µg/L) for drinking water while 53% exceeded the NEQS of 50 ug/L. Mean concentration of As from all the villages of understudy area exceeds WHO and NEQS limits except AK 44.2 ug/L which is greater than WHO limit but within permissible NEQS Table 1. When validated against laboratory values, the EQ kit correctly identified As contamination relative to the guideline in 85% of groundwater.

Previous work in the area has shown that the spatial distribution of As in handpumped water of the Indus plain is spatially highly variable, even within a single village (Farooqi et al., 2007; Farooqi, 2009). In the present study the blanket testing conducted along the bank of the Ravi River also suggest that in highly contaminated areas still some wells meet the WHO standards and is a viable and immediately available option to reduce the As and F exposure in the affected areas. The safer wells can potentially be used of well-switching as the main mitigation option in the short to medium term.

3.2 Cancer risk

Cancer risk was assessed in all the 10 villages of study area through health risk assessment model derived from the USEPA (Integrated Risk Information System (IRIS): As CASRN 7440–38–2, 1998). Coputational results displayed that the residents study areas had toxic risk indices (HQ) ranging from 4 to 6 while cancer risk CR ranging from 1.04E-7.71E (Table 1).

Table 1. Arsenic testing and other basic parameters in the study.

Village	n	Depth (m)	pH	EC (mS/cm)	As (ug/l)			% Wells > WHO	% Wells > NEQS	HQ	CR
					Min	Max	Mean				
AK	61	48	7.6	3069	<10	300	44.3	54	18	4	1.67E-03
BH	94	57	7.5	538	10	300	116.6	99	74	11	4.48E-03
CF	190	36	8	1337	<10	25000	174	49	19	15	6.68E-03
GR	451	45	7.4	1547	<10	1000	197.3	92	74	17	7.71E-03
KAS	127	43	7.9	1739	<10	1000	275.8	98	84	21	1.03E-03
KM	212	36	6.9	2317	<10	1000	157.7	88	57	14	6.33E-03
KP	207	31	7.3	1130	<10	1000	126.7	67	37	12	1.21E-03
KW	30	39	7.9	1817	10	500	292	97	90	23	1.04E-03
SC	215	40	7	3370	<10	1000	61.3	45	20	6	2.04E-03
SD	177	44	8.1	1541	<10	1000	157.3	80	63	14	4.85E-03

3.3 Arsenic enrichment mechanism

So far the results show that the geochemistry for enrichment of As on both sides of the border are entirely different even though both are part of the Indus Basin. Evaporative as well as oxidative mechanisms seems to control As in Pakistani part of Indus basin. Baseline water quality analysis of the untreated groundwater samples has been completed, with 25 percent having nitrate greater than 45 mg/L compared to WHO guidelines, alkaline pH and > 60% samples with positive ORP values.

4 CONCLUSIONS

In outskirts of Lahore drinking water shows widespread As contamination, exceeding the WHO As guideline. This poses a public health threat requiring further investigation and action. For groundwater samples, the EQ kit performed well relative to the WHO As limit and therefore could provide a vital tool for water As surveillance.

ACKNOWLEDGEMENTS

The research work is funded by TWAS, Itay and HEC-USAID under Pak-USAID project.

REFERENCES

Farooqi, A., Masuda, H., Kusakabe, M., Naseem, M. and Firdous, N. 2007. Distribution of highly arsenic and fluoride contaminated groundwater from east Punjab, Pakistan, and controlling role of anthropogenic pollutants in natural hydrological cycle. Geochem. J. 41: 213–234.

Farooqi, A. 2009. Arsenic and Fluoride Contamination- A Pakistan Perspective. Springer, New Delhi, India..145p.

UNICEF/BBS. 2009. Bangladesh National Drinking Water Quality Survey of 2009. UNICEF and Bangladesh. Bureau of Statistics, Dhaka, Bangladesh

van Geen, A., Ahsan, H., Horneman, A.H., Dhar, R.K., Zheng, Y., Hussain, I., Ahmed, K.M., Gelman, A., Stute, M., Simpson, H.J., et al. 2002. Promotion of well-switching to mitigate the current arsenic crisis in Bangladesh. Bull. World Health Organ. 80: 732–737.

van Geen, A., Ahmed, K.M., Seddique, A.A. and Shamsudduha, M. 2003. Community wells to mitigate the arsenic crisis in Bangladesh. Bull. World Health Organ. 81: 632–638.

Arsenic Research and Global Sustainability – Bhattacharya, Vahter, Jarsjö, Kumpiene,
Ahmad, Sparrenbom, Jacks, Donselaar, Bundschuh & Naidu (Eds)
© 2016 Taylor & Francis Group, London, ISBN 978-1-138-02941-5

Arsenic contamination of groundwater in Ghazni and Maidan Wardak provinces, Afghanistan

M.H. Saffi[1] & M.N. Eqrar[2]
[1]DACCAR, Wazirabad, Kabul, Afghanistan
[2]Geosciences Faculty, Kabul University, Kabul, Afghanistan

ABSTRACT: In Afghanistan, arsenic (As) and fluoride (F⁻) contamination are an issue of current drinking water supply systems where users have been using groundwater sources. Arsenic contamination is the major environmental health management concerns especially in Ghazni and Maidan Wardak provinces in WASH sector. Increasing human activities and haphazard urbanization have modified the cycle of heavy metal, non-metal and metalloids. The As contaminated groundwater used for drinking can cause adverse effect of human health of study area. The water quality study with 746 samples from Drinking Water Points (DWPs) have been carried in Khwaja Omari district and center of Ghazni province and Jaghato district of Maydan Wardak province results show that 61% of drinking water points samples exceeded the value of the WHO guideline of 10 µg/L of As, and 38% of analyzed water samples exceeded the Afghanistan Drinking Water Quality Standard (DWQS) of 50 µg/L of As.

1 INTRODUCTION

The inhabitants of study areas are heavily dependent on groundwater containing elevated level of arsenic (As). Therefore, an understanding of the occurrence, behavior, and sources of As along with other water quality parameters in the DWPs is essential to implement drinking-water supply schemes.

The study carried out analyzing physical and chemical parameters integrates data sets from UNICEFE, ECHO funded projects and National Groundwater Monitoring Wells networks to understand the spatial distribution of As concentration along with the hydrochemistry of groundwater in study areas.

The WHO guideline for As in the drinking water is 10 µg/L (WHO, 2004), however the Afghanistan Drinking Water Quality Standard (DQWS) is 50 µg/L (ANSA, 2013). This study focuses the distribution and occurrence of high As concentration in the drinking water points of study areas, but there are no clinical information regarding to the health effect of high As content drinking water.

2 METHODS/EXPERIMENTAL

2.1 Description of study area

The study area is located in the south direction of Kabul and geographically, it is situated between latitude 33.39776–33.84776 and longitude 68.26683–68.61683 (Fig. 1). It covers total area of about 6788 km² with population of about 844,765 (Polhill, 1982)[3]. It has semi-arid climate with major fluctua-

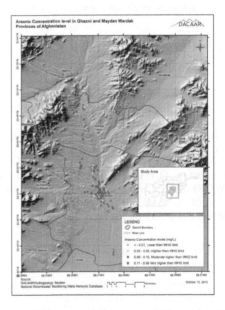

Figure 1. Location of study area and the spatial distribution of arsenic in groundwater samples.

tion in day- and night-time fluctuations. The winter is characterized by low temperatures of less than −20°C while the summer is dominated by high temperatures of more than 35°C. The rainfall and snowfall are the main source of groundwater and surface water, and the area receives an average 200 mm rainfall [4]. There are number of seasonal rivers and abandoned channels which are flowing water in rainy seasons.

Groundwater flow direction is from north mountains front hydrogeolgical boundaries (upstream) to south flood plain (downstream) along the Ghazni seasonal river (Uhl &Tahiri, 2003, US Army, 2009).

2.2 Sampling and field measurements

In total 764, Drinking Water Points (DWPs) including hand pump tube wells and dug wells were collected and tested on-site for measurement of As concentration and physico-chemical parameters like temperature, pH and Electrical Conductivity (EC) using digital Arsenator and pH/conductivity meter (Figure 1). 106 out of 764 water samples were sampled for chemical analyses.

Before collection of samples, each hand-pump was flushed for about 10 minute. The samples were then collected in sterilized 500 mm polyethylene bottle according to the DACAAR's water sample collection procedure. All samples were immediately shipped to the laboratory and stored to the refrigera-tor at 4°C in the dark until analysis. The water sam-ples were analyzed for 34 parameters. The analytical data quality was ensured through col-lection of duplicate samples chemical analysis and comparison. The ionic charge balance of each sample was < 5%.

3 RESULTS AND DISCUSSION

3.1 Distribution pattern of arsenic

Arsenic concentration in the DWPs of study area is mostly geologic occurrence and its spatial distribution is irregular trough the study areas. 61% of drinking water points samples exceeded the value of the WHO guideline of 0.01 mg/L of As, and 38% of analyzed water samples exceeded the Afghanistan Drinking Water Quality Standard (DWQS) of 0.05 mg/L of As. The spatial distribution of As concentration in the drinking water points is shown in the Figure 1.

3.2 Correlation among various water quality parameters

For understanding the hydrogeochemistry of As in the groundwater, the correlation coefficient (r) of As with pH (r = −0.18), Ca^{2+} (r = −0.214), Mg^{2+} (r = −0.176), Na^+ (r = 0.025), EC (r = 0.049), SO_4^{2-} (r = −0162), K^+ (r = 0.152), Cl^- (r = −0.032), HCO_3^- (r = −0.104), Mn^{2+} (r = 0.140), total Fe (r = −0.301), NO_3^- (r = 0.139), NH_4^+ (r = −0.290), SiO_2 (r = 0.156) and F^- (r = 0.058) were plotted by scatter plots using AquaChem 2014.1 software. The observed correlation of As with other water quality parameters indicated complex hydrochemical processes which contribute to mobilization of As in groundwater of the study area. The correlation between SO_4^{2-} and pH is negative, which would be the result of sulfide oxidation. The influencing

Table 1. Water quality statistic analysis results.

No	Elements	Unit	Statistics				Acceptable Limit	
			Count	Min.	Max.	Mean	WHO	NDWQS
1	As	mg/L	106	0.000	0.99	0.037	0.01	0.05
2	Conductivity	µS/cm	106	145	2440	1020	1500	3000
3	pH		106	6.33	8.52	7.43	6.5 − 8.5	6.5 − 8.5
4	ORP	mV	106	0	291	136		
5	Temp	°C	106	6.50	20.8	24.8		
6	Fe++	mg/L	106	0.00	0.30	0.06	0.3	0.3
7	Mn	mg/L	106	0	0.08	0.00	0.4	
8	Cl-	mg/L	106	2.5	500	138.2	250	250
9	PO4 ---	mg/L	106	0.02	1.80	0.51		
10	HCO3-	mg/L	106	115	1170	451		
11	NO3-	mg/L	106	4.20	127.20	41.91	50	50
12	Na+	mg/L	106	24	570	172	200	200
13	K+	mg/L	106	12	90	17		
14	Ca++	mg/L	106	14	200	75		70
15	Mg++	mg/L	106	11	190	43		30
16	Cu	mg/L	106	0.1	0.8	0	2	2
17	SO4--	mg/L	106	3	248	70	250	250
18	F-	mg/L	106	0.02	2.40	0.71	1.5	1.5
19	NH4+	mg/L	106	0.1	0.9	0	1.5 - 3.5	
20	Mn++	mg/L	106	0.0	0.8	0.00	0.05	

hydro-chemical may be dissolution of Fe- and Mn-oxide and sulfide dissolution.

3.3 Hydrochemical statistical analysis

The 106 sampled chemical tested data were analyzed statistically and the result is shown in Table 1.

4 CONCLUSIONS

The As contamination in the DWPs of study area is mostly geologic in origin and its distribution is patchy. Nearly 61% (459 out of 746) of the water samples from the DWPs exceeded the WHO guideline of 10 µg/L of As, and 38% (261 out of 746) of analyzed water samples exceeded the Afghanistan drinking water quality standard of 50 µg/L of As.

ACKNOWLEDGEMENTS

The author gratefully acknowledges from Khalil Rahman the DACAAR water quality analysis Laboratory supervisor for water samples collection and analysis. I also appreciate the cooperation and efforts of Ahmad Jawid, DACAAR Hydro geologist for analyzed water quality data recording and management.

REFERENCES

ANSA, 2013. Afghanistan Drinking Water Quality Standard.

Polhill, R.M. 1982. *Crotalaria in Africa and Madagascar*. Rotterdam: Balkema.

Uhl, V.W. Tahiri, M.Q.. 2003. An overview groundwater resources and challenges of Afghanistan. http://www.vuawater.com/Case-Study-Files/Afghanistan/Afghanistan_Overview_of_GW_Resources_Study-2003.pdf (Accessed on February 3, 2016).

U.S. Army, 2009. Southeast Afghanistan Water Resources Assessment.

WHO 2004. Guidelines for Drinking Water Quality, vol. 1; Recommendations, Third Edition. World Health Organization, Geneva.

Arsenic Research and Global Sustainability – Bhattacharya, Vahter, Jarsjö, Kumpiene,
Ahmad, Sparrenbom, Jacks, Donselaar, Bundschuh & Naidu (Eds)
© *2016 Taylor & Francis Group, London, ISBN 978-1-138-02941-5*

Tracking the fate of arsenic in groundwater discharged to the Meghna River

P.S.K. Knappett[1], K. Myers[1], P. Shuai[1], K. Rhodes[1], K. Jewell[1], J. Peterson[1], N. Dimova[2],
S. Datta[3], M. Berube[3], A. Hossain[4], M. Lipsi[4], S. Hossain[4], A. Hosain[4], K.M. Ahmed[4] &
M.B. Cardenas[5]

[1]*Department of Geology and Geophysics, Texas A&M University, College Station, TX, USA*
[2]*Department of Geological Sciences, University of Alabama, Tuscaloosa, AL, USA*
[3]*Department of Geology, Kansas State University, Manhattan, KS, USA*
[4]*Department of Geology, University of Dhaka, Dhaka, Bangladesh*
[5]*Jackson School of Geosciences, University of Texas at Austin, Austin, TX, USA*

ABSTRACT: The fate of arsenic fluxes to rivers from shallow aquifers is influenced by hydraulic head differences, and the geochemistry and geometry of aquifers and aquitards adjacent to the river. In our study area the eastern side of the Meghna River is generally strongly gaining year-round with some exceptions. The distribution of solid-phase Fe, Mn and As in riverbank sediments was correlated to variations in hydraulic gradients and hydraulic conductivity. The 30 m deep shallow aquifer was mapped 500 m parallel and orthogonal to the river bank at 3 locations, on both sides of the river, using Electrical Resistivity Tomography (ERT). The aquifer dimensions and properties are remarkably consistent between sites. The continuity of the 3–4 m capping clay layer will prevent shallow groundwater from discharging along the shallow river banks. Substantial seasonal fluctuations in dissolved As and Fe concentrations within the aquifer are related to irrigation pumping and natural river level fluctuations.

1 INTRODUCTION

The location, timing, volumetric and chemical fluxes of groundwater discharge to rivers in deltas is not well understood. Rivers that create deltas are influenced by tidal fluctuations. In tropical climates river levels are further influenced by wet and dry seasons. In the Ganges-Brahmaputra-Meghna Delta (GBMD), the Meghna River level and surrounding subsurface water tables fluctuate 3 m annually. The river level peaks at the end of the wet season in September and reaches its lowest point at the end of the dry season in April. Natural tidal and seasonal fluctuations coupled with irrigation pumping during the late dry season have the potential to cause reversals in the direction of water flow between shallow aquifers (<30 m) and the Meghna River. Previous studies have documented the existence of an enriched layer of solid-phase As and Fe within the first 1 m of Meghna River sediments (Datta *et al.*, 2009, Jung *et al.*, 2015). This was postulated to be caused by reducing, high As and Fe groundwater discharging across an oxidized hyporheic zone. The objective of this study is to characterize the hydrogeochemical interactions between the river and the aquifer to understand how the timing and location of groundwater discharge impacts the fate of redox sensitive elements such as As, Mn and Fe.

2 SITE DESCRIPTION AND METHODOLOGY

2.1 Site description

Our site is a 10 km reach of the Meghna River located 30 km east of Dhaka (Fig. 1). At this location the river flows within a single channel 2 km wide during the dry season. It has a natural depth of 10 m but this increases to 30 m in some locations, from local sand harvesting. Twenty kilometers south of the site the Meghna joins with the Padma River and flows out to the Bay of Bengal 200 km south. Tidal fluctuations penetrate 400 km north from the Bay of Bengal up the Meghna River. At our site the typical amplitude of the semi-diurnal tidal fluctuations is 60 cm. These high frequency fluctuations are nested within a 14 day neap-spring tide cycle with an approximate amplitude of 1 m.

2.2 Measuring groundwater discharge to the Meghna river

Three independent, complimentary methods were used to estimate the timing and quantity of groundwater discharge to the river along the 10 km study reach. The first was differential gaging with an Acoustic Doppler Current Profiler (ADCP). The second method employed the natural radioactive tracer, Radon (^{222}Rn). Through ^{222}Rn mass-balance model we were able to evaluate groundwater contri-

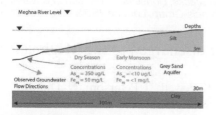

Figure 1. Dimensions of riverbank aquifer and observed seasonal flow directions and concentrations of As and Fe within our 10 km study reach of the Meghna River.

bution to the studied reach of the Meghna. Finally, we estimated groundwater discharge using Darcy's Law within 3 transects of monitoring wells, two on the western side and one on the eastern side.

2.3 Mapping aquifer properties

Groundwater discharge and chemical mass fluxes were up-scaled by mapping river bank aquifers with coring and ERT. Slug tests were performed on all monitoring wells at to measure hydraulic conductivity and one pumping test was performed at one site to estimate aquifer Transmissivity and Storativity.

2.4 Aqueous geochemistry

Nutrients and redox-sensitive elements were measured in the field using Chemetrics test kits on a V-2000 photometer and an alkalinity kit. Dissolved Organic Carbon (DOC) was measured. One 20 mL sample was filtered (0.45 μm) and acidified (HNO_3) for cation and trace metal analysis with an Ion Chromatograph (IC) and an ICP-MS, respectively.

2.5 Numerical modeling

The finite element numerical model COMSOL Multiphysics that solves Richard's equation was used to model the riverbank aquifer. This model allows the calculation of flow within the unsaturated zone, the saturated aquifer and the river allowing us to quantify pore water residence times.

3 RESULTS AND DISCUSSION

3.1 Observed groundwater discharge

Aquifer hydraulic gradients indicate that the river is generally gaining on the eastern side relative to the shallow (<30 m) aquifer. The exception is when an irrigation well is active at that site during the late dry season and a brief period during the early monsoon. This causes a local reversal in flow to away from the river. Hydraulic gradients on the western side of the river indicate the river is losing for more than half the year. The time of peak groundwater discharge is in the early dry period (January).

3.2 Aquifer properties and dimensions

Borehole logging, slug tests and ERT suggests the shallow aquifer is remarkably consistent in its dimensions

and hydraulic conductivity. The aquifer is underlain by a clay aquitard at 30 m depth, and overlain by a 3 m thick clay layer which extends >20 m out into the river from the dry season bank (Fig. 1). This surficial clay layer represents annual floodplain deposits and will greatly limit exchange between the oxic river water and the reducing groundwater. Instead of discharging across the seepage face, which was predicted by our numerical flow model assuming an homogeneous, isotropic aquifer, the presence of this clay cap will cause groundwater to discharge much further from the shoreline in deeper river water than where coring has demonstrated high solid-phase As and Fe.

3.3 Aquifer geochemistry

Aqueous As and Fe concentrations in the aquifer on the eastern bank fluctuated enormously throughout the year under the influence of irrigation pumping and seasonal changes in groundwater flow direction. Under gaining conditions (groundwater flowing towards the river) As and Fe concentrations reached 350 μg/L and 50 mg/L, respectively (Fig. 1). Under losing conditions, induced by irrigation pumping and early monsoon river levels, As and Fe were <10 μg/L and <1 mg/L, respectively. These aqueous geochemical changes were likely caused by induced movement of oxic river water into the aquifer during the late dry season and early monsoon. Such large seasonal geochemical changes have not previously been reported in shallow aquifers in Bangladesh.

4 CONCLUSIONS

Seasonal changes in redox-sensitive elements 100 m into the aquifer suggests a large oxidized hyporheic zone forms during the late dry season and early monsoon causes the subsequent accumulation of As, Fe and Mn during the early dry season when the river becomes gaining again.

ACKNOWLEDGEMENTS

This research is by the National Science Foundation (EAR-1344547) Additional funding to PK was provided by the Texas A&M Foundation. Partial funding for this work was provided to PS by a GSA Student Research Grant.

REFERENCES

Datta, S., Mailloux, B., Jung, H.-B., Hoque, M.A., Stute, M., Ahmed, K.M. and Zheng, Y. 2009. Redox trapping of arsenic during groundwater discharge in sediments from the Meghna riverbank in Bangladesh. *P. Natl. Acad. Sci. USA* 106(40): 16930–16935.

Jung, H.B., Zheng, Y., Rahman, M.W., Rahman, M.M. and Ahmed, K.M. 2015. Redox zonation and oscillation in the hyporheic zone of the Ganges-Brahmaputra Delta: implications for the fate of groundwater arsenic during discharge. *Appl. Geochem.* 63: 647–660.

Arsenic Research and Global Sustainability – Bhattacharya, Vahter, Jarsjö, Kumpiene,
Ahmad, Sparrenbom, Jacks, Donselaar, Bundschuh & Naidu (Eds)
© 2016 Taylor & Francis Group, London, ISBN 978-1-138-02941-5

Hydrogeochemical contrasts across the multi-level aquifers of Bengal basin in Matlab, Bangladesh: Implications for arsenic free and low-manganese drinking water sources

P. Bhattacharya[1], M. Hossain[1], G. Jacks[1], K. Matin Ahmed[2], M.A. Hasan[2],
M. von Brömssen[3] & S.K. Frape[4]

[1]*KTH-International Groundwater Arsenic Research Group, Department of Sustainable Development
Environmental Science and Technology, KTH Royal Institute of Technology, Stockholm, Sweden*
[2]*Department of Geology, Dhaka University, Dhaka, Bangladesh*
[3]*Department of Water Resources, Ramböll Water, Ramböll Sweden AB, Stockholm, Sweden*
[4]*Department of Earth and Environmental Sciences, University of Waterloo, Waterloo, Canada*

ABSTRACT: Targeting shallow, intermediate-deep and deep aquifers, piezometers nests were installed at 15 locations in the Matlab region, an As hot-spot in southeastern Bangladesh. Groundwater levels and water quality were monitored for over a three years period. Stable isotopic composition was used to identify the hydrogeological characteristics of different aquifers, hydraulic connectivity between the contaminated and safe aquifers. Within the shallow depth (up to 100m), two aquifers (Aquifer-1 and Aquifer-2) were identified, and groundwater from Aquifer-1 indicated consistently high As concentration was found to be As-enriched (median As levels upto 714 µg/L). Considerable variability in As concentrations were observed in Aquifer-2 wells (6–30 µg/L) comprising relatively oxidized or less reduced red and off-white sands. The intermediate-deep and deep aquifers were found to contain very low As concentration and these aquifers are hydraulically separated from the shallow aquifers. Groundwater depth and elevation and stable isotope signatures also reflect that intermediate-deep and deep aquifers, in most places belong to the same hydrostratigraphic unit (Aquifer-3).

1 INTRODUCTION

The widespread occurrence of natural arsenic (As) in groundwater through drinking water has drastically reduced the access to safe drinking water in Bangladesh. There has been very little success in mitigation since the discovery of As in the country in 1993 and tens of millions of people are exposed to concentration at levels above the Bangladesh drinking water standard (BDWS; 50 µg/L) and the WHO guideline (10 µg/L). The continued exposure of As through drinking water is challenge, and tubewells have emerged as the accepted option (von Brömssen *et al.*, 2007, Hossain *et al.*, 2015). Local drillers have been recognized as the principal drivers for installation of As-safe tubewells through identifying aquifers based on sediment colour (von Brömssen *et al.*, 2007, Hossain *et al.*, 2014, 2016).

This study deals with a comprehensive and systematic hydrogeological study in the Matlab, Bangladesh. This comprehensive study illustrates the approach included vertical profiling of hydrostratigraphic units (aquifer and aquitard); hydraulic head monitoring using depth-specific piezometers installed to target the shallow, intermediate-deep and deep specific aquifers. The study also considers aquifer sediments; monitoring of As, manganese and other redox parameters along with the major ions for a reasonable time-period; and analysis of groundwater samples for environmental isotopes of oxygen and hydrogen. This kind of characterization based on monitoring is essential for the installation of safe drinking water tubewells for As mitigation.

2 MATERIALS AND METHODS

2.1 Installation of piezometer nests

Matlab region, an As hotspot in Bangladesh (Fig. 1), was chosen for the study of the aquifers. Fifteen piezometer nests were drilled targeting the shallow, intermediate-deep and deep aquifers. In most of the nests four wells were drilled within depths up to 70 m, one well was drilled to a depth of ~110 m and a deep well down to about 235 m (Fig. 1).

For the shallow wells down to hand-percussion was used while the deep well was done using rotary reverse circulation drilling ("donkey drilling") method and sediments samples were collected for characterizing the sediment color (von Brömssen *et al.*, 2007, Hossain *et al.*, 2014).

2.2 Hydraulic head measurements, groundwater sampling and analyses

Groundwater levels were monitored for over a three years period starting pre-monsoon 2009 to post-monsoon 2013. However in the nests 13–17 the monitoring was started one year later including the installation of the P6 piezometers (Fig. 1). Groundwater samples were collected twice annually during

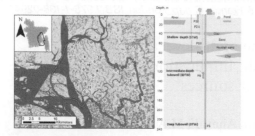

Figure 1. Location map of the study area around Matlab in South-eastern Bangladesh and the distribution of the piezometer nests and their depths.

pre-monsoon and post-monsoon seasons and analysed following the protocol discussed in Bhattacharya *et al.* (2002).

3 RESULTS AND DISCUSSION

3.1 *Hydraulic head monitoring*

The hydraulic heads measured for the piezometers in all 15 nests indicated that the shallow aquifers are separated from the intermediate and deep aquifers (Fig. 2). Groundwater level measurements in shallow piezometers P1 and P2 from all the nests in the study area clearly indicate that they are part of the same hydrostratigraphic unit (Aquifer-1). Shallow piezometers P3 and P4 provided hydraulic head data from Aquifer-2, although variations were observed. Intermediate-deep (P6) and deep (P5) piezometers revealed a similar pattern with considerable overlaps at many places indicating that they mostly belong to the same system (Aquifer-3).

3.2 *Hydrochemical trends, arsenic and redox sensitive parameters*

Distinct variations were observed in the hydrochemical composition of the groundwater samples from the three hydrostratigraphic units. Shallow groundwater abstracted from black colored sands are dominated by $Ca-Mg-HCO_3$ type and with increasing depth it changes to $Ca-Na-Mg-Cl-HCO_3$ to $Na-Cl-HCO_3$ type in off-white and red sediments and these variations reflect generic variations with the colour of the aquifer sediments. Groundwaters abstracted from intermediate-deep and deep aquifers are mostly $Na-Ca-Mg-Cl-HCO_3$ to $Na-Cl-HCO_3$ type.

Shallow P1 and P2 piezometers (Aquifer 1) indicated typically high As with median concentrations ranging between 71 and 646 µg/L, derived from the black sand aquifers. These groundwaters are characterized by elevated DOC, HCO_3, Fe, NH_4-N and PO_4-P and a relatively low concentration of Mn and SO_4 justifying the release of As due to reductive dissolution of Fe—oxyhydroxides (Bhattacharya *et al.*, 2002, von Brömssen *et al.*, 2007). Aquifer-2 monitored through shallow P3 and P4 piezometers revealed wide range of variability with median As

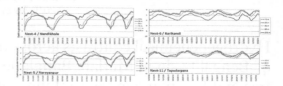

Figure 2. Monitoring groundwater levels (bgl) measured between 2009–2013, in nests 4,5,6 and 11 in Matlab, Bangladesh.

concentrations ranging from 6–30 µg/L and up to 216 µg/L in wells placed in the red and off-white and the black color sediments respectively. These groundwaters were typically low in DOC, HCO_3, Fe, NH_4-N and PO_4-P, but high concentration of Mn and SO_4. The As concentrations in the intermediate-deep (P6) and deep (P5) piezometers in Aquifer 3, were found to be low in As, within the WHO guideline value of 10 µg/L as well as the BDWS, with low concentration of DOC, HCO_3, NH_4-N and PO_4-P.

Sampling and analysis of stable isotopes (δ^3H and δ^{18}O) showed that the groundwater irrespective of depth had a similar origin, being rainwater or river water with a moderate effect of evaporation falling in the Global Meteoric Line.

4 CONCLUSIONS

The study this reveals a distinct hydrogeochemical contrast in the Matlab area in the shallow, intermediate-deep and deep aquifers. The groundwater heads in the shallow aquifers are separated from the intermediate and deep aquifers. The intermediate deep aquifers are connected to the deep aquifers with low As and low Mn to provide safe drinking water to the affected population.

ACKNOWLEDGEMENTS

We acknowledge the Swedish International Development Cooperation Agency (Sida) Global Program (Contribution 75000854).

REFERENCES

Bhattacharya, P., Jacks, G., Ahmed, K.M., et al. 2002. Arsenic in groundwater of the Bengal Delta Plain aquifers in Bangladesh. *B. Env. Contam. Tox.* 69: 538–545.
Hossain, M., Rahman, S.N., Bhattacharya, P., et al. 2015. Sustainability of arsenic mitigation interventions-an evaluation of different alternative safe drinking water options provided in Matlab, an arsenic hot spot in Bangladesh. *Front. Environ. Sci.* 3: 30. doi: 10.3389/fenvs.2015.00030.
Hossain, M., Bhattacharya, P., Jacks, G., et al. 2016. Enhancing the capacity of local drillers for installing arsenic-safe drinking water wells—experience from Matlab, Bangladesh. (Section 5, this Volume).
von Brömssen, M, Jakariya, M., Bhattacharya, P., et al. 2007. Targeting low-arsenic aquifers in groundwater of Matlab Upazila, SE Bangladesh. *Sci. Tot. Environ.* 379: 121–132.

Arsenic Research and Global Sustainability – Bhattacharya, Vahter, Jarsjö, Kumpiene,
Ahmad, Sparrenbom, Jacks, Donselaar, Bundschuh & Naidu (Eds)
© 2016 Taylor & Francis Group, London, ISBN 978-1-138-02941-5

Arsenic in the Bengal Delta Plain: Geochemical complications and potential mitigation option

D. Chatterjee[1], A.K. Kundu[1], S. Barman[1], U. Biswas[1], S. Majumder[2], D. Chatterjee[3]
& P. Bhattacharya[4]
[1]*Department of Chemistry, University of Kalyani, Kalyani, West Bengal, India*
[2]*Department of Environmental Management, International Centre for Ecological Engineering, University of Kalyani, Kalyani, West Bengal, India*
[3]*Heritage Institute of Technology, Chowbaga Road, Anandpur, East Kolkata Township, Kolkata, India*
[4]*KTH-International Groundwater Arsenic Research Group, Department of Sustainable Development, Environmental Science and Engineering, KTH Royal Institute of Technology, Stockholm, Sweden*

ABSTRACT: Groundwaters from the Bengal Delta Plain (BDP) are now significantly enriched with natural arsenic (As), frequently exceeding the WHO guideline value (10 µg/L). The contaminated groundwater is often derived from geologically young sediments (Holocene), low-lying areas and flat terrain where groundwater movement is slow (poorly flushed aquifers). The As content of the aquifer material is not regularly high (3–18 mg/kg), however, the groundwater As content is often exceptionally high (up to 3200 µg/L). The most notable feature of the tubewell groundwater is their predominantly reducing conditions at near-neutral pH values (6.5–7.5) with high redox sensitive species. The issue of deeper aquifer (safe and unsafe) is most challenging in terms of both geological and public health point of view. In this context, deeper aquifer is possibly the most reliable source where remediation technologies are in many cases incapable of yielding As-safe water.

1 INTRODUCTION

The wide-spread occurrence of As-rich groundwater in alluvium is difficult to explain by the low As contents of the Bengal Delta Plain (BDP) sedimentary environment (Chatterjee *et al.*, 2005). In contrast, over the years, it has become gradually evident that instead of absence of a local high As distributive source in the Holocene sediment, several hot-spots (As-rich groundwater > 150 µg/L) have been identified in various parts of BDP (Bhattacharya *et al.*, 2001, Bhattacharyya *et al.*, 2003). One of the major concerns of the BDP groundwater is the unpredictable nature of spatial variation in As concentration (Fendorf *et al.*, 2010). This heterogeneity is a matter of ongoing debate.

Several attempts (mostly hypothesis) have been made to explain the primary (origin) and secondary (mobilization) cause of the incidences of elevated As groundwater of this region and the spatial distribution pattern. In addition, the regional extension of the As distribution anomaly (spatio-temporal heterogeneity of high As patches) has also been significantly noticed in various part of the BDP (Nath *et al.*, 2008a,b, Chatterjee *et al.*, 2004, 2005). Many studies also help to explain the mechanism of As mobilization that primarily deals with release (high vs. low As areas) of As in groundwater (Bhattacharya *et al.*, 1997, 2011, Mukherjee *et al.* 2006). Finally, a model is worked out to accommodate the high release of As in groundwater where sorbed As on secondary phases (principally Fe-/Mn-oxides/hydroxides) that slowly breakdown and release As in groundwater under local anoxic condition (Bhattacharya *et al.*, 2001, 2003b; Islam *et al.*, 2004). However, several questions are remained unanswered in the context of controls of As release and the nature of host sedimentary environment that is responsible for high/low release of As in groundwater.

2 METHODS/EXPERIMENTAL

The Analytical Chemistry division of the University of Kalyani has been involved in groundwater arsenic research for over two decades. For this purpose, we have been collecting groundwater and sediment core samples (mostly focusing on and around the Chakdaha Block, West Bengal (Fig. 1) following the established standard protocols (Nath *et al.*, 2008a,b).

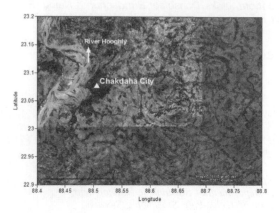

Figure 1. Map of the study area in Chakdaha block, Nadia district, West Bengal, India.

3 RESULTS AND DISCUSSION

The groundwater hydrochemistry study highlights that As is not equivocally present in anoxic environment where bicarbonates and redox species (Fe/Mn) are high. It is likely that an affirmative hydrochemical and/or a biological environment are the basic support to release As in groundwater but not sufficient condition for high As release in groundwater. The widespread high natural distribution of As in groundwater and it's variability does not allow us to isolate a specific class of geochemical process (inorganic redox processes driven by organic matters) rather than a combined complex suite of microbially mediated biogeochemical reaction and processes which appeared to occur simultaneously in very narrow band of water at the redox boundary/zonation and thereby releasing high As in groundwater.

In the deeper aquifer, the As mobilization mechanism is likely to be different from the shallow aquifer. The dissolution of carbonate/mixed carbonate is the major key factor for As release in the deeper aquifer of the BDP. The groundwater quality of the deeper aquifer reveals that the arsenic, alkalinity and manganese contents are relatively high in the deeper aquifer whereas iron content is relatively low. The presence of redox elements along with alkalinity in the deeper aquifers indicates that carbonates act as host environment for As release. The As in deeper aquifer also highlights the enrichment of As in Terminal Pleistocene-Holocene platform that was thought to be free of As (Chatterjee et al., 2005).

Finally, the major challenge is to explain the role of host sedimentary environment (mostly Holocene) and their interaction of their local pattern of slow groundwater flow (sediment-water interaction) to release As in groundwater. Furthermore, the elucidation of the causes for the often observed very heterogeneous As distribution (spatial variability and depth distribution) in the groundwater of contaminated shallow aquifers of the BDP is remained unanswered.

4 CONCLUSIONS

The BDP groundwaters are often reducing in nature with high Fe (II), bicarbonate and As. Reductive dissolution of Fe oxide/hydroxides is interpreted as the principal cause of As release in the environment under local reducing conditions. Recently, a few deeper wells are also found to be contaminated with As where the dissolution of carbonate/mixed carbonate is the key factor for As release and host environment is terminal Pleistocene-Holocene platform. The contamination in the deeper wells is a serious issue from public health point of view because deeper aquifers are mostly exploited for safe/low As drinking water supply. The south-east Asian countries are now working hard to meet up WHO guideline (> 10 µg/L) for their National Standard in a phase manner, and it is important that the remediation technologies should be capable of yielding As-safe water for community supply in anticipation of the future scenario. Until then, deeper aquifer in the BDP is the most reliable source for safe drinking water supply.

ACKNOWLEDGEMENTS

We acknowledge the laboratory support under the DFG-BMZ funded project for Research Co-operation with Developing Countries (Project Stu 169/37-1). We would also like to acknowledge the DST-PURSE fund, Government of India to support the research activities in the Department of Chemistry, University of Kalyani.

REFERENCES

Bhattacharya, P., Chatterjee, D., Jacks, G. 1997. Occurrence of arsenic contaminated groundwater in alluvial aquifers from Delta Plains, Eastern India: Options for safe drinking water supply. Int. J. Water Res. Dev. 13(1): 79–82.

Bhattacharya, P., Jacks G., Jana J., Sracek A., Gustafsson J.P., Chatterjee D. 2001. Geochemistry of the Holocene alluvial sediments of Bengal Delta Plain from West Bengal, India: Implications on arsenic contamination in groundwater. In: G. Jacks, P. Bhattacharya. A.A. Khan (Eds.) Groundwater Arsenic Contamination in the Bengal Delta Plain of Bangladesh. KTH Special Publication, TRITA-AMI Report 3084, pp.21–40.

Bhattacharya, P., Mukherjee, A., Mukherjee, A.B. 2011. Arsenic in groundwater of India. In: J.O. Nriagu (ed.) *Encyclopedia of Environmental Health*. vol. 1, pp. 150–164 Burlington: Elsevier. (DOI: 10.1016/B978-0-444-52272-6.00345-7).

Bhattacharyya, R., Jana, J., Nath, B., Sahu, S. J., Chatterjee, D., Jacks, G. 2003. Groundwater As mobilization in the Bengal Delta Plain, the use of ferralite as a possible remedial measure—A case study. *Appl. Geochem.* 18: 1435–1451.

Chatterjee, D., Chakraborty, S., Nath, B., Jana, J., Mukherjee, P., Sarkar, M. 2004. Geochemistry of arsenic in deltaic sediment. *Geochim. Cosmochim. Acta* 68: A514.

Chatterjee, D., Roy, R.K., Basu, B.B. 2005. Riddle of arsenic in groundwater of Bengal Delta Plain—role of non-inland source and redox traps. *Environ. Geol.* 49: 188–206.

Fendorf, S., Michael, H.A., van Geen, A. 2010. Spatial and temporal variations of groundwater arsenic in south and south East Asia. *Science* 328: 1123.

Islam, F.S., Gault, A.G., Boothman, C., Polya, D.A., Charnock, J.M., Chatterjee, D., Lloyd, J.R. 2004. Role of metal-reducing bacteria in arsenic release from Bengal delta sediments. Nature 430: 68–71.

Nath, B., Stuben, D., Basu Mallik, S., Chatterjee, D., Charlet, L. 2008a. Mobility of arsenic in West Bengal aquifers conducting low and high groundwater arsenic. Part I: Comparative hydrochemical & hydrogeological characteristics. *Appl. Geochem.* 23: 977–995.

Nath, B., Berner, Z., Chatterjee, D., Basu Mallik, S., Stuben, D. 2008b. Mobility of arsenic in West Bengal aquifers conducting low and high groundwater arsenic. Part II: Comparative geochemical profile & leaching study. *Appl. Geochem.* 23: 996–1011.

Arsenic Research and Global Sustainability – Bhattacharya, Vahter, Jarsjö, Kumpiene,
Ahmad, Sparrenbom, Jacks, Donselaar, Bundschuh & Naidu (Eds)
© *2016 Taylor & Francis Group, London, ISBN 978-1-138-02941-5*

Monsoonal influence on arsenic mobilization in groundwater: Geochemical and hydrogeological perspectives

S. Majumder[1,5], S. Datta[2], B. Nath[3], H. Neidhardt[4], Z. Berner[4], M. Hidalgo[5] & D. Chatterjee[6]

[1]*Department of Environmental Management, International Centre for Ecological Engineering,*
University of Kalyani, Kalyani, West Bengal, India
[2]*Department of Geology, Kansas State University, Manhattan, KS, USA*
[3]*Lamont-Doherty Earth Observatory of Columbia University, Palisades, NY, USA*
[4]*Institute of Mineralogy and Geochemistry, Karlsruhe Institute of Technology, KIT, Karlsruhe, Germany*
[5]*Department of Chemistry, University of Girona, Campus Montilivi, Girona, Spain*
[6]*Department of Chemistry, University of Kalyani, Kalyani, West Bengal, India*

ABSTRACT: In this study, we have investigated the potential role of monsoon on the hydrochemical evolution and concomitant As release using stable isotopes ($\delta^{18}O$ and δ^2H). A low-land flood plain and a natural levee have been selected for this purpose. Stable isotope signature depicts recharge of the shallow groundwater by evaporated surface water and Cl/Br molar ratio shows vertical recharge process in the wells within the flood plain area, especially during the post-monsoon season, while influences of both evaporation and vertical mixing are visible in the natural levee wells. Increase in DOC concentrations (from 1.33 to 6.29 mg/L), moving from pre- to post-monsoon season, indicates possible inflow of organic carbon to the aquifer. Simultaneous increase in As_T, Fe(II), and HCO_3^- during the post monsoon season highlights an ongoing reductive dissolution of As-rich Fe-oxyhydroxides. The subsequent increase in As(III) (> 200%), in both the flood plain and natural levee samples, with stable As_T concentration may refer to anaerobic microbial degradation of DOC coupled with the reduction of As(V) to As(III) without triggering additional As release from the aquifer sediments.

1 INTRODUCTION

In the Bengal Delta Plain (BDP), groundwater Arsenic (As) concentrations have been found to frequently exceed the WHO drinking water provisional guideline value of 10 μg/L. Oxygen-18 ($\delta^{18}O$) and deuterium (δ^2H) stable isotopic analysis have been successfully established as methods of choice to delineate hydrogeological environment and recharge processes in cases of groundwater studies, specifically in the case of As-enriched aquifers (Saha *et al.*, 2011). The Cl/Br ratio is another parameter that has been frequently used to delineate the nature of groundwater recharge and mixing (Cartwright & Weaver, 2005, Xie *et al.*, 2012).

The current study uses hydrochemical and isotopic data collected from wells of a high As (flood plain) and a low As (natural levee) area to assess the impact of monsoonal influence on As release and related species distribution in groundwater. This study further shows potential interaction between As and Dissolved Organic Carbon (DOC) in the aquifer.

2 METHODS/EXPERIMENTAL

The study area (latitude 23° 00′ 20″–23° 05′ 20″ N and longitude 88° 31′ 40″–88° 49′ 00″ E) is located in Chakdaha block, Nadia district, approximately 65 km north of Kolkata, West Bengal, India. Groundwater samples have been collected from wells in the flood plain (n = 21 and 22 in the pre- and post-monsoon season, respectively) as well as from wells located on the natural levees (n = 10 and 13 wells during pre- and post-monsoon season, respectively) for hydrochemical and stable isotopic ($\delta^{18}O$ and δ^2H) analyses. The samples for Fe(II) were measured immediately by the 1, 10 – phenanthroline method, using a Lambda 20 UV-VIS spectrophotometer (Perkin Elmer). Cations and As were analyzed using a HR-ICP-MS VG AXIOM, VG Elemental. Anions were determined by IC, DIONEX ICS 1000 equipped with a DIONEX AS 4 SC column. The DOC concentration was determined with a high TOC analyzer (Elementar, Hanau).

3 RESULTS AND DISCUSSION

The stable isotope composition of the ground water (Fig. 1) falls sub-parallel to the Global Meteoric Water Line (GMWL) (as in Craig, 1961). This indicates influences of evaporation on the groundwater samples, reflecting in turn recharge of the shallow groundwater by evaporated surface water, with ponds and infiltrating irrigation water as possible sources. The heavier isotopic composition during the Pre-Monsoon (PRM) season points at a greater influence of mixing with isotopically heavier water sources as compared to the Post-Monsoon (PSTM) season.

Different groundwater samples show two different trends (Fig. 2): 1) in case of major proportions of the natural levee samples and a few flood plain samples, not much change occurs in the Cl/Br ratio with increasing Cl$^-$ concentration and 2) for most of the flood plain samples, a sharp increase in the Cl/Br ratio is accompanied by less variation in respective Cl$^-$ concentrations. The wells in the flood plain area seem to have been influenced by moderate evaporation during the PRM season. On the contrary, the groundwater from the natural levee wells indicates a strong influence of evaporation and less mixing, although lateral mixing cannot be excluded.

Concomitant increase in As$_T$, Fe(II) and HCO$_3^-$ during the post monsoon season highlights an ongoing reductive dissolution of As-rich Fe-oxy-hydroxides. The concentration of DOC increases from 1.33 to 6.29 mg/L (in the flood plain region) in the post-monsoon season, and stimulates the microbial respiration, which drives the system in turn towards more reducing conditions that thermodynamically favor the reduction of As(V) to

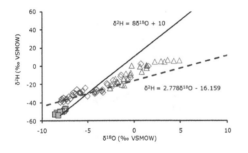

Figure 1. Stable isotope signature of different groundwater samples (Red diamonds—Flood plain PRM; Blue diamonds—Flood plain PSTM; Red triangles—Natural levee PRM; Blue triangles—Natural levee PSTM; White rectangles—Pond water PRM; Green rectangles—Pond water PSTM); Black solid line—GMWL; Blue dotted line—Local Evaporation Line (LEL).

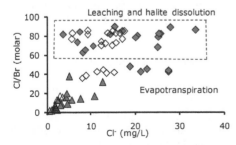

Figure 2. Relationship between Cl- concentration and Cl/Br molar ratio (White diamonds—Flood plain PRM; Blue diamonds—Flood plain PSTM; White triangles—Natural levee PRM; Blue triangles—Natural levee PSTM).

As(III). This supports the sharp increase (> 200%) in As(III) concentration with stable As$_T$ concentration in flood plain and natural levee samples. In that case, As(V) could have been the electron acceptor for DOC oxidation, which results in a conversion of As(V) to As(III) and no (or much less) mobilization of As from the solid phase occurs.

4 CONCLUSIONS

In the case of flood plain wells, the stable isotope signature and Cl/Br molar ratio have shown that vertical mixing could be the major recharge process, especially during the PSTM season. However, in case of natural levee wells, the relationship between As$_T$ concentration and the isotope profile shows a combined effect of both evaporation and mixing. The increase in As(III) proportions relative to the overall concentration of As$_T$ in post-monsoon samples from the flood plain is a major observation of this study. The positive correlation of DOC and HCO$_3^-$ with As(III)/As$_T$ ratios after the monsoon is most likely related to the microbial degradation of DOC, resulting in stronger reducing conditions in the aquifer. This decrease in the redox state is in turn linked to the reduction of dissolved As(V) to As(III), notably without triggering additional As release from the aquifer sediments.

ACKNOWLEDGEMENTS

Authors acknowledge the laboratory support from the DFG-BMZ Programme for Research Cooperation with Developing Countries (Project Stu 169/37–1). SM is thankful to the Erasmus Mundus External Cooperation Window (EMECW-Action II) for doctoral fellowship through EURINDIA programme.

REFERENCES

Cartwright, I. and Weaver, T.R. 2005. Hydrochemistry of the Goulburn Valley region of the Murray Basin, Australia: implications for flow paths and resource vulnerability. *Hydrogeol. J.* 13:752–770.

Craig, H. 1961. Isotopic variations in meteoric water. *Science* 133: 1702–1703.

Saha, D., Sinha, U.K. and Dwivedi, S.N. 2011. Characterization of recharge processes in shallow and deeper aquifers using isotopic signatures and geochemical behavior of groundwater in an arsenic-enriched part of the Ganga Plain. *Appl. Geochem.* 26: 432–443.

Xie, X., Wang, Y., Su, C., Li, J. and Li, M. 2012. Influence of irrigation practices on arsenic mobilization: Evidence from isotope composition and Cl/Br ratios in groundwater from Datong basin, northern China. *J. Hydrol.* 424–425: 37–47.

Arsenic Research and Global Sustainability – Bhattacharya, Vahter, Jarsjö, Kumpiene,
Ahmad, Sparrenbom, Jacks, Donselaar, Bundschuh & Naidu (Eds)
© 2016 Taylor & Francis Group, London, ISBN 978-1-138-02941-5

Variation of arsenic in shallow aquifers of the Bengal Basin: Controlling geochemical processes

A.K. Kundu[1], A. Biswas[1], S. Bhowmick[1], D. Chatterjee[1], A. Mukherjee[2],
H. Neidhardt[3], Z. Berner[3] & P. Bhattacharya[4]

[1]*Department of Chemistry, University of Kalyani, Kalyani, West Bengal, India*
[2]*Department of Geology and Geophysics, Indian Institute of Technology-Kharagpur, Kharagpur, India*
[3]*Institute of Mineralogy and Geochemistry, Karlsruhe Institute of Technology, KIT, Karlsruhe, Germany*
[4]*KTH International Groundwater Arsenic Research Group, Department of Sustainable Development,*
Environmental Science and Engineering, KTH Royal Institute of Technology, Stockholm, Sweden

ABSTRACT: The natural occurrence of high dissolved Arsenic (As) in groundwater is quiet common and has been reported from almost entire globe. Nevertheless the scale of problem is most severe in several countries of South and South East Asia, notably in BDP (Bengal Delta Plain). Different As release mechanism has been put forward to explain As enrichment in groundwater. The groundwater As distribution and their relationship with land-use pattern suggest that As release is influenced by local conditions (e.g. sanitation, presence of surface water, agricultural practice). High rate of groundwater withdrawl can accelerate As mobilization by enhancing the transport of degradable organic matter with recharge water from the surface.

1 INTRODUCTION

The presence of arsenic (As) in groundwater at elevated concentration poses a severe threat to human health worldwide (Bhattacharya *et al.*, 2011, Bhattacharyya *et al.*, 2003, Chatterjee *et al.*, 2005, Biswas *et al.*, 2011, 2014a,b) and the problem is most severe and alarming in groundwater of the Bengal Delta Plain (BDP). In this study, we would like to document a high resolution hydrogeochemical monitoring study from High and Low As site. The current hypothesis is that the reductive dissolution of Fe-oxyhydroxides coupled to the microbially catalyzed mineralization of organic matter releases As in groundwater of the Bengal Basin. The objective of the present study is to develop (or verify) the current understandings of the controlling geochemical process that may regulate the variation of As concentration in groundwater of shallow aquifers of BDP.

2 MATERIALS AND METHODS

Two sets of piezometers (2 × 5), installed at two sites in Nadia district West Bengal, India: i) Site-1 with high As in the vicinity of pond in Sahishpur Village, and ii) Site-2 with low-As close to an irrigation pumping well in Chakudanga village (Fig. 1). Further details on the installations and details are detailed in Biswas *et al.* (2011, 2014a). Groundwater samples were collected from the piezometers over a period of 21 months in acid washed Polyethylene (PE) bottles. On-site measurements were carried out for pH, Eh, conductivity, tempera-

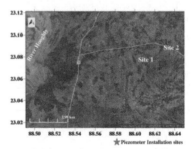

Figure 1. Satellite images of two study sites (Chakdaha block, West Bengal).

ture and alkalinity. Water samples were filtered using 0.45 μm membrane filter) and collected in three separate bottles: one for the analysis of total As (AsT), cations and trace elements (acidified with 7M HNO₃), for analysis using HR-ICP-MS, and anions by UV spectrophotometry and ion chromatography.

3 RESULTS AND DISCUSSION

Groundwater recharge origin at the As rich aquifers of Bengal basin is still a controversial issue. The concentration of major aqueous solutes, dissolved As is higher in Site-1 than Site-2. The higher concentration of chloride, sulphate, δ^2H and As_T are observed for groundwater at the uppermost part of the shallow aquifers (< 21 m) and increased continuously over the monitoring period at both sites. This study supports the view that the reductive dissolution of Fe-oxyhydroxides coupled with competitive PO_4^{3-} sorption reactions in the aquifer sediment enriches As in groundwater of the Bengal Basin. However, the additional Fe released by the weathering of silicate minerals, especially biotite, or the precipitation of Fe as secondary mineral phases such as siderite, vivianite and acid volatile sulfides may result in the decoupling of As and Fe enrichment in groundwater (Biswas et al., 2014b).

The redox zonation within the aquifer possibly regulates the vertical distribution of As in the groundwater in shallowest part of Site-1 as well as recharge of subsurface water. At both sites decreasing trend of DOC (Dissolved Organic Carbon) with increasing depth and enrichment of ammonium ion with increasing depth reflect reducing condition in deeper aquifer. Enrichment of Ca^{2+}, Mg^{2+}, and HCO^{3-} over Na^+ and K^+ in groundwaters of both sites again indicates that dissolution of carbonate minerals are prevailing in aquifers. On the other hand due to huge groundwater abstraction rate of recharge is relatively faster at Site-2, which causes relatively lower value of electrical conductivity i.e. concentration of major aqueous solutes. Groundwater from the shallowest well at Site-1 is more enriched by δ^2H and $\delta^{18}O$ than other deepest wells, which indicate vertical stratification of groundwater in the aquifer.

At Site-2 it is very much likely that the monitoring wells are within the influence of irrigation well, which accelerates the vertical recharge and decreases the groundwater residence time at our study site (viz. Stute et al., 2007). In addition, the anthropogenic input with recharge water possibly increased the concentrations of Cl^- in this part of the aquifer. The vertical layering of groundwater was absent in the aquifer at the low As site (Site-2). The absence of such layering and relatively low major ion concentrations and electrical conductivity could be linked to the enhanced aquifer flushing and decreased water–sediment interactions influenced by local-scale groundwater abstraction. The values of δ^2H and $\delta^{18}O$ in the monitoring samples of two sites and their composition to the Local Meteoric Water Line (LMWL) and Pond Evaporation Line (PEL) indicate that constructed pond do have some role in the groundwater recharge in our study area (Biswas et al., 2014b). The recharge through pond can contribute DOC to the shallow aquifer which would mobilize As either by migrating As plume from the nearest contaminated site or by perturbing the local redox equilibrium through mixing with the DOC (Mukherjee et al., 2008). The seasonal variations of As concentrations in groundwater were observed only in the shallowest part of the aquifers (<30 m).

4 CONCLUSION

The As concentrations in groundwater at the upper-most part of the shallow aquifers (<21 m) increased continuously over the monitoring period

at both sites. This study supports the view that the reductive dissolution of Feoxyhydroxides coupled with competitive PO^{43-} sorption reactions in the aquifer sediment enriches As in groundwater of the Bengal Basin. However, the additional Fe released by the weathering of silicate minerals, especially biotite, or the precipitation of Fe as secondary mineral phases such as siderite, vivianite and acid volatile sulfides may result in the decoupling of As and Fe enrichment in groundwater. The redox zonation within the aquifer possibly regulates the vertical distribution of As in the groundwater.

ACKNOWLEDGEMENTS

We acknowledge the German Research Foundation (DFG) and the German Federal Ministry for Economic Cooperation and Development (BMZ) (Stu 169/37-1) for providing the funding for this research

REFERENCES

Bhattacharya, P., Mukherjee, A., Mukherjee, A.B. 2011. Arsenic in groundwater of India. In: J.O. Nriagu (Ed.) Encyclopedia of Environmental Health. Vol. 1, Elsevier, Burlington. pp. 150–164. (DOI: 10.1016/B978-0-444-52272-6.00345-7).

Bhattacharyya, R., Jana, J., Nath, B., Sahu, S. J., Chatterjee, D., Jacks, G. 2003. Groundwater As mobilization in the Bengal Delta Plain, the use of ferralite as a possible remedial measure—a case study. Appl. Geochem. 18: 1435–1451.

Biswas, A., Majumder, S., Neidhardt, H., Halder, D., Bhowmick, S., Mukherjee-Goswami, A., Kundu, A., Saha, D., Berner, Z., Chatterjee D. 2011. Groundwater chemistry and redox processes: depth dependent arsenic release mechanism. Appl. Geochem. 26: 516–525.

Biswas, A., Gustafsson, J.P., Neidhardt, H., Halder, D., Kundu, A.K., Chatterjee, D., Berner, Z., Bhattacharya, P. 2014a. Role of competing ions in the mobilization of arsenic in groundwater of Bengal Basin: Insight from surface complexation modelling. Water Res. 55: 30–39.

Biswas, A., Neidhardt, H., Kundu, A.K., Halder, D., Chatterjee, D., Berner, Z., Jacks, G., Bhattacharya, P. 2014b. Spatial, vertical and temporal variation of arsenic in the shallow aquifers of Bengal Basin: Controlling geochemical processes. Chem. Geol. 387: 157–169.

Chatterjee, D., Roy, R.K., Basu, B.B. 2005. Riddle of arsenic in groundwater of Bengal Delta Plain – role of noninland source and redox traps. Environ. Geol. 49: 188–206.

Mukherjee, A., von Brömssen, M., Scanlon, B.R., Bhattacharya, P., Fryar, A.E., Hasan, M.A., Ahmed, K.M., Chatterjee, D., Jacks, G., Sracek, O. 2008. Hydrogeochemical comparison and effects of overlapping redox zones on groundwater arsenic near the western (Bhagirathi sub-basin, India) and Eastern (Meghna sub-basin, Bangladesh) margins of the Bengal Basin. J. Cont. Hydrol. 99: 31–48.

Stute, M., Zheng, Y., Schlosser, P., Horneman, A., Dhar, R.K., Hoque, M.A., Seddique, A.A., Shamsudduha, M., Ahmed, K.M., van Geen, A. 2007. Hydrological control of As concentrations in Bangladesh groundwater. Water Resour. Res. 43: W09417.

Arsenic Research and Global Sustainability – Bhattacharya, Vahter, Jarsjö, Kumpiene,
Ahmad, Sparrenbom, Jacks, Donselaar, Bundschuh & Naidu (Eds)
© 2016 Taylor & Francis Group, London, ISBN 978-1-138-02941-5

Low arsenic zones in shallow aquifer system of contaminated areas of middle Ganga Plain, India

D. Saha
Central Ground Water Board, Government of India, New Delhi, India

ABSTRACT: Significant groundwater arsenic (As) contamination exists in the central part of the Gangetic Plains. The present study uses multi-tiered approach; including shallow alluvial stratigraphy, As mobilisation process and geomorphological set up and hydraulic parameters of shallow depth sand aquifers to delineate the low-As conc. areas. These areas can be developed for household scale drinking water supply through hand pumps.

1 INTRODUCTION

The Middle Ganga Plain (MGP), covers the central part of the Gangetic Plains, marked between two subsurface basement ridges, Faizabad in the west and Munger–Saharsa in the east (Saha *et al.*, 2007). Administratively it spread over parts of eastern Uttar Pradesh, central and northern Bihar and north-eastern tip of the Jharkhand states. Though the arsenic (As) contamination is widely discussed for last three decades from the Lower Ganga Plain and the Deltaic areas covering the states of West Bengal in India and Bangladesh (Bhattacharya *et al.*, 1997, Ahmed *et al.*, 2004, Ravenscroft *et al.*, 2005), in the upstream part of the Gangetic Plains (West of Rajmahal Hills) the contamination (> 50 μg/L) was reported later in 2003 (Chakraborti *et al.*, 2003). Subsequently, the contamination has been reported from many parts of the Middle and Upper Ganga Plains.

During the period 2003–2015 significant research has been carried out in the MGP (Sahu & Saha 2015). The contaminated areas are dominantly confined along the Ganga River course, on both the banks (Saha, 2009), though in few places the contamination invades deep inside the flood plains. In Bihar State, it is reported from parts of 15 districts whereas, in Uttar Pradesh and Jharkhand States elevated conc. are reported from eight and one district respectively (Sahu & Saha 2015) the affected districts are shown in Figure 1.

The elevated concentrations of As in groundwater is confined within the 70 m below ground, affecting the potential shallow aquifer system, which is the lifeline for drinking and irrigation water supply in the area (Saha *et al.*, 2011). The present research aims to study the mobilization process, shallow stratigraphy and geomorphology

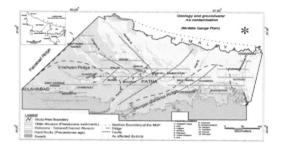

Figure 1. Geology of MGP and contaminated districts.

of the area to understand its spatial variation and ways to delineate the low As areas.

2 METHODS

The research involves interpretation of satellite imagery and Survey of India topographic maps. Field survey conducted to collect data, sediment and water samples, followed by their analyses. Various thematic maps were prepared, geologic sections were drawn and the analytical and collected data were compiled and analyzed in conjunction with the field survey results.

3 RESULTS AND DISCUSSION

The Newer Alluvium belt developed during the Holocene and Recent age is found to be affected in the area (Fig. 1). The lithology of the formations within 70 m below ground, where elevated concentration is confined (Saha & Shukla, 2013), is characterized by fine to very fine sand (micaceous), silt and mud of gray to dark gray clay. At places black mud layers are also seen. By and large, the sand layers within this depth constitute shallow aquifer system. The top part of the formation is often silty or marked by a layer of mud/clay, forming a relatively low permeable zone.

The distribution of As in groundwater has close relation with the geomorphic sub-environments and the sedimentary facies. The contaminated zones (> 50 μg/L) are found to be spatially related with dark gray to black colored organic rich clay sediments, found both in the Active and Older Flood Plain terraces of the Newer Alluvium (NA). These facies assemblages are particularly observed in the abandoned/palaeochannel cut-offs, either filled or in the process of filling under fluvio-lacustrine environments.

The entire NA belt, particularly the Active Flood Plain is flood-prone, where many villages are located on elevated lands. Such settlements, located on anthropogenic fills, particularly covering clay/mud plugged palaeochannels (in ~80% of the tested sources) as well as close to the active channel cut-off lakes (~30% cases), are found to be the hotspots of contamination (max. 987 μg/L). Moderate to high levels of conc. (max. 342 μg/L) are reported (in 50% cases) in areas close to such palaeochannels and the areas traversed by cross-bar channels. The point bars, levees and point bar platforms, predominated by sand facies, are by and large low in As (89–94% cases exhibit < 50 μg/L).

The controlling factors for release of As in the shallow aquifers in MGP are the thickness of the dark gray to black colored clay/mud cover overlying the shallow aquifer sand and the availability of organic carbon are. The spatial variability of conc. is closely associated with the variability of organic matter content in the formation (clay/mud facies). The organic carbon when released into groundwater, creates a reducing front due to consumption and metabolism of the carbon by bacteria at the expense of available dissolved oxygen in groundwater. The result is reductive dissolution of Hydrated Iron Oxide (HFO) coatings on the grains, where As remains adsorbed, and thereby releasing As in groundwater. Lateral and vertical spread of the reducing environment depends upon the volume of organic carbon release, groundwater flow regime, hydraulic conductivity of the formation, and the volume of fresh oxic water recharge that mingles with the spreading anoxic front (Sahu & Saha, 2015). The sandier areas along the scroll bar ridges and levees exhibits low concentration of As as the oxic groundwater in such areas favors stability of hydrated iron oxide.

4 CONCLUSIONS

The research establishes a close relationship between the geomorphological features and the shallow alluvial stratigraphy and the distribution and extent of As concentration in ground water in the MGP. The findings can be used for societal benefits as the shallow aquifers in the low As areas can be used for house-hold scale water supply by hand pumps.

ACKNOWLEDGEMENTS

The author extends thanks to the Chairman, Central Ground Water Board for his critical views and guidance. He also thanks his colleagues for their help and support.

REFERENCES

Ahmed, K.M., Bhattacharya, P., Hasan, M.A., Akhter, S.H., Alam, S.M.M., Bhuyian, M.A.H., Imam, M.B., Khan, A.A., Sracek, O. 2004. Arsenic contamination in groundwater of alluvial aquifers in Bangladesh: An overview. *Appl. Geochem.* 19(2): 181–200.

Bhattacharya, P., Chatterjee, D., Jacks, G. 1997. Occurrence of arsenic-contaminated groundwater in alluvial aquifers from delta plains, eastern India: Options for safe drinking water supply. *Int. J. Water Res. Dev.* 13(1): 79–92.

Sahu S., Saha D. 2015. Role of shallow alluvial stratigraphy and Holocene geomorphology on groundwater arsenic contamination in the Middle Ganga Plain, India. *Environ. Earth Sci.* 73(7): 3523–3536.

Chakraborti, D., Mukherjee S.C., Pati S., Sengupta M.K., Rahman M.M.,Chowdhury U.K., Lodh D., Chanda C.R., Chakraborti A.K., Basu G.K. 2003. Arsenic groundwater contamination in Middle GangaPlain, Bihar, India: a future danger? *Environ. Health Perspect.* 111:1194–1201.

Saha D. 2009. Arsenic groundwater contamination in parts of Middle Ganga Plain, Bihar. *Curr. Sci..* 97(6): 753–755.

Saha D., Upadhyay, S., Dhar, Y.R., Singh, S. 2007. The aquifer system and evaluation of hydraulic parametersin parts of South Ganga Plain, Bihar, Eastern India. *J. Geol. Soc. India* 69: 1031–1041

Saha D., Sahu S., Chandra P.C. 2011. Delineating arsenic-safe deeper aquifers and their hydraulic parameters in parts of Middle Ganga Plain, Eastern India. *Environ. Monit. Assess.* 175(1): 331–348.

Saha D., Shukla R.R. 2013. Genesis of arsenic rich groundwater and the search for alternative safe aquifers in the Gangetic Plain, India. Water Environment Research 85(12): 2254–2264.

Ravenscroft P., Burgess W.G., Ahmed K.M., Burren M., Perrin J. 2005. Arsenic in groundwater of the Bengal Basin, Bangladesh: Distribution, field relations, and hydrogeological setting. *Hydrogeol. J.* 13: 727–751.

Arsenic Research and Global Sustainability – Bhattacharya, Vahter, Jarsjö, Kumpiene,
Ahmad, Sparrenbom, Jacks, Donselaar, Bundschuh & Naidu (Eds)
© 2016 Taylor & Francis Group, London, ISBN 978-1-138-02941-5

Sporadic occurrence of groundwater arsenic: Is it still a mystery?

S. Sahu[1] & D. Saha[2]
[1]Central Ground Water Board, MOWR, RD & GR, SER, Bhujal Bhawan, Khandagiri, Bhubaneswar, Odisha, India
[2]Central Ground Water Board, Government of India, New Delhi, India

ABSTRACT: The distribution of groundwater arsenic (As) in the floodplains of Middle Ganga Plain (MGP) in India, though sporadic, bears specific relation with the floodplain geomorphology and Late Quaternary stratigraphy. The sediments comprise grey to dark grey mud/clay, silt and fine to very fine sand of Holocene age, mainly derived from the Himalaya. A certain correlation is observed between the thickness of the Organic Carbon (OC) rich clay plugs (hanging within the shallow aquifer) and the concentration of the As in groundwater (R^2: ~66%). The floodplain patches rich in sand depict low groundwater As.

1 INTRODUCTION

The sporadic occurrence of elevated concentrations of groundwater arsenic (As) beyond the permissible limits for drinking (WHO, 1993 and BIS, 2012) limits of 10 and 50 µg/L respectively has been a wonder for the researchers, planners and the common people as well. Hand pumps only a few tens of meters apart, tapping the same aquifer at same depth, and even the hand pumps at same location tapping aquifers at different depths may show considerable differences in their As concentrations.

In the Bihar State in MGP, the major part of the contaminated areas falls in the low-lying floodplains of the Ganga River comprising the Newer Alluvium (NA) of Holocene age (Fig. 1). Around ~10 million people (9.6% of state populace) residing in the risk zone (considering the limit of 50 µg/L of As). In the present study we focus to understand the mystery behind such spatial variability in the occurrence of groundwater As in MGP.

2 METHODS

The satellite images (IRS ID, sensor LISS III, months: May and Nov.) were interpreted for main geomorphologic and the morphostratigraphic units

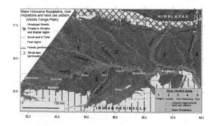

Figure 1. Major floodplains in MGP showing groundwater As contamination (modified after Saha & Sahu, 2015).

in the floodplain. We analyzed ~30 shallow and 91 deeper borehole lithologs (Fig. 2) along with the study of sub-surface alluvial exposures for understanding the sedimentation pattern in different geomorphic environments, the stratigraphic disposition of sediment types and the configuration of aquifers.

The results of As concentration in 285 groundwater samples, collected from hand pumps (depth range: ~10–30 m below ground), were interpreted for their relation with the floodplain geomorphic elements and the sub-surface lithology. The maximum detected As concentrations were related with thickness of Organic Carbon (OC) rich clay plugs for 38 locations, where both the data were available.

3 RESULTS AND DISCUSSION

From last several years, we have field-touch in Bihar State which has been the second major risk zone of groundwater As contamination after the Bengal Basin. The maximum concentration of groundwater As detected in the state stands at 1861 µm/L. Elevated As occurs largely within the depth of ~50 m below ground level (bgl) with the higher incidence within ~30 m bgl.

The NA sediments, deposited in fluvio-lacustrine environment, are grey to dark grey in colour (unoxidized), comprising clay/mud, silt and fine to very fine micaceous sand, derived from the Himalaya by the Ganga and other rivers. Beyond the NA sediments, the craton derived brownish yellow (oxidized) coarser sediments (Pleistocene age) follow (Fig. 2), which depict low groundwater As. The contaminated hand pumps are clustered in villages/settlements located around/over the OC rich clay plugs (thickness: a few centimeters to ~13 m) hanging within the shallow aquifers (Fig. 3). A clay plug originating from the cut off channel of the Ganga can measure up to ~0.21 km³.

A certain correlation is observed between the thickness of the clay plugs and the As levels in

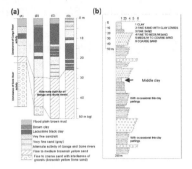

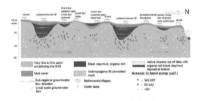

Figure 2. (a) Shallow lithologs from different geomorphic environments in the MGP. A–Point bar platform sequence with cross bar channel, B and C–Channel fill sequence, D–Point bar sequence with buried clay plug. (b) Deeper borehole lithology at Bharauli in the area depicting the disposition of Himalayan and craton derived sediments (modified after Saha *et al.*, 2011; Sahu & Saha, 2015).

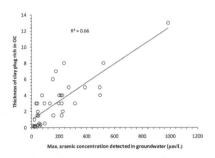

Figure 3. A schematic section showing generalized topography, floodplain morphology and shallow alluvial stratigraphy in MGP. It depicts the concentrations of As in hand pumps in settlements distributed in different geomorphic environments (modified after Sahu & Saha 2014, 2015).

Figure 4. Plot of maximum groundwater As detected in different geomorphic environments in the floodplains of Ganga and the thickness of OC rich clay plug.

groundwater (R^2: ~66%) (Fig. 4). The causative factors of release of As to groundwater (which can also help in predicting As in floodplain) include: (1) availability and content of OC in the stratigraphic column, (2) release of OC to groundwater, status of groundwater development, agitation and fluctuation in water table, (3) hydraulic conductivity in the aquifer, (4) groundwater flow (direction), and (5) the volume of fresh oxic water recharge.

The OC in groundwater is consumed by the microbes present in the shallow aquifer, which in turn consumes the dissolved oxygen in ground-

water. It causes dissolution of HFO, releasing As in groundwater. The dissolution of HFO in As affected areas of MGP is evident by high concentration of dissolved iron in groundwater also.

2 CONCLUSIONS

The groundwater As contamination is associated with OC rich clay plugs present in stratigraphic columns. Release of As to groundwater depends on the availability of OC and its release to groundwater. The spatial variability of groundwater As follows the spatial variability of such the plugs in the stratigraphic column

ACKNOWLEDGEMENTS

The authors express sincere thanks to K. B. Biswas and K.C. Naik for their help. Thanks are also extended to D.P. Pati and V. Srivastava for their suggestions. The opinions expressed in this paper are of authors of their own.

REFERENCES

Bureau of Indian Standard (BIS), 2012. Indian standard specification for drinking water- 10500. Government of India.
Nickson, R.T., McArthur, J.M., Burgess, W., Ahmed, K.M., Ravenscroft, P., Rahman, M. 1998. Arsenic poisoning of groundwater in Bangladesh. Nature 395:338.
Saha, D. 2016. Low arsenic zones in shallow aquifer system of contaminated areas of Middle Ganga Plain, India. In: P. Bhattacharya, M. Vahter, J. Jarsjö, J. Kumpiene, A. Ahmad, C. Sparrenbom, G. Jacks, M.E. Donselaar, J. Bundschuh, R. Naidu (Eds.) *Arsenic Research and Global Sustainability* (As 2016). CRC Press (This Volume).
Saha, D. Sahu, S. 2016. A decade of investigations on groundwater arsenic contamination in Middle Ganga Plain, India. Environ. Geochem. Health Vol 38, pp. 315–337.
Saha, D., Sahu, S., Chandra, P.C. 2011. Arsenic-safe deeper aquifers and their hydraulic parameters in parts of Middle Ganga Plain, Eastern India. *Environ. Monit. Assess.* Vol 175 (1), pp. 331–348.
Sahu, S. 2013. Hydrogeological Conditions and Geogenic Pollution in parts of Western Bihar. *PhD Thesis,* Banaras Hindu University, Varanasi. Uttar Pradesh, India.
Sahu, S. Saha, D. 2014. Geomorphologic, stratigraphic and sedimentologic evidences of tectonic activity in Sone-Ganga alluvial tract in Middle Ganga Plain, India. *J. Earth System Sci.* 123(6): 1335–1347.
Sahu, S. Saha, D. 2015. Role of shallow alluvial stratigraphy, and Holocene geomorphology on groundwater arsenic contamination in Middle Ganga Plain, India. *Environ. Earth Sci.* 73(7): 3523–3536.
Shah, B.A. 2008. Role of quaternary stratigraphy on arsenic contaminated groundwater from parts of Middle Ganga Plain, UP–Bihar, India. *Environ. Geol.* 53: 1553–1561.
van Geen, A., Zheng, Y., Versteeg, R., Stute, M, Horneman, A., Dhar, R., Steckler, M., Gelman, A., Small, C., Ahsan, H., Graziano, J., Hussein, I., Ahmed, K.M. 2003. Spatial variability of arsenic in 6000 tube wells in a 25 km2 area of Bangladesh. *Water Resour. Res.* 39:1140–1155. doi:10.1029/2002WR001617.
WHO, 1993. *Guidelines for Drinking Water Quality.* 2nd edn, v 1, WHO, Geneva.

Arsenic contaminated groundwater of the Varanasi environs in the middle Ganga Plain, India: Source and distribution

N.J. Raju & S. Singh
School of Environmental Sciences, Jawaharlal Nehru University, New Delhi, India

ABSTRACT: To identify the arsenic (As) contamination zones in the Varanasi environs, 95 groundwater samples on either side of Ganga River and four borehole litho-logs at every 3 m interval have been collected and analyzed for various parameters. The As concentration in the groundwater samples of the study area varies from 1 to 80 μg/L. The average As and Fe concentration in the selective sediment samples of litho-logs were 5.8 mg/kg and 529 mg/kg, respectively. The high As concentration is found in the north-eastern part of the meandering Ganga River in the Varanasi environs. In the study area, the western side of Ganga River is free from the As contamination in groundwater though high concentrations of As and iron content in borehole sediment samples are found in between 20–70 m depths.

1 INTRODUCTION

Arsenic (As) contamination of groundwater, exceeding permissible limit of 10 μg/L, is in the global forefront of public health issues for last few decades. In Middle Ganga Plain, in general, groundwater As contamination zones are located mainly along the river Ganga, particularly in its abandoned meander belts. Arsenic contaminated aquifers are confined within lowland organic rich clayey deltaic sediments in the Bengal basin and locally within similar facies in narrow entrenched river channels and isolated tracts of flood plains in Middle Ganga Plain (MGP) (Acharyya, 2005). Holocene sediments, deposited by the Ganga River system forming the major alluvial deposit in Ganga alluvial plain reported to be responsible for As Contamination (Chakraborti *et al.*, 2004). An attempt was made to identify sources and distribution of As contamination zones in and around the Varanasi environs.

1.1 Study area

The study area (~200 km²) situated in the Middle Gangetic Plain of Indian sub-continent and lies in between the latitude 25°15'00"-25°21'03"N and 82°57'00"-83° 07'42"E. The area is mainly underlain by Gangetic alluvium which consists of inter-bedded layers of sand, silt and clay with kankar. Ninety five groundwater samples (Fig. 1) from dugwells, hand pumps and deep borewells on either side of the Ganga River have been collected. Raju *et al.* (2011) have studied hydrogeological characters of the Pleistocene deposits in the middle Ganga plain.

2 METHODS AND MATERIALS

A total of ninety five groundwater samples (68 samples—May 2007 and 27 samples—May 2012) from dug wells hand pumps and deep borewells (Fig. 1)

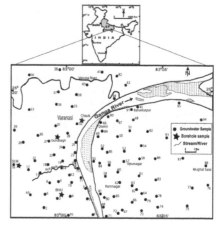

Figure 1. Physiographic and sample location map of the Varanasi environs.

of the study area were collected and analyzed to understand the chemical variations of groundwater. Physico-chemical characteristics of groundwater samples were determined using the standard analytical methods (APHA 2005). Total As in groundwater was determined by FI-HG-AAS.

3 RESULTS AND DISCUSSION

3.1 Chemical quality of groundwater

The range of hydrochemical parameters and their comparison with WHO (2011) are summarized in Table 1. It is observed that most of hydrochemical parameters of the groundwater samples are within the permissible limits of WHO, but 2%, 6%, 12%, 7%, 15% and 6% of Ca^{2+}, Na^+, K^+, HCO_3^-, NO_3^- and F^- are exceeding permissible limits, respectively.

Table 1. Range of chemical parameters and their comparison with WHO (2011) standards for drinking water.

Parameter	Concentration (All values are in mg/L except As is μg/L)			Permissible limits	% exceeding permissible limits
	Min.	Max.	Average		
Ca^{2+}	10	260	76.6	200	2
Mg^{2+}	0.4	127	39.3	150	–
Na^+	13.9	288	90	200	6
K^+	0.6	109	8.9	12	12
HCO_3^-	145	720	386	600	7
SO_4^{2-}	2.5	208	46	600	–
Cl^-	12	494	116	600	–
NO_3^-	0.82	106	19.8	50	15
F^-	0.2	2	0.8	1.5	6
pH	6.9	8.2	7.52	9.2	–
TDS	232	1175	517	1500	–
Fe	0.02	6.9	1.08	0.3	76
As	1	80	10	10	20

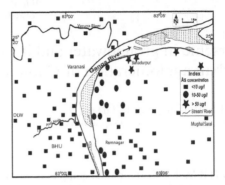

Figure 2. Spatial distribution of arsenic in the Varanasi environs.

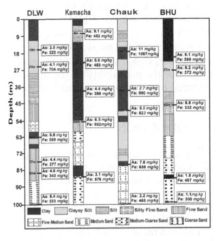

Figure 3. Lithologs and selective depth-wise As and Fe content in bore hole sediments.

3.2 Distribution of As and Fe in groundwater and borehole sediment sample

The chemical analysis of groundwater indicated that As and Fe content varies from 1 to 80 μg/L with an average of 10 μg/L and 0.02 to 6.9 mg/L with an average of 1.08 mg/L, respectively. Fe content of the groundwater samples show that 76% of the samples exceeding permissible limit. Arsenic content in groundwater samples shows that 20% of the samples are exceeding 10 μg/L and 4% of the samples exceeding 50 μg/L (Fig. 2). The high As concentration is found in the northeastern (concave side of river Ganga) parts of Varanasi environs.

Lithofacies sequence of the sediments and vertical distribution of As and Fe in boreholes were prepared from exploratory borehole lithologs (Fig. 3). In the study area, As and Fe concentration in the selective sediment samples of bore hole lithologs were ranges from 1.8 mg/kg to 11.9 mg/kg with an average of 5.8 mg/kg and 222 mg/kg to 1097 mg/kg with an average of 529 mg/kg, respectively.

4 CONCLUSIONS

Groundwater in Varanasi city is virtually As safe (<10 μg/L) due to its position in Pleistocene older alluvium upland surfaces (western side of Ganga) even though borehole sediments shows high As and Fe content in shallow depths whereas villages located in Holocene newer alluvium sediments (eastern side of Ganga) in enriched channels and floodplains of Ganga River have As contaminated groundwater.

ACKNOWLEDGEMENTS

The author is thankful to the Department of Science and Technology (DST), New Delhi for financial support under the research project (SR/S4/ES-160/2005) during 2006–2008.

REFERENCES

Acharyya, S.K. 2005. Arsenic levels in groundwater from Quaternary alluvium in the Ganga plain and the Bengal basin, Indian sub-continent: insight into influence of stratigraphy. Gondwana Research 8: 1–12.

APHA, 2005. Standard Methods for the Examination of Water and Wastewater. 20th Edn., Washington.

Chakraborti, D., Sengupta, M.K., Rahman, M.M., Ahmed, S., Chowdhury, U.K., Hossain, M.A., Mukherjee, S.C., Pati, S., Saha, K.C., Dutta, R.N., Quanmruzzaman, Q. 2004. Groundwater arsenic contaminationand its health effects in the Ganga-Meghna-Brahmaputra plain. J. Environ. Monit. 6: 64–74.

Raju, N.J., Shukla, U.K., Ram, P. 2011. Hydrogeochemistry for the assessment of groundwater quality in Varanasi: a fast urbanizing center in Uttar Pradesh, India. Environ. Monit. Assess. 173: 279–300.

World Health Organization. 2011. Guideline for Drinking Water Quality. 4th Edn. WHO, Geneva, 340p.

Arsenic Research and Global Sustainability – Bhattacharya, Vahter, Jarsjö, Kumpiene,
Ahmad, Sparrenbom, Jacks, Donselaar, Bundschuh & Naidu (Eds)
© 2016 Taylor & Francis Group, London, ISBN 978-1-138-02941-5

Arsenic in groundwater and its potential health risk in a fast growing urban agglomeration of Chota Nagpur Plateau, India

T. Bhattacharya & P. Tirkey

Department of Civil and Environmental Engineering, Birla Institute of Technology, Mesra, Ranchi, Jharkhand, India

ABSTRACT: Groundwater samples from 44 locations of Ranchi urban agglomeration, a fast growing urban center in Chotanagpur plateau of eastern India were collected to assess the drinking water quality with respect to arsenic and trace metals important from public health point of view. The results show that arsenic (As) concentration ranged from bdl – 200 µg/L and bdl – 150 µg/L in monsoon and pre-monsoon season respectively, but mostly below detection limit during post-monsoon season. showed Arsenic concentration were above the WHO drinking water guideline as well as the BIS acceptable limit of 10 µg/L (WHO/BIS), in 84% of the wells sampled during the monsoon season. Arsenic did not show strong correlation with any of the physico-chemical parameters. Principal component analysis showed water quality is dominated by dissolved solids and suggest similar origin for most of the parameters. Health risk assessment signifies the probability of the non-carcinogenic effect due to the presence of arsenic and selenium as compared to carcinogenic effect is more for the residents of the study area.

1 INTRODUCTION

The presence of metals in water can cause intellectual malfunction, lung and kidney problem, gastro-intestinal distress, pulmonary fibrosis and skin dermatitis and even cancer (Dieter *et al.*, 2005). However, the most alarming contaminant and threat to public health in the north-eastern part of India is due to the occurrence of arsenic (As) in groundwater. The source of As in groundwater can be either of geological origin from the local bedrock or from manmade products like paints, rat poisoning, fungicides, and wood preservatives. Leachate from sewage sludge, coal burning, use of pesticides and industrial discharge can also be potential sources of As in water (Garelick *et al.*, 2008). Arsenic contamination poses a serious threat to public health hence need to be detected at the early level of contamination.

2 METHODS

The study area comprises of Ranchi city, capital of the state Jharkhand. Ranchi is a plateau region which is the largest part of the Chotanagpur plateau. 44 water sampling sites were selected according to different land use zones such as rural, peri-urban, urban, industrial and commercial areas (Fig. 1). Samples were collected from dug wells, bore wells and hand pumps in the months of August and September, 2014. The physico-chemical parameters were analysed as per standard methods (APHA, 1999). Heavy Metals analysed using Perkin Elmer Optical 2100DV, Inductively Coupled Plasma-Optical Emission Spectroscopy

(ICP-OES). Health risk assessment of metals was done following the method described by Sam *et al.* (2015). Statistical analysis was done using Excel, 2003 and SPSS version 21. Sampling location map was prepared by using ArcMap, version 9.3.

3 RESULTS AND DISCUSSION

3.1 *Physico chemical parameters and metals*

Total alkalinity, total hardness and total dissolved solids values ranged from 28–340 mg/L, 32–508 mg/L

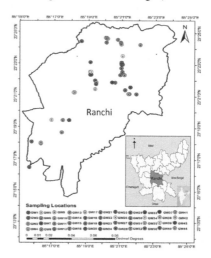

Figure 1. Map of Ranchi showing the location of groundwater sampling. Inset: Location of the study area Ranchi located at the east-central part of Jharkhand, India.

and 51–772 mg/L respectively. In all the samples chloride and fluoride varied from 2–200 mg/L and 0–2.19 mg/L respectively. Sulphate, phosphate and nitrate values ranged upto 0–268 mg/L, 0.004–0.16 mg/L and 0.53–19.01 mg/L respectively. Arsenic was found above the acceptable limit of BIS (Bureau of Indian standards) which is 0.01 mg/L at almost all the sampling locations (Table 1). In this season maximum arsenic (200 µg/L) was found at GW20 namely at Tunkitoli which is in the peri-urban area near Kokar industrial area in monsoon season. Manganese varied from below detection limit (bdl) – 4.199 mg/L. Maximum concentration of 4.199 mg/L Mn was found at GW21, Barhitoli, which is a peri-urban residential area. Nickel (Ni) and Selenium (Se) concentration varied from bdl – 0.077 and 0.032–0.14 mg/L respectively. In the pre-monsoon season As and Mn varied from bdl – 0.015 and bdl – 0.604 mg/L respectively. Ni and Fe were found below detection level and Se varied from bdl – 0.029 mg/L. In all the three season four locations GW4, GW7, GW8 of sector 2 Dhurwa, and GW24 (Near Kanke dam) showed high concentration of various metals. However, according to ANOVA the zonal variation of metal concentrations at 0.05 level of significance for monsoon season were non-significant except Fe ($F = 10.59$, $F_{crit} = 2.6$) and Se ($F = 5.42$). In post monsoon season As ($F = 23.46$) and Se ($F = 58.52$) showed significant zone wise variation. The seasonal variations of metal concentrations were quite significant except Mn.

3.2 Health risk assessment

Health risk assessment of heavy metals revealed that the residents of the study area viz. adults and children are probably at a risk of non-carcinogenic/chronic risk as compared to the carcinogenic risk. The values also signify that the residents are at low risk to carcinogenic effect and medium to high risk to non-carcinogenic effects.

3.3 Principal Component Analysis (PCA)

The first component explained 45.02% of data, whereas the second, third, fourth, fifth and sixth component explained 10.54%, 7.98%, 6.76%, 5.69% and 4.42% of data variance respectively. However, first three components showed more variation as rest of the three showed almost similar variation. So, first three components were used for the rotated space component plot. From the PCA it was evident that the water quality is mostly influenced by the 'solid component' and 'solubility component'. It can also be concluded that these two components are due to same reason that is may be due to percolation and dissolution. The clubbing of number of parameters shows that there may be same source of pollution.

4 CONCLUSIONS

Finally, from the result it can be inferred that groundwater of Ranchi city is contaminated with arsenic, selenium and manganese, as they were found well above the acceptable limits for drinking water. Metal contamination was prominent in monsoon and pre monsoon season. Probable source for contamination can be the weathering of rocks. But detailed investigation is necessary for source identification. Principal component analysis showed water quality is dominated by dissolved solids and suggest similar origin for most of the parameters. Health risk assessment signifies the probability of the non-carcinogenic effect due to the presence of arsenic and selenium as compared to carcinogenic effect is more for the residents of the study area. The presence of elements likes As and Se in toxic amounts is a matter of serious concern and needs immediate government attention.

ACKNOWLEDGEMENTS

Poonam Tirkey is thankful to Technical Education Quality Improvement Programme (TEQIP), India for providing fellowship during the study period.

REFERENCES

Dieter, H.H., Bayer, T.A., Multhaup, G. 2005. Environmental copper and manganese in the pathophysiology of neurologic diseases (Alzheimer's disease and Manganism). *Actahydroch. Hydrob.* 33: 72–78.
Garelick, H., Jones, H., Dybowska, A., Valsami-Jones, E. 2008. Arsenic pollution sources. *Rev. Environ. Contam. Toxicol.*, 197: 17–60.
Sam, N.B., Nakayama, S.M.M., Ikenaka, Y., Akoto, O., Baidoo, E., Mizukawa, H., Ishizuka, M. 2015. Health risk assessment of heavy metals and metalloid in drinking water from communities near gold mines in Tarkwa, Ghana. Environ. Monit. Assess. 187: 397 (DOI 10.1007/s10661-015–4630-3).
APHA. 1999. Standard methods for the examination of water and wastewater (20th Ed.). Washington, DC: APHA, AWWA.

Table 1. Arsenic concentration in groundwater samples.

	Rural/ residential ($n = 11$)	Peri-urban/ residential ($n = 12$)	Urban/ residential ($n = 8$)	Indus-trial ($n = 5$)	Com-mercial ($n = 8$)
As (µg/L)					
Monsoon					
Range	20–100	20–200	bdl-10	30–50	bdl-80
Standard deviation	30	50	40	10	30
Pre-Monsoon					
Range	bdl-10	bdl-10	bdl-10	bdl-20	bdl-10
Standard Deviation	1.0	2.0	2.0	1.0	1.0

Arsenic Research and Global Sustainability – Bhattacharya, Vahter, Jarsjö, Kumpiene,
Ahmad, Sparrenbom, Jacks, Donselaar, Bundschuh & Naidu (Eds)
© 2016 Taylor & Francis Group, London, ISBN 978-1-138-02941-5

Solute chemistry and groundwater arsenic enrichment in southern part of Brahmaputra River basin, India, adjacent to Indo-Burmese ranges

S. Verma[1], A. Mukherjee[1], C. Mahanta[2], R. Choudhury[2] & P. Bhattacharya[3]

[1]*Department of Geology and Geophysics, Indian Institute of Technology (IIT), Kharagpur, India*
[2]*Department of Civil Engineering, Indian Institute of Technology Guwahati, India*
[3]*KTH-International Groundwater Arsenic Research Group, Department of Sustainable Development,*
Environmental Science and Engineering, KTH Royal Institute of Technology, Stockholm, Sweden

ABSTRACT: The present study examines the groundwater chemistry, hydrogeochemical evolution and Arsenic (As) enrichment in shallow aquifers in the southern Brahmaputra river basin situated close to Naga-thrust belt. The major-ion composition dominated by a Na–Ca–HCO3 and Ca–Na–HCO3 hydrochemical facies. Groundwater composition influenced by silicates weathering in S-region of Brahmaputra basin aquifers. The aquifers of S-region are severely contaminated with As (max 0.45 mg/L), nearly 92% collected groundwater sample are enriched with As. As show poor and negative correlation with various redox—sensitive solutes. It suggests that not a single process is controlling factor, although multiple biogeochemical mechanisms might influence As liberation and fate in groundwater of S-region. The geologic explanation for high arsenic in the southern region of Brahmaputra basin (upper Assam) is probably the crustal recycling of arsenic as an incompatible element during tectonic activity.

1 INTRODUCTION

Elevated concentration of dissolved Arsenic (As) in groundwater, used as drinking water in many countries, has threatened the health of more than 100 million people worldwide (Smedley & Kinniburgh, 2002). Of these, the most extensive, As enriched aquifers are present in Southeast Asia, which may be tectonically defined as foreland basins related to the Himalayan orogenic (Mukherjee *et al.*, 2014). The knowledge and understanding of As behavior in the aquifers of river Brahmaputra basin, mostly located in the Indian state of Assam, has been largely undocumented, and unexplored. The present study focuses on the primary solute interaction and groundwater evolution through different weathering mechanisms. Arsenic enrichment and mobilization processes have been discussed with the geological explanation in the aquifer of the southern part of Brahmaputra basin. The study area situated in the southern parts of Brahmaputra River basin of Assam consists of braided alluvial floodplain deposits that are transported by the Brahmaputra River and its tributaries from the adjoining Naga-thrust belt.

2 METHODS

Fifty two (52) groundwater samples collected from southern part of Brahmaputra river basin which located along or close to Naga-Hills. All groundwater samples were collected from hand-pumped and public water supply wells following procedures of Mukherjee & Fryar (2008) in January 2014. The groundwater samples were taken from various depths ranging from 4–62 m below ground level (bgl) but normally <40 m bgl. All of the samples represent the shallow aquifer system in the study area. All field parameters like E_H, temperature, pH, Dissolved O_2 (DO) and specific Conductance (EC) were measured by using a multiparameter probe (Hanna 9282) after stabilization of field parameters. Samples for cations (major and trace) were collected and preserved by acidification with 6N HNO_3 in the field to pH~2. Samples for HCO_3 analysis were collected without adding any preservative and $CHCl_3$ used as preservative for other anions in water samples. Major anions were measured by ion chromatography. The major cations and trace metals were measured by Inductively Coupled Plasma with Optical Emission Spectrophotometer (ICP-OES, Thermo Fisher ICAP 6000).

3 RESULTS AND DISCUSSION

3.1 Groundwater chemistry

The groundwater is circum-neutral to alkaline with measured pH values ranging from 6.5 to 8.6. The dominant cations are Ca and Na in most of the ground water samples. The dominant anion is HCO_3 in groundwater of Brahmaputra basin contributing ~90% to the anion budget. The concentrations of SO_4 and Cl are much lower than HCO_3 in the groundwater samples. Piper plot shows that groundwater in the southern part is dominated by the Na–Ca–HCO_3 Ca–Na–HCO_3 hydrochemical facies (Fig. 1).

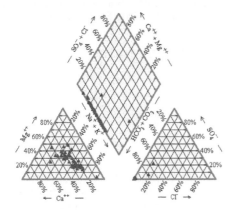

Figure 1. Piper Plots of groundwater samples collected from the study area.

3.2 *Hydogeochemical evolution*

Groundwater evolution is mainly governed by the chemistry of the recharging water, water–aquifer matrix interactions or both, as well as groundwater residence time within the aquifer. In bivariate mixing diagrams, we used Na normalized molar ratios (Ca/Na, Mg/Na, and HCO_3/Na) because the absolute concentrations are affected by dilution and evaporation processes. These correlations indicate that groundwater composition is influenced by both carbonate- and silicate mineral weathering, but most of the samples fall in the global-average silicate weathering domain. It suggests incongruent leaching of argillaceous rock. The shale, siliceous volcanic and ophiolites in the Naga Hills adjoining the Brahmaputra basin are the major controls on hydrogeochemistry of this region. Mineral weathering is assumed to be the governing mechanism controlling the groundwater cation composition of groundwater in the study area.

3.3 *Arsenic in groundwater and tectonics*

About 92% of groundwater samples in southern part are enriched with As (maximum 0.45 mg/L). However, bivariate correlations of As with various redox-sensitive species and solutes in the current dataset were observed by nonparametric Spearman's rho (ρ) calculation. Results suggested that As mobilization in aquifers of study areas is controlled by several hydrogeochemical processes. Arsenic concentrations in southern aquifers show negative correlations with other solutes (e.g., with Fe [$r^2 = -30$], Mn [$r^2 = -0.35$], SO4 [$r^2 = -0.33$], HCO_3 [$r^2 = -0.03$] (Table 1). This poor and negative correlation indicates that multiple reactions and hydrogeochemical processes control As mobilization and fate in the study area.

Mineralogical composition of the aquifers and groundwater As in southern part of Brahmaputra river basin might be influenced by tectonic activity along the northern margin of Indian plate (Indo-Burma Range). The proposed mechanism of arsenic

Table 1. Results of nonparametric Spearman rho calculation for arsenic, where available, are marked in bold.

	E_H	pH	HCO_3	SO_4	Fe	Mn	As
E_H	1	–0.41	0.45	0.03	0.57	0.44	–0.23
pH	–0.41	1	–0.10	–0.11	–0.35	–0.33	0.14
HCO_3	0.45	–0.10	1	0.16	0.40	0.17	–0.03
SO_4	0.02	–0.11	0.16	1	0.17	0.23	–0.33
Fe	0.57	–0.35	0.40	0.17	1	0.72	–0.29
Mn	0.44	–0.33	0.17	0.25	0.72	1	–0.36
As	–0.23	0.14	–0.03	–0.33	–0.29	–0.36	1

enrichment in the Indo-Burmese Mountain belt is consistent with shallow crustal recycling of arsenic as an incompatible element due to a collision between Indian plate and Burmese microcontinents. Subsequently, intense deformation developed thrust belts (Naga-thrust belt) and large foreland basin (Brahmaputra basin). This foreland basin received As enriched sediments through many southern tributaries of Brahmaputra river from adjacent ophiolite belt and arc-related volcanic rocks.

4 CONCLUSIONS

The present study reveals that silicate weathering is the dominant process in the aquifers of S region however cation exchange also contributes to the groundwater chemistry of the study area. The lack of a consistent correlation between As and any other single component indicates that there is no single factor controlling the concentration of dissolved As in aquifers of the southern part. The occurrence of the high arsenic in the aquifers of the southern region of Brahmaputra basin is well explained through crustal recycling of arsenic as an incompatible element during tectonic activity.

ACKNOWLEDGEMENTS

SV acknowledges CSIR (Government of India) for providing CSIR-SR Fellowship. The author acknowledges support of IIT Kharagpur for providing laboratory analyses facilities.

REFERENCES

Mukherjee, A., Bhattacharya, P., Savage, K., Foster, A., Bundschuh, J. 2008. Distribution of geogenic arsenic in hydrologic systems: controls and challenges. *J. Contam. Hydrol.* 99: 1–7.
Mukherjee, A., Verma, S., Gupta, S., Henke, K.R., Bhattacharya, P. 2014. Influence of tectonics, sedimentation and aqueous flow cycles on the origin of global groundwater arsenic: Paradigms from three continents. *J. Hydrol.* 518: 284–299.
Smedley, P.L., Kinniburgh, D.G. 2002. A review of the source, behavior and distribution of arsenic in natural waters. *Appl. Geochem.* 17: 517–568.

Arsenic Research and Global Sustainability – Bhattacharya, Vahter, Jarsjö, Kumpiene,
Ahmad, Sparrenbom, Jacks, Donselaar, Bundschuh & Naidu (Eds)
© *2016 Taylor & Francis Group, London, ISBN 978-1-138-02941-5*

Co-occurrence of arsenic and fluoride in the Brahmaputra floodplains, Assam, India

M. Kumar, N. Das, A. Das & K.P. Sarma
Department of Environmental Sciences, Tezpur Central University, Assam, India

ABSTRACT: Groundwater and grab sediment sampling was conducted in the Brahmaputra River (BR) flood plains from 2011 to 2014 in order to study the co-occurrence of Arsenic (As) and Fluoride (F⁻). Sampling was designed in a way that covers full stretch of BR with the maximum proximity of 10km from the river to understand its influence. Arsenic was found to be mobilized under a predominantly reducing condition due to dissolution of Fe (hydr)oxides, while the sources of F⁻ could be mainly F⁻ bearing minerals. Low levels of F⁻ suggest that alluvial plains where recharge and precipitation rates are high are not suitable for its release. Desorption experiments with raw and Fe (hydr)oxide removed sediment samples at acidic (pH 5) and alkaline (pH 10) conditions reveal that Fe (hydr)oxide played a very important role in the co-release of As and F⁻ at alkaline condition. Thus As and F⁻ have very less probability of co-existence in an alluvial environment with reducing conditions. Isolated aquifers with low recharge and oxidizing conditions could be potential regions where high concentrations of both the contaminants may be observed.

1 INTRODUCTION

Arsenic (As) and Fluoride (F⁻) are important inorganic toxicants which affect the groundwater around the world. Co-occurrence of both the pollutants in groundwater can be a matter of serious concern because of the health hazards involved, As leading to cancer and F⁻ leading to dental and skeletal fluorosis in the long run. While addressing the issue of co-occurrence, it is noteworthy to report that very few workers have researched on this topic. The aforementioned pollutants have also been detected in the Brahmaputra Floodplains (BFP) recently, As has been reported from Jorhat, Dhemaji, Barpeta, Bongaigaon etc. districts of Assam (Singh, 2004), while F⁻ was found to have a more localized distribution in places like Karbi Anglong, Nagaon etc. districts (Das *et al.*, 2003). This study aims to investigate the extent and the distribution of the two pollutants under alluvial settings in the BFP.

2 MATERIALS AND METHODS

2.1 Study area

The BFP is of tectonic origin, formed by the collision of the Indian and the Eurasian plate followed by episodes of continuous sedimentation. It is dominated by undifferentiated alluvial sediments. The region is divided into two distinct hydrogeological units: (i). The dissected alluvial plains which extend from the south of the lower Himalayan piedmont fan zones to the "main rock promontory" of the Garo hills and Shillong plateau in the south; and (ii) the inselberg or the isolated hillock regions of the south.

2.2 Sampling, chemical analyses and statistical techniques

Groundwater and grab sediment samples were collected along Brahmaputra River from Guijan, in the upstream to Dhubri, in the downstream (Fig. 1).

Sampling coordinates were recorded by a handheld GPS set (Garmin GPS map 76CSX). pH, EC, ORP and TDS were measured in situ using the HANNA HI9128 multi parameter water quality portable meter. The parameters analyzed for the groundwater samples were: pH, EC, ORP, TDS, Na^+, K^+, Ca^{2+}, Mg^{2+}, HCO_3^-, SO_4^{2-}, PO_4^{3-}, NO_3^-, Fe, As and TOC.

Parameters analyzed for sediments were: pH, EC, TOC, mineralogy and CEC. All chemical analyses were performed in conformation with the guidelines prescribed by the American Public Health Association (APHA, 2005). Statistical analyses were performed consisted of regression analysis, correlation analysis, principal components analysis

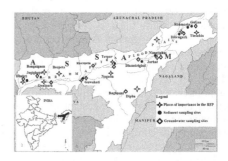

Figure 1. Map of the BFP showing the groundwater samples and the sediment sampling sites.

and hierarchical cluster analysis. Desorption studies were performed on sediments to study the behaviour of As and F- co-occurrence in the BFP based on the method prescribed by (Kim *et al.*, 2012).

3 RESULTS AND DISCUSSION

The BFP was found to be characterized by a mixed aquifer system based on the presence of both oxidising and reducing conditions in the groundwater. Overall reducing conditions were found to be more prevalent than oxidising conditions. This reducing condition was found to be the main driver for the release of As by reductive dissolution of Fe (hydr)oxides. The aforementioned statement is also supported by the fact that As and Fe showed good interrelation in our study area.

Detection of As bearing minerals like arsenopyrite (FeAsS) and walpurgite [$(BiO)_4UO_2(AsO_4)_2$·H_2O] in the soils and sediments by XRD and EDX however indicated towards their role as the primary sources of As. The water type was detected to mostly belong to HCO_3^- type with mixed cations Na^+, K^+, Ca^{2+} and Mg^{2+}.

Various plot of As and F with key parameters are depicted in Figure 2. Depth was found to inversely affect the As level, while no effect was observed on F-. PCA and HCA revealed that Fe (hydr)oxides could also act as a secondary source of anions like F- and SO_4^{2-}. The probable cause could be that metal oxides and hydroxides act as adsorptive surfaces for anions like F-, SO_4^{2-} as well arsenate. The extremely low values of F- in our groundwater samples suggested that the alluvial environment with better recharge and discharge conditions was not suitable for its release. The aforementioned observation indicated towards the existence of alternative modes of As release like sulphide oxidation, albeit on much smaller scale in isolated drier aquifers.

The overall low correlation between As and F- under the natural conditions prompted us to conduct lab based simulation experiments. Based on the sequential extraction data, Fe-(hydr)oxides associated As phase was found to be the most dominant fraction in the soil/sediments.

Batch desorption experiment on raw and Fe (hydr) oxides removed soil at acidic and alkaline conditions (pH 5 and 10 respectively) showed that Fe (hydr) oxide played the most important role in the release of both As and F- at alkaline conditions. Iron (hydr) oxides were removed from the soils by treatment with CBD solution (Na-citrate + Na-bicarbonate + Na-dithionite) based on the method described by Wenzel *et al.* (2001). Overall speciation modeling by MINTEQA2 v3.1 showed that the mineral phases associated with As and F- were under saturated indicating that their levels may increase in the near future due to weathering and dissolution.

4 CONCLUSIONS

The correlation between As and F- in BFP was found to be very poor under the natural settings. Arsenic could mobilize due to reductive hydrolysis of Fe (hydr) oxides, where as the overall alluvial conditions of the BFP with high recharge and precipitation conditions is not very suitable for F- release. Isolated aquifers with low recharge rates in the BFP could however have high F- levels as reported by many workers. Such aquifers could also result in co-occurrence of As with F- due to development of oxidising conditions as revealed by the desorption experiments. The concentration of As and F- however can increase in the near future as SI values of the associated minerals phases indicate under-saturation.

REFERENCES

APHA. 2005. *Standard Methods for the Examination of Water and Wastewater*. (19th Ed.). American Public Health Association, Washington, D.C. 124p.

Bhattacharya, P., Frisbie, S.H., Smith, E., Naidu, R., Jacks, G., Sarkar, B. 2002. Arsenic in the environment: a global perspective. In: Sarkar, B. (Ed.), *Handbook of Heavy Metals in the Environment*. Marcell Dekker Inc., New York, pp. 147–215.

Das, B., Talukdar, J., Sarma, S., Gohain, B., Dutta, R.K., Das, H.B., Das, S.C. 2003. Fluoride and other inorganic constituents in groundwater of Guwahati, Assam, India. *Curr. Sci.* 85(5): 657–661.

Kim, S.H., Kim, K., Ko, K.S., Kim, Y., Lee, K.S. 2012. Co-contamination of arsenic and fluoride in the groundwater of unconsolidated aquifers under reducing environments. *Chemosphere*. 87(8): 851–856.

Kumar, M., Kumar, P., Ramanathan, AL., Bhattacharya, P., Thunvik, R., Singh, U.K., Tsujimura, M., Sracek, O. 2010. Arsenic enrichment in groundwater in the middle Gangetic Plain of Ghazipur District in Uttar Pradesh, India, *J. Geochem. Explor.* 105(3): 83–94.

Singh, A.K. 2004. Arsenic Contamination in Groundwater of North Eastern India. Proceedings of the National Seminar on Hydrology with Focal Theme on "Water Quality", National Institute of Hydrology, Roorkee.

Wenzel, W.W., Kirchbaumer, N., Prohaska T., Stingeder, G., Lombi, E., Adriano, D.C. 2001. Arsenic fractionation in soils using an improved sequential extraction procedure. *Anal. Chim. Acta.* 436(2): 309–323.

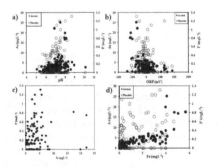

Figure 2. Scatter plot of As and F versus (a) pH (b) ORP (c) F and (d) Fe in the BFP.

Arsenic Research and Global Sustainability – Bhattacharya, Vahter, Jarsjö, Kumpiene,
Ahmad, Sparrenbom, Jacks, Donselaar, Bundschuh & Naidu (Eds)
© 2016 Taylor & Francis Group, London, ISBN 978-1-138-02941-5

Morphological and mineralogical evidences of arsenic release and mobilization in some large floodplain aquifers

C. Mahanta[1], S.S. Sathe[1] & A. Mahagaonkar[2]
[1]*Department of Civil Engineering, Indian Institute of Technology Guwahati, India*
[2]*Center for the Environment, Indian Institute of Technology Guwahati, India*

ABSTRACT: Holocene Brahmaputra floodplain sediments comprising sediments deposited by meandering channels was investigated for deciphering mineralogical evidences for the arsenic (As) from the sediments. Vertical succession of nine core drilled sediment samples were classified based on grain size and color. Samples were analyzed to understand mineralogy, crystallography and chemical composition by XRD, FESEM, TEM and EDS analysis. Morphological and mineralogical analysis revealed presence of crystalline phases of Orpiment, Realgar and Chalcopyrite and other secondary Arsenic bearing minerals. In total 192 groundwater samples were also analyzed to make out the concentration of As and other trace elements. The observed average concentrations of As, Fe, Mn, SO_4^{2-}, PO_4^{3-} and Cl^- were found to be 89µg/L, 76 mg/L, 2.38 mg/L, 1.7 mg/L, 0.7 mg/L and 9.7 mg/L respectively. The results explain how hydrogeology of the aquifer is integrated with dissolution and precipitation of As bearing minerals.

1 INTRODUCTION

Many aquifers in the Brahmaputra floodplain is contaminated with As (Mahanta *et al.*, 2015). Published studies have shown that elevated As concentration is catalyzed by anoxic groundwater and mobilized by reductive dissolution mechanism (Bhattacharya *et al.*, 2011). It is by now understood that high HCO_3^- concentrations were conceivably mediated by oxidation of organic matters mediated by microbiological activities during reduction of Fe oxy-hydroxides minerals (Mahanta *et al.*, 2015). Chemical weathering of phyllosilicate minerals, such as biotite (Seddique *et al.*, 2008) and reduction of magnetite are generally considered the sources of As in a Ganga-Meghna flood plain. However, a complete mineralogical understanding involving primary and secondary arsenic bearing minerals is yet to evolve.

2 METHODS

2.1 *Groundwater and sediment sampling*

Essential groundwater parameters were measured on field using YSI multi probe kit (YSI Inc. 550A, Sondes). Acidified samples were used for cation analysis and non-acidified samples for anion analysis. Core sediments were collected using conventional house-hold technique & reverse rotary drilling method. Classification of sediments was done according to Munsell color chart and based on the morphology, samples were classified (Mahanta *et al.*, 2015).

2.2 *Minerals identification*

Crystalline phase and mineral composition was identified using XRD. This analysis was performed on

a Bruker (Model-D8- Advance) powder diffractometer with *CuK*α radiation ($\lambda = 1.5406$). A step scanning technique at 6 s/0.02° with a fixed time step was applied and data were compiled from 5° to 75° 2θ angle (Matera *et al.*, 2003).

3 RESULTS AND DISCUSSION

3.1 *Hydrogeochemistry*

Hydrogeochemistry endorsed the affinity of dissolved As with Fe, Mn and major cations possibly linked to mineral dissolution. The high alkalinity in individual wells was due to the localized zones with high oxidation of organic matter and the reducing conditions (Mahanta *et al.*, 2015).

The insignificant concentration of PO_4^{3-} (average 0.7 mg/L) and SO_4^{2-} (average 1.7 mg/L) indicates prevailing reducing condition and provenance of mineral precipitation by reductive dissolution in reducing environment.

3.2 *XRD study*

Results of powder XRD were interpreted using PANalytical high score software. XRD analysis represents that the preliminary Fe, As and SO_4^{2-} bearing minerals observed at respective d-spacing values such as orpiment (4.96, 2.23 & 1.37), realgar (4.23, 3.73, 2.98 & 1.99), arsenopyrite (1.82), argentopyrite (1.81) and chalcopyrite (3.35) were mainly observed in silty-clayey and fine sand aquifer at shallow depths. Secondary As bearing arsenolite (3.19 & 1.66) and claudetite (4.49 & 3.66) mineral observed at d-spacing, often in brownish aquifer. Mica group minerals namely biotite

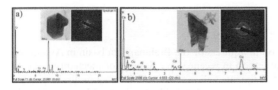

Figure 1. FESEM images for a) Realgar (Monoclinic) b) Pyrite (Isometric) c) Hematite (Hexagonal), d) Biotite (Monoclinic).

Figure 2. TEM, EDS and SAED images representing a) chalcopyrite adhered to chromite, and b) biotite.

(9.88 & 2.27), muscovite (2.56 & 1.35) observed, most of the minerals are enriched in clays than the sandy sediments.

The brownish-yellow (10YR 6/8) soil (with Fe-rich mottle) observed at shallow depths mainly comprises silicates [quartz (3.34 & 2.45)], mica group [muscovite (2.56 & 2.53) and biotite (9.88 & 2.27)] and K-feldspar (4.48 & 4.23) minerals. Crystalline Fe oxy/hydroxides minerals such as magnetite (4.87), goethite (1.5) and hematite (1.45) were significant, but jacobsite and birnessite were less detected probably due to their amorphous nature.

3.3 Morphological study

The grain morphological study by FESEM & TEM and diffraction information from SAED revealed the amorphous, poly-crystalline and crystalline phases of single mineral. The EDS results suggested As affinity with Fe, Mn, Si, Cr and Cu bearing minerals.

Crystals of realgar and pyrite were diagnosed in reduced aquifer sediments (7.5GY 7/1) (Fig. 1a,b), while hematite (hexagonal) (Fig. 1c) of Fe minerals observed in yellow-orange colored (10YR 7/8) medium sand aquifer sediments.

TEM analysis revealed the presence of crystalline chalcopyrite grains with authigenic growths of chromite (Fig. 2a). The aggregate of biotite with platy texture was also present in these sediments (Fig. 2b). The shallow depth (i.e. 10.7 meter) aquifer sediment samples indicated the detrital source of minerals.

4 CONCLUSIONS

The results suggested that the crystalline/polycrystalline minerals are mainly characterized as detritus, enriched mainly with Al, As, Ca, Cr, Cu, Fe, Mn, Mg Na and S bearing minerals. Aquifers are enriched with organic content as well facilitating bacteriological utilization resulting in anoxic conditions.

Based upon lithology and colour, we could understand the oxidized and reduced condition of aquifer enriched with Fe and As bearing minerals respectively.

Groundwater chemistry primarily suggested mobilization of As in shallow aquifer and Fe-oxy/hydroxide reduction as the primary process controlling hydro-geochemistry. The dissolved concentration of Ca (mean 43 mg/L) and HCO_3^- (mean 220 mg/L) in groundwater are due to carbonate dissolution and silicate weathering/organic matter oxidation respectively (Mukherjee & Fryar, 2008). Mineralogical characterization revealed aquifer differences with respect to weathered/ un-weathered minerals and varied sediment provenance. The amorphous and polycrystalline minerals indicate mineral recrystallization and reduction mechanism by hydrogeochemical reactions often accompanied by microbiological activities. The other partly weathered and crystalline minerals (Fig. 1) indicated aquifer predominantly of Holocene age.

The study reinforced that primary and secondary minerals are mainly derived by pedogenic processes from metasediments originated from Himalayan ranges (Mukherjee et al., 2014) and this is validated in case of As minerals as well.

ACKNOWLEDGEMENTS

Departments of Civil, Mechanical and Chemical Engineering, Centre for Nanotechnology, Physics, Chemistry and Central Instrumentation Facility of IIT Guwahati are thanked for performing analyses.

REFERENCES

Bhattacharya, P., Mukherjee, A., Mukherjee, A.B. 2011. Arsenic in groundwater of India. In: J.O. Nriagu (ed.) Encyclopedia of Environmental Health, vol. 1: 150–164 Burlington: Elsevier.

Mahanta, C., Enmark, G., Nordborg, D., Sracek, O., Nath, B., Nickson, R.T., Herbert, R., Jacks, G., Mukherjee, A., Ramanathan, A.L., Choudhury, R. and Bhattacharya, P. 2015. Hydrogeochemical controls on mobilization of arsenic in groundwater of a part of Brahmaputra river floodplain, India. J. Hydrol. Regional Studies 4: 154–171.

Matera, V., Le Hécho, I., Laboudigue, A., Thomas, P., Tellier, S., Astruc, M. 2003. A methodological approach for the identification of arsenic bearing phases in polluted soils. Environ. Pollut. 126(1): 51–64.

Mukherjee, A., Fryar, A.E. 2008. Deeper groundwater chemistry and geochemical modeling of the arsenic affected western Bengal basin, West Bengal, India. Appl. Geochem. 23(4): 863–894.

Mukherjee, A., Verma, S., Gupta, S., Henke, K.R., Bhattacharya, P. 2014. Influence of tectonics, sedimentation and aqueous flow cycles on the origin of global groundwater arsenic: Paradigms from three continents. J. Hydrol. 518: 284–299.

Seddique, A.A., Masuda, H., Mitamura, M., Shinoda, K., Yamanaka, T., Itai, T., Maruoka, T., Uesugi, K., Ahmed, K.M. Biswas, D.K. 2008. Arsenic release from biotite into a Holocene groundwater aquifer in Bangladesh. Appl. Geochem. 23(8): 2236–2248.

Sedimentary controls on arsenic mobilization in groundwater of aquifers in the Brahmaputra River Valley in Assam

R. Choudhury, C. Mahanta & S.S. Sathe
Department of Civil Engineering, Indian Institute of Technology Guwahati, India

ABSTRACT: Groundwater Arsenic (As) contamination in aquifers of the Brahmaputra basin in Assam is a relatively recent discovery and has become a focus of attention in recent years. Sources of mechanisms of As release were assessed from analysis of groundwater and sediments collected from drilled borewells. Sequential extraction depicted Mn- and Fe-oxyhydroxide as the source of major leachable As phases. Mean total leachable solid phase As was 10 mg/kg, while groundwater As ranged from bdl (below detection limit) to 134 μ/L from adjoining tubewell sources. Geochemical studies using sequential-extraction results and scanning electron microscopy, XRD and petrographic study on sediment samples demonstrate that Fe- and Mn-oxyhydroxides and Fe- and Mn-oxyhydroxide-coated quartz and feldspars are the dominant carriers of As in the sediments. Based on results of sequential extraction and mineralogical study, it may be suggested that the reductive dissolution of Fe and Mn oxyhydroxides is the probable mechanism of As release in the study area.

1 INTRODUCTION

Arsenic (As) is widely recognized as atoxic drinking water contaminant and naturally occurring carcinogen (Smith *et al.*, 1992). Tens of millions of people are currently exposed to dangerous levels of As (Bhattacharya *et al.*, 2011). While the reductive dissolution of Fe-oxyhydroxides due to microbial metabolism is generally accepted as the main mechanism of As release into shallow groundwater (Nickson *et al.*, 2000, Bhattacharya *et al.*, 2011) many others have suggested that the release can be attributed to other processes such as chemical weathering and dissolution of phyllosilicate minerals (Pal & Mukherjee, 2009). Uddin *et al.* (2011) reported that detrital heavy mineral phases such as magnetite, biotite, olivine, and apatite along with secondary Fe secondary Fe-(oxy) hydroxide minerals (authigenic goethite, ferrihydrite, and amorphous Fe-(oxy) hydroxides) are some potential sources of As in shallow aquifers. Other mechanisms for the release As into groundwater include oxidation of pyrite, and desorption by competitive anions (Bhattacharya *et al.,* 2011 and references there in).

2 METHODS/EXPERIMENTAL

2.1 *Groundwater sampling and analysis*

Prior to sampling, wells were purged for few minutes to discharge the standing volume of groundwater to obtain fresh water samples from the aquifer. Field parameters like Eh, temperature, pH, Dissolved Oxygen (DO) and Electrical Conductivity (EC) were measured using a multiparameter meter (Hanna 9282). All shallow groundwater samples were filtered by 0.45 μm filters. Samples for cations (major and trace) were collected and preserved by acidification with 6 N HNO_3 in the field to

pH~2. Major cations and trace elements were analyzed by Inductively Coupled Plasma Optical Emission Spectrophotometer (ICP-OES, Thermo Fisher ICAP 6000).

2.2 *Sediment sampling and analysis*

Sediment samples were collected from twelve boreholes (from Darrang district, located on the northern bank of the river Brahmaputra in Assam) using a conventional house-hold technique (hand percussion and reversed circulation). Sediment samples were collected as disturbed samples coming out of the drilling pipe and the depth was measured by knowing the increments of pipes used for drilling. During drilling, color and texture of sediment was noted and lithologs were prepared. Prior to analysis, the sediment samples were dried in a hot air oven. Dried samples were grounded gently by hand with an agate mortar and pestle and the powdered samples were analyzed for their physical, morphological and chemical properties using X-ray diffractometer, Scanning Electron Microscope (SEM), petrographic and sequential extractions study.

3 RESULTS AND DISCUSSION

3.1 *Groundwater chemical compositions*

Groundwater samples show pH values ranging from 6 to 7.8 (mean 6.8) indicating circum-neutral groundwater. Calcium (Ca^{2+}) is the dominant cation, with maximum value of Ca^{2+} (60.5 mg/L) while median concentrations of Ca^{2+} and Na^+ are almost identical (11.2 mg/L and 11 mg/L respectively) and Mg^{2+} ranged between 0.2 mg/L to 24.6 mg/L. Anions are predominated by HCO_3^- with values ranging from 62.7 to 268.5 mg/L (median 120 mg/L) whereas $SO4^{2-}$

and Cl⁻ showed low concentrations in all groundwater samples (median: 3.7 and 4 mg/L, respectively).

Dissolved Oxygen (DO) concentrations in groundwater range from 0.1 to 2.8 mg/L (median 0.6 mg/L). Measured Eh values range between 28.1 and 197.8 mV and generally exhibit higher values near the recharge zone (near Bhutan foothills) compared to the site near Brahmaputra river channel. NO_3^- concentrations (range bdl–9.1 mg/L, median 1.4 mg/L) remain relatively low throughout the study area. Fe concentrations range from bdl to 41.02 mg/L (mean 13.9) and Mn range from bdl to 5.9 mg/L (mean 0.9 mg/L). There are no distinct variation in the concentration of Fe and Mn from the recharge (near Bhutan foothills) to discharge zone (close to river bank). Arsenic concentration range from bdl to 130 µg/L (mean 20 µg/L). ICP measurements identified a total of 65 sources that met the WHO guideline for As of 10 µg/L, another 76 sources with As concentrations up to the national standard of 50 µg/L and 11 sources with concentration >50 µg/L. As concentrations and their distribution as a function of distance from the recharge area indicated a gradual increase from Bhutan foothills to floodplain area. Low As concentrations (<10 µg/L) in foothills regions coincided with high SO_4^{2-} concentrations compared to the floodplain regions where high As concentrations coincided with low SO_4^{2-} concentrations. Depth variation of As, Fe and SO_4^{2-}, indicated that zones of high As correlated with high Fe and low SO_4^{2-} concentrations.

3.2 Sediment geochemistry

Mineralogy of aquifer sediments was dominated by alumino-silicates of Na, K, Mg Al and Fe. The sediment grains were subangular to rounded, suggesting greater distance of transport from the source prior to their deposition (to produce rounding of the sediment grains) and have undergone extensive chemical weathering, leaving sediments dominated by stable mineral quartz (Rowland et al., 2008). XRD peak patterns confirmed the presence of phyllosilicates (17–20–30), quartz (26.95, 26.85, 27), feldspar (20–30, 27.5), chalcopyrite (50), arsenolite (59–60) and clay minerals (illite), with the bulk composition being dominated by quartz, phyllosilicates, feldspars and clay minerals. Phyllosilicates, composed of tetrahedral and octahedral nets, offer available reactive surfaces for As adsorption, presumably via iron hydroxide molecular bonding on the edge of unit layers of the minerals (Nakano et al., 2014).

Total As in the sediment samples range from 7.77 to 13.93 mg/kg with an average concentration of 10.08 mg/kg. Considering the world baseline concentrations of As in sediments of 5–10 mg/kg (Smedley and Kinniburgh, 2002), the average levels of 10.08 mg/kg of As measured in the digested sediment samples are nearly within the natural range (Smedley & Kinniburgh, 2002). However, following Hering and Kneebone (2002), who reported that only a minor fraction (0.09%) of the total As from soils with porosity of 0.30 and an As concentrations near crustal abundance (i.e., 1.8 mg/kg) is necessary to produce dissolved As concentrations that exceed the WHO drinking water guideline 10 µg/L (Bhattacharya et al., 2011), it can be argued that the observed As concentration levels in sediments are of concern. Concentrations of total Mn and Fe range from 61.90 to 114 mg/kg and 1703 to 5129 mg/kg.

4 CONCLUSIONS

Geological, mineralogical, and geochemical investigation of sediment and groundwater data indicates that the reductive dissolution of Fe-oxyhydroxide is the main mechanism of As-mobilization in groundwater of study area. Data from sequential leaching also supports the above contention for As-mobilization in groundwater of the study area. Moreover, the presence of As-bearing minerals such as arsenolite, together with abundant percentage of Fe/Mn in the core sediment samples confirms reductive dissolution mechanism for As release.

ACKNOWLEDGEMENTS

RC acknowledges the institute research fellowship from IIT Guwahati, RC also acknowledges the Central Instrumentation Facilities (CIF) at IIT Guwahati for SEM/ XRD analysis.

REFERENCES

Bhattacharya, P., Mukherjee, A., Mukherjee, A.B. 2011. Arsenic in groundwater of India. In: J.O. Nriagu (Ed.) Encyclopedia of Environmental Health, vol. 1, Elsevier, Burlington, pp. 150–164.

Hering, J.G., Kneebone, P.E. 2002: Biogeochemical controls on arsenic occurrence and mobility in water supplies. In: W.T. Frankenberger, Jr. (Ed.) Environmental Chemistry of Arsenic. Marcel Dekker, Inc., New York, N.Y. pp.155–181.

Nakano, A., Kurosawa, K., et al. 2014. Geochemical assessment of arsenic contamination in well water and sediments from several communities in the Nawalparasi District of Nepal. Environ. Earth Sci. 72(9): 3269–3280.

Nickson, R.T., McArthur, J.M., Ravenscroft, P., Burgess, W.G., Ahmed, K.M., 2000. Mechanism of arsenic release to groundwater, Bangladesh and West Bengal. Appl. Geochem. 15 (4), 403–413.

Pal, T., Mukherjee, P.K., 2009.Study of subsurface gelogy in locating arsenic free groundwater in Bengal Delta, West Bengal, India. Environ. Geol. 56(6): 1211–1225.

Rowland, H.A.L., et al. 2008. Geochemistry of aquifer sediments and arsenic-rich groundwaters from Kandal Province, Cambodia. Appl. Geochem. 23(11): 3029–3046.

Smith, A., Hopenhayn-Rich, C., et al. 1992. Cancer risks from arsenic in drinking water. Environ. Health Perspect. 97: 259–267.

Smedley, P., Kinniburgh, D.G. 2002. A review of the source, behaviour and distribution of arsenic in natural waters. Appl. Geochem.17(5): 517–568.

Uddin, A., Shamsuddha, M., et al. 2011. Mineralogical profiling of alluvial sediments from arsenic-affected Ganges–Brahmaputra floodplain in central Bangladesh. Appl. Geochem. 26(4): 470–483.

*Arsenic Research and Global Sustainability – Bhattacharya, Vahter, Jarsjö, Kumpiene,
Ahmad, Sparrenbom, Jacks, Donselaar, Bundschuh & Naidu (Eds)
© 2016 Taylor & Francis Group, London, ISBN 978-1-138-02941-5*

Groundwater arsenic and extent of contamination along the Mekong River, Vietnam

T. Ito[1], H. Sato[2] & N. Shibasaki[2]
[1]*Kokusai Kogyo Co., Ltd. and Mekong Group, Tokyo, Japan*
[2]*Faculty of Symbiotic Systems Science, Fukushima University and Mekong Group, Fukushima, Japan*

ABSTRACT: The Research Group for Groundwater Contamination by arsenic (As) in the Mekong River Delta (Mekong Group) has studied the actual As contamination and the mechanism of As contamination in the Mekong River Delta since 2008. We have conducted measurements of As concentrations in groundwater, and continuously monitored groundwater levels, the water level of the Mekong River, and rainfall. Based on our study, we published a monograph titled "Groundwater Arsenic Contamination in the Mekong River Delta" and we delivered the monograph to the local relevant agencies/persons. We found serious groundwater contamination by As in Tay Island, which was selected as our pilot study area. After our study, the local government constructed water supply systems and the systems have been operated to supply treated surface water to the entire Tay Island since 2014. While in the inland areas of the Mekong River Delta, most water supply systems have pumped deep groundwater that is slightly contaminated by As. It is predicted that the deep groundwater may be contaminated by As due to induced recharge from upper/lower aquifers by the excessive pumping to meet the increasing demand of water use.

1 INTRODUCTION

We used arsenic (As) field test kits developed by Dr. Hironaka of Asia Arsenic Network (AAN) to grasp As concentrations of groundwater in the Mekong River Delta (MRD). The number of surveyed points is 259 during a period from 2008 to 2015.

The maximum concentration of As is 2400 µg/L, that was detected in Tay Island and Gieng Island located along the Mekong River near the border of Cambodia. While lower concentrations of As around the Vietnamese drinking standard (10 µg/L) were detected from deep aquifers in the inland areas and the middle to downstream areas in the MRD. At present, the economy is rapidly growing in southern Vietnam so that it is concerned that the exploitation of deep groundwater may cause vertical leakage and squeeze of As contaminated water.

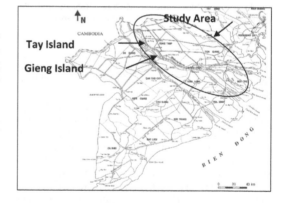

Figure 1. Location map of the study areas, Tay Island and Gieng Island along the Mekong River.

2 MATERIALS AND METHODS

2.1 Geological survey

Two continuous core borings were drilled in Tay Island in 2008 and 2009. Based on the analyses of core samples including X-ray radiographs and ^{14}C dating, characteristics of sedimentary facies were identified.

2.2 Groundwater quality survey

In the field, the As field test kit developed by Dr. Hironaka was employed to analyze pumped groundwater from target wells. At the same time we analyzed Mn^{2+}, Fe^{2+}, NH_4^+, NO_2^-, NO_3^- and F^- by Pack Test (Kyoritsu Chemical-Check Lab.) at the sites. The field groundwater quality survey has been performed at 259 sites with 348 samples during 7 years since 2008.

2.3 Continuous water level monitoring

Groundwater observation holes were constructed at 8 locations in northern Tay Island. The groundwater levels and the level of the Mekong River have been monitored by the automatic water level sensors.

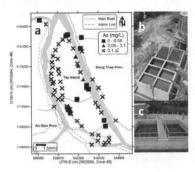

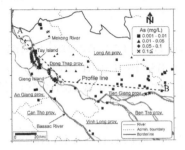

Figure 4. Distribution of arsenic concentration.

Figure 2. a) Distribution of arsenic concentration; b) our purification plant; and c) local government's water supply system.

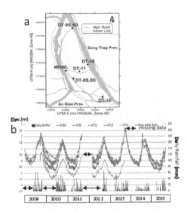

Figure 3. a) Location of observation points; and b) groundwater levels and river water level.

3 RESULTS AND DISCUSSION

3.1 Arsenic concentration in Tay Island

We found serious groundwater contamination by As in Tay Island so that we have selected the island as our pilot study area. As shown in Figure 2, most samples showed more than 10 μg/L of As concentration in Tay Island. As a result, we made some household-level As removal devices and constructed the purification plant. After our study, the local government constructed water supply systems and the systems have been operated to supply treated surface water to the entire Tay Island since 2014. All people can access As-free water now.

3.2 Groundwater level and river water level

Based on the continuous monitoring data of groundwater levels and river water level, daily average values were obtained as and plotted shown in Figure 3. The groundwater levels in Tay Island show that groundwater levels are low in the dry season from March to June, while the levels are high from July to November. The fluctuation patters are

almost similar every year. According to the hearing survey and existing literatures, the groundwater levels have declined in the middle to downstream areas in the MRD. It is reported that the maximum decline is nearly 10 m in the past 10 years.

4 CONCLUSIONS

The distribution of As concentration measured by the Mekong Group from 2008 to 2015 is shown in Fig. 4. In case several measurements were performed at the same well, the maximum concentration value was used. It is understood that elevated As concentrations are found from Tay Island and Gieng Island near the Cambodian border. On the other hand, As concentrations are lower in the rest areas. The depths of highly contaminated wells in Tay and Gieng Islands are 60 m or less, while the well depths in the rest areas range from 100 to 300 m. The latter wells are used as water sources of water supply systems.

The actual results of the field investigations is clearly shown that the shallow groundwater in the upstream areas in the MRD has elevated As concentrations. While As concentrations in the deep groundwater in the downstream areas are lower.

In recent years the groundwater levels in the deep aquifers have declined sharply. Therefore, vertical leakage from shallow aquifers to the deep aquifers may be enhanced. If the current situation continues in the future, the shallow contaminated groundwater by As may reach to the deep aquifers where production wells of water supply systems are pumping deep groundwater.

ACKNOWLEDGEMENTS

We deeply appreciate ERCA, NIA, VAST, and Mekong Group.

REFERENCE

AAN., Mekong Group. 2012. *Groundwater Arsenic Contamination in the Mekong River Delta*. Monograph 58. The Association for the Geological Collaboration in Japan, 151p.

Arsenic Research and Global Sustainability – Bhattacharya, Vahter, Jarsjö, Kumpiene,
Ahmad, Sparrenbom, Jacks, Donselaar, Bundschuh & Naidu (Eds)
© 2016 Taylor & Francis Group, London, ISBN 978-1-138-02941-5

Recent flow regime and sedimentological evolution of a fluvial system as the main factors controlling spatial distribution of arsenic in groundwater (Red River, Vietnam)

J. Kazmierczak[1], F. Larsen[1], R. Jakobsen[1], D. Postma[1], H.U. Sø[1], H.V. Hoan[2], D.T. Trung[2],
P.Q. Nhan[2], A.E. Hass[3], A.H. Hoffmann[3], P.T.K. Trang[4], V.M. Lan[4] & P.H. Viet[4]
[1]*Department of Geochemistry, Geological Survey of Denmark and Greenland (GEUS), Copenhagen, Denmark*
[2]*Hanoi University of Mining and Geology, Hanoi, Vietnam*
[3]*Department of Geosciences and Natural Resource Management, University of Copenhagen, Copenhagen, Denmark*
[4]*CETASD, Hanoi University of Science, Hanoi, Vietnam*

ABSTRACT: We investigate a relationship between geological history, groundwater flow paths and the spatial distribution of arsenic in aquifers of the upper part of the Red River delta in Vietnam. Hydrogeological conditions in the research area are complex. The fining upward sequence of Pleistocene alluvial sediments was partially eroded during the Holocene and covered by sand and clay deposited in fluvial environments. Sedimentary processes lead to the development of two flow systems. Shallow groundwater discharges either to the local surface water bodies or, in the areas where low permeable sediments isolating Pleistocene and Holocene aquifers were eroded, to the deep groundwater flow system discharging to Red River. Previously reported pattern of arsenic groundwater concentrations decreasing with an increasing sediment age is modified by the observed flow regime. Connection of the younger and older river channels resulted in a transport of high arsenic concentrations towards the Pleistocene aquifer, where low arsenic concentrations were expected.

1 INTRODUCTION

Arsenic contamination is a widespread problem for the management of the drinking water resources in SE Asia. It is expected that the groundwater arsenic concentration decreases with increasing sediment age due to hydrogeochemical processes in an aquifer. The relation is clearly visible in the sand bodies characterized by the local flow systems (Postma *et al.*, 2012) and could probably be disturbed under the influence of more complex flow regimes. Thus, our objective was to check if and how the sedimentary processes in a dynamic meandering river system and the recent flow paths modify the spatial distribution of groundwater arsenic concentrations.

2 METHODS

Lithology of the research area and location and geometry of the Holocene river channels and floodplains were delineated along five cross-sections based on ~50 km of Continuously Vertical Electrical Soundings (CVES) profiles. During the CVES survey we used the Wenner method with a 5-m-spacing between the electrodes and a penetration depth up to 70 m. The geophysical data were calibrated against borehole information consisting of geological descriptions and gamma-logging profiles.

Courses of the Holocene rivers were traced on satellite images and Digital Elevation Model (DEM) as visible deformational scars in the landscape, gently bending topographical highs and/or residual water bodies with the use of multiple ground truth observations and Object Based Image Analysis (OBIA).

Obtained geological cross-sections and maps of the buried and recent fluvial structures were combined in GeoScene 3D into a 3D geological model.

A conceptual flow model was developed based on: (i) the 3D geological model, (ii) measurements of the Red River stage and hydraulic heads and properties in the studied aquifers, (iii) flow patterns and rates in the Red River floodplain described by Larsen *et al.* (2008) and Postma *et al.* (2012) and (iv) spatial variations in groundwater chemistry.

Groundwater arsenic concentrations in the Pleistocene and Holocene aquifers were compared with the modeled geological conditions and conceptual flow paths using GeoScene 3D.

3 RESULTS AND DISCUSSION

3.1 *Holocene evolution of the Red River floodplain*

The Red River floodplain was developed under the complex deposition and erosion processes taking place in the dynamic environment of meandering rivers. Along the geological cross-sections we distinguished three major buried meandering river belts of early to late Holocene age partially cutting through the fining upward sequence (gravel-clay) of the Pleistocene deposits or directly overlying the fractured bedrock. Based on the sediment dating (Postma et al., 2012), migration and sedimentary processes within a single river belt can cover a time period of over 2000 years. It leads to the development of the mosaic of younger river channels cutting through the older fluvial sediments. Early to late Holocene river belts are locally separated by fine grained floodplain sediments and overlain by clay and/or sand deposited along the recent (<1000 years) rivers network. Changes in the location of the fluvial channels were additionally triggered by fluctuations of the sea level during Holocene (Tanabe et al., 2006).

In such a dynamic system, connection of the Holocene sand bodies, detected either with the point measurements (borehole data) or linear measurements (CVES survey), into meandering rivers belts is a challenging task. Using OBIA for river reconstruction is a promising methodology (Addink & Kleinhans, 2008). Distribution of the delineated paleo fluvial structures corresponds with the location of the buried river channels interpreted from the geophysical data. Joint analysis of the OBIA results and geological cross-sections allowed for a more certain extrapolation of the Holocene fluvial system.

3.2 Groundwater flow paths

Uniform groundwater age (<40 years) using the $^3H/^3He$ method was detected in the Holocene aquifer in the western part of the research area, along the transect running from the regional recharge zone in the mountains to the regional discharge zone at the Red River. It implies the occurrence of the local flow systems discharging to the nearest surface water bodies (Postma et al., 2012), while the deeper groundwater of the regional flow system discharges to the Red River.

Intense erosion of the fine grained floodplain sediments and Pleistocene clay in the central part of the research area created hydraulic windows between recent and buried river channels and coarse grained Pleistocene deposits. In this area shallow groundwater flows towards the deeper aquifers. Flow reversals from/to the local surface water bodies are additionally dependent on the annual precipitation cycle and changes in the surface water stages (Larsen et al., 2008). The connection between Holocene and Pleistocene aquifers is confirmed by locally observed similar groundwater chemistry.

3.3 Spatial distribution of arsenic in groundwater

The lowest arsenic concentrations were observed along the river belt filled up with sediments

3600–5900 years of age and the highest concentrations—in recent sediments of the Red River. Arsenic is released to the groundwater during oxidation of the organic matter coupled to the reduction of iron hydroxides. Reactivity of the organic matter decreases over time. Therefore, groundwater arsenic concentration in general decreases with increasing sediment age (Postma et al., 2012). However, this seems to be true only for the buried channels which have no connection with the younger fluvial deposits and/or are characterized by the local flow system discharging in the nearest surface water bodies. Flow reversals and hydraulic connection between Holocene and Pleistocene aquifers in the central part of the research area cause transport of high arsenic concentrations towards the older buried river channels and Pleistocene sediments.

4 CONCLUSIONS

Joint analysis of geophysical, borehole and remote sensing data is an invaluable tool for development of 3D geological models of the complex meandering rivers systems.

Previously described pattern of groundwater arsenic concentration decreasing with increasing sediment age (Postma et al., 2012) is modified by the groundwater flow paths. Erosional processes lead to the local hydraulic connection between recent and buried river channels and flow reversals. In these areas, shallow groundwater carrying high arsenic concentration flows towards the Pleistocene aquifer.

ACKNOWLEDGEMENTS

This research has been funded by the European Research Council under the ERC Advanced Grant ERG-2013-ADG. Grant Agreement Number 338972.

REFERENCES

Addink, E.A., Kleinhans, M. 2008. Recognizing meanders to reconstruct river dynamics of the Ganges. ISPRS J Photogramm, ISPRS International Archives XXXVIII-4/C1.
Larsen, F., Pham, N.Q., Dang, N.D., Postma, D., Jessen, S., Pham, V.H., Nguyen, T.B., Trieu, H.D., Tran, L.T., Nguyen, H., Chambon, J., Nguyen, H.V., Ha, D.H., Hue, N.T., Duc, M.T., Refsgaard, J.C. 2008. Controlling geological and hydrogeological processes in an arsenic contaminated aquifer on the Red River floodplain, Vietnam. Appl Geochem. 23: 3099–3115.
Postma, D., Larsen, F., Thai, N.T., Trang, P.T.K., Jakobsen, R., Nhan, P.Q., Long, T.V., Viet, P.H., Murray, A.S. 2012. Groundwater arsenic concentrations in Vietnam controlled by sediment age. Nature Geosci. 5: 656–661.
Tanabe, S., Saito, Y., Vu, Q.L., Hanebuth, T.J.J., Ngo, Q.L., Kitamura, A. 2006. Holocene evolution of the Song Hong (Red River) delta system, northern Vietnam. Sediment. Geol. 187(1): 29–61.

Arsenic Research and Global Sustainability – Bhattacharya, Vahter, Jarsjö, Kumpiene,
Ahmad, Sparrenbom, Jacks, Donselaar, Bundschuh & Naidu (Eds)
© 2016 Taylor & Francis Group, London, ISBN 978-1-138-02941-5

Age and provenance of groundwater in a shallow arsenic-affected aquifer in the lower Mekong Basin, Kandal Province, Cambodia

L.A. Richards[1], J. Sültenfuß[2], D. Magnone[1], A. Boyce[3], C. Sovann[4],
M.J. Casanueva-Marenco[1], C.J. Ballentine[5], B.E. van Dongen[1] & D.A. Polya[1]

[1]*School of Earth, Atmospheric and Environmental Sciences and Williamson Research Centre for Molecular Environmental Science, University of Manchester, Manchester, UK*
[2]*Institute of Environmental Physics, University of Bremen, Bremen, Germany*
[3]*Scottish Universities Environmental Research Centre, East Kilbride, UK*
[4]*Department of Environmental Sciences, Royal University of Phnom Penh; Cambodia*
[5]*Department of Earth Sciences, University of Oxford, Oxford, UK*

ABSTRACT: Arsenic (As) contamination of groundwaters in South/Southeast Asia is a major threat to public health. Understanding the source and age of the groundwaters is critically important to understanding the controls on As mobilization in these aquifers. Utilizing a suite of isotopic tracers, we have determined various site-specific recharge mechanisms, and estimated recharge source contributions. Tritium-helium age determinations show that the groundwater in this area is modern, indicating relatively fast recharge even in the absence of large-scale groundwater abstraction. The relationship between apparent age and depth indicates a strong vertical control on hydrology, which may influence preferential flow paths and ultimately As release in these aquifers.

1 INTRODUCTION

Millions of people in South/Southeast Asia face chronic exposure to groundwater containing dangerous concentrations of naturally-occurring Arsenic (As). As release in these shallow aquifers is widely thought to require metal reducing bacteria (Islam *et al.*, 2004) and bioavailable organic matter, however the nature of the organic matter and the possible anthropogenic exacerbation of As release via large-scale groundwater abstraction (Harvey *et al.*, 2002) remains contentious. Clarifying these controls is essential in determining how future groundwater As hazard may change (Polya & Charlet, 2009). The aim of this work is to provenance As-affected groundwater, in an area unaffected by large-scale groundwater abstraction.

2 METHODS AND MATERIALS

2.1 *Field site description and sample collection*

The field sites are located in Kandal Province, Cambodia, an area heavily affected by As (Lawson *et al.*, 2013, Sovann & Polya, 2014). Groundwater samples were taken along two distinct transects, T-Sand and T-Clay, dominated by sand and clay respectively, and oriented broadly parallel with major inferred groundwater flow paths. Surface and groundwater samples were collected pre-monsoon (May—June 2014) and post-monsoon (Nov—Dec 2014). Samples for stable isotopes (δD and $\delta^{18}O$) were refrigerated in 60 mL amber glass bottles. Tritium samples were collected in

1 L argon filled amber glass bottles and stored with ~4 cm head of argon gas. Noble gas samples were collected in copper tubes and clamped, using back pressure to suppress potential degassing.

2.2 *Stable Hydrogen and Oxygen isotopes*

ΔD and $\delta^{18}O$ analysis was conducted at the Isotope Community Support Facility (ICSF) at SUERC. Recharge source proportions were estimated using the stable isotope mixing model IsoSource Version 1.3.1 (Phillips & Gregg, 2003), using inputs of the isotopic composition of sources (volume-weighted mean for precipitation; annual mean for ponds and rivers) and groundwater, increment and tolerance.

2.3 *Tritium-Helium dating*

^3H, He isotopes and Ne were analyzed at the Institute of Environmental Physics (University of Bremen) (Sültenfuß *et al.*, 2009).

3 RESULTS AND DISCUSSION

3.1 *Stable Oxygen and Hydrogen isotopes*

Precipitation and surface waters show large seasonal isotopic variations. Precipitation peaks in evaporative enrichment around August. Rivers are more isotopically enriched post-monsoon, consistent with the onset of the dry season. Isolated ponds show the opposite trend. Variations in isotopic signatures were used to quantitatively

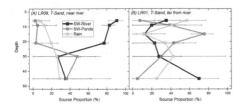

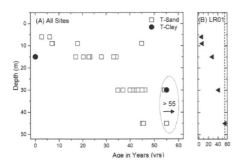

Figure 1. Modelled estimates of recharge source proportions. Errors represent the range of mathematically feasible solutions.

Figure 2. Apparent groundwater ^{3}H-^{3}He age versus depth for (A) all sites and (B) site LR01.

estimate contributions of recharge sources via mixing models (Fig. 1), with sites near the river showing a large contribution of river-similar recharge. The source end-members are not independent and modelled contributions reflect the relative degree of evaporation prior to recharge (ponds representing the most evaporated source). Because the river has both precipitation-based and evaporative components, the model underestimates the river water contribution. Model assumptions/limitations will be further discussed.

3.2 Tritium-Helium dating

Pre- and post-monsoon ^{3}H concentrations are generally similar and broadly consistent with previous work (Lawson et al., 2013). ^{3}H-^{3}He ages increase with depth (Fig. 2), suggesting a major vertical flow control in these aquifers. For all samples except site LR09, the sum of ^{3}H and tritogenic ^{3}He follows the ^{3}H input function for Bangkok precipitation, clearly indicating that no old ^{3}H-free water is admixed to the young groundwater. The largely modern groundwaters provide strong evidence of fast timescales of recharge and challenge previous models which instead simulated flowpath travel times on the order of hundreds of years (Polizzotto et al., 2008). Implications of results on potential surface-groundwater interactions and As mobilization (Lawson et al., 2016) will be discussed.

4 CONCLUSIONS

The provenance of As-affected groundwater was determined using isotopic tracers. Stable isotope data indicated site-specific and varying degrees of surface-groundwater interaction. Estimated recharge source contributions were consistent with lithology and locality. ^{3}H-^{3}He ages showed that most groundwaters are modern (< 55 years), with a strong relationship between age and depth, suggesting a dominant vertical hydrological control. Such hydrological controls may affect As mobilization and subsequent transport.

ACKNOWLEDGEMENTS

This research was funded by a NERC Standard Research Grant (NE/J023833/1) to DP, BvD and CB, a NERC PhD studentship (NE/L501591/1) to DM and NERC Isotope Geosciences Facilities Access IP-1505-1114. We are grateful to Chivuth Kong, Pheary Meas and Yut Yann (all Royal University of Agriculture, Phnom Penh), Chhengngunn Aing, Zongta Sang and Teyden Sok (all RUPP), local land-owners and the local drilling team led by Hok Meas. Ann Hall, Marc Hall, Lori Frees and Lori Allen (Resources Development International—Cambodia) are thanked for logistic support.

REFERENCES

Harvey, C.F., Swartz, C.H., Badruzzaman, A.B.M., Keon-Blute, N., Yu, W., Ali, M.A., Jay, J., Beckie, R., Niedan, V., Brabander, D., Oates, P.M., Ashfaque, K.N., Islam, S., Hemond, H.F., Ahmed, M.F. 2002. Arsenic mobility and groundwater extraction in Bangladesh. *Science* 298: 1602–1606.

Islam, F.S., Gault, A.G., Boothman, C., Polya, D.A., Charnock, J.M., Chatterjee, D., Lloyd, J.R. 2004. Role of metal-reducing bacteria in arsenic release from Bengal delta sediments. *Nature* 430: 68–71.

Lawson, M., Polya, D.A., Boyce, A.J., Bryant C., Ballentine, C.J. 2016. Tracing organic matter composition and distribution and its role on arsenic release in shallow Cambodian groundwaters. *Geochim. Cosmochim. Acta* 178: 160–177.

Lawson, M., Polya, D.A., Boyce, A.J., Bryant C., Mondal, D., Shantz, A., Ballentine, C.J. 2013. Pond-derived organic carbon driving changes in arsenic hazard found in Asian groundwaters. *Environ. Sci. Technol.* 47: 7085–7094.

Phillips, D.L., Gregg, J.W. 2003. Source partitioning using stable isotopes: coping with too many sources. *Oecologia* 136: 261–269.

Polizzotto, M.L., Kocar, B.D., Benner, S.G., Sampson, M., Fendorf, S. 2008. Near-surface wetland sediments as a source of arsenic release to ground water in Asia. *Nature* 454: 505–508.

Polya, D.A., Charlet, L. 2009. Rising arsenic risk? *Nature Geoscience* 2(6): 383–384.

Sovann, C., Polya, D.A. 2014. Improved groundwater geogenic arsenic hazard map for Cambodia. *Environ. Chem.* 11(5): 595–607.

Sültenfuß, J., Roether, W., Rhein, M. 2009. The Bremen mass spectrometric facility for the measurement of helium isotopes, neon, and tritium in water. *Isot. Environ. Health. S.* 45(2): 83–95.

*Arsenic Research and Global Sustainability – Bhattacharya, Vahter, Jarsjö, Kumpiene,
Ahmad, Sparrenbom, Jacks, Donselaar, Bundschuh & Naidu (Eds)
© 2016 Taylor & Francis Group, London, ISBN 978-1-138-02941-5*

Selective chemical extractions of Cambodian aquifer sediments—evidence for sorption processes controlling groundwater arsenic

M.J. Casanueva-Marenco, D. Magnone, L.A. Richards & D.A. Polya
*School of Earth, Atmospheric and Environmental Sciences and Williamson Research Centre
for Molecular Environmental Science, University of Manchester, Manchester, UK*

ABSTRACT: Selective chemical extractions of both sand-rich and clay-rich Cambodian arsenic-prone aquifers have been undertaken to explore the potential role of partial re-equilibration through sorption of arsenic between groundwater and the solid aquifer matrix. Groundwater arsenic was found to be associated with lower Eh, higher pH (particularly at low Eh) and with higher weakly sorbed Fe in clay-rich aquifers but lower weakly sorbed Fe in sand-rich aquifers. The results are suggestive of As(III) sorption on Fe-bearing phases being an important process to consider, particularly in clay-rich aquifers.

1 INTRODUCTION

High arsenic (As) concentrations in shallow reducing groundwaters, that are widely used across the world as a source of drinking, cooking and/or irrigation water, may be controlled by several (bio-)geochemical processes, notably adsorption/desorption, inorganic redox reactions and/or microbiologically mediated reactions (Smedley and Kinniburgh, 2002, Islam *et al.*, 2004).

Irrespective of the processes leading to As mobilization, subsequent movement of As-bearing groundwaters through different aquifer materials may result in changes in concentration of As and other components as a result of (partial) re-equilibration through sorption/desorption (e.g. Radloff *et al.*, 2015). Such processes may thus mask geochemical evidence from groundwater for the nature of previous As mobilization processes.

The aim of this study was, for well-studied (Polizzotto *et al.*, 2008, Lawson *et al.*, 2016 and references therein) arsenic-prone shallow reducing aquifers in Cambodia, to assess the importance or otherwise of such partial re-equilibration in re-setting groundwater arsenic concentrations.

The quantitative importance of these processes might reasonably be expected to depend in part on groundwater pH and Eh and sediment grain size (as a proxy for specific surface area) as well as on the concentrations of weakly and more strongly bound As and related components, notably Fe, Mn, and P in the solid aquifer materials.

Accordingly, the objectives of the present study were to determine the concentrations of weakly and strongly bound As, Fe, Mn and P in aquifer sediments with a view to determining whether or to what extent arsenic concentrations (cf. Richards *et al.*, 2015) in co-existing groundwaters are associated with these solid phase concentrations and other plausibly relevant groundwater/sediment parameters.

2 METHODS/EXPERIMENTAL

2.1 Study area and sampling

The study area was located south of Phnom Penh between the Mekong and Bassac rivers in Kandal Province Cambodia and the study focused on two Holocene sedimentologically contrasting transects, viz., "T-Clay" including floodplain deposits with a relatively high clay content; and "T-Sand" comprising largely sand-dominated scroll bar deposits.

Sediment samples were collected during well construction (Richards *et al.*, 2015). In brief, wells were drilled using manual rotary methods with sediment samples collected at 3 m intervals by manually hammering a custom made sampler with an internal acrylic tube designed to capture the sediment sample. Sediment samples were immediately removed from the sampler at the field site, placed in an aluminum foil bag, placed within a polythene bag which was flushed with nitrogen and then stored in cool box with ice packs in the field. Within 6 hours all samples were re-flushed with nitrogen in a field laboratory and stored in a freezer before transportation to Manchester for analysis. Water sample collection and preservation protocols are outlined in Richards *et al.* (2015).

2.2 Methodology

Two separate single extraction procedures were applied in order to assess: (i) weakly sorbed As (As$_w$), Fe (Fe$_w$) and Mn (Mn$_w$) (v$_0$ = 40 mL of CH$_3$COOH, 0.11 M) and (ii) strongly bound As (As$_s$), Fe (Fe$_s$) and Mn (Mn$_s$) (v$_0$ = 20 mL of NaH$_2$PO$_4$, 0.5 M) (cf. Eiche *et al.*, 2008). The appropriate volume of extractant (v$_0$) was added to ~ 1 g of precisely weighed wet sediment. The mixture was mechanically shaken for 16 hours at a speed of 150 rpm at room temperature and centrifuged at 2700 rpm for 20 min. The supernatant

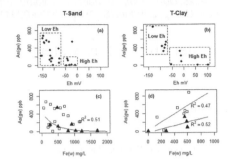

Figure 1. Groundwater arsenic in T-Sand and T-Clay aquifers as a function of Eh (a) & (b) and Fe$_w$ (c) & (d), in which "High Eh" groundwaters are indicated by triangles and "Low Eh" groundwaters by squares.

liquid was transferred to a new tube and stored at 4°C until analysis. Final samples were diluted (20 times) and acidified with high purity HNO$_3$ (at 1%) before analyses by ICP-MS (Agilent 7500cx). Groundwater analysis was carried out using field (pH, Eh) and laboratory (As (As$_{gw}$), Fe (Fe$_{gw}$) and Mn (Mn$_{gw}$)) procedures outlined by Richards *et al.* (2015) with data reduction methods as outlined by Polya & Watts (2016) and Polya *et al.* (2016). Multiple linear regression (MLR) models were built for As$_{gw}$ as a function of the variables measured, viz. pH, Eh, Fe$_{gw}$, Mn$_{gw}$, As$_w$, Fe$_w$, Mn$_w$, As$_s$, Fe$_s$, Mn$_s$. Statistical analysis was conducted mostly using R.

3 RESULTS AND DISCUSSION

Groundwater arsenic, As$_{gw}$ varied between 10 and 800 µg/L and was strongly anti-correlated with Eh in both T-Sand and T-Clay aquifers (Fig. 1(a) & (b)). Whilst As$_{gw}$ was strongly correlated with pH for "Low-Eh" (<−50 mV) groundwater, this was less evident for "High-Eh" (>−50 mV) groundwaters, which also exhibited lower As$_{gw}$ concentrations (Figure 1(a) & (b)). Within T-Sand in the High-Eh water there is an inverse negative correlation between As$_{gw}$ and Fe$_w$ however in the Low-Eh samples there is no relationship between As$_{gw}$ and Fe$_w$ (Figure 1(c)). This suggests that in T-Sand sorption of the sediments may only become important at High-Eh. In T-Clay at both Low-Eh and High-Eh there is a positive correlation between As$_{gw}$ and Fe$_w$ (Figure 1(d)) which suggests that the T-Clay sediments exert a greater control over As$_{gw}$ than T-Sand sediments—a relationship consistent with strongly surface area dependent sorption processes.

4 CONCLUSION

In both the sand and clay dominated aquifers, Eh and pH, particularly at low Eh, were the predominant controls on groundwater arsenic (As$_{gw}$) concentra-

tion. These associations, together with a positive correlation of (As$_{gw}$) with weakly bound Fe (Fe$_w$) in clay dominated aquifers but a negative correlation in sand-dominated aquifers is suggestive of the importance of sorption of aqueous As(III) to iron-bearing minerals phases, particularly in fine-grained high specific surface area clay-rich lithologies. Work is ongoing to quantify these processes and those involving Mn-, and P-bearing and other phases in these aquifers.

ACKNOWLEDGEMENTS

This research was funded by a NERC Standard Research Grant (NE/J023833/1) to DP, Bart van Dongen and Christopher Ballentine and a NERC PhD studentship (NE/L501591/1) to DM. MJCM acknowledges receipt of a University of Cadiz (UCA) Postdoctoral Bridge Contract award.

REFERENCES

Eiche, E., Neumann, T., Berg, M., Weinman, B., van Geen, A. 2008. Geochemical processes underlying a sharp contrast in groundwater arsenic concentrations in a village on the Red River delta, Vietnam. *Appl. Geochem.* 23: 3143–3154.

Islam, F.S., Gault, A.G., Boothman, C., Polya, D.A., Charnock, J.M., Chatterjee, D., Lloyd, J.R. 2004. Role of metal-reducing bacteria in arsenic release from Bengal delta sediments. *Nature* 430: 68–71.

Lawson, M., Polya, D.A., Boyce, A.J., Bryant, C., Ballentine, C.J. (2016). Tracing organic matter composition and distribution and its role on arsenic release in shallow Cambodian groundwaters. *Geochimica et Cosmochimica Acta* 178: 160–177.

Polizzotto, M.L., Kocar, B.D., Benner, S.G., Sampson, M., Fendorf, S. 2008. Near-surface wetland sediments as a source of arsenic release to ground water in Asia. *Nature* 454: 505–508.

Polya, D.A., Watts, M. 2016. Sampling and analysis for monitoring arsenic in drinking water. In: P. Bhattacharya, D.A. Polya, D. Jovanovic. (Eds.) *Best Practice Guide for the Control of Arsenic in Drinking Water*. (Chapter 5), IWA Publishing, London, UK (ISBN13: 9781843393856).

Polya, D.A., Richards, L.A., Al Bualy, A.N., Magnone, D., Lythgoe, P.R. 2016. Groundwater sampling, arsenic analysis and risk communication: Cambodia Case Study. In: P. Bhattacharya, D.A. Polya, D. Jovanovic. (Eds.) *Best Practice Guide for the Control of Arsenic in Drinking Water*. (Chapter 5), IWA Publishing, London, UK (ISBN13: 9781843393856).

Radloff, K.A., Zheng, Y., Stute, M., Weinman, B., Bostick, B., Mihajlov, I., Bounds, M., Rahman, M.M., Huq, M.R., Ahmed, K.M., Schlosser, P., van Geen, A. 2016. Reversible adsorption and flushing of arsenic in a shallow, Holocene aquifer of Bangladesh. *Appl. Geochem.* (doi:10.1016/j.apgeochem.2015.11.003).

Richards, L.A., Magnone, D., van Dongen, B.E., Ballentine, C.J., Polya, D.A. 2015. Use of lithium tracers to quantify drilling fluid contamination for groundwater monitoring in Southeast Asia. *Appl. Geochem.* 63: 190–202.

Smedley, P.L , Kinniburgh, D.G. 2002. A review of the source, behaviour and distribution of arsenic in natural waters. *Appl. Geochem.* 17: 517–568.

Arsenic Research and Global Sustainability – Bhattacharya, Vahter, Jarsjö, Kumpiene,
Ahmad, Sparrenbom, Jacks, Donselaar, Bundschuh & Naidu (Eds)
© 2016 Taylor & Francis Group, London, ISBN 978-1-138-02941-5

Relationship between high arsenic groundwater and surface water distribution in the Hetao basin, China

H. Guo[1,2], Y. Cao[1,2], S. Li[1,2], D. Zhang[1,2], Y. Jia[1,2] & Y. Jiang[1,2]
[1]State Key Laboratory of Biogeology and Environmental Geology, China University of Geosciences,
Beijing, P.R. China
[2]School of Water Resources and Environment, China University of Geosciences, Beijing, P.R. China

ABSTRACT: High As groundwater has been found in the Hetao basin, which is patchily distributed
and challenges the safe drinking water supply. Remote sensing image interpretation and geochemical
investigation were carried out at the basin scale and the local/site scale, respectively. Results showed that
high As groundwater is consistent with the evolution of surface water body. Arsenic concentrations of
porewater near the dry lagoon were generally lower than those near the wet lagoon, although As concentrations increased with the sampling depth in both areas. Organic matter in the recharged surface water
near the wet lagoon would be labile and responsible for As mobilization. Mixing of recharged water and
groundwater also controlled groundwater As concentration.

1 INTRODUCTION

Although distribution of groundwater As was patchy
in the Hetao basin (Guo *et al.*, 2008, 2011, 2013), surface water affected groundwater As concentration. It
was found that interaction between surface water
and groundwater near the irrigation channels and
drainage channels (artificial surface water bodies) led
to relatively low As concentration within the range
of around 1 km (Guo *et al.*, 2011). However, high As
concentrations were usually found near the natural
lakes or lagoons (Jia *et al.*, 2014, Guo *et al.*, 2016).
Both recharge of dissolved organic matter from surface water (Harvey *et al.*, 2002), and sediment organic
matter below surface water body (Neumann *et al.*,
2010, Polizzotto *et al.*, 2008) play important roles in
controlling groundwater As. Besides, sediment lithological conditions, which are normally characterized
by small grain size particles (clay, silty clay and silt),
affected As distribution of shallow groundwater
(Aziz *et al.*, 2008, Guo *et al.*, 2012). However, it is
unclear how the exact relationship between surface
water distribution and high As groundwater is, and
what affects this relationship. This study aims at
investigation of the relation between surface water
and groundwater As distribution by using the remote
sensing technology and delineation of geochemical
factors controlling the relationship.

2 METHODS/EXPERIMENTAL

The study area is located in the Hetao basin, which
lies to the north of the Yellow river and to the south
of the Langshan Mountains (Guo *et al.*, 2008).

Thick sediments have been deposited during the
Pleistocene and Holocene Epoch. Surface water
bodies, including drainage channels, irrigation
channels and natural lagoons, are widely distributed in the basin (Fig. 1), which would potentially
affect As distribution in shallow groundwater. Both
remote sensing image interpretation and geochemical monitoring were carried out to delineate their
relationship at the basin scale and the lagoon scale.

Remote sensing images were interpreted to
extract the distribution of surface water body during last 40 years in the Hetao Basin. Variation of
surface water distribution was analyzed by using
statistical approaches.

Boreholes (around 10 m below land surface,
bls) were drilled near typical dried natural lagoon
and permanent natural lagoon to take sediment
samples and porewater samples. The dried natural
lagoon has been no water for more than 15 years
due to the decline of groundwater table. Groundwater samples were also taken near the lagoons.
Sediment samples and water samples were analyzed for geochemical and isotopic parameters.

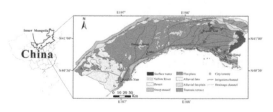

Figure 1. Distribution of surface water body in the
study area.

3 RESULTS AND DISCUSSION

Interpretation of remote sensing images during last 40 year showed variation in surface water body. In the surface water-irrigation areas, area of surface water body increased, while decreased in the groundwater irrigation areas. The increasing trend in surface water body were mainly observed in the areas with high As groundwater. The consistency is possibly due to the presence of clay layers with high organic matter contents near the subsurface of natural lagoons. The other possibility would be the recharge of surface dissolved organic matter into aquifers, which would be bioavailable and labile for microbial metabolism.

Arsenic concentrations of porewater samples from the wet lagoon were higher (0.6–98.6 µg/L; median 7.8 µg/L) than those from the dry lagoon (<0.1 to 47.3 µg/L; median 3.8 µg/L). The results are consistent with interpretation of remote sensing images. Generally, As concentrations increased with increasing the sampling depth in both the dry lagoon and the wet lagoon. From the borehole near the wet lagoon, porewater As concentration was 20.5 µg/L at depth of 0.5 m bls, which increased to 98.4 µg/L at depth of 10.0 m bls. The increase would be related to the gradual release of Fe oxides-adsorbed As in the presence of organic matter during groundwater flow (Neumann *et al.*, 2010, Polizzotto *et al.*, 2008).

Laterally, porewater generally increased its As concentration away from the wet lagoon. However, the highest As concentrations were observed in sediments below the lagoon, which was possibly due to the presence of labile organic matter at the bottom sediments. This trend was consistent with As concentrations in the dry lagoon, where porewaters below the center of the lagoon had higher As concentrations than those near the shore. It indicated that the wetlands would be related to the elevated As concentrations of shallow groundwater via introduction of organic matter. Oxygen and hydrogen isotopes of surface water, porewater and groundwater showed that surface water had been experienced more intensive evaporation than porewater and groundwater. During the recharge of surface water into groundwater, mixing between recharged water (usually with high As concentration) and groundwater may lead to the increase in groundwater As concentrations.

4 CONCLUSIONS

Remote sensing images during the past 40 years showed that the variation in surface water body was consistent with distribution of high arsenic groundwater in the Hetao basin. Near the dry lagoon, porewater and groundwater had relatively lower As concentration than those near the permanent lagoon. Generally, groundwater increased As concentrations along the flow path from the lagoon to the downgradient. Mixture between the recharged water and groundwater also controlled dissolved As concentrations. It suggested that organic matter in the surface water would play an important role in mobilizing As in the aquifer systems.

ACKNOWLEDGEMENTS

The study has been financially supported by National Natural Science Foundation of China (Nos. 41222020 and 41172224), the program of China Geology Survey (No. 12120113103700), the Fundamental Research Funds for the Central Universities (Nos. 2652013028 and 2652015334), and the Fok Ying-Tung Education Foundation, China (Grant No. 131017).

REFERENCES

Aziz, Z., van Geen, A., Stute, M., et al. 2008. Impact of local recharge on arsenic concentrations in shallow aquifers inferred from the electromagnetic conductivity of soils in Araihazar, Bangladesh. *Water Resour. Res.* 44: W07416.

Guo, H.M., Jia, Y.F., Wanty, R., Jiang, Y.X., Zhao, W.G., Xiu, W., Shen, J.X., Li, Y., Cao, Y.S., Wu, Y., Zhang, D., Wei, C., Zhang, Y.L., Cao, W.G., Foster, A. 2016. Contrasting distributions of groundwater arsenic and uranium in the western Hetao basin, Inner Mongolia: Implication for origins and fate controls. *Sci. Total Environ.* 541: 1172–1190.

Guo, H.M., Liu, C., Lu, H., Wanty, R., Wang, J., Zhou, Y.Z. 2013. Pathways of coupled arsenic and iron cycling in high arsenic groundwater of the Hetao basin, Inner Mongolia, China: An iron isotope approach. *Geochim. Cosmochim. Acta* 112: 130–145.

Guo, H.M., Zhang, Y., Xing, L.N., Jia, Y.F. 2012. Spatial variation in arsenic and fluoride concentrations of shallow groundwater from the Shahai town of the Hetao basin, Inner Mongolia. *Appl. Geochem.* 2012, 27: 2187–2196.

Guo, H.M., Zhang, B., Li, Y., Berner, Z., Tang, X.H., Norra, S. 2011. Hydrogeological and biogeochemical constrains of arsenic mobilization in shallow aquifers from the Hetao basin, Inner Mongolia. *Environ. Pollut.* 159: 876–883.

Guo, H.M., Yang, S.Z., Tang, X.H., Li, Y., Shen, Z.L. 2008. Groundwater geochemistry and its implications for arsenic mobilization in shallow aquifers of the Hetao Basin, Inner Mongolia. *Sci. Total Environ.* 393: 131–144.

Harvey, C.F., Swartz, C.H., Badruzzaman, A.B.M., Keon-Blute, N., Yu, W., Ali, M.A., Jay, J., Beckie, R., Niedan, V., Brabander, D., Oates, P.M., Ashfaque, K.N., Islam, S., Hemond, H.F., Ahmed, M.F., 2002. Arsenic mobility and groundwater extraction in Bangladesh. *Science* 298: 1602–1606.

Jia, Y.F., Guo, H.M., Jiang, Y.X., Wu, Y., Zhou, Y.Z. 2014. Hydrogeochemical zonation and its implication for arsenic mobilization in deep groundwaters near the Langshan mountains of the Hetao Basin, Inner Mongolia. *J. Hydrol.* 518: 410–420.

Neumann, R.B., Ashfaque, K.N., Badruzzaman, A.B.M., Ashraf Ali, M., Shoemaker, J.K., Harvey, C.F. 2010. Anthropogenic influences on groundwater arsenic concentrations in Bangladesh. *Nature Geosci.* 3: 46–52.

Polizzotto, M.L., Kocar, B.D., Benner, S.G., et al. 2008. Near-surface wetland sediments as a source of arsenic release to ground water in Asia. *Nature* 454: 505–508.

Arsenic Research and Global Sustainability – Bhattacharya, Vahter, Jarsjö, Kumpiene,
Ahmad, Sparrenbom, Jacks, Donselaar, Bundschuh & Naidu (Eds)
© 2016 Taylor & Francis Group, London, ISBN 978-1-138-02941-5

Relation between sediment salinity and leached arsenic in the Hetao Basin, Inner Mongolia

R. Yuan[1,2], H. Guo[1,2], Y. Jia[1,2]

[1]State Key Laboratory of Biogeology and Environmental Geology, China University of Geosciences, Beijing, P.R. China
[2]School of Water Resources and Environment, China University of Geosciences (Beijing), Beijing, P.R. China

ABSTRACT: Forty-one sediment samples collected from the Hetao Basin were extracted by deionized water for characterizing distribution of sediment salinity and As. Evaporation intensely improved EC of sediments in near-surface stratum. Clay and silty clay generated higher EC, soluble As and HCO_3 than sand. Grain size of sediments, reducing conditions and groundwater flushing were considered to be the main factors influencing soluble components and As in sediments, which may control groundwater chemistry in the aquifers.

1   INTRODUCTION

High As groundwater in inland basins usually has high Total Dissolved Solid (TDS) (Guo *et al.*, 2014), which may either originate from dissolution of sediment minerals or evaporation. In the Hetao Basin, TDS in groundwater was up to 8000 mg/L (Guo *et al.*, 2008), and low flow rate results in a relatively intensive water-rock interaction and accumulation of As into groundwater (Jia *et al.*, 2014). However, distribution of sediment salinity and its role in As mobilization are still unknown.

The objectives of this study are to 1) characterize vertical distribution of sediment salinity; 2) decipher the relation of soluble components to lithology and sampling depth; 3) assess the relation between soluble components and leached As.

2   METHODS/EXPERIMENTAL

2.1   Study area

The Hetao Basin is located in western part of Inner Mongolia, a flat fan shape between the Langshan Mountain and the Yellow River. In the Hetao Basin, the shallow aquifer consists of Quaternary alluvial-lacustrine sediments and hosts high As groundwater (0.6–572 μg/L As). Dark-fine-sand sediments were considered to be the result of reductive dissolution of Fe oxide-minerals (Guo *et al.*, 2008). In the central part of the basin, silty clay and clay rich in organic matter and the alluvial–lacustrine aquifers mainly occur at depths between 10 and 100 m, which formed clay interlayer and leaky-confined or semi-confined.

2.2   Sampling and processing methods

One borehole was drilled below land surface up to 90 m for collecting sediment samples in September 2010. Forty one samples were collected and stored in −80°C within tin foil package.

Four grams of sample were mixed with deionized water by a solid and liquid ratio of 1:5 in PE centrifuge tubes and shaken at 150 r/min in 25°C for 1 h. Supernatant was filtered through 0.22 μm filter membranes. Filtrate was measured promptly by conductivity meter (DDS-307A, SHKY). Major cations and anions were determined by ICP-AES and ICS-600, respectively. For pH measurement, one gram sample was mixed with 10 mL 0.01 M $CaCl_2$ and measured by HANNA pH meters. Saturation indices of soluble components were calculated by PHREEQC.

3   RESULTS AND DISCUSSION

3.1   Vertical distribution of sediment salinity

EC values of near-surface sediment samples showed a decreasing trend along depth (0–20 m) (784–247 μS/cm) (Fig. 1a) due to intense evaporation in arid or semi-arid climate (Guo and Wang, 2005). At depths between 20 and 95 m, EC of sediments kept relatively constant (around 113–252 μS/cm), with the exceptions at depths of 66 and 78 m, where high EC values were observed (657 and 336 μS/cm, respectively). The high values were related to the fine-grained sediments at those depths (clay and silty clay). Relatively low and constant EC value occurring in sandy sediments of aquifer may be due to flush of groundwater.

3.2   Distribution of soluble components

Soluble Ca, Mg, Na, and Cl presented similar trends to EC values. Sodium was the dominant cation (0.052–0.473 mg/g, average 0.112 mg/g) (Fig. 1b, d), which indicated that Na-minerals were more soluble than others. In the near-surface stratum, soluble anions Cl, SO_4, and HCO_3 (Fig. 1c) also showed decreasing trends along depth (0–10 m) (0.395–0.097 mg/g, 0.421–0.221 mg/g and 0.479–0.116 mg/g, respectively). At depths between 10 and 95 m, Cl was relatively constant (around 0.021–0.097 mg/g). At 66 m, where high EC value was observed, extreme high soluble Cl and Ca may be due to dissolution of $CaCl_2$ minerals. However, significant fluctuations of pH, SO_4, HCO_3 versus depth were observed. In clay layers, HCO_3 increased distinctly (0.375–0.61 mg/g) and kept relatively low value in sand layers (around 0.047–0.231 mg/g). Negative correlation between SO_4 and HCO_3 was observed at depths 15–95 m, which was possibly due to effects of corresponding groundwater components.

Soluble HCO_3 was positively correlated with TIC of sediments, indicating that carbonates were mostly soluble. Most samples lay along 1:2 line of mole ratio of $(Ca+Mg)/HCO_3$ and 1:1 line of mole ratio of Na/HCO_3, which indicated that those soluble components originated from dissolution of carbonate minerals. Furthermore, plot of $(Na-Cl)/(HCO_3+2SO_4-2Mg-2Ca)$ implied that soluble minerals have relatively greater effects on groundwater chemistry than cation exchange.

Negative values of $SI_{calcite}$, $SI_{dolomite}$, and SI_{gypsum} revealed that all leachates were unsaturated with respect to calcite, dolomite, and gypsum. Dissolution of gypsum led to a decrease in sediment pH.

3.3 Arsenic leaching

Samples near land surface (0–15 m) with high EC value had low leached As (0.002–0.018 µg/g), which was associated with intense evaporation and oxic-suboxic conditions (Fig. 1e). At depths between 15–95 m, relatively high contents of soluble As were leached from clay and silty clay around 21 m, 63 m,

77 m, and 88 m (0.06–0.147 µg/g). Sand had low contents of soluble As (around 0.005–0.04 µg/g). However, at depths between 47 and 53 m, an increasing trend of leached As was observed (0.018–0.123 µg/g) in sand layer, which may be associated with fine-grained size. Leached As was positively correlated with HCO_3 and negatively correlated with SO_4. The concurrent high contents of leached Fe (1.7–5.2 µg/g) and As (Fig. 1f) in clay or silty clay layers revealed that As may be originated from reductive dissolution of Fe-minerals. Compared with sand layers (0–0.79 µg/g), clay or silty clay had more soluble Fe and As, which was likely due to reducing conditions and weakly flushing of groundwater.

4 CONCLUSIONS

EC values of sediment samples showed a increasing trend (784–247 µS/cm) near surface between 0–20 m due to evaporation, and showed constant distribution (around 113–252 µS/cm) between 20–95 m in sand layer. Clay and silty clay had higher values of EC (366–657 µS/cm) and soluble As (0.06–0.147 µg/g). Fine-grained sediments would have more soluble salt. Meanwhile, flushing of groundwater and large grain size may decrease the sediment salinity of sand in aquifer. Similar trends of Fe, HCO_3 and As revealed that soluble As mainly originated from reductive dissolution of Fe-minerals.

ACKNOWLEDGEMENTS

The study was financially supported by National Natural Science Foundation of China (Nos. 41222020 and 41172224), the program of China Geology Survey (No. 12120113103700), the Fundamental Research Funds for the Central Universities (No. 2652013028). We thank China University of Geosciences (Beijing) Geological Microbiology Laboratory for allowing us to perform the ICS-600 and conductively meter analysis.

REFERENCES

Guo H.M., Yang S.Z., Tang X.H., Li Y., Shen Z.L. 2008. Groundwater geochemistry and its implications for arsenic mobilization in shallow aquifers of the Hetao Basin, Inner Mongolia. Sci. Total Environ. 393: 131–144.
Guo H.M., Wen D.G., Liu Z.Y., Jia Y.F., Guo Q. 2014. A review of high arsenic groundwater in Mainland and Taiwan, China: Distribution, characteristics and geochemical processes. Appl. Geochem. 41: 196–2177.
Guo, H., Wang, Y. 2005. Geochemical characteristics of shallow groundwater in Datong basin, northwestern China: J. Geochem. Explor. 87(3): 109–120.
Jia, Y., Guo, H., Jiang, Y., Wu, Y., Zhou, Y. 2014. Hydrogeochemical zonation and its implication for arsenic mobilization in deep groundwaters near alluvial fans in the Hetao Basin, Inner Mongolia. J. Hydrol. 518: 410–420.

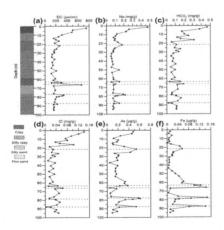

Figure 1. Plots of lithology, EC, Na, HCO_3, Cl, As, Fe versus depth.

Arsenic Research and Global Sustainability – Bhattacharya, Vahter, Jarsjö, Kumpiene,
Ahmad, Sparrenbom, Jacks, Donselaar, Bundschuh & Naidu (Eds)
© 2016 Taylor & Francis Group, London, ISBN 978-1-138-02941-5

Enrichment of high arsenic groundwater and controlling hydrogeochemical processes in the Hetao basin

W. Cao, Y. Zhang & Q. Dong

Institute of Hydrogeology and Environmental Geology, Chinese Academy of Geological Sciences, Zhengding, Hebei, P.R. China

ABSTRACT: In this study we apply cumulative frequency distribution graphs to characterize the distribution of ionic ratios throughout the Hetao basin. All hydrogeochemical processes can basically be classified as recharge intensity of groundwater, evaporation concentration intensity and reductive degree that controlling the spatial distribution of arsenic. We found that the concentration of arsenic is more than 10 µg/l when the (HCO_3 + CO_3)/SO_4 ratio more than 4.1 (strong reductive area). As the evaporation concentration intensity increases, the median value of arsenic rises from 10.74 µg/L to 382.7 µg/L in median reductive area, and rising from 89.11 µg/L to 461.45 µg/L rapidly in strong reductive area. As the river recharge intensity increases (with intensity index from 0 to 5), the median value of arsenic drop from 40.2 µg/L to 6.8 µg/L in median reductive area, and drop markedly from 219.85 µg/L to 23.73 µg/L in strong reductive area. The results provide new insight into the mechanism of As enrichment in groundwater.

1 INTRODUCTION

To quantitatively understand the mechanisms that control As mobilization in groundwater systems, it is essential to determine its hydrogeochemical processes in the aquifer system. Guo et al. (2012, 2015) indicate that hydrogeochemical process is the main factor in controlling groundwater As in the Hetao Basin, Inner Mongolia. Many Ionic ratios (Cl/Br, SO_4/Cl, Na/Cl and so on) have been jointly used to distinguishing hydrogeochemical processes, such as redox condition and evaporation concentration (Cartwright & Weaver, 2005, Cartwright et al., 2006; George Breit, 2015), identify redox condition, recharge and quantitatively evaluate groundwater systems in arid and semi-arid regions (Smedley & Kinniburgh, 2002; Guo, 2015), determine the impact of irrigation return flow on arsenic mobilization in the groundwater system. There are some relations between the concentration of arsenic and the hydrogeochemical processes, such as reductive degree, recharge intensity from ionic ratio and evaporation concentraon intensity. In this study, We apply cumulative frequency distribution graphs to characterize the distribution of ionic ratios throughout the Hetao basin, and thereby gaining the hydrogeochemical indicators that distinguish the suite of hydrogeochemical processes that govern groundwater composition. The result can be used to constrain the hydrogeochemical processes that control As mobilization and enrichment in groundwater.

2 METHODS/EXPERIMENTAL

In this study, we collected 447 samples from the shallow aquifer (10–40 m) which is composed of late Pleistocene and Holocene alluvial and lacustrine deposits. Of those, 80% wells are domestic water wells and the rest are the agricultural supply wells. The sampling campaign was undertaken during the period from June 10th to September 27th, 2010 (Fig.1) across the Hetao basin north to the foot of langshan, south to the Yellow River, west to the wulanbuhe desert and east to the Ulansuhai Nur.

2.1 *Ionic relationships*

The relationships between major ions (Na, Ca, Mg, K, Cl, HCO_3, CO_3 and SO_4) can provide information on the sources of groundwater and the processes that led to their composition. All units are expressed in milliequivalents per litre (meq/L), which converts concentrations to chemical "equivalents". This assists with understanding water-rock processes that may control groundwater composition. Because of the log-normal distribution of data, all axes are presented in logarithmic scale.

3 RESULTS AND DISCUSSION

3.1 *Spatial distribution of arsenic*

In the Hetao basin, 53.5% of shallow groundwater samples were found to contain As concentration exceeding 10 µg/L. Among them, 134 samples were above 50 µg/L, accounting for 30.0% of all samples, and remaining 56.1% were above the WHO drining water guide line value. There are two arsenic anomaly belt: first is located in the connected zone between the two main geomorphic units of northern Hetao basin along the Total Drainage Channel, namely the piedmont alluvial–pluvial plain and the Yellow River alluvial lacustrine plain. This region is the groundwater discharge area of Hetao basin with high pH, TDS, Cl, HCO_3^-, and low SO_4^{2-}. The second is distributed in the several large crevasse splay from Dengkou country to Urad Front Banner along the north bank of Yellow River. Comparing with the first high arsenic belt, the arsenic content of this region was on a low level with a range of 10~50 µg/L.

3.2 Hydrogeochemical process by Ionic ratio

Ionic ratio, including Na/Cl, $(HCO_3 + CO_3)$/Cl, Ca/$(HCO_3 + CO_3)$, Na/HCO_3, Cl/SO_4, $(HCO_3 + CO_3)$/SO_4, were assessed using cumulative frequency distribution graphs to identify hydrogeochemical populations, and demarcation values established that can be used to characterise hydrogeochemical processes based on ionic relationships. In Hetao basin, the processes identified and constrained by this method include three main processes: reductive degree, recharge intensity and evaporation concentration intensity. Hydrogeochemical indicators were developed that define the critical values for differentiation of each process. Overlay analysis of the three main processes of As-contaminated groundwater can be used to constrain the hydrogeochemical processes that control As mobilization and enrichment in groundwater.

Table 1. Hydrogeochemical indicators for various chemical process controlling groundwater composition in the Hetao basin, based off analysis of cumulative frequency distribustion curves.

Ionic ratio	Hydrogeochemical indicator (meq/L ionic ratio)	Hydrogeochemical process
Na/Cl	0~1	Saline groundwater, Evaporative concentration of NaCl
	1~1.8	Background
	>1.8	Mineral hydrolysis or ion-exchange reaction
$(HCO_3 + CO_3)$/Cl	0~1.3	Weak runoff area, evaporation concentration of salt
	>1.3	River recharge, fresher groundwater
Ca/$(HCO_3 + CO_3)$	0~0.15	Strong reducing, evaporation concentration
	0.15~0.6	Intermediate process
	>0.6	Recharge by Ca-HCO_3 water (river recharge), fresher groundwater
Na/HCO_3	0.4~1	Irrigation of Yellow river
	1.1~3.8	Intermediate condition
	>3.8	Halite dissolution or deep fracture brine intrusion
Cl/HCO_4	0~1	Lateral flow recharge from Yinshan mountain or Yellow river, renew faster
	>3.3	High-salty, strong reduce, evaporation concentration
$(HCO_3 + CO_3)$/SO_4	0~1.4	Weak reductive
	1.4~4.1	Intermediate condition
	>4.1	Strong reductive, discharge area with high TDS

3.3 Hydrogeochemical control on As distribution

The success of this assessment was variable for different ion-pairs, but mostly it was possible to assign hydrogeochemical indicators that distinguish the suite of hydrogeochemical processes that govern groundwater composition. The key indicator are defined in Table 1 and illustrate some commonality in between various tested ratio.

The factors of the high arsenic groundwater in Hetao basin are very complex, it is not single hydrogeochemical process that explain the distribution characteristics. According to the previously mentioned, combine the evaporation concentration, reductive condition and the recharge intensity of groundwater

4 CONCLUSIONS

We found that the concentration of arsenic is more than 10 µg/l when the (HCO_3+CO_3)/SO_4 ratio more than 4.1(strong reductive area). As the evaporation concentration intensity increases, the median value of arsenic rises from 10.74 µg/L to 382.7 µg/L in median reductive area, and rising from 89.11 µg/L to 461.45 µg/L rapidly in strong reductive area. As the river recharge intensity increases (with intensity index from 0 to 5), the median value of arsenic drop from 40.2 µg/L to 6.8 µg/L in median reductive area, and drop markedly from 219.85 µg/L to 23.73 µg/L in strong reductive area. The results provide new insight into the mechanism of As enrichment in groundwater.

ACKNOWLEDGEMENTS

This study was supported by National Natural Science Foundation of China (No. 41402234) and the Geological Survey Projects Foundation of China (No. 1212010913010). We thank Drs. Huaming Guo for discussion and comment. We thank two anonymous reviewers for their comment and guest editor for greatly improving the manuscript.

REFERENCES

Breit, G., Guo, H. 2012. Geochemistry of arsenic during low-temperature water–rock interaction. *Appl. Geochem.* 27: 2157–2159.
Cartwright, T.R. Weaver, L.K, 2005. Hydrochemistry of the Goulburn Valley region of the Murray Basin, Australia: implications for flow paths and resource vulnerability *Hydrogeol. J.* 13: 752–770.
Cartwright, T.R. Weaver, L.K. Fifield, 2006. Cl/Br ratios and environmental isotopes indicators of recharge variability and groundwater flow: An example from the southeast Murray Basin: Australia. *Chem. Geol.* 231: 38–56.
Guo, H. Zhang, Y., Xing, L., Jia, Y. 2012. Spatial variation in arsenic and fluoride concentrations of shallow groundwater from the town of Shahai in the Hetao basin, Inner Mongolia. *Appl. Geochem.* 27: 2187–2196.
Guo, H., Liu, Ding, Z.S., Hao, C., Xiu, W., Hou, W. 2015. Arsenate reduction and mobilization in the presence of indigenous aerobic bacteria obtained from high arsenic aquifers of the Hetao basin, Inner Mongolia. *Environ. Pollut.* 203: 50–59.
Smedley, P.L., Kinniburgh, D.G. 2002. A review of the source, behaviours and distribution of arsenic in natural waters. *Appl. Geochem.* 17: 517–568.

Arsenic Research and Global Sustainability – Bhattacharya, Vahter, Jarsjö, Kumpiene,
Ahmad, Sparrenbom, Jacks, Donselaar, Bundschuh & Naidu (Eds)
© 2016 Taylor & Francis Group, London, ISBN 978-1-138-02941-5

Investigation of arsenic contamination from geothermal water in different geological settings of Taiwan: Hydrogeochemical and microbial signatures

J.P. Maity[1,2], C.-Y. Chen[2], J. Bundschuh[3] & P. Bhattacharya[1,4]

[1]*International Center for Applied Climate Studies, University of Southern Queensland, Toowomba, QLD, Australia*
[2]*Department of Earth and Environmental Sciences, National Chung Cheng University, Chiayi, Taiwan*
[3]*Deputy Vice Chancellor's Office (Research and Innovation) & Faculty of Health, Engineering and Sciences, University of Southern Queensland, Toowoomba, Australia.*
[4]*KTH-International Groundwater Arsenic Research Group, Department of Sustainable Development, Environmental Science and Engineering, KTH Royal Institute of Technology, Stockholm, Sweden*

ABSTRACT: The dissemination of dissolved elements, including arsenic (As(V)/As(III)) and microbial diversity was studied in hydrothermal systems of Taiwan considering three different principal geological settings such as Igneous rock terrains, Metamorphic terrains and Sedimentary terrains to understand the cycling, fate and transport and potential impact of As on hydrological systems. The results were indicated as strongly acidic (pH< 3), sulfate-enriched waters of $H-SO_4$-type in igneous-sedimentary rock terrains, slightly alkaline waters (pH: 8–8.95) in metamorphic terrains, and circum-neutral waters (pH 6.47–7.41) of $Na-HCO_3$/$Na-Cl-HCO_3$-type in metamorphic-sedimentary terrains. The geothermal waters were enriched with As in igneous terrains (Beitou: 1.46 mg/L) as compared to sedimentary (0.16 ± 0.14 mg/L) or metamorphic (0.06 ± 0.02 mg/L) terrains. The 16S rRNA gene sequence of bacterial diversity indicates prevalence of mesophilic sulfur- and thiosulfate-oxidizing bacterium in Taipu (igneous rock terrains). The discharge from geothermal springs with significant levels of As and other toxic element contaminate the surface and groundwater of environment.

1 INTRODUCTION

Hydrothermal systems, with geothermal fluids discharging as springs, fumaroles or steam at or close to the earth's surface of the earth, and those are found at many several places of the world (Bundschuh *et al.*, 2013). The hydrothermal systems show various specific hydrochemical characteristic, with suites of enriched trace elements, which are considered common contaminants in surface hydrological and groundwater systems (Bundschuh & Maity, 2015). Arsenic is present at elevated concentrations, compared to average groundwater, in most of the hydrothermal systems throughout the world. Because of its high toxicity, As concentrations in water exceeding the 10 µg/L WHO guideline. Last few decades, hot spring spas and resorts have gained enormous popularity in Taiwan. However, the enrichment of these of hot spring waters with toxic elements such as As, Se, Cr and B can have severe human health and environmental impacts and may affect the natural drinking water and irrigation water sources located at or near hydrothermal fields. Furthermore, the temperature dependent microbial activities of the geothermal system are executed the precipitation, adsorption and mobilization reactions in the presence of mineral and organic matter that drives reducing conditions (Maity *et al.*, 2011).

The present study is a first attempt of an investigation of the water chemistry and microbial diversity of the major and most popular hydrothermal spring systems of Taiwan in different geological.

2 METHODS/EXPERIMENTAL

2.1 *Sampling, chemical and biological analysis*

The water samples were collected from geothermal springs indicated in Figure 1 and stored at 4°C for chemical and biological analysis. EC, TDS, DO, pH, Eh, temperature, salinity, and alkalinity were measured using a portable conductivity meter, a multimeter (SP701, Suntex, Taiwan) and HI-9143 Microprocessor Dissolved Oxygen Meter (HANNA, Taiwan). Alkalinity was measured by titration method. Arsenic speciation was determined by using the HG-AFS (Millenium Excallibur, PS Analytical Ltd, Kent, UK), coupled with a HPLC by checking with reference (SRM 3103a, NIST, USA). The concentrations of trace element were measured with the ICP-MS (Hewlett-Packard 4500, Yamanashi-Ken, Japan) by checking with high purity standards (TMDW; Lot # 623609, USA).

Bacterial isolate were identified by 16S rRNA gene sequence process using forward and reverse primers:

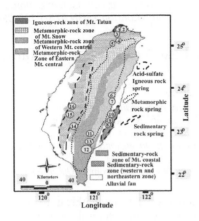

Figure 1. Hydrothermal springs in different lithological terrains of Taiwan: *Igneous rock terrains:* 1.Yangmingshan; 2. Taipu; *Metamorphic terrains:* 3. Beitou; 4. Wulai-1; 5. Wulai-2; 6. Honye (Hualien); 7. Reisuei; 8. Antung-1; 9 Antung-2; 10. Antung-3; 11. Honngye (Taitung); 12. Jiben-1; 13. Jiben-2; 14. *Sedimentary terrains:* Biolai; 15. Kung-Tzu-Ling; 16. Chung-Lun.

16S 27F (forward): 5'-AGAGTTTGATCCTGGCTC AG-3' (genomic positions 8–27, Escherichia coli numbering) and 16S 1492R (reverse): 5'-GGTTAC CTT-GTTACGACTT-3' (genomic positions 1488–1511, Escherichia coli numbering).

3 RESULTS AND DISCUSSION

Highest water temperatures of > 80°C were found in Jiben (12: Jiben-1; 13: Jiben-2) and Kuantzeling (15) hot springs. The temperatures at Yangmingshan (1), Beitou (3) and Hongye (11: Taitung; 6: Hualien) were ranging from 70–80°C, and those from Wulai (4: Wulai-1; 5: Wulai-2), Antung (8: Antung-1; 9: Antung-2; 10: Antung-3) and Chung-lLun (16) ranged from 60–70°C. However, relatively low temperature waters (range 38–50°C) were observed in Taipu (2), Rueisuei (7) and Biolai hot springs (14).

Three groups of hydrothermal spring water samples are distinguished as (i) strongly acidic (pH< 3), sulfate-enriched waters of H-SO$_4$-type (Igneous-Sedimentary rock terrains), (ii) slightly 1kaline waters (pH: 8–8.95) (Metamorphic terrains: Jiben, Antung and Kung-Tzu-Ling), and (iii) circum-neutral waters (pH 6.47–7.41) of Na-HCO$_3$/Na-Cl-HCO$_3$ type (Metamorphic-Sedimentary terrains: Wulai, Hongye, Rueisuei, Chung-Lun and Biolai). The waters are enriched with alkali and alkali earth metals compared to drinking water. Similarly, the water of most of the geothermal spring were found to be enriched with As (highest concentration at Beitou: 1.456 mg/L in Igneous rock terrains) with As (III) being the principal As species. Arsenic concentrations of hydrothermal spring waters in igneous rock terrains exhibit highest concen-

trations (0.69 ± 0.71 mg/L) followed by those of sedimentary (0.16 ± 0.14 mg/L) and metamorphic (0.06 ± 0.02 mg/L) terrains.

The 16S rRNA gene sequence of bacterial diversity shows that the *Sulfurovum lithotrophicum* is Gram-negative, non-motile and coccoid to oval-shaped mesophilic sulfur- and thiosulfate-oxidizing bacterium, and grew chemolithoautotrophically with elemental sulfur or thiosulfate as a sole electron donor and oxygen or nitrate as an electron acceptor in Taipu geothermal water (Igneous rock terrains) in Taiwan. *Limnobacter thiooxidans* (strain Wu-1–2), a thiosulfate-oxidizing bacterium was noticed from Wulai hot geothermal spring (Metamorphic terrains). The *Clostridium sulfidigenes* (Biolai) and *Desulfovibrio psychrotolerans* (Chung-Lun) (moderately alkaliphilic, psychrotolerant and Chemo-organoheterotrophy) are active to reduced sulfur compounds or As (III) (Sedimentary terrains). The discharged geothermal springs water contaminate the surface and groundwater (including drinking and irrigation water), where significant levels of As and other toxic element have detected and hence being a significant risk for environment and health.

4 CONCLUSIONS

The geothermal waters in the igneous rock zone were acidic with higher concentration of As in compare to metamorphic (basic) and sedimentary rock zones (near-neutral pH). Arsenic concentrations of geothermal water of Taiwan are ranging between 0.036 and 1.456 mg/L, where As (III) is the dominant species. The geothermal waters in igneous rock zone were exhibited pH dependent (acidic water) release of higher concentration of As in comparted to metamorphic or sedimentary rock zone (alkaline water).

ACKNOWLEDGEMENTS

We would like to thank NSC, Taiwan.

REFERENCES

Bundschuh, J., Maity J.P., Nath, B., Baba, A., Gunduz, O., Kulp, T.R., Jean, J.S., Kar, S., Tseng, Y.J., Bhattacharya, P., Chen, C.Y. 2013. Naturally occurring arsenic in terrestrial geothermal systems of western Anatolia, Turkey: potential role in contamination of freshwater resources. *J. Hazard. Mater.* 262: 951–959.

Bundschuh, J., Maity J.P. 2015. Geothermal arsenic: Occurrence, mobility and environmental implications. *Renew. Sust. Energ. Rev.* 42: 1214–1222.

Maity, J.P., Liu, C.C., Nath, B., Bundschuh, J., Kar, S., Jean, J.S., Bhattacharya, P., Liu, J.H., Atla, S.B., Chen, C.-Y. 2011. Biogeochemical characteristics of Kuan-Tzu-Ling, Chung-Lun and Bao-Lai hot springs in southern Taiwan. *J. Environ. Sci. Heal. A* 46(11): 1207–1217.

Arsenic Research and Global Sustainability – Bhattacharya, Vahter, Jarsjö, Kumpiene,
Ahmad, Sparrenbom, Jacks, Donselaar, Bundschuh & Naidu (Eds)
© *2016 Taylor & Francis Group, London, ISBN 978-1-138-02941-5*

Arsenic occurrence in groundwater sources of Lake Victoria basin in Tanzania

J. Ijumulana[1,2], F. Mtalo[2], P. Bhattacharya[1,3] & J. Bundschuh[4]

[1]*KTH-International Groundwater Arsenic Research Group, Department of Sustainable Development Environmental Science and Technology, KTH Royal Institute of Technology, Stockholm, Sweden*
[2]*Department of Water Resource Engineering, University of Dar es Salaam, Dar es Salaam, Tanzania*
[3]*International Center for Applied Climate Science, University of Southern Queensland, Toowoomba, Queensland, Australia*
[4]*Faculty of Health, Engineering and Sciences, University of Southern Queensland, Toowoomba, Queensland, Australia*

ABSTRACT: Increasing naturally occurring toxic substances are becoming threat to many safer groundwater sources worldwide. Their detection, scale of occurrence and removal procedures requires sophisticated tools. This study employs geostatistical methods to study the extent of arsenic occurrence over Lake Victoria basin in Tanzania. Water samples are randomly taken at village level from available drinking water source, tested at water quality laboratory to quantify levels of As concentrations. The study further uses fuzzy logics to identify areas with no, low, moderate and high As concentrations based on the World Health Organization (WHO) guidelines and Tanzania Bureau of Standards (TBS) standards of 0.01 mg/L and 0.05 mg/L As respectively. The result of this study is vital to understanding severity of As contamination and prediction of potential human health effects dwelling within these regions.

1 INTRODUCTION

Despite of efforts made by many governments to ensure adequate and accessible drinking water to all citizens, safe drinking water source remain a global problem. Several studies indicate that the levels of geogenic pollutants in drinking water sources are above the WHO drinking water guideline values (WHO, 2011) which may have severe negative human health outcomes among the population (Mandal & Suzuki, 2002, Bhattacharya *et al.*, 2002, Kassenga & Mato, 2008).

Arsenic (As), the 20th most abundant element in the earth's crust, is among the top geogenic pollutants in many groundwater resources and environment worldwide (Bhattacharya *et al.*, 2002, Smedley & Kinniburgh, 2002). In Africa, As concentrations in groundwater range between 0.02 and 1760 µg/L (Ahoulé *et al.*, 2015). Adverse health effects of skin disorder and cancer due to an elevated arsenic concentration were recently reported around the North Mara gold and Geita mining area within Lake Victoria basin (Almas *et al.*, 2012). Apart from reported cases of low As contamination in northern parts of Tanzania (Kassenga & Mato, 2008), the extent of As pollution in groundwater is poorly investigated.

Many naturally occurring elements including As, show spatial variability and thereby exhibit spatial uncertainty (Chiles & Delfiner, 2009). In this study geostatistical models are developed and used to assess spatial variability of As within Lake Victoria basin regions which include Mara, Mwanza, Simiyu, Shinyanga, Geita and Kagera regions. The model development is based on the WHO drinking water guideline of 10 µg/L As and TBS standard of 50 µg/L As. Since As occurs as an oxyanionic species, regression kriging shall be adopted in which statistical significance of other variables such as iron, calcium, manganese shall be taken into consideration.

The present study gives a regional-prediction model of As contaminated site using GIS and remote sensing tools. The results will provide insight to the extent of As contamination in groundwater resources that is an important step towards As determination in areas with minimal data records.

2 CURRENT STATUS OF DRINKING WATER SOURCES AND THEIR CHARACTERIZATION

2.1 *Drinking water sources in Lake Victoria Basin*

During 2013, the Ministry of Water (MoW) conducted a water point mapping project by recording preliminary drinking water quality. Of the 64,020 drinking water sources mapped, 23.4% were located within Lake Victoria basin. Table 1 summarizes the current data on water quality, with a multitude of pollutants in each water source.

2.2 *Approach for delineating the areas affected by arsenic in groundwater in Lake Victoria Basin*

Geostatistical models are used to monitor the extent of As concentrations in the study area. The study samples are developed at village level and analyzed at ward level by adopting ward centroids as our spatial support. Samples in the radius of 1000 m from each centroid are taken and tested to identify water pollution severity. Using the results of tests at each water point a regression kriging model is developed

Table 1. Summary of reported water quality.

Water quality	Water sources (%)	Remarks
Coloured water	0.5	Operational
Fluoride	0.1	Operational
Turbid water	4.4	Operational
Saline water	11.1	Operational
Unknown	4.7	Operational
Soft water	79.0	Operational

Source: (MoW, 2013).

and used to predict As concentrations at unsampled locations. In addition, uncertainty map will be developed using residuals of prediction and compared to WHO guideline and the Tanzanian Bureau of Standards (TBS) for As to identify areas with no, low, moderate and high human health risk.

3 RESULTS AND DISCUSSION

3.1 Prediction of arsenic around Lake Victoria Basin

This study uses the available data on groundwater As from the pioneering study made by Kassenga & Mato (2008) within Lake Victoria Basin to develop As contamination spatial distribution maps. Ordinary regression kriging is being used to predict As at unsampled locations. Figure 1 A and 1B show predicted map and variance map respectively, where it is evidenced that As concentrations vary in space and the variation is location independent as envisaged in Figure 1B.

3.2 Spatial uncertainty modeling

From literature, many naturally occurring water pollutants such as As are uncertain in space or in time or both. Understanding their spatial distribution requires sophisticated models that combine precise deterministic models and probabilistic procedures to quantify them at every location. Fuzzy logics are used in which membership functions are developed based on WHO drinking water guideline value of 10 µg/L for As (WHO, 2011). In this study, we used standard deviations of the data from Kassenga & Mato (2008) to develop uncertainty map of As (Fig. 1B). The predictions have been done at District level though our interest is to develop uncertainty map at local level. The preliminary results indicate that As concentration is independent of the locations, and requires more closer sampling to monitor As variability within Lake Victoria Basin.

3.3 Arsenic prediction model validation

Cross-validation approach has been adopted in this study. Random values at 1 km by 1 km prediction grid have been correlated with estimated values. A correlation value of 0.997 has been obtained indicating that the GIS prediction model developed can perform well in characterizing spatial distribution of As within Lake Victoria Basin.

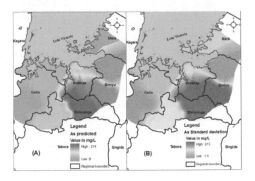

Figure 1. Arsenic prediction map (A) and variance map (B) in Lake Victoria basin. Data source: Kassenga & Mato (2008).

4 CONCLUSIONS

Limited studies on As occurrence within Lake Victoria basin indicate that As exhibits spatial variability. Prolonged ingestion of As within this region may result into negative human health effects. This desktop study thus hold promise for a large scale characterization of spatial distribution of As contamination in drinking water sources and predicting potential population requiring immediate interventions for safe water source.

ACKNOWLEDGEMENTS

We acknowledge the Sida-Tanzania cooperation program No. 51170072 for supporting the research.

REFERENCES

Ahoulé, D.G., Lalanne, F., Mendret, J., Brosillon, S., Maïga, A.H. 2015. Arsenic in African waters: A review. *Water, Air, Soil Pollut.* 26(9): 1–13.

Almås, Å.R., Manoko, M.L. (2012). Trace element concentrations in soil, sediments, and waters in the vicinity of Geita Gold Mines and North Mara Gold Mines in northwest Tanzania. *Soil Sediment Contam.* 21(2): 135–159.

Bhattacharya, P., Frisbie, S.H., Smith, E., Naidu, R., Jacks, G., Sarkar, B. 2002. Arsenic in the Environment: A Global Perspective. In: B. Sarkar (Ed.) *Handbook of Heavy Metals in the Environment*. Marcell Dekker Inc., New York, 147–215.

Chiles, J.P., Delfiner, P. 2009. *Geostatistics: Modeling Spatial Uncertainty* Vol. 497. John Wiley & Sons.

Kassenga, G.R., Mato, R.R. 2008 Arsenic contamination levels in drinking water sources in mining areas in Lake Victoria Basin, Tanzania, and its removal using stabilized ferralsols. *Int. J. Biol. Chem. Sci.* 2(4): 389–400.

Mandal, B.K., Suzuki, K.T. (2002). Arsenic round the world: a review. *Talanta* 58(1): 201–235.

MoW 2013. Water Point Mapping Project. Ministry of Water Resources (MoW), Government of the United Republic of Tanzania. (http://wpm.maji.go.tz/?x = sw6KahIP8RvRm OKWF7 KdFg) Accessed on 29th July, 2015.

Smedley, P.L., Kinniburgh, D.G. 2013. Arsenic in groundwater and the environment In: Selinus, O. (Ed.) *Essentials of Medical Geology*. Springer Netherlands, pp. 279–310.

WHO. 2011. *Guidelines for Drinking-Water Quality*. 4th ed. World Health Organization, Geneva, p. 541.

Arsenic Research and Global Sustainability – Bhattacharya, Vahter, Jarsjö, Kumpiene,
Ahmad, Sparrenbom, Jacks, Donselaar, Bundschuh & Naidu (Eds)
© 2016 Taylor & Francis Group, London, ISBN 978-1-138-02941-5

Occurrence of arsenic in groundwater, soil and sediments in Tanzania

F.J. Ligate[1,2], J. Ijumulana[1,2], F. Mtalo[2], P. Bhattacharya[1,3] & J. Bundschuh[4]

[1]*KTH-International Groundwater Arsenic Research Group, Department of Sustainable Development Environmental Science and Technology, KTH Royal Institute of Technology, Stockholm, Sweden*
[2]*Department of Water Resource Engineering, University of Dar es Salaam, Dar es Salaam, Tanzania*
[3]*International Center for Applied Climate Science, University of Southern Queensland, Toowoomba, Queensland, Australia*
[4]*Faculty of Health, Engineering and Sciences, University of Southern Queensland, Toowoomba, Queensland, Australia*

ABSTRACT: Tanzania is among the sub-Sahara African countries where most of the rural populations use ground water for drinking and other household activities. The presence of elevated concentrations of arsenic in some parts of Tanzania is possibly related to both geogenic and anthropogenic sources. From its origin in the bedrock, inorganic As enters into soil where its concentration depends on the parent rock, climate, forms of As, redox conditions of the soils as well as anthropogenic activities. The interactions of water with rocks and sediments together with environmental deposition have resulted into high concentration in groundwater. Microbial activities in soil and sediments may produce the methylated form of As which are volatile in nature and can re-enter the atmosphere and ultimately be re converted back to inorganic forms. Exposure to elevated concentration of arsenic causes arsenicosis. Further investigations are needed in order to come up with a comprehensive report concerning the occurrence and distribution of As in Tanzania.

1 INTRODUCTION

Arsenic (As) is a trace element which is found at variable concentrations in soils, rocks, natural water, atmosphere and organisms. Arsenic mobilization in the environment is mainly through natural processes such as weathering of rocks, biological activity, geothermal activity and volcanic emissions as well as through human activities such as mining, industrial and agriculture (Bhattacharya et al., 2002, Guo et al., 2014, Herath et al., 2016). Groundwater are especially vulnerable for elevated concentrations of As due to water-rock interactions. Arsenic has profound effects to human health at even low concentrations (Bhattacharya et al., 2016). The WHO provisional guideline value for arsenic in drinking water is 10 µg/L (WHO, 2011). Arsenic has also been reported to be rich in rice and vegetables because of its bio-accumulation tendency. Health problems associated with arsenic results from exposure to contaminated water, food, soil and air. Some of the health effects are; cancer of the skin, liver, lung bladder and other organs (Rubab et al., 2014), hypertension, cardiovascular and cerebrovascular diseases, diabetes mellitus, keratosis and neuro toxicity (IPCS, 2001, Kapaj et al., 2006). Studies on the occurrence of As in groundwater has not received attention so far, although the recent studies have reported an elevated concentration of As in mining areas in lake Victoria basin (Kassenga & Mato, 2008). In view of human health problems associated with As, it is very important to investigate the occurrence of arsenic in different parts of Tanzania especially along the Great Rift Valley.

According to the British Geological Survey report (2008), a number of remediation techniques for groundwater have been reported such as identification of safe tube wells through rigorous water testing and monitoring, groundwater treatment by coagulants such as alum and ferric chloride, addition of potassium permanganate, adsorption using various adsorbents. The sources of arsenic in groundwater depends on the local geology, hydrology and geochemical characteristics of the aquifers (Yan-Chu, 1994). The overall objective of the present study will focus to improve the general understanding and to enhance the present knowledge on occurrence, distribution and health effects of As, specifically this study aims to assess the spatial distribution and extent of contamination of As in ground water surface water soils and sediments, identify the chemical contaminants present in ground water and surface, investigate the effects of seasonal variation on the spatial distribution of geogenic As in groundwater and surface water.

2 GEOLOGIAL AND HYDROGEOLOGICAL CHARACTERISTICS OF THE TANZANIAN RIFT VALLEY

The United Republic of Tanzania is located on the 60°00' South, 35°00' East, with a total area of about 945000 km². It is bordered to the north by Kenya and Uganda, In the east by Indian ocean, in the south by Mozambique in the west by Ruanda, Burundi, Democratic Republic of Congo and Zambia, Its climate varies from tropical along the coast to temperate in the

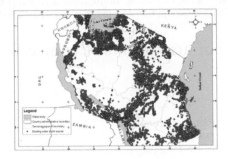

Figure 1. Map of Tanzania and neighboring countries showing the country wide location of the drinking water sources.

highlands. The general geology is comprised of the Precambrian and Phanerozoic formations. Groundwater occurrence is largely influenced by geological conditions, about 75% of Tanzania is underlain by crystalline rocks, dominated by the Precambrian basement complex of varying age and composition forming the basement aquifers (Kongola *et al.*, 1999). According to the BGS (1989), the hydrology of Tanzanian rift valley consists of lateral groundwater flow from the rift escarpment to the rift flow and axial groundwater flows away from the rift flow culmination. The permeability of the volcanic rocks underlying the rift valley are generally poor, Aquifers are normally found in fractured or reworked volcanics and are usually confined.

3   NEED FOR ENHANCING THE REGIONAL DATABASE

The natural As groundwater problem areas are found all over the world but they tend to be confined in several types of environment such as the inland or closed basins in arid or semi-arid areas, the strong reducing aquifers within the sedimentary basins, geothermal areas and in areas of mining activities or where sulfide oxidation has occurred (Bhattacharya *et al.*, 2007, Guo *et al.*, 2014, Herath *et al.*, 2016). Thus the enhancement of regional database for the origin, distribution of As (and other contaminants) in groundwater, soil and sediments, transport and fate in aquifers in Tanzania is relatively less explored. Figure 1 shows the map of Tanzania with the location of the drinking water sources where a very limited information is known. It is there important to investigate and develop the baseline groundwater quality database for As that would enable to identify the potential risk area from public health perspective as well as to develop technologies for control elevated concentration of As.

4   CONCLUSION

Despite the fact that geogenic and anthropogenic As has been extensively investigated in most parts of the world,

there is limited comprehensive study on the contamination of groundwater, soils and sediments from both geogenic and anthropogenic sources in Tanzania. Extensive investigations are required in order to determine the occurrence of As in groundwater, soils and sediments.

ACKNOWLEDGEMENT

The authors thank Sida for financial support (Sida Contribution No. 51170072) through the ongoing DAFWAT research project 2235.

REFERENCES

Allen, D.J. Daring, W.G., Burgess, W.G. 1989. Geothermics and hydrogeology of the southern part of Kenya rift valley with emphasis on the Magadi-Nakuru area. Research report SD/89/1, British Geological Survey, UK.

BGS 2008. Water Quality Fact Sheet: Arsenic. British Geological Survey,Wallingford, U.K.

Bhattacharya, P. Jovanovic, D., Polya, D. 2016. *Best Practices Guide on Control of Arsenic in Drinking. Water.* IWA Publishing, London. (ISBN13:9781843393856) (In press).

Bhattacharya, P., Frisbie, S.H., Smith, E., Naidu, R., et al. 2002. Arsenic in the Environment: A Global Perspective. In: B. Sarkar (Ed.) *Handbook of Heavy Metals in the Environment.* Marcell Dekker Inc., New York, pp. 147–215.

British Geological Survey 1989. The basement aquifer research project, 1984-89. British Geological survey technical report WD/89/15.

Guo, H. Wen, D. et al. 2014. A review of high As groundwater in mainland and Taiwan, China; distribution, characteristics and geochemical processes. *J. Appl. Chem..* 41: 196–217.

Gupta, S.P. 2009 Statistical Methods. 37th Edition. Sultan Chand and Sons. New Delhi, India.

Herath, I., Vithanage, M., et al. 2016. Natural arsenic in global groundwaters: distribution and geochemical triggers for mobilization. *Curr. Pollut. Rep.* 2(1): 68–89.

IPCS 2001. Environmental Health Criteria 224 Arsenic and arsenic compounds. World Health Organization, Geneva, 521p. Available from http://www.who.int./ipcs/publications/ehc_224/en/index.html.

Kapaj, S., Peterson, H., Liber, K., Bhattacharya, P. 2006. Human health effects from chronic arsenic poisoning—a review. *J. Environ. Sci. Hlth A* 41: 2399–2428.

Kassenga, G.R., Mato, R.R. 2008. Arsenic contamination levels in drinking water sources in mining areas in lake Victoria basin, Tanzania, and its removal using stabilized ferralsols. *Int. J. Biol. Chem. Sci.* 2(4): 389–400.

Kongola, L.R.E., Nsanya, G., Sadick, H. 1999. *Groundwater Resources: Development and Management. An Input to the Water Resources Management Policy Review* (Draft). Dar es Salaam, Tanzania.

Rubab, G., Naseem, S., et al. 2014. Distribution and sources of arsenic contaminated groundwater in parts of Thatta district, Sindh. *J. Himalayan Earth Sci.* 47(2): 175–183.

WHO 2011. Guidelines for Drinking-Water Quality. 4th ed. World Health Organization, Geneva, 2011; 541p.

Yan-Chu H. 1994. Arsenic distribution in soils. In: Nriagu, J.O. (Ed.) Arsenic in the Rnvironment. John Wiley and Sons Inc. New York, NY.

Arsenic Research and Global Sustainability – Bhattacharya, Vahter, Jarsjö, Kumpiene,
Ahmad, Sparrenbom, Jacks, Donselaar, Bundschuh & Naidu (Eds)
© 2016 Taylor & Francis Group, London, ISBN 978-1-138-02941-5

Sedimentological controls in distribution of arsenic in the Claromecó basin, Argentina

N.N. Sosa[1], M. Zarate[2] & S. Datta[3]
[1]*Centro de Investigaciones Geológicas (CONICET-UNLP), La Plata, Argentina*
[2]*INCITAP (CONICET-UNLPam), Santa Rosa, La Pampa, Argentina*
[3]*Department of Geology, Kansas State University. Manhattan, KS, USA*

ABSTRACT: The aim of the present work is to demonstrate the preliminary results of the study of the arsenic concentrations in the sediments of the Claromecó river basin. A variation of As content related to lithogenic factors is presented in the study: the Miocene-Pliocene fluvial facies have a reduced arsenic concentration (~2.6 mg/kg) compared with the Late Pleistocene fluvial facies (~11.6 mg/kg). These differences are probably attributed to a major hydraulic gradient during the Pliocene, which is reflected in grain size and in fluvial structures that probably washed out the sediments. Volcanic glass shards probably represent one of the main arsenic sources, especially in the fine silt grain size. On the other hand, different paleosols show different As concentrations due to the presence of amorphous Fe-Mn oxides-hydroxides. Topography is controlling the As concentrations as shown by the increasing of this element in sediments and groundwater in the downgradient of the basin.

1 INTRODUCTION

The Chaco-Pampean plain is an extensive flatland covering 1,000,000 km^2 of central and northern Argentina. It consists of a large mantle of Neogene and Quaternary aeolian (loess) deposits locally reworked by fluvial systems. The region shows one of the highest groundwater As concentrations in the world, causing serious problems to human health. The main possible source of As can be relatead to volcanic ash layers and to volcanic glass shards dispersed in the Pampean loess (Bundshuh *et al.*, 2004; Bhattacharya *et al.*, 2006) that reaches 60 wt.% in the silt fraction (Teruggi et al., 1957). Other authors emphasize the importance of Al, Mn and Fe oxy-hydroxides as another potential source (Smedley *et al.*, 2002; Bhattacharya *et al.*, 2006). In both cases, the oxidising and high pH nature of the groundwaters leads to the release and the mobilization of As through different hydrochemical processes (dissolution, adsorption-desorption).

The Claromecó fluvial basin represents a poorly studied area with high As concentrations in groundwater. The basin is composed of two hydrogeological zones. The upper zone consists of a moderately undulating landscape of low relief (interfluves) composed of Mio-Pliocene silty sandstones capped by a 1.5 m thick pedogenic calcrete (Zarate, 2005). The lower zone is a flatland consisting of a several meters thick aeolian cover including some dune fields close to the Atlantic coast. The valleys are filled with Late Pleistocene fluvial sandy siltstones caped by redoximorphic paleosols. Close to the Atlantic coastline, the Quaternary succession comprises the Late Pleistocene fluvial facies followed by Holocene marshy to estuarine facies.

2 METHODS

2.1 *Geochemical analysis*

The Claromecó basin has been taken as a case study within an on-going project to infer the lithogenic source of As. With this purpose, 34 samples of soils and sediments were collected following chronostratigraphical, sedimentological, pedological and geomorphological criteria and then analyzed via ICP-MS in the Centro de Investigaciones Geológicas Laboratories.

2.2 *Hydrochemical analysis*

A total of 24 water samples were sampled from the Claromecó river basin; 3 surface water samples from the river and its main tributaries, 21 groundwater samples from wells distributed throughout the basin. All the samples were analyzed with IC (major anions) and a HR-ICP-MS (trace and majors elements) at Kansas State University.

3 RESULTS AND DISCUSSION

3.1 *Sediment As concentrations*

A relation between the dissolved As concentrations and the lithogenic factors is evident from the study: the oldest Mio-Pliocene Fluvial Facies (MPFF) have reduced As content (~2.6 mg/kg) compared with the most recent Late Pleistocene Fluvial Facies (LPFF)

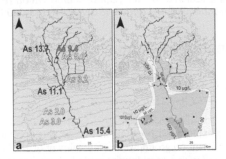

Figure 1. a) Concentrations of arsenic in sediments in the Claromecó basin (mg/kg); b) Concentrations of arsenic in groundwater in the basin. The black dots represent the sample points.

that shows ~11.6 mg/kg of bulk As. Furthermore, the pedogenic calcrete and the redoximorphic paleosols developed in both fluvial facies present very different As contents. While the pedogenic calcrete shows a low content of As (~3.9 mg/kg), the redoximorphic paleosols show the higher As content (~16.5 mg/kg). Close to the coastline, the Holocene estuarine facies show high As content (~18.5 mg/Kg). On the other hand, present day soils show the highest As content, especially in the illuvial horizons (Bt horizon: ~23.3 mg/kg).

The mineralogy is quite similar in all the continental facies with abundant quartz and volcanic glass shards, especially in the fine silt grain size.

The preliminary results so far demonstrate a sedimentological control marking differences in As concentrations between both fluvial facies (MPFF and LPFF). These differences are probably attributed to a major hydraulic gradient during the Mio-Pliocene, which is reflected in grain size and in fluvial structures that probably washed out sediments. A geomorphological control was observed through an increase in As concentrations from the interfluves (MPFF) to the valleys (LPFF) as well as from the upper to the lower basin zone within the LPFF. The Holocene estuarine facies demonstrate high values probably related to the low topographic and hydraulic gradients. Pedogenic calcrete and redoximorphic paleosols developed in MPFF and LPFF respectively. They reflect different pedological processes under different geomorphological and climatic conditions. As content reflect these variations showing the highest values associated with the LPFF paleosols, probably related to the presence of pedofeatures as Fe-nodules and iron rich concretions. Ammonium oxalate extractions on LPFF sediments confirm a strong contribution of As via Fe amorphous oxy-hydroxides. Finally, present day soils show high As concentrations in B and Bt horizons probably due to the grain size and clay mineralogy which is currently under study.

3.2 Water As concentrations

Arsenic concentrations in the groundwater and in the surface waters of the Claromecó basin exceed

by 100% the WHO guideline value of 10 µg/L, and by 80% the Argentine National Standard (CAA) of 50 µg/L. Arsenic concentrations range from 32 and 1010 µg/L, with a mean value of 122 µg/L. Concentration trends coincide with the stratigraphic variations. As concentrations are higher downgradient in the basin probably associated with the lower hydraulic gradient. Other trace elements (Mo, V, F and U) show positive correlations with As (> 0.8) as observed in other areas of the Chaco-pampean plain. These geochemical correlations, accompanied by high pH conditions, suggest that the high As concentrations can be related to the dissolution of the volcanic glass shards dispersed in the sediments.

4 CONCLUSIONS

For the Claromecó basin here-by studied, a primary control on the As concentration was identified based on the hydrological properties reflected by the sedimentary facies. A secondary geomorphological control was observed for the As distributions within the basin: from the interfluves (MPFF) toward the valleys (LPFF) and from the upper to the lower zones of the basin. Groundwater As concentrations follows this regional trend with increasing As downgradient. Different kinds of paleosols show different As values, probably associated with adsorption processes on the surfaces of Fe oxides and hydroxides.

ACKNOWLEDGEMENTS

The author thanks the CONICET (Argentina) for the doctoral fellowship, CIG and INCITAP for giving support facilities and the Kansas State University for the hydrochemical analysis.

REFERENCES

Bhattacharya, P., Claesson, M., Bundschuh, J., Sracek, O., Fagerberg, J., Jacks, G., Martin, R.A., Storniolo, A.R., Thir, J.M.. 2006. Distribution and mobility of arsenic in the Rio Dulce alluvial aquifers in Santiago del Estero Province, Argentina. Sci. Total Environ. 358: 97–120.

Bundschuh, J., Farias, B., Martin, R., Storniolo, A., Bhattacharya, P., Cortes, J., Bonorino, G., Alboury, R. 2004. Groundwater arsenic in the Chaco–Pampean Plain, Argentina: case study from Robles County, Santiago del Estero Province. Appl Geochem. 19(2): 231–243.

Smedley P.L., Kinniburgh D.G. 2002. A review of the source, behavior and distribution of arsenic in natural waters. Appl. Geochem. 17: 517–568.

Teruggi, M.E. 1957. The nature and origin of Argentine loess. J. Sediment. Petrol. 27: 322–332.

Zárate, M., 2005. El Cenozoico tardío continental de la provincia de Buenos Aires. In R.E. de Barrio, R.O. Etcheverry, M.F. Caballé, E. Llambías (Eds.) Relatorio de la Geología y Recursos Minerales de la Provincia de Buenos Aires, 16° Congreso Geológico Argentino, La Plata, Argentina. Relatorio 4: 139–158.

Arsenic Research and Global Sustainability – Bhattacharya, Vahter, Jarsjö, Kumpiene,
Ahmad, Sparrenbom, Jacks, Donselaar, Bundschuh & Naidu (Eds)
© 2016 Taylor & Francis Group, London, ISBN 978-1-138-02941-5

Arsenic in groundwater and sediments in a loessic aquifer, Argentina

L. Sierra, L. Cacciabue, S. Dietrich, P.A. Weinzettel & S.A. Bea
Instituto de Hidrología de Llanuras "Dr. Eduardo J. Usunoff" (IHLLA), Azul, Argentina

ABSTRACT: This work presents the results of a 70 m deep exploration borehole in Buenos Aires province, Argentina. The objective of this work was to establish the As content on groundwater and sediment of the Pampeano aquifer. Groundwater and sediment samples were collected with a sampling interval of about one meter. Thirty representative groundwater and sediment samples were analyzed. Bailed groundwater samples show As concentrations in the range of 8.3–69 (median 32 µg/L). In contrast, pumped groundwater samples showed a higher As content of up to 163 µg/L. Sediment analysis showed typical association of loessic sandy silts with several calcrete horizons. The sediments show As contents in the range 3.5–6.3 mg/kg, and it shows a good correlation with Fe-Mn, which is interpreted as associated to oxy-hydroxides.

1 INTRODUCTION

The Chaco-Pampean Plain is one of the largest known regions (1×10^6 km^2) with high groundwater As concentrations and includes the most populated centers of Argentina (Nicolli *et al.*, 2012). However, a few number of studies have included a detailed aquifer stratigraphy (Smedley *et al.*, 2005, Bhattacharya *et al.*, 2006). Auge *et al.* (2013) determined at a regional scale that As concentration in groundwater across 87% of the Buenos Aires province (267,000 km^2) exceeds 50 µg/L. The groundwater is mainly oxidizing-alkaline and hosted in the Pampeano aquifer which is composed of Tertiary-Quaternary aeolian sediments (loess) and Holocene fluvial reworked material (Smedley *et al.*, 2002, Blanco *et al.*, 2006).

The objective of this work was to analyze the As occurrence in groundwater and sediments of the Pampeano aquifer.

2 MATERIALS AND METHODS

2.1 *Study area*

This study was developed in a 10 km^2 rural area neighboring Tres Arroyos city, Buenos Aires province, Argentina. The mean annual precipitation is 766 mm (1940–2001), with 84% actual evapotranspiration, 13% of recharge and 3% runoff. The study comprises a 120-m thick section of the Pampeano aquifer based on geophysical studies (Weinzettel *et al.*, 2005). Groundwater in the study area is alkaline Na-HCO$_3$ type with electrical conductivity (EC) from 700 to 2750 µS/cm. Arsenic values in groundwater varies from 10 to 234 µg/L. An exploratory borehole was drilled (38°21′54.4″ S; 60°14′39.3″ O) next to the location where highest As value was measured.

2.2 *Sampling and analysis*

A 70 m deep borehole was drilled by percussion without fluid injection and sediment and groundwater sampling were taken at each meter. Drilling sediments were removed from the borehole with a 30 L siphon tool and collected in plastic bags. Following sediment sampling, groundwater samples were bailed with a 2.5 L bailer allowing the borehole to fill back up to the water table at 4.7 meters below surface (mbs). On-site analysis included water pH, EC (25 °C) and alkalinity. Samples were preserved by acidification with HNO$_3$ for cations and trace metals analysis. Except for anions aliquots, all samples were filtered (0.45 µm). The borehole was fully cased with 125 mm diameter PVC casing and fully screened with 1.5 mm screen slot size. Additionally, pumped groundwater sample was obtained at 35 mbs at 20 m^3/h flow rate for 3 hours.

Thirty groundwater and sediments samples were selected and analyzed using standard methodologies (APHA, 2005). Arsenic concentrations were determined by AAS-hydride generation (SM-3114) and F by SPADNS method (SM-4500D). Selected sediment samples were analyzed for Particle Size Distribution (PSD), X-Ray Diffraction (XRD), Loss on Ignition (LOI) and trace element concentrations.

3 RESULTS AND DISCUSSION

3.1 *Borehole groundwater chemistry*

Groundwater samples are alkaline (pH range between 7.57 and 8.12), and chemical composition differences between shallow and deeper waters were observed (Fig. 1). Na-Cl water type dominates down to 24 mbs, with higher EC (in the range of 4500–7390 µS/cm), variable HCO$_3$ (537–1027 mg/L, median 561 mg/L), As (20–69 µg/L, median 40 µg/L) and F (5–20 mg/L, median 8 mg/L). Instead, deeper groundwater compositions are dominated by a Na-mixed-anion type (Cl-SO$_4$) with HCO$_3$ (653–697 mg/L, median 671 mg/L), As (8–42 µg/L, median 29 µg/L) and F (6–25 mg/L, median 15 mg/L). The source of the observed salinity may be associated to evapo-concentration in the upper part of the aquifer. However, since geological conceptual model is not fully resolved yet, mixing processes are not disregarded as contributor to salinity.

Arsenic correlations in the shallower groundwater samples are low for pH ($R^2 = 0.35$), HCO_3 ($R^2 = 0.21$) and F ($R^2 = 0.19$). However, the same correlations in the deeper groundwater samples are lower than the shallow ones. Pumped samples from 35 mbs registered higher As concentration (163 µg/L).

3.2 Sediment description

Collected sediments were predominantly brown, silty, very fine sands with minor clay and coarse sands, in the same way as was described to other loess deposits in Argentina (e.g., Teruggi, 1975).

PSD indicated that sand is the main fraction, in the range 30–83%, followed by silts in the range 14–61%. Clay content is low (3–11%), with a maximum content observed at 20 mbs. It is worth noting that the sand contents increase in the depth intervals of 29–43 and 59–70 mbs (Fig. 1). Organic carbon concentrations (OC = 1/2 LOI 550 °C) were low (1–2%). Black massive nodules and coatings over minerals, aggregates and calcretes were conspicuous. These were interpreted as Fe-Mn oxy-hydroxides based on loess description and the Fe-Mn concentrations. Calcrete layers were very abundant (see LOI 950 °C in Figure 1). Arsenic concentrations were in the range 3.5–6.3 mg/kg (median 4.6 mg/kg), and they are the lowest contents reported in the region (Blanco *et al.*, 2006).

Despite the fact that the bailer sampling method may induce the mixing of waters from different depths, silty sand horizons with elevated Mn sediment contents show high As concentrations in the bailed groundwater samples. On the other hand, pumped groundwater samples shown the highest As concentration in the first

silty sand interval. The low As content in the sediments suggests that the high As concentrations in the pumping water has its origin in the sediment-bound As that can be easily mobilized in the oxidizing groundwater conditions. This observation is consistent with the elevated concentrations of oxy-hydroxides in the interval.

4 CONCLUSIONS

Comparatively low total As concentrations were observed in sediments and bailed groundwater samples from specific horizons. In contrast, pumped groundwater samples showed high As contents, and they could be associated to mixing with As enriched high-productive horizons, and sediment-bound As that are readily mobilized.

ACKNOWLEDGEMENTS

We thank for support at PID75/2011 (ANPCyT-COHIFE), IHLLA Laboratory (Lic. F. Altolaguirre, N. De Libano, Ing. D. Arias, Lic. O. Floriani). and Dr. D. Poiré (CIG—UNLP) for the XRD analyses. Special thanks to the reviewers.

REFERENCES

APHA, 2005. *Standard Methods for the Examination of Water and Wastewater.* 21st ed. AWWA-WEF, Washington, D.C.

Auge, M., Viale, G.E., Sierra, L. 2013. Arsénico en el agua subterránea de la Provincia de Buenos Aires. In N. González & E. Kruse (eds.), *Proc. VIII Argentine Hydrogeologic Congress, La Plata* (2): 58–63.

Bhattacharya, P., Claesson, M., Bundschuh, J., Sracek, O., Fagerberg, J., Jacks, G., Martin, R.A., Storniolo, A.R., Thir, J.M. 2006. Distribution and mobility of arsenic in the Río Dulce Alluvial aquifers in Santiago del Estero Province, Argentina. *Sci. Total Environ.* 358: 97–120.

Blanco, M.D.C., Paoloni, J.D., Morrás, H.J., Fiorentino, C.E., Sequeira, M. 2006. Content and distribution of arsenic in soils, sediments and groundwater environments of the Southern Pampa Region, Argentina. *Env. Toxicol.* 21(6): 561–574.

Nicolli, H.B., Bundschuh, J., Blanco, M., Tujchneider, O., Panarello, H.O., Dapeña, C., Rusansky, J.E. 2012. Arsenic and associated trace-elements in groundwater from the Chaco-Pampeano plain, Argentina: Results from 100 years of research. *Sci. Total Environ.* 429: 36–56.

Smedley, P.L., Kinniburgh, D.G., Macdonald, D.M., Nicolli, H.B., Barros, A.J., Tullio, J.O., Pearce, J.M., Alonso, M.S. 2005. Arsenic associations in sediments from the loess aquifer of La Pampa, Argentina. *Appl. Geochem.* 20(5): 989–1016.

Teruggi, M.E. 1957. The nature and origin of Argentine loess. *Jour. Sed. Res.* 27(3): 322–332.

Weinzettel, P., Varni, M., Usunoff, E. 2005. Caracterización hidrogeológica del área urbana y periurbana de la ciudad de Tres Arroyos, provincia de Buenos Aires. In M. Blarasin, A. Cabrera, & E. Matteoda (eds.), *Proc. IV Argentine Hydrogeologic Congress, Rio Cuarto*, 171–180.

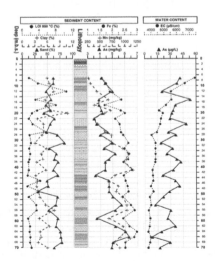

Figure 1. Sand-Clay-LOI 950 °C (left) and As-Fe-Mn (center) sediment logs. EC and As groundwater logs (right).

Arsenic Research and Global Sustainability – Bhattacharya, Vahter, Jarsjö, Kumpiene,
Ahmad, Sparrenbom, Jacks, Donselaar, Bundschuh & Naidu (Eds)
© 2016 Taylor & Francis Group, London, ISBN 978-1-138-02941-5

Distinguishing potential sources for As in groundwater in Pozuelos Basin, Puna region Argentina

J. Murray[1], D.K. Nordstrom[2], B. Dold[3] & A.M. Kirschbaum[1]
[1]*IBIGEO-CONICET, Rosario de Lerma, Salta, Argentina*
[2]*United States Geological Survey, Boulder, CO, USA*
[3]*SUMIRCO (Sustainable Mining Research & Consult EIRL), San Pedro de la Paz, Chile*

ABSTRACT: Groundwater is the main source of drinking water for the inhabitants in Pozuelos Basin, in most cases the As concentrations are above WHO guidelines (8.2–113 µg/L). This study considers three potential sources for As in groundwater i) oxidation of As sulfides in Pan de Azúcar wastes, and acid mine drainage discharge on the basin (0.7–44 mg/L As); ii) weathering and erosion of rich gold mineralized shale's and iii) weathering of volcanic eruptive non-mineralized rocks located in the headwater of some of the streams. High evaporation rates that characterize a prolonged dry season in the basin most probably favors As concentration increase in groundwater and in Pozuelos Lagoon.

1 INTRODUCTION

Argentina was the first Latin American country where Arsenic (As) occurrence in groundwater was reported and the first in the world to document As poisoning from natural sources (Bundschuh *et al.*, 2012). Most of the studies in this country have been focused in the Chaco-Pampean plain region which is known as one of the most extensive areas in the world with high As concentration in superficial and groundwater (Nicolli *et al.*, 2012). In the Puna region of NW Argentina, Andean communities with health problems associated with arsenic-contaminated water were detected (Farías *et al.*, 2009, Concha *et al.*, 2010). Here the arsenic source is associated with volcanic eruptive rocks, volcanic ash, and hydrothermal activity (Satre *et al.*, 1992; Hudson Edwards & Archer, 2012). However extensive studies related to the origin of arsenic in this region are scarce. There is even less knowledge about As anthropogenic sources like mining, which increase the availability and mobility of As by oxidative weathering of mine excavations, waste rocks and tailings (Bowell *et al.*, 2014). This study considers potential sources for high As concentration in groundwater in the Pozuelos Basin, one of the most important closed basins for ecosystem conservation in the Puna region of Argentina.

2 METHODS

2.1 *Water sampling and analysis*

Acid mine waters, groundwater and rivers were sampled between 2011 and 2013 periods. Con-ductivity and pH values were determined in the field with multi-parametric Hanna equipment (Code HI991300N). The samples were filtered in situ with 0.2 Millex HV filtration units by using a syringe, and were stored at 4°C in 15 mL polyethylene centrifuge tubes. Samples for cations and total As analysis were acidified to pH < 2 with ultrapure HNO_3 and were analyzed by Inductively Coupled Plasma-Mass Spectrometry (ICP-MS) and Inductively Coupled Plasma-Optical Emission Spectrometry (ICP-OES). For anions, samples were analyzed by Ion Chromatography (IC).

3 RESULTS AND DISCUSSION

3.1 *Arsenic in groundwater and rivers*

Groundwater is the main source of drinking water for the inhabitants in the basin, and in most of the cases the As concentration are above WHO guidelines (8.22–113 µg/L) (Table 1). In rivers the As concentration range is lower 4.57–17.8 µg/L (Table 1). In both cases the values increases from upstream to downstream in the basin. Both groundwater and rivers discharge in Pozuelos Lagoon that contains As concentration of 47 µg/L (Table 1).

3.2 *Sources of arsenic in Pozuelos Basin*

Pan de Azúcar Mine (Pb-Ag-Zn) was exploited since colonial times until 1990, when the operation ceased without setting any closure plan. Near 70,000 m² of high sulfide content tailings and waste surrounding the mine undergoes high sulfide oxidation and acid waters generation (Murray *et al.*, 2014).

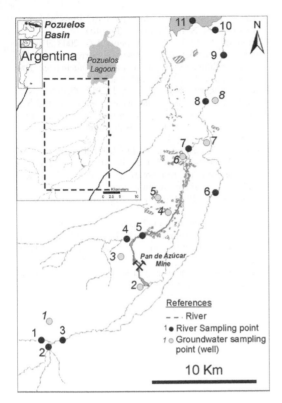

Figure 1. Water sampling stations in Pozuelos Basin.

Table 1. Arsenic content in Pozuelos basin waters.

Groundwater		River and PL	
Sample	As (μg/L)	Sample	As (μg/L)
1	16.1	1	4.6
2	8.86	2	11.3
3	8.22	3	5.11
4	21.4	4	22.8
5	29.7	5	27.0
6	54.7	6	7.5
7	113	7	5.1
8	16.1	8	13.9
		9	15.1
		10	17.8
		11*	47.5

*Pozuelos Lagoon Sample

Pyrite-marcasite are the most abundant sulfide in tailings (9.5 wt%). However, rich As fulfide phases such as arsenopyrite (FeAsS) and freibergite [(Ag,Cu,Fe)$_{12}$(SbAs)$_4$S$_{13}$] are also present. In the primary zone of tailings, the As concentration is 623 mg/kg. In the oxidation zone of tailings As-bearing jarosite [(Na, K, Pb)Fe$_3$(SO$_4$; AsO$_4$)$_2$(OH)$_6$] is the main secondary Fe^{3+} source, but the concentration of As decrease to 92–236 mg/kg. Acid waters pH = 2.1–3.44 with high As and metals (e.g. 0.7–44 mg/L As; 3.2–99.7 mg/L Cd; 0.4–1.7 mg/L Cr; 10–21.3 mg/L Cu; 1.3–47 mg/L Fe; 0.1–1.4 mg/L Pb; 388–8,960 mg/L Zn) drains from the tailings during dry and wet season to the aquifer and rivers in Pozuelos basin.

A diverse composition and age lithology crops out in Pozuelos Basin, several of these rocks contain sulfide metal bearing deposits of Au and Pb-Ag-Zn. Ordovician marine shale are the most abundant rocks in the basin and contain many epithermal and alluvial Au deposits. Epithermal Au is hosted in small size quartz veins which also are rich in arsenopyrite and pyrite. The Au mineralization is structurally controlled and occurs typically along large anticline hinges. Geochemical data indicate that the hydrothermal systems are rich in Au-As and Sb together with subordinate quantities of base metals (Pb, Zn, Cu, and Mo).

Another possible source for As could be related to the presence of effusive volcanic rocks in the surrounding area of Pozuelos Basin, such as Coranzulí ignimbrite volcanic rocks located in the headwater of some of the streams.

Pozuelos Basin (3,500 and 4,500 m. a.s.l.) is a cold and arid region, exposed to intense solar radiation. High evaporation rates characterize a prolonged dry season since scarce rainfall occurs in the austral summer. The annual average rainfall in Pozuelos is 320–350 mm (Garreaud et al., 2003). High evaporation most probably favors As concentration increase in groundwater and in the Pozuelos Lagoon.

4 CONCLUSIONS

In Pozuelos Basin, the source of As could be directly related to the influence of three main different sources i) Oxidation of As sulfides in Pan de Azúcar wastes, and acid mine drainage ii) rich gold mineralized shale's and iii) volcanic eruptive not mineralized rocks.

ACKNOWLEDGEMENTS

We are grateful to the inhabitants of Pozuelos Basin that allowed the water sampling and Pozuelos APN. Jesica Murray received a Conicet postdoc grant and a Fulbright Special Scholarship was awarded to DKN.

REFERENCES

Bundschuh, J., Litter, M.I., Parvez, F., Roman-Ross, G., Nicolli, H.B. 2012. One century of arsenic exposure in Latin America: A review of history and occurrence from 14 countries. *Sci. Total Environ.* 429: 3–36.

Concha, G., Broberg, K., Grandér, M., Cardozo, A., Palm, B., Vahter, M. 2010. High-level exposure to lithium, boron, cesium, and arsenic via drinking water in the Andes of Northern Argentina. *Environ. Sci. Technol.* 44 (17): 6875–6880.

de Sastre, M.S.R., Varillas, A., Kirschbaum, P. 1992. Arsenic content in water in the Northwest area of Argentina. International Seminar Proceedings: Arsenic in the Environment and its Incidence on Health. Universidad de Chile, Santiago, A.M. Sanca (Ed.)

Farías, S.S., Bianco de Salas, G., Servant, R.E., Bovi Mitre, G., Escalante, J., Ponce, R.I. 2009. Survey of arsenic in drinking water and assessment of the intake of arsenic from water in Argentine Puna. In: J. Bundschuh, M.A. Armienta, P. Birkle, P. Bhattacharya, J. Matschullat, A.B. Mukherjee (Eds.) *Natural Arsenic in Groundwater of Latin America.* CRC Press/Balkema, Leiden, The Netherlands. pp. 397–407.

Garreaud, R., Vuille, M., Clement, A.C. 2003. The climate of the Altiplano: Observed current conditions and mechanisms of past changes. *Palaeogeogr. Palaeoclimatol. Palaeoecol.* 194: 5–22.

Hudson-Edwards, K.A., Archer, J. 2012. Geochemistry of As-, F—and B-bearing waters in and around San Antonio de los Cobres, Argentina, and implications for drinking and irrigation water quality. *J. Geochem. Explor.* 112: 276–284.

Murray, J., Kirschbaum, A., Dold, B., Mendes Guimaraes, E., Pannunzio Miner, E. 2014. Jarosite versus soluble iron-sulfate formation and their role in acid mine drainage formation at the Pan de Azúcar mine tailings (Zn-Pb-Ag), NW Argentina. *Minerals* 4: 477–502.

Nicolli, H.B., Bundschuh, J., del C. Blanco, M., Tujchneider, O.C., Panarello, H.O., Dapeña, C., Rusansky, J.E. 2012. Arsenic and associated trace-elements in groundwater from the Chaco-Pampean plain, Argentina: Results from 100 years of research. *Sci. Total Environ.* 42: 936–956.

1.2   Biogeochemical processes

*Arsenic Research and Global Sustainability – Bhattacharya, Vahter, Jarsjö, Kumpiene,
Ahmad, Sparrenbom, Jacks, Donselaar, Bundschuh & Naidu (Eds)*
© 2016 Taylor & Francis Group, London, ISBN 978-1-138-02941-5

Microbial controls on arsenic release and mitigation in aquifer sediments

J.R. Lloyd

School of Earth, Atmospheric and Environmental Sciences, The University of Manchester, Manchester, UK

ABSTRACT: The contamination of groundwaters, abstracted for drinking and irrigation, by sediment-derived arsenic, threatens the health of tens of millions worldwide. Using the techniques of microbiology and molecular ecology, in combination with aqueous and solid phase speciation analysis of arsenic, we have used microcosm and axenic culture-based approaches to provide evidence that anaerobic metal-reducing bacteria can play a key role in the reductive mobilization of arsenic in sediments collected from aquifers in West Bengal, Cambodia, Vietnam and Bangladesh. The critical controls on these activities that are likely to play a role in arsenic mobilization have been described, including the role of organic matter in promoting arsenic-mobilizing respiratory processes, alongside the diversity of organisms involved in mediating these transformations. The impact of other competing anaerobic processes, including sulfate reduction and nitrate-dependent metal oxidation, on arsenic speciation and solubility have also been explored and discussed in the context of mitigating arsenic mobilization.

1 MICROBIALLY DRIVEN MOBILIZATION OF ARSENIC IN AQUIFERS

Contamination of groundwater from naturally occurring arsenic (As) in the subsurface poses a global public health crisis in countries including Mexico, China, Hungary, Argentina, Chile, Cambodia, India (West Bengal), and Bangladesh (Smith *et al.*, 2000). In West Bengal and Bangladesh where the problem has received the most attention, the aquifer sediments are derived from weathered materials from the Himalayas. The mechanism of arsenic release from these sediments has been a topic of intense debate. However, microbially mediated reduction of assemblages comprising arsenic (most likely as arsenate) sorbed to ferric oxyhydroxides has gained consensus as the dominant mechanism for the mobilization of arsenic into these groundwaters (Akai *et al.*, 2004, Islam *et al.*, 2004; van Geen *et al.*, 2004), and the acceptance of this mechanism of arsenic release has paralleled advances in our knowledge of the microorganisms that can respire Fe(III) and sorbed As(V).

For example, an early microcosm-based study from our group in Manchester provided direct evidence for the role of indigenous metal-reducing bacteria in the formation of toxic, mobile As(III) in sediments from the Ganges Delta (Islam *et al.*, 2004). This study showed that the addition of acetate to anaerobic sediments, as a proxy for organic matter and a potential electron donor for metal reduction, resulted in stimulation of microbial reduction of Fe(III) followed by As(V) reduction and the subsequent release of As(III), presumably by As(V)-respiring bacteria that were previously respiring Fe(III). Microbial communities responsible for metal reduction and As(III) mobilization in the stimulated anaerobic sediment were analyzed using molecular (PCR) and cultivation-dependent techniques. Both approaches confirmed an increase in numbers of metal-reducing bacteria, principally *Geobacter* species.

2 MICROBIAL ECOLOGY OF ARSENIC-IMPACTED AQUIFERS: HUNTING FOR THE ORGANISMS THAT MOBILIZE ARSENIC

Although the biogeochemical conditions that promote microbial arsenic mobilization are becoming clearer, it remains a major challenge to identify the organisms that have the potential to cause the reduction and mobilization of the metalloid among the complex microbial communities that exist in the subsurface. One approach that has proved useful is the application of Stable Isotope Probing (SIP), which can link the active fraction of a microbial community to a particular biogeochemical process. Here, sediments are supplemented with a 13C-labeled substrate, and the components of the microbial community that assimilate the substrate are identified by PCR-based analysis of the "heavy" labeled DNA or RNA separated from unlabeled "light" nucleic acids by ultracentrifugation. This technique has been used to identify active As(V)-respiring bacteria in Cambodian aquifer sediments (Lear *et al.*, 2007) implicated in the reductive mobilization of arsenic (Rowland *et al.*, 2007). With the addition of 13C-labeled acetate and As(V), most of which was associated with the mineral phases in the microcosms, an organism closely related to the arsenate-reducing organism *Sulfurospirillum* strain NP4 was identified. Functional gene analysis targeted the As(V) respiratory reductase gene (*arrA*) using highly specific primers, and identified gene sequences most closely related to those found in *S. barnesii* and *G. uraniireducens*.

Most arsenic-related microbial ecology work has focused on Holocene sediments, but a more recent SIP study (Hery *et al.*, 2014) compared the potential role of As(V)-reducing bacteria in the release of arsenic from both Holocene and Pleistocene sediments collected from a Cambodian aquifer. In the Holocene sediment, the metabolically active bacteria stimulated by ^{13}C-acetate and ^{13}C-lactate were again dominated by *Geobacter* and *Sulfurospirillum* spp. (from16S rRNA gene sequencing). The addition of ^{13}C-acetate selected for organisms carrying *arrA* genes most commonly associated with *Geobacter* spp., whereas 13C-lactate selected for *arrA* genes which were not closely related to those in any previously cultivated organism. In contrast, incubation of the Pleistocene sediment with lactate favoured a 16S rRNA-phylotype related to the sulphate-reducing *Desulfovibrio oxamicus* DSM1925, whereas the *arrA* sequences detected clustered with environmental sequences distinct from those identified in the Holocene sediment, suggesting that a distinct microbial ecology was stimulated. A key finding from this study was that in the presence of reactive organic carbon, microbially-mediated As(III) mobilization can occur in Pleistocene sediments, which has clear implications for strategies that aim to reduce arsenic contamination by accessing deeper aquifers containing Pleistocene sediments. Although most SIP studies have focused on "biostimulation" by short chain organic acids (such as acetate and lactate), Rizoulis *et al.* (2014) also used ^{13}C-hexadecane as a model for potentially bioavailable long chain n-alkanes (implicated in stimulating As(V) reduction in the work of Rowland *et al.* (2009)), and a ^{13}C-kerogen analogue used as a proxy for non-extractable organic matter. The data generated were consistent with the utilisation of long chain n-alkanes (but not kerogen) as electron donors for anaerobic processes, potentially including Fe(III) and As(V) reduction in the subsurface.

The ability of *Sulfurospirillum* species to respire As(V), leading to As(III) mobilization, has been well documented in laboratory cultures (Zobrist *et al.*, 2000), while more recent studies using pure cultures of *G. uraniireducens* have confirmed that *Geobacter* species also share this potential to form As(III) from the dissimilatory reduction of As(V) (Lloyd *et al.*, 2011). Further genetic studies are needed to confirm the role of these genes in arsenic metabolism in *G. uraniireducens*, and to explore the diversity of As(V)-reducing *Geobacter* species and the ecophysiology of these organisms in arsenic impacted subsurface sediments. The recent isolation of an organism closely affiliated with the Family *Geobacteraceae*, (and designated strain WB3) from arsenic-rich sediments in Bangladesh (Osboourne *et al.*, 2015) gives a very useful model organism for further complementary laboratory investigations. A broad range of approaches are being used to understand the mechanisms that support reductive mobilization of arsenic from Fe-rich sediments by *Geobacter* and related species, including transcriptomic and proteomic approaches to identify the underpinning physiology, and state of the art imaging techniques to identify the nano-scale impact of this potentially glo-bally important form of microbial metabolism. Studies are also ongoing that build on the successful removal of soluble As(III) from contaminated waters via the targeted stimulation of competing microbial processes that include Fe(II)/As(III) oxidation or sulfate reduction (Omoregie *et al.*, 2013).

ACKNOWLEDGEMENTS

Funding from NERC and the EU are gratefully acknowledged, and also the collaborative support of many colleagues at the University of Manchester including DA Polya, DJ Vaughan, B van Dongen, F Islam, AG Gault, HAL Rowland, G Lear, T Rizoulis, C Boothman, E Omoregie and E Gnanaprakasam.

REFERENCES

Akai, J. et al. 2004. Mineralogical and geomicrobiological investigations on groundwater arsenic enrichment in Bangladesh. *Appl. Geochem.* 19: 215–230.

Héry, M. et al. 2008. Molecular and cultivation-dependent analysis of metal-reducing bacteria implicated in arsenic mobilisation in South East Asian aquifers. *Appl. Geochem.* 23: 3215–3223.

Héry, M. et al. 2014. Microbial ecology of arsenic-mobilizing Cambodian sediments; lithological controls uncovered by stable isotope probing. Environ. Microbiol. DOI:10.1111/1462–2920.12412.

Islam, F.S. et al. 2004. Role of metal-reducing bacteria in arsenic release from Bengal Delta sediments. *Nature* 430: 68–71.

Lear, G. et al. 2007. Molecular analysis of arsenate-reducing bacteria within cambodian sediments following amendment with acetate. *Appl. Environ. Microbiol.* 73 1041–1048.

Lloyd, J.R. et al. 2011. Microbial transformations of arsenic in the subsurface. In: J.F. Stolz, R.S. Oremland. (Eds.) *Environmental Microbe-Metal Interactions II.* ASM Press Washington. pp. 77–90.

Omoregie, E. et al. 2013. Arsenic bioremediation by biogenic iron oxides and sulfides. *Appl. Environ. Microbiol.* 79: 4325–4335.

Osborne, T.H. et al. 2015. Isolation of an arsenate-respiring bacterium from a redox front in an arsenic-polluted aquifer in West Bengal, Bengal Basin. *Environ. Sci. Technol.* 49: 4193–4199.

Rizoulis, A. et al. 2014. Microbially-mediated reduction of Fe(III) and As(V) in Cambodian sediments amended with ^{13}C-labelled hexadecane and kerogen. *Environ. Chem.* 11 538–546.

Rowland, H.A.L. et al. 2007. Organic matter as a critical control in arsenic contamination of drinking water from shallow aquifers. *Geobiology* 5: 281–292.

Rowland, H.A.L. et al. 2009. Microbiological transformations of petroleum and metals leading to arsenic release from West Bengal aquifer sediments. J. Environ. Qual. 38 1598–1607

Smith, A.H. et al. 2000. Contamination of drinking-water by arsenic in Bangladesh: a public health emergency. *Bull. WHO* 78: 1093–1103.

van Geen, A. et al. 2004. Decoupling of As and Fe release to Bangladesh groundwater under reducing conditions. Part II: Evidence from sediment incubations. *Geochim. Cosmochim. Acta* 68: 3475–3486.

Zobrist, J. et al. 2000. Mobilization of arsenite by dissimilatory reduction of adsorbed arsenate. *Environ. Sci. Technol.* 34: 4747–4753.

Arsenic Research and Global Sustainability – Bhattacharya, Vahter, Jarsjö, Kumpiene,
Ahmad, Sparrenbom, Jacks, Donselaar, Bundschuh & Naidu (Eds)
© 2016 Taylor & Francis Group, London, ISBN 978-1-138-02941-5

Mechanism of arsenic release in sulfate rich sediment during microbial sulfate reduction

T.H.V. Phan[1], D. Tisserand[1], F. Bardelli[1], L. Charlet[2], M. Frutschi[3] & R. Bernier-Latmani[3]
[1]*Earth and Planetary Science Department (LGIT-OSUG), University of Grenoble-I, Grenoble, France*
[2]*Institut des Sciences de La Terre (ISTerre), Universite Grenoble Alpes and CNRS, France*
[3]*Ecole Polytechnique Federale de Lausanne, Lausanne, Switzerland*

ABSTRACT: Arsenic contamination of drinking water is a major problem in An Giang, one of the Southwestern Vietnamese provinces. To simulate the natural redox cycles to which natural sediments are subjected, batch redox oscillation bioreactor experiments were conducted on arsenic and sulfate doped natural sediments. Eh oscillation in the range between -300 mV and $+500$ mV was implemented by modulating the influx gas mixture between N_2/CO_2 and compressed air automatically. Cellobiose was added at the beginning of reducing cycles to stimulate metabolism of a natively present microbial community. Results showed that repetitive redox cycling could decrease arsenic mobility significantly during reducing conditions up to 92%. Phylogenetic and functional analyses of 16S rRNA genes from metagenomic sequencing revealed the dominance of sulfur-cycling and iron-cycling bacteria, indicating that sulfate and iron reducing is a key driver of As immobilization during the reducing cycles.

1 INTRODUCTION

Arsenic contamination of drinking water is a major problem in An Giang province of Southwest Vietnam (Hanh *et al*, 2011, Erban, 2013). The studied site, An Phu commune, is located close to the Hau River also called the Bassac River across the border. During the monsoon season (June-October), floodwaters from the Mekong River recharge the aquifer in oxygen rich water, which flows back into the Mekong by normal drainage from November to March. This results in a seasonal wet-dry cycle and generates redox oscillations in the soil and the subsoil, which may lead to the release of arsenic from sediment. Microbial sulfate reduction is an energy-yielding metabolic process during which sulfate is reduced to sulfide and is coupled with the oxidation of organic matter (Barton, 1995). In nature, sulfide produced during microbial sulfate reduction reacts mostly with metal ions, especially Fe^{2+}, and largely precipitating into solids such as pyrite (FeS_2) or mackinawite (FeS_m), but also with As to produce As sulfide minerals such as orpiment (As_2S_3) and realgar (AsS) (Root *et al.*, 2009). Iron sulfide minerals can also sorb or co-precipitate with As to form As-rich pyrite or arsenopyrite (FeAsS) (Charlet *et al.*, 2011). Accordingly, microbial sulfate reduction is generally thought to cause decreased As mobility in subsurface environments (Fendorf *et al.*, 2010).

2 METHODS/EXPERIMENTAL

2.1 *Experimental set-up*

The sediment was sampled from An Phu. Using XRD, mica, chlorite, vivianite, siderite and pyrite were identified. The sediment was suspended in ultrapure water at the concentration of approx. 100g/L. Lab-scale experiments were performed using two redox cycling bioreactors containing suspensions with the same final concentration 50 µM of $As^{(III)}$, and 0.1 mM of SO_4^{2-} or 1 mM of SO_4^{2-} for R1 and R2, respectively, toward two main goals: (1) to investigate the As, Fe, S species generated by oscillating biogeochemical conditions and (2) to pinpoint and describe in detail the role of bacteria in the sediment via DNA mapping.

2.2 *Analytical methods*

Extensive characterization of aqueous phase parameter such as Eh, pH, DOC, S, Fe and As species were performed using appropriate analytical methods. Additionally, solid phase characterization was carried out by XRD and XAS. The 16S rRNA was extracted from slurry samples and amplified, then DNA sequencing was performed to monitor changes in the composition of the microbial community and to probe for the presence of sulfate-reducing bacteria during the experimental cycles.

3 RESULTS AND DISCUSSION

3.1 *Aqueous chemistry*

Eh and pH continuous measurements are given in Figure 1, where reducing and oxidizing steps can be clearly identified. The pH values vary within the range of 5.2 to 7.8. In the reductive process, Eh decrease is driven by the consumption of successive terminal electron acceptors (Fe^{3+}, SO_4^{2-}) and DOC by

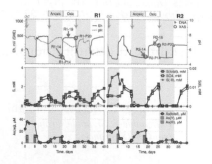

Figure 1. Aqueous phase data during two reactor experiments with R1 (0.1 mM) and R2 (1 mM) of SO_4^{2-}.

the microbial community (Essington, 2004), which results in OH^- production leading to the pH rise. The consumption of sulfate in the reducing process can be attributed to sulfate reduction, coupled to an incomplete oxidation of lactate (Canfield et al., 1998)

The presence of sulfide and acetate in the reducing cycles was associated with lower concentrations of DOC and sulfate in the reactor (Fig. 1). Concentration of SO_4^{2-} decreased during the anoxic half cycle and increased during the oxic half cycle. In contrast, sulfide concentration decreased during the oxidation cycle and rose during the reducing cycle. This indicates that a reversible sulfate reduction occurred during the reactor experiments.

Intra-cycle release of arsenic was measured during each reducing half-cycle. Total concentrations of As decreased during the oxic cycles and increased during the anoxic cycles (Fig. 1). At high values of Eh, $As^{(V)}$ was the major dissolved As species and, upon reduction, $As^{(III)}$ became the major dissolved As species (Masscheleyn, 1991). These ferrous minerals such as FeS, FeS_2 and $FeCO_3$ have been shown to limit arsenic mobility in a reducing environment (Charlet et al., 2011). Moreover, the difference between As(tot) and the sum of As species may be due to the presence of another As form such as thioarsenate which may be absorbed on FeS_2 and FeS (Couture et al., 2013) in the subsequent anoxic half-cycles. Therefore, the concentration of total As decreased significantly after three anoxic cycles.

3.2 Active microbial community

Many identified heterotrophic bacteria are capable of operating under both aerobic and anaerobic conditions. These bacteria can use organic compounds as a source of energy and carbon (Fig. 2). Phylotypes related to S-cycling (SRB) and Fe-cycling bacteria (IRB), including *Desulfobulbus*, *Desulfobulbaceae*, *Desulfobacterales*, *Desulfomicrobium sp*, *Desulfovibrio sp* and *Geobacter sp* were detected in both reactors and in reducing and oxidizing processes, demonstrating the importance of sulfate and iron reduction. Microbial sulfate reduction produces sulfide (Fig. 1) during the anoxic cycles, which can react to precipitate As as As sulfide minerals or form iron sulfide which

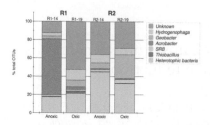

Figure 2. Distribution of microbial 16S rRNA gene sequences at the four samples in R1 (0.1 mM) and R2 (1 mM) of SO_4^{2-}.

sequestered As in the solid phase and decreased As in the aqueous phase (O'Day et al., 2004).

4 CONCLUSIONS

The redox oscillation in natural environment due to flooding and draining has the potential to stabilize As in the solid phase and limit aqueous concentrations. We report experimental evidences that allow to propose mechanisms for As mobility reduction in sulfate rich sediment based on iron and sulfate reducing bacteria, sorption and precipitation process.

REFERENCES

Barton, L.L. 1995. Biotechnology Handbooks 8 - Sulfate-Reducing Bacteria. Springer Science.
Canfield, D.E. 1998. Isotope fractionation and sulfur metabolism by pure and enrichment cultures of elemental sulfur-disproportionating bacteria. *Limnol. Oceanogr.* 43(2): 253–264.
Charlet, L. 2011. Reactivity at (nano)particle-water interfaces, redox processes, and arsenic transport in the environment. *CR Geosci.* 343(2–3): 123–139.
Couture, R.-M. 2013. Sorption of arsenite, arsenate and thioarsenate to iron oxides and iron sulfides: a kinetic and spectroscopic investigation. *Environ. Sci. Technol.* 47(11):5652–9
Erban, L.E., 2013. Release of arsenic to deep groundwater in the Mekong Delta, Vietnam, linked to pumping-induced land subsidence. *Proc. Nat. Acad. Sci.* 110(34): 13751–6.
Essington, M. 2004. Soil and Water Chemistry An Integrative Approach. CRC Press, Boca Raton.
Fendorf, S. 2010. *Arsenic chemistry in soils and sediments.* Lawrence Berkeley National Laboratory.
Hanh, H.T. 2011. Community exposure to arsenic in the Mekong river delta, Southern Vietnam. *Env. Monitor.* 13(7): 2025–2032.
Masscheleyn, P.H. 1991. Effect of Redox Potential and pH on arsenic speciation and solubility in a contaminated soil. *Environ. Sci. Technol.* 25(18): 1414–1419.
O'Day, P. 2004. The influence of sulfur and iron on dissolved arsenic concentrations in the shallow subsurface under changing redox conditions. *Proc. Natl. Acad. Sci. USA* 101(38): 13703–13708.
Root, R.A. 2009. Speciation and natural attenuation of arsenic and iron in a tidally influenced shallow aquifer. *Geochim. Cosmochim. Acta* 73(19): 5528–5553.

Arsenic Research and Global Sustainability – Bhattacharya, Vahter, Jarsjö, Kumpiene,
Ahmad, Sparrenbom, Jacks, Donselaar, Bundschuh & Naidu (Eds)
© 2016 Taylor & Francis Group, London, ISBN 978-1-138-02941-5

The phosphate transporters impart different affinity to arsenate

X.-M. Xue, Y. Yan, Y.Q. Guo, J. Ye & Y.-G. Zhu
Key Lab of Urban Environment and Health, Institute of Urban Environment, Chinese Academy of Sciences,
Xiamen, P.R. China

ABSTRACT: Some researches testing the effects of phosphate on arsenate transport have demonstrated that microorganisms absorb arsenate by phosphate transporters. In this study, four genes (*SphX*, *PstS1*, *sll0540* and *PstS2*) involved in phosphate transportation from the cyanobacterium *Synechocystis* sp. PCC 6803 were disrupted, and the arsenate influx by the mutants was measured. Moreover, the heterologous expression of four genes in *E. coli* was used to test whether arsenic absorption was affected by different phosphate transporters. All of the four phosphate transportation systems were involved in transporting arsenate. *E. coli* Transetta expressing *SphX*, *PstS1* or *sll0540* increased obiviously the absorption of arsenate, while Transetta expressing PstS2 absorbed almost the same amount of arsenate compared to Transetta bearing pET-22b. These results implied that arsenate was transported through phosphate transporters with different affinity to arsenate.

1 INTRODUCTION

Arsenate (As(V)), as an analogue of phosphate (Pi), is taken up via the same transport system with Pi in some organisms. In the freshwater cyanobacterium *Synechocystis* sp. Strain PCC 6803, four associated Phosphate-Binding Proteins (PBPs) have been identified, and characterized to be responsible for Pi uptake (Frances *et al.*, 2010). SphX and PstS1 belong to Pst1 which acts as a low-affinity, high-velocity system for Pi, PstS2 belongs to the second system Pst2 which exhibits high-affinity, low-velocity for Pi. Sll0540 is the fourth PBP, and colocalized with neither pst1 nor pst2. Herein, we assessed the roles of these four phosphate transporters in As(V) uptake by both creating mutants with disruptions of the associated *PstS*, *SphX*, and *Sll0540* and heterologously expressing them in *E. coli*. Our data demonstrated that the four associated PBPs imparted different affinities to As(V).

2 METHODS/EXPERIMENTAL

2.1 Construction of mutants

The genes *PstS1* (GI 951859), *PstS2* (GI 954489), *sll0540* (GI 951967), *SphX* (GI 953314) and *PstS1+SphX* from *Synechocystis* sp. PCC 6803 were deleted as follows: i) The above genes were amplified; ii) The PCR products were cloned into pGEM-T easy vector or pMD19T simple vector to generate plasmid pG-PstS1, pM-PstS2, pM-sll0540, pM-SphX, and pG-(PstS1+SphX); iii) The kanamycin resistance gene was inserted into the above plasmids by using different restriction enzyme sites to yield plasmids pG-PstS1 Kan, pM-PstS2 Kan, pM-sll0540 Kan, pM-SphXKan, and pG-(PstS1+SphX)Kan; iv) These plasmids were transformed into *Synechocystis* sp. PCC 6803 according to previously described (Zang

et al., 2007); v) Mutant was selected. vi) Confirmed gene deletion mutants were cultivated in 100 mL of BG11 medium with 50 µg/mL kanamycin until used.

2.2 The construction of e. coli expressing phosphate transporter protein

To heterologously express the phosphate transporter protein from *Synechocystis* in *E. coli*, each fragment containing ATG start codon and excluding the stop codon was amplified from *Synechocystis* genome DNA. The *PstS1*, *PstS2*, *sll0540* or *SphX* fragment was cloned into *Nde*I and *Xho*I-digested pET-22b vector or *Nco*I and *Xho*I-digested pET-28a vector to generate plasmid pET22b-*PstS1*, pET28a-*PstS2*, pET28a-*sll0540*, or pET22b-*SphX*.

2.3 Bacteria transport assay

When *Synechocystis* Wild Type (WT) and various mutants were grown to stationary phase, approximately 400 mL of each cultured cells were harvested by centrifuging at 5000 g for 10 min at 4°C. The cells was washed thrice with 400 mL volumes of sterilized Pi-free BG-11 medium, prior to resuspension in 400 mL fresh BG-11-Pi medium. After phosphorus starvation for 48 hours, the cells were divided into 50 mL polypropylene tubes with screw-caps. For short-time transport assay, *Synechocystis* was cultured on tripicates in each condition: (A) -Pi-As(V), (B) -Pi+100 µM As(V), (C) +175 µM Pi-As(V), and (D) +175 µM Pi+100 µM As(V). After 30 minutes incubation at 30°C, cells were centrifuged, and followed by washing thrice with 30 mL volumes of ice-cold Pi-free BG-11 medium. In long-time transport assay, As(V) concentration remained constant (10 µM) while K_2HPO_4 concentration was added at 0 or 175 µM. Cells were incubated at 30°C and 96 rpm under lights for 48 hours. The collection method was as described above.

A 25 mL culture of *E. coli* strain Transetta bearing pET22b, pET22b-*PstS1*, pET28a-*PstS2*, pET28a-*sll0540*, or pET22b-*SphX* was grown overnight at 37°C in LB medium containing 100 μg/mL ampicillin or 50 μg/mL kanamycin, and inoculated into 1 L of low Pi medium (Yang *et al.*, 2005) with small modifications by replacing glucose to glycerol in the next day. After incubation for 3 hours with lactose, cells were harvested and followed by washing cells thrice with cold Pi-free buffer containing 75 mM HEPES-KOH (pH7.5), 150 mM KCl and 1 mM MgSO$_4$. The concentrated cell suspension (10^9) was diluted to 2 mL with the same buffer containing gradient sodium As(V) at room temperature in order to initiate the transport assay. The total arsenic content was determined with a Agilent 7500cx inductively coupled plasma-mass spectrometer.

3 RESULTS AND DISCUSSION

3.1 *Short-term uptake of arsenate by Synechocystis*

Our results demonstrated that the As(V) uptake by *Synechocystis* cells competed with Pi uptake. After phosphorus starvation of 48 hours, the background of Pi in *Synechocystis* WT and mutants varied between 10 g/kg (*Synechocystis* ΔSphX)to 15 g/kg (*Synechocystis* WT). All *Synechocystis* strains co-treated with As(V) and Pi absorbed less Pi than only incubated with Pi.

The absorption of As(V) was investigated in *Synechocystis* after cells were treated with As(V) for 30 minutes. When there was no Pi in the medium, all of *Synechocystis* absorbed substantial As(V). *Synechocystis* WT accumulated up to 305 mg/kg, *Synechocystis* ΔSphX absorbed the lowest As as 215 mg/kg.When Pi was added in medium, the amout of arsenic accumulated in *Synechocystis* was greatly reduced. In addition, all of the *Synechocystis* mutant strains except *Synechocystis* ΔPstS2 absorbed arsenic significantly less than *Synechocystis* WT.

3.2 *Long-term accumulation of arsenic by Synechocystis*

The long-term experiments showed that Pi did not affect the accumulation of arsenic in *Synechocystis*. The concentrations of arsenic and phosphorus in cells were measured after *Synechocystis* was treated with As(V) with or without Pi for 48 hours. The phosphorous concentration in *Synechocystis* cells was between 11 g/kg (*Synechocystis* WT) and 8 g/kg (*Synechocystis* Δ(*PstS1*+SphX)) after culturing without Pi for 48 hours. At the same time, *Synechocystis* WT and *Synechocystis* Δ(*PstS1*+SphX) accumulated 159 mg/kg and 114 mg/kg of arsenic respectively. While *Synechocystis* was treated with Pi for 48 hours, each of *Synechocystis* built up large amounts of phosphorus. Arsenic accumulation was increased by *Synechocystis* WT and *Synechocystis* ΔPstS2 when Pi was added, while that was decreased by *Synechocystis* Δsll0540, *Synechocystis* ΔSphX, and *Synechocystis* Δ(*PstS1*+SphX).

3.3 *The uptake of arsenate by E. coli expressing phosphate transporter*

The heterologous expression of phosphate transpoters in *E. coli* Transetta showed that phosphate transpoters had dramatic difference in their affinity to As(V). Transetta expressing SphX absorbed the most As(V) (2012.8 mg/kg) than others (Fig. 1). Transetta expressing PstS2 which acts as a high-affinity, low-velocity transporter to Pi acquired the lowest As(V) (1184.4 mg/kg).

4 CONCLUSIONS

The uptake processes of As(V) in *Synechocystis* sp. PCC 6803 were examined in this paper. The analysis of arsenic concentrations in cyanobacterial extracts showed that Pi was the powerful inhibitor of As(V) uptake, and Pi uptake by *Synechocystis* sp. PCC 6803 was also greatly affected by As(V). Arsenic accumulation was not decrease obviously in *Synechocystis* when Pi was added. The possible reason is that Pi is a key growth-limiting nutrient of *Synechocystis*, the better growth of *Synechocystis* with Pi may result more arsenic transformation and accumulation. Furthermore, *Synechocystis* sp. PCC 6803 has multiple arsenic transformation pathways that As(V) can be converted to other arsenic species and then be accumulated in other ways (Xue *et al.*, 2004).

The phosphate transporters showed different affinities to As(V). From the results, PstS2 with high affinity to Pi has little impact to uptake As(V), and SphX that acts as a low-affinity protein for Pi shown the highest affinity to As(V). This might indicate that As(V) is absorbed through low-affinity systems for Pi by *Synechocystis* sp. PCC 6803.

ACKNOWLEDGEMENTS

Our research is supported by the National Natural Science foundation of China (21507125).

REFERENCES

Pitt, F.D., Mazard, S., Humphreys, L., Scanlan, D.J. 2010. Functional characterization of *Synechocystis* sp. strain PCC 6803 pst1 and pst2 gene clusters reveals a novel strategy for phosphate uptake in a freshwater cyanobacterium. *J. Bacteriol.* 192(13): 3512–3523.

Zang, X., Liu, B., Liu, S., Arunakumara, K.K.I.U., Zhang, X. 2007. Optimum conditions for transformation of *Synechocystis* sp. PCC 6803. *J. Microbiol. (Seoul, Korea)* 45(3): 241–257.

Yang, H.C., Cheng, J.J., Finan, T.M., Rosen, B.P.. Bhattacharjee, H.. 2005. Novel pathway for arsenic detoxification in the legume symbiont *Sinorhizobium meliloti*. *J. Bacteriol.* 187(20): 6991–6997.

Xue, X.M., Raber, G., Foster, S., Chen, S.C., Francesconi, K.A., Zhu, Y.G. 2014. Biosynthesis of arsenolipids by the cyanobacterium *Synechocystis* sp. PCC 6803. *Environ. Chem.* 11(5): 506–513.

Arsenic Research and Global Sustainability – Bhattacharya, Vahter, Jarsjö, Kumpiene,
Ahmad, Sparrenbom, Jacks, Donselaar, Bundschuh & Naidu (Eds)
© 2016 Taylor & Francis Group, London, ISBN 978-1-138-02941-5

Isolation of bacterial strains tolerant to arsenic groundwater

E.E. Pellizzari, L.G. Marinich, S.A. Flores Cabrera, F.E. Cabrera & M.C. Giménez
Departamento de Ciencias Básicas y Aplicadas, Universidad Nacional del Chaco Austral, Chaco, Argentina

ABSTRACT: The main objective of the research was to study the bacterial diversity of groundwater in the department Comandante Fernandez, Chaco, Argentina, contaminated with arsenic (As) and their tolerance to the metalloid. The strains were obtained from 3 wells. Morphological tests included cellular and colonial shape and reaction to Gram coloration. The isolated, purified and tolerant strains were grown in nutrient broth prepared with a solution of arsenic at a final concentration of 1 mg/L arsenic (CNAs). Fifty-one bacterial strains were isolated, of which sixteen strains were tolerant to high concentrations of arsenic. They were identified based on their biochemical properties.

1 INTRODUCTION

Arsenic (As) is a natural geological contaminant of groundwater in the Chaco. The presence of As (V) and As (III) depends on different factors; In the biological system is where many bacteria are able to metabolize As, interaction can beneficially influence the environment, showing that can be a potent source of biotechnologically important enzymes involved in bioremediation processes (Pellizzari *et al.*, 2014). These prokaryotes, play a key role, as they have the ability to transform As, using this defense mechanisms highly developed they have to survive adverse conditions, between systems that cells use is the ability to form biofilms, this cell segregation is a strategy for survival, which occurs in oligotrophic aquatic environments. These cellular systems benefit the development of other bacteria that are not tolerant to As, but its development is important for all natural biota, playing a role as protagonists in biogeochemical cycles.

The main objective of the research was to study the bacterial diversity of groundwater, located in the Departament Comandante Fernandez, Chaco, Argentina, contaminated with As and tolerance to metalloid.

2 METHODS/EXPERIMENTAL

2.1 Materials

The experiments were performed using bacteria isolated from groundwater containing 0.10–0.25 mgAs/L, in Presidencia Roque Sáenz Peña, Chaco province. The nutrient broth was prepared using a solution of 1 mgAs/L, were prepared with distiller water and As(V) stock solutions by dissolving sodium arsenate ($Na_2HAsO_4 7H_2O$) in deionizer water. All prepared solutions were sterilized by autoclave at 120°C for 15 min. The analytical determinations of As in the extracted samples from the reactor were performed by HGAAS (Hydride Generation Atomic Absorption Spectroscopy) using hollow cathode lamps at 193.7 nm wavelength. The instrument quantification limit was 5µg/L; an intermediate precision of less than ±10% was achieved. They were collected and analyzed six samples of water contaminated with As of three wells, located in the urban area were sampled, in order to determine and identify As tolerant bacteria.

2.2 Methods

The isolated and purified samples were seeded into nutrient broth prepared with a solution of As at a final concentration of 1mgAs/L As (CNAs). All culture media were sterilized. The incubation was performed at 37°C for 48 hours to obtain growth.

One mL of water sample was dissolved in 9 mL of 0.85% sodium chloride solution and shaken. Samples were diluted 10–1000 fold with sterile 0.85% sodium chloride solution and plated on nutrient agar plates, pH7.0 all w/v, supplemented with sodium arsenate. These plates were incubated for 48 hours at 37°C. Different colonies were collected and isolated after purification process successfully on the same medium. The colonies showing resistance to arsenate was selected for characterization. The obtained pure isolates were characterized in terms of their morphological, physical, biological and biochemical nature.

Biochemical tests were performed to determine salt tolerance, carbohydrate fermentation, amino acid utilization, H_2S production, citrate utilization, nitrate reduction ability and presence of enzymes like oxidase, catalase, lipase, gelatinase, urease, amylase, etc. Biochemical properties of the isolates were tested according to Bergey's Manual of Systematic Bacteriology

3 RESULTS AND DISCUSSION

3.1 *Results*

Fifty-one colonies were isolated from water samples of 3 wells. All strains were purified and cultured in agar solidified with 1 mg/L As. In the Well 1, 20 strains were isolated, of which 5 were tolerant to As. In the well 2, 17 strains were isolated, resulting six-tolerant. In the well 3, 14 bacterial strains was isolated, 5 strains were tolerant (Table 1). Total bacterial strains tolerant resulting were 16. They were identified based on their biochemical properties. Total tolerant bacterial strains were 16. The cellular morphology and Gram's natureof the isolates were determined by bright field microscopy. They were identified based on their biochemical properties.

3.2 *Discussion*

It is known that the genetic determinants for resistance to heavy metals are widespread microbes (Silver, 1996) without selective pressure that are often present in bacterial genomes (Drewniak *et al.*, 2008).

With the significant growth of different genresof bacteria isolated from groundwater contaminated with As, it allows us to deduce a large presence of microorganisms, ensuring potential bioremediadora as has been shown in previous studies. (Banerjee

Table 1.   Strains isolated from groundwater wells.

Isolates	Strains tolerant	Bacterial genera*
		Well 1
20 Strains	5 Strains	*Cellulomonas uda*
		Oligella urethralis
		Haemophilus ducreyi
		Eikenella
		Pseudomonas aeruginosa
		Well 2
17 Strains	6 Strains	*Garnerella*
		Acidiphilium facilis
		Azotobacter nigricant
		Actinobacillus hominis
		Pasteurella caballi
		Well 3
14 Strains	5 Strains	*Garnerella*
		Moraxella moraxella
		Lacunata lacunata
		Lacunata liquefaciens
		Moraxella phenylpiruica
		Volcaniella eurihalina

* Tolerant bacterial genera identified.

et al., 2011), showing that many microorganisms have the ability to possibly use As in their metabolic systems and confirms the results obtained using native cultures of microorganisms in contact with high concentrations of As, such as *Pseudomonas aeruginosa* (Pellizzari *et al.*, 2015) and *Gluconobacteroxydans* (Marinich *et al.*, 2014), whose extraction capacity of this metalloid is no longer disputed.

4 CONCLUSIONS

The following genera were isolated: Cellulomonas, Oligella, Haemopjilus, Eikenella, Pseudomonas, Garnerella, Acidiphilium, Azotobacter, Actinobacillus, Pasteurella, Gluconobacter, Moraxella, Lacunata and Volcaniella. Therefore, these bacterial strains can be used for bioremediation of water/soil/effluents containing As. These strains are not reported to be isolated from groundwater contaminated with As in urban areas in Chaco, Argentina.

ACKNOWLEDGEMENTS

This work was supported with funds from PI 036/00041/14UNCAus. Universidad Nacional del Chaco Austral, Argentina.

REFERENCES

Banerjee, S., Datta, S., Chattyopadhyay, D., Sarkar, P. 2011. Arsenic accumulating and transforming bacteria isolated from contaminated soil for potential use in bioremediation. *J. Environ. Sci. Heal. A.*, 46: 1736–1747.

Drewniak, L.,Styczek, A., Majder-Lopatka, M., Sklodowska, A. 2008. Bacteria, hypertolerant to arsenic in the rocks of an ancient gold mine, and their potential role in dissemination of arsenic pollution. *Environ. Pollut.* 156: 1069–1074.

Marinich, L.G., Flores Cabrera, S.A., Giménez, M.C., Pellizzari, E.E. 2014. Bioremediación de arsenic presente en agua por *Gluconobacteroxydans*. III simposio argentino de procesos biotecnológicos (saprobio). Universidad Nacional del Litoral (UNL). Santa Fe, Argentina.

Pellizzari, E.E., Giménez, M.C. 2014. Arsenic decrease using *Pseudomonas aeruginosa* on mineral matrix for bioremediation. In: M.I. Litter, H.B. Nicolli, M. Meichtry, N. Quici, J. Bundschuh, P. Bhattacharya & R. Naidu (eds.) *"One Century of the Discovery of Arsenicosis in Latin America (1914–2014) As 2014".* Interdisciplinary Book Series: "Arsenic in the Environment—Proceedings".CRC Press/ Balkema, Leiden, The Netherlands. pp. 762–763.

Pellizzari, E.E., Marinich, L.G., Flores Cabrera, S.A., Giménez, M.C. 2015.Degradación de arsenic por *Pseudomonas aeruginosa* para bioremediación de agua. Estudio preliminar. *Revista Avances en Ciencias e Ingeniería (La Serena-Chile)* 6(1): 1–5.

Silver, S. 1996. Bacterial resistances to toxic metals–A review. *Gene* 179: 9–19.

Arsenic Research and Global Sustainability – Bhattacharya, Vahter, Jarsjö, Kumpiene,
Ahmad, Sparrenbom, Jacks, Donselaar, Bundschuh & Naidu (Eds)
© 2016 Taylor & Francis Group, London, ISBN 978-1-138-02941-5

Arsenic biotransformation by a filamentous cyanobacterium *Nostoc* sp. PCC 7120

Y. Yan, J. Ye, X.-M. Xue & Y.-G. Zhu
Key Lab of Urban Environment and Health, Institute of Urban Environment, Chinese Academy of Sciences, Xiamen, P.R. China

ABSTRACT: Arsenic is a prevalent environmental toxin and carcinogen, and its toxicity and mobility are determined by arsenic speciation. Microbial biotransformation of arsenic plays an important role in the arsenic biogeochemical cycle. Cyanobacteria with a long evolutionary history are widespread in nature. In this paper, *Nostoc* sp. PCC 7120 was chosen to thoroughly study biotransformation pathways of arsenic. *Nostoc* sp. PCC 7120 undergoes multiple processes to cope with arsenic. The reduction-oxidation reactions that interconvert arsenate [As(V)] and arsenite [As(III)] form an arsenic redox cycle in *Nostoc* sp. PCC 7120, and As(III) is then excreted from cells. In addition, *Nostoc* sp. PCC 7120 has the ability to methylate As(III) and further synthesize arsenosugar, and even demethylate both monomethylarsenate [MAs(V)] and monomethylarsenite [MAs(III)] to As(III). The genes of *Nostoc* sp. PCC 7120 involved in As(III) methylation and MAs(III) demethylation have also been identified.

1   INTRODUCTION

Arsenic (As) is a ubiquitous toxic element and occurs primarily as −3, 0, +3, or +5 states with different physicochemical properties. The arsenic toxicity is largely determined by its speciation that is changed by geochemical or biological processes. Microorganisms play a fundamental role in the global As biogeochemical cycle, in particular the redox and methylation cycles, through multiple biotransformation pathways of As (Zhu *et al.*, 2014). The redox cycle that consists of As(V) reduction and As(III) oxidation influences the inorganic forms of this element. Furthermore, the methylation cycle composed of As(III) methylation and methylarsenic demethylation maintains a balance between organic and inorganic As in nature. An arsenite methyltransferase (ArsM) that carried out As(III) methylation has been characterized (Qin *et al.*, 2006). Recently, a C·As lyase, named as ArsI, was identified to be responsible for MAs(III) demethylation (Yoshinaga & Rosen, 2014).

Cyanobacteria are widely distributed in both aquatic and terrestrial environments. Previous studies showed that cyanobacteria are able to accumulate and bio-transform arsenic (Ye *et al.*, 2012), while it remains unclear which genes are involved in the arsenic metabolism. Herein, we choose *Nostoc* sp. PCC 7120 (hereafter as *Nostoc*) to study the As biotransformation pathways.

2   METHODS/EXPERIMENTAL

2.1   *Cyanobacterial culture and arsenic speciation analysis*

Nostoc was axenically grown in BG11 medium without nitrate. To investigate As transformation in *Nostoc*, As(III) was supplied to the medium at the initial concentrations of 10 and 100 µM. The volatile arsenic released from *Nostoc* when treated with As(V) was trapped (Yin *et al.*, 2011). In addition, *Nostoc* was incubated with 1 µM MAs(V) or MAs(III) in triplicate to investigate methylarsenic demethylation. As species in the cells and medium were determined by a high performance liquid chromatography system (HPLC) coupled to an Inductively Coupled Plasma Mass Spectrometer (ICP-MS).

2.2   *Cloning of NsarsM and NsarsI from Nostoc and in vivo assay*

The *NsarsM* and *NsarsI* were amplified from *Nostoc* genomic DNA, respectively. *NsarsM*-pET28a plasmid was constructed according to Yin's method (Yin *et al.*, 2011), and *NsarsI* was inserted into pET22b to generate *NsarsI*-pET22b plasmid (Yan *et al.*, 2015). The predicted molecular mass of NsArsM (323 residues) and NsArsI (150 residues) encoded by *NsarsM* and *NsarsI* were 35.26 and 17.35 kDa. By monitoring $OD_{600 nm}$ of arsenic-hypersensitive strain *E. coli* AW3110 (Δ*arsRBC*) bearing either *NsarsM*-pET28a or *NsarsI*-pET22b, arsenic resistance assays of *NsarsM* and *NsarsI* were performed as previously descried (Yin *et al.*, 2011; Yoshinaga & Rosen, 2014). The As(III)-methylation and MAs(III)-demethylation *in vivo* were analyzed by determining arsenic species in the medium in which *E. coli* AW3110 bearing *NsarsM*-pET28a or *NsarsI*-pET22b was cultured.

2.3   *Purification of NsArsM and NsArsI and in vitro assay*

NsArsM and NsArsI with his-tags were overexpressed in *E. coli* Rosetta (DE3), and purified by Ni-NTA agarose column. The As(III) methylation *in vitro* with 5 µM purified NsArsM was performed in a K_2HPO_4 buffer (pH 7.4) containing 8 mM GSH, 0.3 mM SAM, and 10 µM As(III), at 37°C for 12 h (Yin *et al.*, 2011). The MAs(III) demethylation *in vitro* was performed in a MOPS buffer at pH 7.0, which contained 3 mM TCEP, 1 mM cysteine, 0.1 mM Fe^{2+}, 10 µM NsArsI, and 10 µM MAs(III), at 30°C for 1 h (Yan *et al.*, 2015).

3 RESULTS AND DISCUSSION

3.1 *Arsenic methylation in nostoc*

As(V) was the predominant intracellular species after exposure to 100 µM As(III), DMAs(V) accounted for 5% of the total arsenic. However, only As(III) and As(V) were detected when *Nostoc* was treated with 10 µM As(III). The results demonstrated that *Nostoc* has the ability to methylate As(III) into methylated As, and oxidize As(III) into As(V). *Nostoc* produced volatile arsenical, TMAs(III), when treated with As(V) for 6 weeks. *NsarsM* conferred resistance to As(III) by biomethylating As(III) to DMAs(V) and TMAsO *in vivo*. The purified NsArsM also methylated As(III) *in vitro* with TMAs(III) as the end product (Yin *et al.*, 2011).

3.2 *Arsenic demethylation in nostoc*

Our results demonstrated that *Nostoc* can demethylate MAs(III) to As(III) rapidly and demethylate MAs(V) to As(III) slowly. Almost all of the 1 µM MAs(III) in the medium was demethylated into As(III) by *Nostoc* in 12 hours. However, only 50% MAs(V) was demethylated to As(III) by *Nostoc* during 3 weeks. The heterologous expression of *NsarsI* in AW3110 (*ΔarsRBC*) conferred MAs(III) resistance by MAs(III) demethylation *in vivo*. The purified NsArsI was further characterized to catalyze the formation of As(III) from MAs(III) *in vitro* (Yan *et al.*, 2015).

3.3 *Arsenic transformation in nostoc*

As an ancient species in the earth, cyanobacteria such as *Nostoc* faced multiple selective pressure from different arsenic forms in the environment, so it had to independently evolve versatile strategies to handle them (Fig. 1).

When *Nostoc* lived in the early Earth that contained little oxygen, As(III) was able to transport into *Nostoc* via aquaglyceroporins. The genes *alr1097* and *asr1102* in *Nostoc* encode membrane proteins that possibly catalyze As(III) efflux. At the same time, As(III) may be oxidized to As(V) by unknown oxidases (Yin et al. 2011). Furthermore, As(III) was methylated into DMAs(V) and volatile species by NsArsM when *Nostoc* was exposed to high concentration of As(III) (Yin *et al.*, 2011). MMA(III) and MAs(V) are the predicted intermediates in the pathway of methylation, *Nostoc* can demethylate them, forming an arsenic methylation cycle. When the atmosphere became oxidizing, As(III) was mostly oxidized to As(V) in nature.

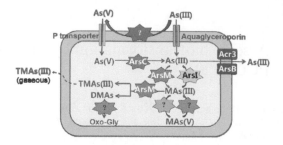

Figure 1. Arsenic metabolism and potential biotransformation pathways in *Nostoc*.

As(V) can absorbed into *Nostoc* cells via phosphate transporters, reduced into As(III), and subsequently detoxified As(III) by undergoing different processes as described above. Moreover, *Nostoc* has the ability to transform inorganic As into oxo-arsenosugar-glycerol (Oxo-Gly) (Miyashita *et al.*, 2012).

4 CONCLUSIONS

In conclusion, *Nostoc* provides a comprehensive view of the molecular mechanism of As redox and methylation cycles, as well as other arsenic biotransformations.

ACKNOWLEDGEMENTS

This work is supported by the National Natural Science foundation of China (21507125 and 31270161).

REFERENCES

Miyashita, S.I., Fujiwara, S., Tsuzuki, M., Kaise, T. 2012. Cyanobacteria produce arsenosugars. *Environ. Chem.* 9(5): 474–484.

Qin, J., Rosen, B.P., Zhang, Y., Wang, G., Franke, S., Rensing, C. 2006. Arsenic detoxification and evolution of trimethylarsine gas by a microbial arsenite S-adenosylmethionine methyltransferase. *Proc. Natl. Acad. Sci. U.S.A.* 103 (7):2075–2080.

Ye, J., Rensing, C., Rosen, B.P., Zhu, Y.G. 2012. Arsenic biomethylation by photosynthetic organisms. *Trends Plant Sci.* 17 (3):155–162.

Yin, X.X., Chen, J., Qin, J., Sun, G.X., Rosen, B.P., Zhu Y.G. 2011. Biotransformation and volatilization of arsenic by three photosynthetic cyanobacteria. *Plant Physiol.* 156 (3): 1631–1638.

Yoshinaga, M., Rosen, B.P. 2014. A C As lyase for degradation of environmental organoarsenical herbicides and animal husbandry growth promoters. *Proc. Natl. Acad. Sci. U.S.A.* 111 (21):7701–7706.

Yan, Y., Ye J., Xue, X.M., Zhu, Y.G, 2015. Arsenic demethylation by a C As lyase in cyanobacterium *Nostoc* sp. PCC 7120. *Environ. Sci. Technol.* 49, (24), 14350–14358.

Zhu, Y.G., Yoshinaga, M., Zhao, F.J., Rosen, B.P. 2014. Earth abides arsenic biotransformations. *An. Rev. Earth Planet. Sci.* 42:443–467.

*Arsenic Research and Global Sustainability – Bhattacharya, Vahter, Jarsjö, Kumpiene,
Ahmad, Sparrenbom, Jacks, Donselaar, Bundschuh & Naidu (Eds)*
© 2016 Taylor & Francis Group, London, ISBN 978-1-138-02941-5

Study on the arsenic accumulation and speciation of arbuscular mycorrhizal symbiont under arsenic contamination

X. Zhang, B.D. Chen & B.H. Ren
*State Key Laboratory of Urban and Regional Ecology, Research Center for Eco-Environmental Sciences,
Chinese Academy of Sciences, Beijing, P.R. China*

ABSTRACT: Wild type and a non-mycorrhizal mutant (TR25:3-1) of *Medicago truncatula* were grown in arsenic (As)-contaminated soil to investigate the influences of Arbuscular Mycorrhizal Fungi (AMF) on As accumulation and speciation in host plants. The results indicated that the plant biomass of *M. truncatula* was dramatically increased by AM symbiosis. Mycorrhizal colonization significantly increased phosphorus concentrations and decreased As concentrations in plants. Moreover, mycorrhizal colonization generally increased the percentage of arsenite in total As both in shoots and roots, while Dimethylarsenic Acid (DMA) was only detected in shoots of mycorrhizal plants. The results suggested that AMF are most likely to get involved in the methylating of inorganic As into less toxic organic DMA and also in the reduction of arsenate to arsenite. The study allowed a deeper insight into the As detoxification mechanisms in AM associations.

1 INTRODUCTION

Excessive As in the soil not only affects plant growth, but also poses a great threat to human health and ecological safety. Arbuscular Mycorrhizal Fungi (AMF), as an important group of soil fungi, can form symbiotic associations with more than 80% of the land plant families. Recent studies show that the arbuscular mycorrhizas naturally occur in As-contaminated soils (Smith *et al.*, 2010) and mycorrhizal inoculation can improve the As tolerance of plants (Orłowska *et al.*, 2012). Nevertheless, most previous experiments investigating the interactions of AMF with host plants under As contaminations were performed under sterilized conditions. Such experiments might have neglected the impacts of other soil microorganisms and thus failed to reveal the significance of AM fungi under natural conditions. Therefore, in the present study wild type and the non-mycorrhizal mutant TR25:3-1 of *M. truncatula* were grown in unsterilized As-contaminated soil to reveal the role of AMF in alleviation of As phytotoxicity under natural conditions.

2 METHODS/EXPERIMENTAL

2.1 Host plants

Seeds of medic plants (*Medicago truncatula* L., wild type cv. Jemalong A17 and the mutant TR25:3-1) were obtained from the Institute of Subtropical Agriculture, Chinese Academy of Sciences. The seeds were surface sterilized in 10% (v/v) H_2O_2 solution for 10 min, then immersed in deionized water for 10 h. They were then pre-germinated on moist filter paper for about 48 h at 27°C till emergence of radicles. The seeds were selected for uniformity before sowing.

2.2 Experimental Procedure

The experimental soil was collected from an As contaminated site (N25°35.22″, E113°00.21″) in Chenzhou City, Hunan Province, China. The extractable As and total As in soil were 4.54 mg/kg and 93.53 mg/kg, respectively.

Wild type and the non-mycorrhizal mutant (which cannot be colonized by AM fungi) of *M. truncatula* were grown in unsterilized As contaminated soil. The experiment was conducted in a controlled environment chamber with 16 h/25°C day, 8 h/18°C night, and a light intensity of 700 μmol m^{-2}s^{-1} provided by supplementary illumination. 16 weeks later, plant shoots and roots were harvested separately.

Percentage root colonization were determined by the grid-intersect method. Approximately 0.2 g freeze-dried samples were weighed and digested by 10 mL HNO_3 using a microwave accelerated reduction system (Mars 5, CEM Co. Ltd, USA). The dissolved samples were analyzed for P by ICP-OES and for As by inductively coupled plasma-mass spectroscopy (ICP-MS) (Agilent7500, Agilent Technology, USA). Different As species in the extracts was determined by high performance Liquid Chromatography-Inductively Coupled Plasma-Mass Spectrometry (HPLC-ICP-MS) (Agilent 7500, Agilent Technology, USA).

Data were analyzed by T-test to compare differences in assayed variables between wild type and mutant. All data analysis was performed using windows-based SPSS 13.0 software package.

3 RESULTS AND DISCUSSION

3.1 Results

No root colonization was observed in the mutant *M. truncatula*, whereas the wild type plants had

Table 1. Dry weights, P and As concentrations of the mutant (TR25:3-1) and wild type of *M. truncatula* grown in an unsterilized arsenic contaminated soil.

	DW (g pot⁻¹)		P conc. (mg/kg)		As conc. (mg/kg)	
	Shoots	Roots	Shoots	Roots	Shoots	Roots
Mutant	0.15	0.07	0.4	0.8	1.27	6.92
Wild type	3.55	0.42	2.22	2.61	0.65	3.63
Significance[a]	***	***	***	***	***	*

[a]By T-test, ***, $P < 0.001$; *, $P < 0.05$.

57% of root length colonized by AMF. The biomass as well as P concentrations in both shoots and roots of the wild type *M. truncatula* were significantly higher than those of the mutant ($P < 0.001$). On the other side, As concentrations in shoots ($P < 0.001$) and roots ($P < 0.05$) tended to be lower in the wild type(Table 1). Small amounts of DMA were found only in the shoots of wild type plants. As(V) and As(III) concentrations were consistently lower in shoots and roots of the wild type compared with the mutant (Fig. 1).

3.2 Discussion

A key finding from the present study was that AM inoculation could alleviate As phytotoxicity potentially by influencing As speciation and transformation in host plants. It was most interesting that DMA was only found in shoots of mycorrhizal plants (Fig. 1). As suggested by Lomax et al. (2012), methylated As species in plants could only originate from soil microorganisms. Plants are unable to methylate inorganic As, but instead take up methylated As. Arsenic methylation has been well demonstrated in a wide range of bacteria, fungi, yeasts and algae. Considering that DMA was only found in mycorrhizal plants and in the mycorrhizosphere soil (Ultra Jr et al., 2007), we could propose that AMF are possibly involved in methylating inorganic As into less toxic DMA.

The use of TR25:3-1, the non-mycorrhizal mutant of *M. trunctula*, provided further support to the ecological significance of mycorrhizal symbiosis in plant adaptation to As contaminated environments. Previous experiments performed under sterilized conditions or unsterilized conditions might inevitably have exaggerated the importance of AMF in As detoxification in plants, as most experiments failed to authentically reflect the natural situations. By growing wild type and the mutant (TR25:3-1) of *M. truncatula* in unsterilized As-contaminated soil, we were able to evaluate the importance of indigenous AMF in plant performance without disturbance of soil microbial communities. All in all, compared with the mutant, the wild type *M. truncatula* exhibited higher plant dry weights, P concentrations and proportion of As(III), but lower As concentrations, suggesting

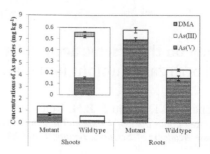

Figure 1. As speciation in the non-mycorrhizal mutant (TR25:3-1) and wild type of *M. truncatula* grown in an unsterilized arsenic contaminated soil (mean ± SE, n = 3). Arsenic speciation in shoots of wild type *M. truncatula* was shown in a small figure with reduced scale of Y-axis.

that AM fungi do play an important role in alleviation of As phytotoxicity under natural conditions.

4 CONCLUSIONS

This study demonstrated that AM fungi could enhance As tolerance of *M. truncatula* by improving plant P nutrition and influencing As accumulation and speciation. The most important finding from this study was that AMF potentially played an important role in the reduction of As(V) to As(III) and also methylation of inorganic As into less toxic organic DMA. The use of non-mycorrhizal mutant of *M. truncatula* demonstrated the ecological importance of AMF in plant tolerance to As under natural conditions. The study allowed a deeper insight into the As detoxification mechanisms in mycorrhizal plants, and also supported the potential use of AMF for bioremediation of As contaminated soils.

ACKNOWLEDGEMENTS

This study was financially supported by National Natural Science Foundation of China (41101246, 41471219).

REFERENCES

Lomax, C., Liu, W.-J., Wu, L., Xue, K., Xiong, J., Zhou, J., McGrath, S.P., Meharg, A.A., Miller, A.J., Zhao, F.-J. 2012. Methylated arsenic species in plants originate from soil microorganisms. New Phytol. 193: 665–672.

Orłowska, E., Godzik, B., Turnau, K. 2012. Effect of different arbuscular mycorrhizal fungal isolates on growth and arsenic accumulation in Plantago lanceolata L. Environ. Pollut. 168: 121–130.

Smith, S.E., Christophersen, H.M., Pope, S., Smith, F.A. 2010. Arsenic uptake and toxicity in plants: integrating mycorrhizal influences. Plant Soil 327: 1–21.

Ultra Jr, V.U., Tanaka, S., Sakurai, K., Iwasaki, K. 2007. Effects of arbuscular mycorrhiza and phosphorus application on arsenic toxicity in sunflower (Helianthus annuus L.) and on the transformation of arsenic in the rhizosphere. Plant Soil 290: 29–41.

Detection and quantification of As(III)-oxidizing microbes in soils highly polluted by breaking-down of old chemical ammunition during inter-war

H. Thouin[1], L. Le Forestier[1], P. Gautret[1], S. Dupraz[2], D. Hube[2] & F. Battaglia-Brunet[2]
[1]*Université d'Orléans, CNRS, BRGM, ISTO, UMR 7327, Orléans, France*
[2]*BRGM, ISTO, UMR 7327, Orléans, France*

ABSTRACT: The open-burning of organo-arsenical compounds present in chemical ammunitions from the First World War was responsible for locally high concentrations of arsenic in top-soil of a highly polluted site from the region of Verdun (France). In order to understand the biogeochemistry of arsenic in this type of environment, quantitative and qualitative characteristics of microbial communities were determined in soil samples with differing As pollution levels. The total concentration of micro-organisms was negatively affected by the pollution level. However the proportion of heterotrophic As(III)-oxidizing organisms and the As(III)-oxidizing rate were higher in the most contaminated than in the less contaminated samples. These results suggest that pollutants, including arsenic, exerted a selective pressure on composition and/or activity of microbial communities.

1 INTRODUCTION

After the Great War, the French military authority was challenged with disposal of large amounts of German chemical ammunitions that had not been fired or when fired, had not detonated. Open burning is one way to break these projectiles down. The burning of Blue Cross shells loaded with solid vomiting and emetic warfare agents, diphenylchloroarsine and diphenylcyanoarsine, resulted in locally intense soil contamination by arsenic and heavy metals. Biogeochemical behavior of arsenic is poorly documented in this type of environment. In the framework of the extended characterization of a highly polluted soil from the region of Verdun (France), characteristics of the microbial communities of samples presenting variable As concentration were examined.

2 METHODS

2.1 *Sampling and analyses*

The study site known as "Place-à-gaz" is located in the Spincourt forest 20 km northeast of Verdun. Between 1926 and 1928, more than 200,000 German chemical shells (especially Blue Cross shells) were open-burned in tranches and rows in the center of this area [1] following a method developed by M. Kostevitch, operating for the British company Pickett & Fils. The severe contamination of this area is visible because of the lack of vegetation. Total concentrations of As, Pb, Cu and Zn were determined in situ using a X-ray fluorescence field portable apparatus,

NITON©. Four soils presenting a gradient of contamination were sampled from the surface, non-saturated black layer (0–10 cm). After size separation at 70 μm, the raw soils were analyzed by ICP-OES for major elements and ICP-MS for trace elements.

2.2 *Microbiological analyses*

Total bacteria were extracted from soils using a Nycodenz gradient separation method. Total microbes were enumerated after fluorescent DAPI staining. Arsenic(III)-oxidizing oligotrophic organisms were enumerated by the Most Probable Number method (MPN). The soil (as wet soil, equivalent to 0.2 g dry soil) was placed in a sterile glass erlenmeyer flask with 10 mL of sterile physiologic water, agitated for 30 min at 25°C, then sonicated 2×20 s at 45 kHz. The soil suspension was serially diluted in sterile physiologic water. Mineral medium containing 100 mg/L As(III) was distributed in Microtest TM Tissue culture plates (96 wells), 250 μL by well. Each well was inoculated with 25 μL of soil suspension dilution. Five wells were inoculated with each dilution. Culture plates were incubated at 25°C for 10 days. The presence of As(III) in the wells was revealed by the formation of the insoluble white complex AsIII-Pyrrolidine DithioCarbamate (PDC). The same method but with 1 g/L yeast extract was applied to determine the MPN of copiotrophic As(III)-oxidizing microbes. The MPN of copiotrophic microbes able to grow in the presence of 100 mg/L As(III) was revealed by the development of turbidity in the wells. As(III)-oxidizing activities were determined as detailed in [2].

3 RESULTS AND DISCUSSION

3.1 Concentrations in heavy metals and As

The level of As, Cu and Pb pollution decreased from soil 1 to soil 4 (Table 1). Soils 1 and 2 were characterized by particularly high concentrations in As and Zn (more than 3%). Copper concentration was also close to 1% in soil 1.

3.2 Microbial concentrations

The total microbes concentration was significantly lower in the most polluted soils 1 and 2 (10X less) than in the less polluted soils 3 and 4 (Fig. 1).

The concentration of oligotrophic As(III)-oxidizing microbes (growing in mineral medium without yeast extract) was lower in the most polluted soils 1 and 2 than in the less polluted soils 3 and 4 (Fig. 2). This result may be linked to a protective effect of yeast extract against toxicity of pollutants. Conversely, the concentration of microbes oxidizing arsenic in presence of 1 g/L yeast extract was the highest in the most polluted soil 1. Remarkably, the same tendency was observed with total microbes concentration growing in presence of both As(III) at 100 mg/L and yeast extract at 1 g/L. Moreover, in the most polluted soil 1, As(III) oxidation was observed in all wells where microbial growth was detected: the concentration in copiotrophic As(III)-oxidizing microbes was equivalent to that of total microbes growing in presence of both As(III) and yeast extract.

3.3 As(III) oxidizing activities

The rates of microbial As(III) oxidation were in the range 0.5 to 2 mg/L/h. The highest rates, close

Table 1. ICP-MS analyses of the four samples of polluted soils.

Soils	As (%)	Pb (%)	Cu (%)	Zn (%)
1	7.28	0.38	0.91	9.02
2	3.08	0.58	0.52	3.73
3	0.63	0.25	0.16	1.07
4	0.19	0.098	0.15	1.33

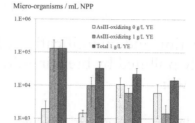

Figure 2. Concentrations of living microorganisms in the four soil samples.

to 2 mg/L/h, was obtained with the most polluted soil 1, whereas the lowest rate, close to 0.5 mg/L/h, was observed with the lowest polluted soil 4.

4 CONCLUSIONS

The presence of As and heavy metals in high concentrations seems to exert a strong selective pressure on the microbial communities of soils polluted by the destruction of chemical weapons from the First World War. The high level of As and metals concentration negatively affected the total microbial concentration, but induced an increase of copiotrophic As(III)-oxidizing microbes concentration. The proportion of As(III)-oxidizing bacteria in the global community and the As(III)-oxidizing activities were clearly positively influenced by the pollution level. These results support the hypothesis of the adaptation of the microbial community, with increased proportions and efficiency of organisms involved in the cycle of arsenic, in highly polluted soils.

ACKNOWLEDGEMENTS

This work was supported by the Région Centre Val de Loire (convention 00087485) and the Labex Vol-taire (ANR-10-LABX-100-01).

REFERENCES

Bausinger, T., Bonnaire, E., Preuß, J. 2007. Exposure as-sessment of a burning ground for chemical ammunition on the Great War battlefields of Verdun. *Sci. Total Environ.* 382: 259–271.

Lescure, T., Moreau, J., Charles, C., Ben Ali Saanda, T., Thouin, H., Pillas, N., Bauda, P., Lamy, I., Battaglia-Brunet, F. 2015. Influence of organic matters on As(III) oxidation by the microflora of polluted soils. *Environ. Geochem. Hlth.* (doi: 10.1007/s10653-015-9771-3).

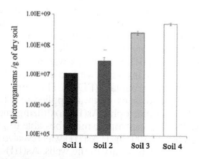

Figure 1. Concentrations of total microorganisms in the four soil samples.

Arsenic Research and Global Sustainability – Bhattacharya, Vahter, Jarsjö, Kumpiene,
Ahmad, Sparrenbom, Jacks, Donselaar, Bundschuh & Naidu (Eds)
© *2016 Taylor & Francis Group, London, ISBN 978-1-138-02941-5*

Diversity of arsenic resistant bacteria from Lonar lake: A meteorite impact alkaline crater lake in India

A.V. Bagade[1], D. Paul[2], T. Rikame[1], A.P. Giri[3], D. Dhotre[2], S. Pawar[2] & K. Kodam[1]
[1]*Department of Chemistry, Savitribai Phule Pune University, Pune, India*
[2]*Microbial Culture Collection, National Centre for Cell Science, Pune, India*
[3]*Division of Biochemical Sciences, CSIR-National Chemical Laboratory, Pune, India*

ABSTRACT: Lonar lake known for its meteorite impact origin and highly alkaline environment harbors a plethora of diverse organisms. Arsenic transforming microbe diversity from Lonar remains unexplored. We attempted to explore the microorganisms causing arsenic transformation using culture-dependent and independent approaches. Amongst the 67 microbes isolated, *Bacillus infantis* L4-18 *and Bacillus solimangrovi* L4-7b could oxidise 15 mM arsenite in 8 days. None of the cultures could reduce arsenate. All the other isolates resisted 2 mM arsenic. In the culture independent approach, microbial diversity revealed *Bacteroides* (41.9%), followed by *Proteobacteria* (17.9%), *Firmicutes* (14.2%), *Actinobacteria* (13.9%), occurring in the Lonar sediment sample. This study provides foundation to study microbial arsenic bio-geochemical cycle along with other biochemical cycles and microbial function in Lonar lake eco system.

1 INTRODUCTION

Extreme environments are considered as mine for diverse microbial community. Arsenic is 20th most abundant element in the earth's crust, even then its toxicity is problematic in many parts of the world. Exploring microbial diversity in saline environment is important for following reasons: i) possibility of finding novel arsenic transforming bacteria from unexplored Lonar Lake, ii) during the early life on earth, there were hyper halophilic conditions and high organic compounds; studying the early evolution of life, how the microbes could adapt and survive this extreme environment. The 3rd reason is that presence of hypersaline conditions on mars gives us the possibility of extinct and/or extant life on Mars (Hongchen *et al.*, 2006).

2 METHODS

2.1 *Culture dependent method*

2.1.1 *Isolation of arsenic tolerant bacteria*
Sediment sample was collected into sterile containers from Lonar Lake, Buldhana district, India. Arsenic content was determined Inductively Coupled Plasma-Atomic Emission Spectroscopy (ICP-AES, Germany). 3 g soil was inoculated in Tris-Mineral Medium (TMM), (Mergeay *et al.*, 1985), supplemented with 0.04% yeast extract and 2 mM of As(III) or As(V). Flasks were incubated on shaker for As(III) and static for As(V) at 30°C for 2 days and 3 ml was inoculated in fresh medium. This procedure was repeated twice.

2.1.2 *Identifying isolates and aio A amplification*
Sequencing of 16S rRNA gene was performed using genomic DNA isolated by alkaline lysis method (Wilson, 2001). Bacterial As(III) oxidase subunit gene (*aio A*) was amplified using primers *aroA* 1F and *aroA* 1R (Inskeep *et al.*, 2007).

2.1.3 *Screening of arsenic transforming bacteria*
The isolates were grown on TMM upto 20 mM As (III or V) for 8 days at 30°C on shaker and checked for arsenic oxidation or reduction ability by using molybdenum blue method (Bachate *et al.*, 2012).

2.2 *Culture independent method*

2.2.1 *DNA extraction, V3 region amplification*
Community DNA was extracted using PowerMax Soil DNA Isolation Kit (Mobio, USA) and V3 region of 16s rRNA was amplified, with fusion Primer 5'C CATCTCATCCCTGCGTGTCTCCGACTCATGA GCGAAGTACTCTAGGGAGGCACA' and universal reverse fusion primer. Reaction mixture containing Taq buffer, MgCl$_2$ 1.75 mM, dNTP mix 0.2 mM, primer 0.08 pM and Taq polymerase 1.5 U, BSA 8 μg, template DNA (100 ng). Programme: 94°C for 4 min; (35x) of 94°C, 30 sec; 69°C, 25 sec and 72°C, 30 sec. The final elongation at 72°C for 5 min. Reactions was processed on Ion Torrent platform PGM.

2.2.2 *Bioinformatics analysis*
The raw data from Ion Torrent PGM sequencing was analyzed in QIIME v1.7 (Caporaso et al. 2010). This project was submitted to NCBI Bio Project ID PRJNA315182; SRP072202.

3 RESULTS AND DISCUSSION

3.1 Culture dependent method

3.1.1 Identification of arsenic oxidizing strains

Arsenic concentration in Lonar lake was found to be 12 µg/L. Arsenic resistant bacteria were isolated by enrichment culture method. Out of 21 bacterial isolates, two (*Bacillus infantis* L4-18 and *Bacillus solimangrovi* L4-7b) showed positive results for arsenic oxidation. Both could oxidize 15 mM As(III) in 8 days. Other isolates namely, *Amphibacillus* sp L-27, *B. cibi* L-18, *B. formantis* L-10, *B. horneckiae* L-25, *B. timonensis* L-3, *B. solii* L-22a, *B. kochii* L-2, *B. paraflexsus* L-15, *Planococcus* L-9, *Sterptobacillus* sp L-25, *Staphylococcus* sp L-32, *Enteractinococcus* sp L-4b, *Methylocaldom* sp L-36, *Pausobacillus* sp L-3, *Methylobacterium* sp L-1, *Candidatus* sp L-4b1, *Neisseria* sp L-12; neither reduced As(V) nor oxidized As(III) but were resistant to 2 mM arsenic. Bachate et al. (2012) showed that *B. firmus* TE7 oxidized 2 mM As(III) in 12 h; the culture simultaneously biotransformed Cr(VI) and As(III).

3.1.2 Amplification of arsenite oxidase gene

AioA gene (500 bp) was amplified from the DNA samples of *Methylobacterium* sp L-1, *B. cibi* L-18a, *B. horneckiae* L-13, *B. infantis* L-4-18, *B. solimangrovi* L4-7b. Even though arsenite oxidase gene was amplified, few bacteria did not show arsenic oxidation, for unknown reasons.

3.2 Culture independent method

3.2.1 Bacterial community analyses

A total of 17 bacterial phyla were detected in the Lonar Lake sediment sample. Taxonomic analysis revealed *Bacteroides* (41.9%), followed by *Proteobacteria* (17.9%), *Firmicutes* (14.2%), *Actinobacteria* (13.9%) (Fig. 1a). Bacterial phyla *Acidobacteria, Tenericutes,* MVP-21, BD1-5, Candidate division TM6, *Elusimicrobia,* and Candidate division SR1 were detected as relatively minor population in Lonar Lake ecosystem. Within *Proteobacteria,* α followed by γ and δ-*proteobacteria* were observed as major groups. Most bacterial phyla *Proteobacteria, Firmicutes* and *Bacteriodates,* which are well known for the metabolic versatility, have ability to thrive and maintain major biogeochemical cycles of the extreme environment (Paul *et al.*, 2016). At the family level taxonomic classification, a total 156 OTUs were found. Within this 99 OTUs were classified at the family level. Bacterial members *Flavobacteriaceae* (21.2%), *ML602J-37* (11.1%), *Rhodobacteraceae* (4.5%), *Bacillaceae* (1.9%), *Halomonadaceae* (1.3%) were found as a predominating groups in this ecosystem (Fig. 1b). Within *Proteobacteria,* members of the *Halomonadaceae, Oceanospirillaceae,* and *Pseudomonadaceae* were prevalent in our taxonomic survey. These bacteria are known for their ubiquity and metabolic flexibility which includes their ability to tolerate extreme and/ or oligotrophic environments, utilize diverse carbon

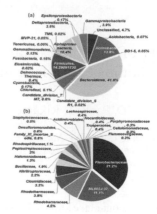

Figure 1. Distribution of major phylogenetic groups of bacteria at (a) phylum and (b) family level.

compounds, and to maintain aerobic and anaerobic lifestyles.

4 CONCLUSION

On comparing the culture dependent and independent aproach *Bacilli, Proteobacteria, Methylobacteria* were seen to be common. Arsenic oxidising and resistant microbes are present in Lonar, which should be further explored for identification of potential arsenic oxidising bacteria for effective bioremediation measures in arsenic contaminated site.

ACKNOWLEDGMENT

SAIF IIT-B, Mumbai for ICP-AES analysis and University Grants Commission, New Delhi, India for funding.

REFERENCES

Hongchen, J., Hailiang, D., et al. 2006. Microbial diversity in water and sediment of lake chaka, an athalassohaline lake in northwestern China. *Appl. Environ. Microbiol.* 72(6): 3832–3845.

Mergeay, M., Nies, D., et al. 1985. *Alcaligenes eutrophus* CH34 is a facultative chemolithotroph with plasmid-bound resistance to heavy metals. *J. Bacteriol.* 162(1): 328–334.

Wilson, K. 2001. Preparation of genomic DNA from bacteria. *Curr. Protoc. Mol. Biol.* 38(1): 2.4.1–2.4.5.

Inskeep, W., Macur, R., et al. 2007. Detection, diversity and expression of aerobic bacterial arsenite oxidase genes. *Environ. Microbiol.* 9(4):934–943.

Bachate, S., Khapare, R., et al. 2012. Oxidation of arsenite by two beta proteobacteria isolated from soil. *Appl. Microbiol. Biotechnol.* 93(5): 2135–2145.Caporaso, J., Kuczynski, J., et al. 2010. QIIME allows analysis of highthroughput community sequencing data. *Nat. Methods* 7(5): 335–336.

Paul, D., Kumbhare, S., et al. 2016. Exploration of microbial diversity and community structure in Lonar Lake. *Front. Microbiol.* 6: 1–12.

Arsenic Research and Global Sustainability – Bhattacharya, Vahter, Jarsjö, Kumpiene,
Ahmad, Sparrenbom, Jacks, Donselaar, Bundschuh & Naidu (Eds)
© 2016 Taylor & Francis Group, London, ISBN 978-1-138-02941-5

Evidence of microbiological control of arsenic release and mobilization in aquifers of Brahmaputra flood plain

S.S. Sathe[1], C. Mahanta[1] & A. Mahagaonkar[2]

[1]*Department of Civil Engineering, Indian Institute of Technology Guwahati, India*
[2]*Center for the Environment, Indian Institute of Technology Guwahati, India*

ABSTRACT: Brahmaputra flood plain is one of the severely arsenic (As) affected areas in India. Published results suggest groundwater is enriched with geogenic As but the mobilization process of As in shallow aquifers remain less understood. Studies have shown that reductive dissolution was a key factor affecting mobility of As between sediments and groundwater systems. Here, we suggest that the mobility of As is greatly affected by microbial organisms, and this remains unstudied so far. The results from this study suggested that As reducing bacteria *Pseudomonas aeruginosa* (ArsC), found at the study area, helps in conversion of As(V) to As(III) and As(III) gets dissolved easily in aqueous phase helping in easier mobilization of the compound. The phylogenetic relationship obtained by comparing 16S rDNA sequences showed the abundance of *P. aeruginosa*, a facultative-aerobic bacteria helping in As(V)-reduction. This facilitates the presence of As in high concentration in the groundwater of the study region.

1 INTRODUCTION

Anoxic condition in an alluvial aquifer is a key factor affecting dissolution of arsenic from sediments to groundwater (Mahanta *et al.*, 2015). Previous studies have established that under oxic condition As^{5+} is predominant and is adsorbed strongly to solid minerals (i.e., ferri-oxy-hydroxide, ferrihydrite, apatite, alumina) (Paul *et al.*, 2015). Conversely, As^{3+} is poorly adsorbed to such minerals in anoxic environment. The objective of our study was to explore As resistant bacteria that may help in mobilization of As in aquifers.

2 MATERIALS AND METHODS

2.1 *Isolation and characterization As(V)-reducing bacteria*

For evaluating As mobilizing bacteria's presence, 100 g of aquifer's sediment sample was taken in 1l ddH_2O with $NaAsO_2$ (Sodium arsenite) and Na_3AsO_4 (Sodium arsenate) at a final concentration of 500 mg/L and kept for 1 week at 27°C. As resistant bacteria were isolated by adding 10 g sediment (triplicates) to 100 ml of 0.85% NaCl solution and were shaken. After serially diluting the solution was plated onto CDM plates (Weeger *et al.*, 1999) containing 100 mg/l $NaAsO_2$. The obtained bacteria were tested for their abilities to oxidize As(III) ($NaAsO_2$) and reduce As(V)(Na_3AsO_4) using a qualitative $KMnO_4$ screening method (Salmassi *et al.*, 2002).

2.2 *16S rDNA sequencing and PCR*

Quality of obtained culture was evaluated on 0.8% Agarose gel. Fragment of 16S rDNA was amplified by PCR using 8F and 1492R. A single discrete PCR amplicon band of 1500 bp was observed, which was purified and used for sequencing. Forward and Reverse DNA sequencing reaction of PCR amplicon was carried out with 704F and 907R primers using BDT v3.1 Cycle sequencing kit on ABI 3730 × l Genetic Analyzer. The 16S rDNA sequence was used to carry out BLAST alignment. Based on maximum identity score first fifteen sequences were selected and aligned using Culture W. Distance matrix was generated using RDP database and the phylogenetic tree was constructed using MEGA5.

2.3 *Morphological and bioaccumulation study*

The obtained cultures were characterized using FESEM and TEM. Bioaccumulation of As in respective culture was determined by semi quantification methods as EDS analysis. These cultures were used for bioaccumulation analysis. They were centrifuged and washed several times by ddH_2O. The accumulated cells at the bottom of centrifuged tubes were used for EDS and morphological studies.

3 RESULTS AND DISCUSSION

3.1 *Power plate and screening test*

Mainly three colored colonies were observed such as reddish (R), white (W) and transparent white (TW) in power plate method (Fig. 1a). The OD (Optimal Density) values obtained from the bacterial cultures represented all phases of growth (Fig. 1b). As the culture grew old, the duration of lag phase decreased while the exponential growth phase was observed within a very short span of time (Fig. 1c). Growth of 0.4 nm OD was obtained quickly after which qualitative screening test was performed as explained by (Salmassi *et al.*, 2002).

3.2 16S rDNA sequencing

The genera of the bacterial strains were identified by analyzing their 16S rDNA genes. The bootstrap consensus tree inferred from 1000 replicates is taken to represent the evolutionary history of the taxa analyzed (Felsenstein, 1985). The culture (W1) showed similarity with *Pseudomonas aeruginosa* strain *P* and closely (by 99.9%) matched with GenBank Accession Number: *KR080311.1* strain, based on nucleotide homology and phylogenetic analysis.

3.3 FESEM, TEM and EDS analysis

The morphological analysis also supported presence of bacillus bacteria, shown in Figure 2a and b.

The semi-quantified EDS analysis of the bacterial culture (Figure 3) showed presence of Mn, As, S and Fe as dominant elements, which indicated arsenate reduction mechanism by detoxification method. The results of TEM signify the similar morphology of the bacterium as seen through FESEM analysis.

4 CONCLUSIONS

The *P. aeruginosa strain P.* (Ars C), is often known to be a facultative-aerobic and As(V)-reducing bacteria. The performed study ground finding to understand the important aspect of As mobilization particularly in deeper aquifers. The finding of this study could establish the role of *P. aeruginosa* (Ars C) bacteria where As(V) reduces to As(III) which is more mobile and toxic form in aqueous phase. Moreover, the increasing OD culture growths in presence of As(V) attribute its characteristic.

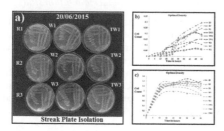

Figure 1. a) Pure culture of three colored colonies, b) & c) Bacterial growth curve (OD at 600 nm), decipher pure culture growth at initial and successive days of Na_3AsO_4 at 100 mg/L concentration.

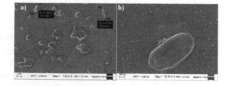

Figure 2. a) Morphology of arsenate reductase pure agglomerated; b) single cell bacteria.

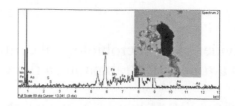

Figure 3. TEM Result showing morphology and chemical composition of the bacterial culture.

This micro-organism based analysis, which is new for this region, opens up a new aspect for understanding the spreading and coverage of As in groundwater aquifers. This also can help in explaining the high As concentrations in the GBM flood plains which receive huge sediment load from Brahmaputra river system, which is proven to be As predominant, also attributed to microbial metabolism by As mobilizing bacteria.

ACKNOWLEDGEMENTS

We thank the Department of Civil Engineering and Center Instrumentation Facility of IIT Guwahati for having provided us with all resources for performing this study. We also acknowledge Xcelris Genomics for helping in sequencing of cultures.

REFERENCES

Felsenstein, J. 1985. Phylogenies and the comparative method. *The American Naturalist* 125(1): 1–15.

Mahanta, C., Enmark, G., Nordborg, D., Sracek, O., Nath, B., Nickson, R.T., Herbert, R., Jacks, G., Mukherjee, A., Ramanathan, A.L., Choudhury, R., Bhattacharya, P. 2015. Hydrogeochemical controls on mobilization of arsenic in groundwater of a part of Brahmaputra river floodplain, India. *J. Hydrol. Regional Studies* 4: 154–171.

Mahanta, C., Sathe, S.S., Mahagaonkar, A. 2016. Morphological and mineralogical evidences of arsenic release and mobilization in some large floodplain aquifers. In: P. Bhattacharya, M. Vahter, J. Jarsjö, J. Kumpiene, A. Ahmad, C. Sparrenbom, G. Jacks, M.E. Donselaar, J. Bundschuh, R. Naidu (Eds.) *Arsenic Research and Global Sustainability (As 2016)*. CRC Press (This Volume).

Malasarn, D., Saltikov, C.W., Campbell, K.M., Santini, J.M., Hering, J.G., Newman, D.K. 2004. arrA is a reliable marker for As(V) respiration. *Science* 306(5695): 455.

Paul, D., Kazy, S.K., Banerjee, T.D., Gupta, A.K., Pal, T., Sar, P. 2015. Arsenic biotransformation and release by bacteria indigenous to arsenic contaminated groundwater. *Bioresource Technol.* 188: 14–23.

Salmassi, T.M., Venkateswaren, K., Satomi, M., Newman, D.K., Hering, J.G. 2002. Oxidation of Arsenite by Agrobacterium albertimagni, AOL15, sp. nov., Isolated from Hot Creek, California. *Geomicrobiol. J.* 19(1): 53–66.

Silver, S., Phung, L.T. 2005. Genes and Enzymes Involved in Bacterial Oxidation and Reduction of Inorganic Arsenic. *Appl. Environ. Microb.* 71(2): 599–608.

Weeger, W., Lièvremont, D., Perret, M., Lagarde, F., Hubert, J.C., Leroy, M., Lett, M.-C. 1999. Oxidation of arsenite to arsenate by a bacterium isolated from an aquatic environment. *Biometals* 12(2): 141–149.

Arsenic Research and Global Sustainability – Bhattacharya, Vahter, Jarsjö, Kumpiene,
Ahmad, Sparrenbom, Jacks, Donselaar, Bundschuh & Naidu (Eds)
© 2016 Taylor & Francis Group, London, ISBN 978-1-138-02941-5

Metagenomic insights into microbial community structure in arsenic-rich shallow and deep groundwater

J.-S. Jean & S. Das

Department of Earth Sciences, National Cheng Kung University, Tainan, Taiwan

ABSTRACT: Microorganisms dwelling in arsenic-rich groundwater have a pivotal role in biogeochemical cycles and are expected to have significant influence on the cycle of arsenic, a metalloid responsible for severe water pollution and the causing agent of the worst mass poisoning in human history. To gain insight into the indigenous bacterial population in arsenic-rich shallow and deep groundwater and to cover a greater genetic diversity, we conducted a 454 pyrosequencing study targeting nine defined hyper-variable regions of 16S rDNAs. The results revealed the presence of diverse and unequal bacterial communities mostly represented by the genera *Acinetobacter, Herbaspirillum, Bacillus, Flavisolibacter, Massilia, Nitrospira, Symbiobacterium, Bellilinea, Arthrobacter, Undibacterium, Sphingomonas, Geobacter, Delftia, Pseudomonas, Hydrogenophaga, Paenibacillus, Gemmatimonas, Anaeromyxobacter, Rheinheimera* and *Rhodococcus*. Overall observations suggest that the indigenous bacteria in arsenic-rich groundwater possess adequate catabolic ability to mobilize arsenic by a cascade of reactions mostly linked to bacterial nutrient acquisitions and detoxification.

1 INTRODUCTION

Arsenic (As) enrichment has frequently been implicated in altering microbial community structure (Oremland & Stolz, 2005). Exploring the microbial community structure and its dynamics leads to a better understanding of As cycling as well as the natural attenuation of As pollution in the As-contaminated environments (Islam *et al.*, 2004). Elevated concentration of As in groundwater of Chianan Plain in southwestern Taiwan is historically associated with endemic blackfoot disease (BFD) (i.e., gangrene). Arsenic concentration in the solid phase of Chianan Plain is mostly represented by As^{+5} and is found to be co-precipitated or adsorbed on various Fe and Mn rich mineral phases (Sengupta *et al.*, 2014). Aqueous As, in contrast, is dominated by As^{+3} and its concentration showed high spatial and depth variations (Sengupta *et al.*, 2014). Biogeochemical activities of indigenous bacteria surviving under As-enriched environment can control the mobility of this metalloid by dissolution of host minerals and/or weathering of rocks (Islam *et al.*, 2004). In spite of several confirmatory lines of evidences, our understanding about the identity of the indigenous bacteria and their probable role in As mobilization in As-rich groundwater remains mostly elusive.

2 METHODS/EXPERIMENTAL

2.1 *Groundwater sampling and analysis*

Groundwater samples were collected from Yenshu 3 (N 23° 18' 6.7" / E 120° 15' 11.1"), Budai-Shinwen (N 23° 20' 22" / E 120° 7' 57.9") and Budai 4 (N 23° 19' 37.8" / E 120° 9' 3.2") at depths of 23 m, 313 m and 300

m respectively. The physicochemical parameters and As concentration of the groundwater reported by Das *et al.* (2015) revealed that groundwaters of Yenshui 3, Budai-Shinwen, and Budai 4 were in anaerobic reducing environments with high As concentrations. The concentrations of As^{+3} were higher than those of As^{+5} in groundwaters of Yenshui 3 (954.1 ± 57.8 µg/L *vs.* 177.1 ± 25.3 µg/L) and Budai-Shinwen (604.5 ± 198.7 µg/L *vs.* 100.2 ± 3.3 µg/L), whereas the concentrations of As^{+3} were lower than those of As^{+5} in groundwater of Budai 4 (24.5 ± 1.3 µg/L *vs.* 285.3 ± 29.4 µg/L).

2.2 *DNA extraction, amplification and pyrosequencing*

Total genomic DNAs were extracted from groundwater samples using water master DNA extraction kit (Epicenter, USA) following the manufacturer's instruction. Bacterial 16S rDNAs at hypervariable regions 1 to 9 were amplified using 6 primers sets to construct pyrosequencing library (Nossa *et al.*, 2010). Pyrosequencing was done on a Roche GX-FLX Titanium 454 pyrosequencer (Roche, Mannheim, Germany).

3 RESULTS AND DISCUSSION

3.1 *Physicochemical properties of groundwater*

The dissolved As in groundwater from Yenshui-3 and Budai-Shinwen were predominantly As^{+3}, whereas it was As^{+5} from Budai-4. The As concentrations in groundwater were much above both World Health Organization (WHO) standard and Taiwan Drinking Water Standard (TDWS) of 10 µg/L. The groundwaters were near neutral to mildly alkaline and reduced as evidenced from the pH and Oxidation Reduction Potential (ORP) values. The groundwaters were rich in Fe and Mn but poor in Total Organic Carbon (TOC) concentrations. The sulfate concentration of groundwater from Budai-Shinwen was much higher compared to that of Yenshui-3 and Budai-4 (Das *et al.*, 2015).

3.2 *Bacterial community structure*

Pyrosequencing analysis revealed that Firmicutes (45.3%) dominated in highly As-rich groundwater of Yenshui 3, whereas Proteobacteria were the dominant phylum in comparatively less As-rich groundwater of Budai-Shinwen and Budai 4 comprising 38.8% and 92%, respectively (Das *et al.*, 2015). It is possible that the endospore forming capability of Firmicutes enabled them to survive under unfavorable conditions. Within Firmicutes, *Bacilli* (32.2%) were the most dominant class found in highly As-enriched groundwater of Yenshui 3, whereas Beta-proteobacteria (21.8%) and Gammaproteobacteria (72.3%) dominated in groundwater of Budai-Shinwen and Budai 4, respectively. Deltaproteobacteria, well recognized as a key player in As cycle (Islam *et al.*, 2004) dominated in highly As-enriched groundwater of Yenshui 3, whereas its presence in less As-enriched groundwater of Budai 4 was negligible. Member of the genera (*Acinetobacter, Herbaspirillum, Bacillus, Nitrospira, Arthrobacter, Sphingomonas, Pseudomonas, Hydrogeno phaga, Paenibacillus, Brevundimonas, Rhizobium*) detected in As-rich groundwater of Chianan Plain are well known for their chemolithoautotrophic mode of metabolism (Oremland & Stolz, 2003). Although often isolated as heterotrophs, these bacteria can use energy and reducing power from oxidation of various inorganic elements during CO_2 fixation or other anaerobic reactions and grow under aerobic or anaerobic oligotrophic environments (Oremland & Stolz, 2003, Das *et al.*, 2015). Several bacterial strains affiliated to these genera have been reported from As-contaminated environment (Oremland & Stolz, 2003, Das *et al.*, 2015) and most of them exhibited weathering activity, which in turn influences As mobility (Drewniak *et al.*, 2013). In addition, bacterial genera like *Acinetobacter, Herbaspirillum, Bacillus, Massilia, Pseudomonas, Hydrogenophaga* are well known for their capacity to withstand elevated concentration of As (Oremland & Stolz, 2003). Metabolic robustness of these genera to grow aerobically and anaerobically utilizing diverse electron/carbon donors and acceptors allow these bacteria not only to play important roles in As biogeochemistry but also create reducing conditions within the As-rich environments. Presence of iron-reducing bacterium (i.e., *Geobacter*) in highly As-rich groundwater of Yenshui 3 and Budai-Shinwen suggests the important role of iron—reducing bacteria in As mobilization in these As-rich aquifers. This could be a microbe mediated Fe(III) reduction utilizing As-bearing Fe(III) minerals causing dissolution and release of As (Islam *et al.*, 2004). Presence of iron-oxidizing bacteria (i.e., *Lysobacter, Thiobacillus, Magnetospirillum* etc.) and sulfur-oxidizing bacteria (i.e., *Rheinheimera, Ignavibacterium, Limnobacter, Thiobacillus*, etc.) in As-rich groundwater signified the potential role of these oxidizers in As mobilization by oxidative dissolution of As-bearing minerals, which leads to As release into the groundwater. Presence of sulfur-reducing bacteria (i.e., *Syntrophobacter, Desulfosporosinus, Desulfotomaculum, Desulfuromonas* etc.) coupled with high level of sulfide in As-rich groundwater of Budai-Shinwen (Das *et al.*, 2015) suggested that the reductive processes involving sulfide could conceivably release toxic level of As into the groundwater of this site (Drewniak & Shlodowska, 2013).

4 CONCLUSIONS

We could envisage an array of microbes mediated processes leading towards As release in shallow and deep groundwater aquifers of Chianan Plain. Unlike preponderance of any particular event, release of sediment bound As could possibly occurred due to a cascade of microbial processes such as dissolution of host minerals, followed by bacterial As^{+5} and/or Fe^{+3} reducing activities or direct action of bacterial reductive machinery on host minerals resulting elevated As^{3+} in groundwater of Chianan Plain. These processes could influence biogeochemical cycling of As therein and are not mutually exclusive but may incorporate each other.

ACKNOWLEDGEMENTS

This work was supported by the National Science Council of Taiwan (NSC 100-2116-M-006-009).

REFERENCES

Das, S., Liu, C.C., Jean, J.S., Liu, T. 2016. Dissimilatory arsenate reduction and in situ microbial activities and diversity in arsenic-rich groundwater of Chianan plain, southwestern Taiwan. *Microb. Ecol.* 71: 365–374.

Drewniak, L., Shlodowska, A. 2013. Arsenic-transforming microbes and their role in biomining processes. *Environ. Sci. Pollut. Res.* 20: 7728–7739.

Islam, F.S., Gault, A.G., Boothman, C., Polya, D.A., Charnock, J.M., Chatterjee, D., Lloyd, J.R. 2004. Role of metal-reducing bacteria in arsenic release from Bengal delta sediments. *Nature* 430: 68–71.

Nossa, C.W., Oberdorf,W.E., Yang, L., Aas, J.A., Paster, B.J., DeSantis, T.Z., Brodie, E.L. Malamud, D., Poles, M.A., Pei, Z. 2010. Design of 16S rRNA gene primers for 454 pyrosequencing of the human foregut microbiome. *World J. Gastroenterol.* 16: 4135–4144.

Oremland, R.S., Stolz, J.F. 2003. Ecology of arsenic. *Science* 300: 939–944.

Oremland, R.S., Stolz, J.F. 2005. Arsenic, microbes and contaminated aquifers. *Trends Microbiol.* 13: 45–49.

Sengupta, S., Sracek, O., Jean, J.-S., Lu, H.-Y., Wang, C.-H., Palcsu, L., Jen, C.-H., Bhattacharya, P. 2014. Spatial variation of groundwater arsenic distribution in the Chianan plain, SW Taiwan: role of local hydrogeological factors and geothermal sources. *J. Hydrol.* 518: 393–409.

Arsenic Research and Global Sustainability – Bhattacharya, Vahter, Jarsjö, Kumpiene,
Ahmad, Sparrenbom, Jacks, Donselaar, Bundschuh & Naidu (Eds)
© *2016 Taylor & Francis Group, London, ISBN 978-1-138-02941-5*

Oxidation of arsenite by using aerobic bacterial granules: A comparison with single bacterial culture

S. Tapase, S. Patki & K. Kodam
Department of Chemistry, Savitribai Phule Pune University, Pune, India

ABSTRACT: Oxidation of arsenite using single culture is known through ages. The aim of present study was to examine the efficiency of bacterial isolate and aerobic bacterial granules (consortium) towards As(III) oxidation. The bacterial granules have high efficiency of oxidizing the As(III) than single culture. The present study showed that *Achromobacter xylosoxidans* MI16 and aerobic bacterial granules effectively oxidize 10 and 50 mM of As(III) concentration respectively in 12 h. Bacterial granules can resist up to 150 mM As(III) under aerobic conditions. This result concludes that aerobic granules have higher oxidation potential than *Achromobacter* sp. represents its important application of bioremediation of As(III) contaminated sites.

1 INTRODUCTION

Arsenic (As) is a metalloid having atomic number 33, exists as pure crystal in many allotropic forms, also existing with various metals. Arsenite oxidation is a potential detoxification mechanism which allows microorganisms to tolerate high levels of arsenite. The oxidation of arsenite to arsenate is used both for detoxification and for energy generation which can also impact the mobility and speciation of arsenic in environment (Stolz *et al.*, 2002). Bacterial granules/consortium is a two or more bacteria living symbiotically. This symbiosis either may be endosymbiotic such as nitrogen fixing bacteria, or ectosymbiosis i.e., living on surface. In consortia, bacteria secrete exopolysaccharides that entraps other bacteria into it and forms granules. It is a very interesting system in which bacteria communicate with each other via chemical signals. This system also helps to survive as in the case some bacteria may lack some enzymes and others might produce them.

Arsenic biotransformation by single organism is known. Hence the aim of this study is to compare the efficacy of bacterial consortia over single bacterial culture for arsenic transformation.

2 METHODS

2.1 Enrichment and isolation of culture

Soil was collected from metal industry at Pirangut, Pune, India (18.51 N 73.60 E). Soil sample (1 g) was inoculated in 30 ml Tris-mineral medium at low phosphate content (Mergeay *et al.*, 1985). Flasks were incubated on shaker at 37°C for 6 days and 1 ml of this enrichment culture was inoculated in fresh medium after 2 days. Arsenite oxidizing bacteria were isolated by plating enrichment cultures on Tryptic Soya Agar (TSA) 0.1X containing 5 mM of As(III).

2.2 Screening for culture showing activity of arsenite oxidation

Screening was done for each individual culture for 24 h with 5 mM As(III) and was checked by molybdenum blue method (Cummings *et. al.*, 1999).

2.3 Identification and phylogenetic analysis

Identification was done by sequencing 16 s RNA gene sequencing (Jiang *et al.*, 2006). Phylogenetic analysis was performed using MEGA version 4 software. Phylogenetic tree was constructed using the neighbor-joining distance method based on p-distance. A total of 500 bootstrap replications were calculated.

2.4 Development of aerobic granules

Aerobic bacterial granules were developed in reactor by passing the air with the help of aerator with different type of microbes which were compatible with each other. Every after 12 h the reactor was fed with 4 g/l TSB medium along with 0.1 mM As(III).

2.5 Arsenic transformation by isolates and granules

Arsenic transformation was checked with varying concentration up to 10 mM As(III) for MI 16 and up to 150 mM As (III) for granules and arsenic transformation was checked by molybdenum blue method (Cummings et al., 1999).

2.6 Amplification of Aro gene responsible for As(III) oxidation in both MI 16 and aerobic bacterial granules

Amplification of gene responsible for As(III) oxidation, *Aro* was performed by PCR technique by using Aro F and Aro R primers (Inskeep et al., 2009).

2.7 Morphological studies

Morphological study of MI 16 and bacterial granules were carried out by using Scanning Electron Microscopy (SEM). Accumulation of arsenic on culture was evaluated by EDS.

3 RESULTS AND DISCUSSION

3.1 Isolation and identification of As(III) oxidizing microorganism

Out of 6 organisms screened, one potential isolates namely MI 16 was chosen for further studies. It was chosen on the basis of oxidation rate as compared to other isolates. After 16 sRNA sequencing, phylogenetic analysis showed that MI 16 formed a clade with *Achromobacter xylosoxidans* NBRC 15126T.

3.2 Development of bacterial granules

Consortium of different bacteria was formed due to exopolysaccharide produced by them. Spherical bodies were formed with light brown in color. The aerobic bacterial granules developed in reactor having very high settling ability and compact structure with average size ranging between 5–8 mm.

3.3 Arsenic transformation by isolates and granules

The isolate *Achromobacter xylosoxidans* MI 16 oxidized As(III) up to 10 mM, while aerobic bacterial granules effectively oxidized As(III) up to 50 mM in 12 h. This results confirmed that bacterial granules have very good potential towards high concentration of As(III).

3.4 Morphological studies

The SEM study showed that in presence of As(III) significant changes were found in morphology of aerobic granules. The porosity of aerobic granules

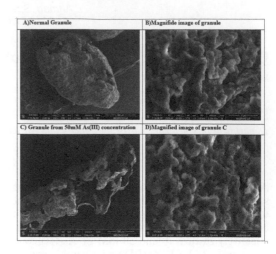

Figure 1. SEM images of bacterial granules.

was increased in presence of As(III), may due to reduction exopolysaccharides present on it (Fig. 1).

4 CONCLUSION

The aerobic bacterial granules has high efficiency of oxidizing As(III) than single culture of *Achromobacter xylosoxidans* NBRC 15126T. The bioreactor development will provide advanced removal of arsenic from contaminated water or to help in the purification system with aid of biological way.

ACKNOWLEDGEMENTS

The author (ST) would like to thank University Grants Commission for BSR fellowship and UPE phase II program for financial assistance.

REFERENCES

Carapito, C., Muller, D., Turlin, E., Koechler, S., Danchin, A., van Dorsselaer, A., Leize-Wagner, E., Bertin, P.N., Lett, M.C. 2006. Identification of genes and proteins involved in the pleitropic response to arsenic stress in *Caenibacter arsenoxydans*, a mettaloresistant betaproteobacterium with an unsequenced genome. *Biochime* 88: 595–606.
Oremland, R.S., Stolz, J.F. 2003. The ecology of arsenic. *Science* 300: 939–944.
Smedley, P.L., Kinniburgh, D.G. 2002. A review of the source, behaviour and distribution of arsenic in natural waters. *Appl. Geochem.* 17: 517–568.

Arsenic Research and Global Sustainability – Bhattacharya, Vahter, Jarsjö, Kumpiene,
Ahmad, Sparrenbom, Jacks, Donselaar, Bundschuh & Naidu (Eds)
© 2016 Taylor & Francis Group, London, ISBN 978-1-138-02941-5

Microbial diversity of arsenic-related bacteria in high arsenic groundwater of Inner Mongolia, China

Y. Wang & P. Li
China University of Geosciences, Wuhan, P.R. China

ABSTRACT: Microbial communities and arsenic-related populations in high arsenic groundwater of Hetao Plain were investigated using illumina sequencing and clone library analysis. Results showed that microbial communities in high As groundwater were predominated by *Psedomonas, Acinetobacter, Alishewanella, Psychrobacter, Methylotenera* and *Crenothrix* which were highly similar with those arsenic-related, nitrogen and sulfur cycling, and organic matter degradation microbes. Arsenic-oxidizing bacteria were mainly composed of *Rhodoferax ferrireducens, Leptothnix, Acidovorax, Pseudomonas* and *Acinetobacter*. Arsenic-reducing populations were dominated by *Geobacter, Desulfosporosinus, Chrysiogenes* and *Desulfurispirillum*. These results expand our current understanding of microbial ecology in high arsenic groundwater of Hetao Plain.

1 INTRODUCTION

Arsenic (As)-contaminated groundwater has impact the health of more than 140 million people all over the world (Ravenscroft et al., 2009). Previous studies have indicated that the release and mobilization of As in groundwater aquifers are controlled by series of microbial mediated reactions and geochemical processes (Oremland et al., 2003). However, little is known about the *in situ* microbial community composition of As metabolizing bacteria. The objectives of this study were to (1) investigate the overall community composition in groundwater samples with different geochemical characteristics, and (2) evaluate the functional microbial groups capable of As-oxidizing and As-reducing in high As groundwater aquifers in Hetao Plain.

2 MATERIAL AND METHODS

2.1 Site description and sample collection

Hetao Plain is one of the Cenozoic rift basins located in the western part of Inner Mongolia. Our case study area Hangjinhouqi County is the most serious As poisoning area in Hetao Plain, with concentrations of As in groundwater up to 1.74 mg/L (Deng, 2008). Samples were collected from domestic wells with approximately 10 L groundwater through 0.2 μm filters (Millipore). The filters were immediately packed into a 50 ml sterile tube, and then frozen in dry ice. All samples were transported to the laboratory on dry ice and then maintained at −80 °C until further analysis.

2.2 Geochemical analysis

The geochemical parameters including temperature, pH, conductivity, Dissolved Oxygen (DO), and Oxidation-Reduction Potential (ORP) were measured *in situ* using a multiple parameter water quality meter (Horiba, Japan). NH_4^+ and Fe(II) were determined using a portable spectrophotometer (HACH, DR890) according to the protocols. As species were separated by anion exchange cartridges and determined using liquid chromatography-atomic fluorescence spectrometry (LC-AFS-9700, Haiguang, China). TOC was measured using TOC analyzer (TOC-V CPH, Shimadzu Corporation, Japan).

2.3 DNA extraction, illumina sequencing and clone library construction

In the laboratory, DNA extractions were conducted using the FastDNA SPIN Kit for Soil (MP Bio, USA). The barcoded universal primer set 515F and 806R were used to amplify the microbial 16S rRNA V4 region. PCR products were purified and quantified using Quint-iT™ PicoGreen dsDNA assay kit (Invitrogen, USA). Equimolar concentrations were combined adequately and then subjected to sequencing by an Illumina MiSeq 2000 instrument. Clone libraries for the functional groups were performed using primer sets As1f (As2f)-As1r for *arrA* gene and aioA95f-aioA 599r for *aioA* gene. Sequencing data was analyzed using QIIME. Phylogenetic trees were constructed by the neighbor-joining method using MEGA program.

3 RESULTS AND DISCUSSION

3.1 Geochemistry

Twenty one groundwater samples were chosen for geochemistry and microbial ecology analyses. Arsenic concentrations ranged from 82 to 991 μg/L and AsIII dominated in most of the high As samples.

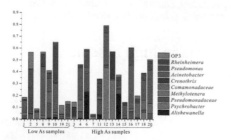

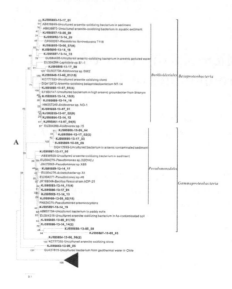

Figure 1. Overall microbial community composition in high As groundwater aquifers of Hetao Plain.

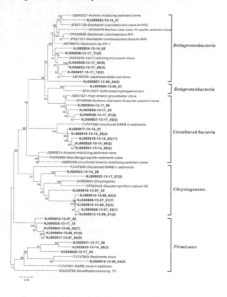

Figure 2. Neighbor-joining tree of arsenic-reducing populations based on amino acid sequences of the *arrA* gene in high arsenic groundwater. The numbers in parenthesis are the number of clones in each OUT.

3.2 *Overall community diversity*

Illumina results demonstrated that the bacterial communities were composed of 62 phyla, showing diverse array of microbes existed in high As groundwater aquifers (Fig. 1). Of these, only nine phyla dominated each community and among which *Proteobacteria* was the most dominant one. The mostly dominant populations in high As groundwater in Hetao Plain included *Pseudomonas, Acinetobacter, Alishewanella, Psychrobacter, Methylotenera* and *Crenothrix*. These populations were highly similar with those microbes capable of As resistant, reduction or oxidation, carbon, sulfur and nitrogen metabolisms in previous studies.

3.3 *Functional arsenic-oxidizing and reducing-communities*

Functional gene analysis of arsenate reducing and arsenite oxidizing community showed that bacteria

Figure 3. Neighbor-joining tree of arsenic-oxidizing populations based on amino acid sequences of the *aioA* gene in high arsenic groundwater. (only shown part of the figure).

related to As redox co-existed in high As groundwaters of Hetao Plain (Figures 2 and 3).

4 CONCLUSIONS

Results indicated that arsenite-oxidizing bacteria were dominated by several populations including *Rhodoferax ferrireducens, Leptothrix, Acidovorax, Pseudomonas* and *Acinetobacter*. The arsenic-reducing populations were dominated by several species including *Geobacter, Desulfosporosinus, Chrysiogenes* and *Desulfurispirillum*. Some of these dominant As related populations were also found dominantly in high arsenic samples by the illumina sequencing, indicating that As functional populations might play key roles in the As release and mobilization processes.

ACKNOWLEDGEMENTS

This research was financially supported by National Natural Science Foundation of China (Grant Nos. 41372348 and 41521001) and Research Fund for the Doctoral Program of Higher Education (Grant No. 2015M572221).

REFERENCES

Deng, Y. 2008. *Geochemical processes of high arsenic groundwater system at Western Hetao Basin*. PhD Thesis. China University of Geosciences, Wuhan, China.
Oremland, R.S., Stolz, J.F. 2003. The ecology of arsenic. *Science*. 300: 939–944.
Ravenscroft, H., Brammer, H., Richards, K. 2009. *Arsenic pollution: A global synthesis*. Wiley, UK.

Arsenic Research and Global Sustainability – Bhattacharya, Vahter, Jarsjö, Kumpiene, Ahmad, Sparrenbom, Jacks, Donselaar, Bundschuh & Naidu (Eds)
© 2016 Taylor & Francis Group, London, ISBN 978-1-138-02941-5

Arsenic biogeochemistry in hot springs in Tengchong geothermal area, China

Z. Jiang, P. Li & Y. Wang
China University of Geosciences, Wuhan, P.R. China

ABSTRACT: Arsenic (As) biogeochemistry in acid and alkaline hot springs of Tengchong geothermal area was investigated using Illumina MiSeq sequencing, functional gene clone library and geochemistry analyses. Results showed As(III) could be oxidized to As(V) in the source water of acid hot spring by some putative microbial populations, such as *Aquificae* and *Pseudomonas*, and then co-deposited with iron at its outflow channel. In sulfide-rich alkaline hot spring, As(III) was firstly derived from thioarsenic transformation probably mediated by *Thermocrinis* and then oxidized downstream by *Thermus* and *Hydrogenobacter*. These results improve our understanding of microbially-mediated As transformation in hot springs.

1 INTRODUCTION

Hot springs provide unique environments for the evolution and establishment of microbial communities and their response to various biogeochemical and metabolic processes involving hydrogen (H_2), sulfur (S), iron (Fe) and arsenic (As). Arsenic biogeochemistry has not been studied in acid sulfate hot springs with low chloride. Besides, the different processes on microbially mediated As species transformation in acid and alkaline hot springs had not been fully understood (Jackson *et al.*, 2001, Planer-Friedrich *et al.*, 2009). Therefore, the objectives of this study were to (1) investigate As geochemistry and microbial community structures in both water and sediments along acid and sulfide-rich alkaline hot springs; (2) evaluate the potential microbially-mediated As transformation processes in Tengchong geothermal area.

2 METHODS/EXPERIMENTAL

2.1 Site description and sample collection

Tengchong, located in southwestern of China, is a typical volcanic geothermal area and has abundant geothermal resources. In the Rehai geothermal field of Tengchong, Zhenzhuquan is a representative acid sulfate hot spring low in Cl, with 128.2 mg/L sulfate, 39.2 mg/L Cl and 71.1 µg/L As. Zimeiquan is a representative alkaline sulfide-rich hot spring, with a pH of 9.0, sulfide concentrations up to 4.8 mg/L, and As concentrations up to 655.6 µg/L. Water and sediment samples were collected through a 0.22 µm syringe polyethersulfone membrane filters and sterile spoons respectively and immediately packed into a 50 mL sterile tube, and then frozen in dry ice. All samples were transported to the laboratory on dry ice and then maintained at −80°C until further analyses.

2.2 Geochemical measurements

The geochemical parameters including temperature, pH and Dissolved Oxygen (DO) were measured *in situ* using a hand-held meter. Sulfide, ammonium, ferrous iron (Fe(II)) and total iron (Fe_{Tot}) were determined using a portable spectrophotometer (HACH, DR890). Arsenic species were separated by anion exchange cartridges (Supelco) and determined using liquid chromatography-atomic fluorescence spectrometry (LC-AFS-9700, Haiguang, China). TOC was measured using TOC analyzer (TOC-V CPH, Shimadzu Corporation, Japan).

2.3 DNA extraction, gene clone library, sequencing, data process and statistical analyses

DNA extractions were conducted using the FastDNA SPIN Kit for Soil (MP Bio, USA). The barcoded, universal primer set 515F and 806R were used to amplify the microbial 16S rRNA V4 region. Equimolar concentrations were combined adequately and then subjected to sequencing by an Illumina MiSeq system. Sequencing raw data were processed through a Galaxy-based pipeline. Over 0.56 million 16S rRNA sequence reads were obtained from 11-paired parallel water and sediment samples along the hot spring's outflow channel. Arsenic oxidization gene *aioA*-based clone libraries were constructed using primers aioA95f and aioA599r. All statistical analyses were performed based on genus-level OTUs at a 97% similarity level with the Vegan package in R.

3 RESULTS AND DISCUSSION

3.1 Arsenic biogeochemistry in acid Zhenzhuquan

In the pool of Zhenzhuquan, the remarkably high ratios of As(V)/As_{Tot} (0.73–0.86) suggested that As(III) oxidation occurred at the discharge source, which

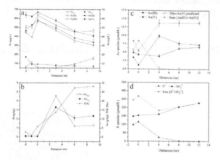

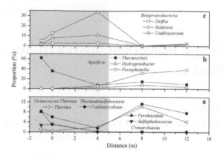

Figure 1. Distribution of selected geochemical data in water and sediment samples along the outflow channel of acid Zhenzhuquan (a, b) and alkaline Zimeiquan (c, d).

Figure 2. Distribution of dominant genus in water samples along Zimeiquan outflow channel.

were distinctly different with previous studies that As(III) was predominate in the source water of those acid sulfate-Cl hot springs. Compared to previous acid sulfate-Cl hot springs with sulfide concentrations of 2.02–12.6 mg/L, the distinctly low concentrations of sulfide (0.03–0.05 mg/L) in the pool provided a prerequisite for microbial As(III) oxidation. Clone libraries of *aioA* gene demonstrated the presence of several groups of As(III)-oxidizing microorganisms in the pool, including a few unidentified families of *Aquificae* and some postulated archaea. Furthermore, based on high 16S sequences similarity with As(III) oxidizing bacteria from geothermal and other environments, *Pseudomonas* and *Ralstonia* inhabiting the pool might be also related to As oxidation.

Coupled with Fe(II) and S oxidation along the outflow channel, arsenic was substantially accumulated in downstream sediments, with As concentrations up to 16.44 g/kg and As/Fe mole ratios up to 6.72 (Figure 1). These were significantly higher that those in previous acid sulfate-Cl hot springs (As/Fe mole ratios: 0.60 to 0.74). Previous studies documented the clay minerals in geothermal area, such as smectite and kaolinite, could host As concentrations up to 4 g/ kg (Pascua *et al.*, 2009). Coincidentally, smectite and kaolinite were also detected in the downstream sediments of Zhenzhuquan, which suggested that As in downstream sediments might also be adsorbed on the clay minerals, except for iron oxides.

3.2 *Arsenic biogeochemistry in alkaline Zimeiquan*

As$_{Sum}$ (combined concentrations of As(III) and As(V)) concentrations significantly increased from 5.45 to 13.86 µmol/L along Zimeiquan outflow channel (Figure 1c). Besides, elevated As(III) from 0 m to 4 m and subsequent decreased with a corresponding increased in As(V) from oxidation after 4 m, strongly suggested that the thioarsenate at this site was first converted to As(III) and then oxidized to As(V) after the thioarsenate disappears, as has been observed in previous studies (Planer-Friedrich *et al.*, 2009). The reduced sulfur generated from thioarsenate transformation

was oxidized to sulfate as DO increase after 4 m, which led to the increase of S$_{sum}$ (combined concentrations of sulfide and sulfate) from 4 m to 12 m (Figure 1d). Based on the maximum of As concentrations observed in three downstream sampling sites and the detection of thioarsenate in the pools, we calculated that thioarsenate concentrations in the Zimeiquan pools should be 5.5–8.4 µmol/L, and accounted for 39.9–60.7% of the total As.

The predominant *Thermocrinis* in upstream samples was probably responsible for transforming the thioarsenate to As(III), as has been demonstrated in previous studies (Figure 2) (Hartig *et al.*, 2014). In contrast, the As(III) oxidation observed downstream might be attributed to the significant appearance of *Thermus* and *Hydrogenobacter*, which are well-known As(III)-oxidizing bacteria inhibiting geothermal environments.

ACKNOWLEDGEMENTS

This research was financially supported by National Natural Science Foundation of China (Grant No. 41120124003, 41521001, 41372348).

REFERENCES

Hartig, C., Lohmayer, R., Kolb, S., Horn, M.A., Inskeep, W.P. Planer-Friedricht, B. 2014. Chemolithotrophic growth of the aerobic hyperthermophilic bacterium *Thermocrinis ruber* OC 14/7/2 on monothioarsenate and arsenite. *FEMS Microbiol. Ecol.* 90 (3): 747–760.

Jackson, C.R., Langner, H.W., Donahoe-Christiansen, J., Inskeep, W.P., McDermott, T.R. 2001. Molecular analysis of microbial community structure in an arsenite-oxidizing acidic thermal spring. *Environ. Microbiol.* 3(8): 532–542.

Pascua, C., Charnock, J., Polya, D., Sato, T., Yokoyama, S., Minato, M. 2005. Arsenic-bearing smectite from the geothermal environment. *Mineral. Mag.* 69(5): 897–906.

Planer-Friedrich, B., Fisher, J.C., Hollibaugh, J.T., Suess, E., Wallschlaeger, D. 2009. Oxidative transformation of trithioarsenate along alkaline geothermal drainages: abiotic versus microbially mediated processes. *Geomicrobiol. J.* 26(5): 339–350.

Arsenic Research and Global Sustainability – Bhattacharya, Vahter, Jarsjö, Kumpiene,
Ahmad, Sparrenbom, Jacks, Donselaar, Bundschuh & Naidu (Eds)
© 2016 Taylor & Francis Group, London, ISBN 978-1-138-02941-5

Characterization of lipid biomarkers and trace elements in Holocene arsenic contaminated aquifers of the Bengal Delta Plains, India

D. Ghosh[1,2], J. Routh[2] & P. Bhadury[1]

[1]*Department of Biological Sciences, Indian Institute of Science Education and Research Kolkata, Mohanpur, India*
[2]*Department of Thematic Studies—Environmental Change, Linköping University, Linköping, Sweden*

ABSTRACT: Earlier studies showed the contrasting properties in the hydrochemistry of grey and brown sand aquifers in the Bengal delta plains. However seldom anything is reported about the organic characteristics, which is vital for driving the As biogeochemical reactions in the sub-surface. In this study, the down core distribution of trace elements and organic matter in two grey sand aquifers are reported. The trace element profiles in these aquifers are similar, but the organic matter characteristics show differences. This is perhaps due to their geographical location. Interestingly, in both types of aquifers, signatures of petroleum-derived hydrocarbons were found, which were reportedly absent in brown sand aquifers indicating the role of the complex hydrocarbons in shaping the indigenous microbial community.

1 INTRODUCTION

Arsenic (As) contamination in the Bengal Delta Plains (BDP) aquifers has been largely reported for the past two decades. Various hypotheses explaining As mobilization have been postulated, of which microbial mediated oxidation of aquifer derived organic matter coupling with reductive dissolution of As bearing Fe(III) minerals is the most widely accepted view (Bhattacharya *et al.*, 1997, Nickson *et al.*, 1998). Diverse groups of arsenite {As(III)} oxidizing bacteria were detected based on *aioA* gene sequencing, only in groundwater from two Grey Sand Aquifers (GSA) (wells 28 and 204) in Nadia district, West Bengal. However, such bacterial signals were not detected in the Pleistocene Brown Sand Aquifers (BSA) in Haringhata (Ghosh *et al.*, 2014). Thus it is very important to determine the key differences in the organic carbon available in these two types of aquifers controlling of overall microbial community structure in these aquifers.

2 METHODS

2.1 Sampling

To study the sedimentary organic matter (SOM) sustaining those bacterial groups reported in GSAs, two boreholes were made next to wells 28 and 204 (Ghosh *et al.*, 2014). Sediments were retrieved at 2 m intervals in the wells (up to 50 m depth). The sections were logged and analyzed for grain size and organic C content. Geochemical analyses included sequential extraction of trace metals and extraction of different lipid classes.

2.2 Trace metal extraction

Freeze dried sediment samples (0.5 g) were used to extract sedimentary trace elements following four step BCR (formerly the Community Bureau of Reference) protocol of sequential extraction (Rauret *et al.*, 2000). The extracts were further analyzed on a Perkin Elmer NexION 300D ICP-MS.

2.3 Total Lipid Extraction

About 10 g of all freeze dried sediment samples were used for extraction of sedimentary lipids following the protocol outlined in Ghosh *et al.* (2015b). Extraction was done on a Dionex 300 automated solvent extractor and the extracts were separated into alkane, alkanol and alkanoic acid fraction using Solid Phase Extractor (Büchi) and analyzed in an Agilent 6890 N GC interfaced to an Agilent 5973 MSD mass spectrometer.

3 RESULTS AND DISCUSSION

3.1 Distribution of trace metals

Arsenic concentration ranged from 2 to 47 mg/kg, and is abundant in clay-rich sediments associated with crystalline oxides and silicate minerals. Arsenic showed significant correlation with Fe ($p = 0.00$; $r = 0.95$), and also with Mn ($p = 0.00$; $r = 0.93$) suggesting presence of Fe and Mn bound As minerals. Mn showed significant correlation with Ca in carbonate fraction ($p = 0.00$; $r = 0.93$), indicating incorporation of Mn into carbonate minerals like rhodochrosite ($MnCO_3$) and $CaCO_3$-$MnCO_3$.

3.2 Distribution of sedimentary lipids

In lipids extracted from these sediments, *n*-alkanoic acids were the dominant lipid monomers (Fig. 1). Alkanoic acid concentrations ranged between 670 to 94,535 ng/mg dry weight in sediments. The alkanoic acids *n*-C16:0 and *n*-C18:0 were most abundant followed by High Molecular Weight (HMW) *n*-alkanoic acids (*n*-C24:0, *n*-C26:0, *n*-C28:0 and *n*-C30:0). The *n*-alkanol concentrations varied from 0.23 to 18.8 ng/mg dry weight. *n*-Alkane concentration varied from 0.16 to 28.5 ng/mg dry weight. Well 28 had a unimodal distribution of *n*-alkanes with the predominance of HMW *n*-alkanes. However, in well 204 a bimodal distribution was observed with equi-dominance of low and high molecular weight *n*-alkanes. Presence of petroleum-derived hydrocarbons was detected in both wells, which were reported in other GSA wells from BDP in previous studies (Rowland *et al.*, 2006). Such hydrocarbons were also detected while characterizing the Dissolved Organic Carbon (DOC) in groundwater (Ghosh *et al.*, 2015a). The detection of hydrocarbon degrading bacterial molecular signatures suggests that soil organic matter plays an important role in sustaining the indigenous microbial flora (Ghosh *et al.*, 2014). Moreover, indigenous bacterial isolates were able to degrade hydrocarbons present in groundwater (Ghosh *et al.*, 2015a). These isolates were also able to oxidize As(III). There was preferential preservation of n-alkanes over n-alkanols in both wells. Along with alkanols, sterols such as brassicasterol, 5α-brassicastanol, 5α-stigmastanol and situstanol were also detected in well 204, indicating terrigenous inputs of organic matter from the soil zone.

4 CONCLUSIONS

Overall, the characteristics and bioavailability of soil organic matter in GSAs are of direct importance to microbial respiratory processes occurring within aquifers, and controlling the flux of As, Fe and other elements. Absence of particular group of organic compounds in BSAs perhaps kept their niche unsuitable for the different microbial groups involved in arsenic cycling, and thus keep them As-safe in comparison to the GSAs in this region.

ACKNOWLEDGEMENTS

DG thanks DST (Govt. of India) for providing the Ph.D. INSPIRE fellowship. The study was financed by the Swedish Research Link-Asia Program.

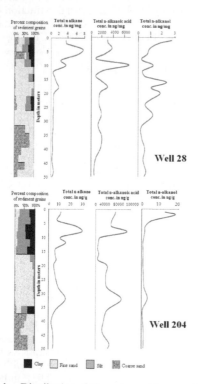

Figure 1. Distribution of biomarkers with respect to sediment grain size in the two GSA aquifers of BDP, India.

REFERENCES

Bhattacharya, P., Chatterjee, D., Jacks, G. 1997. Occurrence of arsenic contaminated groundwater in alluvial aquifers from Delta Plains, Eastern India: Options for safe drinking water supply. *Int. J. Water Res. Manag.* 13: 79–82.

Ghosh, D., Bhadury, P., Routh, J. 2014. Diversity of arsenite oxidizing bacterial communities in arsenic-rich deltaic aquifers in West Bengal, India. *Frontiers in Microbiology,* 5:604. doi: 10.3389/fmicb.2014.00602.

Ghosh D., Routh J., Bhadury P. 2015a. Characterization and microbial utilization of dissolved lipid organic fraction in arsenic impacted aquifers (India). *J. Hydrol.* 527: 221–223.

Ghosh D, Routh J, Dário M, Bhadury P. 2015b. Elemental and biomarker characteristics in a Pleistocene aquifer vulnerable to arsenic contamination in the Bengal Delta Plain, India. *Appl. Geochem.* DOI: 10.1016/j.apgeochem. 2015.05.007.

Nickson, R.T., McArthur, J., Burgess, W., Ahmed, K.M., Ravenscroft, P., Rahman, M. 1998. Arsenic poisoning of Bangladesh groundwater. *Nature*, 395: 338.

Quevallier, P., Rauret, G., López-Sánchez, J.F., Rubio, R., Ure, A., Muntau, H. 1997. Certification of trace metal extractable contents in a sediment reference material (CRM 601) following a three-step sequential extraction procedure. *Sci. Total Environ.* 205: 223–234.

Rowland, H.A.L., Polya, D.A., Lloyd, J.R., Pancost, R.D. 2006. Characterization of organic matter in a shallow, reducing, arsenic-rich aquifer, West Bengal. *Org. Geochem.* 37: 1101–1114.

Arsenic Research and Global Sustainability – Bhattacharya, Vahter, Jarsjö, Kumpiene,
Ahmad, Sparrenbom, Jacks, Donselaar, Bundschuh & Naidu (Eds)
© 2016 Taylor & Francis Group, London, ISBN 978-1-138-02941-5

Mobility of arsenic in two black shale areas in Sweden

G. Jacks[1], B. Nilsson[1] & A. Frank[2]

[1]*Department of Sustainable Development, Environmental Science and Engineering, KTH Royal Institute*
of Technology, Stockholm, Sweden
[2]*Swedish University of Agricultural Sciences, Uppsala, Sweden*

ABSTRACT: This study compares two black shale areas in Central and Northern Sweden with sulphidic black shales with elevated contents of trace metals and arsenic. In the Precambrian area arsenic is mobilized in wetlands and discharged into drains and streams where it is re-adsorbed onto ferric precipitates. The bottom fauna in streams is only pointwise enriched in arsenic while fish shows little or no accumulation of arsenic. contents. In the Cambrian area the arsenic released by oxidation is found in the soil profiles attached to ferric precipitates. Almost no arsenic is seen in ground- and surface water nor in plants. The reason seems to be differences in topography.

1 INTRODUCTION

Black shales formed in marine environments are enriched in trace metals including arsenic. Arsenic is a redox sensitive element that may be mobilized under reducing conditions and then appear in groundwater and in plants. Two areas with respectively Precambrian and Cambrian black shales have been studied and the mobilization of arsenic as well as other trace metals turn out to be different. A partial analysis of the shales is presented in Table 1.

There are considerable differences between the two shales, thus the content of organic carbon is higher in the Cambrian shale and likely to be more reactive.

2 METHODS/EXPERIMENTAL

Sampling of ssoils, groundwater and surface water have been done in both areas. Speciation of the water has been done by in situ dialysis with membranes having pore sizes of 1 kD and 10 kD. In addition samples were filtered by 0.2 µm filters. Water samples were analyzed by ICP-MS and ion chromatography. Soil samples were treated by sequential extraction using the BCR method

(Zemberyová *et al.* 2006), yielding fractions of trace elements soluble in dilute acetic acid, trace elements mobilizable by reduction with hydroxylamine hydrochloride and a residual fraction mostly contained in silicates. Plants were sampled, dried at 50 degree centigrade and dissolved with four acid digestion before analysis by ICP-MS. Arsenic in bottom fauna and fish was speciated into inorganic and organic arsenic (Slejkovec *et al.* 2004).

3 RESULTS AND DISCUSSION

3.1 *Precambrian area*

In the Precambrian area the soils built up by black shales were podzolised and acidic with a pH of 5–6. The B-horizons with its ferric precipitates were enriched in arsenic. In wetland reducing conditions were found with dissolved ferrous iron in the groundwater and arsenic concentration up to around 100 µg/l. Even in some lakes the concentration of arsenic was up to 20 µg/l but the speciation by dialysis showed that the arsenic was to a large extent tied up to ferric precipitates and not likely to be bioavailable (Table 2). The ferric particles and colloids are likely to be tied up to humic matter.

Table 1. Partial analysis of the black shales.

Bedrock	Org. C%	S%	As mg/kg	Zn mg/kg
Precambrian Shale*	3	1	40	440
Cambrian Shale**	12	5	60	450

*Svensson (1980); **Leventhal (1991).

Table 2. Speciation of As and Fe in lake water.

Sample fraction	As µg/l	Fe µg/l
Unfiltered	8.2	2450
Filtered 0.2 µm	5.2	1340
Dialysis 10 kD	3.2	325
Dialysis 1 kD	2.1	49

In discharge areas in drains ferric precipitates were found to contain up to 0.5% As while sandy sediments in the streams contained 200–500 mg/kg of arsenic (Jacks *et al.*, 2013). The bottom fauna showed some accumulation of inorganic arsenic locally in connection to groundwater discharge manifested by ferric precipitates. However, the fish did not show any excess inorganic arsenic. Some plants in discharge areas showed elevated arsenic contents, Salix spp to some extent but *Equisetum spp* to a larger extent. The highest recorded value was 30 mg/kg for *Equisetum fluviatile*. The *Equisetum spp* are silica rich and tend like rice to take up arsenic (Halder *et al.*, 2014). On the whole arsenic did not seem to be a seriouis environmental threat due to the efficient adsorption onto ferric precipitates. However some groundwater wells, notably drilled wells did show high contents of arsenic.

3.2 Cambrian area

The soils in the Cambrian area were not podzolised and had a pH just below. This is due to the the Cambrian shale is overlain to the north and west by limestone which is mixed into the till (Andersson *et al.*, 1985, Snäll, 1988). The content of arsenic in soil profiles were of the same order as that in the black shale itself and showed little variation with depth. The major portion of the arsenic was found in the reducible fraction after BCR sequential extraction. This indicates that the arsenic after oxidation is adsorbed on the site in the soil profiles. The arsenic in groundwater and surface water was low, below or just above the detection limit at 0.1 µg/l. Plants showed little arsenic, in the order of >1 mg/kg. Only *Equisetum pratense* showed 2 mg/kg.

3.3 Discussion

The arsenic behavior is different in the two black shale areas, it seems that arsenic is oxidized in soil profiles in the Cambrian area and adsorbed on site in the soil profiles. As the pH of the soils is well above Zero Point of Charge at pH ~8.2 this is feasible. While the organic matter is higher in the Cambrian black shale than in the Precambrian and likely to have a higher reactivity there is not mobilization of arsenic by reduction. The fresh organic matter in terms of plants is likely to be more reactive in the Cambrian area as well. as the vegetation is far lush there. The main reason that no local mobilization takes place in the Cambrian areas seems to be the topography with no wetlands opposite to the Precambrian area which is flatter and have abundant wetlands in terms of swamps and bogs. The variation on topography in the Cambrian area is of the order of 100 m over a distance of 1 km while it is 30 m over the same distance in the Precambrian area. In neither area the arsenic seems to be a serious environmental problem except for in some drilled groundwater wells in the Precambrian area (Bhattacharya *et al.*, 2010, Jacks *et al.*, 2013).

4 CONCLUSIONS

Arsenic is after oxidation of sulphides in the Cambrian area locally adsorbed onto ferric precipitates and no mobilization to water, fauna and plants occur. In the Precambrian area arsenic is mobilized in wetland under reducing conditions but after discharge of the reduced groundwater, the arsenic is readsorbed onto ferric precipitates. The arsenic found in filtered surface water seems to be tied up to humic-ferric precipitates and not bioavailable and transferred in food chains to fish. Deep drilled groundwater wells could have excessive arsenic concentrations and showed be monitored.

ACKNOWLEDGEMENTS

The authors are thankful for support from ÅForsk and Carl Trygger foundations.

REFERENCES

Andersson, A., Dahlman, B., Gee, D.G., Snäll, S. 1985. The Scandinavian alum shales. *Swed. Geol. Survey* Series Ca 56. 50 pp.

Bhattacharya, P., Jacks, G., von Brömssen, M., Svensson, M. 2010. *Arsenic in Swedish groundwater—Mobility and risk for naturally elevated concentrations.* SGU Research Project—Final Report. TRITA-LWR Report 3030: 25p. (DOI: 10.13140/RG.2.1.2975.7926).

Halder, D., Biswas, A., Slejkovec, Z., Chatterjee, D., Jacks, G., Bhattacharya, P. 2014. Arsenic species in raw and cooke rice: implications for human health in rural Bengal. *Sci. Total Environ.* 497–498: 200–208.

Jacks, G. Slejkovec, Z., Mörth, M., Bhattacharya, P. 2013. Redoc cycling along the water pathways in sulfidic metasediment areas in northern Sweden. *Appl. Geochem.* 35: 35–43.

Leventhal, J.S. 1991. Comparison of organic geochemistry and metal enrichementin two black shales: Cambrian Alum Shale of Sweden and Devonian Chattanoga Shale of United States. *Miner. Deposita* 26: 104–112.

Slejkovec, Z., Bajc, Z., Doganoc, D.Z. Arsenic speciation patterns in freshwater fish. *Talanta* 62: 931–935.

Snäll, S. 1988. Mineralogy and maturity of Alum Shales of south-central Jämtland, Sweden. *Swed. Geol Survey* C 818. 46 pp.

Svensson, U. 1980. Geochemical investigation of minor elements of the principal Precambrian rocks of Västerbotten county, Sweden. *Swed. Geol. Survey* Series C 764.

Zemberyová, M., Barteková, J., Hagarová, J. 2006. The utilization of modified BCR three-step sequential extraction procedure for the fractionation of Cd, Cr, Cu, Ni, Pb and Zn in soil reference materials of different origins. *Talanta* 70: 973–978.

Development of a passive bioremediation process based on sulfate-reduction to treat arsenic-containing acidic mine water

F. Battaglia-Brunet[1], C. Joulian[1] & C. Casiot[2]

[1]*BRGM, Water, Environment and Ecotechnology Division, Environmental Biogeochemistry and Water Quality Unit, 3, Orléans, France*
[2]*HydroSciences Montpellier, Université de Montpellier, Montpellier, France*

ABSTRACT: Arsenic (As) is one of the priority pollutants commonly associated with mine tailings and Acid Mine Drainages (AMD). A bioprocess based on the activity of acido-tolerant Sulfate-Reducing Bacteria (SRB) was studied at laboratory scale in order to precipitate As sulfide. Enrichments of SRB were obtained from sediments of Carnoulès mining site in France. The enriched SRB-containing community was used to inoculate a bioreactor subsequently fed with a synthetic solution, its composition miming that of the Carnoulès AMD. The substrates were immobilized inside the bioreactor in order to treat water in a passive way. Results show the feasibility of selective precipitation of As versus iron in a passive bioreactor fed with AMD.

1 INTRODUCTION

Mining operations trigger significant environmental problems worldwide. A worldwide estimate assumes that about 20,000 km of river and 70,000 ha of lake and reservoir area are seriously damaged by acidic mine effluent. AMD is characterized by low pH and high concentrations of sulfate, iron, heavy metals and metalloids, suspended particulate matter whose impact on downstream watershed can be very dramatic. AMD contamination causes reduction of biodiversity and contamination of water resources, which preclude their use for domestic and agricultural purpose. Arsenic (As) is one of the priority pollutants commonly associated with mine tailings and AMD. At the abandoned Pb–Zn Carnoulès mine (South of France), 1.5 MT of spoil material containing sulfide minerals, metals (Pb, Zn, Tl…) and As was deposited over the former sources of Reigous Creek. The concentrations of As at the Reigous source are among the highest ever reported for AMD. In the framework of the IngECOST-DMA project, passive water treatment processes based on biological reaction are developed. Here, the precipitation of As as sulfide minerals was studied with sulfate-reducing bacteria enriched from the site and a synthetic mine water, its composition miming that of the Carnoulès mine water.

2 METHODS

2.1 Characteristics of the mine water

At Carnoulès mine, AMD is generated at the base of the tailings stock where O_2-rich water drains the tailings material continuously. The water, which emerges at a flow rate of 0.8–1.7 L/s is acid (pH 2–3), rich in sulfate (1–7 g/L), dissolved Fe (0.5 g/L) and As (50–350 mg/L), predominantly in the reduced forms.

2.2 Sulfate-reducing enrichments

Surface sediments were collected at three locations: just at the emergence of the drainage near the tailing dump (Nancucheo & Johnson, 2014), and in the Reigous 100 m (Battaglia-Brunet et al., 2012) and 200 m (Walters et al., 2009) down flow. They were inoculated into anaerobic enrichment medium for sulfate reducers containing glycerol, yeast extract, sulfate, ammonium chloride, potassium phosphate, zinc and As, and prepared at different pH ranging between 3.0 and 5.5. Active cultures precipitating ZnS and As_2S_3 were obtained at all pH values. These enrichments were used for optimizing a passive anaerobic bioreactor treatment of AMD containing As (Nancucheo & Johnson, 2014). DNA extractions, dsrAB genes amplification, cloning and sequencing, and phylogenetic analyses of deduced aminoacids DsrAB sequences were performed as described in Battaglia-Brunet et al. (2012).

2.3 Passive treatment experiment

The bioreactor consisted of a 320 mm high glass column with an internal diameter of 35 mm. It was equipped with a water jacket for temperature regulation and with five rubber-stopped sampling ports. The column was filled with 2–5 mm pieces of pozzolana mixed with agar gel enriched with nutrients (glycerol, yeast extract, NH_4Cl, K_2HPO_4), and biocompounds manufactured by Vertum GmbH (Battaglia-Brunet et al., 2012), then down-flow fed in continuous mode with a synthetic mine water miming the Carnoulès AMD (As 100 mg/L, Zn 20 mg/L, Fe 1 g/L, SO_4 3.6 g/L). Its pH was progressively decreased from 4.5 to 3. Concentrations in As and metals were analyzed by AAS.

3 RESULTS AND DISCUSSION

3.1 Identification of acido-tolerant SRB in the enrichments

Active cultures precipitating ZnS and As_2S_3 were obtained at all pH values. Inventory of *dsr*AB functional gene specific to SRB were performed for two enrichments. The enrichment inoculated with sediment from the Reigous river (location 3, 200 m down flow) with an initial pH of 3.0 contained diverse SRB owing *dsr*AB genes affiliated to *Desulfosporosinus* genes, some species of this genus being isolated from acidic environments. The enrichment obtained at initial pH 3.5 from the sediment of the water emergence (location 1, near the tailing dump) gave *dsr*AB sequences more distant from known sequences, the closest being also affiliated to that of *Desulfosporosinus* species.

3.2 Passive treatment in bioreactor

The bioreactor has worked without any substrate in the feed solution for 200 days. SRB remained active while pH of the feed was decreased from 4.5 to 3.0 (Fig. 1).

The pH was always higher in the outlet than in the feed solution. Some As precipitated in the bioreactor (Fig. 2), whereas iron was not precipitated. This phenomenon can be explained by the low value of pH in the bioreactor that was lower than the optimal pH for FeS precipitation. Conversely, precipitation of As sulfides is favored in acidic conditions. The percentage of precipitated As reached 80% at the beginning of experiment, when the feed

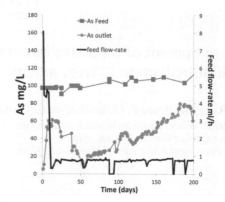

Figure 2. Passive SRB bioreactor experiment. Arsenic concentration in the feed and outlet and feed flow-rate.

pH was fixed at 4.5, then decreased down to 60% with the pH 3.5 feed, and 20% with the pH 3.0 feed. The latter condition corresponds to the composition and pH of the real AMD water of Carnoulès. The decrease of the bioreactor efficiency may be due to (1) a lower SRB activity due to a lower pH or (2) the exhausting of substrates initially placed into the column. The experiment was further pursued with the real AMD water amended with glycerol, and recent results confirmed that SRB activity was rather limited by energetic substrate than by acidity.

4 CONCLUSIONS

Passive bio-precipitation of As sulfide by acido-tolerant SRB has been performed for 200 days in a laboratory bioreactor. Despite a high residence time (10 days) and a decrease of efficiency, this experiment demonstrated for the first time that As can be selectively precipitated by SRB in AMD containing As and Fe in a passive bioreactor device.

ACKNOWLEDGEMENTS

This work was supported by the French National Research Agency (ANR 2013-ECOT-0009).

REFERENCES

Battaglia-Brunet, F., Crouzet, C., Burnol, A., Coulon, S., Morin, D., Joulian, C. (2012) Precipitation of arsenic sulphide from acidic water in a fixed-film bioreactor. *Water Res.* 46: 3923–3933.
Nancucheo, I., Johnson, D.B. 2014. Removal of sulfate from extremely acidic mine waters using low pH sulfidogenic bioreactors. *Hydrometallurgy* 150: 222–226.
Walters, E., Hill, A., Hea, M., Ochmann, C., Horn, H. 2009. Simultaneous nitrification/denitrification in a biofilm airlift suspension (BAS) reactor with biodegradable carrier material. *Water Res.* 43: 4461–4468.

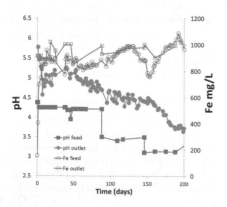

Figure 1. Passive SRB bioreactor experiment, pH and iron concentration of feed and outlet solutions.

Arsenic Research and Global Sustainability – Bhattacharya, Vahter, Jarsjö, Kumpiene,
Ahmad, Sparrenbom, Jacks, Donselaar, Bundschuh & Naidu (Eds)
© *2016 Taylor & Francis Group, London, ISBN 978-1-138-02941-5*

Pollution, degradation and microbial response of roxarsone in vadose zone

Y. Liu, Y. Li & Z. Zhang
Institute of Hydrogeology and Environmental Geology, Chinese Academy of Geological Sciences,
Shijiazhuang, P.R. China

ABSTRACT: This work was to investigate the threat of roxarsone and its principal control factors of transformation in vadose zone. Through field survey and sampling, the arsenic pollution status of study area was clear. The experimental results showed that both biotic processes and abiotic processes acted on roxarsone degradation, in which the effect of microorganisms was 69.6% and the effect of photodegradation was 5.5% during the roxarsone degradation illustrating that the effect of microorganisms was much larger than the effect of photodegradation. In addition, compared with the original chicken manure, the microbial community composition after incubation would become more abundant, but the main microbial community composition had remained unchanged.

1 INTRODUCTION

Roxarsone, 3-nitro-4-hydroxyphenylarsonic acid, is used extensively in the feed of broiler poultry to control coccidial intestinal parasites, improve feed efficiency, and promote rapid growth (Anderson *et al.*, 1983). But most of the roxarsone in the feed is excreted unchanged in the manure and introduced into the environment (Brown *et al.* 2005, Fisher *et al.*, 2015). Roxarsone in the environment can be converted into inorganic arsenic species through biotic and abiotic pathways (Garbarino *et al.* 2003). Inorganic arsenic, such as arsenate and arsenite, is more toxic than roxarsone. Therefore, roxarsone in manure is a hazardous chemical compound (Nachman *et al.*, 2013).

This work was to investigate the threat of roxarsone in livestock and poultry breeding areas, the principal control factors of roxarsone transformation in vadose zone and the microbial response to the roxarsone.

2 METHODS

2.1 *Field survey and sampling*

In this study, 37 groups of groundwater samples in livestock and poultry breeding areas of North Shandong were collected (Fig. 1). The groundwater samples were sent to laboratory to determine arsenic concentration.

According to the survey, a chicken farm in Liaocheng of Shandong Province of southern China was chose as a further study area (Fig. 1), in which the chicken manure was sampled and reserved in pre-sterilized aluminum containers and stored at 4°C.

2.2 *Laboratory experiment*

In the laboratory, the chicken manure was added to the liquid medium with roxarsone (50 mg/L) and the microcosms were incubated under light or dark, respectively. Meanwhile, the chicken manure and 0.2% $HgCl_2$ were added to the liquid medium with roxarsone (50 mg/L) and incubated under dark to determine the effect of microbial action on roxarsone degradation.

At the termination of the incubation, the microbial community were harvested and the DNA were extracted for community diversity analysis by using next-generation DNA sequencing techniques.

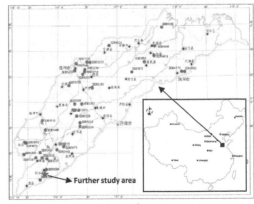

● Groundwater samples ★ Further study area

Figure 1. Sampling locations of groundwater in livestock and poultry breeding areas of Shandong and the further study area location.

3 RESULTS AND DISCUSSION

3.1 Arsenic pollution status in the study area

The groundwater samples in livestock and poultry breeding areas of North Shandong were collected to determine arsenic concentration, in which 64.9% of the samples were detected arsenic and 50% were out of limits and the highest arsenic concentration was 0.136 mg/L. The roxarsone concentration in the feed of broiler poultry was up to 34 mg/kg. The survey results showed that the livestock and poultry breeding areas were seriously polluted by arsenic which might be caused by the feed additives roxarsone.

3.2 Different degradation rate in the dark or light

The change of roxarsone concentration of different treatments during 12 days of incubation was showed in Figure 2. After 12 days of incubation, the degradation rate in the dark or in the light was 89.1% and 94.6%, and the degradation rate in the sterilized microcosms was 19.5%, which illustrated that the effect of photodegradation on the roxarsone degradation was 5.5% and the effect of microorganisms on the roxarsone degradation was 69.6%. The experimental results showed that both biotic processes and abiotic processes acted on roxarsone degradation, and the effect of microorganisms was much larger than the effect of photodegradation.

3.3 The community diversity in different reatments

Molecular characterization revealed the presence of different bacterial groups under different treatments (Fig. 3), in which *Proteobacteria* and *Bacteroidetes* were the main phyla in dark treatment (BG), light treatment (GROX) and original chicken manure (YGJF). The original chicken manure had a simple phylum composition which was made of only three phyla. After incubation in dark or light, the microbial composition became

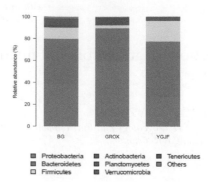

Figure 3. The microbial community structure and composition of different treatments. BG represents chicken manure and roxarsone in dark. GROX represents chicken manure and roxarsone in light. YGJF represents original chicken manure without any treatments.

more abundant and the main phyla of sample BG and sample GROX were similar.

4 CONCLUSIONS

The open accumulation without impervious disposal of litter was the direct way that roxarsone and its derivatives were introduced into environment. Roxarsone could be rapidly degraded by biodegradation and photodegradation, which may result in increased arsenic in surface and groundwater and increased uptake by plants. The effect of microorganisms was much larger than the effect of photodegradation during the roxarsone degradation. In addition, the microbial community diversity after incubation would become more abundant than original chicken manure.

REFERENCES

Anderson, C. 1983. Arsenicals as feed additives for poultry and swine. In: W. Lederer, R. Fensterheim (Eds.) *Arsenic Industrial, Biomedical, Environmental Perspectives.* Van Nostrand Reinhold Co., New York

Brown, B.L., Slaughter, A.D., Schreiber, M.E. 2005. Controls on roxarsone transport in agricultural watersheds. *Appl. Geochem.* 20(1): 123–133.

Fisher, D.J., Yonkos, L.T. & Staver, K.W. 2015. Environmental concerns of roxarsone in broiler poultry feed and litter in Maryland, USA. *Environ. Sci. Technol.* 49(4): 1999–2012.

Garbarino, J.R., Bednar, A.J., Rutherford D.W., Beyer R.S., Wershaw R.L. 2003. Environmental fate of roxarsone in poultry litter. I. Degradation of roxarsone during composting. *Environ. Sci. Technol.* 37 (8): 1509–1514.

Nachman K.E., Baron P.A., Raber G., Francesconi K.A., Navas-Acien A., Love D.C. 2013. Roxarsone, inorganic arsenic, and other arsenic species in chicken: a US-based market basket sample. *Environ. Health Persp.* 121(7): 818–824.

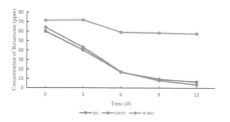

Figure 2. The roxarsone concentration change in different treatments. BG represents chicken manure and roxarsone in dark. GROX represents chicken manure and roxarsone in light. WJBG represent chicken manure, HgCl$_2$ and roxarsone in dark.

*1.3   Geochemical modelling of arsenic and
water-solid phase interactions*

Arsenic Research and Global Sustainability – Bhattacharya, Vahter, Jarsjö, Kumpiene,
Ahmad, Sparrenbom, Jacks, Donselaar, Bundschuh & Naidu (Eds)
© *2016 Taylor & Francis Group, London, ISBN 978-1-138-02941-5*

Geochemical modeling and thermodynamic properties of arsenic species

D.K. Nordstrom[1], X. Zhu[2], R.B. McCleskey[1], L.C. Königsberger[3] & E. Königsberger[3]
[1]*United States Geological Survey, Boulder, CO, USA*
[2]*School of Earth Sciences and Engineering, Nanjing University, Nanjing, P.R. China*
[3]*Chemical and Metallurgical Engineering and Chemistry, Murdoch University, Murdoch, WA, Australia*

ABSTRACT: An internally consistent thermodynamic network for arsenic minerals and aqueous species is being developed based on a high quality starting point: arsenolite and arsenolite solubility. Current research is refining the properties of scorodite solubility as a function of temperature along with improved aqueous species stability constants and activity coefficients for ions and ion pairs based on potentiometric measurements of redox, pH, and conductivity for 5–90°C.

1 INTRODUCTION

Geochemical modeling of water-rock interactions, speciation, and toxicity studies require knowledge of thermodynamic and kinetic properties. Thermodynamic data for many common minerals and aqueous species have reached a tolerable level of inconsistency for practical purposes but data for trace elements is often poorly known or unknown. Arsenic data are well known for a few species and poorly known or unknown for most species. In this paper, we summarize recent evaluations of arsenic thermodynamic data.

2 EVALUATION METHODS

Complete and comprehensive internal consistency is an unachievable goal but something to strive towards. It includes consistency with the basic thermodynamic equations, a consistent chemical model, a consistent mathematical model, and consistency with the fundamental physical constants and atomic weights. One of the criteria for internal consistency is an appropriate choice of starting point in the network. If a highly reliable property is found, that property would have the highest weight in the fitting procedure and could be considered an anchor or cornerstone for the network.

3 RESULTS AND DISCUSSION

3.1 *2014 evaluation: the anchor*

Nordstrom *et al.* (2014) compiled, reviewed, and evaluated thermodynamic data for major aqueous species using data for arsenolite and its solubility as an anchor. This review paper was used as a starting point for interpreting our current experimental data. Data collected from the current experiments could be used to improve the reviewed literature values. An example is shown in Figure 1 with val-

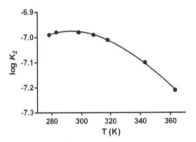

Figure 1.   Values of log K_2 for arsenic acid dissociation as a function of temperature from pH measurements.

Table 1.   Preliminary equilibrium constants for arsenic acids at 25°C (Nordstrom *et al.*, 2014 and this study).

	$H_2AsO_4^- \rightleftharpoons$ $HAsO_4^{2-} + H^+$		$HAsO_4^{2-} \rightleftharpoons$ $AsO_4^{3-} + H^+$	
$\Delta_r C_p$ (J/K/mol)	−207.8	(± 5.8)	−991.9	(± 94.4)
$\Delta_r H°$ (kJ mol⁻¹)	−1.08	(± 0.1)	34.1	(± 1.7)
$\Delta_r S°$ (J K⁻¹ mol⁻¹)	−137.4	(± 0.5)	−111.1	(± 0.01)
log K	−6.98	(± 0.01)	−11.59	(± 0.01)

ues of the logarithm of the 2nd hydrolysis constant for arsenic acid derived from careful pH measurements, an evaluation of activity coefficients using the Hückel equation, and the mean salt method (based on the MacInnes assumption).

The preliminary standard-state properties for arsenic acid dissociation are shown in Table 1.

These values were used in conjunction with solubility data for scorodite and the calcium arsenate minerals to improve the internal consistency of data for these minerals.

4 CURRENT RESEARCH

Improvements in the thermodynamics of the Fe-arsenate system is being obtained by using four

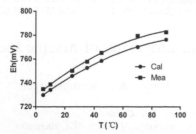

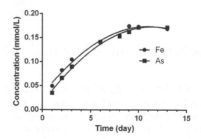

Figure 2. Redox potential measurements of the Fe(II/III) couple in HCl at a pH of 1.62, Fe(II) = Fe(III) = 1 mM compared to redox calculated using evaluated Fe data from Lemire *et al.* (2013) for temperatures of 5–90°C.

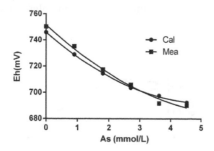

Figure 3. Redox potential measurements of the Fe(II/III) couple in HCl as shown in Fig. 1 but with increasing concentrations of aqueous arsenate at 25°C. Data for ion pairs in this system were used from the evaluation of Nordstrom *et al.* (2014).

different methods: (1) conductivity measurements in the Na_2HAsO_4-H_2O system, (2) pH measurements in the same system, (3) redox potential measurements in the acid Fe(II/III)-arsenate system, and (4) solubility measurements of scorodite, $FeAsO_4 \cdot 2H_2O$.

4.1 Redox potentials

We began with the replication of data from Lemire *et al.* (2013) as shown in Figure 2. The objective was to refine thermodynamic properties of aqueous Fe-arsenate species through redox potential measurements of Fe(II/III).

When an H_3AsO_4 solution is added to this solution, the redox potential changes as shown in Figure 3.

The close agreement shown in Figure 3 confirms the use of the selected stability constant data (Nordstrom *et al.*, 2014).

4.2 Conductivity and pH measurements

Two stock solutions were made, one at pH near 9 where the $HAsO_4^{2-}$ ion is dominant and the other at pH near 4 where $H_2AsO_4^-$ is dominant. A concentrated As_2O_5 solution was added to the Na_2HAsO_4 solution to lower the pH. Measurements of pH and

Figure 4. Scorodite dissolution kinetics at 50°C showing both Fe and As determinations.

conductivity have been reduced to obtain ionic molal conductivities for the $HAsO_4^{2-}$ and $H_2AsO_4^-$ ions.

4.3 Scorodite solubility

The first measurements of scorodite solubility at 50°C in two different laboratories using the same synthesis technique but using different acids for the solubility (HCl vs. H_2SO_4) show dissimilar results. Equilibrium in HCl solution can be reached within about two weeks as shown in Fig. 4. Slight differences in solubility are noted when different starting preparation is used. Studies on pharmacolite and haidingerite solubility are in progress.

5 CONCLUSIONS

Advancements in the thermodynamic properties of arsenate minerals and aqueous species are progressing with multiple measurement techniques, including redox potentials, conductivity, and pH. The temperature-dependence of 5–90°C along with confirmatory checks with PHREEQC computations is providing a much firmer foundation on which to derive thermodynamic properties.

ACKNOWLEDGEMENTS

XZ acknowledges financial support from the China Scholarship Council, DKN and RBM acknowledge the support of the National Research Program of the US Geological Survey and a Sir Walter Murdoch Adjunct Professor award, and LCK and EK acknowledge the support of Murdoch University and the Australian Research Council (Linkage project LP130100991).

REFERENCES

Lemire, R.J., Berner, U., Musikas, C., Palmer, D.A., Taylor, P., Tochiyama, O. 2013. *Chemical Thermodynamics of Iron. Part I*, OECD/NEA.
Nordstrom, D.K., Majzlan, J., Königsberger, E. 2014. Thermodynamic properties for arsenic minerals and aqueous species. In: *Arsenic: Environmental Geochemistry, Mineralogy and Microbiology. Rev. Mineral. Geochem.* 79: 217–255.

Arsenic Research and Global Sustainability – Bhattacharya, Vahter, Jarsjö, Kumpiene,
Ahmad, Sparrenbom, Jacks, Donselaar, Bundschuh & Naidu (Eds)
© 2016 Taylor & Francis Group, London, ISBN 978-1-138-02941-5

Solubility product constants for the calcium ferric arsenate mineral, Yukonite: Fitting possible formulas with PhreePlot

J.J. Mahoney[1], M.T. Bohan[2] & G.P. Demopoulos[2]
[1] Mahoney Geochemical Consulting LLC, Lakewood, Colorado, USA
[2] Department of Mining and Materials Engineering, McGill University, Montreal, Canada

ABSTRACT: Solubility product constants, as log K_{sp} values, were estimated for several possible formulas for the calcium ferric arsenate mineral yukonite. These values used a data set of solution compositions composited from work performed at McGill University over the past 10 years. The fitting routines were performed using the program PhreePlot. Three formulas have been selected and the resultant solubility product constants were estimated based upon a review of a wide range of possible formulas. A fourth formula based upon the simultaneous fitting of the formula and log K_{sp} was also estimated. Analytical uncertainties particularly with the iron concentrations limit the application of this data set. The PhreePlot based results are comparable to other log K_{sp} values estimated using an alternative regression methodology.

1 INTRODUCTION

It has been over 100 years since Tyrell & Graham (1913) identified the ferric calcium arsenate mineral yukonite in mining wastes in northern Canada. In the intervening century, researchers have postulated that the phase may control arsenic concentrations at near neutral and high pH values. Unfortunately, these earlier researchers did not define a generally accepted formula for this phase. The lack of a formula has prevented the definition of thermodynamic properties such as a free energy of formation or a solubility product constant. Previously conducted laboratory work (Bluteau et al., 2009) as well as recently completed experiments provided a robust data set to estimate the log K_{sp} values for likely formulas.

2 EXPERIMENTAL DATA AND SETUP

A composite data set was developed based upon experimental efforts conducted at McGill University over the past 10 years. PhreePlot (Kinniburgh & Cooper, 2011) was selected because it can simultaneously fit multiple parameters (Mahoney, 2015). Yukonite was synthesized by neutralizing solutions of $Fe_2(SO_4)_3 \cdot xH_2O$ and $As_2O_5 \cdot xH_2O$ with NaOH in the presence of $CaSO_4 \cdot 2H_2O$, followed by accelerated ageing for 24 hours at 95°C. Solubility measurements of synthetic yukonite were finalized upon consecutive stable readings of pH controlled equilibration (132–458 days) in either gypsum-free or gypsum-saturated solutions. Figure 1 summarizes the 23 solutions used in this evaluation. Most of the reported iron values were below the reporting limit of 0.05 or 0.001 mg/L. Therefore, ferrihydrite was set as a fixed phase in all samples. To cover a broad range of conditions the

data set was compiled from experiments that included gypsum and some that did not. Samples with gypsum had calcium concentrations greater than 400 mg/L; sulfate ranged from 1400 to 3100 mg/L.

PhreePlot was set up to estimate of the log K_{sp} values. An observation file was set to represent the desired saturation index values of zero for all samples. PhreePlot adjusts the user defined fitting parameters and prepares a PHREEQC based model that compares the observations with the model calculated values. The metric is the Residual Sum of Squares (RSS) and PhreePlot iterates until the fitted parameters minimize the RSS value. The approach and the database that defined the Ca and Fe arsenate complexes are described in Mahoney (2015).

3 RESULTS AND DISCUSSION

One of advantages of this approach is that there is a straightforward means to determine what is not present in the experimental data sets. Different calcium arsenate minerals including tri-calcium arsenate,

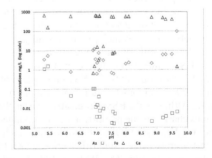

Figure 1. Concentrations of arsenic, iron and calcium for samples in equilibrium with yukonite.

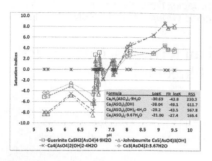

Figure 2. PhreePlot fits comparing calcium arsenate phases to the experimental data. The lined x's represent the zero saturation index values.

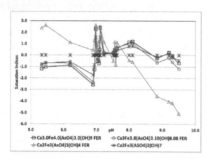

Figure 3. Comparison of saturation indices for yukonite formulas. The lined x's represent the zero saturation index values.

tetra-calcium arsenate, guerinite and johnbaumite were defined and PhreePlot was set up to adjust their log K_{sp} values. Figure 2 shows the results of these fits, which are characterized as poor. Several features demonstrate the poor fits. These include large RSS values, a trend in the fitted functions when compared to the zero saturation index values, and a large range of saturation indices to define the y axis.

These optimized fits required significant changes to the defined solubility constants for these phases. Compared to the yukonite formulas shown in Figure 3 and Table 1, the RSS values for the calcium arsenate phases are too large for an acceptable fit.

Several approaches were used, including two different fitting approaches available in PhreePlot. Fixing the activity of the Fe^{+3} component with ferrihydrite (or goethite) rendered the RSS values insensitive to the iron stoichiometry in the formula. Consequently, the RSS values reflect the ratio of Ca to AsO_4 formula units. The smallest RSS values appear where the Ca:AsO_4 ratio is close to 1:1. Contour plots of RSS as a function of log K_{sp} and AsO_4 stoichiometry (not shown) support these 1:1 ratios. Although the RSS is insensitive to iron, the log K_{sp} values change as the proportion of iron changes.

The $Ca_2Fe_3(AsO_4)_3(OH)_4$ formula was included as it had been selected using other criteria as a likely formula for yukonite. The RSS at 108.75 and the strong downward trend indicate a poor fit. An alternative fit based upon the simultaneous fitting of a

Table 1. Summary of PhreePlot estimated log K_{sp} values for possible yukonite formulas.

Formula*	RSS	Log K_{sp}
$Ca_2Fe_3(AsO_4)_3(OH)_4$	108.75	−112.6
$Ca_3Fe_4(AsO_4)_3(OH)_9$	27.35	−165.5
$Ca_2Fe_3(AsO_4)_2(OH)_7$	12.11	−122.7
Multiparameter Fit		
$Ca_3Fe_{3.8}(AsO_4)_{3.1}(OH)_{8.1}$	25.88	−157.0

* Waters of hydration not included in formulas

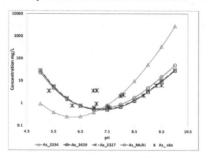

Figure 4. Comparison of modeled and measured As concentration for yukonite formulas. The lined x's represent the measured As concentration in the experiments.

formula and log K_{sp} value is also included in Table 1. The $Ca_3Fe_4(AsO_4)_3(OH)_9$ phase provided the starting conditions. Slight adjustments to the starting formula and improvements in the RSS are noted.

4 CONCLUSIONS

The formulas and log K_{sp} values are still preliminary, but it is expected that this method can be used in an approach that alternates between experimental data and fitting over several cycles to define the data set and the subsequent formulas and log K_{sp} values. With the exception of the $Ca_2Fe_3(AsO_4)_3(OH)_4$ phase, the other formulas provided good fits to the arsenic concentrations in the experiments in models that include ferrihdyrite and gypsum (Fig. 4).

REFERENCES

Bluteau, M.-C., Becze, L., Demopoulos. G.P. 2009. The dissolution of scorodite in gypsum-saturated waters: Evidence of Ca–Fe–AsO4 mineral formation and its impact on arsenic retention. *Hydrometallurgy* 97(3–4): 221–227.

Kinniburgh, D.G., Cooper, D.M. 2011. PhreePlot—creating graphical output with PHREEQC. Available at http://www.phreeplot.org/.

Mahoney, J.J. 2015. Using PhreePlot to calibrate mining related geochemical models: a user's perspective. *10th International Conference on Acid Rock Drainage & IMWA Annual Conference*, p. 736–746. Santiago, Chile.

Tyrrell. J.B., Graham, R.P.D. 1913. Yukonite, a new hydrous arsenate of iron and calcium, from Tagish Lake, Yukon Territory, Canada: with a note on the associated symplesite. *Trans. Roy. Soc. Canada* 7, Sect. 14: 13–18.

*Arsenic Research and Global Sustainability – Bhattacharya, Vahter, Jarsjö, Kumpiene,
Ahmad, Sparrenbom, Jacks, Donselaar, Bundschuh & Naidu (Eds)
© 2016 Taylor & Francis Group, London, ISBN 978-1-138-02941-5*

Revised best-fit parameters for arsenate adsorption to ferrihydrite

J.P. Gustafsson & C. Sjöstedt
Department of Soil and Environment, Swedish University of Agricultural Sciences, Uppsala, Sweden

ABSTRACT: Ferrihydrite is an important environmental sorbent for arsenate, Modelling the chemistry of arsenate therefore requires the use of databases with generic arsenate adsorption parameters for ferrihydrite. Here we present an update to the current best-fit parameters of arsenate adsorption for two models, the diffuse layer model and the CD-MUSIC model. We also investigated whether the variability between different studies could be due to the different equilibration times used. However, it was concluded that no such systematic effects can be seen in the data.

1 INTRODUCTION

The use of surface complexation models for simulating arsenic chemistry in the field can offer new insights into how arsenic is mobilized and transported (e.g. Biswas *et al.*, 2014). To do this successfully it is essential that there are databases with critically reviewed complexation constants. Gustafsson & Bhattacharya (2007) presented such a database for arsenate and arsenite adsorption to ferrihydrite and goethite, using the Diffuse Layer Model (DLM) of Dzombak and Morel (1990) and a now outdated version of a CD-MUSIC model for ferrihydrite; the former has since been used and discussed in a number of studies (e.g. Biswas *et al.*, 2014, Swedlund *et al.*, 2014). Swedlund *et al.* (2014) casted doubt on the relevance of such databases as they showed that slow intra-particle diffusion causes equilibrium to be reached only beyond the timescale of most experiments that form the basis of the databases. Moreover, as the equilibration times of the studies considered in the ferrihydrite database range from 2 to 200 h, there is reason to suspect that slow diffusion may cause variability in the complexation constants.

The aim of this work was threefold: to update the DLM database for arsenate adsorption to ferrihydrite, to derive a new database of generic complexation constants for the current ferrihydrite CD-MUSIC model (Tiberg *et al.*, 2013), and thirdly to investigate whether the consideration of slow intraparticle diffusion can lead to a higher confidence of the derived surface complexation constants.

2 METHODS

2.1 *Selection of data sets*

The strategy for selecting arsenic binding data is described in detail by Gustafsson and Bhattacharya (2007). We extended this data set with arsenate complexation data of Sjöstedt *et al.* (in prep.) In all the data set consists of 20 adsorption edges from 8 different studies, with 210 data points.

2.2 *Data optimization*

We used Visual MINTEQ version 3.1 for all calculations (Gustafsson, 2015). This software contains both the DLM and CD-MUSIC models as well as relevant solution acid-base and complexation constants. The data optimization was done in a different way compared to Gustafsson and Bhattacharya (2007). In the current work the PEST software was used (Doherty, 2004). The complexation constants were optimized by minimizing the weighted sum of squared differences between observed and modelled percentages of adsorbed arsenate. During the optimization, equal weight was given to all data points.

2.3 *Accessible surface area*

As diffusion rates are not possible to include in the present version of Visual MINTEQ, we defined the 'Accessible' Surface Area (ASA) as the area available for arsenate complexation at a given point of time and assumed this to reflect the effect of diffusion. The ASA was calculated for every data point in the kinetic experiments of Swedlund *et al.* (2014). By averaging the results from all six data sets of Swedlund *et al.* (2014) we arrived at:

$$ASA = a \times \ln(t) + b \qquad (1)$$

where ASA is given in $m^2\,g^{-1}$, the equilibration time t is in h, the coefficient a was determined to 36.4 for the DLM and 34.7 for the CD-MUSIC models, and b was selected so that the 'standard' surface areas of the models (600 and 650 m^2/g, respectively) were reached after 6 h. ASA:s relevant for the different sources of the data of the database were then used instead of the standard surface areas to investigate whether this affected the optimization results.

3 RESULTS AND DISCUSSION

All optimizations resulted in very good and consistent model fits to almost all adsorption edges analyzed (see example in Fig. 1). The fits for the DLM and for the CD-MUSIC model were similar in quality. No clear effects were observed when the ASA was used instead of a fixed surface area. For the DLM a small improvement was noted (PEST R correlation coefficient = 0.979 compared to 0.976 with fixed areas); for the CD-MUSIC model the reverse effect was found 0.977 and 0.979, respectively.

Alternative optimization parameters, such as log K_d and log dissolved concentrations were also tested. Still, however, there was no consistent effect of ASA. Two reasons for this may be that (i) most edges were determined after 4 and 24 h, which represent a rather small variation in $\ln(t)$, and (ii) there are other variations in e.g. solid-solution separation, ferrihydrite preparation etc. that may have overshadowed the effect of a variable ASA. Hence the data from the 8 different studies do not show evidence for a variability caused by the different equilibration times used. If slow diffusion effects are to be considered when applying the models, the use of equation 1 or similar approaches will need to be tested in further research.

The revised surface complexation constants are shown in Table 1 and Table 2. They have a smaller confidence interval than the 'old' constants of Gustafsson & Bhattacharya (2007), partly because of

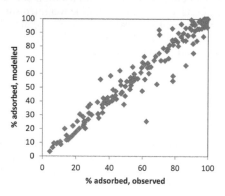

Figure 1. Correspondence between observed and modelled arsenate adsorption for the CD-MUSIC model with a fixed surface area at 650 m²/g (n = 210).

Table 1. Revised best-fit parameters for arsenate adsorption to ferrihydrite with the DLM of Dzombak & Morel (1990).

Species[1]	Δz_o^2	log K	95% CI
FeH$_2$AsO$_4$	0	30.94	30.80–31.08
FeHAsO$_4^-$	−1	25.83	25.62–26.04
FeAsO$_4^{2-}$	−2	19.72	19.55–19.88
FeOHAsO$_3^{3-}$	−3	11.62	11.52–11.72

[1]See Gustafsson & Bhattacharya (2007) for the stoichiometry.
[2]The change of charge in the o-plane.

Table 2. Best-fit parameters for arsenate adsorption to ferrihydrite with the CD-MUSIC model of Tiberg et al. (2013).

Species[1]	$\Delta z_o, \Delta z_b, \Delta z_d^2$	log K	95% CI
FeOAsO$_3$H$_2^{½-}$	0.5,−0.5,0	28.9	28.63–29.17
Fe$_2$AsOOH$^-$	0.58,−0.58,0	32.04	31.86–32.21
Fe$_2$AsO$_2^{2-}$	0.47,−1.47,0	27.40	27.34–27.47

[1]See Gustafsson et al. (2009) for the stoichiometry.
[2]The change of charge in the o-, b and d-planes, respectively.

the use of another optimization method, but also because of the extension of the database, and of correction of some previous errors in the database.

4 CONCLUSIONS

The DLM and CD-MUSIC models could successfully reproduce arsenate adsorption to ferrihydrite from 20 adsorption edges of 8 critically reviewed studies. The optimized constants are recommended for future use of the DLM of Dzombak and Morel (1990) and the CD-MUSIC model of Tiberg et al. (2013). Consideration of slow diffusion through variable time-dependent surface areas did not improve the model performance significantly.

REFERENCES

Biswas, A., Gustafsson, J.P., Neidhardt, H., Halder, D., Kundu, A.K., Chatterjee, D., Berner, Z., Bhattacharya, P. 2014. Role of competing ions in the mobilization of arsenic in groundwater in Bengal basin: insight from surface complexation modeling. *Water Res.* 55: 30–39.

Doherty, J. 2010. PEST—Model-independent parameter estimation. User Manual, 5th Edition. Watermark Numerical Computing, Web: http://www.pesthomepage.org.

Dzombak, D.A., Morel, F.M.M. 1990. *Surface Complexation Modeling*. Wiley, New York.

Gustafsson, J.P., Bhattacharya, P. 2007. Geochemical modelling of arsenic adsorption to oxide surfaces. In: P. Bhattacharya, A.B. Mukherjee, J. Bundschuh, R. Zevenhoven, R.H. Loeppert (Eds.) *Arsenic in Soil and Groundwater Environment: Biogeochemical Interactions, Health Effects and Remediation, Trace Metals and other Contaminants in the Environment*, Volume 9, Elsevier, Amsterdam, The Netherlands: 153–200 (doi 10.1016/S0927-5215(06)09006-1).

Gustafsson, J.P., Dässman, E., Bäckström, M. 2009. Towards a consistent geochemical model for prediction of uranium(VI) removal from groundwater by ferrihydrite. *Appl, Geochem.* 24: 454–462.

Swedlund, P.J., Holtkamp, H., Song, Y., Daughney, C.J. 2014. Arsenate-ferrihydrite systems from minutes to months: a macroscopic and IR spectroscopic study of an elusive equilibrium. *Environ. Sci. Technol.* 48: 2759–2765.

Tiberg, C., Sjöstedt, C., Persson, I., Gustafsson, J.P. 2013. Phosphate effects on copper(II) and lead(II) sorption to ferrihydrite. *Geochim. Cosmochim. Acta* 120: 140–157.

Arsenic Research and Global Sustainability – Bhattacharya, Vahter, Jarsjö, Kumpiene,
Ahmad, Sparrenbom, Jacks, Donselaar, Bundschuh & Naidu (Eds)
© *2016 Taylor & Francis Group, London, ISBN 978-1-138-02941-5*

Co-adsorption of arsenate and copper on amorphous Al(OH)₃ and kaolinite

M. Gräfe[1], D.A. Beattie[2] & J.-F. Lee[3]

[1]*Departamento del Manejo Suelos y Aguas, Instituto Nacional de Investigaciones Agropecuarias, Quito, Ecuador*
[2]*Future Industries Institute, University of South Australia, Mawson Lakes, South Australia*
[3]*Beamline 17C1, National Synchrotron Radiation Research Centre, Hsinchu, Taiwan*

ABSTRACT: We investigated the distribution and speciation of arsenate (As(V)) and copper (Cu(II)) in mineral suspensions of either kaolinite or amorphous (am-) Al(OH)₃. Copper and arsenate are pervasive contaminants in auriferous mining areas with highly antagonistic surface and hydrolysis chemistry, which makes their simultaneous stabilization/ remediation highly challenging. Our results show that even at acidic pH (5.2) the two ions will co-react on the mineral surfaces forming ternary A-type complexes on am-Al(OH)₃ as well as hydrated complexes or highly hydrated precipitates that take on the forms of euchroite (am-Al(OH)₃) or clinoclase (kaolinite). The implications from the speciation results are that the solubility of either Cu(II) or AsO₄ is strongly codependent on the speciation of the co-reacting counterion.

1 INTRODUCTION

Copper (Cu(II)) and arsenate (AsO₄) are pervasive contaminants near copper and gold mines. Strategies to mitigate the risk from metals and metalloids are complicated by the antagonistic surface and hydrolysis chemistry of Cu^{2+} versus AsO_4. Remediation strategies therefore must overcome this antagonism and account for the fundamentally different surface speciation of the metal and the oxyanion when reacting simultaneously at the same soil mineral surface (Jiang *et al.*, 2013). The formation of (surface) copper-arsenate precipitates is a promising pathway for remediation of metal and oxyanion contaminated sites and has been demonstrated to operate at the goethite- and jarosite-water interfaces (Gräfe *et al.*, 2008).The aim of this work was to introduce novel information on sequestration mechanisms between AsO₄ and Cu(II) ions on two common soil minerals, kaolinite and amorphous (am-)Al(OH)₃. The central objective was to determine the molecular scale mechanisms of Cu(II) and AsO₄ interaction on these surfaces.

2 EXPERIMENTAL

2.1 *Sample preparation*

Kaolinite and am-Al(OH)₃ suspensions were prepared for sorption experiments by equilibrating ~ 10 m²/L (200 mg/L am-Al(OH)₃ and 690 mg/L kaolinite) suspensions overnight at a pH 5.2 or 5.5 in 0.01 M NaNO₃. The suspensions were subsequently reacted with either 200 or 250 μM NaH₂AsO₄ and Cu(NO₃)₂ to infer the effects of degree of solution (under-)saturation at pH 5.2 Single ion control experiments were conducted at pH 5.5. All reactions were undersaturated with respect to known Cu(II) and copper-arsenate mineral phases. All reactions were allowed to equilibrate over a 12 day period at 150 rpm with daily adjustment of pH. Solution concentrations of As and Cu were determined using

atomic absorbance spectrometry; the amount of Cu and As in the solid phase was determined by difference to the known initial concentrations.

2.2 *EXAFS spectroscopy*

EXAFS data were collected on the hard XAS, 20 cm period length wiggler beamline, 17C1, of the National Synchrotron Radiation Research Centre (Hsinchu, Taiwan) in fluorescence mode. Spectra were collected in triplicate at the As(V) (11.874 keV) and Cu (8.979 keV) K-edges. Several copper-arsenate mineral standards were diluted in BN and recorded simultaneously in transmission and fluorescence mode. All data reduction and interpretation was performed with WinXAS 3.1 (Ressler, 1998) using standard procedures (Bunker, 2003).

3 RESULTS AND DISCUSSION

3.1 *Solid-liquid phase partitioning*

The solid phase partitioning of AsO₄ and Cu(II) was significantly greater in the am-Al(OH)₃ than in the kaolinite suspensions (Table 1). The solubility of AsO₄ and Cu(II) generally increased in each other's presence in am-Al(OH)₃ suspensions, whereas in the kaolinite suspensions, AsO₄ solubility decreased for a near steady Cu(II) solubility. The Cu:As molar surface ratios on am-Al(OH)₃ were approximately constant (0.24 vs. 0.26), while in kaolinite suspensions, the ratio decreased from 2.0 to 2.5 as the initial AsO₄ and Cu(II) concentrations were lowered from 250 to 200 μM.

3.2 *Summary of the pertinent EXAFS results*

A preliminary analysis of the EXAFS data revealed that the bonding environments of AsO₄ and Cu(II) were very heterogeneous. A linear combination fit (LCF) approach was chosen to identify possible bonding environments using spectra of single ion

Table 1. Percent AsO_4 and Cu(II) removal from solution and surface loadings (µmol/g).

Sample	pH	% removal from solution		Surface loadings µmol g^{-1}	
		As	Cu	As	Cu
AsAlHOx_250	5.5	87		1085.8	
CuAlHOx_250	5.5		33		417.9
CuAsAlHOx_250	5.2	83	20	1035.9	252.1
CuAsAlHOx_200	5.2	78	20	782.5	200.7
AsKaol_250	5.5	11		38.2	
CuKaol_250	5.5		14		52.2
CuAsKaol_250	5.2	27	14	98.9	49.5
CuAsKaol_200	5.2	34	14	98.1	40.5

Table 2. Linear combination fit analysis of As and Cu K-edge EXAFS data. The relevant K-edge is mentioned first and underlined. Arsenic results are shaded in grey.

Sample	C1%	C2%	Σ	Res
	am-Al(OH)$_3$			
CuAsAlHOx_200 Cu:As = 0.26	CuAlHOx, 72	Euchroite, 36	108	26.4
AsCuAlHOx_200 Cu:As = 0.26	AsAlHOx, 90	Euchroite, 11	101	7.5
CuAsAlHOx_250 Cu:As = 0.24	CuAlHOx, 73	Euchroite, 31	104	24.9
AsCuAlHOx_250 Cu:As = 0.26	AsAlHOx, 90	Euchroite, 9	99	7.6
	Kaolinite			
CuKaol_250	Cu(II) (aq), 63	CuAlHOx, 35	98	19.7
AsKaol_250	AsAlHOx, 82	AsO$_4$ (aq), 16	98	10.7
CuAsKaol_200 Cu:As = 0.40	Cu(II) (aq), 92	Clinoclase, 11	103	16.0
AsCuKaol_200 Cu:As = 0.40	AsKaol 84	AsO$_4$ (aq), 17	101	8.6
CuAsKaol_250 Cu:As = 0.50	Cu(II) (aq), 87	Clinoclase, 16	103	15.8
AsCuKaol_250 Cu:As = 050	AsKaol 102		102	11.5

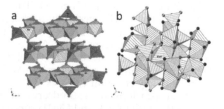

Figure 1. a) Proposed ternary A-type complex of Cu(II) and AsO_4 ions on an amorphous $Al(OH)_3$ surface. Blue octahedra = Al; striped octahedra = Cu(II); green tetrahedra = AsO_4; red spheres = OH/O ligands, b) proposed highly hydrated clinoclase complex forming at or near the surface of kaolinite. The fourth unique Cu(II) site was chosen to display an 8-Å cluster. Green striped tetrahedra = AsO_4, blue striped octahedra and square pyramids = Cu(II); red spheres = oxides (O^{2-}); solid blue spheres = H_2O, and blue-white spheres = OH^-. (Both figures are generated with CrystalMaker v.9.2.3).

sorption experiments and standard copper-arsenate minerals (Table 2).

Between 10 and 30 percent euchroite ($Cu_2[AsO_4]$ $(OH)\cdot 3H_2O$) formed in the am-$Al(OH)_3$ suspensions, whereas in the kaolinite suspensions, only the Cu K-edge suggested the formation of some clinoclase ($Cu_3 AsO_4(OH)_3 \leq 11$–16%). Copper K-edge XANES reflected the highly hydrated state of Cu(II) in the kaolinite system. Least square shell fits suggested that Cu(II) filled vacant octahedral sites in am-$Al(OH)_3$ similar to the complexes Cu(II) formed in the absence of AsO_4. Arsenate adsorbed to the am-$Al(OH)_3$ surface as an inner-sphere complex with Cu(II) and Al next nearest neighbours (Fig. 1a,b).

4 CONCLUSIONS

Copper and AsO_4 formed ternary A-type complexes and a euchroite-like precipitate on am-$Al(OH)_3$ (Fig. 1a) On kaolinite, highly hydrated clinoclase formed as a highly hydrated complex (Fig. 1b). In both mineral systems AsO_4 and Cu(II) interact on or at the surface. The formation of surface-induced precipitates under the given reaction conditions could not be ruled out. The highly hydrated state of the initial complexes suggests that copper-arsenate precipitates may mature at kaolinite and/ or am-$Al(OH)_3$ surfaces over extended periods of time. The hydrolysis of Cu(II) certainly appears to be a key factor.

ACKNOWLEDGEMENTS

This work was supported in part under the Australian Research Council's Discovery funding scheme project number DP0558332 (Faculty of Agriculture, Food and Natural Resources, The University of Sydney) and a University of South Australia Collaborative Research Grant. The work was further supported by the Australian Synchrotron Research Program, which was funded by the Commonwealth of Australia under the Major National Research Facilities Program. This work was conducted by MG as part of post-doctoral fellowships at the University of South Australia and The University of Sydney.

REFERENCES

Bunker, G. 2003. *Overview of the standard XAFS data analysis procedure* [Online]. Bunker, G. Available: http://gbx-afs.iit.edu/training/tutorials.html [Accessed 04/2004].
Gräfe, M., Beattie, D.A., Smith, E., Skinner, W.M., Singh, B. 2008. Copper and arsenate co-sorption at the mineral-water interfaces of goethite and jarosite. *J. Colloid Interf. Sci.* 322: 399–413.
Jiang, W., Jitao, L., Luo, L., Yang, K., Yongfeng, L., Hu, F., Zhang, J., Zhang, S. 2013. Arsenate and cadmium co-adsorption and co-precipitation on goethite. *J. Hazard. Mater.* 262: 55–63.
Ressler, T. 1998. Winxas: A program for x-ray absorption spectroscopy data analysis under MS-windows. *J. Synchr. Rad.* 5: 118–122.

Arsenic Research and Global Sustainability – Bhattacharya, Vahter, Jarsjö, Kumpiene,
Ahmad, Sparrenbom, Jacks, Donselaar, Bundschuh & Naidu (Eds)
© 2016 Taylor & Francis Group, London, ISBN 978-1-138-02941-5

Contrasting arsenic mobility in topsoil and subsoil: Influence of Fe- and Mn- oxyhydroxide minerals

L. Charlet[1], E. Markelova[2,3] & R.M. Couture[3,4]
[1]*Institut des Sciences de La Terre (ISTerre), Universite Grenoble Alpes and CNRS, France*
[2]*Earth and Planetary Science Department (LGIT-OSUG), University of Grenoble-I, Grenoble, France*
[3]*Ecohydrology Research Group, Department of Earth and Environmental Sciences, University of Waterloo,*
Waterloo, Canada
[4]*Norwegian Institute for Water Research-NIVA, Oslo, Norway*

ABSTRACT: Experiments in bioreactors were conducted to investigate As mobility in natural substrates under dynamic redox conditions. In topsoil suspensions the im/mobilization of aqueous As was reversible during oxic/anoxic oscillations indicating high chemical reactivity and bioavailability. In contrast, in subsoil clayey suspensions, As was not susceptible to re-oxidation and remained in the aqueous phase during the experiment. We attribute the differences in As behavior between the two systems to the differences in Fe- and Mn- oxyhydroxide mineral content, which may act as sorbents and as catalysts for As oxidative sequestration.

1 INTRODUCTION

Since the Silurian period (444 Ma ago), mass extinctions coincide with Anoxic Events (AO) and the exposure of organisms to high As, Fe and Mn concentrations (Vandenbroucke *et al.*, 2015). Nowadays, anoxic events are encountered with cyclic periodicity in continental surface environments (floods, paddy irrigation). Here, we study the impact on arsenic mobility of such redox oscillations within contrasted soil and clay-rich subsoil environments.

2 EXPERIMENTAL

2.1 *Experimental conditions*

Two natural substrates were chosen to represent topsoil and subsoil environments: a Mollic Fluvisol 0–10 cm horizon from the Saône River floodplain (France), suspended in river water, and a 22.5 m deep argillaceous Gault formation, from northeastern France, suspended in synthetic pore water.

Batch experiments in bioreactors were performed under oscillating redox conditions, the suspensions being bubbled with gas mixtures either depleted in oxygen (anoxic) or including 0.2 atm pO_2 (oxic) (Couture, 2015). CO_2 concentrations were adjusted to represent near-surface and subsurface conditions (0.03% and 1% CO_2, respectively). Redox oscillations were induced by successive oxic/anoxic periods of 7 days each for 7 weeks. The pH and E_H electrodes were installed into the reactor vessel and a water jacket ensured temperature control (25°C).

Substrate suspensions were spiked with dissolved $Na_2HAsO_4 \cdot 7H_2O$ (≥ 98%, SIGMA) and 20 mM organic carbon as ethanol (≥ 99.8%, SIGMA-ALDRICH). Arsenic was added to achieve initial aqueous concentration of 50 µmol/L in the topsoil or 500 µmol/L in the subsoil suspensions.

2.2 *Analytical methods*

Total aqueous concentrations of As, Mn, and Fe were measured by Inductively Coupled Plasma Optical Emission Spectrometry (ICP-OES). Speciation of As(V) and As(III) was analyzed by High Performance Liquid Chromatography (HPLC) coupled to ICP-OES using an anion exchange column (Hamilton PRP-X100).

3 RESULTS AND DISCUSSION

3.1 *Instantaneous uptake of As*

As(V) sequestration was observed during the first minutes of the topsoil experiment removing >95% of initial contaminant concentration (Fig. 1a). Such rapid As uptake is common due to its strong association with poorly crystalline oxyhydroxide minerals. In contrast, less than 10% As(V) sequestration occurred in the subsoil suspension (Fig. 1c). The relatively weak arsenic uptake by clayey sample can be interpreted in terms of electrostatic repulsion between negatively charged $HAsO_4^{2-}$ and mainly deprotonated basal clay surface at neutral pH (7.2).

The interaction between Point of Zero Charge (PZC) of clay minerals, which is mainly below pH 7, and pK_a of As(V) ($pK_1 = 2.20$, $pK_2 = 6.97$, $pK_3 = 11.53$) determines the adsorption envelope. That is, non-specific arsenate sorption should be favored on positively charged clay edges. However, competitive sorption with other ligands present in the pore water (e.g., SO_4^{2-} and HCO_3^-) may exert a strong influence on this adsorption envelope.

3.2 *Mobility of As during redox oscillations*

As it was shown by Couture et al. (2015) release of As(III) during anoxic conditions was coupled

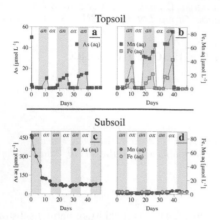

Figure 1. a) and b) Reversible im/mobilization of As coupled to Fe and Mn cycling during oxic/anoxic oscillations in the topsoil suspension; c) and d) Stable As mobility in the absence of Fe and Mn oxyhydroxide minerals in saturated subsoil environment.

Table 1. Metal content of reducible oxide minerals measured after CBD (citrate-bicarbonate-dithionite) extraction.

Formation	Fe (mg/g)	Mn (mg/g)	Al (mg/g)
Topsoil	9.04 ± 0.4	0.42 ± 0.0	0.93 ± 0.0
Subsoil	1.60 ± 0.1	0.04 ± 0.0	0.24 ± 0.0

to the reductive dissolution of Fe(III) and Mn(IV) oxyhydroxides (Fig. 1b) in the topsoil matrix (1.8% by weight) (Table 1). The subsequent oxic cycle resulted in sequestration of As(V) at a rate similar to initial As removal. The im/mobilization of As was fast and repetitive in the topsoil suspension under successive oscillating redox conditions.

In the subsoil suspension the initial decrease of aqueous As concentration was coincident with the reduction of As(V) to As(III) under anaerobic conditions. This reduction was inhibited during the oxic period from day 8 to day 14 (Fig. 1c) and completed during the second reducing period. Resultant As(III)(aq) was not subsequently reoxidised during oxic periods. We suggest that the lack of Fe- and Mn- oxyhydroxide minerals in the subsoil (Table 1) prevented As oxidation and led to high As mobility.

3.3 Mechanism of As reduction and oxidation

Under our experimental conditions (pH~7, T = 25°C) As undergoes reduction via microbial respiration coupled to the oxidation of organic carbon. This process was inhibited in both suspensions during oxic conditions, as aerobic respiration is more energetically efficient than As reduction. Therefore, microbial As reduction (Eqn. 1) was kinetically controlled. An organic carbon availability controlled kinetic rate, which was also inhibited in the presence of terminal electron acceptors of higher redox potential (e.g., O_2).

$$6HAsO_4^{2-} + C_2H_5OH + 10H^+ \rightarrow H_3AsO_3$$
$$+ 2HCO_3^- + H_2O \qquad (1)$$

Chemical oxidation of As by atmospheric oxygen is kinetically hindered (Leuz et al., 2006) as was observed here in the subsoil experiment. In contrast, oxidation of Fe(II) by O_2 at neutral pH is thermodynamically favored and abiotically controlled (Nordstrom et al., 2015). It has been shown by Hug & Leupin (2003) that As oxidation is catalyzed by reactive oxygen species (Eqn. 2), which might be produced during Fe(II) oxidation. This interaction may explain our observations (Fig. 1a) that As oxidation (Eqn. 3) was synchronous to oxidative precipitation of Fe and Mn following oxic and anoxic oscillations.

$$Fe^{2+}_{(aq)} + O_2 \rightarrow Fe(III) + ROS \qquad (2)$$
$$H_3AsO_3 + ROS \rightarrow HAsO_4^{2-} \qquad (3)$$

where, ROS are reactive oxygen species, such as, •OH, O_2•−, or H_2O_2.

4 CONCLUSIONS

We observed a contrast in As behavior in topsoil and subsoil suspensions, which we ascribe to differing content of Fe- and Mn- oxyhydroxide minerals. In the topsoil suspension the im/mobilization of aqueous As was reversible during oxic/anoxic oscillations. Whereas in clay dominated subsoil suspensions, As was not susceptible to re-oxidation and remained in the aqueous phase during the experiment.

ACKNOWLEDGEMENTS

We acknowledge funding from the French National Radioactive Waste Management Agency (ANDRA).

REFERENCES

Couture, R.M., Charlet, L., Markelova, E., Madé, B.T., Parsons, C.T. 2015. On–off mobilization of contaminants in soils during redox oscillations. Environ. Sci. Technol. 49: 3015–3023.

Nordstrom, D.K., Blowes, D.W., Ptacek, C.J. 2015. Hydrogeochemistry and microbiology of mine drainage: An update. Appl. Geochem. 57: 3–16.

Hug, S.J., Leupin, O. 2003. Iron-catalyzed oxidation of arsenic(III) by oxygen and by hydrogen peroxide: pH-dependent formation of oxidants in the Fenton Reaction. Environ. Sci. Technol. 37(12): 2734–2742.

Leuz, A.K., Hug, S.J., Wehrli, B., Johnson, C.A.. 2006. Iron-mediated oxidation of antimony(III) by oxygen and hydrogen peroxide compared to arsenic(III) oxidation. Environ. Sci. Technol. 40: 2565–2571.

Vandenbroucke, T.R., Emsbo, P., Munnecke, A., Nuns, N., Duponchel, L., Lepot, K., Quijada, M., Paris, F., Servais, T., Kiessling, W. 2015. Metal induced malformations in early Palaeozoic plankton are harbinger of mass extinction. Nature Comm. 6: 7966.

Arsenic Research and Global Sustainability – Bhattacharya, Vahter, Jarsjö, Kumpiene,
Ahmad, Sparrenbom, Jacks, Donselaar, Bundschuh & Naidu (Eds)
© 2016 Taylor & Francis Group, London, ISBN 978-1-138-02941-5

Phosphate effects on arsenate binding to soil hydroxides

C. Tiberg, C. Sjöstedt, A.K. Eriksson & J.P. Gustafsson
Department of Soil and Environment, Swedish University of Agricultural Sciences, Uppsala, Sweden

ABSTRACT: Oxyanions may compete with arsenate for sorption sites. This study investigates how phosphate affects the sorption of arsenate on poorly crystalline iron and aluminum hydroxides. A combination of batch experiments, XAS measurements and surface complexation modeling was used. A preliminary interpretation of the results indicated that arsenate was more strongly bound to poorly crystalline aluminum hydroxide than to poorly crystalline iron hydroxide. Phosphate, however, had an even stronger preference for aluminum hydroxide and consequently competed more strongly with arsenate for sorption sites on aluminum than iron hydroxide (preliminary results).

1 INTRODUCTION

Arsenate is strongly bound as inner-sphere surface complexes to both iron and aluminum hydroxides (Arai *et al.*, 2001, Manceau, 1995, Sherman & Randall, 2003). Other oxyanions, such as phosphate, compete with arsenate for sorption sites (Manning & Goldberg, 1996). This competition can greatly affect the sorption-desorption of arsenate in contaminated soils (Tiberg *et al.*, 2016). This study investigates the competition between arsenate and phosphate on combinations of poorly crystalline iron and aluminum hydroxides. The aims were to identify possible differences in the competitive effect on poorly crystalline iron and aluminum hydroxides, identify binding mechanisms and model the competition with a surface complexation model.

2 METHODS/EXPERIMENTAL

2.1 Batch experiments

Arsenate and phosphate were added to batches with different proportions of Ferrihydrite (Fh) and poorly crystalline aluminum hydroxide (Alhox) (Table 1).

The pH was adjusted to about 6.5 and the batches were equilibrated for 24 h before measuring pH and concentration of arsenate and phosphate in the supernatant.

2.2 XAS spectroscopy

XAS measurements were made of the Fh and Alhox from the batch experiments. EXAFS spectra at the arsenic K edge at 11,867 eV were collected at the

Table 1. Composition of batch-series.

Sample	Alhox mM	Fh mM	AsO$_4$ µM	PO$_4$[1] µM
Series 1	1	0	58	0, 50, 100, 200
Series 2	0.75	0.25	58	0, 50, 100, 200
Series 3	0.50	0.50	58	0, 50, 100, 200
Series 4	0.25	0.75	58	0, 50, 100, 200
Series 5	0	1	58	0, 50, 100, 200
Series 6	[2]	[2]	0	100

[1]The PO$_4$ concentration is varied.
[2]One sample with 1 mM Alhox, one with 1 mM Fh and one with 0.50 mM Alhox + 0.50 mM Fh.

wiggler beam line I811, MAX-Lab, Lund, Sweden. XANES spectra at the phosphorus K edge at 2,145 eV were collected at beamline 8, Synchrotron Light Research Institute, Nakhon Ratchasima, Thailand. All XAS spectra were treated in the Athena software (version 0.9.20) (Ravel & Newville, 2005) and final data treatment of the arsenic EXAFS spectra was made in the Artemis software (version 0.0.012) (Ravel & Newville, 2005).

2.3 Geochemical modeling

A surface complexation model with one ferrihydrite surface and one poorly crystalline aluminum hydroxide surface was set up in Visual MINTEQ 3.1 (Gustafsson 2013) with Three Plane CD-MUSIC models. The CD-MUSIC model for ferrihydrite used the surface charging parameters of Tiberg *et al.*, (2013) and the surface charging parameters for poorly crystalline aluminum hydroxide were determined within the study.

3 RESULTS AND DISCUSSION

3.1 *Sorption of arsenate and phosphate*

Preliminary results of arsenate and phosphate concentrations in the supernatant (not shown) suggest that arsenate adsorbs to a greater extent to the Alhox than the Fh but also that phosphate competes more strongly for sorption sites on Alhox.

3.2 *XAS measurements*

According to a preliminary interpretation of the arsenic EXAFS spectra, arsenate seems to prefer Alhox to some extent although it adsorbs strongly also to Fh. This can be illustrated by the Fourier Transforms of some of the EXAFS measurements (Fig. 1). The samples with both Fh and Alhox are combinations of the endmembers with only Fh or Alhox but do follow the Alhox-spectra more closely.

XANES measurements at the phosphorus K-edge showed that a larger part of the phosphate was bound to Alhox than Fh when the concentrations of the sorbents were similar (Series 3). In samples where both arsenate and phosphate were added, the As:P ratio had an effect on the sorption. Arsenic competed with phosphate more strongly on Fh (preliminary interpretation).

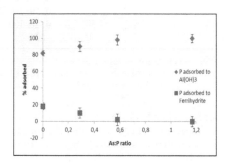

Figure 2.   Preliminary analysis of results from LCF of phosphorus K-edge XANES measurements. Samples with 50% Fh and 50% Alhox.

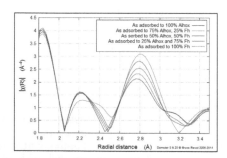

Figure 1.   Stacked Fourier Transforms of k^3-weighted K-edge EXAFS spectra for arsenic.

4 CONCLUSIONS

Preliminary results and interpretations indicate that arsenate adsorbs more strongly to poorly crystalline aluminum hydroxide than ferrihydrite. However, phosphate has an even stronger preference for aluminum hydroxide. As a result, phosphate competes more strongly with arsenate for sorption sites on poorly crystalline aluminum hydroxide than on ferrihydrite.

ACKNOWLEDGEMENTS

Part of this research was carried out at beamline I811, MAX-lab synchrotron radiation source, Lund University, Sweden. Funding for the beamline I811 project was kindly provided by The Swedish Research Council and the "Knut och Alice Wallenbergs Stiftelse". Part of this work was carried out at beamline 8, Synchrotron Light Research Institute, Nakhon Ratchasima, Thailand.

REFERENCES

Arai, Y., Elzinga, E.J., Sparks, D.L. 2001. X-ray Absorption Spectroscopic Investigation of Arsenite and Arsenate Adsorption at the Aluminum Oxide-Water Interface. *J. Colloid Interf. Sci.* 235: 80–88.

Gustafsson, J.P. 2013. Visual MINTEQ 3.1, http://vminteq.lwr.kth.se/.

Manceau, A. 1995. The mechanism of anion adsorption on iron oxides: Evidence for the bonding of arsenate tetrahedra on free Fe(O, OH)6 edges. *Geochim. Cosmochim. Acta* 59: 3647–3653.

Manning, B.A., Goldberg, S. 1996. Modeling competitive adsorption of arsenate with phosphate and molybdate on oxide minerals. *Soil Sci. Soc. Am. J.* 60: 121–131.

Ravel, B., Newville, M. 2005. ATHENA, ARTEMIS, HEPHAESTUS: data analysis for X-ray absorption spectroscopy using IFEFFIT. *J. Synchrotron Radiat.* 12: 537–541.

Sherman, D.M., Randall, S.R. 2003. Surface complexation of arsenic(V) to iron(III) (hydr)oxides: structural mechanism from ab initio molecular geometries and EXAFS spectroscopy. *Geochim. Cosmochim. Acta* 67: 4223–4230.

Tiberg, C., Kumpiene, J., Gustafsson, J.P., Marsz, A., Persson, I., Mench, M.J., Kleja, D.B. 2016. Immobilization of Cu and As in two contaminated soils with zero-valent iron—Long-term performance and mechanisms. *Appl. Geochem.* 67: 144–152.

Tiberg, C., Sjöstedt, C., Persson, I., Gustafsson, J.P., 2013. Phosphate effects on copper(II) and lead(II) sorption to ferrihydrite. *Geochim. Cosmochim. Acta* 120: 140–157.

Arsenic Research and Global Sustainability – Bhattacharya, Vahter, Jarsjö, Kumpiene,
Ahmad, Sparrenbom, Jacks, Donselaar, Bundschuh & Naidu (Eds)
© 2016 Taylor & Francis Group, London, ISBN 978-1-138-02941-5

Effect of organic matter and colloid particle size on arsenic and antimony stability in soils

B. Dousova[1], M. Lhotka[1], E. Duchkova[1], V. Machovic[1], F. Buzek[2] & B. Cejkova[2]
[1] *University of Chemistry and Technology Prague, Prague, Czech Republic*
[2] *Republic Czech Geological Survey, Prague, Czech Republic*

ABSTRACT: The complexation of toxic oxyanions with organic matter and iron particles can affect their stability and binding mechanisms in soils. The adsorption/desorption properties of As^V and Sb^V were tested with soils with a variable content of Natural Organic Matter (NOM) and Fe. Soil samples were investigated from the organic O and mineral B horizons. Arsenic was strongly adsorbed in B horizons on Fe phases, while antimony was predominantly bound in O horizons. The variability of Fe/As ratio in soil solutions and waters was related with surface properties of colloids and explained by model calculation. Most of Fe remained in large colloids with a low surface area, whereas maximum As was included in small particles with a high surface area. The release of As and Sb to distilled water and 0.1M KCl illustrated higher stability of As and Sb in ionic solutions, where available ions prevented the aggregation of OM.

1 INTRODUCTION

The similarities and differences in the geochemical properties of arsenic (As) and antimony (Sb) in the environment result from the same valency and binding characteristics on the one hand, and variable structural forms on the other. Arsenic is predominantly found as pentavalent arsenate, AsO_4^{3-} in the oxidic surface zone (0–3 cm), while the trivalent arsenite, AsO_3^{3-} prevails under slightly reducing conditions (9–12 cm) (Wilson *et al.*, 2010). Antimony as a pentavalent hydrated antimonate, $Sb(OH)_6^-$ represents the major oxidation state over a wide Eh range in the soil profile (Mitsunobu *et al.*, 2008). It has been recently found that the mechanism of As/Sb binding in soils is primarily controlled by the formation of ternary As/Sb complexes with Natural Organic Matter (NOM) and iron (Fe^{III}) (Mikutta & Kretzschmar, 2011, Sharma *et al.*, 2010), with a higher Sb affinity to NOM (Mitsunobu *et al.*, 2008).

The Fe/As and/or Fe/Sb ratios varied significantly depending on the different size of colloids (Buzek *et al.*, 2013). With decreasing particle size, adsorbed amounts and adsorption capacities of As^V and Sb^V have increased.

The aim of this study was to describe the influence of organic matter and colloid size on adsorption/desorption of As^V/Sb^V oxyanions from aqueous solution to soil matrix.

2 EXPERIMENTAL PART

2.1 *Soil and water samples*

Two soil samples from woodland stands in the Czech Republic (similar in soil type and chemical composition and different in the degree of anthropogenic pollution) were taken as four replicate samples of the organic (O, 0–5 to 8 cm) and mineral (B, 10–30 cm) horizons by an auger corer. The beech stand 'Jezeri' (JEZ, locality I) lies in the Ore Mountains close to a brown-coal power plant (exposed to much atmospheric pollution), while the spruce stand 'Na Lizu' (LIZ, locality II) is located in an uncontaminated forest area without direct anthropogenic impact. Samples were dried at 50°C, homogenized by sieving to 2 mm, and used for further analyses.

To evaluate the effect of colloid size on As/Sb content in solution, surface water samples from the Elbe river catchment, Czech Republic, were filtered by micro-sieves and filters for colloid sizes between 80 and 0.45 μm and monitored for As content originating from former acid deposition (Buzek *et al.*, 2013).

2.2 *As^V/Sb^V adsorption*

Model solutions of As^V/Sb^V were prepared from $KH_2AsO_4/NaSbO_4 \cdot 3H_2O$ (of analytical quality) and distilled water, in the concentration range of 2×10^{-5} to 2.5×10^{-4} and pH between 5 and 6. A suspension of the model solution and soil sample (10 g/L) was shaken in a batch procedure at 20°C for 24 hours (Dousova *et al.*, 2012). The filtrate was analysed for residual As/Sb and the saturated soil for the specific surface area (S_{BET}) and by IR spectroscopy.

3 RESULTS AND DISCUSSION

3.1 *Adsorption of As^V/Sb^V oxyanions*

The properties of JEZ and LIZ soils are summarised in Table 1. Arsenic was adsorbed more effectively in the B horizons, and decreased statistically in the O horizons. Conversely, Sb was adsorbed better in the O horizons, with poor adsorption affinity in the B horizons (Fig. 1). The different adsorption behaviour of As and Sb can be understood by the different structural ordering of As^V/Sb^V oxyanions. Arsenates mostly occur as single tetrahedrals (AsO_4^{3-}), forming inner-sphere surface complexes with Fe oxyhydroxides in the mineral horizon. Antimonates such $Sb(OH)_6^-$ with a smaller charge density tend to create multiple hydrated particles with a specific charge distribution and steric characterization, which support their affinity to different active sites, for example, organic matter accumulated in organic horizon.

Langmuir constants K_L, which characterized the binding energy of the adsorption-desorption process, indicate a stronger adsorption affinity of As^V to B horizons than that of Sb^V to O horizons, which correspond to the decreased order of As/Sb binding stability:

$As^V – Fe^{III} \gg Sb^V – OM > As^V – OM > Sb^V – Fe^{III}$* (*adsorption did not run according to the Langmuir isotherm model).

Table 1. Chemical composition, pH and surface properties of tested soils.

Sample	JEZ-O	JEZ-B	LIZ-O	LIZ-B
/ % wt				
Al_2O_3	17.6	20.1	15.9	20.9
SiO_2	65.2	65.3	67.5	64.3
P_2O_5	1.0	0.5	0.6	0.3
Fe_2O_3	6.9	5.4	7.0	6.6
MnO	< 0.1	0.1	< 0.1	< 0.1
TiO_2	1.8	1.1	1.3	1.0
Na_2O/K_2O	0.6/3.4	1.4/4.3	1.4/3.8	1.5/3.3
CaO/MgO	1.5/0.7	0.6/0.9	0.9/0.7	0.6/1.0
TOC / % wt	26.9	4.0	22.7	4.0
CO_2 / % wt	0.1	0.2	0.1	0.3
N / % wt	1.5	0.3	0.9	0.3
TOC : N / % wt	17.9	13.3	25.2	13.3
pH/pH_{KCl}	3.8/3.5	4.9/3.9	4.4/3.0	5.3/4.3
pH_{ZPC}	5.2	8.5	4.0	8.0
$pH_{KCl} : pH_{ZPC}$	0.7	0.5	0.8	0.5
S_{BET} / $m^2 g^{-1}$	4.7	7.6	2.3	12.3

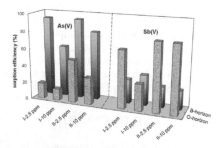

Figure 1. Percentage adsorption efficiency of As^V/Sb^V adsorption on JEZ (I) and LIZ (II) soils at different initial concentrations (2.5 and 10 mg/L).

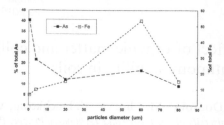

Figure 2. Fe and As content in water filtrates of different particle size fractions—in percent of weight of the element (sampled in the Elbe river catchment).

3.2 Stability of As^V/Sb^V in saturated soils

Desorption experiments were carried out with the As/Sb saturated soils (with mineral horizons for As and organic horizons for Sb, resp.). The saturated samples were leached with distilled water and 0.1 M KCl at a solid-liquid ratio of 1:10. The increased ionic strength of the KCl solution inhibited the release of As/Sb markedly.

3.3 Effect of colloid particle size

According to Figure 2, the distribution of Fe and As between particles is completely different. Most of the Fe is concentrated in a fraction between 80 and 60 μm, while most of the As content is linked to the fraction between 4 and 0.45 μm. The sole association of As with the micro particles indicated a high mobility of As in soil complexes. A similar binding mechanism could be expected with soil antimony.

ACKNOWLEDGEMENTS

This work was part of projects 13-24155S and P210/10/0938 (Grant Agency of Czech Republic).

REFERENCES

Buzek, F. Cejkova, B., Dousova, B., et al. 2013. Mobilization of arsenic from acid deposition after twenty years. Appl. Geochem. 33: 281–293.

Dousova, B. Buzek, F., Rothwell, J., et al. 2012. Adsorption behaviour of arsenic relating to different natural solids: Soils, stream sediments and peats. Sci. Total Environ. 433: 456–461.

Mikutta, C., Kretzschmar, R. 2011. Spectroscopic evidence for terory complex formation between arsenate and ferric iron complexes of humic substances. Environ. Sci. Technol. 45: 9550–9557.

Mitsunobu, S., Takahashi,Y., Sakai, Y. 2008. Abiotic reduction of antimony(V) by green rust (Fe₄(II) $Fe_2(III)(OH)_{12}SO_4 . 3H_2O$). Chemosphere 70: 942–947.

Sharma, P., Ofner, J., Kappler, A. 2010. Formation of binary and ternary colloids and dissolved complexes of organic matter, Fe and As. Environ. Sci. Technol. 44: 4479–4485.

Wilson, S.C., Lockwood, P.V., Ashley, P.M., Tighe, M. 2010. The chemistry and behaviour of antimony in the soil environment with comparisons to arsenic: a critical review. Environ. Pollut. 158: 1169–1181.

Arsenic Research and Global Sustainability – Bhattacharya, Vahter, Jarsjö, Kumpiene,
Ahmad, Sparrenbom, Jacks, Donselaar, Bundschuh & Naidu (Eds)
© *2016 Taylor & Francis Group, London, ISBN 978-1-138-02941-5*

Temporal effects on kinetics of arsenic sorption on composite zeolite: nZVI

D. Muñoz-Lira[1], P. Sepúlveda[1], J. Suazo[1], N. Arancibia-Miranda[1],
M.A. Rubio[2] & A. Galdámez[3]

[1]*Department of Materials Chemistry, Faculty of Chemistry and Biology, University of Santiago, Chile*
[2]*Department of Environment, Faculty of Chemistry and Biology Sciences, University of Santiago, Chile*
[3]*Department of Chemistry, Faculty of Sciences, University of Chile, Chile*

ABSTRACT: This study evaluated the chemical and structural changes in the composite ZN-50%, generated by prolonged effect of removal of As. To study these changes were developed sorption kinetics of long-term (5 min to 6 months) considering an initial concentration As of 1000 mg·L^{-1}, identifying 3 points P1, P2 and P3. The kinetic results showed that from the point P1 to P2 increase material removal almost 6 times for As, remaining constant to the point P5. Desorption studies indicated that the material has greater stabilizing ability As 90%, with a percentage of minimum desorption (<6%) in relation to total retained analyte. This behavior was attributed to the stabilization of new chemical forms generated on the surface of the substrate during the kinetics of removal of As.

1 INTRODUCTION

The interest in the study of nanoparticles of Zero Valent Iron (nZVI) has grown strongly in recent years, due to its ability to react with a variety of contaminants in aqueous systems, either of organic origin (pesticides, organochlorines, etc.) and inorganic, (Pb^{2+}, Cu^{2+}, Cd^{2+}, As) (Sun *et al.*, 2011, Zhang *et al.*, 2015).

The removal mechanisms of contaminants, such as As, by nZVI are mainly attributed to surface corrosion phenomena Fe^{0} and complex formation (Zhang *et al.*, 2015). However, the removal process also can be affected by chemical and structural changes affecting the material during the removal of As, related to the surface formation of a mixture phase of oxides of iron (Kanel *et al.*, 2005, Sun *et al.*, 2011). Due to aggregation processes experienced by nZVI attributed to its high magnetization level, this material has been functionalized using crystalline aluminosilicate such as zeolite, which has increased the sorption capacity and stability of nZVI.

The objective of this study was to evaluate the chemical and structural changes experienced by the composite Zeolite-nZVI (Z – 50%) during the removal process of As due to prolonged stirring time. It was also determined whether these changes intervened the removal capacity of the material and conditioned the desorption of As, avoiding the generation of potentially harmful wastes to people and the environment.

2 METHODS/EXPERIMENTAL

2.1 *Composite synthesis*

The synthesis of compound ZN-50% was developed according to the procedure described by Arancibia-Miranda *et al*, 2016 where the functionalization of the porous material with nZVI, held in proportions 1:0.5 fora theoretical coating 50% (Zeolite:nZVI) using a solution of FeCl$_3$·6H$_2$O (1.0 M) and NaBH4 (1.6 M) as reducing agent.

2.2 *Kinetics of adsorption and desorption studies*

To study the chemical and structural changes of ZN-50% during the sorption of As, a batch type kinetic was performed. The procedure consisted in to mix 40 mL of an aqueous solution containing 1000 mg·L^{-1} AsIII (NaAsO$_2$) and 0.01M NaCl (Electrolyte Support) at pH 3.0 with 100 mg of ZN-50%. The samples were shaken for a time range between 5 minutes and 6 months and were then centrifuged, filtered and stored for the quantication of As using EAA after that, was measured pH and Eh of the resulting solution. The obtained solids were used for the desorption tests using the following extractants agents: H2O (Milli-Q), Sulphate (SO$_4$$^{2-}$) and phosphate (PO$_4$$^{3-}$), which were applied sequentially considering a volume of 40 mL. The concentration of ionic extractants used were of 1000 mg·L^{-1} and 0.01 M NaCl at pH 5.0. The stirring time applied was 24 hours.

2.3 Characterization of materials

The compound ZN-50% and materials obtained post-adsorption (P1, P2 and P3) were characterized structurally by techniques of X-Ray Diffraction (XRD), Scanning Electron Microscopy (SEM) and Isoelectric Point (IEP).

3 RESULTS AND DISCUSSION

3.1 Sorption kinetics

Figure 1 shows the kinetics of removal of As obtained for ZN-50%. In order to evaluate the chemical and structural changes experienced by the composite during removal, three points were identified in the kinetic: P1 (5 min), P2 (1 week) and P3 (6 months).

The kinetic study indicated that during the first five minutes of stirring (P1) the composite achieved As removal 35.2 mg/g, concentration that increased almost 6 times when the shaking it spread for 1 week (P2). This behavior can be attributed to the formation of a surface layer of oxides and hydroxides of Fe^{2+}/Fe^{3+} with a highly affinity for As, which is generated as a result of corrosion of Fe0, a process facilitated by extending the agitation time. Towards the end of the kinetic, the material removal capacity fell slightly, but after 6 months of shaking it reached a similar removal to point P2 corresponding to 56%, which could be explained by the possible stabilization of the new chemical forms on the surface composite, that reduce its affinity for As and remained unchanged removal capacity of the material in point P3.

Sensitive parameters to the changes experienced by the material were pH and Eh, considering that the rise in the capacity of removal of the ZN-50% registered from the point P1 to P2, it generated an increase in the pH of 5.1, decreasing oxidative system capacity by nearly 3 units. At point P3, the values of the parameters pH and Eh reached 4.4 and 257.2 mV respectively.

Table 1. Desorption study using solids obtained from the kinetic points P1 (5 min) and P3 (6 months).

Sample	Total sorbed As (mg/g)	Total sorbed As (post-desorption) (mg/g)	Desorption (%)
P1	35.2 (2.0)	33.3 (0.2)	5.3 (0.5)
P3	222.7 (3.1)	212.1 (0.7)	4.9 (0.7)

3.2 Desorption study

In Table 1 the results of the desorption studies are summarized.

The total absorption, considering the extracting agents (H_2O, SO_4^{2-} and PO_4^{3-}) not exceeded 6% in the points P1 and P5, in relation to the total concentration of analyte sorbed. This behavior indicates that the highest percentage of As removed is in a highly stable phase in the material, which significantly reduces its potential return to the environment. The stabilization process of the contaminant can be conditioned to the formation of different Fe oxides generated during surface corrosion of ZN-50% and the amount of As removed.

4 CONCLUSIONS

According to the above results, prolonged time of removal As had an effect that conditioned in the composite chemical transformations in Fe^0 and subsequent desorption of the contaminant. This phenomenon can still be studied, considering the effect of pH, presence of competing ions, etc.

ACKNOWLEDGEMENTS

This research was conducted with input from CORFO 12IDL2-16251 and Project Basal CEDENNA FB-0807.

REFERENCES

Arancibia-Miranda, N., Baltazar, S., García, A., Muñoz-Lira, D., Sepúlveda, P., Altbir, D. 2016. Nanoscale zero valent supported by zeolite and montmorillonite: template effect on the removal of Pb from aqueous solution. *J. Hazard. Mater.* 301: 371–380.

Kanel, S.R., Manning, B., Charlet, L., Choi, H. 2005. Removal of arsenic(III) from groundwater by Nano cale Zero-Valent Iron. *Environ. Sci. Technol.* 39: 1291–1298.

Zhang, Y., Chen, W., Dai, C., Zhou, C. and Zhou, X. 2015. Structural volution of nanoscale zero-valent iron (nZVI) in anoxic Co^{2+} solution: Interational performance and mechanism. *Sci. Rep.* 5: 13966 (doi:10.1038/srep 13966).

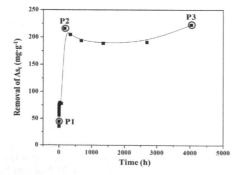

Figure 1. Arsenic removal kinetics for materials ZN-50%.

Arsenic Research and Global Sustainability – Bhattacharya, Vahter, Jarsjö, Kumpiene,
Ahmad, Sparrenbom, Jacks, Donselaar, Bundschuh & Naidu (Eds)
© *2016 Taylor & Francis Group, London, ISBN 978-1-138-02941-5*

Adsorption of arsenic in aquifers of the Red River floodplain controlled by Fe-oxides

H.U. Sø, D. Postma & R. Jakobsen
Department of Geochemistry, Geological Survey of Denmark and Greenland (GEUS), Copenhagen, Denmark

ABSTRACT: Adsorption of As(III) to aquifer sediments from the Red River floodplain, Vietnam, has been investigated. The adsorption affinity differs between the sediments indicating that it is not related to bulk components of the sediment but rather that certain properties of the sediments are important for the adsorption. Normalized to the pool of Fe-oxides in the sediments the adsorption isotherms are almost alike, indicating that Fe-oxides are the main absorbents for As(III) in these three sediments. This has implications for the mobility of As(III) in the groundwater aquifers, as it can vary spatially and also over time, if the Fe-oxides in the sediment become reduced.

1 INTRODUCTION

Adsorption of arsenic is controlling the arsenic mobility in the groundwater aquifers of SE Asia (Radloff *et al.*, 2011). A data compilation of As(III) adsorption isotherms on both oxidized and reduced aquifer sediments from Bangladesh and Vietnam showed very similar adsorption isotherms for four of the five datasets, indicating that As(III) adsorption is controlled by a bulk sediment component (Nguyen *et al.*, 2014).

To further investigate the controlling parameters for As(III) adsorption to aquifer sediments, new As(III) adsorption experiments have been carried out with reduced aquifer sediments from the Red River floodplain in Vietnam. The new results are compared to the results in the above mentioned data compilation, especially the results from Nam Du, Vietnam (Nguyen *et al.*, 2014).

2 MATERIALS AND METHODS

2.1 *Sediment characterization*

Sediment cores were taken in stainless steel tubing using a piston corer and stored frozen within the tubing. The sediments were obtained from Phu Kim and the H-transect north west of Hanoi, Vietnam (Postma *et al.*, 2012). Both aquifer sediments were obtained from the reduced part of the aquifer and are described as grey fine grained sand containing silt and clay. The sediments where characterized by parallel chemical extractions with a) 1 mM HCl at pH 3 with an automatic titrator to dissolve carbonates and phosphates and b) 0.1 M ascorbic acid + 0.2 M ammonium oxalate at pH 3 to dissolve Fe-oxides including crystalline oxides as goethite and hematite, as described in Postma *et al.* (2012).

2.2 *Batch adsorption isotherms*

Batch adsorption experiments were carried out in closed batches within an anaerobic glove box (Coy Laboratory Products, Inc.) with an N_2:H_2 96:4% gas mixture and Pd catalysts. To obtain strict anoxic conditions within the batches, the batches and solutions used for the batches were bubbled with pure N_2 gas pumped through a series of gas washing bottles filled with a buffered Fe(II) solution.

Batch adsorption isotherm experiments were carried out as described in Nguyen *et al.* (2014) using a background solution of 10 mM PIPES + 50 mM NaCl adjusted to pH 7. An As(III) stock solution was added stepwise to the batches (two batches with sediment from the H-transect and two batches with sediment from Phu Kim) and samples were taken 24 h after each addition. The batches were kept inside the glovebox throughout the experiment and shaken manually several times per day.

In contrast to Nguyen *et al.* (2014) the sediments were not leached prior to adsorption.

Aqueous analyses were carried out as described in Nguyen *et al.* (2014). In short, As(V) and As(III) were separated using disposable anion exchange cartridges within the glove box and measured by ICP-MS. The concentration of Fe(II) and PO4 was measured spectrophotometrically in each sample throughout the experiment.

3 RESULTS AND DISCUSSION

3.1 *Sediment extractions*

The content of Fe(II) extracted from the sediments with the two different chemical extraction methods is listed in Table 1. A similar extraction scheme was applied for the Nam Du sediment (Nguyen *et al.*, 2014), why those values are included in Table 1.

The pool of Fe-oxides in the sediment is calculated as the difference in Fe(II) extracted with the two methods in Table 1. The largest pool of Fe-oxides is seen in the Nam Du sediment (75.2 µmol/g), the pool in the H-transect is slightly lower (63.4 µmol/g), and substantially lover in Phu Kim (20.1 µmol/g).

A wide span can also be noted between the sediments in the amount of Fe(II) extracted with 1 mM HCl (here referred to as the secondary Fe(II)). The highest content of secondary Fe(II) is found in the H-transect, while there is only a small amount of secondary Fe(II) in the Nam Du sediment.

3.2 *As(III) adsorption isotherms*

As(III) adsorbs to both the sediment from Phu Kim and the H-transect, but the adsorption of As(III) is notably stronger in the H-transect than the Phu Kim sediment. Compared to previous results from Nam Du, As(III) adsorbs slightly stronger to the Nam Du sediment than the H-transect sediment.

Based on the variation of adsorption affinity for As(III) among the different sediments it does not seem that adsorption is related to a common bulk component of the sediments despite the similarity of adsorption isotherms shown in Nguyen *et al.* (2014). Rather, the variation indicates certain properties of the sediment to be important for the adsorption.

What this property might be appears when normalizing the adsorption of As(III) to the pool of Fe-oxide in the sediment. This results in quite similar adsorption isotherms for the three sediments studied. This indicates that the Fe-oxides are the dominant absorbent in the sediments. This is further investigated at present.

If the Fe-oxides are the main adsorbent for As(III) it has implications for its mobility in aquifers. The mobility of As(III) will vary spatially if there is a spatial variation in Fe-oxide concentration and it will vary in time—as the Fe-oxides in the sediment are reduced, the sorption capacity of the sediment is also reduced.

ACKNOWLEDGEMENTS

This research has been funded by the European Research Council under the ERC Advanced Grant ERG-2013-ADG. Grant Agreement Number 338972.

Table 1. Sediment extracted Fe(II) [µmol/g].

	H-transect	Phu Kim	Nam Du*
1 mM HCl	51.0	12.3	2.7
0.1 M ascorbic acid + 0.2 NH$_4$-oxalate	114.4	32.7	77.9
Sample [m b.g.]	13.25–13.45	7.6–7.85	4.2–4.9

*Data from Nguyen *et al.* (2014). The extraction with 0.5 M formic acid at pH 3 replaced the 1 mM HCl at pH 3. The two methods have been compared and yield identical results.

REFERENCES

Nguyen T.H.M., Postma, D., Pham, T.K.T., Jessen, S., Pham, H.V., Larsen, F. 2014.Adsorption and desorption of arsenic to aquifer sediment on the Red River floodplain at Nam Du, Vietnam. *Geochim. Cosmochim. Acta*. 142: 587–600.

Postma, D., Larsen, F., Nguyen, T.T., Pham, T.K.T., Jakobsen, R., Pham, Q.N., Tran, V.L., Pham, H.V., Murray, A.S. 2012. Groundwater arsenic concentration in Vietnam controlled by sediment age. *Nat. Geosci.* 5: 656–661.

Radloff, K.A., Zheng, Y., Michael, H.A., Stute, M., Bostick, B.C., Mihajlov, I., Bounds, M., Huq, M.R., Choudhury, I., Rahman, M.W., Schlosser, P., Ahmed, K.M., van Geen, A. 2011. Arsenic migration to deep groundwater in Bangladesh influenced by adsorption and water demand. *Nat. Geosci.* 4: 793–798.

Arsenic Research and Global Sustainability – Bhattacharya, Vahter, Jarsjö, Kumpiene,
Ahmad, Sparrenbom, Jacks, Donselaar, Bundschuh & Naidu (Eds)
© 2016 Taylor & Francis Group, London, ISBN 978-1-138-02941-5

Geochemical reactive transport modeling in "4D" of groundwater arsenic distribution in a non-static developing fluvial sediment aquifer system—feasibility study based on the upper part of the Red River, Vietnam

R. Jakobsen[1], D. Postma[1], J. Kazmierczak[1], H.U. Sø[1], F. Larsen[1], P.T.K. Trang[2], V.M. Lan[2], P.H. Viet[2], H.V. Hoan[3], D.T. Trung[3] & P.Q. Nhan[3]

[1]*Department of Geochemistry, Geological Survey of Denmark and Greenland (GEUS), Copenhagen, Denmark*
[2]*Research Centre for Environmental Technology and Sustainable Development (CETASD), Hanoi University of Science, Hanoi, Vietnam*
[3]*Hanoi University of Mining and Geology, Hanoi, Vietnam*

ABSTRACT: We examine the feasibility of combining the complexity of the developing sedimentology of the floodplain aquifer system with the complexity of the network of the most probable controlling geochemical reactions derived from a 1D PHREEQC model—in a "4D" model of the geochemical evolution of an aquifer that changes geometry over time—using PHAST as the modeling tool. It is shown that the combination is feasible in simplified narrow 3D sections and the approach is capable of generating distributions in groundwater arsenic concentrations that resemble the observed very heterogeneous spatial distribution in arsenic concentrations in flood plain sediments.

1 INTRODUCTION

Natural contamination of groundwater by geogenic arsenic is widespread in many recent delta sediments where reactive minerals and organic matter have accumulated. The fact that these sediments are recent implies that the reactivity of these sediments changes relatively rapidly over time and causes the arsenic concentration of relatively older sediments to be lower, around 0.5 μmoles/L, than the concentrations of around 6 μmoles/L found in relatively younger sediments (Postma *et al.*, 2012). We also know that the sedimentation and erosion of the sediments likewise takes place rapidly in these fluvial systems, covering wells near the river within years. Together this forms an interesting challenge for reactive transport modeling of a groundwater system.

As the aquifer system develops physically through sedimentation and erosion, the geochemical characteristics, such as reactivity and sorption capacity, of the components in the sediments responsible for the release of arsenic also change and the water chemistry changes with it. This implies that a model that aims at generating a geochemistry resembling what we observe now needs to include both the physical and the geochemical development of the aquifer system from the onset.

This contribution describes the first trials in this quest by extending the 1D model described in Postma *et al.* (2016) into a changing pseudo 3D model and examining if the results generated resemble our observations of the present Red River aquifer system in terms of the spatially very heterogeneous groundwater arsenic distribution.

2 METHODS

2.1 *Model tools and modeling the developing aquifer system*

The model tool used in this study is PHAST 3, which has the same functionality as PHAST 2 (Parkhurst *et al.*, 2010), but makes use of the PHREEQC 3 derived PhreeqcRM. In general terms PHAST combines a non-density version of HST3D and the general geochemical model PHREEQC. In PHAST 3 a set of solutions and geochemical parameters can be transferred from one model to another using a restart file with the results from a preceding model. The geometry of the preceding model can be different, because the transferred geochemical parameters are only retained in the part of the model that is not overwritten by the next model. So, a model can be run and the resulting sediment and water chemistry from the volume that is not eroded can be used as initial conditions for a model expanding on and eroding into the previous model. This provides the needed tool to model the "4D" development of the geochemistry of the aquifer system.

2.2 *Model setup*

The geochemical model used in the sequence of pseudo 3D PHAST models is a slightly modified version of the 1D PHREEQC model described in Postma et al. (2015 (*this volume*)). Apart from the general groundwater chemistry, the 1D PHREEQC model contains a description of how the reactivity of the organic matter and the arsenic-bearing Fe-oxides change over time as well as a kinetic description of

153

calcite dissolution. A 1D PHAST model gave very similar results to the 1D PHREEQC model. The PHAST flow models were set up to give a vertical flow rate close to the currently observed approximate vertical flow rate of 0.5 m/yr, based on $^3H/^3He$ dating, used in the 1D PHREEQC model, making it possible to compare. The aquifer geometries used in these initial steps are simplifications of the geology that we have derived from borehole information, geophysical modeling and remote sensing. This means that length and depth scales of the aquifer system and features such as canals are comparable with the observations.

2.3 Model setup

The kinetic reactions in the geochemical descriptions are coupled. The Fe-oxides are reduced by the organic matter and the rate of Fe-oxide reduction is a function of the saturation state, strongly dependent on pH, which again is a function of the carbonate dissolution and precipitation as well as the methanogenesis, which also uses the organic matter. These coupled kinetics make the geochemical calculations last rather long. Fortunately the geochemical calculations in PHAST can run in parallel using MPI, exploiting that PHAST uses operator splitting, where the transport and the chemistry equations are solved sequentially without iteration between the two, making parallelization quite efficient. When the number of parallel processes goes beyond 20 the decrease in calculation time per added parallel process is generally small, so there are limits to how large a model can be solved within a reasonable time, even with a very large computer system. The model steps used in this initial study took 24–72 hours using 24 parallel processes.

3 RESULTS AND DISCUSSION

3.1 Simulated complexity

These first trials of combining the complexity of the aquifer sedimentology with the complexity of the geochemical reaction network, show that the approach is feasible and using PHAST it is also relatively simple to go from one step to the next. The combination is capable of producing a distribution of arsenic in the model system that resembles the erratic spatial distribution that we observe in the natural aquifer system in the Red River floodplain.

This first attempt makes use of a few rather large intervals of constant hydrogeology, though in reality the sedimentation of the floodplain is a more continuous process. The sedimentation in a floodplain system is however only continuous to a certain extent. The meanders making up the river may shift suddenly, for example when horse-shoe bends are short circuited by erosion. The hydrogeological model that we are currently constructing for the area seems to confirm an episodic rather than continuous development of the aquifer system.

3.2 Obtaining a simulation of the actual system

To model the actual Red River floodplain system, or more specifically the upper part west of Hanoi that we are focusing on, we will build up a detailed geological and palaeogeographical model in appropriate time steps covering the last 8,000 yr, based on borehole information and geophysical data, supplemented by dating of the sandy aquifer units using core material. Based on this we intend to construct several detailed hydrogeological flow models for different times in the development of the floodplain. The model of the current situation will be calibrated on observed heads and $^3H/^3He$ dating of groundwater ages. Models of preceding stages will need to be calibrated by other means, e.g. on current observations of the depth to the groundwater table. Due to constraints on the duration of the geochemical calculations, the geochemical simulations will be run as pseudo 3D models, meaning narrow 2D or 3D models defined along and including the direction of the flow paths of the set of 3D flow models covering the chosen area over all of the time steps. The geochemical models will use the results of the detailed 3D flow models covering the area to define the boundary conditions to control the flow through the geochemical model.

4 CONCLUSIONS

The model approach appears to be feasible, at least for the simple aquifer system used in these first trials. To what extent the approach is also applicable when we turn to modeling of a more complex more natural system remains to be seen.

ACKNOWLEDGEMENTS

This research has been funded by the European Research Council under the ERC Advanced Grant ERG-2013-ADG. Grant Agreement Number 338972.

REFERENCES

Parkhurst, D.L., Kipp, K.L., and Charlton, S.R. 2010. PHAST Version 2—A program for simulating groundwater flow, solute transport, and multicomponent geochemical reactions: U.S. Geological Survey Techniques and Methods 6–A35, 235 p.

Postma, D., Larsen, F., Thai, N.T., Trang, P.T.K., Jacobsen, R., Nhan, P.Q., Long, T.V., Hung, P., Murray, A.S. 2012. Groundwater arsenic concentrations in Vietnam controlled by sediment age. *Nature Geosci.* 5: 656–661.

Postma, D. Larsen, F., Jakobsen, R., Sø, H.U., Kazmierzak, J. et al. 2016. On the spatial variation of arsenic in groundwater of the Red River floodplain, Vietnam (*This Volume*).

Arsenic Research and Global Sustainability – Bhattacharya, Vahter, Jarsjö, Kumpiene,
Ahmad, Sparrenbom, Jacks, Donselaar, Bundschuh & Naidu (Eds)
© 2016 Taylor & Francis Group, London, ISBN 978-1-138-02941-5

Reactive transport model for predicting arsenic transport in groundwater system in Datong Basin

Q. Yu & H.P. Ye

College of Resources and Environmental Science, South-central University for Nationalities, Wuhan, P.R. China

ABSTRACT: High Arsenic (As) concentration in groundwater of Datong Basin has emerged as an issue of great concern in the past decade because of its serious impact on the health of many people. A 1D reactive transport model (PHREEQC) was employed to investigate the As behavior during surface water and groundwater interaction. The model results revealed that the redox processes, like the reductive dissolution of iron oxyhydroxides, FeS(g), sulfate, nitrate and arsenate, were critical to the transport of As in groundwater, especially for which close to the surface water body.

1 INTRODUCTION

Arsenic (As) behavior can be modeled by reactive transport modeling, which is commonly applied to reconstruct geochemical process of groundwater along the flow tube, based on the advection-dispersion-reaction equations (Parkhurst *et al.*, 1999). Many studies including sediment geochemistry, hydrochemistry and biogeochemistry have been conducted to demonstrate the occurrence of high As groundwater at Datong, however, reactive transport modeling has rarely been applied to illustrate the mechanism of As enrichment in groundwater. Consequently, this paper aim to figure out the controlling processes on As transport during surface-groundwater interaction involving with a 1D reactive transport model.

2 METHODS

The 75-m long and 30-m wide Shanyin Field Site (SYFS) adjacent to the Sanggan River located in Shanyin country of the central part of Datong basin (Fig. 1a). A series of boreholes was drilled and multilevel piezometer nests installed along a transect perpendicular to the river channel (Fig. 1b). The details about the SYFS was presented in Yu (2015).

The 1D reactive transport model was constructed with PHREEQC version 3.0 based on the previous study of Yu (2015). A 10-m 1D column, or flow tube, was defined by 10 cells. All the cells have the same hydraulic and physical properties, with the bulk density of 2.31 kg/L, the effective porosity of 0.22, the pore water velocity of 0.3 m/d and the longitudinal dispersivity of 0.1 m (Yu, 2014). Well 1–2S and Well 2–2S were included in this model. The initial aqueous chemistry and infiltrating water were shown in Table 1. The reactive net of this model was shown in Table 2. A series of nine shifts was performed, and compositions graphed as computed after 30 (i.e. 9×3.3) days. Residence time in one cell was obtained considering

a water flow velocity of 0.3 m/d and a total flow path of 10 m. The contents of calcite, dolomite and quartz were 26.3 mol/L, 2.9 mol/L and 56.0 mol/L, respectively (Yu, 2014). Ion exchange was also included in the 1D model and the concentration of ion exchanger was referred to Parkhurst (1999) with 1.0 mol/L.

3 RESULTS AND DISCUSSION

3.1 *1D reactive transport model*

The calibration results of the 1D model were presented in Fig. 2. A good consistency between calculated values and monitored values in Fe, SO_4, HCO_3 contents of Well 1–2S and Well 2–2S can be observed, indicating the reliability of the 1D model. The monitoring results displayed that there was a slight increase of As concentration of Well 2–2S when surface water infiltrated into the aquifer. However, the model calculations show that there was almost no variation of As concentrations. The inconsistency illustrated that redox processes might have a slight influence on As transport of Well 2–2S. It may be related to the short time mixing of surface-ground water, which could only produce a significant change of As contents for wells locating close to the river other than that locating farer. Therefore, redox is an important process controlling As transport in groundwater system during the flooding period, especially for the wells locating close to the surface water body.

3.2 *Mechanism of arsenic transport*

Redox conditions are critical in controlling As mobilization in aquifer systems (Höhn *et al.*, 2006). The previous research reported that the reductive dissolution of iron oxyhydroxides, sulfate, nitrate and arsenate were the major geochemical processes controlling As transport in the aquifer system (Xie *et al.*, 2008). The increase of Fe(II), As(III), HS^- and NH_4^+ concentrations in groundwater from pre-flood to post-flood also suggested the redox conditions variations induced by

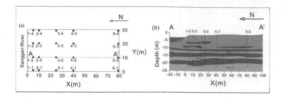

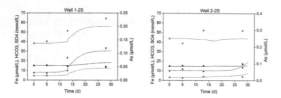

Figure 1. (a) plan view of the Shanyin field Experimental Plot and monitoring wells; (b) hydrogeologic cross section across Sanggan River.

Figure 2. Calibration results of the 1D reactive transport model.

Table 1. Initial aqueous chemistry and aqueous concentrations of infiltrating water (surface water) in the 1D reactive transport model (mg/L, the concentration unit of As is µg/L).

	Well 1–2S	Well 2–2S	Well 3–2S	Well 4–2S	Well 5–2S	Ground water	Surface water
pH	8.13	8.07	8.17	7.94	7.73	8:00 AM	10.01
T (°C)	12.4	12.9	12.8	12.7	13.1	12.8	
pe (mV)	0.15	–0.75	0.3	–0.49	–3.08	–0.8	
Na	1175	1062	440	1112	1229	1003.6 [a]	110.6
Ca	98	126	135	55	229	128.6 [a]	69.8
Mg	218	151	108	193	324	198.8 [a]	40.5
Cl	2199	1935	733	1901	2261	1805.8 [a]	106.9
HCO$_3^-$	762	750	315	875	546	649.6 [b]	214.3
SO$_4$	433	308	551	220	1151	532.6 [b]	258.3
Fe(II)	0.27	0.38	0.47	0.23	0.78	0.4 [b]	0.02
As	14.9	22.1	18.5	47.8	17.8	24.2 [a]	15

[a] Mean value of Wells 1–2S ~ 5–2S, SYFS.
[b] Fitted values for Wells1–2S ~ 5–2S, SYFS.

Table 2. Geochemical reactions included in the 1D Reactive Transport Model.

Redox model	Log K	Reference
$SO_4^{2-} + 10H^+ + 8e^- = H_2S + 4H_2O$	40.644	Archer and Nordstrom, 2003
$H_3AsO_4 + 2H^+ + 2e^- = H_3AsO_3 + H_2O$	18.89	Archer and Nordstrom, 2003
$NO_3^- + 10H^+ + 8e^- = NH_4^+ + 3H_2O$	119.077	Archer and Nordstrom, 2003
$FeOOH + 3H^+ = Fe^{3+} + 2H_2O$	–1.0	Archer and Nordstrom, 2003
$Fe^{3+} + e^- = Fe^{2+}$	13.02	Archer and Nordstrom, 2003
$FeS_{2(S)} + 2H^+ + 2e^- = Fe^{2+} + 2HS^-$	–18.479	Davison, 1991

artificial flooding (Yu et al., 2015). The results of the 1D model further illustrated the important role of redox geochemical processes on As transport, e.g. As(V), SO$_4^{2-}$, NO$_3^-$ were reduced into As(III), HS$^-$ and NH$_4^+$, respectively, together with the reduction dissolution of iron oxyhydroxides and FeS$_{(g)}$.

4 CONCLUSIONS

The 1D reactive transport model was employed to figure out the mechanisms of As transport in groundwater system during the surface water and groundwater interaction in Datong Basin. The numerical modeling calculation indicated that the redox processes, like the reductive dissolution of iron oxyhydroxides, FeS$_{(g)}$, sulfate, nitrate and arsenate, were critical to the transport of As in groundwater, especially for which close to the surface water.

ACKNOWLEDGEMENTS

This research was financially supported by the Fundamental Research Funds for the Central Universities, South-Central University for Nationalities (No. CZQ15006).

REFERENCES

Archer, D.G, Nordstrom, D.K. 2003. Thermodynamic properties of some arsenic compounds of import to groundwater and other applications. *Journal of Chemical & Engineering Data*.

Davison W. 1991. The solubility of iron sulphides in synthetic and natural waters at ambient temperature. *Aquatic Science* 53(4): 309–329.

Höhn, R., Isenbeck-Schröter, M., Kent, D.B., Davis, J.A., Jakobsen, R., Jann, S., Niedan, V., Scholz, C., Stadler, S., Tretnter, A. 2006. Tracer test with As(V) under variable redox conditions controlling arsenic transport in the presence of elevated ferrous iron concentrations. *Journal of Contaminant Hydrology* 88: 36–54.

Parkhurst, D.L., Appelo, C.A.J. 1999. User's Guide to Phreeqc (Version 3)- A Computer Program for Speciation, Batch-reaction, One-Dimensional Transport, And Inverse Geochemical Calculation. U.S. Geological Survey, Denver.

Parkhurst, D.L., Appelo, C.A.J. 1999. *PHREEQC. USGS-WRI Report*: 99–4259.

Yu, Q. 2014. Monitoring and coupled modeling of the hydraulic and hydrogeochemical processes of arsenic transport in hyporheic zone. *China Doctoral Dissertation China University of Geosciences* (in Chinese with English abstract), Chapter 5.

Yu, Q., Wang, Y.X., Xie, X.J. Matthew, C. Pi, K.F. Yu, M. 2015. Effects of short-term flooding on arsenic transport in groundwater system: A case study of the Datong basin. *Journal of Geochemical Exploration* 158: 1–9.

Xie X J, Wang Y X, Duan M Y, Xie, Z.M. 2008. Geochemical and environmental magnetic characteristics of high arsenic aquifer sediments from Datong Basin, northern China. *Environmental Geology* 58(1): 45–52.

Arsenic Research and Global Sustainability – Bhattacharya, Vahter, Jarsjö, Kumpiene,
Ahmad, Sparrenbom, Jacks, Donselaar, Bundschuh & Naidu (Eds)
© 2016 Taylor & Francis Group, London, ISBN 978-1-138-02941-5

Arsenic leaching potential from excavated rock: Sequential Leaching Test (SLT) and Rapid Small-Scale Column Test (RSSCT)—a case study

J.N. Li, T. Kosugi, S. Riya, A. Terada & M. Hosomi
Department of Chemical Engineering, Tokyo University of Agriculture and Technology, Koganei, Tokyo, Japan

ABSTRACT: Arsenic (As) leaching potential from one excavated rock was evaluated by Sequential Leaching Test (SLT) and Rapid Small Scale Column Test (RSSCT) combined with the Sequential Extraction Procedure (SEP). Although the total As content in the rock is low, its leaching concentration was higher than the Japanese environment standard (10 µg/L). Drying treatment can increase the leaching of As. Long-term water extraction in SLT and RSSCT can remove almost all the non-specifically-sorbed As and most of the specifically-sorbed As and can also transform some As to more crystalline phases. The concept of Potential Pollution Leaching Index (PPLI) was proposed in this study and was considered to be possibly used in the risk assessment and land management.

1 INTRODUCTION

Projects such as a tunnel, subway, groundwork and embankment require excavation of huge amounts of soils or rocks. Although in many cases the arsenic (As) concentration levels in these materials were naturally induced, their leaching concentrations were higher than the standards (Tabelin *et al.*, 2014). In Japan, efforts are now directed at the potential risk of excavated soils or rocks, and the Soil Contamination Countermeasures Law of Japan was amended and now covers these solids. In order to better assess the contamination of As in these solids and to develop effective and economical remediation techniques, more detailed knowledge about the As leaching behavior is required. Previous studies have compared the batch and column tests for the release of As from some soils or wastes (Lopez Meza *et al.*, 2008; Kim *et al.*, 2014), but few studies compare them to investigate the quantitative information regarding the potential leaching of As from excavated rocks. The objective of this study was to compare the potential water leaching of As from one excavated rock using batch test and column test.

2 METHODS

2.1 *Material and treatments*

The excavated rock used in our study was obtained from one project in Tokyo, Japan. Aliquot of the sample was air-dried an ambient temperature (AD) for 7 days; aliquot of the sample was used directly without pretreatment (ND). Soil moisture content

was 32.5%; soil pH was 10.2 (1:5 of air-dried sample: water); total As was 8.81 mg/kg.

2.2 *Leaching experiments*

Sequential leaching test was conducted based on the Soil Leachate Standard (SLS) in Japan (Ministry of the Environment, Government of Japan). 3.0 g of sample (dry weight basis) and 30 mL of deionized water were mixed. The mixture was shaken back and forth for 6 h at a rate of 200 rpm/min under ambient conditions. The slurry was centrifuged at 6000 rpm for 15 min, and the supernatant was filtered through a 0.45 µm membrane filter before ICP-MS analysis. These procedures were repeated until the As in the leachates were below detection limit (0.05 µg/L).

Rapid Small-Scale Column Test (RSSCT) was used according to Westerhoff *et al.* (2008). A series of stand-alone RSSCT apparatuses (internal diameter, 1.7 cm; length, 24 cm) were prepared and one slight change is the bottom-to-top flow direction. In order to compare with those results in SLS, 3.0 g of samples were also packed in the columns. The column was first slowly saturated from bottom to top with water, and then the pump flow rate was controlled to 1.0 mL/min. Column effluent samples were collected as a function of volume. The sampling continued until the As concentrations in the leachates reached below its detection limit.

The fractionations of arsenic in the rock before and after SLT and RSSCT were investigated by a five step sequential extraction procedure based on the method described by Wenzel *et al.* (2001).

3 RESULTS AND DISCUSSION

3.1 Arsenic release in SLT and RSSCT

The As in the leachates during the SLT and RSSCT were plotted against the L/S ratio (Fig. 1).

For RSSCT, the curves dropped sharply at the beginning and then slowly to a stable value. But for SLT, the curves showed an increasing trend at first and then a gradual decrease. The total released As in SLT (AD: 34.4%; ND: 31.3%) was higher than those in RSSCT (AD: 24.2%; ND: 23.1%). Besides, the cumulative As releases after AD treatment were slightly higher than those after ND treatment.

Releases of As (L/S from 10 to 150) in SLT and RSSCT are plotted in Fig. 2. At first with L/S of 10, the As releases were higher in RSSCT than that in SLT. But the As concentrations in SLT when L/S above 20 were generally higher than those in RSSCT.

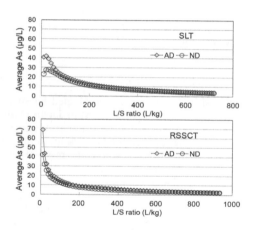

Figure 1. As concentration as a function of the L/S ratio.

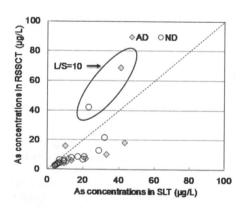

Figure 2. Comparison of SLT versus RSSCT,

3.2 Arsenic fractionations

After SLT and RSSCT finished, it is found that water extraction can remove almost all the non-specifically-sorbed As (F1) and most of the specifically-sorbed As (F2). The amount of amorphous oxides associated As (F3) decreased slightly but the crystalline oxides associated As (F4) increased. Besides, the amount of residual As (F5) also decreased (Fig. 3).

3.3 Potential Pollution Leaching Index (PPLI)

According to Cappuyns and Swennen (2008), the release of As with L/S from 10 to 50 can be fitted to the equation:

$$C_x = a \ln x + b \qquad (1)$$

where x is the L/S ratio (L/kg), C_x is the As released at L/S = x (mg/kg), a and b are two constants. We proposed the concept of Potential Pollution Leaching Index (PPLI). The cumulative As release can be obtained according to the above-mentioned equation; at a certain value of L/S, the ratio of cumulative As to the cumulative volume will reach the Japanese environment standard (10 μg/L). The value of L/S at this time is proposed as the Potential Pollution Leaching Index (PPLI). The approximate PPLI in the following equation can be solved by using Goal seek function in Microsoft Excel 2007.

$$1000C_{PPLI}/PPLI = 1000(a \ln PPLI + b)/PPLI = 10 \qquad (2)$$

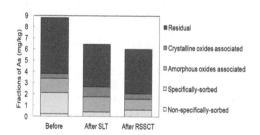

Figure 3. Fractionations of As before and after water extraction.

Table 1. Cumulative release of As fitted to Eq. (1) and Potential Pollution Leaching Index (PPLI).

	Method	R^2	PPLI$_{Calculated}$	PPLI$_{Measured}$
SLT	A D	0.997	272	260 ~ 270
	N D	0.983	208	230 ~ 240
RSSCT	A D	0.999	136	167 ~ 183
	N D	0.999	120	140 ~ 150

The calculated PPLI (Table 1) were generally in accord with those PPLI values measured in the experiment.

4 CONCLUSIONS

Sequential Leaching Test (SLT) and Rapid Small-Scale Column Test (RSSCT) showed different leaching behaviors of As and the total amounts of released As were considerable. We think the Potential Pollution Leaching Index (PPLI) can be used to estimate the potential leaching of arsenic and then to assess its risk. Certainly, whether this method can be expanded to other excavated rocks still needs further investigation.

REFERENCES

Cappuyns, V., Swennen, R. 2008. The use of leaching tests to study the potential mobilization of heavy metals from soils and sediments: a comparison. *Water Air Soil Pollut*. 191: 95–111.

Kim, E.J., Yoo, J.C., Baek, K., 2014. Arsenic speciation and bioaccessibility in arsenic contaminated soils: sequential extraction and mineralogical investigation. *Environ. Poll*. 186: 29–35.

Lopez Meza. S., Garrabrants, A.C., van der Sloot, H.A., Kosson, D.S. 2008. Comparison of the release of constituents from granular materials under batch and column testing. *Waste Manag*. 28: 1853–1867.

Tabelin, C.B., Hashimoto, A., Igarashi, T., Yoneda, T., 2014. Leaching of boron, arsenic and selenium from sedimentary rocks: I. Effects of contact time, mixing speed and liquid-to-solid ratio. *Sci. Tot. Environ*. 472: 620–629.

Wenzel, W.W., Kirchbaumer, N., Prohaska, T., Stingeder, G., Lombi, E., Adriano, D.C. 2001. Arsenic fractionation in soils using an improved sequential extraction procedure. *Anal. Chim. Acta*.436: 309–323.

Westerhoff, P., Benn, T., Chen, A., Wang, L., Cumming, L., 2008. Assessing Arsenic Removal by Metal (Hydr) oxide Adsorptive Media Using Rapid Small Scale Column Tests EPA document # EPA/600/R-08/051.

*1.4   Mobility of arsenic in contaminated soils and sediments,
climate change impacts*

Arsenic Research and Global Sustainability – Bhattacharya, Vahter, Jarsjö, Kumpiene,
Ahmad, Sparrenbom, Jacks, Donselaar, Bundschuh & Naidu (Eds)
© 2016 Taylor & Francis Group, London, ISBN 978-1-138-02941-5

Arsenic concentrations in floodplain soils

V. Cappuyns
Centre for Economics and Corporate Sustainability, KU Leuven, Belgium

ABSTRACT: Arsenic concentrations in a Belgian floodplain affected by industrial activities were compared with As concentrations in European floodplain soils from the FOREGS database. The upper 30 cm of the Belgian floodplain soil showed a clear enrichment with As, related to anthropogenic activities, while single extractions indicated that the mobility of As was significantly higher in the upper 12 cm of the soil. Fe was the independent variable that most significantly explained the total content of As in floodplain soils in the FOREGS database. A regression equation with major elements as independent variables allowed to predict As concentrations in the floodplain soils, and enable d to detect potential anomalies.

1 INTRODUCTION

Floodplain soil samples are often used for global scale geochemical mapping, since they are representative of a large area (Salminen *et al.*, 2005). Arsenic concentrations in a Belgian floodplain soil were compared with As data of floodplain soils across Europe. By analysis of both the Belgian and European data, the relationship between physicochemical soil characteristics and total As concentrations was quantified and areas affected by As contamination were detected. Regression equations, in which As concentrations in floodplain soils were expressed as a function of major element composition allowed to predict anomalous As concentrations. Additionally, single extractions released more As in the anthropogenically affected part of the floodplain.

2 METHODS/EXPERIMENTAL

2.1 *Sampling and sample characterization*

For the European floodplain soils (746 samples), data of the FOREGS Geochemical atlas were used. Details about the sampling and analytical methods are provided in Salminen *et al.* (2005). For the Belgian case, a floodplain along a small river in Central Belgium, located in a region characterized by sandy soils and underlain by a glauconite-rich substrate, was investigated. Organic- and iron-rich wetland soils have developed along this river, that is also affected by wastewater discharge from industrial activities in the upstream part of the river. A soil cylinder with a length of 1 m and a diameter of 3.5 cm was taken at a distance of 25 m from the river, and subdivided into 1 to 5 cm thick slices (30 subsamples in total). Air-dried samples were, among other parameters, analyzed for pseudo-total element concentrations after destruction with a mixture of HNO_3, HCl and HF. Single extractions with $CaCl_2$ 0.01 mol/L, CH_3COOH 0.43 mol/L and ammonium-EDTA 0.01 mol/L were performed as described in Quevauviller *et al.* (1998). Concentrations of major elements (Fe, Al, Ca, K, and

Mg) were measured with flame atomic absorption spectrometry (FAAS, Varian AA6). Arsenic was measured with a HP 4500 ICP-MS. The spectroscopic interference of ArCl, which has the same m/z as As (75) was corrected according to the recommendations of the EPA (Method 200.8, Brockhoff *et al.*, 1999).

2.2 *Statistical analysis*

Statistical analysis was performed with the software package SPSS 22.0 for Windows. Descriptive statistics were calculated for each variable. The 90th-percentile of element concentrations in the upper soil layer of non-contaminated soils is often considered representative for the background value in soils (Carlon *et al.*, 2007). By comparing average and median values, soils with enrichment in As can be detected (De Temmerman *et al.*, 2003). Multiple linear regression according to the stepwise method was performed to deduce possible causal relationships between the variables. Different assumptions of the linear regression (normality of the de residues, autocorrelation, quasi-multicollinearity and heteroscedasticity) were tested.

3 RESULTS AND DISCUSSION

3.1 *(Pseudo) total As concentrations*

Different methods to determine 'total' or 'pseudo-total' concentrations in soils often result in very different results. For As, however, X-Ray Fluorescence (XRF) analysis (giving 'total concentrations') and ICP-MS analysis after Aqua Regia (AR) destruction (giving pseudo-total As-concentrations) yielded comparable results (Fig. 1). Therefore, the results of As in the Belgian samples, in which As was analyzed with a method that results in a more complete sample dissolution compared to AR, are comparable with XRF data. For Fe and other major elements, AR results were systematically lower than XRF results (Fig. 1), and data are not compared. The total As concentrations in the floodplain soil show a median value of 6 mg/kg (both

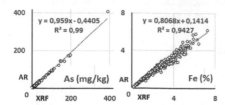

Figure 1.   Scatterplots of Aqua Regia (AR) versus total (determined with XRF) content of As and Fe in floodplain soils from the FOREGS database.

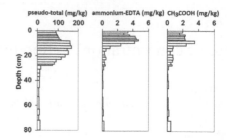

Figure 2.   As concentrations in the Belgian soil profile (a) pseudo-total concentration, (b) concentration extracted with ammonium-EDTA, (c) concentration extracted with acetic acid. Only the upper 80 cm of the core is shown.

for XRF and AR determination). Some high total As values in floodplain soils of the FOREGS database (Fig. 1) are likely due to mineralization and industrial pollution. Average As concentrations in these soils were 11.2 (AR) and 12.2 (XRF) mg/kg, whereas the average As-content in the Belgian soil core was 68 mg/kg.

The 90th percentile of As concentrations was 20 and 23 mg/kg for AR and XRF determination respectively, which already gives an indication that the Belgian soil samples are enriched with As. Figure 2 illustrates that this enrichment mainly occurs in the upper 30 cm of the core, reaching 'normal' background concentrations (< 20 mg/kg) below 30 cm depth. Average As-concentrations of 30 mg/ kg in (uncontaminated) glauconite bearing soils have been reported (Dooley, 2001). The enrichment in the present core is much higher (up to 160 mg/kg), due to industrial activities upstream.

3.2   Single extractions

Both acetic acid and ammonium-EDTA can only extract a small amount of As from the soil samples. There is a distinctly higher extractability of As in the upper 30 cm of the core (Fig. 2), which most likely indicates the anthropogenic origin of this As enrichment. More As is extracted by ammonium-EDTA than with acetic acid, due to the lower final pH of the acetic acid extract (around 3.5) and mobilization of As with increasing pH. The inefficiency of As extraction by ammonium-EDTA is explained by the fact that EDTA does not form stable complexes with As (Tokunaga and Hakuta, 2002). As-concentrations extracted with a 0.01 mol/L $CaCl_2$ solution were in the range 0.5–21 µg/L, with values above 10 µg/L (WHO drinking water guideline) in the upper 12 cm of the core.

3.3   Prediction of as concentrations

Stepwise linear regression with major elements (XRF determination) and organic matter content as independent variables was performed in order to predict the As content in the floodplain soils. A regression equation with log transformed concentrations of Fe, Mn, Ca, K, Na, Si and Ti as independent variables allows to predict As concentrations in the floodplain soils. Fe

concentrations and clay mineral-associated elements explain 44% of the variability in As-concentrations.

4   CONCLUSIONS

Single extractions, as well as the construction of regression equations with major elements and TOC as independent variables enabled to detect potential anomalous As concentrations in the floodplain soils. An approach combining single extractions and statistical data analysis (using major element compositions) is useful to screen for As contamination in soils, taking into account local variations in geological substrate and soil composition.

REFERENCES

Brockhoff, C.A., Creed, J.T., Martin, T.D., Martin, E.R., Long, S.E. 1999. EPA Method 200.8, Revision 5.5: Determination of trace metals in waters and wastes by ICP-MS, EPA-821R-99–017, 61 p.

Carlon, C. (Ed.), 2007. Derivation methods of soil screening values in Europe. A review and evaluation of national procedures towards harmonization. European Commission, Joint Research Centre, Ispra, EUR 22805-EN, 306 p.

De Temmerman, L., Vanongeval, L., Boon, W., Hoenig, M., Geypens, M., 2003. Heavy metal content of arable soils in N Belgium. Water Air Soil Poll. 148: 61–76.

Dooley J. 2001. Baseline Concentrations of Arsenic, Beryllium and Associated Elements in Glauconite and Glauconitic Soils in the New Jersey Coastal Plain. New Jersey Geological Survey, 239 p.

Quevaullier, P., 1998. Operationally defined extraction procedures for soil and sediment analysis I. Standardization. Trends in Analytical Chemistry 17: 289–298.

Salminen, R. (chief-editor) & Batista, M.J., Bidovec, M., Demetriades, A. et al. 2005. Geochemical Atlas of Europe. Part 1: Background Information, Methodology and Maps. Espoo, Finland: EuroGeosurveys & Foregs.

Tokunaga, T., Hakuta, T., 2002. Acid washing and stabilization of an artificial arsenic-contaminated soil. Chemosphere 46: 31–38.

Arsenic Research and Global Sustainability – Bhattacharya, Vahter, Jarsjö, Kumpiene,
Ahmad, Sparrenbom, Jacks, Donselaar, Bundschuh & Naidu (Eds)
© 2016 Taylor & Francis Group, London, ISBN 978-1-138-02941-5

Solubility and transport processes of As(V) in sandy soils from historically contaminated sites at different rainfall intensities

Å. Löv[1], I. Persson[2], M. Larsbo[1], J.P. Gustafsson[1], G. Cornelis[1], C. Sjöstedt[1] & D.B. Kleja[1]

[1]*Department of Soil and Environment, Swedish University of Agricultural Sciences, Uppsala, Sweden*
[2]*Department of Chemistry and Biotechnology, Swedish University of Agricultural Sciences, Uppsala, Sweden*

ABSTRACT: We aim to investigate how different rainfall intensities might affect the mobilization of As in two historically contaminated sandy soils by performing column tests on intact soil columns; to evaluate the laboratory methods used in standard risk assessments; and to characterize possible carriers of As. The result suggests that As predominates in the dissolved phase regardless of site or rainfall intensity and that concentrations were fairly insensitive to rainfall intensity. The solubility in the batch test was lower than in the column test, which could result in a risk of underestimating the amount of As leaching from a contaminated site using standardized methods. Preliminary EXAFS results indicate adsorption of As(V) to Fe in the solid soil and particulate phase and possibly adsorption to Al in the colloidal phase.

1 INTRODUCTION

Climate change induced increased precipitation is expected in Sweden in the near future. Such a scenario will likely increase the role of particulate and colloidal transport. In a sandy soil at extreme rainfall intensities (102 mm/h) at pH 5.6, the importance of particle- and colloidal mobilization has been shown (Hu *et al.*, 2008). Fe-(hydr)oxides are important carriers for As(III) (Goldberg & Johnston, 2001) as well as for As(V) (Slowey *et al.*, 2007; Kumar *et al.*, 2014) and the adsorption of As(V) to ferrihydrite is rapid (Mähler *et al.*, 2013). For As(V) adsorption to Al-hydroxides has also been verified (Kumar *et al.*, 2014).

Our aim was to study the transport processes at different rainfall intensities in contaminated sandy soils. The results were used to evaluate the accuracy of batch tests frequently used in standardized risk assessments. Additionally, possible carriers for As were characterized by using EXAFS (Extended X-ray Absorption Fine Structure).

2 METHODS/EXPERIMENTAL

2.1 *Column test and batch test*

Intact soil columns (20 cm in diameter * 30 cm deep) were collected at an impregnation site and a glass work site. Unsaturated flow was simulated using three different rainfall intensities in a sequence: 2, 10 and 20 mm h^{-1} (three pore volumes per intensity). The leachate was separated in particulate (> 0.45 μm), colloidal (0.45 μm – 10 kDa) and dissolved (< 10 kDa) fractions using micro- and ultrafiltration.

Batch tests for solubility studies were performed (2.5 g soil/50 mL, dry weight) using a 10 mM NaNO$_3$ matrix and pH between 2.4 and natural pH (6.7), equilibrated for 5 days. The kinetics was studied by equilibration for 5, 33, 61 and 90 days.

ISO-standard batch test (SIS-CEN ISO/TS 21268-2:2010) using 1 mM CaCl$_2$ (24 h) were also performed.

2.2 *Speciation*

As and Fe in particulate and colloidal fractions, as well as in the solid soil phase for the impregnation site, were characterized using EXAFS. Concentrations were too low at the glass work site for EXAFS measurements on As, thus only Fe spectra was collected.

The speciation of As was also determined using a filter device which adsorbs all As(V).

3 RESULTS AND DISCUSSION

3.1 *Speciation*

As(V) was the predominating redox state of As at both sites; 95–97% and 84% at the impregnation site and the glass work site, respectively and regardless of rain intensity. Preliminary EXAFS results indicate that As(V) was adsorbed to Fe in the solid phase and particulate phase and possibly to Al in the colloidal phase at the impregnation site.

3.2 *Solubility and kinetics*

The kinetics for As(V) was fairly fast and equilibrium was reached after five days (Figure 1). However, the concentrations in the ISO-standard batch test were considerably lower, indicating a kinetically constrained release in 24 hrs. equilibrations.

3.3 *Column test and batch test*

For the glass work site there was no effect of the change in rainfall intensity for the concentration of dissolved As (Fig. 2). Thus dissolved As seems to be in equilibrium with the solid phase.

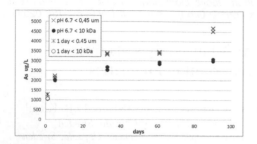

Figure 1. Kinetics of As at natural pH at the impregnation site. The data for 1 day is taken from the standardized batch test.

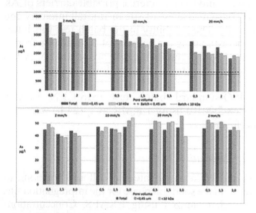

Figure 2. Fractionation of As in column tests. Top graph: impregnation site, bottom graph: glass work site.

At rainfall intensities of 2 mm/h and 10 mm/h for the impregnation site, dissolved As concentrations were fairly constant, indicating equilibrium. In contrast, when the irrigation increased from 10 mm/h to 20 mm/h, the concentration decreased slightly, indicating a kinetically constrained release. As(V) predominates in the dissolved fraction regardless of site. The impregnation site, rich in OM, has a particulate mobilization at the onset of a higher irrigation. However, the glass work site, with low concentration of OM, all As(V) mobilized could be considered to be dissolved. At both sites Al and Fe predominate in the particulate and colloidal fraction (data not shown); only in the dissolved phase can a correlation between Fe, Al and As be identified, indicating potential to mobilize As(V).

The concentration of As(V) in leachates from the column tests was approximately twice as high as from the ISO-standard batch test. As a consequence the calculated Kd differed by a factor two. In standardized risk assessments the Kd value is used as an indicator of the leachability of metal(liod)s, and using the Kd value from the batch test would thus underestimate the amount of As leaching from the impregnation site.

Table 1. Kd values for As calculated from the batch test and the column tests. The Kd values from the column test are calculated based on an average concentration from all rainfall intensities.

Method/Fraction	Impregnation Kd (l/kg)	Glass work Kd (l/kg)
Batch test < 0.45 um	2600	N.A.
Batch test < 10 kDa	3100	N.A.
Total	1100	800
<0.45 um	1200	750
<10 kDa	1300	750
pH leachate, column	≈7.2	≈8
pH leachate, std. batch	≈6.5	N.A.
pH leachate, kinet. batch	≈6.7	N.A.

4 CONCLUSIONS

Mobilization of As during unsaturated flow conditions seems to occur mainly as dissolved As(V), even at extreme rainfall intensities (20 mm/h).

Dissolved As(V) seems to equilibrate with the sorbed pool at moderately to high flow conditions (2 and 10 mm/h). At extreme rainfall conditions (20 mm/h) desorption of As(V) might be kinetically constrained. In soils with high concentrations of OM there might be a fraction which is mobilized by particles (~20%), and the amount of As mobilized could be affected by rainfall intensity. The short equilibration time (24 hrs.) used in the ISO-standard batch test might be too short to be representative for the situation in a percolating solution; thus underestimating the risk of leaching.

Speciation of Fe and As as well as results from three different standard leaching tests (SS-EN 12457-2, ISO/TS 21268-2 and SIS-CEN/TS 14405) for both sites will be presented at the conference, and compared with the results of the column study.

REFERENCES

Goldberg, S., Johnston, C.T. 2001. Mechanisms of Arsenic Adsorption on Amorphous Oxides Evaluated Using Macroscopic Measurements, Vibrational Spectroscopy, and Surface Complexation Modeling. J. Colloid Interf. Sci. 234: 204–216.

Hu, S., Chen, X., Shi, J., Chen, Y., Lin, Q. 2008. Particle-facilitated lead and arsenic ransport in abandoned mine sites soil influenced by simulated acid rain. Chemosphere 71: 2091–2097.

Kumar, P.S., Flores, Q.F., Sjöstedt, C., Önnby, L. 2016. Arsenic adsorption by iron–aluminium hydroxide coated ontomacroporous supports: insights from X-ray absorption spectroscopyand comparison with granular ferric hydroxides. J. Hazard. Mater. 302: 166–174.

Mähler, J., Persson, I., Herbert, R. 2013. Hydration of arsenic oxyacid species. Dalton Trans. 42: 1364–1377.

Slowey, A.J., Johnson, S.B., Newville, M., Brown, G.B. Jr. 2007. Speciation and colloid transport of arsenic from mine tailings. Appl. Geochem. 22: 1884–1898.

Arsenic Research and Global Sustainability – Bhattacharya, Vahter, Jarsjö, Kumpiene,
Ahmad, Sparrenbom, Jacks, Donselaar, Bundschuh & Naidu (Eds)
© 2016 Taylor & Francis Group, London, ISBN 978-1-138-02941-5

Study of arsenic availability in Pampean loess sediments using a sequential extraction procedure

L. Cacciabue[1], S. Dietrich[1], L. Sierra[2], S.A. Bea[2], P.A. Weinzettel[1] & M.G. García[3]
[1]*Instituto de Hidrología de Llanuras, Azul, Buenos Aires, Argentina*
[2]*Instituto de Hidrología de Llanuras "Dr. Eduardo J. Usunoff" (IHLLA), Azul, Argentina*
[3]*Centro de Investigaciones en Ciencias de la Tierra (CICTERRA), CONICET and Universidad
Nacional de Córdoba, Argentina*

ABSTRACT: The aim of this work was to present the preliminary results of a Sequential Extraction
Procedure (SEP) applied to the loess sediments from two study sites in the Pampean plain of Argentina.
It consists in a combined SEP methodology based on Keon *et al.* (2001) and Dold (2003) procedures.
Since mobile arsenic is associated only with some mineral phases, this study was particularly focused on
water soluble minerals, exchangeable ligands and strongly adsorbed arsenic phases. Results showed that
the "mobile" arsenic may come mostly from adsorption complexes formed on Fe oxyhydroxides surfaces.

1 INTRODUCTION

Pampean plain (Argentina) is a large region with
high arsenic (As) in their aquifers, which provide
water for the most numerous population of the
country (Buenos Aires province). Several studies
were conducted to determine the groundwater As
sources in this region (Smedley *et al.*, 2002, Bhattacharya *et al.*, 2006). It is known that volcanic glass,
is an important component of loess sediments,
which contributes As into groundwater (Bhattacharya *et al.*, 2006, Bia *et al.*, 2015). Volcanic glass
is considered as the primary sources of As in this
region, but secondary sources are also present. As
sorbed onto Fe-hydroxides has been assigned as the
most important secondary source of this element,
as it may be released from this phase by desorption
under alkaline conditions (Appelo *et al.*, 2002).

Sequential Extraction Procedures (SEP) have
been developed to study the association of elements with some specific mineral phases. Only a
few previous studies have applied SEP methodology to analyze As associations in the Pampean
loessic sediments (Smedley *et al.*, 2000, Dietrich
et al., 2016) by using procedures that were not specifically developed for natural sediments. Therefore, the aim of this work is: (1) to identify and
quantify As associations by using a SEP particularly designed to study As in natural sediments
(Keon *et al.*, 2001) and; (2) to assess the amount
of As that corresponds to the "mobile fraction"
which is defined as the sum of As concentrations
extracted along with the soluble, exchangeable and
strongly adsorbed arsenic fractions.

2 MATERIALS AND METHODS

Samples used in this work belong to the Pampean
loessic sediments from Buenos Aires province,
Argentina (Fig. 1). One of them was collected near
the city of Azul (S 36° 46'; W 59° 53') whereas the

Figure 1. Location of the two loess samples.

other one was collected from Tres Arroyos city
(S 38° 22'; W 60° 14'). Arsenic contents measured
in shallow groundwater of the study sites are 70 and
234 µg/l in Azul and Tres Arroyos, respectively.

Before analysis, samples were air-dried and sieved
through a 230 µm mesh; the fraction <63 µm, which
is considered as the most reactive fraction, was
used for the analysis. The SEP proposed by Keon
et al. (2001) was adapted in this study (Table 1) in
order to identify the following As-bearing phases
present in the loess: 1) water soluble (SOL) (Dold,
2003); 2) weakly adsorbed As (EXCH); 3) strongly
adsorbed As (ADS); 4) carbonates and amorphous
Fe(III) oxyhydroxides 5) Fe (III) oxyhydroxides
(AM); 6) crystalline Fe(III) oxides (CRYS) (Dold
2003); 7) primary sulfides (SULF).

After each extraction, samples were centrifuged
at 4000 rpm for 15 min. The obtained supernatants
were filtered using 0.22 µm cellulose membranes
and the chemical composition was determined by
ICP-MS and ICP-OES. Detection limit was 1 µg/l
for As. In addition, BET surface area was determined for both samples (Brunauer *et al.*, 1938).

Table 1. SEP used in this work.

Step	Extracted phase	Extractant
1*	Soluble (SOL)	Deionized water
2	Exchangeable (EXCH)	$1M$ $MgCl_2$, pH8
3	Strongly adsorbed (ADS)	$1M$ NaH_2PO_4, pH5
4	Carbonate, Fe-oxyhydroxides (am.)	$1N$ HCl
5	Fe-oxyhydroxides (AM)	NH_4-oxalate, pH3
6*	Crystalline Fe oxides (CRYS)	NH_4-oxalate, 80°C, pH3
7	Pyrite and As-sulfides (am.) (SULF)	$16N$ HNO_3

am.: amorphous, Steps 2, 3, 4, 5 and 7 (after Keon *et al.*, 2001)
* after Dold (2003).

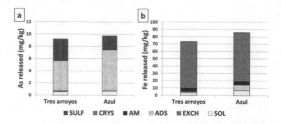

Figure 2. Results of As-Fe samples contents from six steps of the SEP.

3 RESULTS AND DISCUSSION

Results of SEP are shown in Figure 2. The total extracted As concentrations in the sediments were 9.0 and 9.5 mg/kg for Tres Arroyos and Azul samples, respectively. According to SEP results, As is mostly associated with the strongly adsorbed phases. In this sense, Tres Arroyos and Azul samples released 4.9 (55%) and 6.7 mg/kg (70%) in the ADS step, respectively, which is expected due to the affinity of As for the adsorption sites. In addition, 0.56 and 0.67 mg/kg of As were released in SOL step, whereas 0.18 and 0.07 mg/kg of As were released in EXCH step for Tres Arroyos an Azul samples, respectively.

The "mobile" As for these loess samples may be assessed from the sum of SOL, EXCH and ADS steps of the SEP. The Azul loess sample has a higher proportion of the As mobile fraction than Tres Arroyos sample. On the other hand, in AM step 0.04 mg/kg of As was released for both samples. This As is coprecipitated mainly with Fe (III) hydroxides but part of it could be an adsorbed remaining of ADS step. This As should be considered as "potentially mobile", as it could be released due to an eventual pH increase. Finally, 3.9 and 2.3 mg/kg of As were extracted from Tres Arroyos and Azul samples respectively on CRYS and SULF steps. Arsenic associated with these fractions are generally "immobile" as they arer incorporated in crystal lattices.

BET analyses revealed surface area of 24.12 and 26.35 m²/g for Tres Arroyos and Azul samples, respectively. These results suggest that Azul sample may have slightly more capacity of adsorbing ions onto Fe (III) oxy-hydroxides, consistent with the higher Fe contents released in AM and CRYS steps (Fig. 2b).

4 CONCLUSIONS

Loess sediments of the Pampean plain have the highest percentage of As hosted in the adsorption complex (55% and 70% for Tres Arroyos and Azul samples respectively), whereas only the 5% of As is contained in soluble phases and less than 2% is contained in exchangeable phases in both study sites. SEP method

applied in this work is considered an appropriated method to study Pampean loess sediments because it allows distinguishing exchangeable phases from adsorbed phases, which are the As "mobile" phases.

ACKNOWLEDGEMENTS

This work was funded by research project PID2011–075. Special thanks to IHLLA Laboratory (Lic. Fatima Altolaguirre, Lic. Ornella Floriani and Natalia De Libano); CICTERRA Laboratory, Córdoba, Argentina (Lic. Andrea Lojo and Dr. Laura Borgnino). Special thanks to Dr. Carlos Ayora.

REFERENCES

Appelo, C.A.J., Van der Weiden, M.J.J., Tournassat, C., Charlet, L. 2002. Surface complexation of ferrous iron and carbonate on ferrihydrite and the mobilization of arsenic. *Environ. Sci. Technol.* 36(14): 3096–3103.
Bia, G., Borgnino, L., Gaiero, D., García, M.G. 2015. Arsenic-bearing phases in South Andean volcanic ashes: Implications for As mobility in aquatic environments. *Chem. Geol.* 393: 26–35.
Bhattacharya, P., Claesson, M., Bundschuh, J., Sracek, O., Fagerberg, J., Jacks, G., Martin, R.A., Storniolo, A.R.,Thir, J.M. 2006. Distribution and mobility of arsenic in the Río Dulce alluvial aquifers in Santiago del Estero province, Argentina. *Sci. Total Environ.* 358(1–3): 97–120.
Blanco, M.D.C., Paoloni, J.D., Morrás, H., Fiorentino, C., Sequeira, M.E., Amiotti, N.N., Bravo, O., Diaz, S., Espósito, M. 2012. Partition of arsenic in soils sediments and the origin of naturally elevated concentrations in groundwater of the southern Pampa Region (Argentina). *Environ. Earth Sci.* 66(7): 2075–2084.
Brunauer, S., Emmett, P., Teller, E. 1938. Adsorption of gases in multimolecular layers. *J. Amer. Chem. Soc.* 60: 309–319.
Dietrich, S., Bea, S., Weinzettel, P., Torres, E., Ayora, C. 2016. Occurrence and distribution of arsenic in the sediments of a carbonate-rich unsaturated zone. *Environ. Earth Sci.*75: 90 (doi:.10.1007/s12665–015–4892–7)
Dold, B. 2003. Speciation of the most soluble phases in a sequential extraction procedure adapted for geochemical studies of copper sulfide mine waste. *J. Geochem. Explor.* 80: 55–68.
Keon, N.E., Swartz, C.H., Brabander, D.J., Harvey, C. & Hemond, H.F. 2001. Validation of an arsenic sequential extraction method for evaluating mobility in sediments. *Environ. Sci. Technol.* 35: 2778–2784.
Smedley, P.L., Nicolli, H.B., Macdonald, D.M.J; Barros, A.J., Tullio, J.O. 2002. Hydrogeochemistry of arsenic and other inorganic constituents in groundwaters from La Pampa, Argentina. *Appl. Geochem.* 17 (3): 259–284.

Arsenic Research and Global Sustainability – Bhattacharya, Vahter, Jarsjö, Kumpiene,
Ahmad, Sparrenbom, Jacks, Donselaar, Bundschuh & Naidu (Eds)
© 2016 Taylor & Francis Group, London, ISBN 978-1-138-02941-5

Arsenic in Brazilian tropical coastal zone

N. Mirlean & P. Baisch
Geological Oceanography Laboratory, Institute of Oceanography, Federal University of Rio Grande,
Rio Grande RS, Brazil

ABSTRACT: High concentration of arsenic (As) has been found in broad areas of the Brazilian tropical coastal zone. In this region, some carbonate tropical beaches have been enriched with As. High levels of As in sediments correlate with detritus material of calcareous algae involved in diagenetic processes. The As anomalies can be generated by the burial of marsh-lacustrine sediments containing As. Source of As is not clear, but oxyhydroxides layers of Tertiary sediments on the coastline (Barreiras Group) play an significant role. High concentrations of As were detected in groundwater in the Paraiba do Sul delta. The deltaic shallow groundwater aquifer is enriched with As fixed by authigenic sulfides. Arsenic speciation in the marine biota (Açu, Rio de Janeiro) shows that its organic fraction predominates in all samples, representing more than 95% of total As and the main compound is arsenobetaine.

1 INTRODUCTION

Arsenic (As), a toxic metalloid, has attracted environmental researchers' attention worldwide over the last decades. In Brazil, environmental problems concerning As have mostly been described in past and current mining activities and related to metallurgical industries.

Mirlean *et al.* (2011) were the first researchers to identify areas with natural enrichment of As in the Aracruz port terminal, located in Espírito Santo state, on the Brazilian coastline. Afterwards, new studies were carried out and other areas with As enrichment were also identified on the coastline.

The Brazilian environmental act on marine sediment quality control (CONAMA, 2012) has become more interested in the As distribution on the Brazilian coast, mainly due to environmental limitations that are imposed on dredging operations in port areas.

Areas enriched with As may pose health effects to populations that consume groundwater, besides fish, mussels and crustaceous which may have been contaminated.

This study aimed at summarizing the studies of As which have been carried out by the Laboratory of Geological Oceanography at the Federal University of Rio Grande-Brazil, over the last years and at describing the first data on As speciation in the marine biota in Açu, Rio de Janeiro.

2 METHODS/EXPERIMENTAL

2.1 *Study area*

The study area was the Brazilian coastline between 7°38' S and 22°53' S, which comprises 11 Brazilian states. However, the main studies were carried out in Espírito Santo and Rio de Janeiro (in the region of Açu Port).

2.2 *Chemical analyses*

Total As was determined by electrothermal atomic absorption spectrometry that employed a Perkin-Elmer 800 instrument with a Zeeman background corrector with pyrolytically-coated tubes on a platform. Ca and Fe levels were determined by flame (acetylene—air) atomic-absorption spectrometry. Arsenic species of biota were determined by Liquid Chromatography–Inductively Coupled Plasma–Mass Spectrometry (LC–ICP–MS). Separation of As species was performed by an anion exchange column with ammonium phosphate solution as mobile phase. Sediment total sulfide was measured by a Ion-Selective Electrode (ISE). Organic matter content and particle size distribution were determined by common methods (Pansu and Gautheyrou, 2006).

3 RESULTS AND DISCUSSION

3.1 *Arsenic concentrations in coastal zone*

Waters in 4 lagoons monitored in the Açu region (Rio de Janeiro state) has concentrations between 0.1 and 10 μg/g in dissolved As. The monitoring wells of groundwater had irregular behavior, i. e., a group had low As contents whereas the other had high contents, which could reach around 250 μg/g. The total As content in the sands of the beaches under study ranged from 1.1 mg/kg to 119.6 mg/kg while its average in the beach sediments was 18.1 mg/kg. The Threshold Effect Level [TEL]: 19 mg/kg for As was above 33% in all beach sand samples under study, and, in two cases, the Probable Effect Level [PEL]: 70 mg/kg was also exceeded.

Arsenic concentrations in the beach sand in Espírito Santo state averaged about 30 times higher than the background value for sandy sediments and about 18 times the background value for calcareous rocks (1.7 mg/kg). In local marine sediments, the average As concentration is 10-fold the background value for near shore mud (5 mg/kg).

3.2 Geochemical control of As distribution

The enrichment of beach sands with As seems to be the successive result of the diagenetic redistribution in near-shore sediment. Calcareous clastic material is the one that previously composed the upper oxidized layer of near-shore sediment, the skeleton particles of calcareous algae (*Corallina pamizzoi*) and corals on the uppermost stratum of the local sediments. Deposition of Fe(III) hydroxides occurs in this stratum since they adsorb the As(III) and/or As(V) ions diffusing upward from the suboxic sediment horizon. Surfaces of calcareous particles serve as the substrate for the fixation of Fe(III) hydroxides enriched with As.

Analyses of sediment and water sampled in a borewell 3.5 km away from the coastline in the inner part of the beach ridge in the Açu region are shown in Fig. 1. It may be observed that the peak of As concentration was registered at the depth of 15 m, which coincides with the end of the swampy period of the profile development and the beginning of the sediment overlapping eolian sands with consequent drought of the environment.

Elevated concentration of As in the lowest part of the profile coincided with the remarkable presence of sulfides in all sediment samples collected from 15 to 20 m interval (Fig. 1). High concentrations of As most likely appear in groundwater as a result of sulfide complex decomposition through oxidation. The pumping of water from wells can contribute to the penetration of oxygen into the deepest levels of the aquifer and lead to the increase in As mobility.

3.3 Biota arsenic speciation

Seven inorganic and organic As species were identified: Arsenobetaine (AsB), Monomethylarsonic Acid (MMA), Dimethylarsinic Acid (DMA), Monomethyl-Arsonic Acid (MMA), Arsenite [As(III)], Arsenate [As(V)] and p-Arsanilic Acid (p-ASA) in the muscle and liver of three species of marine fish (Bagre bagre, Isopisthus parvipinnis, Larimus breviceps) sampled in the coastal Açu region, Rio de Janeiro. Total As in muscle tissue of fish had contents between 2.3 µg/g and 80.7 µg/g, above the guidelines issued by the Brazilian Health Surveillance Agency (1.0 µg/g). The organic-As species (AsB, MMA, DMA, p-ASA) were prevalent in all liver and muscle samples, representing more than 95% of total As. Arsenobetaine (AsB) was the dominant compound in all samples. Hazards to human health due to fish consumption seem to be low, since the As organic fraction is little toxic and AsB is considered a non-toxic species.

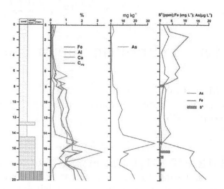

Figure 1. Grain size; Fe, Al, Ca, Corg, As in sediments; Fe, As in water extracts and total sulfide (S^{2-}) along the core of Açu coastal area.

4 CONCLUSIONS

The Brazilian tropical coast has high contents of As in marine sediments, beaches and groundwater due to several geochemical processes, such as early diagenetic, and the process of coastal sediment evolution. High concentrations of As in shallow groundwater in the Paraiba do Sul delta seems to be the result of the release of As fixed by authigenic sulfides.

However, neither some aspects of processes which lead to As enrichment nor the extent of the affected areas have been thoroughly understood. Besides, the source of As has not been identified, despite the fact that the Barreiras formation may be acting as a source to enrich carbonate sand beaches.

There is wide domain of organic As in fish found in the coastal Açu region, Rio de Janeiro, evidence of low risk to human health when the fish is consumed.

High As contents in marine sediments and groundwater cause environmental limitations to the construction of industrial and port facilities, besides dredging operations and may pose hazards to human health.

ACKNOWLEDGEMENTS

The research was funded by Brazilian National Research Council through grant 470541/2010-5.

REFERENCES

CONAMA 2012. Ministério do Meio Ambiente, Conselho Nacional do Meio Ambiente. Resolução nº 454, 01 novembro 2012 (in Portuguese).
Mirlean, N., Baisch, P., Travassos, M.P., Nassar, C. 2011. Calcareous algae bioclast contribution to sediment enrichment by arsenic on the Brazilian subtropical coast. Geo-Marine Lett. 31: 65–73.
Pansu M, Gautheyrou J. 2006. Handbook of Soil Analysis: Mineralogical, Organic and Inorganic Methods. New-York: Springer-Verlag.

Arsenic Research and Global Sustainability – Bhattacharya, Vahter, Jarsjö, Kumpiene,
Ahmad, Sparrenbom, Jacks, Donselaar, Bundschuh & Naidu (Eds)
© 2016 Taylor & Francis Group, London, ISBN 978-1-138-02941-5

Arsenic pollution in soils from the mining district of Linares, Spain

J. Rey[1], M.C. Hidalgo[1], U. Cortada[1] & J. Martínez[2]

[1]*Department of Geology, Campus Científico Tecnológico de Linares, University of Jaen, Spain*
[2]*Department of Mechanical and Mining Engineering, Campus Científico Tecnológico de Linares,*
University of Jaen, Spain

ABSTRACT: Intense mining of lead and copper sulfide deposits has been carried out for centuries in the district of Linares (Jaen, South Spain). These old mining works, abandoned today, generated large accumulations of wastes which were deposited without any prior corrective action. In this study, a geochemical characterization of the soils associated with a stream draining this mining district has been carried out. All the soil samples analyzed present high concentrations of metals and metalloids (Pb, Zn, Cu, Cd, Mn, As, Sb, and Ba). Of the 8 elements analyzed, As and especially Pb show average contents exceeding the limit established by the guidelines of the regional government for soils. The spatial distribution of As is intimately linked to a smelter slag, with a drastic decrease of the concentration values downstream this source of pollution.

1 INTRODUCTION

The study area is located in the abandoned mining district of Linares (southern Spain), which is characterized by the presence of galena ore veins (Azcárate, 1977). The abundance of this sulfide ore enabled the development of an important mining activity, which produced large numbers of wastes (Contreras & Dueñas, 2010). In parallel, an important mineralurgical (gravimetric and flotation processes) and metallurgical industry were developed, generating their own residues (smelter slags). All wastes were deposited on the ground without any pre-conditioning, being exposed to weathering and mobilization processes. Recent studies in this area (Martínez *et al.*, 2008a, b), thirty years after the mine closure, indicate high concentrations of metal(loid)s in soils (Pb, Zn, Cu, Cd, Mn, As, Sb, Ba), well above the limits permitted by regional and European regulations. This work focuses on the analysis of sediments associated with the main stream of this mining area (Baños creek, Fig. 1), paying special attention to As content in soils. In this sense, there are many potential sources of contamination that exist in the drainage basin of this stream, highlighting the presence of smelter slags, tailings and waste-rock heaps or even the industry associated with the city of Linares.

2 SAMPLING AND ANALYSIS

The soil sampling was conducted in 28 sites along the Baños creek, downstream the La Cruz smelter

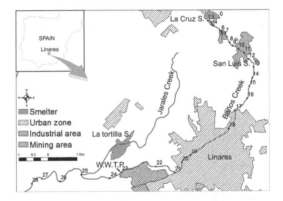

Figure 1. Location of the study region. The position of the 28 sampling sites along the Baños creek is indicated.

(Fig. 1). In each site, soils were sampled at two different positions across the stream section: the stream channel and the floodplain. At the laboratory, samples were dried, homogenized, and crushed to pass through a 2 mm sieve. pH, carbonate content, particle size distribution and organic matter content were analyzed according to UNE 10390:2012, UNE 103200: 1993, 103101:1995, and UNE 103204:1993 rules, respectively. Then a 1 g subsample was ground to < 63 µm for trace element determination, following the procedure described in Ordoñez *et al.* (2003) and Ferreira-Baptista & De Miguel (2005). The total metal content in the 56 soil extracted analytes were determined by inductively coupled plasma mass spectrometry (ICP-MS) analysis.

3 DISCUSSION AND CONCLUSION

High contents of Cu, Pb, Zn, As, Cd, and Sb have been detected in the Baños stream bed (Table 1). It is worth to highlight the concentrations for Pb (10363 mg/kg) and As (166 mg/kg), with average values far exceeding the threshold established by the regional government (275 and 36 mg/kg, respectively). In the case of As, a very high concentration is obtained close to the former La Cruz smelter, where the total content in As exceeded more than 20 times the permissible levels established by the

Table 1. Trace element contents in the Baños Creek samples.

Samples	Cu	Pb	Zn	As	Cd	Sb
1.1	59	2149	171	1921	4,01	139
2.1	38	1094	89	226	2,56	13
3.1	754	25013	3195	319	31,5	147
4.1	149	7546	770	774	6,67	73
5.1	80	1805	192	478	6,58	30
6.1	194	9027	148	345	2,07	24
7.1	441	7691	174	39	1,03	21
8.1	480	14830	144	42	0,85	19
9.1	582	16532	197	41	0,83	24
10.1	426	14244	178	47	1,28	31
11.1	831	17090	302	55	1,37	22
12.1	1236	18365	149	53	0,69	28
13.1	440	9473	97	34	0,35	24
14.1	242	7077	2514	17	0,19	130
15.1	183	5516	1529	23	0,19	115
16.1	232	9202	779	27	0,35	65
17.1	127	3733	423	16	0,22	23
18.1	148	4574	104	17	0,42	6
19.1	171	8198	157	22	0,25	13
21.1	73	1622	215	12	0,32	4
22.1	472	75912	84	56	0,17	72
23.1	77	5096	757	15	0,28	9
24.1	52	5482	94	14	0,09	7
25.1	142	9256	271	20	0,31	28
26.1	45	2145	137	13	0,24	5
27.1	48	3452	122	10	0,22	6
28.1	33	1906	109	10	0,21	6
1.2	40	852	123	1172	2,32	117
2.2	829	8761	6263	1824	38,6	131
3.2	110	1437	67	19	0,98	8
4.2	95	2835	156	599	2,37	39
5.2	201	3656	254	426	4,79	33
6.2	354	16785	822	683	7,21	63
7.2	332	10866	229	37	0,87	20
8.2	621	25027	180	47	1,28	32
9.2	1161	38693	264	51	1,42	39
10.2	647	27408	201	62	1,61	53
11.2	683	15376	308	56	1,24	34
12.2	609	23879	146	59	0,7	30
13.2	917	37355	196	93	1,02	42
14.2	557	22446	178	47	0,7	32
15.2	185	8080	62	20	0,19	11
16.2	203	7019	236	21	0,26	18
17.2	152	5543	133	21	0,34	11
18.2	113	1683	71	17	0,26	3
19.2	198	9496	192	22	0,47	12
20,2	33	1374	73	21	0,2	5
21.2	109	8252	217	20	0,32	10
22.2	76	4358	192	16	0,3	6
23.2	118	4335	801	20	1,57	9
24.2	89	4882	183	16	0,3	7
25.2	114	6517	209	16	0,37	11
26.2	50	2182	323	13	0,18	4
27.2	78	3321	272	11	0,59	5
28.2	123	7845	326	20	0,94	15

Channel (mg.kg) for samples 1.1–28.1; Flood plain (mg.kg) for samples 1.2–28.2.

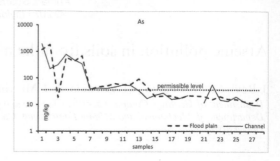

Figure 2. Arsenic distribution across the Baños Creek.

regional government for agricultural soils. Taking into account that concentrations rapidly decrease with the distance to the smelter pile (Fig. 2), it could be considered that tailings ponds located downstream in the vicinity of the riverbed do not constitute sources of As pollution (although this is not the case for other elements such as Pb). Both floodplain and stream channel samples have shown similar distribution for trace elements. Although the maximum values do not match spatially, they seem related to the same pollutant factor.

ACKNOWLEDGEMENTS

This research was funded by the Spanish Ministry of Science and Innovation (Project CGL2013–45485-R) and by the Government of Junta de Andalucía (Project RNM 05959).

REFERENCES

Azcárate, J.E. 1977. Mapa geológico y memoria explicativa de la hoja 905 (Linares), escala 1:50.000. Instituto Geológico y Minero de España, p. 35.

Contreras, F., Dueñas, J. 2010. La minería y la metalurgia en el Alto Guadalquivir: Desde sus orígenes hasta nuestros días. Instituto de Estudios Giennenses. Diputación de Jaén.

Ferreira-Baptista, L., De Miguel, E., 2005. Geochemistry and risk assessment of street dust in Luanda, Angola: A tropical urban environment. *Atmos. Environ.* 39: 4501–4512.

Martínez, J., Llamas, J.F., De Miguel, E., Rey, J., Hidalgo, M.C. 2008a. Soil contamination from urban and industrial activity: example of the mining district of Linares (southern Spain). *Environ. Geol.* 54:669–677.

Martínez, J., Llamas, J.F., De Miguel, E., Rey, J., Hidalgo, M.C. 2008b. Multivariate analysis of contamination in the mining district of Linares (Jaen, Spain). *Appl. Geochem.* 23: 2324–2336.

Ordoñez, A., Loredo, J., De Miguel, E., Charlesworth, S., 2003. Distribution of heavy metals in street dust and soils of an industrial city in Northern Spain. *Arch. Environ. Con. Tox.* 44: 160–170.

Arsenic Research and Global Sustainability – Bhattacharya, Vahter, Jarsjö, Kumpiene,
Ahmad, Sparrenbom, Jacks, Donselaar, Bundschuh & Naidu (Eds)
© 2016 Taylor & Francis Group, London, ISBN 978-1-138-02941-5

Geomorphic regulation of arsenic in soils of Bangladesh

M.T.A. Chowdhury[1,2], C. Deacon[1], P.N. Williams[3], A.A. Meharg[3], A.H. Price[1] & G.J. Norton[1]

[1]*Institute of Biological and Environmental Sciences, University of Aberdeen, Aberdeen, UK*
[2]*Department of Soil, Water and Environment, University of Dhaka, Dhaka, Bangladesh*
[3]*Institute for Global Food Security, Queen's University, Belfast, Belfast, Ireland*

ABSTRACT: Paddy and adjacent non-paddy soil samples from different physiographic regions of Bangladesh were analyzed for total arsenic (As) and a number of chemical elements (Al, Ca, Cd, Co, Cr, Cu, Fe, K, Mg, Mn, Mo, Na, Ni, P, Pb and Zn) to assess the geomorphological and biogeochemical regulation of As in Bangladeshi soils. The mobility and interplay of As in paddy soils from different physiographic regions were also assessed by using diffusive gradients in thin-films (DGT) under anaerobic conditions. This study reveals that As concentrations in Bangladeshi soils vary across the physiographic regions and are potentially linked to inputs of As contaminated groundwater. Significant biogeochemical relationships of As in the soils indicate a biogeochemical control on its occurrence in the soils.

1 INTRODUCTION

Arsenic (As) contamination of groundwater and soil is a serious environmental problem in Bangladesh. Arsenic in the groundwater is unevenly distributed within the physiographic regions and its occurrence shows a close relationship with the surface geology and geomorphology of the country (BGS & DPHE, 2001). Hence, it is crucial to identify the variations in As concentrations along with the underlying biogeochemistry within the physiographic regions to better identify the parameter(s) affecting the concentration and mobilization of As in the environment. In addition, there is very limited data on As in the alluvial sediments as well as in paddy field soils of Bangladesh, that would allow a local and regional picture (Brammer, 2012). This study aims to generate a comprehensive database of Bangladeshi soil As and to assess the geomorphic regulation of As in Bangladeshi soils by understanding its biogeochemical cycling and dynamics within the different geomorphological settings.

2 METHODS

Paddy (n = 1216) and adjacent non-paddy (n = 241) soil samples (0–15 cm) from 11 physiographic regions of Bangladesh were analyzed, using ICP-MS and MP-AES, for total concentrations of Al, As, Ca, Cd, Co, Cr, Cu, Fe, K, Mg, Mn, Mo, Na, Ni, P, Pb and Zn. Prior to analysis, all the samples were oven-dried, ball-milled and digested with a mixture of nitric acid and hydrogen peroxide using a block digester. Diffusive gradients in thin-films (DGT, *Fe-oxide gel*) were used under controlled anaerobic conditions to assess the porewater dynamics and the mobility and interplay of As in paddy field soils from 6 physiographic regions.

3 RESULTS AND DISCUSSION

3.1 *Soil arsenic concentrations*

Arsenic concentrations in the paddy and non-paddy soils from different physiographic regions ranged from 0.6–88 (mean 7.7) and 1.8–24 (mean 5.4) mg/kg, respectively. Arsenic concentrations in the paddy soils were found to be significantly ($p < 0.001$) different from the non-paddy soils, the concentrations being on average higher in the paddy soils compared to the adjacent non-paddy soils. Since the concentration of As varied in the soils, this suggests a natural source of As in the soils with an external source of As being applied to the soils cultivated with rice which is in agreement with Huq & Shoaib (2013) and Meharg & Rahman (2003).

3.2 *Geomorphic variations in soil arsenic*

Arsenic concentrations in the paddy and non-paddy soils showed significant variations ($F = 83.5$; $p < 0.001$ and $F = 5.51$; $p < 0.001$, respectively) between the physiographic regions (Figure 1). Ganges Tidal Floodplain and Ganges River Floodplain showed wider variability in paddy soil As, whereas the paddy soils of Madhupur Tract and Barind Tract had lower As concentrations. Similar geomorphic distribution pattern has also been reported for groundwater As concentrations in Bangladesh (BGS & DPHE, 2001). Statistical analysis of the data indicated that Arial Bil, Gopalganj-Khulna Bils, Ganges River Floodplain and Ganges Tidal Floodplain (A) were significantly higher than the other regions in having As in the paddy soils, while Brahmaputra Floodplain (B) differed significantly from the other regions. Meghna Estuarine Floodplain (C) showed similarity in paddy soil As with Karatoya-Bangali Floodplain and Madhupur tract, the latter being

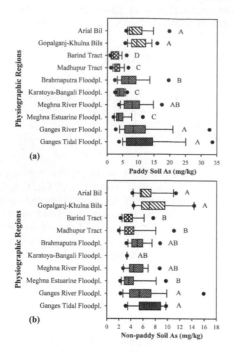

(a)

(b)

Figure 1. The range of arsenic in the paddy (a) and non-paddy (b) soils from the physiographic regions. The no. of samples at each of the physiographic regions, in the order from the top: 42, 63, 67, 118, 207, 15, 166, 224, 263, 49 (a); 10, 8, 15, 15, 65, 1, 28, 28, 59, 11 (b). The letters indicate groupings with significant differences (paddy soils and non-paddy soils were analyzed separately).

Table 1. Soil and DGT characteristics.

Parameters	Physiographic regions*					
	1	2	3	4	5	6
Soil As (mg/kg)	1.4	9.6	2.5	4.7	8.5	8.0
Porewater As (μg/L)	31	23	64	65	21	21
DOC (mg/L)	143	192	149	105	40	42
$^{As}C_{DGT}$ (μg As/L)	14	17	48	43	15	11
^{As}R (ratio)	0.46	0.69	0.73	0.65	0.71	0.53
$^{As}K_d$ (L/kg)	52	2235	100	102	586	1057

*1 = Barind Tract; 2 = Gopalganj-Khulna Bils; 3 = Brahmaputra Floodplain; 4 = Karatoya-Bangali Floodplain; 5 = Ganges River Floodplain; 6 = Ganges Tidal Floodplain.

low in As were found to have a strong potential to release and mobilize As in the porewaters with a dynamic resupply of the element from solid phase, which was also reported by Williams et al. (2011).

4 CONCLUSIONS

Soils of the paddy fields in Bangladesh contain high concentrations of As compared to the soils of the surrounding non-paddy areas. Arsenic concentrations in Bangladeshi soils vary across the physiographic regions and are potentially linked to inputs of As-contaminated groundwater. The soils from different physiographic regions have different potentials to release and mobilize As in the paddy fields.

ACKNOWLEDGEMENTS

The authors acknowledge the Commonwealth Scholarship Commission for a doctoral scholarship and Bangladesh-Australia Centre for Environmental Research of the University of Dhaka for the support during soil collection and sample processing.

different from Barind Tract (D) which was the lowest. The Tracts (B) were similar in non-paddy soil As and were significantly lower than most other regions, except the Brahmaputra floodplain.

3.3 Geochemical Relationships of Arsenic in Soils

Paddy soil As showed significant correlations ($p < 0.001$; $r \geq 0.27$) to Al, Ca, Co, Cu, Fe, K, Mg, Mn, Ni, P, Pb and Zn. The non-paddy soil As had significant relationships ($p < 0.001$; $r \geq 0.29$) with Al, Ca, Co, Cu, Fe, K, Mg, Mn, Ni, Pb, and Zn.

3.4 Porewater and DGT Measured Arsenic

Porewater As concentrations in the paddy soils were variable, the highest concentrations were in the soils from Brahmaputra Floodplain and Karatoya-Bangali Floodplain, which also had the maximum labile As, measured by DGT ($^{As}C_{DGT}$), compared to the other regions during the anaerobic incubation (Table 1). High ^{As}R (ratio, $^{As}C_{DGT}$/ porewater As) and $^{As}K_d$ (distribution coefficient, total soil As/porewater As) values indicated considerable resupply of As and a large pool of As available for resupply in the soils. DOC concentrations in the paddy soils varied significantly ($F = 5.77$; $p = 0.001$) among the regions. The soils

REFERENCES

BGS and DPHE (British Geological Survey and Department of Public Health Engineering) 2001. Arsenic contamination of groundwater in Bangladesh. In D.G. Kinniburg & P.L. Smedley (eds), Final Report: Vol. 2. BGS: Keyworth.

Brammer, H. 2012. The Geography of the Soils of Bangladesh. The University Press Ltd.: Dhaka, Bangladesh.

Huq, S.M.I., Shoaib, J.U.M. 2013. The soils of Bangladesh. Dordrecht: Springer.

Meharg, A.A., Rahman, M. 2003. Arsenic contamination of Bangladesh paddy field soils: implications for rice contribution to arsenic consumption. Environ. Sci. Technol. 37: 229–234.

Williams, P.N., Zhang, H., Davison, W., Meharg, A.A., Hossain, M., Norton, G.J., Brammer, H., Islam, M.R. 2011. Organic matter-solid phase interactions are critical for predicting arsenic release and plant uptake in Bangladesh paddy soils. Environ. Sci. Technol. 45: 6080–6087.

Arsenic Research and Global Sustainability – Bhattacharya, Vahter, Jarsjö, Kumpiene,
Ahmad, Sparrenbom, Jacks, Donselaar, Bundschuh & Naidu (Eds)
© *2016 Taylor & Francis Group, London, ISBN 978-1-138-02941-5*

Enrichment of arsenic and trace metals in the surface sediments of the Indus River, Pakistan

J.A. Khattak, A. Farooqi, S. Bano & A. Raza
Environmental Geochemistry Laboratory, Department of Environmental Sciences, Faculty of Biological Sciences, Quaid-i-Azam University, Islamabad, Pakistan

ABSTRACT: The study was conducted to evaluate the concentration, distribution and sources of arsenic and trace metals in sediments of the Indus River, Pakistan. Sediment samples were collected from different locations along the Indus River and analyzed for major physicochemical properties. XRD results showed that the major minerals present were of quartz, zinnwaldite, albite, calcite and clinochlore. XRF results have confirmed the presence of metal oxides SiO_2, Fe_2O_3, Al_2O_3 in major quantities. The arsenic content in sediment samples range from 3.8 to 115 mg/kg. Sample locations at Kot Mithan and Sukkar Barrages have higher As contamination due to the presence of sugar mills and minerals such as phyllosilicates; kaolinite, mica, and chlorite of these two sites. All the trace metals were found to be within the permissible limits of Scottish Environmental Protection Agency (SEPA) except Ni and Mn at some sites. Pollution load index values of all sites revealed that Sukkar Barrage was the most polluted site in terms of trace metal pollution and arsenic.

1 INTRODUCTION

Contaminants from agricultural, municipal, and industrial sources enter into the water bodies and eventually accumulate in the sediments. Among these environmental pollutants heavy metals have been of great concern because of their abundance, persistence, toxicity in aquatic environments (Beg *et al.*, 2001). Heavy metal contamination and arsenic has been reported in river sediments (Yousafzai *et al.*, 2008). The Indus River is under severe threat of pollution from industries, therefore, it is very important to assess the pollution level of the river sediment (Nickson *et al.*, 2005). The objectives of the study were to evaluate the distribution pattern of arsenic and other heavy metals in surface sediments of the Indus River, to assess the pollution load of heavy metals, and to determine the natural and/or anthropogenic sources of these heavy metals.

2 METHODS/EXPERIMENTAL

2.1 *Study area and sampling*

Figure 1 shows the sampling sites: Taunsa Barrage (Site A), Kot Mithan (Site B), Guddu Barrage (Site C), Sukkar Barrage (Site D) along the Indus River. These sites were selected because of their proximity to the industrial area and agricultural land (Fig.1). A total of 33 sediment samples were collected from these barrages including upstream, downstream, left and right banks and middle of river covering all the directions. The collected samples were then placed in to polyethylene bags, air dried and transferred to the laboratory.

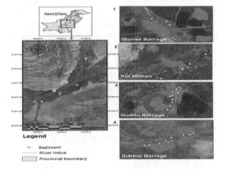

Figure 1. Map of the study area.

2.2 *Data analysis*

The samples were analyzed for trace metals (Ni, Mn, Zn, Cu, Pb, Fe, Cd and Co) using Sequential Flame Atomic absorption Spectrophotometer (Varian Spectra AA-240) under standard operating conditions having r > 0.99. Data were evaluated through the use of field blank, calibration blank, spike samples and replicates for quality control. Similarly, accuracy of the analytical method for arsenic and trace elements was verified by analyzing geological standard Reference Materials (JSL-1 & JSD-1). During the entire experimental process all reagents used were certified and all equipment was soaked in 10% HNO_3 for at least 48 hours prior to their use. Geoaccumulation index, CF and PLI was determined by the following equations; $I_{geo} = \log_2 (Cn/1.5 \text{ Bn})$; $CF = C_{heavy\ metal}/C_{background}$ and $PLI = (CF_1 \times CF_2 \times CF_3...xCFn)^{1/n}$.

3 RESULTS AND DISCUSSION

3.1 Mineralogy and geochemical composition of surface sediments

XRD analysis of sediment samples reveals that the most abundant mineral in each sample is quartz (SiO_2), followed by zinnwaldite KLiFeAl $(AlSi_3)O_{10}$ $(OH, F)_2$, albite $(NaAlSi_3O_8)$, calcite and clinochlore whereas, clay minerals such as Albite and calcite are mostly observed in fine-grained sediments. Silica concentrations (wt%) observed in these four sites are 54.49%, 52.81%, 51.92%, 53.95%, respectively. The high concentration of silica (SiO_2) manifests to the quartz-rich nature of the sediments as well as enrichment of clay minerals. The average concentration (wt%) of Al_2O_3 in sediments is 10.45, 11.22, 11.1 and 10.11, respectively, which may be due to the presence of clay minerals. The average concentration (wt%) of Fe_2O_3 for sediments at site A is 5.7, at site B and C it was 6.6 and 6.7 respectively while at site D it is higher 11.0. These trends of Fe_2O_3 concentration are similar to that of Al_2O_3 (Table 1). MnO_2 content is low followed by P_2O_5 throughout all sediment and soil samples. The P_2O_5 content in four sites is generally low and correlate well with CaO content, suggesting that P in the sediments is fixed in apatite or other Ca-bearing minerals. The concentration of MgO at these four sites are 3.14, 3.23, 3.75 and 0.12 (wt%) respectively. The concentration of CaO is almost similar in first three sites (8.9, 7.9, 8.2 wt% respectively) while for site D it is lower (3.45 wt%) as compared to other sites. At site D the concentration of Na_2O is higher (7.77 wt%) as compared to other sites where the concentration is 2.46, 2.88 and 2.64 (wt%) respectively. The concentration of K_2O is higher at sites B and C (3 wt%) as compared to other sites, probably due to the association with muscovite and biotite. Similarly, the presence of large amount of phyllosilicates such as kaolinite, mica, and chlorite can enhance the concentration of these elements in sediment samples.

3.2 Arsenic and trace metals in the sediments

Table 2 shows that the mean concentration of As in mg/kg at site A is 14.34 followed by 41.11, 10.32 and 24.78 at site B, C and D respectively.

The elevated level of As at site B could be due to effluent of sugar mills which contains heavy metals like arsenic, cadmium, copper, lead, mercury, zinc, chromium. The concentration of Ni crossed the permissible limits of Scottish Environmental Protection Agency (SEPA) i.e. 60 mg/kg in all the sediment samples. The high level of Ni is due to open disposal of solid waste of ghee industries. At site A all of the samples for Mn lie within the permissible limit for Mn whereas at site B and C, 33.33% of the samples exceed the permissible SEPA limit, which could be due to the presence of agrochemical and sugar industries. The Cu, Zn and Pb concentrations in all the sediment samples were found to be within the permissible SEPA limit.

Table 2. Concentration of As and trace elements in the sediments of River in mg/kg.

Sites	A	B	C	D
As	14.34	41.11	10.32	24.78
Ni	157.21	189.31	160.1	187.3
Mn	287.02	229.83	369.18	314.49
Cr	7.59	5.61	11.31	13.57
Cu	13.79	20.36	26.52	30.99
Co	107.49	113.72	149.24	159.24
Pb	46.63	86.63	122.63	159.29
Fe	719.46	3191.46	3115.12	4816.46
Zn	28.87	32.62	39.53	38.17

3.3 The Contamination Factor (CF) and Pollution Load Index (PLI)

The mean CF values for all metals range from 0.0008 to 5.65 for site A, 0.003 to 5.98 for site B, 0.003 to 7.85 for site C and 0.005 to 8.38 for site D. The higher CF values were observed for Co and lowest for Fe. The decreasing order of CF values for trace metals are Co > Pb > Ni > As > Cu > Mn > Zn. PLI of all metals are: 5.4 for site A, 8.3 for site B, 11.3 for site C, and 32 for site D. The high PLI value for site D indicates that it was the most polluted site.

4 CONCLUSIONS

In the study area SiO_2, Fe_2O_3, Al_2O_3 metal oxides were found in major quantities. The Site B and Site C were found to have higher As contamination due to presence of sugar mills and minerals such as phyllosilicates; kaolinite, mica, and chlorite of these two sites. Ni concentrations were high for all the sites due to waste of ghee industries. Overall, the concentrations of trace metal fall within the compositional range of other intertidal sediments throughout the world. High PLI indicates that Site D was most polluted site.

ACKNOWLEDGEMENTS

The authors acknowledge the Geological Survey of Pakistan (GSP) for analytical facilities.

REFERENCES

Bahloul, M., Al-Matrouk, K., Al-Obaid, T., Kurian A. 2001. Chemical contamination and toxicity of sediment from a coastal area receiving industrial effluents in Kuwait. Arch. Environ. Cont. Toxicol. 41: 289–297

Nickson, R., McArthur, J., Shrestha, B., Kyaw-Myint, T., Lowry, D. 2005. Arsenic and other drinking water quality issues, Muzaffargarh District, Pakistan. Appl. Geochem. 20: 55–68.

Yousafzai, A.M., Khan, A.R., Shakoori, A., 2008. Heavy metal pollution in River Kabul affecting the inhabitant fish population. Pakistan Journal of Zoology 40: 331–339.

Arsenic Research and Global Sustainability – Bhattacharya, Vahter, Jarsjö, Kumpiene,
Ahmad, Sparrenbom, Jacks, Donselaar, Bundschuh & Naidu (Eds)
© 2016 Taylor & Francis Group, London, ISBN 978-1-138-02941-5

Seasonal impact on arsenic and trace elements dispersal in agricultural soil of Gangetic Delta region of India

A. Barla, A. Majumdar, A. Shrivastava, S. Singh & S. Bose
Earth and Environmental Science Research Laboratory, Department of Earth Sciences,
Indian Institute of Science Education and Research Kolkata, Mohanpur, Nadia, West Bengal, India

ABSTRACT: Enumeration of the amount of arsenic and other selected trace metals (Fe, Cu, Zn, Pb and Cd) in soil samples of Chakdaha Block in Nadia district of West Bengal during the cultivation of two different rice varieties, Boro and Amon grown in different season's viz. winter-spring and monsoon, with a correlation of metal dispersion and temporal variation. Soil samples from control, and other experimental fields were processed following digestion-dilution by treating with strong acids and analysed to detect trace elements by ICP-MS. Statistical indices of pollution are analysed as Contamination Factor (CF), Pollution Load Index (PLI), Geo-accumulation Index (I_{geo}) and Enrichment Factor (EF) comparing with background concentration. ANOVA shows sampling variation in seasonal respect for same metals. The positive relation between temporal variation and increment of metal dispersion is quite clear. Surrounding anthropogenic contamination and status of contaminated irrigation water with respective parameters plays an important role here.

1 INTRODUCTION

The extensive and prolonged use of agro-chemicals in agricultural field is one of the major causes of accumulation of trace metals in soil. Distribution of arsenic and heavy metals in soil and determining their status of relative abundance in soil and plants provide a valuable access to the knowledge of soil pollution and soil health (Kuhad *et al.*, 1989, Pendias & Pendias, 1992). Contamination by high concentration of arsenic in groundwater increases total content of arsenic in agricultural field and this results into crops uptaking those metals depending on the availability rather than total content of metal in soil (Ahsan *et al.*, 2008, Dudka *et al.*, 1996, Garrett *et al.*, 1998, Moral *et al.*, 2002). Presence of high level of arsenic and heavy metals in agricultural fields of lower region has an adverse effect on crop productivity and quality being a potential health risk factor too.

2 METHODS/EXPERIMENTAL

2.1 *Sampling area*

Nadia district of West Bengal has a potential risk to be contaminated with high quantity of arsenic where our selected study area Chakdaha block has a coordination of 23°01'15.31"N latitude and 88°38'36.86"E longitude.

2.2 *Processing of soil samples*

Soil digestion and dilution process was followed as outlined by Bhattacharya *et al.* (2009). Next those samples were analysed by ICP-MS and graphical analysis along with statistical parameters like ANOVA were done. Indices of pollution, based on background data, were determined (Reddy *et al.*, 2004).

3 RESULTS AND DISCUSSION

The results show the concentration of selected metals in two respective years on the same sampling sites (Table 1). The study tried to find out the concentration (mg/kg) of arsenic (As) and iron (Fe) of soil samples along with four other trace metals (Cu, Zn, Pb and Cd). Significant variation in One-way ANOVA was observed as the $P < 0.0001$ was analyzed. A mix trend has been observed in both type of irrigation viz monsoonal and groundwater irrigation. During analysis of pollution indices, Contamination Factor (CF), Pollution Load Index (PLI), geo-accumulation index (I_{geo}), Enrichment Factor (EF), all shows the range of low to moderate to considerable amount of contamination. Among various types of soil pollutions, contamination with trace metals is an important factor due to their non-biodegradability and prolongs half-life (Shrivastava *et al.*, 2015).

Table 1. Metal concentration in soil with different sampling period in control field (CF) and experimental field (EF).

Metals	Sampling fields	Boro 2013 (n = 40)	Boro 2014 (n = 60)	Amon 2013 (n = 18)	Amon 2014 (n = 45)
Fe	C F	5246 ± 7	6747 ± 6.2	11841 ± 28	32722 ± 117
	E F	20277 ± 12	11379 ± 5.8	12089 ± 39	37785 ± 92
As	C F	7.8 ± 0.3	7.9 ± 0.33	4.4 ± 0.2	11.1 ± 0.3
	E F	47.5 ± 0.6	57.9 ± 0.4	24.5 ± 0.4	36.8 ± 0.3
Zn	C F	27.0 ± 0.38	80.3 ± 0.2	46.0 ± 0.1	113.81 ± 0.9
	E F	41.3 ± 0.37	118.8 ± 0.2	40.3 ± 0.1	112.3 ± 0.4
Pb	C F	4.76 ± 0.1	2.42 ± 0.1	5.63 ± 0.1	4.03 ± 0.03
	E F	16.9 ± 0.1	9.8 ± 0.13	4.83 ± 0.2	4.92 ± 0.03
Cd	C F	0.15 ± 0.01	0.15 ± 0.07	0.35 ± 0.02	0.32 ± 0.01
	E F	0.36 ± 0.02	0.41 ± 0.06	0.32 ± 0.02	0.57 ± 0.02
Cu	C F	38.6 ± 0.3	46.28 ± 0.12	24.0 ± 0.1	64.2 ± 0.41
	E F	68.0 ± 0.4	84.65 ± 0.15	28.9 ± 0.1	63.29 ± 0.12

4 CONCLUSIONS

This study is quite important to show the movement of trace metals through water to soil accumulating in a higher degree. The concentration (mg/kg) of studied metals of soil samples help to assume the situation of agricultural soil in respect of increasing metal(loid)s changing with time due to the continuous use of contaminated groundwater. Groundwater in the area is quite enriched in trace metal(loid)s which enhance the dispersion of elements in agricultural soil.

ACKNOWLEDGEMENTS

SB is thankful to DST Government of India for providing Ramanujan Fellowship Research Grant, (SR/S2/RJN-09/2011) to carry this research.

REFERENCES

Ahsan, D.A., Del Valls, A.C., Balsko, J. 2008. Distribution of arsenic and trace metals in floodplain agricultural soil of Bangladesh. B. Environ. Contam. Tox. 82: 11–15.

Bhattacharya, P., Samal, A.C., Majumdar, J., Santra, S.C. 2009. Transfer of arsenic from groundwater and paddy soil to rice plant (Oryza sativa L.): a micro level study in West Bengal, India. World J. Agric. Sci. 5: 425–431.

Dudka, S., Piotrowskab, M., Terelakbet, H. 1996. Transfer of cadmium, lead and zinc from industrially contaminated soil to crop plants: a field study. Environ. Pollut. 94(2):181–188.

Garrett, R.G., MacLaurin, A.I., Gawalko, E.J., Tkachuk, R. (1998). A prediction model for estimating the cadmium content of durum wheat from soil chemistry. J. Geochem. Explor. 64(1–3): 101–110.

Kabata-Pendias, A., Pendias, H. 1992. Trace Elements in Soil and Plants. 2nd Edition. CRC Press, Boca Raton, FL.

Kuhad, M.S., Mallick, R.S., Shingh, S., Daihiya, I.S. 1989. Background level of heavy metals in agricultural soil of Indo-Gangetic plains of Haryana. J. Ind. Soc. Soil Sci. 37: 700–705.

Moral, R., Gilkes, R.J., Moreno-Caselles, J. 2002. A comparison of extractants for heavy metals in contaminated soils from Spain. Commun. Soil Sci. Plan. 33(15–18): 2781–2791.

Reddy, M.S., Basha, S., Kumar, V.G.S., Joshi, H.V., Ramachandraiah, G. 2004. Distribution, enrichment and accumulation of heavy metals in coastal sediments of Alang-Sosiya Ship Scrapping Yard, India. Mar. Pollut. Bull. 48: 1055–1059.

Shrivastava, A., Barla, A., Yadav, H., Bose, S., 2014. Arsenic contamination in shallow groundwater and agricultural soil of Chakdaha block, West Bengal, India. Front. Environ. Sci. 2: 50 (DOI 10.3389/fenvs.2014.00050).

Arsenic Research and Global Sustainability – Bhattacharya, Vahter, Jarsjö, Kumpiene,
Ahmad, Sparrenbom, Jacks, Donselaar, Bundschuh & Naidu (Eds)
© 2016 Taylor & Francis Group, London, ISBN 978-1-138-02941-5

Arsenic behavior in deep-sea sediments from Nankai Trough

H. Masuda[1], H. Yoshinishi[1], E. Even[1] & S. Fuchida[2]

[1]*Department of Geosciences, Osaka City University, Osaka, Japan*
[2]*National Institute of Environmental Science, Tsukuba, Japan*

ABSTRACT: Arsenic (As) behavior in between interstitial water and sediments of deep-sea cores from Nankai Trough, located at off-shore Japan, were studied to understand the relation to biological activity at the early stage of diagenesis. High As concentration of sediments was recorded at about 350 m depth, where the most of As would be originated from detrital minerals supplied via turibidite from Honshu Island. Pelagic sediments in accretionary prism contained rather smaller As compared with the turbidite layers. Those facts indicate that the subareal insoluble materials are the major source of As in the marine sediments. While, the peak maximum of interstitial water As was at 200 meters below seafloor probably due to decomposition of algae/plant, suggesting that the biogenic As is mobile in the sediments.

1 INTRODUCTION

To understand the Arsenic (As) cycle in the Earth's crust, it must be important to document the accumulation and releasing mechanisms of As in sediments. Here, the As behavior was studied on the sediment column taken from Nankai Trough by a series of IODP drillings. The Nankai Trough is a modern subduction zone located at the off-shore of Honshu Island, Japan on Pacific side. The cored sediments were taken from Kumano frontal basin down to 2200 m from 4000 m depth seafloor to underlying accretionary prism. The site is below the stream of Kuroshio, and biological productivity is very high. The sediment column is good to document the primary source(s) of As and diagenetic reaction related to the biological reactions during early diagenesis.

2 METHODS

2.1 Samples

The samples were taken from Site C0002 at Kumano Basin in Nankai Trough during a series of IODP program, Expeditions 315 (0–200 and 470–1050 mbsf, meters below seafloor), 338 (200–500 and 1100 mbsf) and 348 (2170–2217 mbsf). Interstitial water was squeezed by the traditional way of IODP onboard analyses. The squeezed cakes taken in Exps. 338 and 348 were kept in a freezer until freeze-dry before the analyses. The samples taken in Exp. 315 were requested from the preserved samples in the refrigerated vault at Kochi Core Center.

2.2 Arsenic analyses

Arsenic concentration of interstitial water was determined by ICPMS. For the sediment samples, the concentration was determined by ICPMS after alkaline fusion and acid decomposition. The obtained values were calibrated using the standard solutions prepared with standard sedimentary rock samples (distributed by GSJ-AIST) in the same manner as the sediment samples.

Arsenic in the different phases of sediments was determined by BCR differential chemical extraction (Rauret *et al.*, 1999).

3 RESULTS AND DISCUSSION

3.1 Arsenic profiles in the sediment column

Figure 1 gives the depth profiles of some interstitial water chemistry (Kinoshita et al., 2009; Strasser *et al.*, 2014) and As concentrations of sediments and interstitial water. In this sediment column, methane gas hydrate layer was observed at 200–500 mbsf, where the chlorinity lowered to 120 mM. Bromide concentration changes in accordance with chlorinity below 200 mbsf, however, it increases with depth in the interstitial water in the uppermost 200 m depths. In that depth interval, ammonium ion concentration similarly increases with bromide concentration, suggesting that algae and/or plant is decomposed to release those components into the interstitial water.

Average As concentrations of interstitial water and sediment are 50 µg/L and 5.0 mg/kg respectively. The profiles of those concentrations with depth have two peaks. The highest peak of sediment As is at 350–400 mbsf, where methane hydrate occurs. The second peak appeared at 200 mbsf, consistently the peak of bromide and ammonium concentration maxima. The second peak is rather obvious in the interstitial water than in the sediment, implying that the host of As, of which concentration is not high in the sediments, is easy to breakdown.

No reliable As concentration of interstitial water was obtained for the deep sediment samples, while that of sediments (2173–2217 mbsf) were 1–6 ppm (average: 2.9 ppm), and lower than those from the shallower sediments described above. Thus, the pelagic sediments in the accretionary prism do not contain high amount of As compared with the coastal sediments.

Based on the BCR analysis, a half of As in the sediments was as insoluble phases such as silicates and sulfides, while another half was as adsorbed phases onto the mineral surface.

3.2 Interaction of arsenic in sediment column

The profiles of As concentrations in the sediment column are indicative of the As sources; most of the As is in the detrital mineral(s)/volcanic glass originated from Japan Island. Algae and/or plants would be another As source, which supply rather mobile As than the detrital minerals.

Figure 2 shows the relationship of ratio of As concentrations of interstitial water and host sediment with depth. The ratios are obviously large at 200 and 400 mbsf (A and B in Figure 2), where the peaks of As concentrations were observed as described above. The As was rather mobile and released via decomposition in the uppermost 200 m depth interval.

In this sediment column, methane and other hydrocarbons, which comprise the methane hydrate layer, were mostly biogenic origin (Moore et al., 2013). Thus, it is plausible that the microbial decomposition of As-bearing organic matters of algae and/or plants is responsible for the high concentration in the interstitial water.

Below 200 mbsf, the As concentration of interstitial water would be controlled by the adsorption-desorption equilibrium. It is not clear why the ratio becomes high at that depth, although maturation of organic matters would play an important role to release and fix the As in the sediments. The maximum decomposition temperature of hydrocarbon in the sediment, commonly used as the indicator of maturity of organic matters, increased drastically at the depth of methane hydrate layer, suggesting that a part of As was expelled when the hydrocarbon polymerized.

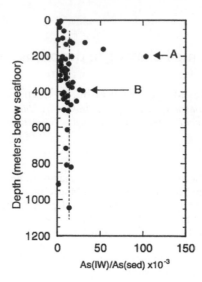

Figure 2. Relationship between ratio of As of sediment and interstitial water and depth.

4 CONCLUSIONS

The studied sediment column provides a good example that the decomposition and maturation of organic matters would be primary important factors to release As into the interstitial water from sediment. Such an As release would cause the risk when the methane hydrate will be economically recovered.

ACKNOWLEDGEMENTS

We thank to Clues of drilling vessel "Chikyu" and Shipboard Scientific Parties of IODP Expeditions 338 and 348. The preserved drilled samples of Exp. 315 was supplied by Kochi Core Center. This work is financially supported by Post-cruise studies by J-DESC.

REFERENCES

Kinoshita, M., Tobin, H., Ashi, J., Kimura, G., Lallemant, S., Screaton, E.J., Curewitz, D., Masago, H., Moe, K.T., and the Expedition 314/315/316 Scientists, 2009. Proceedings of the Integrated Ocean Drilling Program, Volume 314/315/316. Integrated Ocean Drilling Program.

Rauret, G., López-Sánchez, J.F., Sahuquillo, A., Rubio, R., Davidson, C., Ure, A., Quevauviller, Ph. 1999. Improvement of the BCR three stepsequential extraction procedure prior to the certification of new sediment and soil reference materials. J. Environ. Monit. 1: 57–61.

Strasser, M., Dugan, B., Kanagawa, K., Moore, G.F., Toczko S., Maeda, L. and Expedition 338 Scientists. 2014. Proceedings of the Integrated Ocean Drilling Program, Volume 338 Expedition reports, NantoroSEIZE Stage 3: NantoroSEIZE Plate Boundary Deep Riser. Integrated Ocean Drilling Program.

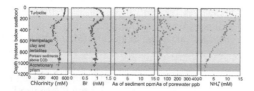

Figure 1. Depth profiles of interstitial water chemistry Data of chlorinity, Br⁻ and NH₄⁺ are from Shipboard Scientific Party IODP Exp. 338.

Arsenic Research and Global Sustainability – Bhattacharya, Vahter, Jarsjö, Kumpiene,
Ahmad, Sparrenbom, Jacks, Donselaar, Bundschuh & Naidu (Eds)
© *2016 Taylor & Francis Group, London, ISBN 978-1-138-02941-5*

Chemical and mineralogical characterization of arsenic, lead, cadmium and zinc in a smelter contaminated soil

S. Rathnayake & A.P. Schwab
Texas A&M University, College Station, TX, USA

ABSTRACT: Soil samples collected from a former smelting site in Stockton, Utah, contained elevated concentrations of potentially toxic elements. Total metal analysis after acid digestion revealed 7,520 mg/kg Arsenic (As), 66,400 mg/kg lead (Pb), 156 mg/kg cadmium (cd), and 10,600 mg/kg Zinc (Zn). These concentrations raise human and ecological health concerns. Metal oxides, sulfides and sulfates were identified, including massicot (PbO), galena (PbS), and arseno-pyrite (FeAsS). Arsenic and lead were associated with iron minerals such as goethite, observed by scanning electron microscopy and energy dispersive X-ray spectroscopy. Metal phases were mainly associated with the silt fraction (2–53 μm) in the soil. The complex mineralogy and metal distribution are critical in assessing risk and designing in-situ remediation strategies.

1 INTRODUCTION

Mining, milling, and smelting are important industrial and economical processes that extract minerals from ores to manufacture consumer products. Additionally, this activity is responsible for releasing large amounts of heavy metals into the environment (Chopin & Alloway, 2007; Dudka & Adriano, 1997; Li & Thornton, 2001). Geochemical weathering redistributes these metals to surrounding soils, sediments, surface water, and ground water affecting crops, animals, and humans (Dudka & Adriano, 1997).

Arsenic is a frequent pollutant found in smelter-contaminated soil (Basta & McGowen, 2004) and cause threats to human health (Järup, 2003). Because dissolved arsenic poses a serious health risk upon exposure, offsite transport of arsenic from contaminated soils is a major environmental concern and requires implementation of appropriate remedial measures (Yang *et al.*, 2007).

2 METHODS/EXPERIMENTAL

2.1 Site description

Soil samples were collected from a contaminated smelter site near Stockton, Utah, which is approximately eight square miles in area including the towns of Stockton and Rush Lake (40.4511° N, 112.3619° W). The risks posed by the site are derived from mining and smelting activities that occurred primarily in 1860's. Contaminated waste is reported to exist in several places, and lead and arsenic are the primary pollutants.

2.2 Chemical and mineralogical analysis

Field soil samples were obtained from a depth of 0 to 30 cm. A portable hand held X-ray fluorescence Analyzer was used to find the pockets of soil with high metal content.

Total metal content was quantified with a strong acid digestion. Soluble metal was extracted using de-ionized water, and plant available arsenic in soil was extracted using $NaHCO_3$ (Woolson *et al.*, 1971). The potentially bioaccessible metal fraction was assessed using a method developed by Ruby (Ruby *et al.*, 1996) to estimate gastrointestinal bioavailability though the ingestion of soil. As, Pb, Cd and Zn concentrations were measured using inductive coupled plasma mass spectrometer (ICP-MS).

Soil samples were analyzed by X-Ray Diffraction (XRD), Scanning Electron Microscope (SEM), and Transmission Electron Microscope (TEM) for mineral detection.

3 RESULTS AND DISCUSSION

3.1 Chemical analysis

Total metal content by acid digestion revealed 7,520 mg As/kg, 66,400 mg Pb/kg, 156 mg Cd/kg, and 10,600 mg Zn/kg. These concentrations raise human and ecological health concerns.

Elemental concentrations determined by different extractions indicate considerable amount of As, Pb, Cd and Zn are readily available.

3.2 Mineralogical analysis

Bulk powder XRD patterns were obtained to identify the crystalline distribution of major min-

Table 1. Metal concentrations determined by different extractions.

	Water Soluble	Plant available	Bioaccessible
		mg/kg	
Zn	7.44	8.65	4433
As	1.83	20.2	246
Pb	17.5	226	20912
Cd	0.11	0.42	138

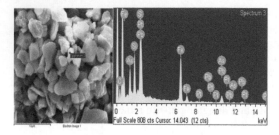

Figure 1. SEM picture and EDS spectrum of arsenopyrite.

erals. XRD patterns indicate presence of quartz (SiO_2), and calcite ($CaCO_3$), some crystalline Fe phases such as jarosite ($KFe_3(OH)_6(SO_4)_2$), magnetite (Fe_3O_4) and common clay minerals kaolinite, montmorillonite, and muscovite.

Additional mineral phases were identified from the electron microscopy data, coupled with energy dispersive X-ray elemental spectrometry (McLean & Bledsoe). Out of the Fe minerals, the most common minerals found were jarosite and pyrite. In addition to pyrite arsenopyrite (FeAsS) was identified. Arsenopyrite was not identified by the XRD technique; therefore, it may be present as non-crystalline form or in minor amounts in the soil. Goethite (α-FeOOH) was present with a dominant Fe peak and most of the time observed As and Pb peaks in the EDS data. Some other metal oxides and sulfides were identified, including massicot (PbO) and galena (PbS).

Based on the SEM data, Pb and As were found associated with Fe minerals, possibly by forming inner-sphere surface complexes. Metal phases were associated mainly with the silt fraction (2–53 µm) and could be primary or secondary mineral phases formed from dissolution-precipitation, oxidation—reduction and adsorption-desorption reactions.

Pockets of contaminated surface soil indicate that sediment might have been deposited during the smelter operation. High concentrations of As, Pb, Cd and Zn in active phases (water soluble and plant available extractions) as well as in the bioaccessible fraction emphasize the necessity of implementing remediation measures.

4 CONCLUSIONS

Contaminated soil consists of several metal oxides, sulfides, and sulfate phases such as massicot, galena, and arseno-pyrite. All these metal phases are distributed in the silt size fraction of the soil (2–53 µm). Because the majority of the soil consists of crystalline Fe oxide, there is a possibility to use non-crystalline Fe source such as Fe rich sludge from drinking water purification plants to stabilize arsenic in soil.

ACKNOWLEDGEMENTS

The authors gratefully acknowledge Agri Life research and Schlumberger foundation for funding the project as well as Y. Deng (Texas A & M University), Lisa Lloyd (US EPA), Tom Daniels (UDEQ), Mark Whitney (Mayor Stockton) and Bryan Tomlin (Texas A & M University).

REFERENCES

Basta, N., McGowen, S. 2004. Evaluation of chemical immobilization treatments for reducing heavy metal transport in a smelter-contaminated soil. *Environ. Pollut.* 127: 73–82.

Chopin, E.I.B., Alloway, B.J. 2007. Trace element partitioning and soil particle characterisation around mining and smelting areas at Tharsis, Ríotinto and Huelva, SW Spain. *Sci. Total Environ.* 373: 488–500.

Dudka, S., Adriano, D.C. 1997. Environmental Impacts of Metal Ore Mining and Processing: A Review. *J. Environ. Qual.* 26: 590–602.

Järup, L. 2003. Hazards of heavy metal contamination. *British Med. Bull.* 68: 167–182.

Li, X., Thornton, I. 2001. Chemical partitioning of trace and major elements in soils contaminated by mining and smelting activities. *Appl. Geochem.* 16: 1693–1706.

Ruby, M.V., Davis, A., Schoof, R., Eberle, S., Sellstone, C.M. 1996. Estimation of lead and arsenic bioavailability using a physiologically based extraction test. *Environ. Sci. Technol.* 30: 422–430.

Woolson, E., Axley, J., Kearney, P. 1971. Correlation between available soil arsenic, estimated by six methods, and response of corn (Zea mays L.). *Soil Sci. Soc. Am. J.* 35:101–105.

Yang, L., Donahoe, R.J., Redwine, J.C. 2007. In situ chemical fixation of arsenic-contaminated soils: An experimental study. *Sci. Total Environ.* 387: 28–41.

Arsenic Research and Global Sustainability – Bhattacharya, Vahter, Jarsjö, Kumpiene,
Ahmad, Sparrenbom, Jacks, Donselaar, Bundschuh & Naidu (Eds)
© *2016 Taylor & Francis Group, London, ISBN 978-1-138-02941-5*

Arsenic contamination of soils and morbidity prevalence in Racha District of Georgia

A.A. Chirakadze[1,2], V.G. Gvakharia[3] & Z.E. Buachidze[1,2]

[1]*Georgian Technical University, Institute for Problems of Engineering Physics, Tbilisi, Georgia*
[2]*Ecological Awareness and Waste Management, Project "Clean Up Georgia-Phase III", Tbilisi, Georgia*
[3]*St. Andrew the First Called Georgian University, Tbilisi, Georgia*

ABSTRACT: Georgia may be regarded as a natural "testing ground" because of the huge amount of poorly managed mining/processing and metallurgical industrial waste. Arsenic is one of the main strategic mineral resources of Georgia and, at the same time, one of the most hazardous pollutants. Arsenic mines and processing facilities are located in the mountainous areas of the regions of Western Georgia having high potential for agricultural production, tourism and health resort. Thus, the problem has not only an environmental, but also socio-economical importance. Because of uniquely high level of soil contamination and relatively low level of water pollution, the examined areas are of special interest for research of health-risks caused by arsenic in soil and food-chain under conditions of clean drinking and irrigating water. The main limitations of investigations carried out in the above areas were related with the insufficiently detailed study of pollution and relative health-risks and lack of the corresponding statistics, yielding in relatively low reliability and accuracy of the obtained results. This work was aimed to overcome the limitations. More detailed studies in a limited set of settlements allowed to increase significantly the accuracy and to establish a clear link between soil pollution and morbidity prevalence.

1 INTRODUCTION

More than 100 thousands of tons of arsenic containing hazardous waste is located in the two mountainous regions of Racha and Lower Svaneti in Western Georgia (Chirakadze, 2010). Content of arsenic in the different kinds of waste varies from tenths of a percent up to dozens of percent. Toxic waste is poorly managed and stored in collapsing buildings (storages and sarcophagus) or even scattered in the open. Three obvious facts determine particular interest in the study of these wastes: extremely high level of soil contamination at relatively low concentrations of arsenic in surface water; full lack of protection of the population and the environment (there were no even light barriers and warnings); and permanently increasing levels of contamination in soil, despite that mining and processing of arsenic has been fully stopped more than 20 years ago. The threats and risks associated with heavy contamination of soils with arsenic were recently investigated by Gvakharia (2014, 2016). The main sources of "secondary" pollution are the above mentioned caved storehouses and poorly managed waste disposals. It has been suggested that the overexposure to arsenic via ingestion of soil and contaminated food can lead to the increase of morbidity, mortality and relevant health-risks. The exposures could be the cause of high prevalence of 10 among the 19 investigated diseases (Gvakharia,

2016). At the same time, the values of 95% confidence intervals were in many cases significantly higher than the determined relative risks of morbidity, indicating insufficient reliability of results. The findings of these studies became the basis for the Dutch PSO Environment project, which involves investigations on arsenic containing mine wastes in Georgia. A detailed study and assessment of the environmental impact of the executed project is also necessary. The main goal of the present study is to overcome the limitations of the previous studies caused by insufficient research on the pollution and relative health-risks and lack of the corresponding statistics, yielding in relatively low reliability and accuracy of the obtained results. Study of a reduced number of pathologies (4 instead of 19), settlements (2 instead of 52) and increased number of analyzed samples (100 additional samples) and interviewed families (for 2- times) in contaminated (Uravi) and control (Oni) settlements was carried out to achieve the posed aims.

2 METHODS/EXPERIMENTAL

2.1 *Sampling, labeling, transportation and storage of samples and soil chemical analyses*

The experimental methods of the research (sampling, analysis of the chemical content of samples, analysis of existing information, questionnaires,

Table 1. New data on the relative disease burden in the region.

Pathologies	Gvakharia (2016)		This study	
	R/R	95% CI*	R/R	95% CI*
Skin cancer	8.7	1.1–66.6.	8.3	4.9–12.1
Hereditary anomalies	2.2	0.9–10.7	2.1	1.4–3.1
Tumors	2.9	1.4–6.0	2.7	1.8–3.9
Mental/ behavioral disorders	1.5	0.7–3.1	1.9	1.3–2.5

R/R: relative risk; *95% confidence interval

interviewing, statistical processing of obtained data) were the same as in Gvakharia (2014, 2016). Sampling, labeling, transporting and storage of samples were performed using the standard US EPA methods. Fifty samples were collected in Uravi, while 50 samples were collected in Oni. Samples were delivered to the laboratory and analyzed within 24 hours after sampling, dried and chemically analyzed using a "RIGAKU" X-ray-fluorescent analyzer. Samples were also analyzed using atom-absorption Perkin-Elmer spectrophotometer and obtained results were compared with the results of roentgen-fluorescent analysis. Newly obtained data were processed together with the data given in Gvakharia (2016).

2.2 Interviewing of population and processing of obtained data

Interviewing of population and processing of obtained data was carried out for10 clusters in Uravi and 20 clusters in Oni using cohort method and EPINFO software.

3 RESULTS AND DISCUSSION

3.1 The new data on the morbidity relative risk and 95% confidence interval

More accurate and reliable data, obtained as a result of additional research carried out in Uravi and Oni settlements, are given in Table 1.

The obtained results clearly show that even in case of safe drinking and irrigation water high level of soil contamination can cause significantly increased morbidity including a wide group of diseases (e.g. skin cancer, tumors, hereditary anomalies, mental and behavioral disorders).

4 CONCLUSIONS

The corresponding data on the morbidity relative risk and 95% confidence interval show that the morbidity prevalence in Racha region of Georgia is caused by heavy pollution of soils with arsenic. The used methods can provide reliable and accurate data on the morbidity relative risk and 95% confidence interval also for Lower-Svaneti region of Georgia

ACKNOWLEDGEMENTS

The research was supported by the Science and Technology Centre in Ukraine (STCU) and the Shota Rustaveli Georgian National Science Foundation grant project #5246.

REFERENCES

Chirakadze A. 2010. Development and bench-scale testing of soft decontamination-remediation methods to be used in the highland of Georgia. In: *Proceedings of the 9th International Symposium on Remediation in Jena, 4–5 October, 1996*, Jena, Book of Abstracts, pp. 50–51.

Gvakharia V.G. 2014. Arsenic pollution of soils and morbidity prevalence in Racha-Lower Svaneti district of Georgia. In: *Proceedings of the 13th International Conference on Clean Energy, 8–12 June, 2014*, Istanbul. pp. 1415–1424.

Gvakharia V.G. 2016. Arsenic pollution of soils and morbidity prevalence in Racha-Lower Svaneti district of Georgia. *International Journal of Global Warming*10(1) (in print).

Arsenic Research and Global Sustainability – Bhattacharya, Vahter, Jarsjö, Kumpiene,
Ahmad, Sparrenbom, Jacks, Donselaar, Bundschuh & Naidu (Eds)
© 2016 Taylor & Francis Group, London, ISBN 978-1-138-02941-5

Impacts of hydroclimate and mining on arsenic contamination of groundwater and surface water systems in Central Asia

J. Thorslund[1], S. Chalov[2], D. Karthe[3] & J. Jarsjö[1]

[1]*Department of Physical Geography and Quaternary Geology and the Bolin Centre for Climate Research,*
Stockholm University, Stockholm, Sweden
[2]*Lomonosov Moscow State University, Faculty of Geography, Moscow, Russia.*
[3]*Department of Aquatic Ecosystem Analysis and Management, Helmholtz Centre for Environmental Research*
Magdeburg, Germany.

ABSTRACT: Arsenic (As) contamination of groundwater due to natural processes (weathering of soil and bedrock) is a well-known problem in Asia. However, anthropogenic activities, such as mining, agriculture and other industries, may considerably increase As inputs to aqueous systems. Arsenic contamination is further coupled to climate change, which may alter hydrological conditions and change the As transport pathways and As concentrations of aqueous systems. We study the impact of mining and hydroclimatic changes on As contamination within the Lake Baikal Drainage basin. Results show that As concentrations, in both ground and surface waters, often exceed WHO permissible levels. Additionally, net increases in loads to riverine systems are seen as a result of mining activities. Considering also pressures related to projected hydroclimatic changes, these results call for continued monitoring and modelling of As spreading to this unique ecosystem and freshwater resource.

1 INTRODUCTION

AArsenic (As) is naturally mobilized due to weather-ing of As rich soil and bedrock and have become a health issue in many Asian countries. In addition to natural processes, 'anthropogenic industrial age' As contamination have produced more than 4.5 million tonnes of As, with mining identified as the number one producer, followed by coal and petro-leum pro-duction (Han et al., 2003). Coupled to these impacts are also hydroclimatic change, which may change hydrological conditions and transport pathways of this contaminant.

We here consider As contamination in Mongo-lia in central Asia, which is known for its extensive mining industry. The impact of climate change is also of par-ticular concern here, since sub-arctic regions have been seen to respond stronger and faster to climate change than other regions (Törn-qvist et al., 2014). We synthesize and analyze results on As spreading under conditions of hydrocli-matic change, with em-phasis on the highly mining impacted region of Zaamar in Mongolia, which is a potential source re-gion for observed metal con-tamination of the unique ecosystem and water resource of Lake Baikal (Rus-sia) (Thorslund et al., 2012, Shinkareva et al., 2015, Pfeiffer et al., 2015). Understanding processes con-trolling metal spread-ing and fate within this basin is of key significance for management.

2 METHODS

2.1 Field measurements and analyses

Water samples were collected in September 2013. Locations included: (i) the Tuul river (both upstream, at Zaamar, including groundwater and waste ponds and downstream) (ii) in the Sharyn-gol river (coal mining), and (iii) the Selenga delta. Arsenic concen-trations were determined by induc-tive coupled plas-ma optical emission spectrometry (ICP-OES). Dis-charge was also estimated along the Tuul River, for estimations of mass loads.

3 RESULTS AND DISCUSSION

3.1 Arsenic concentrations and loads

Concentrations from locations associated with the mining activities are shown in Fig. 1. Both dissolved and total As concentrations exceed maximum per-missible drinking water levels for all locations except the Selenga delta (Fig. 1), with highest con-centra-tions for waste ponds. Further, dissolved As concen-trations, reach almost as high levels as total concen-trations, due to high pH conditions of waters in this region that keeps more toxic As in solution.

Other industrial activities in Mongolia, at the Bor Tologi gold mine and at a coal power plant, showed even higher As concentrations; up to 2820 μg/L re-

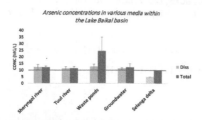

Figure 1. Arsenic measurements from 2013 at various locations in a coal mining region (Sheryngol river), a gold mining region (Tuul river; waste ponds; groundwater) and just before the Lake Baikal (Selenga delta). The patterned bar indicate a value < DL. The red line indicates the World Health Organization (WHO) maximum permissible health risk value of 10 µg/L.

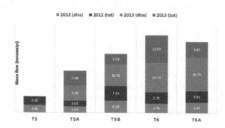

Figure 2. Estimated mass loads of arsenic from 2013 in: upstream (T5-T5A), along site (T5B-T6) and just downstream (T6A) of the Zaamar placer goldmining site, in the Mongolian part of the Lake Baikal basin.

spectively 1170 µg/L (Pfeiffer et al., 2015). They al-so highlight contamination problems and concentrations exceeding permissible levels in groundwater at several locations. Given that placer mining is execut-ed directly in the river valley and waste ponds are often of inadequate construction (Thorslund et al., 2012, Pfeiffer et al., 2015), local river water quality is greatly impacted by these activities.

Arsenic loads along the Tuul river show clear sig-nals from mining activities, with net increases in loads across the Zaamar Goldfield (Fig. 2). Previous quantifications here have shown even higher loads of As, with exports of up to 45 tonnes/year from the Zaamar site to downstream regions and estimated net input of about 30 tonnes/year to this river due to mining activities (Thorslund et al., 2012).

3.2 Impacts of hydroclimatic change

The large riverine export of As within this system will certainly also be influenced by future hydroclimatic changes. For example, particularly fast warming and permafrost degradation of this region com-pared to other regions globally have been seen al-ready and future projections indicate a continued warming trend (Törnqvist et al., 2014). This could lead to changed hydro-periods, altered transport pathways and changes to both loading pattern and magnitudes, either increasing absolute loads on aver-age and/or increasing peak loads due to more

ex-treme events. It could also decrease discharge, po-tentially causing less dilution of As input from in-dustry. Overall, uncertain projections of hydroclimatic changes, especially of discharge trends (Törn-qvist et al., 2014) highlight the need for con-tinued research efforts to improve representations and associated prediction of hydroclimatic changes.

4 CONCLUSIONS

Taken together, the results presented here indicate that there is contamination of both ground and surface waters at several locations of the Lake Baikal basin as a consequence of anthropogenic industries, particularly gold mining. The future uncertainties from combined human inputs of As to the aqueous system and hydroclimatic change, call for more research efforts regarding both monitoring and model-ing of As contamination. These efforts are needed both locally at the anthropogenic source zones and regionally in the vicinity of the Lake Baikal.

ACKNOWLEDGEMENTS

This research was funded by the Swedish Research Council Formas (project nr. 2012-790); the project Hydroclimatic and ecohydrological changes of the Lake Baikal drainage basin, supported by a travel grant from the Faculty of Science, Stockholm Uni-versity, the Russian-Mongolian biological expedition RAS-MAS with the Russian geographical society grant from the Russian Foundation for Basic Re-search (project nr. 12-05-00069-a, 12-05-33090) and by the Russian Scientific Foundation (project nr. 14-17-00155).

REFERENCES

Han, F.X., Su, Y., Monts, D.L., Plodinec, M.J., Banin, A., Tri-plett, G.E. 2003. Assessment of global industrial-age an-thropogenic arsenic contamination. Naturwissen-schaften 90: 395–401.

Pfeiffer, M., Batbayar, G., Hofmann, J., Siegfried, K., Karthe, D., Hahn-Tomer, S., 2014. Investigating arsenic (As) occur-rence and sources in ground, surface, waste and drinking water in northern Mongolia. Environ. Earth Sci. 73: 649–662.

Shinkareva, G.L., Kasimov, N.S., Lychagin, M.Y. 2015. Heavy Metal Fluxes in the Rivers of the Selenga Basin. In: D. Karthe, S. Chalov, N. Kasimov, M. Kappas (Eds.) Water and Environment in the Selenga-Baikal Basin: Interna-tional Research Cooperation for an Ecoregion of Global Relevance, ibidem-Verlag, Stuttgart, Germany, pp. 87-100. (ISSN 1614-4716).

Thorslund, J., Jarsjö, J., Chalov, S.R., Belozerova, E.V. 2012. Gold mining impact on riverine heavy metal transport in a sparsely monitored region: the upper Lake Baikal Basin case. J. Environ. Monit. 14: 2780–2792.

Törnqvist, R., Jarsjö, J., Pietroń, J., Bring, A., Rogberg, P., Aso-kan, S.M., Destouni, G. 2014. Evolution of the hydro-climate system in the Lake Baikal basin. J. Hydrol. 519: Part B, 1953–1962.

1.5   Arsenic speciation and mobility from mine waste and tailings

Arsenic Research and Global Sustainability – Bhattacharya, Vahter, Jarsjö, Kumpiene,
Ahmad, Sparrenbom, Jacks, Donselaar, Bundschuh & Naidu (Eds)
© 2016 Taylor & Francis Group, London, ISBN 978-1-138-02941-5

Arsenic fate following mining of sulfide ore at mine sites and significance of the reduced state

Barry N. Noller[1], Jack C. Ng[2,3], Jade B. Aitken[4] & Hugh H. Harris[5]

[1]*Centre for Mined Land Rehabilitation, The University of Queensland, Brisbane, Queensland, Australia*
[2]*National Research Centre for Environmental Toxicology (Entox), The University of Queensland, Brisbane, Queensland, Australia*
[3]*Cooperative Research Centre for Contamination Assessment and Remediation of the Environment (CRC CARE), Callaghan, NSW, Australia*
[4]*School of Chemistry, The University of Sydney, Sydney, NSW, Australia*
[5]*Department of Chemistry, University of Adelaide, Adelaide, South Australia, Australia*

ABSTRACT: Arsenic fate at mine sites is influenced by the reduced state of the ore. Comparison of arsenic speciation using XANES spectra with known arsenic compounds of ore from various lead and copper sulfidic mines with tailings from their processing shows that there is a significant difference when comparing near surface or shallow ore with deep mined ore. Subsequent comparison with tailings from different mined sulfidic ores, using XANES scan fitting is useful for understanding the chemical form of As in such samples.

1 INTRODUCTION

Arsenic (As) is commonly found in association with base metals including gold and iron in sulfide mineralisation. Historically base metal ore deposits have been mined from surface deposits which become oxidised over geological time. When surface deposits are depleted, deposits are mined from depths of hundreds to thousands of metres corresponding to the reduced As state. Following extraction of metals from mined ore having economic value, the remaining tailings and waste rocks may be stored above or below ground and are rehabilitated. As arsenic has no economic values, it remains in various chemical states and may be remobilised by changes in redox conditions and degree of water saturation.

This study uses synchrotron-induced X-ray Absorption Near Edge Spectroscopy (XANES) to examine chemical structure of As in a range of ore, processed mine waste and other samples. The aim is to identify differences in As speciation of surface and deep-mined sulfide deposits and following processing that may influence As mobility in rehabilitated land forms.

2 METHODS/EXPERIMENTAL

Samples of ores and processed tailings were collected from copper and lead mines in northern Australia and other deposits in Australia, Fiji and Taiwan. Samples initially sieved to <2 mm were ground to <20 µm for X-ray analysis. Total As concentrations were determined by ICP-MS on aqua regia digests of the samples.

Arsenic K-edge XANES spectra were recorded from model compounds and samples at Beamline 20B Australian National Beamline Facility (ANBF) Photon Factory, KEK, Tsukuba, Japan. The spectra were recorded at room temperature in fluorescence mode using a thirty-six-element germanium-array detector. BL-20B was equipped with a channel-cut Si(111) monochromator which was detuned 50% to reject harmonics. Sodium arsenate was used as an internal standard for energy calibration, with the first peak of the first derivative assumed to be 11,873.6 eV. Solid models diluted in boron nitride (ground to <20 µm) were pressed into aluminium spacers then covered between two 63.5 µm Kapton tape windows (window size, 2*10 mm). Data analysis, including calibration, averaging and background subtraction of all spectra and Principal Component Analysis (PCA), target and linear regression analyses of XANES spectra were performed using the EXAFSPAK software package (George and Pickering 2000). The precision of fitting was determined to be ~10% based on analyses of control model compound mixtures.

3 RESULTS AND DISCUSSION

3.1 *Arsenic concentrations in samples*

Table 1 gives the total As concentrations in samples.

3.2 *Observations from XANES scans*

Figure 1 shows XANES As K-edge scans for: A. model compounds (solid) in boron nitride (1000 mg/kg)

Table 1. Total As concentrations in samples.

Sample	Total arsenic (mg/kg)
Natural lead ore at surface	1060
Upper catchment sediment	9
Lead Mine 1 deep ore	14
Lead Mine 2 deep ore	113
Lead Mine 3 deep ore	112
Lead Mine 4 deep ore	9250
Tailings (near surface ore)	229
Tailings (deep ore)	404
Copper mine 1 deep ore	1154
Copper mine 1 deep ore	3700
Copper mine 3 deep ore	181
Queensland arsenic mine surface ore	>10,000
Udu copper mine Fiji ore near surface	85
S. Taiwan shale with pyrite at surface	10

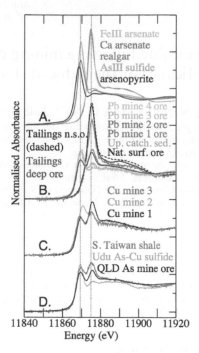

Figure 1. XANES As K-edge scans for: A. model compounds; B. lead mines, natural surface ore and upper catchment sediment; C. copper mines; D. South (S) Taiwan surface erosion of shale, Udu copper mine Fiji and Queensland (QLD) As mine ore from surface; vertical lines are at 11,869.5 eV (As-I) and 11,875.5 eV (AsV).

ferric arsenate (FeIIIAsV), calcium arsenate (CaAsV), Realgar (As$_4$S$_4$), arsenic III sulfide (AsIIIS) and arsenopyrite (FeAsS); B. Lead (Pb) mines 1, 2, 3 and 4 ores, upper catchment sediment and natural surface ore (n.s.o.); C. Copper (Cu) mines 1, 2 and 3 mines ores; and D. South (S) Taiwan surface erosion of shale, Udu copper mine Fiji and Queensland (QLD) As mine ore.

Figure 1B indicates that natural surface ore, upper catchment sediment, lead mine 1 ore and tailings from near surface ore (n.s.o.) all show presence of AsV, primarily FeIIIAs but may also include other metal arsenates; there is little or no reduced As present in these samples. Comparison of Fig. 1B lead mine ores 2, 3 and 4 and tailings deep ore (from these deep mined ores), Fig. 1C copper mine ores 1, 2 and 3 and Fig. 1D South (S) Taiwan surface erosion of shale, Udu copper mine Fiji and Queensland (QLD) As mine ore, all show a combination of AsV and reduced As(-I) which may be arsenopyrite or AsIII sulfide.

3.3 Significance of As speciation

Basu & Schreiber (2013) identify that As mineralogy partially controls release via weathering reactions. In host rocks arsenopyrite was the most abundant mineral and higher As concentrations were found in smaller grain size fractions in tailings and stream sediments (Basu & Schreiber, 2013). This is similar to the observations of data in Fig. 1 where tailings in particular are finely ground (to <200 µm). There is a clear distinction from the XANES spectra in Fig. 1 between near surface and deep-mined ore.

4 CONCLUSIONS

This study shows that comparison of molecular structure from XANES spectra for difference naturally-occurring sulfide-containing ore and tailings samples gives a clear distinction between near surface ore and deep-mined ore. This data may be useful in understanding mobilisation of As in rehabilitated mine waste structures.

ACKNOWLEDGEMENTS

This work was performed at the BL20B Beamline Facility Photon Factory (PF), KEK Tsukuba-Japan with support from the Australian Synchrotron Research Program, which is funded by the Commonwealth of Australia under the major National Research Facilities Program. Entox is a partnership between Queensland Health and the University of Queensland.

REFERENCES

Basu, A., Schreiber, M.E. 2013. Arsenic release from arsenopyrite weathering: Insights from sequential extraction and microscopic studies. *J. Hazard. Mater.* 262: 869–904.
George, G.N., Pickering, I.J.. 2000. *EXAFSPAK: A Suite of Computer Programs for Analysis of X-ray Absorption Spectra.* Stanford Synchrotron Radiation Laboratory: Stanford, CA.

Arsenic Research and Global Sustainability – Bhattacharya, Vahter, Jarsjö, Kumpiene,
Ahmad, Sparrenbom, Jacks, Donselaar, Bundschuh & Naidu (Eds)
© 2016 Taylor & Francis Group, London, ISBN 978-1-138-02941-5

Behaviour and mobility of arsenic in a Mexican hydrosystem impacted by past mining activities

A. Barats[1], A.M. Orani[1,2], C. Renac[1], J.P. Goudour[1], G. Durrieu[3], H. Saint-Martin[4],
M.V. Esteller[5] & S.E. Garrido Hoyos[6]

[1]*Université Nice Sophia Antipolis, CNRS, IRD, Observatoire de la Côte d'Azur, Géoazur, Campus Azur CNRS,*
Valbonne, France
[2]*International Atomic Energy Agency, Environment Laboratories, Monaco*
[3]*Université de Toulon, PROTEE, La Garde, France*
[4]*Universidad Nacional Autónoma de México, Instituto de Ciencias Físicas, Cuernavaca, México*
[5]*Universidad Autonoma del Estado de Mexico, Centro Interamericano de Recursos del Agua, Cerro Coatepec*
S/N Ciudad Universitaria, Toluca, México
[6]*Instituto Mexicano de Tecnologia del Agua, Paseo Cuauhnahuac, Jiutepec, México*

ABSTRACT: This study focusses on arsenic (As) because this metalloid presents a great health issue in a region of Morelos state in Mexico (Huautla district), due to past mining activities. The main aim of this study is to better constrain As behavior and transport in this hydrosystem. During two contrasting hydrological periods, total and dissolved concentrations of As were determined in samples of river water, springs, and drinking waters. Arsenic concentrations were also determined in suspended particulate matter (SPM) collected from water samples and in sediments from the rivers (bulk samples, and different size and magnetic fractions). In-situ physicochemical parameters were also measured. In waters, arsenic occurs mainly as dissolved species at high concentrations, above the WHO limits of 10 μg/L. SPM, sediments and their different fractions also contain large concentrations of As. This study reveals here that not only the water is As-contaminated, but there is also a significant As-stock in solids which may provide supplementary As inputs to waters during eventual transfer events.

1 INTRODUCTION

Arsenic is the prime naturally-occurring carcinogen found in the environment (Nriagu, 2001). This metalloid is a threat due to its presence in many waters used for domestic consumption (Smedley & Kinniburgh, 2002). In many areas of the world, As concentrations are higher than the WHO drinking water limit of 10 μg/L. Arsenic is mobile over a wide range of pH and redox conditions (Cullen *et al.*, 1989). Its occurrence depends also on water/mineral interactions, themselves influenced by the chemistry of the aqueous phase (pH, E_h, ionic strength, Dissolved Organic Carbon (DOC)) and by the abundance and characteristics of the solid phases (Bissen & Frimmel, 2003, Grosbois *et al.*, 2011, Haque *et al.*, 2008). The aim of this study is to better understand the physical and geochemical processes governing As transport and mobility in a natural hydrosystem.

2 MATERIALS AND METHODS

2.1 Sampling sites

The study area is located in the southern region of Morelos state, central Mexico, between 18°26–18°32 N latitude and 98°58–99°10 W longitude, at about 900 m of altitude, geologically described in previous studies (Esteller *et al.*, 2015, Gonzales *et al.*, 2013). Water samples from two mines, three from the Nexpa River, three drinking waters, three springs, and river sediments were collected in June 2013 (dry season) and in September 2014 (rainy season).

2.2 Sampling and analytical methods

The following in-situ measurements were performed: temperature, electrical conductivity (EC), pH, E_h, dissolved oxygen (diss. O_2), alkalinity, and turbidity. Sampling and analytical methods were previously described for the water (Barats *et al.*, 2014), and for major ions (Esteller *et al.*, 2015). Water was filtered using a Nalgene Polysulfone filtration unit and 0.45 μm filters. Field collected wet filters were dried under a laminar hood in a clean room, and wet sediments were oven dried at 40°C.

Sediments collected during water sampling were sieved and weighed to characterize mineralogical compositions (SEM-EDS and XRD). Two size fractions, the 100–400 μm and the clay-size (<2 μm) fractions, were analyzed in details. In order to characterize metallic source within sediments, the 100–400 μm particles were magnetically separate into paramagnetic and diamagnetic particles. Bulk sediments and the 100–400 μm fractions were ground. Grounded powders and SPM on filters were mineralized by an acid digestion assisted by microwaves. The digested materials were then diluted before ICPMS analyses.

3 RESULTS AND DISCUSSION

3.1 *Physicochemical conditions and water nature*

All water samples are slightly basic (pH = 7.8 ± 0.6) and oxygenated (diss. O_2 = $63 \pm 31\%$; E_h = 150 ± 45 mV). EC, alkalinity and turbidity are variable with time, exhibiting high values during the wet period. Diss. O_2 displays low contents (20–50%) in some drinking waters and springs even during the rainy period.

Major ion compositions reveal calcareous- and sulfate-rich waters, except for the America mining site with sodium- and sulfate-rich waters. Whatever the sampling period, DOC concentrations were high with 5.9 ± 0.6 mg/L and 16 ± 2 mg/L in the river and in drinking waters, respectively. In spring or mine waters, DOC values strongly depend on the sampling period (higher during rainy period).

3.2 *Arsenic in waters*

In river and some drinking water samples, total As concentrations (unfiltrated water samples) remain lower or close to the WHO limit. But, springs and mine waters can display high total As concentrations, above the Mexican limit (25 µg/L), which may be higher during the dry period, especially in waters from the America mine (from 82 to 127 µg/L).The situation is alarming because one of the mine waters with an 15 µg/L of As, irrespective of the sampling period, is used as drinking water. Dissolved As concentrations are similar to total concentrations, at least 88% of total arsenic.

3.3 *Arsenic in particles and sediments*

In river, As concentrations in SPM were higher during the dry season than during the wet season: 6–12 and 1–3 mg/kg, respectively. In springs, mines or drinking waters, they are higher, varying from 25–90 and 2–20 mg/kg, during the dry and wet seasons, respectively. In sediments, As concentrations are slightly lower than in SPM, averaging 4 mg/kg whatever the sampling period. In sediment fractions, As concentrations seem homogeneous, ranging from 3–6 mg/kg.

3.4 *Arsenic behavior*

Such as previously suggested (Esteller *et al.*, 2015), the occurrence of dissolved As species in water samples was confirmed. Accounting in-situ measurements of pH and E_h, arsenic is suggested to be under arsenate species, $HAsO_4^-$, one of the most toxic species (Hughes *et al.*, 2002). This study highlights thus the huge mobility of this metalloid in this hydrosystem. Sediments and SPM contain also high As concentrations (from few to ten mg/L), 500–1200 times higher than those in waters. These high As concentrations may be related to the occurrence of As-rich minerals (such as residual sulfides) or of minerals with high particle specific surface areas such as ferruginous clays (Fe-Mg-smectite). Springs, mines or drinking waters collected during this study contain fortunately low SPM concentrations (1–30 mg/L). But, these sediments and SPM threaten the water quality because they constitute a passive As stock which may be remobilized. These results on solids allow also the previous hypothetic origin of As in waters—induced by water/rock interactions—to be confirmed (Esteller *et al.*, 2015).

4 CONCLUSIONS

In this area of Mexico, the environmental and human threats reside not only on the occurrence of As-contaminated waters, but also of As-rich minerals in SPM and sediments. Transfer processes from the solid to aqueous phase can eventually lead to higher concentrations in waters.

ACKNOWLEDGEMENTS

This research was supported by ECOS Nord in France and by CONACYT in Mexico. We are grateful to A. Leclerc, H. Bali, S. El Hamaoui, students helping in water sampling and analyses.

REFERENCES

Barats, A., Féraud, G., Potot, C., Philippini, V., Travi, Y., Durrieu, G., Dubar M., Simler R. 2014. Naturally dissolved arsenic concentrations in the Alpine/Mediterranean Var River watershed (France). *Sci. Total Environ.* 473–474: 422–436.

Bissen, M., Frimmel, F.H. 2003. Arsenic—a review. Part I: occurrence, toxicity, speciation, mobility. *Acta Hydrochim. Hydrobiol.* 31(1): 9–18.

Cullen, W.R., Reimer K.J. 1989. Arsenic speciation in the environment. *Chem. Rev.* 89(4): 713–764.

Esteller, M.V., Domínguez-Mariani, E., Garrido, S.E., Avilés, M. 2015. Groundwater pollution by arsenic and other toxic elements in an abandoned silver mine, Mexico. *Environ. Earth Sci.* 74(4): 2893–2906.

Grosbois, C., Courtin-Nomade, A., Robin, E., Bril, H., Tamura, N., Schafer J., Blanc G. 2011. Fate of arsenic-bearing phases during the suspended transport in a gold mining district (Isle river Basin, France). *Sci. Total Environ.* 409: 4986–4999.

Haque, S., Ji, J., Johannesson K.H. 2008. Evaluating mobilization and transport of arsenic in sediments and groundwaters of Aquia aquifer, Maryland, USA. *J. Contam. Hydrol.* 99: 68–84.

Hughes, M.F. 2002. Arsenic toxicity and potential mechanisms of action. *Toxicol. Lett.* 133: 1–16.

Nriagu, J.O. 2001. Arsenic poisoning through the ages. New York, (Frankenberger, W.T., Jr., Ed.), Marcel Dekker.

Smedley, P.L., Kinniburgh D.G. 2002. A review of the source, behaviour and distribution of arsenic in natural waters." *Appl. Geochem.* 17(5): 517–568.

Arsenic Research and Global Sustainability – Bhattacharya, Vahter, Jarsjö, Kumpiene,
Ahmad, Sparrenbom, Jacks, Donselaar, Bundschuh & Naidu (Eds)
© 2016 Taylor & Francis Group, London, ISBN 978-1-138-02941-5

Arsenic release from gold extraction tailings in Ecuador

J. Van Daele[1] & V. Cappuyns[2]
[1]*Department of Earth and Environmental Sciences, KU Leuven, Belgium*
[2]*Centre for Economics and Corporate Sustainability, KU Leuven, Belgium*

ABSTRACT: Major environmental problems associated with gold mining and extraction are acid mine drainage and high levels of arsenic and other elements in the mine tailings. A mine tailing paste sample from Zaruma, Ecuador was subjected to different types of leaching tests, analyzing the influence of pH, grain size, L/S ratio and time on As release. Despite elevated total concentrations (541 mg/kg), As showed low solubility, and the release in different leaching tests was very low, even at extreme pH values (pH 2 and 11). Since pyrite in this sample showed low reactivity, the risk of acid mine drainage is limited. The release of As from the tailing material under changing environmental conditions (e.g. co-disposal with acid or alkaline waste) is not of particular concern, but dispersal of fine particles by water erosion should be avoided.

1 INTRODUCTION

Most studies focus on the mechanical and physical properties of mine tailing paste, while chemical properties are only reported in a few studies (Coussy *et al.*, 2012). Mine tailings resulting from gold deposits usually contain elevated concentrations of potentially toxic elements, among which arsenic (As) is a major pollutant. In the present study, mine tailing paste (MT) from a gold extraction plant using CN^- was analyzed for chemical and mineralogical properties. pH_{stat} leaching tests and cascade batch leaching tests were performed on this mine tailing sample in order to assess the release of As as function of pH, time, Liquid/Solid (L/S) ratio and grain size distribution.

2 METHODS

2.1 *Sampling and sample characterization*

Mine tailings (MT) were sampled at the processing plant of a gold mine in Zaruma, southern Ecuador (Barie *et al.*, 2015). A fresh sample of mine tailing paste obtained after gold extraction was taken and washed thoroughly to remove remaining cyanide from the solids. Total concentrations of major elements (Al_2O_3, CaO, FeO, K_2O, MgO, MnO, Na_2O) were determined after lithium metaborate fusion (LB). Aqua regia (AR) destruction was used to determine elements not incorporated in the silicate matrix. The resulting solutions were measured with ICP-OES (Varian 720-ES equipment). All analyses were performed on triplicate samples. The mineralogical composition of the samples was determined with quantitative X-ray powder diffraction as described in Barie *et al.* (2015).

2.2 *Leaching tests*

PH_{stat} leaching tests (at pH 2, 4, 5, 7, 9 and 11) and geochemical modelling (VisualMinteq) of the pH-dependent leaching behavior were performed as described in

Cappuyns *et al.* (2014). During the pH_{stat} leaching test, the kinetics of element release were monitored for 48 h. The influence of L/S ratio was investigated with a serial batch test in which the material was successively extracted five times with water at L/S ratio of 20, resulting in cumulative L/S ranges from 20 to 100. This serial batch test was performed on 2 subsamples (3 replicates each) of different grain size (<45 μm and 45–100 μm resp.). Single extractions with acetic acid and ammonium-EDTA (Cappuyns, 2012) were performed according to standardized methods.

3 RESULTS AND DISCUSSION

3.1 *Chemical and mineralogical characterization*

The mine tailing paste was composed of quartz (45%), chlorite (14%), K-feldspar (12%), muscovite and illite (12%), plagioclase (11%), and pyrite (6%) (composition not corrected for the amorphous content). Important differences exist between the major element concentrations determined with the LB and AR method, (Table 1), since only 13% of the MT sample was dissolved by the AR method (insoluble residue = 87%). The pH of the tailing material was 9.1 ± 0.1, and the medium grain size (d_{50}) was 22 μm.

3.2 *Influence of pH, time, L/S ratio, and grain size on As release*

The cumulative release of As during the serial batch test amounted to 0.3 mg/kg for the 45–100 μm sample, while it was lower (0.17 mg/kg) for the sample with the smaller grainsize (< 45 μm). This was likely caused by the lower pH in the < 45 μm sample (pH = 8.9 ± 0.3) compared to the 45–100 μm sample (pH 9.5 ± 0.2), which tends to promote As surface retention. The release of As as a function of pH (Figure 1) followed a V-shaped pattern as is usually observed for heavy metals. Maximal release of As occurred at pH 2, but remained very limited (1 mg/kg) (Fig. 1).

Table 1. Concentrations of As (in mg/kg) and major elements (in%) in the MT sample (mean ± stdv of 3 replicates) determined with the Lithium metaborate (LB) and Aqua Regia (AR) methods.

	LB	AR		LB	AR
As	–	541 ± 35	Na_2O	1.02 ± 0.02	0.02 ± 0.00
Al_2O_3	11.7 ± 0.13	2.18 ± 0.05	SiO_2	68.8 ± 0.03	–
CaO	1.35 ± 0.01	0.41 ± 0.01	S	–	3.13 ± 0.01
FeO	7.09 ± 0.10	6.25 ± 0.16	TiO_2	0.46 ± 0.01	–
K_2O	2.50 ± 0.03	0.76 ± 0.03			
MgO	2.31 ± 0.01	1.77 ± 0.04	IR	–	86.9 ± 1.24
MnO	0.17 ± 0.01	0.14 ± 0.00	LOI	4.53	–

* IR = insoluble residue (in%), LOI = Loss on ignition (in%).

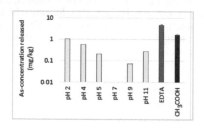

Figure 1. Release of As as a function of pH (48 h pH_{stat} leach-ing, compared to As extracted with EDTA and acetic acid.

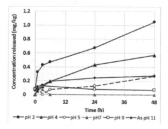

Figure 2. release of arsenic as a function of time during pH_{stat} leaching at pH 2, 4, 5, 7, 9 and 11.

At low pH, the release of As is most likely explained by the dissolution of mineral surfaces, whereas at alkaline pH, desorption of arsenate species is likely to occur. The MT was characterized by a low ANC (62 mmol/kg at pH 2). Acetic acid (pH~3) extracted more As than ammonium-EDTA (pH~7), despite the lower pH of the acetic acid solution. EDTA does not form stable complexes with arsenic, whereas it has been shown that acetic acid is able to extract consid-erable amounts of As (Cappuyns, 2012).

Reaction kinetics during pH_{stat} leaching at pH 2, 4 and 5 show that equilibrium was not yet reached after 48 h (Fig. 2). Quantitative XRD analysis of the MT before and after pH_{stat} leaching at pH 2 indicated that pyrite was not dissolved after 48 h of leaching. Geochemical modeling of the dissolution of pyrite as a function of pH indicated that, from a thermo-dynamic point of view, pyrite should completely dis-solve at pH 2. Slow reaction kinetics (Fig. 2), and a highly stable mineral phase (pyrite) might explain the

very low release of As, even at extreme pH values (pH 2 and pH 11). Besides the sufficient supply of electron acceptors (Fe^{3+}, O_2), the reactivity of pyrite depends on ore mineralogy, including the effects of associated sulfide impurities (Cruz *et al.*, 2001).

4 CONCLUSIONS

Despite elevated total concentrations (541 mg/kg), As showed a low mobility, and the release in different leaching tests was very limited, even at extreme pH values of 2 and 11. Since pyrite shows a low reactivity, the risk of acid mine drainage is limited. The release of As from the tailing under changing environmen-tal conditions (e.g. co-disposal with acid or alkaline waste) is not of particular concern, but dispersal of the very fine particles (d_{50} of the tailing material = 22 μm) by water erosion is a main concern. Priority should be given to the implementation of a sustainable tailings disposal system.

ACKNOWLEDGEMENTS

Grateful acknowledgements are made to Lieven Machiels for providing the mine tailing sample, and to Elvira Vassilieva for her assistance with the labora-tory analyses.

REFERENCES

Barrie E., Cappuyns V. et al. 2015. Potential of inorganic polymers (geopolymers) made of halloysite and volcanic glass for the immobilisation of tailings from gold extrac-tion in Ecuador. *Appl. Clay Sci.* 109–110: 95–106.

Cappuyns V. 2012. A critical evaluation of single extractions from the SMT program to determine trace element mobil-ity in sediments. *Applied and Environmental Soil Science* Article ID 672914: 1–15, doi:10.1155/2012/672914

Cappuyns V., Alian V., Vassilieva E., Swennen R. 2014. pH dependent leaching behavior of Zn, Cd, Pb, Cu and As from mining wastes and slags: kinetics and mineralogical control. *Waste and Biomass Valorization* 5(3): 355–368.

Coussy S., Paktunc D., Rose J., Benzaazoua M. 2012. Arsenic speciation in cemented paste backfills and synthetic calci-um–silicate–hydrates. *Minerals Engineering* 39: 51–61

Cruz R., Bertrand V., Monroy M., González, I. 2001. Effect of sulfide impurities on the reactivity of pyrite and pyritic con-centrates: a multi-tool approach. *Appl. Geochem.* 16: 803–819.

Arsenic Research and Global Sustainability – Bhattacharya, Vahter, Jarsjö, Kumpiene,
Ahmad, Sparrenbom, Jacks, Donselaar, Bundschuh & Naidu (Eds)
© *2016 Taylor & Francis Group, London, ISBN 978-1-138-02941-5*

Arsenic mobilization from an abandoned smelting slag dump in the Linares mining district, Spain

M.C. Hidalgo[1], J. Rey[1], J. Martínez[2], M.J. de la Torre[1] & D. Rojas[1]
[1]*Department of Geology, Campus Científico Tecnológico de Linares, University of Jaen, Spain*
[2]*Department of Mechanical and Mining Engineering, Campus Científico Tecnológico de Linares, University of Jaen, Spain*

ABSTRACT: The slags from the smelting dump of La Cruz present very high total metal(loid) contents (4400, 51000, 23000 and 300 mg kg^{-1} for Cu, Zn, Pb, and As, respectively). The leachates generated are sodium sulfate type, with pH > 8, electrical conductivity up to 15 mS/cm, several mg/L for dissolved Cu, Fe, Pb, and As concentrations that exceed 40 mg/L. The soils analyzed downstream the smelter also featured high concentrations for Cu (240 mg/kg), Zn (1800 mg/kg), and Pb (9900 mg/kg), with total As reaching 500 mg/kg. Despite the high total metal(loid) contents, exchangeable fractions in soils were low and this circumstance may contribute to prevent the mobility of these elements in the environment.

1 INTRODUCTION

The Linares mining district is located in Andalusia, S Spain, to the N of the province of Jaen. These metal mines are hosted in a small granitoid massif (extension: 80 km^2) emplaced at the end of the Hercynian orogenic cycle. Galena was the main ore and the exploitation of the veins by underground mining is known since Roman times. At the beginning of the 20th century, this mining district became the world's first supplier of Pb, and an important minerallurgical and metallurgical industry was developed. Despite the wealthy of the ores, in 1991, the last exploitation ceased and many mine waste facilities were abandoned without applying any corrective action. The La Cruz smelting dump is composed of slags created during the metallurgical processing of lead. These are vitreous deposits, with a surface area of 4.7 ha and an accumulated volume of 1,345,000 m^3. Geologically, this structure lies directly on Triassic materials (Rey *et al.*, 2013). Important surface erosion by water, along with excavated slopes (by extraction of the material for cement production), exudations and leachates at the foot of the deposit, were detected. In this study, we analyze the mobilization of metal(loid)s in water and soils near this abandoned smelter.

2 MATERIAL AND METHODS

A geochemical characterization of the smelting slags and the surrounded soils was carried out. A microwave-assisted total acid digestion, using nitric acid (HNO$_3$) and Hydrochloric acid (HCl), was performed on the solid samples for multielement acid extraction. The total concentration of 28 elements was determined by Inductively Coupled Plasma Mass Spectrometry (ICP-MS) analysis. In order to evaluate the mobility of these elements, a five-step sequential extraction procedure has been carried out to differentiate, from more to less mobile fractions: water soluble fraction, exchangeable fraction, carbonated bound fraction, fraction linked to sulfates and organic matter, and linked to amorphous oxides. In addition, the mineralogy of the soils was studied by X-Ray Diffraction (XRD) and Scanning Electron Microscopy (SEM). Precipitates and crusts were also obtained from a seasonal pond downstream the smelting dump. Leachate samples were collected for the analysis of major constituents and trace elements. The temperature, Electrical Conductivity (EC) and pH were determined in situ for all the water samples.

3 RESULTS AND DISCUSSION

The leachates from the smelting dump are characterized by pH values >8, with electric conductivities ranging from 2 to 15 mS/cm and up to 1.5, 2.3, 10, 43 and 4400 mg/L of dissolved Cu, Fe, Pb, As, and SO$_4^{2-}$, respectively. These lixiviates of Na-sulfate type are temporarily accumulated in small ponds and flushed during the rainy periods. During the dry season, white salts and reddish crusts are deposited in its borders (Figure 1). The SEM results indicate that detrital minerals are predominant in the soils (quartz and phyllosilicates), with galena and cerussite in low proportion, as well as neoformed phases rich in Pb, Fe, and As. In addition, the precipitates are primarily constituted by Ca-Na sulfates and carbonates, accompanied by Na arsenates (Fig. 1).

In the slag of the La Cruz smelter, a slightly acidic pH (6.4) and an electrical conductivity of

Figure 1. SEM image of precipitates sampled downstream the smelting slag dump.

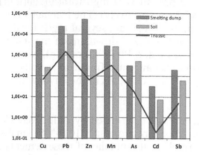

Figure 2. Total concentrations for selected elements in the smelting slag and surrounded soils. Local geochemical background value for Triassic materials is also shown.

110 µS/cm were measured. These smelting wastes have very high total metal(loid) contents (4400, 51000, 23000 and 300 mg/kg for Cu, Zn, Pb, and As, respectively) and are arranged directly on the Triassic deposits. Martínez et al. (2012) observed a direct relationship between the slag dump and the substratum on which it lies. This implies that the leachates generated by rainfall infiltration can reach the subsoil and affect the subsurface flow.

The soils sampled downstream the smelter, in a lower area of olive groves receiving the polluted leachates, also had elevated trace elements concentrations for Cu (240 mg/kg), Zn (1800 mg/kg), and Pb (9900 mg/kg), with total As reaching 500 mg/kg. These figures are very alarming, even if we compared them with the enriched local geochemical background (Martínez et al., 2008) for Triassic materials (Fig. 2).

The results obtained for element distributions indicated that the residual fraction was more important at the smelting dump than in the soil samples. Trace element partitioning by sequential extraction showed greater As potential mobilization and bioavailability at the smelting site than in downstream soils, although Mn, Zn, and Pb distribution were similar in both sites. Considering the percentage distribution of the studied fractions in the soils (Fig. 3), the predominant fraction for Cu was the oxide form, and carbonate bound fraction

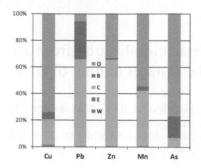

Figure 3. Trace element partitioning (% of the sum of the different fractions) in soils downstream the smelting dump: W, water soluble; E, exchangeable; C, carbonate bound; B, sulfate and organic matter; O, amorphous oxides.

dominated for Pb and Zn. Arsenic was largely contained in the organic matter and oxides bound fraction. Despite the high total trace element concentrations, it should be noted that the water soluble and exchangeable fractions were very low. This circumstance could contribute to prevent the mobility of these elements in the environment.

4 CONCLUSIONS

The smelting slag dump represents long-term source of pollutants, particularly of As, to local surface waters and soils. Considering our results, we can conclude that there is an important Pb and As mobilization from the smelting dump in the form of alkaline waters of Na-SO$_4$ type. These leachates contribute downstream to the soil pollution and the contamination level makes it necessary to consider a water treatment coupled with remediation of the soils near the smelter.

ACKNOWLEDGEMENTS

This research was funded by the Spanish Ministry of Science and Innovation (Project CGL2013-45485-R) and by the Government of Junta de Andalucía (Project RNM 05959).

REFERENCES

Martínez, J., Llamas, J.F., De Miguel, E., Rey, J., Hidalgo, M.C. 2008. Multivariate analysis of contamination in the mining district of Linares (Jaen, Spain). Appl. Geochem. 23: 2324–2336.

Martínez, J., Rey, J., Hidalgo, M.C., Benavente, J. 2012. Characterizing abandoned mining dams by geophysical and geochemical methods: The Linares-La Carolina district (Southern Spain). Water Air Soil Poll. 223: 2955–2968.

Rey, J., Martínez, J., Hidalgo, M.C., Rojas, D. 2013. Heavy metal pollution in the Quaternary Garza basin: A multidisciplinary study of the environmental risks posed by mining (Linares, southern Spain). Catena 110, 234–242.

Arsenic Research and Global Sustainability – Bhattacharya, Vahter, Jarsjö, Kumpiene,
Ahmad, Sparrenbom, Jacks, Donselaar, Bundschuh & Naidu (Eds)
© 2016 Taylor & Francis Group, London, ISBN 978-1-138-02941-5

Influence of limestone drains and seasonality on arsenic speciation in a stream surrounding a gold mineralization area

R.W. Veloso[1], J.W.V. de Mello[1,2], C.B. Rosado[1] & S. Glasauer[3]

[1]*Universidade Federal de Viçosa, Department of Soils, Viçosa-MG, Brazil*
[2]*National Institute of Science and Technology on Mineral Resources, Water and Biodiversity—INCT-Acqua, Brazil*
[3]*School of Environmental Sciences, University of Guelph, Guelph, ON, Canada*

ABSTRACT: Arsenic is a trace element that often occurs in gold mineralization areas associated with sulfide minerals. Acid Mining Drainage (AMD) caused by sulfide and ferrous iron oxidation, as well as biomethylation are important processes related to As mobility and toxicity in mining areas. This work focuses on the influence of limestone drains on As mobility and speciation in a stream surrounding a gold mineralization area. Total arsenic contents and speciation were mainly influenced by the proximity to the mineralized area, but limestone drains efficiently decreased soluble As. Arsenate (As^V) was the dominant species downstream the limestone drain. For organic arsenicals, only DMA was detected close to the mineralization area, upstream the limestone drain, which was related to the presence of algal mats and gel flocs.

1 INTRODUCTION

Arsenic is often associated to gold mineralization, where it occurs typically as arsenopyrite (FeAsS) or impurities in pyrites (FeS_2). Arsenic is mobilized when sulfide minerals are exposed to oxygen and water, causing Acidic Mine Drainage (AMD) (Evangelou, 1995). Neutralization of AMD is often carried out by the use of alkaline agents, such as Ca and Mg carbonate minerals. AMD neutralization can also promote precipitation of Fe (hydr)oxides, thereby increasing the immobilization of As (Simón et al., 2005). Biological mediation of As mobility, including biomethylation, is considered an important mechanism to detoxify the environment. Therefore it has been considered in remediation of contaminated areas (Mello et al., 2007; Rahman and Hassler, 2014). This work evaluated the influence of limestone drains and seasonality on As speciation and dispersion in a water stream surrounding a gold mineralization area.

2 EXPERIMENTAL

2.1 Samples collection

Water samples and sediments were collected from a creek, close to a gold mineralization area, located northwest of Minas Gerais State, Brazil (Figure 1), during the dry (November, 2011) and rainy (February 2012) seasons. There were four sampling sites, along the creek: two sites (B.1 and B.1.1), closer to the mineralized area and close to springs, upstream the limestone drain, and other two sites (B.2 and B.3) downstream this drain.

2.2 Samples pretreatments

Water and sediments samples were preserved at 4°C and stored in the dark just after collection. Water

Figure 1. Sketch map of the sampling sites location.

samples were filtered through a Millipore membrane (<0.45 µm). Sediment samples were freeze-dried, passed through a 2 mm sieve and stored in polyethylene bottles. Subsequently, sediments were analyzed for pH (1:2.5 in aqueous suspension). Total As and Fe contents in sediments were determined by ICP-OES after acid digestion. Arsenic speciation was performed in filtered water samples by HG-AFS coupled to a HPLC column. Total soluble As was determined by the sum of As species.

3 RESULTS AND DISCUSSION

3.1 Sediments characterization

Values for pH in sediments did not show evidence of acid drainage, ranging from 5.6 to 7.0, for both seasons. Lower pH values were observed in samples B.1.1 which presented also the highest As contents, during both dry and rainy seasons. This sample presented a thin gelatinous orange layer, named gelfloc, suggesting accumulation of Fe (hydr)oxides on biogenic material (Figure 2). This is confirmed by the highest Fe contents in B.1.1 (Table 1). In sites B2 and B3, located downstream

Figure 2. Iron (hydr)oxides gel flocs in water streams close to the gold mineralization area.

Table 1. Values of pH, As and Fe contents in sediments during the dry (November 2011) and rainy (February 2012) seasons.

	pH		As*		Fe	
	Dry	Rainy	Dry	Rainy	Dry	Rainy
			g kg⁻¹			
B.1	5.9	6.5	2.5	0.1	61.2	37.7
B.1.1	5.6	6.1	8.7	5.1	110.9	84.8
B.2	7.0	7.0	1.5	0.3	66.1	35.9
B.3	6.9	7.0	1.1	1.0	107.1	55.8

*As and Fe content in sediment by EPA 3052.
Detection Limits: 15 mg/kg for As and 19 mg/kg for Fe.

limestone drain, arsenic contents were smaller than at B.1.1 (Table 1).

3.2 Water characterization

Soluble As decreased downstream, immediately after the limestone drain (Table 2). It was observed the presence of Fe (hydr)oxides coatings on the gravel limestone. Then these (hydr)oxides, especially the poorly crystalline phases, can be responsible for As retention in the limestone drain.

The decrease of soluble As in water could not be due to dilution effects because groundwater is below the depth of these shallow creek. Therefore, arsenic may be associated with sediments, in particular the composites of Fe (hydr)oxides, clay minerals and algae mats that coat the stream beds (Table 1). Identification of chemical forms of As in stream sediments is warranted to assess their long term stability against As release.

Inorganic As species (As^V and As^{III}) dominated in the water samples. High As concentrations were obtained close to the mineralization area, in B.1.1 (Table 2). The highest concentration for As^{III} was during the dry season, but for As^V it was during the rainy season. These results reveal the influence of seasonality on As speciation. In spite of the high As^{III} contents upstream the limestone drain, As^V was the main species downstream, for both seasons (Table 2). These results are indicative of the effectiveness of limestone drain in controlling As mobility, as well as a progressive oxidation of As

Table 2. As speciation in water samples, collected during the dry (November 2011) and rainy (February 2012) seasons.

	As^V		As^{III}		DMA		Sun	
	Dry	Rainy	Dry	Rainy	Dry	Rainy	Dry	Rainy
				µg L⁻¹				
B.1	43	21	LD	123	LD	3.8	34	148
B.1.1	LD	237	265	0.5	LD	5.9	265	234.5
B.2	32	16	LD	LD	LD	LD	32	16
B.3	22	13	LD	LD	LD	LD	22	13

* D–Dry season and R–rainy season
# Detection limits, DL = 0.45 µg/L (As^V); 0.28 µg/L (As^{III}); and 0.39 µg/L (DMA).

downstream. Organic species were detected only in minor concentrations, just as DMA during the rainy season, close to the mineralization area (Table 2). Interestingly DMA detection was associated to the occurrence of gel flocs. This suggests As biotransformation by living organisms, as algae or cyanobacteria, and is in line with findings by Mello et al. (2007) in waterlogged soils from the same area.

4 CONCLUSIONS

The limestone drain proved to be efficient to remove arsenic from water. Seasonality affected both total arsenic contents in sediments and speciation in water samples. Organic species were detected only as DMA in minor concentrations during the rainy season. It was associated to the presence of gel flocs.

ACKNOWLEDGEMENTS

To the funding Brazillian agencies CNPq (Conselho Nacional de Desenvolvimento Científico e Tecnológico), CAPES (Coordenação de Aperfeiçoamento de Pessoal de Nível Superior) and FAPEMIG (Fundação de Amparo a Pesquisa do Estado de Minas Gerais).

REFERENCES

Evangelou, V.P. 1995. *Pyrite Oxidation and its Control.* Boca Raton: CRC Press. 293p.

Mello, J.W.V., Talbott, J.L., Scott, J., Roy, W.R., Stucki, J.W. 2007. Arsenic speciation in arsenic-rich brazilian soils from gold mining sites under anaerobic incubation. *J. Soils Sediments* 6: 1–9.

Simón, M., Martín, F., García, I., Bouza, P., Dorronsoro, C., Aguilar, J. 2005. Interaction of limestone grains and acidic solutions from the oxidation of pyrite tailings. *Environ. Pollut.* 135: 65–67.

Rahman, M.A. Hassler, C. 2014. Is arsenic biotransformation a detoxification mechanism for microorganisms? *Aquat. Toxicol.* 146:212–219.

Arsenic Research and Global Sustainability – Bhattacharya, Vahter, Jarsjö, Kumpiene,
Ahmad, Sparrenbom, Jacks, Donselaar, Bundschuh & Naidu (Eds)
© 2016 Taylor & Francis Group, London, ISBN 978-1-138-02941-5

Redistribution of arsenic in two lakes affected by historical mining activities, Stollberg, Sweden

M. Bäckström
Man-Technology-Environment Research Centre, Örebro University, Örebro, Sweden

ABSTRACT: Milling occurred between 1898 and 1931 and tailings containing As (2 000 mg/kg dw), Pb (15 000 mg/kg dw) and Zn (20 000 mg/kg dw) were dumped directly into two lakes. Surface sediments contain very high trace element concentrations; 2 000 mg/kg dw As, 8 000 mg/kg dw Pb and 12 000 mg/kg dw Zn. High trace element concentrations were also found in suspended matter (1 700–2 400 mg/kg dw As, 1 000–1 800 mg/kg dw Pb and 2 400–3 900 mg/kg dw Zn). As/Zr ratios for suspended matter indicate that a significant portion of the resuspended materials consist of tailings. Both lakes are still working as sinks for trace elements even though diffusion of As into the surface waters is significant. Results indicate that tailings are still redistributed within the sediments of the two lakes more than 80 years after closure of the milling activities at the lake shores.

1 INTRODUCTION

In the historical Stollberg area mining has been ongoing from at least the Middle Ages, even though today there is no active mine. Historical enrichment plants were often located close to running waters and the waste product was often disposed of in or close to the water.

The main objective of this study is to investigate the redistribution processes of primarily As in two lakes in order to determine unknown sources of As to the lakes and assess the sediments as a potential secondary source for As to the downstream recipient.

2 MATERIALS AND METHODS

2.1 Study area

In the area (Fig. 1) milling activities took place between 1898 and 1931. At Silvhyttan 1, a mill producing jig tailings was located between 1898 and 1919. At Silvhyttan 2, a gravity based mill was in operation between 1913 and 1919. In 1931 all milling operations ceased.

Present sulphide minerals in the area are chalcopyrite ($CuFeS_2$), pyrite (FeS_2), pyrrhotite (FeS), arsenopyrite (FeAsS), galena (PbS) and sphalerite (ZnS).

2.2 Sampling and analysis

Sediment cores were sampled and analysed throughout both lakes. Sediment pore waters were extracted through centrifugation. Bottom water were collected in order to be able to calculate the diffusion flux.

Suspended matter was sampled using both sediment traps and pumping through filters.

Suspended matter was analyzed for major and trace elements using alkaline fusion. Surface waters were sampled and analysed 4 times during one year. Water flow was also measured during each sampling.

Figure 1. Map of the lakes Staren and Plogen and the connecting creek Silvhyttebäcken. Green areas indicate mining waste in or very close to the surface waters.

3 RESULTS AND DISCUSSION

3.1 Jig tailings on shore

Tailings on the lake shores lack vegetation and are totally bare. During rain events tailings are likely being flushed into the lake. Wave erosion is also a likely mechanism. The tailings contain very high concentrations of trace elements (average 2 000 mg/kg dw As, 15 000 mg/kg dw Pb and 20 000 mg/kg dw Zn).

3.2 Sediments

Sometimes more than 60 cm of tailings were found in the sediments and sometimes tailings were clearly noted visually more than 150 m from the shore. In the southern part of lake Staren an approximately 5 cm thick layer of silty material was found in almost every sampling point at somewhat varying depths. This material contained very high concentrations of trace elements (on average 1 700 mg/ dw As, 31 000 mg/kg dw Pb and 34 000 mg/kg dw Zn).

Maximum As concentrations for both lakes are around 3 000 mg/kg dw at a depth around 5–10 cm into the sediments. It is also clear that the As concentrations in the surface sediments are a lot lower even though the concentrations are still very high.

Average sediment concentrations are 1 080 mg/ kg dw As, 8 970 mg/kg dw Pb and 10 800 mg/kg dw Zn for lake Staren. For lake Plogen the average sediment concentrations are 1 560 mg/kg dw As, 32 500 mg/kg dw Pb and 48 200 mg/kg dw Zn.

3.3 Suspended matter

Arsenic concentrations in suspended matter was around 200 mg/kg dw in lake Plogen. In lake Staren, however, the concentrations at the bottom was around 2 100 mg/kg dw compared to around 210 mg/kg dw in the surface water. Resuspension of sediments was estimated by comparing the concentrations of organic matter (as LOI), Zr, TiO_2 and SiO_2. In Staren the estimated resuspension was between 10 and 50% and in Plogen around 15%.

3.4 Pore water and diffusion

Diffusion over the sediment/bottom water interface was calculated according to Chaillou et al. (2003). Results indicate that the only element with significant flux from the sediments in both lakes to the water body is As with around 10 kg/year.

3.5 Mass balance calculations

A mass balance approach was used in order to determine if any source or sink for trace elements had been missed. Results from the calculations for Lake Staren are found in Table 1.

It is clear that there is at least one trace element source lacking in the calculations (negative net values for all trace elements). It is likely that the missing trace element source is leaching and physical erosion from the on-shore tailings.

Table 1. Mass balance calculations for Lake Staren for As, Pb and Zn (kg/year). Net balance is calculated using amounts from sediment traps. Positive values indicate an addition to the water and negative values indicate a loss from the water body. Negative net values indicate that a source for trace elements to the lake is missing.

	As	Pb	Zn
To Staren	+4.4	+21	+88
From Staren	−21.9	−36.8	−650
Sum surface waters	−17	−15	−560
Diffusion	+10	0	0
Sedimentation	−30	−110	−390
Resuspension	+15	+56	+195
Net	−22	−69	−760

The results for Lake Plogen are somewhat different. For As and Pb there is a missing source and Zn there is a missing sink. A likely source is the on-shore tailings.

4 CONCLUSIONS

The original contamination of the sediments in Lake Staren was due to the mills. Tailings were released almost directly into the lake whereby the heavier particles settled immediately while the finer particles were transported further and contaminated the entire lake. The sediments in both Lake Staren and Lake Plogen are still a trap for trace elements, even though the diffusion of As from the sediments is significant. Today the sediments are buried with sediments containing lower trace element concentrations, but resuspension and diffusion keep contaminating the sediments.

In Lake Plogen the source of trace elements is similar to the source for Lake Staren. In addition a lot of dissolved and suspended trace elements are entering the lake through the inlet. Results indicate that the surface sediments still today contain a large portion of tailings (around 25%). A large part of this is probably redistributed to the surface sediments through resuspension.

ACKNOWLEDGEMENTS

Swedish Geological Survey (SGU) is acknowledged for providing funding for the project.

REFERENCE

Chaillou, G., Schäfer, J., Anschutz, P., Lavaux, G., Blanc, G. 2003. The behavior of arsenic in muddy sediments of the Bay of Biscay (France). Geochim. Cosmochim. Acta 67(16): 2993–3003.

Arsenic Research and Global Sustainability – Bhattacharya, Vahter, Jarsjö, Kumpiene,
Ahmad, Sparrenbom, Jacks, Donselaar, Bundschuh & Naidu (Eds)
© 2016 Taylor & Francis Group, London, ISBN 978-1-138-02941-5

The release of arsenic from cyanidation tailings

R. Hamberg, L. Alakangas & C. Maurice
Department of Civil, Environmental and Natural Resources Engineering, Division of Geosciences and
Environmental Engineering, Luleå University of Technology, Luleå, Sweden

ABSTRACT: Tailings from a gold mine containing 1000 mg/kg of As were used to predict the release of As over an extended period of time. Post-cyanide mine processes were aiming to form arsenates and Fe-hydrates for effective As-immobilization. Speciation of the As in ore and tailings samples revealed that mining processes have dissolved the majority of the arsenopyrite in the ore, causing secondary As phases to co-precipitate with newly formed Fe-hydrates. Weathering Cell Tests (WCT) were conducted to assess the effect of weathering on the stability of As in the tailings. As-bearing Fe-hydrates remained intact during the early stages of the WCT. During later stages of the WCT, the release of As, Fe and S increased due to pyrrhotite oxidation and the destabilization of As-bearing Fe-hydrates. Low proportions of As was released in WCT, but additional pyrrhotite oxidation as pH falling to < 3 could further destabilize As-bearing Fe-hydrates.

1 INTRODUCTION

At a gold mine, in northern Sweden, gold was extracted from inclusions in arsenopyrite by cyanide leaching. The benchmark of As-release (15 µg/l) into surrounding waters has been exceeded on a regular basis during these years. Effluents from the cyanide leaching process were treated with $Fe_2(SO_4)_3$, $CuSO_4$ and lime for the immobilization of As and the detoxification of cyanides before deposition into tailings dams. Iron sulfate ($Fe_2(SO_4)_3$) was added to process water deriving from the detox-process where toxic cyanides are destroyed in an alkaline medium (pH 9.5–10). Cyanidation and tailings treatment processes ($Fe_2(SO_4)_3$ and lime addition could alter the distribution of As. The ore and tailings were subjected to static and kinetic leaching and chemical composition to evaluate the speciation and distribution of As. The knowledge of geochemical processes controlling As mobility, will lead to a better management of As-bearing tailings to avoid high As concentrations in the mine drainage. The specific objects for this study were to:

- Perform chemical characterization and speciation of As phases in ore and tailings
- Determine the mobility of As phases in oxidizing conditions by kinetic leaching

2 MATERIALS AND METHODS

2.1 Materials

Tailings (TA) originating from a gold mine, were sampled at depths of 0–30 cm on approx. ten differ-

Table 1. Chemical composition of the tailings and ore ($n = 3, \pm SD$) in terms of selected elements and inorganic C.

Element	Unit	Tailings	Ore
As	mg/kg TS	2011070 ± 30	2014703 ± 781
Cr	”	166 ± 3	247 ± 45
Cu	”	147 ± 7	82.0 ± 12.1
Ni	”	63.8 ± 2.1	126 ± 12
Pb	”	4.56 ± 0.46	7.71 ± 2.58
S	”	20933 ± 493	25367 ± 2363
Zn	”	25.0 ± 0.4	10.6 ± 1.5
$C_{inorganic}$	%	0.19 ± 0.03	N.D

ent locations on the tailings dam and mixed to a bulk sample of approx. 15 kg. Tailings and crushed ore were provided by the mining company (approx. 3 kg).

2.2 Methods

2.2.1 Accelerated weathering—Weathering Cell Test (WCT)

A modified version of the WCT-test developed by Cruz et al. (2001) has been used to accelerate weathering. Approx. 70 g of tailings was placed on a paper filter placed in a Büchner-type funnel. The samples were exposed to weekly cycles consisted of: one day of leaching, three days of ambient air exposure, another day of leaching and finally two days of air exposure. Leaching of the samples was conducted with 50 ml of deionized water over approximately 2 h, and then the leachates were recovered with the help of vacuum. Test was conducted over a period of 32 cycles (217 days).

2.2.2 Sequential extraction test

Arsenic speciation was assessed using the modified sequential extraction scheme described by Dold (2003). In each extraction sequence, 2 g of material were extracted with seven different solutions in succession. That in order to specify As into seven different phases: Water-extractable As, exchangeable As, As bound to Fe (III) oxy-hydroxides, As bound to Fe (III) oxides, secondary sulfides, primary sulfides and the residual fraction. Triplicate samples were stored cold (–4°C) in darkness until analysis.

3 RESULTS AND DISCUSSION

3.1 Sequential extraction test

The sequential extraction results showed that most of the As in the ore was associated with the primary sulphide fraction (98%) and approx. 1% was associated with the residual fraction (silicates).

The sequential extractions of the tailings suggested that a majority of the As content was associated with Fe (III) oxy-hydroxides (94.6%), approx. 2.5% with primary sulfides (arsenopyrite). This indicates that some of the arsenopyrite remained unoxidized. Less than 1% of As was associated with the Adsorbed-Exchangeable-Carbonate (AEC), phase. The solid-phase speciation of As in Svartliden tailings suggested that As has been released from sulphide minerals in the ore and subsequently adsorbed onto Fe-(oxy)-hydroxides (Fig. 1 and 2). Alkaline, As-rich (< 2500 mg/kg) tailings slurries from the cyanidation process were treated by the addition of $Fe_2(SO_4)_3$ aiming to form Fe-hydroxides amenable for As-adsorption. Adsorption of As onto Fe-oxides is more efficient if As is present as As (V). As (V) is assumed to be the dominant As-specie in tailings due to pre-treatment methods including lime, oxidization (by H_2O_2 and O_2).

3.2 Weathering cell test

Weathering cell tests were conducted for 217 days, at the end of the WCT-test, pH decreased to < 3.5 and concentrations of As, Fe and S increased in the leachate. The Fe/S molar ratio in the leachate increased to 0.8 by the end of the WCT, suggesting

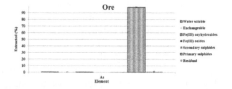

Figure 1. Distributions of As across different phases in unprocessed ores from the Svartliden gold mine based on the results of sequential extraction tests (Error bars, n = 3, ± SD).

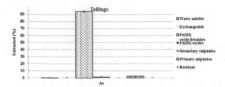

Figure 2. Distributions of elements across different phases in tailings from the Svartliden gold mine based on the results of sequential extraction tests (Error bars, n = 3, ± SD).

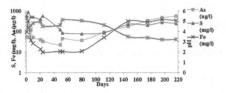

Figure 3. Evolution of pH and release of Fe, S and As from gold mine tailings over time during the WCT.

that pyrrhotite had been oxidized over the course of the experiment. The release of As increased during pyrrhotite oxidation. Moreover, when the pH falls below 3, As-bearing Fe-precipitates become unstable and more As is released (Fig. 3).

4 CONCLUSIONS

The solid-phase speciation of As in ore and tailings suggested that mining processes has re-distributed As from sulphide minerals in the ore to subsequently be adsorbed onto Fe-(oxy)-hydroxides in the tailings. Results from the weathering cell tests shows that As-release increased during pyrrhotite oxidation. Moreover, when the pH falls below 3, As-bearing Fe-precipitates become unstable and more As is released.

ACKNOWLEDGEMENTS

Financial support from Ramböll Sverige AB, Ramböll Foundation, SUSMIN—Tools for sustainable gold mining in EU, and the Center of Advanced Mining and Metallurgy (CAMM) at Luleå University of Technology are gratefully acknowledged.

REFERENCES

Cruz, R., Bertrand, V., Monroy, M., González, I. 2001. Effect of sulfide impurities on the reactivity of pyrite and pyritic concentrates: a multi-tool approach. *Appl. Geochem.* 16(7–8): 803–819.

Dold, B. 2003. Speciation of the most soluble phases in a sequential extraction procedure adapted for geochemical studies of copper sulfide mine waste. *J. Geochem. Explor.* 80(1): 55–68.

Arsenic Research and Global Sustainability – Bhattacharya, Vahter, Jarsjö, Kumpiene,
Ahmad, Sparrenbom, Jacks, Donselaar, Bundschuh & Naidu (Eds)
© 2016 Taylor & Francis Group, London, ISBN 978-1-138-02941-5

Characterization of arsenic-resistant endophytic bacteria associated with plant grown on mine tailings

B. Román Ponce, R. Rodríguez-Díaz, I. Arroyo-Herrera, M.S. Vásquez-Murrieta & E.T. Wang

Departamento de Microbiología, Escuela Nacional de Ciencias Biológicas, Instituto Politécnico Nacional,
México D. F., México

ABSTRACT: The As resistant plants may harbor microbial communities able to detoxify As. Until now, there is no report about the diversity and As transforming ability of the endophytic bacteria in Mexico. Thirty As resistant bacteria were isolated from the roots of *P. laevigata* and *S. angustifolia* grown on the mine tailings in Villa de la Paz, SLP, Mexico. These isolates were identified into eight genera within three phyla (mainly Firmicutes and Actinobacteria). These bacteria showed a high resistance to As: 100 mM to As^{5+} and 50 mM to As^{3+}. Some of them presented PGP characteristics. Twenty-one strains were able to transform As on CDM. Therefore, the two sampled plants harbored endophytic bacteria with high As resistance, which could transform As and produced siderophore at the presence of As^{3+} and/or As^{5+}. These bacteria have potential in the remediation of the As contaminated sites and worthy for further study.

1 INTRODUCTION

Microorganisms have an important role in the transformation of arsenic (As) in the environment and affect the speciation and mobility of As; therefore alter its bioavailability and toxicity in soils (Oremland *et al.*, 2005, Borch *et al.*, 2010). Up till now, little information is available about the arsenic-resistant endophytic bacteria (AEB) associated with the arsenic hyperaccumulator plants; and plant growth promoting features have been reported for these bacteria (Zhu *et al.*, 2014, Xu *et al.*, 2016).

Nowadays, there is no report about the AEB associated with plant grown on mine tailings in Mexico. For this reason, the aims of this study were to isolate and characterize AEB, evaluate their ability in As tolerance, As transformation and As induced siderophore production. Our findings substantiate the potential application of native bacteria for the detoxification of As in contaminated soils.

2 METHODS

2.1 *Sampling site, isolation, identification and characterization of endophytic bacteria*

The sampling site is located in Villa de la Paz in the state of San Luis Potosí (23.7 N, 178.7 W). The mine tailing have As concentration above 8420 mg kg^{-1} (Franco-Hernández *et al.*, 2010). Roots of two endemic plants, *Prosopis laevigata* and *Spharealcea angustifolia*, were sampled. The AEB were isolated from roots according to Márquez-Santacruz *et al.* (2010) and Barzanti *et al.* (2007). Genomic DNA was extracted from each isolate using the protocol of Zhou *et al.* (1995). The 16S rRNA genes were amplified using the genomic DNA as template and the primers fD1/rD1 (Weisburg *et al.*, 1990). The amplicons (1400 pb) were sequenced following Sun *et al.* (2010) and the acquired sequences were compared with those in the GenBank database using the BLAST program. A neighbour-joining phylogenetic tree was constructed and the isolates were identified according to their phylogenetic relationships as described by Román-Ponce *et al.* (2015). Plant growth promoting characters, including siderophore production, Idolacetic Acid (IAA), phosphate solubilization and nitrogen fixation were assayed following the methods of Schwyn & Neilands (1987), Perez-Miranda *et al.* (2007), Kuklinsky-Sobral *et al.* (2004), Bric *et al.* (1991) and Rodriguez-Caceres (1982).

2.2 *Arsenic resistance, transformation and organic ligands synthesis*

All isolates were tested for their As resistance by growth in the medium proposed by Rathnayake *et al.* (2013) supplied with 5–20 mM As^{3+} or 5–100 mM As^{5+}. To examine As transformation, all isolates were cultivated in Chemical Defined Medium (CDM) (Weeger *et al.*, 1999) amended with 2 mM of NaAsO$_2$ or NaH$_2$AsO$_4 \cdot$7H$_2$O; The method of silver nitrate was used for screening the biotransformation; and the arsenic biotransformation was further quantitatively verified using the molybdenum blue method (Hu *et al.*, 2012). The siderophore production was evaluated for the isolates according to Nair *et al.* (2007).

3 RESULTS AND DISCUSSION

3.1 Low diversity of As-resistant endophytic bacteria

In the present study, thirty AEB resistant 5 mM of As^{5+} or As^{3+} were isolated from the two sampled plants. A low diversity was observed among the isolates that were identified into 3 phyla belonging to 8 genera: *Bacillus, Staphylococccus, Pseudomonas, Kocuria, Micrococcus, Microbacterium, Leucobacter,* and *Arthrobacter* (Fig. 1). Among these genera, *Microbacterium, Micrococcus, Arthrobacter, Kocuria, Bacillus* and *Pseudomonas* have been reported as AEB (Zhu *et al.*, 2014; Xu *et al.*, 2016). These results suggested that Gram-positive bacteria in Firmicutes and Actinobacteria were dominant AEB; while the Gram-negative bacteria in Proteobacteria were the minor group in AEB associated with the sampled plants.

3.2 Limited potential in AEB for plant growth promotion

In particular, strains *Bacillus* spp. NP1E3, NP2E5, CS3E3 and *Arthrobacter* sp. CS3E2 showed positive for at least a half of the evaluated plant growth promoting traits. The proportion of isolates with plant growth promoting features in this study was lower than that presented in the AEB of the arsenic hyperaccumulator *Pteris vittata* (Zhu *et al.*, 2014; Xu *et al.*, 2016).

3.3 Arsenic transformation and siderophore production

Among the isolated AEB, 51% were capable of transforming arsenate to arsenite, and 50% oxidized

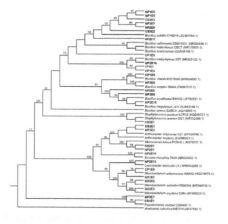

Figure 1. Neighbor-joining (NJ) tree of 16S rRNA genes reconstructed with the Jukes–Cantor distance. Bootstrap values greater than 50% were presented at the nods. *Anabaena cylindrica* NIES 19 was used as an out group.

arsenite in the CDM medium. Only 6 isolates were unable to transform arsenic under the test condition. Evaluated with the molybdenum blue method, only 3 isolates could reduce As^{5+}: NE2E1 (*Micrococcus luteus*) reduced 94.17% ± 0.36; NM2E15 (*Bacillus* sp.) reduced 69.16% ± 1.14; and CE3E2 (*Bacillus* sp.) reduced 24.9% ± 4.69. With respect to the arsenite oxidation, only isolate NM2E7 (*Pseudomonas stutzeri*) oxidized 45.91% ± 2.92. Arsenic biotransformation by members of the genera *Micrococcus, Bacillus* and *Pseudomonas* have been reported in previous reports (Majumder *et al.*, 2013; Zhu *et al.*, 2014; Guo *et al.*, 2015; Xu *et al.*, 2016). Over 80% (20 isolates) and 54% (14 isolates) of the AEB showed ability to produce siderophores against As^{5+} and As^{3+}, respectively; and the isolates with siderophore secretion belonged to *Microbacterium, Bacillus, Staphyloccoccus, Brevibacterium, Micrococcus, Pseudomonas, Kocuria* and *Staphylococcus*.

4 CONCLUSIONS

Our study characterized AEB associated with *P. laevigata* and *S. angustifolia* growth in the mine tailings contaminated with high arsenic concentrations. The isolated endophytic bacteria had an innate capability of arsenic resistance and some plant growth-promoting characteristics. *Bacillus* was the most dominant AEB group in the isolates. More than fifty percent of the isolates could reduce As^{5+}. Further studies should be carried out to characterize their mechanisms in As tolerance, transformation and siderophore secretion. Characterization of arsenic-resistant endophytic bacteria reported here is the first step to use them in bioremediation of As-contaminated sites.

ACKNOWLEDGEMENTS

This research was funded with Projects SIP- IPN 20130722 and 20130828. B.R.P. received Doctoral scholarships support from the CONACyT.

REFERENCES

Hu, S., Lu, J., Jing, C. 2012. A novel colorimetric method for fields arsenic spetiation analysis. *J. Environ. Sci.* 24: 1341–1346.
Nair, A., Juwarkar, A.A., Singh, S.K. 2007. Production and characterization of siderophores and its application in arsenic removal from contaminated soil. *Water Air Soil Poll.* 180: 199–212.
Xu, J.Y., Han, Y.H., Chen, Y., Zhu, L.J. & Ma, L.Q. 2016. Arsenic transformation and plant promotion characteristics of As-resistant endophytic bacteria from As-hyperaccumulator *Pteris vittata. Chemosphere* 144: 1233–1240.

Arsenic Research and Global Sustainability – Bhattacharya, Vahter, Jarsjö, Kumpiene,
Ahmad, Sparrenbom, Jacks, Donselaar, Bundschuh & Naidu (Eds)
© 2016 Taylor & Francis Group, London, ISBN 978-1-138-02941-5

Arsenic removal by acid tolerant sulfate reducing bacteria in a semiarid system

E.D. Leiva[1,2,3], I.T. Vargas[2,3] & P.A. Pasten[2,3]

[1]*Departament of Inorganic Chemistry, Faculty of Chemistry, Pontificia Universidad Católica de Chile, Santiago, Chile*
[2]*Department of Hydraulic and Environmental Engineering, Pontificia Universidad Católica de Chile, Santiago, Chile*
[3]*CEDEUS, Centro de Desarrollo Urbano Sustentable, Chile*

ABSTRACT: Rivers in northern Chile are enriched in arsenic (As) derived from natural and anthropogenic sources. Mining activity in the area, enhances the generation of Acid Mine Drainage (AMD), promoting the mobilization of As from the solid phase. AMD poses a threat to waterways due to low pH, high contents of toxic metals and dissolved salts. Sulfate Reducing Bacteria (SRB) can be used for biological treatment of the contaminated waters, because SRB allow the reduction of sulfate that precipitate metals and metalloids (i.e. As) as insoluble sulfides. Therefore, bioremediation systems based on SRB could be used to remove As from acidic waters indirectly. This study was conducted to determine the presence of sulfate reduction processes under acidic conditions and their potential for As removal from AMD waters. Batch reactors were inoculated with SRB enriched from anaerobic acid sediments. Our results show that As concentrations in aqueous phase were attenuated (> 70%) due to stabilization processes in Fe-rich sediments. In addition, microbiological analysis confirmed the presence of SRB. Thus, processes of sulfate reduction mediated by acid-tolerant SRB can achieve neutralization of pH and a significant As removal from AMD waters, and would be useful in the development of more effective AMD remediation systems.

1 INTRODUCTION

Northern Chile is known for mining and water scarcity. This area also has high concentrations of arsenic (As) in aqueous and solid phase, seriously affecting the quality of water resources and limiting the use of water for different purposes (Leiva *et al.*, 2014). In addition, arid and semiarid climate strongly contributes to the generation of As-rich waters, due to high evaporation rates.

Arsenic is widely distributed in natural aqueous environmental systems, where interaction between iron (Fe) minerals and As can have a significant role in the As release and mobilization. The environmental fate of As is linked to the environmental fate of Fe, because As is mainly sorbed onto ferric Fe(III)-oxyhydroxides due to its high affinity and abundance (Dixit & Hering, 2003). The main source of As in the environment is the release from As-rich minerals (e.g. arsenopyrite (FeAsS) (Smedley & Kinniburgh, 2002), but mining operations can accelerate their release to the aqueous phase. Likewise, AMD associated with mining has a negative impact on the mobilization of As and can promote the release of As directly through the generation of AMD or by dissolution of As-rich mineral phases.

AMD remediation is a major focus of interest for the mining industry worldwide. Unfortunately, conventional treatment technologies for acid neutralization, metal or salts removal have several limitations, both economic and operational. In this sense, bioremediation systems based on acid tolerant Sulfate Reducing Bacteria (SRB) can be used for treatment of these contaminated waters. Sulfate reduction processes mediated by SRB allow the reduction of sulfate concentrations, precipitation of metals and acidity consumption (Neculita *et al.*, 2007). Neutralization of pH can facilitate precipitation of Fe(III) oxyhydroxides and the As removal by adsorption. This study investigates the removal of As from AMD waters by biological sulfate reduction processes under acidic conditions. The relevance of these findings lies in the possibility of optimizing As removal in AMD treatment systems.

2 METHODS

2.1 *Study site: Aguas Calientes area in upper section of Azufre River sub-basin*

The Aguas Calientes area is located in the Azufre River sub-basin near the Tacora volcano, in the Lluta River Watershed in northern Chile (18°00′–18°30′S and 70°20′–69°22′W). In this area the As concentrations in hydrothermal and surface waters exceeds 0.9 mg/L, Fe concentrations exceed 70 mg/L and sulfate concentrations are higher than 1000 mg/L. A wetland grows in the Aguas Calientes area and sulfate reduction processes can occur continuously. To assess this, five sampling points in the wetland area were selected and the sediment samples were used to inoculate batch bioreactors.

2.2 *Batch experiments*

Batch bioreactors were assessed during a 80-day batch experiment. Five batch reactors in 250 mL glass bottles were inoculated with anaerobic sediments from the five sampling points. Natural water from the Azufre River with high As (> 3 mg/L) and Fe concentrations (> 80 mg/L) and low pH (< 2.0) was then added to bioreactors and As removal and pH neutralization were evaluated over time.

2.3 Hydrogeochemical analyses

Water samples from the bioreactors evolution were taken each 7 days and the bioreactor performance was assessed by monitoring pH, temperature, electrical conductivity and by the measurement of total dissolved metals (As, Fe). Measurements of the total dissolved metal concentrations were carried out by ICP-OES.

3 RESULTS AND DISCUSSION

3.1 pH neutralization in batch bioreactors

Batch tests were conducted to evaluate the presence of biological sulfate reduction processes. The pH neutralization was observed over time in five bioreactors (Fig. 1).

The pH increases rapidly from 1.7 to 6.0–7.5. Additionally, an increase in the total alkalinity was observed (from 5 mg/L to 500 mg/L $CaCO_3$), suggesting that the processes of acid neutralization may be due to biological sulfate reduction. Interestingly, R5 bioreactor shows a constant pH above 7.0, suggesting that the microbial consortia of the R5 sediment can be enriched in SRB. The presence of SRB was confirmed by characterization of the microbial community and by decrease of sulfate concentrations in aqueous phase (data not shown).

The optimal reduction of sulfate occurs at pH in the range of 5.0 to 8.0 (Willow & Cohen, 2003). Thus, the pH values reported during the batch experiments suggest that SRB are active during operation and tolerate the initial acidity.

3.2 Arsenic and iron concentrations are removed in sulfate reduction bioreactors

To reveal if pH neutralization and sulfate reduction process promote the As removal, we evaluate the change in Fe and As concentrations in batch bioreactors. The results showed a higher removal of Fe (~90%) and As (~80%). Figure 2 shows that the higher removal of As and Fe occurs in correlation with the increase of pH (Figure 1), suggesting an initial mechanism pH-dependent.

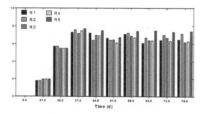

Figure 1. pH evolution in batch bioreactors inoculated with sulfate reduction sediments. The graphs show the average of data.

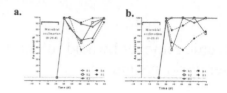

Figure 2. Total dissolved Fe (a) and As (b) evolution during batch bioreactor experiments. A marked removal of As and Fe was observed.

The pH increase in Fe-rich waters favors the precipitation of Fe(III)-oxyhydroxides and subsequent adsorption of As on these minerals. Thus, our data support the idea that the removal of Fe is directly controlled by a pH-dependent mechanism, while As removal is associated with a mechanism of adsorption onto Fe oxides.

4 CONCLUSIONS

Our results show that the Fe and As removal are influenced by the pH increase mediated by acid tolerant SRB. The removal of As is controlled by pH increase and stabilization onto Fe-minerals. pH values remain in circum-neutral ranges due to increase in alkalinity during sulfate reduction and this promotes the precipitation of metal carbonates and hydroxides in oxic microenvironments. These results are a step forward in understanding sulfate reduction systems under acidic conditions, and to increase its potential as a strategy for AMD remediation and As removal.

ACKNOWLEDGEMENTS

The authors acknowledge a FONDECYT Postdoctorado 3140515, FONDECYT 1130936/2013, and CONICYT/FONDAP 15110020 grant.

REFERENCES

Dixit, S., Hering, J.G. 2003. Comparison of arsenic(V) and arsenic(Iii) sorption onto iron oxide minerals: implications for arsenic mobility. Environ. Sci. Technol. 37: 4182–4189.

Leiva, E.D, Ramila, C.d.P., Vargas, I.T., Escauriaza, C.R., Bonilla, C.A., Regan, J.M., Pizarro, G.E, Pasten, P.A. 2014. Natural attenuation process by microbial oxidation of arsenic in a high Andean watershed. Sci. Total Environ. 466–467: 490–502.

Neculita, C.M., Zagury, G.J., Bussiere, B. 2007. Passive treatment of acid mine drainage in bioreactors using sulfate-reducing bacteria: critical review and research needs. J. Environ. Qual. 36(1): 1–16.

Smedley, P.L., Kinniburgh, D.G. 2002. A review of the source, behaviour and distribution of arsenic in natural waters. Appl. Geochem. 17: 517–568.

Willow, M.A., Cohen, R.R.H. 2003. pH, dissolved oxygen, and adsorption effects on metal removal in anaerobic bioreactors. J. Environ. Qual. 32: 1212–1221.

Arsenic Research and Global Sustainability – Bhattacharya, Vahter, Jarsjö, Kumpiene, Ahmad, Sparrenbom, Jacks, Donselaar, Bundschuh & Naidu (Eds)
© 2016 Taylor & Francis Group, London, ISBN 978-1-138-02941-5

The distribution of arsenic and arsenic-bearing minerals in basins of the Northern Atacama region, Chile

J. Tapia[1,2], A. Menzies[3], C. Dorador[4] & D. Cornejo[4]

[1]*Universidad Austral de Chile (UACh), School of Geology, Campus Isla Teja, Valdivia, Chile*
[2]*Universidad de Antofagasta (UANTOF), Mines Engineering Department, Campus Coloso, Antofagasta, Chile*
[3]*Universidad Católica del Norte (UCN), Geology Department, Antofagasta, Chile*
[4]*Universidad de Antofagasta, Instituto Antofagasta, Biotechnology Department, Campus Coloso, Antofagasta, Chile*

ABSTRACT: Northern Chile has a long history of arsenic (As) contamination. Examples include arsenic found in Chinchorro mummies' hair and high rates of lung and bladder cancer in the city of Antofagasta. The natural environment of the northern Atacama Region is comparable to areas associated with high natural arsenic concentrations in Chile. To quantify natural concentrations of arsenic of this region, the northern Atacama Region basins were sampled at the El Salado river, the Pedernales salt-flat, and the Río Negro and Laguna Verde hot-springs. Our results show that: (1) to the east of Diego de Almagro, all water samples contain dissolved-As values above 700 μg/L; (2) As bearing phases were not observed in the colloidal fraction; (3) As-bearing minerals were found at the Pedernales salt-flat (silicate minerals with ~20% As), the Potrerillos site (apatite and Fe-oxides), and the El Salado river close to the Potrerillos site (iron oxides); (4) in 80% of the studied sediments there are enzymes that metabolize As, however this phenomenon is nearly absent in the Pedernales salt-flat indicating a high solubility of the As-bearing silicates.

1 INTRODUCTION

Northern Chile has a long history As pollution (Fraser, 2012), from Chinchorro mummies (Byrne *et al.*, 2010) to current Antofagasta inhabitants who have the highest rates of bladder and lung cancer in Chile due to a chronic consumption of As between 1958–1970 (Fraser, 2012). This pollution is believed to be the result of natural sources (e.g. volcanic rocks, hot-springs and mineral deposits; Romero *et al.*, 2003) and anthropogenic sources, e.g. mineral deposits exploitation (Romero *et al.*, 2003).

The northern Atacama Region is located at the southern limit of the Atacama Desert and is generally characterized by the presence of volcanic rocks, hot-springs, and porphyry and epithermal ore deposits. The main superficial water source is the non-perennial El Saldo river, which flows close to two National Parks (Pan de Azúcar on the coast and Nevado de Tres Cruces in the Andes). Since the beginning of the 20th Century this region has been affected by mining activities through the exploitation of the Potrerillos and El Salvador porphyry copper deposits. Subsequently, due to mining activities between 1938 and 1975, all mine tailings were discharged directly into the El Salado River without any treatment (Paskoff & Petiot, 1990). The unique historical background and geological features of this area motivate this study in which the As distribution in surficial water and As-bearing minerals are analyzed.

2 METHODS

2.1 Sampling

Samples from the Andes were taken in SE-NW transverse sections. Within the El Salado River, water samples were collected in an EW transverse orientation (Fig. 1, RS01 to RS16). A total of 25 water and filter (0.45 μm) samples, 10 sediment samples, and 1 precipitate sample (SP02) were collected. Water samples were filtered in acid cleansed polypropylene vials and were acidified in the field. These samples were kept at 4°C until analysis. Filters were sealed and sediments were kept in sterile 125 ml sampling bags.

2.2 Water analysis

The elemental matrix was quantified using the ICP-MS at the Géosciences Environnement Toulouse (GET) Laboratory, and the data quality was confirmed against the SLRS-5 standard.

2.3 Sediment analysis

Colloids were identified by Scanning Electronic Microscope (SEM) and Energy Dispersive Spectrometer (EDS) in the Geology Department of the UCN. Sediment mineralogy was characterized by a QEMSCAN® E430, based on a ZEISS EVO 50 SEM combined with Bruker Series 4 EDS detectors.

Routine analysis was performed using a beam size of less than 1 μm with an operative voltage of 25 kV and beam current of 5 nA. Samples were analyzed in Fieldscan mode (field size 1500, ~50x). Standard X-ray counting of 1000 photons at each point was acquired at a pixel spacing of 10 μm. Determinations and data reductions were performed using iMeasure v5.3.2 and iDiscover v5.3.2, respectively.

Genomic ADN and ArsC gen were amplified to identify arsenate reductase enzymes. These analyses were performed in the Complejidad Microbiana and Ecología Funcional Laboratory.

3 RESULTS AND DISCUSSION

3.1 Water

All water samples located to the east of Diego de Almagro contained dissolved As concentrations over 700 μg/L (Fig. 1). In a separate stream to the west, As concentrations were well below 10 μg/L. In addition, pH was circumneutral in all studied sites (Fig. 1).

The elementary matrix obtained by ICP-MS was subjected to a multivariate statistical analysis (Li, B, Mg, Sc, Ti, V, Cr, Mn, Fe, Co, Ni, Cu, Zn, As, Sn, Mo, Cd, Sb, Ba, Pb, U). Data obtained was grouped into drinkable and non-drinkable As. However, 3 samples were outside of this regional trend: SP02 (Pedernales salt-flat), RS02 (El Salado river, close to Potrerillos), and PO01 (Potrerillos); these difference could be related to epithermal mineralization or anthropogenic sources.

3.2 Sediments

SEM imagery coupled with EDS spectra showed that the predominant minerals retained in 0.45 μm filters were halite and gypsum, yet As-bearing colloids were not found in this fraction. QEMSCAN® analysis showed that mineralogy of these basins

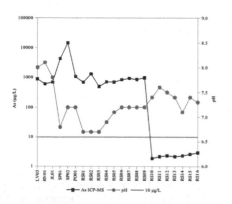

Figure 1. Dissolved As distribution in an E-W transect in Northern Atacama Region. The maximum dissolved As concentration allowed in drinking water is displayed in red (US EPA and WHO).

corresponds to quartz, k-feldspar, micas, carbonates, and clays. Only three sites revealed the presence of As-bearing minerals: (1) Pedernales salt-flat (SP02), where As is a notable constituent of an undetermined silicate (20% As), (2) Potrerillos (PO01), where As is tied to apatite and iron oxides, and (3) El Salado river close to Potrerillos (RS02), where As-bearing minerals are iron oxides.

Universal primers that codify ArsC were positive for 80% of the studied sediment samples. This enzyme performs the first steps of arsenate metabolism before being expelled, reducing arsenate [As(V)] to arsenite [As(III)] (Silver et al., 2001). One of the negative samples was SP02, implying that this unknown silicate might be highly soluble (14,637 μg/L in water).

4 CONCLUSIONS

The highest dissolved As values are located to the east of Diego de Almagro. In water, this element is distributed regionally as drinkable As and non-drinkable As. The Pedernales salt-flat and Potrerillos area do not follow this trend. Arsenic bearing minerals correspond to highly soluble silicates, apatite, and iron oxides.

ACKNOWLEDGEMENTS

This study was funded by the UANTOF and was supported in the field by Stéphane Audry and Sergio Villagrán in addition to CISEM staff in the UCN who prepared and analyzed the samples using QEMSCAN® (Marina Vargas, Monserrat Barraza, Pamela Fonseca, and Carolina Ossandon).

REFERENCES

Byrne, S., Amarasiriwardena, D., Bandak, B., Bartkus, L., Kane, J., Jones, J., et al. 2010. Were Chinchorros exposed to arsenic? Arsenic determination in Chinchorro mummies' hair by laser ablation inductively coupled plasma-mass spectrometry (LA-ICP-MS). Microchem. J. 94: 28–35.

Fraser, B., 2012. Cancer cluster in Chile linked to arsenic contamination. The Lancet 379: 603.

Paskoff, R., Petiot, R., 1990. Coastal progradation as a by-product of human activity: an example from Chañaral Bay, Atacama Desert, Chile. J. Coast. Res. 6: 91–102.

Romero, L., Alonso, H., Campano, P., Fanfani, L., Cidu, R., Dadea, C., et al. 2003. Arsenic enrichment in waters and sediments of the Rio Loa (Second Region, Chile). Appl. Geochem. 18:1399–1416.

Silver, S., Phung, L., Rosen, B., 2001. Arsenic metabolism: resistance, reduction and oxidation. In: W.T. Frankenberger Jr. (Ed.) Environmental Chemistry of Arsenic Marcel Dekker, New York. pp. 247–272.

US EPA, O., 2009. http://water.epa.gov/drink/contaminants/ (accessed 26.1.2015).

WHO, 2011. http://www.who.int/water_sanitation_health/publications/2011/dwq_guidelines/en/ (accessed 27.1.2015).

Arsenic Research and Global Sustainability – Bhattacharya, Vahter, Jarsjö, Kumpiene,
Ahmad, Sparrenbom, Jacks, Donselaar, Bundschuh & Naidu (Eds)
© 2016 Taylor & Francis Group, London, ISBN 978-1-138-02941-5

Leaching of arsenic from waste deposits at La Aurora mine, Xichú mining district, Guanajuato, Mexico: Characterization and remediation

A. Guerrero Aguilar[1] & Y.R. Ramos Arroyo[2]

[1]*University of Guanajuato, Engineering Division, Environmental Engineering Department, Guanajuato, Mexico*
[2]*University of Guanajuato, Engineering Division, Hydraulic Engineering Department, Guanajuato, Mexico*

ABSTRACT: The present work focuses on the leachate characterization, the dimensioning of the issue of arsenic release into the Xichú River and the proposal of low-cost remediation strategies for mining waste at La Aurora mine located in the Xichú Mining District, Guanajuato, Mexico. Approximately 1 million tons of sulfur-rich mining waste were deposited on the banks of the Xichú River during the mining of metamorphic hydrothermal deposits. A total of 47 leachate samples containing high concentrations of potentially toxic elements (As, Cd, Cu, Fe, Mn, Pb and Zn) were analyzed by atomic absorption spectroscopy. Chemical composition diagrams (Piper and Stiff) were created with the obtained data with the software AQUACHEM to visualize the impact of arsenic leaching in the Laja-Xichú basin.

1 INTRODUCTION

The mining district of Xichú in Guanajuato district was a silver producer until 1950. Unfortunately the generation and disposal of mine waste was neglected. Four structures of sulfidic mine waste where disposed at the riverside of the Xichú river and they represent approximately 1 million tons. The exposure of these mine tailings to water oxidizes the sulfidic waste and releases potentially toxic elements in leachate.

This work focuses on the characterization of leachate and the study of how it affects the Laja-Xichú basin as well as the proposal of remediation strategies.

2 METHODS/EXPERIMENTAL

2.1 Sampling strategy

A sampling strategy was paramount for a proper comprehension of how arsenic is distributed in the Laja-Xichú basin. This was achieved as follows:

1. Delimitation of the study area in the Rio Laja.
2. Identification of key locations with presence of arsenic where different types of waters were found.
3. Sample collection and measure of field parameters.

2.2 Sample analysis

Samples were analyzed by atomic absorption spectroscopy (AAnalyst 100 Spectrometer), volumetric and photometric techniques. Chemical composition diagrams (Piper and Stiff) were created with the resulting data with the AQUACHEM software and geochemical simulations were made with the PHREEQC software to visualize the possible scenarios of arsenic leaching.

3 RESULTS AND DISCUSSION

3.1 Variation of natural waters

For most of sampled natural waters in the Laja-Xichú basin it is common to present a carbonate-calcic composition. This due to the presence of fractured rhyolite in the area where fresh water springs can appear.

Figure 1 shows the variation in the major anions and cations in water samples that are related to mining waste leachate at La Aurora mine. Only the most representative sites are highlighted. The bicarbonate-calcic character stream "Charco Azul", located above the gradient of the waste deposits, is represented as water unaffected by leachate.

Leachate 1, 2 and "White Gels" are sulfated product of the oxidation of sulphides in tailings. There are mostly magnesium and calcium, the leachate 1 resulted with very high concentrations of chlorides. The variation of the character of the elements in place "after tailings 2" in the Xichú river shown on this site samples were collected during four seasons in the field (black diamond). It shows that the influence of leachate is higher in the dry season (December 2014) that during the rainy season (September 2013), this influence is reflected in the sulfated character.

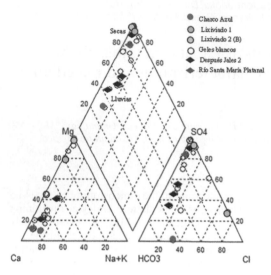

Figure 1. Tailings leachate samples and how they affect the Xichú River.

3.2 Chemical variability of water

Except for the spring "El Platanal" showing a sulfated character, every other sampled site has the dominant anion bicarbonate. The reason for this is that the hydrogeological base in El Platanal is shales, rocks that may contain FeS_2 and releasing SO_4^{-2} when rusted. The prevalent cation in all samples is calcium.

The spring "Casitas" also shows a calcium-sulfated as the other sites, the presence of chloride and sodium suggests anthropogenic involvement. This spring is located in fractured rhyolite and is below the local cemetery, it remains highly vulnerable to contamination.

Salinity that is appreciated, is greater in the spring "El Ojo de Agua" which is the output of the hydrological Laja-Xichú basin, a tributary of the Santa Maria River basin. This spring has up to 20 mEq Ca^{2+}, when the others are in the order of 5 mEq. Apparently significant amounts of sulphides come out of the spring El Ojo de Agua, this was not quantified but at the Santa Maria River after Laja significantly increases the concentration of sulfates by the oxidation of sulphides.

3.3 Comparison with NOM-127-SSA1-1994

As shown in Table 1, nearly all samples have some kind of effect on its purity either anthropogenic or naturally. Significantly, 93.5% of the analyzed samples exceed the maximum permissible limits of Mn. This comes from a rhodochrosite Mn ore with potential to precipitate given site conditions. 49% exceeds the concentration range of sulfates and 34% has a concentration of more than 0.025 mg/L of arsenic,

Table 1. Percentage of samples that exceed the Maximum Permissible Limit in NOM-127-SSA1-1994.

Parameter	MPL	% over
pH	6.5–8.5	29.79
As	0.025	34.04
Cd	0.005	8.51
Cu	2.0	12.77
Cl	250.0	12.77
Fe	0.3	23.40
Mn	0.2	93.62
Pb	0.01	6.38
SO_4	400.0	48.94
Zn	5.0	12.77

which is the limit established in NOM-SSA1-1994. The high amount of As is partly due to the acidic drainage (Smedley & Kinniburgh, 2002, Nordstrom, 2011) and heavy metals leaching released into the environment and partly due to the mineral springs of arsenic as FeAsS, typical of the region. This suggests that deposits of tailings from the Aurora influence the composition of the water in the hydrologic system of the basin of the Laja-Xichú.

4 CONCLUSIONS

Arsenic was present in the 47 analyzed samples. 16% of the samples exceeded the maximum permissible limit for arsenic concentrations in water in the Mexican regulation (0.025 mg/L) wile 8% of the samples exceed a concentration of 1.0 mg/L of arsenic. The leachate that comes from the mine tailings evidently affects the Laja-Xichú basin by releasing potentially toxic elements to the environment. The covering of the tailings with limestone is considered to be the best option for remediation, this because of the available limestone in the area and because it represent the less harmful alternative for the environment at the Laja-Xichú basin.

ACKNOWLEDGEMENTS

We acknowledge the University of Guanajuato, the Engineering Division and the authorities of Xichú, Guanajuato for their support.

REFERENCES

Nordstrom, D.K. 2011. Hydrogeochemical processes governing the origin, transport and fate of major and trace elements from mine wastes and mineralized rock to surface waters. *Appl. Geochem.* 26: 1777–1791.
Smedley, P.L., Kinniburgh, D. 2002. A review of the source, behaviour and distribution of arsenic in natural waters. *Appl. Geochem.* 17: 517–568.

Arsenic Research and Global Sustainability – Bhattacharya, Vahter, Jarsjö, Kumpiene,
Ahmad, Sparrenbom, Jacks, Donselaar, Bundschuh & Naidu (Eds)
© *2016 Taylor & Francis Group, London, ISBN 978-1-138-02941-5*

Simulation of arsenic retention in constructed wetlands

M.C. Valles-Aragón[1] & M.T. Alarcón-Herrera[2]
[1]*Autonomous University of Chihuahua, Faculty of Agrotechnological Science, Chihuahua, Mexico*
[2]*Advanced Materials Research Center (CIMAV), Durango, Mexico*

ABSTRACT: The software RCB-Arsenic was developed to simulate the water flow and reactive transport, contemplating the processes of As retention in Constructed Wetlands (CW). The objective of this research was to validate the RCB-Arsenic model simulating the behavior of a CW for As removal from water. The validation was made using data from a 122-day experiment, with two CW-prototypes, one planted with Eleocharis macrostachya (CWA) and another one unplanted (CWB). For simulation, the data addition was in 2 stages that considered the system mechanisms: 1) aqueous complexation, precipitation/dissolution, adsorption on granular media, 2) uptake and rhizofiltration by plants. Stage 1 was compared with CWB, the As means were 40.79 ± 7.76 and 39.96 ± 6.32, respectively. Stage 2 was compared with CWA, the As means were 9.34 ± 4.80 and 5.14 ± 0.72, respectively. The mass-balance simulated and experimental had a similar As retention rate (94 and 91%). The RCB-Arsenic model, adequately simulated the As retention in a CW.

1 INTRODUCTION

Constructed Wetlands (CWs) represent a feasible and inexpensive alternative for As removal from water (Zurita *et al.*, 2012). For an environmental assessment associated with CWs, it is essential to predict pollutant transport under the combined effects of advection, mass dispersion, and biological reaction (Bin 2012).

The RCB-Arsenic, which uses the Retraso-Code-Bright (RCB) software, was developed in previously to simulate CWs treating groundwater contaminated with As (Llorens *et al.*, 2013). The model simulates flow and reactive transport by contemplating As retention processes occurring inside CWs such as: mineral precipitation, adsorption on granular media and absorption by plants (Rahman *et al.*, 2011). The model was calibrated according to experimental data from previous studies, which were obtained from CW prototype-scaled (Olmos-Marquez *et al.*, 2012).

2 METHODS/EXPERIMENTAL

2.1 *Experimental validation*

The validation was performed using data from a 122-day experiment. There were used CW prototypes measuring 1.5×0.5 m length-width (ratio 3:1) and a 2.5% slope. The granular media was silty sand ($\rho = 1.4$ g cm^{-3}, porosity: 38%, hydraulic conductivity: 18.53 cm/h). The wetland prototypes were filled to a height of 0.35 m. The CWA was planted with *Eleocharis macrostachya* whereas the CWB remained unplanted as a control. The wetland prototypes were operated in parallel with subsuperficial flow, feeding of synthetic water prepared with well water and sodium arsenite ($NaAsO_2$) with a water column of 0.30 m. Samples were taken every week at the water inlet and outlet. The analytical determination of As was made via atomic absorption spectrometry with hydride generation in a GBC Avanta Sigma equipment.

2.2 *Simulation of arsenic retention*

A CW prototype was represented with a 2D mesh sized in accordance to the experimental dimensions. The following experimental factors were considered: intrinsic permeability of 1.2×10^{-10} m^2, porosity of 38%, and water flow of 54 L/day. The data addition for the simulation of As retention in CWs was made in two stages: Stage 1 Aqueous complexation, precipitation/dissolution and adsorption in the granular media. As adsorption by mechanisms performed in the soil. Stage 2 Aqueous complexation, precipitation/dissolution and adsorption on the granular media, uptake (absorption) and rhizofiltration (adsorption) by the plants. As adsorption by mechanisms effected in the soil and plant. The reactive transport model was calibrated in a quasi-steady state due to the RCB limitations (constant flow and composition) (Llorens *et al.*, 2013). A concentration of 0.058 mg Fe/L, pH of 7.98, and a temperature of 25°C (average experimental values) were indicated as water conditions. The simulation results in the stage 1 were compared with the unplanted wetland (CWB) and in the stage 2 were done with the planted wetland (CWA). A general linear model (PROC GLM) in the SAS 9.1.3 software was considered.

3 RESULTS AND DISCUSSION

3.1 Arsenic retention

Arsenic concentration decreased from the inlet to the outlet (Schiwindaman *et al.*, 2014). In Figure 1, As concentration in the influent and the contrast of CWA and CWB in the effluent, is shown. Between the 30 and 37 days of operation, CWB exposed As concentrations below the maximum permissible level (MPL) (NOM 127 1994). CWA recorded As concentrations in the effluent below the MPL until day 122. Arsenic retention during the 122 days of operation in CWA was 91%, whereas in CWB was 57%. The planted wetland showed greater retention than reported: 69–79% (Schiwindaman *et al.*, 2014) and 80–85% (Rahman *et al.*, 2014). Arsenic retention in the planted wetland was greater than in the unplanted (42%), which is higher than the reported: 15% (Rahman *et al.*, 2011), 20% (Zurita *et al.*, 2012), and 32–37% (Rahman *et al.*, 2014).

3.2 Simulation of arsenic retention

Stage 1: Comparing As concentrations at the outlet of the unplanted wetland, the simulated As outflow was higher than the experimental, from day 1 until 37 and from day 70 till 81 (Figure 2). The experimental mean of As concentration was 40.79 ± 7.76 and the simulated 39.96 ± 6.32. Stage 2: Arsenic concentrations from the simulation remained below of the experimental during the entire stage, except at day 45. The experimental mean of As concentration was 9.34 ± 4.80 and the simulated 5.14 ± 0.72. Stage 1 showed a reduced As retention than Stage 2. This is in agreement to the experimental results, where the activity of the plant roots contributes to a potentially high As immobilization in the planted wetland (Rahman *et al.*, 2011). Hence, the model simulated the assertion that plants play an important role in the immobilization and retention of As in the porous media (Schiwindaman *et al.*, 2014). Furthermore, the mass balances of experimental (CWA) and simulated (Stage 2) As retention had similar percentages 91 and 94%, respectively.

The statistical analysis indicates a non-significant difference (P>0.5) between experimental (CWB) and simulated values of the unplanted wetland (Stage 1). There was a significant difference between

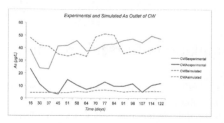

Figure 2. Arsenic concentration at the outlet of CW prototypes against simulated with RCB-Arsenic.

experimental (CWA) and simulated values of the planted wetland (Stage 2). Besides, the contrast between planted and unplanted experimental wetlands established a significant difference (P < 0.5).

4 CONCLUSIONS

The calibrated RCB-Arsenic model is a useful tool to predict the response of As retention by a CW system with submerged flow. The validity of the system can be modified depending on the input data like: composition of the porous media, the plant, water characteristics, and operation method. Further research is needed, an adequate implementation of the main processes of As retention within CW into numerical models. CW are an effective alternative technology for As removal from water.

REFERENCES

Bin, C. 2012. Transport of biocomponent contaminant in free-surface wetland flow. *J. Hydrodyn.* 24(6): 925–929.

Llorens, E., Obradors, J., Alarcón-Herrera, M., Poch, M. 2013. Modelling of arsenic retention in constructed wetlands. *Bioresource Technol.* 147: 221–227.

NOM 127 SSA1. 1994. Agua Para Uso y Consumo Humano-Límites Permisibles de Calidad y Tratamientos a que Debe Someterse el Agua para su Potabilizacion. Mexico.

Olmos-Márquez, M., Alarcón-Herrera, M., Martín-Dominguez, I. 2012. Performance of *Eleocharis macrostachya* and its importance for arsenic retention in constructed wetlands. *Environ. Sci. Pollut. Res.* 19: 763–771.

Rahman, K., Wiessner, A., Kuschk, P., Afferden, M., Mattusch, J., Müller, R. 2014. Removal and fate of arsenic in the rhizosphere of juncus effusus treating artificial wastewater in laboratory-scale constructed wetlands. *Ecol. Eng.* 69: 93–105.

Rahman, K., Wiessner, A., Kuschk, P., Afferden, M., Mattusche, J., Müllera, R. 2011. Fate and distribution of arsenic in laboratory-scale subsurface horizontal-flow constructed wetlands treating an artificial wastewater. *Ecol. Eng.* 37(8): 1214–1224.

Schiwindaman, J., Castle, J., Rodgers, J. 2014. Fate and Distribution of Arsenic in a Process-Desing Pilot-Scale Constructed Wetland Treatment System. *Ecol. Eng.* 68: 251–259.

Zurita, F., Del Toro-Sánchez, C., Gutierrez-Lomelí, M., Rodriguez-Sahagún, A., Castellanos-Hernandez, O., Ramírez-Martínez, G., White, J. 2012. Preliminary study on the potential of arsenic removal by subsurface flow constructed mesocosms. *Ecol. Eng.* 47: 101–104.

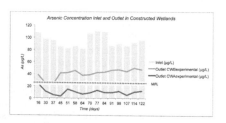

Figure 1. Arsenic retention in constructed wetlands.

*1.6   Arsenic in airborne particulates from
natural and anthropogenic sources*

Arsenic Research and Global Sustainability – Bhattacharya, Vahter, Jarsjö, Kumpiene,
Ahmad, Sparrenbom, Jacks, Donselaar, Bundschuh & Naidu (Eds)
© 2016 Taylor & Francis Group, London, ISBN 978-1-138-02941-5

Global atmospheric transport of arsenic

S. Wu, K.M. Wai & X. Li
Michigan Technological University, Houghton, MI, USA

ABSTRACT: Arsenic and many of it compounds are toxic pollutants in the global environment. They can be transported long distance in the atmosphere before depositing to the surface, but the global source-receptor relationships between various regions have not been studied yet. We develop the first global model for arsenic to better understand and quantify the inter-continental transport of arsenic. Our model reproduces the observed arsenic concentrations in surface air for various sites around the world. Arsenic emissions from Asia and South America are found to be the dominant sources for arsenic in the Northern and Southern Hemisphere, respectively. Asian emissions are calculated to contribute 39% and 38% of the total arsenic deposition over the Arctic and Northern America, respectively. Another 14% of the arsenic deposition to the Arctic region is attributed to European emissions. Our results indicate that the reduction of anthropogenic arsenic emissions in Asia and South America can significantly reduce arsenic pollution not only locally but also globally.

1 INTRODUCTION

Arsenic is a ubiquitous metalloid in the global environment. Elemental arsenic and many of its compounds have high toxicity and have been listed by the International Agency for Research on Cancer (IARC) as Group 1 carcinogens (IARC, 2013). There have been many studies showing increased lung cancer risk for people living or working near arsenic-emitting industrial plants such as smelting facilities (Cordier *et al.*, 1983, Pershagen, 1985). They, even at relatively low exposure levels, can also cause many other adverse health effects related to the brain and nervous system, digestive system, and skin (Navas-Acien *et al.*, 2008, Ettinger *et al.*, 2009).

There are large spatial variations for the atmospheric concentrations of arsenic, which can vary by several orders of magnitudes from less than 0.1 ng/m^3 in remote sites to more than 10 ng/m^3 in urban/industrial areas, presumably reflecting the impacts from anthropogenic activities. The arsenic concentrations were reported to be less than 41 pg/m^3 in the south polar atmosphere (Maenhaut *et al.*, 1979). In China and Chile, the dominant arsenic source regions in the Northern and Southern Hemispheres, respectively, the arsenic concentrations were found to reach 15 ng/m^3 or higher (Gidhagen *et al.*, 2002, Li *et al.*, 2010). The typical residence time of arsenic in the atmosphere is several days (Pacyna, 1987, DET, 2000), making it capable of long-range transport. This implies that arsenic emissions from one region can significantly affect other regions downwind. However, the global source-receptor relationship between various regions has not been quantified so far, in contrast to the extensively studied source-receptor relationship for other anthropogenic pollutants and dust.

2 METHODS/EXPERIMENTAL

In this study, we develop the first-ever global gridded emission inventory for arsenic and implement it in a global atmospheric chemical transport model to examine the global transport and source-receptor relationships for arsenic. The global arsenic model is based on the GEOS-Chem chemical transport model (http://geos-chem.org) v9-01-01. The GEOS-Chem model has been applied to a wide range of research related to atmospheric trace gases, aerosols and mercury (Bey *et al.*, 2001, Huang *et al.*, 2013). It is driven by assimilated meteorological fields from NASA GMAO.

We develop the global arsenic emission inventories by combining various data sources available, including existing regional arsenic emission inventories, industrial activities (in particular coal combustion and metal smelting which are major processes leading to anthropogenic arsenic emissions), and volcanic emissions. The arsenic emissions are processed and gridded into spatial resolution of 4° latitude by 5° longitude with a base year of 2005.

In order to better examine the source-receptor relationships between various regions in terms of arsenic concentration and deposition, we carry out a suite of sensitivity simulations where anthropogenic arsenic emissions from a certain region are turned off in the model. For example, we shut off emissions from Asia in the sensitivity model run and then compare the calculated atmospheric arsenic deposition (Dno_Asia) with those from the base run (Dbase) to derive the percentage contribution of Asian emissions to atmospheric arsenic in the receptor region: ContributionAsia = (Dbase – Dno_Asia) / Dbase × 100%. Figure 3 shows the contribution to total (wet + dry) deposition from each continental-scale source region.

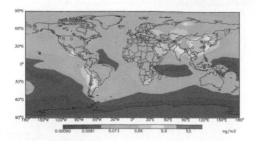

Figure 1. Annual mean arsenic concentrations in ambient surface air (in ng/m³). Model-simulation results (background) are compared with measurement data at various sites (circles) around the world.

3 RESULTS AND DISCUSSION

The global arsenic emissions are calculated to be 30.7 Gg yr-1 (with 15.8 and 4.4 Gg/yr from East Asia and South America, respectively). The model-simulated annual mean concentrations of atmospheric arsenic are compared with available measurement data in Figure 1. We find very good agreement between model results and observations with a high correlation ($r^2 = 0.98$).

High arsenic concentrations (10 ng/m³ or higher) are found over large areas in eastern China and northern Chile (Fig. 1), which are at least one order of magnitude higher than those in the United States and Europe. Figure 1 also illustrates the outflow of arsenic plumes from Asia, which are transported over the North Pacific and North America following the Westerlies. Similarly the arsenic plumes from North America are transported across the North Atlantic towards Europe. In the Southern Hemisphere, the major arsenic source is Chile. The arsenic plumes at lower latitudes are transported towards the tropical Pacific following the trade winds, and those at higher latitudes are transported towards the Southern Atlantic following the Westerlies.

Anthropogenic arsenic emissions from Asia are found to make the largest contributions to atmospheric arsenic deposition over the North Pacific Ocean and western North America. About 10–60% of atmospheric arsenic concentration and 30–70% of total arsenic deposition over the western part of North America are attributed to Asian emissions. Significant contributions to the Arctic region (up to 60% for atmospheric concentration and 70% for total arsenic deposition) are calculated for Asian emissions. The contribution of anthropogenic arsenic emissions from South America is found to dominate over the Southern Hemisphere except for Southern Africa and Australia. Up to 90% of arsenic deposition over the Antarctic is attributed to emissions from South America.

4 CONCLUSIONS

On average, about 39% of the total arsenic deposition over the Arctic region is attributed to Asian anthropogenic emissions, reflecting the strong arsenic emissions from Asia. The European anthropogenic emissions are calculated to contribute almost 14% of the total arsenic deposition to the Arctic. The North American contribution to arsenic in the Arctic (about 4%) is found to be much less than those from Asia or Europe, reflecting both the lower anthropogenic emission strengths and the lower latitudes of the sources. The Asian anthropogenic emissions are found to contribute to the total arsenic deposition in North America by 38%.

The inter-continental transport of arsenic, especially the significant global impacts associated with arsenic emissions from certain source regions as shown by our results, highlights the benefits of international cooperation to reduce arsenic pollution around the world. These source-receptor relationships should be considered by researchers and policymakers in designing mitigation strategies for arsenic pollution.

ACKNOWLEDGEMENTS

The authors thank the EMEP data providers for making the arsenic data available. This study was supported by U.S. EPA (grant 83518901) and NSF (grant 1313755).

REFERENCES

Bey, I., et al. 2001. Global modeling of tropospheric chemistry with assimilated meteorology: Model description and evaluation. *J. Geophys. Res.* 106: 23,073–23,095.

Cordier, S., Thériault, G., Iturra, H. 1983. Mortality pattern in a population living near a copper smelter. *Environ. Res.* 31: 311–322.

DET. 2000. A review of arsenic in ambient air in the UK. Department of the Environment, Transport and the Regions, Scottish Executive, The National Assembly for Wales.

Ettinger, A.S., et al. 2009. Maternal arsenic exposure and impaired glucose tolerance during pregnancy. *Environ. Health Perspect.* 117: 1059–1064.

Gidhagen, L., Kahelin, H., Schmidt-Thomé, P., Johansson, C. 2002. Anthropogenic and natural levels of arsenic in PM10 in Central and Northern Chile. *Atmos. Environ.* 36: 3803–3817.

Huang, Y., Wu, S., Dubey, M.K., French, N.H.F. 2013. Impact of aging mechanism on model simulated carbonaceous aerosols. *Atmos. Chem. Phys.* 13: 6329–6343.

IARC. 2013. IARC Monographs on the evaluation of carcinogenic risks to humans. International Agency for Research on Cancer.

Li, C., et al. 2010. Concentrations and origins of atmospheric lead and other trace species at a rural site in northern China. *J. Geophys. Res.* 115: D00 K23

Maenhaut, W., Zoller, W.H., Duce, R.A., Hoffman, G.L. 1979. Concentration and size distribution of particulate trace elements in the south polar atmosphere. *J. Geophys. Res.* 84: 2421–2431.

Navas-Acien, A., Silbergeld, E.K., Pastor-Barriuso, R., Guallar, E. 2008. Arsenic exposure and prevalence of type 2 diabetes in US adults. *JAMA* 300: 814–822.

Pacyna, J.M. 1987. In Lead, Mercury, Cadmium and Arsenic in the Environment SCOPE 31, T.C. Hutchinson, K.M. Meema, Eds. (Wiley, Chichester, 1987) Chap. 7.

Pershagen, G. 1985. Lung cancer mortality among men living near an arsenic-emitting smelter. *Am. J. Epidemiol.* 122: 684–694.

Arsenic Research and Global Sustainability – Bhattacharya, Vahter, Jarsjö, Kumpiene,
Ahmad, Sparrenbom, Jacks, Donselaar, Bundschuh & Naidu (Eds)
© 2016 Taylor & Francis Group, London, ISBN 978-1-138-02941-5

Detection and analysis of arsenic-bearing particles in atmospheric dust using Mineral Liberation Analysis

M. Gasparon[1,2], I. Delbem[3], M. Elmes[1] & V.S.T. Ciminelli[2,4]

[1] *The University of Queensland, School of Earth Sciences, Brisbane, Australia*
[2] *National Institute of Science and Technology on Mineral Resources, Water and Biodiversity—INCT-Acqua, Brazil*
[3] *Universidade Federal de Minas Gerais, Center of Microscopy, Department of Chemistry, Belo Horizonte, Brazil*
[4] *Universidade Federal de Minas Gerais, Department of Metallurgical and Materials Engineering, Belo Horizonte, Brazil*

ABSTRACT: A new method was developed for the detection and analysis of arsenic-bearing particles in atmospheric dust using Mineral Liberation Analysis. This method, originally developed for metallurgical and geological samples, was modified for the analysis of atmospheric dust and is suitable for TSP and PM_{10} low-volume air samples. Arsenic-bearing particles with an average diameter as small as 1 micrometer can be detected and measured with an analytical throughput of over 10,000 particles per hour. The method provides single particle data on particle size, morphology, elemental composition and mineralogy, and is suitable for both crystalline and non-crystalline phases. The first application of this method to a Brazilian mining town confirmed that the arsenic detected in air is not associated to arsenic-bearing mineral phases excavated at the mine site. This new technique provides crucial data on the source, distribution and potential toxicity of arsenic-bearing atmospheric particles in urban, rural and mining environments.

1 INTRODUCTION

Air pollution is an ever-increasing environmental challenge facing both industrialized and developing nations. A major component of air pollution is Atmospheric Particulate Matter (APM), which is considered a key parameter in air quality monitoring due to its negative impacts on human health, the environment, and atmospheric chemistry (IARC, 2013). Unlike other environmental media (soil and water), APM transport of pollutants, such as arsenic, presents the most significant potential to cross regional boundaries and spread over large areas because air masses have minimal topographic boundaries for confinement (Csavina *et al.*, 2012).

Following the global growing consensus on the correlation between negative health impacts and air pollution, various monitoring programs have been established to assess the level of particles in the air and their spatial variability and temporal trends. Monitoring programs typically include the measurement of particle mass using a range of sampling devices and spatial and temporal sampling schedules. Less commonly, the elemental concentrations in the bulk samples are measured using a range of methods that include X-ray fluorescence or Inductively Coupled Plasma Mass Spectrometry (ICP-MS). Although these methods are effective in providing information on bulk particle size and composition, they cannot identify different sources, nor provide quantitative information on the

elemental concentrations in individual sources. This is a serious limitation because in any environment there are typically different sources that contribute to the total particle load (such as agriculture, transport, construction and industrial activities, mining, natural erosion, and forest fires). As a result, environmental regulators are often forced to establish stringent dust suppression procedures that fail to achieve a decrease in total particle load, because the sources have not been correctly identified.

2 EXPERIMENTAL

The Mineral Liberation Analysis (MLA) is an automated digital image analyser coupled to a scanning electron microscope equipped with an energy dispersive X-ray spectrometer (SEM/EDS) that was developed to examine fine grained mineral mixtures in order to determine liberation potential and improve the efficiency of mineral processing plants (Gu, 2003). For this project, the method was modified to allow the analysis of particles deposited on membranes used for the collection of TSP and PM_{10} particles.

The instrument used was the MLA 650 system coupled to a FEI Quanta 650 SEM-FEG, equipped with a Bruker Quantax EDS analyzer. The SEM was set to operate in a high vacuum with an accelerating voltage of 25kV, a spot size of 5.5 and a 10 mm working distance. The MLA was configured to process 20,000 particles per sample. Analyses were performed at the Federal University of Minas Gerais (UFMG) Microscopy Center (CM-UFMG). SKC 'AirChek 52' low-volume active pumps fitted with IOM (Institute of Occupational Medicine) inhalable samplers and polycarbonate membranes were used for sample collection. The IOM sampler meets a range of international standards, including the Australian standard for inhalable particulate matter (SKC, 2015). The spectral reference library used for particle identification was constructed using mineral phases and materials collected from local lithologies and environmental sources.

3 RESULTS AND DISCUSSION

The method relies on segmentation of particle regions based on Back-Scattered Electron (BSE) brightness, followed by X-ray analysis. Once individual particles have been identified, the first stage identifies all distinct mineral phases and defines their boundaries. This process is called phase segmentation and is performed on each individual particle. Phase segmentation outlines regions of homogenous BSE values and assigns a specific colour to each identified region (Figure 1), with the average BSE greyscale value of each defined region directly related to a "mineral" of unique "average atomic number" determined by its composition.

Mineral classification compares the sample information with a pre-defined list of mineral spectra. This library is constructed before analysis and involves the collection of high quality X-ray spectra for each mineral in the sample with each mineral phase typically characterized by 2–5 spectra, depending on their complexity. Mineral phases are classified based on user-defined criteria and a false colour is assigned to each phase/composition to produce a mineral map of the particles which allows for the direct visual comparison of samples.

Results of the first pilot study carried out in the Paracatu area (Minas Gerais, Brazil), where gold mineralization is associated with arsenopyrite, show that at this specific mine site atmospheric arsenic is primarily associated with organic combustion residues of forest fires, and not with the dispersion of arsenopyrite from the local mine site as previously postulated.

4 CONCLUSIONS

This study outlines the successful development of a new method based on MLA analysis for single particle characterization of atmospheric particulate matter. Arsenic-bearing particles with an average diameter as small as 1 micrometer can be detected and measured with an analytical throughput of over 10,000 particles per hour. The analysis provides single particle data on particle size, morphology, elemental composition and mineralogy, and is suitable for both crystalline and

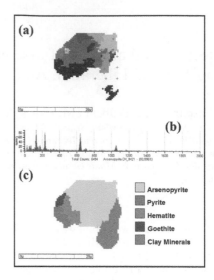

Figure 1. (a) Example of a composite particle, showing the phase segmentation (areas with different colour indicating different BSE brightness) and individual EDS analysis points (red crosses). (b) Example of the arsenopyrite EDS spectrum obtained for one point. (c) BSE data and EDS spectra are combined and matched to the spectral library to produce a phase map of each particle.

non-crystalline phases. This method can be applied in principle to trace the source of most elements in APM, and can therefore be used to identify and manage activities and sources responsible for the release of potentially toxic elements into the atmosphere.

ACKNOWLEDGEMENTS

The authors are grateful to the Brazilian government agencies (CNPq, Fapemig, CAPES and FINEP) for financial support and the PVE fellowship to M. Gasparon, and to the UFMG Center for Microscopy for the infrastructure used to carry out the analyses. Kinross Brasil Mineração is gratefully acknowledged for assistance with sample collection.

REFERENCES

Csavina, J., Field, J., Taylor, M.P., Gao, S., Landázuri, A., Betterton, E.A., Sáez, A.E. 2012. A review on the importance of metals and metalloids in atmospheric dust and aerosol from mining operations. *Sci. Total Environ.* 433: 58–73.

Gu, Y. 2003. Automated scanning electron microscope based mineral liberation analysis. *Journal of Minerals and Materials Characterization and Engineering* 2: 33–41.

IARC (International Agency for Research on Cancer). 2013. *Outdoor Air Pollution a Leading Environmental Cause of Cancer Deaths.* World Health Organisation, Lyon, France.

SKC 2015. https://www.skcinc.com, accessed 12 November 2015.

Reactivity of As-rich smelter dust in contrasting soils—a two year *in situ* experimental study

A. Jarošíková[1], V. Ettler[1], M. Mihaljevič[1], V. Penížek[2] & T. Matoušek[3]

[1]*Institute of Geochemistry, Mineralogy and Mineral Resources, Faculty of Science, Charles University in Prague, Praha, Czech Republic*
[2]*Department of Soil Science and Soil Protection, Faculty of Environmental Sciences, Czech University of Life Sciences Prague, Praha, Suchdol, Czech Republic*
[3]*Institute of Analytical Chemistry of the ASCR, v. v. i., Brno, Czech Republic*

ABSTRACT: The reactivity of copper smelter dust enriched in arsenic (53% As), predominantly composed of arsenolite (As_2O_3), was investigated by an in situ weathering experimental approach. The dust placed in perforated experimental bags (1 μm mesh) was incubated for 6, 12, 18 and 24 months in different depths of four contrasting soils developed under different vegetation covers. Total concentrations and speciation of As were measured in soils and soil pore water. Mass losses increased over time from 4 to 42% and As concentration in soil increased to 849 mg/kg. Soil pore water contained up to 13 mg/l and As(V) was predominant species. SEM and XRD investigation indicated that arsenolite was highly weathered especially in Cambisols developed under the beech cover and complex Ca-Pb-Cu-Fe-Zn-As-S phase formed as a secondary alteration product.

1 INTRODUCTION

Smelting industry is an important source of local contamination. Especially surrounding soils represent an important sink for smelter emissions. To understand the fate of smelter particulates when deposited in soils and depicting subsequent dynamics of smelter-related contaminants in soil systems interacting with water and biota, an *in situ* experimental approach can be used (Birkefeld *et al.*, 2006; Pareuil *et al.*, 2010). The goal of this study is to investigate the long-term transformation of As-rich dust from copper smelter in four contrasting soils.

2 MATERIALS AND METHODS

2.1 *Smelter dust*

Dust collected in bag house filters of a Cu smelter is predominantly composed of arsenolite (As_2O_3), galena (PbS), and gypsum ($CaSO_4 \cdot 2H_2O$). Total As content in the dust is 533 ± 16 g/kg.

2.2 *In situ experiment*

According to Ettler et al. (2012), double experimental polyamide bags (1 μm mesh) filled with 1 g of the dust were horizontally placed for incubation into different depths (5 to 55 cm) of four contrasting soils in the Czech Republic: (i) neutral-to-alkaline Chernozem on loess (grass cover; mean annual precipitation 490 mm), (ii) neutral-to-slightly acidic Cambisol (grass cover; annual precipitation 560 mm), (iii) and (iv) acidic Cambisols developed under beech and spruce forests (annual precipitation 600 mm), respectively. Moreover, experimental polyamide bags (1 μm mesh) with cellulose were placed in each horizon to determine proxy microbial activity in soils using litterbag method. Each 6 months, bags were collected and soils were vertically sampled (each 5 cm of depth) by physical rings. When possible, soil pore water was collected using rhizon pore water samplers. Arsenic concentrations in soils and pore water were determined using Inductively Coupled Plasma Mass Spectrometry (ICP-MS; ThermoScientific XSeries[II]). Arsenic speciation analysis was performed by anion exchange HPLC-ICP-MS (Agilent 7700x ICP MS spectrometer and Agilent 1200 Series LC pump). Changes in phase composition of dust was observed using X-ray diffraction analysis (XRD; PANanalytical X'Pert Pro diffractometer) and scanning electron microscope (SEM; TESCAN VEGA 3XM) with X-ray microanalysis (EDS; Bruker Quantax 200 X-Flash 5010).

3 RESULTS AND DISCUSSION

3.1 *Arsenic in soil and pore water*

Transport of released As in soils was observed within the closest few centimeters below the experimental bag's position, as shown in Figure 1. Arsenic leaching was the most significant in the beech forest Cambisol both after 6 and 12 months (590 mg/kg and 849 mg/kg, respectively). This is probably caused by low pH and specific seepage conditions leading to more rapid flush regime under the beech. Soils developed under the spruce stand exhibited the lowest values of As released from dust: 164 mg/kg (6 months) and 484 mg/kg (12 months).

Pore water sampled in the beech forest contained 9 mg/l As (6 months) and up to 13 mg/l (12 months) indicating significant As leaching of the dust. Under the spruce, As in soil pore water was only up to 47 µg/l. Arsenic speciation analysis of pore water showed that almost all of As (98–100%) has been oxidized to As(V).

A proxy microbial activity ranged between 2–81% during winter period of year (October—April) and 7–98% during the summer period (April—October), with the highest values in soils of the agricultural area. The lowest microbial activity was observed in soils developed under the spruce.

3.2 *Dust transformation*

The overall mass losses from the dust tend to be increasing over time: 4–28% (6 months), 19–32% (12 months), 20–39% (18 months), 25–42% (24 months). X-ray diffraction analysis indicated that arsenolite and gypsum were significantly dissolved. The highest dissolution of arsenolite

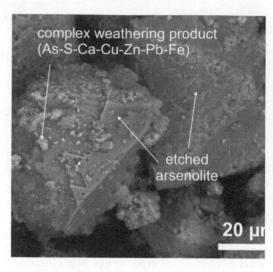

Figure 2. SEM image of the smelter dust after 12 months of incubation in Cambisol developed under the beech forest.

occurred at the beech forest locality and SEM demonstrated highly etched arsenolite surfaces and secondary formation of a complex weathering product, probably metal arsenate (Fig. 2).

4 CONCLUSIONS

Copper smelter dust particles enriched in As are highly reactive in soil systems and its dissolution increased over time. The highest dissolution leading to As release into the soil and pore water was observed in Cambisol developed under the beech forest due to lowest pH and specific seepage conditions leading to more rapid flush regime.

ACKNOWLEDGEMENTS

This study was supported by Czech Science Foundation (GAČR 13-17501S), Grant Agency of Charles University (GAUK 259521) and Czech Academy of Sciences (RVO:68081715).

REFERENCES

Birkefeld, A., Schulin, R., Nowack, B. 2006. In situ investigation of dissolution of heavy metal containing mineral particles in an acidic forest soil. *Geochim. Cosmochim. Acta* 70: 2726–2736.

Ettler, V., Mihaljevič, M., Šebek, O., Grygar, T., Klementová, M. 2012. Experimental in situ transformation of Pb smelter fly ash in acidic soils. *Environ. Sci. Technol.* 46: 10539–10548.

Pareuil, P., Bordas, F., Joussein, E., Bollinger, J.C. 2010. Alteration of a Mn-rich slag in contact with soil: In-situ experiment during one year. *Environ. Pollut.* 158: 1311–1318.

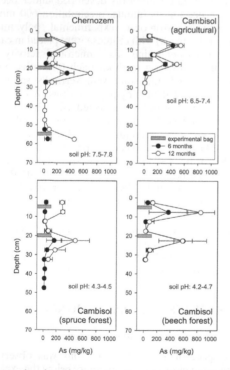

Figure 1. Arsenic concentrations in four contrasting soils after 6 and 12 months of smelter dust incubation.

Arsenic Research and Global Sustainability – Bhattacharya, Vahter, Jarsjö, Kumpiene,
Ahmad, Sparrenbom, Jacks, Donselaar, Bundschuh & Naidu (Eds)
© 2016 Taylor & Francis Group, London, ISBN 978-1-138-02941-5

Arsenic in airborne particulates in the vicinity of anthropogenic source

O.V. Shuvaeva[1], V.V. Kokovkin[1] & V.F. Raputa[2]

[1]*Nikolaev Institute of Inorganic Chemistry, Siberian Branch of Russian Academy of Sciences, Novosibirsk State
University, Novosibirsk, Russia*
[2]*Institute of Computational Mathematics and Mathematical Geophysics, Siberian Branch of Russian Academy
of Sciences, Novosibirsk State University, Novosibirsk, Russia*

ABSTRACT: In present work the quantitative regularities at the regional scale from the point-type
contamination source have been investigated using experimental study and numerical analysis. It has been
shown with the use of snow cover as an indicator of long-term pollution that the concentration of aerosol
constituents monotonically decreases with the distance from the contamination source. Application of
mathematical model demonstrated a satisfactory agreement between calculated and experimental results.

1 INTRODUCTION

Novosibirsk Tin Plant (NTP) is located on the left
bank of the Ob River in the industrial zone of the
city. The main activity of the company is production of commercial tin alloys, solders, and babbitt.
The feedstocks used are tin-containing minerals
which also may include lead, arsenic, sulfur, zinc,
indium, iron and others. Processing of raw materials at the plant is carried out under the scheme: fin-
ishing concentrates – roasting – reduction smelting
– refining. Poor materials, slags and wastes are
processed by fuming and electrofusion. The main
toxicants released into environment are arsenic,
lead, zinc and sulfur oxides. The main emission
of harmful substances into the atmosphere occurs
through a hundred-meter chimney which height,
on the one hand, allows an effective scattering of
emitted harmful impurities, but on the other hand,
can lead to significant gas-aerosol pollution of an
urban area. There are a lot of examples of air pollution investigations at the vicinities of the anthropogenic sources but the quantitative regularities of
these processes are not very well studied.

In frames of the present investigation the model
of long-term contamination from anthropogenic
source of point type with the use of a small amount
of necessary parameters was developed. The model
has been successfully applied to describe the contamination picture in the vicinity of tin plant.

2 METHODS/EXPERIMENTAL

2.1 Mathematical modeling

The snow cower was used as the indicator of aerosol pollution. It is known that during winter period
it accumulates large amounts of aerosol deposi-
tions and may serve as an indicator of territory
contamination.

The levels of pollutants contents have been estimated taking into account the winter winds directions. To assess the long-term (month, season, year)
contamination of the area from a "point type"
source of observational data has been proposed
and approved the following regression dependence
(Raputa *et al.*, 1997, Kokovkin *et al.*, 1999):

$$p\left(r,\varphi,\overline{\Theta}\right)=g\left(\varphi+180°\right)\cdot f(r,\Theta_1,\Theta_2) \qquad (1)$$

$$f(r,\Theta_1,\Theta_2)=\Theta_1 r^{\Theta_2}\exp\left(-c/r\right), \qquad (2)$$

where p is specific content of impurity in snow
cover (soil, air); r, φ - the polar coordinates of the
calculated point at the beginning of the source
position; $g(\varphi)$ - climate repeatability of wind directions for the given period of time; c- the value
which depends on the height of the source, temperature and volume of the ejected gas and wind
speed; $\Theta=(\Theta_1,\Theta_2)$ - vector of unknown parameters. Θ_1 is proportional to the power of emission
and depends on wind speed, turbulence, the height
of the source and the aerosol sedimentation rate.

$$\Theta_2=-2-w\big/\left[k_1\left(n+1\right)\right], \qquad (3)$$

where w - the rate of aerosol particles sedimentation; k_1- coefficient of vertical turbulent diffusion
at the height of 1 m; n is exponent of the approximation of the horizontal component of the wind
velocity. When $\Theta_2\to-2$ then $w\to0$ which is true
for poorly settling impurities, $c=2r_{max}$ in accordance with the earlier study by Raputa *et al.* (1997).

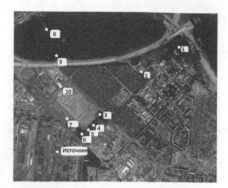

Figure 1. The scheme of snow sampling.

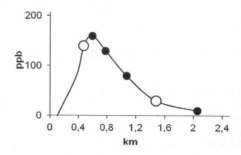

Figure 2. Regression model of As spatial distribution at the rate of aerosol particles monodispersity.

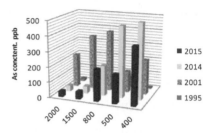

The distance from the source of emission, m

Figure 3. A comparison of the data concerning snow cover contamination with arsenic.

Unknown parameters Θ_1, Θ_2 can be found using the observation data. r_m parameter can be estimated from the geometric characteristics of the source. To estimate the parameters, it is sufficient to use a limited set of reference points of observation (in this study two reference points were used). This circumstance provides an additional opportunity to check the adequacy of the model on the remaining points of the experimental sampling, which act as control ones.

2.2 *Sampling and analysis*

Sampling was carried out on the snow full height with the help of snow-gauge standard way in March 2015 according to the rout shown in Figure 1. The sample preparation procedure included snow melting following by filtering. Suspended matter collected on the filter was subjected to microwave mineralization procedures, Arsenic determination in both water soluble and particulate fractions was performed by ICP-AES (iCAP 6500). As(III) species in water-soluble fraction was detected using whole-cell bacterial biosensor on the base of E-coli (Daunert *et al.*, 2000).

3 RESULTS AND DISCUSSION

The results of the study are presented in Figure 2. It is seen that maximum arsenic concentration corresponds to the point of 800 meters away from the source of contamination (the chimney of tin plant). It is also clear that calculated curve (open circles) is in a good agreement with observations (black circles).

A comparison of these results with the data of the previous years shows that arsenic emissions decreased, however, the maximum of particles deposition moved closer to the source indicating an increase in their size. Figure 3 shows the distribution of As in snow cover during the period of 1995–2015.

4 CONCLUSIONS

The experimental research and numerical simulation allows us to conclude the following:

- Recovery models adequately describe snow cover contamination with arsenic in the vicinity of the tin plant.
- Analysis of the data demonstrated that emissions of arsenic by NTP did not change significantly compared to the previous years.
- These patterns of distribution of arsenic in the snow cover in the vicinity of the NTP can be used to analyze the data of aerosol pollution of air and soil.

REFERENCES

Berlyand, M.E. 1975. Modern problems of atmospheric diffusion and air pollution, Hydrometeoizdat, Leningrad.

Daunert, S., Barrett, G., Feliciano, J.S., Shetty, R.S., Shrestha, S., Smith-Spencer, W. 2000. Genetically engineered whole-cell sensing systems: coupling biological recognition with reporter genes. *Chem. Rev.* 100(7): 2705–2738.

Kokovkin, V.V., Raputa, V.F., Shuvaeva, O.V. 1999. *Chemistry for Sustainable Development.* 477p.

Raputa, V.F., Sadovsky, A.P., Olkin, S.E. 1997. The reconstruction of benzapyrene fallouts in the vicinities of the Novosibirsk Electrode Plant. *Meteorology and Hydrology* 2: 33–41.

1.7 Analytical advancements and challenges in measuring and monitoring of arsenic in solid and aqueous media

Arsenic Research and Global Sustainability – Bhattacharya, Vahter, Jarsjö, Kumpiene,
Ahmad, Sparrenbom, Jacks, Donselaar, Bundschuh & Naidu (Eds)
© 2016 Taylor & Francis Group, London, ISBN 978-1-138-02941-5

New insights in the mechanisms of arsenic association with iron oxides in the environment

V.S.T. Ciminelli[1,2], E. Freitas[1,2,3], L.A. Montoro[4], J.W.V. de Mello[2,5] & M. Gasparon[6]

[1]*Department of Metallurgical and Materials Engineering, Universidade Federal de Minas Gerais-UFMG,*
 Belo Horizonte, Brazil
[2]*National Institute of Science and Technology on Mineral Resources, Water and Biodiversity,*
 INCT-Acqua, Brazil
[3]*Center of Microscopy, Universidade Federal de Minas Gerais-UFMG, Belo Horizonte, Brazil*
[4]*Department of Chemistry, Universidade Federal de Minas Gerais-UFMG, Belo Horizonte, Brazil*
[5]*Department of Soils, Universidade Federal de Viçosa, Viçosa, Brazil*
[6]*The University of Queensland, School of Earth Sciences, Queensland, Australia*

ABSTRACT: A mechanism responsible for long-term As fixation in the environment is described here for the first time. Arsenic (1.6 ± 0.5 wt%) was unambiguously identified in mesocrystals of Al-hematite with an As/Fe molar ratio of 0.026 ± 0.006 in environmental samples. The As-bearing Al-hematite is interpreted as a secondary phase formed from iron (hydr)oxides. Experimental evidence supports a mechanism whereby arsenic adsorption onto Al-ferrihydrite Nanoparticles (NPs) is followed by Al-ferrihydrite aggregation in a process of self-assembly Oriented Attachment (OA) to produce nanosized Al-hematite mesocrystals, with goethite as a possible intermediate phase. Arsenic species are incorporated and thus irreversibly trapped in the aggregates. Structural Al was identified in all the As-bearing iron (hydr)oxides. Aluminum plays a major role in increasing As adsorption onto ferrihydrite NPs, controlling the NPs size, and favoring crystal aggregation by decreasing the NPs surface energy to promote attachment events.

1 INTRODUCTION

The key role of iron or aluminum (hydr)oxides in the mobility and bioavailability of As in the environment is well established. More recently, As-bearing mixed Al-Fe-compounds have been investigated, and Al has been shown to increase the sorption capacity of arsenate (Adra *et al.*, 2016, Violante *et al.*, 2009). However, the detailed ageing/recrystallization mechanisms taking place following the initial adsorption/co-precipitation of As, and the role of Al in these processes remain to be fully understood.

The structural characterization of As-bearing phases in environmental samples is not trivial due to particles heterogeneity, small grain size and low concentrations. Synchrotron-based analytical techniques combined with theoretical molecular modeling or with other spectroscopic techniques have been increasingly applied to investigate As distribution, speciation, and bonding characteristics. The spatial resolution limit of the aforementioned techniques, however, can be overcome by Transmission Electron Microscopy (TEM). This method allows for the characterization of solid phases at the nanoscale, with better spatial resolution than any other technique.

Spectroscopy and X-ray diffraction (for bulk analysis) were combined with high resolution TEM (for phase characterization at nanoscale) to investigate As-bearing phases in environmental samples. The present work provides new insights into the mechanism of As incorporation in the structure of Al-Fe-(hydr)oxides such as Al—substituted goethite and hematite.

2 EXPERIMENTAL

The samples consisted of Fe-Al-enriched oxisols that have been in contact with concentrated As-sulphide waste produced from a gold cyanidation plant. After more than a decade of exposure the impoundments were removed and the oxisol liner samples were recovered for analyses. A sample from the local oxisol, representing the material before exposure to the As-sulfide tailings, was also investigated. The main constituents of the samples were identified by XRD analysis as quartz, kaolinite, muscovite, siderite, albite, gibbsite, hematite and goethite (Freitas *et al.*, 2015).

Arsenic concentrations were measured in triplicate using a Perkin Elmer 7300 Inductively Coupled Plasma with Optical Emission Spectrometer (ICP-OES). Precision and accuracy measured against international reference materials were better than 95%.

High Resolution TEM (HR-TEM), Nano-Beam Electron Diffraction (NBD), Energy Dispersive X-Ray Spectroscopy (EDS), and Electron Energy-Loss Spectroscopy (EELS) were used for samples characterization. The HRTEM, NBD and punctual EDS analyses were performed in the Center of Microscopy at the Universidade Federal de Minas Gerais (CM-UFMG), Brazil, using a TEM-LaB6

Tecnai G2-20 (FEI), operating at 200kV, equipped with a Si(Li) EDS detector (EDAX) with a 30 mm² window. The EDS mapping and EELS analyses were performed in the Brazilian Nanotechnology National Laboratory (LNNano) using a TEM-FEG JEM 2100F (JEOL), operating at 200kV, equipped with EDS (Noran), HAADF (High Angle Annular Dark Field) detector and a Gatan Imaging Filter GIF Tridiem system. The JEMS software was used for the electron diffraction analyses. Sample preparation is described in detail elsewhere (Freitas et al., 2015).

3 RESULTS AND DISCUSSION

The As content was 12 ± 7 mg/kg in the natural oxisol (NOX) and ranged from 1,886 mg kg^{-1} to 6,375 mg. kg^{-1} in the oxisol liners (OXL). TEM analysis showed the presence of ferrihydrite, goethite and hematite. Most arsenic (60–69 wt%) in the OXL samples was found in association with crystalline nanoparticle aggregates of Fe-Al-(hydr)oxides. Structural Al was identified in all As-bearing phases by EDS analysis. Arsenic was homogeneously distributed in the structure of nanocrystals of Al-substituted hematite (α-(Fe,Al)$_2$O$_3$) (Fig. 1) and goethite (α-(Fe,Al)OOH). The arsenic content found by EDS in Al-hematite ranged from 0.9 to 3.23 wt%, e.g. from 9,000 to 32,300 mg/kg. In Al-goethite, the As content ranged from 1.2 to 2.4 wt% (12,000 to 24,000 mg/kg). As hematite and goethite are the most stable phases among Fe-(hydr)oxides, the large amount of As incorporated in the crystal structure of such phases implies low As mobility in the environment.

A mechanism is proposed for the incorporation of As into hematite, in which Al-ferrihydrite nanoparticles (NPs) act as a template. According to this mechanism, As is adsorbed onto existing or newly-formed Al-ferrihydrite, thus modifying its surface properties. The Al-ferrihydrite NPs then aggregate through a self-assembling Oriented Attachment (OA) process, resulting in hematite mesocrystals by ageing. The for-

mation of goethite in an intermediate step cannot be excluded. It is known that Al in the system increases the surface area of ferrihydrite, thus improving its adsorption capacity, slowing down phase transformation into hematite, and giving more stability to these phases (Schwertmann et al., 1979; Violante et al., 2009). The presence of Al has been shown to favor the OA-based crystal growth, and therefore to increase the amount of As immobilized in the OA upon attachment of the Al-Fe-(hydr)oxide nanoparticles. Then, arsenic primarily sorbed onto the NPs may be trapped in the aggregates as they grow, and finally be irreversibly incorporated in the Al-hematite phase.

4 CONCLUSIONS

A natural remediation process for long-term As immobilization in Fe-Al-(hydr)oxides present in oxisols is described. The mechanism involves As adsorption onto Al-ferrihydrite nanoparticles, followed by aggregation in a process of self-assembly oriented attachment and ageing to produce Al-hematite mesocrystals.

ACKNOWLEDGEMENTS

The authors are grateful to the Brazilian sponsor agencies–CNPq (Conselho Nacional de Desenvolvimento Científico e Tecnológico), Fapemig (Fundação de Amparo a Pesquisa do Estado de Minas Gerais) and CAPES (Coordenação de Aperfeiçoamento de Pessoal de Nível Superior)/PROEX for financial support. Kinross Brasil Mineração is gratefully acknowledged for the samples. The authors are also grateful to the Science Without Borders fellowship to M. Gasparon, and to the Brazilian Nanotechnology National Laboratory (LNNano/LNLS) and to the Center of Microscopy (CM-UFMG).

REFERENCES

Adra, A., Morin, G., Ona-Nguema, G., Brest, J. 2016. Arsenate and arsenite adsorption onto Al-containing ferrihydrites. Implications for arsenic immobilization after neutralization of acid mine drainage. *Appl. Geochem.* 64: 2–9.

Freitas, E.T.F., Montoro, L.A., Gasparon, M., Ciminelli, V.S.T. 2015. Natural attenuation of arsenic in the environment by immobilization in nanostructured hematite. *Chemosphere* 138: 340–347.

Schwertmann, U., Fitzpatrick, R.W., Taylor, R.M., Lewis, D.G. 1979. The influence of aluminum on iron oxides. Part II. Preparation and properties of Al-substituted hematites. *Clays Clay Miner.* 27(2): 105–112.

Violante, A., Pigna, M., Del Gaudio, S., Cozzolino, V., Banerjee, D. 2009. Coprecipitation of arsenate with metal oxides. 3. Nature, mineralogy and reactivity of iron(II)-aluminum precipitates. *Environ. Sci. Technol.* 43: 1515–1521.

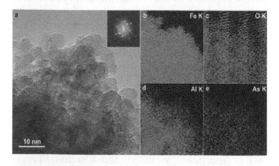

Figure 1. (a) HRTEM image of an aggregate of crystalline hematite NPs. The inset shows the fast Fourier Transform of the whole aggregate; (b-e) EDS elemental maps of Fe, O, Al and As of the aggregate shown in (a). The periodicity in image (c) is due to some noise during measurements.

Arsenic Research and Global Sustainability – Bhattacharya, Vahter, Jarsjö, Kumpiene,
Ahmad, Sparrenbom, Jacks, Donselaar, Bundschuh & Naidu (Eds)
© 2016 Taylor & Francis Group, London, ISBN 978-1-138-02941-5

Solid phase extraction for thioarsenate stabilization in iron-rich waters

B. Planer-Friedrich & M.K. Ullrich
Environmental Geochemistry Group, University of Bayreuth, Bayreuth, Germany

ABSTRACT: Thioarsenates play an important role in nature. However, due to difficulties in stabilisation, there are still too few studies considering their occurrence. In iron-rich waters, stabilization is especially problematic, since flash-freezing leads to precipitation of iron oxyhydroxides, acidification to precipitation of As-S-minerals. We show here that arsenate and thioarsenates sorb on the anion exchanger Bio-Rad AG2-X8 (retention 97–100%) and can be eluted quantitatively and species-conserving using 0.5 M alkalinized salicylate (recovery 90–102%). Trithioarsenate is the species most difficult to mobilize and requires up to three elution steps. Arsenite and cationic iron pass the resin. Neither presence of iron nor high chloride concentrations interfere with sorption or stability of thioarsenates on the resin. Resin storage does not change speciation.

1 INTRODUCTION

Arsenite and arsenate are often assumed to be the only relevant species when investigating inorganic arsenic speciation. Under sulfate-reducing conditions, this is wrong. The high affinity between arsenic and reduced sulfur leads to formation of thioarsenites ($[As^{III}S^{-II}_nO_{3-n}]^{3-}$, n = 1–3) and thioarsenates ($[HAs^VS^{-II}_nO_{4-n}]^{2-}$, n = 1–4) which differ in mobility (Burton *et al.*, 2013, Couture *et al.*, 2013), microbial transformation (Edwardson *et al.*, 2014, Planer-Friedrich *et al.*, 2015), and toxicity (Hinrichsen *et al.*, 2014, 2015) from the known behavior of their respective oxyanions. Due to analytical difficulties, the existence of thioarsenites has only been confirmed in synthetic solutions. Thioarsenates can be separated routinely by ion chromatography, but need a very strong eluent (100 mM NaOH) for desorption (Planer-Friedrich *et al.*, 2007).

The reason why still only few researchers consider thioarsenic species in their studies, is the tedious on-site sample preservation. Acidification leads to transformation to arsenite and precipitation of As-S-minerals. Filtration and flash-freezing is the currently best available method in iron-free systems (Planer-Friedrich *et al.*, 2007). However, it is inconvenient to organize in remote locations and fails in iron-rich waters. Iron oxyhydroxides precipitate upon thawing and lead to adsorption of (thio)arsenic species. We have previously investigated use of pH-buffered EDTA for Fe complexation or addition of ethanol (Suess *et al.*, 2015). However, all these preservation methods eventually lead to some artefact formation and required short turnover times.

In this study, we tested solid phase extraction which had previously been used for arsenite/arsenate and sulfate/thiosulfate separation. The idea was that arsenate and thioarsenates could be stabilized on an anion exchanger while non-charged arsenite and cationic iron pass without retention, thereby elimi-

nating iron interferences on thioarsenate stability. Based on the known strong sorption of especially higher thiolated arsenates on the chromatographic column, we already suspected that a strong eluent would be required for complete desorption. Species stability during sorption and elution was a further point of concern. Finally, we wanted to test the potential of matrix interference/elimination (e.g. in marine waters) and preconcentration for waters where matrix to target species ratios are high.

2 METHODS

Solid phase extraction cartridges were prepared from AG2-X8 (Bio-Rad) resin material (chloride form) and conditioned with 0.1 M KOH or deionized water. All experiments were conducted inside an oxygen-free glovebox (5%H_2/95%N_2). Up to 5 times 50 mL test solution was loaded on the columns at a flow rate of 2–3 mL/min using a vacuum pump. Test solutions contained sulfide, thiosulfate, sulfate, arsenite, arsenate, mono- and trithioarsenate individually or as mixtures in the presence and absence of iron (5 mg/L) and chloride (140 mg/L). After sample application, 4 mL deionized water was used to wash off non-retained species. Elution was done at a flow rate of 1 mL/min using up to 3 times up to 15 mL for complete desorption. Different soaking times (up to 1h) were tested. For eluents, KCl (0.5, 3 M), citrate (0.5 M), and salicylate (0.5 M) were tested both for preserving the original speciation in solution and, if species integrity was confirmed, for completeness of elution. Arsenic speciation and total concentrations were determined in the initial solution, the passages, wash solution, and eluates as described previously (Planer-Friedrich *et al.*, 2007). For all samples, retention was calculated as ($n_{initial\ solution} - n_{passage}$)/ $n_{initial\ solution}$*100% and recovery as $n_{eluate1-5}$/($n_{initial\ solution} - n_{passage} - n_{wash}$) *100%. To test species stability during storage, exchangers were loaded with natural water from a mineral spring at Frantiskovy Lazne, Czech Republic, that was known to contain thioarsenates (Suess *et al.*, 2015). Cartridges were stored cold and dark until elution and analysis after 1, 2, and 6 days.

3 RESULTS AND DISCUSSION

Preconditioning with KOH lead to sorption of arsenite, presumably due to deprotonation or oxidation at the high pH of remaining conditioning liquid or, in the presence of iron, due to iron oxyhydroxide precipitation and co-sorption. Deionized water was used for conditioning in all further experiments. Complete retention of the initially applied amount on the resin was confirmed for sulfate, thiosulfate, arsenate, monothioarsenate, and trithioarsenate with 100%, 100%, 97%, 99%, and 100%, respectively. Arsenite passed the resin with <3% retention. Arsenic concentrations in the wash solutions were below detection limit. Use of KCl as eluent was sufficient for full recovery of sulfate, thiosulfate, arsenate, and monothioarsenate, but trithioarsenate was not recovered quantitatively. Citrate impaired chromatographic separation of arsenate and was not tested in elution experiments. Salicylate in concentrations below 0.05 M had no negative effects on chromatographic separation. We found, however, an increase in ICP-MS sensitivity that had to be compensated by using a matrix-matched calibration. For elution, 0.5 M salicylate enabled full and species-conserving recovery of 102%, 102%, 96%, 92%, and 90% for sulfate, thiosulfate, arsenate, monothioarsenate, and trithioarsenate, respectively. Only trithioarsenate was not recovered in a single elution step. Longer soaking times or up to 3 elution steps were required for full recovery. The necessity of multiple analyses or dilution of concentration is a serious drawback here. An alternative could be removing the resin material from the cartridge and stirring it in salicylate to increase surface reaction kinetics compared to the flow-through version applied in our experiments.

In solutions with free sulfide we observed unexpected arsenite binding to the resin, obviously via sulfide bridging to organic resin material. Upon elution in salicylate, arsenite desorbed as arsenite but reacted with eluting sulfide to form thioarsenate artefacts. We also observed that the salicylate matrix accelerated thioarsenate formation from arsenite and sulfide as has been described for other high ionic strength or buffered solutions before (Planer-Friedrich *et al.*, 2010). To avoid this artefact thioarsenate formation, salicylate with excess NaOH was used. At OH:SH ratios >1, arsenite and sulfide do not react to form thioarsenates via thioarsenites (Planer-Friedrich *et al.*, 2010).

The optimized method worked well also in the presence of iron. As expected, iron passed the resin without changing thioarsenate speciation. The passage was acidified to keep arsenite and iron in solu-

tion. Chloride had no effect on sorption efficiency. Eluates of natural samples loaded on the resins showed no speciation changes from 1 to 2 or 6 days.

4 CONCLUSIONS

Thioarsenates are difficult to preserve in natural waters, but their importance requires them to be considered in more studies. AG2-X8 anion exchangers are presented as an economic alternative compared to flash-freezing that also works well in iron-rich waters. Loaded resins can be stored without species transformations and elution with alkalinized salicylate recovers (thio)arsenates quantitatively and species-conserving for subsequent IC-ICP-MS analysis.

ACKNOWLEDGEMENTS

We acknowledge generous funding by the German Research Foundation within the Emmy Noether program (Grant No. PL 302/3-1) and help with experiments from V. Misiari, A. Gayout, and S. Will.

REFERENCES

Burton, E.D., Johnston, S.G., Planer-Friedrich, B. 2013. Coupling of arsenic mobility to sulfur transformations during microbial sulfate reduction in the presence and absence of humic acid. *Chem. Geol.* 343: 12–24.

Couture, R.M., Rose, J., Kumar, N., Mitchell, K., Wallschlager, D., Van Cappellen, P. 2013. Sorption of Arsenite, Arsenate, and Thioarsenates to Iron Oxides and Iron Sulfides: A Kinetic and Spectroscopic Investigation. *Environ. Sci. Technol.* 47(11): 5652–5659.

Edwardson, C., Planer-Friedrich, B., Hollibaugh, J. 2014. Transformation of monothioarsenate by haloalkaliphilic, anoxygenic photosynthetic purple sulfur bacteria. *FEMS Microbiol. Ecol.* 90: 858–868.

Hinrichsen, S., Geist, F., Planer-Friedrich, B. 2015. Inorganic and methylated thioarsenates pass the gastrointestinal barrier. *Chem. Res. Toxicol.* 28(9): 1678–1680.

Hinrichsen, S., Lohmayer, R., Zdrenka, R., Dopp, E., Planer-Friedrich, B. 2014. Effect of sulfide on the cytotoxicity of arsenite and arsenate in human hepatocytes (HepG2) and human urothelial cells (UROtsa). *Environ. Sci. Pollut. R.* 121(17): 10151–10162.

Planer-Friedrich, B., Härtig, C., Lohmayer, R., Suess, E., McCann, S.H., Oremland, R. 2015. Anaerobic chemolithotrophic growth of the haloalkaliphilic bacterium strain MLMS1 by disproportionation of monothioarsenate. *Environ. Sci. Technol.* 49(11): 6554–6563.

Planer-Friedrich, B., London, J., McCleskey, R.B., Nordstrom, D.K., Wallschläger, D. 2007. Thioarsenates in geothermal waters of Yellowstone National Park: determination, preservation, and geochemical role. *Environ. Sci. Technol.* 41(15): 5245–5251.

Planer-Friedrich, B., Suess, E., Scheinost, A.C., Wallschlaeger, D. 2010. Arsenic speciation in sulfidic waters: Reconciling contradictory spectroscopic and chromatographic evidence. *Anal. Chem.* 82: 10228–10235.

Suess, E., Mehlhorn, J., Planer-Friedrich, B. 2015. Anoxic, ethanolic, and cool—an improved method for thioarsenate preservation in iron-rich waters. *Appl. Geochem.* 62: 224–233.

Arsenic Research and Global Sustainability – Bhattacharya, Vahter, Jarsjö, Kumpiene,
Ahmad, Sparrenbom, Jacks, Donselaar, Bundschuh & Naidu (Eds)
© 2016 Taylor & Francis Group, London, ISBN 978-1-138-02941-5

Simultaneous determination of inorganic arsenic species in a pore water, paddy soil, and rice by using ion chromatography with inductively coupled plasma mass spectrometry

S.-H. Nam, M.-Y. Park, S.-H. Son & S.-W. Kwon
Department of Chemistry, Mokpo National University Muan-Gun, Chonnam, South Korea

ABSTRACT: This study focused on the determination of inorganic arsenic species in rice which were produced in several places in Korea. If existed, to identify the source of inorganic arsenics, the surrounding paddy soil and pore water were analyzed. The developed method for the quantitative determination of As species included the selective extraction method, the optimum separation and detection method, and all the important instrumental and methodological parameters. The prevention of loss and transformation of As species was confirmed with the developed extraction and analytical method.

1 INTRODUCTION

Chemical transformation of elements can result in various chemical species that differ considerably in properties and in quantity. The toxic and biological effects of arsenic species depend on the different chemical forms in a sample. In particular, the determination of inorganic arsenic in a rice is important because rice is a staple food in many countries including Korea. This study focus on the determination of inorganic arsenic species in a rice which was produced in several places in Korea. If existed, to identify the source of inorganic arsenics, the surrounding soil and water were analyzed.

2 METHODS/EXPERIMENTAL

2.1 Instrumentation

An inductively coupled plasma—mass spectrometry (SPECTRO, Germany) was used for the determination of As. GS-50 gradient pump (DIONEX, USA) equipped with an injection valve was used as the sample and eluents delivery system. A Hamilton PRP-X100 analytical column (Hamilton Company, USA) was used for the separation of the As species.

2.2 Extraction method

Water: Taking a 10 mL of the sample and centrifuge for 10 min at 4000 rpm in order to filter out impurities. It was treated with 0.45 μm syringe filter, then used as the final sample.

Soil: The extraction solution was the mixture of 1 M Phosphoric acid and 0.5 M ascorbic acid. 0.2000 g of soil and 7.5 mL of the extraction solution were transferred to a 50 mL falcon tube. Then, the mixture was sonicated for 30 min using a vibra cell to extract the asenic species. It was centrifuged for 20 minutes at 3200 rpm. The supernatant was filtered through a 0.45 μm nylon syringe filter to a 50 mL falcon tube. The remaining samples were repeated with the extraction process. The final sample is diluted with a pure water to 50 mL.

Rice: A 1.0000 g of a dry sample and 10 mL of 5 mM Malonic acid were placed into a 50 mL falcon tube. Then, the mixture was heated for 30 min over 80°C water and ultra-sonicated for 5 min. This extraction procedure was repeated three times, and finally the ultrasonic extraction was done for 10 minutes. It was cool down for 2 hours. The extracted solutions were centrifuged at 3,000 rpm for 10 min and filtered through a 0.45 μm nylon syringe filter to a 50 mL falcon tube.

3 RESULTS AND DISCUSSION

3.1 Pore water

We analyzed more than 30 pore water samples. Arsenic in most samples existed in very low levels. Only the inorganic arsenic species were in pore water samples. The organic arsenic species did not exist in any sample.

As (III) primarily existed which its concentration ranged from 0.3 to 8.5 μg/kg while the As (V) rarely existed.

3.2 Paddy soil

Paddy soils were extracted by the mixture of 1 M Phosphoric acid and 0.5 M ascorbic acid using sonicator. Arsenic in most paddy soils were the inorganic arsenic species, but extremely low levels of DMA existed in the samples. The major arsenic species was As (V) in paddy soil, its concentration range was from

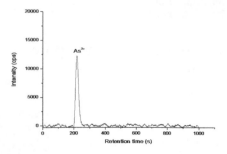

Figure 1. Arsenic chromatogram of pore water.

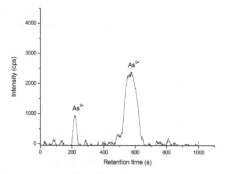

Figure 2. Arsenic chromatogram of a paddy soil.

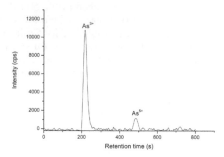

Figure 3. Arsenic chromatogram of a rice.

0.5 to 165 mg/kg. The concentrations of As (III) in paddy soils were in the range of 0.3–8.6 mg/kg.

3.3 *Rice*

Both inorganic arsenic species and organic arsenic species existed in a rice. In a rice, the main arsenic species was As (III) which its concentration range was from 8 to 65 μg/kg. The concentrations of As (V) in rices were in the range of 1.4–6.8 μg/kg. The major organic arsenic species in a rice was DMA. But DMA existed in a few rice samples.

The Figure 4 showed the ratio of inorganic arsenic species to total arsenic species in a rice for a certain region.

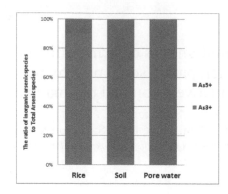

Figure 4. The ratios of the inorganic arsenic to the total arsenic in a certain region.

4 CONCLUSIONS

The concentration of As (V) is much higher than that of As (III) in paddy soil, but the concentrations of As (III) in rice and pore water were higher than that of As (V). It was described that the mobility and uptake of arsenic relyed on the arsenic species. The mobility and uptake rate of As (III) were higher than that of As(V). The high content of arsenic species in paddy soil were closely related with the total concentration of arsenic species in rice.

ACKNOWLEDGEMENTS

This work was funded by Ministry of Environment in Korea.

REFERENCES

Abedin, M.J., Feldman, J., Meharg, A.A. 2002. Uptake kinetics of arsenic species in rice plants. *Plant Physiol.* 128: 1120–1189.

Chen, Z., Akter, K.F., Rahman, M.M., Naidu, R. 2008. The separation of arsenic species in soils and plant tissues by anion-exchange chromatography with inductively coupled mass spectrometry using various mobile phases. *Microchem. J.* 89: 20–28.

Giral, M., Zagury, G.J., Deschenes, L., Blouin, J.-P. 2010. Comparison of four extraction procedures to assess arsenate and arsenite species in contaminated soils. *Environ. Pollut.* 158: 1890–1898.

Raber, G., Stock, N., Hanel, P., Murko, M., Navratilova, J., Francesconi, K.A. 2012. An improved HPLC–ICPMS method for determining inorganic arsenic in food: Application to rice, wheat and tuna fish. *Food Chem.* 134: 524–532.

Sanz, E., Muñoz-Olivas, R., Cámara, C., Sengupta, M.K., Ahamed, S. 2007. Arsenic speciation in rice, straw, soil, hair and nails samples from the arsenic-affected areas of Middle and Lower Ganga plain, *J. Environ. Sci. Heal. A* 42, 1–11.

Arsenic Research and Global Sustainability – Bhattacharya, Vahter, Jarsjö, Kumpiene,
Ahmad, Sparrenbom, Jacks, Donselaar, Bundschuh & Naidu (Eds)
© 2016 Taylor & Francis Group, London, ISBN 978-1-138-02941-5

A new analytical method for rapid analysis and speciation of arsenic in rice

O.G.B. Nambiar

Former Scientist, National Chemical Laboratory, Pune, India

ABSTRACT: A new analytical method for the direct quantitative determination of arsenic (As) in rice is presented in this paper. This procedure is capable of performing speciation of As (III) and As (V). Thiamine present in the sample can also be measured by this procedure. Filter-plug inserted inside the reactor cap fixes the hydrogen sulfide gas generated in the reactor vessel. Arsenic along with the hydrogen sulfide is measured by substituting the filter plug with cotton plug without the sulfide absorbent. The hydrogen sulfide also produces pink stain in the detector tube. Arsenic and thiamine are found to be in the molar ratio of 1:1. For determining total As, the arsenate is converted to arsenite by a catalyst incorporated in the reagent, which is converted to arsine by sodium borohydride. The main As species detected in rice samples was inorganic As (III), which is consistent with the reported results. It takes only about 20 minutes to analyze one sample of rice. The present procedure can also analyze biological samples like urine, hair etc.

1 INTRODUCTION

Exposure to elevated levels of arsenic (As) lead to a host of illness like cancer, diabetes, bronchitis, hypertension, skin hypo and hyperpigmentation, skin, bladder and lung cancers. A lot of attention is focused on the analytical method for quantization of As and its speciation in rice samples (Williams *et al.*, 2005). The analytical protocol presently practiced for the analysis of rice sample is quite elaborate and time consuming. Analytical procedure involves digestion of powered sample nitric acid or TFA and finished by ICP-MS (Zavala *et al.*, 2008a,b, Narukawa *et al.*, 2008, Huang *et al.*, 2012). Further As speciation is carried out by HPLC system. There is also certain uncertainty regarding the inter conversion of As species during nitric acid/ TFA extract. Costly equipment are also required for the analysis. A rapid, sensitive and reliable method is reported in this paper. This methodology is an extension of the procedure used for the determination of As in water samples including their speciation.

2 METHODS/EXPERIMENTAL

2.1 *Materials*

Rice and paddy samples including different types of basmati varieties, polished, aromatic, long, medium or short grains, collected from wholesale market were analyzed.

2.2 *Arsenic-monitoring system*

The Combo kit (UNICEF /NCL Technology) supplied by Chemin-Corporation, Pune India, was used for the study. Fig. 1 Standard Arsenite and Arsenate solutions and thiamine nitrate purchased from Merck were used. Silicon-oil based foam-suppressor was

used in experiments. Average moisture content in all samples were found to be in the range of 10 to 14%.

The equipment consists of a 250 mL capacity polycarbonate reaction vessel with a screw-cap terminating into a tube fitted with a silicon rubber tube (Fig. 1). The cap is fitted with a silicon-rubber washer for a leak-proof fitting. The cavity inside the cap is used for packing the cotton filter-plug. The pre-calibrated detector tube is inserted into the silicon-rubber tube. The cotton filter plug is treated with suitable chemicals to absorb alkaline vapors or hydrogen sulfide, which can interfere with the analysis. The detector tube is calibrated in a scale from 0 to 100 ppb, with marking at every 10 ppb intervals. Arsine gas produces deep pink color to the powder inside the detector-tube. The calibration on the detector tube is for 50 g/50 mL sample. If the sample taken for analysis is 10 g the reading is multiplied with factor 5. For example if the reading

Figure 1. Equipment for arsenic analysis.

is 25 on the detector tube for a sample of 10 g then the actual concentration is $25 \times 5 = 125$ ppb.

2.3 Determination of arsenic in rice

Arsenic (III): 10 g powdered rice samples was weighed into the polycarbonate reaction vessel, followed by 50 mL of distilled water. Five drops of foam-suppressor was added. The reactor cap with filter plug and washer placed in the groove and the detector-tube inserted in the silicon rubber tube is kept ready. The reagent A from vial is added to the reactor followed by the reagent B. The reactor bottle is closed and tightened. The reactor kept aside undisturbed for 15 minutes. Arsine gas produced by the reaction produces pink color in the detector tube.

For the determination of total As (As_{total}) in the sample, the reagent (A) is substituted by (A*). Reagent (A*) contains catalyst which converts all As(V) to As (III). The reading gives the As_{total} in the sample. In the third experiment on the sample the filter plug is changed, which allows hydrogen sulfide to pass through to the detector tube. Total quantity of As along with thiamine is obtained in this experiment. Calibration of thiamine on the detector tube was carried out using standard thiamine nitrate solutions.

3 RESULTS AND DISCUSSION

3.1 Results

Results obtained for As content along with thiamine values are presented in Table 1.

Calibration of the detector tubes were checked by using standard As (III), As (V) and thiamine solutions. Reproducibility and recoveries were tested at the entire range of concentrations. Standard thiamine solutions were prepared using thiamine nitrate equated to molar concentrations of As. Standard addition method was used for recovery and validation of method. Recovery of thiamine was studied by synthetic standard solutions. Recoveries were found to be 100%. Reduction of As (V) to As (III) by thiamine; As (V) is reduced to As (III) by thiamine quantitatively. This method was used for the determination of free thiamine in the samples.

3.2 Analysis of grain samples

Grain samples are analyzed in three steps: i) determination of As (III) in which powdered sample 5 to 10 g added to the reaction vessel containing 50 mL distilled water. Reagents A and B are added and closed with the cap fitted with detector tube. Sulfide filter is used in the filter cavity. The pink stain in the detector tube gives the As(III) concentration., ii) the above experiment is repeated by addition of reagent A* in place of reagent A. The reading gives the value for As (V) and As (III); iii) in the third step of the experiment, the filter plug is changed to the normal cotton without the sulfide filter. The reading indicates As (III) plus thiamine.

Table 1. Concentration of arsenic species (μg/kg).

Sample	As_{total}	As(III)	As(V)	Thiamine* (equated to As
Basmati	125	125	ND	125
Basmati II	75	75	ND	75
Basmati III	125	125	ND	125
Silky Kolam	200	200	ND	200
Rice Powder (fried)	225	225	ND	225
Beaten Rice (Flakes)	250	250	ND	250
Broken Rice brown (old)	140	120	20	120
Rice Paddy	125	125	ND	125
Rice Paddy (culture 1)	150	150	ND	150
Rice Paddy (parboiled)		125	ND	125
Wheat Flour	Traces	Traces	ND	650

3.3 Stability of As (III) - thiamine compound

Potato grown in certain region in India were found to contain considerable amounts of As. Potato chips fried in oil at temperature around 180 degree C were found to retain As (III) as thiamine compound. Parboiled paddy is also found to retain As (III)-thiamine complex. Rice powder even after dehydrating at elevated temperatures retains As in the lower oxidation state.

4 CONCLUSIONS

Arsenic in rice samples is present as the thiamine compound. Arsenic in the rice grain is in its lower oxidation state and is very stable. Even at high temperatures As remains as As (III).

ACKNOWLEDGEMENTS

Author is thankful to the Agricultural University, Kerala for providing paddy samples.

REFERENCES

Huang, J.H., Fecher, P., Ilgen, G., Hu, K.N., Yang, J. 2012. Speciation of arsenite and arsenate in rice grain-verification of nitric acid based extraction method and mass sample survey. Food Chem. 130: 453–459.

Narukawa, T., Inagaki,, K., Kuroiwa,. T., Chiba, K. 2008. The extraction and speciation of arsenic in rice flour by HPLC-ICP-MS. Talanta 77: 427–432.

Williams, P.N., Price, A.H., et al. 2005. Variation in arsenic speciation and concentration in paddy rice related to dietary exposure. Environ. Sci. Technol. 39: 5531–5540.

Zavala, Y.J., Duxbury, J.M. 2008. Arsenic in rice: I. Estimating normal levels of total arsenic in rice grain. Environ. Sci. Technol. 42: 3856–3860.

Zavala, Y.J., Gerads, R, .Gorleyok H, Duxbury J.M. 2008. Arsenic in rice: II. Arsenic speciation in USA grain and implication for human health. Environ. Sci. Technol. 42: 3861–3866.

Arsenic Research and Global Sustainability – Bhattacharya, Vahter, Jarsjö, Kumpiene,
Ahmad, Sparrenbom, Jacks, Donselaar, Bundschuh & Naidu (Eds)
© *2016 Taylor & Francis Group, London, ISBN 978-1-138-02941-5*

Arsenic speciation in environmental samples by a new specific tri-isobutyl phosphate polymer material pre-concentration and HPLC-ICP-MS determination

X. Jia[1,2], J. Wang[1,2], Q.Q. Chi[1,2] & X. Zhang[1,2]
[1]*Key Lab of Urban Environment and Health, Institute of Urban Environment, Chinese Academy of Sciences,*
Xiamen, P.R. China
[2]*Ningbo Urban Environment Observation and Research Station, Chinese Academy of Sciences, Ningbo,*
P.R. China

ABSTRACT: A new specific tri-isobutyl phosphate functionalized poly (chloromethyl styrene-co-styrene) was synthesized, characterized and used as the adsorbent to preconcentrate arsenic species in environmental waters by on line Solid-Phase Extraction (SPE). The trace speciation analysis of arsenic in water samples has been used by HPLC-ICP-MS after on-line SPE. Avoiding external reagent, As(III), MMA, DMA and As(V) have been adsorbed on the home-made mini-column with large number of positively charged adsorption groups, and then eluted rapidly (within seconds) with a mixed solution of ammonium nitrate and ammonium dihydrogen phosphate. This work provides an effective tri-isobutyl phosphate functionalized polymer material, and a reliable on-line of the preconcentration method and detection coupled method. Under the optimized experimental conditions, the enrichment factors obtained for As(III) and DMA with 25 mL sample solution were 28, while for MMA and As(V) reached 30. The low detection limits of 1.8 ng/L, 1.0 ng/L, 0.9 ng/L and 1.3 ng/L, with the Relative Standard Deviations (RSDs) of 3.9%, 5.6%, 3.2% and 4.5% were obtained for As(III), DMA, MMA and As(V), respectively. At the same time, the established method was successfully applied to analyze three environmental samples of lake water, river water and seawater, respectively.

1 INTRODUCTION

Arsenic (As) is a toxic trace element that is naturally present in all terrestrial and aquatic environments, but it is also an important environmental pollutant. There is increasing evidence of cancer risk associated with chronic exposure to low levels of arsenic, e.g., through drinking water (Huang et al., 2010). Thus, analyzing their individual species in the environmental water samples is extremely important for adequately revealing their toxicity and bioavailability, monitoring mutual transformation eventually evaluating their impact on the environment and health of human body.

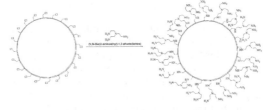

Figure 1. Synthesis process and chemical structure of the functional polymer.

into the stainless steel hollow mini-column. Prior to use, ethanol and ultrapure water were passed through the column in sequence in order to clean it. The column was then conditioned to the desired pH with buffer solution.

2 EXPERIMENTAL

2.1 Synthesis of functionalized organic polymer material and minicolumn preparation

The materials for preconcentration of arsenic species were prepared by the emulsion copolymerization of styrene and p-chloromethyl styrene in presence of crosslinker divinyl benzene with Benzoyl Peroxide (BPO) as initiator, followed by the functionalized with tri-isobutyl phophate (structure in Figure 1). The functionalized material precipitates was centrifuged, washed with methanol and finally dried to a constant weight (5.3 g).

A total of 430 mg of the functional organic polymer anion-exchange resin adsorbent was filled

2.2 Procedure for preconcentration and elution

In the current work, the prepared minicolumn was fixed on the rotary injection valve (6 channel) to substitute the primary sample loop. Under the pressure of a pump, three replicates of 25 mL sample solution containing the analytes were loaded on to the column, then all of the arsenic species were adsorbed on the functional organic polymer material with positively charged adsorption site quickly due to the strong electrostatic affinity.

After preconcentration, the system was washed by ultra-pure water with a flow rate of 5 mL min⁻¹, and 1 min was sufficient to avoid residue in the pipeline.

Then the valve was switched, the analytes carried by eluent (20 mM mixed solution of NH_4NO_3 and $NH_4H_2PO_4$) were directly transported into the HPLC-ICP-MS for separation and determination.

3 RESULTS AND DISCUSSION

3.1 Optimization of the pH value for samples

The effect of pH on the retention of arsenic species on functionalized anion-exchange minicolumn was investigated. For this purpose, 25 mL sample solutions containing 0.5 ng/mL each target species were prepared separately and pH values of the sample solutions were adjusted to range of 7–11 with $NH_3 \cdot H_2O$ solution, while other experimental conditions is set. The results indicated that, no significant changes of the adsorption percentage were observed for arsenic species except DMA, as shown in Figure 2. The reason might be that the steric hindrance of DMA molecular structure is relatively large.

Since more than 90% adsorption percentage were achieved for all the analytes at pH 7, finally, in order to avoid the exogenous contamination and consequently minimize the background, the experiment was performed at neutral pH value without additional reagents. This shows that the applicability of the synthesized material is promising.

3.2 Effect of flow rate and sample volume

The influence of the sample flow rate on the Enrichment Factors (EFs) and recoveries (>90%) of As(III), DMA, MMA and As(V) were also investigated in the ranges of 1–9 mL/min, while keeping the other conditions constant (the maximal flow rate is depended on the maximal tolerable pressure of the pump and the minicolumn). The results in Figure 3 depicted that the adsorption percentage of arsenic species were almost kept constant till 5 mL/min, then began to decrease with further increase of flow rate. That may because the analytes and the adsorption sites can't contact completely at high flow rate. Thus, all the experiments were carried out at a flow rate of 5 mL/min in this study for shortening the analytical time and maintaining the high adsorption efficiency.

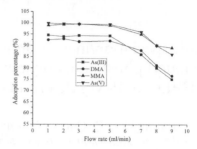

Figure 3. Effects of flow rate on the analytes.

It is also necessary to examine the maximum applicable sample volume when the practical samples containing a very low level of analytes were analyzed. The influence of sample volume on the recoveries of a mix standard solution of arsenic species was investigated in the range of 5–30 mL. The results shown that the recoveries were found to be stable only till 25 mL, then decreased with further increasing of volume. It might be due to the limited adsorptive capacity of minicolumn and the loss of adsorbed analytes rinsed by the larger volume of sample. Therefore, 25 mL was selected as the breakthrough volume for this work.

4 CONCLUSIONS

A novel, simple and rapid method has been developed for the speciation of arsenic in water samples by HPLC-ICP-MS after on-line SPE preconcentration by employment of a new specific isobutyl phosphate functionalized polystyrene. The features of the synthesized ploymer material as a new solid phase sorbent allowed fast separation and enrichment of As species in one step using only a single mini-column without the need for any excess oxidation or reduction reagent. Thanks to the high recognition of developed SPE, high EFs were obtained and low LOD were achieved within only 5 min for enrichment. This technique has distinctive advantages over the conventional methods with respect to very short sample preparation time and free of toxic organic extraction solvent.

ACKNOWLEDGMENTS

This work was financially supported by Knowledge Innovation Program of Institute of Urban Environment, Chinese Academy of Sciences (IUEMS201409), Science and Technology of Fujian Province (2015H0045), and the National Natural Science Fund of China (21507126).

REFERENCE

Huang, G.H., Kretzschmar, R. 2010. Sequential extraction method for speciation of arsenate and arsenite in mineral soils. Anal. Chem. 82(13): 5534–5555.

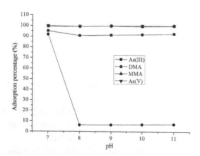

Figure 2. Effect of pH on the adsorption percentage of arsenic species.

Arsenic Research and Global Sustainability – Bhattacharya, Vahter, Jarsjö, Kumpiene,
Ahmad, Sparrenbom, Jacks, Donselaar, Bundschuh & Naidu (Eds)
© 2016 Taylor & Francis Group, London, ISBN 978-1-138-02941-5

Analysis of total arsenic in water samples, soils, sediments and rice by means of Total Reflection X-ray Fluorescence analysis

H. Stosnach
Bruker Nano GmbH, Berlin, Germany

ABSTRACT: Total Reflection X-ray Fluorescence (TXRF) spectroscopy was tested for the analysis of total arsenic in different samples types. In water samples the analysis of total arsenic is possible as well in low concentrated natural waters as in highly concentrated waste waters. Soil and sediment samples can be analyzed without sample digestion or extraction in concentration ranges from low mg/kg up to wt.-%. Here the analysis is limited by the homogeneity of the sample. In rice samples current limit values for total arsenic can be controlled after a very fast and simple sample preparation.

1 INTRODUCTION

Regarding the analysis of total arsenic it can be distinguished between fixed laboratory and portable field assays.

Fixed laboratory assays are applied for accurate analysis of arsenic in liquid and solid samples to µg/L and µg/kg levels, respectively. Several accepted analytical methods are available, including Atomic Fluorescence Spectroscopy (AFS), Graphite Furnace Atomic Absorption (GFAA), Hydride Generation Atomic Absorption Spectroscopy (HGAAS), Inductively Coupled Plasma-Atomic Emission Spectrometry (ICP-AES), and Inductively Coupled Plasma-Mass Spectrometry (ICP-MS).

These methods demand a pretreatment of the samples, either with acidic extraction or acidic oxidation digestion. The instruments itself are bulky, expensive to operate and maintain, and require fully equipped laboratories for operation.

The most common field assays are colorimetric field kits and Anodic Stripping Voltammetry (ASV). If used by trained staff they are acceptable for purposes of sample screening or site surveys, strive for similar detection goals as fixed laboratory methods, are relatively inexpensive, and can produce a large number of screening results in a short time. However, they are restricted to liquid samples and cannot be directly applied to solid sample like soils or food.

Total Reflection X-Ray Fluorescence Spectroscopy (TXRF) is a well-established method for trace element analysis in various sample types (Klockenkaemper & von Bohlen, 2015). Because of the working principles of this method (direct analysis of liquid and solid samples, quantification based on internal standardization only) and instrumentation (no external media necessary for operation, low power consumption) this method is easy to operate at low to moderate operation and maintenance costs.

Commercially available instruments can be operated as fixed laboratory assays, certain spectrometers can also be operated mobile as field assays (Mages *et al.*, 2003).

In this paper the possibilities and restrictions of TXRF for the analysis of total arsenic in waters, soils, sediments and rice are presented.

2 METHODS/EXPERIMENTAL

In a TXRF spectrometer the incident X-ray beam from the X-ray tube passes a monochromator and hits a flat reflecting surface, holding the sample. The monochromatic excitation beam has an incident angle slightly less than the critical angle and the form of a paper strip. The characteristic X-rays, emitted by the sample, enter a solid-state detector and their intensity and energy is measured, applying an amplifier and multichannel analyzer. The detector itself is closely coupled to the sample and a rectangular orientation.

Samples for TXRF should be presented as solutions, suspensions, fine powders or thin films. Depending on the matrix, liquids can be prepared directly or after dilution. Solid samples must be ground as fine as possible with a maximum grain size of 50 µm. The fine powder can then be prepared as slurry, using an aqueous Triton X100 solution (1 volume %) as a suspension agent. Alternatively, solid materials can be prepared as dissolutions, applying suitable solvents or microwave assisted acid digestion.

For quantitative analysis the addition of an internal standard is mandatory. Internal standardization is performed by adding a standard of an element, not present in the sample itself to the sample. A known quantity of this internal standard is added to the sample in order to get concentrations, similar to those of the analyte. After thorough homogenization, a small amount (0.5 to 10 µL) of sample is pipetted onto a clean sample carrier and subsequently dried by heat or in vacuum.

3 APPLICATION EXAMPLES

3.1 *Water samples*

Trace element analysis of water samples by means of TXRF is widely reported and ranges from samples with low concentrations groundwater or landfill leachates (e.g. Cataldo, 2012).

Sample preparation for this sample type is done by adding a defined amount of an internal standard element to the sample, homogenization and subsequent preparation of a 10 µL droplet onto a reflective sample carrier. The sample is placed in the instrument for analysis with typical measurement times of 1000s after drying.

Typical detection limits for total arsenic are between 0.20 µg/L for low concentrated water and 30 µg/L for high concentrated wastewater samples (Table 1).

The only limiting factor for the analysis of total arsenic in water samples is the line interference with lead. But only in case of high lead concentration the accurate quantification of arsenic is compromised.

3.2 *Soil and sediment*

For common analytical techniques solid samples like soils must be prepared as acid digestions or extractions. Although these preparations could also be applied for TXRF, much less expensive and time consuming preparation as slurries is applied.

Arsenic could be analyzed in this sample type with a broad linear range from normal concentration levels in the low mg/kg range (Towett & Shepherd, 2013) up to weight%-ranges in highly contaminated samples (US EPA, 2006).

The only major restriction of this method is that the accuracy and reproducibility is limited by the very small analyzed sample amount and, thus, the homogeneity of the sample.

3.3 *Rice samples*

The analysis of arsenic in rice is a very current topic and a limit value of 0.2 mg/kg came into effect for the total arsenic concentration in certain countries.

With an optimized TXRF method it could be proven that the safe control of total arsenic concen-

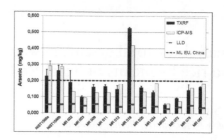

Figure 1. Comparison of TXRF and ICP-MS values for different rice samples.

trations in rice samples after an easy slurry preparation is possible. Figure 1 shows a comparison of As values, measured with ICP-MS and TXRF in several rice samples. The detection limits of TXRF is distinctly lower than the actual legal limit values for As in rice, indicating that a safe control by TXRF is possible.

4 CONCLUSIONS

TXRF spectroscopy is a versatile analytical tool for the analysis of total arsenic in various sample types. Even though it does not offer the same capability regarding species analysis and sensitivity as ICP-MS it is suitable in laboratory environments where the technical requirements for these high-end techniques are not given.

Applied mobile it can support the common field assay techniques for solid samples like soil or plant materials.

ACKNOWLEDGEMENTS

We would like to thank the Queen's University Belfast, Institute for Food Safety for providing the rice samples and ICP-MS data.

REFERENCES

Cataldo, F., 2012, Multielement analysis of a municipal landfill leachate with total reflection X-ray fluorescence (TXRF). A comparison with ICP-OES analytical results, *J. Radioanal. Nuc. Chem.* 293: 119–126.

Klockenkamper, R., von Bohlen, A. 2015. Total reflection X-ray fluorescence analysis and related methods, 2nd edition, Wiley, London.

Mages, M., Woelfl, S., Ovari, M, von. Tumpling, W. 2003. The use of a portable total reflection X-ray fluorescence spectrometer for field investigation. *Spectrochim. Acta Part B* 58: 2129–2138.

Towett, E., Shepherd, K.D. 2013. Quantification of total element concentrations in soils using total X-ray fluorescence spectroscopy (TXRF). *Sci. Total Environ.* 463–464: 374–388.

US EPA 2006. XRF Technologies for Measuring Trace Elements in Soil and Sediment Report EPA/540/R-06/005.

Table 1. ICP-OES and TXRF values for landfill leachates and groundwater samples (data from Cataldo, 2012).

	ICP-OES (µg/L)	TXRF (µg/L)	TXRF LLD (µg/L)
Landfill leachate	223	232	2,3
Groundwater	<1	12	0,65
Groundwater	<1	13	0,9

Arsenic Research and Global Sustainability – Bhattacharya, Vahter, Jarsjö, Kumpiene,
Ahmad, Sparrenbom, Jacks, Donselaar, Bundschuh & Naidu (Eds)
© *2016 Taylor & Francis Group, London, ISBN 978-1-138-02941-5*

Arsenic determination by anodic stripping voltammetry using graphene screen-printed electrode

C. Núñez & V. Arancibia
Chemistry Faculty, Pontificia Universidad Católica de Chile, Santiago, Chile

ABSTRACT: Based in the adsorption of As(III) onto graphene surface electrode, was developed a simple and sensitive method by Anodic Stripping Voltammetry (ASV). The sensitivity of the analytical method was 4.53 Na µg/L (RSD = 5%) with a detection limit of 0.25 µg/L. The lineal response was in the range 0.1 to 5.0 µg/L using buffer phosphate 0.01mol/L pH 6.5. Some common metals were evaluated and did not present significant interference at concentrations lower than 700 µg/L. The method was applied for determination of total arsenic in water samples from the Loa River in the North of Chile, obtaining satisfactory results (RE: 1.26%). Measurements did not require the used of hazardous solvent, offering a more rapid procedure with improved analytical characteristics in comparison with the current ones.

1 INTRODUCTION

The pollution of arsenic in natural water sources is a worldwide problem. The most common form of exposure is by consumption of contaminated drinking water, especially in rural and semi-urban areas where water irrigates food crops, or drinking water is often used without any treatment (Mandal *et al.*, 2002).

The North of Chile is an extremely arid zone; the only permanent superficial water is the Loa River where arsenic concentrations can reach 21800 µg/L (Smedley *et al.*, 2002). This source provides water for cities (tap water), mining activity, and agricultural use. The quality of water from the Loa River is poor due to high salinity and high dissolved arsenic and boron, because its main tributary (El Salado) has its origin in the geothermal field of El Tatio (Romero *et al.*, 2003). Therefore, arsenic detection is critical to the health care of the population in these areas.

Many methods have been developed in order to monitor and control arsenic concentrations. For this aim, the electroanalytical techniques have advantage above the traditional techniques, as spectrometry and chromatography, due to its sensitivity, low cost and direct applicability to determine As(III) and As(V) without previous separation.

On the other hand, graphene is a two-dimensional carbon nanomaterial with many distinctive properties that are electrochemically beneficial. Furthermore, it is used for the simultaneous removal of inorganic species, trivalent and pentavalent, in high concentration (Dongyun *et al.*, 2015). In the present work, graphene modified screen-printed was used in order to develop a method able to detect and to quantify arsenic in natural water samples without any pre-treatments.

2 EXPERIMENTAL

2.1 *Apparatus and equipment*

Anodic Stripping Voltammetry (ASV) measurements were made with a Metrohm 757 VA Computrace Stand with a graphene modified screen-printed electrode. The reference electrode was Ag/AgCl/KCl 3 mol/L, and the auxiliary electrode was a platinum wire. Solutions were stirred during accumulation steps. pH was measured with an Orion model 430 pH meter.

2.2 *Anodic stripping voltammograms*

Measurements were performed in 10 mL of phosphate buffer 0.1M pH 6.5. The calibration curves were obtained and linear regression and detection limits were calculated. In order to eliminate matrix effects the standard addition method was used. The first test to validate the method was carried by determining arsenic in spiked water and was applied to the determination of total arsenic in river water from the arsenic rich zone in the north of Chile (city of Calama). The measurements were performed subsequent to the addition of 80 µL 0.40 mol/L $Na_2S_2O_3$ solution for every 5.0 mL of the sample.

The Detection Limit (DL) was calculated from:

$$y_{DL} = a + 3\sigma_{x/y} \text{ and } y_{DL} = a + bx_{DL} \quad (1)$$

where a is the intercept, $\sigma_{x/y}$ is the random error in x and y, and b is the slope. The Quantitation Limit (QL) was calculated from:

$$y_{DL} = a + 10\sigma_{x/y} \quad (2)$$

3 RESULTS AND DISCUSSION

3.1 Operational parameters optimization

Initially Cyclic Voltammetry (CV) was employed to identify the signal peak corresponding to As(III). It can be observed a peak current in the oxidation zone increasing when As(III) is added. This suggests that a redox reaction occurs in the graphene electrode surface. In order to attain high a sensitivity and precision of the method, the major ASV parameters such as pH, buffer concentration, amplitude pulse, potential and time accumulation, were optimized. The results show that the As(III) peak decreased when the buffer concentration increased 0.01 mol/L. pH was studied in the range 6.0 to 8.0. It was not observed an important variation on arsenic the signal current. In the same way, were evaluated the effect to amplitude pulse, potential and time accumulation. Only the amplitude pulse showed a significant variation on the arsenic peak current increasing at higher voltage. However, at potentials higher that 200 mV, the signal widens thereby increasing the possibility of overlap with any interference. The optimum analytical conditions found for all parameters studied were: Buffer concentration 0.01 mol/L; pH 6.5 and pulse amplitude 200 mV.

3.2 Interference studied and validation method

The interference of some common metals such as Fe, Cu, Bi, Pb, Hg, Cd, Mn, Mg, Se and Cr was studied in terms of selectivity and competition. The selectivity decreases when the interfering concentrations are higher to 1000 µg/L. On the other hand, when the interference was studied in arsenic presence, the competition for active sites on the surface electrode increased when the concentrations of these metals were higher than 700 µg/L. The major interferences in both cases were Cu, Hg Bi and Se.

Under the optimum analytical conditions, calibration curves were obtained. The peak current is proportional to As(III) concentration from 0.1 to 5.0 µg/L, the DL was found to be 0.25 µg/L (Fig. 1). The correlation coefficient and slope were 0.9987 and 4.53 µAL/µg. The repeatability was studied making 20 successive measurements at the lowest level of the method. The calculated relative standard deviation was 5%. The usefulness of the present method was evaluated determining arsenic in water spiked with 50 µg/L of As(III). The value obtained was 50.64 ± 0.10 µg/L (RE 1.26%).

3.3 Real water samples

The proposed method was applied to the determination of total arsenic in surface waters. The sample was collected from the Loa River in the north of Chile. The obtained result was compared with measurement made by an external analytical laboratory (Table 1).

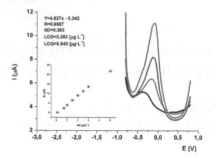

Figure 1. Anodic voltammograms as a function of As(III) concentration and calibration curve. Buffer 0.01M pH 6.5.

Table 1. Analysis of water samples by GPH-SPE Method.

Sample	GPH-Method µg/L	ICP-MS µg/L
Loa River	448 ± 0.22	510

*Loa River: sample collected from a horticultural crop irrigation area.

4 CONCLUSIONS

Graphene has been widely applied in electrochemical sensing due to its unique structure and exceptional chemical and physical properties. In this work, these characteristics were beneficial for arsenic determination. It was possible to develop a sensitive method, capable to obtain total arsenic from natural water sample without any pre-treatment. More studies are required to determine the robustness of the method. The positive results open the possibility of developments in this field.

ACKNOWLEDGEMENTS

Financial support by FONDECYT under Regular Project 1130081, CONICYT through a doctoral Scholarship N°21130492 and *Tesis en el sector productivo* N°7815120004 and Sponsorship Ecometales Limited.

REFERENCES

Mandal, B.K., Suzuki, K.T. 2002. Arsenic around the world: a review. *Talanta* 58(1): 13–16.

Smedley, P.L., Kinniburgh, D.G. 2002. A review of the source, behavior and distribution of arsenic in natural waters. *Appl. Geochem.* 17(5): 517–568.

Romero, L., Alonso, H., Campano, P., Fanfani, L., Cidu, R., Dadea, C., Keegan, T., Thornton, I., Farago, M. 2003. Arsenic enrichment in waters and sediments of the Rio Loa (Second Region, Chile). *Appl. Geochem.* 18(9): 1399–1416.

Dongyun, Z., Hui, H., Xiaojun, L., Shengshui, H. 2015. Application of graphene in electrochemical sensing. *Curr. Opin. Colloid In.* 20(5–6): 383–405.

Arsenic Research and Global Sustainability – Bhattacharya, Vahter, Jarsjö, Kumpiene,
Ahmad, Sparrenbom, Jacks, Donselaar, Bundschuh & Naidu (Eds)
© *2016 Taylor & Francis Group, London, ISBN 978-1-138-02941-5*

Stripping voltammetric determination of arsenic with a multi-walled carbon nanotube screen printed electrode modified with alginic acid from brown algae

V. Arancibia, C. Núñez, C. Muñoz & M. Valderrama
Chemistry Faculty, Pontificia Universidad Católica de Chile, Santiago, Chile

ABSTRACT: We report an electro-analytical methodology for determining As in natural waters using square wave voltammetry with a screen-printed electrode modified with carbon nanotubes and alginic acid from brown algae (CNTAa-SPE). The modified electrode was typically prepared by deposition of Alginic acid at −0.70 V by 900 s (C_{Aa} 2.5 mg/L). Variables like pH, accumulation potential (E_{acc}) and accumulation time (t_{acc}) were optimized. Under the best experimental conditions (pH 2.0; 0.01 mol/L HNO_3), the peak current is proportional to the As concentration up to 10.0 µg/L (R = 0.9961), with a 3σ detection limit of 0.68 µg/L and a sensitivity of 2.42 µA/ µg/L. The relative standard deviation for a As solution (5.0 µg/L) was 3.8% for six successive assays. Finally, the method was validated using synthetic drinking water spiked with As(III) and was applied to the determination of arsenic in water from the Loa River (North of Chile).

1 INTRODUCTION

Arsenic is one of the most toxic elements found in nature, and it constitutes one of the main concerns in relation to human health. Consumption of water with high concentrations of this nonmetal over an extended period of time causes serious diseases, including development of cardiovascular and peripheral vascular disease anomalies, lung fibrosis, hematological disorders and carcinoma (Ng *et al.*, 2003). Sensitive, selective and accurate methods are required for the determination of arsenic in biological and environmental matrices. Stripping voltammetry has important advantages that include speed, high selectivity and sensitivity, low detection limit, and it is suitable for the direct determination of As(III) without previous separation. To improve sensitivity in stripping techniques, in most cases, the electrode surfaces are modified by using carbon in nano-size, ligands or metals. The aim of this study was to optimize stripping voltammetric procedure for As determination using multi-walled carbon nanotube screen printed electrode modified with alginic acid (Aa). Alginate is a polysaccharide produced by some algae and is the common name given to a family of linear polysaccharides containing 1,4-linked β-D-mannuronic and α-L-guluronic acid residues. The ratio of mannuronic and guluronic groups and the affinity for metals ions vary with the type of algae. The advantage of using alginate for entrapment of arsenic is caused by environment provided by the gels for the entrapped material as well as their high porosity; similarly, multiwalled carbon nanotubes electrode was used to provide a more porous surface to facilitate adsorption of Aa and enhanced voltammetry sensitivity (Kauffmann *et al.*, 2015, Polyak *et al.*, 2004).

2 EXPERIMENTAL

2.1 *Apparatus and equipment*

The voltammograms were obtained on a BASi CV50 W with multiwalled carbon nanotube screen-printed electrodes (DRP-110CNT) which were purchased from DropSens (Spain). The reference electrode was Ag/AgCl/KCl 3 mol/L and the auxiliary electrode was a platinum wire externally connected. pH was measured with an Orion model 430 pH-meter.

2.2 *Chemicals and reagents*

Deionized water, HPLC quality, Merck, was used for sample preparation, dilution of the reagents, and rinsing purposes. Standard solutions of arsenic were prepared by diluting a commercial standard solution containing 1000 mg/L (2 wt.% HNO_3, Aldrich). Alginic acid sodium salt from brown algae (CAS 9005-38-3) was obtained from Sigma.

2.3 *Electrode preparation*

The multiwalled Carbon Nanotube Screen Printed Electrode (CNT-SPE) was transferred into the alginic acid solution (2.5 mg/L) and the Aa was deposited to −0.70 V by 15 min. The same electrode was used for a series of measurements (CNTAa-SPE).

2.4 Procedure for obtaining voltammograms

18.0 mL of water, 2.0 mL of 0.1 mol/L HNO_3, and different volumes of 1.0 mg/L As(III) solution were pipetted into the voltammetric cell. Then an accumulation potential (E_{acc}) by accumulation time (t_{acc}) was applied to the freshly prepare modified multiwalled carbon nanotube electrode while the solution was stirred (700 rpm). After a 10 s quiescent period, the potential was scanned between –0.80 and 0.80 V using square wave modulation with 4 mV step amplitude, 25 mV pulse amplitude, and a frequency of 25 Hz. All data were obtained at room temperature (~25°C).

3 RESULTS AND DISCUSSION

The carboxylic, sulfonic and hydroxyl groups presents in Aa interacts with a variety of metallic species which was exploited for As(III) determination by anodic stripping voltammetry at ex situ modified electrodes. Figure 1 shows anodic stripping voltammo-grams for As solution in 0.01 mol/L HNO_3 using a CNT-SPE with and without Aa. Square Wave Anodic Stripping Voltammetry (SWASV) using this modified electrode showed one anodic wave for the oxidation of Aa at 0.03 V and a small well-resolved peak at –0.10 V due probably to the oxidation of As-Aa complex.

3.1 Effect of pH

The first step of this study was to find out the influ-ence of pH on the peak current in the 2.0 to 10.0 range using Britton Robinson buffers. However, it was found that at acid pH adjusted with nitric acid peak current of As-Aa had adequate values.

3.2 Effect of E_{acc} and t_{acc}

The influence of the variation of the accumulation potential on the peak current (As-Aa and free Aa) was examined over the 1.00 to –1.00 V range (30 and 60 s). Both signals are almost constant over the whole potential range. In addition subsequent experiments were performed without accumulation time in order to avoid extending the time of analysis.

3.3 Linear range, detection limit, and reproducibility of the method

Optimal analytical conditions were found to be pH 2.0 (0.01 mol/L HNO_3), step amplitude 4 mV; pulse amplitude 25 mV; and frequency 25 Hz. Under these conditions the peak current was proportional to the concentration of As over the 0.0–10.0 µg/L range (R: 0,9961), with a 3σ detection limit of 0.68 µg/L and a sensitivity of 2.42 µA/ µg/L without an accumulation potential and time. Reproducibility for a 5.0 µg/L As(III) solution was 3.8% (n = 6). On the other hand, Cu(II), Pb(II) and Cd(II) form

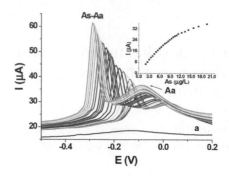

Figure 1. Voltammograms and calibration curves in function of As(III) concentration. Conditions: pH = 2.0 (0.01 mol/L HNO_3). Screen-printed electrode modified with carbon nanotubes and Aa from brown algae (CNTAa-SPE). Curve a: multi-walled carbon nanotube screen printed electrode without Aa.

complexes with Aa which present electrochemical reactions at potentials more negative, so they did not interfere with the determination of As. The usefulness of the present method was also evaluated by determining As in spiked drinking water (5.0 µg/L) and the value obtained was 4.8 ± 0.3 µg/L. Figure 1 illustrates the calibration curve and voltammo-grams as a function of As(III) concentration.

4 CONCLUSIONS

In the present paper a novel arsenic determination was carried out forming an As-Aa complex which is adsorbed on a multi multiwalled carbon nanotube screen printed electrode modified with Aa. The presence of Aa allowed the adsorption of the com-plex on the electrode. The advantages of this new procedure are fast complexation kinetics, high sensi-tivity, simplicity and speed. The method can be ap-plied to both natural water and biological samples where arsenic concentration is higher than 0.7 µg/L.

ACKNOWLEDGEMENTS

The authors acknowledge with thanks the financial support of FONDECYT under Project 1130081.

REFERENCES

Ng, J.C., Wang, J., Shraim A. 2003. A global health problem caused by arsenic from natural sources. *Chemosphere* 52(9): 1353–1359.

Polyak, B., Geresh, S., Marks, R.S. 2004. Synthesis and characterization of abiotin–alginate conjugate and its application in a biosensor construction. *Biomacromolecules* 5: 389–396.

Sakira, A.K., Issa, T., Ziemons, E., Dejaegher, B., Mertens, DHubert, P.; Kauffmann, J.M. 2015. Determination of arsenic(iii) at a nanogold modified solid carbon paste electrode. *Electroanalysis* 27(2): 309–316.

Arsenic Research and Global Sustainability – Bhattacharya, Vahter, Jarsjö, Kumpiene,
Ahmad, Sparrenbom, Jacks, Donselaar, Bundschuh & Naidu (Eds)
© 2016 Taylor & Francis Group, London, ISBN 978-1-138-02941-5

A simple ultrasound-assisted method for the determination of total arsenic in rice using Hydride Generation Atomic Absorption Spectrometry (HG-AAS)

M. Pistón[1], I. Machado[1], M. Belluzzi Muiños[1], F.C. Pinheiro[2] & J.A. Nóbrega[2]

[1]Analytical Chemistry. DEC, Faculty of Chemistry. Universidad de la República (UdelaR), Montevideo, Uruguay
[2]Group of Applied Instrumental Analysis, Department of Chemistry. Federal University of São Carlos. São Carlos, SP, Brazil

ABSTRACT: Uruguay is among the top eight exporters of rice in the world. Assessing the levels of Arsenic (As) in rice grain is a concern for human health. To ensure food safety, the laboratories need simple and rapid validated analytical methodologies to provide fast response, to allow the commercialization of the products. A simple ultrasound-assisted method for sample preparation and determination of total arsenic using Hydride Generation Atomic Absorption Spectrometry was validated. The figures of merit were adequate for control purposes in terms of precision and trueness. Commercial samples of Uruguayan rice were analyzed and the results did not differ significantly with those obtained using ICP-MS. The total arsenic concentrations ranged from 0.10 to 0.24 mg kg^{-1}. These values were below the maximum limit established in the region. The proposed methodology is simple, economic and does not require drastic conditions of sample preparation or expensive techniques.

1 INTRODUCTION

The levels of contaminants in food are controlled by the sanitary authorities in all the countries. In the MERCOSUR region there is a regulation that indicates that the maximum limit of total arsenic (As) in rice is 0.3 mg/kg. Previous studies in Uruguayan aquifers revealed that the levels of total As are of concern. Thus the determination of As in rice is very important to assess the situation in our country. A simple ultrasound-assisted method for the determination of total As in rice using Hydride Generation Atomic Absorption Spectrometry (HG-AAS) was validated. This method can be postulated as an alternative to expensive techniques such as ICP-MS not available in many laboratories in developing countries.

2 EXPERIMENTAL

2.1 Experimental procedure for HG-AAS determinations

All chemicals used were of analytical reagent grade and all the solutions were prepared with ultrapure water of 18.2 MΩ cm resistivity (ASTM Type I). All glassware was soaked overnight in 10% (v/v) HNO3 before use. HCl (37% v/v, Merck) was used for sample preparation. A commercial standard solution (1000 mg/L) of As(V) (Merck) was used for preparation of analytical calibration solutions.

An ultrasonic homogenizer (VC505, Sonics) 750 W; 20 kHz; 23 VAC equipped with a 13 mm titanium alloy probe was employed. For analytical determinations by HG-AAS, a flame atomic absorption spectrometer (AAnalyst 200, Perkin Elmer) coupled to a MHS system (PerkinElmer) was used. Operating conditions were: wavelength 193.7 nm and acetylene-air flame. Nitrogen (99.99%) was used as carrier gas at a flow rate of 60 mL min^{-1}. A quartz T-tube cell with a pathlength of 165 mm and a diameter of 12 mm was heated to approximately 900°C. Five of the more popular brands of rice in Uruguay were analyzed.

Sample preparation: 0.25 g of sample was accurately weighted in a glass vessel and then 10.0 mL of HCl 50% (v/v) were added. The probe was immersed in this suspension for 30 min (35% sonication amplitude). The slurries obtained were then centrifuged at 3700 rpm for 3 min and 5.0 mL of the supernatant were treated with 0.5 mL of 20% (w/v) KI as pre-reductant for 1 h. Then the solution was put into the reaction vessel of the instrument where NaBH4 2% (w/v) (prepared in 1% (w/v) NaOH) was added to form the hydride. Three drops of octanol were added to each sample to avoid foaming. Reagent blanks were also run.

2.2 Experimental procedure for ICP-MS determinations

All solutions were prepared using HNO$_3$ (Merck, Germany) distilled in a sub-boiling apparatus (Milestone, Italy) and ultrapure water obtained from a Milli-Q system (Millipore, USA). Five of the more popular brands of rice in Uruguay were analyzed using an inductively coupled plasma quadrupole mass spectrometer iCAP Q-ICP-MS (Thermo Fisher Scientific, USA) operated in standard mode. The plasma was operated at a RF applied power of 1.55 kW, with a plasma gas, auxiliary gas and nebulizer gas flow rates of 14, 0.8 and 1.05 L/min, respectively. Approximately 0.200 g of sample in triplicate were digested in 5 mL of HNO_3 2 mol/L and 3 mL of H_2O_2 30% m/m using a microwave-assisted heating program (8 min ramp, 15 min. plateau at 155°C; 4 min ramp, 30 min. plateau at 210°C) with closed flasks (Four Wave Speed, Berghof Analytik, Germany) and diluted to a final volume of 50 mL with distilled-deionized water. Later on, digests were diluted and spiked to give added concentrations from 0.05 to 0.2 μg/L As for application of the standard additions method for calibration. Final acid concentration was 0.014 mol/L HNO_3.

3 RESULTS AND DISCUSSION

3.1 Ultrasound-assisted method and determination using HG-AAS (US/HG-AAS)

The main figures of merit are shown in Table 1.

These parameters are adequate for food safety control considering the regional regulations.

3.2 Microwave-assisted method and determination using ICP-MS (MW/ICP-MS)

The efficiency of microwave-assisted digestion of organic samples using diluted nitric acid solutions and hydrogen peroxide were discussed by Bizzi et al. (2011). It was demonstrated that the temperature gradient established in closed vessels promotes the regeneration of nitric acid in the presence of hydrogen peroxide, which acts as a source of oxygen, thus enabling the decomposition of organic

Table 1. Main figures of merit.

Parameter	US/HG-AAS
Linearity (μg/L)	Up to 10.0
LOD (3σ, n = 10) (mg/kg)	0.015
LOQ (10σ, n = 10) (mg/kg)	0.045
Precision (RSD, n = 6)	≤ 10%
Trueness (compared to certified value (n = 6)*	95.5

*Certified Reference Material NIST 1568a Rice Flour.

Table 2. A comparison of arsenic concentration in Uruguayan rice measured by US/HG-AAS and MW/ICP-MS.

Sample	As (mg/kg) US/HG-AAS	As (mg/kg) MW/ICP-MS
Brand A	0.209 ± 0.017	0.236 ± 0.048
Brand B	0.173 ± 0.014	0.200 ± 0.024
Brand C	0.104 ± 0.008	0.101 ± 0.005
Brand D	0.160 ± 0.021	0.205 ± 0.022
Brand E	0.233 ± 0.019	0.221 ± 0.024

(mg kg^{-1}, mean ± standard deviation).

samples at low acid concentrations. The detection limits were calculated as three times the standard deviation of ten analytical blanks (authentic replicates) divided by the slope of the analytical calibration curve obtained for each replicated digested sample. Thus, LOD was expressed as a range (0.006–0.008 mg/kg). Recovery was 98.6% compared with the Certified Reference Material NIST 1568a Rice Flour. The same Uruguayan commercial samples were analyzed in Uruguay and Brazil. Results are shown in Table 2.

Results obtained using both methodologies are in agreement. The levels of total As in all the samples were below the maximum level admitted by the local regulation.

4 CONCLUSIONS

A simple ultrasound-assisted procedure was validated for the determination of total As in rice. This procedure was applied in Uruguayan commercial samples and results were compared with those obtained using ICP-MS. The proposed method resulted adequate as an alternative to expensive techniques still not frequently available in developing countries.

ACKNOWLEDGEMENTS

CSIC for financial support, Thermo Scientific and Analitica for technical support.

REFERENCE

Bizzi, C.A., Barin, J.S., Müller, E.I., Schmidt, L., Nóbrega, J. A.; Flores, E.M. M. 2011. Evaluation of oxygen pressurized microwave-assisted digestion of botanical materials using diluted nitric acid. Talanta 83(5):1324–1328.

Arsenic Research and Global Sustainability – Bhattacharya, Vahter, Jarsjö, Kumpiene, Ahmad, Sparrenbom, Jacks, Donselaar, Bundschuh & Naidu (Eds)
© 2016 Taylor & Francis Group, London, ISBN 978-1-138-02941-5

Evaluation of ultrasound-assisted methods for sample preparation for the determination of total arsenic in globe artichoke (*Cynara cardunculus* L. subsp. *Cardunculus*)

I. Machado[1], E. Rodríguez Arce[1], M. Pistón[1] & M.V. Cesio[2]

[1]*Analytical Chemistry. DEC, Faculty of Chemistry. Universidad de la República (UdelaR), Montevideo, Uruguay.*
[2]*Pharmacognosy and Natural Products. DQO, Faculty of Chemistry. Universidad de la República (UdelaR), Montevideo, Uruguay*

ABSTRACT: The aim of the present study was to evaluate the food safety of globe artichoke crops, in terms of total arsenic content, using simple and rapid analytical methodologies. For this purpose, two ultrasound-assisted methods followed by Hydride Generation Atomic Absorption Spectrometry (HG-AAS) were evaluated and compared with a microwave-assisted digestion followed by Electrothermal Atomic Absorption Spectrometry (ET-AAS). The proposed methods showed to be adequate for the purpose of control of this contaminant according to the regional regulations and in good agreement with the principles of the Green Chemistry. The levels of total arsenic in all analyzed samples were below the maximum level admitted.

1 INTRODUCTION

Globe artichoke is a well known plant in South America, not only as a healthy food (fruits), but also as raw material used in herbal medicine products to relieve indigestion and as hepatoprotective (leaves) (Pistón *et al.*, 2014). In the MERCOSUR region there is a regulation that indicates that the maximum limits of total Arsenic (As) in infusion and edible vegetables are 0.6 and 0.3 mg/kg respectively (MERCOSUR, 2015).

The aim of the present study is to evaluate the food safety of these crops, in terms of total arsenic content, using simple and rapid analytical methodologies. For this purpose, two simple ultrasound-assisted methods for sample preparation followed by HG-AAS and a microwave-assisted digestion followed by ET-AAS were compared. Samples from a family of farmers that are dedicated to grow crops of globe artichokes where analyzed in order to make an assessment.

2 EXPERIMENTAL

2.1 *Instrumentation*

An ultrasonic homogenizer (VC505, Sonics) 750 W; 20 kHz; 230 VAC equipped with a 13 mm titanium alloy probe and an ultrasonic bath (8893, Cole Parmer) 469 W; 47 kHz were used for ultrasound-assisted methods. For total digestion procedure a Microwave oven (Mars6, CEM) 400–1800 W provided with 12 EasyPrep™ vessels was employed.

For analytical determinations by HG-AAS, a flame atomic absorption spectrometer (AAnalyst 200, Perkin Elmer) coupled to a MHS system (Perkin Elmer) was used. Operating conditions were: wavelength 193.7 nm, lamp current 400 mA, slit width 0.5 nm and acetylene-air flame. Nitrogen (99.99%) was used as carrier gas, with a flow rate of 60 mL/min. A quartz T-tube cell with a path-length of 165 mm and a diameter of 12 mm, was heated to approximately 900°C. For those determinations using ET-AAS, an atomic absorption spectrometer (iCE 3500, Thermo) equipped with a graphite furnace atomizer, auto-sampler and Zeeman-based correction was employed. Argon (99.99%) was used as purge gas. Operating conditions were: wavelength 193.7 nm, lamp current 10 mA and spectral resolution 0.5 nm. Extended life graphite tubes treated with permanent modifier were used.

2.2 *Reagents*

All chemicals used were of analytical reagent grade and all the solutions were prepared with ultrapure water of 18.2 MΩcm resistivity (ASTM Type I).

All glassware was soaked overnight in 10% (v/v) HNO_3 before use. HNO_3 (65% v/v, Merck) and HCl (37% v/v, Merck) were used for sample preparation. A commercial standard solution (1000 mg/L) of As(V) (Merck) was used for the calibration curves.

2.3 *Samples*

Globe artichoke leaves (2 kg) and fruits (10 kg) were collected in Montevideo, Uruguay.

The fragments of leaves were identified as *Cynara cardunculus* subsp. *Cardunculus*. Samples were chopped and dried in an oven with forced air circulation (70°C) and stored at 20°C, at light-free conditions.

2.4 *Ultrasound-assisted methods: using a probe and a bath*

For both methods 0.5 g of sample were accurately weighted in a glass vessel and then 10.0 mL of HCl 50% (v/v) were added. The probe was immersed in this suspension for 15 min (35% sonication amplitude); when an ultrasonic bath was employed, the vessel was put into the bath for 90 min. The slurries obtained by both methods were then centrifuged at 3700 rpm for 3 min. Reagent blanks were also run.

2.5 *Microwave-assisted method*

For this method 0.5 g of sample were accurately weighted, and 10.0 mL of HNO_3 were added into the EasyPrep™ vessel. The vessel was sealed with the screw cap and placed inside the microwave oven: power 400–1800 W, 20 min ramp time until 200°C, 10 min hold at 200°C, 800 psi pressure. Reagent blanks were also run.

2.6 *Total arsenic determination*

HG-AAS: 5.0 mL of the solutions obtained in 2.4 were treated with 0.5 mL of 20% (w/v) KI as pre-reductant for 1 hour. Then the solution was put into the reaction vessel of the instrument where 2% (w/v) $NaBH_4$ (prepared in 1% (w/v) NaOH) was added to form the hydride. Three drops of octanol were added to each sample to avoid foaming.

ET-AAS: Graphite tubes were treated with niobium by pipetting 50 µL of a 1000 mg/L $Nb(NO_3)_5$ solution and submitting the tube to the following temperature program: [temperature/ramp time/hold time]: drying (100°C/10 s/60 s), atomization (2700°C/0 s/5 s). The entire procedure was repeated 6 times in order to obtain a 300 µg of modifier on the tube. The temperature program for As determination was: drying (100 °C/10 s/30 s), pyrolysis (1200°C/15 s/15 s), atomization (2200°C/0 s/3 s), cleaning (2600°C/0 s/3 s). Argon flow of 0.2 mL/min was used, except during atomization step. Injection volume was 30 µL. Solutions obtained in 2.5 were directly injected.

3 RESULTS AND DISCUSSION

The main features of merit for both analytical techniques are shown in Table 1.

Table 2 shows the results obtained using the three procedures for samples of artichoke leaves and fruits from the family crop. These results show that the methods are comparable.

The ultrasound-assisted methods where simple and relatively fast and can be postulated as an alternative

Table 1. Main figures of merit obtained.

Parameter	HG-AAS	ET-AAS
Linearity (µg/L)	Up to 10.0	Up to 20.0
LOD (3σ, n = 10) (mg/kg)	0.007	0.007
LOQ (10σ, n = 10) (mg/kg)	0.022	0.024
Precision (RSD, n = 6)	≤ 10%	≤ 10%
Trueness (compared to certified value, n = 6)*	98%	97%

*CRM NIST 1570a Trace Elements in Spinach Leaves.

Table 2. Comparison of the sample treatments.

| Sample | As (mg/kg) | | |
	US-probe	US-bath	MW
Leaves	0.065 ± 0.007	0.062 ± 0.005	0.064 ± 0.005
Fruits	0.025 ± 0.002	0.025 ± 0.003	0.026 ± 0.004

Results expressed as: mean value ± standard deviation, n = 6. US: ultrasound

to those classical methods involving total digestion. All the obtained concentrations were below the maximum admitted values for this kind of food.

4 CONCLUSIONS

Ultrasound-assisted methods show to be simple and in better agreement with the principles of the Green Chemistry since they utilize diluted acid. These procedures showed to be adequate for the purpose of control of this contaminant, according to the regional regulations.

ACKNOWLEDGEMENTS

We thank Bianco family for providing samples. The research that gives rise to the results, received funding from Agencia Nacional de Investigación e Innovación (ANII POS_NAC_2013_1_11407), PEDECIBA-Química, and Comisión Sectorial de Investigación Científica (CSIC-UdelaR).

REFERENCES

MERCOSUR. 2015. "Reglamento técnico MERCO-SUR" (Technical Regulation) MERCOSUR/GMC/RES. N° 12/11 http://www.puntofocal.gov.ar/doc/ (Accessed 18-10-2015).

Pistón, M., Machado, I., Branco, C.S., Cesio, V., Heinzen, H., Ribeiro, D., Fernandes, E. Campos Chisté, R. & Freitas, M. 2014. Infusion, decoction and hydroalcoholic extracts of leaves from artichoke (Cynara cardunculus L. subsp. cardunculus) are effective scavengers of physiologically relevant ROS and RNS. *Food Res. Int.* 64: 150–156.

Arsenic Research and Global Sustainability – Bhattacharya, Vahter, Jarsjö, Kumpiene,
Ahmad, Sparrenbom, Jacks, Donselaar, Bundschuh & Naidu (Eds)
© 2016 Taylor & Francis Group, London, ISBN 978-1-138-02941-5

Waterbox & AQUA-CHECK—A newly developed package for drinking water supply and monitoring

K. Siegfried[1,2], S. Hahn-Tomer[1,2] & S. Heuser[2]
[1]*Aquacheck GmbH, Hof, Germany*
[2]*Helmholtz Centre for Environmental Research–UFZ, Leipzig, Germany*

ABSTRACT: The Waterbox and the AQUA-CHECK multi parameter field test kit are developed as a comprehensive and transportable drinking water supply system. The Waterbox was developed as a compact and portable drinking water treatment facility. It contains compartments for particle filtration, precipitation and disinfection. AQUA-CHECK 2 incorporates a handheld photometric device for measurement of multiple water quality parameters. The newly developed AQUA-CHECK 3 also comprises a luminometer for the measurement of arsenic with the ARSOlux biosensor. Moreover, this device will be equipped with GPS sensors and software for data storage and transfer to web data base systems. The Waterbox—AQUA-CHECK package could be a very practical, efficient and cost-effective combination to monitor and treat decentralized drinking water supply in remote and less developed areas of countries in Asia, Latin America and Africa. The performances of the components of the Waterbox—AQUA-CHECK setup were tested in several research trials in India, Mongolia, Mexico and Argentina.

1 INTRODUCTION

Drinking water pollution problems caused by growing populations, improper treatment of industrial and communal waste and catastrophic events are increasingly contaminating large areas and endangering the environment and health of millions of people (Guo *et al.*, 2007, Klebercz *et al.*, 2012). The most recent catastrophic discharge of highly contaminated waste waters and mud of the Samarco iron ore mine into the Rio Doce river in Brazil (Eisenhammer, 2015) shows that heavy metal and arsenic pollution is no less a problem today than in the past.

2 MATERIALS & METHODS

2.1 *The Waterbox*

The Lavaris Waterbox (Lavaris Technologies GmbH, Hof, Germany), a compact and space-saving construction, which can be purchased as a stand-alone unit at a cost of approximately 20.000 € or a networked operation combined with several units, eliminates hazardous metals and actively influences the lime-carbonic acid system. Moreover, a neutralization of corrosion-chemical parameters takes place. Additionally, the neutral pH-value is restored. Fresh water is transformed into drinking water with a performance of 20 m^3 per day (at 20 h/d) corresponding to the World Health Organization and the EU-Water Guideline 98/83.

CarbonAdd® (Lavaris Technologies GmbH, Hof, Germany) is a mixed, inorganic powder consisting of calcium carbonate, calcium chloride, sodium hydrogen carbonate and sodium carbonate. The most important applications are the treatment of potable water, the remineralization of desalinated sea water and the treatment of surface water. By dosing substances such as calcium carbonate as well as hydrogen carbonate in different kinds of soluble salts, the carbonic acid system of the raw water will be influenced actively. Deacidification, increasing hardness, remineralization and interactions between the water and the piping materials are processes that can be independently adjusted by the use of CarbonAdd®. The enumerated processes take part in one treatment step (co-precipitation). For heavily contaminated waters the product IPTEC which is a mixture of calcium peroxide, calcium chloride and calcium dihydroxide is used in the Lavaris water treatment systems. The Waterbox also contains components for particle filtration, UV disinfection and activated carbon filtration. The system can provide pure drinking water within three minutes and is utilized in the Lavaris Waterbox as well as the Lavaris mobile treatment plants for larger applications, which are able to treat up to 13,000 l/h.

2.2 *The AQUA-CHECK field test kit*

In order to guarantee that all water sources treated by Lavaris Technologies are safe and sustainable, the AQUA-CHECK 2 was developed. This test kit enables the user to measure indicators such as fluoride, ammonia, nitrite, nitrate, phosphate, chlorine, silicon, copper, iron, oxygen and pH at a cost of 840 €. The AQUA-CHECK 3, which is currently under final development, will be a comprehensive, highly sensitive, professional water analysis kit with maximum operational comfort. A mobile handheld photometer and luminometer are brought together in one interchangeable device. The luminometer is used for the measurement of arsenic with the ARSOlux biosensor (Harms *et al.*, 2006, Siegfried *et al.*, 2012, 2015) and it is also going to be applicable for observation of unspecific toxicity.

A built in GPS and camera enable the user to easily locate and record the location of the water sample taken and the USB port guarantees a simple transfer of all data to electronic devices and a web data base. The analytical tool operates independently on a battery system. The functionality of AQUA-CHECK 3 was patented internationally and the development was guided by the work regulations for water analysis as postulated by the German "unified methods for testing water, waste water and sludge".

3 RESULTS AND DISCUSSION

3.1 Waterbox

Currently the Waterbox is operating in Germany, Mexico, Kenya, India, China and Nepal. Results showing the efficiency of the treatment system are presented in Table 1. The system can be optionally equipped with a solar panel for energy independent operations.

3.2 AQUA-CHECK

The AQUA-CHECK 2 multi parameter test kit has been used in Germany, Mongolia and India. Research centers, universities and industries have especially shown interest internationally, where remote water analysis often poses severe challenges. The data shown below are taken from field study sites in India (Table 2) and Argentina (Fig. 1).

Table 1. Raw water (processed water) quality data and data of treated water (IPTEC).

Parameter	Raw water	Treated water mg/L	Threshold*
As	0.023	0.001	0.01
Cd	0.015	0.0005	0.003
Cr	0.27	0.042	0.05
Cu	0.95	0.03	2.0
Zn	11	0.024	–
COD	61,200	926	–
PO_4^{3-}	32	<1.0	–
S^{2-}	50	0.02	–

*Drinking Water Ordinance Germany

Table 2. Results of the AQUA-CHECK 2 test kit and the TWAD (Tamil Nadu Water Supply And Drainage Board) test of water samples from Murshidabad, India.

Parameter	AQUA-CHECK mg/l	TWAD Kit
DO	3.2	n/a
pH	8	8
F^-	0.37	0
Fe	0.74	0.3
NH_4^+	0.05	0.5
NO_2^-	0.7	0.5
NO_3^-	30	45

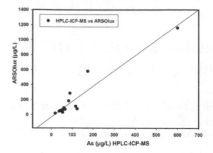

Figure 1. Results of the ARSOlux Biosensor and ICP-MS analysis of arsenic contaminated samples from Rafaela, Santa Fe, Argentina (Siegfried et al., 2015).

4 CONCLUSIONS

The results of first small scale pilot tests of the Waterbox and AQUA-CHECK combination, demonstrated a very good performance of the technologies under difficult conditions in remote areas. Further extensive field campaigns in contaminated areas in Germany and abroad will be executed to show the performance at larger scales.

ACKNOWLEDGEMENTS

This research was funded and supported by the Lavaris Technologies GmbH, Hof and the Helmholtz Centre for Environmental Research, Leipzig, Germany.

REFERENCES

Eisenhammer, S. 2015. Brazil mining flood could devastate environment for years. Reuters. 15.11.2015. Download at: http://www.reuters.com/article/2015/11/15/us-brazil-damburst-environment-idUSKCN0T40PY20151115#FrFPJzPft1hE2zJT.97 (Accessed: 30.11.2015).

Guo, J.X., Hu, L., Yand, P.Z., Tanabe, K., Miyatalre, M., Chen, Y. 2007. Chronic arsenic poisoning in drinking water in Inner Mongolia and its associated health effects. J. Environ. Sci. Health A 42(12):1853–1858.

Harms, H., Wells, M., Van der Meer, J.R. 2006. Whole-cell living biosensors – are they ready for environmental application? Appl. Microbiol. Biot. 70: 273–280.

Klebercz, O., Mayes, W.M., Anton, A.D., Feigl, V., Jarvis, A.P., Gruiz, K. 2012. Ecotoxicity of fluvial sediments downstream of the ajka red mud spill, Hungary. J. Environ. Monitor. 14: 2063–2071.

Siegfried, K., Endes, C., Bhuiyan, A.F., Kuppardt, A., Mattusch, J., van der Meer, J.R., Chatzinotas, A., Harms, H. 2012. Field testing of arsenic in groundwater samples of Bangladesh using a test kit based on lyophilized bioreporter bacteria. Environ. Sci. Technol. 46: 3281–3287.

Siegfried, K., Hahn-Tomer, S., Koelsch, A., Osterwalder, E., Mattusch, J., Staerk, H.-J., Meichtry, J.M., De Seta, G.E., Reina, F.D., Panigatti, C., Litter, M.I., Harms, H. 2015. Introducing simple detection of bioavailable arsenic at rafaela (Santa Fe Province, Argentina) using the ARSOlux biosensor. Int. J. Environ. Res. Pub. Health 12: 5465–5482.

Gold nanoparticles-based chemical sensor for on-site colorimetric detection of arsenic in water samples

K. Shrivas

Department of Chemistry, Guru Ghasidas Vishwavidyalaya, Bilaspur, Chhattisgarh, India

ABSTRACT: We report the Lauryl Sulphate (LS) modified gold nanoparticles (AuNPs) as a Localized Surface Plasmon Resonance (LSPR) based chemical sensor for the colorimetric detection of arsenic in water samples. This colorimetric LSPR based detection of arsenic found to be a simple, selective, sensitive, and can be applied at the sample source. The method was based on the color change of AuNPs from pink to blue with the addition of arsenic to NPs that caused the shift in LSPR band due to the inter-particle coupling effect. The calibration curve was linear over 5–500 μgL^{-1} arsenic with Limit Of Detection (LOD) of 2 $\mu g/L$ and correlation estimation (r^2) of 0.994. The concentration of arsenic found in water samples of central India was in the range 15 to 350 μgL^{-1} which was found higher than WHO Tolerance Limit Value of 10 μgL^{-1} (TLV).

1 INTRODUCTION

Arsenic (As) is a toxic and ubiquitous element presents in environmental samples such as soil, sediment, water, aerosol, rain, aquatics and vegetations. There are four oxidation states (–3, +3, 0, and +5) of arsenic that can exist in the nature and As^{3+} is more toxic than As^{5+}. In India, ground waters of West Bengal and Chhattisgarh states are most commonly contaminated with arsenic that exceeded the World Health Organization (WHO) guidelines for drinking water (10 $\mu g/L$). In recent years, metal Nanoparticles (NPs) based colorimetric sensors have drawn tremendous attention over conventional colorimetric methods for the determination of metal ions and organic molecules due to the optical, electronic and chemical properties of NPs. In the present work, Lauryl Sulphate (LS) modified AuNPs was used as a chemical sensor for on-site colorimetric detection of arsenic in real environmental samples.

2 METHODS/EXPERIMENTAL

2.1 *Synthesis of gold NPs*

AuNPs modified with Lauryl Sulphate (LS) was prepared by reducing the salts of gold ($HAuCl_4$) with $NaBH_4$. Briefly, 100 mL of 0.25×10^{-3} M solution of $HAuCl_4$ was mixed with 0.5 mL of 0.1 M LS and stirred for 20 min. After, 1 mL of 0.1 M $NaBH_4$ was added to the solution mixture and stirred for 30 min. The color of the solution was changed from yellow to pink showing the formation of AuNPs.

2.2 *Apparatus*

A VIS-spectrophotometer type-1800 (Shimadzu, Japan) matched with 1 cm quartz cell was used for the absorbance measurement. The computer software UV probe was used to perform the absorbance measurement.

2.3 *Procedure for AuNPs based LSPR colorimetric method for detection of arsenic*

1 mL aqueous sample containing arsenic was taken in to a 5-mL of glass bottle followed by the addition of same volume of AuNPs, and then the solution mixture was kept for prescribed time at room temperature (25 + 2)°C while maintaining the pH of the solution. The color of the solution mixture was changed from pink to blue that can be observed by our naked eyes. The color intensity of the solution was measured with UV-visible spectrophotometer. Similarly, the collected filtered sample with pH 5.0 was taken in to a glass bottle and 0.1 mL of ascorbic acid (1%) was added to convert the arsenate to arsenite present in the sample. Then, equal volume of AuNPs was added, the color change of solution mixture was obtained followed by the measurement using UV-visible spectrophotometer.

3 RESULTS AND DISCUSSION

3.1 *AuNPs as a LSPR based chemical sensor for colorimetric detection of arsenic*

The metal ions such as K^+, Na^+, Mg^{2+}, Ca^{2+}, Ba^{2+}, Cu^{2+}, Ni^{2+}, Zn^{2+}, Al^{3+}, Co^{3+}, Cr^{3+}, Fe^{3+}, Pb^{2+}, As^{3+}, Hg^{2+} and Cd^{2+} (5 mg/L) were tested for the color change of AuNPs by adding 1:1 ratio of metal ions to AuNPs in to a glass vial. Only the solution of arsenic with NPs showed the color change from pink to blue and other metal ions did not show the color change, shown in Fig. 1A. The UV-visible spectra of AuNPs with different types of metal ions are shown in Fig. 1B.

The color change from pink to blue color was due to the aggregation of NPs after the addition of As(III) as As_2O_3 to the solution mixture. The addition of As_2O_3 to solution mixture causing the replacement of LS capped group from the surface of AuNPs that caused the aggregation followed by the color change due to instability of NPs.

3.2 Optimization of AuNPs-based chemical sensor for colorimetric detection of arsenic

The AuNPs based colorimetric method was optimized in order to obtain the best experimental conditions for the detection of arsenic from water sample. The LSPR band obtained at wavelength of 530 nm and 730 nm was used for the optimization of the present method. The best performance for the detection of arsenic was obtained when the pH of the solution was 5.0, concentration of NPs was 25 μM and the reaction time was 5 min.

3.3 Linear range, limit of detection and precision using AuNPs—based chemical sensor for colorimetric detection of arsenic

The linearity range of arsenic was calculated based on the ratio of maximum absorbance at 730 nm to 530 nm and the linear range obtained was in the range of 5–500 μg/L with a correlation of estimation (r^2) of 0.994. LOD was evaluated based on the minimum quantity of arsenic that could give a change in the absorbance value at three time of standard deviation. The value of LOD obtained was 2 μg/L. Intra-day precision was calculated based on the analysis of 100 μg/L of arsenic in triplicate analyses (n = 3) on same day and found to be 3.5–5.6%. Inter-day precision determined on

Table 1. Determination of arsenic (μg/L) using AuNPs based LSPR colorimetric method from water samples of central India.

Water samples	Arsenic (μg/L), AuNPs-based colorimetric method	Arsenic (μg/L) HG-AAS method
Tube well-1	25	23
Tube well-2	48	43
Well-1	150	157
Well-2	350	339
Pond-1	50	53
Pond-2	65	68

three consecutive days for analyzing the same concentration of arsenic was found to be 3.8–6.6%.

3.4 Application for AuNPs as a localized surface plasmon resonance (LSPR) based chemical sensor for colorimetric detection of arsenic

The feasibility of the proposed method was assessed by analyzing real water samples (tube well, well and pond) from Ambagarh Chowki, Central India, where water, soil and sediment samples are highly contaminated with arsenic. In the present procedure, AuNPs based colorimetric method was used for the detection of arsenic in contaminated water samples. The concentration of arsenic in tube well, well and pond water samples was found in the range 15–65, 105–350 and 50–108 μg/L, respectively, as given in Table 1. The average concentration of arsenic in well water sample was found to be higher than pond and tube well water samples. The reason for higher concentration of arsenic found in well water sample is probably due to the leaching of more arsenic from underground rocks and also has a larger surface area compared to tube well. The results obtained by AuNPs based colorimetric method were also validated with results of HG-AAS and the concentrations of arsenic were found to be in good agreement.

ACKNOWLEDGEMENTS

We would like to thank the Department of Science Technology, New Delhi for awarding Kamlesh Shrivas a fast track project (NO.SB/FT/CS-128/2012).

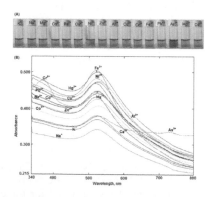

Figure 1. (A) Photographs of glass vial containing solution mixture of different metal ions and AuNPs and (B) UV-visible spectra of AuNPs (25 μM) with 500 μg/L of different concentration of metal ions.

REFERENCES

Shrivas, K., Patel, K.S. 2004. On site deter mination of arsenic in contaminated water samples. *Anal. Lett.* 37: 333–344.
Motl, N.E., A.F. Smith, DeSantisa, C.J. 2014. Engineering plasmonic metal colloids through composition and structural design. *Chem. Soc. Rev.* 43: 3823–3834.

Arsenic Research and Global Sustainability – Bhattacharya, Vahter, Jarsjö, Kumpiene,
Ahmad, Sparrenbom, Jacks, Donselaar, Bundschuh & Naidu (Eds)
© 2016 Taylor & Francis Group, London, ISBN 978-1-138-02941-5

Determination of arsenic in groundwater by Sector-Field ICP-MS (ICP-SFMS)

M.J. Ríos-Lugo[1], H. Hernández-Mendoza[2], B. González-Rodríguez[1] & E.T. Romero Guzmán[2]

[1]*Universidad Autónoma de San Luis Potosí, San Luis Potosí, México*
[2]*Laboratorio Nacional de Investigaciones en Forense Nuclear, Instituto Nacional de Investigaciones Nucleares, La Marquesa, Ocoyoacac, Edo. de México, México*

ABSTRACT: A method for measuring the concentrations of Arsenic (As) in groundwater Used Inductively Coupled Plasma-Sector Field Mass Spectrometry (ICP-SFMS) was to optimize in this work. Method 3015 A was used for removal of organic matter on water samples. The results obtained shown than levels of As are higher at those permitted by Mexican Official Standard NOM-127-SSA1-1994 in 4 wells with groundwater of a total of 25 wells and according by the United States Environmental Protection Agency (USEPA) and World Health Organization (WHO), the wells with concentrations higher of 0.01 mg L^{-1} were 244. The proposed method is fast and ideal for measuring As in groundwater.

1 INTRODUCTION

The major threats to health through exposure to heavy metals have been associated with Lead (Pb), Cadmium (Cd), Mercury (Hg) and Arsenic (As). These metals are studied extensively and it is effects on human health have been reviewed frequently by international bodies like the World Health Organization (WHO) and United States Environmental Protection Agency (USEPA). In Mexico, heavy metals concentration has been detected above the limits set by Mexican Official Standard (NOM-127-SSA1-1994) in the states of Aguascalientes, Coahuila, Chihuahua, Durango, Guanajuato, Hidalgo, Jalisco, San Luis Potosi, Sonora and Zacatecas (Gutiérrez, 1996). These studies concluded that the presence of As in groundwater is high. Other studies have allowed assessing contamination As in other regions of the country, such as are Durango, Sonora, Zacatecas, Chihuahua and Baja California Sur (Gutiérrez, 1996, Carrillo, 1998). Additionally, it is noteworthy that, studies in Matehuala in San Luis Potosi have allowed contamination recognize that As is a major health problem for human population in Mexico (Yáñez, 1997).

The aim of this work was to develop a fast method for quantification As in groundwater from wells at community of San Diego de la Union, Guanajuato.

2 METHODS/EXPERIMENTAL

2.1 Instrumentation

Sector-Field ICP-MS (ICP-SFMS, high resolution ICP-MS, Element 2/XR Thermo Fisher Scientific, Germany) was used to measurements of As. Aqueous samples were introduced with a SC-2 DX autosampler Element Scientific Inc. (ESI), and micro concentric nebulizer coupled to a Twister with a Helix 50 mL cyclonic borosilicate glass spray chamber (Elemental Scientific Inc., USA). The torch of the ICP-SFMS instrument (Elemental Scientific Inc., USA) was shielded with a grounded platinum electrode (GuardElectrode™, Thermo Scientific). A microwave MARS6 (CEM, Matthews, North Carolina) was used for digestion of samples.

2.2 Materials and reagents

The mass calibration of ICP-SFMS was performed using a certified multi-element solution XXIII (Ba, B, Co, Fe, Ga, In, K, Li, Lu, Na, Rh, Sc, Y, Tl, and U) from Merck (Germany). For optimization of instrumental conditions of ICP-SFMS was performed by using aliquot of 10 µg/L from certified solution of 1000 µg/L of As (Merck, Germany), the last solution was used for calibration curve. In addition, 1 µg/L of internal standard solution was added during the measurements for monitoring the instrumental stability. The solutions were prepared using analytical grade reagents from Merck (Germany). High purity water (> 18 MΩ/cm) was obtained from a Milli-Q® Reference (Millipore México). Nitric acid was purified by distillation in a Milestone Duopur (Milestone s.r.l., Italy) sub-boiling system. Certified Ar gas (99.96%) was supplied by INFRA, S.A. de C.V. México.

2.3 Preparation of samples

The samples were collected in the period of May to August 2015 in the community of San Diego de la Union, Guanajuato. The method 3015 A (acid digestion by microwave for aqueous samples and extracts) was used for removal of organic matter on water samples. The samples were evaporation to dryness and resuspended in HNO_3 2% v/v, 10 mL for analysis of As using ICP-SFMS.

Table 1. Results obtained of As in groundwater samples by ICP-SFMS.

Wells	As mg/L	Location
P1	0.010 ± 0.0001	21°23"12.02" N 100°47"18.38" O
P2	0.039 ± 0.0005	21°22"37.70" N 100°46"38.47" O
P4	0.026 ± 0.0007	21°24"58.32" N 100°46"46.96" O
P5	0.034 ± 0.0009	21°19"22.52" N 100°44"07.79" O
P6	0.018 ± 0.0004	21°31"24.28" N 100°46"09.91" O
P7	0.165 ± 0.0014	21°30"10.00" N 100°48"08.81" O
P8	0.027 ± 0.0003	21°22"48.74" N 100°40"08.33" O
P10	0.018 ± 0.0003	21°25"52.25" N 100°44"51" O
P11	0.018 ± 0.0006	21°27"6.20" N 100°50"33.72" O
P12	0.015 ± 0.0009	21°24"27.49" N 100° 53" 14.31" O
P14	0.018 ± 0.0002	21°27"25.0" N 100°51"30.7" O
P15	0.014 ± 0.0003	21°24"33.11" N 100° 53"28.43" O
P16	0.016 ± 0.0010	21°22"23.38" N 100° 55"58.62" O
P17	0.023 ± 0.0010	21°20"59.60" N 100° 58"30.11" O
P18	0.026 ± 0.0007	21°19"45.86" N 101° 1"40.29" O
P19	0.026 ± 0.0007	21°21"47.58" N 100° 49"53.55" O
P20	0.024 ± 0.0006	21°19"44.05" N 100° 48"10.11" O
P21	0.017 ± 0.0002	21°19"44.05" N 100° 49"53.58" O
P22	0.006 ± 0.0001	21°20"03.63" N 100° 51"33.68" O
P23	0.041 ± 0.0010	21°19"51.22" N 100° 58"46.95" O
P24	0.022 ± 0.0005	21°26"57.57" N 100° 47"1.28" O
P25	0.036 ± 0.0006	21°26"19.95" N 100° 54"47.11" O

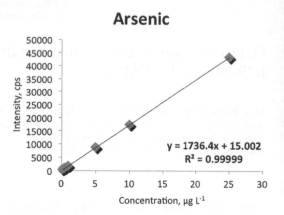

Figure 1. Calibration curve of As.

3 RESULTS AND DISCUSSION

The results obtained for the levels of As are shown in Table 1. Several parameters have been taken into account and evaluated for the optimization of the analytical method for quantitative detection of As, such as; linearity of calibration curve, Limit of Detection (LOD), Limit of Quantification (LOQ) and the precision in terms of relative standard deviation (RDS).

Results obtained in stability test of 7Li, 115 In and 238U using the certified multi-element solution XXIII shown a 2% (RSD). Moreover, LOD and LOQ of As were $0.515 ± 0.032$ µg L^{-1} and $5.15 ± 0.032$ µg L^{-1} respectively, and this values were estimated using the followings equations LOD = 3σ/S and LOQ = 10σ/S. Figure 1 shows the calibration curve obtained in measurements of As using high resolution mode was R2 = 0.99999. RSD was calculated through the total analysis and that was less that 5% for As.

According to the official Mexican Standard NOM-127-SSA1-1994, the results of the concentration levels of As groundwater water, only exceeded (0.025 mg/L) in four wells. Moreover, 24 wells were found with As concentrations greater than 0.01 mg/L according to USEPA standard and WHO drinking water limit values.

4 CONCLUSIONS

Arsenic has been studied extensively in environmental and biological matrices due to its high toxicity on human health, which is why the WHO and USEPA has programs worldwide on monitoring of As in food, drinking water and environmental matrices. ICP-SFMS proved to be powerful technique for determination of As in groundwater sample.

ACKNOWLEDGEMENTS

Acknowledgment at Secretary of Energy (SENER) by financial support obtained in this project and CONACYT for the support granted in the No. 232762 project for the creation of the National Laboratory of Forensic Nuclear Research. (LANAFONU).

REFERENCES

Gutiérrez, P., Rodríguez, R.E, Romero, G., Velázquez, G.A. 1996. *Eliminación de arsénico en agua potable de pozos. Actas INAGEQ* 2: 319–332.

Carrillo, A., Drever, J. 1998. Environmental assessment of the potential for arsenic leaching into groundwater from mine wastes in Baja California Sur, México. *Geofísica Internacional* 37: 35–39.

Yáñez L, Calderón J, Carrizales L y Díaz-Barriga F. 1997. Evaluación del riesgo en sitios contaminados con plomo aplicando un modelo de exposición integral (IEUBK). In: F. Díaz-Barriga (Ed.) *Evaluación de riesgos para la salud en la población expuesta a metales en Bolivia.* Centro Panamericano de Ecología Humana y Salud OPS/OMS.

*Arsenic Research and Global Sustainability – Bhattacharya, Vahter, Jarsjö, Kumpiene,
Ahmad, Sparrenbom, Jacks, Donselaar, Bundschuh & Naidu (Eds)
© 2016 Taylor & Francis Group, London, ISBN 978-1-138-02941-5*

Arsenic determination in agricultural water in Chihuahua, Mexico

M.C. Hermosillo-Muñoz & M.C. Valles-Aragón
Autonomous University of Chihuahua, Faculty of Agrotechnological Science, Chihuahua, Mexico

ABSTRACT: Arsenic-contaminated water has affected the live of many people. In recent years, it's been reported an increasing As contamination of water bodies, soil and crops in many regions of the world. So, the objective of this research was to determinate the As concentration and the physicochemical parameters of water for agricultural use on Chihuahua´s south-center region in Mexico. The study was realized in the municipalities of Rosales, Meoqui, Delicias and Julimes, where their principal economic activity is the agriculture. 65 irrigation wells were sampled. For the samples were determined As, conductivity, total dissolved solids, turbidity, alkalinity, pH, redox potential, and temperature. 48% of the samples showed presence of As in water, 9% exceeded the MPL of 100 µg/L. There was no correlation between physicochemical parameters and As concentration. This region has wells with occurrence of As, therefore is recommendable to analyze soil and crops to evaluate the problem scope.

1 INTRODUCTION

The massive water extraction for agricultural purposes has lead to drilling deep wells up to fossil waters, which often, are already contaminated with As (Ortega García *et al.*, 2010). From an agricultural point of view, there is a problem on areas flooded with arsenic-contaminated water as the crops may contain certain amounts of this metalloid making them dangerous to eat (Moreno-Jimenez, 2010).

The agricultural use of arsenic-contaminated waters necessarily leads to the risk of agricultural products to absorbing and storing this metalloid. Vegetables are the crops that are most exposed to this condition because they remain in direct contact with water and soil containing it (Ortega García *et al.*, 2010). Thus, the objective of this research was to determinate the As concentration and the physicochemical parameters of water for agricultural use on Chihuahua´s south-center region in Mexico.

2 METHODS/EXPERIMENTAL

2.1 Area description

The study was realized on south-center region of Chihuahua´s state, in the municipalities of Rosales, Meoqui, Delicias and Julimes. On this region predominates the planting of chili, lucerne, corn, peanut, sorghum, wheat, beans, oats, onion, among others.

2.2 Sampling

The well water sampling was realized in coordination with the Irrigation District 005 of the region.

Sixty five irrigation wells were sampled: 38 from Meoqui, 3 from Julimes, 9 from Delicias and 15 from Rosales. Two samples in each well of 1 L were collected, one for the physicochemical parameters that was refrigerated, and another for As determination, in which 2 mL of HNO_3 was added for preservation. Likewise, pH; redox potential (ORP) and temperature were analyzed in field with a potentiometer HI 98127 (pHep®4)—HI 98128 (pHep®5) HAN-NA INSTRUMENTS.

Conductivity, Total Dissolved Solids (TDS) and turbidity were determinate in laboratory using an OAKTON Handheld Conductivity/TDS/ Temperature/RS232C Meter CON 11 model. While analysis of Alkalinity was carried out by titration, taking 50 mL per sample which were titrated with HCl to 0.02N using an orange color indicator.

The analytical determination of As was made via atomic absorption spectrometry with hydride generation in a Perker Elmer Analyst Spectrometer. To ensure quality control of the measurements, the analysis was done in duplicate, using targets and standard solutions. The limit of As quantification was 5 µg/L.

The Maximum Permissible Levels (MPL) set by the Mexican law "Ley Federal de Derechos Disposiciones Aplicables en materia de Aguas Nacionales" (CONAGUA, 2013) for agricultural water were considered as a reference and compared with the data obtained.

The physicochemical parameters and As concentration in the wells sampling were used for the generation of a geostatistical model. The Inverse Distance Weighting (IDW) model was used to represent the distribution of the metalloid on the research region and their correlation with water features. A Forward Selection of Explanatory Variables in the SAS 9.1.3 software was considered.

3 RESULTS AND DISCUSSION

3.1 Analytic determination

31 samples (48%) showed presence of As in water, 6 of them (9%) exceeded the MPL of 100 μg of As L^{-1} set by the regulations.

The total of samples exceeding the physicochemical parameters limits were as follows:

- 1200 μS for Conductivity: 24.62%
- 500 ppm for TDS: 35.38%
- 10 NTU for Turbidity: 29.23%
- 400 ppm for Alkalinity: 6.15%
- 6 9.5 permissible pH range: 6.15%
- 500 ppm for hardness: 9.23%

Currently, there is an undergoing draft law in México that includes conductivity regulations for water. Also, there is not set MPL for ORP for agricultural use of water in this country.

The statistical analysis showed that there was not collinearity between physicochemical parameters and As concentration.

3.2 Modeling

The model of As concentration showed the dispersion in the south-center region of Chihuahua´s state. The darker color means greater concentration magnitude and the lighter color the lowest.

The problem of As pollution of water in agricultural wells of the region is evident and relevant. Other authors had demonstrated this but, for wells of human consumption (Olmos-Márquez, 2011). A research conducted by Comisión Nacional del Agua in the Delicias–Meoqui aquifer (the one of study), reported 50% of the wells having As concentrations higher than 50 μg/L (Camacho et al., 2011).

This result is an imperative problem of analysis, due to there are signs that the use of arsenic-contaminated water for irrigation has led to the accumulation of As in soils, also lead bioaccumulation of As in edible plants and crops (Bundschuh et al., 2012). Besides, drinking water, crops irrigated with water contaminated with arsenic can be a contribution to the daily intake of arsenic in food (Dahal, Fuerhacker et al., 2008).

Regretably, rural wells, farms, and small communities as the ones on this region, generally have inadequate, if any, treatment systems, thus are exposed to a higher risk of developing As related health problems (Camacho et al., 2011).

4 CONCLUSIONS

This region has wells with presence of As that can represent a health problem, not just for the intake of contaminated water, also for the possibility of As integrated in the food chain by aliments produced in the territory. Thus is recommendable analyze soil and crops to determinate the magnitude of the problem in the site, especially in the wells that are over the MPL for agricultural water.

ACKNOWLEDGEMENTS

To the National Council of Science and Technology of Mexico (CONACyT) for the financial support granted to the development of this project.

REFERENCES

Bundschuh, J., Nath, B., Bhattacharya, P., Liu, C., Armienta, M.A., Moreno, L., Lopez, D., Jean, J.S., Cornejo, L., Lauer, L., Tenuta, A. 2012. Arsenic in the human food chain: the Latin American perspective. Sci. Total Environ. 429: 92–106.

Camacho, L.M., Gutiérrez, M., Alarcón-Herrera, M.T., Villalba, M.L, Deng S. 2011. Occurrence and treatment of arsenic in groundwater and soil in northern Mexico and southwestern USA. Chemosphere 83(3): 211–225.

CONAGUA. 2013. Ley Federal de Derechos Disposiciones Aplicables en materia de Aguas Nacionales. México: Comisión Nacional del Agua.

Dahal, B.M., Fuerhacker, M. Mentler, A. Karki, K.B. Shrestha, R.R., Blum, W.E.H. 2008. Arsenic contamination of soils and agricultural plants through irrigation water in Nepal. Environ. Pollut. 155: 157–163.

Moreno Jimenez, E. 2010. Recuperación de Suelos Mineros Contaminados con Arsénico Mediante Fitotecnologías. Primera Ed. Madrid: Universidad Autónoma de Madrid.

Olmos-Márquez, M. 2011. Remoción de arsénico en el agua por fitorremediación con Eleocharis macrostachya en humedales construidos de flujo subsuperficial. Chihuahua: CIMAV.

Ortega García, J., Lugo Sepúlveda, R.E., Espinoza Ojeda, E. 2010. Análisis de arsénico en papa usando un arsenómetro portátil. Invurnus 5(1): 28–33.

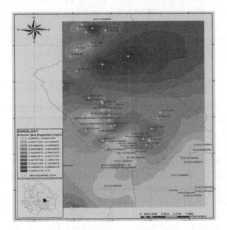

Figure 1.   Arsenic dispersion in the State of Chihuahua.

Arsenic Research and Global Sustainability – Bhattacharya, Vahter, Jarsjö, Kumpiene,
Ahmad, Sparrenbom, Jacks, Donselaar, Bundschuh & Naidu (Eds)
© *2016 Taylor & Francis Group, London, ISBN 978-1-138-02941-5*

Response of bacterial bioreporters to surface-bound arsenic

C.M. van Genuchten[1], A. Finger[1], J. van der Meer[2], & J. Peña[1]
[1]*Institute of Earth Surface Dynamics, University of Lausanne, Lausanne, Switzerland*
[2]*Department of Fundamental Microbiology, University of Lausanne, Lausanne, Switzerland*

ABSTRACT: Previous work shows that the response of bacterial bioreporters to aqueous Arsenic (As) is altered by solution chemistry. However, the influence of solid phase Fe precipitates on the detection of As by these engineered microorganisms is not understood. In this study, we applied bacterial bioreporters to assess the bioavailability of As(V) bound to the surface of a suite of Fe(III) (oxyhydr)oxide minerals. We found that As(V) bound to the surface of crystalline goethite (As:Fe = 1 mol%) was not detected by the bioreporters. By contrast, substantial fractions of As(V) (35–50%) adsorbed to nanocrystalline two-line ferrihydrite (As:Fe = 1–6 mol%) were detected by the bacteria, which suggests that structural properties of the solid phase play a key role in the detection of surface-bound As(V) by bioreporters. Our findings stress the importance of preventing the oxidation of Fe(II), which is common in As-contaminated aquifers, to ensure accurate As measurements when applying bioreporters in the field.

1 INTRODUCTION

Bacterial bioreporters are engineered microorganisms that bioluminesce in response to bioavailable target compounds. These microorganisms are being developed to measure bioavailable arsenic (As) for several applications, including low-cost screening of As-contaminated groundwater wells (e.g. Stocker *et al.*, 2013). Although previous work has shown that co-occurring aqueous species can interfere with the detection of As by bacterial bioreporters (Harms *et al.*, 2005), no study has investigated the response of bioreporters when used in systems containing solid phase iron (Fe)-(oxyhydr)oxides, which are efficient As scavengers. Our objective in this work is to determine the impact of solid phase Fe precipitates on the detection of As(V) by bacterial bioreporters. This knowledge is critical to understand the limitations of the bioreporter technique as an As detection method and provides insight into the impact of mineral surfaces on As bioavailability.

2 METHODS

2.1 *Bacterial bioreporters*

We used the bioreporter strain *E. coli* DH5α-2697 and followed the bioassay procedures outlined in Stocker *et al.* (2013).

2.2 *Adsorption and co-precipitation experiments*

Solid samples with surface-bound As(V) were generated via two methods: i) As(V) adsorption experiments using goethite (Goe, α-FeOOH) and 2-line ferrihydrite (Fh) and ii) co-precipitation of Fe(III) and As(V). In this work, we refer to samples prepared via adsorption and co-precipitation as ADS and CO samples, respectively. Adsorption experiments were performed by equilibrating Goe and Fh suspensions with aqueous As(V) for 24 h at pH 7.5 in the presence or absence of 1 mM Ca^{2+}. Co-precipitation experiments were performed using an Fe(0) Electrocoagulation (EC) cell. The EC cell generates Fe(II), which is oxidized by O_2 to produce Fe(III) precipitates in-situ. For CO samples, the EC electrolyte contained a range of As(V) concentrations in the presence or absence of Ca^{2+}. The structures and solid phase compositions of the CO samples have been described in detail (van Genuchten *et al.*, 2014). In all experiments, solids were separated after the reaction with 0.1 µm filters. The As:Fe surface excess was calculated using As measurements in the filtrate obtained by ICP-OES. The As:Fe surface excess ranged from 1 to 6 mol% for ADS samples and from 10 to 60 mol% for CO samples.

2.3 *Bioreporter measurements of surface-bound As*

To obtain bioreporter measurements of surface-bound As(V), the filtered solids from ADS and CO experiments were first rinsed with DI water. The rinsed solids were then resuspended in fresh DI water and the solution pH was adjusted to 7.5. After resuspension, two unfiltered and two filtered aliquots were taken. The first unfiltered aliquot was used directly in bioreporter experiments and the second was acidified and analyzed by ICP-OES for total Fe and As concentrations. For the filtered aliquots, the first was given to the bioreporters in separate experiments to account for any mobilized (aqueous phase) As in the resuspension. The second filtered aliquot was acidified and analyzed by ICP-OES to verify the presence of any mobilized As(V) that was measured by the bioreporters. The surface-bound As(V) concentration detected by the bioreporters was then calculated by subtracting the bioreporter-measured As(V) concentration in the filtered suspension from the bioreporter-measured As(V) concentration in the unfiltered resuspension.

3 RESULTS AND DISCUSSION

3.1 Bioreporter response to ADS samples

Figure 1A shows the concentration of surface-bound As(V) detected by the bioreporters as a function of total input As(V) concentration for Goe (1 mol%) and Fh (1 and 6 mol%) ADS samples. A shaded grey line illustrates a hypothetical 1:1 response of the bioreporters to input As(V). For each sample, 3–4 solids concentrations were tested, which are indicated by the dotted lines. The response of the bioreporters to As(V) in each sample was less than 1:1, which indicates that Fe(III) minerals inhibits the bioreporter response. However, the bioreporter response was affected by the two Fe(III) phases differently. The bioreporters did not detect As(V) bound to Goe at any solids concentration. By contrast, significant fractions of As(V) bound to Fh were detected by the bioreporters, which is clearly evident when the bioreporter-measured As(V) concentration is normalized by the total As(V) concentration (Fig. 1B). In addition, the fraction of total As(V) detected by the bioreporters increased from ~38 to 50% with an increased surface excess from 1 to 6 mol%. These results imply that the detection of surface-bound As(V) by bioreporters depends on structural properties of the adsorbent and the As:Fe surface excess.

The effect of 1 mM Ca^{2+} present during the generation of the Fh ADS samples is shown in Figure 1B. The presence of Ca^{2+} lead to a decrease in the fraction of bioreporter-detected As(V) from 38 to 21% in the 1 mol% Fh ADS sample and from 50 to 30% in the 6 mol% Fh ADS sample. This significant reduction in As(V) detected by the bioreporters can be explained by Ca^{2+}-induced aggregation of As(V)-laden Fh crystallites into large flocs (Kaegi et al., 2010), which can lead to less surface-bound As(V) exposed to the bulk solution.

3.2 Bioreporter response to CO samples

Figure 2 shows the fraction of As(V) detected by the bioreporters as a function of the As:Fe surface excess for CO samples. For reference, the fraction of As(V) detected in the 6 mol% Fh ADS

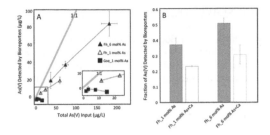

Figure 1. A) response of bioreporters to Goe and Fh ADS samples, B) effect of surface loading and co-occurring Ca^{2+} on the fraction of surface-bound As(V) detected by the bioreporters.

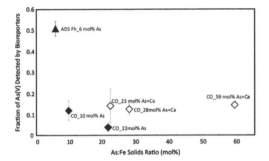

Figure 2. Response of bioreporters to CO samples generated in the presence (empty diamonds) and absence of Ca^{2+} (filled diamonds).

sample is also given. Figure 2 illustrates the different response of the bioreporters to ADS and CO samples. For example, in contrast to ADS samples, there is no strong relationship between the bioreporter-detected As(V) fraction and the As:Fe surface loading. Furthermore, the presence of Ca^{2+} did not produce the same systematic decrease in bioreporter-detected As(V) in the CO samples as was observed in ADS samples. The different bioreporter response to ADS and CO samples is likely due to the different methods of generating each sample series. Whereas the solid phase in ADS samples is pre-formed, the Fe(III) precipitate is produced in-situ in the CO samples, producing sub-nanoscale precipitates with large As:Fe surface excesses (van Genuchten et al., 2014). These sub-nanoscale precipitates rapidly aggregate into sub-spherical particles (Kaegi et al., 2010), which decreases the surface area accessible by the bioreporters and limits the interaction between the bioreporters and sorbed As(V).

Despite the different response of the bioreporters to the ADS and CO samples, surface-bound As(V) was detected by the bioreporters in both sample series. This result might indicate that bioavailable As is not strictly limited to As species in solution.

ACKNOWLEDGEMENTS

We thank Siham Beggah Möller and Vitali Maffenbeier for their assistance with bioreporter experiments and acknowledge funding from the BCV Foundation and the Sandoz Family Foundation.

REFERENCES

Kaegi, R., Voegelin, A., Folini, D., et al. 2010. Effect of phosphate, silicate, and Ca on the morphology, structure, and elemental composition of Fe(III)-precipitates formed in aerated Fe(II) and As(III) containing water. *Geochim. Cosmochim. Acta* 74: 5798–5816.

Harms, H., Rime, J., Leupin, O., et al. 2005. Effect of groundwater composition on arsenic detection by bacterial bioreporters. 2005, *Microchim. Acta* 151: 217–222.

van Genuchten, C.M., Pena, J., Gadgil, A.J. 2014. Fe(III) nucleation in the presence of bivalent cations and oxyanions leads to subnanoscale 7 Å polymers. *Environ. Sci. Technol.* 48: 11828–11836.

1.8   Synchrotron applications and innovative research on arsenic in natural systems

Arsenic Research and Global Sustainability – Bhattacharya, Vahter, Jarsjö, Kumpiene,
Ahmad, Sparrenbom, Jacks, Donselaar, Bundschuh & Naidu (Eds)
© *2016 Taylor & Francis Group, London, ISBN 978-1-138-02941-5*

Arsenic in soil-plant systems: Comparison of speciation by chemical extractions and X-ray absorption spectroscopy

R.B. Herbert Jr.[1], T. Landberg[2], I. Persson[3] & M. Greger[2]
[1]*Department of Earth Sciences, Uppsala University, Uppsala, Sweden*
[2]*Department of Ecology, Environment and Plant Sciences, Stockholm University, Stockholm, Sweden*
[3]*Department of Chemistry and Biotechnology, Swedish University of Agricultural Sciences, Uppsala, Sweden*

ABSTRACT: Cultivation experiments were conducted where the terrestrial plant lettuce (Lactuca sativa) and the emergent plant common cottongrass (Eriophorum angustifolium) were exposed to As during growth in soils and sediments, with the aim of determining the relative changes in As speciation that occurred during uptake from a soil, and oxidized mine sediment, and a reduced mine sediment. The chemical extractions and XANES of soils indicate that As is predominantly present as As(V); uptake and transport in the two plant species appears to result in a change in the As(V)-As(III) ratio, with relatively more As(III) in shoots and roots. This result is confirmed for cottongrass by XANES, but is not observed for lettuce. Plant species and environmental conditions appear to be important factors controlling As speciation.

1 INTRODUCTION

In aerated, terrestrial soils, the primary chemical form of arsenic (As) is the oxidized species arsenate (As(V); AsO_4^{3-}), while reduced forms such as arsenite (As(III); AsO_3^{3-}) can occur in water saturated soils. At circum-neutral pH, the dominating forms are dihydrogen arsenate ($H_2AsO_4^-$) and arsenous acid (H_3AsO_3), respectively. During plant uptake, As(III) and As(V) in soils can be transformed and immobilized in the plant tissues, such as through the binding of As(III) to reduced sulfur in phytochelatins (Moreno-Jiménez et al., 2012).

The aim of this work was to determine the relative changes in As speciation that occurred during uptake in a terrestrial and submerged (water saturated) soil-plant system.

2 METHODS

A series of cultivation experiments were conducted where the terrestrial plant lettuce (*Lactuca sativa*) and the emergent plant common cottongrass (*Eriophorum angustifolium*) were exposed to As during growth in soils and sediments. Initially, lettuce and cottongrass were grown for three weeks and five weeks, respectively, in uncontaminated potting soil before being transferred to the As-bearing soils and sediments, where they were then cultivated for 7 days in a greenhouse at 23°C. At harvest, the plant tissues (shoots and roots) were rinsed from soil particles using 20 mM EDTA and dried at 80°C for 48h.

The unsaturated soils consisted of an agricultural soil developed from alum shale in Kinnekl-eva, Sweden. The sediments were oxidized and reduced sediments acquired under water-saturated conditions from the Adak mine site, Sweden.

Arsenic species were analyzed in soils, sediments and plants. The total As and plant-available As concentrations in soils and sediments, as well as As in plants, were extracted according to procedures provided in Bergqvist et al. (2014). Following extraction, solutions were analyzed for As(III) and As(V) using anion exchange chromatography (PRP X-100 column) coupled to vapor generation-AAS.

Bulk soil and sediment samples, as well as freeze-dried plant shoots and roots, where analyzed using synchrotron-based X-ray absorption spectroscopy. Arsenic K-edge XANES and EXAFS spectra were collected at the wiggler beam-line I811 at the MAX-lab facility at Lund University, Sweden. Beam-line conditions are provided in Bergqvist et al. (2014).

3 RESULTS AND DISCUSSION

3.1 Soils and sediments

The total chemical extractions of soils and sediments (Table 1) indicate that the alum shale soil and oxidized sediment contained almost exclusively As(V), while the reduced mine sediment contained a greater proportion of As(III) relative to As(V). This result is partially confirmed by XANES (Fig. 1) in that the spectrum for the alum soil strongly resembles the spectra for the $HAsO_4^{2-}$ standard. However, the spectrum for the oxidized mine sediment is somewhat shifted to lower energies, suggesting the presence of As(III) component

Table 1. Average concentrations of As(III) and As(V) in soils and plants after cultivation for 7 days. Concentrations in mg/kg. n = 4–7.

	Alum shale soil		Oxidized sediment		Reduced sediment	
	AsIII	AsV	AsIII	AsV	AsIII	AsV
Soils						
Total	1.2	139	6.0	410	14	48
0.1M HCl	0.22	11.9	0.4	24	1.0	3.2
Cotton grass						
Root	51	64	134	167	70	95
Shoot	6.6	6.7	22	56	15	21
Lettuce						
Root	78	33	189	95	94	51
Shoot	3.8	1.5	21	21	15	17

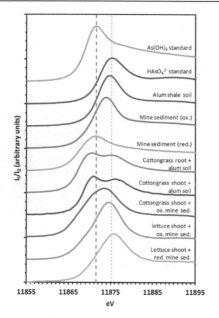

Figure 1. XANES spectra of reference samples, soils, sediments, and plants. Dashed vertical bars indicate peak positions of the $As(OH)_3$ and $HAsO_4^{2-}$ reference spectra.

as well. The reduced mine sediment has a peak centered at 11871.4 eV, indicating a strong As(III) contribution, similar to the $As(OH)_3$ standard.

Plant-available As concentrations (0.1M HCl extractable) are roughly an order of magnitude lower than the total concentrations (Table 1), but exhibit similar relationships between the As(III) and As(V) species.

3.2 Plant roots and shoots

Among lettuce and cottongrass shoots and roots, As concentrations are consistently greater in the roots than in the shoots (Table 1). In cottongrass, As is, to a first approximation, evenly distributed between As(III) and As(V) in roots and shoots, regardless of growth medium. This result is reflected by the XANES analyses where a double peak for the As(III) and As(V) species is observed in roots and shoots grown in alum soil. While this double peak is absent in the spectrum for cottongrass grown in oxidized mine sediments, the shoulder on the lower energy side of the peak is clearly concave, suggesting the presence of a As(III) contribution.

In lettuce, the chemical extractions indicate that there is a greater accumulation of As(III) in the roots, relative to As(V) (Table 1). This is contrary to the behavior exhibited by cottongrass. As(III) and As(V) concentrations are quite similar in lettuce shoots. This speciation is not, however, indicated by XANES (Fig. 1), which instead shows that most arsenic is present as As(V).

EXAFS data (not shown) provide a more detailed picture of As speciation than can be provided by XANES. The EXAFS results indicate that As(V)···O binding environments predominate in oxidized soils and sediments, while a significant As(III) component is present in the reduced sediment. EXAFS spectra of cottongrass roots indicate scattering lengths of ca. 1.77 Å and 2.24 Å, agreeing with the bond lengths for As(III)···O and As(III)···S scattering (e.g. in phytochelatins) respectively. Mixed As(V)···O/As(III)···O binding is common in roots and shoots of lettuce.

4 CONCLUSIONS

The chemical extractions of alum shale soils indicate that As is predominantly present as As(V); uptake and transport in the two plant species results in the accumulation of both As(III) and As(V) in shoots and roots. This result is confirmed for cottongrass by XANES, but is not observed for lettuce. Plant species and environmental conditions appear to be important factors controlling As speciation. A comparison of extraction data and XANES spectra indicates that the two methods more consistently yield similar results for soils than for plant tissues.

ACKNOWLEDGEMENTS

This project has been financed by the Swedish Research Council.

REFERENCES

Bergqvist, C., Herbert, R., Persson, I., Greger, M. 2014. Plants influence on arsenic availability and speciation in the rhizosphere, roots and shoots of three different vegetables. Environ. Pollut. 184: 540–546.
Moreno-Jiménez, E., Esteban, E., Peñalosa, J.M. 2012. The fate of arsenic in soil-plant systems. Rev. Env. Contam. Toxicol. 215: 1–37.

Arsenic Research and Global Sustainability – Bhattacharya, Vahter, Jarsjö, Kumpiene,
Ahmad, Sparrenbom, Jacks, Donselaar, Bundschuh & Naidu (Eds)
© 2016 Taylor & Francis Group, London, ISBN 978-1-138-02941-5

Understanding the relation of As and Mn biogeochemistry within Fe-rich sediments in southeast Asian aquifers

S. Datta[1], M. Vega[1], P. Kempton[1], N. Kumar[2], K. Johannesson[3] & P. Bhattacharya[4,5]

[1]*Department of Geology, Kansas State University, Manhattan, Kansas, USA*
[2]*Department of Geological Sciences, Stanford University, Stanford, USA*
[3]*Department of Earth and Environmental Sciences, Tulane University, New Orleans, USA*
[4]*KTH-International Groundwater Arsenic Research Group, Department of Sustainable Development, Environmental Science and Engineering, KTH Royal Institute of Technology, Stockholm, Sweden*
[5]*International Center for Applied Climate Science, University of Southern Queensland, Toowoomba, Queensland, Australia*

ABSTRACT: Aquifer sediments were collected from varying depths from selected villages in Murshidabad, India. Hydrological and physico-chemical parameters of groundwater were measured and aquifer cores were collected. Solid phase analysis includes X-Ray Diffraction (XRD), X-ray Absorption Spectroscopy (XAS) and x-ray imaging. Primary focus of this work was to understand the inter-relation of As and Mn biogeochemistry in presence of Fe-rich minerals. Mechanism of As and Mn release is controlled by the oxidation-reduction of Fe bearing minerals in these subsurface aquifers. With our spectroscopy data, we have observed reduced As and Mn species in aquifer sediment as the depth increased relating directly to elevated concentration of As and Mn observed in groundwater. These observations are vital in understanding the inter-depending behavior of As and Mn in subsurface sediments.

1 INTRODUCTION

Elevated concentrations of both manganese (Mn) and arsenic (As) have been observed in the groundwaters of Murshidabad (West Bengal), in eastern India. Manganese, is known to cause neuromuscular problems and inhibition of neurological developments particularly in children. The health impacts of bioavailable As are well known with As being a Class-I carcinogen. The presence of Mn in the already As afflicted regions of SE Asia poses serious implications for millions of inhabitants (von Brömssen *et al.*, 2007). The concomitant release of As and Mn in these groundwaters has inflicted more questions to the inter-relation of these oxy-anions with Fe-oxides in the sediments. The prevalence of Fe-bearing hyporheic zones, also known as 'iron curtains' that sequester Fe-minerals to a great extent, which in turn acts both as a sink or source of various oxy-anions. Series of studies have identified these reactive zones for attenuation of Mn and As at Waquoit Bay and also along the Meghna river banks in Bangladesh (Datta *et al.*, 2009, Jung *et al.*, 2012). However, the role of geochemical processes in constraining the chemical fluxes in groundwater discharge was less significant in these studies. In the Meghna river studies, it has been established that enrichment of As in shallow subsurface riverbank sediments along the river can be attributed to redox trapping of groundwater As by Fe[III]-(oxyhydr)oxides precipitated in the hyporheic zone during dry season, when the hydraulic gradient

drives groundwater flow from the aquifer to the river (Jung *et al.*, 2012).

The current study aims to address two objectives in understanding biogeochemical cycling of Mn and As in groundwaters where sediment chemistry is mainly associated with Fe-(oxyhydr)oxides: (i) the occurrence and overall distribution (lateral and temporal) of groundwater Mn and As; (ii) the relation between Mn, As and Fe-minerals in the sediments and their impact on the subsequent release of Mn and As from these sediments.

Several recent studies have suggested that temporal redox fluctuation is a key dynamic event within this natural reactive barrier (NRB) consisting of Fe[III]-(oxyhydr)oxides at these river banks (Jung *et al.*, 2012). The sand-silty riverbank sediments may undergo redox oscillation with time due to seasonal fluctuation in the groundwater table and river water level between dry/rainy seasons as well as tidal effects, that may affect oxygen flux to the subsurface. The focus of the present work is detail understanding of the sediment chemistry of Fe-minerals that effectively controls the release and adsorption of both As and Mn concurrently.

2 METHODS/EXPERIMENTAL

2.1 Sample collection and analysis

Aquifer sediment samples were collected from 5 villages near the Bhagirathi river in Murshidabad. Needle samplers coupled with conventional drilling

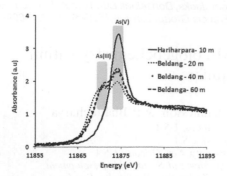

Figure 1. As-K-edge XANES spectra collected for aquifer sediment from different depths in two sites confirming As speciation in solid phase. Sediments show enrichment of As^{III} with depth in this area.

techniques were used to collect shallow to intermediate depth sediment cores, which were kept intact under N_2 as frozen samples until being shipped to Kansas State. Total metal concentrations were obtained by using bulk X-Ray Florescence (XRF) and total acid digestion. A series of selective extraction procedures were followed to decipher As, Mn and Fe phases in the sediments (Keon *et al.*, 2011). Sediment composition and mineralogy were examined by combining total acid digestions, quantitative x-ray diffraction, and petrographic analysis to provide a quantitative assessment of the major mineral composition of sediments.

2.2 X-ray Absorption Spectroscopy

Aquifer sediment samples were analyzed for x-ray imaging and X-Ray Absorption Spectroscopy (XANES and EXAFS) at Stanford Synchrotron Radiation Lightsource (SSRL) beamlines 2–3 and 11–2 respectively and spectra were collected for As, Mn, and Fe-K-edges. On an average 10–12 spectra were collected for each sample.

3 RESULTS AND DISCUSSION

Two distinct scenarios govern the distribution of Mn in Murshidabad: i) Reductive dissolution of Mn-Fe-(oxyhydr)oxide-As rich minerals in the presence of dissolved organic matter and (ii) Mn reduction and subsequent mobilization in aquifers with positive redox potentials because Mn is a strong oxidant. This could also be contributing to the oxidation of As from As^{III} to As^{V} (Fig. 1). 83% of the total wells surveyed contain Mn levels that exceed the recommended WHO limit of 0.4 mg/L. Dissolved As at the same locations show a range from <10μg/L to ~4000 μg/L. Bulk Mn concentrations in sediment do not show significant variation between the oxidizing/reducing aquifers. Inference is that, Mn sediment availability does not play a key role in groundwater Mn distribution/release.

During reducing conditions assisted by microbial processes, reductive dissolution of Fe occurs and results in release of harmful level of dissolved As and also Fe^{II} into groundwater. Mn-oxides may also reductively dissolve (resulting in high dissolved Mn in water) during this process. However, where sufficiently abundant, the highly oxidising nature of the Mn-oxides limits reductive dissolution of the As bearing Fe-oxides. Any As that is desorbed from Feoxides will be oxidised by Mn-oxides from As^{III} to As^{V}, which has greater sorption affinity to Fe oxides than As^{III}. So, As^{V} rapidly re-sorb to any remaining Fe oxides and precipitates, thus restricting mobility to groundwater.

4 CONCLUSIONS

Our results suggest three different mechanisms responsible for controlling Mn and As release and accumulation in groundwater (i) Microbially mediated reduction of Mn-Fe-(oxyhydr)oxides containing sorbed As complexes, as we found positive correlation between Mn, As, and Fe. (ii) Mn reduction and subsequent release as a result of other species being oxidized by Mn in aquifer with positive redox potential. Elemental speciation in aquifer sediment in this area indicates co-existence of reduced Mn with oxidized As and (iii) Possible inhibition of Mn release due to the presence of protein-like dissolved OM. Probing into the natural release mechanisms of Mn in separate ways provide a multi-variate approach to understanding how Mn behaves in the aqueous environment.

ACKNOWLEDGEMENTS

Financial assistance for this work has been provided by National Science Foundation and synchrotron work was executed at SSRL.

REFERENCES

Datta, S., Mailloux, B., Jung, H.B., Hoque, M.A., Stute, M., Ahmed, K.M., Zheng, Y. 2009. Redox trapping of arsenic during groundwater discharge in sediments from the Meghna riverbank in Bangladesh. *Proc. Natl. Acad. Sci. USA* 106(40): 16930–16935.

Jung, H.B., Bostick, B.C., Zheng, Y. 2012. Field, experimental, and modeling study of arsenic partitioning across a redox transition in a Bangladesh aquifer. *Environ. Sci. Technol.* 46(3): 1388–1395.

Quicksall, A.N., Bostick, B.C., Sampson, M.L. 2008. Linking organic matter deposition and iron mineral transformations to groundwater arsenic levels in the Mekong delta, Cambodia. *Appl. Geochem.* 23(11): 3088–3098.

Keon, N.E., Swartz, C.H., Brabander, D.J., Harvey, C.F., Hemond, H.F. 2001. Validation of an arsenic sequential extraction method for evaluating mobility in sediments. *Environ. Sci. Technol.* 35(13): 2778–2784.

von Brömssen, M., Jakariya, M., Bhattacharya, P., Ahmed, K.M., Hasan, A., Sracek, O., Jonsson, L., Lundell, L., Jacks, G. 2007. Targetting low-arsenic aquifers in Matlab upazila, southeastern Bangladesh. *Sci. Total Environ.* 379: 121–132.

Arsenic Research and Global Sustainability – Bhattacharya, Vahter, Jarsjö, Kumpiene,
Ahmad, Sparrenbom, Jacks, Donselaar, Bundschuh & Naidu (Eds)
© 2016 Taylor & Francis Group, London, ISBN 978-1-138-02941-5

Structural characterization of arsenic compounds in volcanic ashes and loessic sediments

G.L. Bia, M.G. García & L. Borgnino
Centro de Investigaciones en Ciencias de la Tierra (CICTERRA).
CONICET, and FCEFyN Universidad Nacional de Córdoba, Córdoba, Argentina

ABSTRACT: X-ray Absorption Fine Structure spectroscopy (XAFS) was used to determine the oxidation state of Arsenic (As) and its local chemical coordination in recent and ancient volcanic ashes and Chacopampean loess sediments. XANES analysis allowed to discriminate the oxidation states of arsenic in the studied samples: As^{5+} is dominant in loess sediments while in volcanic ashes As^{-1} and As^{3+} were identified. The proposed EXAFS models fit well with the experimental data, suggesting that in loess sediments, As^{5+} could be in the form of arsenate ions adsorbed onto ferric oxyhydroxides. In recent volcanic ashes, the identified As^{-1} species is likely associated with arsenopyrite or arsenical pyrite. Besides, As^{3+} is likely related to As atoms present as impurities within the glass structure, in the form of oxy-hydroxide complexes. In ancient volcanic ashes As^{3+} is the dominant oxidation state but As^{5+} was also identified.

1 INTRODUCTION

The Chacopampean region (Central and Northern Argentina), is affected by the presence of high concentrations of Arsenic (As) in groundwater. The origin of these As have traditionally been assigned to the alteration of volcanic glass spread in the loessic sediments that blanket the entire region (i.e., Nicolli *et al.*, 2012). Bia *et al.* (2015) identified by XPS the presence of As (III)-S and As (V)-O nanocoatings, deposited on the surface of volcanic glass particles. The As contained in these coatings is easily released in contact with water, but a higher proportion of As remains within the aluminosilicate glass structure as an impurity.

The aim of this work is to determine the oxidation state of As and its local chemical coordination (to a radius of ~ 4 Å Around As) in recent and ancient volcanic ashes and loess sediments.

2 METHODS AND EXPERIMENTAL

2.1 Sampling

Two volcanic ash samples were collected immediately after the eruptions of the Chaitén (2008) and Puyehue (2011) volcanoes in nearby regions. The ancient volcanic ash sample (T1) was collected from a sedimentary column in the southern part of Chacopampean plain. Sample L10 correspond to loessic sediments collected from two different sections in the northern Chacopampean plain. The general chemical and mineralogical characteristics of the samples are described in detail in Borgnino *et al.* (2013) and Bia *et al.* (2015).

2.2 XAFS data collection and analysis

Arsenic K-edge spectra (11867 eV) were collected at beamline XAFS1 at the Brazilian Synchrotron Light Laboratory (LNLS), in Campinas, Brazil. Structural information was obtained by fitting Fourier Transformed (FT) EXAFS data, with the ARTEMIS software (Ravel & Newville, 2005). The FT of the EXAFS was fit with the predicted function by varying the number of coordinating atoms, their distance, mean square displacement and passive electron reduction factor in order to obtain the best fit between the experimental and predicted spectra.

3 RESULTS AND DISCUSSION

3.1 XANES analysis: Oxidation state of As

The As XANES spectra with the reference materials are presented in (Fig 1a). According to the results, the recent and ancient volcanic ashes contain mainly As^{3+}. Besides, for the Chaitén sample, XANES spectrum is broad indicating the presence of the more reduced As^{-1} species. In ancient ashes, a minor proportion of As^{5+} species were also identified. On the contrary, in loess sediments As^{5+} is the dominant oxidation state.

3.2 EXAFS analysis: As structure

Figure 1b and c shows the EXAFS analysis and the schematic illustration of the As coordination environment of the studied samples. In recent volcanic ashes, the As first shell is composed of ~3 oxygen atoms at a distance of 1.79 Å around the central atom (Table 1), similar to the corresponding values for the sodium arsenite tetrahydrate reference material. This indicates that the main As structural unit in the silicate glass is As^{3+}-O. However, the degree of disorder revealed by the σ^2 values (Table 1), suggests the presence of the As(OH)$_3$ and AsO(OH)$_2^-$ species, as reported in similar

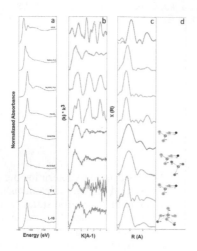

a) Normalized Absorbance — Energy (eV)
b) $(k)\cdot k^3$ — K(Å-1)
c) X (R) — R (Å)
d)

Figure 1. a) Normalized arsenic K-edge X-ray absorption near edge structure (XANES), b) k^3-weighted EXAFS spectra, and c) Fourier transform magnitudes at As K-edge of the samples and As-bearing reference materials. d) Schematic illustrations of the As coordination environment of the studied samples. Solid lines represent experimental data; dashed lines correspond to fits obtained with the parameters indicated in Table 1.

Table 1. Local structure of arsenic in the studied samples derived from fitting EXAFS spectra at As K-edge.

Samples	Shell	N	R (Å)	σ^2(Å^2)	Δe (eV)	χ^2
Reference solids						
FeAsS	As-Fe	3	2.37	0.00120	7.02	0.789
	As-S	1	2.33	0.00134		
	As-As	1	3.07	0.00109		
	As-As	2	3.19	0.00183		
	As-S	3	3.31	0.00182		
NaAsO$_2$·4H$_2$O	As-O	3	1.79	0.00167	6.80	0.809
	As-As	2	3.24	0.00153		
Na$_3$HAsO$_4$·7H$_2$O	As-O	4	1.69	0.00101	6.56	0.767
	As-MS	12	4.12	0.00100		
FeAsO4	As-O	4	1.68	0.00089	6.79	0.832
	As-Fe	14	3.05	0.00132		
	As-As	4	3.35	0.00152		
Recent volcanic ash						
Chaitén	As-O	3.00	1.79	0.00256	11.4	0.452
	As-Fe	3.00	2.42	0.00234		
	As-X	1.00	3.16	0.03670		
Puyehue	As-O	3.00	1.79	0.00239	11.19	0.421
	As-x	1.00	3.17	0.03120		
Ancient volcanic ash						
T1	As-O	3.2	1.79	0.00204	12.45	0.673
	As-x	1.1	3.21	0.03780		
Loessic sediments						
L-10	As-O	4	1.69	0.00336	6.74	0.523
	As-Fe	2	3.40	0.00161		
	As-Fe	2	3.54	0.00122		

samples (Borisova *et al.*, 2010). In the ancient volcanic ash sample, the results reveal the presence of an As first shell at 1.75 Å with a coordination of ~3.2, which suggests the occurrence of As^{3+}-O and As^{5+}-O compounds. The second shell of the recent and ancient volcanic ashes at ~3.17–21 Å with ~1 neighbors around As, is associated with Na/K/Si/Al atoms (Borisova *et al.*, 2010). Therefore, the As^{3+} hydroxide species, may be bound via one of its oxygen atoms to the Al-Si network or coordinated to the major cations K/Na via electrostatic interaction in the disordered structure of the glass. In the Chaitén sample, the distance at 2.42 Å is assigned to the coordination first-shell of As-Fe in

arsenopyrite (FeAsS; 2.40 Å, this study) or arsenical pyrite ($FeS_{2-2x}As_x$; 2.31 Å).

In contrast, the As first shell in loess sediments is found at ~1.69 Å with coordination numbers of ~4 which is assigned to As^{5+}-O in arsenates (Table 1). The second shell at a distance of ~3.4 Å is similar to the As-Fe distance in scorodite (3.35 Å, this study) and to As^{5+}-O adsorbed in goethite. However the fourth shell of As-As in scorodite, that contains a single As atom at 4.21 Å is not identified. This indicates that scorodite is likely not present in the samples. Therefore, the identified second shell distance suggests that AsO_4^{3-} is adsorbed via bidentate double corner sharing between the apices of adjacent $Fe(III)O_6$ octahedra.

4 CONCLUSIONS

This study reports the transformations in the redox state of As from its primary sources in volcanic ashes to the final phases found in loess sediments, using XAFS analysis. While As^{3+} and As^{-} species predominate in the primary sources found in the recent volcanic ashes, As^{5+} is the dominant oxidation state in loess samples. In these sediments As^{5+} is mostly found as arsenate adsorbed onto ferric oxyhydroxides. Speciation in the ancient volcanic ashes indicates the presence of both redox states, representing an intermediate stage between primary and secondary sources.

ACKNOWLEDGEMENTS

Authors wish to acknowledge the assistance of LNLS (Campinas-Brasil), CONICET and UNC for their support and the facilities used in this investigation. G. Bia acknowledges a doctoral fellowship from CONICET.

REFERENCES

Bia, G., Borgnino, L., Gaiero, D., Garcia, M.G. 2015. Arsenic-bearing phases in South Andean volcanic ashes: Implications for As mobility in aquatic environments. *Chem. Geol.* 393–394: 26–35.

Borgnino, L., Garcia, M.G., Bia, G., Stupar, Y., Le Coustumer, Ph., Depetris, P.J. 2013. Mechanisms of fluoride release in sediments of Argentina's central region. *Sci. Total Environ.* 443: 245–255.

Borisova, A.Y., Pokrovski, G.S., Pichavant, M., Freydier, R., Candaudap, F. 2010. Arsenic enrichment in hydrous peraluminous melts: Insights from femtosecond laser ablationinductively coupled plasma-quadrupole mass spectrometry, and in situ X-ray absorptionfine structure spectroscopy. *Am. Mineral.* 95: 1095–1104.

Nicolli, H.B., Bundschuh, J., Blanco, M. del C., Tujchneider, O.C., Panarello, H.O., Dapeña, C., Rusansky, J.E. 2012. Arsenic and associated trace-elements in groundwater from the Chaco-Pampean plain, Argentina: results from 100 years of research. *Sci. Total Environ.* 429: 36–56.

Ravel, B., Newville, M. 2005. ATHENA, ARTEMIS, HEPHAESTUS: data analysis for X-ray absorption spectroscopy using IFEFFIT. *J. Synchrotron Rad.* 12, 537–541.

Arsenic Research and Global Sustainability – Bhattacharya, Vahter, Jarsjö, Kumpiene,
Ahmad, Sparrenbom, Jacks, Donselaar, Bundschuh & Naidu (Eds)
© 2016 Taylor & Francis Group, London, ISBN 978-1-138-02941-5

Solid-phase speciation of arsenic in abandoned mine wastes from the northern Puna of Argentina using X-ray absorption spectroscopy

N.E. Nieva, L. Borgnino & M.G. García
Centro de Investigaciones en Ciencias de la Tierra (CICTERRA). CONICET, and FCEFyN
Universidad Nacional de Córdoba, Córdoba, Argentina

ABSTRACT: This study investigates the solid-phase speciation of arsenic (As) in the abandoned sulfide mine wastes accumulated in La Concordia mine, northern Puna of Argentina. Samples were collected from the exposed walls of an oxidation profile formed in one of the tailing dams. Mineralogy was characterized by XRD and SEM/EDS. X-ray absorption fine structure spectroscopy (XAFS) was used to determine the arsenic oxidation state. The primary sources are As-bearing sulfides, such as arsenian pyrite and arsenopyrite. As-jarosite and Fe (hydroxy)sulfates are the more abundant secondary As-bearing minerals. X-ray absorption near edge structure (XANES) analysis indicates that As^{+5} is the dominant oxidation state in the entire oxidation profile but minor proportions of As^{+3} are also observed in the bottom saturated layers. The in-depth distribution of the As redox species reveals that after 30 years of exposure, oxidation has affected almost the entire sedimentary column, eventhough signs of anoxic conditions are macroscopically observed.

1 INTRODUCTION

When sulfidic mine wastes are exposed to the weathering agents and to the action of microorganisms, a highly acidic drainage, rich in dissolved metal(oid)s and sulfate is generated (Lottermoser, 2010). The main process involves the oxidation of

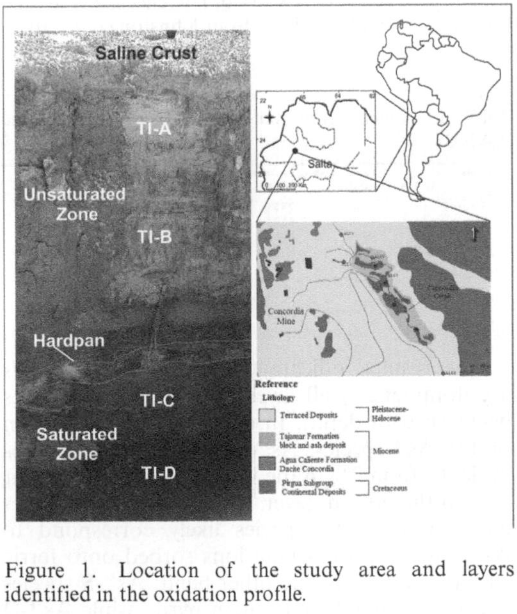

Figure 1. Location of the study area and layers identified in the oxidation profile.

the parent sulfide and the subsequent precipitation of secondary minerals. During such process, elevated concentrations of arsenic (As) are released to the water and then incorporated into a cycle that includes the precipitation/dissolution of As-bearing minerals as well as adsorption/desorption from Fe or Al (hydr)oxide sites (Lottermoser, 2010). The release of As from these phases strongly depends on the compound solubility. Once in solution, this element is highly mobile over a wide range of pH (Smedley & Kinniburgh, 2002).

The aim of this work is to characterize the solid-phase speciation of As in sediments accumulated in the oxidized profiles of tailing dams that were abandoned 30 years ago without an appropriate closure plan in La Concordia mine in northern Puna of Argentina (Fig. 1).

2 METHODS/EXPERIMENTAL

2.1 Sampling and chemical/mineralogical characterization

Samples were collected from five layers identified in a 200 cm depth oxidation profile formed in one of the mine tailing dams. The profile was divided into two different zones: 1) unsaturated oxidized zone; and 2) saturated unoxidized zone (Figure 1). In the uppermost zone two layers TI-A and TI-B, were identified. A laterally discontinuous layer (TI-H) was also recognized at the base of layer

TI-B that acts as a hardpan barrier. The saturated unoxidized zone encompasses two layers (TI-C and TI-D).

After collection, samples were air-dried and stored in plastic zip-lock bags until analysis. Mineralogical characterization was performed by XRD and SEM/EDS. The pseudo-total chemical composition was determined in sediment extracts obtained after acid digestion and analysed by ICP-OES.

2.2 *X-Ray absorption fine structure spectroscopy (XAFS)*

Arsenic K-edge spectra (11867 eV) were collected at beamline XAFS1 at the Brazilian Synchrotron Light Laboratory (LNLS) in Campinas, Brazil. Air-dried, finely ground tailing samples and As reference material (As_2S_3, $NaAsO_2.4H_2O$; As(V)-FeOOH) were mounted behind Kapton tapes in a Teflon holder and analyzed in transmission mode. Five scans were collected for each sample, and then merged. The data were examined using the X-ray absorption near edge structure (XANES) spectra. The XANES spectra were analyzed using Athena 0.8.056 for linear least-squares combination fits using well-characterized As reference materials.

3 RESULTS AND DISCUSSION

3.1 *Mineralogical /chemical composition of mine wastes*

Concentrations of As in the study sediments vary between 890 and 3100 mg/kg. The highest concentrations were measured in the hardpan layer (TI-H). Layers in the unsaturated zone show higher As concentrations than the bottom saturated layers. The main primary minerals determined by XRD and SEM/EDS are quartz, K-feldspar, zircon, and illite, as well as sulfides such as arsenical pyrite, arsenopyrite, some scarce grains of galena and polymetallic sulfides. The main secondary As minerals are As-jarosite and Fe (hydrous)sulfates, which are more abundant in the uppermost layers. A detailed description of the chemical and mineralogical composition of these wastes can be found in Nieva *et al.*, (2016).

3.2 *As-XANES analysis*

The As XANES spectra of the samples and reference materials are shown in Figure 2. According to the obtained results, all samples show characteristic edge features that correspond As^{5+} compounds. However, samples from the bottom layers show pre-edge shoulders in the range of As^{+3} species. The linear combination fits are presented in Table 1.

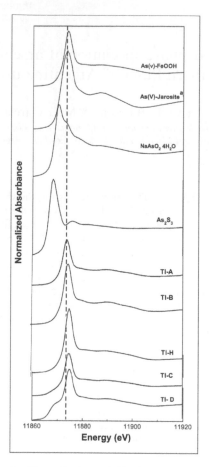

Figure 2. Normalized arsenic K-edge X-ray absorption near edge structure (XANES) for reference materials and mine waste samples. a) Data from Johnston *et al.*, 2012.

Table 1. Linear combination fit results for arsenic XANES.

Samples	Linear Combination – XANES				Statistical Parameters	
	As^{3+}-S	As^{3+}-O	As^{5+}-O	As^{5+}-Jrs	R factor	Reduced X^2
TI-A			78.5%	21.5%	0.00150	0.00217
TI-B			83.0%	17.0%	0.00091	0.00144
TI-H		6.3%	81.5%	12.2%	0.00122	0.00175
TI-C	2.6%	11.5%	73.3%	12.6%	0.00132	0.00211
TI-D	12.7%	17.2%	47.0%	23.1%	0.00015	0.00023

The results indicate that As^{5+}-O compounds are dominant in all layers, but their percentages decrease with depth. In contrast, the proportion of the As^{3+} species, As^{3+}-S and As^{3+}-O, increases in depth from 0% in the uppermost layers to near 30% in the bottom layer. Considering the samples mineralogy, As(V) species likely correspond to As-jarosite, and arsenate ions sorbed onto ferric oxyhydroxides. On the other hand, As^{3+}-S species likely correspond to arsenian pyrite while As^{3+}-O could be assigned to arsenite ions sorbed onto ferric oxyhydroxides.

4 CONCLUSIONS

The obtained results indicate that As^{5+} compounds predominate all along the tailing profile, with minor contributions of As^{3+} species. The latter are more abundant in the bottom layers, in coincidence with the mineralogical composition observed by XRD and SEM/EDS analysis. Eventhough, a partial oxidation of As^{3+} could have occurred during treatment of the bottom layer's samples, the results suggest that after 30 years of exposure, oxidation has affected almost the entire profile. This has important environmental implications, because As is mostly associated with the more labile phases identified in the sediments.

ACKNOWLEDGEMENTS

Authors wish to acknowledge the assistance of CONICET and UNC whose support facilities and funds were used in this investigation. The Brazilian Synchrotron Light Laboratory (LNLS, Brazil) facilities were used in this investigation. Special thanks to Dr Scott Johnston for sharing As-jarosite XANES data with us. N.E. Nieva acknowledges a doctoral fellowship from CONICET.

REFERENCES

Lottermoser, B.G 2010. *Mine Wastes: Characterization, Treatment and Environmental Impacts*. 3rd Edition. Springer; p. 400.

Nieva, N.E; Borgnino, L.; Locati F.; García M.G. 2016. Mineralogical control on arsenic release during sediment-water interaction in abandoned mine wastes from the Argentina Puna. *Sci. Total Environ*. (DOI: 10.1016/j.scitotenv.2016.01.147).

Johnston, S.G.; Burton, E.D.; Keene, A.F.; Planer-Friedrich, B.; Voegelin, A.; Blackford, M.G.; Lumpkin, G.R. 2012. Arsenic mobilization and iron transformations during sulfidization of As(V)-bearing jarosite. *Chem. Geol*. 334: 9–24.

Smedley, P.L.; Kinniburgh, D.G. 2002. A review of the source, behaviour and distribution of arsenic in natural water. *Appl. Geochem*. 17(2): 517–568.

Arsenic Research and Global Sustainability – Bhattacharya, Vahter, Jarsjö, Kumpiene, Ahmad, Sparrenbom, Jacks, Donselaar, Bundschuh & Naidu (Eds)
© 2016 Taylor & Francis Group, London, ISBN 978-1-138-02941-5

Arsenic mobilization from synthesized Fe(III) oxides under natural conditions

D. Zhang[1], H. Guo[1,2] & W. Xiu[2]
[1]*State Key Laboratory of Biogeology and Environmental Geology, China University of Geosciences, Beijing, P.R. China*
[2]*School of Water Resources and Environment, China University of Geosciences, Beijing, P.R. China*

ABSTRACT: Reductive dissolution of iron (hydr)oxides has been accepted as a major formation mechanism for high arsenic groundwater. However, few researches are available for illustrating arsenic release from different iron oxides in natural aquifer. An in-situ method was applied to elucidate mobilization of arsenate adsorbed on two-line-ferrihydrite, goethite and hematite. During the 80 d experiment, desorption of arsenate was observed in these iron oxides, but the reductive dissolution only occurred in ferrihydrite which was transformed into lepidocrocite and goethite. The redox potential and bicarbonate and DOC concentrations were the key factor controlling arsenic mobilization during desorption. At the end of the experiment, more arsenic was released from goethite than hematite.

1 INTRODUCTION

Elevated arsenic groundwaters have been widely found in north-west of China, where high As groundwater severely endangers the health of local people (Guo *et al.*, 2014). Although, reductive dissolution of iron oxide minerals has been generally accepted as a main mechanism of arsenic release to groundwater, few in-situ experiments focus on the release of adsorbed arsenic from different iron (hydr)oxides. Ferrihydrite, goethite, and hematite has been found in sediments hosting high As groundwater, which are probably the most important adsorbents for As in sandy aquifers due to their great abundance and strong affinity.

This investigation was designed to evaluate the key environmental factors affecting As mobilization and transportation, and to explain the relation of As release to Fe oxide mineralogy, and finally to assess the potential capacity of three iron oxide minerals releasing As into groundwater.

2 EXPERIMENTAL

2.1 Study area and materials

The study area, located in the western Hetao basin, Inner Mongolia, is a typical region where shallow aquifers host high As groundwater (Fig. 1). Two wells 2–2 and 2–4, which are setup at the same site with different depths (10 m and 20 m), are located in the district of Hangjinhouqi in the Hetao basin. Screened intervals were 1 m for each well. Well 2–2 has a relatively higher ORP with lower concentrations of NH_4^+, Fe^{2+} (average of 0.63 and 0.3 mg/L) and As (<10 µg/L) than Well 2–4 with lower ORP (−140 mV) and high concentrations of NH_4^+, Fe^{2+} (average of 3.67 and 1.96 mg/L) and As (103 µg/L).

Two-line ferrihydrite, goethite, and hematite were prepared using the procedure of Schwertmann & Cornell (2003). Fe(III) oxides-coat quartz sand was saturated with As(V) (Arsenate Loaded-Iron Mineral Coated Quartz Sand was referred as ALIMCQS hereafter), which was deployed into bottom of wells sealed inside a piece of 150 mesh nylon bag and sampled at different time intervals. A peristaltic pump was used to purge each well for 1 h before sample deployment.

On retrieval, samples were kept in a 5 mL amber anaerobic bottle filled with pure nitrogen (N_2 > 99.999%), and preserved at 4°C during transportation.

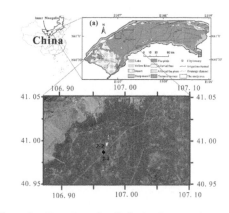

Figure 1. Location of wells for in-situ experiments.

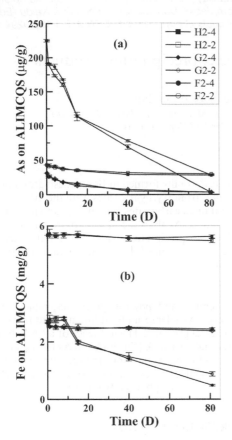

Figure 2. Changes of As and Fe contents in ALIMCQS from wells 2–2 and 2–4.

2.2 Analytical methods

Back to laboratory, samples were determined by powder X-Ray Diffraction (XRD), X-Ray Absorption Near Edge Structure (XANES), Scanning Electron Microscopy (SEM), Qualitative Energy Dispersive Spectrum (EDS), and chemical digestion.

3 RESULTS AND DISCUSSION

3.1 Changes of As and Fe in ALIMCQS

Desorption of arsenate occurred in all ALIMCQS (Fig. 2). Content of As decreased, while no change of Fe content was observed in 8 d for f-ALIMCQS, and in 80 d for g-ALIMCQS and h-ALIMCQS. However, there was a strong correlation between the lost As and lost Fe in f-ALIMCQS after 8 d, indicating that reductive dissolution of ferrihydrite was coupled with As mobilization (Fig. 2). Chemical digestion showed

more As and Fe retaining on ALIMCQS in 2–2 than in 2–4, indicating that redox potential was the key factor controlling arsenic mobility.

At 80 d, XANES showed that As(V) was partly changed to As(III) in f-ALIMCQS, but no As(III) was detected in g-ALIMCQS and h-ALIMCQS.

3.2 Mineralogy change

At 80 d, XANES showed that As(V) was partly changed to As(III) in f-ALIMCQS, but no As(III) was detected in g-ALIMCQS and h-ALIMCQS. Mineralogy change Results of SEM and XRD indicated that transformation of ferrihydrite to lepidocrocite and goethite after 80 d. This agreed with the previous research that Fe(II) with low concentrations catalyzed transformation of ferrihydrite to crystalline minerals.[3]

3.3 Release of As in three different Fe-oxides

The proportion of lost As in ferrihydrite, goethite and hematite in 80 d were about 87%, 90% and 33%, respectively. It demonstrated that more As was released from ferrihydrite and goethite than from hematite.

4 CONCLUSIONS

Two processes were observed to release As from ferrihydrite to groundwater: desorption and reductive dissolution. Transformation of ferrihydrite to lepidocrocite and goethite would partly release adsorbed As(V). Desorption was the major process for As mobility in goethite and hematite, although more arsenic was released from goethite than hematite.

ACKNOWLEDGEMENTS

National Natural Science Foundation of China (Nos. 41222020 and 41172224).

REFERENCES

Edward, D.B., Scott, G.J., Richard, T.B. 2011. Microbial sulfidogenesis in ferryidrite-rich environments: effects on iron mineralogy and arsenic mobility. *Geochim. Cosmochim. Acta* 75: 3072–3087.

Guo, H., Wen, D., Liu, Z., Jia, Y., Guo, Q. 2014. A review of high arsenic groundwater in Mainland and Taiwan, China: Distribution, characteristics and geochemical processes. *Appl. Geochem.* 41 (1): 196–217.

Schwertmann, U., Cornell, R. M. 2003. *Iron Oxides in the Laboratory*. Wiley-VCH.

Arsenic Research and Global Sustainability – Bhattacharya, Vahter, Jarsjö, Kumpiene,
Ahmad, Sparrenbom, Jacks, Donselaar, Bundschuh & Naidu (Eds)
© 2016 Taylor & Francis Group, London, ISBN 978-1-138-02941-5

Arsenic mobility in natural and synthetic coprecipitation products

M. Martin[1], R. Balint[1], R. Gorra[1], L. Celi[1], E. Barberis[1] & A. Violante[2]

[1]*Department of Agricultural, Forest and Food Sciences, University of Torino, Grugliasco, Torino, Italy*
[2]*Department of Agriculture, University of Naples Federico II, Portici (Napoli), Italy*

ABSTRACT: The As release from coprecipitation products naturally formed in Bangladesh groundwaters was compared with that from synthetic Fe-As precipitates. Iron oxy(hydr)oxides were the main component of all materials, but the natural coprecpitates also contained carbonates, phosphates and microbial organic matter. Synthetic coprecipitation products released more As than the natural ones with several diluted extractants. Kinetics of As release from synthetic Fe-As coprecipitates, evidenced an hindrance in As mobilization when a coverage of Ca-polygalacturonate was introduced. Hence, the organic interface could be a key-factor in affecting the stability of such systems.

1 INTRODUCTION

The introduction of As in the food chain largely depends on its mobility in the soil-water system. Sorption/desorption reactions mostly control As mobility in oxic environments, while under alternating redox conditions, such as in aquifers or temporarily flooded soils, precipitation/dissolution reactions may play a major role. These transformations can involve both abiotic or microbially-mediated reactions, resulting in rather variable, complex products, which nature and stability, still not fully understood, depend on the environmental conditions, kind and activity of the microbial communities. The parallel study of natural systems and in-vitro simplified models was aimed to understand the role of different components of the complex systems forming in nature and the mechanisms involved in As mobility.

2 METHODS/EXPERIMENTAL

2.1 Sampling of natural coprecipitation products

Arsenic-enriched coprecipitation products naturally formed were collected from different groundwater storage systems in Bangladesh. They were analyzed for chemical and mineralogical composition and stability, potential As release under different conditions, and composition of the biofilm-forming microbial communities.

2.2 Synthesis of coprecipitation products

Synthetic Fe-As(III) coprecipitation products were prepared at different As/Fe ratios. Part of each sample was immediately freeze-dried, part was aged for one year. All the samples were characterized for chemical and physical properties, mineralogy, potential As release under different conditions, as the natural products. The influence of a Ca-Polygalacturonate network (Ca-PGA) on As release kinetics from the synthetic coprecipitation products was also evaluated by extractions with inorganic (Pi) and organic phosphate (IHP) solutions.

3 RESULTS AND DISCUSSION

3.1 Characterization of of natural coprecipitation products

The natural co-precipitates were heterogeneous materials, mainly consisting in amorphous and poorly-crystalline phases, containing 30–40% Fe (w/w), 5–10% Ca, 2.5–6% P, 2.5–4% organic C (2–4%) and inorganic C (1–4%). Arsenic (0.2–0.5%) was present in both tri- and pentavalent forms, diffusely distributed within the amorphous matrix. The crystalline phases consisted of calcite, Ca- and Fe-phosphates. The complex distribution of P, As, Fe and Ca within the co-precipitate components suggested the adsorption of P and As on the iron oxides surfaces, but did not exclude the possible nucleation of Ca/Fe arseno-phosphates. The organic matter was mostly from microbial origin. The microbial community contained methylotrophic *Betaproteobacteria*, Gallionella-like iron oxidizers, methanogens, methylotrophs, and sulphate reducers while no phylotypes known to be directly involved in As(V) respiration or As(III) oxidation were found (Gorra et al., 2012).

The crystallinity of these coprecipitation products did not increase with ageing. The As extractability with different anions potentially occurring in soils and waters was moderate (Fig. 1). Inositol phosphate was often the best As extractant, particularly

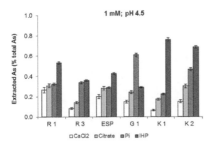

Figure 1. Percentage of total As extracted with 1 mM CaCl₂, citrate, inorganic phosphate (Pi), inositol hexaphosphate (IHP) from the natural coprecipitation products.

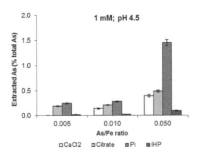

Figure 2. Percentage of total As extracted with 1 mM CaCl₂, citrate, inorganic phosphate (Pi), inositol hexaphosphate (IHP) from synthetic coprecipitation products.

when the organic matter was low, probably because of its competition with As for adsorption as well as for its Fe-complexing ability (Celi et al., 2003).

3.2 Characterization of synthetic coprecipitation products

The synthetic As/Fe coprecipitation products were poorly crystalline and their specific surface area increased with As content. After one year of aging some goethite formed in the 0.005 and 0.010 As/Fe molar ratio products, while the 0.050 one remained poorly crystalline (Violante et al., 2007). The proportion of extractable As increased with As saturation (Fig. 2) and, differently from the natural coprecipitates, inorganic P always extracted more As than IHP. This was attributed to the microporous structure of the materials, hampering the large organic molecules to access the surfaces (Celi et al., 2003).

The ageing of the coprecipitation products often resulted in an increased As release, attributed to the crystallization process inducing a decrease in the specific surface area.

3.3 Discussion

When the coprecipitation products were embedded in a Ca-polygalacturonate network, the release of As with both inorganic P (Fig. 3) and IHP (not shown) was slower, probably because of the diffu-

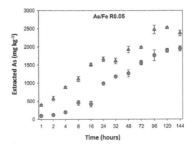

Figure 3. Kinetic of As extraction with inorganic phosphate from uncovered (triangles) or Ca-PGA covered (circles) coprecipitation product with 0.05 As/Fe molar ratio.

sion of the solutes was hampered by the organic network. The presence of the PGA network was even more effective in preventing As release when the larger IHP molecule was involved (not shown).

4 CONCLUSIONS

The natural coprecipitation products were complex materials, formed in anoxic/oxic niches supporting a phylogenetically and metabolically diverse group of biofilm-forming prokariotes. Although these materials remained in poorly crystalline form, hence potentially highly reactive, their potential As release into the environment was probably limited also by the presence of microbial organic matter embedding the mineral phases. The synthetic, pure As/Fe coprecipitates released more As than the natural ones when treated with diluted extractants. The inclusion of the synthetic coprecipitation products within a Ca-polygalacturonate network resulted in a slower and overall scarcer As release, approaching the behavior obtained with the naturally formed ones.

ACKNOWLEDGEMENTS

This study was funded by the Italian Ministry of University and Research (MIUR), PRIN Project 2010–11.

REFERENCES

Celi, L., De Luca, G., Barberis, E. 2003. Effects of interaction of organic and inorganic P with ferrihydrite and kaolinite-iron oxide systems on iron release. Soil Sci. 168: 479–488.
Gorra, R., Webster, G., Martin, M., Celi, L., Mapelli, F., Weightman, A. 2012. Dynamic microbial community associated with iron-arsenic co-precipitation products from a growndwater storage system in Bangladesh. Microbial Ecology, 64: 171–186.
Violante, A., Gaudio, S.D., Pigna, M., Ricciardella, M., Banerjee, D. 2007. Coprecipitation of arsenate with metal oxides. 2. Nature, mineralogy, and reactivity of iron(III) precipitates. Environ. Sci. Technol. 41: 8275–8280.

Section 2: Arsenic in the food chain

2.1   Arsenic in rice

Arsenic Research and Global Sustainability – Bhattacharya, Vahter, Jarsjö, Kumpiene,
Ahmad, Sparrenbom, Jacks, Donselaar, Bundschuh & Naidu (Eds)
© 2016 Taylor & Francis Group, London, ISBN 978-1-138-02941-5

People eat rice and drink water—not soil—so why is arsenic in rice and water so under regulated in the UK compared to arsenic in soil?

D.A. Polya[1], A.A.N. Al Bualy[1], A. Speak[1], P.R. Lythgoe[1], A. Bewsher[1], K. Theis[1],
G. Chana[1] & A.K. Giri[2]

[1]*School of Earth, Atmospheric and Environmental Sciences and Williamson Research Centre for Molecular Environmental Science, University of Manchester, Manchester, UK*
[2]*Molecular Genetics Division, CSIR—Indian Institute of Chemical Biology, Kolkata, India*

ABSTRACT: Over 3,000,000,000 people around the world consume rice as a staple at 100 s g/day in contrast to soil, which other than for those engaging in geophagia, typically of clay, is typically only inadvertently consumed at or below the 100 mg/day level. Furthermore, the bio-accessibility of inorganic arsenic in rice is much higher than that in many soils. Notwithstanding this, in the European Union, even newly introduced regulation for arsenic in polished rice is much less demonstrably protective of public health than is regulation of arsenic in soils. A somewhat smaller discrepancy exists between regulatory values for water and soil. Possible reasons and implications for this discrepancy are explored and discussed.

1 INTRODUCTION

Over 3 billion people around world consume rice as a staple (Meharg & Zhao, 2012) whereas for the majority of people soil is only consumed inadvertently at or below the 100 mg/day level (von Lindern *et al.*, 2016). Taking into account typical consumption rates and bioaccessibilities for rice, water and soil consumption by a typical "standard" adult of these media with arsenic concentrations at their regulatory values gives rise to internal doses of 0.6, 0.3 and 0.004 µg/kg-bw/day (Table 1) compared to a reported lower bound for the lung cancer $BMDL_{01}$ of 0.3 µg/kg-bw/day (EFSA, 2014).

Table 1. Inorganic arsenic internal doses from consumption of rice, water and soil with (a) As concentrations at UK regulatory values and (b) typical consumption rates for "standard" adult consuming rice as a staple.

Parameter	Rice	Water	Soil
Regulatory value (µg/kg)	200[a]	10[b]	32,000[c]
Intake (g/day)	250[d]	2000[e]	0.1[f]
As External Dose (µg/day)[g]	50	20	3.0
Bioaccessibility	80%[h]	100%[i]	10%[j]
As Internal Dose ((µg/day)[g]	40	20	0.3
As Internal Dose (µg/kg-bw/day)[k]	0.6	0.3	0.004

[a] EC (2015) to supersede from 1/1/2016 subject to any deferment the 1000 µg/kg value for food (HMSO, 1959); [b] WHO (2004) provisional guide value; [c] C4SL for residential soils (DEFRA, 2013); [d] Mean UK Asian-Bangladeshi rice purchase rate (Meharg & Zhao, 2012); [e] EA (2009); [f] von Lindern *et al.* (2016); [g] Calculated; [h] Based on high (90%) i-As rice (cf. Polya & Lawson (2015); [i] Assuming 100% i-As; [j] Cave *et al.* (2013) for UK (Northampton) soils; [k] Assuming adult body-weight of 70 kg.

Thus, there is: (i) little or no safety factor evident in the regulation of water or rice; and (ii) there is a factor of around two orders of magnitude between the intakes from rice or water consumption and that from typical inadvertent soil consumption. Why? And what are the implications?

2 DISCUSSION—REASONS—RICE

Direct evidence of detrimental health outcomes arising from consumption of high inorganic-arsenic rice is lacking although Banerjee *et al.* (2013) reported an association between micronuclei (MN) frequency in urothelial cells and arsenic in consumed cooked rice for a cohort of West Bengal villagers not otherwise substantially exposed to arsenic. This work received criticism from Prof. Samuel Cohen, University of Nebraska Medical Center, Nebraska, USA who stated that the Giemsa stained objects identified as MN could instead be arsenic-bearing intracellular inclusions (Cohen, S., Comment posted at http://www.nature.com/articles/srep02195; accessed 20/3/16). A subsequent independent pilot study by Dr Ashok K. Giri using a more DNA-specific DAPI stain, suggests, however, that around 80–90% of the microscopic objects identified as MN by Banerjee *et al.* (2013) were correctly identified (Fig. 1) lending support to the safety of the conclusions in that paper that the same doses of inorganic arsenic from rice or water suitably corrected for bioaccesssibility are broadly equivalent in terms of health risks. This suggests that the apparently less protective regulations of inorganic arsenic in rice compared to arsenic in drinking water cannot be justified other than perhaps on economic grounds.

Given the global rice production that could be impacted if regulatory values for inorganic arsenic in rice were con-

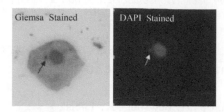

Figure 1. Micronuclei from Urothelial Cells stained in Giemsa followed by de-staining and counter staining with DAPI. (Photographs: AKG).

Table 2. Total arsenic and inorganic arsenic content (dry weight basis) in rice products on sale in Manchester, UK in 2014.

Product	Total Arsenic μg/kg (range)	Inorganic Arsenic μg/kg (range)
Rice cereal (n = 3)	180 (120–285)	109 (75–157)
Baby rice (n = 2)	135 (98–172)	93 (68–118)
Rice cakes (n = 5)	182 (134–233)	161 (125–199)

*Preparation and analysis by PRL & AB (MAGU); recovery of ΣAs and i-As in SRM1568a was 103% & 95% respectively.

sistent with those for drinking water (cf. Meharg & Raab, 2010, Meharg & Zhao, 2012) (let alone those for soil), such regulation could have large socio-economic impacts, and implicitly argued as a reason for slowing policy change given the levels of people's willingness to pay to avoid arsenic-attributable risks (cf. Gibson et al., 2012).

3 DISCUSSION—REASONS—SOIL

Whilst regulatory values for arsenic in soils in the UK have been derived from epidemiological data and with an explicit consideration of soil uptake rates, age and behavioural dependence of individuals, and bioaccessibility, the values reflect rather conservative treatment of particularly of uncertainties in soil ingestion rates (EA, 2009, and cf. von Lindern et al., 2016 and references therein).

4 IMPLICATIONS

Arguably conservative C4SL or SGV values may give rise to unnecessary and substantial remediation costs, on the order of £50 K to £1765 K per hectare (HCA, 2015) depending upon the intended use and the scale/nature of the contamination, as well as delays in development of land. Might less strict regulation of arsenic in soil be indicated?

In contrast, even the new EC (2015) regulatory values for rice permit human exposures (cf. Table 2) which are close to the lower limit for $BMDL_{01}$ suggesting that there is a *prima facie* case for tightening these regulatory values further.

Lastly, we note that tighter regulation of arsenic in drinking water may be merited—something that should be of interest to the estimated greater than 100,000 UK people (calculated from DWI, 2015) whose drinking water, either from private or from public water supplies, contains between 5 and 10 μg/L of inorganic arsenic.

ACKNOWLEDGEMENTS

Funding for various aspects of this work was provided by NERC Standard Research Grant (NE/J023833/1) to DAP (PI), NERC IAA Award (Uni. Manchester NERC IAA 05) to DAP (PI), British Council UKIERI Award (PRAMA; SA07-09) to DAP and AKG. The comparison study of DAPI and Giemsa stained urothelial cell micronuclei was conducted with ethical approval at IICB by AKG. We thank Louise Ander, Stefano Bonassi and Samuel Cohen for discussions. The views expressed here do not necessarily reflect those of any of the funders or individuals who assistance we acknowledge here.

REFERENCES

Banerjee, M., Banerjee, N., Bhattacharjee, P, Mondal, D, et al. 2013. High arsenic in rice is associated with elevated genotoxic effects in humans. *Scientific Reports* 3: 2195.

Cave, M.R., Wragg, J., Harrison, H. 2013. Measurement modelling and mapping of arsenic bioaccessibility in Northampton, United Kingdom. *J. Environ. Sci. Heal. A* 48: 629–640.

Gibson, J., Rigby, D., Polya, D.A., Russell, N. 2015. Discrete choice experiments in developing countries: Willingness to Pay vs Willingness to Work. *Environ. Resource Econ.* (DOI 10.1007/s10640-015-9919-8).

DEFRA 2013. Development of Category 4 Screening Levels for assessment of land affected by contamination. SP1010.

DWI 2015. Annual Reports for 2014. HMSO.

EA 2009. Contaminants in soil: updated collation of toxicological data and intake values for humans. Inorganic arsenic *Environment Agency. Science Report*: SC050021/TOX 1.

EC 2015. Commission Regulation (EU) 2015/2016 of 25 June 2015 amending Regulation (EC) No 1881/2006 as regards maximum levels of inorganic arsenic in foodstuff. *Official Journal of the European Union.* L161/14 26/6/2015.

EFSA 2014. Dietary exposure to inorganic arsenic in the European population. *EFSA Journal* 12(3): 3597.

HMSO 1959. The Arsenic in Food Regulations 1959 (and as amended in 1960 and 1973).

Homes and Communities Agency 2015. Guidance on dereliction, demolition and remediation costs. HMSO.

Meharg A.A., Raab, A. 2010. Getting to the bottom of arsenic standards and guidelines. *Environ. Sci. Technol.* 44: 4395–4399.

Meharg, A.A., Zhao, F.-J. 2012. *Arsenic & Rice.* Springer.

Mondal, D., Polya, D.A. 2008. Rice is a major exposure route for arsenic in Chakdha Block, West Bengal. *Appl. Geochem.* 23: 2986–2997.

Mondal, D., Banerjee, M., Kundu, M. Banerjee, N. et al. 2010. Comparison of drinking water, raw rice and cooking of rice as arsenic exposure routes in three contrasting areas of West Bengal, India. *Environ. Geochem. Hlth* 32: 463–477.

Polya, D.A., Lawson, M. 2015. Geogenic and anthropogenic arsenic hazards in groundwater and soils: distribution, nature, origin and human exposure routes. In: J.C. States (Ed.) *Arsenic: Exposure Sources, Health Risks and Mechanisms of Toxicity.* John Wiley & Sons.

Von Lindern, I., Spalinger, S., Stifelman, M.L., Stanek, L.W. Bartrem, C. 2016. Estimating children's soil/dust ingestion rates through retrospective analyses of blood lead biomonitoring from the Bunker Hill Superfund Site in Idaho. *Environ. Health Perspect.* (doi: 10.1289/ehp.1510144).

*Arsenic Research and Global Sustainability – Bhattacharya, Vahter, Jarsjö, Kumpiene,
Ahmad, Sparrenbom, Jacks, Donselaar, Bundschuh & Naidu (Eds)
© 2016 Taylor & Francis Group, London, ISBN 978-1-138-02941-5*

Inorganic arsenic in rice and rice products on the Swedish market 2015 Part 1: Survey

B. Kollander, B. Sundström, V. Öhrvik & L. Abramsson
National Food Agency, Uppsala, Sweden

ABSTRACT: Inorganic arsenic (iAs) in 63 rice and 36 other rice products including rice cakes and rice drinks, was determined using HPLC-ICP-MS (prEN 16802). None of the rice nor the rice cakes had an inorganic arsenic content that exceeded the maximum levels that will come into force in the EU with effect from 1 January 2016. The concentration of iAs varied between < 1 µg/kg up to 322 µg/kg. There was no significant difference in iAs content between organic and conventionally produced products. A study of the effect of cooking the rice in an excess of tap water and discarded, showed a decrease of iAs with 40–70%. In addition composite samples from a "Market Basket" were analyzed. The food group Fish and Cereals contained in average around 10 µg iAs/kg while the other food groups contained around 5 µg/kg or lower.

1 INTRODUCTION

The Swedish National Food Agency (NFA) has been working for many years mapping the sources of consumers' intake of arsenic. Rice and rice products represent one third of the total exposure to arsenic in Sweden. In 2013, the Swedish NFA investigated the total arsenic content in a selection of products intended for children (Kollander & Sundström, 2015). The results of the investigation also led to several companies subsequently working to reduce the arsenic content in their products. This project is part of the Swedish NFA work to map the occurrence of arsenic in various foods and to investigate the intake of arsenic from various types of food. It is also part of work on a more long-term objective, to induce rice producers to work more actively to ensure that the rice raw material has a lower arsenic content and in this way reduce consumers' intake of arsenic.

This presentation describes the sampling, analysis and the results from the survey of inorganic arsenic in rice and rice products on the Swedish market 2015, and it represents Part 1 of three. The subsequent parts are *Part 2. Risk Assessment*, and *Part 3. Risk Management*.

2 METHODS/EXPERIMENTAL

2.1 Sampling

A selection of rice and rice products was bought in various food stores in Uppsala, Halmstad and Västerås and via the internet during March-April 2015. The selection was made with the aim of covering the large supermarket chains' own brands, as well as other well-known brands on sale in many different stores and also organic products. A number of individual random samples of rice were also bought from smaller specialist shops. In the larger stores, rice products were also bought that were intended for consumers with food allergies/coeliac disease.

A total of 102 different products were bought, made up of 63 rice (basmati, jasmine, long-grain, round-grain, whole grain), 11 rice cakes, 9 fresh rice porridges, 6 breakfast cereals, 5 rice drinks, 4 gluten-free breads, 3 noodles and 1 gluten-free pasta. At least 1 kg of all products was bought from at least 2 packs of the same product. This is in accordance with the commission's regulation (EC 333/2007) for controls pursuant to current legislation.

2.2 Analytical methods

Analysis of inorganic arsenic was performed by HPLC-ICP-MS (High Performance Liquid Chromatography-Inductively Coupled Plasma Mass Spectrometry) in the Swedish NFA laboratory. An HPLC from Agilent (1260) with a strong anion exchange column (Dionex Ionpac AS7 and precolumn Dionex Ionpac AG7 were used to separate the different arsenic compounds in the sample. The analytical method (prEN 16082) is accredited in accordance with ISO/IEC 17025 by SWEDAC for inorganic arsenic for rice, rice products and other foodstuffs within the range 1–25 000 µg/kg. The Limit Of Detection (LOD) is between 1 and 3 µg/kg depending on the dilution of the sample before analysis, and the measurement uncertainty is +/– 26%.

The total content of arsenic in the samples was analysed by ALS Scandinavia AB, Luleå, using high resolution ICP-MS (HR-ICP-MS, ELEMENT XR, Thermo Scientific). All samples were analysed using two different instruments to safeguard the results. To increase sensitivity to arsenic, methane gas was added to the sample flow. The LOD for arsenic was 1.7 µg/kg, calculated as 3 times the standard deviation for blank sample (n = 11).

3 RESULTS AND DISCUSSION

3.1 Levels of inorganic arsenic in rice and rice products on the Swedish market

Among the products tested, rice cakes (n = 11) contained the highest levels of inorganic arsenic, with an average of 152 µg/kg (range 86–322 µg/kg). Whole grain rice and raw rice (n = 9) had the next-highest level with an average of 117 µg/kg (range 75–177 µg/kg). Basmati rice (n = 17, average 63 µg/kg) and jasmine rice (n = 18, average 69 µg/kg) had a significantly lower inorganic arsenic content than other types of rice (range 30–107 µg/kg). The gluten-free breads (n = 4) contained lower levels of arsenic than rice cakes, with an average of 42 µg/kg (range 22–56 µg/kg).

For the fresh rice porridges (n = 9), which apart from the rice itself had a water content of 60–90 per cent, the average content was 14 µg/kg (range 10–17 µg/kg), and for the rice drinks (n = 6) the average was 8 µg/kg (range 5–10 µg/kg).

3.2 Organic products

The study included 18 organic products. The results showed that there was no significant difference in arsenic content between organic and conventionally produced products. Neither could any difference be detected on the basis of country of origin.

3.3 Effect of rinsing and cooking in tap water

To investigate whether preparation and cooking affected the inorganic arsenic content, a further six different types of rice were analysed before cooking, after rinsing, after cooking where all the water was absorbed and after cooking with an excess of tap water, where cooking water was discarded. The tap water used (Uppsala municipality) contained < 1 µg iAs/L. Rinsing before cooking did not reduce the inorganic arsenic content. On the other hand, the content was reduced by between 40 and 70 per cent if the rice was cooked with an excess of water, compared with when all the cooking water was absorbed.

3.4 Market basket samples

In order to estimate the general intake of inorganic arsenic in the Swedish population, analyses were also made of food samples that were included in the Swedish NFA earlier study, Market Basket 2010, which were earlier only analyzed for total arsenic (NFA, 2012). The highest levels of inorganic arsenic were found in the following food groups (average

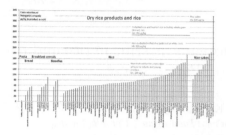

Figure 1. Content of inorganic arsenic in the dry rice and rice products included in the survey. The products are sorted by group and listed in order of their inorganic arsenic content. The Maximum Level (ML) for inorganic arsenic is marked with red lines (applies from 1 January 2016).

content (range) in µg/kg): Fish 13 (10–21), Cereals 11 (4–15), Sugar and similar 5 (< 2–12) and Fruit 3 (< 2–7). In the food groups Meat, Egg, Dairy, Cooking fat, Bakery, Soft drinks, Vegetables and Potatoes, the level of inorganic arsenic was in most cases below the detection limit of 1–2 µg/kg (in wet and dry samples respectively).

4 CONCLUSIONS

Arsenic is a element that is carcinogenic and should be avoided as far as possible. The purpose of the introduction of maximum levels for inorganic arsenic, is to decrease the consumers' exposure of inorganic arsenic. At the same time, the introduction of maximum levels set on the ALARA (as low as reasonably achievable) principle should not affect trade on global markets to any great extent (maximum around 10 per cent). According to this work, and others, regarding levels of inorganic arsenic in rice and rice products on the European market, it can be concluded that the maximum levels that begin to apply in the EU with effect from 1 January 2016 could be a third to a half lower according to the ALARA principle.

REFERENCES

Kollander, B., Sundström, B. 2015. Inorganic arsenic in rice and rice products on the Swedish Market—Part 1. A Survey of inorganic arsenic in rice and rice products. Swedish National Food Agency Report: 16/2015, Uppsala, Sweden

NFA. 2012. Market Basket 2010 – chemical analysis, exposure estimation and health-related assessment of nutrients and toxic compounds in Swedish food basket, Swedish National Food Agency Report: 7/2012, Uppsala, Sweden

prEN 16802 Foodstuffs—Determination of elements and their chemical species—Determination of inorganic arsenic in food of marine and plant origin by anion exchange HPLC-ICP-MS following waterbath extraction. CEN TC275/WG10, 2015.

Arsenic Research and Global Sustainability – Bhattacharya, Vahter, Jarsjö, Kumpiene, Ahmad, Sparrenbom, Jacks, Donselaar, Bundschuh & Naidu (Eds)
© *2016 Taylor & Francis Group, London, ISBN 978-1-138-02941-5*

A comparative assessment of arsenic distribution in rice produced in Pakistan and other geographical regions

H. Rasheed, R. Slack & P. Kay
water@leeds, School of Geography, University of Leeds, UK

ABSTRACT: Variability in total arsenic (As) in rice for human consumption was evaluated using 250 commercial rice samples obtained from cereal markets and retail stores in the primary rice growing area of Punjab, Pakistan. Twelve of the 250 rice samples were found to exceed the recently set advisory limit of 0.2 mg/kg of As in rice, with a maximum detected concentration of 0.356 mg/kg. This study demonstrates the need to quantify the relationship between As in water, soil and paddy crop. It is thus hypothesized that the origin of As in rice grains is from As contaminated ground water used for paddy field irrigation. The data set produced from 250 rice samples is compared with the literature values to derive a global As distribution of As in rice. The order of ranking with respect to distribution and intensity of As in rice grain is revealed as China > Bangladesh > USA > EU > Spain > Taiwan > Thaiand > India > Sweden > Pakistan.

1 INTRODUCTION

Arsenic (As) contamination of water supplies has been reported to affect 137 million people in more than 70 countries (Ravenscroft *et al.*, 2009). Arsenic associated health implications include skin lesions, cancer, developmental toxicity, neurotoxicity, cardiovascular diseases and diabetes. Arsenic contaminated groundwater is used for drinking, washing and irrigation purposes. In particular, crops requiring large volume of water to grow such as rice, might be a lethal source of As.

Rice is a staple food for more than half of the world's population. A number of studies have reported the presence of As in rice (e.g. Halder *et al.*, 2014, Williams *et al.*, 2006) suggesting a link to As contaminated groundwater used extensively for irrigation or cooking of rice. However, sufficient data could not be provided to set a global Maximum Contaminant Level (MCL) of As in rice and is a current issue under debate. In July 2014, the WHO-Codex Alimentarius Commission recommended 0.2 mg/kg as the advisory level of inorganic As in polished (*i.e.* white) rice.

Arsenic in groundwater has been reported in Pakistan (Tahir & Rasheed, 2014). However; no systematic study has been conducted on As levels in the food chain. Pakistan is the 4th largest rice producer in the world after China, India and Indonesia. Rice is cultivated over 10% of the total cropped area and is an important part of the country's exports. This study aimed to investigate As exposure via consumption of rice grown in primary rice growing areas of Pakistan. The specific objectives are: 1) to investigate risks of As exposure from rice; 2) to understand the links between groundwater and As in food.

2 METHODS/EXPERIMENTAL

2.1 *Sampling*

Two hundred and fifty rice grain samples were collected from cereal markets or godowns, wholesalers and open markets of ten primary rice producing districts (Table 1). A one kilogram sample was collected from randomly selected storage sacks, and sealed in ultraviolet (UV) sterilized zipped lock plastic bag, labeled and transported at ambient temperature to the National Water Quality Laboratory, Islamabad.

2.2 *Laboratory analysis*

Official Methods of Analysis of AOAC (Association of Analytical Chemists) were used to digest and ana-

Table 1. Arsenic concentrations in rice samples.

District	n	Mean (mg/kg)	SD	Min (mg/kg)	Max (mg/kg)	No. of samples ≥ 0.2 mg/kg
Kasur	24	0.079	0.038	0.000	0.204	0
Sialkot	22	0.066	0.035	0.000	0.162	0
Okara	27	0.126	0.070	0.047	0.329	4
Sheikhupura	28	0.063	0.026	0.029	0.141	0
Gujranwala	24	0.080	0.050	0.009	0.223	2
Narowal	24	0.050	0.022	0.014	0.085	0
Gujrat	24	0.069	0.032	0.024	0.150	0
Mandi Bahauddin	23	0.129	0.076	0.048	0.276	5
Lahore	27	0.056	0.020	0.026	0.094	0
Hafizabad	27	0.117	0.093	0.037	0.356	1
Overall	250	0.082	0.054	0.009	0.356	12

*Permissible limit of Codex Alimentarius Committee (March, 2014).

lyse the samples. For quality control, method blank, previously analyzed sample and NIST traceable certified reference material (1000 ± 0.2 mg/L of As) were analyzed with percent recovery of 95%.

2.3 Statistical analysis

Descriptive statistics and regression analysis was performed using Microsoft Excel, 2010 and Minitab v.14 (State College, PA).

3 RESULTS AND DISCUSSION

3.1 Arsenic distribution in rice grains

The mean concentration of As in the 250 rice samples was 0.054 mg/kg, calculated on the regional basis, with a min-max range of 0.085–0.356 mg/kg As in rice grains (Table 1). Twelve samples exceeded the advisory limit of 0.2 mg As/kg.

The R^2 (coefficient of determination) between maximum level of As in rice and previously reported maximum As level of ground water sources of these districts (Tahir & Rasheed, 2014) is 0.92. Data were compared with previously published studies (Codex Alimentarius Report, 2014; FDA, 2013; EFSA, 2009; Meharg et al., 2009) to ascertain Pakistan's position in the global issue of rice contaminated by As and to determine the scale of the problem posed to rice consumers (Fig. 1).

3.2 Discussion

The low level As risk in rice grown in Pakistan (Fig. 1) may be attributed to the As contaminated groundwater used for the irrigation of paddy fields. These findings are comparable with other studies by Halder et al. (2014) and Williams et al. (2006) who found a strong correlation between As in rice and irrigation or cooking water. The average water requirement for land preparation, intentional drainage and deep percolation of paddy field in the study area is reported to be around 1600 mm (Bhatti & Kijne, 1992). Deep percolation restores a more favorable salt level for the subsequent crop and may also cause As leaching from surface to the root zone for uptake by the plant. Long term irrigation with As contaminated groundwater may result in the gradual increase of As level of soil and eventually in rice plants.

A comparison with the advisory limit of 0.2 mg/kg has shown (Figure 1) that rice samples from Bangladesh (0.3–0.9 mg/kg) has a wider variability with 4.5 folds higher than the acceptable limit, whereas; rice from India (0.08–0.4) and Thailand (0.1–0.4 mg/kg) are not significantly different. However, mean As concentrations for rice from Pakistan (0.1–0.35 mg/kg) and rice from Swedish retail market (0.1–0.24) are comparatively lower. In all these countries of the world, rice as a staple food may be a major source of As exposure. For such areas, it is important to study the levels of As in irrigation water and paddy field soils, uptake of As by different parts of rice plant, As absorption by raw and cooked rice grains, bioavail-

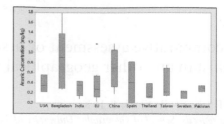

Figure 1. Total arsenic concentration in rice grain reported for major rice producing and marketing countries.

ability, impact of different rice varieties and flow of As through rice export to other regions of the world.

4 CONCLUSIONS

Prevalence of excessive As in 12 rice samples (0.085–0.356 mg/kg) does not seem to be an established risk at present. A review of global data shows that rice contaminated with As might be a problem in other countries around the world. This suggests that a statutory limit should be applied globally without restriction to countries having As issues.

ACKNOWLEDGEMENTS

This research was supported in part by the United Nations Industrial Development Organization (UNIDO).

REFERENCES

Bhatti, M.A. Kijne, J.W. 1992. Irrigation management potential of paddy/rice production in Punjab in Pakistan. Proceedings Intern. Workshop, AIT, Bangkok, 355–366p.

Codex Alimentarius Commission, 2014. FAO-Â News Article, Geneva 14–18 July 2014.

EFSA. 2009. EFSA Panel on contaminants in the food chain (contam):Scientific opinion on arsenic in food. EFSA J 7: 60–71.

FDA. 2013. Arsenic in Rice and Rice Products. http://www.fda.gov/Food/FoodborneIllnessContaminants/Metals/ucm319870.htm. [Accessed October 02, 2015].

Halder, D., Biswas, A., Šlejkovec, Z., Chatterjee, D., Nriagu, J., Jacks, G, Bhattacharya, P. 2014. Arsenic species in raw and cooked rice: Implications for human health in rural Bengal. Sci. Total Environ. 497–498: 200–208

Meharg AA, Williams, P.N., Adomako, E., Lawgali, Y.Y., Deacon, C., Villada, A., Feldmann, J., Raab, A., Zhao, F.J., Islam, R., Hossain, S., Yanai, J. 2009. Geographical variation in total and inorganic arsenic content of polished (white) rice. Environ. Sci. Technol.43:1612–17

Ravenscroft, P., Brammer, H., Richards, K.S. 2009. Arsenic Pollution: A Global Synthesis. Wiley Blackwell, UK, 588p.

Tahir, M.A, Rasheed, H. 2014. Arsenic Monitoring and Mitigation in Pakistan. Technical Report, PCRWR, Islamabad, Pakistan.

Williams, P. Islam, M.R., Adomako, E.E., Raab, A., Hossain, S.A., Zhu, Y.G., et al. 2006. Increase in rice grain arsenic for regions of Bangladesh irrigating paddies with elevated arsenic in groundwaters. Environ. Sci. Technol. 40:4903–4908.

Arsenic Research and Global Sustainability – Bhattacharya, Vahter, Jarsjö, Kumpiene,
Ahmad, Sparrenbom, Jacks, Donselaar, Bundschuh & Naidu (Eds)
© 2016 Taylor & Francis Group, London, ISBN 978-1-138-02941-5

Monitoring of the arsenic content of white rice and brown rice produced in the Republic of Korea

K.-W. Lee & S.G. Lee
Department of Biotechnology, Korea University, Seoul, South Korea

ABSTRACT: We have monitored Arsenic (As) content of 200 white rice samples and 100 brown rice samples produced in Korea. The total arsenic content in samples was analyzed by Inductively Coupled Plasma-Mass Spectrometry ICP-MS after microwave assisted digestion. Arsenic speciation i.e. As^{III}, As^V, DMA, MMA was determined by wet-sonication and HPLC-ICP-MS. The major compound among As species in rice is an inorganic arsenic (As^{III}, As^V). The proportion of inorganic As/total As is respectively 68.45% in white rice and 72.46% in brown rice. We confirmed that all white rice samples were lower than CODEX standard (0.2 mg/kg).

1 INTRODUCTION

Arsenic (As) is widely known to cause harmful effects for human beings and animals (Manju *et al.*, 1998) and exists as various chemical forms. Among them, Inorganic Arsenic species (iAs) containing As^{III} and As^V are far more toxic than organic ones (Edmonds & Francesconi, 1993). Normally, it is known that the toxicity of As^{III} is higher than that of As^V (Edmonds & Francesconi, 1993). According to International Agency for Research on Cancer (IARC), Arsenic and arsenic compounds are classified as a Group 1, carcinogenic to humans (Rousseau *et al.*, 2005). Previous research reported that iAs causes detrimental risk of urinary bladder, lung and skin lesion (Rousseau *et al.*, 2005).

Contaminated soil and drinking water have been recognized as the major source of As from the past. Recently rice and rice-based foods have been considered as an important dietary source of iAs because they contain relatively high levels. Therefore, many countries where rice is a staple food are concerned about dietary ingestion of iAs from rice. However, there's not many latest data about iAs contents of rice products. Accordingly, the importance of As speciation is gradually highlighted.

There are many kinds of literature for As speciation by a diverse combination of the analytical instrument. Coupling of High-Performance Liquid Chromatography (HPLC) and Inductively Coupled Plasma-Mass Spectrometry (ICP-MS) technique has many advantages such as high level of sensitivity, excellent separation ability and high elemental specificity (Ronkart *et al.*, 2007).

In the current study, the purpose was to provide information about quantitative analysis of white rice distributed in the market and the percentage of inorganic As/total As.

2 METHODS/EXPERIMENTAL

2.1 Samples collection

A total number of 300 rice samples that used in this study was randomly collected from the supermarkets and national online shopping mall. All the samples were produced in Korea in 2014 and were stated product's origin including province. The areas of production were a total of 8 provinces of Korea as follows: Gyeonggi, South Chungcheong, North Chungcheong, South Gyeongsang, North Gyeongsang, South Jeolla, North Jeolla, and Gangwon province.

2.2 Instrument and reagents

For quantitative analysis, An ICP-MS 7700x (Agilent. Palo Alto, CA, USA) collision mode and HPLC 1100 series with the quaternary pump were used. Separations were achieved by Hamilton PRP-X100 anion exchange column (250 mm × 4.1 mm, 10 µm, Teknokroma, Barcelona, Spain). The isocratic mobile phase was 5 mM malonic acid (Wako, Japan) which are adjusted with ammonia solution (Samchun, Korea). Sodium arsenite (Fisher Scientific, USA), Arsenic standard solution for ICP, sodium arsenate dibasic heptahydrate and Cacodylic acid (Sigma-Aldrich, USA), Disodium methyl arsonate hexahydrate (Chemservice, USA), were used for analysis.

2.3 Total As determination

Around 0.5 g powdered rice sample was pre-digested with 4 mL HNO_3 (70%) at the hot plate of 80–100°C for 90 min. After pre-digestion, 3 mL HNO_3 (70%) and HNO_3 (70%) were added in the vessel. Then it was digested using Microwave (Milestone ETHOS EASY, Italy). The digested solution was cooled to

Table 1. Method validation followed via linearity, LOD, LOQ and precision for tAs, As speciation.

	Analyte				
	tAs	As^{III}	As^V	DMA	MMA
Linearity(R^2)	0.9999	0.9998	1.000	0.9998	0.9996
LOD^a	0.06	1.24	0.73	0.74	0.58
LOQ^b	0.20	3.77	2.21	2.25	1.75
Precision (CV%)	6.68	4.03	3.64	2.55	3.82

^{a}LOD: Limit of detection (3.3 × σ [y intercept] /slope[average] of calibration curve; n= 3); ^{b}LOQ: Limit of quantification (10 × σ [y intercept]/slope[average] of calibration curve; n = 3)

Table 2. Method validation followed via analysis of CRM (mean ± SD, n = 5).

CRM 1568b	Certified value^c	Observed value^c	Recovery (%)
tAs	285 ± 14	282 ± 5	99.11 ± 2
^{d}iAs	92 ± 10	84.43 ± 1.30	91.77 ± 1.41
DMA	180 ± 12	159.83 ± 2.04	88.79 ± 0.99
MMA	11.6 ± 3.5	10.03 ± 0.18	86.49 ± 1.33

^{c}The values are expressed in μg/kg; ^{d}iAs: Sum of As^{III} and As^V; CRM(Certified Reference Materials), NIST SRM 1568b white rice flour.

room temperature and was made up to 25 g with deionized water. The total arsenic concentrations in the digest were determined by ICP-MS.

2.4 Arsenic speciation

About 1 g of the powdered rice samples was weighed into a 15 mL polypropylene tube, and a solution (10 mL) of 5 mM Malonic acid (pH 5.6) was added. After shaking fully, wetted samples were heated by a water bath at 80°C for 30 min and then was sonicated for 1 min. The heating and sonication procedure were repeated four times. The extracts were centrifuged for 10 min at 3000 g. The supernatant was filtered using 0.45 μm nylon membrane filter and was used for HPLC-ICP-MS analysis.

3 RESULTS AND DISCUSSION

3.1 Validation of analytical method

Validation of analytical method has been formerly implemented before monitoring. Results are shown in Tables 1 and 2.

3.2 Determination of tAs, As species in white rice and brown rice

The data on total arsenic and arsenic species in 300 rice samples (Table 3) showed that the As^{III} content

Table 3. Total arsenic and speciation results in white and brown rice samples (μg/kg).

	White rice (n = 200)		Brown rice (n = 100)	
	Conc.	Range	Conc.	Range
tAs	86.31±19.98	30.95~135.49	158.63±41.51	80.33~282.12
iAs	59.95±12.73	21.36~101.22	113.78±28.41	49.93~192.51
As^{III}	57.05±12.44	21.04~99.95	105.03±25.31	46.62~172.94
As^V	1.43±0.59	N.D~3.38	8.75±4.96	3.20~39.01
DMA	1.88±1.19	N.D~9.57	2.54±1.63	N.D~9.30
MMA	0.63±0.45	N.D~2.07	1.26±0.54	N.D~2.46
iAs/tAs	68.45±9.82	43.47~90.62	72.46±8.63	55.13~91.34

was highest among arsenic species in both white rice and brown rice.

4 CONCLUSIONS

Results of this study, reporting As levels in white rice and brown rice produced in Korea confirmed both are high the proportion of iAs (43.47~91.34%) which is major compound. Brown rice showed that high content than white rice compared with all As species. The content of As in white rice (0.086 mg/kg) was low compared with the CODEX standard (0.2 mg/kg). However, the Korea Standard of arsenic in rice is not established. Thus, this monitoring data will help the establishment of Korean standard of As (or inorganic As) in brown rice and rice products.

ACKNOWLEDGEMENTS

This research was supported by a grant (15162MFDS077) from Ministry of Food and Drug Safety in 2015.

REFERENCES

Edmonds, J. S., & Francesconi, K. A. (1993). Arsenic in seafoods: human health aspects and regulations. Mar. Pollut. Bull. 26(12): 665–674.
IARC. 2004. Some drinking-water disinfectants and contaminants, including arsenic. Working Group on the Evaluation of Carcinogenic Risks to Humans, World Health Organization, & International Agency for Research on Cancer. Vol. 84, IARC.
Manju, G.N., Raji, C., Anirudhan, T.S. 1998. Evaluation of coconut husk carbon for the removal of arsenic from water. Water Res. 32(10): 3063–3070.
Rousseau, M.-C., Straif, K., Siemiatycki, J. 2005. IARC carcinogen update. Environ. Health Persp. 113(9) A580–A581.
Ronkart, S. N., Laurent, V., Carbonnelle, P., Mabon, N., Copin, A., Barthélemy, J.P. 2007. Speciation of five arsenic species (arsenite, arsenate, MMAA^V, DMAA^V and AsBet in different kind of water by HPLC-ICP-MS. Chemosphere 66(4): 738–745.

Arsenic Research and Global Sustainability – Bhattacharya, Vahter, Jarsjö, Kumpiene,
Ahmad, Sparrenbom, Jacks, Donselaar, Bundschuh & Naidu (Eds)
© *2016 Taylor & Francis Group, London, ISBN 978-1-138-02941-5*

Can irrigation practice for rice cultivation reduce the risk of arsenic to human?

S. Islam[1,2,3], M.M. Rahman[1,2], M.R. Islam[3], M. Nuruzzaman[1,2] & R. Naidu[1,2]

[1]*Global Centre for Environmental Remediation (GCER), Faculty of Science and Information Technology,*
The University of Newcastle, Callaghan NSW, Australia
[2]*Cooperative Research Centre for Contamination Assessment and Remediation of the Environment*
(CRC CARE), Australia
[3]*Department of Soil Science, Bangladesh Agricultural University, Mymensingh, Bangladesh*

ABSTRACT: Arsenic bioaccumulation in rice grain has been identified as a major problem in Bangladesh and many parts of the world. Rice is one of the crops affected by arsenic due to its semiaquatic nature. A field study was conducted to investigate the effect of variety and water management on the bioaccumulation of arsenic within the rice plants in different rice cultivars. Ten of the most popularly grown BRRI, BINA and local rice cultivars were screened for susceptibility to arsenic under varying irrigation options. Total grain arsenic accumulation was higher in the plants grown in high soil arsenic in combination with conventional irrigation practice. Results showed that appropriate water management practice and suitable variety resulted in a reduction of grain arsenic level around 39% in addition to increase grain yield around 38%.

1 INTRODUCTION

Arsenic (As) intake from rice represents an important route of exposure for human being, especially for people consuming substantial amount of rice in their diet. Paddy rice is more efficient in As uptake than other cereal crops (Williams *et al.*, 2007). Rice grown under flooded conditions was found to accumulate much more As than that grown under aerobic condition and this difference was attributed to the higher bioavailability of As under flooded soil conditions (Williams *et al.*, 2007).

A significant genetic variability also found among the cultivars for grain As bioaccumulation. In Bangladesh, As-contaminated groundwater extracted through shallow tube-wells is widely used to irrigate rice crops during the dry season (Saha & Ali, 2007), which has resulted in elevated As concentrations in soils and rice grains (Islam *et al.*, 2007), and significant yield losses due to As phytotoxicity (Khan *et al.*, 2009). Water management techniques and rice cultivars dramatically affect the concentration of As in rice grains. Selecting appropriate irrigation practice and rice cultivars in As contaminated soils will benefit high yields and low As in grain.

2 MATERIALS AND METHODS

2.1 Experimentation

A field experiment was carried out using ten of the most popularly grown BRRI, BINA and local cultivars at highly As contaminated soils of Faridpur district in Bangladesh, during the winter season (January–May) of 2014. Rice varieties tested were local, high yielding and aromatic. Two water management options i.e. Continuous Flooding (CF) and Alternate Wetting Drying (AWD) was followed with 4 replications. Each plot had received recommended dosed of N, P, K and S fertilizers. The soil was clay loam in texture, with pH 7.6, organic matter 1.98%, total N 0.13%, available P 13.7 mg/kg, available S 13.7 mg/kg, exchangeable K 0.11 meq/100g, cation exchange capacity 10.1 meq/100g soil and total As 15.69 mg/kg. The chemical properties of soil and irrigation water was presented in Table 1.

2.2 Sample digestion and analysis

Trace analytical grade (70% HNO_3) was used for the digestion of the rice samples by the procedure of Rahman *et al.* (2009). An Agilent 7500c (Agilent

Table 1. Chemical element concentration in soil and irrigation water.

Chemical Element	Concentration	
	Soil (mg/kg)	Irrigation water (µg/L)
Arsenic	15.69	255.4
Cadmium	0.09	ND
Lead	7.30	0.4
Zinc	24.35	1.4
Selenium	0.19	1.3
Nickle	ND	1.4
Chromium	ND	0.2

ND = Not Detectable.

Technologies, Tokyo, Japan) Inductively Coupled Plasma Mass Spectrometry (ICP-MS) was used for the determination of As in digested rice grain samples. For quality control a standard reference material (SRM1568b rice flour) obtained from the National Institute of Standard and Technology (NIST), USA was used to verify the analytical results. The nature of As speciation in rice grains was determined by using Trifluoroacetic Acid (TFA) followed by the extraction technique of Abedin et al. (2002).

3 RESULTS AND DISCUSSION

3.1 Grain yield

Arsenic levels in the surface paddy soils and irrigation water in the field site was 15.69 mg/kg and 255.4 µg/L, respectively indicating high As contamination of soil and ground water (Table 1). This high As contamination reduced the grain and straw yields of rice. There was a significant yield difference among the cultivars under the continuous flooding and alternate wetting and drying irrigation practices. Alternate wetting and drying irrigation treatment significantly increased grain yields for around 38% over continuous flooding practice (Fig. 1). The variety with the highest yields was BRRI dhan58 (7.48 ton/ha) followed by BRRI dhan47 (6.89 ton/ha), Binadhan10 (6.89 ton/ha) and Binadhan8 (6.89 ton/ha).

There was also a significant difference in grain As concentrations among the cultivars under different water management. Arsenic concentrations in different rice varieties were markedly higher in the CF treatments than in the AWD treatments (Fig. 2). The As levels in rice grains under AWD and CF irrigation conditions ranged from 172–453 µg/kg and 221–580 µg/kg, respectively. Overall, there was a 9 to 39% difference in grain As concentrations between the CF and AWD irrigation practices among the rice cultivars. Speciation of grain As were found not to differ among the varieties in respect to inorganic As around 88 to 99%.

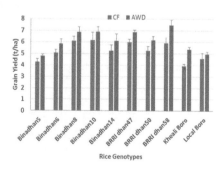

Figure 1. Grain yield differences under CF and AWD irrigation practices (n = 4).

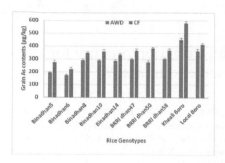

Figure 2. Total As contents varies under CF and AWD irrigation practices (n = 4).

4 CONCLUSIONS

This study demonstrates that suitable rice genotypes with AWD irrigation option not only significantly increase rice grain yields, but also reduce the concentration of As in rice plants.

ACKNOWLEDGEMENTS

The author thank The University of Newcastle for the postgraduate scholarship (UNIPRS). Financial support from CRC CARE is greatly acknowledged.

REFERENCES

Abedin, M.J., Cresser, M.S., Meharg, A.A., Feldmann, J., Cotter-Howells, J. 2002. Arsenic accumulation and metabolism in rice (Oryza sativa L.). Env. Sci. Technol. 36: 962–968.

Islam, M., Jahiruddin, M., Islam, S. 2007. Arsenic linkage in the irrigation water-soil-rice plant systems. Pak. J. Sci. Indust. Res. 50: 85–90.

Khan, M.A., Islam, M.R., Panaullah, G.M., Duxbury, J., Jahiruddin, M., Loeppert, R. 2009. Fate of irrigation-water arsenic in rice soils of Bangladesh. Plant and Soil 32: 263–277.

Meharg, A.A., Rahman, M.M. 2003. Arsenic Contamination of Bangladesh Paddy Field Soils: Implications for Rice Contribution to Arsenic Consumption. Env. Sci. Technol. 37: 229–234.

Rahman, M., Owens, G., Naidu, R. 2009. Arsenic levels in rice grain and assessment of daily dietary intake of arsenic from rice in arsenic-contaminated regions of Bangladesh—implications to groundwater irrigation. Environ. Geochem. Hlth 31: 179–187.

Saha, G.C., Ali, M.A. 2007. Dynamics of arsenic in agricultural soils irrigated with arsenic contaminated groundwater in Bangladesh. Sci. Total Environ. 379: 180–189.

Williams, P.N., Villada, A., Deacon, C., Raab, A., Figuerola, J., Green, A.J., Feldmann, J., Meharg, A.A. 2007. Greatly Enhanced Arsenic Shoot Assimilation in Rice Leads to Elevated Grain Levels Compared to Wheat and Barley. Env. Sci. Technol. 41: 6854–6859.

Arsenic Research and Global Sustainability – Bhattacharya, Vahter, Jarsjö, Kumpiene,
Ahmad, Sparrenbom, Jacks, Donselaar, Bundschuh & Naidu (Eds)
© 2016 Taylor & Francis Group, London, ISBN 978-1-138-02941-5

Continuous use of arsenic contaminated irrigation water: A future threat to sustainable agriculture in Pakistan

A. Javed[1,2], B. Afzal[2], I. Hussain[2] & A. Farooqi[2]

[1]*Department of Earth and Environmental Sciences, Bahria University, Islamabad, Pakistan*
[2]*Environmental Geochemistry Laboratory, Department of Environmental Sciences, Faculty of Biological Sciences, Quaid-i-Azam University, Islamabad, Pakistan*

ABSTRACT: Research work was conducted in Waziarabad District to investigate the concentrations of arsenic in irrigation water, soil and plant parts and to assess the future threat to the economy. The level of arsenic in irrigation water was much above the WHO permissible limit (0.01 mg/l) for drinking water, but within the FAO permissible limit of 0.10 mg/l for irrigation water. Results indicate that the arsenic is transferred from irrigation water and paddy soil to various parts of rice plants. In none of the studied samples the concentration of arsenic in soil and in rice exceeded the permissible limit (25 mg/kg for soil and 0.2 mg/kg for rice grain) which means that consumption of rice does not pose a significant health threat to population; however, the future projections of soil As possess threats to the economy of the country in future To avoid the future problems due to arsenic contamination, there is a need to take appropriate steps, such as development of new irrigation strategies and continuous monitoring of irrigation water, soil and grain.

1 INTRODUCTION

Arsenic (As) rich groundwater is used extensively for crop irrigation, particularly for the paddy rice which may result in yield reduction (Dittmar *et al.*, 2010). In Pakistan, rice meets more than 2 million ton of nation's food requirement and earns about US $ 933 annually from foreign exchange (Manzoor *et al.*, 2006). District Wazirabad is well-known for its high quality export rice where ground water is extensively used for irrigation. The main objective of this study was to investigate the concentrations of As in irrigation water, soil and plant parts, its distribution and mobility in the soil of a rice paddy and to assess the threat to the economy in coming years.

2 METHODS/EXPERIMENTAL

2.1 *Study area and sampling*

Wazirabad is situated on the bank of the Chenab Rivernearly 100 kilometers north of Lahore on the Grand Trunk Road. Samples of groundwater, soil, and rice were taken from agricultural land of five different villages, i.e, Dohnkal (Site A), Begowali (site B), Mansoorwali (Site C), Sohdra (Site D) and Ojlankalan (Site E) of Wazirabad. From each site, five shallow tube wells at depths ranging from ca. 20–50 m were selected as sampling points. 25 irrigation water samples (collected in pre-washed polyethylene 100 ml bottles), five composite soil samples (20–25 g at depth 0–25 cm were collected with the help of PVC pipe sampler), and five composite rice samples (total 25 samples collected in polyethylene bag), from each tube well and its surrounding cultivated land in 2014. The soil and plant samples were dried, stored at 4°C and transported to laboratory.

2.2 *Data analysis*

Hydride Generation (HG) AAS was used for the analysis of total As in all of the water and soil samples (detection limit 5 µg/L). Total As in rice plants were analyzed by ion hydride generator atomic absorption spectroscopy. The precision of the analysis was also checked by certified standard geological reference materials from Japan, such as JSD1 and JSD2 were used for the quality assurance of the results for As determinations in soil.

3 RESULTS AND DISCUSSION

3.1 *Arsenic in water, soil and plant parts*

Figure 1 shows the mean As concentration in water, soil, and different parts of plant. The results reveals that the mean As concentrations in the irrigation water are well within the FAO limits of 100 (µg/l) for irrigation water, however, the mean concentration of all water samples exceeds WHO limits of 10 µg/l for drinking water. Analysis of Irrigated rice paddy soil revealed that As concentration in soil was found to be in the range of 2.0–6.9 mg/kg, which was below the maximum acceptable limit for agricultural soil (25 mg/kg). The present study indicates that the

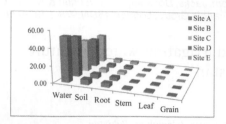

Figure 1.   Mean concentration of As in water (μg/L), soil, root, stem, leaf and grain in mg/kg) in study area.

agricultural soil of the study area accumulate As due to excessive use of As contaminated irrigation water; this can been seen by high As concentration in irrigation water and soil in both Site A and Site B as compared to other sites (Fig. 1).

The total As in grain was found to have the following range; 0.001 to 0.06 mg/kg in grain, which is less than global normal distribution range of 0.08 to 0.20 mg/kg and is also lower as compared to Bengal of 0.028 to 0.961 mg/kg (Halder *et al.*, 2014) suggesting that population is not at risk by the direct consumption of rice.

3.2   *Arsenic transfer from water to soil and plant*

The correlation coefficients (r^2) among As concentrations in irrigation water, soil and in different parts (root, stem, leaf and grain) revealed that the As content of the soil is significantly correlated with the As content of irrigation water ($r = 0.582$). This indicates that the usage of As rich irrigation water in the study area have a potential to contaminate the agricultural soil with As. High significant correlation is obtained between As concentrations in irrigation water and root ($r = 0.601$). Thus from the results it can be concluded that the As contaminated irrigation water and the agricultural field soil are highly responsible for the transfer and uptake of As in rice plant. Moreover, As uptake by plant is also affected by the rice variety/grain size which is well demonstrated by Halder *et al.* (2012) who concluded that average accumulation of As in rice grain increases with decrease of grain size.

3.3   *Future threats to sustainable agriculture*

On basis of results of As in irrigation water and soil in present study we have assessed the potential effect of As concentrations in irrigation water on soils for the coming years up to 2045 using Linear regression model. A correlation factor of As addition to soil per 10 years was calculated from the study of Brammer & Revenscroft (2009). We have estimated an increase in soil As concentration for year 2035 (52.5 mg/kg) and for 2045 (147 mg/kg). Similar trend of increasing As in soil is reported by Dittmar *et al.* (2010) that As concentration top in paddy soil will increases by a factor of 1.5–2 by the year 2050.

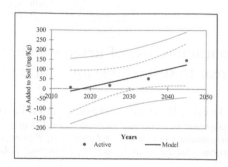

Figure 2.   Regression of As added to soil (mg/kg) by year ($R^2 = 0.8534$).

Thus, future simulation of As in paddy soil poses a threat to the economy of the country in future.

4   CONCLUSIONS

The level of As in irrigation water, soil and grain of the study area were found to be with the permissible limits. Arsenic is transferred from irrigation water and paddy soil to various parts of rice plants. We also found that consumption of rice does not pose a significant health threat to population; however, the future projections of soil As possess threats to the economy of the country in future. To avoid the future disturbance due to As contamination, there is a need to take appropriate steps.

ACKNOWLEDGEMENTS

The author acknowledges the Geological Survey of Pakistan (GSP) for analytical facilities.

REFERENCES

Brammer, H., Ravenscroft, P. 2009. Arsenic in groundwater: a threat to sustainable agriculture in South and South-east Asia. *Environ. Internat.* 35(3): 647–654.

Dittmar, J., Voegelin, A., Roberts, L.C., Hug, S.J., Saha, G.C., Ali, M.A., Badruzzaman, A.B.M., Kretzschmar, R. 2010. Arsenic accumulation in a paddy field in Bangladesh: seasonal dynamics and trends over a three-year monitoring period. *Environ. Sci. Technol.* 44: 2925–2931.

Halder, D., Biswas, A., Slekjovec, Z., Chattarjee, D., Nriagu, J., Jacks, G., Bhattacharya, P. 2014. Arsenic species in raw and cooked rice: Implications for human health in rural Bengal. *Sci. Tot. Environ.* 497: 200–208.

Halder, D., Bhowmick, S., Biwas, A., Mandal, U., Nriagu, J., Guha Mazumder, D.N., Chatterjee, D., Bhattacharya, P. 2012. Consumption of brown rice: a potential pathway for exposure in rural Bengal. *Environ. Sci. Technol.* 46: 4142–4148.

Manzoor, Z., Awan, T. H., Zahid, M. A., & Faiz, F. A. 2006. Response of rice crop (super basmati) to different nitrogen levels. *J. Animal Plant Sci.* 16(1–2): 52–55.

Arsenic Research and Global Sustainability – Bhattacharya, Vahter, Jarsjö, Kumpiene,
Ahmad, Sparrenbom, Jacks, Donselaar, Bundschuh & Naidu (Eds)
© 2016 Taylor & Francis Group, London, ISBN 978-1-138-02941-5

Dissolved ferrous iron concentration optimal for simultaneous reduction of arsenic and cadmium concentrations in rice grain

T. Honma[1], T. Tsuchida[1], H. Ohba[1], T. Makino[2], K. Nakamura[2] & H. Katou[2]
[1]*Niigata Agricultural Research Institute, Nagaoka, Japan*
[2]*National Institute for Agro-Environmental Sciences, Tsukuba, Japan*

ABSTRACT: Arsenic (As) and cadmium (Cd) in rice grain pose threats to human health. Their concentrations in rice grain are sensitive to water managements such as flooding and drainage, and resultant changes in the soil redox potential. However, there has been a paucity of reports on the soil redox conditions under which uptake of As and Cd by rice and their concentrations in grain may be simultaneously suppressed. We carried out field experiments to find the optimal water management and soil conditions, in terms of dissolved ferrous iron (Fe(II)) concentration, for simultaneous reduction of As and Cd in the soil solution and rice grain. Dissolved As and Cd concentrations during pre-heading 3 weeks and post-heading 3 weeks, which exerted strong influence on the As and Cd concentrations in grain, showed distinct relationships with the dissolved Fe(II) concentration. In a given paddy field, the dissolved Fe(II) serves as an indicator of the soil redox condition, and the concentration within the range of 20–30 mg/L was found to be optimum for simultaneous reduction of dissolved As and Cd concentrations in the paddy field tested.

1 INTRODUCTION

Rice is a staple food for the Asian population, but in some cases, it contains high levels of arsenic (As) and cadmium (Cd). Codex Alimentarius Commission adopted a maximum level of inorganic As at 0.2 mg/kg and Cd concentration at 0.4 mg/kg, respectively (Codex, 2006, Codex, 2014). Water management during the filling period is known to strongly affect the grain As and Cd concentrations. Arao *et al.* (2009) revealed the trade-off relationship between the As and Cd concentrations in rice grain. Intermittent irrigation has been shown to be a useful technique for simultaneous reduction of As and Cd concentrations (Hu *et al.*, 2013). However, there has been a paucity of data on the soil redox conditions under which uptake of As and Cd by rice and their concentrations in grain may be simultaneously suppressed. We carried out field experiments to find the optimal water management and soil conditions, in terms of the dissolved ferrous iron (Fe(II)) concentration, for simultaneous reduction of As and Cd in the soil solution and rice grain.

2 MATERIAL AND METHODS

2.1 *Experimental field*

Field experiments were conducted in 2014 in a paddy field developed on an alluvial plain in central Japan. The soil was classified as a Typic Hydraquent. The topsoil had a total carbon content of 16.2 g/kg, a total nitrogen content of 1.53 g/kg, a 1 M HCl-extractable As concentration of 2.49 mg/kg, and a 0.1 M HCl-extractable Cd concentration of 0.84 mg/kg. The textural composition of the soil was 52% sand (0.02–2 mm), 30% silt (2 µm–0.02 mm), and 18% clay (< 2 µm). The soil pH at a soil:water ratio of 1:2.5 was of 5.8.

2.2 *Field experiments*

Seedlings of rice (*Oryza. sativa* L. cv. Koshihikari) were transplanted on 19 May 2014. The field was kept flooded for 36 days after transplanting. After the subsequent midseason drainage for 14 days, five different water managements were practiced in the field plots for pre-heading 3 weeks and post-heading 3 weeks. The water management strategies included Prolonged Flooding (PF); intermittent irrigation with irrigation intervals of 6, 8, and 10 days (Int-6, Int-8, and Int-10); No Irrigation (NI). For the intermittent irrigation, the plots were kept flooded for 3 days after irrigation followed by 3-day drainage (Int-6), 5-day drainage (Int-8), or 7 day-drainage (Int-10). Each treatment had three replicated plots in the field. Rice was harvested on 17 September 2014.

2.3 *Soil and plant analysis*

Soil redox potential (Eh) was measured with platinum electrodes installed at 15 cm depth, using an Eh meter. Soil solutions were sampled at the same depth during the growth period at intervals of 1–2 weeks for determination of dissolved As, Cd, and ferrous iron (Fe(II)) concentrations. Dissolved Fe(II) concentration was determined colorimetrically from the absorbance due to the Fe(II) 2,2'-bipyridyl complex. Total dissolved As and Cd concentrations were determined by flow Injection-Inductively Coupled Plasma Mass Spectroscopy (ICP-MS). Rice grain samples were powdered and digested with HNO_3. The total As and Cd concentrations in grain were measured using ICP-MS. As(III), As(V), MMA(V) and DMA(V) in the rice grains were determined according to the methods of Baba et al. (2014) with minor modifications.

3 RESULTS AND DISCUSSION

3.1 Effects of water management on the grain As and Cd concentrations

The grain inorganic As (= As(III) + As(V)), and Cd concentrations were strongly affected by water management during pre-heading 3 weeks and post-heading 3 weeks. A distinct trade-off relationship between inorganic As and Cd concentrations in rice grain were found among the different water management plots. The grain inorganic As concentrations among the plots were in the order of PF > Int-6 > Int-8 > Int-10 > NI, of which the concentration in the PF plot exceeded the maximum level adopted by Codex. The grain Cd concentrations were in the order of NI > Int-10 > Int-8 > Int-6 > PF, of which the concentration in NI plot exceeded the allowance maximum level adopted by Codex. In the present study, the water managements Int-8 and Int-10 were found promising for simultaneous suppression of the grain As and Cd.

3.2 Relationships among dissolved As, Cd, and Fe(II) concentrations

Our previous study (Honma et al., unpublished data) showed that the inorganic As and Cd concentrations in rice grain were, respectively, strongly correlated with the averaged concentrations of dissolved As and Cd during post-heading three weeks. Figure 1 shows the relationships between the dissolved As and Cd concentrations and the dissolved Fe(II) concentration during growth period. A linear relation between the dissolved As and Fe(II) is suggestive of As solubilization by reductive dissolution of As-containing Fe (hydr)oxides. Contrastingly, the dissolved Cd was appreciable only when the dissolved Fe(II) concentration was below 10 mg/L. Both of the dissolved As and Cd concentrations were relatively low when the dissolved Fe(II) concentration was within the range of from 20–30 mg/L, corresponding to an Eh of −71 to −50 mV. The results suggest that water management which leads to the dissolved Fe(II) concentration within the above range could be effective for simultaneous reduction of grain As and Cd concentrations in the paddy field in the present study. We expect the optimum dissolved Fe(II) concentration may vary with the poorly crystalline Fe hydroxide content, the total As and Cd contents in soil as well as other soil properties such as the cation exchange capacity. In view of the paucity of data on the soil redox conditions conductive to simultaneous

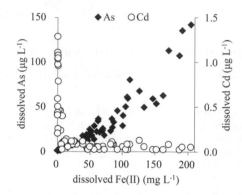

Figure 1. Relationship between the dissolved As and Cd concentrations and the dissolved Fe(II) concentration.

reduction of As and Cd uptake by rice, the optimum level of dissolved Fe(II) for suppression of dissolved As and Cd should be determined for a wider range of soil types and climate conditions.

4 CONCLUSIONS

Dissolved As and Cd concentrations at 15 cm depth of a paddy soil showed a linear relationship and a highly skewed negative correlation, respectively, with the dissolved Fe(II) concentration. For the paddy field tested in the present study, dissolved Fe(II) concentration of 20–30 mg/L was found to be optimal for simultaneous reduction of dissolved As and Cd, and their concentrations in rice.

ACKNOWLEDGEMENTS

This work was supported by a grant from the Ministry of Agriculture, Forestry, and Fisheries of the Japanese Government (Research project for improving food safety and animal health As-240).

REFERENCES

Arao, T., Kawasaki, A., Baba, K., Mori, S., Matsumoto, S. 2009. Effects of water management on cadmium and arsenic accumulation and dimethylarsinic acid concentrations in Japanese rice. Environ. Sci. Technol. 43: 9361–9367.
Baba, K., Arao, T., Yamaguchi, N., Watanabe, E., Eun, H., Ishizaka, M., 2014. Chromatographic separation of arsenic species with pentafluorophenyl column and application to rice. J. Chromatogr. A 1354: 109–116.
Codex Alimentarius Commission 2006. Report of the 29th session of the CODEX Alimentarius Commission. CODEX Alimentarius Commission, Alinorm 06/29/41.
Codex Alimentarius Commission 2014. Joint FAO/WHO food standards program, codex alimentarius commission, Thirty-seventh Session. CICG, Geneva, Switzerland 14–18 July 2014.
Hu, P., Huang, J., Ouyang, Y., Wu, L., Song, J., Wang, S., Li, Z., Han, C., Zhou, L., Huang, Y., Luo, Y., Christie, P., 2013. Water management affects arsenic and cadmium accumulation in different rice cultivars. Environ. Geochem. Hlth. 35: 767–778.

Arsenic Research and Global Sustainability – Bhattacharya, Vahter, Jarsjö, Kumpiene,
Ahmad, Sparrenbom, Jacks, Donselaar, Bundschuh & Naidu (Eds)
© *2016 Taylor & Francis Group, London, ISBN 978-1-138-02941-5*

Utilizing genetic variation and water management for cultivating low grain arsenic rice

G.J. Norton[1], T.J. Travis[1], A. Douglas[1], A.H. Price[1], M.H. Sumon[2] & M.R. Islam[2]
[1]*The Institute of Biological and Environmental Sciences, University of Aberdeen, Aberdeen, Scotland*
[2]*Department of Soil Science, Bangladesh Agricultural University, Mymensingh, Bangladesh*

ABSTRACT: Arsenic in rice grains poses a threat to human health for people that consume large quantities of rice. Numerous ways have been proposed to mitigate the accumulation of arsenic in rice grains. In this study a large number of rice varieties from the aus subgroup were screened for arsenic accumulation under conventional flooding as well as using the water management technique Alternate Wetting and Drying (AWD). These varieties were then sequenced to identify Single Nucleotide Polymorphisms (SNPs) and a SNP database was created to conduct genome wide association mapping to identify loci that control natural variation for arsenic accumulation. The AWD technique on average significantly reduced grain arsenic by ~15% when compared to rice plants grown under flooded conditions. The genetic mapping study identified a number of loci involved in the accumulation of arsenic. These results indicate that AWD and breeding are potential ways to decrease rice grain arsenic.

1 INTRODUCTION

Rice (*Oryza sativa* L.) is perhaps the most important crop plant, given that it is estimated to be the staple food of half the world's population. However, the amount of arsenic in rice grains can be a risk to human health where rice is a dietary staple, for example populations in South and South-East Asia.

Rice accumulates high concentrations of arsenic due to a number of factors, including being grown under anaerobic cultivation conditions. This method of cultivation facilitates the mobilization of arsenic into soil solution. Reducing water usage during rice cultivation can decrease arsenic accumulation by 10-fold in grain (Xu *et al.*, 2008). However, growing rice under aerobic conditions on average reduces yield.

A way to decrease water use, avoid cultivation under consistent anaerobic systems, and also not negatively affect yield is to employ a technique called Alternate Wetting and Drying (AWD). AWD combines the beneficial aspects of both aerobic and anaerobic cultivation. AWD can lower water use for irrigated rice by ~35% (Zhang *et al.*, 2009), and increase rice yield by ~10% relative to permanent flooding (Zhang *et al.*, 2009).

A number of studies have highlighted genetic variation for arsenic accumulation in rice grains. This genetic variation has the potential to be used in breeding programs to decrease arsenic in rice grains. Genome Wide Association (GWA) mapping is a technique used to identify the genomic loci which are responsible for traits. In this study 280 rice genotypes (predominantly from the *aus* subgroup of rice) were grown under both AWD and continually flooded conditions. The concentration of arsenic and other elements was determined in the shoots and grains. The rice genotypes were sequenced and a rice Single Nucleotide Polymorphism (SNP) database created. The trait data was used to conduct GWA mapping. This paper explores the effect of AWD on different elements in the rice plants, as well as genomic loci identified in the GWA study.

2 METHODS

2.1 *Field conditions*

A field trial was conducted at the Bangladesh Agricultural University, Mymensingh (Figure 1). Two different irrigation techniques were tested; for each treatment four replicate blocks in a complete random design were used. The water irrigation techniques used were continually flooded and Alternate Wetting and Drying (AWD). The field under AWD underwent 4 AWD cycles. The continually flooded field was kept flooded (2–5 cm of standing water) through the experiment until the start of flowering.

Prior to the first AWD cycle the AWD field was kept flooded to the same depth as the continually flooded field. The first AWD cycle started ~ 20 days after transplanting. A cycle is when the water level in the field is allowed to drop to the point where the soil is saturated at a depth of 15 cm below the soil surface; once the water drops to this point the field is reflooded and the cycle repeated. The water depth is monitored by perforated tubes in the field. As with the continually flooded treatment, at the start of flowering the AWD cycling was stopped.

The experiment consisted of 280 rice genotypes (260 identified as *aus* rice genotypes and 20 check

Figure 1. Field experiment at the Bangladesh Agricultural University, Mymensingh.

genotypes). Prior to this experiment these rice genotypes had been sequenced and a SNP database, consisting of ~2 million SNPs, had been developed.

2.2 Elemental analysis

Once the plants had flowered and the grain reached maturity, the plants were hand harvested and the grains separated from the shoot material. The grains were then dehusked and the shoot material powderised, prior to being acid digested. Total elemental analysis was conducted using inductively coupled plasma mass spectrometry. Rice flour (NIST 1568b) was used as a certified reference material.

2.3 Genome wide association mapping

GWA mapping was conducted on individual phenotypes using EMMAX with a kinship matrix and the IBS (Identity By State) method to correct for stratification in the population of rice cultivars. The GWAS results were visualized in R as 'Manhattan' plots using "QQplot" (Turner, 2014).

3 RESULTS AND DISCUSSION

3.1 Effect of AWD on grain arsenic concentrations

There were both AWD treatment effects and genotypic differences for the grain elements measured, as well as genotype by treatment interactions for some elements (Table 1). For grain arsenic the average arsenic concentration across all the genotypes grown under the continually flooded conditions was 0.476 mg/kg, while under AWD it was significantly less, with an average concentration of 0.401 mg/kg. Therefore, growing the genotypes under AWD reduced grain arsenic by 15.8% on average. There was also a wide range in grain arsenic between the genotypes; grain arsenic ranged from 0.231–0.675 mg/kg for the plants grown under AWD, while for the plants grown under continually flooded conditions the grain arsenic concentrations were from 0.287–0.764 mg/kg.

3.2 Genome wide association mapping of grain arsenic

GWA mapping revealed a large number of significant SNPs for grain arsenic. These map to various locations across the rice genome. One region that is of interest is on the top of chromosome 3. A number of SNPs identified as being significantly associated with grain arsenic were located at approximately 0.46 Mb.

Table 1. Summary of genotypic and treatment effects on selected elements in rice grains. The values is the f-value from the ANOVA.

	ANOVA results			
Element	Genotypic (G) difference	Treatment (T) effect	Direction of treatment effect	G x T
Fe	2.0***	46.8***	↓	NS
Cu	2.0***	2804.5***	↑	NS
Zn	5.7***	72.5***	↑	NS
As	2.73***	174.2***	↓	1.3***
Cd	15.0***	889.2***	↑	7.6***

NS = not significant, *p <0.05, **p <0.01 and ***p <0.001. ↑ = AWD increased element concentration (compared to continually flooded), ↓ = AWD decreased element concentration (compared to continually flooded).

In this region are two potential candidate genes that could affect grain arsenic. One is Lsi2, a gene identified to be responsible for arsenic transport (Ma et al., 2007) and the other a gene which is annotated as an arsenate reductase gene (Chao et al., 2014). These loci are currently undergoing further analysis.

4 CONCLUSIONS

This study highlights the potential of both field management and genetics in reducing grain arsenic. Appropriate use of water management can reduce grain arsenic without reducing grain yield (data not shown). The genetic mapping highlights a number of genomic regions that could be used in breeding to reduce grain arsenic, however these still need further investigation.

ACKNOWLEDGMENTS

This work was funded by BBSRC research project BB/J00336/1.

REFERENCES

Chao, D.Y., Chen, Y., Chen, J., Shi, S., Chen, Z., Wang, C., Danku, J.M., Zhao, F.-J., Salt, D.E. 2014. Genome-wide association mapping identifies a new arsenate reductase enzyme critical for limiting arsenic accumulation in plants. PLoS Biol 12(12): e1002009.

Ma, J, Yamaji, N., Mitani, N., Tamai, K., Konishi, S., Fujiwara, T., Katsuhara, M., Yano, M. 2007. An efflux transporter of silicon in rice. Nature 448:209–212.

Turner, S.D. 2014. Qqman: An R Package for Visualizing GWAS Results Using Q-Q and Manhattan Plots. bioRxiv. Cold Spring Harbor Labs Journals. doi:10.1101/005165.

Xu, X.Y., McGrath. S.P., Meharg, A.A., Zhao, F.J. 2008. Growing rice aerobically markedly decreases arsenic accumulation. Environ. Sci. Technol. 42:5574–79.

Zhang, H., Xue, Y., Wang, Z., Yang, J., Zhang, J. 2009. Alternate wetting and moderate soil drying improves root and shoot growth in rice. Crop Sci. 49: 2246–2260.

Arsenic Research and Global Sustainability – Bhattacharya, Vahter, Jarsjö, Kumpiene,
Ahmad, Sparrenbom, Jacks, Donselaar, Bundschuh & Naidu (Eds)
© *2016 Taylor & Francis Group, London, ISBN 978-1-138-02941-5*

Replacement of arsenic-contaminated soil for improved rice yields in Bangladesh

B.L. Huhmann[1], C.F. Harvey[1], J. Duxbury[2], A. Uddin[3], I. Choudhury[3], K.M. Ahmed[3],
B. Bostick[4] & A. van Geen[4]

[1]*Department of Civil and Environmental Engineering, MIT, Cambridge, Massachusetts, USA*
[2]*School of Integrative Plant Science, Cornell University, Ithaca, New York, USA*
[3]*Department of Geology, University of Dhaka, Dhaka, Bangladesh*
[4]*Lamont-Doherty Earth Observatory, Columbia University, New York, USA*

ABSTRACT: Winter season (boro) rice in Bangladesh is often irrigated with high-arsenic groundwater. This arsenic builds up in soil and reduces rice yield. To investigate the yield effects of arsenic, we exchanged the top fifteen centimeters of soil between high-arsenic (near the irrigation inlet) and low-arsenic (far from the irrigation inlet) plots in Faridpur district, Bangladesh. We measured soil arsenic and rice yield for thirteen plots where the soil was exchanged and thirteen adjacent plots where the soil was not exchanged. The average yield during these experiments was 5.9 t/ha and the average soil As was 20.6 mg/kg. We observed a negative correlation between difference in yield (ranging from +2.3 to −1.9 t/ha) and difference in soil As (ranging from −8.8 to +19.1 mg/kg) between the plots. This indicates that replacing high As soil with low As soil has the potential to improve rice yields.

1 INTRODUCTION

Much of the groundwater in Bangladesh is contaminated with high levels of arsenic, posing a risk to food security when this water is used for irrigation. Rice comprises 71% of caloric consumption in Bangladesh (FAOSTAT), and winter season (boro) rice is grown under flooded conditions sustained by groundwater irrigation. Arsenic from irrigation water builds up in soil over time (Meharg & Rahman, 2003) and can be taken up by rice plants, leading to decreased yields (Panaullah *et al.*, 2009).

2 METHODS

The study was conducted in fields irrigated by high-arsenic wells in Faridpur district, Bangladesh. The wells drew water from 40 to 100 m in depth and had arsenic concentrations of 200–1000 µg/L as measured with the ITS Econo-Quick field kit.

In January before the fields were transplanted with boro rice, we exchanged the top fifteen centimeters of soil between thirteen 5 × 5 m high-arsenic (near the irrigation inlet) and low-arsenic (far from the irrigation inlet) plots. We measured soil arsenic for the twenty-six plots where soil was exchanged ("soil replacement plots") and twenty-six adjacent plots where soil was not exchanged ("control

plots"). We also measured rice yields for thirteen pairs of soil replacement and control plots.

We measured soil arsenic using the ITS Econo-Quick field kit based on the Gutzeit method, with 10 minute extractions of 0.5 g of soil in 50 mL of water. We validated this method by comparing kit As measurements with X-Ray Fluorescence (XRF) spectrometer measurements on 102 soil samples and found a good correlation ($R^2 = 0.64$) across the 1–120 mg/kg range of concentrations.

Soil arsenic was measured in 5 cm increments from 0 to 20 cm on a single soil core taken from each plot during months two, three, and four after transplanting and on four sets of three 20 cm cores collected from each plot at harvest. EC, pH, organic carbon, N, P, K, S, and Zn for all cores were measured at the BRAC soil laboratory in Gazipur, Bangladesh.

Rice yields were measured for a 3 × 3 m area in the center of each 5 × 5 m plot. The rice was threshed immediately after harvest, its weight and moisture content were recorded, and yield values were adjusted to 14% moisture content.

3 RESULTS AND DISCUSSION

Kit soil arsenic concentrations across the top 20 cm averaged 5.8 ± 0.7 mg/kg in the thirteen low arsenic control plots, and 9.7 ± 1.4 mg/kg in the adjacent originally low arsenic soil replacement plots, indicating that replacing the top 15 cm with high arsenic soil increased soil arsenic concentrations, as expected. Similarly, soil arsenic concentrations were 17.9 ± 2.0 mg/kg in the high arsenic control plots and 12.6 ± 2.2 mg/kg in the originally high arsenic soil replacement plots, indicating a decrease in arsenic with soil replacement (Fig. 1). However, the shift in soil arsenic concentrations was not complete, with soil arsenic values in the replacement plots intermediate to those in the control plots, possibly because of insufficient soil removal and/or subsequent mixing of the soil by tilling.

To investigate the effects of soil arsenic on rice yield, we computed the difference in soil arsenic measured at harvest and difference in rice yield between thirteen soil replacement and adjacent control plots. Taking the differences between pairs of adjacent plots allowed us to control for other factors that may affect yield, such as fertilizer and pesticide use, farmer care (e.g. weeding), transplanting and harvesting dates, irrigation water source, rice variety, and other variations in local conditions (e.g. shade). We also verified that soil chemistry and nutrient content (EC, pH, organic carbon, N, P, K, S, and Zn) were similar between the paired plots.

We observed a negative correlation between difference in rice yield and difference in soil As for both rice varieties planted in our study plots (Fig. 2). The difference in rice yield ranged from +2.3 to −1.9 t/ha (average yield of 5.9 t/ha) and the difference in soil arsenic ranged from −8.8 to +19.1 mg/kg (average soil As of 20.6 mg/kg). In addition to the negative slope, the relationship between yield difference and soil arsenic difference

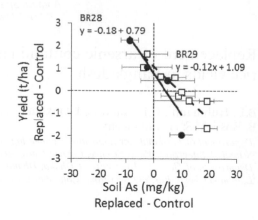

Figure 2. Yield difference as a function of soil arsenic difference between replacement and control plots for rice varieties BR28 (filled circles) and BR29 (unfilled squares).

has a positive intercept, indicating that the process of swapping the soil itself (regardless of change in soil arsenic) provided an increase in yield of about 1 t/ha. This may be due to physical and/or chemical changes in the soil introduced during the soil swap, such as breaking up or aerating the soil.

4 CONCLUSIONS

We have shown that soil arsenic negatively impacts boro rice yield and that replacement of arsenic-contaminated soil can restore yield. We have also shown that the field kit provides a promising method for farmers to rapidly and affordably identify their high-arsenic soils, and thus could be used to determine areas suitable for mitigation. Given the demonstrated negative effects of soil arsenic on rice yields, and the fact that soil arsenic will continue to build up unless farmers find a source of low-arsenic irrigation water, soil remediation should be seriously considered to improve agricultural productivity in Bangladesh.

REFERENCES

Food and Agriculture Organization of the United Nations. FAOSTAT http://faostat3.fao.org
Meharg, A.A., Rahman, M.M. 2003. Arsenic contamination of bangladesh paddy field soils: implications for rice contribution to arsenic consumption. *Environ. Sci. Technol.* 37(2): 229–234.
Panaullah, G.M., Alam, T., Hossain, M.B., Loeppert, R.H., Lauren, J.G., Meisner, C.A., Ahmed, Z.U., Duxbury, J.M. 2009. Arsenic toxicity to rice (Oryza sativa L.) in Bangladesh. *Plant Soil* 317(1): 31–39.

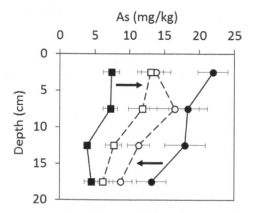

Figure 1. Kit soil arsenic depth profiles averaged across the thirteen low arsenic control (filled squares), low arsenic replacement (unfilled squares), high arsenic replacement (unfilled circles), and high arsenic control (filled circles) plots.

Arsenic Research and Global Sustainability – Bhattacharya, Vahter, Jarsjö, Kumpiene,
Ahmad, Sparrenbom, Jacks, Donselaar, Bundschuh & Naidu (Eds)
© *2016 Taylor & Francis Group, London, ISBN 978-1-138-02941-5*

Countermeasures against arsenic and cadmium contamination of rice

T. Arao[1], T. Makino[2], S. Ishikawa[2], A. Kawasaki[2], K. Baba[2] & S. Matsumoto[3]
[1]*Central Region Agricultural Research Center, NARO, Tsukuba, Ibaraki, Japan*
[2]*Institute for Agro-Environmental Sciences, NARO, Ibaraki, Japan*
[3]*Shimane University, Shimane, Japan*

ABSTRACT: The Codex Alimentarius Commission has decided a Maximum Level for inorganic As in polished rice in 2014 to protect consumers' health risk. Flooding cultivation increases As concentration in rice grains, whereas aerobic treatment increases the concentration of Cd. Our strategy is a combination of flooded cultivation to decrease Cd and the application of iron materials to decrease As. Applications of materials lowered the As concentration in rice grain under flooded conditions. Using nontransgenic rice mutants which had nearly non-detectable levels of Cd in the grains, we have been testing the hypothesis that growing the low Cd cultivar aerobically simultaneously reduce As and Cd.

1 INTRODUCTION

Inorganic Arsenic (As) has been identified as a human carcinogen. Rice tends to be a major source of inorganic arsenic from food, particularly in Asia and other countries where it is a staple food. In some cases, human intake of As from the consumption of rice exceeds that from drinking water. A market-basket survey indicated that rice and rice cake contributed about 60% to total daily intake of inorganic As in the Japanese population (Oguri. *et al.*, 2014). To protect consumers from excessive exposure, the Codex Alimentarius Commission agreed a Maximum Level (ML) for inorganic As in polished rice at 0.2 mg/kg (2014). On the other hand, the Codex ML for cadmium (Cd) is 0.4 mg/kg for polished rice (Ref). Since rice is a staple crop in Japan and Asian countries, reducing the concentration of As and Cd in rice is an urgent matter.

Flooding increased As concentrations in rice grains, whereas aerobic treatment increased the concentration of Cd. Flooding for both 3 weeks before and after heading was most effective in reducing grain Cd concentrations, but this treatment increased the As concentration considerably, whereas aerobic treatment during the same period was effective in reducing As concentrations but increased the Cd concentration markedly (Arao *et al.*, 2009). Concentrations of dimethylarsinic acid (DMA) in grain were very low under aerobic conditions but increased under flooded conditions. DMA accounted for 3–52% of the total As concentration in grain grown in soil with a lower As concentration and 10–80% in soil with a higher As concentration.

Prospective As adsorbents under flooded anaerobic conditions were selected (Suda *et al.*, 2015). The objective of this study is to investigate the simultaneous effects on As and Cd levels in rice grains induced by water management and soil amendment of paddy field.

2 METHODS/EXPERIMENTAL

2.1 *Pot experiments*

Pot experiments with three replications each were performed in a greenhouse at ambient temperatures under sunlight. Wagner pots (0.02 m²) were filled with 3 kg of soil B as stated in Arao *et al.* (2009). Various soil amendments were added for rice cultivation. Materials a, c and e are $FeSO_4$, metal Fe powder and Fe oxide material respectively. Material b is a silicates and material d is an aluminum oxide. Rice (*Oryza sativa* L. cv. Koshihikari) seedlings were germinated on perlite and then transplanted to 1/5000a pots. Eh was measured at a depth of 10 cm. A soil-water sampler was buried in the middle of the soil of each pot for collecting soil solution. Th soil solution was sampled 33, 54, 70, 83 and 90 days after transplanting and diluted with 10% HNO_3 at a ratio of 9:1 immediately after collection and filtered through a sterilized 0.45-µm filter. The rice was grown under flooded conditions for five weeks from transplanting, followed by intermediate irrigation for three weeks. For intermediate irrigation planted pots were irrigated daily. Irrigation volume was decided according to the daily weather so that Eh value would not be below zero. During heading stage of rice, 4 water-management treatments were examined in the experiment: Treatment 1 involved flooding for five weeks; treatment 2, flooding for two weeks until heading time followed by under intermediate irrigation for three weeks; treatment 3, under intermediate irrigation for two weeks until heading time followed by flooding for three weeks; treatment 4, under intermediate irrigation for five weeks. Then the rice was grown under drainage conditions before harvest.

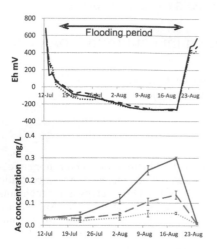

Figure 1. Effects of material e on soil redox potential and As concentrations in soil solution.

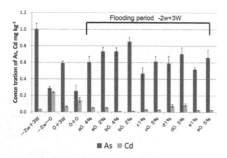

Figure 2. Effects of materials on concentration of As and Cd in brown rice.

2.2 Analysis of arsenicals

The grains and straw samples were each dried and ground in a mill to a fine powder, and samples were digested in 5:1 (v/v) HNO_3/H_2O_2 in a microwave oven. Concentrations of Cd and As in the digested samples and soil solution were determined by ICP-MS. To determine As speciation in rice grain, a powdered sample of ground rice was extracted with HNO_3 (0.15 M) at 80°C for 2 h. The As speciation in the grain of rice were determined by HPLC/ICP-MS (Baba et al., 2014). Accuracy was evaluated by using a certified reference material (rice flour, NMIJ CRM 7503-a, National Institute of Advanced Industrial Science and Technology, Japan).

3 RESULTS AND DISCUSSION

3.1 Soil eh and as concentration in soil solution

Although applications of the material e did not affect the change of Eh, it lowered As concentration in soil solution under flooded condition. Material e is an iron-containing material and

arsenic in the soil solution might be adsorbed by this material.

3.2 As and Cd concentration in brown rice

Applications of materials lowered the As concentration in brown rice grain. Although the effect of drainage after heading stage on As concentration in rice grain is larger than the effect of application of materials, Cd concentration increased under drained condition. But application of materials b, c and e did not increase Cd concentration in rice grain.

3.3 Ongoing studies

Paddy field experiments to investigate the effects of iron containing materials on mobility and availability of As have been conducted at five regions in Japan. We developed nontransgenic rice mutants with mutations of OsNramp5 gene, which had nearly nondetectable levels of Cd in the grains (Ishikawa et al., 2012). Using this cultivar, we have been testing the simultaneous effects of water managements on As and Cd in the grains.

4 CONCLUSIONS

The application of iron containing materials is effective to reduce As concentration in soil solution and rice grain.

ACKNOWLEDGEMENTS

This work was supported by a grant from the MAFF of Japan (Research Project for Ensuring Food Safety from Farm to Table [AC-1122] and Improving Food Safety and Animal Health As-110).

REFERENCES

Arao, T., Kawasaki, A., Baba, K., Mori, S., Matsumoto, S. 2009. Effects of water management on cadmium and arsenic accumulation and dimethylarsinic acid concentrations in Japanese rice. Environ. Sci. Technol. 43: 9361–9367.

Baba, K., Arao, T., Yamaguchi, N., Watanabe, E., Eun, H., Ishizaka, M. 2014. Chromatographic separation of arsenic species with pentafluorophenyl column and application to rice. J. Chromatogr. A, 1354: 109–116.

Ishikawa, S., Ishimaru, Y., Igura, M., Kuramata, M., Abe, T., Senoura, T., Hase, Y., Arao, T., Nishizawa, N., Nakanishi, H. 2012. Ion-beam irradiation, gene identification, and marker-assisted breeding in the development of low-cadmium rice. Proc. Natl. Acad. Sci. U.S.A. 109: 19166–19171.

Oguri T. Yoshinaga J. Nakazato T. 2014. Inorganic Arsenic in the Japanese Diet: Daily Intake and Source. Arch. Environ. Contam. Toxicol. 66: 100–112.

Suda A. Baba K. Yamaguchi N. Akahane I. Makino T. 2015. The effects of soil amendments on arsenic concentrations in soil solutions after long-term flooded incubation. Soil Sci Plant Nutri. 61, 592–602.

Arsenic Research and Global Sustainability – Bhattacharya, Vahter, Jarsjö, Kumpiene, Ahmad, Sparrenbom, Jacks, Donselaar, Bundschuh & Naidu (Eds)
© 2016 Taylor & Francis Group, London, ISBN 978-1-138-02941-5

Water management for arsenic and cadmium mitigation in rice grains

W.-I. Kim[1], A. Kunhikrishnan[1], M.-K. Paik[1], J.-H. Yoo[1], N. Cho[1] & J.-Y. Kim[2]
[1]*Chemical Safety Division, National Institute of Agricultural Science, Wanju, Republic of Korea*
[2]*Hazardous Substance Analysis Division, Gwangju Regional FDA, Gwangju, Republic of Korea*

ABSTRACT: Managing arsenic (As) and cadmium (Cd) together in rice (*Oryza sativa* L.) plants is challenging and different strategies are being developed for mitigating As and Cd loading into the rice grains. This study investigated the effect of water management on As and Cd accumulation in brown rice. A field plot experiment was conducted with five water management regimes [Flooded control, alternate wetting drying (AWD – 60 and 40) and row (R-60 and R-40)] using two rice cultivars (Indica and Japonica). Compared to the flooded control, all the four treatments significantly reduced the concentration of As in brown rice with R-40 showing the least concentration. AWD and row treatments reduced As levels by 45–60% and by 32–55% in Indica and Japonica cultivars, respectively. However, increased Cd concentrations were noticed in row and AWD treatments. AWD-60 treatment for As and Cd in Indica cultivar reduced As without greatly increasing Cd concentration in brown rice. AWD water management offers some promising solutions, however, additional field studies and As bioaccessibility research are required to control both As and Cd in paddy soils and rice grains.

1 INTRODUCTION

Managing low concentrations of arsenic (As) and cadmium (Cd) together in rice plants has always been a serious issue and there is an urgent need to develop strategies for reducing As and Cd loading into the rice grains. Numerous studies have reported that water management affects Cd and As (Hu *et al.*, 2013a, b, Linquist *et al.*, 2015) bioavailability in soils and their subsequent uptake by rice. When a paddy field is flooded and the soil has a low redox potential, any Cd present in the soil combines with sulfur (S) to form CdS which has a low solubility in water. Thus, flooding during the growing season, especially during later stages of plant growth, can effectively reduce Cd concentrations in rice grains (Arao *et al.*, 2009). In contrast, anaerobic conditions in paddy soil lead to the reduction in As(V) to As(III) which enhances the bioavailability of As to rice plants. Therefore, growing rice aerobically results in a significant decline in As accumulation in rice (Xu *et al.*, 2008). However, both Cd and As can occur together as contaminants in paddy fields, and they can accumulate simultaneously in rice plants (Williams *et al.*, 2009).

To help reduce water consumption during rice cultivation there has been considerable interest in expanding the aerobic cultivation practices employed in upland rice to lowland environments where anaerobic paddy cultivation is traditional. However, the reduced yields and pest control problems associated with aerobic cultivation need to be addressed. One major recent advance in rice water management is termed Alternate Wetting and Drying (AWD) (Price *et al.*, 2013). AWD combines the beneficial aspects of both aerobic and anaerobic cultivation. With all these points taken into consideration, our study investigated the effect of water management on As and Cd accumulation in brown rice in two rice cultivars grown in Korea.

2 METHODS/EXPERIMENTAL

Arsenic- and Cd-contaminated paddy fields near a mining area were selected for the field experiments. The water management experiment was conducted in Seosan city in Korea. The total As concentrations in Seosan plots was 182.4 mg/kg. The Seosan field was severely contaminated with Cd (total −28.2 mg/kg). Two rice cultivars Indica and Japonica, grown in Korea, were used for the water management experiment. Five water management treatments were laid out in a randomized complete block design and replicated three times. Treatments were: (i) Flood (continuously flooded control), (ii) AWD/60, (iii) AWD/40, (iv) row/60 (R-60) and (v) row/40 (R-40), where AWD represents alternate wetting and drying. For the AWD water treatments, the plots were irrigated to a flood depth of 10 cm and the water was allowed to subside via evapotranspiration and percolation until soil moisture reached the critical moisture level for that treatment (60 and 40% of saturated volumetric water) when the fields were reflooded. For the row treatments, rice plants were planted on beds and watering by furrows until soil in beds reaches the critical moisture level (60% and 40% of field capacity). In the flood treatment, water was maintained at 10 cm. The total As and Cd concentrations in brown rice samples were analysed using ICP-MS after microwave digestion.

3 RESULTS AND DISCUSSION

Compared to the flooded treatment, AWD and row treatments significantly reduced the concentration of As in brown rice with R-40 showing the least concentration of As. AWD and row treatments reduced As levels by 45–60% and by 32–55% in Indica and Japonica cultivars, respectively. In the case of Cd, the trend was opposite; increased Cd concentrations were noticed in row and AWD treatments compared to the flooded treatment. Compared to the flooded control, Cd levels increased by 1.3–1.8 and 1.1–1.6 times in Indica and Japonica cultivars, respectively, with R-40 treatments showing a significant increase. AWD-60 treatments for As and Cd in Indica cultivars indicate reduction in As without greatly increasing Cd concentration in brown rice.

Uptake of Cd and As has been shown to differ with rice cultivars (Hu *et al.*, 2013a). With increasing irrigation from aerobic to flooded conditions, the As concentrations increased significantly in the straw, husk and brown rice, whereas the Cd concentrations decreased (Hu *et al.*, 2013a, b). They also observed that the intermittent and conventional treatments produced higher grain yields than the aerobic and flooded treatments. In this study, different water regimes influenced the grain yields of both cultivars. In the AWD and row treatments, only a slight decrease in grain yield was noticed when compared to control and Japonica displayed a slight increase in yield compared to Indica. Yang *et al.* (2009) observed that alternate wetting and moderate soil drying reduces Cd in rice grains and increases grain yield. Recently, Linquist *et al.* (2015) noticed that, relative to the flooded control treatment and depending on the AWD treatment, yields were reduced by < 1–13% but the grain As concentrations reduced by up to 64%.

4 CONCLUSIONS

Arsenic and Cd contamination threatens food security, safety and quality, as well as the long-term agricultural sustainability of rice crops.

Managing low concentrations of As and Cd together in rice plants is challenging. Water management affects the bioavailability of As and Cd in the soil and hence their accumulation in rice grains and grain yields. While As is immobilized under oxidizing conditions and solubilized under reducing conditions these trends are the opposite in Cd. This study indicates that AWD water managements offer some promising solutions to mitigate As and Cd, however, additional field studies using different rice cultivars are required to control both As and Cd in paddy soils and their uptake in rice grains.

REFERENCES

Arao, T., Kawasaki, A., Baba, K., Mori, S., Matsumoto, S., 2009. Effects of water management on cadmium and arsenic accumulation and dimethylarsinic acid concentrations in Japanese rice. *Environ. Sci. Technol.* 43: 9361–9367.

Hu, P., Huang, J., Ouyang, Y., Wu, L., Song, J., Wang, S., Li, Zhu., Han, C., Zhou, L., Huang, Y., Luo, Y., Christie, P., 2013a. Water management affects arsenic and cadmium accumulation in different rice cultivars. *Environ. Geochem. Heal.* 35: 767–778.

Hu, P., Li, Zhu., Yuan, C., Ouyang, Y., Zhou, L., Huang, J., Huang, Y., Luo, Y., Christie, P., Wu, L., 2013b. Effect of water management on cadmium and arsenic accumulation by rice (*Oryza sativa* L.) with different metal accumulation capacities. *J. Soil Sed.* 13: 916–924.

Linquist, B.A., Anders, M.M., Adviento-Borbe, M.A.A., Chaney, R.L., Nalley, L.L., da Rosa, E.F.F., van Kessel, C., 2015. Reducing greenhouse gas emissions, water use, and grain arsenic levels in rice systems. *Glob. Change Biol.* 21(1): 407–17.

Price, A.H., Norton, G.J., Salt, D.E., Ebenhoeh, O., Meharg, A.A., Meharg, C., Islam, R., Sarma, R.N., Dasgupta, T., Ismail, A.M., McNally, K.L., Zhang, H., Dodd, I.C., Davies, W.J., 2013. alternate wetting and drying irrigation for rice in Bangladesh: Is it sustainable and has plant breeding something to offer? *Food and Energy Security* 2(2): 120–129.

Williams, P.N., Lei, M., Sun, G.X., Huang, Q., Lu, Y., Deacon, C., Meharg, A.A., Zhu, Y.G., 2009. Occurrence and partitioning of cadmium, arsenic and lead in mine impacted paddy rice: Hunan, China. *Environ. Sci. Technol.* 43: 637–642.

Xu, X.Y., McGrath, S.P., Meharg, A.A., Zhao, F.J., 2008. Growing rice aerobically markedly decreases arsenic accumulation. *Environ. Sci. Technol.* 42: 5574–5579.

Yang, J., Huang, D., Duan, H., Tan, G., Zhang, J., 2009. Alternate wetting and moderate soil drying increases grain yield and reduces cadmium accumulation in rice grains. *J. Sci. Food Agr.* 89: 1728–1736.

2.2 Arsenic in marine species

Arsenic Research and Global Sustainability – Bhattacharya, Vahter, Jarsjö, Kumpiene,
Ahmad, Sparrenbom, Jacks, Donselaar, Bundschuh & Naidu (Eds)
© *2016 Taylor & Francis Group, London, ISBN 978-1-138-02941-5*

Study on the analytical method for arsenic species in marine samples using ion chromatography coupled with mass spectrometry

S. Cui[1], J.H. Lee[2] & C.K. Kim[1,2]
[1]*Neo Environmental Business Company, Korea*
[2]*Korea Research Institute of Standards and Science, Daejeon, Korea*

ABSTRACT: The chemical speciation of element has been important because its toxic and biological effects cannot be determined only by the information of the total element, but they depend on the particular chemical forms of an element. This study was focused on the analytical methods for the separation and quantitation of arsenic species (As^{3+}, As^{5+}, MMA, DMA and AsB) in marine samples. Arsenic species in samples were extracted by the solvent of methanol: water (1:1,v:v) and separated using a Hamilton PRP-X100 analytical column. And then the concentrations of each species were selectively detected by mass spectrometry. The recovery efficiency of this method was 86.4%, and good reproducibility was also obtained. Due to the high content of chloride in the marine samples, we will focus on the interference caused by ArCl at mass 75 in the future study. We will try to detect arsenic at mass 91 by combining arsenic with oxygen (AsO).

1 INTRODUCTION

The quantification of the total arsenic in environment samples has routinely been performed. But, it has been very well known that toxicity and biological effect depend on the chemical species of arsenic in samples. Of the chemical species of arsenic, inorganic arsenic As^{3+} and As^{5+} are highly toxic. Therefore, the importance on research of arsenic species has greatly increased. Speciation of arsenic species in environmental samples is essential in order to provide a meaningful assessment of exposure due to differences in toxicities of the chemical forms.

2 METHODS/EXPERIMENTAL

2.1 Experimental

A Thermo Xseries II mass spectrometry was used for the determination of As. Operating conditions were shown as Table 1. Arsenic was detected at the mass of 75. A DIONEX ICS-5000+ ion chromatography coupled with a Hamilton PRP-X100 analytical column was used for the separation of the arsenic species. The operating conditions were shown in Table 2.

2.2 Methods

The freeze-drying seafood samples used in this study were crushed and homogenized. A CRM (Oyster Powder, KRISS) was also used for the validation of our analytical method. A 0.4000 g of a dry powder sample and 20 mL of 50% methanol (1:1, v/v) were added to a 50 mL falcon tube, and

Table 1. Operating conditions for mass spectrometry.

Plasma gas flow rate	Ar, 15 L/min
Carrier gas flow rate	Ar, 1.0 L/min
Auxiliary gas flow rate	Ar, 0.7 L/min
CCT gas flow rate	H_2 7%, He 5.5 mL/min
Pump flow rate	1.0 mL/min
ICP-RF Power	1400 W
Torch	Standard torch
Nebulizer	Meinhard concentric type
Spray chamber	Single pass type
Sampler cone	Nickel
Skimmer cone	Nickel

Table 2. Operating conditions for ion chromatography.

Analytical column	Hamilton PRP-X100 analytical column
Guard colmn	Hamilton PRP-X100 guard column
Flow rate	1.0 mL/min
Sample loop	100 μL
Mobile phase A	20 mM ammonium carbonate
Mobile phase B	50 mM ammonium carbonate

sonicated for 20 min. After sonicating, the sample and extraction solution were separated using a centrifuge. The extracted solution was filtered through a 0.2 μm filter to a 50 mL beaker. The solvent was removed by a thermostatic bath to about 1 mL (eliminate the methanol). The beaker was washed with DI water, and the solution was transfer to a 50 mL falcon tube making the final volume to 20 mL. Five arsenic species (As^{3+}, As^{5+}, MMA, DMA and AsB) in the sample solution was detected by IC-ICP-MS.

3 RESULTS AND DISCUSSION

3.1 *Validation of the analytical method*

A CRM (Oyster Powder, KRISS) was used for the validation of our analytical method. The CRM was determined using the analytical. The chromatogram was shown in Figure 1. The recovery efficiency and reproducibility were also determined in this study. The results were shown in Tables 3 and 4.

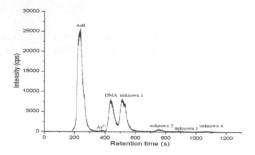

Figure 1.   Chromatogram of arsenic species in the CRM.

Table 3.   Recovery efficiency of the method.

	AsB
Theoretical value (mg/kg)	6.6 ± 0.31
Measure value (mg/kg)	5.71 ± 0.05
Recovery efficiency (%)	86.4

Table 4.   Reproducibility of the method (mg/kg).

Runs	AsB
First time	5.77
Second time	5.70
Third time	5.66
Fourth time	5.68
Fifth time	5.81
Average	5.72
SD	0.06
RSD	0.01

Table 5.   Results for marine samples (in mg/kg).

Runs	AsB	As^{3+}	DMA	MMA	As^{5+}
Konosirus punctatus	1.18 ± 0.04	ND	1.35 ± 0.07	ND	ND
Chelon haematocheilus	1.44 ± 0.02	ND	ND	ND	ND
Synechogobius hasta	6.79 ± 0.48	ND	ND	ND	ND
Sebastes schlegeli	3.09 ± 0.12	ND	ND	ND	ND
Sebastes koreanus	3.80 ± 0.05	ND	ND	ND	ND
Lateolabrax maculatus	7.44 ± 0.03	ND	ND	ND	ND
Zoarces gilli	10.2 ± 0.49	ND	ND	ND	ND
Tapes philippinarum	5.48 ± 0.23	ND	ND	ND	ND
Mactra veneriformis	4.20 ± 0.20	ND	ND	ND	ND

ND: Not Detected.

3.2 *Application*

The analytical method was applied to various marine samples. The results were shown in Table 5.

3.3 *Future study*

At present the methods generally used for removing the interference of ArCl in the detection of arsenic could not completely eliminate the interference. Therefore, we will combine Arsenic with Oxygen (AsO) and detect arsenic at mass 91 in order to avoid the interference. A certain percentage of oxygen (mixture of oxygen and Helium) will be used as the CCT gas. And the parameters will be adjusted to obtain the best results.

4 CONCLUSIONS

The quantitative determination of arsenic species in marine samples was performed by IC-ICP-MS. The 50% methanol (1:1, v/v) was used as the extraction solvent for arsenic species. Five arsenic species studied in this study were separated by a Hamilton PRP—X100 analytical column and the peaks of the species occurred within 1200 s. The recovery efficiency of this method validated with the certified reference material was 86.4%. And good reproducibility was also obtained from the method. The method was applied to various marine samples. The results showed that the main arsenic species determined was AsB.

ACKNOWLEDGEMENTS

This research was a part of the project titled 'Development of techniques for assessment and management of hazardous chemicals in the marine environment', funded by the Ministry of Oceans and Fisheries, Korea.

REFERENCES

Pizarro I. 2003. Arsenic speciation in environmental and biological samples Extraction and stability studies, *Anal. Chim. Acta* 495: 85–98.

Wenhua Geng. 2009. Arsenic speciation in marine product samples: Comparison of extraction–HPLC method and digestion–cryogenic trap method, *Talanta* 79: 369–375.

Nam, S.-H. 2010. A study on the extraction and quantitation of total arsenic and arsenic species in seafood by HPLC-ICP-MS. *Microchem. J.* 95: 20–24.

Leufroy, A.. 2011. Determination of seven arsenic species in seafood by ion exchange chromatography coupled to inductively coupled plasma-mass spectrometry following microwave assisted extraction: Method validation and occurrence data. *Talanta* 83: 770–779.

Arsenic Research and Global Sustainability – Bhattacharya, Vahter, Jarsjö, Kumpiene,
Ahmad, Sparrenbom, Jacks, Donselaar, Bundschuh & Naidu (Eds)
© 2016 Taylor & Francis Group, London, ISBN 978-1-138-02941-5

Arsenobetaine a possible methyl donor in the one carbon cycle?

M. Bergmann & W. Goessler
Institute of Chemistry, Analytical Chemistry, University of Graz, Graz, Austria

ABSTRACT: It is well known that arsenobetaine is present in high concentration in marine animals and ingested in high concentrations via seafood. Whether it is metabolized in the same way as glycine betaine is not known until now. Therefore, we administered orally a dose of 2 mg of deuterated arsenobetaine to five volunteers. With a newly developed HPLC-ESMS method we determined possible transformation products in the urine samples that have been collected for almost four days. From our results we can conclude that less than 0.05% of the ingested deuterated arsenobetaine was metabolized. It is worth mentioning that we have observed that one of the five volunteers is a low arsenobetaine excreter. Whereas four volunteers excreted the ingested arsenobetaine with a half-life of ~10 hours, it took around 30 hours for one female volunteer before 50% of the arsenobetaine was eliminated from her body.

1 INTRODUCTION

Since its discovery in 1977 arsenobetaine (AB) has been reported in many marine animals at high concentrations (Edmonds et al., 1977, Francesconi & Edmonds, 1997). In fishes up to 170 mg arsenic per kg wet mass with up to >95% of AB have been reported (Francesconi & Edmonds, 1997). Arsenobetaine, is the arsenic analogue to glycine betaine (GlyB), which is an important osmolyte and methyl-donor. It has been suggested that high concentrations of AB in marine animals might also function as an osmolyte (Gailer et al., 1995). E. coli accumulates AB with the transporters of GlyB (Ratriyanto et al., 2009). Clowes and Francesconi have shown that the uptake and elimination of AB by the mussel *Mytilus edulis* is correlated with the salinity of the water (Clowes & Francesconi, 2004). In the present work we investigated the possibility that AB can act as a methyl-donor in the one carbon cycle (see Scheme 1).

In the one-carbon cycle a methyl group is transfered from glycine betaine to homocysteine to form methionine. In return gycine betaine is converted to dimethylglycine. Betaine-homocysteine-methyl transferase catalyzes this reaction.

Upon consumption of seafood, especially marine animals, we ingest high amounts of AB but there are no known toxic effects. The fast excretion of AB via the kidneys might be with a possible explanation. A half-life of ~7–11 hours has been reported (Lehmann et al., 2001, Johnson & Farmer, 1991). In 1983 Vahter used ^{73}As-labelled AB which was orally given to mice, rats and rabbits to investigate the metabolism of AB in the mammalian body. In the urine and tissue extracts of these animals AB was the only radiolabelled arsenical detectable (Vahter, 1983). With the design of this study it was not possible to answer the question whether the methyl groups of AB might be exchanged in the mammalian body. Therefore, we studied the possible exchange of the methyl groups with deuterated AB and molecule-selective mass spectrometry.

2 EXPERIMENTAL

2.1 Chemicals

Deuterated arsenobetaine bromide (d3AB) trimethylarsine-d3 (using CD₃I, 99.5 atom% D, Sigma Aldrich) and bromoacetic acid, synthetically grade 97%, Merck KGaA (McShane, 1982).

2.2 Urine collection

Five volunteers (2 male and 3 female) in the age of 17 to 48 years participated in this study. The volunteers did not consume any seafood for seven days before and during the time of the study to ensure low AB concentrations in the urine. One day before the ingestion of the d3AB (2 mg As) they started to collect and determine the amount of urine volumetrically for 3.5 days. An aliquot (40 mL) of all urine samples (113) was stored at −20°C until the analysis. Before freezing the samples, their specific gravity was measured with a total solids refractometer (TS400, Serum Protein 6.54, Leica Microsystems Inc.).

Before the analysis the urine sample were filtered through Nylon syringe filters (0.2 μm, Markus Bruckner Analysentechnik). For the analysis we used an Agilent 6120 quadrupole LC/ESIMS (Agilent Technologies, Waldbronn Germany). For the separation of the d3AB we chromatographed the urine samples on a Zorbax 300-SCX (150 × 4.6 mm, 5 μm) cation-exchange column with a 20 mM ammonium formate buffer (≥95% ammonium formate, Carl Roth GmbH+Co.

301

KG with ultrapure water) at pH 2.5 (adjusted with formic acid ≥98% p.a., Rotipuran®, Carl Roth GmbH+Co.KG) and 65% of methanol (Rotisolv® HPLC, ≥99.5%, Carl Roth GmbH+Co.KG) (v/v) as mobile phase A flow rate was 1.5 mL/min and a temperature of 30°C was used. The injection volume was 20 μL. To avoid the overloading of the ESI source we have split the flow post column 1:1. For ESMS we used a positive polarity with a drying gas flow of 13 L/min, a drying gas temperature of 350°C, a nebulizer pressure of 35–60 psig, a capillary voltage of 4000 V, and a fragmentor voltage of 100 V. We monitored m/z 50–210 in the scan mode and m/z 179, 182, 185 and 188 in the SIM mode representing AB, dAB, d2AB or d3AB. To overcome possible matrix effects we spiked the urine samples with AB. Quantitative arsenic excretion profiles were determined with HPLC-ICPMS.

3 RESULTS AND DISCUSSION

3.1 Excretion of AB

The determined total arsenic concentrations agreed well with the sum of the arsenic species determined with HPLC-ICPMS (90 and 110%). Before ingestion of d3AB different arsenic species (inorganic As, AB, DMA and MA) were found at low concentrations in the urine. After ingestion d3AB accounted more than 96% of the arsenic in the urine. Excretion profiles of the five volunteers are shown in Figure 1.

Four of the five volunteers excreted the ingested AB at a half-live of ~10 hours. One female volunteer showed a much longer half live of almost 30 hours. Even when we repeated the ingestion experiment a half-live of more than 20 hours was observed.

3.2 Metabolism of AB

The analysis of the urine samples with HPLC-ESMS revealed the methyl groups of AB are not exchanged in the human body as it is the case with glycine betaine. With the experimental setup used in the study we can state that the transformed d3AB is less than 0.5 μg/L in the urine samples.

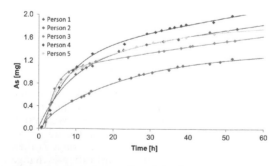

Figure 1.   Excretion profiles of the five volunteers.

Table 1.   Results of the AB spike experiments of the urine samples with the highest concentration of d3AB from all 5 volunteers, spiked with AB, measured with HPLC-ESMS.

	USG	Added AB [μg/L]	% of d3AB
Person 1	1.015	0.1	0.01
Person 2	1.014	0.5	0.04
Person 3	1.003	0.1	0.02
Person 4	1.024	0.5	0.01
Person 5	1.015	0.5	0.03

To exclude possible matrix effects AB was spiked at a concentration of less 0.1 to 0.5 μg/L to the urine with the highest d3AB concentration (Table 1).

4 CONCLUSIONS

With the developed methods it was possible to show that the methyl group of arsenobetaine is not used in the one carbon cycle. Differences in the molecule-size of AB to GlyB and the huge excess of GlyB could be reasons for the observations made in the present study. In vitro experiments with betaine-homocysteine-methyltransferase in the absence of GB could help to enlighten this situation.

REFERENCES

Clowes, L., Francesconi, K. 2004. Uptake and elimination of arsenobetaine by the mussle Mytilus edulis is related to salinity. Comp. Biochem. Phys. C 137: 35–42.

Edmonds, J., Francesconi, K., Cannon, J., Raston, C., Skelton, B., White, A. 1977. Isolation, crystal structure and synthesis of arsenobetaine, the arsenical constituent of the western rock lobster Panulirus longipes cygnus george. Tetrahedron Lett. 18: 1543–1546.

Francesconi, K., Edmonds, J. 1997. Arsenic and marine organisms. Adv. Inorg. Chem. 44: 147–189.

Gailer, J., Francesconi, K., Edmonds, J., Irgolic, K. 1995. Metabolism of arsenic compounds by the blue mussel Mytilus edulis after accumulation from seawater spiked with arsenic compounds. Appl. Organomet. Chem. 9: 341–355.

Johnson, L., Farmer, J. 1991. Use of human metabolic studies and urinary arsenic speciation in assessing arsenic exposure. Bull. Environ. Contam. Toxicol. 46: 53–61.

Lehmann, B., Ebeling, E., Alsen-Hinrichs, C. 2001. Kinetics of arsenic in human blood after ingestion of fish. Gesundheitswesen 63 (1): 42–48.

McShane, W. 1982. The Synthesis and Characterisation of Arsenocholine and Related Compounds. PhD Dissertation Texas. A & M University, College Station, Texas.

Ratriyanto, A., Mosenthin, R., Bauer, E., Eklund, M. 2009. Metabolic, osmoregulatory and nutritional functions of betaine in monogastric animals. Asian. Austral. J. Anim. 22(10): 1461–1476.

Vahter, M. 1983. Metabolism of arsenobetaine in mice, rats and rabbits. Sci. Total Environ. 30: 197–211.

Arsenic Research and Global Sustainability – Bhattacharya, Vahter, Jarsjö, Kumpiene,
Ahmad, Sparrenbom, Jacks, Donselaar, Bundschuh & Naidu (Eds)
© 2016 Taylor & Francis Group, London, ISBN 978-1-138-02941-5

Speciation and toxicity of arsenic in shellfish seafood from Fujian, China

G.-D. Yang, Z.-Q. Qiu, Z.-M. Lv, P.-Y. Shuai & W.-X. Lin
Fujian Provincial Key Laboratory of Agroecological Processing and Safety Monitoring, College of Life Sciences,
Fujian Agriculture and Forestry University, Fuzhou, Fujian, P.R. China

ABSTRACT: A Microwave Assisted Extraction (MAE) and Ion Chromatography Inductively Coupled Plasma Mass Spectrometry (IC-ICP-MS) coupling technology for the analysis of arsenic (As) speciation in shellfish seafood from Fujian province of China was developed in this study. The results showed that the optimal extracting solvent was Distilled Deionized (DDI) water and the optimal ratio of extracting solvent volume to shellfish seafood powder mass was 100 mL/g. Arsenic species could be well separated under the gradient elution by 1.0 mmol/L $(NH_4)_2CO_3$ before 2 min and 10.0 mmol/L NH_4HPO_4 after 2 min. More than 90% of As species was organic arsenic in the 42 shellfish seafood from Fujian province of China. The concentration of inorganic As was the highest in oyster. However, its concentration was much lower than the Maximum Residue Limits (MRLs), which meant that there existed little As intake risk from shellfish seafood from Fujian province of China.

1 INTRODUCTION

Shellfish is the popular seafood in the cuisine of many countries. Arsenic (As) concentration is very high in the shellfish seafood (based on its dried weight) due to the pollution from sea water and its high accumulation ability for As element (Yang *et al.*, 2009). However, the As toxicity is close related to its speciation (Bissen & Frimmel, 2003, Caruso *et al.*, 2003). Therefore, it is very important to accurately determine As speciation in the shellfish seafood to evaluate its uptake safety for human health.

2 METHODS/EXPERIMENTAL

2.1 *Sampling site and samples*

Forty two shellfish seafoods were collected from the coastal waters of Fujian province in southeastern China (see Figure 1).

Samples were collected from seven sites: S1 in Lianjiang (119.53 °E, 26.20 °N), S2 in Fuzhou (119.3 °E, 26.08 °N), S3 in Ningde (119.52 °E, 26.65 °N), S4 in Zhangzhou (117.35 °E, 24.52 °N), S5 in Quanzhou (118.58 °E, 24.93 °N), S6 in Xia-Men (118.10 °E, 24.46 °N) and S7 in Putian (119.00 °E, 25.44 °N) in autumn 2014. The samples of Venerupis philippinarum, Oyster and Sinonovacula were collected from site S1~S7, the sample of Tegillarca granosa was collected from site S1~S5, and the other 4 clam samples were collected from S1, S2, S4 and S5 (the dried weight of each sample was about 500 g). All fresh shellfish seafood samples were deshelled, homogenated, vacuum freeze-

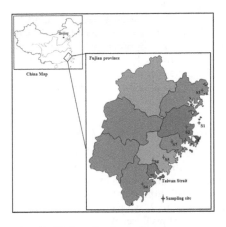

Figure 1. Shellfish seafood sampling sites.

dried and stored in clean valve bags, the valve bags with dried shellfish seafood samples were put in the desiccator.

2.2 *Optimization for the extraction of As speciation in shellfish seafood*

A microwave assisted extraction (MAE) program was used to extract the As speciation in shellfish seafood. Dried shellfish seafood powder from S1 site was chosen as the material to investigate the effects of the extracting solvent composition and the ratio of extracting solvent volume to shellfish seafood powder mass on the extraction of As speciation (3 replicates each).

2.3 Speciation and toxicity of As in shellfish seafood

A coupling system of Ion Chromatography and Inductively Coupled Plasma Mass Spectrometry (IC-ICP-MS) was used to separate arsenobetaine (AsB), arsenite [As (III)], dimethylarsinic acid [DMA (V)], monomethylarsonic acid [MMA (V)] and arsenate [As (V)] and detect the As speciation in shellfish seafood. The toxicity and uptake safety of As from the shellfish seafood was evaluated according to the Maximum Residue Limits (MRLs) in GB2762–2012 and NY 5073–2006.

3 RESULTS AND DISCUSSION

3.1 Optimization for the extraction of As speciation in shellfish seafood

The effects of the extracting solvent composition, the ratio of extracting solvent volume to shellfish seafood powder mass, microwave heating temperature and heating time on the extraction of As speciation were investigated, we found that the optimal extracting solvent was Distilled-Deionized (DDI) water, the optimal ratio of extracting solvent volume to shellfish seafood powder mass was 100 mL/g, the optimal microwave heating temperature was 120 °C and the optimal microwave heating time was 15 min.

3.2 Speciation of As in shellfish seafood

Under the optimal MAE and IC-ICP-MS conditions, the Chromatograms of mixed As standard solutions, Lianjiang clam extraction and its spiked extraction solution are shown in Figure 2.

From Figure 2, we could see that five As species could be well separated under the optimal IC-ICP-MS parameters. The RSDs (n = 5) of migration time and peak area were 0.20%~1.27% and 0.85%~4.85% at the concentration of 2 µg/L AsB, 10 µg/L As (III), 5 µg/L [DMA (V), MMA (V), As (V)] and the lowest detected concentration was 0.1 µg/L AsB, 0.2 µg/L As (III), 0.1 µg/L DMA (V), 0.1 µg/L MMA (V), 0.1 µg/L As (V), respectively.

From Figure 2, we could also see that AsB was the predominant As species in Lianjiang clam. Its spiked extraction recovery was (100.6±2.31)%, (110.3±4.57)%, (106.2±3.28)%, (95.45±3.47)%, (100.3±2.98)%, respectively, for AsB, As (III), DMA (V), MMA, As (V). The spiked recovery of 95.45~110.3% meant that the method developed by us could be used to determine As species in shellfish seafoods.

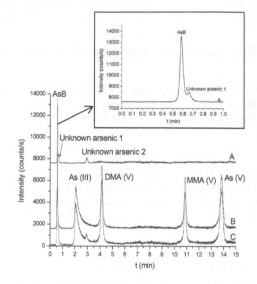

Figure 2. Chromatograms of IC-ICP-MS. A. Lianjiang clam, B. Standard solution of 2 µg/L AsB; 10 µg/L As (III); 5 µg/L [DMA (V); MMA (V); As (V)], C. Lianjiang clam spiked by 2.0 mg/kg AsB; 10.0 mg/kg As (III); 5.0 mg/kg [DMA (V); MMA (V); As (V)].

Table 1. Concentration of As species in Fujian shellfish seafood samples.

Samples	AsB	As (III)	As (V)	DMA	Unknown As
1**	0.25–0.91	0.00–0.02	0.00–0.02	0.02–0.06	0.21–0.63
2	0.35–0.99	0.00–0.04	0.00–0.03	0.06–0.12	0.14–0.30
3	0.13–0.60	0.00–0.00	0.00–0.01	0.01–0.06	0.16–0.73
4	0.48–0.98	0.00–0.02	0.00–0.01	0.03–0.05	0.05–0.30
5	0.15–0.44	0.00–0.01	0.01–0.01	0.03–0.05	0.23–0.41
6	0.19–0.41	0.00–0.00	0.00–0.01	0.02–0.04	0.10–0.20
7	0.67–1.52	0.01–0.04	0.00–0.03	0.01–0.02	0.08–0.32
8	0.23–0.47	0.01–0.02	0.01–0.02	0.03–0.05	0.05–0.08

*Concentration of As speciation in different clams (mg/kg)

*based on wet weight. **1-Venerupis philippinarum, 2-Oyster, 3-Sinonovacula, 4-Tegillarca granosa, 5-Cyclina sinensis, 6-Trigonia, 7-Clam, 8-Oil clam

3.3 Toxicity analysis of as in shellfish seafood

Under the optimal MAE and IC-ICP-MS conditions, the As species in 42 Shellfish seafoods from 7 sampling site was determined, the results were shown in Table 1. We found that AsB was the predominant As species in Fujian shellfish seafood samples, accounting for 24.58%~86.57% of the total As species. The concentration of As (III) and As (V) was 0~0.04 mg/kg and 0~0.03 mg/kg, much lower than MRLs in GB2762-2012 and NY 5073-2006, which meant that the risk for As intake from shellfish seafood from Fujian, China is minimal.

4 CONCLUSIONS

Arsenic speciation in shellfish seafood from Fujian province of China was determined in this study. The results showed that 24.58%~86.57% of the total As species was AsB, the concentration of high toxic As (III) and As (V) was much lower than MRLs in GB2762-2012 and NY 5073-2006, which meant that there existed little As intake risk in shellfish seafood from Fujian province of China.

ACKNOWLEDGEMENTS

This work was financially supported by Natural Science Foundation of China (31271670), Fujian Provincial Department of Science and Technology (2013Y0005), Fujian Provincial Department of Education (201510389108) and Key Subject of Ecology, Fujian Province, China (6112C0600).

REFERENCES

Bissen, M., Frimmel, F.H. 2003. Arsenic—a review. Part I: occurrence, toxicity, speciation, mobility. *Acta Hydroch. Hydrob*. 31(1): 9–18.

Caruso, J.A., Klaue, B., Michalke, B., Rocke, D.M. 2003. Group assessment: elemental speciation. *Ecotox. Environ. Safe*. 56(1): 32–44.

Yang, G.D., Xu, J.H., Zheng, J.P., Xu, X.Q., Wang, W., Xu, L.J., Chen, G.N., Fu, F.F. 2009. Speciation analysis of arsenic in *Mya arenaria Linnaeus* and *Shrimp* with capillary electrophoresis-inductively coupled plasma mass spectrometry. *Talanta* 78(2): 471–476.

Arsenic Research and Global Sustainability – Bhattacharya, Vahter, Jarsjö, Kumpiene,
Ahmad, Sparrenbom, Jacks, Donselaar, Bundschuh & Naidu (Eds)
© 2016 Taylor & Francis Group, London, ISBN 978-1-138-02941-5

Organoarsenicals in seaweed are they toxic or beneficial: Their analysis, their toxicity and their biosynthesis

J. Feldmann, J.F. Kopp, A. Raab & E. Krupp
Department of Chemistry, TESLA (Trace Element Speciation Laboratory), University of Aberdeen, Aberdeen, UK

ABSTRACT: Lipid-soluble arsenic compounds have been the focus of the last decade on arsenic containing natural products due to the advances of analytical chemistry. At least 50 new compounds have been identified. The identification and quantification using HPLC-ICPMS/ESI-MS will be presented, which are key to study biosynthetic pathways and how their synthesis respond to different environmental conditions.

1 INTRODUCTION

Arsenic interaction with biological systems are dependent on its speciation. Hence, the identification of the different molecular forms are key to understand how arsenic is taken up, biotransformed and accumulated in biota. Since the arsenic is in the low or sub ppm level in biota, it is challenging to identify the different organoarsenic compounds. However in recent years the combination of ICP-MS as arsenic selective detector and ESI-MS molecular detector can run simultaneously when the species are separated by HPLC methods. This methodology was used to determine more than 50 different lipid-soluble arsenic compounds. Due to the lack of arsenolipid standards, the use of stable water-soluble arsenic standard in combination of the elemental instead of molecular response of the ICP-MS can be used for reliable quantification. Here in this overview the different classes of arsenolipids identified so far will be shown and the analytical methodology illustrated by the analysis of different organs of pilot whales In the second part the methods will be applied to Ectocarpus exposed to different environmental conditions in order to study the biological response in terms of biosynthesis of arsenolipids.

2 METHODS/EXPERIMENTAL

2.1 Samples and experiments

Cod liver oils, which is well-characterized has been analysed for their arsenic speciation using HPLC-ICPMS/ESI-MS since no arsenolipids standards are available. While NIST standard reference seaweed samples were used for the identification of arsenosugars and arsenolipids. The individual arsenic species were analysed using accurate mass, MS/MS fragmentation pattern and the arsenic response from the ICP-MS (Raab *et al.*, 2013). Their quantification was done by using dimethylarsinic acid and by correcting the response with the C-enhancement effect from the methanol gradient according to Amayo *et al.* (2011). This analysis was applied to the different extracts from seaweed and other marine organisms.

3 RESULTS AND DISCUSSION

3.1 The analysis for arsenolipids

The usual analysis consists of the extraction. Most arsenolpids were found in the methanol/DCM fraction, hence, there are in the fraction of polar lipids. This is due to the dimethylarsinoyl moiety.

The extracts were cleaned-up and derivatised using acetylation and phospholipase to identify those especially those complex arsenic containing phosphorlipids (see Figure 1). In fish tissues and oils a series of AsHC and AsFA were identified, while seaweeds contain in addition to those classes of arsenolipids also some AsPLs.

3.2 Arsenolipid response to environmental changes

In a microcosm experiment using different strains of filamentous brown algae (Ectocarpus), in which different environmental conditions such as nutrient deficiency with regards to nitrate and phosphate as well as oxidative stress was tested (Pétursdóttir *et al.*, 2016). The hydrophilic and the lipophilic fractions of the Ectocarpus were identified and quantified

Figure 1. Four different classes of arsenolipids: Arsenic containing hydrocarbons (AsHC), arsenic containing fatty acids (AsFA), arsenic containing fatty alcohols (AsFAl) and arsenic containing phospholipids (AsPL).

and compared to their controls. The arsenolipids and arsenosugar profiles differed depending on the experimental conditions. Under low phosphate conditions, a significant reduction of phosphorus-containing arsenosugars was noticed, and a significant increase of phosphate-containing AsPL was found when compared with the controls. Oxidative stress increased the amounts of arsenolipids in the algae.

3.3 Discussion

The arsenolipids biosynthesis was influenced by the environmental conditions. It is however unclear why these compounds are formed and how they are arsenic been transformed from arsenate into arsenolipids. Would the introduction of a redox-active element, which is covalently bound to the lipids in a cell membrane, be beneficial for the algae when exposed to oxidative stress? This will be discussed in the lecture in more detail.

Additionally it should be pointed out that some of the arsenolipids such as AsHC and to a lower extent AsFA had shown comparable toxic effects to human cell lines as arsenite (Meyer et al., 2014a,b). This aspect is concerning especially when marine food such as seaweed or fish are communed by us. Do we have an emerging arsenic pollutant in addition to that of inorganic arsenic?

In order to address the food monitoring issue for seaweed, a new field test kit, which was developed for inorganic arsenic in rice (Bralatei et al., 2015) was applied for the analysis of inorganic arsenic in seaweed and compared to a speciation analysis using HPLC-ICPMS and it has been found that some seaweed species like Laminaria digitata contain large fraction of the arsenic in the form of inorganic arsenic.

Hence, not only contains seaweed large amounts of cytotoxic arsenolipids but also significant amount of inorganic arsenic and this is not only restricted to hijiki.

4 CONCLUSIONS

Complex analytical procedures are necessary for the identification and quantification of arsenolipids, while simple methods can measure inorganic arsenic. Arsenolipids are maybe beneficial for the algae, not only as a detoxification product of inorganic arsenic, but it maybe also toxic for consumers from a higher trophic level.

ACKNOWLEDGEMENTS

We thank the TESLA research fund and the College for Physical Sciences for funding.

REFERENCES

Amayo, K.O., Petursdottir, A., Newcombe, C., Gunnlaugsdottir, H., Raab, A., Krupp, E.M., Feldmann, J. 2011. Identification and quantification of arsenolipids using reversed-phase HPLC coupled simultaneously to high-resolution ICPMS and high-resolution electrospray MS without species-specific standards. Anal. Chem. 83(9), pp. 3589–95.

Bralatei, E., Lacan, S., Krupp, E.M., Feldmann, J. 2015. Detection of inorganic arsenic in rice using a field kit: a screening method. Anal. Chem. 87(22): 11271–11276.

Meyer, S., Matissek, M., Müller, S.M., Taleshi M.S., Ebert, F., Francesconi, K.A., Schwerdtle, T. 2014a. In vitro toxicological characterisation of three arsenic-containing hydrocarbons. Metallomics: Integrated Biometal Science 6(5): 1023–1033.

Meyer, S., Schulz, J., Jeibmann, A., Taleshi, M.S., Ebert, F., Francesconi, K.A., Schwerdtle, T. 2014b. Arsenic-containing hydrocarbons are toxic in the in vivo model Drosophila melanogaster. Metallomics 6: 2010–2014.

Pétursdóttir, Á.H., Fletcher, K., Gunnlaugsdottir, H., Krupp, E.M., Kuepper, F.C., Feldmann, J. 2016. Environmental effects on arsenosugars and arsenolipids in Ectocarpus (Phaeophyta). Environ. Chem. 13(1): 21–33.

Raab, A., Newcombe, C., Pitton, D., Ebel, R., Feldmann J. 2013. Comprehensive analysis of lipophilic arsenic species in a brown alga (Saccharina latissima). Anal. Chem. 85(5): 2817–2824.

Arsenic Research and Global Sustainability – Bhattacharya, Vahter, Jarsjö, Kumpiene,
Ahmad, Sparrenbom, Jacks, Donselaar, Bundschuh & Naidu (Eds)
© *2016 Taylor & Francis Group, London, ISBN 978-1-138-02941-5*

Arsenolipids—the underestimated threat in marine food sources?

J.F. Kopp, Z. Gajdosechova, A. Raab, E. Krupp & J. Feldmann
Department of Chemistry, TESLA (Trace Element Speciation Laboratory), University of Aberdeen, Aberdeen, UK

ABSTRACT: The term arsenolipids describes a class of arsenic species that are not water-, but lipid-soluble. So far, there are 4 subcategories known: Arsenohydrocarbons (AsHC), Arseno Fatty Acids (AsFA), Trimethylarsino Fatty Alcohols (TMAsFOH), and Arsenosugar Phospholipids (AsPL). They have been found mainly in the marine environment (algae, various fish), but recent studies show their accumulation in land-based organisms after ingestion as well. Due to the lack of pure compounds, their toxicological potential was unclear until the synthesis was published (Taleshi *et al.*, 2014) which made cytotoxicological studies possible. These studies quickly showed, that they are able to exert a similar cytotoxicity as inorganic As (Meyer *et al.*, 2014). A study on *Globicephala melas* (long-finned pilot whale) regarding accumulation and metabolism of these compounds is being conducted in order to be able to assess the threat arsenolipids pose to humans and the environment.

1 ARSENOLIPIDS AND THEIR TOXICOLOGY

Arsenic is a ubiquitous pollutant at trace level. To this date, only inorganic arsenic is addressed as toxic by the World Health Organization (WHO, 2012), whereas organic arsenic species are mostly regarded as non-toxic. Organic arsenic species can be subdivided into water-soluble and lipid-soluble species. The main environmentally occurring organic water-soluble species are dimethylarsinic acid, arsenobetaine and various arsenosugars.

Lipid-soluble species have first been reported by Lunde *et al.* (1968) in 1968 (Lunde, 1968). After their discovery, arsenolipid-research has been put aside, because relatively low concentrations and complex structures made analysis and structure elucidation very difficult. Eventually, in the early 2000s, the developments in accurate mass-MS and ICP-MS had lowered detection limits significantly and made structural determination without complicated derivatisation reactions possible. Today, about 60 lipid soluble species have been reported comprising of arsenohydrocarbons (AsHC), Arseno Fatty Acids (AsFA), Trimethylarsino Fatty Alcohols (TMAsFOH), and Arsenosugar Phospholipids (AsPL).

In early 2015 a toxicological study by Meyer *et al.* was published showing that arsenohydrocarbons exert a similar cytotoxicity as inorganic As^{III} (Meyer *et al.*, 2014a), which was supported shortly after by a study on the *in vivo* model *Drosophila melanogaster* (Meyer *et al.*, 2014b).

Based on these studies, questions about the production, bioavailability and accumulation of these compounds in other organisms arise.

One study on environmental effects influencing the production of arsenolipids (and—sugars) in 3 species of the brown algae *Ectocarpus* by Pétursdóttir *et al.* (2015) showed increased arsenolipid-levels under low phosphate- and oxidative stress-conditions. Since various fish species in the marine food chain have been shown to contain arsenolipids, it would be no surprise to find them in marine mammals.

A sample set consisting of 21 long-finned pilot whales (*Globicephala melas*) originating from a mass stranding in 2012 near Anstruther on the Scottish coast represents a good opportunity to study accumulation and expression patterns of arsenolipids.

Due to disclosure conflicts this abstract will only discuss the methodologies and possible drawbacks when dealing with this type of matrix. Results will be presented at the actual conference.

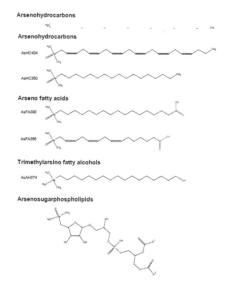

Figure 1.   Exemplary structures for the 4 different types of arsenolipids.

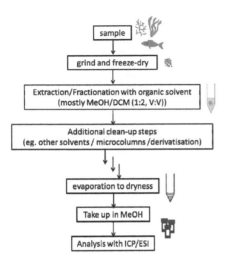

Figure 2. Typical extraction schematic for the speciation of arsenolipids.

2 METHODOLOGY AND ISSUES

2.1 *Separation, identification and quantification of arsenolipids*

Due to their lipophilic character, arsenolipids are best separated on a reversed phase column (C8/C18) using a gradient (mostly MeOH/H$_2$O). For the identification of arsenic-containing peaks this is coupled with an ICP-MS detector. Molecular information is gained by splitting the eluent post-column and infusing parts of it into ESI-MS (Amayo *et al.* 2011). This allows targeted search for intact molecules at the retention times of arsenic containing species as well as MS/MS experiments.

Due to the use of HPLC-gradients, for arsenolipid-analysis, the carbon enhancement effect in ICP-MS needs to be corrected for. There are three major approaches to correct for this: First, the introduction of a 2nd HPLC, running a mirrored gradient, which is mixed with the eluent from the separating HPLC in a 1:1 ratio (Pröfrock & Prange, 2009). Second, a mathematical correction using a response factor (Amayo *et al.*, 2011). Lastly, the addition of organic solvent to the spraychamber which leads to carbon-saturation in the gas phase, rendering gradient-based fluctuations negligible (Raber *et al.*, 2010).

2.2 *Extraction procedure for lipophilic arsenic species*

A typical extraction procedure for arsenolipid-analysis is depicted in Figure 2. Depending on the type of sample, samples are ground and freeze-dried and then extracted in an organic solvent. The most commonly used/effective mixture is methanol and dichloromethane in a 1:2 (v/v) ratio. The following steps are strongly dependent on the matrix and the aim of the

experiment. For quantitative purposes (especially in samples with low As-concentrations) it is advisable not to implement too many clean-up steps, since too much analyte might be lost in the process. On the other hand, when trying to identify the compounds in the first place, the clean-up procedure should be as thorough as possible in order to remove any matrix components that reduce ionization capabilities in the ESI-MS.

As the organs of pilot whales have a high natural fat content, the extraction of lipophilic molecules present in comparably low concentrations requires a sophisticated extraction procedure.

ACKNOWLEDGEMENTS

The authors would like to thank the School of Natural and Computing Sciences at University of Aberdeen for funding. Further acknowledgement goes to the entire group of Prof. Kevin Francesconi at the Karl-Franzens-University in Graz for very fruitful discussions and student-exchanges.

REFERENCES

Amayo, K.O. Petursdottir, A., Newcombe, C., Gunnlaugsdottir, H., Raab, A., Krupp, E.M., Feldmann, J. 2011. Identification and quantification of arsenolipids using reversed-phase HPLC coupled simultaneously to high-resolution ICPMS and high-resolution electrospray MS without species-specific standards. *Anal. Chem.* 83(9): 3589–3595.

Lunde, G. 1968. Analysis of arsenic in marine oils by neutron activation. Evidence of arseno organic compounds. *J. Am. Oil Chemists Soc.* 45(5): 331–332.

Meyer, S., Matissek, M., Müller, S.M., Taleshi M.S., Ebert, F., Francesconi, K.A., Schwerdtle, T. 2014a. In vitro toxicological characterisation of three arsenic-containing hydrocarbons. *Metallomics* 6(5): 1023–1033.

Meyer, S., Schulz, J., Jeibmann, A., Taleshi, M.S., Ebert, F., Francesconi, K.A., Schwerdtle, T. 2014b. Arsenic-containing hydrocarbons are toxic in the in vivo model Drosophila melanogaster. *Metallomics* 6: 2010–2014.

Pétursdóttir, Á.H. Fletcher, K., Gunnlaugsdottir, H., Krupp, E.M., Kuepper, F.C., Feldmann, J. 2015. Environmental effects on arsenosugars and arsenolipids in Ectocarpus (Phaeophyta). *Environ. Chem.* 13(1): 21–33.

Pröfrock, D., Prange, A. 2009. Compensation of gradient related effects when using capillary liquid chromatography and inductively coupled plasma mass spectrometry for the absolute quantification of phosphorylated peptides. *J. Chromatogr. A* 1216(39): 6706–6715.

Raber, G. Raml, R., Goessler, W., Francesconi, K.A. 2010. Quantitative speciation of arsenic compounds when using organic solvent gradients in HPLC-ICPMS. *J. Anal. At. Spectrom.* 25(4): 570–576 25.

Taleshi, M.S. Seidler-Egdal, R.K., Jensen, K.B., Schwerdtle, T., Francesconi, K.A. 2014. Synthesis and character-ization of arsenolipids: naturally occurring arsenic compounds in fish and algae. *Organometallics* 33(6): 1397–1403.

WHO, 2012. Arsenic—Fact Sheet N°372. Available at: http://www.who.int/mediacentre/factsheets/fs372/en/ [Accessed December 12, 2015].

2.3 Arsenic in food and beverages

Arsenic Research and Global Sustainability – Bhattacharya, Vahter, Jarsjö, Kumpiene,
Ahmad, Sparrenbom, Jacks, Donselaar, Bundschuh & Naidu (Eds)
© 2016 Taylor & Francis Group, London, ISBN 978-1-138-02941-5

Arsenic speciation and bioavailability in vegetables

R.A. Schoof & E. Handziuk
Ramboll Environ, Seattle, WA, USA

ABSTRACT: Arsenic is naturally present in most foods. Historically dietary studies and food surveys have estimated intake of total arsenic (totAs), but only the fraction that is inorganic is of primary health concern. Average daily inorganic arsenic (iAs) intake has been reported to range from 2 to 8 μg/d, with vegetables contributing up to one quarter of that amount, and rice and fruits contributing another 30%. In many cases the cost of analysing iAs is cost prohibitive compared with cost of analysing totAs. For that reason, an approach to predicting the contribution of vegetables to dietary iAs intake is needed. This paper presents a review of the fraction of totAs that is iAs, the effects of cooking on As concentrations and the bioavailability of iAs in prepared foods. Current understanding of these factors is reviewed and the implication for assessment of dietary arsenic exposure is discussed.

1 INTRODUCTION

Arsenic is naturally present in most foods. Historically dietary studies and food surveys have estimated intake of totAs, but only the fraction that is inorganic is of primary health concern. Daily iAs intake has been reported to average 3.2 μg/d for U.S. adults and children (Schoof *et al.*, 1999 a, b, Yost *et al.*, 2004). A probabilistic model of dietary iAs exposure in the U.S. using 2003–2004 National Health and Nutrition Examination Survey (NHANES) data estimated mean intake to be 2.0 μg/d (Xue *et al.*, 2010). Kurzius-Spencer *et al.* (2004) estimated geometric mean daily intake of iAs ranging from 5.8 to 8 μg/d using food consumption data from the 2003–2004 NHANES and other databases. Baeyens *et al.* (2009) reported an average dietary iAs intake for Belgian consumers of 5.8 μg/d.

2 TOTAL ARSENIC IN VEGETABLES

Studies of arsenic in food most commonly report totAs concentrations. The most comprehensive datasets of totAs concentrations in vegetables are from government organizations compiling data on contaminant concentrations in market foods (EFSA, 2009, Health Canada, 2014, FDA, 2014). All of these datasets prepare the foods as consumed.

The largest single dataset for arsenic in foods is from the European Food Safety Authority (EFSA, 2009). A majority of the samples (66%) were below Limits of Detection (LOD). The mean arsenic concentration from all samples in the category "vegetables, nuts, and pulses" was 0.0367 mg/kg wet wt. for the upper bound (i.e., using LOD for nondetects) and 0.0262 mg/kg wet wt. for the lower bound (using zero for the nondetects). Means were derived by adjusting the means for each vegetable type based on actual dietary contribution. The highest concentrations were found in dried vegetables, oil seeds, mushrooms, and fresh herbs. Of the categories consumed in greatest quantities, the highest mean arsenic was found in leafy vegetables.

Each year the U.S. Food and Drug Administration (FDA) analyzes market foods. Data from 2006 through 2011 were combined to calculate the mean arsenic concentration from selected vegetables (giving equal weight to each kind of vegetable) of 0.0097 mg/kg wet wt. (upper bound) and 0.00081 mg/kg wet wt. (lower bound). Variation in concentration among kinds of vegetables could not be discerned due to the high proportion of nondetects.

Health Canada Total Diet Studies from 2005 through 2007sampled vegetables from three to four different markets in Toronto, Halifax, and Vancouver. Arsenic concentrations were measured in each vegetable as prepared (cooked or peeled if applicable) and the average for typical garden vegetables (giving equal weight to each vegetable) was 0.011 mg/kg wet wt., comparable to what was found in the FDA total diet study dataset. Herbs had the highest arsenic concentration, mean 0.15 mg/kg wet wt., while tomatoes and peppers had very low concentrations.

3 INORGANIC ARSENIC IN VEGETABLES

Xue *et al.* (2010) found that the major food category contributing to U.S. dietary iAs intake was vegetables, contributing nearly one quarter of intake (24%). Fruit juices and fruits contributed 18% of intake, while rice contributed another 17%. Beer and wine (12%) and other grains including flour, corn, and wheat (11%) were also important exposure sources. Poultry, meat, and seafood contributed smaller amounts to intake. Few studies have reported both totAs and iAs concentrations in vegetables and fruit. Schoof et al. (1999a, b) found iAs was an average of 51% of totAs in vegetables purchased at U.S. markets, with a range from 9% to 104%. For fruits, the mean and range were 47% and 21% to 103%. Font-cuberta et al. (2011) reported a mean estimate that 23% of totAs in market vegetables and fruits from Spain was iAs, with 13% of spinach arsenic being iAs and 44% of cabbage arsenic being iAs. JECFA (2011) reports the proportion of iAs to totAs in 36 vegetable samples to vary from 33% to 74%. Higher proportions of iAs have been reported in other studies, frequently reflecting very high soil and water arsenic concentrations (Diaz et al., 2004, Helgesson & Larsen, 1998, Smith et al., 2006). EFSA (2009) concluded that 70% reflected the best overall average iAs content for food communities other than fish and seafood.

iAs in rice has now been characterized in a large number of studies, including some systematic surveys of a large number of samples (Fontcuberta et al., 2011, FDA, 2013, Lynch et al., 2014). These studies demonstrate that the relative proportion of organic arsenic and iAs in rice is highly variable.

4 COOKING IMPACTS

Evidence is accumulating that cooking may reduce both totAs and iAs concentrations in vegetables and rice. Cooking losses may be significant from leafy greens. Diaz et al. (2004) report that for exposed vegetables (chard, spinach, asparagus, beans, cauliflower) iAs when cooked, was 22–44% of the raw concentrations. Root vegetables also had significant reductions, with iAs in cooked carrot, potatoes and beetroot 17–29% of the raw concentrations.

5 BIOAVAILABILITY

It has long been assumed that dietary arsenic is almost completely absorbed. A recent review by Yager et al. (2015) indicates that while absorption is generally high (typically 70% or higher), dietary arsenic absorption may be reduced compared with absorption of arsenic in drinking water. Many available studies use in vitro digestion assays to measure arsenic bioaccessibility, but seldom include measures of iAs both before and after the

digestion procedure, thus preventing an assessment of the relative bioavailability of iAs.

6 CONCLUSIONS

Vegetables contribute the greatest amount to dietary intake of iAs; however, there is considerable variation in concentrations both within and across kinds of vegetables. Detection limits have been too high to allow for reliable determination of distribution of concentrations. Accurate estimates of dietary iAs intakes should reflect concentrations in foods prepared as they are consumed and should also consider relative bioavailability. Application of the EFSA (2009) assumption that 70% of totAs is iAs to vegetables may substantially overestimate iAs intake.

REFERENCES

Baeyens, W., Gao, Y., De Galan, S., Bilau, M., Van Larebeke, N., Leermakers, M. 2009. Dietary exposure to total and toxic arsenic in Belgium: Importance of arsenic speciation in North Sea fish. Mol. Nutr. Food Res. 53: 558–565.

Diaz, O.P., Leyton, I., Munoz, O., Nunez, N., Devesa, V., Suner, M.A., Velez, D., Montoro, R. 2004. Contribution of water, bread, and vegetables (raw and cooked) to dietary intake of inorganic arsenic in a rural village of Northern Chile. J. Agr. Food Chem. 52: 1773–1779.

EFSA (European Food Safety Authority). 2009. Scientific Opinion on Arsenic in Food. EFSA J. 7(10): doi:10.2903/j.efsa.2009.1351.

FDA (U.S. Food and Drug Administration). 2013. FDA Statement on Testing and Analysis of Arsenic in Rice and Rice Products. September 6. http://www.fda.gov/Food/Food borneIllnessContaminants/Metals/ucm367263.htm

FDA (U.S. Food and Drug Administration). Total Diet Study. http://www.fda.gov/Food/FoodScienceResearch/TotalDietStudy/default.htm. Accessed August 2015.

Helgesen, H., Larsen, E.H. 1998. Bioavailability and speciation of arsenic in carrots grown in contaminated soil. Analyst 123(5): 791–796.

JECFA (Joint FAO/WHO Expert Committee on Food Additives). 2011. Safety evaluation of certain contaminants in food: report of the seventy-second joint FAO/WHO Expert Committee on Food Additives (JECFA). World Health Organization (WHO) WHO Food Additives Series 63, 8. FAO JECFA Monographs.

Kurzius-Spencer, M., Burgess, J.L., Harris, R.B., Hartz, V., Roberge, J., Huang, S., Hsu, C.-H., O'Rourke, M.K. 2014. Contribution of diet to aggregate arsenic exposures—An analysis across populations. J Expo Sci Environ Epid 24(2): 156–162.

Schoof, R.A., Yost, L.J., Crecelius, E., Irgolic, K., Guo, H.-R., Greene, H.L. 1998. Dietary arsenic intake in

Taiwanese districts with elevated arsenic in drinking water. *Hum. Ecol. Risk Assess.* 4(1): 117–136.

Schoof, R.A., Yost, L.J., Eickhoff, J., Crecelius, E.A., Cragin, D.W., Meacher, D.M., Menzel D.B. 1999a. A market basket survey of inorganic arsenic in food. *Food Chem. Toxicol.* 37: 839–846.

Schoof, R.A., Eickhoff, J., Yost, L.J., Crecelius, E.A., Cragin, D.W., Meacher, D.M., Menzel, D.B. 1999b. Dietary exposure to inorganic arsenic. In: W.R. Chappell, C.O. Abernathy, R.L. Calderon (Eds.) *Proc. Third International Conference on Arsenic Exposure and Health Effects.* Elsevier Science Ltd., pp. 81–88.

Schoof, R.A. Yager, J.W. 2007. Variation of total and speciated arsenic in commonly consumed fish and seafood. *Hum. Ecol. Risk Assess.* 13: 946–965.

Smith, N.M., Lee, R., Heitkemper, D.T., DeNicola Cafferky, K., Haque, A., Henderson, A.K. 2006. Inorganic arsenic in cooked rice and vegetables from Bangladeshi households. *Sci. Total Environ.* 370: 294–301.

Xue, J., Zartarian, V., Wang, S.-W., Liu, S.V., Georgopolous, P. 2010. Probabilistic modeling of dietary arsenic exposure and dose and evaluation with 2003–2004 NHANES data. *Environ. Health Perspect.* 118: 345–350.

Yager J.W., Greene, T., Schoof, R.A. 2015. Arsenic relative bioavailability from diet and airborne exposures: Implications for risk assessment. *Sci. Total Environ.* 536: 368–381.

Yost, L.J., Tao, S.-H., Egan, S.K., Barraj, L.M., Smith, K.M., Tsuji, J.S., Lowney, Y.W., Schoof, R.A., Rachman, N.J. 2004. Estimation of dietary intake of inorganic arsenic in U.S. children. *Hum. Ecol. Risk Assess.* 10: 473–483.

Arsenic Research and Global Sustainability – Bhattacharya, Vahter, Jarsjö, Kumpiene,
Ahmad, Sparrenbom, Jacks, Donselaar, Bundschuh & Naidu (Eds)
© 2016 Taylor & Francis Group, London, ISBN 978-1-138-02941-5

Intake of inorganic arsenic from food in Hungary, Romania and Slovakia

G.S. Leonardi[1], P. Gnagnarella[2], T. Fletcher[1] & Other Members of ASHRAM Study Group
[1]*London School of Hygiene and Tropical Medicine, London, UK*
[2]*European Oncology Institute, Milan, Italy*

ABSTRACT: The report presents secondary analysis findings from the "ArSenic Health Risk Assessment and Molecular epidemiology" (ASHRAM) study. Aim: to quantify exposure from ingestion of inorganic arsenic (iAs) from food. Methods: Participants (n = 540) were controls selected for a case-control study, residents of Slovakia, Hungary and Romania aged 30–79, and were general hospital surgery in-patients (appendicitis, abdominal hernia, duodenal ulcer, or cholelithiasis) and orthopedic and traumatology patients (fractures). Dietary habits were assessed using a validated food frequency questionnaire, and these were linked to iAs contents in rice and potatoes to estimate iAs intake from food at individual level. Results: (The total iAs intake from food for the general population in areas of Central Europe was estimated as 10.14 µg/day (±6.83), and 19.80 µg/day (±20.78) after adding intake from cooking water. Conclusions: The main contributor to daily iAs intake in this population was rice.

1 INTRODUCTION

The main inorganic arsenic (iAs) exposure for humans is via food and water, but methods for quantification of iAs exposure from food at population level have not been widely used. Increased concentrations of As in food may arise from irrigation or industrial pollution. In 1993 World Health Organization (WHO) reduced the recommended water standard from 50 to 10 µg/L, however in contrast to iAs legislative regulations in drinking water in the US and European Union, there are no food standards for either total or iAs for foods. This strengthens the need for accurate information on dietary intake of the toxicologically potent iAs, and evidence for associated health risks, in order to set guidelines for iAs in food products. The objective of the present work was to develop an exposure model to estimate the intake of iAs from food from direct observation at population level in Hungary, Romania and Slovakia.

2 METHODS

2.1 Study population

The study selected and analysed data collected from the case-control "ArSenic Health Risk Assessment and Molecular epidemiology" (ASHRAM) study, conducted in Hungary, Slovakia and Romania. Dietary exposure measurements were evaluated among controls, as these were selected to be representative of the population. Controls (n = 540) were general hospital surgery in-patients (appendicitis, abdominal hernia, duodenal ulcer, or cholelithiasis), and orthopedic and traumatology patients (fractures).

2.2 Questionnaires

A 'General Questionnaire' and 'Food Frequency Questionnaire' (FFQ) were applied to the recruited participants by an interviewer in the native language. The general questionnaire obtained information regarding demographic (age, gender, body weight, height) and socioeconomic characteristics (income, education, profession), medical history, smoking habits, skin type and history of exposure to sunlight and location of residential water sources consumed. The FFQ is an effective instrument for determining the dietary pattern and the consumption of distinct food items. In the present study, a FFQ was administered to determine fluid and nutritional consumption particularly before 1989, and after in consideration of the expected dietary differences before and after political, social changes and interventions in 1980 in Hungary and 1989 in Slovakia. The nutrition questionnaire was detailed and had four main sections assessing: general aspects of eating habits, fat-intake pattern, 112 questions aimed to assess past food intake and questions assessing present food intake. Food sampling was conducted only in Slovakia, and the information produced was adapted for the use in Hungarian and Romanian participants.

The present study focused on the questions associated with potatoes and rice consumption and recipes containing these ingredients. The exposure model was applied for 'Potatoes—boiled or puree', 'Potatoes—fried or baked', 'Goulash soup', 'Vegetable soup', 'Ragu rice' and 'Vegetable rice'.

2.3 Dietary iAs intake calculations

For each food item the intake was calculated based on the consumption frequency (coefficients for the frequency of consumption were assigned according to a protocol developed for ASHRAM), along with serving sizes (g) (the quantity was estimated based on the recipes) and iAs content ($\mu g/100g$) as measured in ASHRAM study. The iAs content for peeled potatoes was found 0.023 $\mu g/100g$, for unpeeled potatoes 0.33 $\mu g/100g$ and for rice 15.83 $\mu g/100g$. Therefore, the average contaminant intake ($\mu g/day$) = $EDI_{iAs} \times$ Frequency. The estimation of complete iAs consumption was obtained by summing iAs ingested from various food items identified in the FFQ. Seasonal variation was not included as the consumption of investigated food items was considered stable throughout the year.

For potatoes the iAs daily intake from four recipes was calculated three times—using iAs concentration either for peeled potatoes, unpeeled potatoes and from the general knowledge considering that 5% of the population consumed unpeeled potatoes, while the remaining 95% ate peeled potatoes. Similarly, the iAs intake from rice was calculated. Finally, the total iAs daily intake from food was calculated by summing the iAs intake from potatoes and rice examined in the current study.

3 RESULTS AND DISCUSSION

3.1 Intake from rice and potatoes

The average iAs intake from potatoes was 0.09 $\mu g/day$ (± 0.05), assuming that 95% of eaten potatoes were peeled while only 5% were eaten unpeeled. Unpeeled potatoes contributed much more to the total daily iAs intake compared to peeled potatoes. Intake from rice averaged 10.12 $\mu g/day$ ± 7.01 The total iAs ingestion from food (considering all meals that included potatoes and rice) was 10.14 $\mu g/day$, ± 6.83, largely from rice.

3.2 Intake from cooking water

Adding cooking water to the exposure model, nearly doubled the dietary intake at 19.80 $\mu g/day$

(± 20.78) (the maximum range value increased from 37.01 to 205.37 $\mu g/day$).

4 CONCLUSIONS

The present study estimated the level of daily dietary iAs intake from staple foods rice and potatoes in the general Hungarian, Slovakian and Romanian population, with and without contribution by cooking water. It is a unique finding as a validated FFQ was applied in a large case-control study to identify dietary patterns and estimate iAs daily intake.

ACKNOWLEDGEMENTS

We thank the participants for their contributions to this study. Financial support provided by European Commission project No. QLK4-CT-2001–00264 (ASHRAM) is gratefully acknowledged. The authors state that they have no competing financial interests. Members of the ASHRAM study group were (in alphabetical order): Tony Fletcher, Walter Goessler, Eugen Gurzau, Kari Hemminki, Rupert Hough, Kvetoslava Koppova, Rajiv Kumar, Giovanni Leonardi, Peter Rudnai, Marie Vahter.

REFERENCES

Decarli, A., Franceschi, S., Ferraroni, M., Gnagnarella, P., Parpinel, M.T., Vecchia, C.L., et al. 1996. Validation of a food-frequency questionnaire to assess dietary intakes in cancer studies in Italy results for specific nutrients. Ann. Epidemiol. (2): 110–118.

Hough, R.L., Fletcher, T., Leonardi, G.S., Goessler, W., Gnagnarella, P., Clemens, F., et al. 2010. Lifetime exposure to arsenic in residential drinking water in Central Europe. Int. Arch. Occup. Environ. Health 83(5):471–481.

Leonardi, G., Vahter, M., Clemens, F., Goessler, W., Gurzau, E., Hemminki, K., et al. 2012. Inorganic arsenic and basal cell carcinoma in areas of Hungary, Romania, and Slovakia: A case-control study. Environ. Health Perspect. 120(5):721–726.

London School of Hygiene and Tropical Medicine. 2005. ASHRAM. Arsenic Health Risk Assessment and Molecular Epidemiology. Final Report for European Commission Framework Programme 5. August 2005.

Arsenic Research and Global Sustainability – Bhattacharya, Vahter, Jarsjö, Kumpiene, Ahmad, Sparrenbom, Jacks, Donselaar, Bundschuh & Naidu (Eds)
© 2016 Taylor & Francis Group, London, ISBN 978-1-138-02941-5

Occurrence of arsenic, vanadium and uranium in powdered milk from Argentina

A.L. Pérez Carrera[1,2], F.E. Arellano[1,2], W. Goessler[3], S. Braeuer[3] & A. Fernández Cirelli[1,2]

[1]*Instituto de Investigaciones en Producción Animal (UBA-CONICET), Universidad de Buenos Aires, Buenos Aires, Argentina*
[2]*Centro de Estudios Transdisciplinarios del Agua, Facultad de Ciencias Veterinarias, Universidad de Buenos Aires, Buenos Aires, Argentina*
[3]*Institute of Chemistry—Analytical Chemistry, University of Graz, Graz, Austria*

ABSTRACT: In Latin America the information related to trace elements content in milk and milk products is scarce. In this region, arsenic naturally occurring in ground water, it can reach production areas. We studied the correlation of As with other trace elements, like vanadium and uranium, and also calculated the daily As intake rate. The mean As concentration in all samples was 20 ± 11 µg/kg. A correlation between As and V (r: 0.74, $p < 0.05$) and also between As and U (r: 0.70, $p < 0.05$) was found in all samples. The daily intake rate was below the maximum limit of intake (0.015 mg/kg bw or 2.1 µg/kg/d). This work shows the importance of the characterization of trace elements in milk and its products.

1 INTRODUCTION

In Argentina, arsenic (As) is naturally present in the groundwater, used as animal drinking water in very high concentrations. The As affected region coincides with the main dairy production areas (Smedley & Kinniburgh, 2002). This element and others that are present in the Argentinean water in higher concentrations, like uranium (U) or vanadium (V), could be transferred to milk. Their presence could affect the quality of milk products (Ayar *et al.*, 2009) and represent a risk for human health. In spite of the importance of milk for food, there are few studies in Latin America about the composition of trace elements in milk and other dairy products. The aim of this study was to quantify the levels of As in milk powder, to study whether there is any correlation with the U and V levels in these samples, and estimate the risk from consumption through As daily intake rate.

2 MATERIALS AND METHODS

A number of 26 commercial powdered bovine milk samples (5 infant formula, 11 skimmed milks and 10 whole milks from different trademarks) were obtained from Argentinean supermarkets and stores. About 0,5 g of each sample was digested with 5 mL of nitric acid (65%, Carl-Roth, Karlsruhe, Germany, subboiled) in a microwave digestion system (Ultraclave IV, MLS GmBH, Germany). Then they were diluted with ultrapure water (18.2 MΩ*cm) to an acidity of 10% v/v. As, V and U levels were measured with Inductively Coupled Plasma Mass Spectrometry (Agilent ICP-MS 7500ce, Waldbronn, Germany). NIST SRM 1640a and IAEA 153 were used as reference materials. All samples were prepared in triplicate. For the statistical analysis we used Infostat® software.

The risk of As toxicity was estimated using the USEPA equation (USEPA, 1992, 2004). Standardized values were used for this calculation, taking into account a balanced diet of 0.06 kg/day of powdered milk for a 70 kg body weight for an adult, and 0.07 kg/day of infant formula and 11.25 kg average weight for a baby/child from 0 to 4 years old.

The following equation was used to calculate the intake of toxic substances (USEPA, 1992, 2004):

$$ADDI = C* IR* EF*ED*CF /BW * AT \qquad (1)$$

where:
ADDI average daily intake dose (in mg/kg·d)
C concentration of the dangerous substance (in mg/kg)
IR daily intake rate (in mg/d)
EF exposure frequency (in d/year)
ED exposure duration (in year)
CF conversion factor (1.10^{-6} kg/mg)
BW body weight of the exposed person (in kg)
AT average time correction factor (ED * 365 days for non carcinogenetic substances; statistical duration of human life (70) * 365 days for carcinogenetic substances).

Table 1. Summary statistics for total samples, babies and child formula, skimmed and whole powdered milk.

Powdered Milk	Trace element	N	Mean	SD	Min	Max
Total		26	20	11	3	44
Infant formula	As	5	20	14	3	36
Skimmed	(µg/kg)	10	26	8	10	38
whole		11	16	10	4	44
Total		26	31	22	5	95
Infant formula	V	5	34	27	5	73
skimmed	(µg/kg)	10	41	26	18	95
whole		11	21	11	12	46
Total		26	6	6	1,3	30
Infant formula	U	5	5	4	1,3	10
skimmed	(µg/kg)	10	4	2	1,9	8
whole		11	7	7	2,6	30

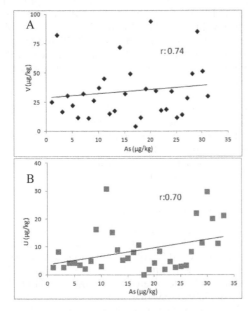

Figure 1. A. Correlation diagram of A) As and V; B) Dispersion diagram of As and U.

3 RESULTS AND DISCUSSION

The mean As concentration in samples was 20 ± 11 µg/kg, with levels from 16 ± 10 µg/kg in whole milk to 26 ± 8 µg/kg in skimmed milk (Table 1). In the case of V the mean was 31 ± 22 µg/kg, with levels between 21 ± 11 µg/kg (whole milk) and 41 ± 26 (skimmed milk). The mean level for U was 6 ± 6 µg/kg, showing high variability (Table 1). No significant differences between concentrations of different elements were found in all types of powdered milk.

Correlations between As and V and between As and U were calculated using Sperman's correlation coefficients. A correlation between As and V was found (r: 0.74 p < 0.05), and also between As and U (r: 0.70 p < 0.05) in all samples (Fig. 1).

An arsenic intake rate (AsIR) of 0.014 µg/kg/d was calculated for whole milk powder. The AsIR value for skimmed milk was 0.022 µg/kg/d and finally for the infant formula 0.124 µg/kg/d.

4 CONCLUSIONS

Considering the Maximum tolerate level (LMR) of 0.05 mg/kg of As (MERCOSUR, 2013) all samples present levels below this value. Furthermore, the present results show that the daily intake rates for both are below the maximum recommended intake level (0.015 mg/kg bw or 2.1 µg/kg/d) described by by the World Health Organization (Carignan et al., 2015). Studies on health effects of U and V are scarce in Argentina. A positive correlation between As and V, and between As and U was observed. This work shows the importance of the characterization of trace elements in milk and its products.

ACKNOWLEDGEMENTS

Authors are indebted to the University of Buenos Aires and to CONICET (National Research Council) for financial support and UBA (Universidad de Buenos Aires). The authors also want to thank OEAD WTZ (Österreich Argentinien, Projekt 08/2013 2013–2015) for financial support.

REFERENCES

Ayar, A., Sert, D., Akın, N. 2009. The trace metal levels in milk and dairy products consumed in middle Anatolia—Turkey. Environ. Monit. Assess. 152: 1–12.

Carignan, C.C., Cottingham, K.L., Jackson, B.P., Farzan, S.F., Gandolfi, A.J., Punshon, T., Folt, C.L., Karagas, M.R.. 2015. Estimated exposure to arsenic in breast-fed and formula-fed infants in a United States cohort. Environ. Health Perspect. 123: 500–506.

Pérez Carrera, A.L., Álvarez Gonçalvez, C.V., Fernández Cirelli, A. 2015. Vanadio en agua de bebida animal de tambos del sudeste de Córdoba, Argentina. In Vet. 16 (1): 39–47.

Reglamento Técnico MERCOSUR. 2013. Ministerio de Salud Pública. Decreto N° 14. www.aladi.org/nsfaladi/... nsf/.../$ FILE/Decreto%20N°%2014–2013.pdf

Smedley, P.L., Kinniburgh, D. 2002. A review of the source, behaviour and distribution of arsenic in natural waters. Appl. Geochem. 17: 517–568.

United States of Environmental Protection Agency (USEPA). 1992. Guidelines for exposure assessment, Environmental Protection Agency, Risk Assessment Forum, Washington (DC).

United States of Environmental Protection Agency (USEPA). 2004. RAGS. Volume I: HHEM (Part E, Supplemental Guidance for Dermal Risk Assessment), Interim Guidance. EPA/540/R-99/005. PB99-963312. OSWER 9285.7-02EP.

Arsenic Research and Global Sustainability – Bhattacharya, Vahter, Jarsjö, Kumpiene,
Ahmad, Sparrenbom, Jacks, Donselaar, Bundschuh & Naidu (Eds)
© 2016 Taylor & Francis Group, London, ISBN 978-1-138-02941-5

Arsenic and trace elements in groundwater, vegetables and selected food grains from middle Gangetic plain—human health perspective

Manoj Kumar[1], AL. Ramanathan[1], M.M. Rahman[2], R. Naidu[2] & P. Bhattacharya[3,4]

[1]School of Environmental Sciences, Jawaharlal Nehru University, New Delhi, India
[2]Global Centre for Environmental Remediation (GCER), University of Newcastle, Callaghan Campus,
Callaghan, NSW, Australia; Cooperative Research Centre for Contamination Assessment and Remediation
of the Environment (CRC CARE), Australia
[3]KTH-International Groundwater Arsenic Research Group, Department of Sustainable Development,
Environmental Science and Engineering, KTH Royal Institute of Technology, Stockholm, Sweden
[4]International Center for Applied Climate Science, University of Southern Queensland, Toowoomba, Queensland, Australia

ABSTRACT: Very limited efforts have been directed to determine the degree of Arsenic (As) and other trace elements contamination in food grains and understanding the associated risk to human through consumption of food in middle Gangetic plain of Bihar, India. Arsenic concentration in groundwater (>80%) samples found above BIS and WHO permissible limit, while 28% samples exceeded the previous WHO drinking water guideline (400 µg/L) for manganese (Mn). In dietary food grains As followed the sequence as rice > wheat > maize. The estimated daily intake by individual of As, Mn, Ni, Cd, Co, Pb, Zn and Cr from drinking water and dietary food grains were 169,14582, 474, 19, 26, 1449, 12955 and 882 µg/kg. The Health Risk Index (HRI) were >1 for As in drinking water, vegetables and rice indicated the potential health risk to the residents of the study area. However, HRI <1 for wheat and maize indicate a relative absence of health risks associated with their ingestion.

1 INTRODUCTION

Geogenic As contamination of groundwater in middle Gangetic plain is known for more than a decade (Bhattacharya et al., 2011). The problem was reported in 2002 in Semaria Ojha Patti village, Sahapur block in the Bhojpur district, India (Chakraborti et al., 2003). The area amid flood prone belt of the Kosi and the Gandak fans (Saha & Shukla, 2013). Arsenic in low concentration can stimulate plant growth but it can be accumulated above WHO threshold concentration for safe ingestion of food crops (Rahman & Naidu, 2009) which may pose serious risks to the local residing population (Rahman & Naidu, 2009). Therefore it is of prime need to evaluate As exposure via commonly consumed food grains and vegetables. The As concentration >50 µg/L has been reported in Samastipur district (Saha & Shukla, 2013). From the health perspective it is equally important to study the trace elements other than As in drinking water, vegetables and commonly consumed food grains. The objective of this study was to estimate the total exposure to As and other trace elements via drinking water and food. An attempt was also made to understand the potential health risk caused by consumption of contaminated water and food.

2 METHODS/EXPERIMENTAL

2.1 Study area and sample collection

Two blocks (Mohiuddinagar and Mohanpur) of Samastipur district, Bihar, India were selected for the current study. Figure 1 shows the location of the study area in Bihar, India (middle Gangetic plain) and the corresponding sampling points. Tubewell water samples (n = 23), vegetables (n = 34), rice (n = 15), wheat (n = 35) and maize (n = 31) which are commonly consumed by residents of this area were collected.

2.2 Sample digestion and analysis

A microwave digester (CEM, MARS 6) inbuilt 42 digester vessels was used to digest all food samples. Approximately 0.5 g of ground food samples were weighed directly into a Teflon vessel followed by adding 5 ml concentrated HNO_3 and 2 ml H_2O_2. An Agilent7500ce (Agilent Technologies, Tokyo, Japan) ICP-MS with detection limits of 0.01, 0.03, 0.05, 0.03, 0.02, 0.01, 0.10,0.05, 0.2 and 0.01 µg/l for As, Cd, Co, Cr, Cu, Mn, Ni, Pb, Se and Zn respectively.

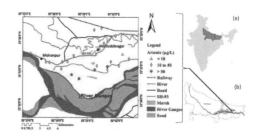

Figure 1. Map of the study area with sample location. a) India, b) middle Gangetic plain of Uttar Pradesh and Bihar (modified from Kumar et al., 2016).

Table 1. Mean concentration (µg/kg) of As and trace elements in Vegetables and food grains samples of Samastipur district.

Parameters	As	Cd	Co	Cr	Cu	Mn	Ni	Pb	Zn
Vegetable	452	102	173	5823	17253	24040	2921	717	35413
Rice	51	19	10	294	4255	5928	378	2277	9123
Wheat	27	18.8	33	1354	5459	31821	449	1395	20474
Maize	13	1.70	3.71	1087	2554	5664	187	1636	12373
Green gram	23	17	65	728	10990	16857	894	1345	28707

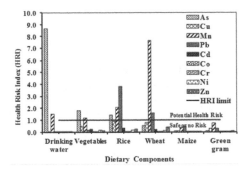

Figure 2. Health risk index for As, Cu, Mn, Pb, Cd, Co, Cr, Ni and Zn. (Line shows the HRI limit i.e. 1) (modified from Kumar et al., 2016).

3 RESULTS AND DISCUSSION

3.1 Arsenic and trace elements concentration in groundwater

The concentration of As in groundwater ranged from 6.25 to 135 µg/L. Approximately (> 80%) of groundwater samples exceeded for As and 28% samples exceeded for Mn to their respective WHO limits of 10 and 400 µg/L respectively. Groundwater were below to the selected reference drinking water standards of BIS and WHO for other trace elements like Ni, Pb, Se, Cu, Co, Cd, Cr, and Zn.

3.2 Arsenic and trace elements in vegetables and food grains

Mean concentration of As and trace elements shown in Table 1.

Table 1. Mean concentration (µg/kg) of As and trace elements in Vegetables and food grains samples of Samastipur district.

The concentration of As followed a sequence with in vegetables with their mean values as luffa (800) > brinjal (492) > cucumber (399) > ladyfinger (375) > gourd (268) > green gram (174 µg/kg). In food grains the trend followed as rice > wheat > maize with their mean concentration of 51, 27 and 13 µg/kg respectively. Vegetables were found enriched with Cd (102 µg/kg) than dietary food grains.

3.3 Chronic daily intake of As and trace elements and health risk assessment

The estimated chronic daily intake of As from drinking water, rice, wheat, vegetables, green gram and maize were 116.67, 19.59, 7.09, 23.96, 1.11 and 0.96 µg. It was observed that an adult consumes 169 µg of As from drinking water and dietary food grains. The daily Mn take from vegetable, drinking water, wheat, rice, maize and green gram were 1274, 1637, 8241, 2240, 396 and 792 µg respectively. It appears that HRI values exceed safe limit1 for As in drinking water, rice and vegetables (Fig. 2).

4 CONCLUSION

In this study > 88% samples were found to have As > 10 µg/L and 28% samples having Mn > 400 µg/L which are their respective BIS and WHO permissible limits. The concentration of As and other trace elements varied extensively. Drinking water contributed major (~67%) of daily chronic intake of As.

Rice contributed for Pb, Cu, Ni and Cd. It was also found that potential health risk for As in vegetables, drinking water and rice for Mn in drinking water, rice, wheat and vegetables, for Pb in wheat and rice. The risk from other estimated trace elements were negligible for remaining food components.

ACKNOWLEDGEMENTS

The author is thankful to the JNU and SIDA, Sweden for partial financial support. Manoj Kumar got Crawford Fund, Australia and Council of Scientific and Industrial Research (CSIR), India fellowship to do this work.

REFERENCES

Bhattacharya, P., Mukherjee, A., Mukherjee, A.B. 2011. Arsenic in groundwater of India. In: J.O. Nriagu (ed.) Encyclopedia of Environmental Health, vol. 1: 150–164 Burlington: Elsevier.

Chakraborti, D., Mukherjee, S.C., Pati, S., Sengupta, M.K., Rahman, M.M., Chowdhury, U.K., Lodh, D., Chanda, C.R., Chakraborti, A.K., Basu, G.K. 2003. Arsenic groundwater contamination in Middle Ganga Plain, Bihar, India: a future danger? *Environ. Health Perspect.* 111: 1194–1201.

Kumar, M., Rahman, M.M., Ramanathan, AL., Naidu, R. 2016. Arsenic and other elements in drinking water and dietary components from the middle Gangetic plain of Bihar, India: health risk index. *Sci. Total Environ.* 539: 125–134.

Rahman, F., Naidu, R. 2009. The influence of arsenic speciation (AsIII & AsV) and concentration on the growth, uptake and translocation of arsenic in vegetable crops (silverbeet and amaranth): greenhouse study. *Environ. Geochem. Hlth.* 31: 115–124.

Saha, D., Shukla, R.R. 2013. Genesis of arsenic-rich groundwater and the search for alternative safe aquifers in the Gangetic Plain, India. *Water Environ. Res.* 85: 2254–2264.

Arsenic Research and Global Sustainability – Bhattacharya, Vahter, Jarsjö, Kumpiene,
Ahmad, Sparrenbom, Jacks, Donselaar, Bundschuh & Naidu (Eds)
© 2016 Taylor & Francis Group, London, ISBN 978-1-138-02941-5

Arsenic in the urine of two Giant Pandas and their food

S. Braeuer[1], W. Goessler[1], W. Hoffmann[2], E. Dungl[2], T. Chunxiang[3] & L. Desheng[3]

[1]*Institute of Chemistry—Analytical Chemistry, University of Graz, Graz, Austria*
[2]*Vienna Zoo, Austria*
[3]*China Conservation and Research Centre for the Giant Panda, China*

ABSTRACT: The arsenic concentrations in urine samples of the Giant Pandas at Vienna Zoo and in their food (bamboo) were determined with Inductively Coupled Plasma Mass Spectrometry. In the urine samples, the arsenic levels were exceedingly high, namely up to 180 µg/L, although the animals were not exposed to high arsenic concentrations via their drinking water. However, the bamboo samples, especially the leaves, contained somewhat high amounts of arsenic (up to 920 µg/kg dry mass). The reasons for the occurrence of such high arsenic concentrations in the urine and the bamboo samples remain to be investigated.

1 INTRODUCTION

Arsenic has been investigated with much interest in various environmental systems. Besides water and soil, living organisms have been studied as well. However, when looking at terrestrial mammals, the available data is restricted mostly to humans, monkeys, apes, livestock and small animals like rats, mice, guinea pigs, etc. (National Research Council, 2005). Therefore we decided to take a look on the arsenic status of a more exotic and endangered mammal, the Giant Panda. Since this animal almost exclusively feeds on bamboo, we also determined the arsenic concentrations in samples of this plant.

2 EXPERIMENTAL

2.1 Sample preparation

29 urine samples of the two Giant Pandas (*Ailuropoda melanoleuca*) living at Vienna Zoo were collected with syringes from the tiled ground. Animal, date and floor type were recorded, and the samples were stored in polypropylene tubes at 4°C until analysis. For arsenic determination, samples were filtered through 0.2 µm Nylon® filters (La-Pha-Pack®GmbH, Langerwehe, Germany) and then diluted with ultrapure water (18.2 MΩ*cm, Millipore, Bedford, USA) and nitric acid (≥ 65% p.a., subboiled, Carl Roth GmbH + Co.KG, Karlsruhe, Germany) to a final acidity of 10% v/v.

Bamboo samples were taken directly from the plants that were intended as feed for the animals at the zoo. In total we collected 13 plant samples of 9 different species (Table 1). They were washed,
divided into leaves and trunk/branches and then dried. Aliquots were digested with nitric acid in a microwave heated autoclave at 250°C. Afterwards, the samples were diluted with ultrapure water to a final acidity of 10% v/v.

2.2 Measurement

For total arsenic quantification we used Inductively Coupled Plasma Mass Spectrometry (ICPMS, 7700ce, Agilent, Waldbronn, Germany). The analysis was performed in collision gas mode with helium at 4 mL/min as collision gas. Arsenic was quantified *via* external calibration. For quality assessment, germanium was used as internal standard. It was added online to the samples. Further on, we used the Standard Reference Material® (SRM) 1640a (Trace Elements in Natural Water, NIST, Gaithersburg, USA) and ClinChek® – Urine

Table 1. List of the investigated bamboo samples.

Bamboo species	number of samples
Phyllostachys viridis mitis	2
Phyllostachys viridiglaucescens	2
Phyllostachys nigra boryana	2
Phyllostachys edulis	2
Phyllostachys viridis sulphurea	1
Phyllostachys aureosulcata spectabilis	1
Phyllostachys pubesces edulis	1
Phyllostachys bissetii	1
Pseudosasa japonica	1

Table 2. Arsenic in the urine of the Giant Pandas at Vienna zoo (n = 29), not normalized to specific gravity.

	Arsenic [µg/L]
Mean ± standard deviation	70 ± 40
Range	5–180
First percentile	35
Median	60
Third percentile	100

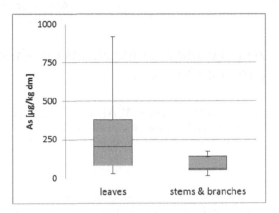

Figure 1. Arsenic concentrations in leaves and stems & branches of the bamboo samples as box plots.

Control (Recipe Chemicals + Instruments GmbH, Munich, Germany) for control of trueness.

3 RESULTS AND DISCUSSION

3.1 Quality control

The results for both reference materials (SRM® 1640a and ClinChek®) were in good agreement with the certified values (100 ± 4% and 104 ± 3%).

3.2 Urine samples

The arsenic concentrations in the Giant Pandas' urine samples were surprisingly high, with a median of 60 µg/L. More information about the results is given in Table 2.

Usually, the arsenic concentration in the urine of terrestrial mammals is between 5 and 10 µg/L (Lindberg et al., 2006). Of the 29 investigated Panda urine samples, only two contained less than 10 µg/L. The drinking water of the Giant Pandas was of good quality and contained less than 1 µg As/L. Hence, this cannot be the (only) source of these high amounts of arsenic that the animals are taking up.

3.3 Bamboo samples

Besides the drinking water, bamboo is the only other possible candidate for providing the Giant Pandas with high amounts of arsenic.

In general, the arsenic concentrations were higher in the leaves than in the corresponding stems and branches (Fig. 1). While the latter ones contained 15–170 (median: 62) µg As/kg dry mass (dm), the leaves had 29–920 (median: 205) µg As/kg dm.

These results are significantly higher than those reported for bamboo shoots (Zhao et al., 2006). There, the authors found between 27 and 94 µg/kg dm in 15 different samples from China. Only three of these samples contained more than 50 µg/kg dm. To the best of our knowledge, no other data

about arsenic concentrations in bamboo exists. Therefore, more samples from different origin have to be investigated to find out if our results are typical for bamboo plants or not.

4 CONCLUSIONS

The Giant Pandas at Vienna Zoo appear to excrete unusually high amounts of arsenic via their urine. One possible source is the bamboo, which is their main feed, where some samples also contained high levels of arsenic. To elucidate this phenomenon, it will be important to determine the chemical forms of the arsenic that are present in the animals' urine and in the plants.

ACKNOWLEDGEMENTS

The authors want to thank the coordinators and animal keepers of Vienna Zoo for collecting and providing the samples.

REFERENCES

Lindberg, A.-L., Goessler, W., Gurzau, E., Koppova, K., Rudnai, P., Kumar, R., Fletcher, T., Leonardi, G., Slotova, K., Gheorghiu, E., and Vahter, M. 2006. Arsenic exposure in Hungary, Romania and Slovakia. *J. Environ. Monit.* 8 (1): 203–208.

National Research Council. 2005. Mineral tolerance of animals. National Academy Press, Washington, D.C., USA.

Zhao, R., Zhao, M., Wang, H., Taneike, Y., Zhang, X. 2006. Arsenic speciation in moso bamboo shoot—A terrestrial plant that contains organoarsenic species. *Sci. Total Environ.* 371: 293–303.

Arsenic Research and Global Sustainability – Bhattacharya, Vahter, Jarsjö, Kumpiene,
Ahmad, Sparrenbom, Jacks, Donselaar, Bundschuh & Naidu (Eds)
© 2016 Taylor & Francis Group, London, ISBN 978-1-138-02941-5

Arsenic and fluoride effects on white clover (*Trifollium repens*) seeds early development

A. Fernández Cirelli, C.V. Alvarez Gonçalvez & A. Pérez Carrera
*Centro de Estudios Transdisciplinarios del Agua (CETA—UBA), Facultad de Ciencias Veterinarias, Instituto de
Investigaciones en Producción Animal (INPA) UBA—CONICET, Universidad de Buenos Aires, Argentina*

ABSTRACT: In several areas of the Chacopampean plain (Argentina) white clover fields are irrigated
with groundwater, in which arsenic (As) and other inorganic trace elements like fluoride (F) are naturally
present. In this study we analyzed the effect of a combination of As and F on the germination and growth
parameters of white clover seeds. A non significant decrease in germination percentage regard to control
was observed for all treatments. Nevertheless, a significant negative effect on both, radicle and hypocotyl
elongation was observed when seeds were exposed to As or As-F mixture. The present results denote that
more investigation on the effect of As on forage plants and the interaction with several factor like other
ions present in water is needed.

1 INTRODUCTION

White clover (*Trifollium repens*) is one of the main
forage species used for cattle in Argentine due to
its high digestibility and palatability (Carrillo
et al., 2003). In several areas of the Chacopampean
plain (Argentina) white clover fields are irrigated
with groundwater, where arsenic (As) and fluoride (F⁻) are present naturally. Several studies have
reported the effects of As and F on the germination and growth parameters of plants. It is known
that As is non-essential for plants, and it is generally
toxic. Arsenic inhibits root elongation, cell proliferation, biomass production and damage by oxidative stress (Garg & Singla, 2011, Finnegan & Chen,
2012). Fluoride is another non-essential element for
plants. It was seen that F retards the rate of seedling
root growth during germination, reduces the viability and the levels of soluble proteins of germinating
seeds, decreases DNA synthesis, causes a deficiency
of calcium in the plants and it is involved in the
inhibition of enzymes (Barbier *et al.*, 2010, Panda
et al., 2015), among other deleterious effects.

Little information is available regarding the
effects of some trace element present in irrigation
water on the forage germination and biomass production. It is important to know that usually plants
are not exposed to a single element but rather to
a combination of several elements; and often, the
exposure to these chemical combinations has different effects than would have an independent exposure to them. Although the biological effects of
arsenic and fluoride administered alone have been
well studied, the effect of the simultaneous exposure
has been poorly studied and the results are controversial. Therefore, the aim of this study is to analyze
the effect of As, F and As + F on the germination
and growth parameters of white clover seeds.

2 MATERIAL AND METHODS

This study was carried out on white clover seeds
(*Trifolium repens*). These were supplied by a pro-
vider of commercial seeds. Their viability was
confirmed before the beginning of the experiment, they presented up to a 90% of germination. Ideal stocking conditions were guaranteed
(low temperatures and relative humidity). Seeds
were placed on sterile plastic Petri boxes with
filter paper of 80 g/m², 250 μm thick, pores of
14 μm and a permeability of 14 l/s/m² (measured
by method DIN 53887). The seeds were irrigated
with: As, F or As + F. All of them were at a final
concentration of 1mg/L. The As used was sodium
arsenate 7 hydrate (NaHAsO₄. 7H₂O) from Bio-
pack (CAS#10048-95-0). The fluoride solution
was prepared from sodium fluoride (NaF) from
Merk (CAS#7681-49-4). The experiment was
carried out for five days under lab conditions,
in darkness and at a constant temperature of
$22 \pm 2°C$. A modified version of an EPA protocol
was used. Impact of As and F on germination and
plant elongation were measured and statistically
analyzed with InfoStat®.

3 RESULTS AND DISCUSSION

A non significant decrease in germination percentage ($p > 0.05$) regarding control was observed for
all treatments (Fig. 1).

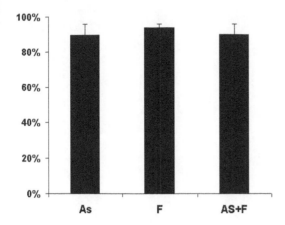

Figure 1. Normalized percentage of germination of white clover seeds regard to control. There are not significant differences ($p < 0.05$) between treatments.

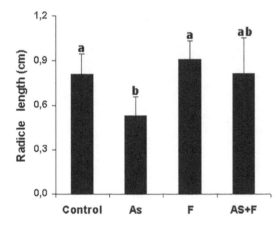

Figure 2. White clover hypocotyls length for different treatments. Bars show confidence intervals. Different letters denote significant differences ($p < 0.05$) between treatments.

It was observed a significant negative effect ($p < 0.05$) on radicle elongation (Fig. 2) when seeds were exposed to As solutions in regard to control, however this effect was not significant when seeds were exposed to F + As or F solutions. Considering hypocotyls elongation, white clover showed a significant and negative variation ($p < 0.05$) on hypocotyl length (Fig. 3) when seeds were exposed to a combination of As and F, but there are not significant differences regard to control when seeds were exposed to As or F alone.

4 CONCLUSIONS

These results show that irrigation with As or As + F negatively affect seedling growth parameters.

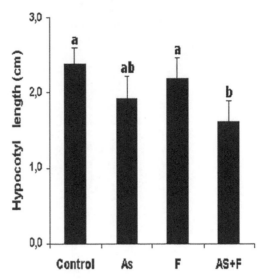

Figure 3. White clover hypocotyls length for different treatments. Bars show confidence intervals. Different letters denote significant differences ($p < 0.05$) between treatments.

This study also suggests that radicle and hypocotyl are affected differently by both elements, isolated or combined. More studies are needed to establish the conjoined effects of both elements on forages. The present results denote that more investigation on the effect of As on forage plants and the interaction with several factors (soil, other water ions and compounds, soil microorganisms, etc.) is necessary to clarify the role of As in forage production.

ACKNOWLEDGEMENTS

The authors are grateful to the University of Buenos Aires and CONICET, MINCyT for financial support.

REFERENCES

Barbier, O., Arriola-Mendoza, L., Del Razo, L.M. 2010. Molecular mechanisms of fluoride toxicity. *Chem.-Biol. Interact.* 188: 319–333.

Carrillo, J. 2003. *Manejo de Pasturas.* EEA INTA Balcarce. Balcarce, Buenos Aires Argentina. pp 458.

Finnegan, P.M., Chen, W. 2012. Arsenic toxicity: the effects on plant metabolism. *Front. Physiol.* 3: Article 192: 1–12.

Garg, N., Singla, P. 2011. Arsenic toxicity in crop plats: physiological effects and tolerance mechanism. Environ. Chem. Lett. 9: 303–321.

Panda, D. 2015. Fluoride toxicity stress: physiological and biochemical consequences on plants. *Int. J. Biores. Env. Agril. Sci.* 1: 70–84.

2.4   Bioavailability

Fungal bioaugmentation of the rice root-zone to reduce arsenic uptake by rice from soils

P.K. Srivastava

Environmental Sciences, CSIR—National Botanical Research Institute, Lucknow, India

ABSTRACT: Arsenic (As) is affecting people at risk upon consumption of contaminated water and food produces. To reduce the bioavailable soil As and uptake by rice, the soil fungal bioaugmentation was attempted by using consortia of four fungal stains. In a soil-column study, four fungal consortia treatments namely, non-inoculated soil (T1), spore suspension (T2), fungal biomass (T3), and talc-formulation (T4) were used. Rice seedling transplanted into each column after fungal treatments to soils. After the experiment, the bioavailable As was reduced by 50% in T4 compared to T1. The bioaugmentation increased fungal cfu from 3.01 to 4.74 log units in the rice root-zone soils upon treatments. Soil ergosterol content increased under different fungal treatments compared to 0.78 in T1. The grain arsenic content (μg/kg) in rice with T4 was lowered by 76% compared to T1. Among treatments, T4 was effective and can be applied for arsenic mycoremediation for safe rice cultivation in As-contaminated soils.

1 INTRODUCTION

Arsenic (As) is a food chain contaminant. The accumulation of As in soils and crops due to the use of As-contaminated groundwater for irrigation has created immediate concern worldwide. Building of soil As level in rice fields may lead to higher As uptake in rice grains and provide more As for ingestion to the inhabitants of the affected region. Assessment of soil arsenic bioavailability may help the extent of remediation required at contaminated sites by providing actual toxic exposure estimates. The impact of As accumulation in soil on plants and the transfer of As to the food-chain depend on As bioavailability in soils. Rhizospheric manipulation of plant roots using microbes with As bioremediation capabilities can be one of the strategies to mitigate As contamination in soil-plant interface. Fungi may serve as a sink to bioaccumulate/sorb As along with biomethylation and bioconversion of As for remediation purposes. The objective of the study was to reduce the bioavailable fraction of soil As using consortia of four novel promising fungal strains having ability to remove and accumulate As (Srivastava *et al.*, 2011, 2012). It was hypothesized that this approach may lower down the As uptake by rice upon bioaugmentation of As removing fungal consortia in the soil-plant root zone.

2 METHODS/EXPERIMENTAL

2.1 *Treatment and experiment*

The pure cultures of four soil fungal strains were maintained according to Srivastava *et al.* (2012). The soil core samples (0–30 cm depth, n = 5) were collected using PVC columns (10 cm dia. and 35 cm height) from As-contaminated (mean total As 15 mg/kg) rice fields from Ballia district of Uttar Pradesh, India (Srivastava *et al.*, 2015). The consortia treatments were prepared as spore suspension (T2); fungal biomass (T3), and talc-based formulation (T4). The soil without any fungal consortia application was set as the control (T1). For the experiment, fungal treatments were applied into the upper layer (0–6 cm) of soil column and kept for incubation for 7d. Further, nursery-grown 10d old rice seedlings transplanted into each soil column. Soil columns irrigated with an As solution (10 mg/L) as per rice agronomy practice for 3 months. No extra fertilizer and weedicide were used.

2.2 *Analysis of soil properties, soil arsenic fractionation and total arsenic in rice*

After harvesting of entire rice plant from soil column (upon 135d), the root-zone soil was collected. The soil samples were analyzed for different physico-chemical properties and enzyme activities. The fungal cfu, ergosterol, live fungal biomass contents were also measured in soil samples. Arsenic fractionation in <2 mm soil samples was carried out using sequential extraction procedure (Wenzel *et al.*, 2001). Total As content in soil and rice plant parts (root, shoot and dehusked grain) were determined using ICP–MS after microwave digestion of samples.

3 RESULTS AND DISCUSSION

3.1 *Soil characteristics*

Upon different fungal treatments, physico-chemical properties (except bulk density and pH) and enzyme activities have changed, following the pattern of T1 > T2> T3> T4,

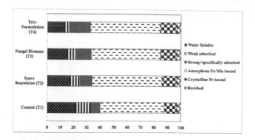

Figure 1. Percent contribution of different arsenic fractions to total soil arsenic content in different treatments.

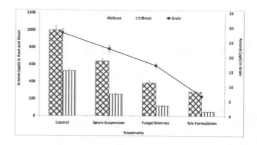

Figure 2. Arsenic (µg/kg) content in various plant parts of rice under different treatments.

with their respective highest values in T4. The soil fungal cfu increased from 3.01 in T1 to 4.74 in T4. The soil ergosterol content (µg/g), an effective biochemical marker of live fungal biomass, was also increased exponentially from 0.78 in T1 to 6.5 in T4. Results of soil properties clearly indicate that the T4 treatment had the strongest effects on fungal population growth and presumably colonization of root-zone soils under study.

3.2 Soil arsenic fractionation

The total soil As contents (µg/g) were 16.99, 12.65, 9.52 and 7.90 in the T1, T2, T3 and T4, respectively. The maximum reduction in the total As content was 53% in T4 compared to T1. The water soluble and weakly adsorbed soil As fraction were reduced by 72 and 84% in T4 compared to T1. While, strongly/specifically adsorbed soil As fraction was increased by 5.5% in T4 compared to T1. As shown in Figure 1, the water soluble and weakly adsorbed fractions of soil As were reduced from 21 and 11% share of the total soil As content in the case of T1 to 12 and 4% share of the total soil As content in T4. Whereas, the strongly/specifically adsorbed fraction of soil As was increased from 7% share of the total soil As content in T1 to 16% share of the same in T4. The percent shares of the remaining fractions of soil As were not significantly changed among all treatments. The water soluble and weakly adsorbed fractions of soil As represent potentially bioavailable soil As content. It was observed that this potentially bioavailable soil As fraction was significantly reduced by 50% upon T4 compared to T1, along with an increase of 128% in specifically adsorbed soil As fraction in the same comparison. This increase in specifically adsorbed soil As might be due to increase in fungal biomass, followed by subsequent enhanced sorption (chemisorption to various functional groups on fungal cell wall, and passive adsorption) of soil As onto fungal biomass/cell wall, and possibly As bioaccumulation by fungal biomass.

3.3 Arsenic in rice plant

Most As accumulated in the roots, followed by shoots and grains under different treatments (Fig. 2). The highest As values in different plant parts of rice were observed in T1, whereas the lowest As values were found in T4. The accumulation of As (µg/kg) in rice grains ranged 6.68

in T4 to 28.25 in T1. The grain As content was reduced by 19, 40 and 76% upon the treatment of T2, T3 and T4, respectively compared to T1. This reduction in grain As was significantly and positively correlated (p <0.05, $r^2 = 0.95$) with the reduction in the potentially bioavailable soil As fraction under study. The sharp decline in grain As content may also be due to bioaccumulation and bio-volatilization of soil As by increased population of tested fungal strains (Srivastava et al., 2011, 2012).

4 CONCLUSIONS

The study demonstrated that the talc-formulation consortia of novel soil fungal strains having capabilities to remediate As can be exploited to lower down the potentially bioavailable soil As fraction in the rice root-zone. This reduction of bioavailable soil As may facilitate reduction in grain As content and may lead to safe rice cultivation in As-contaminated rice fields under in-situ conditions.

ACKNOWLEDGEMENTS

Author is thankful to the Director, CSIR—NBRI for institutional support and the financial support from the Department of Biotechnology, Government of India through grant BT/PR12764/BCE/8/1119/2015 to carry out this study.

REFERENCES

Srivastava, P.K., Vaish, A., Dwivedi, S., Chakraborthy, D., Singh, N., Tripathi, R.D. 2011. Biological removal of arsenic pollution by soil fungi. Sci. Total Environ. 409(12): 2430–2442.

Srivastava, P.K., Shenoy, B.D., Gupta, M., Vaish, A., Mannan, S., Singh, N., Tewari, S.K., Tripathi, R.D. 2012. Stimulatory effects of arsenic-tolerant soil fungi on plant growth promotion and soil properties. Microbes Environ. 27(4): 477–482.

Srivastava, P.K., Singh, M., Gupta, M., Singh, N., Kharwar, R.N., Tripathi, R.D., Nautiyal, C.S. 2015. Mapping of arsenic pollution with reference to paddy cultivation in the middle Indo-Gangetic Plains. Environ. Monit. Assess. 187(4): 198.

Wenzel, W.W., Kirchbaumer, N., Prohaska, T., Stingeder, G., Lombi, E., Adriano, D.C. 2001. Arsenic fractionation in soils using an improved sequential extraction procedure. Anal. Chim. Acta 436: 309–323.

Arsenic Research and Global Sustainability – Bhattacharya, Vahter, Jarsjö, Kumpiene,
Ahmad, Sparrenbom, Jacks, Donselaar, Bundschuh & Naidu (Eds)
© *2016 Taylor & Francis Group, London, ISBN 978-1-138-02941-5*

Metal interaction on arsenic toxicity in both *in vivo* and *in vitro* biological systems including human cells

C. Peng[1,3], Q. Xia[1,3], S. Muthusamy[1,3], V. Lal[1,3], J.C. Ng[1,3], D. Lamb[2,3], M. Kader[2,3],
M. Mallavarapu[2,3] & R. Naidu[2,3]
[1]*National Research Centre for Environmental Toxicology (Entox), The University of Queensland, Brisbane, Queensland, Australia*
[2]*Global Centre for Environmental Remediation (GCER), The University of Newcastle, Newcastle, New South Wales, Australia*
[3]*CRC for Contamination Assessment and Remediation of the Environment (CRC CARE), University of Newcastle, Newcastle, New South Wales, Australia*

ABSTRACT: Cadmium and lead are often co-exist with arsenic at elevated levels in the environment. They are of significant environmental and health concern. However, there is a scarcity of data for the interaction effect of cadmium and lead on arsenic. Previously, we have demonstrated cadmium or lead can reduce the bioavailability of arsenic when co-administered in solutions orally to rats. This binary mixture interaction effect could only explain 50% of the variation when soil was dosed to rats implicating other metals or factors may influence the bioavailability and kinetics of arsenic. In order to explore further and extend to ternary mixtures, in vitro systems including the United Barge Method (UBM) for bioaccessibility and human liver cells (HepG2 cell line) for toxicity evaluation were employed in this study. Results indicated that concentration addition may be appropriate for the risk assessment of these elements. However, interaction effects varied with dosages and soil properties.

1 INTRODUCTION

Arsenic (As) may co-exist with cadmium (Cd) and lead (Pb) in contaminated sites. For assessment of the risk posed by these contaminants, it is vital that the bioavailability and toxicity of contaminants are adequately understood. Current toxicity assessments are generally restricted to individual contaminants and rarely mixtures. This study fills data gaps needed for the risk assessment of mixed contamination by generating scientifically based toxicological knowledge and also improve the current risk assessment methods by integrating bioaccessibility data and *in vitro* toxicity assays to investigate potential interaction effect of mixtures containing As, Cd and Pb (see Figure 1).

2 METHODS

2.1 *Bioaccessibility of spiked- and aged soils*

Seven different variants of soils were collected from Victoria and South Australia for spiking with sodium arsenate, cadmium nitrate or lead acetate at levels near or above the current Health Investigation Levels (NEPC, 2013) and aged for 12 months, then dried and sieved to ≤250 μm for analyses. For mixture interaction experiments, portions of the same soil type containing As, Cd or Pb were mixed. United BARGE Method (UBM)

(Denys *et al.*, 2012) was used to measure bioaccessibility (BAC) of As, Cd and Pb individually and in combinations of binary and ternary mixtures. Soil properties included elemental concentrations, pH, total C, N_2, S, total organic carbon, cation exchange capacity, amorphous oxides of Al, Fe and Mn. Sand, clay and silt contents were also determined (data not shown).

2.2 *Effect of Cd and Pb on As toxicity*

Toxicity of individual elements and in mixtures were studied using human liver cells (HepG2) for various end points including cytotoxicity (MTS assay),

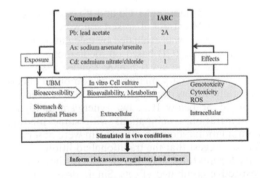

Figure 1. A conceptual diagram summarizing the study approaches and objectives. IARC = International Agency for Research on Cancer.

oxidative stress (Nrf2) and genotoxicity (micronucleus assay). The study included As, Cd and Pb in water and UBM extracts of spiked- and aged-soils.

3 RESULTS AND DISCUSSION

3.1 *Effect of soil properties on bioaccessibility*

We have previously shown that the bioavailability and pharmacokinetics of As in rats are influenced by co-exposure to Pb (Diacomanolis *et al.*, 2013) and Cd (Diacomanolis *et al.*, 2014).

In this study, results show that As has the highest BAC followed by Cd and Pb in both the stomach phase (acid pH) and intestinal phase (near neutral pH). Organic carbon, iron oxide and aluminium oxide were the most important parameters influencing the BAC of As, Cd and Pb. No interactions between As, Cd and Pb in binary or ternary mixtures during BAC test were observed in any of the seven different types of soils, which indicates As, Cd and Pb aged independently did not interact while being solubilized during the BAC test contrasting to in vivo data (Diacomanolis *et al.*, 2013, 2014). For soil spiked and aged separately as in this current study, additive effect may be proposed when estimating the BAC of As, Cd and Pb mixtures.

UBM extracts of As, Cd and Pb were dosed to HepG2 cells (human hepatocellular carcinoma cells) for the uptake study. Results demonstrate the uptake of As in pure solution increased in a dose-response manner while the uptake of As extracted in UBM solution was significantly promoted. The uptake of Cd extracted from different soils by UBM solution showed no significant difference but was two times as high as that of Cd in pure solution. Due to Pb precipitating with inorganic salts in the culture medium, few data of Pb was obtained. HepG2 cells may be a useful *in vitro* model to predict the uptake of As and other metals by correlating with their respective BAC data. UBM extracts of soils were also used in the toxicity study.

3.2 *Mixture toxicity*

Cytotoxicity data of As, Cd and Pb either in solution or UBM extract of spiked soil were obtained. Mixture effect was evaluated by calculating Combination Index (CI) based on median effect principle mass-action law (Chou & Talalay, 1984). For binary and ternary mixtures the combined effects generally follow toxicity addition for concentrations at or below IC_{50}, but synergism at higher concentration combinations with some mixture combinations showing antagonism were observed. These suggest that the traditional addition of toxicity for mixtures may over or underestimate their respective combined effects. Single dose response curves for As, Cd and Pb were determined in all seven soil samples by exposing UBM soil extracts to HepG2 cells. Their respective toxicity to liver cells was also variable depending on soil types with the exception of arsenic.

For oxidative stress, individual dose response for As, Cd and Pb as well as the concentration that induces response induction ratio of 1.5 ($EC_{IR1.5}$) were determined. Generally, the results showed that mixtures of As, Cd and Pb induced oxidative stress adaptive response in ARE reporter-HepG2 cells and that Concentration Addition (CA) is an appropriate model to predict interaction effect of these selected mixtures.

Micronucleus formation was measured in liver cells. As (0.3–32 µm) and Pb (0.3–32 µm) showed clear dose response and positive induction of micronucleus. Cd (0.06–1µm) did not show positive response and the maximum dose tested is limited by its high cytotoxicity to HepG2 cells. Therefore, only interaction effect between Pb and As was examined. Lead exerted variable effects from antagonism to near addition on induction potential of As.

4 CONCLUSIONS

Organic carbon, Fe oxide and Al oxide were found to be key soil properties influencing As bioaccessibility. However, these key soil properties were not significant for the interaction between As and Cd or Pb during UBM extraction. Interaction effect of Cd and Pb on As is quite complicated. Mixture effect is dose and biological end point dependent. The results have also shown that traditional addition method for risk assessment of mixtures may over or underestimate the risk in some scenarios.

ACKNOWLEDGEMENTS

This research was funded by CRC CARE (project number 3.1.01.11-2), the Chinese Science Council and UQ. Entex is a partnership between Queensland Health and the University of Queensland.

REFERENCES

Denys, S., Caboche, J., Tack, K., Rychen, G., Wragg, J., Cave, M., Jondreville, C., Feidt, C. 2012. In vivo validation of the unified BARGE method to assess the bioaccessibility of arsenic, antimony, cadmium, and lead in soils. *Environ. Sci. Technol.* 46: 6252–6260.

Diacomanolis, V., Noller, B.N., Ng J.C. 2014. Bioavailability and pharmacokinetics of arsenic are influenced by the presence of cadmium. *Chemosphere* 112: 203–209.

Diacomanolis, V., Noller, B.N., Ng, J.C. 2013. Interaction effects of lead on bioavailability and pharmacokinetics of arsenic in the rat. *Environ. Geochem. Hlth.* 35: 757–766.

Chou, T.C., Talalay, P. 1984. Quantitative analysis of dose-effect relationships: the combined effects of multiple drugs or enzyme inhibitors. *Adv. Enzyme Regul.* 22: 27–55.

NEPC 2013. Guideline on health-based investigation levels. Schedule B7 in National Environment Protection (Assessment of Site Contamination) Measure. – *Guideline on Investigation Levels for Soil and Groundwater,* National Environmental Protection Council, Canberra.

Arsenic Research and Global Sustainability – Bhattacharya, Vahter, Jarsjö, Kumpiene,
Ahmad, Sparrenbom, Jacks, Donselaar, Bundschuh & Naidu (Eds)
© *2016 Taylor & Francis Group, London, ISBN 978-1-138-02941-5*

Structural models of arsenic and metals bonding in rat stomach and small intestine to understand *in vivo* toxicological interactions

R.B. Taga[1], B.N. Noller[2], J.C. Ng[1,3] & H.H. Harris[4]

[1]*National Research Centre for Environmental Toxicology (Entox), The University of Queensland, Brisbane, Queensland, Australia*
[2]*Centre for Mined Land Rehabilitation, The University of Queensland, Brisbane, Queensland, Australia*
[3]*Cooperative Research Centre for Contamination Assessment and Remediation of the Environment (CRC CARE), Callaghan, NSW, Australia*
[4]*Department of Chemistry, Adelaide University, Adelaide, South Australia, Australia*

ABSTRACT: Structural models are developed for *in vivo* bonding of As with Cd, Cu or Pb to compare toxicological interactions in rat stomach and small intestine for short term dosing (1–3 h) with solutions. Comparison of molecular structure from XANES spectra for tissues recorded at ~10 K from oral gavage of solutions to rats enables validation of predicted metal arsenate or arsenite formation, demonstrating the applicability of XANES to study As speciation in the stomach and intestinal phases, including with time using post oral gavage sampling. The *in vivo* method provides a means for validating *in vitro* bioaccessibility testing.

1 INTRODUCTION

Synchrotron-induced X-ray Absorption Near Edge Spectroscopy (XANES) is used for *in-vivo* bonding of arsenic (As) respectively with Cadmium (Cd), Copper (Cu) or lead (Pb) from short (1–3 h) rat dosing experiments (Diacomanolis *et al.*, 2014). The rat As bioavailability based on blood or urine ratio from soil ingestion against a pure As compound is cancelled by metabolism. XANES spectra of tissue samples held at −80 °C can provide direct structural information from metal-As interactions in rat tissues (Diacomanolis *et al.*, 2010). This study aims to develop molecular models from rat uptake studies for validating *in-vitro* tests for human gastro-intestinal bioaccessibility following As and metal ingestion.

2 METHODS/EXPERIMENTAL

2.1 *Rat uptake studies*

Rat dosing *in vivo* experiments had animal ethics approvals from the University of Queensland (Ethics No. 006/13) to provide stomach and small intestine tissue samples. Doses (mg/kg bw to rat) by oral gavage were: As (2.5), Cd (3), Cu (25) and Pb (20). Analysis of rat tissues by ICP-MS showed that the doses of metal and As solutions by oral gavage could be detected in XANES spectra; the combined tissues data (range mg/kg wet wt.) was: As (0.48–28), Cd (0.2–850), Cu (0.4–1190), and Pb (0.1–1500). Samples were stored at −80°C and transported in a Cryoshipper for synchrotron experiments.

2.2 *XANES scans*

Arsenic K-edge XANES spectra were recorded from model compounds and rat tissues at Beamline 20B Australian National Beamline Facility (ANBF) Photon Factory, KEK, Tsukuba, Japan and at the XAS beamline Australian Synchrotron (AS), Melbourne (Diacomanolis *et al.*, 2015). Spectra were recorded at ~10 K in fluorescence mode with sample held in a He-expansion cryostat to minimise photo damage. Solid models were diluted in boron nitride or cellulose (ground to <20 μm) and pressed into aluminium spacers covered with Kapton tape windows; free aqueous forms of Na arsenate were recorded. Data analysis of XANES spectra was performed using EXAFSPAK software (Diacomanolis *et al.*, 2015).

3 RESULTS AND DISCUSSION

3.1 *Observations from XANES scans*

Figure 1 shows XANES As K-edge scans for: A. model compounds (solid) in cellulose (1000 mg/kg) Na arsenate (NaAsV), Pb arsenate (PbAsV), Fe arsenate (FeAsV), Cu arsenate (CuAsV), Cd arsenate (CdAsV) and Ca arsenate (CaAsV); B. Model compounds As tris-glutathione in solution $(As(GS)_3)$; and solid in cellulose (1000 mg/kg) Na arsenite (NaAsIII), Cd arsenite (CdAsIII), Ca Arsenite (CaAsIII) and Orpiment (AsIIIS); C. Stomach (stom.) and small intestine (S.I.) collected 1 h and 3 h post oral gavage with Cd chloride and

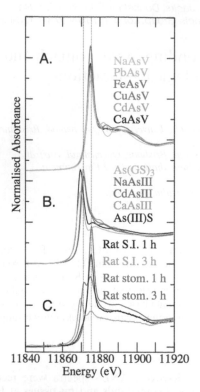

Figure 1. XANES As K-edge scans for: A and B model compounds; and C stomach (stom.) [1 h red and 3 h pink] and small intestine (S.I.) [1 h black and 3 h green] tissues of rats collected post oral gavage with cadmium chloride and sodium arsenate solutions; vertical lines are at 11,869.5 eV (As-I) and 11,875.5 eV (AsV) (recorded at AS-XAS).

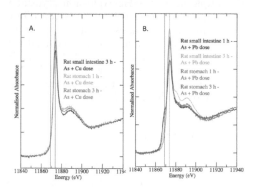

Figure 2. XANES As K-edge scans for stomach and small intestine tissues of rats collected 1 h and 3 h post oral gavage with A copper sulfate and sodium arsenate solutions and B lead nitrate and sodium arsenate solutions, respectively; vertical lines are at 11,869.5 eV (As-I) and 11,875.5 eV (AsV) (Recorded at BL20B).

Na arsenate solutions. Figure 2 shows XANES As K-edge scans for: A. Stomach and small intestine collected 1 h (stomach only) and 3 h post oral gavage with As (Na arsenate) and Cu (sulfate),

respectively in stomach and small intestine both showing AsV (E_{max} 11,874 eV), indicating arsenate cf. the vertical lines at 11,869.5 eV and 11,875.5 eV; and B. Stomach and small intestine collected 1 h and 3 h post oral gavage with Pb acetate and Na arsenate solutions for stomach 1 h and 3 h and small intestine 1 h and 3 h exposures showed AsV and some As-I in the XANES scans for As+Pb cf. the vertical lines at 11,869.5 eV and 11,875.5 eV. The latter is more likely sulfur bound AsIII having lower peak energies at the edge than real arsenites.

3.2 Observations from rat uptake

The XANES spectra for As in rat tissues co-administered with Cu (Fig. 2A showing AsV) and Pb (Fig. 2B shows AsV with As-1), respectively in stomach and small intestine both show AsV (E_{max} 11,874 ev), implying arsenate. These observations relating chemical interaction of Cu and Pb respectively with As in the stomach and intestine following oral gavage and the molecular form of As are different compared to liver and kidney). Similarly As and Cd may form arsenate prior to absorption in the gastrointestinal tract or post absorption (Fig. 1.C) which showed both As-I and AsV in both stomach (gastric) and via progression of feed to the intestine. Thus As-I and AsV may occur in the gastric phase for in vivo and show subtle features of As speciation change. The CdAsV XANES (Fig. IA) was similar to PbAsV (Fig. 2B) but different from Cu (Fig. 2A) showing only AsV (without As-I).

The CdAsV XANES (Fig. 1A) data also support the change of pharmacokinetic parameters in rat with or without co-administration of Cd relating chemical interaction of As with Cd or other ligands in stomach and intestine. Decreased As in blood may result from co-administration of Cd giving decreased As solubility in the gastro-intestinal tract due to chemical interaction forming less soluble compounds with Cd and/or other chemical forms of phosphorus or sulfur. Formation of Cd-arsenate in the rat stomach is plausible based on the dosing regime. Similar interaction may also occur in other mammalians.

4 CONCLUSIONS

This paper demonstrates the applicability of XANES to study As speciation in both solution and solid phase for the unresolved question of As speciation in the stomach and intestinal phases, including with time post oral gavage and provides a method for validating in vitro bioaccessibility techniques.

ACKNOWLEDGEMENTS

This work was performed at the BL20B Beamline Facility Photon Factory (PF), KEK Tsukuba-

Japan (November 2012) with support from the Australian Synchrotron Research Program, which is funded by the Commonwealth of Australia under the major National Research Facilities Program, and on the XAS beamline at the Australian Synchrotron, Victoria, Australia (March 2015). Entox is a partnership between Queensland Health and the University of Queensland.

REFERENCES

Diacomanolis, V., Ng, J.C., Sadler, R., Nomura, M., Noller, B.N., Harris, H.H. 2010. Consistent Chemical Form of Cd in Liver and Kidney Tissues in Rats Dosed with a Range of Cd Treatments: XAS of Intact Tissues. *Chem. Res. Toxicol.* 23(11): 1646–1649.

Diacomanolis, V., Noller, B.N, Ng, J.C. 2013. Interaction effects of lead on bioavailability and pharmacokinetics of arsenic in the rat. *Environ. Geochem. Hlth* 35(6): 757–766.

Diacomanolis, V., Noller, B., Ng, J.C. 2014. Bioavailability and pharmacokinetics of arsenic are influenced by the presence of cadmium. *Chemosphere* 112: 203–209.

Diacomanolis, V., Noller, B.N., Taga, R., Harris, H.H, Aitken, J.B., Ng, J.C. 2015. Relationship of arsenic speciation and bioavailability in mine wastes for human health risk assessment. *Environ. Chem.* (doi: 10.1071/EN14152).

Arsenic Research and Global Sustainability – Bhattacharya, Vahter, Jarsjö, Kumpiene,
Ahmad, Sparrenbom, Jacks, Donselaar, Bundschuh & Naidu (Eds)
© 2016 Taylor & Francis Group, London, ISBN 978-1-138-02941-5

Intestinal arsenic absorption is increased by colon suspension from IBD patients

M. Calatayud[1], A. Geirnaert[1], C. Grootaert[2], G. Du Laing[3] & T. Van de Wiele[1]

[1]*Laboratory of Microbial Ecology and Technology (LabMET), Ghent University, Ghent, Belgium*
[2]*Laboratory of Food Chemistry and Human Nutrition, Ghent University, Ghent, Belgium*
[3]*Department of Applied Analytical and Physical Chemistry, Ghent, Belgium*

ABSTRACT: Arsenic (As) is a metalloid that brings about substantial health risks. Since human As exposure mainly occurs through the intake of contaminated water and food, gut epithelial barrier function is a key determinant of As toxicokinetics. If gut barrier function is compromised, as is the case for patients with Inflammatory Bowel Disease (IBD), As may be increasingly absorbed. Using a combination of semi-continuous gut simulators and Caco-2 cell cultures, we describe how simulated colon suspension derived from IBD patients displays a higher apparent permeability coefficient for As ($4.2 \pm 1.3 \times 10^{-5}$ cm/s) than colon suspension from healthy individuals ($0.7 \pm 0.1 \cdot 10^{-5}$ cm/s). Using a mixture of microorganisms (butyrate producing Clostridia cluster IV and XIVa), the epithelial barrier function was strengthen and thereby lower the epithelial As absorption rates ($20 \pm 7\%$), also in the presence of simulated colon fluids derived from IBD patients. Thus, individuals with compromised gut barrier functioning may be more vulnerable to pollutant exposure in general and inorganic As more specifically. Yet, our *in vitro* data show that probiotic strategies may alleviate these symptoms.

1 INTRODUCTION

Arsenic (As) is an environmental contaminant widely distributed through the Earth crust. As can be present in organic and inorganic forms. Inorganic As (iAs) is highly toxic to humans and it is considered as a human carcinogen by the International Agency for Research on Cancer. As toxicity mechanisms are constantly being updated and recently, new advances on gut microbiome and As interactions suggest significant influence of As on gut microbiome structure and activity (Lu *et al.*, 2014). The risk assessment related to As exposure is usually performed in healthy individuals, however some populations with of pre-existent diseases could be in a higher risk of As internal exposure. Inflammatory Bowel Disease (IBD) patients have a compromised intestinal epithelial barrier function (Bernstein *et al.*, 2009, Hering *et al.*, 2012) and therefore the risk of internal exposure could be increased, because As intestinal absorption occurs mainly by paracellular route (Calatayud *et al.*, 2010). This study assessed *in vitro* As absorption in IBD patients and the effect of the treatment of gut microbiota from IBD patients with butyrate-producing bacteria to reduce its intestinal absorption and toxicity.

2 METHODOLOGY

2.1 Simulated gut suspension and Caco-2 cell

Faecal samples from 9 individuals (3 IBD patients in active stage, 3 IBD patients in remissive condition and 3 healthy individuals) were incubated in semi-continuous anaerobic gut simulators (Geirnaert, 2015). At the start of the incubation the faecal microbiota were supplemented with a mix of butyrate-producing bacteria (*Butyricicoccus pullicaecorum 25–3, Faecalibacterium prausnitzii, Roseburia hominis, Roseburia inulinivorans, Eubacterium hallii, Anaerostipes caccae;* MIX). Supernatants were taken after 42 hours and filtered (0.2 μm), thus obtaining a sterile simulated intestinal water, which was added to the apical compartment of Transwell® system with colon adenocarcinoma cells (Caco-2, ATCC® HTB-37™ATCC) developed during 7 days in DMEM supplemented with fetal bovine serum (10%). Every two days until 18 days post seeding (11 days of exposure to intestinal water), the apical and basal medium were replaced. At day 18 post seeding, the apparent permeability coefficient (Papp) of As^{V} (1μM; 2 hours) was calculated (Calatayud *et al.*, 2010).

2.2 Intracellular redox status

The intracellular redox status in Caco-2 cells exposed to intestinal water from healthy individuals and IBD patients amended or not with MIX was evaluated by assessing the intracellular ratio between oxidized and reduced glutathione (GSH/GSSG) through the fluorometric method described by Hissin et al. (1976).

3 RESULTS AND DISCUSSION

3.1 Apparent permeability coefficient of As

Papp values of As after exposure to intestinal water from healthy individuals were statistically significant lower $(0.7 \pm 0.06 \times 10^{-6}$ cm/s) than those obtained after exposure of cells to intestinal water from IBD patients (1.9 to 5.2×10^{-6} cm/s). As Papp in Caco-2 cells exposed to intestinal water, both from healthy and IBD patients, was higher than Papp values obtained from cells exposed to cell culture media $(0.13 \pm 0.05 \times 10^{-5}$ cm/s). This results suggest that As ingested from contaminated water and food can reach a higher intestinal absorption and therefore internal exposure in IBD patients.

3.2 Intracellular redox status

Caco-2 cells exposed to intestinal water from healthy individuals showed similar GSH/GSSG ratio that control cells (cell culture media) (Table 1).

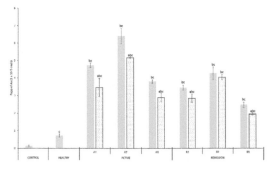

Figure 1. Apparent permeability of As. Bars represent the mean ± SEM (n ≥ 3) of As Papp in Caco-2 cell monolayers exposed 11 days to cell culture media (control), intestinal water from healthy individuals or intestinal water from IBD patients supplemented (dotted bars) or no supplemented (solid bars) with a mix of butyrate producing bacteria (MIX). Statistically significant differences were marked by uppercase letters: a, comparison between intestinal water from IBD patients supplemented or not with MIX; treatments and control treated with cell culture media; b, comparison between intestinal water from IBD patients and healthy individuals; c, comparison between intestinal water from IBD patients and cell culture media.

Table 1. Reduced and oxidized glutathione ratio (GSH/GSSG) in Caco-2 cells exposed to intestinal water from healthy individuals and IBD patients amended or not with a mix of butyrate producing bacteria (mean ± SEM; n ≥ 3).Statistically significant differences between cells exposed to intestinal water from IBD individuals with and without amendment with MIX are marked by an asterisk (*).

			GSH/GSSG ratio
	Control		56 ± 6
	Healthy		48 ± 6
Active	A1		28 ± 6
		MIX	39 ± 2*
	A2		23 ± 5
		MIX	51 ± 3*
	A3		49 ± 2
		MIX	44 ± 4
Remission	R1		32 ± 6
		MIX	44 ± 6
	R2		40 ± 2
		MIX	60 ± 5*
	R3		37 ± 3
		MIX	52 ± 1*

The cell monolayers exposed to intestinal water from IBD patients without amendment with MIX presented lower GSH/GSSG rations than cells exposed to control or intestinal water from healthy individuals.

Cells exposed to intestinal water from 5 IBD patients amended with MIX exhibited an increase in GSH/GSSG ratio (1.4–1.6 fold-increase) compared with those no supplemented with MIX.

Reactive Oxygen Species (ROS)-mediated oxidative damage is a common mechanism of As cellular toxicity. The imbalance caused both by pre-existent diseases (e.g. IBD) and As exposure could affect the epithelial barrier function and worsen the consequences of As exposure in high risk individuals.

The in vitro amendment with MIX could, not only reduce As absorption by enhancing the tight junction integrity, but also improve the antioxidant defenses of the intestinal epithelium.

4 CONCLUSIONS

Individuals with compromised gut barrier functioning may be more vulnerable to inorganic As. Thus, the internal dose assessment for IBD individuals should differently evaluated that those from healthy individuals, and therefore a new high risk group of As exposure is proposed.

The *in vitro* data show that probiotic strategies could alleviate iAs absorption at intestinal level and restore the oxidative stress through.

ACKNOWLEDGEMENTS

This work was supported by Ghent University (Special Research Fund, BOFPDO2014000401; GOA, BOF12/GOA/008 and Strategisch Basisonderzoek–SBO) and the Institute for the Promotion of Innovation by Science and Technology in Flanders (IWT-Vlaanderen, project nr. 100016).

REFERENCES

Bernstein, C.N., Nugent, Z., et al. 2009. Isotretinoin is not associated with inflammatory bowel disease: a population-based case-control study. *Am. J. Gastroenterol.* 104(11): 2774–2778.

Calatayud, M., Gimeno, J., Vélez, D., Devesa, V., Montoro, R. 2010. Characterization of the intestinal absorption of arsenate, monomethylarsonic acid, and dimethylarsinic acid using the Caco-2 cell line. *Chem. Res. Toxicol.* 23(3): 547–556.

Geirnaert, A. 2015. Probiotic potency of butyrate-producing bacteria for modulating the microbiome and epithelial barrier in inflammatory bowel disease. Ph.D. Thesis, Ghent University.

Hering,, N.A., Fromm, M., Schulzke, J.-D. 2012.Determinants of colonic barrier function in inflammatory bowel disease and potential therapeutics. *J. Physiol.* 2012. 590(Pt 5): 1035–1044.

Hissin, P.J., Hilf, R. 1976. A fluorometric method for determination of oxidized and reduced glutathione in tissues. *Anal. Biochem.* 74: 214–226.

Lu, K., Abo, R.P., Schlieper, K.A., Graffam, M.E., Levine, S., Wishnok, J.S., Swenberg, J.A., Tannenbaum, S.R., Fox, J.G. 2014. Arsenic exposure perturbs the gut microbiome and its metabolic profile in mice: an integrated metagenomics and metabolomics analysis. *Environ Health Perspect* 122: 284–291.

Arsenic Research and Global Sustainability – Bhattacharya, Vahter, Jarsjö, Kumpiene,
Ahmad, Sparrenbom, Jacks, Donselaar, Bundschuh & Naidu (Eds)
© *2016 Taylor & Francis Group, London, ISBN 978-1-138-02941-5*

Influence of silicon on uptake and speciation of arsenic in lettuce

M. Greger[1], T. Landberg[1] & R.B. Herbert Jr.[2]
[1]*Department of Ecology, Environment and Plant Sciences, Stockholm University, Stockholm, Sweden*
[2]*Department of Earth Sciences, Uppsala University, Uppsala, Sweden*

ABSTRACT: The aim of this work was to investigate if silicon (Si) decreases arsenic (As) uptake and influences As speciation in lettuce. Therefore, lettuce was grown for one and four days in a 1 and 10 µM arsenate or arsenite solution, with and without 1 mM Si added. Silicon decreased both arsenite and arsenate uptake in lettuce. Silicon decreased the arsenate and arsenite concentration in shoots to highest extent, by 50%, when grown in an arsenate solution. Silicon increased the bound fraction of As in the shoot when grown in an arsenite solution. Both decreased arsenate and arsenate in shoots and the increased percentage of firmly-bound As in shoots might be a manner for the plant to tolerate higher As levels in the presence of Si. For human health, it is less critical to find high arsenate and not high arsenite concentration in the lettuce shoots, since arsenite is more toxic than arsenate.

1 INTRODUCTION

Arsenic (As) is an element detrimental to health. The intake of As is mainly via food and drinking water. Lettuce accumulates high As concentrations in the shoot (McBride, 2013, Bergqvist *et al.*, 2014). Arsenic exists in various inorganic and organic forms, species; organic As species are less common in plants. Inorganic As is more toxic than organic As and, of the inorganic As species, arsenate is considered less toxic to organisms than arsenite (Meharg & Hartley-Whitaker, 2002). Arsenic speciation is thus as important as the total As content when dealing with As in food crops.

Silicon (Si) is the second most common element in soil but the availability is very low, about 1/5000 of the total Si content is available for plant uptake (Greger & Landberg 2015). There is increasing interest in using Si as a fertilizer of crops since Si increases the biomass production and the resistance against abiotic and biotic stresses (Ma 2004). In a field experiment (Greger & Landberg, 2015), Si decreased the accumulation of As by 25–40% in the edible parts of several crops cultivated in the field, while there was not a significant change in the availability of As in soil by Si addition.

The aim of this study was to investigate if Si influences the uptake and speciation of As in lettuce.

2 MATERIALS AND METHODS

In order to study the effect of Si on As uptake and speciation in lettuce (*Lactuca sativa cv.* Americanischer brauner), plants were cultivated in: 1)

nutrient medium containing 1 µM arsenite or arsenate in the presence or absence of 1 mM potassium silicate for 24 hours; 2) nutrient medium containing 10 µM arsenite or arsenate in the presence or absence of 1 mM potassium silicate for four days. Plants were either used intact or divided in roots and shoot. Total As was analysed by Atomic Absorption Spectrophotometry (AAS) and arsenite and arsenate by HPLC-AAS according to Greger *et al.* (2015).

3 RESULTS AND DISCUSSION

Arsenic was taken up by lettuce, both in the presence and absence of Si and when added as arsenite or arsenate. Arsenate uptake was not significantly different from arsenite uptake (Table 1). When Si was added the total As uptake decreased, both when arsenite and arsenate was added (Table 1).

Plants contain more arsenate than arsenite, even though arsenite was added (Table 2). This was the

Table 1. Average total arsenic content in whole plants of lettuce after treatment for 24 hrs in nutrient medium containing 1 µM arsenate or arsenite with or without 1 mM Si. n = 4, ±SE.

Treatment	Arsenite added	
	µgAs(gFW plant)$^{-1}$	Arsenate added
– Si	14.7 ± 0.8	16.1 ± 1.1"
+ Si	11.6 ± 0.7*	14.1 ± 0.8*"

*Indicates significant difference from data without Si in each column. "Indicates significant difference between arsenite and arsenate treatment.

Table 2. Arsenite and arsenate in roots and shoots of lettuce after treatment for four days with arsenite or arsenate, with or without 1 mM Si. Concentration and % difference between Si treatment and no treatment is given n = 4.

Treatment	Arsenite, μg gDW^{-1}				Arsenate, μg gDW^{-1}			
	Root		Shoot		Root		Shoot	
	conc	%	conc	%	conc	%	conc	%
Arsenate	41		43		1226		180	
Arsenate + Si	30*	73	21*	49	1109	91	92*	51
Arsenite	57		44		1145		70	
Arsenite + Si	43*	75	40	91	928*	81	61	87

*Indicates significant difference between Si treatment and non-Si treatment.

Table 3. Firmly bound As in percentage of total As in lettuce tissue after lettuce was treated for four days with arsenite or arsenate, with or without 1mM Si. n = 4.

Treatment	Bound As,%	
	Root	Shoot
Arsenate	7	46
Arsenate + Si	22	31
Arsenite	8	30
Arsenite + Si	5	84*

*Indicates significant difference between Si treatment and non-Si treatment.

case in both roots and shoot. This does not agree with the findins of Bergqvist *et al.* (2014), in which As is transformed to arsenite in roots. Such a transformation of arsenate to arsenite in roots would result in the binding of arsenite by phytochelatins in the root cells and thereby detoxify As (Meharg & Hartley-Whitaker, 2002). Otherwise the arsenate is translocated to the shoot and there can influence photosynthesis negatively. Less arsenite than arsenate was however, found in the shoots, which would prevent the toxic effects of arsenite on the photosynthetic apparatus (Uroic *et al.*, 2012).

In lettuce, a decrease of arsenate was shown in the presence of Si and the largest effect was found in the shoots when arsenate was added (Table 2). Also, arsenite decreased in both roots and shoots when Si was added (Table 2), as a cause of the lowered total uptake of As (Table 1). Adding Si did not result in any changes in the ratio between arsenite and arsenate, neither in roots nor shoots compared with the non-silicon treatments (Table 2). Therefore, enzymes responsible for arsenate/arsenite metabolism (Moreno-Jiménez et al., 2012) were most likely unaffected by Si.

More As is firmly bound in the shoot than in the roots (Table 3). This may be a mechanism to tolerate high As levels in shoot tissues. In the presence of Si, more As is bound up in the shoot when arsenite was added (Table 3). The binding is most likely an As‐O binding to an organic compound (Greger et al., 2014).

4 CONCLUSIONS

We conclude that Si decreases the uptake of As in lettuce and makes lettuce more tolerant to this metalloid by binding up As in the shoot. Silicon might therefore act in several ways to prevent toxic effects by As in the plants. Silicon decreases the uptake of As in plant leaves and also helps to decrease As intake via lettuce.

ACKNOWLEDGEMENTS

This work was funded by the Swedish Research Council and the Knut and Alice Wallenberg foundation. Lillvor Wikander is acknowledged for her laboratory assistance.

REFERENCES

Bergqvist, C., Herbert R., Persson I., Greger M. 2014. Plants influence on arsenic availability and speciation in the rhizosphere, roots and shoots of three different vegetables. *Environ. Pollut.* 184: 540–546.

Greger, M., Landberg, T. 2015. Silicon reduces cadmium and arsenic levels in field-grown crops. *Silicon* (DOI 10.1007/s12633-015-9338-z (Published online: 27 November, 2015)

Greger, M., Landberg, T., Herbert, R. & Persson I. 2014. Arsenic speciation in submerged and terrestrial soil plant systems. *Proceedings of the 5th International Congress on Arsenic in the Environment, May 11–16, 2014, Buenos Aires, Argentina.* CRC Press/Balkema, Leiden.

Greger M, Bergqvist C, Sandhi, A, Landberg T. 2015. Influence of silicon on arsenic uptake and toxicity in lettuce. *J. Appl. Bot. Food Qual.* 88: 234–240.

Ma, J.F. 2004. Role of silicon in enhancing the resistance of plants to biotic and abiotic stresses. *Soil Sci. Plant Nutr.* 50: 11–18.

McBride, M.B. 2013. Arsenic and lead uptake by vegetable crops grown on historically contaminated orchard soils. *Applied Environmental Soil Science* 2013: Article ID 283472, 8p.

Meharg, A.A., Hartley-Whitaker, J. 2002. Arsenic uptake and metabolism in arsenic resistant and nonresistant plant species. *New Phytol.* 154: 29–43.

Moreno-Jiménez, E., Esteban, E., Peñalosa, J.M. 2012. The fate of arsenic in soil-plant systems. *Rev. Environ. Contam. T.* 215: 1–37.

Uroic, M.K., Salaün, P., Raab, A., Feldmann, J. 2012. Arsenate impact on the metabolite profile, production, and arsenic loading of xylem sap in cucumbers (*Cucumis sativus* L.). *Front. Physiol.* 3: 55: 1–23.

Arsenic Research and Global Sustainability – Bhattacharya, Vahter, Jarsjö, Kumpiene,
Ahmad, Sparrenbom, Jacks, Donselaar, Bundschuh & Naidu (Eds)
© *2016 Taylor & Francis Group, London, ISBN 978-1-138-02941-5*

Bioavailability of arsenic in two Italian industrial contaminated soils

C. Porfido, I. Allegretta, O. Panzarino, R. Terzano, E. de Lillo & M. Spagnuolo
Department of Soil, Plant and Food Sciences, University of Bari "Aldo Moro", Bari, Italy

ABSTRACT: In the present work, the bioavailability of As in soils from two Italian dismissed mining and industrial sites respectively of Valle Anzasca (Piedmont) and Scarlino (Tuscany) is studied coupling biological assays and chemical analyses. The mineralogy and As total content were estimated respectively via XRPD and WDXRF. Sequential extractions coupled with TXRF were used to assess the potential bioaccessibility of As in soils. Mortality and reproduction tests, TXRF and μXRF were performed on earthworms (Eisenia andrei) exposed to contaminated soils in order to assess the eco-toxicity and the bioavailability of As. These preliminary results show that when As is associated with amorphous Fe-hydrous-oxides, the concentration of As in the coelomic fluids is directly related to the As concentration in soil. On the contrary, when As is associated with well crystallized Fe-hydrous-oxides, the concentration of As in the coelomic fluids is comparable with that observed in earthworm exposed to non-contaminated soils.

1 INTRODUCTION

Arsenic (As) is a metalloid element, naturally associated with gold, sulphur, iron and heavy metals and is often found in soils and waste around former mines and industrial sites treating As-bare minerals. The assessment of the bioavailability of As in these soils is very important in order to protect human and ecosystems health. Earthworms are often used to assess the bioavailability of As in soils (Langdon *et al.*, 2003). In this work, *Eisenia andrei* was exposed to As-polluted soils from Valle Anzasca and Scarlino (Italy) in order to evaluate the bioavailability of arsenic. Different X-ray based techniques were used to evaluate the concentration and the distribution of As both in soils and earthworms.

2 EXPERIMENTAL

2.1 *Soils analysis*

Three soils samples per polluted site were collected, sieved (2 mm) and dried. The mineralogical characterization of the soils was carried out by XRPD (Miniflex II, Rigaku). The total As was estimated via portable XRF (Niton XL3t GOLDD+, Thermo Scientific), while sequential extraction (Wenzel *et al.*, 2001) coupled with TXRF (S2 PCOFOX, Bruker) were used to study the solubility of As. Soils elemental maps were acquired using μXRF (M4 TORNADO, Bruker) in order to evaluate the As distribution and its correlation with other elements. All the above analyses were conducted also on two control soils, OECD standard soil and cattle manure.

2.2 *Earthworms analysis*

Ten sexually mature earthworms were exposed to each contaminated soil and controls. The mortality was assessed after 14 days of exposure (acute toxicity). Oxidative stress was estimated by measuring H_2O_2, catalase, phenoloxidase, glutathione S-transferase and malondialdehyde. The effect of As on the reproduction (chronic toxicity), was assessed after 28 days of exposure. Earthworms were embedded in epoxidic resin and sections (100 μm of thickness) were analyzed via μXRF in order to localize As accumulation. Since detoxification mechanisms seem to act mainly inside the coelom (e.g. for Cd, Panzarino *et al.*, 2016), coelomic fluids (few μl) were electrically extruded from worms and analysed via TXRF in order to quantify the As concentration.

3 RESULTS AND DISCUSSION

3.1 *Soil characterization*

The mineralogy of soil from Valle Anzasca (V1, V2, V3) is uniform and characterized by the presence of illite, quartz, albite and orthoclase. Two Scarlino soils are characterized by the presence of illite, kaolinite and quartz (S1 and S2) while gypsum and magnetite have been also detected in S3. No As-baring minerals was detected by XRD in all the six soils. However, sequential extraction results (Table 1) showed that As is in mainly associated with amorphous Fe-hydrous-oxides and this fraction increases with the increase in As concentration in Valle Anzasca soils. Differently, in Scarlino soils As is mainly associated with well-crystallized

Table 1. Relative percentage of As in each extraction step and total As concentration estimated with portable XRF.

Extraction step	S1	S2	S3	V1	V2	V3
1. Non-specif. sorbed	0	0	1	2	1	0
2. Specif. sorbed	15	4	12	11	25	12
3. Amorphous h-ox of Fe	46	3	41	50	67	85
4. Well-cryst. h-ox of Fe	0	90	42	28	2	1
5. Residue	38	3	4	9	6	1
Total As (mg/kg)	41	224	736	134	3174	9135

Fe hydrous-oxides, except in S1 where the total As concentration is low (41 mg/kg). This is confirmed by μXRF maps which show an overlapping between Fe and As, probably precipitates, around quartz or feldspar grains.

3.2 Bioavailability of As

Biological tests showed that no organism dies after a period of 14 days of exposition to As-contaminated soils. However, reproduction tests evidenced a reduction of the number of new born earthworms with the increase of As concentration in the soil. An oxidative stress was recorded in all the earthworms exposed to contaminated soils, without evidence of a concentration effect, while no stress was detected in organisms exposed to control soils.

Elemental maps acquired on earthworm sections with μXRF (Fig. 1) showed that As is found in the coelomic cavity together with S. No As signal is recorded in other parts of the earthworms grown in contaminated soils. The co-presence of Fe and As in the intestine is due to the presence of soil residues.

3.3 Bioavailability of As

Biological tests showed that no organism dies after a period of 14 days of exposition to As-contaminated soils. However, reproduction tests evidenced a reduction of the number of new born earthworms with the increase of As concentration in the soil. An oxidative stress was recorded in all the earthworms exposed to contaminated soils, without evidence of a concentration effect, while no stress was detected in organisms exposed to control soils.

Elemental maps acquired on earthworm sections with μXRF (Figure 1) showed that As is found in the coelomic cavity together with S. No As signal is recorded in other parts of the earthworms grown in contaminated soils. The co-presence of Fe and As in the intestine is due to the presence of soil residues.

The concentrations of As in coelomic fluids (Fig. 2) increases with the amount of arsenic in

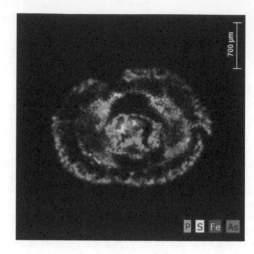

Figure 1. Distribution of P, S, Fe and As in a cross section of an earthworm grown in V3 for 28 days.

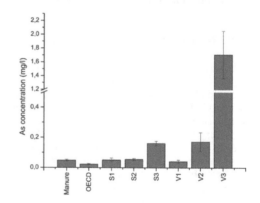

Figure 2. Concentration of As in coelomic fluids determined by TXRF.

soils, although the type of association between As and Fe-hydrous oxides seems to be crucial. When a consistent part of the As is associated with well crystallized Fe-hydrous oxides the concentration of As in the coelomic fluids is comparable to that found in the earthworm exposed to control soils. On the contrary, when the As is associated with amorphous hydrous oxides of iron, the concentration of As in coelom increases considerably.

4 CONCLUSIONS

Some results on the bioavailability of As in mine and industrial soils from Valle Anzasca and Scarlino are presented. As is compartmentalizeded mainly in the ceolomic cavity of earthworms and its concentration increases with the amount of As in soils. Arsenic associated with amorphous

Fe-hydrous oxides seems to influence more than As associated with well crystallized Fe-hydrous oxides its concentration in the coelom, appearing therefore more bioavailable. However, biological tests suggest that, after short exposure period, As does not exert an acute toxicity on earthworms but rather affects their reproduction capacityand causes oxidative stress. Further biological analyses will be performed in order to better assess the ecotoxicity of As.

ACKNOWLEDGEMENTS

Analysis were carried out at the Micro X-ray Lab financed by Regione Puglia (FESR 2000-2006—Risorse Liberate—Obiettivo Convergenza).

REFERENCES

Langdon, C.J., Piearce, T.G., Meharg, A.A., Semple, K.T. 2003. Interactions between earthworms and arsenic in the soil environment: a review. *Environ. Pollut.* 124: 361–373.

Panzarino O., Hyršl P., Dobeš P., Vojtek L., Vernile P., Bari G., Terzano R., Spagnuolo M., de Lillo E. 2016. Rank-based biomarker index to asses cadmium ecotoxicity on the earthworm Eisenia andrei. *Chemosphere* 145: 480–486.

Wenzel, W.W., Kirchbaumer, N., Prohaska, T., Stingeder, G., Lombi, E., Adriano, D.C. 2001. Arsenic fractionation in soils using an improved sequential extraction procedure. *Anal. Chim. Acta* 436(2): 309–323.

Arsenic Research and Global Sustainability – Bhattacharya, Vahter, Jarsjö, Kumpiene,
Ahmad, Sparrenbom, Jacks, Donselaar, Bundschuh & Naidu (Eds)
© *2016 Taylor & Francis Group, London, ISBN 978-1-138-02941-5*

Arsenic in groundwater, agricultural soil and crops of Sahibganj in the middle Gangetic plain: A potential threat to food security and health

S. Chakraborty, O. Alam & W.A. Shaikh
Department of Civil and Environmental Engineering, Birla Institute of Technology, Mesra,
Ranchi, Jharkhand, India

ABSTRACT: Groundwater, agricultural soil and crops of Sahibganj district of Jharkhand, on the middle Gangetic basin of India was analysed for arsenic (As) concentrations to study the soil to plant transfer and associated health risks in human. Physical and chemical characteristics of groundwater and soil were studied to understand the probable release mechanism. Results indicated groundwater to be Ca-HCO_3 type with low redox potential and high TDS. Arsenic varied between 1–133 µg/L in groundwater and 33 to 902 µg/kg in soil. Positive correlation was observed between As with depth, PO_4^{3-} and Fe in groundwater and Fe and clay fraction of soil suggesting reductive dissolution mechanism of Fe (hydr)oxides the probable mechanism of As release. Considerable accumulation of As was seen in spinach (0.35 µg/kg), onion (0.32 µg/kg) and wheat grains (0.30 µg/kg). Health risk assessment revealed high carcinogenic and non-carcinogenic risk for both adults and children.

1 INTRODUCTION

Sahibganj district in Jharkhand, India, is situated on the western bank of river Ganges. It has been reported to be arsenic (As) affected by UNICEF, Central Groundwater Board of India and some researchers (Bhattacharjee *et al.*, 2005, Bhattacharya *et al.*, 2011). Relatively few studies have been carried out in this area addressing the health effects of local communities due to exposure to As contaminated groundwater. However detailed study of As contamination in groundwater, agricultural soil and crops were not undertaken till date. The present investigation was done with the objective to: 1) determine the spatial variation of total As and physicochemical parameters in groundwater and agricultural surface soils to understand the source, probable release mechanisms and As enrichment in the soil, 2) study the uptake and translocation of As in crops including grains, vegetables and pulses to assess the potential health risk associated with consumption of groundwater and locally grown crops.

2 METHODS/EXPERIMENTAL

2.1 Sampling and analysis

In the present study, twenty As affected villages in three blocks (Sahibganj, Rajmahal and Dudhwa) of Sahibganj district of Jharkhand has been selected for detailed investigation. The study area is basically an alluvial plain on which agriculture is practised irrigated with groundwater. Collection of groundwater was done following standard protocol and preserved with 14 M HNO_3 for trace element (As, Fe and Mn) analysis. Geographical coordinates of the sampling locations were recorded using Garmin Etrex-30 GPS. Onsite analysis of physicochemical parameters was measured with MultiParameter Analyser (YSI) handheld instrument and rest brought to lab for lab analysis. Surface soils collected from agricultural fields were analysed for trace elements, texture and major ions. Crop samples were collected during the harvesting season, brought to the lab in polythene bags, washed, segregated into different parts, dried till constant weight and analysed for total As concentration after standard digestion methods. Trace metal analysis was done with the help of Perkin Elmer Optima 2100 DV ICP-OES.

2.2 Statistical analysis and health risk assesment

Statistical analysis including Pearson's correlation, Single factor ANOVA, regression, was performed using Microsoft Excel. PCA (Principal component analysis) was performed in SPSS 17 software. Piper diagram or trilinear diagram was constructed by Groundwater chart software of USGS. Carcinogenic and non-carcinogenic health risk was calculated in children and adults due to consumption of groundwater according to Bortey-Sam *et al.* (2015). Bioaccumulation factor, Translocation factor and Enrichment factor were calculated according to Ghosh & Singh (2005).

3 RESULTS AND DISCUSSION

3.1 Physicochemical characteristics and groundwater arsenic occurrences

Groundwater analysis revealed the water to be circum neutral with low redox potential, varying between −1 mV to −199 mV and high TDS load (mean 559 ± 172 mg/L). The concentration of major anions and cations in the present study area was found in the order of $HCO_3^- > Cl^- > SO_4^{2-} > NO^{3-} > PO_4^{3-}$ and Ca > Na > Mg > K respectively. The groundwater was found to be $Ca-HCO_3$ type. Mean HCO_3 concentration was found to be 368 ± 84 mg/L. A positive correlation was noted between As-Fe suggesting reductive dissolution of Fe (hydr)oxides as the mechanism of As mobilization in groundwater, and correlation amongst As and PO_4^{3-} tends to suggest that PO_4^{3-} which competes with As as oxyanions for the adsorption sites on Fe (hydr)oxides are also mobilized in the process (Biswas et al., 2014). Depth-wise heterogeneity of As concentration suggested incremental trend with significant positive correlation (0.82; $p < 0.05$).

3.2 Arsenic concentration in agricultural soil

Arsenic concentration in surface soil varied between 33 to 902 µg/kg with a mean value of 487 ± 186 µg/kg. Fe and Mn ranged between 265 to 819 mg/kg and 18–29 mg/kg. Texture analysis revealed varying percentage of sand, silt and clay. Mean values varied between 36–62% sand, 7–35% silt and 13–42% clay. N, P and K showed an average concentration of 506 ± 120 kg/ha, 114 ± 44 kg/ha and 511 ± 163 kg/ha. TOC was found in very less concentration (range 0–1% with mean value of 0.42 ± 0.20%. Arsenic showed significant positive correlation with Fe (0.77; $p < 0.05$) and moderate positive correlation with clay fraction (0.48; $p < 0.05$). These results are similar to the previous studies in the Bengal Basin, where the fine grained Gangetic alluvial sediments are relatively enriched in As (Mukherjee & Bhattacharya, 2001, Bhattacharya et al., 2011)

3.3 Arsenic concentration in crops

Arsenic analysis in the aerial as well as below ground parts of the crops and vegetables exhibited considerable variation. In bulb, tuber and roots, As varied between 0.11 to 0.32 mg/kg and 0.01 to 0.35 mg/kg in leaf, grain and fruits. Among the aerial parts of the plant highest As concentration was recorded in garlic leaf (0.24 mg/kg), followed by 0.20 mg/kg in wheat grains and spinach leaves. These are all edible parts of the plants and thus pose significant risk of consumption. While among the roots and bulbs, highest As concentration was recorded in pigeon pea (0.88 mg/kg), followed by 0.76 mg/kg in paddy roots and 0.70 mg/kg in ladyfinger.

Bioaccumulation Factor (BAF) was found to be maximum in mustard seeds (4.07) followed by maize (3.18) and onion (2.13) indicating high rate of accumulation in edible parts of the plants. Translocation factor ranged between 0.02 and 1.61 with the highest TF in garlic and lowest in sesame.

3.4 Health risk assessment

Health risk assessment calculation due to groundwater consumption showed very high As induced noncarcinogenic health risk for both adults and children in all the sampling locations. HQ values was as high as 26.270 (adults) and 25.200 (children) at Dehari village and as low as 1.284 (adults) and 1.230 (children) at English. The cancer risk from drinking As laden water for both adults and children were found to be equally very high in the study area, ranging from 5.78E-04 (high risk) to 1.18E-02 (very high risk). Mean cancer risk for the exposed communities suggested very high risk of 4.63E-03 in adults and 4.45E-03 in children.

4 CONCLUSIONS

Overall observation of the present study indicated that arsenic is present in considerable amounts in groundwater above the WHO guideline value and the BIS standards and probable release mechanism can be reductive dissolution of Fe (hydr)oxides. Arsenic also accumulated in agricultural surface soil due to irrigation of groundwater. Sources of As in soil can be from the alluvium as well as synthetic fertilizers. Crops accumulated fair amount of As as evident by their BAF and TF, which can pose serious health risk. Carcinogenic and non-carcinogenic hazard due to water consumption was found to be alarming and thus require proper management on an immediate basis.

ACKNOWLEDGEMENTS

Department of Science and Technology, New Delhi is highly acknowledged for funding the SERB project SR/FTP/ES-2/2013 for this study.

REFERENCES

Bhattacharjee, S., Chakravarty, S., Maity, S., Dureja, S., Gupta, K.K..2005. Metal contents in the groundwater of Sahebganj district, Jharkhand, India, with special reference to arsenic. Chemosphere 2005; 58: 1203–1217.

Bhattacharya, P., Mukherjee, A.,Mukherjee, A.B. 2011. Arsenic in Groundwater of India. In: J.O. Nriagu (ed.) Encyclopedia of Environmental Health, vol. 1, pp. 150–164 Burlington: Elsevier.

Biswas, A., Bhattacharya, P. Mukherjee, A., Nath, B., Alexanderson, H., Kundu, A.K., Chatterjee, D., Jacks, G. 2014. Shallow hydrostratigraphy in an arsenic affected region of Bengal Basin: Implication for targeting safe aquifers for drinking water supply. Sci Total Environ. 485–486: 12–22.

Bortey-Sam, N. Nakayama, S.M.M., Ikenaka, Y., Akoto, O., Baidoo, E., Yohannes, Y.B., Mizukawa, H., Ishizuka, M. 2015. Human health risks from metals and metalloid viaconsumption of food animals near gold mines in Tarkwa, Ghana: estimation of the daily intakes and target hazard quotients (THQs). Ecotox. Env. Safety. 111: 160–167.

Ghosh, M., Singh, S.P. 2005. A review on phytoremediation of heavy metals and utilization of it's by products. Appl. Ecol. Environ. Res. 3(1): 1–18.

Mukherjee, A.B., Bhattacharya, P. 2001. Arsenic in Groundwater in the Bengal Delta Plain: slow poisoning in Bangladesh. Environ. Rev. 9: 189–220.

Arsenic Research and Global Sustainability – Bhattacharya, Vahter, Jarsjö, Kumpiene,
Ahmad, Sparrenbom, Jacks, Donselaar, Bundschuh & Naidu (Eds)
© 2016 Taylor & Francis Group, London, ISBN 978-1-138-02941-5

A comparative study of arsenic accumulation in agricultural fields

S.H. Farooq[1], W. Dhanachandra[2], D. Chandrasekharam[2], A.K. Chandrashekhar[2],
S. Norra[3], Z. Berner[3] & D. Stüben[3]

[1]*School of Earth, Ocean and Climate Sciences, IIT Bhubaneswar, Bhubaneswar, India*
[2]*Department of Earth Sciences, IIT Bombay, Mumbai, India*
[3]*Institute of Mineralogy and Geochemistry, Karlsruhe Institute of Technology, KIT, Karlsruhe, Germany*

ABSTRACT: A study has been carried out to determine the fate of Arsenic (As) supplied to agriculture fields (through contaminated irrigation water), and the relationship between As accumulation in agriculture soils and the crop pattern. A paddy and a wheat cultivating agriculture field irrigated by contaminated water having As concentration of 137 µg/L and 67.3 µg/L were selected. Results clearly indicate that in comparison to wheat field though higher quality of contaminated water was used in paddy field but still paddy field soils show lesser As accumulation. Such an accumulation indicate the existence of an effective As removal mechanism in paddy soils. Absence of any such mechanism in wheat fields is responsible for build-up of As in wheat soils. Different harvesting methods applied in paddy and wheat cultivation are responsible for different accumulation of As in agriculture soils.

1 INTRODUCTION

Arsenic contamination of groundwater is a major global problem. Higher concentrations of As in groundwater of Bengal delta (West Bengal, India and Bangladesh) have been reported by several workers (Bhattacharya *et al.*, 1997, 2002 and references there in). A dense network of surface water bodies (streams and ponds) exist in Bengal delta. The irrigation water demand can easily be met by the surface water resources, however farmers in Bengal delta still relies on groundwater for irrigational needs. It is observed that in many cases the groundwater used for irrigation contains high concentrations of As. Studies have demonstrated that As brought along with irrigation water accumulates in agriculture soils and different parts of plant matter. In Bangladesh alone around 1360 tons of As are introduced into paddy fields annually through contaminated irrigation water (Ali *et al.*, 2003). A correlation between As concentrations in irrigation water, its accumulation in agriculture soils and its relation with the cultivation practices has not yet been investigated. In the present study an attempt is made to highlight the differential accumulation of As in agriculture fields as a function of harvesting techniques implied to cultivate different crops.

2 METHODS

Kaliachak block of Malda district of West Bengal (India) has been selected for the study. Two agricultural fields, one cultivating different varieties of paddy throughout the year, and the other cultivating wheat as main crop along with some seasonal vegetables, have been selected for the study. Groundwater used to irrigate these fields was collected in Polyethylene (PE) bottles after filtering with 0.45 µm cellulose nitrate filter. The samples were acidified by adding few drops of ultrapure HNO_3 (14 M concentration) and stored at 4^0C until further analysis. The As concentration in water samples were analyzed by high resolution ICP-MS (Axiom, Thermo/VG Elemental, UK). Soil samples were collected by digging 110 cm deep tranches in both agricultural fields. Soil samples were collected at every 5 cm interval for first 30 cm depth and then at an interval of 10 cm till the end of the profile. Collected soil samples were stored immediately in plastic bags. In the laboratory, the samples were freeze dried and stored at 4^0C in the nitrogen atmosphere until further analysis. Soil samples were processed to perform XRF (Spectra 5000, Atomica) and XRD (Kristalloflex D500, Siemens). Soil samples were extracted with Phosphate and Dithionite Citrate to understand the association of As with different mineral components (As pools) present in the sediments.

3 RESULTS AND DISCUSSION

Arsenic concentrations in the groundwater used for irrigation in paddy and wheat fields were found to be 137 µg/L and 67.3 µg/L, respectively. This means, in general As concentrations in paddy irrigation water were two times higher than in wheat irrigation water.

The soil profiles have been divided into two horizons; the upper soil horizon (0–15 cm) and the lower soil horizon (15–110 cm) based on the fact that the roots of paddy and wheat plants generally remain within top 15 cm of soil profile. The upper soil horizon of both (paddy and wheat) profiles show higher As concentrations as compared to the lower soil horizon (Fig. 1). Further, it is interesting to note that though paddy field is irrigated with higher As containing water (2X) and almost four times higher quantity of water is pumped into the paddy field but still As concentration in paddy soils up to a depth of 80 cm are lower than that of the wheat soils. Figure 1 also shows that as compared to paddy soils, significantly higher As accumulation occurs in upper horizon of wheat soils.

XRD results show that there is no significant difference in mineralogy of upper soil horizons of both profiles. The clay minerals content in upper soil horizon is 37.2% and 35.5%, respectively for paddy and wheat soils. The concentrations of most other minerals are also comparable. Thus, the higher As concentrations in upper wheat soil horizon or lower As concentrations in upper paddy soil horizon cannot be attributed to mineralogical differences.

The amount of As extracted by phosphate represents fraction of As adsorbed on mineral surfaces, while dithionite citrate extraction indicate fraction of As bound to poorly crystalline and well crystallized iron-oxide/hydroxide phases. The results of these extractions are given in Table. 1. It is clearly evident from the extraction results (Table 1) that the upper soil horizon of the paddy profile contains lesser amounts of adsorbed As as compared to the lower soil horizon. Such a trend indicates existence of a mechanism in upper paddy soil horizon (i.e. surficial environment) that hinders the adsorption of As on soil particle surfaces, however with increasing depth the effect of this mechanism diminishes. This finding is in-line with the earlier findings demonstrating that the reducing conditions generated in paddy fields due to decay of remains of previous crop under flooded condition are responsible for lesser As concentration in upper

Table 1. Percentage of total As (mg/kg) extracted by various extractants.

Extractant	Phosphate		Dithionite Citrate	
	Paddy	Wheat	Paddy	Wheat
Upper Horizon (0–15 cm)	21.6	33.1	56.3	44.4
Lower Horizon (15–110 cm)	24.8	31.9	58.2	54.0

soil (Farooq et al. 2010). It has been observed that the conditions in wheat field are completely in contrast to the paddy field. Due to lesser pumping of irrigation water, flooded conditions do not exit, and due to burning of remains of previous crop lesser amount of organic matter remains available for decay in the wheat field. Under such circumstances (i.e. non-flooded conditions and low organic matter) reducing conditions strong enough to keep As in mobilized phase does not exists thus a significant quantity of As supplied with irrigation water gets adsorbed on the surface of sediment particles which is evident from the results shown in Table 1.

4 CONCLUSIONS

The study demonstrates that As build-up in agricultural soils does not depend solely on As concentration in irrigation water. Agriculture practices (i.e. flooded/non-flooded conditions; and presence of organic matter in agriculture field) also play an important role in modulating As concentrations in agriculture soils.

ACKNOWLEDGEMENTS

The authors gratefully acknowledge the support from DST, Govt. of India and German Academic Exchange Program (DAAD) to conduct this study.

REFERENCES

Ali, M.A., Badruzzaman, A.B.M., Jalil, M.A., Hossain, M.D., Ahmed, M.F., Masud, A.A., Kamruzzaman., M., Rahman, M.A. 2003. Fate of arsenic extracted with groundwater. In: M.F. Ahmed (Ed.) Fate of Arsenic in the Environment. ITN International Training Network, Dhaka. pp. 7–20.

Bhattacharya, P., Chatterjee, D., Jacks, G. 1997. Occurrence of arsenic-contaminated groundwater in alluvial aquifers from delta plains, eastern India: Options for safe drinking water supply. Int. J. Water Res. Dev. 13(1): 79–92.

Farooq, S.H., Chandrasekharam, D., Berner, Z., Norra, S., Stüben, D. 2010. Influence of traditional agricultural practices on mobilization of arsenic from sediments to groundwater in Bengal delta. Water Res. 44(19): 5575–5588.

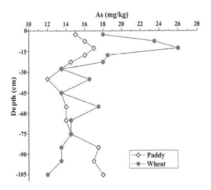

Figure 1. Depth wise variation of As concentration in paddy and wheat soils.

Section 3: Arsenic and health

3.1 Epidemiology

Arsenic Research and Global Sustainability – Bhattacharya, Vahter, Jarsjö, Kumpiene,
Ahmad, Sparrenbom, Jacks, Donselaar, Bundschuh & Naidu (Eds)
© 2016 Taylor & Francis Group, London, ISBN 978-1-138-02941-5

Urinary arsenic metabolites and risk of Breast Cancer molecular subtypes in northern Mexican women

L. López-Carrillo[1], B. Gamboa-Loira[1], A. García-Martínez[1],
R.U. Hernández-Ramírez[1] & M.E. Cebrián[2]
[1]*Instituto Nacional de Salud Pública, Cuernavaca, México*
[2]*Departamento de Toxicología, Centro de Investigación y Estudios Avanzados del Instituto Politécnico Nacional, D.F., México*

ABSTRACT: Epidemiological evidence suggests that Breast Cancer (BC) risk factors (i.e breastfeeding, parity, menarche, etc.) may vary by tumor pathology. We assessed if BC risk related to urinary monomethyl arsenic acid percentage (%MMA) vary by molecular subtype. A total of 499 patients were assigned to one of three tumor marker categories: HR+ (ER+ and/or PR+ and HER2–), HER+ (regardless of ER or PR status) and TNBC (ER–, PR–, and HER2–). Controls were healthy women, with no history of cancer, matched 1:1 by age.%MMA remained as a significant risk factor for BC only for HR+. HER2+ and TNBC tumors were not significantly associated with urinary arsenic (As) metabolites. Our results suggest that increased BC risk related to As exposure is not mediated by HER2 expression. This is the first report on As exposure and BC molecular subtypes which requires further confirmation.

1 INTRODUCTION

For many years, the presence of arsenic (As) in drinking water of some areas in Northern Mexico has been a major concern (Cebrián *et al.*, 1994). As levels have ranged in Sonora from 71 to 305 µg/L (Wyatt *et al.*, 1998), Chihuahua from 15 to 300 µg/L (Camacho *et al.*, 2011) and Región Lagunera from 7 to 600 µg/L (Del Razo *et al.*, 1990). We have recently reported that women residing in Northern Mexico, with higher urinary monomethyl arsenic acid percentage (%MMA) had an increased breast cancer (BC) risk (MMA% ORQ5vs.Q1 = 2.63; 95%CI: 1.89, 3.66; p for trend <0.001) (López-Carrillo *et al.*, 2014).

Breast tumor subtypes with distinctive biology and treatment responses are defined by the immunohistochemical expression of estrogen and progesterone receptors (HR) and human epidermal growth factor receptor 2 (HER2). Epidemiological evidence suggests that BC risk factors (i.e. breastfeeding, parity, menarche, etc.) may vary by tumor pathology (Yang *et al.*, 2007). We assessed if%MMA is associated with a particular BC molecular subtype.

2 METHODS

2.1 *Study design*

A population based case–control study was performed from 2007 to 2011 in five states of Northern Mexico. Incident cases were identified from main public tertiary hospitals in the study area. A total of 499 patients with histopathologically confirmed BC and molecular subtype status available information were identified. Controls were healthy women, with no history of cancer, matched 1:1 by age.

2.2 *Breast Cancer molecular subtypes*

Tumor marker information for Estrogen Receptor (ER), Progesterone Receptor (PR) and HER2 was obtained from medical records. ER+ and PR+ was based on > = 1% cell staining. HER2+ status was determined when at least 30% of protein overexpression was observed in the tumor cell (Hammond *et al.*, 2010). Cases were assigned to one of three tumor marker categories: HR+ (ER+ and/or PR+ and HER2–), HER+ (regardless of ER or PR status) and TNBC (ER-, PR-, and HER2–).

2.3 *Arsenic determination*

Urinary concentrations (µg/L) of species As^{3+}, As^{5+}, MMA^{5+}, DMA^{5+}, and arsenobetaine (AsB) were determined by high-performance liquid chromatography ICP-MS (Gilbert-Diamond *et al.*, 2011). Detection limits were As^{3+}: 0.12; As^{5+}: 0.20; MMA^{5+}: 0.12; DMA^{5+}: 0.08; AsB: 0.08. Results were reported by g of creatinine. Total As was calculated by using both the sum of iAs, MMA,

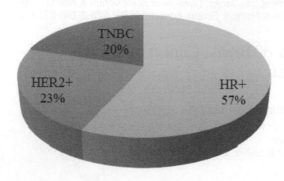

TNBC 20%

HER2+ 23%

HR+ 57%

Figure 1. Breast Cancer molecular subtypes among cases.

DMA plus or minus AsB concentration (TAs or TAs-AsB, respectively). iAs was calculated as the sum of trivalent (As^{3+}) and pentavalent (As^{5+}). %iAs, %MMA and %DMA species were estimated using TAs-AsB as denominator.

2.4 Statistical analysis

Using conditional logistic regression models, the associations between BC molecular subtypes and %MMA and %DMA was estimated according to the observed quintile distribution among controls. Based on the median value of each quintile a lineal trend was assessed accordingly. All analyses were performed with the software Stata 13 (StataCorp, College Station, TX, USA).

Table 1. Breast Cancer molecular subtypes odds ratios in relation to percentages of urinary arsenic metabolites.

| Arsenic species percentages[a] | Odds ratios (95% CI) | | | | | |
| | Quintiles | | | | | |
	1st	2nd	3rd	4th	5th	p for trend[e]
MMA						
All subtypes (N)	67/110	84/98	103/101	105/94	140/96	
	1.00	1.52	2.03	2.11	2.98	<0.001
		(0.90, 2.57)	(1.19, 3.47)	(1.22, 3.64)	(1.72, 5.14)	
HR (N)	33/61	51/51	60/60	63/57	77/55	
	1.00	2.72	2.88	3.70	5.66	<0.001
		(1.27, 5.84)	(1.30, 6.40)	(1.63, 8.41)	(2.45, 13.08)	
HER2 (N)	17/27	15/27	27/22	25/18	30/20	
	1.00	0.41	1.51	2.26	1.31	0.233
		(0.12, 1.46)	(0.45, 5.10)	(0.64, 7.98)	(0.37, 4.63)	
TNBC (N)	17/22	18/20	16/19	17/19	33/21	
	1.00	1.08	1.93	1.12	2.14	0.168
		(0.32, 3.69)	(0.56, 6.65)	(0.36, 3.47)	(0.73, 6.27)	
DMA						
All subtypes (N)	120/100	117/88	93/102	79/113	90/96	
	1.00	1.31	1.03	0.68	1.15	0.480
		(0.80, 2.15)	(0.62, 1.69)	(0.42, 1.10)	(0.68, 1.95)	
HR (N)	65/51	72/51	52/65	46/67	49/50	
	1.00	0.98	0.62	0.48	0.90	0.152
		(0.48, 1.99)	(0.30, 1.25)	(0.23, 0.96)	(0.42, 1.93)	
HER2 (N)	32/25	21/16	23/22	16/25	22/26	
	1.00	1.66	1.89	0.55	0.86	0.552
		(0.51, 5.33)	(0.59, 6.06)	(0.18, 1.68)	(0.23, 3.26)	
TNBC (N)	23/24	24/21	18/15	17/21	19/20	
	1.00	2.19	1.95	1.27	1.85	0.393
		(0.74, 6.50)	(0.61, 6.20)	(0.43, 3.74)	(0.59, 5.82)	

a. Using Tas-AsB in the denominator.
b. Model adjusted for log transformed Tas-AsB (μg/L), AsB (μg/L), BMI (kg/m2), total breastfeeding (months), alcohol (g per week) and creatinine (mg/dL), and original raw values for smoking (pack-years) and age at first pregnancy.
c. Log transformed variables.
d. Based on control group distribution, quintile cut-offs were as follows: MMA% (≤7.01, >7.01–8.92, >8.92–10.79, >10.79–13.30 and >13.30), DMA% (≤72.37, >72.37–77.81, >77.81–81.38, >81.38–84.96, and >84.96).
e. p for trend across quintile medians.

352

3 RESULTS AND DISCUSSION

3.1 Characteristics of study population

The mean age of cases and controls was around 54 years with 51 years of residence in the study zone. In the entire population, TAs ranged from 0.47 to 303.29 µg/L (0.45 to 1683.83 µg/g creatinine) (data not shown). HR+ BC patients were 57%, HER2 23% and TNBC 20% (Figure 1). These figures are similar to those reported among other groups of Mexican cancer patients (Lara-Medina et al., 2011; Martinez et al., 2013, Pérez-Rodríguez, 2015).

3.2 Breast Cancer molecular subtypes risk in relation to urinary arsenic metabolites

After controlling by key BC factors,%MMA remained as a significant risk factor for BC only for HR+. HER2+ and TNBC tumors were not significantly associated with urinary As metabolites (Table 1).

4 CONCLUSIONS

Our results suggest that increased BC risk related to As exposure is not mediated by HER2 expression. This is the first report on As exposure and breast cancer subtypes which requires further confirmation.

REFERENCES

Camacho, L.M., Gutiérrez, M., Alarcón-Herrera, M.T., Villalba, M.D.L. and Deng, S. 2011. Occurrence and treatment of arsenic in groundwater and soil in northern Mexico and southwestern USA. Chemosphere 83: 211–25.

Cebrián, M.E., Albores, A., García-Vargas, G., Del Razo, L.M. and Ostrosky-Wegman, P. 1994. Chronic arsenic poisoning in humans: the case of Mexico. In: Nriagu, J.O. (Ed.), Arsenic in the Environment, Part II: Human Health and Ecosystem Effects. John Wiley & Sons, Inc, New York, pp. 93–107.

Del Razo, L., Arellano, M. and Cebrián, M. 1990. The oxidation states of arsenic in well-water from a chronic arsenicism area of northern Mexico. Environ. Pollut. 64: 143–53.

Gilbert-Diamond, D., Cottingham, K.L., Gruber, J.F., Punshon, T., Sayarath, V., Gandolfi, A.J., Baker, E.R., Jackson, B.P., Folt, C.L. and Karagas, M.R. 2011. Rice consumption contributes to arsenic exposure in US women. Proc. Natl. Acad. Sci. U.S.A. 108: 20656–20660.

Hammond, M.E.H., Hayes, D.F., Dowsett, M., Allred, D.C., Hagerty, K.L., Badve, S., Fitzgibbons, P.L., Francis, G., Goldstein, N.S., Hayes, M., Hicks, D.G., Lester, S., Love, R., Mangu, P.B., McShane, L., Miller, K., Osborne, C.K., Paik, S., Perlmutter, J., Rhodes, A., Sasano, H., Schwartz, J.N., Sweep, F.C.G., Taube, S., Torlakovic, E.E., Valenstein, P., Viale, G., Visscher, D., Wheeler, T., Williams, R.B., Wittliff, J.L. and Wolff, A.C. 2010. American Society of Clinical Oncology/ College Of American Pathologists guideline recommendations for immunohistochemical testing of estrogen and progesterone receptors in breast cancer. J. Clin. Oncol. 28: 2784–2795.

Lara-Medina, F., Pérez-Sánchez, V., Saavedra-Pérez, D., Blake-Cerda, M., Arce, C., Motola-Kuba, D., Villarreal-Garza, C., González-Angulo, A.M., Bargalló, E., Aguilar, J.L., Mohar, A. and Arrieta, Ó., 2011. Triple-negative breast cancer in Hispanic patients: high prevalence, poor prognosis, and association with menopausal status, body mass index, and parity. Cancer 117: 3658–69.

López-Carrillo, L., Hernández-Ramírez, R.U., Gandolfi, A.J., Ornelas-Aguirre, J.M., Torres-Sánchez, L. and Cebrian, M.E. 2014. Arsenic methylation capacity is associated with breast cancer in northern Mexico. Toxicol. Appl. Pharmacol. 280: 53–59.

Martinez, M.E., Wertheim, B.C., Natarajan, L., Schwab, R., Bondy, M., Daneri-Navarro, A., Meza-Montenegro, M.M., Gutierrez-Millan, L.E., Brewster, A., Komenaka, I.K. and Thompson, P.A. 2013. Reproductive factors, heterogeneity, and breast tumor subtypes in women of Mexican descent. Cancer Epidemiol. Biomarkers Prev. 22: 1853–1861.

Pérez-Rodríguez, G. 2015. Prevalencia de subtipos por inmunohistoquímica del cáncer de mama en pacientes del Hospital General Regional 72. Instituto Mexicano del Seguro Social. Cir. Cir. 83: 193–198.

Wyatt, C.J., Fimbres, C., Romo, L., Méndez, R.O. and Grijalva, M. 1998. Incidence of heavy metal contamination in water supplies in northern Mexico. Environ. Res. 76: 114–119.

Yang, X.R., Sherman, M.E., Rimm, D.L., Lissowska, J., Brinton, L.A., Peplonska, B., Hewitt, S.M., Anderson, W.F., Szeszenia-Dabrowska, N., Bardin-Mikolajczak, A., Zatonski, W., Cartun, R., Mandich, D., Rymkiewicz, G., Ligaj, M., Lukaszek, S., Kordek, R. and García-Closas, M. 2007. Differences in risk factors for breast cancer molecular subtypes in a population-based study. Cancer Epidemiol. Biomarkers Prev. 16: 439–443.

Arsenic Research and Global Sustainability – Bhattacharya, Vahter, Jarsjö, Kumpiene,
Ahmad, Sparrenbom, Jacks, Donselaar, Bundschuh & Naidu (Eds)
© 2016 Taylor & Francis Group, London, ISBN 978-1-138-02941-5

Arsenic metabolism and cancer risk. A meta-analysis

B. Gamboa-Loira[1], L. López-Carrillo[1] & M.E. Cebrián[2]
[1]Instituto Nacional de Salud Pública, Cuernavaca, México
[2]Departamento de Toxicología, Centro de Investigación y Estudios Avanzados del Instituto
Politécnico Nacional, D.F., México

ABSTRACT: Epidemiologic evidence is inconsistent regarding the association between cancer and
arsenic (As) metabolism pattern, measured though its urinary metabolites. We performed a meta-analysis
of 13 articles published up to August 2015. Summary risk estimates were obtained with the DerSimonian
and Laird method for random effects model. The Q-statistic was used to identify heterogeneity in the
outcome variable across studies. The potential for publication bias was evaluated through the Begg's test.
We estimated a significant summary measure for cancer in relation to Monomethyl Arsenic Acid (MMA),
2.04 (95%CI: 1.70, 2.45) and Dimethyl Arsenic Acid (DMA) percentage, 0.55 (95%CI: 0.37, 0.82). We
found no evidence of significant heterogeneity in the summary association measure for %MMA but the
test was significative for %DMA. No evidence of publication bias was observed (p = 0.961). Overall, these
results should be regarded as a strong evidence for the relationship between As metabolism pattern and
cancer.

1 INTRODUCTION

Inorganic Arsenic (iAs) has been classified as a
human carcinogen (IARC, 2004), based not only
on epidemiologic evidence showing a relationship
between As in drinking water and various types of
cancer but also on results suggesting the biotransformation of iAs, which produces monomethylated
and dimethylated (MMA and DMA, respectively)
metabolites plays an important role on cancer risk.

Some but not all epidemiologic studies have
shown a positive association between %MMA
and cancer risk and, a negative relationship with
%DMA. We estimated summary cancer risks in
relation to urinary As metabolites.

2 METHODS

2.1 Studies search and identification

Epidemiologic studies on As and cancer risk were
identified in PubMed and MEDLINE databases
using the MeSH terms: arsenic, methylation,
metabolism, cancer, carcinoma and epidemiologic
study. Inclusion criteria were: epidemiologic study
with histopathologically confirmed cancer cases,
providing values for urinary %MMA and %DMA
as well as measures of association and Confidence
Intervals (CI) between cancer and % DMA and/
or %MMA, respectively. A total of 23 studies were
eligible but only 13 met the inclusion criteria from

which we extracted: author(s), publication year,
study place and year, epidemiologic design, sample
size, cancer anatomic site, total As exposure, measures of association and CI, covariates considered
in the analysis and general observations.

2.2 Statistical analysis

We estimated summary measures and CI between
cancer and the extreme categories of urinary %DMA
and/or %MMA, performing a meta-analysis by the
inverse of variance method for fixed-effects model
and the DerSimonian and Laird method for random-
effects model. Heterogeneity was tested though the
Q-statistic, stratifying by epidemiologic design and
anatomic site of cancer. We assessed the potential
for publication bias using the Begg's test. Statistical
analyses were performed using the software Stata 13
(StataCorp, College Station, TX, USA).

3 RESULTS AND DISCUSSION

3.1 Characteristics of selected studies

A total of 13 studies were included in this meta-
analysis, 2 cohorts and 11 case-control studies.
Two of the studies were considered twice because
they gave information of different populations
(Steinmaus et al., 2006) and anatomic sites (Melak
et al., 2014). These studies included a total of
8285 participants (2409 cases/5876 controls).

Six evaluated urothelial (Huang *et al.*, 2008a,b, Melak *et al.*, 2014; Pu *et al.*, 2007; Steinmaus *et al.*, 2006; Wu *et al.*, 2013), 3 skin (Chen *et al.*, 2003; Hsueh *et al.*, 1997; Yu *et al.*, 2000), 2 lung (Steinmaus *et al.*, 2010, 2006), 1 renal (Huang *et al.*, 2012), 1 breast (López-Carrillo *et al.*, 2014) and 1 several anatomic sites of cancer (Chung *et al.*, 2009). Urinary total As concentration ranged 0.26–303.29 in studies that reported in μg/L and <13.09–38.70 in those that reported in μg/g creatinine. Covariates frequently controlled for were: age (all studies) and sex (12 studies).

3.2 Quantitative synthesis

Random effects model summary association measure for %MMA was 2.04 (95% CI: 1.70, 2.45) and for %DMA was 0.55 (95% CI: 0.37, 0.82) (Figure 1). We found evidence of significant heterogeneity in the summary measure for %DMA ($\chi^2 = 50.59$; df = 10; p < 0.001) but not for %MMA ($\chi^2 = 17.94$; df = 14; p = 0.209). All sources of heterogeneity evaluated (Table 1) were significant for %DMA, except within lung ($\chi^2 = 0.34$; df = 1; p = 0.560) and skin ($\chi^2 = 1.64$; df = 1; p = 0.200) cancer studies. No evidence of publication bias was observed (p = 0.961).

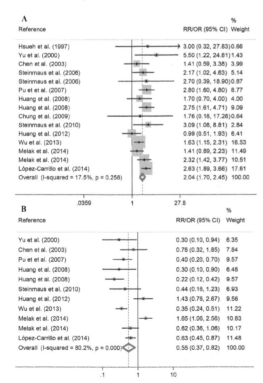

Figure 1. Meta-analysis of the association between urinary arsenic metabolites and cancer. A, %MMA; B, %DMA.

Table 1. Summary association measures for urinary arsenic and cancer.

Studies includes in the analysis	%MMA		%DMA	
	No. of studies	RR/OR (95%CI)	No. of studies	RR/OR (95%CI)
Cohort	2	1.71 (0.76–3.85)	1	–
Case-control	13	2.06 (1.68–2.52)	10	0.57 (0.38–0.87)
Nested case-control	1	–	0	–
Population based	7	2.24 (1.77–2.85)	5	0.68 (0.40–1.18)
Hospital based	5	1.84 (1.28–2.65)	5	0.49 (0.27–0.91)
Anatomic site of cancer[a]				
Bladder	6	1.97 (1.52, 2.54)	4	0.48 (0.20, 1.18)
Lung	2	2.44 (1.57, 3.80)	2	0.58 (0.36, 0.93)
Skin	3	2.26 (0.95, 5.34)	2	0.51 (0.21, 1.26)

a–These results exclude: breast (1) and kidney cancer (1) and cohorts (2).

4 CONCLUSIONS

This meta-analysis provides strong evidence for the relationship between As exposure, altered As metabolism, particularly a high urinary % MMA, and cancer.

REFERENCES

Chen, Y.-C., Guo, Y.-L.L., Su, H.-J.J., Hsueh, Y.-M., Smith, T.J., Ryan, L.M., Lee, M.-S., Chao, S.-C., Lee, J.Y.-Y. and Christiani, D.C. 2003. Arsenic Methylation and Skin Cancer Risk in Southwestern Taiwan. *J. Occup. Environ. Med.* 45: 241–248.

Chung, C.-J., Hsueh, Y.-M., Bai, C.-H., Huang, Y.-K., Huang, Y.-L., Yang, M.-H. and Chen, C.-J., 2009. Polymorphisms in arsenic metabolism genes, urinary arsenic methylation profile and cancer. *Cancer Cause. Control* 20: 1653–1661.

Hsueh, Y., Chiou, H. and Huang, Y., 1997. Serum beta-carotene level, arsenic methylation capability, and incidence of skin cancer. *Cancer Epidemiol. Biomarkers Prev.* 6: 589–596.

Huang, C.-Y., Su, C.-T., Chung, C.-J., Pu, Y.S., Chu, J.-S., Yang, H.-Y., Wu, C.-C. and Hsueh, Y.-M., 2012. Urinary total arsenic and 8-hydroxydeoxyguanosine are associated with renal cell carcinoma in an area without obvious arsenic exposure. *Toxicol. Appl. Pharmacol.* 262: 349–354.

Huang, Y.-K., Huang, Y.-L., Hsueh, Y.-M., Yang, M.-H., Wu, M.-M., Chen, S.-Y., Hsu, L.-I. and Chen, C.-J.,

2008a. Arsenic exposure, urinary arsenic speciation, and the incidence of urothelial carcinoma: a twelve-year follow-up study. *Cancer Cause. Control* 19: 829–839.

Huang, Y.-K., Pu, Y.-S., Chung, C.-J., Shiue, H.-S., Yang, M.H., Chen, C.J. and Hsueh, Y.M. 2008b. Plasma folate level, urinary arsenic methylation profiles, and urothelial carcinoma susceptibility. *Food Chem. Toxicol.* 46: 929–938.

López-Carrillo, L., Hernández-Ramírez, R.U., Gandolfi, A.J., Ornelas-Aguirre, J.M., Torres-Sánchez, L. and Cebrian, M.E., 2014. Arsenic methylation capacity is associated with breast cancer in northern Mexico. *Toxicol. Appl. Pharmacol.* 280, 53–59.

Melak, D., Ferreccio, C., Kalman, D., Parra, R., Acevedo, J., Pérez, L., Cortés, S., Smith, A.H., Yuan, Y., Liaw, J. and Steinmaus, C. 2014. Arsenic methylation and lung and bladder cancer in a case-control study in northern Chile. *Toxicol. Appl. Pharmacol.* 274: 225–231.

Pu, Y.S., Yang, S.M., Huang, Y.K., Chung, C.J., Huang, S.K., Chiu, A.W.H., Yang, M.H., Chen, C.J. and Hsueh, Y.M., 2007. Urinary arsenic profile affects the risk of urothelial carcinoma even at low arsenic exposure. Toxicol. Appl. Pharmacol. 218: 99–106.

Steinmaus, C., Bates, M.N., Yuan, Y., Kalman, D., Atallah, R., Rey, O. a, Biggs, M.L., Hopenhayn, C., Moore, L.E., Hoang, B.K. and Smith, A.H., 2006. Arsenic methylation and bladder cancer risk in case-control studies in Argentina and the United States. *J. Occup. Environ. Med.* 48: 478–488.

Steinmaus, C., Yuan, Y., Kalman, D., Rey, O. a, Skibola, C.F., Dauphine, D., Basu, A., Porter, K.E., Hubbard, A., Bates, M.N., Smith, M.T. and Smith, A.H., 2010. Individual differences in arsenic metabolism and lung cancer in a case-control study in Cordoba, Argentina. *Toxicol. Appl. Pharmacol.* 247: 138–45.

Wu, C.-C., Chen, M.-C., Huang, Y.-K., Huang, C.-Y., Lai, L.-A., Chung, C.-J., Shiue, H.-S., Pu, Y.-S., Lin, Y.-C., Han, B.-C., Wang, Y.-H., Hsueh and Y.-M., 2013. Environmental tobacco smoke and arsenic methylation capacity are associated with urothelial carcinoma. *J. Formos. Med. Assoc.* 112: 554–560.

Yu, R., Hsu, K., Chen, C. and Froines, J., 2000. Arsenic methylation capacity and skin cancer. *Cancer Epidemiol. Biomarkers Prev.* 9: 1259–1262.

Smoking and metabolism phenotype interact with inorganic arsenic in causing bladder cancer

T. Fletcher[1], G.S. Leonardi[1], M. Vahter[1] & Other Members of ASHRAM Study Group
[1]*London School of Hygiene and Tropical Medicine, London, UK*
[2]*Institute of Environmental Medicine, Karolinska Institutet, Stockholm, Sweden*

ABSTRACT: This study aimed to quantify bladder cancer relative risks in relation to inorganic arsenic (As) exposure below 100 μg/L in drinking water and the potential modifying effects of As metabolism and cigarette smoking. A case-control study was conducted in areas of Hungary, Romania, and Slovakia. Cases were histologically confirmed cases of bladder cancer and controls were general surgery, orthopedic, and trauma patients. Exposure indices were constructed based on information on As intake over lifetime of participants. The study recruited 185 cases of histologically verified bladder cancer and 540 controls. The estimated effect of As on cancer was stronger in participants with urinary markers indicating inefficient metabolism of As. By smoking category moderate smokers showed lower As-related risks than either never smokers or heavy smokers. We found a positive association between urothelial cancer and exposure to iAs through drinking water at concentrations below 100 μg/L.

1 INTRODUCTION

Inorganic As is a recognized carcinogen and toxicant, commonly present in groundwater. Causal relationships between long-term elevated inorganic arsenic (As) exposure and cancers of skin, lung, and bladder are well estalished. Although certain populations, e.g. in Bangladesh, West Bengal, Taiwan, China, Argentina, and northern Chile, have had areas with exposure to very high concentrations of As in drinking water (several hundred μg/L), there is widespread exposure world-wide to lower concentrations of inorganic As, i.e. in the range of 5–50 μg/L in drinking water and 5–500 μg/kg in food, especially cereals and vegetables. The National Research Council risk assessment indicated a comparatively high cancer risk even at concentrations as low as 10 μg/L in drinking water, however, these risk estimates, as well as current WHO, EU and US drinking water guidelines are based on linear extrapolation of cancer risks at low doses in studies with relatively high As exposure, mainly in Taiwan. Concerns have been raised about the validity of such extrapolation, in part because several modes of action, which do not include direct DNA mutations, could be expected to result in a threshold dose-response. Cigarette smoking is a known cause of bladder cancer and suspected to interact with iAs at either biological or statistical level.

The Arsenic Health Risk Assessment and Molecular Epidemiology (ASHRAM) study aimed to quantify the risks of several cancer types in relation to long-term low-level As exposure via drinking water in Hungary, Romania, and Slovakia. The associations with bladder cancer are presented here.

2 METHODS

2.1 Study design and population

The study areas included districts in central Slovakia where drinking water is derived from a cracked hard rock aquifer, as well as districts in eastern Hungary and western Romania located on the Great Hungarian Plain, an alluvial basin that straddles the Hungarian-Romanian border. Routine monitoring of sources indicated that approximately 1,100,000 individuals had used water with >10 μg As/L, but generally less than 100 μg/L, at some point during the past 30 years. The study was conducted among individuals who provided written informed consent and were part of the majority Caucasian population, excluding the minority Roma population. In the absence of population registers, we worked with public health agencies to identify incident cases of bladder cancer (ICD-10 code C67) and controls diagnosed among those aged 30–79 years in the same areas. Controls were general surgery in-patients (appendicitis, abdominal hernia, duodenal ulcer, cholelithiasis; with diagnostic codes ICD-10: K35-K37, K40-K46, K26, K80) and orthopedic and trauma patients (fractures, with diagnostic codes ICD-10: SO2, S12, S22, S32, S42, S52, S62, S72, S82, S92, TO2, TO8, T10, T12) aged 30–79. We recruited controls diagnosed with a wide variety of conditions from two distinct hospital departments to reduce the possibility that geographic variation for a specific diagnosis or clinical practice might introduce systematic differences between control and base population distributions. Recruitment of cases and controls was carried out over 21 months (January 2003–September 2004). Pathologists at the public hospitals were responsible for histological confirmation of all bladder cases in the study areas.

2.2 Information collected

A face-to-face interview was conducted with each participant to obtain a detailed residential history focused on identification of drinking water sources at home and information on potential confounders including smoking habits, categorised into number of pack-years. For each individual, the concentrations of As in drinking water at home addresses over their lifetime were derived from measurements at the time of the study and historical data provided by national water authorities. Based on lifetime concentration profiles and individuals' fluid intakes, three exposure indices were constructed: the time-weighted average concentration (μg As/L) over the lifetime of each individual, the lifetime cumulative dose (g As) and the peak daily dose rate (μg As/day) at the participant's residence with the highest water As concentration. Each participant was asked to provide an early morning urine sample. Urinary metabolites of inorganic As (iAs), Methylarsonic Acid (MA) and Dimethylarsinic Acid (DMA) were measured with HPLC on-line with Inductively Coupled Plasma-Mass Spectrometry (ICPMS).

2.3 Statistical analyses

Data were analyzed in STATA version 10 (Statacorp, 2007) using separate multivariable logistic regression models for each As exposure index that included the potential confounders defined, giving adjusted Odds Ratios (ORs) per unit of exposure. Stratified analyses were conducted to evaluate effect modification by methylation efficiency and by smoking status on the association between bladder cancer and As exposure.

3 RESULTS AND DISCUSSION

In general, ORs increased with increasing As exposure by each measure. All three As exposure indices were associated with bladder cancer when exposures were modeled as continuous variables. Associations with all three indices of As exposure were stronger in participants with low DMA% or high MA% in urine, with little or no evidence of associations among participants with high DMA% or low MA%. Associations were stronger among non smokers and heavy smokers compared to light smokers.

Table 1 shows the ORs per daily arsenic intake of 10 μg As/day (at the residence with highest As drinking water concentration), stratified by urinary DMA and smoking status. Similar patterns were evident for the other exposure metrics.

Table 1. Bladder cancer risk stratified by smoking and arsenic metabolism status. ORs and 95% Confidence Intervals per 10 μg As/Day peak daily intake.

Smoking category	Urinary DMA below median of 76.6%		Urinary DMA above median of 76.6%	
	Arsenic OR (95% CI)		Arsenic OR (95% CI)	
Non	1.34	(1.13–1.58)	1.01	(0.86–1.18)
Light	1.12	(0.87–1.42)	0.84	(0.63–1.11)
Medium	1.01	(0.84–1.21)	0.76	(0.58–0.98)
Heavy	1.13	(0.91–1.40)	0.85	(0.64–1.11)
V. heavy	1.58	(1.06–2.35)	1.19	(0.81–1.75)

4 CONCLUSIONS

The ASHRAM study demonstrates evidence of an association between long-term low-level exposure to As in drinking water and bladder cancer, most clearly in participants with urinary markers indicating inefficient metabolism of As. A non-linear interaction with smoking intensity was indicated, which warrants further research.

ACKNOWLEDGEMENTS

We thank the participants for their contributions to this study. Financial support provided by European Commission project No. QLK4-CT-2001-00264 (ASHRAM) is gratefully acknowledged. The authors state that they have no competing financial interests. Members of the ASHRAM study group were (in alphabetical order): Tony Fletcher, Walter Goessler, Eugen Gurzau, Kari Hemminki, Rupert Hough, Kvetoslava Koppova, Rajiv Kumar, Giovanni Leonardi, Peter Rudnai, Marie Vahter.

REFERENCES

Hough, R.L., Fletcher, T., Leonardi, G.S., Goessler, W., Gnagnarella, P., Clemens, F., et al. 2010. Lifetime exposure to arsenic in residential drinking water in Central Europe. *Int. Arch. Occup. Environ. Health* 83(5): 471–481.

Leonardi, G., Vahter, M., Clemens, F., Goessler, W., Gurzau, E, Hemminki, K., et al. 2012. Inorganic arsenic and basal cell carcinoma in areas of Hungary, Romania, and Slovakia: a case-control study. *Environ. Health Perspect.* 120(5): 721–726.

Lindberg, A., Sohel, N., Rahman, M., Persson, L.A. and Vahter, M. 2010. Impact of smoking and chewing tobacco on arsenic-induced skin lesions. *Environ. Health Perspect.* 118: 533–538.

Statacorp (2007) Stata statistical software: Release 10. Statacorp LP, College Station, TX, USA.

Arsenic Research and Global Sustainability – Bhattacharya, Vahter, Jarsjö, Kumpiene, Ahmad, Sparrenbom, Jacks, Donselaar, Bundschuh & Naidu (Eds)
© 2016 Taylor & Francis Group, London, ISBN 978-1-138-02941-5

The joint effects of arsenic exposure, cigarette smoking and *TNF-alpha* polymorphism on Urothelial Carcinoma risk

C.-C. Wu[1,2], Y.-H. Wang[3] & Y.-M. Hsueh[4]

[1]*Department of Urology, School of Medicine, College of Medicine, Taipei Medical University, Taipei, Taiwan*
[2]*Department of Urology, Shuang Ho Hospital, Taipei Medical University, New Taipei City, Taiwan*
[3]*Graduate Institute of Clinical Medicine, College of Medicine, Taipei Medical University, Taipei, Taiwan*
[4]*School of Public Health, College of Public Health and Nutrition, Taipei Medical University, Taipei, Taiwan*

ABSTRACT: The aim of the present study is to investigate the combined effects of urinary total arsenic level, cigarette smoking and secondhand smoke on Urothelial Carcinoma (UC) risk. We conducted a hospital-based case-control study consisting of 584 UC cases and 1098 cancer-free controls and found that ever smokers who exposed to secondhand smoke and carried the G/A and A/A genotypes of the TNF-α -308 G/A polymorphism had a significantly higher UC risk (OR = 5.17, 95% CI = 2.26–11.86). Ever smokers who exposed to higher total arsenic level ($\geq$ 16.38 µg/g creatinine) and secondhand smoke had a significantly increased UC risk (OR = 13.18, 95% CI = 6.48–26.83). In conclusion, our findings suggest that the joint effects of urinary total arsenic level, cigarette smoking, secondhand smoke and the TNF-α -308 G/A and A/A genotypes on UC risk were statistically significant and showed a dose-response relationship.

1 INTRODUCTION

Urothelial Carcinoma (UC) includes cancers of the bladder, ureter, and renal pelvis. Cigarette smoking is the major risk factor for bladder cancer. Secondhand smoke (also known as environmental tobacco smoke) contained several chemical carcinogens and has been thought to be significantly associated with bladder cancer (Tao *et al.*, 2010). Other risk factors including chronic arsenic exposure through drinking well water, occupational exposure to carcinogenic chemicals, and inflammation are known risk factors for bladder cancer.

Arsenic exposure, cigarette smoking and secondhand smoke can generate Reactive Oxidative Stress (ROS) and chronic inflammation. Oxidative stress can induce several inflammatory cytokines such as tumor necrosis factor alpha (TNF-α) (Zhou *et al.*, 2011). Several single nucleotide polymorphisms (SNPs) have been identified in the promoter region of the *TNF-α* gene. Many studies have focused on the -308 G/A polymorphism (rs1800629) of *TNF-α* gene. The -308 G/A single nucleotide transition has been shown to influence TNF-α expression. *TNF-α* -308 G/A polymorphism has also been investigated to explore its effect on several cancers such as bladder cancer (Jeong *et al.*, 2004).

To investigate the effect of the -308 G/A polymorphism of the *TNF-α* gene on UC risk in Taiwan, we conducted a hospital-based case–control study to address this issue. Furthermore, we examined the combined effects of cigarette smoking, secondhand smoke exposure, urinary total arsenic level and the

TNF-α -308 G/A polymorphism on the development of UC.

2 METHODS

2.1 *Study population*

A total of 584 UC cases diagnosed with histopathological confirmation. A total of 1,098 cancer-free controls recruited from a hospital-based pool, including those who received a general health examination. All participants provided informed consents before with the interview and then donated bio-specimen. The Research Ethics Committee approved this study, and this study also complied with the World Medical Association Declaration of Helsinki.

2.2 *Questionnaire and biospecimen*

Basic characteristics and lifestyle such as cigarette smoking, secondhand smoke exposure and history of disease were collected using a structured questionnaire. A 6–8 ml of vein blood was drawn from each participant for genotype determination. A 50 ml of spot urine was collected at the time of recruitment and immediately stored at a −20°C freezer for the detection of urinary total arsenic level.

2.3 *Determination of urinary total arsenic*

The detailed procedures have been described previously (Hung *et al.*, 2008). Briefly, a spot urine of 200 µl was used for identifying urinary arsenic species by High-Performance Liquid Chromatography (HPLC) (Waters 501, Waters Associates, MA). The contents of inorganic arsenic and their metabolites were quantified by hydride generator atomic absorption spectrometry (HG-AAS) (Hsueh *et al.*, 1998). Urinary concentration of the sum of inorganic arsenic, MMA and DMA was normalized against urinary creatinine levels (μg/g creatinine).

2.4 Genotyping of TNF-α -308 G/A polymorphism

Genotyping was detected by a Polymerase Chain Reaction-Restriction Fragment Length Polymorphism (PCR-RFLP). The primers were: 5'-TCCTC CCTG CTCCGATTCCG-3' (sense) and 5'-AGGCAATAG GTTTTGAGGGCCAT-3' (antisense). The resulting DNA fragments which represented *TNF-α* -308 G/A polymorphism were determined.

2.5 Statistical analysis

The joint effects of cigarette smoking, secondhand smoke exposure, urinary total arsenic level and *TNF-α* -308 G/A polymorphism on UC risk were estimated by Odds Ratios (ORs) and 95% Confidence Intervals (CIs) using a multivariate-adjusted logistic regression. All data were analyzed using Statistical Analysis Software for Windows, version 9.2 (SAS Institute, Cary, NC). The *P*-values of < 0.05 were considered statistically significant.

3 RESULTS AND DISCUSSION

3.1 Demographic characteristics of subjects and their association with UC

Ever smokers had a significantly increased UC risk (OR = 2.01, 95% CI = 1.56–2.58) compared those who had never smoked. Those with secondhand smoke exposure also had a significantly higher UC risk (OR = 1.75, 95% CI = 1.37–2.23) compared those who had never exposed to secondhand smoke. Subjects carrying the G/A and A/A genotypes of the *TNF-α* -308 G/A polymorphism had a increased UC risk of 1.02 (95% CI = 0.74–1.40) compared to those with the G/G genotype.

3.2 Urinary total arsenic level, cigarette smoking and secondhand smoke on UC risk

Ever smokers exposed to a lower total arsenic level (< 16.38 μg/g creatinine) and secondhand smoke had a significantly increased UC risk of 3.16 (95% CI = 1.39–7.32) compared to never smokers who exposed to lower total arsenic level but not secondhand smoke. Ever smokers who exposed to a higher total arsenic level (≥ 16.38 μg/g creatinine) and secondhand smoke had a significantly greater UC risk of 13.18 (95% CI = 6.48–26.83).

3.3 The combination analysis

Never smokers who exposed to secondhand smoke and carried the G/A and A/A genotypes had a significantly increased UC risk of 2.53 (95% CI = 1.18–5.43) compared to never smokers who unexposed to secondhand smoke carrying the G/G genotype as the reference group. Furthermore, ever smokers who exposed to secondhand smoke and carried the G/A and A/A genotypes had a greater UC risk of 5.17 (95% CI = 2.26–11.86) compared to never smokers who unexposed to secondhand smoke carrying the G/G genotype as the reference group.

4 CONCLUSIONS

Individuals who had exposed to cigarette smoking, higher total arsenic level (≥ 16.38 μg/g creatinine) and secondhand smoke had a greater risk of UC. Moreover, the joint effects of these three environmental risk factors and the G/A and A/A genotypes of the *TNF-α* -308 G/A polymorphism on UC were significant and showed a dose-response relationship.

ACKNOWLEDGEMENTS

This study was supported by grants from National Science Council (NSC-97-2314-B-038-015-MY3 (1–3), NSC-97-2314-B-038-015-MY3 (2–3), NSC-97-2314-B-038-015-MY3 (3–3)), NSC 100-2314-B-038-026, MOST 104–2314-B-038–079 and MOST 104-2314-B-038-024.

REFERENCES

Hsueh, Y.M., Huang, Y.L., Huang, C.C., Wu, W.L., Chen, H.M., Yang, M.H., Lue, L.C. and Chen, C.J. 1998. Urinary levels of inorganic and organic arsenic metabolites among residents in an arseniasis-hyperendemic area in Taiwan. *J Toxicol. Environ. Health Part A* 54: 431–44.

Huang, Y.K., Huang, Y.L., Hsueh, Y.M., Yang, M.H., Wu, M.M., Chen, S.Y., Hsu, L.I. and Chen, C.J. 2008. Arsenic exposure, urinary arsenic speciation, and the incidence of urothelial carcinoma: a twelve-year follow-up study. *Cancer Causes Control* 19: 829–839.

Jeong, P., Kim, E.J., Kim, E.G., Byun, S.S., Kim, C.S. and Kim, W.J. 2004. Association of bladder tumors and GA genotype of -308 nucleotide in tumor necrosis factor-alpha promoter with greater tumor necrosis factor-alpha expression. *Urology* 64: 1052–6.

Tao, L., Xiang, Y.B., Wang, R., Nelson, H.H., Gao, Y.T., Chan, K.K., Yu, M.C. and Yuan, J.M. 2010. Environmental tobacco smoke in relation to bladder cancer risk--the Shanghai bladder cancer study. *Cancer Epidemiol. Biomarkers Prev.* 19: 3087–3095.

Zhou, P., Lv, G.Q., Wang, J.Z., Li, C.W., Du, L.F., Zhang, C. and Li, J.P. 2011. The TNF-alpha-238 polymorphism and cancer risk: a meta-analysis. *PLoS ONE* 6: e22092.

Arsenic Research and Global Sustainability – Bhattacharya, Vahter, Jarsjö, Kumpiene,
Ahmad, Sparrenbom, Jacks, Donselaar, Bundschuh & Naidu (Eds)
© 2016 Taylor & Francis Group, London, ISBN 978-1-138-02941-5

Dose-response relationship between arsenic exposure from drinking water and chronic kidney disease in low-to-moderate exposed area in Taiwan—a 14-year perspective study

L.-I. Hsu[1], K.-H. Hsu[2], H.-Y. Chiou[3] & C.-J. Chen[1]
[1]*Genomics Research Center, Academia Sinica, Taipei, Taiwan*
[2]*Department of Health Care Management, Chang-Gung University, Taoyuan, Taiwan*
[3]*School of Public Health, Taipei Medical University, Taipei, Taiwan*

ABSTRACT: The association of arsenic exposure with Chronic Kidney Disease (CKD) has never been reported in low-to-moderate exposed area in Taiwan. The aim of this study is to examine the dose-response relationship between ingested arsenic and CKD in northeastern Taiwan where well arsenic ranged from undetectable to 3590 µg/L. A total of 6,153 participants were followed 1998–2011. The disease status was ascertained through linkage with National Health Insurance database. Cox regression analysis was used to determine the hazard ratio of the disease associated with arsenic exposure. The dose-response relationship between two arsenic exposure indices and CKD was observed. Hazard ratio was 1.00, 0.92 (95% confidence interval (CI), 0.66–1.28), 1.20 (95%CI, 0.90–1.60), 1.46 (95%CI,1.09–1.95), 1.08 (95%CI,0.80–1.49) and 1.58 (95%CI,1.14–2.18) in relation to well water arsenic concentration of <10.0, 10.0–24.9, 25.0–49.9, 50.0–99.9, 100.0–499.9 and ≥ 500.0 µg/L (P_{trend}:0.0099). Significantly increased CKD risk was observed among people consuming contaminated water with arsenic concentration >50 µg/L.

1 INTRODUCTION

Arsenic exposure is an important public health issue worldwide. Chronic arsenic ingestion was significantly associated with adverse health effects including various cancers, cardiovascular diseases, hypertension and diabetes in a dose-response manner. The epidemiological studies have shown the association between high arsenic exposure and adverse kidney disease outcomes (Zheng et al., 2014). Previous studies in Taiwan have shown that higher urinary arsenic was associated with an abnormal β2 microglobulin as well as a decreased estimated glomerular filtration rate in normal population (Hsueh et al., 2009, Chen et al., 2011). However, the effect of low-to-moderate arsenic levels on renal dysfunction remains inconclusive.

The aim of this study is to examine the dose-response relationship between ingested arsenic and Chronic Kidney Disease (CKD) in low-to-moderate exposed area in Taiwan.

2 MATERIAL AND METHODS

The northeastern cohort was recruited from 18 villages of Lanyang Basin during 1991–1994. A total of 8,088 residents aged 40 or more years old were recruited. The residents in Lanyang Basin had consumed well water (<40 m deep) for more than 50 years since 1940s. The water in the wells was found to have an arsenic concentration ranging from undetectable to 3590 µg/L.

During the recruitment, detailed histories of residency and duration of drinking artesian well water of each participant were obtained. At the same time, a total of 3,901 well-water samples (one sample from each household) were collected and their arsenic concentrations were estimated by hydride generation combined with flame atomic absorption spectrometry. The Cumulative Arsenic Exposure (CAE) from the artesian well water was defined as $(C \times D)$, where C was the arsenic level of the well water from individual's household and D was the duration of drinking the artesian water.

The disease status of each subject was ascertained through linkage with computerized Taiwan National Health Insurance (NHI) database. In Taiwan, 99% of residents were insured for NHI at the end of 2008. Identification of CKD patients was defined including: (1) at least one inpatient claim record for ICD-9 code = 585; or (2) at least two outpatient records for ICD-9 code = 585; (3) the record from Registry for catastrophic illness patients. The participants were followed from 1998 through 2011. The participants who died before 1998, or who had missing well arsenic concentration, or the prevalent CKD cases diagnosed before

1998/07/01 were excluded for the analysis. Finally, a total of 6,153 participants were included for the analysis.

The CKD incidence of each group was calculated by dividing the number of newly-diagnosed CKD by the total number of follow-up person-years observed in a given group. Cox regression analysis was used to determine the hazard ratio of the disease associated with arsenic exposure indices including well arsenic concentration and cumulative arsenic exposure. We categorized the well arsenic concentration into six categories (<10.0, 10.0–24.9, 25.0–49.9 50.0–99.9, 100.0–499.9 and ≥ 500.0 µg/L), and divided cumulative arsenic level into <1000.0, 1000.0–4999.9, 5000.0–9999.9, 10000.0–19999.9 and ≥ 20000.0 µg/L*year) using and well arsenic concentration <10.0 µg/L as referent group.

3 RESULTS

There were 444 newly diagnosed chronic kidney disease cases during a follow-up period of 70,795 person-years (mean time: 11.5 years). Old age, high body mass index (BMI > 25), analgesic use were significant risk factors for CKD. Disease status such as hypertension, diabetes, dyslipidemia or cirrhosis were associated with increased CKD risk. The crude incidence rate (per 100,000 person-years) was 548.1, 613.7, 666.8, 717.7, 590.7, 876.1, respectively in relation to well water arsenic concentrations of < 10.0, 10.0–24.9, 25.0–49.9, 50.0–99.9, 100.0–499.9 and ≥ 500.0 µg/L; and 578.8, 791.4, 703.7, 718.7, 810.3 per 100,000 person-years in relation to CAE <1000.0, 1000.0–4999.9, 5000.0–9999.9, 10000.0–19999.9 and ≥ 20000.0 µg/L*years, respectively. Dose-response relationship was observed between two arsenic exposure indices and the CKD risk.

With the adjustment of age, sex, education level, BMI, analgesic use, cigarette smoking, hypertension, diabetes and dyslipidemia, the Hazard Ratio (HR) was 1.00, 0.92 (95% confidence interval (CI),0.66–1.28), 1.20 (0.90–1.60), 1.46 (1.09–1.95), 1.08 (0.80–1.49) and 1.58 (1.14–2.18) in relation to categories from ≦ 10.0 to ≧ 500.0 µg/L (Ptrend:0.0099); HR was 1.00, 1.12 (0.88–1.42). 1.42 (1.08–1.87), 1.24 (0.86–1.78), 1.26 (0.81–1.96) and 1.76 (1.13–2.75) in relation to the categories from reference group to CAE ≥ 20000.0 µg/L*year (P_{trend}: 0.0053).

4 CONCLUSIONS

There was a significant dose-response trend between ingested arsenic exposure and chronic kidney disease. Significantly increased CKD risk was observed among people consuming contaminated water with arsenic concentration >50 µg/L.

REFERENCES

Chen, J.W., Chen, H.Y., Li, W.F., Liou, S.H., Chen, C.J., Wu, J.H. and Wang, S.L. 2011. The association between total urinary arsenic concentration and renal dysfunction in a community-based population from central Taiwan. *Chemosphere* 84: 17–24.

Hsueh, Y.M., Chung, C.J., Shiue, H.S., Chen, J.B., Chiang, S.S., Yang, M.H., Tai, C.W. and Su, C.T. 2009. Urinary arsenic species and CKD in a Taiwanese population: A case-control study. *Am. J. Kidney Dis.* 54: 859–870.

Zheng, L.Y., Kuo, C.C., Fadrowski, J., Agnew, J., Weaver, V.M. and Navas-Acien, A. 2014. Arsenic and chronic kidney disease: A systematic review. *Curr. Envir. Health Rpt.* 1: 192–207.

Arsenic Research and Global Sustainability – Bhattacharya, Vahter, Jarsjö, Kumpiene,
Ahmad, Sparrenbom, Jacks, Donselaar, Bundschuh & Naidu (Eds)
© *2016 Taylor & Francis Group, London, ISBN 978-1-138-02941-5*

Arsenic exposure, non-malignant respiratory outcomes and immune modulation in the Health Effects of Arsenic Longitudinal Study (HEALS) cohort

F. Parvez[1], V. Slavkovich[1], J.H. Graziano[1], Y. Chen[2], F. Wu[2], C. Olopade[3], M. Argos[3],
H. Ahsan[3], M. Eunus[4], R. Hasan[4], A. Ahmed[4], T. Islam[4], G. Sarwar[4]

[1]*Department of Environmental Health Sciences, Mailman School of Public Health, Columbia University,*
New York City, NY, USA
[2]*Departments of Environmental Medicine, New York University School of Medicine, New York, NY, USA*
[3]*Departments of Health Studies, Medicine and Human Genetics and Cancer Research Center, The University*
of Chicago, Chicago, IL, USA
[4]*Columbia University Arsenic Research Project, Dhaka, Bangladesh*

ABSTRACT: Limited evidence exists on the effects of arsenic (As) exposure on non-malignant pulmonary outcomes, particularly among those exposed to low-to-moderate levels of As or without skin lesions. In a population-based study, using follow-up data in the Health Effects of Arsenic Longitudinal Study (HEALS) cohort, we found that As exposure from drinking water was significantly related to increased incidence of chronic obstructive lung diseases, and pulmonary tuberculosis and respiratory infections in children. We have also observed As induced immune modulation in the same population.

1 INTRODUCTION

Epidemiologic research has demonstrated associations of exposure to chronic As on non-malignant respiratory outcomes including obstructive lung disease, impaired lung function and also with respiratory infections (Mazumder *et al.*, 2005, Rahman *et al.*, 2011, Parvez *et al.*, 2013). A few studies from an As endemic area of Chile have also reported increased mortality from pulmonary tuberculosis and chronic obstructive lung disease including chronic bronchitis and bronchiectasis (Smith *et al.*, 2006, Smith *et al.*, 2011). However, evidence on non-malignant respiratory effects of As are largely based on studies with methodological limitations including ecological study design, retrospectively collected exposure and outcome data with limited sample size. While the physiological mechanisms by which As induces non-malignant lung diseases remain largely unknown, emerging evidence from animal and a few human studies suggest disruption of the immune system as a plausible mode of action (Biswas *et al.*, 2008, Burchiel *et al.*, 2014).

We prospectively evaluated the effects of As exposure on respiratory outcomes in a population exposed to wide range of water with As concentrations (0.1–1,517 µg/L) and includes a large number of study participants exposed to low-to-moderate levels of exposure in the Health Effects of Arsenic Longitudinal Study (HEALS) cohort.

2 MATERIALS AND METHODS

The HEALS cohort was established in Araihazar in 2000, Bangladesh, to prospectively examine health effects of As exposure from drinking water among ~35,000 adults and their children (Ahsan *et al.*, 2006). Since recruitment, we have been following HEALS participants for their health status using a number of procedures including active quarterly in person visits by research assistants, bi-annual visits by physicians, and regular surveillance by village health workers. During the follow-up visits all cohort members were evaluated by trained physicians using standard protocol to ascertain lung disorders.

We conducted population-based analyses to evaluate the association between As exposure, measured in well water and urine samples, incident of chronic obstructive lung diseases, asthma, tuberculosis and immune function in the HEALS participants.

2.1 *Assessment of exposure and study outcomes*

Assessment of exposure: Water, urine and urinary As metabolites, namely Monomethylarsonic Acid (MMA), Dimethylarsinic Acid (DMA) and inorganic As (InAs), were measured by High-Performance Liquid Chromatography Coupled Plasma Mass Spectrometry (HPLC-ICP-MS) using the method as described earlier (Parvez *et al.*, 2008).

Assessment of study outcomes: Obstructive lung disease, Pulmonary Tuberculosis (PTB) and asthma cases were ascertained by standard clinical protocol including pulmonary function tests, chest x-ray, medical records and medication use. T-cell function was determined by activating T-cell with Phytohemagglutinin (PHA) in Peripheral Blood Mononuclear Cell (PBMC) samples.

3 RESULTS AND DISCUSSION

We have been actively following the HEALS and ascertained Chronic Obstructive Pulmonary Diseases (COPD) and asthma. As of October 2015, we have identified 403 and 225 incident cases of COPD and asthma respectively as part of our ongoing follow-up procedure. The ceases were confirmed by standardized clinical and diagnostic protocols. We used cox proportional hazard regression models to estimate Hazard Ratios (HRs) for COPD and asthma with levels of As exposure assessed at baseline, adjusting for age, gender, smoking, body mass index, education and arsenic-related skin lesion status. HRs (95% CI) were: 1.14 (1.04–1.25) and 1.13 (1.01–1.27) for COPD and asthma respectively, in relation to every 98 µg/L increase in water As, in interim analyses. However, no association was observed with emphysema. The findings will be presented at the meeting in June. In a similar analysis with 228 tuberculosis cases, we also have observed a dose-response effect of As exposure. As compared to those at the lowest quartile of water As ($< = 5.5$ µg/L), the HRs of tuberculosis were 1.08 (95% CI: 0.75–1.62), 1.17 (95% CI: 0.81–1.71), and 1.27 (95% CI: 0.84–1.83), for the 2nd–4th quintiles of baseline water As concentration (5.5–43.2, 43.2–116, >116 µg/L), respectively in adjusted model. Similar relationship was observed in relation to baseline urinary As. The association between PTB and MMA% were slightly stronger. Males were found to be at higher risk.

Among children less than 5 years of age living in households with no smokers, we observed an increased risk of clinically confirmed acute Respiratory Tract Infection (ARI). The associations were [*trend test*]: OR (odds ratio): 2.64 (95% CI: 1.19–5.85), p < 0.01 and 1.99 (95% CI: 0.97–4.05), p < 0.05 with *in utero* and early life water and urinary As exposure respectively in models adjusted for age and gender. Association was stronger among boys as observed in earlier studies.

In an ongoing study among never-smokes, we observed dose dependent suppression of PHA stimulated PBMC T-cell proliferation associated with urinary arsenic ($\beta = -0.36$, p = 0.1) and MMA ($\beta = -0.42$, p = 0.2) and a number of cytokines including (TNFα, IL4, IL6, and IL1β) particularly significantly negatively with IL17 ($r = -0.92$, p = 0.02) as observed in animal study.

4 CONCLUSIONS

To our knowledge, our study is the first to report findings on low-to-moderate levels of As exposure in relation to incidence of COPD asthma and tuberculosis. The study has several strengths including a large sample size, prospectively-collected data on As exposure measured in water and urine samples, and clinically confirmed cases of obstructive lung disease, asthma and tuberculosis. We also have found that the associations for obstructive lung disease were stronger among smokers. Our analysis also reveals an As associated immune modulation which may explain part of As related non-malignant respiratory outcomes.

ACKNOWLEDGEMENTS

This work was supported by US NIEHS grants P42 ES10349, P30 ES 09089, R01 ES019968. 02S1 and R01 ES023888. Thanks to our field staff and the study participants.

REFERENCES

Ahsan, H., Chen, Y., Parvez, F., Argos, M., Hussain, A.I., Momotaj, H., Levy, D., van Geen, A., Howe, G. and Graziano, J. 2006. Health effects of arsenic longitudinal study (HEALS): description of a multidisciplinary epidemiologic investigation. *J. Expo. Sci. Env. Epid.* 16(2): 191–205.

Biswas, R., Ghosh, P., Banerjee, N., Das, J.K., Sau, T., Banerjee, A., et al. 2008. Analysis of T-cell proliferation and cytokine secretion in the individuals exposed to arsenic. *Hum. Exp. Toxicol.* 27(5): 381–386.

Burchiel, S.W., Lauer, F.T., Beswick, E.J., Gandolfi, A.J., Parvez, F., Liu, K.J. and Hudson, L.G. 2014. Differential susceptibility of human peripheral blood T cells to suppression by environmental levels of sodium arsenite and monomethylarsonous acid. *PLoS ONE* 9(10): e109192.

Mazumder, D.N., Steinmaus, C., Bhattacharya, P., von Ehrenstein, O.S., Ghosh, N., Gotway, M., Sil, A., Balmes, J.R., Haque, R., Hira-Smith, M.M. and Smith, A.H. 2005. Bronchiectasis in persons with skin lesions resulting from arsenic in drinking water. *Epidemiology* 16(6): 760–765.

Parvez, F., Chen, Y., Yunus, M., Olopade, C., Segers, S., Slavkovich, V., Argos, M., Hasan, R., Ahmed, A., Islam, T., Akter, M.M., Graziano, J.H. and Ahsan, H. 2013. Arsenic Exposure and Impaired Lung Function: Findings from a Large Population-based Prospective Cohort Study. *Am. J. Resp. Crit. Care* 188(7): 813–819.

Rahman, A., Vahter, M., Ekström, E.C. and Persson, L.Å. 2011. Arsenic exposure in pregnancy increases the risk of lower respiratory tract infection and diarrhea during infancy in Bangladesh. *Environ. Health Persp.* 119(5): 719–724.

Smith, A.H., Marshall, G., Yuan, Y., Ferreccio, C., Liaw, J., von Ehrenstein, O., Steimaus, C., Bates, M.N. and Selvin, S. 2006. Increased mortality from lung cancer and bronchiectasis in young adults after exposure to arsenic in utero and in early childhood. *Environ. Health Persp.* 114(8): 1293–1296.

Smith, A.H., Marshall, G., Yuan, Y., Liaw, J., Ferreccio, C. and Steinmaus, C. 2011. Evidence from Chile that arsenic in drinking water may increase mortality from pulmonary tuberculosis. *Am. J. Epidemiol.* 173(4): 414–420.

Arsenic Research and Global Sustainability – Bhattacharya, Vahter, Jarsjö, Kumpiene,
Ahmad, Sparrenbom, Jacks, Donselaar, Bundschuh & Naidu (Eds)
© *2016 Taylor & Francis Group, London, ISBN 978-1-138-02941-5*

Effect of drinking arsenic safe water for ten years in an arsenic exposed population: Study in West Bengal, India

D.N. Guha Mazumder[1], A. Saha[2], N. Ghosh[1] & K.K. Majumder[3]
[1]*DNGM Research Foundation, Kolkata, India*
[2]*Regional Occupational Health Center (Eastern), (ICMR), Kolkata, India*
[3]*Department of Community Medicine, KPC Medical College & Hospital, Jadavpur, Kolkata, India*

ABSTRACT: A study was done in 2010–2011 on the cohort population of 2620 of previous epidemiological study done during 1995, assessing various levels and duration of arsenic exposure in the past and its impact on arsenical skin lesion following consumption of arsenic safe water in West Bengal, India. Following drinking of arsenic safe water 36 out of 131 mild cases of pigmentation (out of 2620 participants) had cleared the lesion, while 65 cases had mild (9 new appearance) and 24 had moderate pigmentation (1 new appearance) and 7 cases had severe pigmentation during present examination. Further, 17 out of 46 mild cases of keratosis cleared the lesion, 17 remained mild (1 new appearance), 11 cases became moderate keratosis and one case severe keratosis during the present examination. Increased severity of skin lesion even following taking safe water was found to be significantly associated with higher initial arsenic level and dose of arsenic exposure.

1 INTRODUCTION

Reports are scanty in the literature on long term effect of chronic arsenic toxicity after stoppage of drinking arsenic contaminated water on skin manifestations. The current report is based on a study done in 2010–2011 on the cohort population of previous epidemiological study carried out in south 24 Parganas, West Bengal during 1995, assessing various levels of arsenic exposure in the past and its impact on arsenical skin lesion following consumption of arsenic safe water. The object of the study is to ascertain the natural history of arsenical skin lesion following drinking of arsenic safe water in an arsenic affected population.

2 METHODS

A follow up study was done on a population of 5,562 residing in 947 households during 2010–11 that was a subgroup of group of population studied during 1995 (surveyed in South 24 Parganas). The details of method of selection of the previous 1995 epidemiological study done and analysis of their drinking water sources for arsenic level has been described earlier (GuhaMazumder *et al.*, 1998).

Each participant was questioned briefly about his or her sources of drinking water, diet and water intake and clinical symptoms; a general medical examination was done, including a careful inspection for arsenic skin lesions. Water samples were collected from private and public tube wells used for drinking and cooking purposes by each recruited household. Arsenic levels were measured by flow-injection hydride generation atomic absorption spectrophotometer. The arsenic concentration in tube well water in the villages ranged up to 3400 µg/L. Supply of arsenic safe drinking water scheme had been completed by PHED, Govt. of West Bengal through deep tube wells from spot sources and supply of filtered surface water through pipe line system covering the nine arsenic affected blocks of the district of South 24 Parganas in the state since 2001 (PHED, 2013).

Here, in this present paper; however we have analyzed 2620 subjects because of the fact that they were drinking safe water 2 for at least 10 years after the stoppage of their earlier drinking of unsafe water. Hence this constituted a cohort that was first examined in 1995 and was kept under drinking of safe water afterwards and finally reexamined in 2010–11.

Each participant was questioned briefly about his or her current sources of drinking and cooking water, and duration of water use from the source. Demographic characteristics and socio economic condition of the participant was recorded in a proforma. All patients were examined in the field by the physicians who have had long years experience in diagnosing arsenic-caused skin lesions.

Evidence of drinking of relatively arsenic safe water (<0.05 mg/L) from current drinking water source was found in 2620 participants only during the follow up period. The present paper is hence based on outcome analysis of clinical status of arsenical skin lesion following drinking of arsenic safe water with correlation of arsenic exposure data in the past on these 2620 participants.

Arsenic level in drinking water source, duration of intake of arsenic contaminated water and cumulative arsenic dose of arsenic intake in the past was considered for correlation with severity of arsenical skin lesion in the past and during current examination.

3 RESULTS AND DISCUSSION

Data of previous study showed that mean arsenic level in drinking water in the past was Mean ± SD 0.22 ± 0.24 mg/L. During the past examination 2488 subjects were found to have no pigmentation, while 131 cases had mild pigmentation and 1case had moderate pigmentation. Following drinking of arsenic safe water 36 (27.48%) out of 131 mild cases of pigmentation (past arsenic level in drinking water was Mean ± SE, 0.21 ± 0.005 mg/L) had clearance of the lesion, while 65 cases were found to have mild pigmentation (Including 9 subjects who had new appearance from subjects without any lesion in the past), 24 were found to have moderate pigmentation (1 case new appearance), and 7 cases were found to have severe pigmentation during present examination. In regard to keratosis during the past examination, 2574 subjects were found to have no lesion, while 46 cases had mild keratosis. Following drinking of arsenic safe water 17 (36.95%) out of 46 mild cases of keratosis (past arsenic level in drinking water, Mean ± SE, 0.22 ± 0.005 mg/L) cleared the lesion, 17 cases were found to have mild keratosis (with one case of new appearance from subjects without lesion), 11 cases were found to have moderate pigmentation and one case was found to have severe keratosis during the present examination. It was further observed that higher the level of arsenic intake in the past, higher was the severity of pigmentation and keratosis observed during current examination and the dose response relationship was found to be statistically significant.

It was evident from the study that significant improvement had taken place in regard to status of both pigmentation and keratosis after intake of arsenic safe drinking water for a prolonged period (2001–2010)[2] in an arsenic exposed population. However increment of severity and new appearance of pigmentation and keratosis were observed in significant number of cases in spite of drinking arsenic safe water for a prolonged period. It was further observed that higher the level of arsenic intake and cumulative dose of arsenic exposure in the past, higher was the progression of severity of pigmentation and keratosis observed during current examination and the dose response relationship was found to be statistically significant. It was observed that only male gender was more prone to be associated with non improvement of both skin lesions after drinking arsenic safe water.

A few studies conducted earlier investigated the effect of drinking arsenic free water on arsenical skin lesion caused by ground water arsenic contamination highlighting either improvement of skin lesion or new appearance in a few (Tseng et al., 1968, Yeh et al., 1973, Guha Mazumder et al., 2001, 2014, Sun et al., 2006). However, the studies were carried out on small number of participants and for a short period of follow up. No study described increment of severity of skin lesion even after drinking arsenic safe water. Further individual exposure data in the past was not considered for assessing the outcome results of follow up study. Thus, a major strength of this study is that it is the first large population-based study with individual exposure data, followed up for a prolonged period following drinking arsenic safe water, which can provide critical information with which to characterize the exposure-response relationship in regard to effect of drinking arsenic safe water.

4 CONCLUSIONS

Intervention with arsenic safe water in West Bengal was found to be associated with clearance of arsenical skin lesion in people drinking water in the past with arsenic level less than Mean ± SE, 0.21 ± 0.005 mg/L. However higher level of arsenic exposure in drinking water in the past was found to be associated with increment of severity or reappearance of new skin lesion in some of the arsenic exposed people in spite of drinking of arsenic safe water.

ACKNOWLEDGEMENTS

This work was supported by funding from Public Health Engineering Department, Govt. of West Bengal; vide Memo No.PHE/IV/985/W-127/10/AO-204 dt 31st March, 2010. The authors express their thanks to Shri Barun Mitra, Soumendu Mondal and Ranjit Das for their help in carrying on the field study and to Shri Kaushiki Chakraborty for his secretarial assistance.

REFERENCES

Guha Mazumder, D.N., Haque, R., Ghosh, N., De, B.K., Santra, A., Chakraborty, D. and Smith, A.H. 1998. Arsenic levels in drinking water and the prevalence of skin lesions in West Bengal, India. *Int. J. Epidemiol.* 27(5):871–877.

Guha Mazumder, D.N., Ghosh, N., De, B.K., Santra, A., Das, S., Lahiri, S., Haque, R., Smith, A.H. and Chakraborti, D. 2001. Epidemiological study on various non carcinomatous manifestations of chronic arsenic toxicity in a district of West Bengal. In: C.O. Abernathy, R.L. Calderon W.R. Chappell (Eds.) *Arsenic Exposure and Health Effects IV*, Oxford, UK: Elsevier Science: pp. 153–164.

Guha Mazumder, D.N., Ghosh, A., Deb, D., Biswas, A. and Guha Mazumder, R.N. 2014. Arsenic exposure, health effects and biomarker and treatment of arseni-cosis - experience in West Bengal, India. In: M. Litter, H.B. Nicolli, M. Meichtry, N. Quici, J. Bundschuh, P. Bhattacharya and R. Naidu. (Eds.) One Century of the Discovery of Arsenicosis in Latin America (1914–2014) (As2014),Balkema.pp 481–484.

PHED 2013. Water Quality Survey Reports. (Access:http://www.wbphed.gov.in/main/index.php/component/content/category/95-water-quality).

Sun, G., Li, X., Pi, J., Sun, Y., Li, B., Jin, Y. and Xu, Y. 2006. Current research problems of chronic arsenico-sis in China. *J. Health Popul. Nutr.* 24:176–181.

Tseng, W.P., Chu, H.M., How, S.W., Fong, J.M., Lin, C.S. and Yeh, S. 1968. Prevalance of skin cancer in an endemic area of chronic arsenicism in Taiwan. *J. Natl..Canc. Inst.* 40: 453–463.

Yeh, S. 1973. Skin cancer in chronic arsenicism. *Human Pathol.* 4: 465–485.

Arsenic Research and Global Sustainability – Bhattacharya, Vahter, Jarsjö, Kumpiene,
Ahmad, Sparrenbom, Jacks, Donselaar, Bundschuh & Naidu (Eds)
© 2016 Taylor & Francis Group, London, ISBN 978-1-138-02941-5

Associations of chronic arsenic exposure with circulating biomarkers of cardiovascular diseases

K. Hossain[1], M. Rahman[1], S. Hossain[1], M.R. Karim[2], M.S. Islam[2] & S. Himeno[3]

[1]*Department of Biochemistry and Molecular Biology, University of Rajshahi, Rajshahi, Bangladesh*
[2]*Department of Applied Nutrition and Food Technology, Islamic University, Kushtia, Bangladesh*
[3]*Laboratory of Molecular Nutrition and Toxicology, Faculty of Pharmaceutical Sciences, Tokushima Bunri University, Tokushima, Japan*

ABSTRACT: Mechanism of arsenic-induced Cardiovascular Diseases (CVDs) is largely unknown. We explored the associations of arsenic exposure with circulating markers of CVDs recruiting human subjects from arsenic-endemic and non-endemic rural areas in Bangladesh. We found that arsenic exposure levels of the study subjects were positively associated with the levels of plasma big Endothelin-1 (Big ET-1). The study subjects in arsenic-endemic areas who were hypertensive had also higher levels of BigET-1 compared to that of normotensive study subjects. We also investigated the several circulating molecules involved in the atherosclerosis. We found that arsenic exposure decreased circulating High Density Lipoprotein Cholesterol (HDL-C) with concomitant increase in Oxidized Low Density Lipoprotein (Ox-LDL), Intercellular Adhesion Molecule-1 (ICAM-1), Vascular Cell Adhesion Molecule-1 (VCAM-1), C-Reactive Protein (CRP), Plasma Uric Acid (PUA), Vascular Endothelial Growth Factor (VEGF) and Matrix Metalloproteinase-2 and -9 (MMP-2 and MMP-9). Therefore, all these findings are important for obtaining novel mechanistic insights into arsenic-induced CVDs.

1 INTRODUCTION

Arsenic is a potent environmental pollutant and human carcinogen. Arsenic poisoning is becoming an emerging epidemic in Asia especially in Bangladesh. A large numbers of toxicity cases have been reported in the different parts of Bangladesh. About 80–100 millions of additional people are currently at risk of toxicity because of consuming arsenic through drinking water at greater than the permissive limit (10 μg/L) set by World Health Organization (WHO). One of the major causes of arsenic-exposure related morbidity and mortality is CVDs. Mechanism of arsenic-induced CVDs is largely unknown. Therefore, we explored the associations of arsenic exposure with circulating markers of CVDs recruiting human subjects from arsenic-endemic and non-endemic rural areas in Bangladesh.

2 METHODS

Study areas and study subjects were selected from the north-west region of rural Bangladesh as we described in our studies (Karim et al., 2010; Hossain et al., 2012; Karim et al., 2013). Arsenic concentrations in the drinking water, hair and nails of the study subjects were determined by inductively coupled plasma mass spectroscopy. HDL-C and PUA were analyzed by commercially available kits with an analyzer, while Ox-LDL, ICAM-1, VCAM-1, CRP, VEGF, MMP-2 and MMP-9 were quantified by immunoassay kits with a micro plate reader.

3 RESULTS AND DISCUSSION

Endothelial dysfunction is thought to be a key event for the formation of atherosclerotic plaque leading to CVDs. BigET-1 is a specific marker of endothelial damage, and a precursor form of well known vasoconstrictor, Endothelin-1 (ET-1). In our study (Hossain et al., 2012), arsenic concentrations in the drinking water, hair and nails of the study subjects were found to be positively associated with the plasma levels of Big ET-1 (Table 1). Hypertension is a very common form of CVDs. Elevated levels of ET-1 are implicated in inducing blood pressure or hypertension possibly because of its vasoconstrictive activity. Intriguingly, we found that the study subjects in arsenic-endemic areas who were hypertensive had also higher levels of plasma BigET-1 as compared to that of normotensive study subjects suggesting the involvement of Big ET-1/ET-1 in arsenic-induced hypertension. In subsequent study (Karim et al. 2013), we investigated several

Table 1. Associations of arsenic exposure with circulating markers of CVDs.

Circulating markers	Water As	Hair As	Nail As
Big ET-1 (n = 304)	$r_s = 0.428^b$	$r_s = 0.441^b$	$r_s = 0.406^b$
HDL-C (n = 324)	$r_s = -0.387^b$	$r_s = -0.368^b$	$r_s = -0.369^b$
Ox-LDL (n = 324)	$r_s = 0.361^b$	$r_s = 0.403^b$	$r_s = 0.327^b$
ICAM-1 (n = 324)	$r_s = 0.371^b$	$r_s = 0.376^b$	$r_s = 0.334^b$
VCAM-1 (n = 316)	$r_s = 0.313^b$	$r_s = 0.372^b$	$r_s = 0.300^b$
CRP (n = 313)	$r_s = 0.354^b$	$r_s = 0.339^b$	$r_s = 0.277^b$
PUA (n = 483)	$r_s = 0.362^b$	$r_s = 0.332^b$	$r_s = 0.260^b$
VEGF (n = 260)	$r_s = 0.363^b$	$r_s = 0.205^a$	$r_s = 0.190^a$
MMP-2 (n = 369)	$r_s = 0.208^b$	$r_s = 0.163^a$	$r_s = 0.160^a$
MMP-9 (n = 373)	$r_s = 0.163^a$	$r_s = 0.173^a$	$r_s = 0.182^b$

As = Arsenic; r_s = Correlation coefficient; [a]$p < 0.01$, [b]$p < 0.001$; r_s and p-values were from Spearman correlation coefficient test; n = Number of study subjects.

circulating anti-atherogenic, pro-oxidative and pro-inflammatory and adhesion molecules involved in the atherosclerosis. We found that arsenic exposure decreased the plasma HDL-C with the concomitant increased in plasma Ox-LDL, ICAM-1, VCAM-1 and CRP levels (Table 1). Recent studies have established PUA as a surrogate marker of CVDs. Uric acid has both pro-oxidant and anti-oxidant activities. Uric acid can act as a pro-oxidant inside the cells to induce inflammatory pathways associated with CVDs through the production of Ox-LDL and CRP. We (Huda et al., 2014) found that chronic arsenic exposure increased the PUA levels (Table 1) suggesting the involvement of uric acid in arsenic-induced CVDs. Furthermore, our recent studies (Rahman et al., 2015; Islam et al., 2015) demonstrated that arsenic exposure increased circulating VEGF, MMP-2 and MMP-9 levels (Table 1). VEGF and MMPs are deeply implicated in cardiovascular pathology through angiogenesis, endothelial dysfunction and atherosclerotic plaque formation and plaque instability. Therefore, findings of our population based studies are important for obtaining novel mechanistic insights into arsenic-induced CVDs.

4 CONCLUSIONS

All the findings observed in our studies may be the hallmark features of arsenic-induced pro-atherogenic events leading to CVDs. The biochemical indicators determined in this study strongly suggest that residents in arsenic-endemic areas are at risk for atherosclerosis and future development of CVDs.

REFERENCES

Hossain, E., Islam. K., Yeasmin, F., Karim, M.R., Rahman, M., Agarwal, S., Hossain, S., Aziz, A., Mamun, A.A., Sheikh, A., Haque, A., Hossain, M.T., Hossain, M., Haris, P.I., Ikemura, N., Inoue, K., Miyataka, H., Himeno, S. & Hossain, K. 2012. Elevated levels of plasma Big endothelin-1 and its relation to hypertension and skin lesions in individuals exposed to arsenic. *Toxicol. Appl. Pharm.* 259(2): 187–194.

Huda, N., Hossain, S., Rahma, M., Karim, M.R., Islam, K., Mamun, A.A., Hossain, M.I., Mohanto, N.C., Alam, S., Aktar, S., Arefin, A., Ali, N., Salam, K.A., Aziz, A., Saud, Z.A., Miyataka, H., Himeno, S. & Hossain, K. 2014. Elevated levels of plasma uric acid and its relation to hypertension in arsenic-endemic human individuals in Bangladesh. *Toxicol. Appl. Pharm.* 281(1): 11–18.

Islam, M.S., Mohanto, N.C., Karim, M.R., Aktar, S., Hoque, M.M., Rahman, A., Jahan, M., Khatun, R., Aziz, A., Salam, K.A., Saud, Z.A., Hossain, M., Rahman, A., Mandal, A., Haque, A., Miyataka, H., Himeno, S. & Hossain, K. 2015. Elevated concentrations of serum matrix metalloproteinase-2 and -9 and their associations with circulating markers of cardiovascular diseases in chronic arsenic-exposed individuals. *Environ. Health* 14 (1): 92.

Karim, M.R., Rahman, M., Islam, K., Mamun, A.A., Hossain, S., Hossain, E., Aziz., A., Yeasmin, F., Agarwal, S., Hossain, M.I., Saud, Z.A., Nikkon, F., Hossain, M., Mandal, A., Jenkins, R.O., Haris, P.I., Miyataka, H., Himeno, S. & Hossain K. 2013. Increases in oxidized low-density lipoprotein and other inflammatory and adhesion molecules with a concomitant decrease in high-density lipoprotein in the individuals exposed to arsenic in Bangladesh. *Toxicol. Sci.* 135(1): 17–25.

Karim, M.R., Salam, K.A., Hossain, E., Islam, K., Ali, N., Haque, A., Saud, Z.A., Yeasmin, T., Hossain, M., Miyataka, H., Himeno, H. & Hossain, K. 2010. Interaction between chronic arsenic exposure via drinking water and plasma lactate dehydrogenase activity. *Sci. Total Environ.* 409(2): 278–83.

Rahman, M., Mamun, A.A., Karim, M.R., Islam, K., Amin, A.A., Hossain, S., Hossain, M.I., Saud, Z.A., Noman, A.S.M., Miyataka, H., Himeno, S. & Hossain, K. 2015. Associations of total arsenic in drinking water, hair and nails with serum vascular endothelial growth factor in arsenic-endemic individuals in Bangladesh. *Chemosphere* 120: 336–342.

Arsenic Research and Global Sustainability – Bhattacharya, Vahter, Jarsjö, Kumpiene,
Ahmad, Sparrenbom, Jacks, Donselaar, Bundschuh & Naidu (Eds)
© 2016 Taylor & Francis Group, London, ISBN 978-1-138-02941-5

Levels of arsenic in drinking water and short and long-term health effects in northern Chile, 1958–2010

C. Ferreccio[1] & C. Steinmaus[2]

[1]*School of Medicine, Pontificia Universidad Católica de Chile, Santiago, Chile*
[2]*School of Public Health, University of California, Berkeley, Berkeley, CA, USA*

ABSTRACT: From 1958 to 1970 the 270,000 inhabitants of Antofagasta City received drinking water containing 850 ug/L of arsenic until health effects attributed to arsenic were identified and prompted its control. Beginning in 1970 water treatment plants lowered arsenic concentrations to 110 ug/L, 70 ug/L, 40 ug/L and 10 ug/L in the 1970's, 80's, 90's and 2003 respectively. Arsenic-associated short-term effects first occurred in children 11–15 years old consisted of skin arsenicism (50%) and respiratory symptoms (8%), systemic vascular disease, and infant mortality. Rates of these effects returned to expected levels in the 1970's. Twenty to 40 years after exposure, the rates of skin, bladder, lung, kidney and liver cancers increased. An exception was breast cancer mortality which decreased during the high exposure period, returning to normal by the end of the 1970's. Long term effects of arsenic exhibit a dose-response effect and increase by younger age of exposure, and with exposure to tobacco smoke, occupational dust and obesity. To date, neurological or developmental effects have not been studied in this population.

1 INTRODUCTION

Northern Chile, in particular the Antofagasta Region, contains the driest dessert in the world, and whose only water source are rivers originating in the Andes mountains. In these oxygenated waters most inorganic Arsenic (As) is in pentavalent form. The content of As in these rivers varies from 10 to 1,000 µg/L, and have been very stable over time. The local population resides either in small rural or large urban cities, all served by municipal drinking water companies. There are no private wells in the region. Until the 1990´s water companies were government owned. Arsenic measurements are available since 1930. In 1958 the Toconce river, with levels of As over 800 ug/L, became the new water source for the capital city of Antofagasta and remains so today (Ferreccio & Sancha, 2006).

In response to the health effects in Antofagasta children, health authorities established maximum tolerable limits for As in drinking water in 1970, which were reviewed in 1984 and 2006; corresponding maximum levels of As were 120 ug/L, 50 ug/L and 10 ug/L. From 1970 to 1972 two As removal plants relying on adsorption with aluminum sulphate or iron hydroxide particles were installed.

2 METHODS

We reviewed historical data on As levels in drinking water, population censuses and surveys and relevant studies that measured As levels in food, air and in individuals. We reviewed health effects using publications and reports of clinical casess, case series, and ecological and case control studies. Only selected references are listed due to space limitations.

3 RESULTS AND DISCUSSION

3.1 Arsenic in drinking water

The main cities of northern Chile have had diverse levels of As in drinking water from 1930 to 2010. The lowest levels measured were in Arica (10 ug/L throughout the period); the highest were in Antofagasta (860 ug/L in the high exposure period) (Table 1). Figure 1 shows the evolution of As drinking water concentrations in the city of Antofagasta.

3.2 Arsenic in food during the high exposure period

Arsenic concentrations in beverages during 1968–1969) (N = 134 samples) ranged from 150 ug/L for Orange Crush to 290 ug/L for seltzer-water, from 390–480ug/L in beer, and from 80 ug/L in fresh cow's milk to 350 ug/L in pineapple juice (Zaldívar, 1974). In vegetables, As ranged from 0.01 mg/100 g in tomatoes to 0.1 mg/100 g in radishes. In spaghetti and canned fish, concentrations were 0.01 and 0.02 mg/100 g respectively. Arsenic levels dropped soon after the treatment plant began operation (Borgoño et al., 1977).

Table 1. Arsenic concentration among the northern Chilean cities 1930–2010.

City Population	Average Arsenic Concentration (µg/L)							
	1930-57	1958-70	1971-77	1978-79	1980-87	1988-94	1995-02	2003+
Arica (168,594)	10	10	10	10	10	10	9	9
Iquique (196,941)	60	60	60	60	60	60	10	10
Tocopilla (21,827)	250	250	636	110	110	40	10	10
Calama (125,946)	150	150	287	110	110	40	38	10
San Pedro (4,522)	600	600	600	600	600	600	600	10
Antofagasta (270,184)	90	860	110	110	70	40	50	10

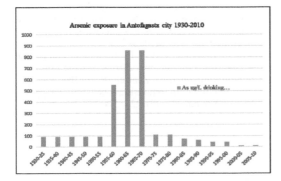

Figure 1. Levels of arsenic exposure in drinking water of Antofagasta city, Northern Chile (1930–2010).

3.3 Internal dose

Studies of As in hair in a population sample in Antofagasta (N = 514) in 1968 averaged 9.2 mg/kg (0.8 to 15 mg/kg). In nails (N = 26) mean As was 29 mg/hg. In 1976 As dropped to 2.7 mg/kg in hair (N = 274) and to 14 mg/kg in nails (N = 273). In fresh human milk from 1968–1970, As concentrations averaged 210 µg/L (Zaldívar, 1974).

3.4 Health effects

Short term health effects: Dermatological and systemic manifestations were first noted two years after the high exposures in Antofagasta began (1960). These included bronchiectasis, Raynaud's syndrome, peripheral vascular phenomena, ischemia and myocardial infarction, and mesenteric thrombosis (Borgoño et al., 1977). In 1962 the first cases of children with cutaneous lesions were identified in Antofagasta. In 1968, over 100 cases seen at the Antofagasta hospital received the diagnoses of arsenicism (Zaldívar, 1974). In 1976, Borgoño found skin lesions in 50% and 2% of children aged 11 to 15 years and < 11 years, respectively (N = 1587) (Borgoño et al., 1977).

Rosenberg (1973) described anatomo-pathological findings for 2 unrelated Antofagasta infants exposed to As in utero who died from myocardial infarction at 2 years of age. Both had occlusive, diffuse, generalized lesions in medium and small size arteries predominantly of the heart, skin, gastrointestinal tract, kidneys, pancreas, and liver. Interestingly, no inflammatory changes were identified (Rosenberg, 1973). Later he studied 3 additional Antofagasta children on autopsy, presenting the same systemic vascular damage, all accompanied by arsenical skin lesions and As in various internal organs (Rosenberg, 1974). The following progression of acute and short term effects was described by Zaldivar among 475 hospitalized cases in Antofagasta from 1968 to 1971 (Zaldivar & Guillier, 1977): 1) weakness and gastric symptoms: anorexia, vomiting and weight loss; 2) respiratory, intestinal, arterial and skin lesions: leucomelanodermia, hyperqueratosis of hands and feet, chronic bronchitis, bronchiectasis, recurrent bronchopneumonia, Raynauds with or without finger gangrene, chronic diarrhea and hepatic cirrhosis; 3) cardiovascular and neural lesions: cardiomegaly, systemic occlusive arterial disease, cerebral thrombosis and cardiac infarct; less frequently, aphasia, palsy and convulsions; 4) malignant growth of skin and liver: skin carcinoma, liver hemangioendotelioma and lung cancer.

Skin, respiratory symptoms and cardiovascular lesions were also reported in children and adults, particularly under 40 years of age. Most As exposed cardiovascular diseases were in subjects who also presented with As-related skin lesions (Zaldivar, 1980).

During the high exposure years, infant mortality increased in Antofagasta compared to the non-exposed Region of Valparaíso with an estimated rate ratio attributed to As exposure of 1.7 (95% CI 1.5–1.9) for late fetal mortality (Hopenhayn-Rich et al., 2000).

Long term health effects: Lung cancer: Two case-control studies of As and lung cancer (cases N = 457, controls N = 1059) found a dose-response relationship for average lifetime As concentrations. In a 1994–96 study, odds ratios (OR) results were 3.6 and 8.5 for water concentrations of 30–50 ug/L and 200–400 ug/L respectively (1). In a 2007–2010study, they reported ORs of 1.7 (80–197 ug/L) and 3.2 (> 197 ug/L) (10). A mortality study from 1989–1993 found a lung cancer SMR in men of 3.8 and 3.2 in women in Antofagasta vs. the rest of Chile (Smith et al., 1998). Bladder cancer: A case control study of 232 cases and 640 controls recruited from 2007–2010 found ORs of 2.6 to 6.0 for exposures from 80–197ug/L to > 197ug/L (Steinmaus et al. 2014). Mortality SMR in men was 6.0 and 8.2 in women in Antofagasta vs the rest of Chile (Smith et al., 1998). Kidney renal pelvis and ureter: Lifetime average As exposure was significantly higher among cases

(N = 24) than controls (N = 640). Mortality SMR was 1.6 for men and 2.7 for women in Antofagasta vs the rest of Chile. Other health effects reported were increased mortality of pulmonary tuberculosis, particularly in men and bronchiectasis, especially among people born during the high exposure period. Finally excess of liver cancer mortality was found among children aged 0–19 years, all exposed to As at young ages.

REFERENCES

Borgoño, J.M., et al. 1977. Arsenic in the drinking water of the city of Antofagasta epidemiological and clinicalstudy before and after the installation of a treatment plant. *Environ. Health Perspect.* 19: 103.

Ferreccio C, Sancha AM. 2006. Exposure to arsenic and its impact on health in Chile. *J. Health Popul. Nutr.* 24 (2): 164–175.

Hopenhayn-Rich, C., et al. 20xx. Chronic arsenic exposure and risk of infant mortality in two areas of Chile. *Environ. Health Persp.* 108(7): 667–673.

Rosenberg, H.G. 1973. Systemic arterial disease with myocardial infarction report on two infants. *Circulation* 47(2): 270–275.

Rosenberg, H.G. 1974. Systemic arterial disease and chronic arsenicism in infants. *Arch. Pathol.* 97(6): 360–365.

Smith, A.H., et al. 1998. Marked increase in bladder and lung cancer mortality in a region of Northern Chile due to arsenic in drinking water. *Amer. J. Epidemiol.* 147(7): 660–669.

Steinmaus, C. et al. 2014. Increased lung and bladder cancer incidence in adults after in utero and early-life arsenic exposure. *Cancer Epidemiol. Biomarkers Prev.* 23(8): 1529–1538.

Zaldívar, R. 1974. Arsenic contamination of drinking water and foodstuffs causing endemic chronic poisoning. *Beiträge zur Pathologie* 151: 384–400.

Zaldivar, R. 1980. A morbid condition involving cardiovascular, bronchopulmonary, digestive and neural lesions in children and young adults after dietary arsenic exposure. Zentralblatt fur Bakteriologie.1.Abt. Originale B, Hygiene, Krankenhaushygiene, *Betriebshygiene, Praventive Medizin* 170(1–2): 44–56.

Zaldivar, R., Guillier, A. 1977. Environmental and clinical investigations on endemic chronic arsenic poisoning in infants and children. Zentralblatt fur Bakteriologie, Parasitenkunde, Infektionskrankheiten und Hygiene, Abt. I (Originale) 165(2): 226–234.

Arsenic Research and Global Sustainability – Bhattacharya, Vahter, Jarsjö, Kumpiene,
Ahmad, Sparrenbom, Jacks, Donselaar, Bundschuh & Naidu (Eds)
© 2016 Taylor & Francis Group, London, ISBN 978-1-138-02941-5

Arsenic toxicity: Who is most susceptible?

C. Steinmaus, A.H. Smith & C. Ferreccio
School of Public Health, University of California, Berkeley, Berkeley, CA, USA

ABSTRACT: Ingested arsenic has been associated with cancer, cardiovascular disease, respiratory disease, skin lesions, infectious disease and other health effects. Susceptibility to these appears to vary greatly from person to person. A number of studies have linked factors such as poor nutrition, inter-individual differences in arsenic metabolism, and genetics to potential increases in arsenic-related risks. Recently expanding evidence also suggests that early-life exposure, obesity, and ancestry might play an important role. For example, in northern Chile high relative risks of lung and bladder cancer were seen in adults who were highly exposed to arsenic in early life but not later. Risks were even higher in subjects who had higher body mass indices. Similar findings were also seen for non-malignant lung disease. Overall, a variety of important susceptibility factors have been identified, and incorporating these into standard setting processes could lead to regulations that help protect major susceptibility groups.

1 INTRODUCTION

Tens of millions of people worldwide are exposed to arsenic in drinking water and food. Ingested arsenic is an established cause of lung, bladder, and skin cancer, and has been linked to a variety of other health effects including cardiovascular disease, non-malignant lung disease, skin lesions, and others (NRC, 2014). A number of studies from Taiwan, Bangladesh, India, the United States, and elsewhere have provided data suggesting that nutrition, genetics, and differences in metabolism may confer marked increases in susceptibility to the health effects of arsenic. Our research group has used a unique exposure situation to study the adverse effects of arsenic and factors conferring susceptibility to these effects. This area is the driest habitable place on earth with relatively few individual water sources, with a range of arsenic water concentrations from <10–860 µg/L. Arsenic records are available for many of these sources for decades in the past, which has allowed us to estimate lifetime exposure with fairly good accuracy. In one of these cities, Antofagasta, exposure levels were especially high from 1958 when two rivers with high arsenic concentrations were diverted to the city for drinking, until 1970 when an arsenic treatment plant was installed.

Using data from this area we found markedly elevated relative risks of cancer and other disease in adults associated with early-life arsenic exposure. We have also identified major increases in risk associated with certain genetics, poor arsenic metabolism, and obesity, including some evidence of increased risk at lower arsenic exposures. In this paper, we review results from Chile suggesting that certain large populations groups are markedly susceptible to arsenic.

2 METHODS

Data from a recently completed cancer case-control study (n = 1301) and a lung function-spirometry study (n = 795) are presented. In the first, incident cases of lung and bladder cancer were ascertained from all pathologists and radiologists in the three most northern Regions of Chile (Steinmaus *et al.*, 2013). Controls were randomly selected from the Chilean Electoral Registry which includes >90% of all adults in this area. In the lung function study, subjects were randomly selected from the Chile Electoral Registry for three of the four largest cities in this area: Arica, Iquique, and Antofagasta, with historical drinking water arsenic concentrations of 10, 60, and 860 µg/L, respectively (Ferreccio *et al.*, 2000). Subjects in both studies were interviewed regarding their lifetime residential history, smoking, water sources, medical history, and work-related exposures. In the lung function study, subjects were also asked about presence of cough, shortness of breath, and other lung symptoms, and lung function was assessed using spirometry. Residential history was linked to arsenic water concentrations so that an estimated drinking water arsenic concentration could be assigned to each year of each subject's life. Typical drinking water intake at the time of interview and 20 years before were used to estimate total arsenic intake from drinking water (Steinmaus *et al.*, 2013). Water intake was not available for proxy subjects, and in an earlier analysis categorization of proxy subjects was based only on arsenic drinking water levels (Steinmaus *et al.*, 2014, 2015). Here, we show results where arsenic intakes in proxy subjects were estimated by assigning them the mean cancer case or control specific drinking water intake from non-proxy subjects. Body Mass Indices (BMI) were assessed by self-reported height and weight at age 20, 40 years, and 10 years preceding interview. Inorganic Arsenic (InAs) and its major metabolites, MMA and DMA, were measured in urine at the time of interview.

Logistic regression was used to calculate Odds Ratios (ORs) for cancer or respiratory symptoms, with adjustments for age, sex, race/ethnicity, occupational exposures, smoking, and other factors. In the lung function study, linear regression was used to calculate age, sex, and height adjusted residuals for Forced Vital Capacity (FVC) and Forced Expiratory Volume in One second (FEV1).

3 RESULTS AND DISCUSSION

As previously reported, relative risks for both lung and bladder cancer were markedly high in those subjects who were born during or just before the high exposure period in Antofagasta. For example, while the lung cancer OR for those people who were adults during the high exposure period was 1.32 (95% Confidence Interval (CI), 0.75–2.34), for those who were *in utero* or children during this period the lung cancer OR was 5.24 (95% CI, 3.05–9.00) (Steinmaus *et al.*, 2014). For lung function and respiratory disease, subjects born in Antofagasta also had elevated prevalence of cough, shortness of breath, and other respiratory symptoms compared to subjects born in the much lower exposure city of Arica. Lung function as measured by spirometry was also decreased in subjects with high early-life exposure.

For both cancer (Steinmaus *et al.*, 2015) and non-malignant lung disease (Fig. 1), arsenic-related associations were greater in those with elevated BMIs than in those with lower BMIs. For cancer, increased associations were also seen in subjects with elevated ratios of MMA in their

Table 1. Lung cancer ORs in subjects who never had known arsenic drinking water concentrations >100 µg/L.

As (µg/day)	Mean	Cases	Cont	OR	90% CI
<13.2	5.6	22	96	1.00	Reference
13.2–54.9	21.7	29	96	1.40	0.78–2.50
>54.9	58.2	41	96	1.89	1.08–3.31

As, estimated arsenic intake from drinking water (categories based on tertiles in controls); Cont, controls; Mean, mean intake in this category; OR, odds ratio adjusted for age, sex, and smoking

urine and in those with certain polymorphisms in arsenic metabolizing genes.

Some evidence of elevated risks for cancer were also seen for arsenic water concentrations <100 µg/L (Table 1). For example, the lung cancer OR in subjects with arsenic water concentrations near 60 µg/L was 1.89 (one-sided p-trend = 0.03) (Table 1). And, in subjects with highest five-year average arsenic water concentrations <100 µg/L, the lung cancer OR in subjects with an elevated % MMA compared to subjects with lower % MMA in urine was 2.23 (90% CI, 1.06–4.68).

4 CONCLUSIONS

Using data on lifetime exposure in Chile, evidence of marked susceptibility to arsenic-related disease was seen for early-life exposure, excess BMI, arsenic metabolism, and other factors. Although the BMI results are novel, these other results are consistent with findings from a number of other studies from Taiwan, Bangladesh, India and elsewhere (NRC, 2014). Accounting for major differences in susceptibility could lead arsenic standards and regulations in food and water that help protect all major groups.

ACKNOWLEDGEMENTS

This research was supported by US National Institute of Environmental Health Sciences grants 5R01ES014032 and P42ES04705.

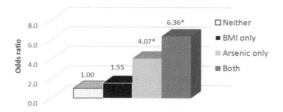

Figure 1. ORs for any respiratory symptom in people with elevated BMI (>27.4 kg/m²) and low arsenic (< 11 µg/L) ("BMI only"); low BMI (≤ 27.4 kg/m²) but elevated arsenic (≥ 11 µg/L) ("Arsenic only"); and with both elevated BMI and arsenic ("Both"), using people with low BMI and low arsenic as the reference. *Two-sided p-values <0.05.

REFERENCES

Ferreccio, C., Gonzalez, C., Milosavjlevic, V., Marshall, G., Sancha, A.M., Smith, A.H. 2000. Lung cancer and arsenic concentrations in drinking water in Chile. *Epidemiology* 11(6): 673–679.
NRC. Critical Aspects of EPA's IRIS Assessment of Inorganic Arsenic: Interim Report. 2014 June 2, 2014]; Available from: www.nap.edu/catalog.php?record_id = 18594.

Steinmaus, C., Castriota, F., Ferreccio, C., Smith, A.H., Yuan, Y., Liaw, J., Acevedo, J., Pérez, L., Meza, R., Calcagno, S., Uauy, R., Smith, M.T. 2015. Obesity and excess weight in early adulthood and high risks of arsenic-related cancer in later life. *Environ. Res.* 142: 594–601.

Steinmaus, C., Ferreccio, C., Smith, A.H. 2015. Three Authors Reply. *Am. J. Epidemiol.* 182(1): 90–92.

Steinmaus, C., Ferreccio, C., Yuan, Y., Acevedo, J., González, F., Perez, L., Cortés, S., Balmes, J.R., Liaw, J., Smith, A.H. 2014. Elevated lung cancer in younger adults and low concentrations of arsenic in water. *Am. J. Epidemiol.* 180(11): 1082–1087.

Steinmaus, C.M., Ferreccio, C., Acevedo Romo, J., Yuan, Y., Cortes, S., Marshall, G., Moore, L.E., Balmes, J.R., Liaw, J., Golden, T., Smith, A.H. 2013. Drinking water arsenic in northern Chile: high cancer risks 40 years after exposure cessation. *Cancer Epidemiol. Biomarkers Prev.* 22: 623–30.

Steinmaus, C., Ferreccio, C., Acevedo, J., Yuan, Y., Liaw, J., Durán,V., Cuevas, S., García, J., Meza, R., Valdés, R., Valdés, G., Benítez, H., VanderLinde, V., Villagra, V., Cantor, K..P., Moore, L.E., Perez, S.G., Steinmaus, S., Smith, A.H. 2014. Increased lung and bladder cancer incidence in adults after in utero and early-life arsenic exposure. *Cancer Epidemiol. Biomarkers Prev.* 23: 1529–38.

Arsenic Research and Global Sustainability – Bhattacharya, Vahter, Jarsjö, Kumpiene,
Ahmad, Sparrenbom, Jacks, Donselaar, Bundschuh & Naidu (Eds)
© *2016 Taylor & Francis Group, London, ISBN 978-1-138-02941-5*

Epidemiological evidences on drinking water arsenic and type 2 diabetes

D.D. Jovanović[1], K. Paunović[2], D.D. Manojlović[3] & Z. Rasic-Milutinović[4]
[1]*Institute of Public Health of Serbia "Dr Milan Jovanović Batut", Belgrade, Serbia*
[2]*Institute of Hygiene and Medical Ecology, Faculty of Medicine, University of Belgrade, Belgrade, Serbia*
[3]*Institute of Chemistry, Technology and Metallurgy, Center of Chemistry, Belgrade, Serbia*
[4]*Departments of Endocrinology, University Hospital Zemun, Belgrade, Serbia*

ABSTRACT: The highest arsenic concentrations in Serbia are measured in drinking water in the Middle Banat region within Vojvodina. The research was designed as a registry-based cross-sectional study comprising two separate populations in Serbia. Mean arsenic concentrations in drinking water from public water supply systems in 383 samples in Middle Banat region was 56.1 µg/L, Median value 38.0 µg/L, range 1.0–349.0 µg/L. This cross-sectional study showed that population from Middle Banat region in Serbia, exposed to low levels of arsenic in drinking water, was at higher risk for the occurrence of type 2 diabetes, in comparison to the unexposed population in Central Serbia.

1 INTRODUCTION

Arsenic is not uniquely present in waters in Serbia. Vojvodina region on the north belongs to the southern part of the Pannonian Basin, which contains high concentrations of naturally occurring arsenic (Varsanyi & Kovacs, 2006). On the other side, the central and southern regions of Serbia belong to the Danube river basin, which contains much lower levels of arsenic (Dangic, 2007). The highest arsenic concentrations in Serbia are measured in drinking water in the Middle Banat region within Vojvodina (Jovanović *et al.*, 2011). Prevalence of type 2 diabetes in Serbia is approximately 600 thousand persons, or 8.2% of the total population, similarly to European prevalence (Sicree *et al.*, 2009). The aim of this study was to explore the association between exposure to arsenic in drinking water and the occurrence of type 2 diabetes in Middle Banat region, Serbia.

2 METHODS/EXPERIMENTAL

2.1 Study design

The research was designed as a registry-based cross-sectional study comprising two separate populations in Serbia. The populations were selected based on arsenic levels in drinking water. The exposed population consuming higher arsenic levels in drinking water was chosen from the Middle Banat region within Vojvodina. The unexposed population was selected from six regions in Central Serbia where arsenic is not present in drinking water. All newly diagnosed cases of type 2 diabetes were obtained from the National Registry of Diabetes in Serbia from 2006 to 2010.

2.2 Epidemiological evidences

The National Registry of Diabetes in Serbia contains person's age, gender, family history of diabetes, smoking habits (current smoker, non-smoker), presence of overweight (defined as body mass index $\geq 25 \, kg/m^2$), central obesity (defined as waist circumference greater than 94 cm for men, and greater than 80 cm for women), and previous diagnosis of arterial hypertension (defined as medically confirmed diagnosis or use of any antihypertensive drug). The number of reported cases was used to calculate standardized incidence rates of diabetes in both areas, and odds ratios for the occurrence of diabetes for the exposed population in comparison to the unexposed population for the period 2006–2010.

3 RESULTS AND DISCUSSION

3.1 Arsenic in drinking water

Mean arsenic concentrations in drinking water from public water supply systems in 383 samples in Middle Banat region was 56.1 µg/L, Median value 38.0 µg/L, range 1.0–349.0 µg/L. Mean arsenic concentration in 525 samples in Central Serbia was 2.0 µg/L, Median value 2.0 µg/L, range 0.5–4.0 µg/L.

3.2 Risk factors for diabetes

Men with type 2 diabetes from the exposed area were more likely to be overweight, have a positive family history of diabetes, low HDL cholesterol, and high LDL cholesterol. Men with type 2 diabetes from Middle Banat were significantly less likely to be current smokers, have central obesity, and have high cholesterol and triglyceride levels than were men from the unexposed area.

Women with type 2 diabetes living in the exposed area were more likely to have a positive family history of diabetes, low HDL cholesterol, and high LDL cholesterol level. They were also less likely to be overweight, with central obesity, and have high cholesterol and triglyceride levels than were women from the unexposed area.

3.3 Standardized incidence rates

The standardized incidence rates of type 2 diabetes were significantly higher for the exposed population, men and women, in comparison to the unexposed population from Central Serbia at any time point from 2006 to 2010 (Table 1).

Odds ratios for type 2 diabetes were significantly higher for the exposed population, both men and women, in the period from 2006 to 2010, when compared with the unexposed population (Table 2).

The presented results are similar to studies conducted in other geographically diverse parts of the world. A positive association between ingested inorganic arsenic and the prevalence of diabetes was found among people from southern Taiwan, residing in villages where arsenic is endemically present in drinking water (Lai et al., 1994). In a recent study in Taiwan, higher prevalence of diabetes in the arsenic-exposed area (7.5% vs. 3.5% in the non-exposed area) across all age groups, and for both men and women. The prevalence odds ratio for diabetes was 2.69 (95% CI = 2.65–2.73) in the endemic area, in comparison to the non-endemic area, after adjustment for age and gender (Wang et al., 2003).

Table 1. Standardized incidence rates of type 2 diabetes in the exposed and the unexposed populations.

Year of registry	Population from Middle Banat		Population from Central Serbia	
	SIR	95% CI	SIR	95% CI
2006	218.6	199.2–239.4	136.4	134.0–138.8
2007	150.3	134.4–167.6	114.1	111.9–116.3
2008	173.4	156.5–191.5	107.9	105.8–110.0
2009	147.5	131.8–164.5	119.2	117.0–121.4
2010	204.9	186.4–224.7	126.7	124.4–129.0

SIR – Standardized incidence rate.
CI – confidence interval.

Table 2. Odds ratio for type 2 diabetes in the exposed compared to the unexposed populations.

Year of registry	OR	95% CI
2006	1.56	1.42–1.70
2007	1.31	1.17–1.46
2008	1.64	1.48–1.81
2009	1.23	1.10–1.37
2010	1.62	1.48–1.78

On the contrary to these studies, the Health Effects of Arsenic Longitudinal Study in Bangladesh found no association between arsenic levels in water and urine samples and type 2 diabetes (Chen et al., 2010). Similarly, a reanalysis of the data from the 2003–2004 National Health and Nutrition Examination Survey in the USA did not indicate that arsenic exposure would increase the risk of diabetes (Steinmaus et al., 2009).

4 CONCLUSIONS

This cross-sectional study showed that population from Middle Banat region in Serbia, exposed to low levels of arsenic in drinking water, was at higher risk for the occurrence of type 2 diabetes, in comparison to the unexposed population in Central Serbia. These results support the hypothesis that exposure to arsenic in drinking water may play a role in the occurrence of type 2 diabetes.

ACKNOWLEDGEMENTS

All authors wish to thank their collaborators in the Institute of Public Health of Zrenjanin, whose supports are greatly acknowledged.

REFERENCES

Chen, Y., Ahsan, H., Slavkovich, V., Loeffler Peltier, G., Gluskin, R.T., Parvez, F., et al. 2010. No association between arsenic exposure from drinking water and diabetes mellitus: a cross-sectional study in Bangladesh. *Environ. Health Perspect.* 118: 1299–1305.

Dangic, A. 2007. Arsenic in surface—and groundwater in central parts of the Balkan Peninsula (SE Europe). In: P. Bhattacharya, A.B. Mukherjee, J. Bundschuh, R. Zevenhoven, R.H. Loeppert (Eds). *Arsenic in Soil and Groundwater Environment—Biogeochemical Interactions, Health Effects and Remediation.* Elsevier, Amsterdam.

Jovanović, D., Jakovljević, B., Rašić-Milutinović, Z., Paunović, K., Peković, G., Knezević, T. 2011. Arsenic occurrence in drinking water supply systems in ten municipalities in Vojvodina region, Serbia. *Environ. Res.* 111: 315–318.

Lai, M.S., Hsueh, Y.M., Chen, C.J., Shyu, M.P., Chen, S.Y., Kuo, T.L., et al. 1994. Ingested inorganic arsenic and prevalence of diabetes mellitus. *Am. J. Epidemiol.* 139: 484–492.

Sicree R., Shaw, J.E., Zimmet, P.Z. 2009. The Global burden of diabetes. In: D. Gan (Ed.). *Diabetes Atlas.* Fourth Ed. International Diabetes Federation, Brussels.

Steinmaus, C., Yuan, Y., Liaw, J., Smith, A.H. 2009. Low-level population exposure to inorganic arsenic in the United States and diabetes mellitus: a reanalysis. *Epidemiology* 20: 807–815.

Varsanyi I., Kovacs, L.O. 2006. Arsenic, iron and organic matter in sediments and groundwater in the Pannonian Basin, Hungary. *Appl. Geochem.* 21: 949–963.

Wang, S.L., Chiou, J.M., Chen, C.J., Tseng, C.H., Chou, W.L., Wang, C.C., et al. 2003. Prevalence of non-insulin-dependent diabetes mellitus and related vascular diseases in southwestern arseniasis-endemic and nonendemic areas in Taiwan. *Environ. Health. Perspect.* 111:155–159.

Arsenic Research and Global Sustainability – Bhattacharya, Vahter, Jarsjö, Kumpiene,
Ahmad, Sparrenbom, Jacks, Donselaar, Bundschuh & Naidu (Eds)
© 2016 Taylor & Francis Group, London, ISBN 978-1-138-02941-5

Groundwater arsenic poisoning in Buxar District, Bihar, India: Health hazards

A. Kumar, R. Kumar, M. Ali, V. Gahlot & A.K. Ghosh
Mahavir Cancer Institute and Research Centre, Patna, Bihar, India

ABSTRACT: In the recent times, arsenic poisoning due to contaminated groundwater in the middle Gangetic plain in India has resulted to lots of health related problems in the population. In Bihar (India), 17 districts have been reported to be affected with arsenic poisoning. In the present study, Simri and Tilak Rai Ka Hatta village, a flood plain Diara region of river Ganga in Buxar district was targeted for the groundwater arsenic assessment and health related problems assessments. The study showed high contamination of arsenic in the groundwater in the entire village where arsenic levels were more than 100 µg/L and the maximum level recorded was 1929 µg/L. The typical symptoms of arsenicosis, cancer incidence and hormonal imbalance in the rural population were highly prevalent. Present study thus concludes that, arsenic poisoning in entire villages has caused severe health hazards in the rural population.

1 INTRODUCTION

Water pollution in the present scenario is causing lots of health hazards to humans. The developing countries have the maximum burden of pollution in comparison to the other countries. Heavy metals as chemical pollutants in water have toxic cumulative toxic properties, carcinogenic potential and cause severe adverse effects to human health. Arsenic (As) is abundant in the crust of the earth and is found in all environments and it is found in soil, minerals, surface and groundwater. Today, it is estimated that more than 5 million people in the Bihar state are drinking water with As concentrations greater than 50 µg/L and presently the groundwater As contamination has spread to 17 districts of the state. Buxar is one of the most As affected district in Bihar (Singh *et al.*, 2014).

Thus, present study deals with the groundwater As assessment and health assessment of the rural population.

2 METHODS/EXPERIMENTAL

2.1 Location

The study was done at Simri village (25°38′17.6″N 84°06′49.4″E) and Tilak Rai Ka Hatta village (25°41′36″N, 84°07′51″E) of Buxar district of Bihar. The Simri village is very large in area and is divided in to 7 strips—Bakullaha Patti, Bhan Bharauli, Khaira Patti, Ramo Patti, Halwa Patti, Doodhi Patti and Gope Bharauli. Among these 7 strips, Doodhi Patti strip acquires the largest part of the village area.

The population of the Simri village is 17,670 while there are 2621 households (Census, 2011 Interim report). The population of the Tilak Rai Ka Hatta village is 5,348 with 340 households (Census, 2011).

2.2 Sample collection and survey

Ethical approval was obtained from the Institutional Ethics Committee (IEC) of the institute before collection of the blood samples of the subject from the targeted site.

Altogether, 322 water samples from Simri village and 110 water samples from Tilak Rai Ka Hatta village while 120 blood samples from both the villages were collected for As assessment through atomic absorption spectrophotometer (Pinnacle 900T, Perkin Elmer, Singapore). All the results were statistically analysed through Graphpad Prism 5 software, USA. Arsenic related health problems among village people were also studied. A survey in the entire area was conducted to estimate the per capita consumption of drinking water through hand pumps by the villagers. For determining the exact location of the hand pump, hand held Global Positioning System (GPS) receivers (Garmin etrex10, of USA) with an estimated accuracy of ≈ 10 m were utilised.

3 RESULTS AND DISCUSSION

3.1 Arsenic assessment

The study shows novel findings ever explored in this area. Among the 7 strips of the village, the Halwa patti and Doodhi patti strips were the most affected strips where As concentration in hand pumps was found to be more than 100 µg/L. In Simri village the maximum As concentration in water sample recorded was 1929 µg/L while in Tilak Rai Ka Hatta village was 1908 µg/L (Fig. 1).

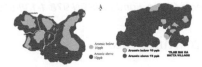

Figure 1. Arsenic map of Simri Village and Tilak Rai Ka Hatta Village of Buxar district of Bihar.

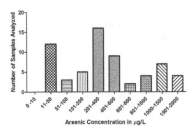

Figure 2. Arsenic concentration in handpumps in Simri and Tilak Rai Ka Hatta villages of Buxar District (n = 62 samples assayed through AAS).

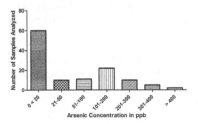

Figure 3. Arsenic concentration in blood samples of village people of Simri and Tilak Rai Ka Hatta villages of Buxar District (n = 120 samples assayed through AAS).

3.2 Health assessment

The rural population exhibited the typical symptoms of arsenicosis like hyperkeratosis in palm and sole, melanosis in palm and sole, blackening of tongue, skin irritation, anaemia, gastritis, constipation, loss of appetite, bronchitis & cough etc. These symptoms were also observed in children of age between 8–14 years. The most important finding of the study is that the blood samples of the village people had high concentration of As which correlates with their arsenicosis symptoms (Fig. 4).

The maximum As concentration in the blood sample recorded in the village was 664.7 µg/L denotes the level of As toxicity. The most fascinating result was that the cancer cases were also reported from these villages i.e. skin cancer, gall bladder cancer and breast cancer cases (Fig. 3).

The other health related assessments showed hormonal imbalance in the population. The rural population exhibited elevated levels of serum estrogen while decreased levels of serum testosterone levels denotes that the As contamination in groundwater and its consumption by the rural population has caused severe health problems to them (Kumar et al., 2015).

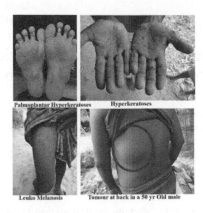

Figure 4. Arsenicosis symptoms in village people.

The entire study thus reveals that high As concentration in groundwater in these two villages and drinking of this contaminated water has led to severe health related problems in the population.

4 CONCLUSIONS

The survey study of these 2 villages of Buxar district, deciphered a unique health related problems in the population. The high As concentration in drinking water as well in their blood has revealed the As related health problems in the rural populations especially the major problems like arsenicosis, hormonal imbalance and few cases of cancer. Thus, As poisoning is not only causing skin diseases or cancer but the entire population is threatened with infertility problem also. The children of age between 8–14 years are at very high risk to get these diseases. Therefore, a proper strategy is immediately required to cater the severity of the As poisoning in these particular areas.

ACKNOWLEDGEMENTS

The authors extend their appreciation to the Department of Science & Technology, Ministry of Science & Technology, Government of India, New Delhi for the financial assistance of this work and to the institute for the entire infrastructural facilities.

REFERENCES

Census. 2011. Interim Report of Population Census of India. (http://www.censusindia.gov.in/2011).

Kumar, A., Ali, M., Rahman, S.M., Iqubal, A.M., Anand, G., Niraj, P.K., et al. 2015: Ground water arsenic poisoning in "Tilak Rai Ka Hatta" village of Buxar District, Bihar, India causing severe health hazards and hormonal imbalance. J. Environ. Anal. Toxicol. 5: 290.

Singh, S.K., Ghosh, A.K., Kumar, A., Kislay, K., Kumar, C., Tiwari, R.R., Parwez, R., Kumar, N., Imam, M. 2014. Groundwater arsenic contamination and associated health risks in Bihar, India. Int. J. Environ. Res. 8(1): 49–60.

*Arsenic Research and Global Sustainability – Bhattacharya, Vahter, Jarsjö, Kumpiene,
Ahmad, Sparrenbom, Jacks, Donselaar, Bundschuh & Naidu (Eds)*
© 2016 Taylor & Francis Group, London, ISBN 978-1-138-02941-5

Arsenic exposure and high incidence of drowning rate among children of Bangladesh

M. Rahman[1], N. Sohel[2] & M. Yunus[3]

[1]*Research and Evaluation Division, BRAC, Mohakhali, Dhaka, Bangladesh*
[2]*Clinical Epidemiology and Biostatistics Department, McMaster University, Hamilton, Canada*
[3]*International Centre for Diarrhoeal Disease Research, Dhaka, Bangladesh*

ABSTRACT: Arsenic contamination manifests various health hazards related to respiratory, skin, gastrointestinal, liver, cardiovascular, nervous system, pregnancy outcome, diabetes, etc. Previous studies showed arsenic poisoning as a risk factor of pregnancy related complications. The present study identifies a potential link between in-utero arsenic exposure and increased mortality rate among their children. A cohort study was undertaken to follow-up the children until they were five years old. Children who were born to 11,414 pregnant women between 2002 and 2004 were included in the follow-up and analyzed. According to the research findings, children who have been exposed to arsenic in-utero could be at an increased risk of drowning due to the effect that it can have on their physical behavior. Reduction of arsenic exposure is required in countries where groundwater arsenic concentrations are high threatening serious public health.

1 INTRODUCTION

In Bangladesh, over 50 million people are at high risk to arsenic exposure exceeding the limits recommended by the WHO guideline (McLellan, 2002). It is evident that arsenic exposure is associated with impaired cognitive function (Hamadani et al., 2011) and fetal development (von Ehrenstein et al., 2007). About 42% deaths among children occur due to drowning mostly in the rural areas of Bangladesh, while 75% deaths occur in natural water bodies adjacent to their homes (Hyder et al., 2008). Around 50 children die every day and 18,000 children every year due to drowning in rural Bangladesh. Most of the studies conducted in arsenic contamination showed the association with exposure risk assessment, morbidity and mortality, but the present study explored a potential link between arsenic exposures during pregnancy with drowning mortality among the offspring. Therefore, the study aimed one step further to find out the potential relationship between arsenic exposure during pregnancy and drowning mortality among the offspring

2 METHODS/EXPERIMENTAL

This cohort study was conducted in Matlab subdistrict, Bangladesh. Children who were born to 11,414 pregnant women between 2002 and 2004 were included. All individuals were prospectively followed-up as part health and demographic system.

Over 13,000 functional tubewells out of 16,430 were identified and screened for arsenic concentrations by using test kits in 2002–3. Water samples from tubewells were collected to test in the laboratory. Individual's drinking water history during pregnancy was collected to measure arsenic exposure. Maternal age at birth, sex, education, asset ownership and arsenic exposure were analyzed to measure the association.

To measure the association between arsenic exposure and drowning deaths, we estimated Relative Risk (RR) adjusted for mother's age, education and parity; children's gender and distance of water bodies by calculating confidence interval, we used bootstrap method. Due to low incidence, we choose 50% sample randomly with replacement for 100 times. The nonparametric bootstrap 95% CI would be 2.5th and 97.5th percentiles of those bootstrap samples.

3 RESULTS AND DISCUSSION

The mean arsenic concentration in drinking water was 139 µg/L (median: 51.1 µg/L). More than half of the pregnant women used drinking water exceeding 10 µg/L arsenic. A significant association was found between prenatal arsenic exposures and drowning among children in Bangladesh (RR 1.74, 95% CI: 1.03, 3.44, lowest vs highest tertile) (Table 1). The women who had more than one children and had no primary education tended to experience higher drowning deaths among their children. However, the underlying reasons behind such association could not be explained due to unavailable information what type of information in this regard. Available literature depicted effects of arsenic exposure on children's cognitive development which might have a relation with increased drowning deaths among them. Exposure to chronic arsenic concentrations was reported to have

Table 1. *In-utero* arsenic exposure and adjusted relative risks of drowning among children.

In-utero As exposure (ug/L)	Adjusted Relative Risk*	95% CI
0.1–4.6	1.0	
4.7–186.6	1.02	(0.49, 2.37)
186.7–3644	1.74	(1.03, 3.44)

*Adjusted with age, education and asset score.

adverse effects on children's intelligence in several studies conducted in China, Taiwan, West Bengal and Bangladesh (Siripitayakunkit *et al.*, 2001; Wang *et al.*, 2007; Tsai *et al.*, 2003; Wasserman *et al.*, 2006).

The children between the ages of 1–4 years (86.3 per 100000) mostly died due to drowning within 20 meters from their homes. The children usually start walking at this age and are often less conscious about the risk factors threatening to life. The UNICEF estimates that about 46 children drown daily, while 16,500 die each year for drowning. In rural Bangladesh available and unprotected water bodies to meet households' daily needs make children's lives even more risky. This might be one of the reasons of higher mortality rate in rural areas than in urban areas. One of the Millennium Development Goals was to reduce mortality by two-thirds among the under-five children. The number of child mortality was declined from 94 per 1000 live births in 1990 to 52 in 2007 (Sayem *et al.*, 2011). The MDG target of child mortality has been renewed to Sustainable Development Goals (SDGs) in recent years. The new commitment aimed at reducing under-five child mortality to 25 deaths per 1000 live births. To reach the goal within the time all the potential risk factors (direct and indirect) endangering children's lives need to be emphasized.

A speculation had been made in this study that prenatal exposure to arsenic might have increased drowning deaths among the children. One of the limitations of this study was that other potential risks factors of drowning were not considered during analysis. However the significant association indicates a new dimension of indirect effects of arsenic contamination in human health beside some direct and obvious health outcomes.

4 CONCLUSIONS

A significant association between prenatal arsenic exposure through drinking water and higher risk of drowning mortality among the under-five children was observed. The indirect effects of arsenic poisoning should be further investigated beyond its direct health outcomes to facilitate national and international goal to reduce under-five child mortality and facilitate children's health and wellbeing.

ACKNOWLEDGEMENTS

The authors would like to thank Professor Joseph H. Graziano, Mailman School of Public Health, Columbia University, New York for reviewing the revised draft and Dr Fahmida Tofail, ICDDRB to critically review the child development part. The AsMat study was conducted at the icddrb with the support of Swedish International Development Agency (Sida) World Health Organization (WHO) and United States of Agency for International Development (USAID).

REFERENCES

Austin, P.C., Chiu, M., Ko, D.T., Goeree, R., Tu, J.V. 2010. Propensity score matching for estimating treatment effects. In: D.E. Faries, A.C. Leon, D.M. Haro, R.L. Obenchain (Eds.) *Analysis of Observational Health Care Data Using SAS. Cary, NC*: SAS Press, pp. 51–84.

Hamadani, J.D., Tofail, F., Nermell, B., Gardner, R., Shiraji, S., Bottai, M., et al. 2011. Critical windows of exposure for asrsenic associated impairment of cognitive function in pre-school girls and boys: a population-based cohort study. *Int. J. Epidemiol.* 40: 1593–604.

Hyder, A.A., Borse, N.N., Blum, L., Khan, R., El Arifeen, S., Baqui, A.H. 2008. Childhood drowning in low- and middle-income countries: urgent need for intervention trials. *J. Paediatr. Child Health* 44: 221–227.

McLellan, F. 2002. Arsenic contamination affects millions in Bangladesh. *Lancet* 359: 1127.

Sayem, A.M., Nury, A.T.M.S., Hossain, A.D. 2011. Achieving the Millennium Development Goal for under-five mortality in Bangladesh: Current status and lessons for issues and challenges for further improvements. *J. Health Popul. Nutr.* 29(2): 92–102.

Siripitayakunkit, U., Lue, S., Choprapawan, C. 2001. Possible effects of arsenic on visual perception and visual-motor integration of children in Thailand. In: W. Chappell, C.O. Abernathy, R.L. Calderon (Eds.) Arsenic Exposure and Health Effects: *Proceedings of the Fourth International Conference on Arsenic Exposure and Health Effects*. Oxford: Elsevier Science, Ltd, San Diego, California; pp. 165_72.

Tsai, S.Y., Chou, H.Y., The, H.W., Chen, C.M., Chen, C.J. 2003. The effects of chronic arsenic exposure from drinking water on the neurobehavioral development in adolescence. *Neurotoxicology* 24: 747_53.

von Ehrenstein, O.S., Poddar, S., Yuan, Y., Mazumder, D.G., Eskenazi, B., Basu, A. et al. 2007. Children's intellectual function in relation to arsenic exposure. *Epidemiology* 18: 44–51.

Wang, S-X., Wang, Z-H., Cheng, X-T., Li, J., Sang, Z-P., Zhang, X-D., et al. 2007. Arsenic and fluoride exposure in drinking water: children's IQ and growth in Shanyin County, Shanxi Province, China. *Environ. Health Perspect.* 115: 643_7.

Wasserman GA, Liu X, Parvez F, Ahsan H, Levy D, Factor-Litvak P, et al. 2006. Water manganese exposure and children's intellectual function in Araihazar, Bangladesh. *Environ. Health Perspect.* 114: 124–129.

Arsenic Research and Global Sustainability – Bhattacharya, Vahter, Jarsjö, Kumpiene,
Ahmad, Sparrenbom, Jacks, Donselaar, Bundschuh & Naidu (Eds)
© 2016 Taylor & Francis Group, London, ISBN 978-1-138-02941-5

Cardiovascular risk in people chronically exposed to low level of arsenic in drinking water

S.M. Htway[1] & Ohnmar[2]

[1]*Department of Physiology, University of Medicine, Magway, Myanmar*
[2]*Department of Physiology, University of Medicine (1), Yangon, Myanmar*

ABSTRACT: The aim of this study was to investigate arsenic-related cardiovascular risks in apparently healthy people chronically exposed to low level of arsenic in drinking water. This cross-sectional study compared cardiovascular risk parameters between arsenic-exposed group (51 µg/L) and non-exposed group (no detectable arsenic). Systemic arterial Blood Pressure (BP), corrected QT interval (QTc) and Ankle Brachial Pressure Index (ABPI) were assessed as cardiovascular risk parameters. All cardiovascular risk parameters were within normal limits in both groups. However, the estimated risk of high normal mean arterial blood pressure, borderline prolongation of QTc interval and borderline decrease in ABPI were 5.3, 1.3 and 2.1 times higher in the exposed group than the non-exposed group, respectively. This study finds increased level of cardiovascular risk in people exposed to the low level of arsenic (51 µg/L) in drinking water.

1 INTRODUCTION

According to World Bank Policy Report (2005), 3.4 million people in Myanmar are exposed to arsenic-contaminated water of 50 µg/L or more. In Myanmar, permissible level of arsenic contamination in drinking water is 50 µg/L (World Bank, 2005). Reviewing world-wide research activities on the pathological effects of chronic exposure to arsenic, recent studies emphasize on non-cancer effects (i.e., ischemic heart disease, hypertension, peripheral vascular disease and diabetes). Recent epidemiological studies have shown that the cardiovascular system is particularly sensitive to chronic arsenic exposure via drinking water (Tseng, 2008). Cardiovascular effects include altered myocardial depolarization such as prolonged QT interval and nonspecific ST segment changes, cardiac arrhythmia, ischemic heart disease, cerebrovascular disease, hypertension, and peripheral vascular diseases (Glazener *et al.*, 1968; Tseng *et al.*, 1977; Goldsmith & From, 1986; Chen *et al.*, 1996; Chiou *et al.*, 1997; Rahman *et al.*, 1999). This study aimed to investigate the arsenic-related cardiovascular risks in apparently healthy people with chronic exposure to low level of arsenic in drinking water.

2 METHODS

2.1 Subjects

Seventy-five subjects (male = 35, female = 40), living in Monpin Village, Sintgaing Township, Mandalay Region (Middle Myanmar), who had been drinking arsenic-contaminated water (>50 µg/L) from a common dug-well for at least 2 years were recruited. Another age and sex matched 75 subjects, living in Shwebonthar Village (about 10 km away from Monpin Village), who consumed water with no detectable arsenic, were also recruited. Arsenic levels of water were measured by the Occupational Health Department (Myanmar) within 6 months before the study. The study subjects were young adult (age range 18–45 years), non-smokers, non-obese and apparently healthy, so that they had low risk of cardiovascular diseases.

2.2 Blood pressure measurement

After 15-minute supine rest, systemic arterial blood pressure was taken from both right and left arms using indirect method by mercury sphygmomanometer (300-Desk, Merino Co., Ltd, Japan). The average of three measurements at one minute interval was taken. Mean Arterial Pressure (MAP) was calculated from Systolic Blood Pressure (SBP) and Diastolic Blood Pressure (DBP). Then, ankle and brachial systolic pressures were taken from both right and left arms and legs using Doppler ultrasound (Bidop ES-100V3, Hadeco Inc., Japan) to calculate Ankle Brachial Pressure Index (ABPI).

2.3 ECG

Routine 12-lead ECG together with 10-second rhythm strip was performed after 15-minute rest (lying quietly on the bed) using digital ECG (DECG-03 A, Schenzhen Mindray Biomedical Electronics Co.,

Ltd, China). The R-R intervals and QT intervals were measured using vernier caliper. Heart rate and corrected QT interval (QTc) were calculated.

3 RESULTS AND DISCUSSION

3.1 *Mean MAP, QTc and ABPI*

Although cardiovascular risk parameters of all subjects were within normal limits, Mean Arterial blood Pressure (MAP) was significantly higher in the exposed group than the non-exposed group ($p < 0.001$). Similarly, corrected QT interval (QTc) was longer in the exposed group than the non-exposed group though statistically insignificant ($p = 0.082$). Regarding Ankle Brachial Pressure Index (ABPI), it was significantly lower ($p < 0.05$) in the exposed group than the non-exposed group (Table 1).

3.2 *Blood pressure*

In the present study, since blood pressures of all subjects were within normal range, it was categorized into high normal blood pressure and normal blood pressure. High normal mean arterial blood pressure was defined as 100–110 mmHg (Luepker et al., 2004). In the exposed group, 5 out of 75 subjects (6.7%) in the exposed group were high normal mean arterial blood pressure, but the only one subject out of 75 subjects (1.3%) in the non-exposed group had high normal BP. The odd ratio was 5.3 (95% CI 0.6–46.4). Therefore, the estimated risk of high normal mean arterial blood pressure was 5.3 times higher in the exposed group than the non-exposed group.

3.3 *QTc*

The QTc more than or equal to 450 ms is defined as the prolonged QTc interval and less than or equal to 430 ms is regarded as normal (Goldenberg et al., 2006). The QTc between 430 ms and 450 ms is considered borderline. We found 10 out of 75 (13.3%) subjects in the exposed group and the 8 out of 75 (10.7%) subjects in the non-exposed group had borderline QTc interval. The odd ratio was 1.3 (95% CI 0.5–3.5).

3.4 *ABPI*

Although normal value of ABPI ranges from 0.9 to 1.3, the ABPI value from 0.91 to 0.99 was defined as borderline ABPI (Pickering et al., 2005). In the present study, 6 out of 75 subjects (8%) in the exposed group were found to be borderline ABPI while 3 out of 75 subjects (4%) in the non-exposed group were borderline ABPI. The odd ratio was 2.1 (95% CI 0.5–8.7).

4 CONCLUSIONS

This study found that chronic arsenic exposure might cause increased systemic arterial blood pressures, prolonged corrected QT interval and low ankle brachial pressure index in apparently healthy people. Therefore, it could be concluded that arsenic-related cardiovascular risks seem to be increased in apparently healthy people exposed to even low level of arsenic contamination (51 µg/L) in drinking water for long-term duration.

ACKNOWLEDGEMENTS

We would like to offer our heartfelt thanks to Dr. Min Than Nyunt, former Director-General, Department of Health, for his expert opinion, excellent advices and methodological supports for the research. We wish to mention our thanks to Dr. Kyi Kyi Mar, former Township Medical Officer, Sintgaing Township Hospital, for her judicious guidance in study site selection and sampling.

REFERENCES

Chen, C.J., Chiou, H.Y., Chiang, M.H., Lin, L.J., Tai, T.Y. 1996. Dose-response relationship between ischaemic heart disease mortality and long-term arsenic exposure. *Arterioscl. Throm. Vas.* 16(4): 504–510.

Chiou, H.Y., Huang, W.I., Su, C.L., Chang, S.F., Hsu, Y.H., Chen, C.J. 1997. Dose-response relationship between prevalence of cerebrovascular disease and ingested inorganic arsenic. *Stroke* 28(9): 1717–1723.

Glazener, F.S., Ellis, J.G., Johnson, P.K. 1968. Electrocardiographic findings with arsenic poisoning. *Calif. Med.* 109(2): 158–162.

Goldenberg, I., Moss, A.J. Zareba, W. 2006. QT interval: How to measure it and what is normal. *J. Cardiovasc. Electr.* 17: 333–336.

Table 1. Comparison of MAP, QTc and ABPI between arsenic-exposed group and non-exposed group.

Parameter	Exposed group (n = 75)	Non-exposed group (n = 75)	p value
MAP	85.20 ± 8.28 mmHg	78.57 ± 7.09 mmHg	0.001
QTc	405.71 ± 20.51 ms	399.44 ± 23.24 ms	0.082
ABPI	1.07 ± 0.07	1.10 ± 0.06	0.05

Goldsmith, S., From, A.H.L. 1986. Arsenic-induced atypical ventricular tachycardia. *New Engl. J. Med.* 303: 1096–1097.

Luepker, R.V., Evans, A., Mckeigue, P., Reddy, K.S. 2004. *Cardiovascular Survey Methods.* World Health Organization, Geneva.

Pickering, T.G., Hall, J.E., Appel, L.J. 2005. AHA Recommendations for blood pressure measurement in humans and experimental animals: Part 1: Blood pressure measurement in humans. *Circulation* 111: 697–716.

Rahman, M.M., Tondel, M., Ahmad, S.A., Chowdhury, I.A., Faruquee, M.H., Axelson, O. 1999. Hypertension and arsenic exposure in Bangladesh. *Hypertension* 33: 74–78.

Tseng, C.H. 2008. Cardiovascular disease in arsenic-exposed subjects living in the arseniasis-hyperendemic areas in Taiwan. *Atherosclerosis* 199:12–18.

Tseng, W.P. 1977. Effects and dose-response relationships of skin cancer and Blackfoot disease with arsenic. *Environ. Health Persp.* 19: 109–119.

World Bank 2005. Towards a more effective operational response: Arsenic contamination of groundwater in South and East Asian Countries. Vol 1 & 2. World Bank Policy Report.

3.2   Biomarkers of exposure and metabolism

Arsenic Research and Global Sustainability – Bhattacharya, Vahter, Jarsjö, Kumpiene,
Ahmad, Sparrenbom, Jacks, Donselaar, Bundschuh & Naidu (Eds)
© 2016 Taylor & Francis Group, London, ISBN 978-1-138-02941-5

Prospective evaluation of arsenic exposure in rural Bangladeshi children shows continuous exposure both via water and rice

M. Kippler[1], H. Skröder[1], S.M. Rahman[1,2], J. Hamadani[2], F. Tofail[2] & M. Vahter[1]
[1]*Institute of Environmental Medicine, Karolinska Institutet, Stockholm, Sweden*
[2]*International Center for Diarrhoeal Disease Research in Bangladesh, Dhaka, Bangladesh*

ABSTRACT: This longitudinal mother-child cohort study (n = 1,017) in rural Bangladesh, an area with a wide range of highly toxic arsenic in the well-water, aims at evaluating the effects of the massive mitigation activities and the health effects of early-life arsenic exposure. Arsenic and other elements, including manganese, in the drinking water used during pregnancy (2002–2003) and when the children were 5 and 10 years of age, were measured using ICP-MS. As marker of the actual arsenic exposure, we measured urinary concentrations of arsenic metabolites using HPLC on line with hydride generation and ICP-MS. Median water arsenic decreased from 23 µg/L (range 0.02–882) during pregnancy to < 2 µg/L (< 0.01–672) at 10 years. The fraction of wells exceeding the national standard (50 µg/L) decreased from 58 to 27%. Still, 1/3 of the children had higher water arsenic than prenatally. Installation of deeper wells (>50 meters) explained much of the lower arsenic concentrations, but increased the manganese concentrations, especially in 50–100 meter wells (median 0.9 mg/L). Low arsenic and manganese were found mainly in >100 m deep wells. Urinary arsenic decreased much less than water arsenic, indicating additional exposure through food, mainly rice.

1 INTRODUCTION

Access to safe drinking water is important for human health and development. With the increasing use of ground water for drinking and agricultural purposes, the often naturally occurring arsenic has become a growing public health concern globally. Inorganic arsenic is a potent toxicant and carcinogen. While adverse health effects in adults are well-documented, consequences of early-life exposure are not well researched.

Our on-going studies in rural Bangladesh have prospectively evaluated children's exposure to arsenic through drinking water during this period of intensive mitigation activities, and associated health effects. Maternal arsenic exposure during pregnancy was found to be associated with impaired growth and increased risk of infant mortality and morbidity, especially in infectious diseases, related to suppressed immune defense. Arsenic exposure also impaired child cognitive function at 5 years of age.

These results prompted us to continue to follow the children's health at 10 years of age, with focus on exposure pattern over time.

2 METHODS

Our longitudinal mother-child cohort (N~2,000) was initiated nested in a large randomized food and micronutrient supplementation trial in pregnancy (Persson *et al.*, 2012) in Matlab, rural Bangladesh. More than 95% of the inhabitants in the study area use drinking water from local wells, in which the arsenic concentrations vary considerably (Vahter *et al.*, 2006). Women were recruited in early pregnancy and the children's exposure to arsenic and other toxic agents were assessed until 10 years of age.

Drinking water used during pregnancy was sampled (acidified) in a parallel water arsenic screening project (Rahman *et al.*, 2006) and matched to the cohort of pregnant women (Vahter *et al.*, 2006). We subsequently collected water samples when the children were 5 and 10 years of age (n = 1,017) (Rahman *et al.*, 2015). We measured arsenic and other elements in the water using ICP-MS (Agilent 7700x; Agilent Technologies, Tokyo) (Kippler *et al.*, 2016). Elevated manganese concentrations were found in the deep wells (Ljung *et al.*, 2009), the main mitigation alternative. Thus, we also focused on this metal, the excess exposure of which may cause neurotoxicity.

For evaluation of the actual arsenic exposure, we measured metabolites of inorganic arsenic in maternal urine during pregnancy, using ion exchange chromatography and AAS (Vahter *et al.*, 2006). Arsenic metabolites in the urine of their children at 5 and 10 of age were measured using HPLC on line with hydride generation and ICP-MS (Fangstrom *et al.*, 2009, Gardner *et al.*, 2011). Concentrations measured by the two methods correlated strongly ($r_s = 0.98$). Urinary concentrations were adjusted for variations in urine dilution by specific gravity (average 1.012).

We evaluated arsenic and manganese exposure in relation to age, depth of the wells, geographical area, as well as socio-economic factors.

3 RESULTS AND DISCUSSION

3.1 Exposure through drinking water over time

The arsenic concentrations in drinking water decreased from a median of 23 µg/L (range 0.02–882 µg/L) in pregnancy to 1.8 µg/L (<0.01–672 µg/L) when the children were 10 years of age. However, still 27% of all the children used water with arsenic concentrations >50 µg/L, the national standard. Further comparison showed that 22% of the children with prenatal water >300 µg/L (8% of all the mothers) had still such high water concentrations at 10 years of age.

In parallel to decreasing water arsenic, the concentrations of manganese increased, from a median of 213 µg/L (range 1.3–6,550 µg/L) during pregnancy to 348 µg/L (range 0.10–6,245 µg/L) at 10 years of age. While 44% of the mothers had used water with manganese concentrations >300 µg/L (the life-time health advisory level of U.S. EPA; www. EPA.gov) during pregnancy, 52% of their children used water with >300 µg/L at 10 years. The correlation between water concentrations of arsenic and manganese was $r_s = -0.47$ ($p < 0.001$). The association was markedly non-linear, and characterized by high manganese concentrations at very low arsenic concentrations and *vice versa* (Ljung *et al.*, 2009).

The changes in the water concentrations of arsenic and manganese over time paralleled the installations of deeper wells. Of all the shallow wells (at 10 years), 26% had water arsenic concentrations <10 µg/L, while 69% of those of 50–100 meters depth and 87% of those deeper than 100 meters had such low concentrations. On the other hand, ~70% of both the shallow and deep wells had water manganese concentrations <300 µg/L, but only 30% of those 50–100 meters had <300 µg/L. Low concentrations of both arsenic and manganese were found in 19% of the wells at 10 years, and out of these, 50% were deeper than 100 meters and 41% 50–100 meters.

3.2 Other sources of exposure

The concentrations of arsenic metabolites in urine decreased less over time than did the drinking water concentrations. Although, the overall association between arsenic concentrations in drinking water and urine was significant ($r_s = 0.64$; $p < 0.001$), it was much weaker ($r_s < 0.18$) at water arsenic concentrations below 10 µg/L, indicating other sources of exposure. The intercept in the linear regression analysis of urinary *vs.* water arsenic at 10 years of age was 61 µg/L, and that for the mothers was 82 µg/L. Based on the found arsenic concentrations in rice from the studied families (range 56–316 µg/kg dry weight), we conclude that rice was the main source of arsenic besides the drinking water. Indeed, recent studies showed alarming exposure through food, especially rice, including rice-based infant foods (Meharg *et al.*, 2008, Ljung *et al.*, 2011).

4 CONCLUSIONS

This population-based, longitudinal study of children's arsenic exposure in rural Bangladesh, from early in utero to 10 years of age (2002–2013), showed a clear decrease in drinking water arsenic concentrations over time; from a median of 23 µg/L in 2002–03 to less than 2 µg/L a decade later. The installation of deeper tube wells explained much of the lower arsenic concentrations. Still, many children had higher water arsenic concentrations at 10 years than prenatally, indicating installation of new wells without testing for arsenic. In spite of the generally successful water arsenic mitigation activities, the children's exposure to the highly toxic and carcinogenic arsenic continues to be elevated, indicating a substantial intake of arsenic through food, mainly rice.

ACKNOWLEDGEMENTS

Support was obtained from the Swedish Research Councils VR and Formas, the Swedish development Cooperation (Sida), and the MINIMat food micronutrient supplementation trial (PIs LA Persson, Uppsala University, and S El Arifeen, icddr,b, Dhaka).

REFERENCES

Fangstrom, B, Hamadani, J., Nermell, B., Grander, M., Palm, B., Vahter, M. 2009. Impaired arsenic metabolism in children during weaning. *Toxicol. Appl. Pharmacol.* 239: 208–214.

Gardner, R., Hamadani, J., Grandér, M., Tofail, F., Nermell, B., Palm, B., Kippler, M., Vahter, M. 2011. Persistent exposure to arsenic via drinking water in rural bangladesh despite major mitigation efforts. *Am. J. Public Health* 101 Suppl 1: S333–338.

Kippler, M., Skröder, H., Rahman, S.M., Tofail, F., Vahter, M. 2016. Elevated childhood exposure to arsenic despite reduced drinking water concentrations—a longitudinal cohort study in rural Bangladesh. *Environ. Int.* 86:119–125.

Ljung, K.S., Kippler, M.J., Goessler, W., Grander, G.M., Nermell, B.M., Vahter, M.E. 2009. Maternal and early life exposure to manganese in rural Bangladesh. *Environ. Sci. Technol.* 43(7): 2595–2601.

Persson, L.A., El Arifeen, S., Ekstrom, E.C., Rasmussen, K.M., Frongillo, E.A., Yunus, M., Team MIS. 2012. Effects of prenatal micronutrient and early food supplementation on maternal hemoglobin, birth weight, and infant mortality among children in Bangladesh: the MINIMat randomized trial. *JAMA* 307(19): 2050–2059.

Rahman, M., Vahter, M., Sohel, N., Yunus, M., Wahed, M.A., Streatfield, P.K., et al. 2006. Arsenic exposure and age and sex-specific risk for skin lesions: a population-based case-referent study in Bangladesh. *Environ. Health Perspect.* 114(12): 1847–1852.

Rahman, S.M., Kippler, M., Ahmed, S., Palm, B., El Arifeen, S., Vahter, M. 2015. Manganese exposure through drinking water during pregnancy and size at birth: A prospective cohort study. *Reprod. Toxicol.* 53: 68–74.

Vahter, M.E., Li, L., Nermell, B., Rahman, A., El Arifeen, S., Rahman, M., Persson, L.A., Ekström, E.C. 2006. Arsenic exposure in pregnancy: a population-based study in Matlab, Bangladesh. *J. Health Popul. Nutr.* 24(2):236–245.

Arsenic Research and Global Sustainability – Bhattacharya, Vahter, Jarsjö, Kumpiene, Ahmad, Sparrenbom, Jacks, Donselaar, Bundschuh & Naidu (Eds)
© 2016 Taylor & Francis Group, London, ISBN 978-1-138-02941-5

Assessment of arsenic exposure among the residents living along the Mekong River in Cambodia

S. Himeno[1], C. Tohmori[1], D. Sumi[1], H. Miyataka[1] & S. Sthiannopkao[2]

[1]*Laboratory of Molecular Nutrition and Toxicology, Faculty of Pharmaceutical Sciences, Tokushima Bunri University, Tokushima, Japan*
[2]*Department of Environmental Engineering, Dong A University, Busan, Korea*

ABSTRACT: Arsenic (As) exposure levels of Cambodian people living along the Mekong River were assessed by determination of As in water, human samples, and fish, and by questionnaire surveys on the use of water resources and fish consumption. Although average concentrations of As in tube well water were over 500 µg/L, only a few patients of arsenicosis were found. Speciation of urinary As and fish consumption survey revealed that total As concentration in urine is influenced by fish consumption pattern in Cambodia where protein nutrition is derived largely from freshwater fish and partly from marine fish. Freshwater fish contained very low levels of arsenobetaine, but small amounts of inorganic As were detected. The intake of inorganic As from freshwater fish was estimated to be far less than those from the water containing 10 µg/L.

1 INTRODUCTION

Arsenic (As) contamination of ground water is evident in Cambodia (Sthiannopkao *et al.*, 2010). However, relatively few patients of arsenicosis with skin lesions have been found in this area compared with other As-polluted areas. In Cambodia, people utilize a variety of water resources such as tube well, river, rain water, and tap water. In addition, Cambodian people consume high amounts of fish, predominantly freshwater fish. Thus, the determination of As concentrations in tube well water or urine samples alone may not provide precise information on the As exposure levels among Cambodian people. Therefore, we collected tube well water samples, hair, nail, and urine samples from the residents, and fish samples consumed by the residents in four villages along the Mekong River in Cambodia. We also conducted a dietary survey for fish consumption to assess the amount of As intake from fish.

2 METHODS

2.1 Sampling of exposure markers

Four villages located along the Mekong River close to the boundary to Vietnam were selected as study areas (Fig. 1). PS, PC and CK are As-polluted villages, and KT is a non-polluted village. 180 individuals (90 males and 90 females) were recruited as study participants by home visiting. Hair, nail, and urine samples were collected, and a questionnaire survey for their use of water resources was conducted. As concentrations in samples were analyzed by using ICP-MS. Speciation of

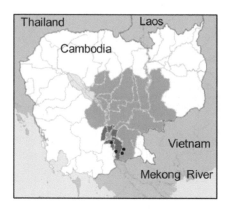

Figure 1. Map of study areas. Circles show the villages for this study. Dark areas are severely polluted by arsenic.

As in urine was conducted with HPLC-ICP-MS. The proposal of this study was approved by the National Ethics Committee for Health Research under the Ministry of Health of the Kingdom of Cambodia.

2.2 Fish consumption survey

40 species of freshwater fish and 20 species of marine fish and other aquatic animals were collected at local markets. The edible parts of fish were digested with nitric acid, and total As concentration was determined by ICP-MS. Certified Reference Material for "cod" was used for validation of the precision of analysis. After freeze-drying and methanol/water extraction, chemical forms of As in fish meat were analyzed by HPLC-ICP-MS.

By using the photos of the 40 freshwater fish and 20 marine fish, we performed a dietary survey for fish consumption among the residents in the same four villages. Average amounts of consumption and As concentrations (both total and each chemical forms) in each species of fish were used for the estimation of average intake of As from each species of fish.

3 RESULTS AND DISCUSSION

3.1 Evaluation of exposure markers

Average As concentrations in tube well water in PS, PC and CK were similar and over 500 µg/L, 50-fold higher than the WHO guideline. The tube well water of KT, non-polluted village, contained less than 10 µg/L As. The questionnaire survey revealed that people in all four villages had been drinking tube well water for more than 10 years. Recently, however, the frequency of using tube well water has been reduced significantly in the severely polluted areas. Chlorine-treated tap water from the Mekong River was introduced several years ago in CK. In PS where a few arsenicosis patients with keratosis were found, the village leader installed a filter for tube well water and began to distribute the filtered water to villagers several years ago. On the contrary, no measures for prevention of As consumption except for painting the tube well in red have been conducted in PC. Thus, most of study participants were exposed to As from tube well water only in the past, but some of them especially those in PC are still exposed to As from tube well water.

We measured total As levels in hair, nail, and urine samples in order to evaluate whether these markers reflect the past and current exposure to As in these areas. Hair and nail As levels in PS, PC, and CK showed higher concentrations than those in non-contaminated KT. There was no significant difference in nail As concentrations among PS, PC, and CK. As nail As levels are known to reflect As exposure for a long time, the similar levels of nail As among the three villages may reflect the past drinking of tube well water.

On the contrary, urinary total As level in PC was the highest among the As-polluted areas, probably reflecting a current exposure to As in PC. Unexpectedly, urinary As level in KT, non-contaminated village, was similar to those in PS and CK. Therefore, we performed speciation analysis of urinary As, and found that only the urine samples of KT contained high concentrations of arsenobetaine, suggesting a consumption of marine fish only in KT. Speciation of urinary As also showed that the residents in PC showed the highest levels of mono- and di-methylated arsenicals. These data suggest that total As levels in urine may not be a good indicator for As exposure especially in Cambodia where people eat both marine and freshwater fish. Speciation of urinary As is highly required.

3.2 Arsenic intake from fish

The results of questionnaire survey showed that the study participants in all four villages eat high amounts of fish. As freshwater fish are known to contain very low levels of As, we next examined the amounts of consumption of freshwater and marine fish among the residents in these areas where a variety of freshwater fish are consumed.

We collected 40 species of freshwater fish which are highly consumed by Cambodian people. 20 species of marine fish and other aquatic animals were also collected. By using the photos of these fish, we performed a questionnaire survey among the residents in the same four villages. The results clearly showed that the frequency of eating marine fish was the highest in KT. Actually, PS, PC, and CK people hardly eat marine fish. These data confirmed the high concentrations of urinary arsenobetaine only in KT.

Average level of total As in freshwater fish was 0.087 µg/g, while that of marine fish and other aquatic animals was 4.24 µg/g. Speciation of As in fish revealed that the concentrations of arsenobetaine were extremely low in freshwater fish, whereas a variety of arsenicals including inorganic arsenicals (iAs) were detected in freshwater fish, though the levels were low. As the consumption amounts of freshwater fish were very high, we calculated the average intake of iAs from each fish species based on the results of fish consumption survey. The consumption of tilapia, snakehead, and some carps contributed to the intake of iAs. To evaluate the risk of freshwater fish for the intake of iAs, we calculated total intake of iAs from all freshwater fish. The calculated value was 14 µg/week/person. However, if a person drinks 2 L/day of water containing 10 µg/L As (WHO guideline), total intake of iAs shall be 140 µg/week/person. Thus, the intake of iAs from freshwater fish among Cambodian people was far less than that from drinking water containing 10 µg/L As.

4 CONCLUSIONS

Current and past exposure to As among the residents in As-polluted areas in Cambodia was assessed by the measurement of multiple exposure markers. Speciation of urinary As is essential in assessing iAs intake among Cambodian people who eat freshwater fish dominantly and marine fish partly.

ACKNOWLEDGEMENTS

This study was supported by a JSPS KAKENHI 24406009.

REFERENCE

Sthiannopkao, S., Kim, K.-W., Cho, K.H., Wantala, K., Sotham, S., Sokuntheara, C., Kim, J.H. 2010. Arsenic levels in human hair, Kandal Province, Cambodia: The influences of groundwater arsenic, consumption period, age and gender. *Appl. Geochem.* 25: 81–90.

Arsenic Research and Global Sustainability – Bhattacharya, Vahter, Jarsjö, Kumpiene,
Ahmad, Sparrenbom, Jacks, Donselaar, Bundschuh & Naidu (Eds)
© 2016 Taylor & Francis Group, London, ISBN 978-1-138-02941-5

Arsenic exposure and metabolism among women in Bolivia

J. Gardon[1], N. Tirado[2], K. Broberg[3] & M. Vahter[3]
[1]*IRD, Hydrosciences, Montpellier, HSM, France*
[2]*Institute of Genetics, School of Medicine, Universidad Mayor de San Andrés, La Paz, Bolivia*
[3]*Institute of Environmental Medicine, Karolinska Institutet, Stockholm, Sweden*

ABSTRACT   Inorganic arsenic (iAs) in drinking water is found in south regions of Bolivia, but human exposure and metabolism efficiency remain unknown. Inefficient methylation of arsenic with a low fraction of DMA in urine is a susceptibility factor for toxic effects. To elucidate arsenic exposure and metabolism in different Bolivian population groups, 74 women from communities (Llapallapani, Pampa-Aullagas, Quillacas) around Lake Poopó were recruited. Arsenic exposure was assessed by the sum of arsenic metabolites (iAs, MMA, DMA) in urine (U-As) using HPLC-HG-ICPMS. The efficiency of arsenic metabolism was expressed as the percentages of the metabolites. The median U-As was 62.2, 62.3 and 83.9 µg/L and the median%DMA was 83.0, 81.3 and 79.1% in Llapallapani, Pampa Aullagas and Quillacas, respectively. Women in Llapallapani had a significantly higher fraction of DMA. The women showed elevated arsenic exposure and an efficient arsenic metabolism, which might influence their risk for arsenic-related disease.

1 INTRODUCTION

The highlands of the Andes Mountains (Altiplano) in northern Chile, Argentina, and Bolivia, constitute a region with high levels of arsenic in many springs and rivers. Several studies in Chile and Argentina have characterized the human exposure to arsenic and arsenic-related diseases, including susceptibility factors. However, very little is known about the situation in Bolivia.

Inorganic arsenic (iAs) is methylated in the body through the one-carbon metabolism, resulting in the formation of Methylarsonic Acid (MMA) and Dimethylarsinic Acid (DMA), which are excreted in the urine. In particular, a high fraction of MMA in urine has been associated with increased risk of various arsenic-related adverse health effects (recently reviewed in Antonelli *et al.*, 2014), while an efficient methylation to DMA has been related to a high rate of arsenic excretion in the urine, thus being beneficial. There is a marked variability in the metabolism of inorganic arsenic within and between population groups (Vahter, 2002). We have previously found a particularly efficient arsenic metabolism in indigenous people in *San Antonio de los Cobres* in the Argentinean Andes (Vahter, 1995).

In this study, we investigated the exposure to arsenic through drinking water and food as well as the metabolism of arsenic in women from south Bolivian Altiplano, with the aim to extend the ongoing evaluation of variations in arsenic methylation efficiency and other susceptibility factors.

2 METHODS

2.1 *Study areas and populations*

The field study took place in the south of the Oruro Department, close to the Poopó Lake, 400 km south of the Titicaca Lake. In this region, previous studies found elevated arsenic concentrations in the drinking water, mainly well water, sometimes exceeding 100 µg/L (Van den Bergh *et al.*, 2010, Ramos *et al.*, 2012). Although mining pollution has been observed east of the Poopó Lake, the water arsenic in the study area is mainly of geogenic origin, released through leaching of volcanic rocks (Van den Bergh *et al.*, 2010, Orma-chea Muñoz *et al.*, 2013). Drinking water samples were collected in each of three villages, Llapalla-pani, Pampa-Aullagas and Quillacas, in order to compare with the arsenic concentrations previously observed.

We recruited women (*n* = 74) with the assistance of medical personnel in the villages. Oral and written informed consent regarding the study was obtained from all participants. The women were interviewed about background characteristics, including diseases, dietary preferences, family origins and different life style factors. Spot midstream urine samples were obtained for assessment of arsenic exposure and potential co-exposure to other toxic trace elements. The samples were tested for pH and potential presence of sugar and proteins, frozen and transported to the trace element laboratory at Karolinska Institutet, Sweden.

2.2 Assessment of arsenic exposure

Exposure to inorganic arsenic (from all sources) was assessed based on the sum of the concentrations of iAs and its metabolites (iAs+MMA+DMA) in urine (U-As), determined by high-performance liquid chromatography, coupled with hydride generation (HG) and inductively coupled plasma mass spectrometry (Agilent 7500ce, Agilent Technologies, Tokyo) as detailed previously (Fängström et al., 2008). Arsenic concentrations were adjusted to the mean specific gravity (1.021). Efficiency of arsenic metabolism was assessed by the fractions (percentages) of iAs, MMA, and DMA in urine.

The concentrations of total arsenic in acid-diluted sub-samples of urine were determined by ICP-MS (Agilent 7700x, Agilent Technologies, Tokyo).

3 RESULTS AND DISCUSSION

3.1 Drinking water concentrations of arsenic

Median arsenic concentrations in drinking water samples were 27.5, 26.6 and 127.5 µg/L in Llapallapani, Pampa Aullagas and Quillacas respectively. Thus, the WHO health-based guideline level of 10 µg/L was clearly exceeded, particularly in Quillacas.

3.2 Human exposure to arsenic

Consistent with the arsenic concentrations observed in drinking water, the women's arsenic exposure (U-As) was found to be higher in Quillacas compared with the two other villages (KWallis test $p<0.001$) (Table 1). However, there was a wide variation within each village.

3.3 Arsenic metabolism

Regarding arsenic metabolites in urines, our preliminary evaluation indicates that the fraction of DMA in urine was fairly high (median 81.3%), indicating efficient arsenic metabolism. The fractions were similar to those previously observed in the Andean population in northern Argentina (Vahter et al. 1995). In addition, we found differences between communities, the %DMA being significantly higher in the Llapallapani village than in the two other villages (KWallis $p = 0.03$). U-As influenced negatively the DMA fraction (spearman $r = -0.21$, $p = 0.07$).

In the same line, the MMA fraction in urine (overall median 8.4%) was lower in Llapallapani (KWallis,

$p = 0.007$) than in the other villages, supporting a high efficiency of the methylation of arsenic. No difference was observed in the fraction of iAs in urine (median 10.0%) between the three villages.

4 CONCLUSIONS

In this region of Bolivia the studied arsenic-exposed women showed efficient metabolism of arsenic, with a low fraction of MMA and a high fraction of DMA in urine. Further studies are needed to identify factors influencing the arsenic metabolism, as well as if their efficient arsenic metabolism protects them from arsenic-related toxicity.

ACKNOWLEDGEMENTS

We thank the women in the communities for participation in the study and local medical teams. The research was funded by IRD-HSM, Eric Philip Sörensens foundation, Kungliga Fysiografiska Sällskapet, and Karolinska Institutet.

REFERENCES

Antonelli, R., Shao, K., Thomas, D.J. Sams, R 2nd, Cowden J. 2014. AS3MT, GSTO, and PNP polymorphisms: impact on arsenic methylation and implications for disease susceptibility. Environ. Res. 132: 156–167.

Fängström, B., Moore, S., Nermell, B., Kunstl, L., Goessler, W., Grandér, M., Kabir, I,. Palm., B., El Arifeen, S., Vahter, M. 2008. Breast-feeding Protects against Arsenic Exposure in Bangladeshi Infants. Environ. Health Perspect. 116: 963–969.

IARC. 2012. A review of human carcinogens: Arsenic, metals, fibres, and dusts. Volume 100C. IARC Monographs on the Evaluation of Carcinogenic Risks to Humans, International Agency for Research on Cancer (World Health Organization, Lyon, France).

Ormachea Muñoz, M., Wern, H., Johnsson, F., Bhattacharya, P., Sracek, O., Thunvik, R., Quintanilla, J., Bundschuh, J. 2013. Geogenic arsenic and other trace elements in the shallow hydrogeologic system of Southern Poopó Basin, Bolivian Altiplano. J. Hazard. Mater. 262: 924–940.

Ramos, O.E., Cáceres, L.F., Ormachea, M., Bhattacharya, P., Quino, I., Quintanilla, J., Sracek, O., Thunvik, R., Bundschuh, J., García, M.E. 2012. Source and behavior of arsenic and trace elements in groundwater and sur-face water in the Poopó Lake Basin, Bolivian Altiplano. Environ. Earth Sci. 66: 793–807.

Vahter, M. 2002. Mechanisms of arsenic biotransformation. Toxicology 181–182: 211–217.

Vahter, M., Concha, G., Nermell, B., Nilsson, R., Dulout, F., Natarajan, A.T. 1995. A unique metabolism of inorganic arsenic in native Andean women. Eur. J. Pharmacol. 293: 455–462.

Van Den Bergh, K., Du Laing, G., Montoya, J.C., De Deckere, E., Tack, F.M. 2010. Arsenic in drinking water wells on the Bolivian high plain: Field monitoring and effect of salinity on removal efficiency of iron-oxides-containing filters. J Environ. Sci. Health A Tox. Hazard. Subst. Environ. Eng. 45: 1741–1749.

Table 1. Mean, median, 5th and 95th percentiles of arsenic concentrations in urine (µg/L, adjusted to mean specific gravity) by village of residence.

Village	Mean (n)	p50 (p5-p95)
Llapallapani	61.3 (22)	62.2 (30.1–104.5)
Pampa Aullagas	56.3 (18)	62.3 (24.4–83.1)
Quillacas	109.3 (34)	83.9 (34.3–244.3)

Arsenic Research and Global Sustainability – Bhattacharya, Vahter, Jarsjö, Kumpiene,
Ahmad, Sparrenbom, Jacks, Donselaar, Bundschuh & Naidu (Eds)
© *2016 Taylor & Francis Group, London, ISBN 978-1-138-02941-5*

Arsenic exposure and methylation efficiency during pregnancy and birth size

F. Harari[1], K. Broberg[1], M. Vahter[1] & E. Casimiro[2]
[1]*Institute of Environmental Medicine, Karolinska Institutet, Stockholm, Sweden*
[2]*Hospital Dr. Nicolás Cayetano Pagano, San Antonio de los Cobres, Salta, Argentina*

ABSTRACT: Variation in inorganic arsenic methylation and toxicity is mainly mediated by AS3MT polymorphisms. Arsenic exposure is associated with adverse pregnancy outcomes but the influence of AS3MT genotype has not been investigated. We aimed at elucidating the impact of early-life arsenic exposure on birth size and the role of AS3MT polymorphisms, in a population-based mother-child cohort in northern Argentina ($n = 148$) exposed to arsenic from drinking water. We measured arsenic in maternal blood (total) and urine (iAs, MMA and DMA) using ICPMS and HPLC-HG-ICPMS, respectively. We inferred haplotypes from 4 AS3MT SNPs. Outcomes were birth weight, length and head circumference. We found inverse associations between arsenic exposure during pregnancy and birth weight in boys and in newborns to mothers carrying haplotypes associated with less efficient arsenic methylation. In conclusion, arsenic exposure during pregnancy seemed to impair birth size, with increased susceptibility in boys and newborns of mothers with poorer arsenic methylation efficiency.

1 INTRODUCTION

Inorganic arsenic is a toxic and carcinogenic metalloid found at elevated concentrations in drinking water in many regions worldwide (IARC, 2012). As a consequence, many people are continuously exposed to inorganic arsenic. In humans, inorganic arsenic is metabolized by a series of methylation reactions, converting it to mono- (MMA) and dimethylated (DMA) metabolites, which are excreted in urine (Vahter, 2002). The methyl transfer is mainly accomplished by the arsenic (+3 oxidation state) methyltransferase (AS3MT), but also by other methyltransferases, and polymorphisms in these genes contribute substantially to the arsenic methylation efficiency (Engstrom *et al.*, 2011; Harari *et al.*, 2013).

Arsenic easily crosses the placenta (Concha *et al.*, 1998), exposing the fetus to considerable concentrations. Exposure to arsenic during pregnancy has been associated with an increased risk of spontaneous abortion, neonatal and infant mortality as well as with a decreased birth weight (Kippler *et al.*, 2012). However, little is known about the role of inorganic arsenic methylating genes on these associations.

The aim of this study is to elucidate the impact of arsenic exposure from drinking water during pregnancy on birth size in a population with a higher prevalence of an efficient arsenic methylating genotype.

2 MATERIALS AND METHODS

2.1 Study population

This study is part of an ongoing population-based mother-child cohort on health effects of early-life exposure to drinking water pollutants (i.e. arsenic, lithium and boron) in Los Andes Department (~3800 m above sea level), Province of Salta, northern Argentina. Pregnant women were recruited ($n = 148$, 88% participation rate) and interviewed about their personal characteristics, including e.g. age, parity, smoking, dietary preferences, diseases and drinking water sources. Gestational age was calculated based on date of the last menstrual period. Blood and spot-urine samples were collected during each visit (1st, 2nd and 3rd trimesters) and kept frozen at $-80°$ C until analysis.

2.2 Arsenic exposure

Arsenic exposure was assessed based on the arsenic concentrations in blood and urine. Total blood arsenic concentrations were measured by Inductively Coupled Plasma Mass Spectrometry (ICPMS), diluting aliquots (0.2 mL) of the blood samples 1:25 with an alkali solution (Lu *et al.*, 2015). The relative concentrations of arsenic metabolites in urine (%iAs, %MMA and %DMA) were measured by high-performance liquid chromatography coupled with hydride generation and ICP-MS (HPLC-HG-ICPMS) (Harari *et al.*,

2013). All measurements were performed at Karolinska Institutet, Sweden.

2.3 AS3MT haplotype

Four SNPs of *AS3MT* were analyzed using Taqman® SNP genotyping assays on a fast real-time PCR System (7900HT, Applied Biosystems, CA, USA) and haplotypes were inferred by PHASE software (Harari *et al.*, 2013).

2.4 Birth outcomes

Birth anthropometric measures included weight, length and head circumference, all obtained within a few hours after birth for all women. Birth weight was measured using a baby balance (Seca 725 Mechanical Beam Baby Scale, Brooklyn, NY, USA), birth length using a portable wood infantometer to the nearest 5 mm with the child in supine position, and head circumference with a soft, non-stretchable plastic tape line.

3 RESULTS AND DISCUSSION

3.1 General characteristics

In total, 194 women were recruited, out of whom 148 had data on exposure in the 3rd trimester (on average GW 33) and birth outcomes. The general characteristics are presented in Table 1. None reported cigarette smoking or alcohol intake.

3.2 Arsenic exposure, birth outcomes and AS3MT

Our preliminary analyses based on linear regression models adjusted for gestational age at birth, maternal age and height, parental monthly income, blood lithium and cesium, serum boron, infant sex

Table 1. Characteristics of the participating pregnant women in the 3rd trimester.

Characteristics	Median	Range
Age (years)	24	13–41
Parity (n)	1	0–12
Height (cm)	153	134–169
Weight (kg)	62	46–93
Years of education	9	0–17
Gestational week	33	27–41
Total urinary arsenic (μg/L)	129	13–1,826
MMA (%)	3.8	1.3–7.4
DMA (%)	89	55–95
iAs (%)	7.2	2.5–43
Blood arsenic (μg/L)	2.2	0.42–14

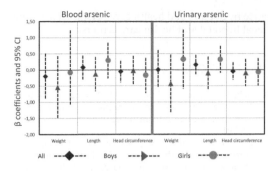

Figure 1. β coefficients and 95% CIs of associations between arsenic exposure (log2-transformed blood and urinary concentrations) in the 3rd trimester of pregnancy and pregnancy outcomes (birth weight, length and head circumference) in all newborns as well as stratified by infant sex.

and *AS3MT* haplotype, showed no statistically significant associations between arsenic exposure in the 3rd trimester of pregnancy and birth weight, length or head circumference. However, when stratifying by infant sex, the associations in boys were inverse for birth weight and length (Fig. 1), although nonsignificant possibly related to the small sample size. Similar results were previously found in a Bangladeshi cohort (Kippler *et al.*, 2012).

In this study, the two major *AS3MT* haplotypes were GCTC (74% of frequency) and TGTT (21% of frequency). After stratifying the multivariable-adjusted linear regression models by the *AS3MT* haplotypes, associations between maternal urinary iAs metabolite and birth weight and length were inverse ($p = 0.09$) for those with a less efficient arsenic metabolism (higher MMA% in urine). No associations were found for those with efficient arsenic metabolism. Analyses concerning the influence of the newborns' *AS3MT* haplotype are ongoing.

4 CONCLUSIONS

In this study, we observed inverse associations between arsenic exposure and birth size in boys as well as in newborns to mothers with less efficient arsenic metabolism. Data regarding the infants' arsenic efficiency and its influence in these associations will be presented.

ACKNOWLEDGEMENTS

We appreciate the participation of the mothers from San Antonio de los Cobres and the assistance by doctors and health care personnel from the local hospital and health care centers.

REFERENCES

Concha, G., Vogler, G., Lezcano, D., Nermell, B., Vahter, M. 1998. Exposure to inorganic arsenic metabolites during early human development. *Toxicol. Sci.* 44(2):185–190.

Engstrom, K., Vahter, M., Mlakar, S.J., Concha, G., Nermell, B., Raqib, R., et al. 2011. Polymorphisms in arsenic(+III oxidation state) methyltransferase (AS3MT) predict gene expression of AS3MT as well as arsenic metabolism. *Environ. Health Perspect.* 119(2):182–188.

Harari, F., Engstrom, K., Concha, G., Colque, G., Vahter, M., Broberg, K. 2013. N-6-adenine-specific DNA methyl-transferase 1 (N6 AMT1) polymorphisms and arsenic methylation in Andean women. *Environ. Health Perspect.* 121(7):797–803.

IARC. 2012. Arsenic, metals, fibres, and dusts. A review of human carcinogens. IARC Monographs, Lyon, France.

Kippler, M., Wagatsuma, Y., Rahman, A., Nermell, B., Persson, L.A., Raqib, R., et al. 2012. Environmental exposure to arsenic and cadmium during pregnancy and fetal size: a longitudinal study in rural Bangladesh. *Reprod. Toxicol.* 34(4): 504–511.

Lu, Y., Kippler, M., Harari, F., Grander, M., Palm, B., Nordqvist, H., et al. 2015. Alkali dilution of blood samples for high throughput ICP-MS analysis-comparison with acid digestion. *Clin. Biochem.* 48(3):140–147.

Vahter, M. 2002. Mechanisms of arsenic biotransformation. *Toxicology* 181–182:211–217.

Arsenic Research and Global Sustainability – Bhattacharya, Vahter, Jarsjö, Kumpiene,
Ahmad, Sparrenbom, Jacks, Donselaar, Bundschuh & Naidu (Eds)
© 2016 Taylor & Francis Group, London, ISBN 978-1-138-02941-5

Factors influencing arsenic metabolism in children

H. Skröder Löveborn[1], M. Kippler[1], Y. Lu[1], S. Ahmed[1,2], R. Raqib[2],
D. Kuehnelt[3] & M. Vahter[1]

[1]*Institute of Environmental Medicine, Karolinska Institutet, Stockholm, Sweden*
[2]*International Center for Diarrhoeal Disease in Bangladesh, Dhaka, Bangladesh*
[3]*Institute of Chemistry, Analytical Chemistry, NAWI Graz, University of Graz, Graz, Austria*

ABSTRACT   Inorganic arsenic (iAs) is metabolized through reduction and methylation. Incomplete methylation to Dimethylarsinic Acid (DMA), with elevated urinary excretion of Methylarsonic Acid (MMA), is a susceptibility factor in adults. However, little is known for children. In our mother-child cohort study in rural Bangladesh, we repeatedly measured urinary arsenic metabolites (U-As) in mothers and children (1.5, 4.5 and 9 years), and evaluated main factors influencing the children's metabolism. We also measured certain toxic and essential elements in children's blood. Children had higher %DMA than their mothers, unrelated to the children's lower exposure. Nutritional status, especially factors related to the one-carbon metabolism, was influential for the children's arsenic methylation. Also selenium status, assessed as erythrocyte selenium, appeared highly influential. Available data indicate that the factors influencing children's urinary arsenic metabolite pattern differ from that in adults. Whether arsenic methylation efficiency is a susceptibility factor for arsenic toxicity in children needs further evaluation.

1 INTRODUCTION

Naturally occurring arsenic in drinking water, mainly well water, is a growing public health concern globally. Also, the presence of arsenic in rice is of increasing concern, considering the number of people exposed. Inorganic arsenic is a potent toxicant and carcinogen. While adverse health effects in adults are well-documented, consequences of early-life exposure are not well researched. Our ongoing studies in rural Bangladesh, based on a large mother-child cohort, have prospectively evaluated children's exposure to arsenic through drinking water and associated health effects. Maternal arsenic exposure during pregnancy was found to be associated with impaired child growth and development and increased risk of infant mortality and morbidity, especially in infectious diseases. In support, arsenic was found to be immunosuppressive in the children.

A well-established susceptibility factor for arsenic-related adverse health effects in adults is inefficient arsenic metabolism (Lindberg *et al.*, 2008a). Arsenic is methylated in the body via the one-carbon metabolism, and the main metabolites excreted in the urine are Methylarsonic Acid (MMA) and Dimethylarsinic Acid (DMA), besides some remaining inorganic arsenic (iAs). Higher fractions of iAs and MMA in the urine have been associated with higher retention of arsenic (Vahter, 2002) and increased risk of adverse health effects

(Lindberg, *et al.*, 2008a). Multiple factors have been shown to influence the metabolism of arsenic in adults, including polymorphisms in the arsenic (+III oxidation state) methyltransferase (AS3MT) gene (Engstrom *et al.*, 2011), gender, nutrition, smoking and pregnancy (Lindberg *et al.*, 2008b, Howe *et al.*, 2014, Lindberg *et al.*, 2010, Gardner *et al.*, 2012).

Because little is known about the metabolism as a susceptibility factor for arsenic toxicity in children, we have evaluated the urinary metabolites of arsenic in children, including influencing factors.

2 METHODS

Our mother-child cohort on arsenic exposure is established in Matlab, 50 km southeast of the capital of Bangladesh, Dhaka, where most people use drinking water from local wells with considerable variations in arsenic concentrations (Vahter *et al.*, 2006). Our longitudinal mother-child cohort was nested in a large randomized food and micronutrient supplementation trial in pregnancy (Vahter *et al.*, 2006), and the children's exposure to arsenic and other toxic agents have been followed from early gestation until 10 years of age.

For evaluation of the exposure to and metabolism of arsenic, and associate health effects, we measured iAs and its metabolites in urine (U-As) of a subsample of the mothers during pregnancy and their children at 1.5, 4.5 and 9 years of age. For analysis of arsenic metabolites in urine, we used HPLC on line with hydride generation and ICPMS (Fängström et al., 2009, Gardner et al., 2011). For compensation of differences in urine dilution, we adjusted the obtained concentrations by the average specific gravity (1.012) of the urine samples. We also measured zinc, iron, magnesium, selenium, and manganese in erythrocytes by ICPMS following dilution in an alkaline solution (Lu et al., 2015).

For evaluation of the efficiency of arsenic methylation we calculated the relative amounts of the arsenicals in urine (%iAs, MMA and DMA). We evaluated these in relation to age, U-As, genetic predisposition, season of sampling, Socio-Economic Factors (SES) and nutritional factors, using linear regression analysis.

3 RESULTS AND DISCUSSION

Our preliminary evaluations show that the children at had lower U-As (at 9 years, median 53, range 9–1,268 µg/L) than their mothers at GW8 (77 µg/L, range 2–2,063 µg/L; $p < 0.001$). This was largely associated with installation of deeper wells with lower arsenic concentrations. The children also had significantly lower %iAs and higher %DMA in urine collected at 9 years, compared to that of their mothers during pregnancy ($p < 0.001$) (Skröder et al., 2016). This was true also when considering the difference in U-As. At 9 years, the children also had slightly lower %MMA compared to their mothers ($p < 0.001$). Similar pattern of urinary arsenic metabolites was observed at 4.5 and 1.5 years of age (Fängström et al., 2009, Skröder et al., 2016).

In the assessment of factors influencing the metabolism of arsenic in the children at 9 years, the factors positively associated with %iAs and inversely associated with %DMA in urine included Ery-Se and season of sampling (Skröder et al., 2016). On the other hand, plasma folate and urinary selenium were inversely associated with%iAs and positively associated with %DMA. Also, the children's %iAs and %DMA were positively associated with the corresponding metabolite fractions in their mothers.

When assessing factors influencing %MMA in children's urine, we found positive associations with family SES, maternal education, maternal %MMA, and child U-As and selenium status. Also, season of sampling seemed influential.

The observed association between the children's urinary pattern of arsenic metabolites and that of their mothers suggests genetic predisposition, most likely through the genetic variation in the As3MT gene (Engstrom et al., 2011). Still, the children

appeared to have more efficient methylation of arsenic to DMA compared to their mothers, in line with other studies (Fangstrom et al., 2009, Hall et al., 2009, Lindberg et al., 2008b).

4 CONCLUSION

In conclusion, nutritional factors, including those involved in one-carbon metabolism, appeared to be influential for arsenic methylation efficiency in Bangladeshi children. Whether arsenic methylation efficiency is a risk factor for arsenic toxicity in children needs further evaluation.

ACKNOWLEDGEMENTS

We acknowledge support from Swedish Research Councils VR and Formas and the Swedish development Cooperation (Sida), as well as the MINI-Mat food micronutrient supplementation trial (PIs LA Persson, Uppsala University, and S El Arifeen, icddr,b, Dhaka).

REFERENCES

Engstrom, K., Vahter, M., Mlakar, S.J., Concha, G., Nermell, B., Raqib, R., Cardozo, A., Broberg, K. 2011. Polymorphisms in arsenic(+III oxidation state) methyltransferase (AS3MT) predict gene expression of AS3MT as well as arsenic metabolism. *Environ. Health Perspect.* 119(2): 182–188.

Fängström, B., Hamadani, J., Nermell, B., Grander, M., Palm, B., Vahter, M. 2009. Impaired arsenic metabolism in children during weaning. *Toxicol. Appl. Pharmacol.* 239(2): 208–214.

Gardner, R., Hamadani, J., Grander, M., Tofail, F., Nermell, B., Palm, B., Kippler, M., Vahter, M. 2011. Persistent exposure to arsenic via drinking water in rural Bangladesh despite major mitigation efforts. *Am. J. Public Health* 101 Suppl 1: S333–338.

Gardner, R.M., Engstrom, K., Bottai, M., Hoque, W.A., Raqib, R., Broberg, K., Vahter, M. 2012. Pregnancy and the methyltransferase genotype independently influence the arsenic methylation phenotype. *Pharmacogenet. Genomics* 22(7): 508–516.

Hall, M.N., Liu, X., Slavkovich, V., Ilievski, V., Pilsner, J.R., Alam, S., Factor-Litvak, P., Graziano, J.H., Gamble, M.V. 2009. Folate, Cobalamin, Cysteine, Homocysteine, and Arsenic Metabolism among Children in Bangladesh. *Environ. Health Perspect.* 117(5): 825–831.

Howe, C.G., Niedzwiecki, M.M., Hall, M.N., Liu, X., Ilievski, V., Slavkovich, V., Alam, S., Siddique, A.B., Graziano, J.H., and Gamble, M.V. (2014). Folate and cobalamin modify associations between S-adenosylmethionine and methylated arsenic metabolites in arsenic-exposed Bangladeshi adults. *J. Nutr.* 144(5): 690–697.

Lindberg, A.L., Ekstrom, E.C., Nermell, B., Rahman, M., Lonnerdal, B., Persson, L.A., Vahter, M. 2008b. Gender and age differences in the metabolism of inorganic arsenic in a highly exposed population in Bangladesh. *Environ. Res.* 106(1): 110–120.

Lindberg, A.L., Rahman, M., Persson, L.A., Vahter, M. 2008a. The risk of arsenic induced skin lesions in Bangladeshi men and women is affected by arsenic metabolism and the age at first exposure. *Toxicol. Appl. Pharmacol.* 230(1): 9–16.

Lindberg, A.L., Sohel, N., Rahman, M., Persson, L.A., Vahter, M. 2010. Impact of smoking and chewing tobacco on arsenic-induced skin lesions. *Environ. Health Perspect.* 118(4): 533–538.

Lu, Y., Kippler, M., Harari, F., Grander, M., Palm, B., Nordqvist, H., Vahter, M. 2015. Alkali dilution of blood samples for high throughput ICP-MS analysis-comparison with acid digestion. *Clin. Biochem.* 48(3): 140–147.

Skröder, H., Kippler, M., Lu, Y., Ahmed, S., Kuehnelt, D., Raqib, R., Vahter, M. 2016. Arsenic metabolism in children differs from that in adults. *Toxicol Sci.* Accepted.

Vahter, M. 2002. Mechanisms of arsenic biotransformation. *Toxicology* 181–182, 211–217.

Vahter, M.E., Li, L., Nermell, B., Rahman, A., El Arifeen, S., Rahman, M., Persson, L.A., Ekstrom, E.C. 2006. Arsenic exposure in pregnancy: a population-based study in Matlab, Bangladesh. *J. Health Popul. Nutr.* 24(2): 236–245.

Arsenic Research and Global Sustainability – Bhattacharya, Vahter, Jarsjö, Kumpiene,
Ahmad, Sparrenbom, Jacks, Donselaar, Bundschuh & Naidu (Eds)
© 2016 Taylor & Francis Group, London, ISBN 978-1-138-02941-5

AS3MT: Mechanism

B.P. Rosen, H. Dong & J.J. Li
Department of Cellular Biology and Pharmacology, Herbert Wertheim College of Medicine, Florida International University, Miami, Florida, USA

ABSTRACT: Arsenic is the most ubiquitous environmental toxin and carcinogen. Human As(III) S-adenosylmethionine (SAM) methyltransferases (hAS3MT) methylates As(III) to trivalent methylated species that are more toxic and potentially more carcinogenic than inorganic arsenic. A synthetic hAS3MT gene was expressed, and the purified enzyme characterized. A new catalytic pathway is proposed in which the products are more toxic trivalent methylarsenicals, but arsenic undergoes oxidation and reduction as enzyme-bound intermediates.

1 INTRODUCTION

In humans, AS3MT catalyzes transfer of methyl groups from SAM to As(III), producing primarily MAs and DMAs (Thomas & Rosen, 2013). AS3MT is a member of the large superfamily of SAM methyltransferases (MTs) and is an example of a "Phase II" enzyme that is simultaneously protective but also activates arsenic into more toxic and potentially carcinogenic organoarsenicals species (Styblo *et al.*, 2000). AS3MT transforms As(III) into trivalent methylated species methylarsenite (MAs(III)) and dimethylarsenite (DMAs(III)) (Dheeman *et al.*, 2014). The trivalent methylarsenicals are excreted in urine, where they oxidize to pentavalent methylarsenate (MAs(V)) and dimethylarsenate (DMAs(V)) (Le *et al.*, 2000).

The pathway of methylation remains controversial. Challenger (Challenger, 1947) proposed a series of alternating oxidative methylations and reductions, generating the pentavalent products. Hayakawa and coworkers (Hayakawa *et al.*, 2005) proposed that the products are the trivalent that rapidly oxidize non-enzymatically to the observed urinary species. Our proposed pathway is consistent with Hayakawa's proposal but includes enzyme-bound pentavalent intermediates similar to those predicted by Challenger. hAS3MT has four conserved cysteine residues, Cys32, Cys61, Cys156 and Cys206. Each is involved in methylating As(III), but only Cys156 and Cys206, which form the arsenical binding site, are required for methylation of MAs(III). Cys32 and Cys61 reduce the pentavalent intermediates, forming a disulfide bond that is re-reduced by Trx to regenerate the active form of the enzyme.

2 METHODS/EXPERIMENTAL

2.1 *AS3MT activity*

A hAS3MT gene corresponding to the sequence of the cDNA clone, which lacks the last nine residues of the hAS3MT sequence, was chemically synthesized. Significant features of the chemical synthesis: 1) there was no change in the amino acid sequence of the protein; 2) the codon usage was optimized for expression in *E. coli*; 3) the GC content was optimized, and stem-loop structures were disrupted to increase ribosome binding and the half-life of the mRNA. AS3MT and its mutants were expressed in *E. coli* BL21(DE3). AS3MT activity was determined in a low volume 384-well microtiter plates in a Synergy H4 Hybrid Multi-Mode Microplate Reader (Dong *et al.*, 2015). The Homogeneous Time-Resolved Fluorescence (HTRF) was calculated from the ratio of emission at 665 nm and 620 nm. The concentration

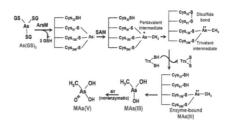

Figure 1. First methylation step in the hAS3MT catalytic cycle. Key features: 1) The substrate is As(GS)₃. 2) There are both penta- and trivalent intermediates. 3) Formation of a disulfide bond reduces the pentavalent intermediate. 4) The first product, MAs(III), slowly dissociates from AS3MT, followed by nonenzymatic oxidation in air to MAs(V). The second methylation step (not shown) forms DMAs(III), which is also nonenzymatically oxidized to MAs(V).

of SAH was calculated with a calibration curve constructed with known concentrations of SAH. HPLC-ICP-MS analysis of arsenic methylation was described (Dong *et al.*, 2015) with a C18 300 A reverse-phase column, with arsenic measured by Inductively Coupled Plasma Mass Spectrometry (ICP-MS) using an ELAN 9000.

3 RESULTS AND DISCUSSION

3.1 *The catalytic mechanism of AS3MT*

All AsMTs have four conserved cysteine residues, which are Cys32, Cys61, Cys156 and Cys206 in hAS3MT. A goal has been to elucidate their role in catalysis. All four cysteines are required for the first methylation step (As(III)→MAs(III)), but Cys32 and Cys61 are not required for the second methylation step (MAs(III)→DMAs(III)). In various crystal structures of the algal ortholog, the first cysteine pair forms a disulfide bond, while the second pair binds trivalent arsenicals. Thus the C-terminal cysteine pair forms the metalloid binding site, while the N-terminal cysteine pair play a different role in catalysis that we postulate is to keep arsenic reduced during the methylation pathway.

We propose a distinctive new hypothesis that explains how arsenic is transformed into more toxic and potentially more carcinogen species (Dheeman *et al.*, 2014, 2015). Figure 1 shows the key features of the first methylation step (As(III)→MAs(III)). Novel features of our model is that the substrates and products are trivalent, but there are transient pentavalent enzyme-bound intermediates in each step that are reduced by conserved cysteine residues, creating a disulfide bond cascade mechanism of alternating oxidations and reductions of the bound arsenic. Human AS3MT has long been considered a special class of enzyme, but it's really not. We predict that all AsMTs employ a basic catalytic mechanism similar to that of O-, N-, C- and S-methyltransferases. What differentiates AsMTs from other members of the superfamily is the necessity to bind trivalent arsenicals and maintain them in reduced forms, for which they evolved and utilize four conserved cysteine residues that are absent in other MTs.

4 CONCLUSIONS

Members of the very large superfamily of methyltransferase enzymes append a methyl group from SAM to acceptor groups by S_N2 displacement mechanisms, involving attack of a nucleophile on the methyl group of SAM with inversion of configuration and concomitant release of SAH. Methyltransferases are categorized based on the electron-rich, methyl accepting atom, usually O, N, C, or S. These enzymes all have a conserved SAM binding fold and use a common enzymatic mechanism of methyltransfer, where SAM serves as a potent alkylating agent.

While Challenger proposed a series of oxidations and reductions along with three rounds of methylation, none of the O-, N-, C- and S-methyltransferases oxidizes their substrates. The evidence that AS3MT directly generates As(V) metabolites is not strong. Lysine and arginine methyltransferases are good models for arsenic methyltransferases as they carry out three very similar methyl transfers without oxidation of the nitrogen. In summary, MTs methylate their substrates without oxidation and without unfavorable leaving groups. Nature is conservative, and there is no reason to consider that AS3MT would use a different enzymatic mechanism from any other member of the superfamily. Yet, a major chemical difference between the metalloid arsenic and the nonmetal atoms (O, N, C, S) acceptors of the methyl group is its redox activity. If, as we propose, AS3MT catalyzes formation of oxidized intermediates, there must be a mechanism to reduce the intermediate to produce the trivalent products. We propose disulfide bonds form during each methylation step and are reduced by conserved cysteine residues. The role of Trx is to reduce those disulfide bonds. This is a novel feature of AS3MT that differentiates it from other members of the methyltransferase superfamily.

ACKNOWLEDGEMENTS

This work was supported by National Institutes of Health grants R01 ES023779 and R37 GM55425.

REFERENCES

Challenger, F. 1947. Biological methylation. *Sci. Prog.* 35(139): 396–416.
Dheeman, D.S., Packianathan, C., Pillai, J.K., Rosen, B.P. 2014. Pathway of human AS3MT arsenic methylation. *Chem. Res. Toxicol.* 27(11): 1979–89.
Dong, H., Xu, W., Pillai, J.K., Packianathan, C., Rosen, B.P. 2015. High-throughput screening-compatible assays of As(III) S-adenosylmethionine methyltransferase activity. *Anal. Biochem.* 480: 67–73.
Hayakawa, T., Kobayashi, Y., Cui, X., Hirano, S. 2005. A new metabolic pathway of arsenite: arsenic-glutathione complexes are substrates for human arsenic methyltransferase Cyt19. *Arch. Toxicol.* 79(4): 183–91.
Le, X.C., Lu, X., Ma, M., Cullen, W.R., Aposhian, H.V., Zheng, B. 2000. Speciation of key arsenic metabolic intermediates in human urine. *Anal. Chem.* 72(21): 5172–7.
Marapakala, K., Packianathan, C., Ajees, A.A., Dheeman, D.S., Sankaran, B., Kandavelu, P., Rosen, B.P. 2015. A disulfide-bond cascade mechanism for As(III) S-adenosylmethionine methyltransferase. *Acta Crystallogr. D. Biol. Crystallogr.* 71(Pt 3): 505–15.
Styblo, M., Del Razo, L.M., Vega, L., Germolec, D.R., LeCluyse, E.L., Hamilton, G.A., Reed, W., Wang, C., Cullen, W.R., Thomas, D.J. 2000. Comparative toxicity of trivalent and pentavalent inorganic and methylated arsenicals in rat and human cells. *Arch. Toxicol.* 74(6): 289–99.
Thomas, D.J., Rosen, B.P. 2013. Arsenic methyltransferases. In: R.H. Kretsinger, V.N. Uversky, E.A. Permyakov (Eds.) *Encyclopedia of Metalloproteins.* New York, Springer New York: 138–143.

Arsenic Research and Global Sustainability – Bhattacharya, Vahter, Jarsjö, Kumpiene, Ahmad, Sparrenbom, Jacks, Donselaar, Bundschuh & Naidu (Eds)
© 2016 Taylor & Francis Group, London, ISBN 978-1-138-02941-5

AS3MT: Inhibitors

B.P. Rosen, H. Dong, C. Packianathan & M. Madegowda
Department of Cellular Biology and Pharmacology, Herbert Wertheim College of Medicine, Florida International University, Miami, Florida, USA

ABSTRACT: Modulators of hAS3MT activity may be useful for prevention or treatment of arsenic-related diseases. Using a new high-throughput assay for hAS3MT, we identified ten novel noncompetitive small molecule inhibitors. *In silico* docking analysis indicate that the inhibitors bind in a cleft distant from the As(III) or SAM binding sites, implying the presence of a regulatory site. Newer bisubstrate inhibitors are being developed that will bridge the SAM and As(III) binding sites. These inhibitors may be useful tools for research in arsenic metabolism and are the starting-point for development of drugs against hAS3MT.

1 INTRODUCTION

We developed a rapid and sensitive microplate assay to identify modulators of AS3MT activity (Dong *et al.*, 2015). We screened a library of more than 30 million small molecules and identified 10 inhibitor of hAS3MT activity with IC_{50} values of 30–50 μM (Dong, Madegowda, Nefzi, Houghten, Giulianotti and Rosen 2015). None inhibits binding of either As(III) or MAs(III), and they do not inhibit catachol o-methyltransferase (COMT), a non-arsenic SAM MT and so appear to be non-competitive inhibitors selective for AS3MT. *In silico* docking analysis of the inhibitors to the crystal structure of an AS3MT ortholog (Ajees *et al.*, 2012) indicates that small molecules bind in a cleft in the enzyme located distant from the As(III) and SAM binding sites. The inhibitor binding location may be an allosteric regulatory site, which implies the existence of physiological regulators of arsenic methylation. These are the first identified selective small molecule inhibitors of hAS3MT. New bisubstrate molecules are being developed that will bind to both the SAM and As(III) binding sites. These inhibitors provide a scaffold for future rational design of clinically-useful drugs that modulate hAS3MT activity and may also be useful probes for research into AS3MT function.

2 METHODS/EXPERIMENTAL

2.1 *Homology modeling and in silico docking analysis of hAS3MT*

A homology model of hAS3MT was built on the structure of the PhAs(III)-bound algal ortholog (4KU9) using the homology modeling server

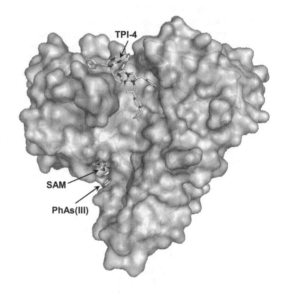

Figure 1. *In silico* binding analysis of inhibitor TPI-4 to AS3MT. TPI-4 was docked with the crystal structure of an AS3MT ortholog (4FS8). The surface plot shows TPI-4, SAM and phenylarsenite in stick form.

SWISS-MODEL (http://swissmodel.expasy.org/). The 3D structure of small molecule inhibitors were calculated as described (Dong *et al.*, 2015). The inhibitors were docked using the ADVina. Predicted molar dissociation constants (K_d) were calculated from the ΔG values using the relationship

$$K_d = exp(\Delta G/RT) \qquad (1)$$

where *R* is 1.9872 and *T* is 298.15 K. Molecular models were generated using PyMol.

3 RESULTS AND DISCUSSION

3.1 *Identification of small molecule inhibitors of AS3MT*

A newly developed TR-FRET assay for AS3MT activity that is both rapid and highly sensitive was used for high throughput screening of potential AS3MT inhibitors (Dong *et al.*, 2015) utilizing the Torrey Pines Institute for Molecular Studies (TPIMS) Scaffold Ranking Library, which contained over 30 million synthetic compounds systematically arranged into 70 samples (Houghten *et al.*, 2008, Wu *et al.*, 2013). A set of individual compounds all containing a bisguanidine pyrrolide core with differing R groups was screened. Ten TPIMS compounds inhibited hAS3MT with IC_{50} values in the range of 30 to 50 µM (Dong *et al.*, 2015). We determined the effect of the small molecule inhibitors on the first and second methylation steps individually. The TR-FRET assay measures primarily the first methylation step, and each of the ten compounds inhibit. By HPLC-ICP-MS the effect of the inhibitors on the second methylation step was determined. When the substrate is As(III), DMAs is the primary final compound, a combination of both the first and second methylation steps. However, when MAs(III) is used as substrate, only the second methylation step occurs. Five of the small molecules inhibited both As(III) and MAs(III) methylation. In contrast, the other five did not inhibit MAs(III) methylation. These results suggest that all 10 TPIMS compounds inhibited the first methylation step (As(III) → MAs(III)), while only five are effectively inhibited the second step (MAs(III) → DMAs(III)).

3.2 *In silico docking analysis inhibitor binding to AS3MT*

A virtual screening approach using Autodock Vina was applied to calculate the lowest free energy binding site for the 10 small molecule inhibitors (Dong *et al.*, 2015). The crystal structure of the AS3MT ortholog from *C. merolae*, was used for the docking studies (Ajees *et al.*, 2012). It has three domains, an N-terminal domain with the SAM binding site, a middle domain with the As(III) binding site, and a C-terminal domain of unknown function. Both the As(III) binding site and C-terminal domain are unique to AS3MTs and are not found in other SAM MTs. All 10 TPIMS compounds bound in a cleft between the first and third domains located on the opposite side of the protein from the As(III) and SAM binding sites (shown with TPI-4 in Figure 2).

4 CONCLUSIONS

A new high-throughput assays and crystal structures have allowed identification of synthetic small molecule inhibitors of AS3MT. The inhibitors bind to a potential allosteric site on the surface of AS3MT distant from the active site. The suggests that there could be physiological regulation of arsenic methylation by cytosolic molecules. Newer bisubstrate inhibitors that span the distance between SAM and As(III) binding sites are being developed. Small molecule inhibitors are the starting point for development of drugs to prevent arsenic-related diseases.

ACKNOWLEDGEMENTS

This work was supported by National Institutes of Health grant R01 ES023779 and R37 GM55425. MM was supported by a Raman Postdoctoral Fellowship from the University Grants Commission, Government of India.

REFERENCES

Ajees, A.A., Marapakala, K., Packianathan, C., Sankaran, B. and Rosen, B.P. 2012. Structure of an As(III) S-adenosylmethionine methyltransferase: insights into the mechanism of arsenic biotransformation. *Biochemistry* 51(27): 5476–85.

Dong, H., Madegowda, M., Nefzi, A., Houghten, R.A., Giulianotti, M.A. and Rosen, B.P. 2015. Identification of small molecule inhibitors of human As(III) S-Adenosylmethionine Methyltransferase (AS3MT). *Chem. Res. Toxicol.* 28(12): 2419–25.

Dong, H., Xu, W., Pillai, J.K., Packianathan, C. and Rosen, B.P. 2015. High-throughput screening-compatible assays of As(III) S-adenosylmethionine methyltransferase activity. *Anal. Biochem.* 480: 67–73.

Houghten, R.A., Pinilla, C., Giulianotti, M.A., Appel, J.R., Dooley, C.T., Nefzi, A., Ostresh, J.M., Yu, Y., Maggiora, G.M., Medina-Franco, J.L., Brunner, D. and Schneider, J. 2008. Strategies for the use of mixture-based synthetic combinatorial libraries: scaffold ranking, direct testing in vivo, and enhanced deconvolution by computational methods. *J Comb. Chem.* 10(1): 3–19.

Wu, J., Zhang, Y., Maida, L.E., Santos, R.G., Welmaker, G.S., LaVoi, T.M., Nefzi, A., Yu, Y., Houghten, R.A., Toll, L. and Giulianotti, M.A. 2013. Scaffold ranking and positional scanning utilized in the discovery of nAChR-selective compounds suitable for optimization studies. *J. Med. Chem.* 56(24): 10103–17.

Arsenic Research and Global Sustainability – Bhattacharya, Vahter, Jarsjö, Kumpiene,
Ahmad, Sparrenbom, Jacks, Donselaar, Bundschuh & Naidu (Eds)
© 2016 Taylor & Francis Group, London, ISBN 978-1-138-02941-5

Advances on the analytical development of feasible techniques and validated methods for the determination of arsenic metabolites in urine in Uruguay

V. Bühl[1], C. Álvarez[2], M.H. Torre[3], M. Pistón[1] & N. Mañay[2]

[1]*Analytical Chemistry. DEC, Faculty of Chemistry, Universidad de la República (UdelaR), Montevideo, Uruguay*
[2]*Toxicology. DEC, Faculty of Chemistry, Universidad de la República (UdelaR), Montevideo, Uruguay*
[3]*Inorganic Chemistry. DEC, Faculty of Chemistry, Universidad de la República (UdelaR), Montevideo, Uruguay*

ABSTRACT: Exposure to arsenic (As) is considered a serious environmental public health problem in many countries, but in Uruguay has only been considered in occupational health. The aim of this work is to present the advances of the analytical development of feasible methodologies for the determination of toxicologically relevant species of As in both occupationally exposed and non- exposed populations, as an alternative for more expensive techniques such as HPLC-HG-ICPMS. These analytical methodologies are useful to have the scientific evidence of population exposure to low doses of arsenic in Uruguay in order to prevent long-term health effects and establish a basis for the development of new scientific research.

1 INTRODUCTION

Exposure to arsenic (As) is considered a serious environmental public health problem in many countries, but in Uruguay has only been considered in occupational health, although this element may be widespread in the environment and its inorganic species may have adverse effects on human health even at low levels of exposure.

In the Faculty of Chemistry in Uruguay, we are developing analytical technologies to assess and control exposure to toxic arsenic species in vulnerable populations. We also have background research and collaboration with national, regional and international leaders in this field (Mañay *et al.*, 2014).

In our country, there are no background studies on arsenic exposure, and the current legislation only takes into account the regular biomonitoring of arsenic in exposed workers. However, the risk of exposure in relation to all other non-occupational variables is not evaluated.

Thus we are now developing a multidisciplinary research project which aims to deepen the knowledge about the different factors that can affect arsenic toxicology, metabolism and speciation in relation to levels of inorganic As and methylated metabolites in urine as well as the assessment of the occupational and natural sources along with other environmental parameters and individual, including genotoxicity.

In this project's framework, we present the advances of the analytical development of feasi-ble techniques and validated methods to fill the gap of the "inorganic arsenic methylation challenges". This will provide analytical tools in order to further understand about the impact of As on exposed Uruguayan workers population health and the possible correlations of arsenic exposure not only in their workplace but also owing to the natural environmental exposure.

2 EXPERIMENTAL

2.1 *Determination of toxicologically relevant species of arsenic in urine using HG-AAS*

Sample preparation consisted of a derivation with L-cysteine of arsenic species. This method is based on the fact that As(III) + As(V) and the methylated species (MMA + DMA) react with L-cysteine in acid medium, generating tioderivated compounds that originate arsines at similar velocities, which allows quantification of the total amount without overestimating or underestimating other species. Thus sum of the four toxicologically relevant species is determined (As(III) + As(V) + MMA + DMA), being this sum essentially the 'hydride-reactive' fraction. To determine the sum of the four species, hydride generation technique based on the reduction of them to arsine (AsH_3) with $NaBH_4$ in acid medium (HCl) was used. This methodology was validated and published in a previous work (Bühl *et al.*, 2015).

2.2 Hyphenated technique: HPLC-HGAAS

The separation conditions of As (III), As (V), DMA and MMA species were optimzed on a conventional Shimadzu HPLC with UV diode array arrangement detector. The working conditions were as follows: λ = 193.7 nm, anion exchange column: Kinetex® PFP, 150 x 4.6 mm, mobile phase: 0.1% (v/v) HCOOH containing 1% (v/v) MeOH. Once the optimizations conditions were achieved, a separate pump module with manual control (Shimadzu) was coupled to the atomic absorption spectrometer (SpectrAA 55B Varian) with HG-AAS detection. The above mentioned mobile phase was used for the separation and for hydride generation the eluent from the column was mixed with 1% $NaBH_4$ and 3M HCl for subsequent determination.

3 RESULTS AND DISCUSSION

3.1 Determination of toxicologically relevant species of arsenic in urine using HG-AAS

A comparison between the validated methodology with the one considered as "gold technique" was made. The "gold technique" involves the coupling of a separation technique, an atomic emission technique and a mass detector (HPLC-ICPMS) for As speciation. This hyphenated instrumentation is very expensive and it is not yet available for routine analysis in Uruguay.

The comparison was performed analyzing 40 samples (from a previous project) by HPLC-ICPMS in the laboratory of Karolinska Institutet, Stockholm, Sweden and using the proposed method. The obtained results using both approaches appear to be statistically equal in the lowest range of exposure, from 5 to 40 µg/L which is adequate for monitoring purposes according the regulations (Bühl et al., 2015).

The validated method was applied for toxicological analysis in urine samples from exposed workers to arsenic to comply with international recommendations (ACGIH) risk prevention health effects.

3.2 Hyphenated technique: HPLC-HG-AAS

Under the optimized conditions described in 2.2 using conventional HPLC-UV, it is possible to evaluate the separation performance at ppm levels of As species. Figure 1 shows a chromatogram obtained in these conditions. The four species were separate in the following order: t: 2.230 min. - As(III), t: 2.447 min. - MMA, t: 2.638 min. DMA, t: 3.238 min As(V). But ppm levels are not adequate for the proposed objectives.

The coupled system developed in our laboratory (HPLC-HG-AAS) using the same conditions than with the conventional HPLC-UV was applied for the determination of As(III) and As(V) at ppb

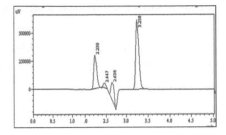

Figure 1. Chromatogram obtained using HPLC-UV detection prior to couple the pump and column to the HG-AAS system.

levels. The total run lasts 5 min with good separation of the inorganic species. The methodology is in development stage for the determinations in urine.

4 CONCLUSIONS

Economic and feasible analytical tools for the determination of the main metabolites to assess arsenic human exposure were developed and validated. The proposed methods has proven to be an alternative for more expensive techniques, such as HPLC-HG-ICP-MS, used for arsenic speciation in urine. These analytical methodologies are useful since through them we will have the scientific evidence of population exposure to low doses of arsenic in Uruguay in order to prevent long-term health effects and establish a basis for the development of new scientific research.

ACKNOWLEDGEMENTS

The research that gives rise to the results, received funding from Agencia Nacional de Investigación e Innovación (ANII FMV_3_2013_1_100439), PEDECIBA-Química.

REFERENCES

ACGIH, 2014. Threshold limit values for chemical substances and physical agents and biological exposure indices. American Conference of Governmental Industrial Hygienists, Cincinnati, OH

Bühl, V., Álvarez, C., Kordas, K., Pistón, M. and Mañay, N. 2015. Development of a simple method for the determination of toxicologically relevant species of arsenic in urine using HG-AAS. *Journal of Environment Pollution and Human Health* 3(2): 46–51.

Mañay, N., Pistón, M. and Goso, C. 2014. Arsenic environmental and health issues in Uruguay: A multidisciplinary approach. In: M.I. Litter, H.B. Nicolli, M. Meichtry, N. Quici, J. Bundschuh, P. Bhattacharya and R. Naidu (Eds.) "*One Century of the Discovery of Arsenicosis in Latin America (1914–2014) As 2014*". CRC Press/Taylor and Francis, 485–487.

Arsenic Research and Global Sustainability – Bhattacharya, Vahter, Jarsjö, Kumpiene,
Ahmad, Sparrenbom, Jacks, Donselaar, Bundschuh & Naidu (Eds)
© 2016 Taylor & Francis Group, London, ISBN 978-1-138-02941-5

A modified creatinine adjustment method to improve urinary biomonitoring of exposure to arsenic in drinking water

D.R.S. Middleton[1,2,3], M.J. Watts[2], T. Fletcher[3] & D.A. Polya[1]

[1]*School of Earth, Atmospheric and Environmental Sciences and Williamson Research Centre for Molecular Environmental Science, University of Manchester, Manchester, UK*
[2]*Inorganic Geochemistry, Centre for Environmental Geochemistry, British Geological Survey, Nicker Hill, Keyworth, Nottinghamshire, UK*
[3]*Centre for Radiation, Chemicals and Environmental Hazards (CRCE), Public Health England, Chilton, Didcot, Oxfordshire, UK*

ABSTRACT: Spot urinary concentrations of arsenic (As) require adjustment for hydration variation. There is no consensus as to the most appropriate method. We compared the performance of creatinine (Cre), specific gravity (SG), osmolality and a modified variant of Cre adjustment on spot urinary As concentrations of 203 volunteers from the UK. Strength in correlation between drinking water total As and urinary non-arsenobetaine As was used to indicate good performance. Performance of the order: Cre (modified) > SG > osmolality > unadjusted > Cre (routine) was observed. Only Cre (modified) adjustment performed significantly better than unadjusted concentrations and both Cre (modified) and SG performed significantly better than Cre (routine). This is the first study to compare the performance of multiple hydration adjustments of urinary As on the basis of reflectance of external exposure and, importantly, allows others to apply this methodology to existing data without further measurements.

1 INTRODUCTION

Urinary biomonitoring is routine for assessing recent exposure to arsenic (As) given its rapid excretion from the body (2–4 days post-ingestion) (Orloff *et al.*, 2009). Twenty-four hour sampling is often not feasible and spot urine samples require adjustment for inter-volunteer differences in hydration i.e. urinary dilution (Aylward *et al.*, 2014). The Google Scholar search term: "*arsenic*; *urine*; (one of) *creatinine, specific gravity, osmolality*; *adjustment OR correction*" yielded 5070, 2840 and 625 results for creatinine (Cre), specific gravity (SG) and osmolality, respectively. This reflects the wide use of Cre adjustment, relative to alternative adjustment methods. Creatinine adjustment has received criticism in the literature due to demographic and nutritionally derived variations in creatinine excretion (Barr *et al.*, 2005). Less attention has been paid to the mathematical application of creatinine adjustment, routinely performed:

$$As_{adjusted} = As_{unadjusted}/Cre_{specimen}, \quad (1)$$

with results commonly expressed as As µg/g Cre (The term "*arsenic*; *"/g cre"* OR *"/g cr"* OR *"/g creatinine"* yielded 3320 results). This assumes a proportional change in the concentration of both As and Cre in response to change in Urinary Flow Rate (UFR). This has been disputed by findings that a range of urinary analytes exhibit different concentration changes in response to UFR and adjustments require the inclusion of a coefficient to account for this relationship (Araki *et al.*, 1990). We, therefore, propose a modification of the Cre adjustment, based on previous work (Vij & Howell, 1998) as follows:

$$As_{adjusted} = As_{unadjusted} \times (Cre_{ref}/Cre_{specimen})^z \quad (2)$$

where: Cre_{ref} is the median Cre concentration of the study group and z is a coefficient, empirically derived by dividing the regression slope (b) of As against UFR by that of Cre against UFR (a) hence $z = b/a$. We test the performance of modified Cre adjustment against alternative methods using the correlation between urinary As and drinking water As, with stronger correlations indicative of better performance.

2 METHODS/EXPERIMENTAL

2.1 Data acquisition

Two-hundred and three volunteers from Cornwall, UK provided paired urine and drinking water samples as per their participation in a wider bio-monitoring study for which ethical approval was granted by the University of Manchester Research Ethics Committee (Ref 13068) and the NHS Health Research Authority National Research Ethics Committee (NRES) (Ref 13/EE/0234). Urinary As speciation was performed by HPLC-ICP-MS and $As^{III} + As^{V} + MA + DMA$ concentrations used as the exposure biomarker: U-As^{IMM}. Urinary Cre was determined by the Jaffe method using a colorimetric liquid assay kit, SG was determined using a digital refractometer and osmolality was measured using a cryoscopic osmometer.

2.2 Adjustment factor performance assessment

Urinary-As^{IMM} concentrations were adjusted for: Cre (routine method—Equation 1); SG and osmolality (Levine-Fahy method) (Levine & Fahy, 1945) and Cre (modified method—Equation 2). The z value used in Equation 2 was derived numerically by fixing the denominator (a) at 0.68 (Araki et al., 1990) and, given that no published values for As against UFR (b) were available, processing multiple correlations for a range of b between 0 and 1.5 at increments of 0.01. The value yielding the strongest correlation was selected and resulted in a z value of 0.49. Pearson correlation coefficients (r_p) between unadjusted and various adjusted U-As^{IMM} concentrations against drinking water total As were calculated. Variables were natural log-transformed prior to analysis due to positively skewed distributions. To test the significance of differences between correlations, Williams' tests were performed (Venables, 2002). All statistical analyses were performed in the R programming environment (R Core Team, 2013).

3 RESULTS AND DISCUSSION

Performance correlations for the range of adjustment factors investigated are presented in Table 1 with their significant differences relative to one-another. The order of performance on the basis of numerical correlations alone was Cre (modified) ($r_p = 0.69$) > SG ($r_p = 0.68$) > osmolality ($r_p = 0.67$) > unadjusted ($r_p = 0.65$) > Cre (routine) ($r_p = 0.62$). Williams' test results indicated that only modified Cre adjustment yielded a statistically stronger correlation than unadjusted concentrations and modified Cre and SG adjustments yielded a statistically stronger correlations than routine Cre adjustment. The failure to detect significantly stronger correlations between SG and osmolality and unadjusted concentrations may be attributed to the sample size and a lack of statistical power. It is possible to modify SG

Table 1. Drinking water total As versus U-As^{IMM} Pearson correlations across the range of adjustment methods. Correlations share a letter when not significantly different from one another based on Williams' tests.

Adjustment method	r_p (95% CI)	Significance
Unadjusted	0.65 (0.56, 0.72)	a,c
Cre[a]	0.62 (0.53, 0.70)	a
SG	0.68 (0.60, 0.75)	b,c
Osmolality	0.67 (0.58, 0.74)	a,b
Cre[b]	0.69 (0.61, 0.75)	b

[a]Routine method—Equation 1.
[b]Modified method—Equation 2, $z = 0.49$.

and osmolality adjustments in the manner which was performed for Cre adjustment (Vij & Howell, 1998) and future efforts may also improve their performance.

4 CONCLUSIONS

While acknowledging the criticisms of the use of Cre adjustment of biomonitoring results due to demographic differences in Cre clearance, we show in this example that an appreciable degree of performance is hindered by the mathematical application of the Cre metric, seldom addressed to-date. These findings reiterate the element-specific nature of urinary hydration adjustments and provide an opportunity for research groups to retrospectively adjust existing biomonitoring data using this modified approach without the requirement of extra measurements. This is the first study to compare Cre, SG and osmolality adjustments of urinary As concentrations against a performance criteria based on external environmental exposure.

ACKNOWLEDGEMENTS

DRSM was funded by NERC via a University of Manchester/BUFI (Centre for Environmental Geochemistry) PhD studentship (GA/125/017, BUFI Ref: S204.2). We thank Elliott Hamilton, Louise Ander, Andrew Marriott and Andrew Dunne (all BGS) and Giovanni Leonardi, Rebecca Close, Helen Crabbe, Karen Exley, Amy Rimell and Mike Studden (all PHE) for their involvement in the wider project. We thank Nigel Kendall and David Gardner from the University of Nottingham Biosciences for facilitating creatinine and osmolality measurements. Murray Lark, Chris Milne, Mark Cave and Simon Chenery from the BGS are thanked for advice.

REFERENCES

Araki, S., Sata, F. and Murata, K. 1990. Adjustment for urinary flow rate: An improved approach to biological monitoring. *Int. Arch. Occ. Env. Hea.* 62: 471–477.

Aylward, L.L., Hays, S.M., Smolders, R., Koch, H.M., Cocker, J., Jones, K., et al. 2014. Sources of variability in biomarker concentrations. *J. Toxicol. Env. Heal. B* 17:45–61.

Barr, D.B, Wilder L.C, Caudill S.P, Gonzalez A.J, Needham L.L, Pirkle J.L. 2005. Urinary creatinine concentrations in the US population: Implications for urinary biologic monitoring measurements. *Environ. Health Perspect.* 113:192.

Levine, L. and Fahy, J.P. 1945. Evaluation of urinary lead concentrations. I. The significance of the specific gravity. *J. Indus. Hyg. Toxicol.* 27: 217–223.

Orloff, K., Mistry, K. and Metcalf, S. 2009. Biomonitoring for environmental exposures to arsenic. *J. Toxicol. Env. Heal. B* 12: 509–524.

R Core Team. 2013. R: A language and environment for statistical computing, r foundation for statistical computing, Vienna, Austria. www.R-project.org.

Venables, W.N.R.B.D. 2002. *Modern Applied Statistics with S.* Fourth edition. Springer, New York. ISBN 0–387–95457–0.

Vij, H.S. and Howell, S. 1998. Improving the specific gravity adjustment method for assessing urinary concentrations of toxic substances. *Am. Ind. Hyg. Assoc. J.* 59:375–380.

Arsenic Research and Global Sustainability – Bhattacharya, Vahter, Jarsjö, Kumpiene,
Ahmad, Sparrenbom, Jacks, Donselaar, Bundschuh & Naidu (Eds)
© 2016 Taylor & Francis Group, London, ISBN 978-1-138-02941-5

The ACGIH Biological Exposure Index for urinary arsenic: To adjust for urinary creatinine or not?

G.P. Kew[1], J.E. Myers[2] & B.D. Johnson[3]

[1]*EOH Workplace Health and Wellness, Cape Town, South Africa*
[2]*Consultant, Dundee Precious Metals, Cape Town, South Africa*
[3]*Vice-President: Environment, Dundee Precious Metals, Toronto, Canada*

ABSTRACT: Employees at a Namibian Copper Smelter are occupationally exposed to inorganic arsenic as a contaminant of the complex ore concentrate being smelted and processed. Arsenic is monitored biologically by means of total urinary arsenic (uAs). The relationship between uncorrected uAs and creatinine corrected uAs (uAs_{Cr}) was examined for 4313 end of shift, end of week measurements with a view to understanding the impact of adjustment by urinary creatinine (uCr) concentration when comparisons are made with the American Conference of Governmental Industrial Hygienists (ACGIH) Biological Exposure Index (BEI). uAs and uAs_{Cr} were positively correlated. uAs_{Cr} provides a higher numerical value than uAs at uAs values below 25 µg/L, and a lower numerical value compared with uAs values above 25 µg/L, complicating comparisons with the uncorrected uAs ACGIH standard. The use of uncorrected uAs as required by the (ACGIH) can be misleading in both high and low exposure settings.

1 INTRODUCTION

Employees at a copper smelter in Tsumeb, Namibia are exposed to inorganic arsenic as a contaminant of metallurgically complex ore concentrates smelted. Mandatory biological monitoring comprises urinary arsenic (uAs). The relationships between uncorrected uAs, urinary creatinine (uCr) and creatinine corrected uAs (uAs_{Cr}) were examined in relation to the ACGIH Biological Exposure Index for arsenic (ACGIH, 2001).

2 METHODS

2.1 *Sampling and analytic methods*

Spot urine samples were obtained from 4313 employees at the end of a working shift at the end of a working week. These were representative of the entire workforce between 2012 to 2015. All company employees and contractors were monitored bi-annually in winter and summer. Samples were analysed for total arsenic (inorganic, methylated (MMA & DMA), and arsenobetaine), and creatinine concentration. Analysis was by ICP/MS in an accredited pathology laboratory (Goulle et al., 2005) and uCr values between 0.34 to 3.4 g/l were considered valid. Employees ranged from those with the lowest exposures e.g. in the administration block, to the highest exposed in the arsenic

plant where arsenic trioxide is recovered from the smelting process, purified and bagged. Intermediate exposures were in smelting. The work involves exposure to heat from the smelting process, and the weather, particularly in the summer months, can be very hot with temperatures reaching 40 degrees Celsius, with low to moderate humidity. The work is physically arduous for many, particularly in hot working environments, where adequate hydration is important. Historically, work was made challenging by the previous owners' use of sub-standard personal protective equipment and engineering exposure controls.

2.2 *Statistical methods and ethics*

STATA 12 was used for statistical analysis. uAs and uCr were correlated using the Pearson correlation coefficient. Graphical exploration of the relationship between uAs and uAs_{Cr} was undertaken with *mrunning*. Multiple linear regression was used to examine the relationship between uAs_{Cr} and uAs at low and high levels of uCr. As biological monitoring was mandatory, both in terms of Namibian law and company policy, the study did not require ethical clearance. All employees are made aware of the company's ongoing program of improving arsenic trioxide management, exposures and stewardship—an important component of which is surveillance of occupational exposures and health status.

3 RESULTS AND DISCUSSION

3.1 *The relationship between uAs and uCr*

Overall there is a highly significant positive correlation between uAs and uCr ($r = 0.34$, $n = 4313$, $p < 0.0001$).

3.2 *Graphical examination of the relationship between uAs_{Cr} and uAs*

The bivariate plot and smoothed red curve (with 95% confidence bands in yellow and green) generated by *mrunning* in Figure 1 shows the correlation between uAs_{Cr} and uAs, while Figure 2 shows that uAs_{Cr} provides a higher numerical value than uAs at uAs values below a crossover point at 25 µg/l, and a lower numerical value compared with uAs above 25 µg/l, complicating comparisons with the uAs ACGIH standard.

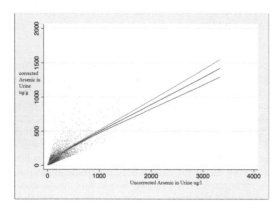

Figure 1. uAs_{Cr} versus uncorrected uAs for all values.

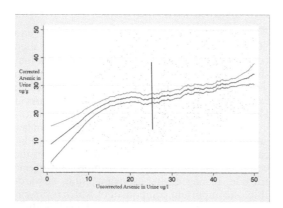

Figure 2. uAs_{Cr} versus uAs for uAs < 50 µg/l.

Table 1. Correlation between uAs and uCr.

uCr (g/L)	β (uAs µg/l)	Constant	R^2	p
Full range	0.48	37	0.64	0.0001
> 1.5	0.41	14.7	0.85	0.0001
< 1.0	1.32	11.5	0.86	0.0001

3.3 *The relationship between uAs_{Cr} and uAs given the value of uCr*

Table 1 shows the results of regression analysis of uAs_{Cr} on uAs. For the overall relationship, correction by uCr substantially reduces the numerical value of the uAs values ($\beta = 0.48$). Restricting analysis to uCr > 1.5 g/l (the relatively dehydrated) the relationship is virtually unchanged ($\beta = 0.41$). Restricting to uCr values below 1 g/l (the relatively well hydrated) the relationship is reversed and uAs_{Cr} values exceed uAs ($\beta = 1.32$) values numerically.

4 CONCLUSIONS

We conclude that the ACGIH BEI documentation of 2001 has not sufficiently characterized the relationship between uAs and uAs_{Cr}. This has particular relevance for high exposure environments typical of a copper smelter on the one hand, and for hot working conditions within warm climatic zones such as Namibia on the other. These environments are typically associated with more concentrated urine. For low level exposure environments (below 25 µg/L uAs) without high heat stress and where urine is not concentrated, correcting for creatinine will result in a more conservative (preventive) exposure estimation.

ACKNOWLEDGEMENTS

We acknowledge the DPM management, employees, and occupational health staff for their cooperation and support.

REFERENCES

StataCorp. 2011. Stata Statistical Software: Release 12. College Station, TX: StataCorp LP.

ACGIH 2001. Arsenic and soluble inorganic compounds. Recommended BEI. American Conference of Governmental Industrial Hygienists.

Goulle, J.P., Mahieu, lL., Castermant, J. and Nevue, N. 2005. Metal and metalloid multi-elemental ICP-MS validation in whole blood, plasma, urine and hair. *Forensic Science International* 153:39–44.

Arsenic Research and Global Sustainability – Bhattacharya, Vahter, Jarsjö, Kumpiene,
Ahmad, Sparrenbom, Jacks, Donselaar, Bundschuh & Naidu (Eds)
© 2016 Taylor & Francis Group, London, ISBN 978-1-138-02941-5

Rapid health risk assessment of discharge of tailings water containing elevated level of arsenic and other metals

J.C. Ng[1], S. Muthusamy[1] & P. Jagals[2]

[1]*The University of Queensland, National Research Centre for Environmental Toxicology (Entox), Brisbane, Queensland, Australia; CRC for Contamination Assessment and Remediation of the Environment (CRC CARE), University of Newcastle, Newcastle, New South Wales, Australia*
[2]*School of Population Health, The University of Queensland, Brisbane, Queensland, Australia*

ABSTRACT: The mine waste water treatment plant was vandalized in a gold mine located in Honiara of the Solomon Islands. With the wet season fast approaching there was a risk the tailings water containing elevated level of arsenic and other metals might spill into the river system downstream. A rapid risk assessment was conducted by testing potential toxicity of mine water against river water samples obtained from both upstream and downstream of the designated discharge point. Three cell-based bioanalytical in vitro assays were employed for the comparative toxicity evaluation. The test results supported a controlled discharge of the tailings water into the river in order to avoid a potential disastrous event which might occur during the big wet season. The short-term discharge plan was allowed to be implemented until a more sustainable waste water treatment plant could be restored.

1 INTRODUCTION

The waste water treatment plant was vandalized in a gold mine in Honiara of the Solomon Islands in 2014. This region is known to have huge rain falls during the wet season. Water in the tailings storage facility is known to have contained elevated levels of arsenic (~30 µg/L) and other metals. Since the tailings dam was near its capacity there was a risk for over spilling large volume of tailings into the river system and villages downstream resulting in a disastrous event. Various stakeholders particularly the local villagers, ministries of Health, and Environment, and WHO were concerning about if it would be safe for the environment and communities to allow the mining company to discharge untreated tailings water directly into the river. In December 2014, we were engaged by the Department of Foreign Affairs via WHO to investigate the potential health risk of such a discharge proposal. Water samples were delivered on Christmas Day and the work was commenced as soon as they were released by the Australian Quarantine Inspection Services on Boxing Day. A rapid risk assessment approach was required to deal with this emergency situation. The approach was to take the water in its entirety and test for "whole" mixture toxicity regardless what were knowns or unknowns in the specimen.

2 METHODS/EXPERIMENTAL

2.1 Sample description

The WHO office in Honiara, the Solomon Islands, coordinated the sample collection. On the 20th Dec 2014, three water samples labeled as GR0, GR7 and GR9 were collected into clean plastic bottles (1 L) were kept cold then hand carried into Australia on the 25th Dec 2014. The samples were kept in a fridge at the Brisbane AQIS office until they were released on the 26th Dec 2014. The samples were kept at 4°C until analysis. GR0 is the influent river water (upstream) into the mine, GR7 is the tailings dam water, and GR9 is downstream river water.

2.2 In vitro comparative toxicity testing

Water pH of GR0, GR7 and GR9 were 6.72, 6.93 and 7.31 respectively. Filtered water samples (0.45 µm) were tested in HepG2 cells for cytotoxicity (MTS assay, 24 h exposure), oxidative stress response (Nrf2 induction in Antioxidant Response Element (ARE)-reporter cells, 24 h exposure) and genotoxicity (micronucleus test, 60 h exposure) (Peng et al., 2015). The filtered water samples were concentrated by 10 fold, then aliquots of the concentrates were used for these three toxicity validation tests.

Each test was conducted in triplicate unless otherwise specified and with varying doses, and with appropriate negative and positive controls to ensure the assays are working. The highest dose at 1000 µL/mL for the oxidative stress and cytotoxicity was discarded because precipitation occurred in the culturing medium and the result was deemed unreliable. This precipitation was probably because the concentrated salts were forming insoluble product(s) with the salts and nutrients present in the culture medium.

3 RESULTS AND DISCUSSION

3.1 *Comparative toxicity data*

It has been proposed, as part of the mine waste management strategy, that controlled dewatering from the tailing dam could prevent disastrous event of overspill and resulting in large amount of sediment (tailings) running into the nearby river system.

The study was focused on how best to minimize potential human health impact, if any, from such dewatering program. Since mine wastewater is a complex mixture we will focus on potential mixture effect by treating the water as an entity and compare its relative toxicity to the upstream control (pristine) and downstream river water. The comparative toxicity results are shown in Tables 1–3. ICP-MS measured As concentrations were 1.4, 29.6 and 1.2 µg/L in GR0, GR7 and GR9 respectively; and Hg were 3.2, 2.3 and 3.3 µg/L respectively. Other toxic heavy metal concentrations of human health importance were relatively low (data not shown).

Table 1. Water sample oxidative stress (NRf2 assay) potential expressed as fold increase (mean ± s.e, triplicate and three independent runs) compared with the negative control measured by fluorescent spectroscopy. TBHQ positive control data are not shown. * µL of sample in 1 mL of culture medium.

Dose* (µL/mL)	GR0	GR7	GR9
0	1.00 ± 0.18	1.00 ± 0.18	1.00 ± 0.18
15.6	1.12 ± 0.11	1.04 ± 0.07	1.05 ± 0.03
31.25	1.14 ± 0.03	1.08 ± 0.13	1.03 ± 0.06
62.5	1.04 ± 0.05	1.09 ± 0.05	1.14 ± 0.04
125	1.21 ± 0.11	1.15 ± 0.02	1.13 ± 0.07
250	1.38 ± 0.06	1.18 ± 0.08	1.22 ± 0.04
500	1.12 ± 0.08	1.25 ± 0.01	1.23 ± 0.08

Table 2. Water sample cytotoxicity expressed as% growth (mean ± s.e, triplicate and three independent runs) compared to the negative control as measured by an MTS assay kit. CdCl$_2$ positive control data are not shown. * µL of sample in 1 mL of culture medium.

Dose* (µL/mL)	GR0	GR7	GR9
0	100	100	100
31.25	103.4 ± 0.06	106.1 ± 0.06	115.9 ± 0.03
62.5	102.8 ± 0.06	101.6 ± 0.02	116.9 ± 0.05
125	100.4 ± 0.06	102.5 ± 0.04	105.8 ± 0.05
250	97.2 ± 0.03	96.6 ± 0.06	109.2 ± 0.02
500	95.9 ± 0.04	94.1 ± 0.04	87.9 ± 0.04

Table 3. Water sample induced micronucleus formation expressed as fold increase (mean ± s.e, triplicate and three independent runs) compared to the negative control (pure H$_2$O). The value is the average of duplicate. Benzo[a]pyrene positive control data are not shown. * µL of sample in 1 mL of culture medium.

Dose* (µL/mL)	GR0	GR7	GR9
0	1	1	1
25	0.56	0.54	0.60
50	0.54	0.64	0.57
100	0.80	0.60	0.75
250	0.68	0.61	0.67
500	0.91	1.13	1.05
1000	0.81	0.85	1.63

3.2 *Interpretation of toxicity data*

Taking into account of the assay variation, it would appear that a dose response relationship was observed for the last high doses compared to the lower doses and the negative control (pure 18 MΩ.cm Milli-Q water) for both the oxidative stress (Nrf2) and cell viability end points.

For the oxidative stress indicator GR7 (tailing dam water) at the highest dose of 500 µL/mL (which is very high dosage in terms of bioassay) seems to have similar level of stress induction as the water from GR9 and both the water samples have about 10% higher stress causing potency.

For the cytotoxicity and at the highest dosing concentration, GR7 seems to have a similar potency as that of GR0 but GR9 (existing river water) seems to have a higher potency.

For the genotoxicity potency as measured by micronucleus formation all three water samples seem to be non-genotoxic.

4 CONCLUSIONS

All three water samples have similar but very limited toxicity potency and only occur at very high dosing concentration (for cytotoxicity and oxidative stress) bearing in mind that the conclusions are limited to acute toxicity assays. The important observation is that water from the tailings dam is no more toxic than either the background upstream water or the receiving water downstream from the intended discharge outlet.

As a precautionary principle, any discharge of this tailings dam water into the river system should take into consideration of the river flow and discharge flow rates, and appropriate dilution should be applied in order to meet environmental and health guidelines, noting that this current report focused on only the health aspect. Ideally the tailings dam water should be treated before discharge. This study is limited by a single sample from each of three sampling points. The sampling regime should be wider and seasonal variation should be monitored. Sampling during discharging period should be taken for further toxicity (or lack of toxicity) confirmation.

ACKNOWLEDGEMENTS

We are grateful for logistic support by Ministry of Health and Ministry of Environment, and WHO for logistic support and facilitating the study and obtaining financial support from DFAT. Entox is a partnership between Queensland Department of Health and the University of Queensland.

REFERENCE

Peng, C., Muthusamy, S., Xia, Q., Lal, V., Denison, M.S. & Ng, J.C. 2015. Micronucleus formation by single and mixed heavy metals/loids and PAH compounds in HepG2 cells. *Mutagenesis* 1–10, doi:10.1093/mutage/gev021.

Arsenic Research and Global Sustainability – Bhattacharya, Vahter, Jarsjö, Kumpiene,
Ahmad, Sparrenbom, Jacks, Donselaar, Bundschuh & Naidu (Eds)
© 2016 Taylor & Francis Group, London, ISBN 978-1-138-02941-5

Challenges on the surveillance of arsenic exposed workers in Uruguay

N. Mañay[1], V. Bühl[2], C. Alvarez[1] & M. Pistón[2]
[1]*Toxicology. DEC, Faculty of Chemistry, Universidad de la República (UdelaR), Montevideo, Uruguay*
[2]*Analytical Chemistry. DEC, Faculty of Chemistry, Universidad de la República (UdelaR),*
Montevideo, Uruguay

ABSTRACT: Urinary arsenic level is the most common biomarker for assessing human exposure risks at workplaces and it is necessary to quantify inorganic arsenic species in urine and its methylated metabolites (arsenate + arsenite + DMA + MMA). Reference limits in Uruguay are less than 35 µg/L of the sum of inorganic arsenic (iAs) and its metabolites in urinemeasured at end of a workweek. Regular controls are performed to comply with national regulations. However,other sources of arsenic exposure besides that from workplaces such as food or drinking water and other non-occupational variables have not being taken into account in the development of toxic effects of arsenic and no background data is available. In this work some related biomonitoring studies and research focused on workers exposed to arsenic in the industries of wood preservation in Uruguay are presented.

1 INTRODUCTION

Urinary arsenic (As) level is the most common biomarker for assessing human exposure risks at workplaces. It is necessary to quantify inorganic arsenic (iAs) species in urine and its methylated metabolites (arsenate + arsenite + DMA + MMA), for biomonitoring assessment and risk management of workers' As exposure in order to prevent adverse effects or medical interventions. Reference limits in Uruguay are regulated by a law adopting those recommended by ACGIH (< 35 µg/L) which considers the sum of iAs and its metabolites in urine-measured at end of a workweek. This issue has represented a major analytical challenge in Uruguay since the determination of the sum of these toxicological relevant species of As were determined using an alternative method. This method was developed using Hydride Generation-Atomic Absorption Spectrometry (HG-AAS) because HPLC-ICPMS is not available for routine analysis (Bühl *et al.*, 2015).

However, other sources of As exposure besides that from workplaces such as food or drinking water and other non-occupational variables have not being taken into account in the development of toxic effects of As and no background data is available.

Our analytical toxicology research group is currently developing a multidisciplinary project. The aim is to establish relationships among observed effects, levels of As species in urine and the influence of non-occupational As exposure in workers along with other environmental parameters and individual susceptibility, including genotoxicity. The final results will be of great importance to prevent long-term health effects exposures to low doses of As.

In this work some related biomonitoring studies and research focused on workers exposed to As in the industries of wood preservation in Uruguay are presented.

2 METHODS/EXPERIMENTAL

2.1 *Arsenic occupational exposed workers in Uruguay: Current situation*

In Uruguay most industries uses Chromated Copper Arsenate (CCA) as wood preservative (solution 60%). The mixture can be expressed as Cr, Cu and As oxides (28.5% Cr_2O_3, 11.1% CuO and 20.4% As_2O_5). This product is registered as a pesticide, with a restricted use, and the only available formulation is an aqueous solution. This formulation minimizes the risk of contamination in case that an accident occurs during the transport of the product to the plant, but industries devoted to wood preservation in our country increased its use, leading to a higher risk for workers. The processes carried out in these industries and occupational exposure are reviewed.

2.2 *Arsenic biomonitoring*

Urine samples from workers presumably exposed to As, in different workplaces are periodically analyzed in our Laboratory, for toxicologically relevant species in order to assess health risks according to national regulations. Results of this routine biomonitoring, in a period of two years, were statistically assessed.

2.3 *New challenges*

Several of ongoing research studies to assess the incidence of low levels of As on exposed workers health and their correlations with different associate parameters focused on toxicological aspects, are presented.

3 RESULTS AND DISCUSSION

3.1 *Exposed workers*

There are over a dozen industries in Uruguay that perform wood preservation with CCA. Many of them have modern plants for the impregnation process. However some of these tasks are still manually made for some particular wooden parts (those of small size), with consequent health risks of workers exposure to As.

3.2 *Arsenic biomonitoring*

176 urine samples were analyzed for the last two years for biomonitoring purposes. The sum of toxicologically relevant species of As (iAs +MMA+DMA) were determined using HG-AAS (Bühl *et al.*, 2015). The results are summarized in Table 1.

Only 16 exposed workers showed higher levels than those established by regulations (As-Urine < 35 µg/L), although half of the population have rates over 5 µg/L. The average result of this routine biomonitoring was 10 µg/L (range 5–82 µg/L). Those workers with high levels of As were taken away from the workplace for at least 15 days, and then the analysis was repeated; all of them showed lower levels after this action.

Table 1. Analytical results of urinary As during the last two years for biomonitoring purposes.

	As-Urine (µg/L) Sum of iAs+DMA+MMA
Number of samples	176
Detection Limit (HG-AAS)	1.5
Quantification Limit	5.1
Mean value	10.2 ± 1.2
Minimum value	< 5.1
Maximum value	82.5 ± 9.9

3.3 *New challenges*

Besides this routine biomonitoring to As exposed workers, the possible factors which may influence on As urine levels and develop adverse effects are being assessed.

Our pilot study is taking special attention on potential occupational factors that are considered together with known or suspected exposure to other chemical or physical agents at work, home or leisure activities. For this purpose, we are processing surveys done to a target group of exposed workers asking about characteristics of particular dietary issues, seafood, rice and tap water consumption, among others. Correlations between time/duration of exposure, and habits, lifestyles, and medical/family history of each worker will be studied. Urine, blood and mouth cells are also being taken for further studies as biomarkers to determine other possible effects and establish correlations with low levels of As exposure. Nutrition factors and water resources also will be taken into account

4 CONCLUSIONS

The results of an ongoing research were reviewed. These results will highlight the impact on health and the consequences of As exposure in exposed workers in their working place and natural environment, considering the influence of other factors that affect the development of adverse effects caused by this element. No background information is available in our country and these As issues have been underestimated until now.

This research will contribute to provide scientific evidence to prevent long-term health effects exposures to low doses of As.

ACKNOWLEDGEMENTS

The research that gives rise to the results, received funding from Agencia Nacional de Investigación e Innovación (ANII FMV_3_2013_1_100439), PEDECIBA-Química.

REFERENCE

Bühl, V., Álvarez, C., Kordas, K., Pistón, M. and Mañay, N. 2015. Development of a simple method for the determination of toxicologically relevant species of arsenic in urine using HG-AAS. *Journal of Environmental Pollution and Human Health* 3(2): 46–51.

Arsenic Research and Global Sustainability – Bhattacharya, Vahter, Jarsjö, Kumpiene,
Ahmad, Sparrenbom, Jacks, Donselaar, Bundschuh & Naidu (Eds)
© 2016 Taylor & Francis Group, London, ISBN 978-1-138-02941-5

Mineral nutrients and toxicants in Bangladesh groundwater—do we need a holistic approach to water quality?

M.A. Hoque & A.P. Butler
Department of Civil and Environmental Engineering, Imperial College, London, UK

ABSTRACT: Groundwater, a drinking water source for > 95% of inhabitants living in Bangladesh, has received much attention because of its naturally occurring Arsenic (As), but the coupling or decoupling of As toxicity by other water constituents remains unexplored. Although food provides nutrients, in rural settings, where people subsist on a low nutrient diet, drinking-water-nutrients may supply quantities significant for human health thereby preventing diseases. Here, we show, using augmented datasets from Bangladesh, that the content of groundwater is such that in some areas individuals obtain up to 50% or more of the Recommended Daily Intake (RDI) of some nutrients (e.g., calcium, magnesium, iron) from just 2 litres of drinking water. We also show the spatial association of groundwater nutrients and health outcomes in Bangladesh. We suggest, therefore, that a holistic approach, i.e., considering the entire range of groundwater constituents, should be considered when undertaking health impact studies of groundwater.

1 INTRODUCTION

Drinking water, often rich in minerals, has rarely been considered as a source of mineral nutrients. Historically, the focus has been on the diseases associated with certain elements (e.g., arsenic, fluoride, manganese etc.), and rarely has there been consideration of the positive health effects of mineral-nutrients (e.g., calcium, magnesium, iron etc.) (Rosborg, 2015).

Populations living in resource-poor settings, who subsist on a low nutrient diet, may suffer from nutrient deficiency and associated health consequences. In such settings drinking water rich in mineral content can provide a significant amount of the recommended daily nutrient requirement, but this contribution remains unrecognised. We postulate that this varied, and often unrecognised, mineral content could help some (where intake via food is insufficient) to reach the recommended daily level of nutrients, and hence avoid nutrient deficiency related consequences (Hoque & Butler, 2016).

Medical hydrogeology should, therefore, also consider the presence of beneficial mineral nutrients including those that influence the toxicity of harmful elements or alter groundwater composition during storage, i.e. the time between collection and consumption. We carried out a preliminary exploratory study in Bangladesh to assess drinking water nutrients and toxicants and the possible health benefits of these groundwater nutrients at a national scale.

2 METHODS AND MATERIALS

The British Geological Survey (BGS) in collaboration with the Department of Public Health and Engineering (DPHE) have made available a database of chemical composition of 3805 tube-wells from across Bangladesh (BGS/DPHE, 2001) which was used for comparison with demographic health (meta) data (e.g., probability of stunting in children, the prevalence of rickets and other leg deformities, anaemia, and arsenicosis patients).

The Recommended Daily Intake (RDI) for different nutrients vary for different age groups, as well as for special groups (e.g., pregnant women), however, for simplicity a fixed conservative value (e.g., FAO and WHO, 2001) is considered here for each mineral nutrient: calcium (Ca, 800 mg/d), fluorine (F, 4 mg/d), iron (Fe, 14 mg/d), iodine (I, 150 µg/d), potassium (K, 3000 mg/d), magnesium (Mg, 300 mg/d), manganese (Mn, 2 mg/d), sodium (Na, 2000 mg/d), selenium (Se, 55 µg/d) and zinc (Zn, 10 mg/d).

3 RESULTS AND DISCUSSION

3.1 *Nutrients and toxicants*

The human body needs relatively large amounts of macro-minerals (Ca, Mg, Na, K etc.), most of which are found in groundwater. Significant amounts, up to 50% of RDI for Ca and Mg, can be derived from groundwater in parts of Bangladesh. However, these are often accompanied with As (Fig. 1) and, in coastal areas, excessive Na (Khan *et al.*, 2014). Furthermore, around 30% of wells in Bangladesh provide more than 50% of RDI for Fe—an important micronutrient that prevents anaemia.

Groundwater generally contains limited amounts of K and in Bangladesh it is almost entirely less than 10% of RDI The presence of Se, F, Zn are insignificant and rarely provide anything close of the RDI. The groundwater is NO_3 safe, with concentrations well below the WHO guideline value of 50 mg/L. The data also indicate that around 20% of wells can provide more than 100% of RDI for Mn, and 43% of wells have As concentrations above the WHO guideline, 10 µg/L, for drinking water.

3.2 *Health impact*

We have assumed nutrients provided by drinking water are entirely absorbed by the body. Although in practice this depends on many factors, nevertheless, previous studies in Bangladesh (Merrill *et al.*, 2011) indicate significant amounts of Fe absorption by the human body derived from drinking water. A preliminary analysis of district wide anaemic patients indicates that the numbers of patients are low in areas where tube wells have a higher Fe concentration (Hoque and Butler, 2016).

From dietary surveys, Ca and Mg intake for much of the Bangladeshi population does not meet daily requirements. For example, Ca intake from food is estimated to be 200 to 400 mg/d (Islam *et al.*, 2003). This is less than 50% of the RDI.

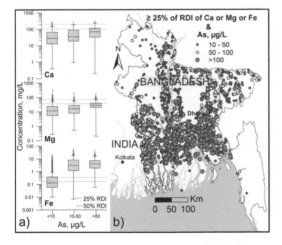

Figure 1. The adverse effects of As in Bangladesh Groundwater are well known, but the same water contains beneficial mineral nutrients (e.g. Ca, Mg, Fe) too. The linkage of arsenic toxicity with other water constituents is unknown.

However, this perceived 'Ca under nutrition' may be less widespread than originally inferred due to the high presence of Ca in some groundwaters. Magnesium deficiency during pregnancy can lead to low birth weight (Takaya *et al.*, 2006). In Bangladesh the distribution of Mg in tubewells and stunting in children under-5 years, assumed to be a proxy for low birth-weight, show good spatial correlation (Hoque and Butler, 2016).

Health impacts of As (Argos *et al.*, 2010) and Na (Khan *et al.*, 2014) are well known but their potential reduction due to the presence of beneficial nutrients in groundwater has not been investigated.

3.3 *Hydrogeochemical controls*

Groundwater acquires its chemical composition during recharge and then from subsequent reactions with aquifer materials. In Bangladesh aquifers with high organic content give rise to microbial reduction (Nickson *et al.*, 1998), which leads to consumption of oxygen and NO_3 from recharging water. This, in turn, releases Mn and Fe, and adsorbed arsenic, from MnOOH and FeOOH minerals, into solution.

4 CONCLUSIONS

- Groundwater in Bangladesh contains significant amount of mineral nutrients that may be beneficial to human health.
- The issue of estimating usual intake distributions of nutrients considering only food sources may be counterfactual, particularly in areas with high levels of dissolved solids in groundwater sources, as the contribution of drinking water to nutrient and mineral supply can be substantial.
- Often the health impacts of groundwater toxicants are studied in isolation, for instance As, F, and Na, but other ions present in the same water, which can potentially reduce the toxicity, indicate the need for a holistic approach.

ACKNOWLEDGEMENTS

The authors acknowledge financial support from Leverhulme Trust, Wellcome Trust and Economic and Social Research Council, UK.

REFERENCES

Argos, M., Kalra, T., et al. 2010. Arsenic exposure from drinking water, and all-cause and chronic-disease mortalities in Bangladesh (HEALS): a prospective cohort study. *The Lancet* 376(9737): 252–258.

BGS/DPHE. 2001. *Arsenic contamination of ground-water in Bangladesh* (No. WC/00/19). Keyworth: Department of Public Health Engineering (DPHE) of Government of Bangladesh and British Geological Survey (BGS).

FAO & WHO. 2001. *Human Vitamin and Mineral Requirements, Report of a joint FAO/WHO expert consultation Bangkok, Thailand*. Food and Nutrition Division, FAO Rome: Food and Agriculture Organization (FAO) of the United Nations, and World Health Organization (WHO).

Hoque, M.A., and Butler, A.P. 2016. Medical hydrogeology of Asian deltas: Status of groundwater toxicants and nutrients, and implications for human health *Int. J. Environ. Res. Public Health* 13, 81 (doi:10.3390/ijerph13010081).

Islam, M.Z., Lamberg-Allardt, C., et al. 2003. Dietary calcium intake in premenopausal Bangladeshi women: do socio-economic or physiological factors play a role? *Eur. J. Clin. Nutr.* 57(5): 674–680.

Khan, A.E., Scheelbeek, P.F.D., et al. 2014. Salinity in drinking water and the risk of (Pre)Eclampsia and gestational hypertension in coastal Bangladesh: A case-control study. *PLoS ONE* 9(9): e108715.

Merrill, R.D., Shamim, A.A., et al. 2011. Iron status of women is associated with the iron concentration of potable groundwater in rural Bangladesh. *J. Nutr.* 141(5): 944–949.

Nickson, R., McArthur, J.M., et al. 1998. Arsenic poisoning in Bangladesh groundwater. *Nature* 395: 338.

Rosborg, I. (Ed.). 2015. Drinking Water Minerals and Mineral Balance: Importance, Health Significance, Safety Precautions. London: Springer International Publishing Switzerland.

Takaya, J., Yamato, F., et al. 2006. Possible relationship between low birth weight and magnesium status: from the standpoint of "fetal origin" hypothesis. *Magnes Res.* 19(1): 63–69.

Arsenic Research and Global Sustainability – Bhattacharya, Vahter, Jarsjö, Kumpiene,
Ahmad, Sparrenbom, Jacks, Donselaar, Bundschuh & Naidu (Eds)
© 2016 Taylor & Francis Group, London, ISBN 978-1-138-02941-5

Groundwater and blood samples assessment for arsenic toxicity in rural population of Darbhanga district of Bihar, India

S. Abhinav[1], S. Navin[1], S.K. Verma[1], R. Kumar[2], M. Ali[2], A. Kumar[2,3] & A.K. Ghosh[3]

[1]*L.N. Mithila University, Darbhanga, Bihar, India*
[2]*Mahavir Cancer Institute and Research Centre, Patna, Bihar, India*
[3]*A.N. College, Patna, Bihar, India*

ABSTRACT: Arsenic poisoning through consumption of ground water is global problem. India is one of the most affected country exposed to arsenic poisoning, where more than 100 million people are at risk. Investigation by Central Ground Water Board (CGWB) of India has confirmed arsenic contaminated groundwater in 86 districts of 10 Indian states. Bihar is the second most groundwater arsenic contaminated state in India. Assessment of groundwater arsenic contamination and associated health-risks in the four villages of Darbhanga districts Bihar, India was undertaken. Paghari and Habidih-East are the two villages under Baheri block and Parri and Bairampur are the two villages under Biraul block of Darbhanga district. Results revealed that maximum contamination of arsenic in the groundwater was found to be 911 μg/L in Paghari village of Baheri block. Strong correlation was observed between elevated level of arsenic in the groundwater and elevated level of arsenic in the blood samples.

1 INTRODUCTION

Groundwater pollution due to heavy metal is one of the most serious problems we are facing today. It has become evident that increasing human activities have modified the global cycle of heavy metals. Among these metals, arsenic exhibits a complex metabolism and is possibly the most abundant pollutant which is linked with carcinogenic, non-carcinogenic health impacts (Chen *et al.*, 2010, Naujokas *et al.*, 2013). In India, Bihar is the only state where rate of groundwater arsenic contamination is much higher compared to other states (Chaurasia *et al.*, 2012, Singh *et al.*, 2014). In a recent report (CGWB, 2015), 17 out of 38 districts of Bihar has been identified with sever groundwater arsenic contamination. A population of more than 5 million people are drinking groundwater with As concentration beyond permissible limit guided by World Health Organization of 10 μg/L (WHO, 2011) and the Bureau of Indian Standard (BIS) of 50 μg/L. Present study was done to estimate arsenic in the groundwater and human blood samples collected from four villages of Darbhanga district of Bihar, India along with the assessment of related health problems.

2 METHODS/EXPERIMENTAL

2.1 *The study area*

The study was done at four different villages under two blocks of Darbhanga district (26.17°N 85.9°E) in the state of Bihar, India.

Block—Baheri	Block—Biraul
Village 1. - Paghari	Village 3. - Parri
Village 2. - Habidih-east	Village 4. - Bairumpur

2.2 *Sample collection and survey*

Altogether, 200 groundwater samples from hand tube wells were collected (50 water samples each villages) for primary screening of arsenic contamination using field test kit (Merck, Germany). After primary screening 48 arsenic contaminated positive (As > 10 μg/L) groundwater samples of hand tube wells and 48 human blood samples belonging to the same households were collected from four villages (12 groundwater and 12 human blood samples from each villages). Net quantity of arsenic in the groundwater and human blood samples were assessed through Graphite Furnace—Atomic Absorption Spectrophotometer (GF-AAS) of PerkinElmer, USA. Arsenic associated health related problems among village people were also studied during the entire survey. Parameters like pH, temperature, depth and GPS location of hand tube wells were also include in this study.

3 RESULTS AND DISCUSSION

3.1 *Arsenic assessment*

Out of 48 groundwater samples from all the four villages only four samples shows arsenic level

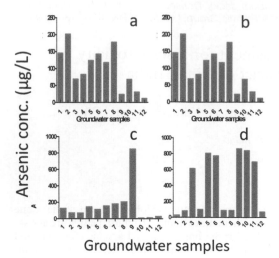

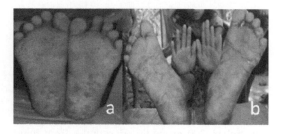

Groundwater samples

Figure 1. Distribution of arsenic in tubewell water samples in the investigated villages: a) Paghari and b) Habidih-east (Baheri Block); c) Parri and d) Bairumpur (Biraul Block).

Figure 2. Arsenicosis symptoms in village people.

below 10 μg/L. Maximum arsenic contamination level was found to be 911 μg/L in Paghari village (Fig. 1a). The average concentration of As was 232 μg/L in Paghari village and 201 μg/L in Habidih east village (Fig. 1b), while the concentrations in wells were as high as 843 μg/L in Parri village (Fig. 1c) and 862 μg/L in Bairampur village (Fig. 1d).

3.2 *Health impacts*

The villagers exhibited the typical symptoms of arsenicosis like hyperkeratosis, melanosis in palm and sole (Fig. 2). A total of 23 persons with visible symptom of arsenic poisoning were detected in the study area. An increased amount of arsenic in blood samples was also found in these populations.

4 CONCLUSIONS

Groundwater arsenic contamination in these four village of Darbhanga distrct was much higher than the standard guideline. Population of these areas are at very higher risk because of regular consumption of arsenic contaminated drinking water. Arsenic in their blood samples act as a biomarker means indicators of exposure. Therefore, there is a need of sustainable technologies in rural areas in developing countries for successfully mitigation of groundwater arsenic contamination.

ACKNOWLEDGEMENTS

The authors extend their appreciation to the INSPIRE DIVISION Department of Science & Technology, Ministry of Science & Technology, Government of India, New Delhi for the financial assistance of this work.

REFERENCES

Chaurasia, N., Mishra, A. and Panday, S.K. 2012. Finger print of arsenic contaminated water in India. A Review. *J. Forensic Res.* 3: 1–4.
CGWB. 2015. *Groundwater Quality Monitoring: Groundwater Quality Scenario in India.* Central Ground Water Board, Ministry of Water Resources, Government of India, Faridabad, India. 117p.
Chen, Y., Ahsan, H., Slavkovich, V., Loeffler Peltier, G., Gluskin, R.T., Parvez, F., et al. 2010. No association between arsenic exposure from drinking water and diabetes mellitus: a cross-sectional study in Bangladesh. *Environ. Health Perspect.* 118: 1299–305.
Naujokas, M.F., Anderson, B., Ahsan, H., Aposhian, H.V., Graziano, J.H., Thomas, C. and Suk, W.A. 2013. The broad scope of health effect from chronic arsenic exposure: Update on a worldwide public health problem. *Environ. Health Perspect.* 121: 295–302.
Singh, S.K., Ghosh, A.K., Kumar, A., Kislay, K., Kumar, C., Tiwari, R., Parwez, R., Kumar, N. and Imam, M. 2014. Groundwater arsenic contamination and associated health risks in Bihar, India. *Int. J. Environ. Res.* 8(1): 49–60.
WHO. 2011. Guideline for Drinking Water Quality. 4th Ed. World Health Organization, Geneva.

Arsenic Research and Global Sustainability – Bhattacharya, Vahter, Jarsjö, Kumpiene,
Ahmad, Sparrenbom, Jacks, Donselaar, Bundschuh & Naidu (Eds)
© 2016 Taylor & Francis Group, London, ISBN 978-1-138-02941-5

Assessing the impact of arsenic in groundwater on public health

J.-H. Chung[1], J.-D. Park[2] & K.-M. Lim[3]
[1]*Research Institute of Pharmaceutical Sciences, Seoul National University, Seoul, Korea*
[2]*College of Medicine, Chung-Ang University, Seoul, Korea*
[3]*College of Pharmacy, Ewha Womans University, Seoul, Korea*

ABSTRACT: The determination of As in groundwater is fundamental to assess its impact on public health. Here, we measured As concentration of groundwaters in 722 sites covering all 6 major provincial areas of Korea. Groundwater was measured As concentration in two occasions (summer and winter) employing highly sensitive ICP-MS. Among targeted 722 sites, seasonal variation of As concentration in groundwater was minimal, while geographical difference, prominent. The concentrations of As in groundwaters ranged from 0.1 to 48.4 µg/L and the majority displayed lower than 10 µg/L. However, some area exceeded 10.0 µg/L (2–3%) and most remarkably, the urinary excretion of As in the population around 3 groundwaters with high arsenic levels was measured to be markedly higher with all stratified analysis yielding higher As levels compared to non-contaminated area (< 5 µg/L). Especially, urinary As excretion of non-smokers was significantly higher in the contaminated area, suggesting the possible contribution of As-contaminated groundwaters to low-dose arsenic exposure.

1 INTRODUCTION

Groundwater is still a major source of drinking water especially in rural areas. Groundwater is also being extensively used for agriculture, food processing and washing, suggesting additional contribution to arsenic exposure via contaminated foods and utensils. Therefore, the contamination of groundwater by arsenic may be a major threat to public health, indicating an urgent need for the accurate monitoring of arsenic levels nationwide ideally with a sensitive quantitative method to meet the current As standard (10 µg/L). Indeed many countries at the risk of arsenic exposure regularly monitor As contamination in drinking water and groundwaters (Mandal & Suzuki 2002, Nordstrom, 2002).

Here, we monitored arsenic concentrations in groundwaters of Korea nationwide and evaluated geographical distribution and seasonal variation of arsenic levels with a highly sensitive analytical method, ICP-MS with the limit of detection of 0.1 µg/L, hundred fold lower than current As standard, 10 µg/L. In addition, we measured and compared the urinary excretion of arsenic of the population around the As-contaminated groundwater sites (As >10 µg/L) to determine the contribution of contaminated groundwaters to the human exposure to arsenic.

2 METHODS/EXPERIMENTAL

2.1 Study area

South Korea is divided into 6 provincial areas as follows, Seoul and Gyeonggi (capital), Gangwon (east), Chungcheong (west), Gyeongsang (south east), Jeolla (south west) and Jeju (volcanic island). Sampling of groundwater was performed by a grab method twice in each during summer and winter. Of 722 sites, 98% and 88% were successfully sampled in the summer and the winter, respectively.

2.2 Analysis of As concentration in groundwater in urine

The concentration of arsenic in groundwater was analyzed by inductively coupled plasma-mass spectrometry (ICP-MS, Perkin Elmer, Elan 6100 DRC plus, Shelton, CT). Selection of sampling population: The target area was selected (Gosung, Gyeongsang, Dang-Jin, Chungcheong and Gok-sung, Jeolla) such that As contamination level exceeded 10 µg/L and the groundwater was the major source of drinking water. Control, i.e., non-contaminated area was selected to match geographical location. Total As concentration urine was determined by using inductively coupled-plasma dynamic reaction cell-mass spectrometry (ICP-DRC-MS) (Perkin Elmer, Elan 6100 DRC plus) after wet digestion with nitric acid.

3 RESULTS AND DISCUSSION

The concentration of arsenic in groundwater sampled from 722 selected sites of 6 areas ranged from 0.1 to 48.4 µg/L (0.12 to 41.9 µg/L in summer, 0.1 to 48.4 µg/L in winter). The levels of arsenic in groundwater are divided into 4 categories; group

I (lower than 2.0 µg/L), group II (2.0–4.9 µg/L), group III (5.0–9.9 µg/L) and group IV (higher than 10 µg/L). In summer, the number of group I was the largest where 624 sites (88.1%) fell, while 49 sites (6.9%) were group II, 20 sites (2.8%), group III and 15 sites (2.1%). Similar patterns of arsenic levels were observed in winter. 566 sites (89.0%) were group I, 41 sites (6.4%), group II, 15 sites (2.4%), group III and 14 sites (2.2%), group IV in winter. These data indicated that the seasonal variation in arsenic levels of groundwater was insignificant ($X2 = 0.43$, $p > 0.1$).

In contrast, geographical variation was noted. In Gangwon area, there was no site belonging to group IV but 4.9% and 9.2% of groundwaters of Gyeongsang area were group IV in summer and winter, respectively. A national distribution of arsenic concentration in groundwater of study sites in summer and winter were presented in Figure 1.

To examine the contribution of As-contaminated drinking waters to systemic exposure, we measured the urinary excretion of As of the population around the As-contaminated groundwaters (As > 10 µg/L) and compared with non-contaminated area (< 5 µg/L). The urinary excretion of As in the population around 3 As-contaminated groundwaters (11.0~42.2 µg/L) was measured to be markedly higher (79.7 ± 5.2 µg/g creatinine vs 68.4 ± 5.4, $p = 0.052$) although a statistical significance was not barely achieved. However, in non-smokers, As excretion were significantly higher in the contaminated area (85.9 ± 12.7 vs 54.0 ± 6.3, $p = 0.030$), suggesting that As-contaminated drinking waters might contribute to the its systemic exposure at least in part.

Recently, public health threat from arsenic contaminated groundwaters in China has attracted a large attention even though the portion of people at stake is around 2% of total population (Rodríguez-Lado et al., 2013). In the present study, we found that 2.2% of sampled sites of Korea exceeded 10 µg/L which was comparable to other countries.

In addition, geographical variation was evident; in Gangwon area, no groundwater sites showed arsenic levels exceeding 10 µg/L, while in Gyeongsang area, 4.9% and 9.2% sites were determined to be higher than 10 µg/L during summer and winter season, respectively, suggesting that local variation in As exposure may exist and management of As contamination may be necessary in some areas.

More than 2% of groundwater sites sampled exceeded 10 µg/L, indicating that potential health problems associated with exposure to arsenic could not be precluded. Strongly supporting this, we could find that compared to non-contaminated area (< 5 µg/L), the urinary excretion of As in the population around 3 groundwaters with high arsenic levels (11.0~42.2 µg/L) was measured to be markedly higher. Actually in all subgroups, average urinary As excretion of people near the contaminated groundwaters was higher than those of non-contaminated area (109~159% of non-contaminated). Especially in non-smokers, As excretion was significantly increased by As-contaminated groundwaters, suggesting that As-contaminated groundwaters could indeed contribute to systemic exposure of As at least in part. This may be from the wide use of groundwaters in agriculture, food processing and washing as well as drinking, which can further contribute to human exposure to arsenic (Navas-Acien et al., 2009).

4 CONCLUSIONS

The nationwide survey in groundwater for arsenic contamination with the highly sensitive analytical method was the first report in Korea although the number of sampled sites were limited. Especially, we could demonstrate that As-contaminated groundwaters even though with minimal extents (10~50 µg/L) may increase the systemic exposure of As indeed. Therefore, more extensive monitoring for As in groundwaters and examination of related health effects are necessary to protect public health from As exposure.

REFERENCES

Mandal, B.K., and Suzuki, K.T. 2002. Arsenic round the world: a review. Talanta 58: 201–235.
Navas-Acien, A., Umans, J.G., Howard, B.V., Goessler, W., Francesconi, K.A., Crainiceanu, C.M., Silbergeld, E.K., and Guallar, E. 2009. Urine arsenic concentrations and species excretion patterns in American Indian communities over a 10-year period: the Strong Heart Study. Environ. Health Perspect. 117: 1428–1433.
Nordstrom, D.K. 2002. Public health. Worldwide occurrences of arsenic in ground water. Science 296: 2143–2145.
Rodríguez-Lado, L., Sun, G., Berg, M., Zhang, Q., Xue, H., Zheng, Q., and Johnson, C.A. 2013. Groundwater arsenic contamination throughout China. S198cience 341(6148): 866–868.

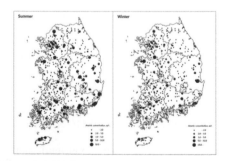

Figure 1. Nationwide distribution of arsenic concentration in the groundwater in summer and winter, respectively.

3.3   Biomarkers of health effects

Arsenic Research and Global Sustainability – Bhattacharya, Vahter, Jarsjö, Kumpiene,
Ahmad, Sparrenbom, Jacks, Donselaar, Bundschuh & Naidu (Eds)
© 2016 Taylor & Francis Group, London, ISBN 978-1-138-02941-5

Metabolomics of arsenic exposure: The Man *vs.* the mouse

M. Stýblo[1], R.C. Fry[1], M.C. Huang[1], E. Martin[1], C. Douillet[1], Z. Drobná[1], M.A. Mendez[1],
C. González-Horta[2], B. Sánchez-Ramírez[2], M.L. Ballinas-Casarrubias[2], L.M. Del Razo[3]
& G.G. García-Vargas[4]

[1]*University of North Carolina at Chapel Hill, Chapel Hill, North Carolina, USA*
[2]*Facultad de Ciencias Químicas, Universidad Autónoma de Chihuahua, Chihuahua, México*
[3]*Departamento de Toxicología, Centro de Investigación y de Estudios Avanzados del IPN,*
  México City DF, Mexico
[4]*Universidad Juárez del Estado de Durango, Gómez Palacio, Durango, México*

ABSTRACT: Mice have been used as laboratory models to study adverse effects associated with human exposures to Arsenic (As). However, differences exist between mice and humans in As metabolism and susceptibility to As toxicity. This paper compares data from two independent studies that examined changes in the mouse and human metabolomes in response to As exposure. In one study, C57BL/6 mice were exposed to 0 or 1 ppm As; the other study involved 86 residents of Chihuahua (Mexico) exposed to 0.1 to 285 ppb As in drinking water. Identical analytical techniques were used to analyze urine and plasma metabolomes in both studies. The numbers of plasma/urinary metabolites associated with As exposure were lower in mice (N = 12/23) as compared to humans (N = 15/61) in spite of the higher exposure level; differences were also found in the pathways enriched for these metabolites. This comparison suggests that the metabolic shifts associated with As exposure differ between mice and humans.

1 INTRODUCTION

The mouse has been the preferred species in laboratory studies examining the adverse effects associated with As exposure in humans. It has been shown, however, that the effects of As in mice may differ from those in humans. For example, the carcinogenic effects of As-exposure found in humans can only be partially reproduced in mice that are genetically modified to increase susceptibility to these effects (Germolec *et al.*, 1998). Similarly, higher levels of As may be needed in mice as compared to humans to develop diabetes (Maull *et al.*, 2012). Although the reported differences in As metabolism may be in part responsible for the different susceptibility of mice and humans to As toxicity (Vahter *et al.*, 1994), more research is needed to justify using the mouse model in studies of As toxicity and to characterize limitations of this model. The present study compares effects of As exposure on mouse and human metabolomes using data from a laboratory and a human population study carried out by our team.

2 METHODS

2.1 *The mouse study*

Seventeen-week old male and female C57BL/6 mice (Jackson Labs) were fed a standard Prolab Isopro RMH 3000 chow (LabDiet) and drank either deionized water or deionized water containing 1 ppm As as arsenite *ad libitum* for 4 weeks. Urine for metabolomics analysis was collected from mice housed individually in metabolic cages for 24 hours without access to food. Plasma was isolated from blood collected prior to sacrifice via submandibular bleeding. All procedures involving mice were approved by the University of North Carolina (UNC) Institutional Animal Care and Use Committee.

2.2 *The population study*

This study (Martin *et al.*, 2015) used samples of fasting plasma and urine collected from 90 diabetic and 86 nondiabetic individuals who were selected from the Chihuahua cohort in Mexico. This cohort was established between 2008 and 2012 to examine the association between As exposure and diabetes (Mendez *et al.*, 2016). Only adults with a minimum of 5-year residency in the study area were recruited for this cohort. Pregnant women and individuals with occupational exposures to As were excluded. The metabolites of As (inorganic As, methyl-As, and dimethyl-As) in urine, and As in drinking water were measured by HG-CT-AAS. For the mouse *vs.* human metabolome comparison, we used only data for the 86 non-diabetic subjects (Table 1). All procedures were approved by the IRBs of UNC and Cinvestav-IPN, and all participants signed a written consent.

Table 1. Basic characteristics of the study participants.

	N (%)	Mean (range)
Females	65 (75.6%)	
Males	21 (24.4%)	
Age (years)		50 (18–78)
BMI[a]		30.1 (19.2–43.6)
As in water (µg/L)		65.8 (0.1–284.7)
≤ 10 µg/L	11 (12.8%)	
≥ 10 µg/L	75 (87.2%)	
≤ 25 µg/L	14 (16.3%)	
≥ 25 µg/L	72 (83.7%)	
As in urine (µg/L)[b]		80.5 (6.11–348.6)

[a]Body mass ratio; [b]sum of As metabolites in urine

2.3 Metabolomic analysis

The metabolic profiling was carried out in the same laboratory using GC-TOF-MS and LC-TOF-MS. A detailed description of the instrumentation, and of data acquisition and analysis is provided in the published human study (Martin et al., 2015).

2.4 Statistical analysis

Multi-variate regression models were used to establish significant associations ($p < 0.05$) between the plasma and urinary metabolites and As exposure in the mouse and the population study (Martin et al., 2015).

3 RESULTS

3.1 Plasma metabolome

The exposure to 1 ppm As had relatively minor effects on the metabolite profiles in mouse plasma. Only 12 metabolites were significantly shifted (all increased), including 11 in plasma of male and 1 in plasma of female mice. The pathway analysis using MetaboAnalyst software determined that 4 pathways of amino acid metabolism and glutathione metabolism were significantly enriched for the metabolites that were changed by As exposure (Table 2). In human plasma, 15 metabolites were associated with As exposure (characterized by sum of As metabolites in urine), but only one pathway in amino acid metabolism was found to be significantly enriched for the exposure-associated metabolites.

3.2 Urine metabolome

The exposure to As had a greater effect on metabolite profiles in urine than in plasma. A total of 23 and 61 urinary metabolites were significantly associated with As exposure in mice and humans,

Table 2. Pathways enriched for the plasma and urinary metabolites associated with As exposure.

	Mice[a]		Humans	
	No. of pathways	No. of metabolites	No. of pathways	No. of metabolites
PLASMA				
Amino acids	4	3	1	1
Other[b]	1	2		
URINE				
Amino acids	3	3	3	10
Energy			5	11
Lipids	2	2[c]		
Vitamins			1	2
Other[d]	1	1		

[a]Both males and females; [b]glutathione metabolism; [c]multiple phosphatidylcholine and acyl-carnitine species were counted as 2 metabolites; [d]nitrogen metabolism.

respectively. Of the 23 exposure-associated metabolites in mouse plasma, 21 were found in plasma of female mice, including 13 phosphatidylcholine and 3 acyl-carnitine species—all increased in response to As exposure. However, the MetaboAnalyst software counted these species only as 2 metabolites and determined that 2 pathways of lipid metabolism were enriched (Table 2). Other pathways enriched for the exposure-associated metabolites were 3 pathways of amino acid metabolism and a pathway of nitrogen metabolism. The pathways enriched for the exposure-associated metabolites found in human urine included 5 pathways of energy metabolism, 3 pathways of amino acid metabolism and 1 pathway of vitamin (riboflavin) metabolism.

4 DISCUSSION

Results of this study suggest that the metabolic responses to As exposure differ between mice and humans. Smaller numbers of metabolites were shifted in mice in spite of the higher exposure level, and different pathways were affected (lipid, glutathione and nitrogen metabolism vs. energy and vitamin metabolism in humans). Notably, the responses in mice were sex-dependent. These findings could help to design future laboratory studies using mice and to guide the translation of data from the mouse studies to humans.

REFERENCES

Germolec, D.R., Spalding, J., Yu, H.S., Chen, G.S., Simeonova, P.P., Humble, M.C., Bruccoleri, A., Boorman, G.A., Foley, J.F., Yoshida, T., Luster, M.I. 1997.

Arsenic enhancement of skin neoplasia by chronic stimulation of growth factors. *Am. J. Pathol.* 153: 1775–1785.

Maull, E.A., Ahsan, H., Edwards, J., Longnecker, M.P., Navas-Acien, A., Pi, J., Silbergeld, E.K., Styblo, M., Tseng, C.H., Thayer, K.A., Loomis, D. 2012. Evaluation of the association between arsenic and diabetes: a National Toxicology Program workshop review. *Environ Health Perspect.* 120: 1658–1670.

Vahter, M. 1994. Species Differences in the metabolism of arsenic compounds. *Appl. Organomet. Chem.* 8: 175–182.

Martin, E., González-Horta, C., Rager, J., Bailey, K.A., Sánchez-Ramírez, B., Ballinas-Casarrubias, L., Ishida, M.C., Gutiérrez-Torres, D.S., Hernández Cerón, R., Viniegra Morales, D., Baeza Terrazas, F.A., Saunders, R.J., Drobná, Z., Mendez, M.A., Buse, J.B., Loomis, D., Jia, W., García-Vargas, G.G., Del Razo, L.M., Stýblo, M., Fry, R. 2015. Metabolomic characteristics of arsenic-associated diabetes in a prospective cohort in Chihuahua, Mexico. *Tox. Sci.* 239: 144: 338–346.

Mendez, M.A., González-Horta, C., Sánchez-Ramírez, B., Ballinas-Casarrubias, L., Hernández, Cerón, R., Viniegra Morales, D., Baeza Terrazas, F.A., Ishida, M.C., Gutiérrez-Torres, D.S., Saunders, R.J., Drobná, Z., Fry, R.C., Buse, J.B., Loomis, D., García-Vargas, G.G., Del Razo, L.M., Stýblo, M. 2016. Chronic exposure to arsenic and markers of cardiometabolic risk-a cross-sectional study in Chihuahua, Mexico. *Environ. Health. Perspect.* 124(1): 104–111.

Arsenic Research and Global Sustainability – Bhattacharya, Vahter, Jarsjö, Kumpiene,
Ahmad, Sparrenbom, Jacks, Donselaar, Bundschuh & Naidu (Eds)
© *2016 Taylor & Francis Group, London, ISBN 978-1-138-02941-5*

Arsenite and its metabolite, methylarsonite, inhibit calcium influx during glucose-stimulated insulin secretion in pancreatic islets

M.C. Huang, C. Douillet & M. Stýblo
University of North Carolina at Chapel Hill, Chapel Hill, North Carolina, USA

ABSTRACT: Numerous epidemiological studies have shown significant associations between chronic exposure to arsenic (As) and an increased risk of diabetes. However, the molecular mechanisms underlying the diabetogenic effects of As exposure have not been sufficiently studied. Previous studies published by our laboratory have shown that inorganic As (iAs) and its trivalent methylated metabolites inhibit Glucose-Stimulated Insulin Secretion (GSIS) in isolated murine pancreatic islets. Here, we investigate the effects of iAs (arsenite) and one of its metabolite, methylarsonite (MMAs), on calcium influx, one of the key steps in GSIS. Our data suggest that iAs and MMAs inhibit calcium influx after glucose stimulation and that MMAs may be more potent than iAs as an inhibitor of this process. Further studies using isolated islets and β-cell lines are warranted to confirm these observations and to identify the molecular targets of iAs and MMAs.

1 INTRODUCTION

Chronic exposure to high levels of inorganic arsenic (iAs) in drinking water have been linked to an increased risk of diabetes (Maull *et al.*, 2012). Studies investigating mechanisms underlying iAs-associated diabetes have addressed insulin signaling and insulin secretion, two pathways that are impaired in diabetes. We and others have shown that *in vitro* exposure to iAs or its methylated trivalent metabolites inhibits glucose-stimulated insulin secretion (GSIS) in pancreatic islets and β-cell lines (Douillet *et al.*, 2013; Lu *et al.*, 2010; Díaz-Villaseñor *et al.*, 2006). The insulin-secreting β-cells are found in the islets of Langerhans in the pancreas. These cells secrete insulin, the primary hormone regulating glucose homeostasis, in response high blood glucose. The metabolism of glucose in β-cells results in increased ATP/ADP ratio which causes closure of ATP-dependent potassium channels, leading to depolarization of the cell membrane and calcium influx via the voltage-gated calcium channels (Rorsman *et al.*, 2012). As such, calcium influx is a key step in GSIS. Here, we present preliminary data from a study examining the effects of arsenite (iAs) and its methylated trivalent metabolite, methylarsonite (MMAs), on calcium influx in isolated murine pancreatic islets.

2 METHODS

2.1 *Pancreatic islet isolation and exposure*

Pancreatic islets were isolated from adult male C57Bl/6 mice (Jackson Laboratories, Bar Harbor, MA) as previously described (Douillet *et al.*, 2013). The isolated islets were incubated overnight in RMPI 1640 at 37°C, 5% CO_2. The islets were cultured in a 6-well plate (40–50 islets per well) in 3 ml of medium containing either iAs (0.5, 1, or 2 µM) or MMAs (0.05, 0.1 or 0.5 µM) for 48 hours. To limit oxidation of iAs and MMAs, the medium was changed after 24 hours.

2.2 *Ratiometric calcium imaging of islets*

Islets exposed to arsenicals and control (unexposed) islets were incubated with 2 µM fura-2 AM (Thermo Fisher Scientific) for 40–50 minutes at 37°C in KRP buffer (114 mM NaCl, 4.7 mM KCl, 1.2 mM KH_2PO_4, 1.16 mM $MgSO_4$, 10 mM HEPES, 2.5 mM $CaCl_2$, 0.1% BSA, 25.5 mM $NaHCO_3$, pH 7.2). The calcium imaging was carried out in the Delta T live-cell imaging flow-chamber (Bioptechs). The islets in the chamber were perfused with KRP buffer containing 2.5 mM glucose (low glucose), followed by KRP buffer containing 16.7 mM glucose (high glucose) at a flow rate of 0.3 ml/min. The fluorescent images were captured using the Olympus IX81 inverted microscope and a Hammamatsu ORCA R2 cooled CCD camera at 340 nm and 380 nm with the Semrock FURA2-C000 dichroic filter sets. Camera exposure times were kept consistent throughout each experiment. Imaging was controlled by Velocity (Improvision) and data was analyzed using ImageJ and in Excel.

3 RESULTS

Examples of the ratiometric signals (340/380 nm) recorded during perfusion of As-exposed and control islets with low- and high-glucose buffers are shown in Figure 1. The data suggest that basal intracellular calcium levels (recorded at low glucose) did not change due to either iAs- or MMAs- exposure. Calcium influx in response to high glucose appeared to be inhibited by both arsenicals, but at different concentrations. A lag in calcium influx was observed in islets exposed to

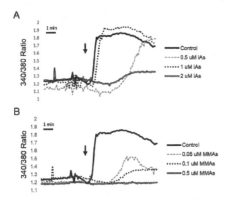

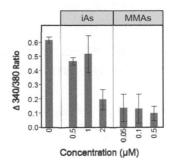

Figure 2. The intracellular calcium signal in control pancreatic islets and in islets exposed to iAs or MMAs for 48 hours and stimulated with 16.7 mM glucose: Mean ±SE, N = 8 for control islets, N = 2–3 for As-treated islets.

Figure 1. Time course of changes in intracellular calcium of control and As-exposed islets: Ratiometric fluorescent signals using fura-2 AM were plotted vs. time to show changes in intracellular calcium of murine islets perfused with low- and high-glucose buffers. Islets were exposed to iAs (A) or MMAs (B) for 48-hours prior to imaging. Data from a single experiment using islets pooled from pancreata of 3 mice is shown. Start of perfusion with the high-glucose (16.7 mM) buffer is indicated by the arrows.

0.5 μM, but not 1 μM iAs; influx was substantially decreased only at 2 μM iAs (Figure 1 A). In contrast, all three concentrations of MMAs, starting at 0.05 μM, inhibited calcium influx (Figure 1B).

Figure 2 summarizes results of several calcium imaging experiments carried out to date. Here, the difference between average of 340/380 ratios recorded during the last three minutes at high- and low-glucose buffers are shown for control islets and islets exposed to iAs or MMAs. These results suggest that iAs is a much weaker inhibitor of glucose-stimulated calcium influx than MMAs, and that MMAs concentrations below 0.05 μM could significantly inhibit this process.

4   DISCUSSION AND CONCLUSIONS

Our previously published data showed that GSIS was significantly decreased in islets exposed for 48 hours to 2 μM iAs and 0.1 μM MMAs and that these exposures were not cytotoxic (Douillet *et al.*, 2013). Preliminary data described here indicate that the same concentrations of iAs and MMAs also inhibit calcium influx after stimulation of islets with high glucose. Previously, iAs has been found to affect calcium transport and signaling in other cells types, including human keratinocytes (Hsu *et. al.*, 2012) or airway epithelial cells (Sherwood *et al.* 2011). However, this is the first report of the effects of iAs and its methylated metabolite MMAs on calcium signaling in pancreatic islets. The molecular mechanisms underlying inhibition of calcium influx in pancreatic islets exposed to iAs and MMAs will be examined in future studies.

ACKNOWLEDGEMENTS

This research was funded by grants 1R01ES 022697–01 A1 to M. S. and 5T32ES007126–32 to M. H. The authors would like to thank Dr. Bagnell and associates at the UNC microscopy facility and Will Ostrom for their assistance in this project.

REFERENCES

Díaz-Villaseñor, A., Sánchez-Soto, M.C., Cebrián, M.E., Ostrosky-Wegman, P. and Hiriart, M. 2006. Sodium arsenite impairs insulin secretion and transcription in pancreatic β-cells. *Toxicol. Appl. Pharmacol.* 214: 30–34.

Douillet, C., Currier, J., Saunders, J., Bodnar, W.M., Matousek, T. and Styblo, M. 2013. Methylated trivalent arsenicals are potent inhibitors of glucose stimulated insulin secretion by murine pancreatic islets. *Toxicol. Appl. Pharmacol.* 267: 11–15.

Fu, J., Woods, C.G., Yehuda-Shnaidman, E., Zhang, Q., Wong, V., Collins, S., Sun, G., Andersen, M.E. and Pi, J. 2010. Low-Level Arsenic Impairs Glucose-Stimulated Insulin Secretion in Pancreatic Beta Cells: Involvement of Cellular Adaptive Response to Oxidative Stress. *Environ. Health Perspect.* 118: 864–870.

Hsu, W.L., Tsai, M.H., Lin, M.W., Chiu, Y.C., Lu, J.H., Chang, C.H., Yu, H.S. and Yoshioka, T. 2012 Differential effects of arsenic on calcium signaling in primary keratinocytes and malignant (HSC-1) cells. *Cell Calcium.* 52: 161–169.

Maull, E.A., Ahsan, H., Edwards, J., Longnecker, M.P., Navas-Acien, A., Pi, J., Silbergeld, E.K., Styblo, M., Tseng, C.-H., Thayer, K.A. and Loomis, D. 2012. Evaluation of the association between arsenic and diabetes: A national toxicology program workshop review. *Environ Health Perspect.* 120: 1658–1670.

Rorsman, P., Braun, M., and Zhang, Q. 2012. Regulation of calcium in pancreatic alpha and beta-cells in health and disease. *Cell Calcium.* 51: 300–308.

Sherwood, C., Lantz, R.C., Burgess, J.L. and Boitano, S. 2011. Arsenic Alters ATP-dependent Ca2+ signaling in human airway epithelial cell wound response. *Tox Sci.* 121 (1): 191–206.

Arsenic Research and Global Sustainability – Bhattacharya, Vahter, Jarsjö, Kumpiene,
Ahmad, Sparrenbom, Jacks, Donselaar, Bundschuh & Naidu (Eds)
© *2016 Taylor & Francis Group, London, ISBN 978-1-138-02941-5*

Assessment of cytogenetic damage among a cohort of children exposed to arsenic through drinking tubewell water in West Bengal, India

A.K. Bandyopadhyay[1], A.K. Giri[1] & S. Adak[2]

[1]*Molecular Genetics Division, CSIR-Indian Institute of Chemical Biology, Kolkata, India*
[2]*Department of Zoology, Seth Anandaram Jaipuria College, Kolkata, India*

ABSTRACT: Arsenic exposure through drinking water and dietary sources mainly from rice has been associated with several health disorders which include cytogenetic damage, cancers and epigenetic alterations, etc. Until date, a comprehensive study on early life exposure in children has not been conducted in arsenic exposed areas in West Bengal. We have assessed the health effects, cytogenetic damage as measured by micronucleus assay in three different cell types in 68 children exposed to arsenic. Clinical examination reported tenderness at the gastric point, with verbal confirmation of gastritis and persistent nausea in exposed children. A significant increase in micronuclei frequencies were observed in all three cell types i.e. lymphocytes, buccal and urothelial cells of exposed children when compared to unexposed children. Our study infers that arsenic exposure through drinking water and dietary sources may lead to induction of cytogenetic damage in children.

1 INTRODUCTION

Arsenic contamination in groundwater is a major problem now in several part of the world. In West Bengal, India about 26 million people are exposed to arsenic through drinking water much above the permissible limit set by the (WHO, 1996). Many possible modes of arsenic toxicity in humans have been proposed, including chromosomal abnormalities, generation of oxidative stress, impaired DNA repair, altered DNA methylation and cell proliferation. Studies on arsenic toxicity suggested association of respiratory symptoms, cardiovascular diseases and certain cancers with arsenic in drinking water (Banerjee *et al.*, 2013a). Recent studies to assess child development and their intelligence have found that both in Bangladesh and US, arsenic exposure through drinking water poses a threat to children (Smith *et al.*, 2013). However, study on an overall clinical manifestation in children due to arsenic is lacking in the population of West Bengal as well as in other affected parts of India. Authors have already reported that rice is a potential source of arsenic exposure in this district (Banerjee *et al.*, 2013b) and most of the children are taking rice as a staple in this district of Murshidabad and rural Bengal. Hence, in this study we attempted to evaluate the health effects and genetic damage in the children exposed to arsenic through drinking water and rice in West Bengal, India.

2 METHODS

2.1 Selection of study cohorts

Study participants includes 68 children aged 5–15 years from the arsenic exposed districts of Murshidabad, while the unexposed subjects were recruited from the East Midnapore district, where arsenic content is within the safe limit laid down by WHO. Selected exposed and unexposed samples were matched for age-sex, nutritional status to eliminate potential confounders of the outcome in relation to arsenic exposure and health effects.

2.2 Sample collection and analysis

Consultant pediatrician thoroughly examined the all participating children. After a well informed, written consent from the guardian of the study participants, blood, urine, buccal samples and drinking water samples were collected for further analysis. The study was in accordance to the Declaration of Helsinki II and approved by the Institutional Ethical Committee on human subjects of CSIR-Indian Institute of Chemical Biology. Estimation of arsenic content in water and the urine samples was estimated using Atomic Absorption Spectrometer (AA-7000), Shimadzu, Japan using GFA-7000A graphite furnace atomizer.

2.3 Micronucleus assay for measurement of cytogenetic damage

Micronucleus assay was performed from three different cell types namely lymphochyte, urothelial and buccal cells from blood, urine and buccal mucosa cells respectively following standard protocol (Basu et al., 2004). All slides were observed under the Nikon Eclipse 80i microscope and scoring was done as described in Basu et al. (2004).

2.4 Statistical analyses

The statistical analysis was done using GraphPad InStat Software. Since this is a matched case-control study pair wise t-test was performed to compare the central tendencies of continuous independent variables (As-content in water, urine) between the arsenic exposed and unexposed groups.

3 RESULTS AND DISCUSSION

3.1 Demographic characteristics and arsenic exposure of the study participants

The arsenic content measured in the urine of the exposed children (av. 75.23 µg/L) is significantly higher than the unexposed children ($p < 0.0001$).

3.2 Induction of cytogenetic damage in arsenic exposed children

The induction of cytogenetic damage as measured by micronucleus assay performed in the three cell types i.e. buccal cells, urothelial cells and lymphocyte are expressed as MN frequency per 1000 cells and is significantly higher ($p < 0.0001$) in the arsenic exposed group than the unexposed group for all the cell types assessed (Table 1). Since this population does not have any other environmental exposure and the children do not consume any seafood during the last 6 months of the sample collection. So

Table 1. Comparison of cytogenetic damage in arsenic exposed and unexposed children.

Study Parameters	Unexposed children	Exposed Children
Lymphocyte MN/ 1000 Cells	2.02 ± 0.67 (N = 44)	3.01 ± 1.18[a] (N = 57)
Buccal MN/1000 Cells	1.22 ± 0.31 (N = 51)	2.76 ± 0.33[a] (N = 68)
Urothelial MN/1000 Cells	1.52 ± 0.52 (N = 47)	2.50 ± 0.66[a] (N = 61)

Paired t test; Pa < 0.0001 when compared with arsenic unexposed children. N = Number of children.

the present results suggests that exposure to arsenic through drinking water brings about considerable cytogenetic damage in children.

3.3 Arsenic exposure possibly leads to gastritis

More than 50% of the arsenic exposed children complained of nausea, refusal to food. On clinical examination it was found that gastric point was highly tender. However these are clinical evaluations, which are yet to be confirmed by extensive investigation.

4 CONCLUSIONS

Hence our study is first report to bring forth relevance of the micronucleus assay as one of the important biomarkers of genetic damage to study arsenic toxicity in children. Also our study unveils the gastric problem in the children exposed to arsenic both through water and rice.

ACKNOWLEDGEMENTS

This work has been supported by Emeritus Project Grant no. 21(0885)/12/EMR-II to A.K.G and University Grants Commission (UGC) grant (PSW-069/13-14 (ERO) #WC2-124) to S.A.

REFERENCES

Banerjee, N., Paul, S., Sau, T.J., Das, J.K., Bandyopadhyay, A., Banerjee, S. and Giri, A.K. 2013a. Epigenetic modifications of DAPK and p16 genes contribute to arsenic-induced skin lesions and nondermatological health effects. Toxicol. Sci. 135:300–8.

Banerjee, M., Banerjee, N., Bhattacharjee, P., Mondal, D., Lythgoe, P.R., Martínez, M., Pan, J., Polya, D.A. and Giri, A.K. 2013b. High arsenic in rice is associated with elevated genotoxic effects in humans. Sci. Rep. 3: 2195. doi: 10.1038/srep02195.

Basu, A., Ghosh, P., Das, J. K., Banarjee, A., Ray, K. and Giri, A.K. 2004. Micronuclei as biomarkers of carcinogen exposure in populations exposed to arsenic through drinking water in West Bengal, India: a comparative study in 3 cell types. Cancer Epidemiol. Biomark. Prev. 13: 820–827.

Smith, A.H., Yunus, M., Khan, A.F., Ercumen, A., Yuan, Y., Smith, M.H., Liaw, J., Balmes, J., von Ehrenstein, O., Raqib, R., Kalman, D., Alam, D.S., Streatfield, P.K. and Steinmaus, C. 2013. Chronic respiratory symptoms in children following in utero and early life exposure to arsenic in drinking water in Bangladesh. Int. J. Epidemiol. 42: 1077–86.

WHO. 1996. Guidelines for Drinking Water Quality. Health Criteria and other supporting information 2. World Health Organisation, Geneva. pp. 940–949.

Arsenic Research and Global Sustainability – Bhattacharya, Vahter, Jarsjö, Kumpiene,
Ahmad, Sparrenbom, Jacks, Donselaar, Bundschuh & Naidu (Eds)
© 2016 Taylor & Francis Group, London, ISBN 978-1-138-02941-5

Pro-atherogenic effects of arsenic and the protective potential of selenium and high-selenium lentils

R.M. Krohn[1], M. Lemaire[1], K.K. Mann[2] & J.E.G. Smits[1]
[1]*Faculty of Veterinary Medicine, University of Calgary, Calgary, Alberta, Canada*
[2]*Lady Davis Institute for Medical Research, McGill University, Montréal, Canada*

ABSTRACT: Cardiovascular Disease (CVD) is the major cause of death worldwide, and arsenic (As) intake through drinking water is a well-known risk factor for CVD. We tested the potential of high-selenium (Se) lentils from Saskatchewan in reducing As-triggered atherosclerosis in an ApoE$^{-/-}$ mouse model and are studying the effects of As in Human Aortic Endothelial Cells (HAoEC). Mice exposed to 200ppb As in their drinking water and control mice received one of three lentil diets: Se-deficient (0.009 mg/kg), Se-adequate (0.16 mg/kg), or Se-fortified (0.3 mg/kg). After 13 weeks, As-induced atherosclerotic lesions in the aortic arch and sinus were substantially reduced in As-exposed mice on the Se-fortified diet (p < 0.05). Arsenic inhibited HAoEC wound healing, and treatment with selenium restored wound healing capacity. Arsenic-induced expression of IL-8 in HAoEc was inhibited by selenite. These data suggest a protective effect of selenium, and especially high-selenium lentils against the pro-atherogenic actions of As.

1 INTRODUCTION

Cardiovascular Disease (CVD) is the number one cause of death in the developed world. While it is established that chronic arsenic (As) exposure is associated with vascular disease in humans, the risk for CVD from low to moderate As exposure (< 300 µg/L) is difficult to evaluate since epidemiological studies have revealed inconsistent finding. More recent studies have emerged, however, that link even low As exposure with CVD. In a previous study it was shown that As exposure as low as 200 µg/L caused increased atherosclerosis in a mouse model (Lemaire *et al.*, 2011).

In this study, we examined the potential of naturally Se-rich lentils from the prairie regions of Canada to counteract As-induced atherosclerosis in apolipoprotein E-deficient (apoE$^{-/-}$) mice with moderate As exposure, as well as to identify the concurrent cost of selenium deficiency. We also investigated the pro-atherogenic effects conferred by As on human aortic endothelial cells (HAoEC) and the potential of different Se species to prevent these effects.

2 METHODS/EXPERIMENTAL

2.1 Lentil diets

Three different rodent diets were formulated, each containing 50% lentils. The high-Se diet (Se = 0.3 mg/kg) contained 50% of Saskatchewan grown high-Se lentils, the Se-adequate diet (0.16 mg/kg) contained a mix of high- and low-Se lentils, and the Se-deficient diet (0.009 mg/kg) contained 50% low-Se lentils. Apart from the Se content, the diets were nutritionally balanced.

2.2 Mouse experiment

Six groups of ApoE$^{-/-}$ male mice received tab water or 200ppb As in their drinking water and one of the three lentil diets.

2.3 Atherosclerotic lesion and plaque composition assessment

Atherosclerotic lesion formation was assessed in paraformaldehyde-fixed aortic arches and sinus sections via Oil-red-O staining. Smooth Muscle Cell (SMC) and macrophage content of the lesions in the aortic sinus were identified by immunofluorescence, using anti-α-SMC actin (α –SMA) or MOMA-2 monoclonal antibodies.

2.4 Serum lipid profile

Serum was analyzed for high density-and low density-lipoprotein (HDL and LDL)-cholesterol, and triglycerides levels by IDEXX Laboratories (Toronto, ON).

2.5 Analysis of hepatic glutathione

Reduced glutathione (GSH) and oxidized GSH (GSSG) levels were determined via enzymatic reaction in a plate reader.

2.6 Wound healing assay

HAoEC were grown to confluence and incubated with 5µM As and selenite or selenomethionine (SeMet) at various concentrations for 24 hrs. Then a scratch wound was made in the monolayer and wound closure was measured after 4hrs.

2.7 IL-8 expression

IL-8 expression in the supernatant of HAoEC was assessed via ELISA after 24 hr incubation of cells with 5μM As and selenite or SeMet.

3 RESULTS AND DISCUSSION

After 13 wks of As-exposure, mice developed plaques in the aorta. Minor plaque development was observed in the control mice. Plaque development was significantly reduced in the aortic sinus, or completely abolished in the aortic arch (Fig. 1) of As-exposed mice on the high-Se lentil diet. Here, the diets had no effect on plaque development in control mice.

Se deficiency plus As resulted in significantly higher Low Density Lipoproteins (LDLs), the so-called 'bad' cholesterol, and Se deficiency appeared to cause a less stable plaque phenotype, indicated by higher macrophage and lower SMC content.

In As-exposed mice only, we observed the activation of the Glutathione Peroxidase (GPX) radical scavenging system, depicted by the decrease in GSH and increase in GSSG. GPX is a selenopro-

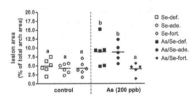

Figure 1. Atherosclerotic lesions in the aortic arch in the three dietary groups. Open symbols are control animals, and closed symbols resemble As-exposed animals. [a,b] Groups marked with *a* are significantly different from groups marked with *b* ($p < 0.05$).

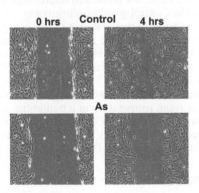

Figure 2. Examples of endothelial wound healing. HAoEC were incubated with or without As for 24hrs prior injury with a pipet tip. Wound healing was measured after 4 hrs.

tein, therefore Se in the diet was required in order to activate this antioxidant system in response to As.

Damaged vascular endothelium is a major driver for atherogenesis, therefore we tested the effect of As on endothelial wound healing. Wound closure of HAoEC was delayed when cells were incubated with 5 μM As prior to the experiment (Fig. 2), and healing was restored with the addition of selenite (at 1 μM) and SeMet (at 0.01 μM). Selenite was chosen as it is commonly used in tablet fomr as antidote for As poisoning. SeMet is the major Se species found in the lentils.

IL-8 is a cytokine in humans, which is released by compromised endothelial cells and plays a role in plaque development. Selenite, but not SeMet, inhibited As-induced IL-8 expression in HAoEC.

4 CONCLUSIONS

The experimental mouse study emphasizes the importance of adequate Se status for cardiovascular health, and suggests that Se-deficiency constitutes a major health risks to As-exposed individuals.

Our findings with mice, suggest that dietary Se intake above the US-FDA recommended daily allowance may be beneficial to reduce the risk of As-related atherosclerosis and warrants human-specific investigation. A possible mechanism by which As disrupts atherogenesis is the interference with endothelial wound healing, as demonstrated on human endothelial cells. Further investigation will shed light on the genes involved in this process.

ACKNOWLEDGEMENTS

We thank Jaime Rodríguez-Estival, and Barry Goertz for technical assistance, and Albert Vandenberg from the University of Saskatchewan (UofS) for producing and providing the two lentil types.

REFERENCES

James, K.A., Byers, T., Hokanson, J.E., Meliker, J.R., Zerbe, G.O., et al. 2015. Association between lifetime exposure to inorganic arsenic in drinking water and coronary heart disease in Colorado residents. *Environ. Health Persp.* 123(2): 128–134.

Krohn, R., Lemaire, M., Negro Silva, L., Mann, K., Smits. J.E. 2016. High-selenium lentil diet protects against arsenic-induced atherosclerosis in mice. *J. Nutr. Biochem.* 27: 9–15.

Lemaire, M., Lemarie, C.A., Molina, M.F., Schiffrin, E.L. Lehoux, S., Mann, K.K. 2011. Exposure to moderate arsenic concentrations increases atherosclerosis in ApoE-/- mouse model. *Toxicol. Sci.* 122(1): 211–221.

Moon, K.A., Guallar, E., Umans JG, Devereux RB, Best LG, et al. 2013. Association between exposure to low to moderate arsenic levels and incident cardiovascular disease. A prospective cohort study. *Ann. Intern. Med..* 159(10): 649–659.

Arsenic Research and Global Sustainability – Bhattacharya, Vahter, Jarsjö, Kumpiene,
Ahmad, Sparrenbom, Jacks, Donselaar, Bundschuh & Naidu (Eds)
© 2016 Taylor & Francis Group, London, ISBN 978-1-138-02941-5

Reciprocal expressions of VEGF and the numbers of dendritic cells in arsenic-induced skin cancer: A plausible cause of impaired dendritic cell activation in arsenic carcinogenesis

C.-H. Lee[1], C.-H. Hong[2] & H.-S. Yu[3]

[1]*Department of Dermatology, Kaohsiung Chang Gung Memorial Hospital, Chang Gung University College of Medicine, Kaohsiung, Taiwan*
[2]*Department of Dermatology, Kaohsiung Veterans General Hospital, Kaohsiung, Taiwan*
Department of Dermatology, Faculty of Medicine, National Yang-Ming University, Taipei, Taiwan
[3]*National Environmental Toxicology Center, National Health Research Institutes, Zhunan, Taiwan*

ABSTRACT: Arsenic remains an important environmental hazard that causes several cancers. Arsenic-induced Bowen's disease (As-BD), a skin carcinoma in situ, is the most common arsenical cancer. Patients with As-BD have an impaired contact hypersensitivity response. We have reported that arsenic paradoxically impairs Dendritic Cell (DC) migration through STAT3 upregulation and VEGF production from epidermal keratinocytes using cell, tissue, and animal models. In this study, we further demonstrated an increased expression of VEGF and decreased numbers of DC in epidermis. More importantly, there are spatial interactions and reciprocal changes in VEGF-DC in epidermis from tissue with As-BD, further validating that immune interactions in the microenvironment play an important role in regulating the disease course of arsenical cancers.

1 INTRODUCTION

Arsenic remains an important environmental threat to human health. Hundreds of millions people in the world, including those in Bangladesh, West Bengal, and Mongolia, remain being exposed to arsenic through drinking well water contaminated with arsenic. Arsenic-induced Bowen's disease (As-BD) is the most common arsenical cancer. It tends to influence multiple areas of the skin. As-BD is usually restricted to the epidermis for several decades, although a few of them may subsequently evolve to invasive carcinoma. In the underlined dermis, there are usually moderate immune cell infiltrates (Lee *et al.*, 2011).

Cell-mediated immunity is dysregulated in patients with As-BD (Yu *et al.*, 1998). In patients with As-BD, contact hypersensitivity responses to 2,4-Dinitrochlorobenzene (*DNCB*) is markedly impaired (Yu *et al.*, 1998). Our previous study showed that epidermal keratinocytes from As-BD preferentially induce the apoptosis of CD4+ cells in the cutaneous environment by Fas-FasL interactions (Liao *et al.*, 2009). For dendritic cells, we demonstrated that arsenic induces STAT3 expressions in cultured keratinocytes, in addition, we showed that Vascular Endothelial Growth Factor (VEGF) production from keratinocytes by arsenic is abolished when STAT3 activations are abrogated by interference RNA (Hong *et al.*, 2015). In fact, VEGF from tumors interferes with DC differentiation and

maturation (Gabrilovich *et al.*, 1996). However, whether VEGF production from epidermal keratinocytes would interact spatially with dendritic cells in As-BD remains to be determined.

2 METHODS/EXPERIMENTAL

2.1 *Immunohistochemical and immunofluorescent exams of arsenic-induced Bowen's disease*

Skin specimens of As-BD were obtained from endemic areas in southwestern Taiwan, where artesian water was contaminated by high concentrations of arsenic. Control group specimens were obtained from the non-sun-exposed skin of age-comparable patients who did not live in the endemic areas and did not have a previous history of cancer. The tissues were fresh-frozen, fixed in acetone at 20°C for 5 minutes, air-dried for 5 minutes, and washed in PBS for 15 min. They were then incubated overnight at room temperature in a humidifier, with or without a mouse anti-VEGF (1:500) antibody and a mouse anti-CD207 antibody (for dendritic cells, 1:500) (eBioscience, San Diego, CA).

3 RESULTS AND DISCUSSION

3.1 *Increased expressions of VEGF in As-BD*

We first examined the expression of VEGF from the tissue with As-BD and compared it from skin tissue of normal controls by immunohistochemistry. The result showed that while expression of VEGF is minimal in normal skin, the expression of VEGF is increased in epidermis of As-BD, with a localization at intercellular space (Fig. 1).

3.2 Decreased dendritic cell numbers in epidermis of As-BD

We then examined the numbers of dendritic cells from tissue with As-BD and compared it from skin tissue of normal controls by immunohistochemistry with CD207 antibody. The result showed that the numbers of dendritic cells are decreased in As-BD than those in normal skin (Fig. 2). The characteristic features of dendritic cells are evident for its featured dendritic expressions of CD207.

3.3 Colocalizations of VEGF from keratinocytes and the CD207 in dendritic cells

Finally, we overlaid the expressions of CD207 and VEGF in skin from normal controls and As-BD.

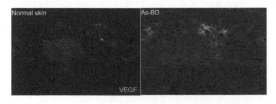

Figure 1. Increased VEGF expressions in As-BD.

Figure 2. Decreased epidermal dendritic cells in As-BD.

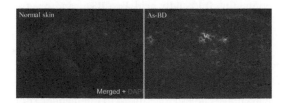

Figure 3. Colocalizations of VEGF from keratinocytes and CD207 in dendritic cells (Merged, VEGF-Green+CD207-Red).

The result showed that expressions of CD207 and VEGF are colocalized (Fig. 3), further validating the spatial interactions and close vicinity between VEGF from keratinocytes and the CD207 dendritic cells.

4 CONCLUSIONS

Dendritic cell migration and the contact hypersensitivity response are impaired in patients with arsenical cancers. The enhanced VEGF production in epidermis from As-BD decreased the dendritic cell numbers in the epidermis, as evidenced by spatial colocalization and the reciprocal expressions of VEGF and CD207. These results may explain why arsenic paradoxically impairs the dendritic cell activation in the microenvironment of arsenical skin cancers.

ACKNOWLEDGEMENTS

This work is supported by grants from Ministry of Science and Technology in Taiwan (MOST103-2314-B-182 A-020, Dr. Lee; MOST102-2314-B-037-044, Dr. Yu; MOST102-2314-B-010-005-MY2, Dr. Hong), National Health Research Institute (NHRI-EO-096-PP-11 and NHRI CN-PD-9611P, Dr. Yu), and Chang Gung Medical Research Program (CMRPG8C0821 and CMRPG8D1571, Dr. Lee).

REFERENCES

Gabrilovich, D.I., Chen, H.L., Girgis, K.R., Cunningham, H.T., Meny, G.M., Nadaf, S., Kavanaugh, D., Carbone, D.P. 1996. Production of vascular endothelial growth factor by human tumors inhibits the functional maturation of dendritic cells. *Nat. Med.* 2: 1096–1103.

Hong, C.H., Lee, C.H., Chen, G.S., et al. 2015. STAT3-dependent VEGF production from keratinocytes abrogates dendritic cell activation and migration by arsenic: a plausible regional mechanism of immunosuppression in arsenical cancers. *Chem Biol Interact.* 227: 96–103.

Lee, C.H., Liao, W.T., Yu, H.S. 2011. Aberrant immune responses in arsenical skin cancers. *Kaohsiung J. Med. Sci.* 27: 396–401.

Liao, W.T., Yu, C.L., Lan, C.C., Lee, C.H., Chang, L.W. et al. 2009. Differential effects of arsenic on cutaneous and systemic immunity: Focusing on CD4+ cell apoptosis in patients with arsenic-induced Bowen's disease. *Carcinogenesis* 30: 1064–1072.

Yu, H.S., Chang, K.L., Yu, C.L., Wu, C.S., Chen, G.S., Ho, J.C. 1998. Defective IL-2 receptor expression in lymphocytes of patients with arsenic-induced Bowen's disease. *Arch. Dermatol. Res.* 290: 681–687.

Arsenic Research and Global Sustainability – Bhattacharya, Vahter, Jarsjö, Kumpiene,
Ahmad, Sparrenbom, Jacks, Donselaar, Bundschuh & Naidu (Eds)
© 2016 Taylor & Francis Group, London, ISBN 978-1-138-02941-5

Arsenic impaired cholesterol efflux by inhibiting ABCA1 and ABCG1

Z. Yuan[1], X.H. Tan[1], X.L.Xu[1], J.X. Zhang[1,2], X. Nan[1,2], Y.H. Li[1,2] & L. Yang[1,2]
[1]*School of Medicine, Hangzhou Normal University, Hangzhou, Zhejiang, P.R. China*
[2]*Ministry of Education Key Laboratory of Xinjiang Endemic and Ethnic Disease, Shihezi University, Shihezi,*
Xinjiang, P.R. China

ABSTRACT: The impaired efflux capacity of macrophages for cholesterol causally linked to the pathogenesis of atherosclerosis. Exposure to arsenic is a risk factor of atherosclerosis. However, the effect of arsenic exposure on cholesterol efflux is still poorly understood. In the present study, we investigated the effect of arsenic treatment on cholesterol efflux and the expression levels of ABCA1 and ABCG1 in macrophage cells. The results indicate that arsenic decreases the expression of ABCA1 and ABCG1 and inhibites the cholesterol efflux rate, in a dose dependent manner both in THP-1 macrophage cell lines and mouse primary macrophage cells. In conclusion, reduced cholesterol efflux by inhibiting ABCA1 and ABCG1 may contribute atherosclerosis by arsenic.

1 INTRODUCTION

Arsenic is a widespread environmental contaminant to which millions of people are exposed worldwide. Epidemiologic evidence indicates that humans exposed to arsenic have an increased risk not only of cancer but also of developing cardiovascular diseases, such as ischemic heart disease and atherosclerosis (Rosenson *et al.*, 2012, Stea *et al.*, 2014). Despite the clear epidemiologic links between arsenic and an increased risk of atherosclerosis, the mechanisms by which arsenic enhances atherosclerosis are unclear. HDL particles are critical acceptors of cholesterol from lipid-laden macrophages and thereby participate in the maintenance of net cholesterol balance in the arterial wall and in the reduction of arterial cholesterol-loaded macrophages (Du *et al.*, 2015). ATP-Binding Cassette Transporter A1 (ABCA1) (Heinecke, 2015; Liu *et al.*, 2012; Wang & Smith, 2014) and ABCG1 (Liu *et al.*, 2012; Yvan-Charvet *et al.*, 2010) are the two important channels for cholesterol-mediated efflux from macrophage foam cells. The effect of arsenic on cholesterol efflux and the expression level of ABCA1 and ABCG1 are still poorly studied. In the present study, the THP-1 cells and the primary mouse macrophage cells obtained from wild type C57B/6 mice were treated with arsenic, and the reverse cholesterol efflux been determined by scintillation counting technique. ABCA1 and ABCG1 mRNA and protein levels were measured by quantitative real time PCR (qPCR) and Western-blot assay.

2 METHODS/EXPERIMENTAL

2.1 Cell lines and chemicals

THP-1 cells and isolate mouse macrophage cells were grown in RPMI 1640 medium containing 10% fetal bovine serum and 1% Penicillin-Streptomycin Solution (HyClone). $NaAsO_2$ (Sigma) was prepared into 0.5 mol/L stock solution in 0.1 N sodium hydroxide and diluted in PBS before use.

2.2 Reverse cholesterol efflux

Oxidized-LDL (50 µL, 12.5 µg) mixed 1,2,3-*N*-[³H] cholesterol (1 µCi) for 15 minutes. THP-1 macrophages (5×10^4 cells per well) were labeled in RPMI 1640 medium containing mixture for 24 hours. Unincorporated ³H was then removed by extensive washing in PBS containing 0.1% fatty acid–free BSA. Cells were subsequently stimulated NaA_sO_2 with or without 22R-OHC/9-*cis*-RA in RPMI medium 1640 containing 0.2% fatty acid–free BSA for 6 hours. Cells were washed in PBS containing 0.1% fatty acid–free BSA, and reverse cholesterol efflux was performed by incubating the cells in RPMI 1640 medium containing 0.1% fatty acid–free BSA and 10 µg/mL of purified human Apo-A1 for 18 hours. ³H was measured in both media and cell extracts, and the percentage of total cholesterol efflux was calculated as follows: percentage cholesterol efflux = [³H in media/(³H in media + ³H in extracts)] × 100.

3 RESULTS AND DISCUSSION

3.1 *Aresnic inhibit ABCA1 and ABCG1 in THP-1 human macrophages*

The ABCA1 expression level was very low in macrophage cells, and Liver X Receptor (LXR) ligand 22R-OHC and 9-cis-RA strong induces ABCA1 expression (Padovani *et al.*, 2010). The THP-1 cells were pretreated with 4 µg/mL of 22R-OHC and 10 µmol/L of 9-cis-RA to increase the expression level of ABCA1. Then, THP-1 macrophages were incubated with different concentrations of $NaAsO_2$ (0.625, 1.25, 2.5 and 5 µmol/L). LXR ligand treatment for 6 hours induced a 3-fold increase the expression of ABCA1, while $NaAsO_2$ treatment significantly inhibited ABCA1 expression (fig1. A and B). $NaAsO_2$ treatment decreased mRNA and protein levels of ABCA1 and ABCG1, in a dose-dependent manner (Fig. 1 C and D).

3.2 *Arsenic inhibits cholesterol efflux in THP-1 human macrophages*

The THP-1 cells were treated with LXR agonist and different dose arsenic as described above. The cholesterol efflux rates were determined by scintillation counting technique. LXR ligand treatment for 6 hours induced a 3-fold increase of total effluxed cholesterol. Arsenic reduced the cholesterol efflux in a dose-dependent manner (Fig. 2).

3.3 *Arsenic inhibit ABCA1 and ABCG1 in primary macrophages cells*

The primary macrophage cells isolated from suckling C57B/6 mice. And the effect of ABCA1 and

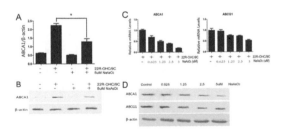

Figure 1. ABCA1 and ABCG1 decreased in arsenic treated THP-1 human macrophage cells. (A) THP-1 human macrophages were induced with 4 µg/mL of 22R-OHC and 10 µmol/L of 9-*cis*-RA, with or without 5 µmol/L of $NaAsO_2$ for 6 hours. The mRNA levels isolated from cells were analyzed by qPCR. And, (B) The protein levels assessed by Western blot. (C) THP-1 human macrophages were treated LXR agonist and different concentrations of arsenic for 6 hours, then the mRNA levels of ABCA1 and ABCG1 were determined by qPCR. And, (D) the protein of ABCA1 and ABCG1 been assessed by Western blot.

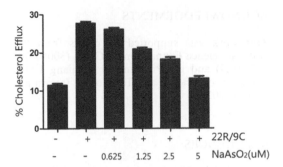

Figure 2. Arsenic inhibits cholesterol efflux from THP-1 human macrophages. THP-1 human macrophages were loaded with [³H]cholesterol, washed, and induced with 4 µg/mL of 22R-OHC and 10 µmol/L of 9-*cis*-RA, with or without 0.625, 1.25, 2.5 or 5 µmol/L of $NaAsO_2$ for 6 hours. After incubating in fatty acid-free medium for18 hours, the amount of radiolabeled cholesterol in medium and cell extracts were determined by scintillation counting technique. And the% cholesterol of efflux expressed as percentage of total cholesterol.

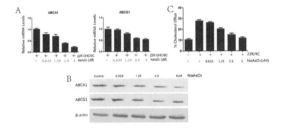

Figure 3. Arsenic decrease ABCA1 and ABCG1 expression, and reduce cholesterol efflux in primary macrophages cells. (A) The primary macrophages obtained from C57B/6 mouse were treated as previously stated, and the mRNA levels of ABCA1 and ABCG1 measured by qPCR. And, (B) The protein levels were analyzed by Western blot. (C) The primary macrophages cells treated with arsenic as described above for THP-1 cells. And, the% cholesterol of efflux were determined.

ABCG1 expression, and cholesterol efflux were tested as above. The results, as we expect, showed that arsenic decreased ABCA1 and ABCG1, and reduced cholesterol efflux from primary mouse macrophage cells (Fig. 3).

4 CONCLUSIONS

Arsenic decreases the expression of ABCA1 and ABCG1, and inhibits cholesterol efflux, both in THP-1 human macrophages and primary mouse macrophages. Inhibition of ABCA1 and ABCG1 expression and reduction of cholesterol efflux rate may contribute increasing the blood lipid levels and the pro-atherogenic effects of Arsenic.

ACKNOWLEDGEMENTS

This work was supported by grants from the Natural Science Foundation of China (30060074; 30560129) and the Program for Zhejiang Leading Team of Science and Technology Innovation (2011R50021).

REFERENCES

Du, X.M., Kim, M.J., Hou, L., *et al.* (2015). HDL particle size is a critical determinant of ABCA1-mediated macrophage cellular cholesterol export. *Circ. Res.* 116: 1133–1142.

Heinecke, J.W. 2015. Small HDL promotes cholesterol efflux by the ABCA1 pathway in macrophages: implications for therapies targeted to HDL. *Circ. Res.* 116: 1101–1103.

Liu, X., Xiong, S.L., and Yi, G.H. 2012. ABCA1, ABCG1, and SR-BI: Transit of HDL-associated sphingosine-1 -phosphate. *Clin. Chim. Acta* 413: 384–390.

Padovani, A.M., Molina, M.F., Mann, K.K. 2010. Inhibition of liver x receptor/retinoid X receptor-mediated transcription contributes to the proatherogenic effects of arsenic in macrophages in vitro. *Arterioscler Thromb. Vasc. Biol.* 30: 1228–1236.

Rosenson, R.S., Brewer, H.B., Jr., Davidson, W.S., *et al.* 2012. Cholesterol efflux and atheroprotection: advancing the concept of reverse cholesterol transport. *Circulation* 125: 1905–1919.

Stea, F., Bianchi, F., Cori, L., *et al.* (2014). Cardiovascular effects of arsenic: clinical and epidemiological findings. *Environ Sci Pollut Res Int* 21, 244–251.

Wang, S., Smith, J.D. 2014. ABCA1 and nascent HDL biogenesis. *Biofactors* 40, 547–554.

Yvan-Charvet, L., Wang, N., Tall, A.R. 2010. Role of HDL, ABCA1, and ABCG1 transporters in cholesterol efflux and immune responses. *Arterioscler Thromb Vasc Biol* 30, 139–143.

3.4   Arsenic and genomic studies

Arsenic Research and Global Sustainability – Bhattacharya, Vahter, Jarsjö, Kumpiene,
Ahmad, Sparrenbom, Jacks, Donselaar, Bundschuh & Naidu (Eds)
© 2016 Taylor & Francis Group, London, ISBN 978-1-138-02941-5

Maternal polymorphisms in arsenic (+3 oxidation state)-methyltransferase AS3MT are associated with arsenic metabolism and newborn birth outcomes: Implications of major risk alleles and fetal health outcomes

R.C. Fry[1], E. Martin[1], K.S. Kim[1], L. Smeester[1], M. Stýblo[1], F. Zou[1], Z. Drobná[1],
M. Rubio-Andrade[2] & G.G. García-Vargas[2]

[1]*University of North Carolina at Chapel Hill, Chapel Hill, North Carolina, USA*
[2]*Universidad Juárez del Estado de Durango, Gómez Palacio, Durango, México*

ABSTRACT: Arsenic (+3 oxidation state) methyltransferase (AS3MT) is the key enzyme in the metabolism of inorganic arsenic (iAs). Polymorphisms of AS3MT have been shown to influence adverse effects associated with exposure to iAs in adults, but little is known about its role in iAs metabolism in pregnant women and newborn birth outcomes. The relationship between seven Single Nucleotide Polymorphisms (SNPs) of AS3MT and concentrations of iAs and its methylated arsenic metabolites were assessed in mother-baby pairs of the Biomarkers of Exposure to ARsenic (BEAR) cohort in Gomez Palacio, Mexico. Comparisons of the genotype distributions among women in Argentina and Bangladesh showed that the BEAR cohort has an allelic frequency that more closely resembles the population in Bangladesh. These data highlight that the major allele in the population in Gómez Palacio represents an at-risk allele for arsenic toxicity.

1 INTRODUCTION

Inorganic exposure during pregnancy has been associated with adverse health and infant outcomes including increased risk of spontaneous abortion, stillbirth, infant mortality, low birth weight, decreased head and chest circumferences, and increased risk of infection in infants (Vahter, 2009). In humans, ingested iAs is biotransformed via multiple consecutive methylation steps into methylated metabolites, Methyl-Arsenic (MMAs) and Dimethyl-Arsenic (DMAs), that has been described as enhancing removal and thus decreasing the overall toxicity of arsenic (Vahter, 2002). In general, individuals who are thought to be efficient in iAs methylation have lower urinary levels of iAs (10–20%) and MMAs (10–20%) and a higher level of DMAs (60–80%) (Vahter, 2002). They have also been shown to have less demonstrable levels of health effects. In the present study, we examined the impact of genotype on arsenic metabolism in the Biomarkers of Exposure to ARsenic (BEAR) Cohort and compared the allelic frequencies to the frequencies observed in women in San Antonio de los Cobres, Argentina and Matlab Bangladesh to examine the impact of allelic frequency on arsenic metabolism.

2 METHODS

2.1 *Ethics statement, study subjects and sample collection*

This study was approved by the Institutional Review Boards of the University of North Carolina at Chapel Hill and Universidad Juárez del Estado de Durango (UJED). A total of 200 pregnant women residing in Gómez Palacio, in the State of Durango, Mexico, were recruited at the General Hospital of Gómez Palacio to participate in the BEAR prospective pregnancy cohort. At birth, maternal and fetal DNA samples and birth outcome measures were collected from all mother-baby pairs as detailed previously (Rager *et al.*, 2014, Laine *et al.*, 2015).

2.2 *DNA isolation and genotyping*

DNA was isolated from the 200 maternal whole blood. Seven polymorphisms in the *AS3MT* gene (ID: 57412; NM_020682.3; NP_065733.2) that have been previously associated with inter-individual differences in iAs metabolism and/or *AS3MT* expression were evaluated. These SNPs represent functionally tested (rs7085104, rs3740400, rs3740390, rs11191439, rs10748835, rs1046778) or validated (rs3740393) TaqMan assays purchased from AB Applied Biosystems (Foster City, CA).

2.3 *Statistical analysis*

SNP genotypes were numerically coded as zero, one or two as the number of copies of the less common/minor allele. To examine the relationships between each SNP and arsenic species multiple linear regression was performed. *A priori* covariates were selected based on known factors that influence birth outcomes as well as metabolism. These factors included: U-tAs (as a measure of exposure), maternal age, smoking status, drinking status, and education level (a surro-

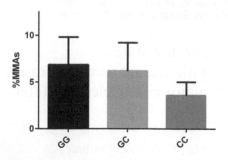

Figure 1. Representative plot showing the associations between %MMAs and genotype of rs3740393. The major allele for this frequency is G (frequency = 0.75) and the less common allele is C (frequency = 0.25).

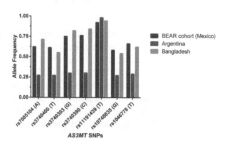

Figure 2. Allele frequency within the BEAR cohort as compared to cohorts in San Antonio de los Cobres, Argentina and Matlab, Bangladesh.

gate used for socioeconomic status). For all models, adjusted p-value thresholds were used to assess for significance.

3 RESULTS

3.1 *Characteristics of the study population*

Selected maternal demographic characteristics, indicators of iAs exposure, pregnancy and birth outcomes of the mother-infant pairs of BEAR cohort are described in detail elsewhere (Laine *et al.*, 2015). The average percent of iAs, MMAs and DMAs with regard to U-tAs were 6.1% iAs, 6.4% MMAs, and 87.6% DMAs, which is in agreement with the higher reported methylation capacity of pregnant women. For this cohort, we have previously reported a significant positive correlation between DW-iAs and U-tAs (r = 0.51; $p < 0.0001$) (Laine *et al.*, 2015).

3.2 *Identification of SNP—metabolite associations*

The women and infants of the cohort were classified into one of three categories depending upon their genotypes as wild type homozygotes, variant homozygotes or heterozygotes. Six of the tested SNPs are located within intronic regions (rs7085104, rs3740400, rs3740393, rs3740390, rs10748835 and rs1046778)

of *AS3MT*. One non-synonymous SNP present in exon 9 is known to alter the amino acid sequence of the AS3MT protein (rs11191439, Met287Thr). For each of the seven SNPs analyzed, multiple linear regression analyses were conducted. Six of the seven SNPs (rs7085104, rs3740400, rs3740393, rs3740390, rs10748835, and rs1046778) had significant ($p < 0.007$) associations with U-MMAs, U-DMAs, %MMAs, or %DMAs while rs11191439 showed no significance at any examined level with iAs or any of the urinary metabolites. Lower levels/proportions of U-MMAs and higher levels/proportions of U-DMAs were associated with the presence of variant (minor) allele for all of these SNPs as shown in Figure 1.

3.3 *Comparison to previously published studies*

The genotype frequencies for the BEAR cohort were compared to genotype frequencies found in two cohorts, San Antonio de los Cobres, Argentina (n = 176) and one in Matlab, Bangladesh (n = 359). These data demonstrate that the BEAR cohort has an allele frequency that more closely resembles women in Matlab, Bangladesh than San Cristobal, Argentina (Figure 2).

4 CONCLUSIONS

We demonstrate that the majority of pregnant women from the BEAR cohort in Gómez Palacio, Mexico are potentially poorer metabolizers of inorganic arsenic producing higher levels of MMAs and comparatively lower levels of DMAs. Comparison of these results to previously published studies on allele frequency shows differences to the Argentinian population and similarities to the Bangladeshi population. The results of this study suggest that the majority of women in Gómez Palacio carry the *AS3MT* risk alleles for elevated MMAs with clear links to potential poorer health outcomes.

REFERENCES

Laine, J.E., Bailey, K.A., Rubio-Andrade, M., Olshan, A.F., Smeester, L., Drobná, Z., Herring, A.H., Stýblo, M., García-Vargas, G.G., Fry, R.C. 2015. Maternal arsenic exposure, arsenic methylation efficiency, and birth outcomes in the Biomarkers of Exposure to ARsenic (BEAR) pregnancy cohort in Mexico. *Environ. Health Perspect.* 123(2): 186–192.

Rager, J.E., Bailey, K.A., Smeester, L., Miller, S.K., Parker, J.S., Laine, J.E., Drobná, Z., Currier, J., Douillet, C., Olshan, A.F., Rubio-Andrade, M., Stýblo, M., García-Vargas, G., Fry, R.C. 2014. Prenatal arsenic exposure and the epigenome: altered microRNAs associated with innate and adaptive immune signaling in newborn cord blood. *Environ. Mol. Mutagen.* 55(3): 196–208.

Vahter, M. 2002. Mechanisms of arsenic biotransformation. *Toxicology* 181–182: 211–217.

Vahter, M. 2009. Effects of arsenic on maternal and fetal health. *Annu. Rev. Nutr.* 29: 381–399.

Arsenic Research and Global Sustainability – Bhattacharya, Vahter, Jarsjö, Kumpiene, Ahmad, Sparrenbom, Jacks, Donselaar, Bundschuh & Naidu (Eds)
© 2016 Taylor & Francis Group, London, ISBN 978-1-138-02941-5

AS3MT, Locus 10q24 and arsenic metabolism biomarkers in American Indians: The Strong Heart Family Study

P. Balakrishnan[1], D. Vaidya[1], N. Franceschini[2], V.S. Voruganti[2], M. Gribble[3], K. Haack[4], S. Laston[4], J. Umans[5], K. Francesconi[6], W. Goessler[6], K. North[2], E. Lee[7], J. Yracheta[8], L. Best[8], J. MacCluer[4], J. Kent, Jr[4], S. Cole[4] & A. Navas-Acien[1]

[1]*Johns Hopkins University, Baltimore, Maryland, USA*
[2]*University of North Carolina, Chapel Hill, North Carolina, USA*
[3]*Emory University, Atlanta, Georgia, USA*
[4]*Texas Biomedical Research Institute, San Antonio, Texas, USA*
[5]*Georgetown University, Washington, DC, USA*
[6]*University of Graz, Graz, Austria*
[7]*University of Oklahoma Health Sciences Center, Oklahoma City, Oklahoma, USA*
[8]*Missouri Breaks Industries Research, Inc, Timber Lake, South Dakota, USA*

ABSTRACT: Genetic determinants may partly explain metabolic differences of inorganic arsenic(iAs) between individuals. We investigated association of arsenic metabolism biomarkers with variants previously associated with cardiometabolic traits (~200,000, Illumina Cardio MetaboChip) or arsenic metabolism (670) among 2,428 American Indian participants in the Strong Heart Family Study. Arsenic traits included percent urine arsenic species, measured by HPLC-ICPMS, (iAs, monomethylarsonate [MMA], dimethylarsinate [DMA], divided by their sum and principal components of the arsenic species. Each phenotype was regressed on allele dosage, accounting for familial relatedness, age, sex, total arsenic levels, and population stratification. SNP associations were stratified by center and meta-analyzed. Multiple testing was accounted using Bonferroni. 10q24 variants were statistically significant for all arsenic phenotypes. Index SNP for percent arsenic species (rs12768205) and principal components (rs3740394, rs3740393) were located near AS3MT, whose gene product catalyzes methylation of iAs to MMA and DMA. Among the candidate arsenic variant associations, functional 10q24 SNPs were most significant (P<9.33e−5).

1 INTRODUCTION

Arsenic toxicity may be better understood through investigating patterns of arsenic metabolism and methylation, which v may be influenced by genetic susceptibility, age, sex, and other risk factors (Concha *et al.*, 2002). The proportion of variation explained by genetic determinants, heritability, range from 50–53% for iAs%, 16–50% for MMA%, and 33–63% for DMA% (Tellez-Plaza *et al.*, 2013). Previous studies have implicated Single Nucleotide Polymorphisms (SNPs) particularly locus 10q24 and arsenic (III) methyltransferase gene *AS3MT*, whose gene product biotransforms iAs to MMA and DMA (Pierce *et al.*, 2012; Schläwicke Engström *et al.* 2009). Our objective was to investigate common variants with markers of arsenic metabolism, highlighting mechanisms of arsenic toxicity and related diseases in the Strong Heart Family Study (SHFS).

2 METHODS

2.1 *Study population*

The SHFS is a large, multigenerational cohort recruited from the Strong Heart Study (SHS), an ongoing population-based study conducted in 13 American Indian tribes/communities in Arizona, Oklahoma and North/South Dakota. Families with a core sibship consisting of 3 original SHS participants and at least 5 additional living family members including 3 original SHS participants were recruited. Our study was restricted to 2,428 participants who were free of diabetes and had arsenic and genotype data.

2.2 *Arsenic measurements*

Total urine arsenic and arsenic species concentrations from spot urine samples was determined by Inductively Coupled Plasma Mass Spectrometry

(ICPMS) and High Performance Liquid Chromatography-ICPMS (HPLC-ICPMS), respectively. For the 9.1% samples below the Limit Of Detection (LOD) of 0.10 µg/L for iAs, 2.6% for MMA, and <0.1% for DMA, the value was imputed as LOD/√2. Percent arsenic species (iAs%, MMA%, DMA%) was calculated as the relative proportion of the species to the sum of all three arsenic species.

2.3 Genotyping

Blood DNA was genotyped using Illumina Cardio MetaboChip and candidate arsenic panel with 120,975 and 670 SNPs, respectively. Quality control included Hardy-Weinberg equilibrium, mismatch between genotyped and reported sex, identity by descent clustering, and outlier in Principal Components Analysis (PCA) of ancestry.

2.4 Statistical analysis

Percent arsenic species were logit transformed to approximate a normal distribution and also used to create orthogonal principal components using PCA. Each trait was regressed on allele dosage, accounting for familial relatedness, age, sex, total arsenic level, and population stratification. SNP associations were stratified by center in SOLAR (Blangero & Almasy, 1996) and meta-analyzed in METAL (Willer et al., 2010). Multiple testing was accounted using Bonferroni and Linkage Disequilibrium (LD).

3 RESULTS AND DISCUSSION

The median (IQR) for sum of inorganic and methylated species was 6.6 (3.9–11.6) µg/L.

PCA resulted in 2 orthogonal principal components (PC1, PC2) of arsenic species.

Table 1. Baseline characteristics.

	Total	Arizona	Oklahoma	South/North Dakota	P-value
No. of Participants	2428	703	819	906	
Mean age, yrs (SD)	35.2 (15.2)	30.9 (13.1)	38.8 (15.7)	35.3 (15.4)	<0.01
No. females (%)	1469 (60.5%)	438 (62.3%)	486 (59.34%)	545 (60.15%)	0.48
Arsenic levels					<0.01
Tertile 1 (0.21–4.73µg/L)	804 (33.1%)	100 (14.2%)	357 (43.5%)	347 (38.3%)	
Tertile 2 (4.74–9.35µg/L)	799 (32.9%)	204 (29.0%)	293 (35.7%)	302 (33.3%)	
Tertile 3 (9.37–176.6µg/L)	825 (33.9%)	399 (56.7%)	169 (20.6%)	257 (28.3%)	
Median iAs% (IQR)	9.78 (6.4–14.0)	10.8 (7.6–15.2)	8.2 (5.4–12.3)	10.1 (6.7–14.5)	<0.01
Median MMA% (IQR)	13.9 (10.5–17.7)	12.8 (10.0–16.2)	13.7 (10.3–17.8)	14.9 (11.5–18.5)	<0.01
Median DMA% (IQR)	75.6 (68.6–81.6)	75.2 (69.0–81.1)	77.2 (70.5–83.3)	74.3 (67.3–80.4)	<0.01
Mean PC1 (SD)	0.04 (12.21)	0.62 (12.05)	-2.20 (11.64)	1.62 (12.54)	<0.01
Mean PC2 (SD)	-0.01 (4.91)	1.59 (4.87)	-0.70 (4.53)	-0.63 (5.00)	<0.01

Table 2. Summary of arsenic principal components.

	PC1	PC2
Variance in arsenic species explained (%)	86.1	13.9
Standard deviation	12.21	4.91
Weight for iAs%	0.49	0.65
Weight for MMA%	0.32	-0.75
Weight for DMA%	-0.81	0.10

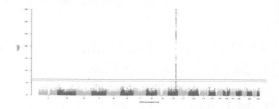

Figure 1. Genomewide association for PC1.

Table 3. Index SNPs for Cardio MetaboChip.

SNP	Chr	Position¹	Allele	MAF	Gene	Location	Trait	iAs%	MMA%	DMA%	PC1	PC2
rs3740393	10	104626645	C/G	0.17	AS3MT	intron	PC2	8.63e-7	1.07e-13*	2.71e-23*	2.58e-34*	1.56e-8*
rs3740394	10	104624464	A/G	0.18	AS3MT	intron	PC1	1.27e-4	4.66e-7*	3.87e-20*	2.19e-38*	5.57e-6
rs12768205	10	104637839	G/A	0.27	AS3MT	intron	iAs%, MMA%, DMA%	8.27e-8*	1.20e-15*	5.90e-24*	1.15e-29*	1.78e-7*

Table 4. Index SNPs for candidate arsenic panel.

SNP	Chr	Position¹	Allele	MAF	Gene	Location	iAs%	MMA%	DMA%	PC1	PC2
rs11191439	10	104638723	T/C	0.18	AS3MT	coding	3.23e-4	2.60e-7*	2.89e-19*	1.12e-36*	7.54e-7*
rs7911488	10	105154089	A/G	0.25	USMG5	UTR	1.04e-6*	6.54e-8*	7.04e-18*	3.57e-31*	6.63e-6*
rs4925	10	106022789	C/A	0.12	GSTO1	coding	0.02	0.11	1.55e-3	4.60e-6*	9.18e-7*
rs2297235	10	106034491	A/G	0.12	GSTO2	UTR	0.08	0.20	4.58e-3	1.67e-5*	4.31e-3

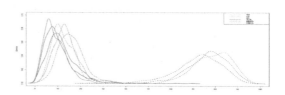

Figure 2. Percent arsenic species by rs12768205.

All arsenic traits were significantly associated with AS3MT SNPs that were in LD ($r^2 > 0.80$). The strongest signal was for principal component 1.

Of the 670 candidate arsenic SNPs, the top significant SNPs were either coding or UTR variants in locus 10q24.

Each copy of G for rs12768205 had a separation in the distribution of percent arsenic species. The variant genotype GG had iAs% and MMA% distributions shifted toward lower and DMA% distribution shifted toward higher percentages.

4 CONCLUSIONS

Association signals in AS3MT and 10q24 are consistently associated with arsenic metabolism biomarkers. Functionally annotated 10q24 variants also show a strong relationship. Associations including AS3MT and GSTO1/2 highlight oxidative stress in arsenic biotransformation and thus arsenic related diseases. Due to high LD at 10q24 in most populations and especially among American Indians in our study, further investigations are needed.

ACKNOWLEDGEMENTS

Work funded by National Institute of Environmental Health Sciences (R01ES021367, T32 ES007141-32).

REFERENCES

Blangero, J., Almasy, L. 1996. *Solar: Sequential oligogenic linkage analysis routines*. Population Genetics Laboratory Technical Report 6.

Concha, G., Vogler, G., Nermell, B., Vahter, M. 2002. Intra-individual variation in the metabolism of inorganic arsenic. *Int. Arch. Occup. Environ. Heal.* 75(8): 576–80.

Pierce. B.L., Kibriya, M.G., Tong, L., Jasmine, F., et al. 2012. Genome-wide association study identifies chromosome 10q24. 32 variants associated with arsenic metabolism and toxicity phenotypes in Bangladesh. *Plos Genetics* 8(2):e1002522.

Schläwicke Engström, K., Nermell, B., Concha, G., Strömberg, U., et al. 2009. Arsenic metabolism is influenced by polymorphisms in genes involved in one-carbon metabolism and reduction reactions. *Mutat. Res.-Fund. Mol. M.* 667(1):4–14.

Tellez-Plaza, M., Gribble, M.O., Voruganti, V.S., Francesconi, K.A., Goessler, W., et al. 2013. Heritability and preliminary genome-wide linkage analysis of arsenic metabolites in urine. *Environ. Health Perspect.* 121(3): 345–351.

Willer, C.J., Li, Y., Abecasis, G.R. 2010. METAL: Fast and efficient meta-analysis of genomewide association scans. *Bioinformatics* 26(17): 2190–2191.

Arsenic Research and Global Sustainability – Bhattacharya, Vahter, Jarsjö, Kumpiene, Ahmad, Sparrenbom, Jacks, Donselaar, Bundschuh & Naidu (Eds)
© 2016 Taylor & Francis Group, London, ISBN 978-1-138-02941-5

Effects on DNA methylation of arsenic exposure *in utero*

K. Engström[1], M. Kippler[1], R. Raqib[2], S. Ahmed[2], M. Vahter[1] & K. Broberg[1]

[1]*Institute of Environmental Medicine, Karolinska Institutet, Stockholm, Sweden*
[2]*International Centre for Diarrhoeal Disease Research, Bangladesh (ICDDR, B), Dhaka, Bangladesh*

ABSTRACT: In a previous study, we evaluated cord blood DNA methylation among children from Bangladesh, and we found that prenatal arsenic exposure appears to decrease DNA methylation of specific, cancer-related genes, especially in boys. In the present study, we followed-up the same children at 9 years of age (N = 113). DNA methylation of venous blood was evaluated using the HumMeth450K from Illumina. Maternal urine samples for prenatal arsenic exposure were collected in gestational weeks 5 to 14 and child urine samples for current arsenic exposure were collected at 9 years of age. We found gender differences in the epigenetic response to arsenic: Based on the number of statistically significant CpGs, the effect of current exposure seemed to be the strongest for boys, while for girls, the effect of prenatal exposure was the strongest. The majority of CpGs significantly associated with arsenic were hypermethylated, suggesting dynamic responses to arsenic in early life.

1 INTRODUCTION

Recent studies suggest that arsenic can alter fetal programming by alteration of DNA methylation (Bailey *et al.*, 2014). Still, data is scarce regarding the persistence of the effects of prenatal arsenic exposure on epigenome-wide DNA methylation. We have previously investigated the association between arsenic exposure *in utero* and cord blood DNA methylation (Broberg *et al.*, 2014), where we found that early prenatal arsenic exposure appears to decrease DNA methylation of specific genes, particularly in boys. In the present study, we follow-up the same children at the age of 9 years. The aim of this study is to investigate how prenatal as well as current arsenic exposure alters DNA methylation in childhood.

2 METHODS/EXPERIMENTAL

2.1 *Study area and design*

Our research on arsenic-related health effects was nested in a large randomized food and micronutrient supplementation trial conducted during pregnancy [Maternal and Infant Nutrition Interventions in Matlab (MINIMat)], in Matlab, about 50 km southeast of Dhaka, Bangladesh (Persson *et al.*, 2012). Of the 127 mother-child pairs included in the study regarding prenatal arsenic exposure and DNA methylation (Broberg *et al.*, 2014), 114 had DNA available for analysis of DNA methylation at 9 years of age. The study was approved by the ethics committees at icddr, b (Dhaka, Bangladesh) and the Karolinska Institutet (Stockholm, Sweden). Consent was obtained from all participants, and participants were free to refrain from any part of the study at any time.

2.2 *Analysis of arsenic*

Arsenic exposure was based on the sum concentration of arsenic metabolites (inorganic arsenic, methylarsonic acid (MMA) and dimethylarsinic acid (DMA) in spot urine samples collected, as previously described (Vahter *et al.*, 2006). Maternal urine samples for prenatal arsenic exposure were collected in early pregnancy between GW 5 and 14 (median GW 8) and child urine samples for current exposure were collected at 9 years of age. All the measured arsenic concentrations in maternal urine and child urine were above the limit of detection, i.e. >1 ug/L. All arsenic concentrations were adjusted for the average variation in urine dilution by specific gravity. The average specific gravity of both maternal urine samples and those of the children at 9 years was 1.012 g/mL.

2.3 *Analysis of DNA methylation*

Venous blood samples were collected from the 9-year-old children in lithium-heparin treated tubes and the lymphocytes were separated by centrifugation. Mononuclear cells were separated by Ficoll (Pharmacia-Upjohn/McNeill Laboratories) density gradient centrifugation. DNA was isolated using QIAamp DNA Blood Midi kit (Qiagen). The DNA quality was evaluated with a NanoDrop spectrophotometer (NanoDrop Products,) and Bioanalyzer 2100 (Agilent) and showed good quality (260/280 nm > 1.80). 250 ng of DNA was bisulfite-treated with the EZ DNA Methylation kit (Zymo Research). For each sample, the bisulfite converted DNA was eluted in 30 µL, which was further evaporated to a volume of <4 µL, and used for methylation analysis using the HumMeth450K Beadchip (Illumina). Image processing, background correction, quality control, filtering, and normalization were performed in the R package minfi. The 450k included a total of 485,512 sites before filtering. All samples performed well, with at least > 98% of the CpGs with detection p-value below 0.01. We removed CpGs for which more than 20% of the samples had a detection p-value above 0.01 (N = 322). Furthermore, the following probes were removed: rs probes and CpH probes (N = 3,091), probes with in silico non-specific binding (N = 29,118) probes on the X and Y chromosomes (N = 10,329), probes with common SNPs (according to the function dropLociWithSnps in minfi; N = 15,424). M-values were adjusted for batch via the ComBat function. In total, 426,936 probes were left for further analysis. For the analyses stratified for sex, X and Y chromosomes were not removed, thus 437,179 probes were included in these analyses.

2.4 Statistical analysis

One individual was removed due to low DNA concentration. In total, 113 children were included in the analysis. Principal component analysis (PCA) was performed to evaluate the influence of technical and biological variables on DNA methylation. Peripheral blood contains different types of cells, and cell composition could impact the DNA methylation. We estimated the cell type proportions using the minfi estimate CellCounts function. We evaluated the influence of: 1) prenatal arsenic (gestational weeks 5–14) and 2) current arsenic (9 years of age) on DNA methylation. Analyses were performed including all individuals and stratifying for gender. Differently methylated positions were evaluated by fitting a robust linear regression model to each CpG using the R package limma with adjustment for cell type proportions, as well as sex of the child in the analyses comprising all individuals. Empirical Bayes smoothing was applied to the standard errors. P-values were adjusted for false discovery rate (FDR) to obtain q-values. A q-value of 0.05 or lower was considered statistically significant.

3 RESULTS AND DISCUSSION

3.1 Results and discussion

Current arsenic exposure was not significantly associated with any estimated cell type proportion. Preliminary results show that there were few statistically significant findings when evaluating DNA methylation for all individuals together (adjusted for sex) (Table 1). However, when stratifying for sex, there were a number of statistically significant findings, especially for boys (Table 1). For boys, DNA methylation at 9 years of age was more strongly associated with current arsenic exposure than with prenatal arsenic exposure, based on the number of statistically significant CpGs. For girls, the prenatal arsenic exposure was more strongly

Table 1. Number of CpGs significantly associated with arsenic.

Study group	Total	Nr. Hypermet.	Nr. Hypomet.
Prenatal As			
– All	14	10 (71%)	4 (29%)
– Boys	436	180 (41%)	256 (59%)
– Girls	108	65 (60%)	43 (40%)
Postnatal As			
– All	1	1 (100%)	–
– Boys	872	665 (76%)	207 (24%)
– Girls	55	34 (62%)	21 (38%)

associated with DNA methylation than with exposure at 9 years of age. For boys, current arsenic exposure was associated with hypermethylation of specific genes: 76% of the statistically significant CpGs were hypermethylated.

4 CONCLUSIONS

This study shows gender differences in the epigenetic response to arsenic. For boys, the effect of current exposure seems to be the strongest, while for girls, the effect of prenatal exposure was the strongest.

ACKNOWLEDGEMENTS

This research was supported by the Swedish Research Council for Health, Working Life and Welfare (FORTE); the Swedish Research Council, the Swedish International Development Cooperation Agency, and the Karolinska Institutet.

REFERENCES

Bailey, K.A., Fry RC. 2014. Arsenic-associated changes to the epigenome: what are the functional consequences? Curr. Environ. Health Rep. 1: 22–34.

Broberg, K., Ahmed, S., Engstrom, K., Hossain, M.B., Jurkovic Mlakar, S., Bottai, M., Grander, M., Raqib, R., Vahter, M. 2014. Arsenic exposure in early pregnancy alters genome-wide DNA methylation in cord blood, particularly in boys. J. Dev. Orig. Health Dis. 5(4): 288–298.

Persson, L.A., El Arifeen, S., Ekstrom, E.C., Rasmussen, K.M., Frongillo, E.A., Yunus, M., Team MIS. 2012. Effects of prenatal micronutrient and early food supplementation on maternal hemoglobin, birth weight, and infant mortality among children in Bangladesh: the MINIMat randomized trial. JAMA 2012; 307(19):2050–2059.

Vahter, M.E., Li, L., Nermell, B., Rahman, A., El Arifeen, S., Rahman, M., Persson, L.A., Ekström, E.C. 2006. Arsenic exposure in pregnancy: a population-based study in Matlab, Bangladesh. J. Health Popul. Nutr. 24(2):236–245.

Arsenic Research and Global Sustainability – Bhattacharya, Vahter, Jarsjö, Kumpiene,
Ahmad, Sparrenbom, Jacks, Donselaar, Bundschuh & Naidu (Eds)
© 2016 Taylor & Francis Group, London, ISBN 978-1-138-02941-5

Exposure to inorganic arsenic and gene expression in peripheral blood

K. Broberg[1], S.S. Ameer[2], G. Concha[3], M. Vahter[1] & K. Engström[1]
[1]Institute of Environmental Medicine, Karolinska Institutet, Stockholm, Sweden
[2]Division of Occupational and Environmental Medicine, Lund University, Lund, Sweden
[3]National Food Agency, Uppsala, Sweden

ABSTRACT: Arsenic is an established carcinogen and a risk factor for several non-malignant diseases. Mechanisms of arsenic toxicity may include interference with gene expression. The impact of arsenic exposure on gene expression was evaluated in women (n = 80) living in the Puna of the northern Argentinian Andes with varying concentrations (10–1251 µg/L) of arsenic in drinking water. DirectHyb HumanHT-12 v4.0 was used for genome wide gene expression analysis. Robust linear regression model was used to each array to evaluate the relations between arsenic exposure, gene expression and with the influence of arsenic metabolism efficiency. Also, Ingenuity Pathway Analysis (IPA) was performed to look for relevant pathways, diseases, networks etc. associated with the genes. In the association between arsenic and gene expression, most of the genes were downregulated. Pathway analyses revealed different expression pattern associated with arsenic exposure between women with high and low urinary %MMA, indicating variation in arsenic susceptibility by arsenic-methylation.

1 INTRODUCTION

Millions of people world-wide are exposed to inorganic arsenic (iAs) through drinking water and certain food. Arsenic is an established carcinogen and a risk factor for several non-malignant diseases, including cardiovascular and lung diseases (IARC, 2004). Inorganic arsenic is Metabolized to Methylarsonic Acid (MMA), and dimethylarsinic acid (DMA). The degree of methylation to DMA varies within and between study groups (Vahter, 2002). Incomplete arsenic metabolism, with higher fractions of iAs and MMA in the urine, is a marker of increased susceptibility to arsenic-related diseases, such as arsenic-related skin lesions, cancer and cardiovascular disease (Antonelli et al., 2014).

Studies have indicated that a mechanism of arsenic toxicity is through interference with gene expression, either by interaction with DNA methylation or with gene-expression regulatory proteins. Mice exposed to iAs prenatally have indicated altered gene expression levels of oncogenes, tumor suppressor genes, and stress-related genes in their livers as adults (Liu et al., 2004, 2006). Maternal exposure to iAs measured by arsenic in toenails was found to be associated with differentially expressed gene expression of inflammatory signalling pathways and transcriptional response in newborn cord blood (Fry et al., 2007). A genome-wide, gene-specific DNA methylation study showed that promoters of 183 genes were differentially methylated in Mexican individuals with arsenic-associated skin lesions, compared to unexposed individuals with no skin lesions (Smeester et al., 2011) and higher arsenic exposure during pregnancy was associated with lower methylation in CpG sites in cord blood mainly in boys (Broberg et al., 2014).

In this study, we evaluated the association between arsenic exposure and gene expression, and possible modifying effects of arsenic metabolism efficiency, in arsenic exposed population in Northern Argentina.

2 MATERIALS AND METHODS

2.1 Study population and arsenic assessment

The study participants were women (n = 80) living in the Puna of the northern Argentinian Andes in the province of Salta. The study participants, recruited in 2008, were mostly from the village San Antonio de los Cobres (water arsenic level ~200 µg/L), and few others were living in the small surrounding villages (water arsenic level 3.5–70 µg/L). For gene expression, blood were collected in PAXgene blood RNA tubes (PreAnalytiX, Qiagen), and stored in -80°C. Exposure to arsenic was determined in urine samples based on the sum of concentrations of iAs and its metabolites (iAs+MMA+DMA) in urine (U-As) by high-performance liquid chromatography coupled with Hydride Generation (HG) and inductively coupled plasma mass spectrometry (Agilent 7500ce, Agilent Technologies). Efficiency of arsenic metabolism was assessed by the percentages of iAs, MMA, and DMA in urine.

2.2 Gene expression

RNA was isolated using PAXgene blood RNA kit (PreAnalytiX, Qiagen). RNA concentration and purity were measured by a Nanodrop spectrophotometer (Wilmington, DE). DirectHyb Human-HT-12 v4.0 (Illumina, CA, USA) was used for gene expression analysis. Gene expression data was quan-

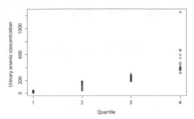

Figure 1. Concentrations of U-As (μg/L) divided into quartiles.

tile normalized. 47323 probes were included in the array, and probes with a detection p-value above 0.1 (i.e. p-value for the difference between the array and the background noise) in 90% or less of the samples were removed, which meant that 14290 (30%) probes remained.

2.3 Statistical analysis

We had DNA methylation data for 80 individuals, and we estimated the cell type proportions based on the method by Houseman et al., (2012) in this data. By evaluating Principal Component Analysis (PCA) we adjusted the associations of arsenic with gene expression for age and coca chewing, as well as the (fractions of) cell types most strongly associated with gene expression (granulocytes, monocytes and CD4+ cells).

Robust linear regression model were performed for each array using the R package limma to evaluate the relations between gene expression and arsenic. We used gene expression as outcome variable, U-As concentrations divided into quartiles (Fig. 1) as independent variable, and age, coca-use, fractions of granulocytes, monocytes and CD4+ cells as covariates. The influence of efficiency of arsenic metabolism on the associations between arsenic exposure and gene expression was evaluated by stratifying for high and low%iAs or%MMA (below and above median) in urine. P-values were adjusted for multiple comparisons by the Benjamini-Hochberg FDR method (Benjamini & Hochberg, 1995) to obtain q-values. A q-value ≤0.05 was considered statistically significant.

Finally, we uploaded both the up and down regulated genes with p value = 0.1 to the Ingenuity Pathway Analysis (IPA), to look for relevant pathways, diseases, networks etc. associated with the genes.

3 RESULTS AND DISCUSSION

3.1 Arsenic exposure and gene expression

Preliminary analyses show that the majority of genes were downregulated in relation to arsenic in urine, among the top 100 genes, about two-third of the genes were downregulated and one-third genes were upregulated.

3.2 Modification by arsenic methylation

When stratifying by high/low urinary arsenic metabolites, pathway analyses identified different expression pattern particularly between high and low%MMA groups. In the high%MMA group, U-As was associated with networks related to cancer, whereas in the low group U-As was associated with networks related to cellular function and maintenance. Further evaluation of the data is ongoing.

4 CONCLUSIONS

This study indicated that that people with inefficient arsenic metabolism (higher% iAs or higher% MMA) may be at high risk since they showed more cancer-related gene expressed than the efficient metabolizers.

ACKNOWLEDGEMENTS

The authors thank Alejandro Cardozo, Plácida Mamaní and Esperanza Casimiro.

REFERENCES

Antonelli, R., Shao, K., Thomas, D.J., Sams, R. 2nd, et al. 2014. AS3MT, GSTO, and PNP polymorphism:impact on arsenic methylation and implications for disease susceptibility. Environ. Res. 132. 156–167.
Benjamini, Y., Hochberg, Y. 1995. Controlling the false discovery rate: A practical and powerful approach to multiple testing. Journal of the Royal Statistical Society Series B (Methodological): 289–300.
Broberg, K., Ahmed, S., Engström, K., Hossain, M.B., et al., 2014. Arsenic exposure in early pregnancy alters genome-wide DNA methylation in cord blood, particularly in boys. J. Dev. Orig. Health Dis. 5:288–298.
Fry, R.C., Navasumrit, P., Valiathan, C., Svensson, J.P., et al. 2007. Activation of inflammation/nf-kappab signaling in infants born to arsenic-exposed mothers. PLoS Genet. 3(11): e207.
Houseman, E.A. Accomando, W.P., Koestler, D.C., Christensen, B.C., et al. 2012. DNA methylation arrays as surrogate measures of cell mixture distribution. BMC Bioinformatics 13: 86.
IARC. 2004. Some drinking-water disinfectants and contaminants, including arsenic. IARC Monogr. Eval. Carcinog. Risks Hum. 84: 41–267.
Liu, J., Xie, Y., Ducharme, D.M.K., Shen, J., et al. 2006. Global gene expression associated with hepatocarcinogenesis in adult male mice induced by in utero arsenic exposure. Environ. Health Perspect. 114: 404–411.
Liu, J., Xie, Y., Ward, J.M., Diwan, B.A., Waalkes, M.P. 2004. Toxicogenomic analysis of aberrant gene expression in liver tumors and nontumorous livers of adult mice exposed in utero to inorganic arsenic. Toxicol. Sci. 77: 249–257.
Smeester, L. et al. 2011. Epigenetic changes in individuals with arsenicosis. Chem. Res. Toxicol. 24: 165–167.
Vahter, M. 2002. Mechanisms of arsenic biotransformation. Toxicology 181–182: 211–217.

Arsenic Research and Global Sustainability – Bhattacharya, Vahter, Jarsjö, Kumpiene,
Ahmad, Sparrenbom, Jacks, Donselaar, Bundschuh & Naidu (Eds)
© *2016 Taylor & Francis Group, London, ISBN 978-1-138-02941-5*

Exposure to inorganic arsenic and mitochondrial DNA copy number and telomere length in peripheral blood

S.S. Ameer[1], Y.Y. Xu[1], K. Engström[1], H. Li[1], P. Tallving[1], A. Boemo[2], L.A. Parada[3],
L. Guadalupe Peñaloza[2], G. Concha[4], F. Harari[5], M. Vahter[5] & K. Broberg[5]

[1]*Division of Occupational and Environmental Medicine, Lund University, Lund, Sweden*
[2]*Facultad de Ciencias Exactas and Consejo de Investigación, Universidad Nacional de Salta, Salta, Argentina*
[3]*Institute of Experimental Pathology, UNSa—CONICET, Argentina*
[4]*National Food Agency, Uppsala, Sweden*
[5]*Institute of Environmental Medicine, Karolinska Institutet, Stockholm, Sweden*

ABSTRACT: Exposure to inorganic arsenic (iAs) is a risk factor for cancer. Alterations in mitochondrial DNA copy number (mtDNAcn) and telomere length (TL) have been associated with cancer risk. Two Argentinean groups were studied: A) Puna area of Andes and B) Chaco. Arsenic exposure was assessed as the sum of arsenic metabolites (iAs, MMA, and DMA) in urine (U-As) using HPLC-HG-ICPMS. MtDNAcn, TL, and genotype of the arsenic-methylating gene AS3MT were determined in blood by real-timePCR. The Chaco participants had less-efficient metabolism, with higher%iAs and%MMA in urine, and lower frequency of the efficient-metabolizing AS3MT haplotype. U-As was associated with increased mtDNAcn in Chaco but not in Andes. U-As was associated with longer TL in Chaco, but less so in Andes. Individuals with%iAs>median showed significantly higher mtDNAcn and TL in both groups. Arsenic was associated with increased mtDNAcn and TL, particularly in individuals with less-efficient arsenic metabolism, who might have increased risk for arsenic-related cancer.

1 INTRODUCTION

Exposure to inorganic arsenic (iAs) is a known risk factor for cancer (IARC, 2012). The susceptibility to arsenic differs between individuals. Incomplete arsenic metabolism, with higher fractions of iAs and MMA in the urine, seems to be a marker of increased susceptibility to arsenic-related diseases, (Antonelli *et al.*, 2014). Genetic variation in *AS3MT*, a major arsenic-methylating enzyme, significantly contributes to arsenic metabolism efficiency and is likely an underlying factor determining susceptibility to arsenic. Alterations in mitochondrial DNA copy number (mtDNAcn) and Telomere Length (TL) in blood have been associated with cancer risk. In this study, we evaluated the effect of arsenic, including modifying effects of the arsenic metabolising capacity, on TL and mtDNAcn in two arsenic-exposed groups in northern Argentina.

2 METHODS AND MATERIALS

2.1 *Study population and arsenic assessment*

Men and women in Puna area of the Andes (n = 264) and in the Chaco region (n = 169) of eastern part of the province of Salta, northern Argentina were recruited for the study. Both the areas had wide range of arsenic concentration in water. Urine samples were collected for the assessment of arsenic exposure and blood samples were collected for genotyping and measurement of mtDNAcn and TL. Exposure to arsenic was determined based on the sum of concentrations of iAs and its metabolites (iAs+ MMA+DMA) in urine (U-As) by high-performance liquid chromatography (Waldbronn) coupled with Hydride Generation (HG) and inductively coupled plasma mass spectrometry (Agilent Technologies). Efficiency of arsenic metabolism was assessed by the fractions (percentages) of iAs, MMA, and DMA in urine.

2.2 *MtDNA, telomere length, genotyping and statistical analyses*

The relative mtDNAcn and relative TL were measured by using real-time PCR (7900HT, Applied Biosystems). Six SNPs were genotyped within or close to the *AS3MT* locus (5' to 3' direction): rs486955, rs9527, rs3740400, rs3740393, rs11191439, and rs1046778, by Taqman allelic discrimination assays (Life Technologies) using real-time PCR. Data from Andes and Chaco were analyzed separately. Multivariable-adjusted linear

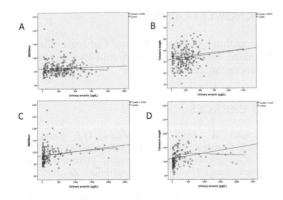

Figure 1. Scatter plots depicting relations between urinary arsenic (sum of inorganic arsenic metabolites) concentrations and mtDNAcn and telomere length in Andes (a, b) and Chaco (c, d).

regression analyses were performed to evaluate the relations between arsenic and mtDNAcn and arsenic and TL. Association between percentages of iAs, MMA and DMA with AS3MT haplotype were analyzed using the general linear model. The influence of efficiency of arsenic metabolism on associations between arsenic and mtDNAcn or TL was analyzed by including an interaction term between total arsenic in urine and fraction of each metabolite (above and below median). We performed linear regression analyses in individuals with below and above median U-As metabolites.

3 RESULTS AND DISCUSSION

3.1 Arsenic, mtDNAcn and telomere length

The median for U-As was 196μg/L and 80μg/L for Andes and Chaco respectively. The metabolite

pattern differed significantly between the study groups: the median%iAs and median%MMA were higher and median%DMA was lower in Chaco compared with Andes.

Increasing U-As was significantly associated with both higher mtDNAcn and longer TL in Chaco, and to a lesser extent in Andes in linear regression analyses. Further, the influence of different cell composition in blood had no significant effect on these associations.

3.2 Study groups differed in arsenic metabolism efficiency

Frequencies of AS3MT haplotype showed striking differences between the Andes and Chaco study groups. Haplotype 1 was the most common haplotype in Chaco (59%, compared with 23% in Andes), while haplotype 2 was the most common haplotype in Andes (69%, compared with 21% in Chaco). Haplotype 1 was associated with significantly higher% iAs, higher%MMA and lower%DMA in Andes, but not in Chaco. Haplotype 2 was associated with significantly lower%iAs and%MMA and higher%DMA in Andes and significantly lower%MMA and higher%DMA in Chaco.

3.3 Modification by arsenic metabolism efficiency

We found significant interactions between%iAs and U-As, on the effect of mtDNAcn in Andes (p = 0.017 for%iAs), but not in Chaco. After stratifying the data at the median values of%iAs, the results showed significant positive associations between U-As and mtDNA among individuals with high%iAs in Andes, but not among individuals with low%iAs. We observed a similar trend for the Chaco study group, though the association was not statistically significant.

Table 1. Multivariable regression analyses of the associations between U-As and mtDNAcn and telomere length stratified for iAs (above and below median).

Groups	mtDNA cn[a]	% iAs	n	β_1 (95% CI)[b]	p[b]	p[c]
Andes	0.56	11	122	−0.012 (−0.041 to 0.017)	0.41	0.017
	0.58	≥11	122	0.036 (0.006 to 0.065)	0.017	
Chaco	0.76	17	80	0.009 (−0.011 to 0.028)	0.40	0.98
	0.93	>17	79	0.015 (−0.003 to 0.033)	0.096	
	TL					
Andes	0.41	11	94	0.003 (−0.009 to 0.014)	0.67	0.35
	0.45	≥11	114	0.009 (0.0004 to 0.017)	0.041	
Chaco	1.01	17	80	0.011 (−0.006 to 0.028)	0.20	0.35
	1.19	>17	79	0.018 (0.002 to 0.034)	0.029	

[a]presented as median relative values of either mtDNAcn or telomere length
[b]mtDNAcn/telomere length = α+β1 × U-As(100μg/L) +β2 × age+β3 × gender.
[c]p-interaction (β4) = α+β1 × U-As(per 100 μg/L)+β2 × age+β3 × gender+β4 × (U-As × and>median%iAs).

There were no significant interactions between arsenic metabolites and U-As on TL. However, statistically significant positive associations between U-As and TL were only found among individuals with high%iAs in both study groups. There was less clear pattern of modification by%MMA or%DMA.

4 CONCLUSIONS

Arsenic was associated with increased mtDNAcn and TL, particularly in individuals with less-efficient arsenic metabolism, who might have increased risk for arsenic-related cancer.

ACKNOWLEDGEMENTS

The authors thank Alejandro Cardozo, Plácida Mamaní, Esperanza Casimiro and Eva Assarsson for recruiting the volunteers.

REFERENCES

Antonelli, R., Shao, K., Thomas, D.J., Sams, R. 2nd, Cowden, J. 2014. *AS3MT*, *GSTO*, and *PNP* polymorphism:impact on arsenic methylation and implications for disease susceptibility. *Environ. Res.* 132. 156–167.
IARC. 2012. Arsenic, metals, fibres and dust. *IARC Monogr. Eval. Carcinog. Risks Hum.* 100: 11.

Arsenic Research and Global Sustainability – Bhattacharya, Vahter, Jarsjö, Kumpiene,
Ahmad, Sparrenbom, Jacks, Donselaar, Bundschuh & Naidu (Eds)
© *2016 Taylor & Francis Group, London, ISBN 978-1-138-02941-5*

Health effects, genetic and epigenetic changes in the population exposed to arsenic in West Bengal, India

A.K. Giri

Molecular Genetics Division, CSIR-Indian Institute of Chemical Biology, Kolkata, India

ABSTRACT: We have evaluated the health effects, genetic damage and epigenetic changes in the arsenic exposed population of West Bengal, India. We observed that the people having risk genotype had also significant genetic damage when compared to non risk genotype. In our arsenic exposure assessment it has been observed arsenic through rice can alone induced genetic damage in human. Epigenetic changes were mainly carried out through DNA methylation study in some tumor suppressor genes and DNA repair genes. Significant hypermethylation and hypomethylation were observed in arsenic exposed individuals when compared to arsenic unexposed individuals. When telomere length was observed, we found that the telomere lengths were significantly increased in arsenic exposed population. In the microRNA profiling in the arsenic exposed population showed a significant increase in mir21, miR99a and miR155 expression. Thus it can be concluded that epigenetic changes may play a key role in arsenic induced cancer in human.

1 INTRODUCTION

In West Bengal, India about 26 million people are exposed to arsenic through drinking water more than the permissible limit. Arsenic induces different types of skin lesions and skin cancers. Arsenic also induced cancers of liver, kidney, bladder, and other internal organs apart from peripheral neuropathy, conjunctivitis and respiratory diseases. (Guha Mazumder *et al.*, 2001). Although a large number of individuals are exposed to arsenic through drinking water but only 15 to 20% individuals showed arsenic induced skin lesions. This indicate that genetic variants plays an important role in arsenic induced toxicity and carcinogenicity. Since arsenic is not mutagenic but carcinogenic in human so it is assumed that epigenetic changes may leads to cancer in human.

2 METHODS

2.1 *Groundwater sampling and analyses*

Study subjects were recruited from three highly arsenic-affected districts of West Bengal i.e. North 24 Parganas, Nadia and Murshidabad. The control subjects were chosen from East Midnapur district of the same state with little or no history of arsenic contamination in the ground water. All the study subjects were age and sex matched with similar socio economic status. The selected study subjects had at least 10 years of arsenic exposure through drinking water. The selection criteria were followed as described in Ghosh *et al.* (2007).

2.2 *Sampling*

Samples were collected from those subjects who provided informed consent to participate in the study. This study was conducted in accord with the Helsinki II Declaration and approved by the Institutional Ethics Committee of CSIR-Indian Institute of Chemical Biology. Blood samples were collected from arsenic exposed individuals with skin lesions individuals without skin lesions from three arsenic affected districts of West Bengal. The control blood samples were collected from the East Midnapur district of the same state. The collected samples included drinking water, urine, nails and hair. The samples were analyzed by Atomic Absorption Spectrometer. Bladder exfoliated urothelial cells were isolated from the urine as well as lymphocyte cultures in RPMI-1640 were used for analysis of the cytogenetic damage by micronucleus assay. Blood lymphocytes were also used to isolate DNA, RNA and protein for subsequent studies.

3 RESULTS AND DISCUSSION

3.1 *Genetic variations and susceptibility in arsenic exposed population*

PCR, sequence-BLAST and RFLP analysis have found that several SNPs of key functional genes of DNA-repair, immunological and tumor suppressor pathways have been associated with increased incidence of arsenic-induced skin lesions. SNPs in important genes that may be responsible towards arsenic induced skin lesions and premalignant form of skin lesions in the population exposed to arsenic through drinking water.

3.2 Arsenic induces epigenetic deregulation

Promoter methylation status was determined by bisulfite conversion of genomic DNA and methylation-specific PCR. Realtime PCR and western blotting determined the expression titer of both the genes; which indicated that significant hypermethylation was found in the promoters of both DAPK and p16 genes in the cases compared with the controls resulting in down regulation of the genes in the cases. There was a 3.4-fold decrease in the expression of death-associated protein kinase and 2.2-fold decrease in gene expression of p16 in the cases compared to the controls, the lowest expression being in the cancer tissues. Promoter hypermethylation of the genes was also associated with higher risk of developing arsenic-induced skin lesions, peripheral neuropathy, ocular and respiratory diseases.

3.3 Arsenic mitigation strategies provide some relief to chronically exposed individuals

In a two-wave cross sectional study we found that significant decrease of arsenic exposure (190.10 µg/L to 37.94 µg/L) resulted in significant amelioration of the severity of dermatological disorders (p < 0.0001) Micronucleus formation in urothelial cells and lymphocytes also decreased significantly (p < 0.001). However, there was a significant (p < 0.001) rise in the incidence of each of the non-dermatological diseases like, peripheral neuropathy, conjunctivitis and respiratory distress over the period (Paul *et al.*, 2013). Thus, a complete amelioration calls for better strategies apart from removal of arsenic through drinking water.

3.4 Rice is a potential source of arsenic exposure to humans

Arsenic in rice is a potential threat to human (Halder *et al.*, 2012; 2013) and cooked rice arsenic content on its own is sufficient to give rise to genotoxic effects and adverse health effects in humans (Banerjee et al. 2013; Halder *et al.*, 2014). We have chosen 417 arsenic exposed individuals from three districts of West Bengal, namely, Murshidabad, Nadia and East Midnapur whose arsenic content in drinking water was < 10 µg/L. The entire study population was divided into 6 exposure groups based on the arsenic content in rice and recorded micronuclei formation in their urothelial cells. Results show that the rice arsenic content of > 200 µg/kg is associated with significant increased genetic damage as is evident from the increased micronuclei formation in the urothelial cells of the arsenic exposed individuals. (Banerjee *et al.*, 2013).

4 CONCLUSIONS

Thus, the facets of arsenic toxicity range from genotypes of an individual (SNPs) to the molecular toxicity (epigenetic alterations) encompassing a wide genre of the molecular machinery within the cells. Arsenic-contaminated groundwater can occur as a result of microbially driven reductive dissolution of naturally occurring Fe(III) oxyhydroxides coupled to reduction of As(V) to As(III) coupled to organic carbon degradation. The organic carbon that fuels these processes can be derived from many sources, including domestic wastewater, petroleum hydrocarbons, and organic-rich sediments in eutrophic lakes. The potential impact of both historical and current intentional or accidental re-lease of produced water from oil and gas development on arsenic contamination in the subsurface poses significant challenges ranging from obtaining samples to obtaining reliable determinations of trace elemental concentrations in samples with very high concentrations of dissolved salts.

REFERENCES

Banerjee, M., Banerjee, N., Bhattacharjee, P, Mondal, D, Lythgoe, P.R., Martínez, M., Pan, J., Polya, D.A., Giri, A.K. 2013. High arsenic in rice is associated with elevated genotoxic effects in humans. *Sci. Rep.* 3: 2195.

Ghosh, P., Banerjee, M., De Chaudhuri, S., Chowdhuri, R., Das, J.K., Mukherjee, A., Sarkar, A.K., Mondal, L.K., Baidya, K., Sau, T.J., Banerjee, A., Basu, A, Chaudhuri, K., Ray, K., Giri, A. K. 2007 Comparison of health effects between individuals with and without skin lesions in the population exposed to arsenic through drinking water in West Bengal, India. *J. Exp. Sci. Env. Epid.* 17: 215–223.

Guha Mazumder, D.N., De, B.K., Santra, A., Ghosh, N., Das, S., Lahiri, S., Das, T. 2001. Randomized placebo-controlled trial of 2,3-dimercapto-1-propanesulfonate (DMPS) in therapy of chronic arsenicosis due to drinking arsenic-contaminated water. *J. Toxicol.-Clin. Toxic.* 39: 665–674.

Halder, D., Bhowmick, S., Biswas, A., Chatterjee, D., Nriagu, J., Guha Mazumder, D.N., Šlejkovec, Z., Jacks, G., Bhattacharya, P. 2013. Risk of arsenic exposure from drinking water and dietary components: Implications for risk management in rural Bengal. *Environ. Sci. Technol.* 47: 1120–1127.

Halder, D., Bhowmick, S., Biswas, A., Mandal, U., Nriagu, J., Guha Mazumder, D.N., Chatterjee, D., Bhattacharya, P. 2012. Consumption of brown rice: A potential pathway for arsenic exposure in rural Bengal. *Environ. Sci. Technol.* 46: 4142–4148.

Halder, D., Biswas, A., Šlejkovec, Z., Chatterjee, D., Nriagu, J., Jacks, G., Bhattacharya, P. 2014. Arsenic species in raw and cooked rice: Implications for human health in rural Bengal. *Sci. Total Environ.* 497–498: 200–208.

Paul, S., Das, N., Bhattacharjee, P., Banerjee, M., Das, J.K., Sarma N, Sarkar A, Bandyopadhyay A K, Sau T J, Basu S, Banerjee, S., Majumder, P., Giri, A.K. 2013. Arsenic-induced toxicity and carcinogenicity: a two-wave cross-sectional study in arsenicosis individuals in West Bengal, India. *J. Exp. Sci. Env. Epid.* 23(2): 156–162.

3.5   Risk assessment of chronic ingestion

Arsenic Research and Global Sustainability – Bhattacharya, Vahter, Jarsjö, Kumpiene, Ahmad, Sparrenbom, Jacks, Donselaar, Bundschuh & Naidu (Eds)
© 2016 Taylor & Francis Group, London, ISBN 978-1-138-02941-5

Development of the IRIS toxicological review of inorganic arsenic

J.S. Lee, J.S. Gift & I. Cote
U.S. Environmental Protection Agency, National Center for Environmental Assessment, Research Triangle Park, USA

ABSTRACT: The U.S. Environmental Protection Agency (U.S. EPA) Integrated Risk Information System (IRIS) Program is developing a state-of-the-science Toxicological Review of Inorganic Arsenic (iAs). The approaches to assess the hazard and dose-response of iAs, and the associated uncertainty, are based on evolving practices in the IRIS Program and are guided by National Research Council (NRC) recommendations. The review is incorporating several new elements, including comprehensive problem formulation and planning involving stakeholders, and explicit quantitative consideration of sensitive sub-population risks and risk modifiers. The review will also integrate thorough analyses of adverse outcome pathways and networks to inform causal determinations and dose-response model choices, categorical and Bayesian regression meta-analyses of studies to examine dose-response, and Bayesian analyses of uncertainties.

1 INTRODUCTION

Inorganic arsenic (iAs) is a naturally occurring element widely distributed throughout the Earth's crust. In addition to natural sources, industrial activities can release iAs. Inorganic arsenic is found in water, food, soil, and air. This prevalence increases the potential for human exposure; therefore, characterization of the human health impacts of iAs exposure is important. U.S. EPA is currently developing a human health assessment for iAs that considers both cancer and noncancer hazards.

2 HAZARD IDENTIFICATION

PubMed, Web of Science, and TOXLINE were searched using the chemical name and CASRN (Chemical Abstracts Service Registry) number. The literature results from databases were combined, and duplicate records were removed. Non-peer-reviewed articles, abstracts, posters, and review articles were separated in the initial screening of the comprehensive list of references. Remaining references were grouped using natural language processing based on text similarities in the titles and abstracts. Following screening by title and abstract, the full text of all epidemiologic and toxicology studies identified was further reviewed to identify characteristics of the study design and the health effects reported in the study to determine if the study would inform the hazard identification for iAs.

Studies considered relevant for hazard identification were subject to risk-of-bias evaluations.

Risk of bias for each study was evaluated using the questions and considerations proposed in the Office of Health Assessment and Translation (OHAT) approach (NTP, 2013). Each study was evaluated independently by two scientists.

Evidence of health effects identified through the literature search will be summarized and integrated as described in the Cancer Guidelines (U.S. EPA, 2005) and the Integrated Science Assessments for lead and ozone (U.S. EPA, 2013a, b). The process is depicted in Figure 1.

Adverse health effects under consideration are shown in Figure 2.

U.S. EPA will collect and analyze Mode of Action (MOA) information for each health endpoint of significant concern. MOA analyses might inform the shape of the dose-response curve beyond the range of the observational data and can be used to inform susceptibility and variability features, integrate mechanistically related outcomes, and help evaluate multiple risk modifiers (e.g., preexisting disease backgrounds, differences in genetic susceptibilities, smoking, alcohol consumption, diet). Risk modifiers will be evaluated using an evidence-of-susceptibility framework adapted from the Integrated Science Assessments (U.S. EPA, 2013).

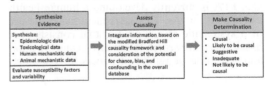

Figure 1. Overview of hazard identification for arsenic.

Adverse Health Effects Under Consideration
• Lung, skin, bladder, prostate, renal, and liver cancer
• Cardiovascular disease
• Skin lesions
• Diabetes
• Nonmalignant respiratory disease
• Pregnancy outcomes
• Neurodevelopmental toxicity
• Immune effects

Figure 2.   Adverse health effects under consideration.

3   DOSE-RESPONSE ANALYSES

Dose-response analyses will be performed on end-points for which iAs is determined as "causal" or "likely causal." Level 1 represents a standardized approach. Levels 2 and 3 are more comprehensive methods that will be applied, if warranted, based on the endpoint and data set under consideration. Levels 2 and 3 extend the Level 1 approach by deriving model-based predictions of risk across ranges of doses, while attempting to quantify model uncertainty, exposure uncertainty, biological considerations, and individual and study population variability. Level 3 extends the Level 2 analysis through the use of Bayesian methods for quantifying uncertainties and, where possible, the use of MOA information to quantify individual variability.

4   RESULTS

Our initial literature search conducted in January 2013 identified over 40,000 references. After removing duplicates and non-peer reviewed literature, about 27,000 references remained. Clustering was performed using 900 seed references identified from other human health assessments. Literature search updates conducted through July 2014 and studies identified from clustering resulted in about 7000 articles to screen. Clustering used to identify studies to support MOA analyses resulted in about 5000 references. Screening for relevance identified about 600 human studies and about 100 toxicology studies.

5   CONCLUSIONS

U.S. EPA is currently developing a human health assessment for iAs. This assessment has made use of a systematic review process that consists of a broad literature search followed by categorization and screening of published papers for relevance.

Risk of bias evaluation of epidemiology and toxicology studies was conducted. Hazard identification considers multiple streams of data to determine whether or not iAs exposure is associated with development of adverse health effects. Mechanistic models will help us better interpret the data in terms of causality, response modifiers, and dose-response considerations. Dose-response analyses will be performed according to a three level approach that reflects increasing complexity.

ACKNOWLEDGEMENTS

We thank Vincent Cogliano, Samantha Jones, Glinda Cooper, Ellen Kirrane, Tom Luben, Ryan Jones, Ingrid Druwe and others at the U.S. EPA. We thank Audrey Turley, Robyn Blain, Bill Mendez, Sorina Eftim, Cara Henning, Michelle Cawley, Dave Burch, Pam Ross, and others at ICF International. We also thank Andy Rooney, Kris Thayer, Bruce Allen, Kan Shao, and Lyle Burgoon.

REFERENCES

NRC (National Research Council). 2013. Critical aspects of EPA's IRIS assessment of inorganic arsenic: Interim report. Washington, D.C: The National Academies Press.

NRC (National Research Council). 2014. Review of EPA's Integrated Risk Information System (IRIS) process. Washington, DC: The National Academies Press. http://www.nap. edu/catalog.php?record_id = 18764.

NTP (National Toxicology Program). 2013. Draft OHAT approach for systematic review and evidence integration for literature-based health assessments—February 2013. National Institute of Environmental Health Sciences, National Institutes of Health. http://ntp.niehs.nih.gov/ntp/ohat/eval uationprocess/drafto-hatapproach_february2013.pdf.

U.S. EPA (U.S. Environmental Protection Agency). (2005). Guidelines for carcinogen risk assessment. (EPA/630/P-03/001F). Washington, DC: U.S. Environmental Protection Agency, Risk Assessment Forum. http://www.epa.gov/ cancerguidelines/.

U.S. EPA (U.S. Environmental Protection Agency). 2013a. Integrated science assessment for lead EPA Report, EPA/600/R-10/075F. Research Triangle Park, NC. http://ofmpub.epa.gov/eims/eimscomm. getfile?p_download_id = 514513.

U.S. EPA (U.S. Environmental Protection Agency). 2013b. Integrated science assessment for ozone and related photochemical oxidants. EPA/600/R-10/076F. Research Triangle Park, NC: U.S. Environmental Protection Agency, National Center for Environmental Assessment.http://cfpub.epa.gov/ w/isa/recordisplay. cfm?deid = 247492.

*Arsenic Research and Global Sustainability – Bhattacharya, Vahter, Jarsjö, Kumpiene,
Ahmad, Sparrenbom, Jacks, Donselaar, Bundschuh & Naidu (Eds)*
© 2016 Taylor & Francis Group, London, ISBN 978-1-138-02941-5

Inorganic arsenic in rice and rice products on the Swedish market 2015 Part 2: Risk assessment

S. Sand, G. Concha, V. Öhrvik & L. Abramsson
National Food Agency, Uppsala, Sweden

ABSTRACT: The exposure to inorganic arsenic in Sweden occurs mainly via certain foods. This assessment shows that rice is the greatest source of exposure to inorganic arsenic (27–31 percent) for the population of Sweden. The median exposure from foods, including rice, is estimated to be approximately 0.07 for adults, 0.10 for 11/12-year-olds, 0.13 for 8/9 year-olds and 0.18 µg/kg b.w./day for 4 year-olds. The Swedish National Food Agency's so-called "Risk Thermometer" has been used to evaluate the risks. The Risk Thermometer has five different risk classes and the estimated exposure to arsenic in food classify, generally speaking, in risk class 3. For children, and especially young children, the exposure is close to or above the limit of what is generally acceptable from a health perspective. The acceptable arsenic exposure is regarded to be approximately 0.15 µg per kilo body weight per day, of which 0.045 µg per kilo body weight per day, or 30 percent, comes from rice.

1 INTRODUCTION

The Swedish National Food Agency has been working for many years mapping the sources of consumers' consumption of arsenic. Rice and rice products represent one third of the total exposure to arsenic in Sweden. In 2013, the Swedish National Food Agency investigated the arsenic content in a selection of products intended for children. The results of the investigation also led to several companies subsequently working to reduce the arsenic content in their products. This project is part of the Swedish National Food Agency's work to map the occurrence of arsenic in various foods and to investigate the intake of arsenic from various types of food. It is also part of a more long-term objective, to induce rice producers to work more actively to ensure that the rice raw material has a lower arsenic content and in this way reduce consumers' intake of arsenic.

The overall project comprises three parts: Part 1—*a survey of inorganic arsenic in rice and rice products*, Part 2—*risk assessment*, and Part 3—*risk management*. The second part is described herein.

2 MATERIALS AND METHODS

2.1 *Consumption data and concentration data*

Consumption data from the 2003 and 2010–11 Swedish National Food Agency dietary surveys were used (Amcoff *et al.*, 2012; Enghardt Barbieri *et al.*, 2006). The 2003survey on children included 590 4-year-olds, 889 school children in the second year (8/9 years old) and 1,016 in the fifth year (11/12 years old). The 2010–11 survey on adults comprised a total of 1,797 individuals.

The concentrations levels of arsenic in foods used in this risk assessment are reported in *Part 1- A Survey of Inorganic Arsenic in Rice and Rice Products* (Kollander & Sundström, 2015).

2.2 *Exposure assessment*

The exposure assessment is based on the part of the population that consumes rice and rice products (consumers only). In the Swedish National Food Agency's dietary surveys, approximately 50% reported some level of rice consumption (Amcoff *et al.*, 2012; Enghardt Barbieri *et al.*, 2006). The median and 95th percentile of exposure to inorganic arsenic from foods was calculated in term of intake per kilo body weight and day (using reported weights). Average inorganic arsenic concentrations in foods have been used in the exposure assessment.

2.3 *Risk thermometer*

The Swedish National Food Agency has developed a tool for risk characterization called the "Risk Thermometer" (Sand *et al.*, 2015). The Risk Thermometer is based on the traditional principle for risk characterization, but differs in the sense that the severity of the critical health effect is also considered in a systematic manner. The underlying risk characterization measure in the Risk Thermometer is therefore called Severity-Adjusted Margin Of Exposure (SAMOE), rather than Margin of Exposure (MOE).

$$SAMOE = RP / (AF_{BMR} \times AF \times SF \times E) \qquad (1)$$

RP is the reference point, in this case corresponding to the $BMDL_{0.05}$ for lung cancer (FAO/WHO 2011).

Table 1. Relation between the SAMOE and the risk classes.

SAMOE	Risk Class	Concern level
0.01	5	high
0.01–0.1	4	moderate-to-high
0.1–1	3	low-to-moderate
1–10	2	no-to-low
> 10	1	no

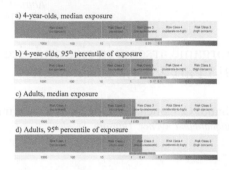

a) 4-year-olds, median exposure

b) 4-year-olds, 95th percentile of exposure

c) Adults, median exposure

d) Adults, 95th percentile of exposure

AF_{BMR} is an assessment factor for response-adjustment (a generalization of the principle of using an assessment factor for LOAEL to NOAEL extrapolation). The standard benchmark response is 10% in the Risk Thermometer. $AF_{BMR} = 1/10$ is used in this study since the RP only corresponds to a 0.5% risk. $AF = 10$ is the default assessment factor for differences in susceptibility within the human population. $SF = 100$ describes the severity of the critical health effect. A health effect classification scheme has been developed as a basis for determining the value of the SF, which can range from 1 to 100 (Sand et al., 2015).

The SAMOE is classified in one of five risk classes describing different levels of health concern (Table 1). Exposures that categorize in risk classes 1 and 2 are not regarded to represent a health risk in a long-term perspective. Risk class 3 is a grey zone. The middle of risk class 3 (SAMOE = 0.316) is used as a reference that in a balanced manner takes into account traditional risk assessment practice where an exposure that is below the health-based reference value is considered to be safe. Exposures that categorize in risk classes 4 and 5 are regarded to represent a potential health risk.

3 RESULTS AND DISCUSSION

3.1 Dietary exposure to inorganic arsenic

The median exposure for 4, 8/9, and 11/12 year-olds was 0.18, 0.13, and 0.1 µg/kg b.w./day, and the 95th percentile of exposure was 0.27, 0.21, and 0.16 µg/kg b.w./day, respectively.

For adults the median and 95th percentile of exposure was 0.065 and 0.11 µg/kg b.w./day, respectively. Among both adults and children, rice represented the largest source of exposure to inorganic arsenic (27–31 percent of the total intake of inorganic arsenic).

3.2 Risk characterization

The median arsenic exposure as well as the 95th percentile of exposure classifies in risk class 3 for all consumer groups. The results are illustrated in Figure 1 for 4-year-olds and adults: these two groups represent the extremes in estimated inorganic arsenic exposure in each direction.

Figure 1. The wide grey bars show the size of the SAMOE that classifies in one of five risk classes that describe different levels of health concern. The thin grey bars show the uncertainty interval for the SAMOE. The ends of the intervals describe the 5th and 95th confidence limit. See also the report on the Risk Thermometer (Sand et al., 2015) for details regarding the uncertainty analysis.

4 CONCLUSIONS

The estimated exposure to inorganic arsenic in food generally classify in risk class 3, and for children (especially younger children) the exposure is close to or above the limit for what is acceptable from a health perspective. Also, accounting for estimated uncertainties, the arsenic exposure from foods may be higher than desirable for a small portion of the adults. The acceptable arsenic exposure (in a lifetime perspective) is assessed to be approximately 0.15 µg per kg body weight per day, of which 0.045 µg per kg body weight per day (or 30 percent) comes from rice and rice products. Based on scenario analyses the acceptable arsenic exposure from rice corresponds to approximately 3–4 portions per week for children and 6 portions per week for adults.

REFERENCES

Amcoff, E., Edberg, A., Enghardt Barbieri, H., Lindroos, A.K., Nälsén, C., Pearson, M., Warensjö Lemming, E. 2012. Riksmaten—vuxna 2010–11. Livsmedels-och näringsintag bland vuxna i Sverige. Swedish National Food Agency, Uppsala, Sweden.

Enghardt Barbieri, H., Pearson, M., Becker, W. 2006. Riksmaten—barn 2003. Livsmedels- och näringsintag bland barn i Sverige. Swedish National Food Agency, Uppsala, Sweden.

Kollander, B., Sundstrom, B. 2015. Inorganic arsenic in rice and rice products on the Swedish Market—Part 1. A Survey of inorganic arsenic in rice and rice products. Swedish National Food Agency Report 16/2015, NFA, Uppsala, Sweden.

Sand S, Bjerselius R, Busk L, Eneroth H, Sanner-Färnstrand J, Lindqvist R. 2015. The Risk Thermometer—a tool for risk comparison. Swedish National Food Agency Report: 8/2015, NFA, Uppsala, Sweden.

Arsenic Research and Global Sustainability – Bhattacharya, Vahter, Jarsjö, Kumpiene,
Ahmad, Sparrenbom, Jacks, Donselaar, Bundschuh & Naidu (Eds)
© *2016 Taylor & Francis Group, London, ISBN 978-1-138-02941-5*

Inorganic arsenic exposure and children at risk in Majuli—the most thickly populated river island of the world

R. Goswami & M. Kumar
Department of Environmental Science, Tezpur University, Tezpur, Assam, India

ABSTRACT   Chronic arsenic (As) toxicity due to drinking of As contaminated water is a major environmental health hazard. Although substantial information is available about chronic As toxicity in adults, knowledge of such health effect on children is scanty. An effort was made in Majuli to understand the extent of the As exposure in children through analyzing 40 groundwater samples from 40 schools (one sample from each school) and 120 biological samples (40 hair, 40 nail, and 40 urine). The results showed significant correlation between As in the drinking water (max = 318µg/L) with As in hair (median = 943 µg/kg, $r = 0.53$), nail (median = 1797 µg/kg, $r = 0.83$) and urine (median = 101.8 µg/L, $r = 0.77$) from the As exposed children. Analysis of biological samples revealed that the children might be sub clinically affected. Further studies on the metabolism and effects of As in children are urgently needed.

1 INTRODUCTION

Many aquifers in various parts of the world includ-ing India have been found to be contaminated with inorganic arsenic, iAs (Bhattacharya *et al.*, 2002). There is sufficient evidence from human studies that chronic ingestion of iAs causes cutaneous and sys-temic manifestations and skin, bladder and lung can-cer in adults (Chakraborti *et al.*, 2004). In Asian countries, where more than 130 million people are exposed to iAs con-taminated groundwater contain-ing more than 10 µg/L (WHO guideline value), at least 20 million children below 11 years are vulner-able (Goswami *et al.*, 2014). Many studies have shown that children are at higher risk of iAs expo-sure (Concha *et al.*, 1998). Though lot of information is available on health effects due to chronic iAs toxicity in adults, study of such effect on children is limited in North eastern region of India. A cross sectional study using environmental and biologi-cal samples was undertaken to assess the relationship between iAs concentration in drinking water and their As concentrations in hair, nail, and urine of children living in Majuli-the most thickly populated river island of the world where the groundwater is naturally As rich. 40 children aged between 5 and 11 years were selected from the schools where groundwater As con-centration was >50 µg/L.

2 METHODS

2.1 *Study area*

Majuli, the study area, is located on the middle of the mighty Brahmaputra River. Geographically, the island is a part of the great alluvial tract of the Brahmaputra River, which is by nature a geo-synclinal basin com-posed of alluvial and more recent Pleistocene sedi-ment deposits. The island is populated by 1,68,000 people in 248 villages (Cen-sus, 2011). Geologically, Majuli is occupied by un-consolidated alluvial sedi-ments of quaternary age. The island comprises both younger and older alluvi-um as well as active flood plain deposits of recent age.

2.2 *Sample collection, digestion and analysis*

Groundwater ($n = 40$) and biological samples includ-ing hair ($n = 40$), nail ($n = 40$) and urine ($n = 40$) were collected (from the schools where groundwater As conc. was >50 µ/L) to evaluate the extent of arsenic burden in the children. Before groundwater sample collection, arsenic field test was performed using field testing kit (Merck).

The modes of biological sample collection, diges-tion and analytical procedures for hair and nail have been reported elsewhere (Samanta *et al.*, 1999). Sam-ples (after digestion) were analyzed for As by FI-HG-AAS.

3 RESULTS AND DISCUSSION

3.1 *Arsenic in tube well water samples*

Of the hand tube-wells sampled from the selected 40 no of schools of Majuli, 50% of samples were found to have As concentrations above 10 µg/L. Nearly 15% of the samples had As concentrations greater than 50 µg/L. Moreover, 3% of samples had As greater than 300 µg/L, the concentration predicting overt arsenical skin lesions (Chakraborti *et al.*, 2004) (from the present study maximum As concentration was found 318 µg/L).

Figure 1. Photograph of children from Majuli who were drinking arsenic contaminated water.

3.2 Arsenic in biological samples from exposed children of Majuli

Arsenic concentration in biological samples as hair, nails and urine plays an important role in evaluating the As body burden. Urinary As is usually considered as the most reliable indicator of recent exposure to As and is used as the main bio-marker of exposure. Levels of total As in urine, hair and nail samples collected from the children of Majuli are shown in Table 1. About 70% of urine samples contained As level above the maximum normal level (50 µg/L). Urinary As concentration is even higher in several cases within 10% of the urine samples containing As concentrations greater than 300 µg/L. All hair samples evaluated contained As above the maximum normal level (200 µg/kg) (NRC, 1999). Five percentage of hair samples were found with As concentrations above 3000 µg/kg from one school where the children consumed very high concentrations of As through drinking water (max: 318 µg/L). Of the analyzed nail samples, 82% of nail samples contained As above the maximum normal level (500 µg/kg) (NRC, 1999); Two percentage of nail samples of children evaluated contained As above 5000 µg/kg. The results showed significant correlation between As in the drinking water with As in hair ($r = 0.53$), nail ($r = 0.83$) and urine ($r = 0.77$) of the children evaluated.

3.3 Excretion of arsenic in urine of children compared to adults

It is generally considered that children are more susceptible to the adverse effects of As than adults. Experimental studies have indicated that the methylation of As is influenced by the dose level, mode of administration, form of As administered, and nutritional status of the subject (Chakraborti et al., 2004). However, the influence of age, particularly among children has been scarcely evaluated. A comparative evaluation of urinary concentration between child and adults drinking water from the same source (> 50 µg As/L) was evaluated in a subgroup conformed by 7 teachers (> 25 years) and 7 children (<12 years). Interestingly the urinary mean As concentration in children was much higher (339 µg/L), compared to the adults (154 µg/L). Therefore, it is evident that the concentrations of As in the urine of the studied children group were much higher than the normal range.

Table 1. Arsenic concentrations in urine, hair and nail samples from 40 children of Majuli Island.

Parameters	As in urine[a]	As in hair[b]	As in nail[c]
Mean	155	1,023	2,500
Median	101.8	943	1,797
Minimum	20.8	224	426
Maximum	695	5,000	10,722
SD	154.9	1,039.8	2,327

[a]Normal urine As ranges from 5 to 40 µg/L.5 L/day (NRC, 1999); [b]Normal hair As is 80–250 µg/kg, with 1,000 µg/kg an index on toxicity (NRC, 1999); [c]Normal As concentration of nail is 430 –1,080 µg/kg.(NRC, 1999).

4 CONCLUSIONS

The present work, for the first time in North East India, reports the As distribution in hair, nail and urine of children of Majuli Island. The studied groups of children have not shown arsenical skin lesions, still high arsenic body burden in their biological samples and thus could be sub-clinically affected. Urinary As concentrations in children were higher than adults from the comparative study in the subsample evaluated. Thus the role of high As exposure in these children at later ages should be evaluated.

ACKNOWLEDGEMENTS

We thank the Department of Science and Technology (DST), under Govt. of India for funding the research project under Fast Track Young Scientist Scheme.

REFERENCES

Bhattacharya, P., Frisbie, S.H., Smith, E., Naidu, R., Jacks, G., Sarkar B. 2002. Arsenic in the Environment: A Global Perspective. In: B. Sarkar (Ed.) Handbook of Heavy Metals in the Environment. Marcell Dekker Inc., New York, pp. 147–215.
Census Report. 201.1 http://censusindia.gov.in/.
Chakraborti, D., Sengupta, M.K., Rahaman, M.M., et al. 2004. Groundwater arsenic contamination and its health effects in the Ganga- Meghna-Brahmaputra plain. J. Environ. Monitor. 6(6): 74–83.
Concha, G., Nermelli, B., Vahter, M. 1998. Metabolism of inorganic arsenic in children with high arsenic exposure in northern Argentina. Environ. Health. Perspect. 106: 355–359.
Goswami, R., Rahman, M.M., Murrill, M., Sarma, K.P., Thakur, R., Chakraborti, D. 2014. Arsenic in the groundwater of Majuli—The largest river island of the Brahmaputra: Magnitude of occurrence and human exposure. J. Hydrol. 518: 354–362.
NRC. 1999. Arsenic in Drinking Water. National Academy of Sciences. Washington, DC.
Samanta, G., Chowdhury, T.R., Mandal, B.K., et al. 1999. Flow injection hydride generation atomic absorption spectrometry for determination of arsenic in water and biological samples from arsenic affected districts of West Bengal, India and Bangladesh. Microchem. J. 62: 174–191.

Arsenic Research and Global Sustainability – Bhattacharya, Vahter, Jarsjö, Kumpiene,
Ahmad, Sparrenbom, Jacks, Donselaar, Bundschuh & Naidu (Eds)
© *2016 Taylor & Francis Group, London, ISBN 978-1-138-02941-5*

Speciation and health risk assessment of arsenic in groundwater of Punjab, Pakistan

M.B. Shakoor[1], N.K. Niazi[1,2], I. Bibi[1,2], M.M. Rahman[3,4], R. Naidu[3,4], M. Shahid[5],
M.F. Nawaz[6] & M. Arshad[1]

[1]*Institute of Soil and Environmental Sciences, University of Agriculture Faisalabad, Faisalabad, Pakistan*
[2]*Southern Cross GeoScience, Southern Cross University, Lismore, NSW, Australia*
[3]*Global Centre for Environmental Remediation, Faculty of Science and Information Technology,*
The University of Newcastle, Callaghan, NSW, Australia
[4]*Cooperative Research Centre for Contamination Assessment and Remediation of the Environment*
(CRC CARE), University of Newcastle, Newcastle, New South Wales, Australia
[5]*Department of Environmental Sciences, COMSATS Institute of Information Technology, Vehari, Pakistan*
[6]*Department of Forestry, Range Management and Wildlife, University of Agriculture Faisalabad, Pakistan*

ABSTRACT: In this study, we examined the total and speciated Arsenic (As) concentrations and other drinking water quality parameters for unraveling the health risk of As from drinking water to humans. Groundwater samples ($n = 62$) were collected from three previously unexplored rural areas (Chichawatni, Vehari, Rahim Yar Khan) of Punjab, Pakistan. The As concentration in the groundwater samples ranged from <10–206 µg/L which was higher than the WHO safe limit of 10 µg/L. Arsenite (As(III)) constituted 13–67% of total As and arsenate (As(V)) ranged from 33–100% in the study area. For As health risk assessment, the hazard quotient (11–18 times) and cancer risk (46–600 times) values were found more than US-EPA recommended values. Various other water quality parameters also enhanced the health risk. The results show that the consumption of As-contaminated groundwater poses an emerging health threat to the communities, thus immediate remedial and management measures are required for providing safe drinking water to the people living in As-affected areas.

1 INTRODUCTION

Arsenic (As) contamination of groundwater is a global health issue because of its carcinogenic nature (Shakoor *et al.*, 2015). Arsenic naturally occurs as inorganic and organic forms in different environmental and biological samples. Inorganic As is mostly observed in groundwater, surface water, soil and various foods. Arsenic contamination of groundwater (up to 5000 µg/L) has been reported worldwide in over 105 countries affecting more than 200 million people, of which nearly 100 million are the resident of south east Asia. Arsenic health risk assessment is fundamentally important to calculate potential health risk (chronic and carcinogenic effects) such as, Average Daily Dose (ADD), Hazard Quotient (HQ), and Carcinogenic Risk (CR) from consumption of As-contaminated water (Sultana *et al.*, 2014). Hence, the objectives of the current study were to (i) determine the total As content and species of As in groundwater from three previously unexplored areas of Punjab (Chichawatni, Vehari and Rahim Yar Khan), Pakistan; (ii) evaluate the concentration of various water quality parameters such as

major cations, anions and trace metals, and (iii) estimate the health risk of As by calculating ADD, HQ and CR values for As-contaminated drinking water by United States-Environmental Protection Agency (US-EPA) model and bootstrap method.

2 MATERIALS AND METHODS

Groundwater samples ($n = 62$) were collected from electric pumps/wells, tubewells and hand pumps of various previously unexplored areas of Punjab (Chichawatni, Vehari and Rahim Yar Khan), Pakistan. An inductively coupled plasma mass spectrometer (ICP-MS) was used for multi-elements analysis such as total As, calcium (Ca), magnesium (Mg), potassium (K), sodium (Na), iron (Fe), manganese (Mn), nickel (Ni), lead (Pb), cadmium (Cd), copper (Cu), silicon (Si), cobalt (Co), aluminum (Al), zinc (Zn), boron (B), chromium (Cr), phosphorus (P) and sulfur (S), For anions analyses including nitrate (NO_3^-), sulfate (SO_4^{2-}), flouride (F^-), chloride (Cl^-) and phosphate (PO_4^{3-}) in water samples, ion chromatography (IC) was used. Arsenic speciation was done by using an IC-ICP-MS. Other anions such as carbonates (CO_3^{2-}) and bicarbonates (HCO_3^-) were determined by acid titration (H_2SO_4) using phenolphthalein and methyl orange as indicators. Geochemical modeling software, PHREEQC, was used to calculate the Saturation Indices (SI) of possible As-bearing minerals phases using groundwater chemical composition data.

3 RESULTS AND DISCUSSION

Arsenic in the water samples ranged from < 10 to 206 µg/L (mean = 37.9 µg/L) (Table 1) which was significantly higher than the safe limit of As in water set by World Health Organisation (WHO) (10 µg/L).

Speciation data of the selected groundwater samples showed that As(III) constituted 13 to 66% of total As and As(V) ranged from 45 to 100%. Figure 1 compares the mean As concentrations of this study with the previous studies conducted in Pakistan, Indonesia, Cambodia, Vietnam, Bangladesh, and India. Mean As concentrations found in our study were higher than the As concentrations reported in Pakistan, Indonesia and Vietnam while less than the concentrations documented in Cambodia, Bangladesh and India.

Saturation indices values indicated the precipitation of iron (Fe) oxide minerals such as goethite (5.3), hematite (13) and magnetite (5), demonstrating that these Fe oxide mineral phases could possibly control sorption/desorption of As in the studied aquatic environments, under the alkaline pH (> 7–8.3). For health risk assessment of As, the ADD, HQ and CR values ranged from 3.6×10^{-5} to

Table 1. Summary statistics of arsenic in groundwater samples collected from three rural areas of Punjab, Pakistan.

Chichawatni (CW) ($n = 8$)		Vehari (Vh) ($n = 28$)		Rahim Yar Khan (RYK) ($n = 26$)	
Mean	Range	Mean	Range	Mean	Range
120 ± 55	23–201	42 ± 38	1.3–144	9 ± 5	1.6–35

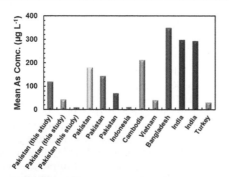

Figure 1. Comparison of mean As concentration in groundwater in this study with other studies.

5.6×10^{-3} mg/kg/day, 0.12 to 18.53 and 5.4×10^{-5} to 8.3×10^{-3}, respectively. Hazard quotient and CR are considered to be present if the values are HQ > 1.00 and CR > 10^{-6} and in the present study the HQ and CR values calculated by both the US-EPA model and bootsrap method were higher (HQ: 11–18 and CR: 46–600 times) than the US-EPA limit suggesting that the people in the study area are at high risk of As-induced carcinogenicity from drinking water. In addition, sodium adsorption ratio (SAR), Cl, NO_3, SO_4, Fe, Na, Mg, Mn and Pb also exceeded than their permissible limits set by WHO and Pakistan Environmental Protection Agency (Pak-EPA).

4 CONCLUSIONS

The results showed that over 60% groundwater samples collected from various sites of CW, Vh and RYK were unfit for drinking purpose having high values of EC, Cl, NO_3, SO_4, Fe, Mn and Pb. High levels of As were detected in groundwater of the study area. Both the US-EPA model and bootstrap methods provided high carcinogenic and non-carcinogenic As risks to the individuals of the study area. The ADD and HQ were in the order of CW > Vh> RYK with the maximum HQ value of 18.5 reported in CW. The CR values were significantly (46–600 times) higher than its default acceptable level (10^{-6}) exposing people to chronic As carcinogenicity. Higher cancer risk values were obtained by bootstrap method compared to US-EPA model. Given the health risk associated with extent of groundwater As contamination, government of Pakistan should reduce current (50 µg/L) safe limit of As to 10 µg/L, so that people should be protected from mass poisoning of As.

ACKNOWLEDGEMENTS

The authors are thankful to the Grand Challenges Canada–Stars in Global Health (GCC Grant No. 0433–01) for providing financial support. We acknowledge financial support from the CRC-CARE (project 3.1.3.11/12), Australia.

REFERENCES

Shakoor, M.B., Niazi, N.K., Bibi I., Murtaza, G., Kunhikrisnan, A., Seshadri, B., Shahid, M., Ali, S., Bolan, N.S., Sik, Y., Abid, M., Ali, F. 2015. Remediation of arsenic-contaminated water using agricultural wastes as biosorbents. *Crit. Rev. Env. Sci. Tec.* 46(5): 467–499.
Sultana, J., Farooqi, A., Ali, U. 2014. Arsenic concentration variability, health risk assessment, and source identification using multivariate analysis in selected villages of public water system, Lahore, Pakistan. *Environ. Monit. Assess.* 186: 1241–1251.

Section 4: Clean water technology for control of arsenic

4.1   Technologies based on adsorption and co-precipitation

Arsenic Research and Global Sustainability – Bhattacharya, Vahter, Jarsjö, Kumpiene,
Ahmad, Sparrenbom, Jacks, Donselaar, Bundschuh & Naidu (Eds)
© *2016 Taylor & Francis Group, London, ISBN 978-1-138-02941-5*

Relationship between composition of Fe-based LDHs and their adsorption performances for arsenate in aqueous solutions

H. Lu[1], Z. Zhu[1], Q. Zhou[1], H. Zhang[1] & Y. Qiu[2]
[1]*State Key Laboratory of Pollution Control and Resource Reuse, Tongji University, Shanghai, P.R. China*
[2]*Key Laboratory of Yangtze River Water Environment, Ministry of Education, Tongji University, Shanghai, P.R. China*

ABSTRACT: A series of iron based Layered Double Hydroxides (LDHs), which contain transition and rare earth metal cations including Ni^{2+}, Cu^{2+}, La^{3+} and Ce^{3+} as the layer ions, with carbonate or sulfate as intercalation anions and Corresponding Calcined Products (CLDHs) were synthesized. These synthesized LDHs have been used to investigate the relationship between the composition and adsorption performances for arsenate in water. The results showed that the sulfate intercalated CuMgFeLa-LDH had the highest adsorption capacity for arsenate. The adsorption isotherm can be well described by Sips isotherm model and the maximum adsorption capacity was 313.48 mg/g. It was also applied to treat practical polluted water sample of Xikuangshan area, and can meet the drinking water standard of WHO and EPA after the adsorption process with a dosage of 0.2 g/L.

1   INTRODUCTION

Arsenic (As) exists widely in water environment as a notoriously toxic element (Smedley & Kinniburgh, 2002). Layered Double Hydroxides (LDHs) as one kind of functional materials have been investigated and applied for As removal due to their excellent anion exchange capacity and large surface area (Goh *et al.*, 2008). Based on our previous researches (Hong *et al.*, 2014; Lu *et al.*, 2015), the objective of this work is to study the adsorption performances of arsenic on different ferric based LDHs and discuss the effect of different laminate metals and interlayer anions in LDHs. Then explore the application possibility of the synthesized LDHs for treating the practical polluted water sample. A series of ferric based Layered Double Hydroxides (LDHs) with different laminate metals and interlayer anions and Corresponding Calcined Products (CLDHs) were synthesized and used to study the removal performance for arsenate in aqueous solutions. The possible mechanisms of arsenate adsorption on these LDHs were discussed. Furthermore, the selected CuMgFeLa-SO$_4$-LDH was applied to treat practical As polluted water sample of Xikuangshan area and gave a satisfactory result.

2   EXPERIMENTAL

2.1   *Synthesis and characterization*

A series of ferric based Layered Double Hydroxides (LDHs), which contain transition and rare earth metal ions including Ni^{2+}, Cu^{2+}, La^{3+} and Ce^{3+}, with carbonate and sulfate intercalation were synthesized by co-precipitation method. The molar ratio of divalent metal ions and trivalent metal ions was 3. The Corresponding Calcined Products (CLDHs) were prepared with the same method of our previous study (Guo *et al.*, 2013). The characterization of materials like X-Ray Diffraction (XRD), Fourier Transfer Infrared (FTIR), surface area, pore size distribution, Scanning Electron Microscope (SEM) and Energy Dispersion Spectrometer (EDS) were same with previous study (Lu *et al.*, 2015). The As(V) concentrations were determined by Inductively Coupled Plasma Optical Emission Spectrometry (ICP-OES) Agilent 720 (Agilent, USA), and Atomic Fluorescence Spectrometry (AFS) FP6-A (PERSEE, China).

2.2   *Batch adsorption experiments*

The effect of initial As(V) concentrations on adsorption were studied to evaluate the adsorption performance of the different synthesized materials. The batch adsorption experiments were carried out at the temperature of 298° K and the initial solutions pH was 7. The experiments were replicated three times and average results were used. The batch adsorption experiments were conducted in 150 mL flasks with 50 mL required concentrations of As(V) solution. The flasks were shaken in a thermostatic shaker at 150 rpm for 24 hours then filtered with 0.45 μm membrane. The practical water sample from Xikuangshan area was taken

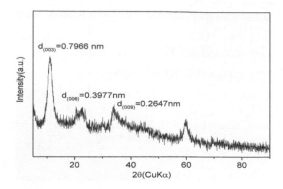

Figure 1. XRD pattern of synthesized CuMgFeLa-SO$_4$-LDH.

Table1. The adsorption capacity of LDHs and CLDHs.

	Adsorption capacity (mg/g^1)				
	Ni/Fe	Mg/Fe	Mg/Fe/La	Mg/Fe/Ce	Cu/Mg/Fe/La
CO$_3$$^{2-}$	18.00	16.99	33.49	21.06	63.39
CLDH	12.49	39.36	17.99	45.22	41.48
SO$_4$$^{2-}$	37.77	45.57	49.93	47.56	70.56

and treated by the selected CuMgFeLa-SO$_4$-LDH material.

3 RESULTS AND DISCUSSION

3.1 Characterization

The XRD patterns of all the synthesized LDHs materials had shown the characteristic peaks of layered double hydroxides. Typical peaks such as (001), (002) and (003) were observed and no other crystalline phases occurred. The XRD pattern of CuMgFeLa-SO$_4$-LDH as an example was presented in Figure 1. The SEM images were consistent with XRD pattern, indicating the layered structures of the synthesized LDHs. The FTIR spectra provided the necessary evidence for the identification of carbonate and sulfate intercalation for the related LDHs.

3.2 Relationship between composition and adsorption capacity of LDHs

The initial As(V) concentration was set at 20 mg/L and the adsorbent dosage was 0.2 g/L. The results of different adsorbents were shown in Table 1.

From Table 1, it can be seen that, for the same metal ion of laminates, the adsorption ability of sulfate intercalated LDHs is greater than that of carbonate intercalated LDHs. The addition of rare earth metal can improve the As(V) adsorption performance of the LDHs, and Lanthanum showed a more significantly positive influence on As(V) adsorption when compared with that of cerium for the carbonate intercalation LDHs. However, under the same conditions for the sulfate intercalated LDHs, it was less affected. The presence of transition metal like Ni and Cu can also improve the arsenic adsorption of the related LDH materials. It was found that the LDHs with the sulfate

intercalation was more efficient than that of carbonate intercalated and CLDH materials on the adsorption removal of As(V). The CuMgFeLa-SO$_4$-LDH showed a highest adsorption capacity and was selected for further treatment of the practical arsenic polluted water samples.

3.3 Adsorption isotherms

Langmuir, Freundlich and Sips models were used to analyze the adsorption experimental data of synthesized LDHs. It can be found that the adsorption of arsenate on the carbonate intercalated LDHs can be described by Langmuir isotherm model. However, the isotherm data of sulfate intercalated LDHs are in good accordance with the Freundlich and Sips isotherm model. The multilayer adsorption and affinities over the heterogeneous surface may occur at surface of the sulfate intercalated LDHs. The maximum adsorption capacity of As(V) on CuMgFeLa-SO$_4$-LDHs can be calculated from Sips isotherm model as 313.48 mg/g.

3.4 Practical water sample treating

When the dosage of the CuMgFeLa-SO$_4$-LDH was 0.2 g/L, the concentration of As in a practical water sample can be reduced from the original value of 800 μg/L to below 10 μg/L. It can meet the standard of limit for arsenic in drinking water of WHO, USEPA and China.

4 CONCLUSIONS

The addition or dropping of transition and rare earth metals can improve the arsenic adsorption capacities of the related LDHs materials. The LDHs with sulfate intercalation is more efficient for arsenic adsorption than that of carbonate intercalated LDHs and the corresponding CLDHs. The CuMgFeLa-SO$_4$-LDH showed highest adsorption capacity of 313.48 mg/g among all the synthesized materials. When the dosage of the CuMgFeLa-SO$_4$-LDH was 0.2 g/L, the concentration of arsenic in practical polluted water sample can be reduced from 800 μg/L to below 10 μg/L. It is a potential efficient adsorbent for decontamination of arsenic in water.

ACKNOWLEDGEMENTS

This work was supported by the National Natural Science Foundation of China (No. 41372241).

REFERENCES

Goh, K.-H., Lim, T.-T., Dong, Z. 2008. Application of layered double hydroxides for removal of oxyanions: A review. *Water Res.* 42: 1343–1368.

Guo, Y., Zhu, Z., Qiu, Y., Zhao, J. 2013. Enhanced adsorption of acid brown 14 dye on calcined Mg/Fe layered double hydroxide with memory effect. *Chem. Eng. J.* 219: 69–77.

Hong, J., Zhu, Z.L., Lu, H.T., Qiu, Y.L. 2014. Effect of metal composition in lanthanum-doped ferric-based layered double hydroxides and their calcined products on adsorption of arsenate. *Rsc Advances* 4: 5156–5164.

Lu, H., Zhu, Z., Zhang, H., Zhu, J., Qiu, Y. 2015. Simultaneous removal of arsenate and antimonate in simulated and practical water samples by adsorption onto Zn/Fe layered double hydroxide. *Chem. Eng. J.* 276: 365–375.

Smedley, P.L., Kinniburgh, D.G. 2002. A review of the source, behaviour and distribution of arsenic in natural waters. *Appl. Geochem.* 17: 517–568.

Arsenic Research and Global Sustainability – Bhattacharya, Vahter, Jarsjö, Kumpiene,
Ahmad, Sparrenbom, Jacks, Donselaar, Bundschuh & Naidu (Eds)
© 2016 Taylor & Francis Group, London, ISBN 978-1-138-02941-5

Fate of low arsenic concentrations during full-scale aeration and rapid filtration

J.C.J. Gude, L.C. Rietveld & D. van Halem
Delft University of Technology, Delft, The Netherlands

SUMMARY: Arsenic (As) mobility during full-scale aeration and rapid filtration was investigated. As(III) remained largely mobile during aeration and as supernatant water and was not efficiently immobilized. In the filter bed however, oxidation of As(III) was complete within 2 min of contact time and As removal efficiency improved. Therefore, the overall conclusion is that not aeration and supernatant storage, but the filter bed is the crucial treatment step for rapid As(III) removal, indicating the importance to control the oxidation sequence of Fe and As for achieving improved As removal efficiencies.

1 INTRODUCTION

In the Netherlands, groundwater treatment commonly consists of aeration, with subsequent sand filtration without using chemical oxidants like chlorine. With arsenic (As) concentrations well below the current guideline of 10 µg As/L, groundwater treatment plants have been exclusively designed for the removal of iron (Fe), manganese (Mn) and ammonium. The aim of this study was to investigate the As removal capacity at two of these ground Water Treatment Plants (WTPs) (13 and 26 µg As/L) in order to identify operational parameters that can contribute to lowering the filtrate As concentration to <1 µg/L. For this purpose a sampling campaign and experiments with supernatant water were executed to identify the key mechanisms controlling As removal (Gude et al., 2016).

2 MATERIALS AND METHODS

2.1 Supernatant water experiments

Jar tests experiments were executed to simulate As behaviour in the supernatant water with extended residence times. 12 L of supernatant water was collected and distributed evenly over six jars containing magnetic stirrers. Three jars were spiked with extra 22 µg/L As(V), consequently the three jars contained both As(III) and As(V). The jars were continuously stirred at 20 rpm and analysed for pH, ORP, conductivity, oxygen and temperature. Samples were taken after 10, 20 and 60 minutes contact time, filtered through 0.45 µm and analyses by ICP-MS.

2.2 WTP sampling campaign

At two WTPs, a water sampling campaign was executed with 10–12 sampling points in the raw water, supernatant water, multiple sample points in the filter bed and filtrate of one filtration step. Before sampling, the rotation of the groundwater well circuit was stopped to ensure a constant flow and water quality during the complete runtime of the sampled filter. All data were collected between a filter runtime of 16–22 hours to ensure optimal working conditions and minimal pore blocking by already retained Hydrous Ferric Oxide (HFO) flocs.

3 RESULTS AND DISCUSSION

3.1 Supernatant water experiments

In order to investigate As removal by precipitating Fe(II) in the supernatant water, adsorption of natural occurring As(III) and a spike of 23 µg/L As(V) onto the formed HFO was measured over 60 min contact time. The results are depicted in Figure 1.

At the start of this experiment both WTPs had similar levels of As(III) and Fe(II) in the aerated supernatant water, namely about 12 µg/l As(III) and 0.7 mg/L Fe(II) for WTP1 and WTP2. HFO flock concentrations were 0.7 mg/L and 3.5 mg/L respectively. The As(III) concentration of both WTPs showed a limited drop (1 µg/L and 2 µg/L) during the 60 minutes. In the same jar, the spiked As(V) was readily immobilized (10 µg/L and 16 µg/L), making oxidation of As(III) imperative for complete As removal.

3.2 As removal in a filter bed

As observed in the supernatant water experiments, most of the As was still mobile before entering

Table 1. Water quality WTPs.

Parameter	Unit	WTP1		WTP2	
		Raw	Filtrate	Raw	Filtrate
As	µg/L	13.2	6.18	26.1	2.44
Fe	mg/L	1.40	0.01	4.33	0.14
Mn	mg/L	0.036	0.001	0.178	0.001

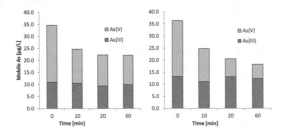

Figure 1. Addition of 23 µg/L As(V) to supernatant water of WTP1 (left) and WTP2 (right) with naturally occurring As(III). Graphs show As species over time, with at t = 0 for WTP2 0.7 mg/L Fe(II) and 0.7 mg/L HFO flocs and for WTP3 0.7 mg/L Fe(II) and 3.5 mg/L HFO flocs.

the filter bed, predominantly consisting of As(III) (>90%). Upon entering the filter bed, the water quality changed rapidly, as depicted for Fe, As, and Mn in Figure 2.

Removal of As in the rapid filters was more efficient than removal of As on Fe flocs in the supernatant water. WTP1 and WTP2 removed 48% and 90% As in the filter bed respectively. As was removed simultaneously with Fe and removal stopped when Fe removal was finished. At WTP2 >97% Fe was removed within the first meter of the filter bed. However, the Fe profile in the WTP2 filter bed was inconsistent with WTP1, which can potentially be explained by the dual filter bed at this location. The HFO flocs (3.5 mg/L) formed in the supernatant water above the filter penetrated deeper into the filter bed because of the larger size of the anthracite in the top layer of the filter bed, subsequently, resulting in more elevated As concentrations higher in the filter bed compared to WTP1.

The mobile As in the filtrate WTP1 and WTP2 was in the As(V) form. This means that As(III) oxidation was completed in the filter bed within 10 min residence time.

Mn was barely found in the backwash water, but abundantly on the filter grains, and must therefore mostly be accumulated on the filter grains as Mn oxides. These Mn oxides are able to oxidize As(III) (Driehaus et al., 1995; Lafferty et al., 2010; Manning et al., 2002), and potentially explain the accelerated As(III) oxidation and subsequent adsorption to the formed HFO in the top layer of the filter bed.

4 CONCLUSIONS

Results showed that As(III) was not efficiently removed in the supernatant water after aeration. Also, after extending residence times during supernatant water experiments, all Fe(II) was precipitated into filterable HFO flocs but most As(III) remained mobile. However, when adding As(V) during these experiments, As(V) was readily adsorbed.

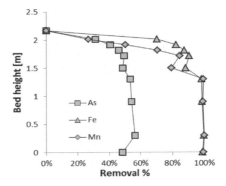

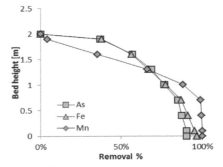

Figure 2. Removal percentage of As, Fe and Mn over filter bed height of WTP1 (top) and WTP2 (bottom).

In the rapid filter, an instant acceleration of As(III) removal was observed compared to the aerated supernatant water, also after 60 minutes residence time. The Mn oxides, formed in the top of the filter bed, are able to oxidize As(III) rapidly and potentially explain the enhanced As removal.

The overall conclusion is that not aeration and supernatant storage, but the filter bed is the crucial treatment step for rapid As(III) removal, indicating the importance to control the oxidation sequence of Fe and As for achieving improved As removal efficiencies.

REFERENCES

Driehaus, W., Seith, R., Jekel, M. 1995. Oxidation of arsenate(III) with manganese oxides in water treatment. *Water Res.* 29: 297–305.
Gude, J.C.J., Rietveld, L.C., van Halem, D. 2016. Fate of low arsenic concentrations during full-scale aeration and rapid filtration. *Water Res.* 88: 566–574.
Lafferty, B.J., Ginder-Vogel, M., Zhu, M., Livi, K.J.T., Sparks, D.L. 2010. Arsenite oxidation by a poorly crystalline manganese-oxide. 2. Results from X-ray absorption spectroscopy and X-ray diffraction. *Environ. Sci. Technol.* 44: 8467–8472.
Manning, B.A., Fendorf, S.E., Bostick, B., Suarez, D.L. 2002. Arsenic(III) oxidation and arsenic(V) adsorption reactions on synthetic birnessite. *Environ. Sci. Technol.* 36: 976–981.

Arsenic Research and Global Sustainability – Bhattacharya, Vahter, Jarsjö, Kumpiene,
Ahmad, Sparrenbom, Jacks, Donselaar, Bundschuh & Naidu (Eds)
© *2016 Taylor & Francis Group, London, ISBN 978-1-138-02941-5*

Arsenic adsorption kinetic studies from the reverse osmosis concentrate onto lateritic soil

C.E. Corroto[1,2], A. Iriel[2,3], E. Calderón[2,3], A. Fernádez Cirelli[2,3] & A. Pérez Carrera[2,3]

[1]*Agua y Saneamientos Argentinos S.A. (AySA S.A.), Buenos Aires, Argentina*
[2]*Centro de Estudios Transdisciplinarios del Agua (Universidad de Buenos Aires) Centro de Estudios Transdisciplinarios del Agua (CETA- UBA), Buenos Aires, Argentina*
[3]*Instituto de Investigaciones en Producción Animal (INPA-UBA/CONICET), Facultad de Ciencias Veterinarias, Buenos Aires, Argentina*

ABSTRACT   Reverse osmosis technology is an efficient option for arsenic removal from drinking water. Its implementation generates an effluent (concentrate) with well defined physicochemical properties such as high conductivity. In this study we analyzed As adsorption using a lateritic soil as a low cost adsorbent to reduce arsenic content in the concentrate. Results showed a reduction in arsenic concentration of 40% in the concentrate and that the adsorption process followed a pseudo-second order kinetics.

1 INTRODUCTION

In Argentina, Arsenic (As) presence in groundwater is a huge environmental and health issue. Some cities in Buenos Aires have used reverse osmosis for arsenic abatement in water for human consumption. This technology produces an effluent (concentrate) that contains high levels of salts and arsenic dissolved. The discard of this concentrate in surface waters can cause a serious environmental impact due to the presence of As and other trace elements. In the literature, the options proposed for the concentrated treatment before the discharge are different and diverse being the dilution the most used. However, this is not the best solution for the receiving body. An alternative solution is to use a low cost material as adsorbent in order to reduce the arsenic concentration in the concentrate. For that reason, the aim of this study is to analyze the kinetic aspects of the adsorption processes of arsenic on a lateritic soil from residual concentrated water obtained by reverse osmosis.

2 EXPERIMENTAL

2.1 Materials

Lateritic soil that was used as adsorbent material was obtained from Misiones province (Argentina) and used without any posterior treatment. Working solutions were obtained from the effluent of the reverse osmosis process (concentrate) with physicochemical parameters: pH = 7.90, conductivity = 320 µS/cm, Cl⁻ = 68.7 mg/L, SO_4^2 = 71.6 mg/L, NO_3^- = 121 mg/L, NO_2^- < 0.01 mg/L, F⁻ = 2,5 mg/L, alkalinity = 1070 mg/L, Na^+ = 420 mg/L, K^+ = 16 mg/L, Fe (total) = 0,08 mg/L, Mg^{2+} = 22 mg/L, total hardness (CO_3Ca) = 183 mg/L, NH_4^+ < 0.05 mg/L, P (total) = 0.45 mg/L.

2.2 Batch experiments

Batch adsorption experiments were performed by contacting 3 g of laterite with 500 mL of arsenic solution (residual effluent of the reverse osmosis) with an initial concentration of 135 µg/L. The experiments were performed upon agitation with a magnetic stirrer for a period of 76 hours at 300 rpm at room temperature (20°C). During this time, several samples were taken, centrifuged and filtered (0.45 µm) for solid removal. Arsenic concentrations were determined in the supernatant and the quantity of As adsorbed in the particles was calculated as following:

$$q_e = \frac{(C_i - C_e)V}{W} \qquad (1)$$

where q_e is amount of As adsorbed per unit mass of adsorbent at equilibrium mg g⁻¹, C_i and C_e are the initial and equilibrium concentrations (µg/L) of arsenic solution respectively. V is the volume (L) and W is the mass (g) of the adsorbent. Arsenic adsorption kinetics was determined by analyzing adsorptive uptake of As from the aqueous solution at different time intervals.

2.3 Arsenic analysis

Arsenic analysis was carried out using ICP-OES (Perkin Elmer, Optima 2000). All measurements were performed at 193.7 nm using an external calibration standard. Reference material from Lake Ontario water was provided by Certified Reference Materials & Quality Assurance.

3 RESULTS AND DISCUSSION

3.1 Kinetics study

As shown in Figure 1 As content diminishes around 40% during the first 24 hours.

Data were fitted to the most often used models such as pseudo-first order and pseudo-second order to obtain kinetics parameters from the adsorption process. The linearized rate equation of pseudo-first order is given as:

$$\log(q_e - q_t) = \log q_e - \frac{K_{ad}}{2,303}t \qquad (2)$$

where q_t (mg/g) is the amount of As adsorbed per unit mass of adsorbent at time t and k_{ad} (min^{-1}) is the rate constant of the pseudo-first-order adsorption process.

Additionally, pseudo-second order kinetics equation is given as:

$$\frac{t}{q_t} = \frac{1}{h} + \frac{1}{q_e}t \qquad (3)$$

where $h=kq_e^2$ (mg/g/min), k (g/mg/min) is the rate constant of pseudo-second-order adsorption process. Obtained parameters are presented in Table 1.

Results presented in Table 1 showed that the model of pseudo-second order fits with the experimental data. Particularly, R^2 value is closer to the unity and theoretical and experimental q_e values are

Table 1. Kinetic parameters from As adsorption on lateritic soil.

Pseudo first order	K_{ad} (min^{-1}) 0.196	q_e (mg/g) 0.038	R^2 0.926
Pseudo second order	k (g/mg/min) 0.390	q_e (mg/g) 0.010	R^2 0.995

Arsenic initial concentration is 135 µg/L and $q_{e\ exp}$ is 0.009 mg/g.

in a good accordance with each other. Therefore, it is possible to suggest that the sorption of As from the concentrate followed a pseudo-second order type reaction kinetics. Finally, a maximum quantity of As adsorbed at the surface of 0.009 mg/g can be determined for the experimental data.

4 CONCLUSIONS

In this study we have studied the possibility to use laterite as a low cost adsorbent to reduce the As concentration in the concentrate of reverse osmosis process. In this sense, it was observed that As adsorption takes place during the first 24 hours and follows a kinetic of pseudo-second order. Further studies are necessary in order to complete the physicochemical characterization of the adsorption process from the concentrated solutions.

ACKNOWLEDGEMENTS

University of Buenos Aires, Consejo Nacional de Investigación Científica y Tecnológica and Agua y Saneamientos Argentinos S.A.

REFERENCES

Litter, M.I., Morgada, M.E. Bundschuh, J. 2010. Possible treatments for arsenic renoval in Latin American waters for human consumption. *Environ. Pollut.* 158: 1105–1118.
Lin, L., Xu, X., Pepelis, C., Cath, T., Xu, P. 2014. Sorption of metals and metalloids from reverse osmosis concentrate on drinking water treatment solids. *Sep. Purif. Technol.* 134: 37–45.
Ranjan, D., Talat, M., Hasan, S.H. 2009. Biosorption of arsenic from aqueous solution using agricultural residue "rice polish". *J. Hazard. Mater.* 166: 1050–1059.
Salameh, Y., Al-Lagtah, N., Ahmad, M.N.M., Allen, S.J., Walker, G.M. 2010. Kinetic and thermodynamic investigations on arsenic adsorption onto dolomitic sorbents. *Chem. Eng. J.* 160: 440–446.

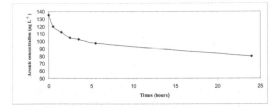

Figure 1. Dependence of the arsenic concentration with the time of contact with laterite.

Arsenic Research and Global Sustainability – Bhattacharya, Vahter, Jarsjö, Kumpiene,
Ahmad, Sparrenbom, Jacks, Donselaar, Bundschuh & Naidu (Eds)
© 2016 Taylor & Francis Group, London, ISBN 978-1-138-02941-5

Bioadsorption studies of arsenate from aqueous solutions using chitosan

M.C. Giménez[1], N.M. Varela[1] & P.S. Blanes[2]

[1]*Departamento de Química Analítica, Universidad Nacional del Chaco Austral, Argentina*
[2]*Área Química General, FCBioyF, IQUIR-CONICET, Universidad Nacional de Rosario, Argentina*

ABSTRACT: This study proposed the use of chitosan (biopolymer with a deacetylation degree of 85%) to remove arsenate [As(V)] ions from aqueous solutions. Batch adsorption experiments were carried out as a function of pH, contact time, initial concentration of the adsorbate, adsorbent dosage, and co-anions were evaluated. Maximum adsorption (82%) has taken place at pH 5.0, while equilibrium was achieved in 60 min. at temperature 25°C, with adsorbent dose (1 mg/g), initial concentration 1 g/L As(V), and shaking speed 200 rpm. The presence of coexisting anions has a negative effect on As(V) adsorption. The equilibrium adsorption data were fitted reasonably well for the Langmuir adsorption isotherm and showing maximum As(V) adsorption capacity of chitosan was 123.7 mg/g. Kinetic study reveals that adsorption of As(V) is fast and follows pseudo-second-order kinetics. The results reveals that chitosan biopolymer (with deacetylation degree of 85%) can be an effective adsorbent for removing As(V) from groundwater.

1 INTRODUCTION

The problem of groundwater arsenic contamination in Argentina is more serious in rural areas. In Chaco province, where 60% of the total water consumption is groundwater, arsenic contamination was detected (Buchhamer *et al.*, 2012). For this reason have increased the need to eliminate/minimize this contaminant from water before its use by human beings.

Various technologies are being used for removal of arsenic from water. However, technologies like nanofiltration, reverse osmosis, and electrodialysis have high removal efficiency but involved high capital and running cost, and high tech maintenance (Mohan & Pittman, 2007). Adsorption is a simple and economical alternative to conventional arsenic removal techniques. Among available different adsorbents for removal of arsenic from water, chitosan (N-deacetylated derived from chitin) has received increased attention as one of the promising renewable polymeric materials (Annaduzzaman *et al.*, 2014).

The main objective of this research was to evaluate the feasibility of using chitosan as an adsorbent for removal of As(V). Effects of pH, adsorbent dose, and initial arsenate concentration, contact time and co-anions were also studied in a batch study. Additionally, the sorption isotherm was explored to describe the experimental data.

2 EXPERIMENTAL

Chitosan used in this study was purchased from Parafarm Laboratory, Buenos Aires, Argentina. All other chemical used were of analytical reagent (AR) grade. Milli-Q quality water (Millipore, USA) was used throughout As(V) solution. Stock solution of As(V) were prepared by dissolving sodium salt heptahydrate ($Na_2HAsO_4.7H_2O$) in Milli-Q water (at a concentration of 1000 mg/L) and the working solutions were obtained by appropriate dilutions of stock solutions. Characterization of chitosan biopolymer was performed by FTIR analysis. The residual As(V) concentration in the solution was analyzed by HG-AAS M4200, (Metrolab Instruments). The pH of solution was measured by using a standard digital pH meter (OAKTON M510 pH/mV/°C). Carbonate and chlorides were determined by titration, fluoride using ion-selective electrode and sulfate by UV-Vis spectrophotometer. The influence of sorption parameters such as the adsorbent dose (0.5–5 g/L), pH solution (2–11), and contact time (10–120 min) were studied thoroughly to evaluate optimum conditions.

3 RESULTS AND DISCUSSION

3.1 *FTIR analysis of biosorbent*

FTIR spectra of the sorbent, before and after As(V) uptake, were recorded. The positions of the major adsorption bands for chitosan are as follows: 3430 cm^{-1}, 1634 cm^{-1}, 1384 cm^{-1} and 1106 cm^{-1}.

This observation indicate the presence of functional groups such as $-NH_2$, $-OH$, and $-CO-$ which are involved in binding As(V) to chitosan (Fig. 1).

3.2 *Adsorption experiment*

The pH of the medium is one of the important parameter, which significantly affects As(V) adsorption. Maximum As(V) adsorption was observed at pH 5, while decreasing with increase in alkalinity. The As(V) removal (%) increased with time up to about 60 minutes and then attainment the adsorption equilibrium. The percent removal of As(V) increased (60–82%) with increase in the adsorbent dose from 0.05 to 1.0 g/L.

3.3 *Effect of coexisting anions*

Arsenic contaminated groundwater contains several other ions which can compete with arsenic in the adsorption process. In the present study, the effects of F^-, Cl^-, SO_4^{2-}, and CO_3^{2-} were investigated separately. The solution was prepared from their sodium salts and concentration of each co-anion was varied in the range of 1–500 mg/L where the initial As(V) concentration was maintained at 1 mg/L with other optimal removal condition. The presence of anions has a deleterious effect on the adsorption of As(V).

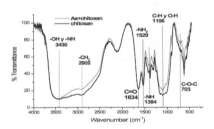

Figure 1. FTIR spectra from chitosan adsorption.

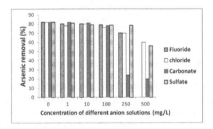

Figure 2. Effect co-anions on removal of arsenate (optimum removal condition).

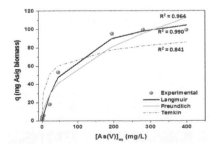

Figure 3. Equilibrium biosorption isotherm for As(V) (optimum removal condition).

However, the experimental result showed that the presence of carbonate significantly affected the As(V) removal process (Fig. 2).

3.4 *Isotherms and adsorption kinetics*

The experimental data were fitted with the selected adsorption isotherm models: Langmuir, Freundlich and Temkin (Fig. 3). Adsorption of As(V) ion was best explained by Langmuir model as the value of regression coefficient ($R^2 = 0.990$). The adsorption capacity for As(V) was found as 123.7 mg/g by Langmuir model and this reveals the monolayer adsorption during solid/liquid phase interaction. The adsorption kinetic data were well described by pseudo second order rate expressions ($r^2 = 0.99$; $q_e = 113$ mg/g), assuming chemisorption as the uptake mechanism.

4 CONCLUSIONS

Chitosan biomass has been used in the present work to adsorb As(V) ions across liquid phase in batch test. Maximum percentage removal As(V) was obtained as 82% at temperature 25°C, 60 min contact time, 1 mg/L initial concentration of As(V), 1 g/L adsorbent dose, pH 5 and shaking speed 200 rpm. The presence of other co-anions, particularly carbonates, in water has negative effects on arsenate uptake. The experimental data of the sorption equilibrium from arsenate ions solution are well fitted with the Langmuir isotherm and pseudo-second order kinetics model.

ACKNOWLEDGEMENTS

This work was supported with funds from PI 041/00005 UNCAus (Universidad Nacional del Chaco Austral).

REFERENCES

Annaduzzaman, M., Bhattacharya, P., Ersoz, M., Lazarova, Z. 2014. Characterization of a chitosan biopolymer and arsenate removal for drinking water treatment. In: M.I. Litter, H.B. Nicolli, M. Meichtry, N. Quici, J. Bundschuh, P. Bhattacharya, R. Naidu (eds.) *"One Century of the Discovery of Arsenicosis in Latin America (1914–2014) As 2014", Interdisciplinary Book Series: "Arsenic in the Environment— Proceedings".* Series Editors: J. Bundschuh & P. Bhattacharya, CRC Press/Taylor and Francis (ISBN 978-1-138–00141–1), pp. 745–747.
Buchhamer, E.E., Blanes, P.S., Osicka R.M, Giménez M.C. 2012. Environmental risk assessment of arsenic and fluoride in the Chaco province, Argentina: research advances, *J. Toxicol. Env. Heal. A* 75(22–23): 1437–1450.
Mohan, D., Pittman Jr., C.U. 2007. Arsenic removal from water/wastewater using adsorbents—A critical review. *J. Hazard. Mater.* 142: 1–53.

Arsenic Research and Global Sustainability – Bhattacharya, Vahter, Jarsjö, Kumpiene, Ahmad, Sparrenbom, Jacks, Donselaar, Bundschuh & Naidu (Eds)
© 2016 Taylor & Francis Group, London, ISBN 978-1-138-02941-5

Adsorptive removal of arsenic from aqueous solutions by iron oxide coated natural materials

S. Arıkan, D. Dölgen & M.N. Alpaslan
Department of Environmental Engineering, Faculty of Engineering, Dokuz Eylül University, Buca, İzmir, Turkey

ABSTRACT: Arsenic contamination in water sources is a global problem. Adsorptive filtration is considered as an alternative emerging technology for arsenic treatment without any side effects or treatment process alterations. In this study, iron-oxide coated sand material was used to remove arsenic ions from aqueous solutions. Effect of pH, contact time, adsorbent dose, initial arsenic concentration, and particle size which are important parameters for adsorption were investigated. Adsorption isotherm and kinetic studies were studied in order to evaluate adsorption capacity and rate. Results showed that 5 g/L iron-oxide coated sand could treat the solution consisting 100 µg/L arsenic below the Turkish standard of 10 µg/L. Isotherm data were fitted into both the Freundlich and the Langmuir isotherms. Batch studies showed that iron-oxide coated sand can be effectively used for arsenic removal due to its availability, cost-effectiveness and high performance.

1 INTRODUCTION

Arsenic is a heavy metal which is both toxic and carcinogenic and even fatal due to its intake to the body. Arsenic can be found in the groundwater with high concentration as a result of both natural and anthropogenic activities. Decreasing in quality and quantity of potable water in recent years caused global arsenic contamination in water sources, as well (Dolgen & Alpaslan, 2009). Because of its negative health effects, WHO recommends its value below 10 µg/L with the provisional guidelines (WHO, 1993). Arsenic removal below this limit by conventional methods is not possible for high arsenic contaminated water. For this reason, a cost efficient and easily operated additional/advanced treatment system is an urgent need. Adsorptive filtration is such a system which can remove arsenic efficiently, with low/no sludge production and low-cost when using available adsorbents in the affected region (Li et al., 2011, Maji et al., 2011).

2 EXPERIMENTAL STUDY

2.1 Preparation of adsorbent and arsenic contaminated aqueous solution

Diluted solutions were prepared from arsenic stock solution to use in the experiments in a particular range with respect to the arsenic contaminated water in Turkey (100–3000 µg/L). 0.1 M NaOH and H_2SO_4 were used to adjust the pH of solution.

In order to make surface of raw sand positively charged for arsenic affinity, metal coating process was applied onto particles. Materials were sieved in the range of 300–425 µm, 425–600 µm and 710–1000 µm. Then, they were washed with deionized water and dried at 105°C. Afterward, 20 g of these materials were mixed with 50 ml of a 0.5 M ferric nitrate solution. The mixture was then dried at 200°C in an oven. The coated materials were washed with deionized water until the runoff was clear, dried at 200°C and stored in capped bottles.

SEM-EDS analyses and XRD analyses were done to investigate the effect of coating. Figure 1 shows the SEM photos of sand and coated sand in 1000x magnification.

2.2 Batch studies

Effect of pH, contact time, adsorbent dose, initial arsenic concentration, and particle size were investigated in order to determine optimum operational and design parameters. The samples were placed on a mechanical shaker and shaken at 100 r/min. All the experiments were conducted at room temperature 23°C ± 2. Experiments were carried out simultaneously using sand and iron-oxide coated sand to compare the performance of the materials

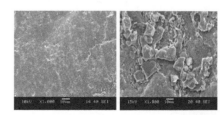

Figure 1. SEM images of sand and iron-oxide coated sand.

and investigate the effect of coating. Samples were taken from inlet and effluent, and arsenic measurements were done by ICP-OES.

3 RESULTS AND DISCUSSION

3.1 *Impacts of pH, contact time, adsorbent dose, initial As concentration, and particle size*

Maximum arsenic removal (76%) obtained at neutral pH range (pH 7) for 2 hours of contact time, 5 g/L adsorbent dose and 100 µg/L initial arsenic concentration. At alkaline conditions, arsenic removal was reduced to 52%.

The effect of contact time was investigated. The results showed that 4 hours of contact time was sufficient to reach 85% removal efficiency.

The effect of adsorbent dose on the adsorption of the arsenic was also examined. In the experiments, adsorbent dose was between 1–50 g/L. The results showed arsenic removal rate increased sharply up to 5 g/L ironoxide coated sand, and approximately 90% removal efficiency was obtained. A further increase in adsorbent dosage did not cause significant improvement in arsenic adsorption. This may be due to the adsorption of almost all arsenic to the adsorbent and the establishment of equilibrium between the arsenic adsorbed to the adsorbent and those remaining unadsorbed in the solution.

In order to determine the impacts of initial arsenic concentration, arsenic solutions were prepared in the range of 100–3000 µg/L at pH 7. For 100 µg/L arsenic solution, removal rate was 90% with 5 g/L adsorbent dose and four hours contact time. Increasing arsenic concentrations resulted with decreased removal rates, for 270 µg/L and 530 µg/L arsenic concentrations removal ratio was 79% and 70%, respectively. For higher concentrations, arsenic removal rapidly decreased and approximately 25% removal was obtained at 3000 µg/L arsenic concentration.

To investigate the effects of particle size, the experiments were carried out with variable size adsorbents (350, 500 and 850 µm, averagely). Optimum particle size was provided as 850 µm yielding 90% efficiency for 100 µg As /L at pH 7 for four hours contact time.

3.2 *Isotherm and kinetic studies*

In order to determine adsorption capacity and rate, isotherm and kinetic studies were performed. Langmuir and Freundlich isotherm equations are the most widely used models to describe the experimental data of adsorption isotherms. Thus, in the study, Langmuir and Freundlich models were studied for coated sand in order to evaluate suitable model, results were summarized at Table 1. Results

Table 1. Isotherm study results for coated sand.

Isotherm constants for coated sand	Langmuir			Freundlich		
	q_{max} (mg/g)	b (L/mg)	R^2	n	k_f (mg/g)	R^2
	0,117	20,7	0,944	2,44	0,142	0,989

of the isotherm studies were fitted into both the Freundlich and the Langmuir isotherms.

The coefficient of determination (R^2) for all the isotherms ranged between 0,94 and 0,99 representing an excellent fit of the observed data. First-order, pseudo first- order, second order and pseudo second-order reaction kinetics were studied to obtain best fit kinetic model. Pseudo second-order reaction kinetic perfectly represented iron-oxide coated sand with 0.998 coefficient of determination.

4 CONCLUSIONS

In general, pH is an important parameter for metal adsorption from aqueous solutions, for iron oxide coated sand material adsorption was favorable at neutral pH. Increased arsenic concentrations were resulted decreasing in the removal efficiency. In the experiments, 5 g/L of coated sand was sufficient to reduce 100 µg/L arsenic nearly to 10 µg/L without pre-treatment in 4 hours (app. 90% arsenic removal). Studies were also performed with raw sand material to compare the performance of coated material. Treatment efficiency was negligible for raw sand material. Thus, modification of sand particles by coating iron oxide was proven as an efficient operation for arsenic removal. It is considered that using iron-oxide coated sand in column systems will be convenient, because quartz sand is a high strength material. Further, quartz sand is quite a lot available all over the world so it is a cost-effective adsorbent to use especially in the small villages at rural areas.

REFERENCES

Dolgen, D., Alpaslan, M.N. 2009. Appropriate arsenic removal strategies for drinking water systems in Turkey. In: *Proceedings of the International Workshop on Urbanization, Land Use, Land Degradation and Environment, Turkey. Workshop II-Environment.* pp. 169–175.

Li Z., Jean J-S., Jiang W-T., Chang P-H., Chen C-.J., Liao L. 2011. Removal of arsenic from water using Fe-exchanged natural zeolite. *J. Hazard. Mater.* 187: 318–323.

Maji S.K., Kao Y.-H., Liu C.-W. 2011. Arsenic removal from real arsenic-bearing groundwater by adsorption on iron-oxide-coated natural rock (IOCNR) *Desalination* 280: 72–79.

WHO, 1993. *Guidelines for Drinking-Water Quality.* Second edition, Volume 1.

Arsenic Research and Global Sustainability – Bhattacharya, Vahter, Jarsjö, Kumpiene,
Ahmad, Sparrenbom, Jacks, Donselaar, Bundschuh & Naidu (Eds)
© 2016 Taylor & Francis Group, London, ISBN 978-1-138-02941-5

Arsenic adsorption by iron-aluminium hydroxide coated onto macroporous supports: Insights from X-ray absorption spectroscopy and comparison with granular ferric hydroxides

P. Suresh Kumar[1,2], R. Quiroga Flores[1,3], L. Önnby[1,5] & C. Sjöstedt[4]

[1]*Department of Biotechnology, Lund University, Lund, Sweden*
[2]*Wetsus, Leeuwarden, The Netherlands*
[3]*Instituto de Investigaciones Fármaco Bioquímicas, Facultad de Ciencias Farmacéuticas y Bioquímicas,*
Universidad Mayor de San Andrés, La Paz, Bolivia
[4]*Department of Soil and Environment, Swedish University of Agricultural Sciences, Uppsala, Sweden*
[5]*Eawag, Swiss Federal Institute of Aquatic Science and Technology, Dübendorf, Switzerland*

ABSTRACT: The arsenic adsorption characteristics of a macroporous polymer coated with coprecipitated iron-aluminium hydroxides (MHCMP) is evaluated in this study. The MHCMP had a maximum adsorption capacity of 82.3 and 49.6 mg As/g adsorbent for As(III) and As(V) ions respectively. Extended X-ray Absorption Fine Structure (EXAFS) confirmed that binding of As(III) ions took place on the iron hydroxides coated on the MHCMP, whereas for As(V) ions the binding specificity could not be attributed to one particular metal hydroxide. Column experiments were run in a well water spiked with 100 µg/L concentration of As(III) and for comparison a commercially available adsorbent (GEH®102) based on granular iron-hydroxide was used. The MHCMP was able to treat seven times more volume of well water as compared to GEH®102, maintaining the threshold concentration of less than 10 µg As/L, indicating that the MHCMP is a superior adsorbent.

1 INTRODUCTION

Adsorption is a method that is especially suited for treating lower concentrations of pollutants such as As in groundwater. We have previously developed a technique which uses cryogels of acrylamide as a backbone and a coating of co-precipitated iron-aluminium hydroxides for adsorption of arsenic (Suresh Kumar et al., 2014). This study presents the results of adsorption studies of both arsenite and arsenate to these Metal-Hydroxide-Coated Macroporous Polymers (MHCMP) (Suresh Kumar et al., 2016). Also, a column experiment was performed where the MHCMP were compared to a commercially available adsorbent (GEH®102) (based on granular ferric hydroxide) where well water was spiked with 100 µg As(III)/L.

2 METHODS/EXPERIMENTAL

2.1 Production of metal-hydroxide-coated macroporous polymers

The MHCMPs were produced according to a method described in detail in Suresh Kumar et al., 2014. In brief, 6% acrylamide cryogel monolith (about 10 mm × 7 mm) was placed in a column and the following solutions were pumped at a flow-rate of 1 mL/min in the mentioned order: aqueous solution (1 mL) containing equimolar concentration (0.4 M each) of iron and aluminium salts, 25% NH₄OH (1 mL), and distilled water (20 mL). The composites containing the Fe-Al hydroxides coated cryogels were dried overnight at 105 °.

2.2 Batch experiments

Batch experiments were performed at 22°C for both As(III) and As(V). A single MHCMP was used in 10 mL of arsenic solution (adsorbent/ Fe-Al hydroxides concentration of 0.7 to 0.8 g/L), and adsorption was allowed to run overnight in test tubes on a rocking table. Studies for kinetics were performed at 1 mg/L and pH 7. Effect of pH was studied by varying the initial pH of the arsenic solutions between pH 3 and pH 11 at 1 mg/L.

2.3 EXAFS experiments

Samples at pH 3, 5, 7 and 9 were studied with EXAFS spectroscopy to reveal the mechanism of the adsorption. The experiments were performed at beamline I-811, at MAX-lab, Lund, Sweden.

2.4 Column experiments

Well water was collected from outside Sjöbo, southern Sweden, and spiked to 100 µg/L of As(III) (pH 8.1). Two MHCMPs (total mass corresponding to

76 mg) were packed together in a column (bed height 20 mm, inner diameter 7 mm, bed volume = 1 ml) and the well water was pumped upflow through the column at a flowrate of 0.26 mL/min. Adsorption using GEH®102 was performed under the same pH and flowrate. The column was packed with 100 mg of GEH®102 to have a comparable mass of adsorbent to that of MHCMP (bed volume = 0.087 ml).

3 RESULTS AND DISCUSSION

3.1 Batch experiments

Both arsenic species (As(III) and As(V)) showed similar kinetic pattern with high removal (about 95% removal for As(III), 88% removal for As(V)) at an initial arsenic concentration of 1 mg/L.

Pseudo first-order, pseudo second-order and Elovich kinetic models were fitted using non-linear regression analysis to investigate the mechanism of adsorption and determine the potential rate-limiting steps. These models are derived based on the assumptions that the adsorbate binds onto one surface site on the adsorbent for pseudo first-order, whereas it binds to two sites for the pseudo second-order (Deliyanni et al., 2007). As(III) was best fitted with the pseudo-second order model ($R^2 = 0.9909$) and As(V) with the pseudo-first order model ($R^2 = 0.9511$). Thus the best fit models for As(III) and As(V) indicate that As(III) and As(V) are bound through two and one sites on the adsorbent particles in the MHCMP, respectively. The effect of pH on As(III) and As(V) adsorption onto MCHMP was negligible throughout the tested pH range, i.e. pH 3 to 11.

3.2 EXAFS experiments

For As(III), the EXAFS spectra at pH 5, 7 and 9 were best fitted with a model using an As--Fe distance of 2.94–2.95 Å, indicating a bidentate mononuclear complex.

For As(V), the spectra at all four pH-values were similar to each other. They were best fitted with a model using an As--Al distance of 3.11–3.16 Å, indicating a bidentate binuclear complex. However, a model using an As--Fe distance of ~3.3 Å could also be modelled. It was therefore not possible to conclude which metal hydroxide As(V) preferred.

3.3 Column experiments

Figure 1 shows the effluent concentration as a function of volume of well water passed through the column for the MHCMP and the benchmark adsorbent GEH®102. Thus 76 mg of MHCMP is able to treat 150 mL of the well water to a recommended As-concentration, which is seven times more treatable volume than achieved by 100 mg of the GEH®102, which could only treat 20 mL of well water before

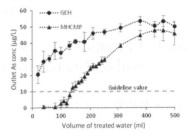

Figure 1. As adsorption from well water in column mode using MHCMP and GEH®102 (pH ≈ 8, T = 22 °C). Guideline value refers to the WHO guideline of 10 μg/L As in drinking water.

reaching the threshold concentration. Possibly some of the As(III) was oxidized during the experiment, but if so that would have lowered total As adsorption since MHCMP was better at removing As(III).

4 CONCLUSIONS

MHMCPs showed a high adsorption capacity for As(III) and As(V). Adsorption for both species was not affected for a wide pH range. EXAFS analysis determined that As(III) and As(V) adsorption occurred via inner-sphere complexes, implying stronger and more selective adsorption as compared to outer-sphere complexes. Column experiments showed that the MHCMP could treat higher volumes of As(III) solutions when compared with a commercially available adsorbent. The MHCMP was found to be a suitable adsorbent for low concentrations of arsenic. The MHCMP could thus be used both as polishing step in a wastewater treatment plant and be considered for drinking water applications.

ACKNOWLEDGEMENTS

LÖ acknowledges the Swedish Research Council for financial support. MAX-lab is acknowledged for beam time and support.

REFERENCES

Deliyanni, E., Peleka, E., Lazaridis, N. 2007. Comparative study of phosphates removal from aqueous solutions by nanocrystalline akaganéite and hybrid surfactant-akaganéite, Sep. Purif. Technol. 52: 478–486.

Suresh Kumar, P., Önnby, L,. Kirsebom, H. 2014. Reversible in situ precipitation: a flow-through approach for coating macroporous supports with metal hydroxides. J. Mater. Chem. A 2: 1076–1084.

Suresh Kumar, P., Quiroga Flores, R., Sjöstedt, C., Önnby, L. 2016. Arsenic adsorption by iron-aluminium hydroxide coated onto macroporous supports: Insights from X-ray absorption spectroscopy and comparison with granular ferric hydroxides. J. Hazard. Mater. 302: 166–174.

Arsenic Research and Global Sustainability – Bhattacharya, Vahter, Jarsjö, Kumpiene,
Ahmad, Sparrenbom, Jacks, Donselaar, Bundschuh & Naidu (Eds)
© *2016 Taylor & Francis Group, London, ISBN 978-1-138-02941-5*

Removal of arsenic (III) and arsenic (V) from water using material based on the natural minerals

M. Szlachta & P. Wójtowicz
Faculty of Environmental Engineering, Wrocław University of Technology, Wrocław, Poland

ABSTRACT: This work focuses on analyzing the efficiency of material derived from natural minerals in arsenic (III) and arsenic (V) removal from aqueous solution. The material used in the study (F7M) is an ion exchange adsorbent composed mainly of magnesium, iron, silica, calcium and aluminium (oxides or carbonates). The performance of the material was examined in batch adsorption system using model solutions. It was confirmed that the arsenite was more difficult to remove from water in comparison to the arsenate and the efficiency of the process depends on different operating parameters including contact time, material dose and solution pH.

1 INTRODUCTION

Arsenic may occur naturally in the aquatic environment or may be released into the water through anthropogenic sources. In the aquatic ecosystem arsenic is typically present in inorganic forms, as oxyanions of trivalent arsenite or pentavalent arsenate. However, the chemistry of the arsenic species strongly depends on environmental conditions. As (III) is dominant in more reducing conditions, whereas As (V) is mostly present in an oxidizing environment (Zhu *et al.*, 2011). Arsenic is classified as one of the most toxic chemical elements and excessive, long-term human intake of arsenic from contaminated water may result in serious health problems. Ion exchange and adsorption have already been proven to be effective methods for the removal of anionic species of inorganic arsenic from waters (Szlachta *et al.*, 2012). There is a large number of studies available in the literature on selective ion exchange and porous materials. However, many of them are not technically applicable or economically feasible in the treatment of arsenic contaminated waters (Mohan & Pittman, 2007).

The main objective of this work was to evaluate the efficiency of material derived from natural minerals in arsenic removal from aqueous solution. The material is produced by Aquaminerals Finland Oy Ltd (represented by Oulu Water Alliance Oy Ltd, Finland). The ion exchange adsorption was examined using model solutions containing arsenic (III) or arsenic (V). Conducted experiments included the equilibrium isotherm and adsorption kinetics tests. The performance of the ion exchange adsorbent in arsenic removal was also analysed using various material doses and solution pH.

2 METHODOLOGY

To investigate the arsenic removal using material F7M, the batch adsorption technique was applied. The tests were carried out in the laboratory scale using model solutions spiked with arsenic (III) or arsenic (V). The mineral material F7M was used as provided and characterized by the morphology studies, BET specific surface area, XRD and point of zero charge. The adsorption equilibrium experiments were conducted using the bottle point method. The solution with an initial concentration of As (III) or As (V) of 1–300 mg/L was added into capped tubes containing ion exchanger in the of 2 g/L. The samples were agitated in a shaker until equilibrium was reached.

In the kinetics tests 2.5 g of material was added to 500 mL of water sample spiked with 1 mg/L and 20 mg/L of As (III) or As (V). The initial pH of the solution was adjusted to 7.0. The test was carried out for 5 h using a magnetic stirrer. Additionally, experiments with various material dose (1–10 g/L) and solution pH (2–11) were carried out with the initial concentration of As(III) or As(V) of 20 mg/L.

3 RESULTS AND DISCUSSION

The F7M material used in this study is an ion exchange adsorbent and can be utilized in a batch system followed by clarification or flotation. Material is derived from natural minerals and consists mainly of magnesium, iron, silica, calcium and aluminium (oxides, carbonates). Apart from arsenic, the ion exchanger is able to effectively remove a wide range of elements including chromium, phosphorus, uranium, nickel, copper, lead, zinc, manganese and cadmium. As presented in Figure 1, the surface of F7M exhibits smooth morphology with low porosity in the micrometre scale. The BET specific surface area of F7M was 11.9 m²/g and an average pore size was 6.47 nm. The point of zero charge of mineral material was found at pH of 10.6.

Adsorption kinetics of arsenite and arsenate on F7M material is presented in Figure 2. The time dependent studies for both anionic species of arsenic showed relatively rapid uptake in the initial stage of the process. Under applied conditions, it was possible to achieve the high rate of arsenic removal. After 5 hours of contact time approximately 98.5% and 84.8% of As (V) were removed from the solution with initial concentration of 1 mg/L and 20 mg/L, respectively. As expected the percentage of As (III) removal was lower compare to As (V). An increase in the initial concentration of arsenite from 1 mg/L to 20 mg/L decreased the efficiency of the process from 78% to 50%, respectively. It was recorded that after the process the solution pH increased over 3 units, from its initial value of 7.0 to final pH of about 10.3–10.4 suggesting that the removal of arsenic species takes place through the exchange of hydroxyl ions.

The influence of solution pH on F7M performance was also analysed and obtained data indicate that the removal of arsenic was affected by pH changes. In the case of As (III), with increasing the solution pH from 2 to 11, the removal of arsenite decreased only about 10%. These slight changes in the process efficiency may be related to the fact that within tested pH range, the fully protonated arsenic (III) ($pK_{a1} = 9.2$) predominates (Goldberg, 2002) in experimental solutions. The stronger effect of pH on the ion exchange adsorption of arsenate onto F7M was observed. The removal of As (V) from alkaline solution was reduced by approximately 20% compared to the efficiency of this process carried out under acidic conditions. The difference in the process performance results from various species of arsenate present in the solutions, i.e. $H_3 AsO_4$ ($pK_{a1} = 2.3$), $H_2 AsO_4^-$ ($pK_{a2} = 6.8$) and $HAsO_4^{2-}$ ($pK_{a3} = 11.6$) – which is the least susceptible to the removal (Goldberg, 2002). The observed pH dependence of arsenic removal is also related to

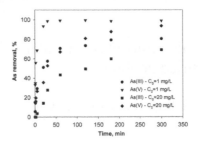

Figure 2. Adsorption kinetics of As (III) and As (V) onto F7M.

the surface charge of ion exchange adsorbent use in the study. Within the tested pH range, the surface of F7M was predominantly positively charged, which overall enhances the removal of anionic forms of arsenic from solutions of pH below 10.6.

4 CONCLUSIONS

An ion exchange adsorbent F7M derived from natural minerals successfully removed the As (III) and As (V) from aqueous solutions. However, the arsenite was more difficult to remove from water compared to the arsenate. Therefore, when considering the water or wastewater treatment by adsorption/ion exchange using F7M, the peroxidation step to convert trivalent arsenic to pentavalent is recommended, if the high removal rate of arsenite is required. The performance of the material was examined in batch adsorption system and it was confirmed that the efficiency of the process depends on different operating parameters, including contact time, material dose and solution pH.

ACKNOWLEDGEMENTS

This research is supported by the National Centre for Research and Development grant (2014–2017) "Tools for sustainable gold mining in EU"— SUSMIN, within the FP7 ERA-NET ERA-MIN program.

REFERENCES

Goldberg, S. 2002. Competitive adsorption of arsenate and arsenite on oxides and clay minerals. *Soil Science Society of America Journal* 66: 413–421.
Mohan, D., Pittman Jr., C.U. 2007. Arsenic removal from water/wastewater using adsorbents—A critical review. *J. Hazard. Mater.* 142: 1–53.
Szlachta, M., Gerda, V., Chubar, N. 2012. Adsorption of arsenate and selenite using an inorganic ion exchanger based on Fe-Mn hydrous oxide. *J. Colloid Interf. Sci.* 365: 213–221.
Zhu, Y., An, F., Tan, J. 2011. Geochemistry of hydrothermal gold deposits: A review. *Geosci. Front.* 1(3): 367–374.

Figure 1. The SEM micrograph of F7M material.

Arsenic Research and Global Sustainability – Bhattacharya, Vahter, Jarsjö, Kumpiene,
Ahmad, Sparrenbom, Jacks, Donselaar, Bundschuh & Naidu (Eds)
© 2016 Taylor & Francis Group, London, ISBN 978-1-138-02941-5

Arsenic removal from water using metal and metal oxide modified zeolites

A.K. Meher, P. Pillewan, S. Rayalu & A. Bansiwal
Environmental Materials Division, CSIR-NEERI, Nehru Marg, Nagpur, Maharashtra, India

ABSTRACT: In this paper, we have discussed arsenic removal from drinking water by using metal and metal oxide incorporated zeolites as an adsorbent. Copper Exchanged Zeolite (CEZ) was prepared by impregnation of copper on Commercial Zeolite-A by ion exchange method. Another adsorbent namely Iron Enriched Aluminosilicate Adsorbent (IEASA) was synthesized by alkali fusion of fly ash followed by ageing and hydrothermal curing. Both the materials were thoroughly characterized by pXRD, SEM and FTIR to understand the nature of the adsorbents and arsenic removal mechanism. For both adsorbents Langmuir model was best fitted and CEZ showed an adsorption capacity of 1.37 and 1.48 mg/g for As(III) and As(V) respectively and IEASA having a capacity of 0.592 mg/g. The adsorbents effectively remove arsenic in a broad range of pH and in presence of various competitive anions. The water quality after treatment with the adsorbents suggests, it's safe for drinking purposes.

1 INTRODUCTION

The most severe cases of Arsenic (As) poisoning due to uptake of contaminated water have been documented in Indo genetic plains. The International Agency for Research on Cancer (IARC) has classified As as a carcinogen to humans. Because of the adverse effects on human health, many regulatory bodies have framed a strict guideline value of 10 µg/L for drinking water. Many reported technologies like precipitation, coagulation, ion exchange, bioremoval, membrane filtration etc. for removal of As from drinking water suffers some limitations. Adsorption seems to be a feasible technique because of its process simplicity and availability of wide range of cheap adsorbents with a considerable adsorption capacity (Ali, 2012, Mohan and Pittman, 2007). For this reason adsorbents like activated alumina, zeolites and activated carbon are widely used for water purification. Zeolites are hydrated aluminosilicate natural materials with exceptional ion-exchange and sorption properties. Natural zeolites have less potential for adsorption of As. However, the hydrophilic/hydrophobic properties can be altered by methods like metal exchange/enrichment in the zeolitic framework, which leads to enhanced As adsorption.

The objective of this study was to evaluate the influence of metal incorporation on As removal efficiency of zeolite and to study the effect of various adsorption parameters. Detailed physicochemical analysis of the untreated and treated drinking water was done to assess the portability of treated water.

2 METHODS

In one of our work, Cu was incorporated on commercial salt-A (CZA) by the ion exchange method. Zeolite-A was mixed with copper nitrate solution at near neutral pH. After 3 h the material was filtered and washed with DI water to remove the excess copper nitrate solution. The modified adsorbent was used after drying in oven.

In another work, zeolite was synthesized using a novel method of alkali fusion of fly ash followed by ageing and hydrothermal curing, which was intentionally enriched with Fe. Iron oxide content in the synthetic adsorbent was found to be 6.76%, as compared to 0.1% of the commercial Zeolite-A. From FTIR results it was confirmed that the material developed Fe-OH group.

Inorganic arsenic species dissociate to some degree in water depending upon pH and Eh. As (V) exists as a charged species in aqueous media in a wide range of pH ranging from 2.5 to 14. However, As(III) exists as a charged species across a much narrower range of pH ranging from 9 to14. Probably due to this reason, As(V) is better adsorbed on most adsorbents as compared to As(III).

Incorporation of metals like Fe, Cu etc. results in the development of functional group like –OH, which has a strong tendency to bind with arsenic.

$$Fe\text{-}OH + H+ = Fe\text{-}OH^{2+} \qquad (1)$$
$$Fe\text{-}OH = H+ + Fe\text{-}O\text{-} \qquad (2)$$

Anions such as $H_2AsO_4^-$ can be either adsorbed through nonspecific columbic interaction according to the scheme:

Table 1. Some metal/metal oxide modified zeolites for arsenic removal from aqueous media.

Type of zeolite	Metal/ Metal oxide	Initial conc. of As	Adsorption capacity	Optimal pH	Ref.*
Natural	Zero-valent Iron	10–200 mg/L.	512 mg/kg	6.5	1
Clinoptilolite	MnO_2	0.5–50 µg/L	2.5 µg/g	6	2
Synthetic	Fe-Mn binary oxides	2 mg/L	296.23 mg/g [As(III)] 201.10 mg/g [As(V)]	7	3
Clinoptilolite	Fe	147 µg/L	100 mg/kg	–	4
Clinoptilolite	FeO	100 mg/L	48 mg/g [As(III)] 68 mg/g [As(V)]	6.5–9	4
Zeolite-A	Fe	1 mg/L	0.592 mg/g	4–10	5
Chabazite	Fe	50 µg/L	3.62 mg/g [As(III)] 0.809 mg/g [As(V)]	7–10	6
Clinoptilolite	Fe	50 µg/L	14.19 mg/g [As(III)] 30.21 mg/g [As(V)]	4–11	
Zeolite-A	Cu	2.0 mg/L	1.37 mg/g [As(III)] 1.48 mg/g [As(V)]	7	7

*1: Andrews (2009); 2: Camacho *et al.* (2011); 3: Kong *et al.* (2014); 4: Li *et al.* (2011); 5: Pilawan *et al.* (2014); 6: Payne *et al.* (2005); 7: Pilewan *et al.* (2014).

$$Fe\text{-}OH^{2+} + H_2\,AsO_4^- = Fe\text{-}OH^{2+}\text{-}O_4H_2\,As \qquad (3)$$

Or they can undergo direct exchange with surface hydroxyl group as shown below:

$$Fe\text{-}OH + H_2\,AsO_4^- = Fe\text{---}H_2\,AsO_4 \qquad (4)$$
$$Fe\text{-}OH^{2+} + H_2\,AsO_4^- = Fe\text{---}H_2\,AsO_4 + H_2O \qquad (5)$$

3 RESULTS AND DISCUSSION

Modification of zeolite with metal/metal oxides leads to enhanced adsorption capacity and better performance in a wide range of pH. The iron enriched zeolite exhibit Langmuir adsorption capacity of 0.592 mg/g at an initial arsenic concentration of 1 mg/L, whereas commercial zeolite showed negligible affinity for arsenic. Similarly CZA showed ap proximately 10 times higher adsorption capacity than the unmodified one with an adsorption capacity of 1.37 mg/g for As(III) and 1.48 mg/g for As(V).

Both the adsorbents showed effective removal of arsenic in a wide range of pH and in presence of competitive anions. Comparison of various metal/metal oxide zeolites reported for arsenic removal is presented in Table1.

4 CONCLUSIONS

The synthesized materials namely CEZ and IEASA were found to be selective for arsenic, as evident from the competitive ion studies. The experiment also proves that, the physico-chemical parameters of treated water meets the guideline values of Indian and WHO standards and is safe for drinking purposes. It can be concluded from the studies that, the metal and metal oxide modified zeolites can be effectively used in the arsenic affected regions to resolve the issue of provision of safe drinking water.

REFERENCES

Ali, I. 2012. New generation adsorbents for water treatment. *Chem. Rev.* 112(10): 5073–5091.

Andrews, J.R. 2009. Arsenic Removal Using Iron-Modified Zeolites. M.Sc. Thesis, New Mexico Institute of Mining and Technology, New Mexico.

Camacho, L.M., Parra, R.R., Deng, S. 2011. Arsenic removal from groundwater by MnO2-modified natural clinoptilolite zeolite: Effects of pH and initial feed concentration. *J. Hazard. Mater.* 189(1–2): 286–293.

Kong, S., Wang, Y., Zhan, H., Yuan, S., Yu, M., Liu M. 2014. Adsorption/Oxidation of arsenic in groundwater by nanoscale Fe-Mn binary oxides loaded on zeolite. *Water Environ. Res.* 86(2): 147–155.

Li, Z., Jean, J.S., Jiang, W.T., Chang, P.H., Chen, C.J., Liao, L. 2011. Removal of arsenic from water using Fe-exchanged natural zeolite. *J. Hazard. Mater.* 187(3): 318–323.

Li, Z., Jiang, W.T., Jean, J.S, Hong, H., Liao, L., Lv, G. 2011. The combination of hydrous iron oxide precipitation with zeolite filtration to remove arsenic from contaminated water. *Desalination* 280(1–3): 203–207.

Meher, A.K., Das, S., Rayalu, S., Bansiwal, A. 2015. Enhanced arsenic removal from drinking water by iron-enriched aluminosilicate adsorbent prepared from fly ash. *Desalination Water Treat.* 1–13.

Mohan, D., Pittman Jr. C.U. 2007. Arsenic removal from water/wastewater using adsorbents-A critical review. *J. Hazard. Mater.* 142(1–2): 1–53.

Payne, K., Abdel-Fattah, T. 2005. Adsorption of Arsenate and Arsenite by Iron-Treated Activated Carbon and Zeolites: Effects of pH, Temperature, and Ionic Strength. *J. Environ. Sci. Hlth A* 40(4): 723–749.

Pillewan, P., Mukherjee, S., Meher, A.K., Rayalu, S., Bansiwal, A. 2014. Removal of asrsenic (III) and arsenic (V) using copper exchange Zeolite-A. *Environmental Progress and Sustainable Energy* 33(4): 1274–1282.

Arsenic Research and Global Sustainability – Bhattacharya, Vahter, Jarsjö, Kumpiene,
Ahmad, Sparrenbom, Jacks, Donselaar, Bundschuh & Naidu (Eds)
© 2016 Taylor & Francis Group, London, ISBN 978-1-138-02941-5

Surface properties of clay sorbents for decontamination of water polluted by arsenic

M. Lhotka & B. Dousova
University of Chemistry and Technology Prague, Prague, Czech Republic

ABSTRACT: Arsenic is of an increasing environmental attention due to their significance to human health. The use of clay materials as selective sorbents of different contaminants belongs to very effective decontamination methods. This paper describes the effect of different temperatures on the rehydration process, surface and physical properties of a kaolinite-based rehydrated metakaolinite. The rehydration of metakaolinite was studied from 150 to 250°C under autogenous pressures. A natural kaolinite was calcined to metakaolinite and then rehydrated at different temperature to a highly porous kaolinite. To improve sorption properties to anionic particles, prepared kaolinite was treated by rehydration in autoclave with Fe^{2+} ions. The rehydrated kaolinites were used for the adsorption of arsenic oxyanions and the adsorption efficiency were investigated.

1 INTRODUCTION

The hydrated surface of clay minerals belongs to very effective sorbents both in natural and in technological processes. In general, aluminosilicates are selective sorbents of cationic contaminants thanks to a low pH_{ZPC} (Jelínek *et al.*, 1999). A simple pretreatment of initial material with Fe (Al, Mn) ions can significantly improve its sorption affinity to oxyanions (Doušová *et al.*, 2006, 2009, Grygar *et al.*, 2007, Gupta *et al.*, 2005).

During the interaction of raw clay and Fe salt solution, reactive ion-exchangeable species and very poorly crystalline hydrated oxide particles in stable oxidation state (Fe^{3+}) were fixed on the clay surface forming active adsorption sites. The substantial variability in growing Fe phases (hydrated Fe_2O_3, non-specific Fe^{3+} species, ferrihydrite) resulted from the different type of aluminosilicate carrier and the treatment conditions (Doušová, 2009). Most of the active sites have been stabilized in surface complexes during the adsorption process, while unoccupied sites tended to be transformed to more stable and/or crystalline mineral Fe phases (goethite).

Rocha *et al.* (1990) studied the rehydration of metakaolinite to kaolinite which was heated at 155–250°C for 1–14 days and concluded that the amorphous metakaolinite can be transformed to crystalline form. The specific surface area highly increased during the rehydration. Lhotka *et al.* (2012) studied the rehydration of metakaolinite to kaolinite which was heated at 150–250°C for 1–14 days and concluded that the rehydration of metakaolinite to kaolinite was strongly dependent on temperature and time of the hydrothermal process. The optimum transformation from the point of view of the surface properties was observed after longer-term autoclaving (4–7 days) at 175°C, when the specific surface S_{BET} of raw kaolinite increased more than three times. Cationic particles were better adsorbed on the longer-term rehydrated raw sorbents in the order of $Cd^{2+} > Zn^{2+} \geq Pb^{2+}$, while oxyanions showed a higher adsorption affinity for Fe-modified sorbents. Pb^{2+} and SeO_3^{2-} particles exhibited the best adsorption properties. During the interaction of raw clay and Fe salt solution, ion-exchangeable Fe^{3+} particles in amorphous and/or poorly crystalline form have been fixed on the clay surface forming active adsorption sites.

The aim of this work was to prepare kaoline-based sorbents using rehydration method, to describe surface properties, and to compare sorption efficiency and sorption capacity of modified sorbents from the point of view of rehydration time and the type of oxyanionic particle.

2 EXPERIMENTAL PART

2.1 *Preparation of sorbents*

The crystalline kaolinite from West Bohemia was used for the preparation of modified sorbents. This kaolinite was calcined at 650°C for 3 h and convert-ed to metakaolinite. For the preparation of rehydrat-ed kaolinite, 8 g of metakaolinite was mixed with 30 g of water. The suspension was stirred for 2 min at room temperature and then inserted into autoclave. The autoclave was heated at 175°C for 1, 4, 7 and 14 days. In next experiments, 30 g of 0.6M $FeCl_2$ was used instead of water. After removal from autoclave, the suspension was filtered off, washed with distilled water and dried at 80°C for 24 h.

Table 1. As(III)/As(V) adsorption on rehydrated kaolinite and Fe-modified rehydrated kaolinite under the various rehydration time.

Sample	Surface area S_{BET} (m^2/g)	As(III) sorption q-As (mmol/g)	ε-As (%)	As (V) sorption q-As (mmol/g)	ε-As (%)
Kaolinite	15.84				
Metakaolinite	17.81				
S1 (1 day)	31.75	0.35	0.5	–	–
S4 (4 days)	98.57	0.65	5.9	–	–
S7 (7 days)	103.10	0.27	1.1	–	–
S1Fe (1 day-Fe)	70.21	5.24	57	10.49	99
S4Fe (4 days-Fe)	73.81	6.00	59	10.49	99
S7Fe (7 days-Fe)	71.05	6.74	56	8.99	100
S14Fe (14 days-Fe)	69.71	6.00	61	6.00	99

q – maximum adsorption capacity, ε – adsorption efficiency.

2.2 Arsenic adsorption and analytical methods

The As(V) and As(III) model solutions were prepared from KH_2AsO_4 and $NaAsO_2$ of analytical quality and distilled water in the concentration about 40 mg/L and pH $\approx$ 6.4–7.0. The suspension of model solution and sorbent (6 g/L) was shaken in sealed polyethylene bottle at room temperature for 24 hours. The product was filtered off; the filtrate was analyzed for residual As concentration and pH value. Equilibrium adsorption isotherms of nitrogen were measured at 77 K using static volumetric adsorption systems (TriFlex analyzer, Micromeritics). The adsorption isotherms were fitted in the BET specific surface area and the pore size distribution by the DFT and BJH method. The concentration of As in aqueous solutions was determined by HG-AFS using PSA 10.055 Millennium Excalibur.

3 RESULTS AND DISCUSSION

3.1 Characterization of sorbents

The samples of sorbents were prepared under different reaction conditions. The specific surface area of newly prepared sorbents was increased compare to raw kaolinite. The comparison of S_{BET} of sorbents without and with Fe ion is shown in Table 1.

The surface area of the newly prepared kaolinites was much larger than raw kaolinite and metakaolinite (from 15.8 to ~103.1 m^2/g). From the IR spectra of samples treated at 175°C the characteristic kaolinite bands have been developed. The Al – OH stretching (3800–3500 cm^{-1}) and bending modes (940 and 910 cm^{-1}), along with Al-O$_6$ stretching, were useful for the monitoring of kaolinite rehydroxylation. The rehydration of metakaolinite to kaolinite strongly depended on the temperature and time of hydrothermal process. The optimum transformation from the point of view of the surface properties was observed after longer-term autoclaving (4–7 days) at 175°C, when the specific surface area S_{BET} of raw kaolinite increased more than three times.

3.2 Sorption experiments

Adsorption of As(III) and As(V) was ineffective in the case of rehydrated kaolinites. The addition of Fe improve the sorption capacity and efficiency significantly (0.5–90.8%), which correspond well with above described adsorption mechanism forming surface Fe-As complexes. The comparison of adsorption efficiency ε and maximum sorption capacities q for investigated oxyanions are summarized in Table 1.

4 CONCLUSIONS

As(III) and As(V) adsorption on Fe-modified kaolin ran almost quantitatively by high sorption capacities, and the procedure according to Langmuir model should be considered. The 4-days rehydrated kaolin demonstrated the highest specific surface and the best adsorption properties for both As(III) and As(V) oxyanions.

ACKNOWLEDGEMENTS

This work was part of projects 13-24155S (Grant Agency of Czech Republic).

REFERENCES

Doušová, B., Grygar, T. Martaus, A., Fuitová, L. Koloušek, D., Machovič, V. 2006. Sorption of AsV on aluminosilicates treated with FeII nanoparticles. J. Colloid Interf. Sci. 302: 424–431.

Doušová, B., Fuitová, L., Grygar, T., Machovič, V., Koloušek, D., Herzogová, L., Lhotka, M. 2009. Modified alumino-silicates as low-cost sorbents of As(III) from anoxic groundwater. J. Hazard. Mater. 165: 134–140.

Grygar, T., Hradil, D., Bezdička, P., Doušová, B., Čapek, L. & Schneeweiss, O. 2007. Fe(III) modified montmorillonite and bentonite: Synthesis, chemical and UV-VIS spectral characterization, arsenic sorption, and catalysis of oxidative dehydrogenation of propane. Clays and Clay Minerals 55(2): 165–176.

Gupta, V.K., Saini, V.K. & Jain, N. 2005. Adsorption of As(III) from aqueous solution by iron oxide-coated sand. J. Colloid Interf. Sci. 288: 55–60.

Jelinek, L.; Inoue, K. & Miyajima, T. 1999. The Effect of Hu-mic Substances on Pb(II) Adsorption on Vermiculite, Chem. Lett. 1: 65.

Lhotka, M., Doušova, B., & Machovič, V. 2012. Preparation of modified sorbents from rehydrated clay minerals. Clay Miner. 47: 251–258.

Rocha, J., Adams, J.M., Klinowski, J. 1990. The rehydration of metakaolinite to kaolinite: Evidence from solid-state NMR and Cognate techniques. J. Solid State Chem. 89: 260–274.

Arsenic Research and Global Sustainability – Bhattacharya, Vahter, Jarsjö, Kumpiene, Ahmad, Sparrenbom, Jacks, Donselaar, Bundschuh & Naidu (Eds)
© 2016 Taylor & Francis Group, London, ISBN 978-1-138-02941-5

Influence of silica on arsenic removal from groundwater by coagulation, adsorption and filtration

A. González, A.M. Ingallinella, G.S. Sanguinetti, H. Quevedo, V.A. Pacini & R.G. Fernández
Center of Sanitary Engineering, Faculty of Exact Sciences, Engineering and Surveying, National University of Rosario, Argentina

ABSTRACT: The objective of this project was to study the influence of silica (SiO_2) on Arsenic (As) removal by coagulation with aluminium salts since, according to literature, silica is one of the ions with major interference such as phosphate (PO_4^{3-}). Jar tests using natural water containing As concentrations from 0.10 to 0.20 mg/L and from 50 to 60 mg/L SiO_2, and model water prepared in the laboratory were performed. The assays were carried out under different doses of PACl, different concentrations of SiO_2 and different pH conditions. The results indicated that while the concentration of SiO_2 increases As removal decreases. In natural water, the highest SiO_2 removal was 22% at pH > 8, while the highest As removal (96%) took place at pH between 5.5 and 6.0 but, in both cases, with higher values of residual Al. An increment in residual Al concentration in treated water was observed as the concentration of SiO_2 increased, which could indicate that a large amount of silica inhibits the formation of Al hydroxide precipitates.

1 INTRODUCTION

The matrix of water to be treated is an important factor in the process of selection of As removal technologies. Despite the importance that the presence of competing ions has, enough attention was not given to the matrix in case where coagulation—adsorption—filtration processes were applied. The bibliographic research shows that the most important interfering ions are silica (SiO_2) and phosphate (PO_4^{3-}) (Meng *et al.*, 2000). Gregory et al. (1998) reported that not only SiO_2 competes with As for the adsorption sites available on the Al hydroxides, but also influences the formation thereof, either positively or negatively depending on concentration. The aim of this project was to analyse the influence of SiO_2 on the As removal by coagulation—adsorption—filtration process using aluminium (Al) salts. Jar Tests were conducted using natural waters and prepared water in order to evaluate the performance of As, SiO_2 and Al under different coagulant dosage, different concentrations of SiO_2 and different pH conditions.

2 EXPERIMENTAL METHODS

Jar tests were conducted using groundwater from Villa Cañas, Santa Fe Province (Argentina), which has As concentrations between 0.10 to 0.20 mg/L, and SiO_2 between 50–60 mg/L. Experiences were also performed using prepared water containing similar concentrations as the above described. Initial pH conditions were achieved by adding sulphuric acid 1 N, H_2SO_4 (96%) or sodium hydroxide 5,25 N, NaOH. The coagulant used was polyaluminium chloride, PACl 18 (Al_2O_3 16.9% m/m). Jar tests were conducted under the following conditions: 1 minute of rapid mixing (G: 300 sec^{-1}),

15 minutes of flocculation (G: 50 sec^{-1}), and 20 minutes of settling. The supernatant of each jar was filtered through cellulose nitrate membranes of nominal pore diameter of 0.45 um. Assays were performed by duplicate. The first series of tests were conducted based on natural waters, at doses of PACl between 40 and 140 mg/L, and different initial pH conditions: natural pH (8.2) and pH 6.9. The second series of tests was conducted using natural water, at a dose of 100 mg/L PACl, and an initial pH range between 3 and 10. The third series of tests was performed also with natural water, at dose of 100 mg/L PACl, pH 6.9 and different concentrations of SiO_2. To increase the concentration of SiO_2 in natural waters Monohydrate Sodium Metasilicate, $Na_2SiO_3 \cdot H_2O$, was added. In the last series of tests prepared water was used, at dose of 100 mg/L PACl and pH 6.9. Model water was prepared using distilled water with the addition of: sodium hydrogen carbonate, $NaHCO_3$ to achieve an alkalinity of about 400 mg/L $CaCO_3$, arsenic acid solution, H_3AsO_4 in order to achieve a final concentration of 0.15 mg/L As(V), and $Na_2SiO_3 \cdot H_2O$ to achieve different concentrations of SiO_2 in water. In filtered samples there were analyzed: pH, Turbidity, As (Hydride Generation/Atomic Absorption Spectrophotometry Method), SiO_2 and Al.

3 RESULTS AND DISCUSSION

3.1 *As and SiO_2 removal vs dose of PACl*

The results of the first series of tests show that As removal increases with increasing dose of PACl and with decreasing pH, which is consistent with previous research work (Ingallinella *et al.*, 2003). As removal was around 80% with doses of 100 mg/L PACl and pH 6.9, while at natural pH, As removal was approximately

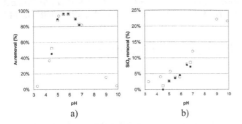

Figure 1. As (a) and SiO₂ (b) removal in natural waters at different pH, with 100 mg/L PACl (○ Test 3, □ Test 4, ■ Dup. Test 4). Initial As concentration: 0.13 mg/L. Initial SiO₂ concentration: 53 mg/L.

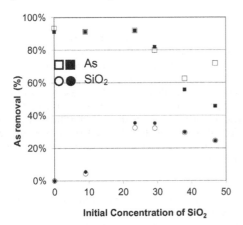

Figure 2. As removal in prepared waters with varying initial concentrations of SiO₂ with 100 mg/L PACl and pH 6.9 (■ Test 8, ● Dup. Test 8, □ Test 9, ○ Dup. Test 9). Initial As concentration: 0.15 mg/L.

55%. Most SiO₂ removal occurred at natural pH, which is consistent with that reported by Meng et al. (2000).

3.2 As and SiO₂ removal vs pH

In Figure 1 As and SiO₂ removal efficiencies in natural waters at different pH is shown.

It is demonstrated that the SiO₂ removal increases with increasing pH and it was stabilized around 22% at pH 9, corroborating the results of Tests 1 and 2, and the studies by Meng et al. (2000). The largest As removal occurred at pH range from 5 to 7. Regarding Al residual concentrations, the optimum value was achieved at pH 6.5–7.0 (0.03 mg/L).

3.3 As and SiO₂ removal vs SiO₂ concentration

Table 1 shows the results of As and SiO₂ removal in natural water, using different initial SiO₂ concentrations.

It can be seen that As removal decreases from 76% to 41% with increasing initial SiO₂ concentration, while SiO₂ removal was about 12% in all tests. For residual Al, it can be seen that with increasing SiO₂ concentration in water, the concentration of residual Al increases from 0.05 mg/L to 0.24 mg/L. It could be also concluded that As removal decreases with increasing concentration of

Table 1. As and SiO₂ removal efficiencies in natural waters for different initial concentrations of SiO₂, at pH 6.9 and 100 mg/L PACl. Initial As concentration: 0.13 mg/L (Test 5, 6 and 7).

Initial SiO₂ (mg/L)	Average removal (%)		Residual Al (mg/L)
	As	SiO₂	
57.0 (± 0.4)	76 (± 4)	12 (± 1)	0.05 (±0.01)
81 (± 2)	59 (± 3)	12 (± 1)	0.11 (±0.03)
106 (±1)	41 (±6)	12 (±2)	0.24 (±0.03)

Number of samples: 6 (Mean ± Standard Deviation)

SiO₂. This could be explained by a strong association of Al precipitates with SiO₂, which could compete with As for available surface sites (Meng et al., 2000). Furthermore, Gregory et al. (1998) showed that the presence of high concentrations of SiO₂ (> 90 mg/L) at pH 7, inhibit the formation of precipitates of Al, implying a lower As removal. The latter justifies the observed increase in the concentration of residual Al with increasing SiO₂ concentration in the test water.

Figure 2 shows the test with prepared water The presence of SiO₂ affected As removal from concentrations > 20 mg/L. In this experiment, As removal was lower than that obtained with natural water and 100 mg/L PACl at pH 6.9 (46%–72% vs 75% -82% removal of As respectively, see results Tests 2, 3, 4 and 5). Conversely, the removal of SiO₂ in a concentration of 47 mg/L (similar to natural water) was 25%, while for trials with natural water, under the same conditions, was in the range between 9% and 13%.

4 CONCLUSIONS

It was demonstrated that in both natural water and prepared water, the increasing initial concentration of SiO₂ decreases the efficiency in As removal.

The higher removal of SiO₂ was 22% at pH greater than 8, and the maximum As removal, 96%, occurred at pH between 5.5 and 6.0, but with higher values of residual Al in both cases.

For treated water, an increase in Al residual concentration, as SiO₂ concentration increased, for the same dose, was observed. This could indicate that the presence of that compound inhibits the formation of Al precipitated and therefore decreases the As removal.

REFERENCES

Gregory J., Duan J., 1998. The Influence of Silicic Acid on Aluminium Hydroxide Precipitation and Flocculation by Aluminium Salts. J. Inorg. Biochem. 69: 193–201.

Ingallinella, A.M., Fernández, R., Stecca, L.M. 2003. Proceso ArCIS-UNR para la remoción de As y F⁻ en aguas subterráneas: una experiencia de aplicación, Revista Ingeniería Sanitaria y Ambiental, Edición N° 66 y 67.

Meng X., Bang S., Korfiatis G.P. 2000. Effects of silicate, sulfate, and carbonate on arsenic removal by ferric chloride [J]. Water Res. 34(4): 1255–1261.

Arsenic Research and Global Sustainability – Bhattacharya, Vahter, Jarsjö, Kumpiene,
Ahmad, Sparrenbom, Jacks, Donselaar, Bundschuh & Naidu (Eds)
© *2016 Taylor & Francis Group, London, ISBN 978-1-138-02941-5*

Evaluation of native sulphate-reducing bacteria for arsenic bioremediation

C.M. Rodríguez[1], E. Castillo[2] & P.F.B. Brandão[1]

[1]*Laboratorio de Microbiología Ambiental y Aplicada, Departamento de Química, Universidad Nacional de Colombia, Bogotá, Colombia*
[2]*Laboratorio de Química Ambiental, Grupo de Estudios para la Remediación y Mitigación de Impactos Negativos al Ambiente (GERMINA), Departamento de Química, Universidad Nacional de Colombia, Bogotá, Colombia*

ABSTRACT: The arsenic precipitation efficiency of two native strains of Colombia was evaluated: a strain previously isolated and adapted in vitro, and another strain isolated from agricultural soils from the Bogota savannah. Desulfovibrio acrylicus strain CMPUJ U361 was used in arsenic precipitation assays at different temperatures, initial pH and sulphate concentration through a full factorial design, where the significance level was determined for each variable. Based on the results, precipitation assays were performed using a native isolated strain (P29), which was selected according to its resistance to the metalloid (up to 100 mg/L). Strain P29 showed a precipitation efficiency of 63% while strain D. acrylicus, adapted in vitro under optimal growth conditions, showed an 83% efficiency. Nonetheless, the first showed higher precipitation efficiency at pH 5.0 and 7.0, and at 20, 40 and 60 °C, suggesting a better versatility of the new isolate to adapt and stabilize the metalloid species formed.

1 INTRODUCTION

Sulphate-reducing bacteria (SRB) have the ability to use sulphate as a terminal electron acceptor and those with no assimilatory metabolism use redox reactions to generate metabolic energy and sulphide expulsion from the cell (Hansen, 1994). Sulphide reacts with Arsenic (As) to generate arsenic sulphide, an insoluble compound that forms an outer layer around the cell wall (Utgikar, 2002). This gives these bacteria the ability to precipitate and decrease the bioavailability of arsenic in the environment. In this work, native SRB from Colombia were studied to evaluate its arsenic sulphide precipitation ability and thus its feasibility to be implemented in an arsenic remediation process.

2 METHODS

SRB are anaerobic or aerotolerant microorganisms, thus experimental setup and procedures were established for their appropriate laboratory manipulation (Hurst *et al.*, 2007). Two native SRB strains (*Desulfovibrio vulgaris* CMPUJ U153 and *Desulfovibrio acrylicus* CMPUJ U361) from the Pontificia Universidad Javeriana Microorganism Collection (Colección de Microorganismos de la Pontificia Universidad Javeriana—CMPUJ), Bogotá, Colombia, were evaluated for use in the experiments. For these two strains, the Minimum Inhibitory Concentration (MIC) was evaluated against arsenic. The strain that showed better growth at higher arsenic concentrations was used to establish the best arsenic precipitation conditions. For this, a full factorial design was implemented by varying the conditions of three variables: temperature (20, 40 and 60°C), initial medium pH (5.0, 7.0 and 9.0) and initial sulphate concentration (10, 30 and 50 mM). These assays were performed in triplicate under anaerobic conditions. After incubation, the cultures were centrifuged to separate precipitate and supernatant and the arsenic concentration was determined in both by Hydride generation atomic absorption spectrometric (HG-AAS). Simultaneously to the assays with the CMPUJ native strains, further native SRB were isolated from agricultural soil with presence of arsenic (1.74 mg/L). This was done to access further the diversity of native sulphate-reducing microorganisms from Colombia and access their arsenic precipitation abilities. Identification studies are underway to sequence its 16S rRNA genes for subsequent phylogenetic analysis. For the strains with higher MIC against As the variables temperature (20, 40 and 60°C) and pH (5.0, 7.0 and 9.0) were evaluated in order to establish their best conditions and precipitation efficiency.

3 RESULTS AND DISCUSSION

D. acrylicus CMPUJ U361 showed better growth at higher As concentrations (up to 5 mg/L) and thus was used to perform a full factorial analysis on the arsenic precipitation assay. The best precipitation

Table 1. Percentage of precipitation efficiency using inorganic sulphide (20°C), *Desulfovibrio acrylicus* CMPUJ U361 and new native strain P29.

| Inorganic sulphide | | *Desulfovibrio acrylicus* | | | | | |
| | | MPUJ U361 | | | New isolate P29 | | |
		20°C	40°C	60°C	20°C	40°C	60°C
5,0	13%	1%	1%	2%	46%	19%	42%
7,0	4%	1%	35%	1%	63%	60%	39%
9,0	26%	2%	83%	16%	28%	6%	4%

conditions found were at pH 9.0, incubation temperature of 40°C and sulphate initial concentration of 50 mM, in which, on average, higher values of arsenic ratio were found in the precipitate and supernatant. Through a statistical analysis of the precipitation assay results with strain *D. acrylicus* CMPUJ U361D, the variables significance was determined. It was found that variables pH and temperature were more influential in the process. The initial sulphate concentration was not a significant variable, since its variation did not involve a significant change in the precipitation efficiency. The precipitation phenomenon is linked to the bacterial growth and the chemical species stability in the medium, and is limited by the arsenic sulphide accumulation phenomenon on the cell surface (Utgikar, 2002).

Several soil samples from the Bogota savannah were analysed for arsenic and the one showing the highest concentration (1.74 mg/L) was selected for native SRB isolation using USBA (Baena, 1998) and Postgate C (Hurst, 2007) selective media. Among the isolates recovered, strain P29 was selected based on its resistance to arsenic (up to 100 mg/L) and the amount of precipitate formed at higher metalloid concentration. The best precipitation conditions with this strain were pH 7.0 and 20°C. Assays with inorganic sulphide were performed as control of precipitation. The precipitation efficiency (Table 1) was determined and it was found it was lower (63%), at 20 °C and pH 7.0, for the native isolate P29 compared with strain *D. acrylicus* CMPUJ U361 (83%), at 40 °C and pH 9.0, which was adapted *in vitro* and had shown less resistance to As.

Under acidic and neutral conditions (pH 5.0 and 7.0) a higher precipitation efficiency was observed with strain P29 at different temperatures (20, 40 and 60°C) compared to *D. acrylicus* CMPUJ U361, thus showing it to be more versatile, able to adapt and to stabilize the chemical species formed to induce the metalloid precipitation. Greater precipitation efficiency was found, as expected, using SRB compared to inorganic sulphur in all media where growth of these microorganisms was reported.

Using SRB in remediation systems can be influenced by their ability to assimilate certain nutrients present in the environment, in addition to structural characteristics that can influence their degree of resilience to a metal or metalloid elements and

their precipitation. Thus, the optimum precipitation conditions will be different by using other microorganisms, which may improve efficiencies or be more resistant to harsher growth conditions in relation to the arsenic concentration and other contaminants.

4 CONCLUSIONS

Previously isolated native SRB strains were evaluated and *D. acrylicus* CMPUJ U361 showed a better response in MIC assays against arsenic and with which metalloid precipitation studies were developed. Through a statistical study of variables interaction, it was established that the most significant variables were temperature and pH. These were evaluated to establish the best precipitation conditions of the new native strain P29, selected among various sulphate-reducing microorganisms isolated from agricultural soils with arsenic. Among all the assays performed, strain *D. acrylicus* CMPUJ U361 showed the greater precipitation efficiency (83%) compared to the native P29 isolate (63%), and both showed higher efficiency than the inorganic sulphide control (26%).

ACKNOWLEDGEMENTS

The Research Division DIB of the Universidad Nacional de Colombia is thanked for financial support to this project (code DIB 27086). We thank Professor Sandra Baena, and her students Carolina, Gina and Marcela, from the Pontificia Universidad Javeriana, Bogotá, Colombia, for access to the native SRB from the CMPUJ used in this study and for their help to establish procedures for microbial anaerobic growth in our laboratory. Finally, we thank the members of the research group GERMINA, Rodrigo Pérez, for his help on the factorial design and statistical analysis, and Diana Tamayo, for her help with the molecular identification of the newly isolated SRB strains.

REFERENCES

Baena, S., Fardeau, M.L., Labat, M., Ollivier, B., García, J.L., Patel, B.K. 1998. *Desulfovibrio aminophilus* sp. nov., novel amino acid degrading and sulfate reducing bacterium from an anaerobic dairy wastewater lagoon. *Systematic and Appl. Microbiol.* 21: 498–504.

Hansen T.A. 1994. Metabolism of sulfate reducing prokaryotes. *Antonie Van Leeuwenhoek* 66(1–3): 165–85.

Hurst, Crawford, Garland, Lipson, Mills, Stetzenbach. 2007. *Manual of Environmental Microbiology* (3rd Edition).

Teclu, D., Tivchev, G., Laing, M., Wallis, M. 2008. Bioremoval of arsenic species from contaminated waters by sulfate-reducing bacteria. *Water Res.* 42: 4885–4893.

Utgikar, V., Harmon, S.M., Chaudhary, N., Tabak, H.H. Govind, R., Haines J.R. 2002. Inhibition of sulfate-reducing bacteria by metal sulfide formation in bioremediation of acid mine drainage. *Environ. Toxicol.* 17(1): 40–48.

Arsenic Research and Global Sustainability – Bhattacharya, Vahter, Jarsjö, Kumpiene,
Ahmad, Sparrenbom, Jacks, Donselaar, Bundschuh & Naidu (Eds)
© *2016 Taylor & Francis Group, London, ISBN 978-1-138-02941-5*

Evaluation of dewatering performance and physical-chemical characteristics of iron chloride sludge

S.E. Garrido[1] & K. García[1,2]
[1]*Instituto Mexicano de Tecnología del Agua, Jiutepec, México*
[2]*Universidad Nacional Autónoma de México, Jiutepec, México*

ABSTRACT: The arsenic removed from drinking water generates a residue which is toxic. The aim of this study was to analyze the physical and chemical factors in the thickening and dewatering of sludge produced in As(V) removal by coagulation-flocculation process. The critical process variables associated with the removal of As(V) from water and sludge production were evaluated on experimental design. The optimum conditions were: pH 7.20; dosage: $FeCl_3$ 34.37 mg/L and polymer 0.89 mg/L, settling solids 7.59 mL/L and final As(V) concentration 0.003 mg/L. Subsequently, the $FeCl_3$ sludge was characterized and the factors studied that influence in the dewatering performance: Specific Resistance to Filtration (SRF) with a minimum value of $6.29.10^{+12}$ g/cm and Filtration time (Ft) of 107 s. Regarding moisture distribution sludge is comprised 75% mainly interstitial and free water.

1 INTRODUCTION

Arsenic (As) removal from drinking water processes currently used are shown in Table 1, all of these technologies generate waste sludge forma with metal hydroxides and reject water, with high concentrations of dissolved solids, dominated As(III) and As(V) ions, whose forms are toxic, so their waste must be treated before disposal. The aim of this study was to analyze the physical and chemical factors in the thickening and dewatering of $FeCl_3$ sludge produced in As(V) removal by coagulation-flocculation process in drinking water.

2 MATERIALS AND METHODS

2.1 *Optimization of the conditions for obtaining As(V) sludge*

In order to study and evaluate efficiently the critical operating parameters of the coagulation-flocculation process to determine the values that maximized the As(V) removal from water and sludge production a central composite design was used, N = 2^3 was performed, in which the factors considered were: A: pH 6.5 to 7.5; B: Dose of $FeCl_3$ 20 to 40 mg/L; C. Dosage polyacrylamine cationic polymer; 0.5 to 1.5 mg/L. Initial As(V) concentration were 0.050–0.100–0.150 mg/L. The tests were carried out in test Jars Phipps & Bird, with velocity gradients, G: Rapidly Mixer: 400 s^{-1}, 10 s, slow mixing: 60, 24, 12, 10 s^{-1} for each 5 min, and sedimentation 20 min. Arsenic was determined photometrically (Garrido

et al., 2013) and physical-chemical parameters according to Mexican standards.

2.2 *Physical and chemical effects in the thickening, and dewatering in $FeCl_3$ sludge with As(V)*

Once optimized to obtain $FeCl_3$ sludge of coagulation-flocculation, they are characterized and studied the factors that influence in the dewatering performance. a) Specific resistance to filtration (SRF) and filtration time (Ft). SRF test involves placing 100 mL of $FeCl_3$ sludge in a Buchner funnel with 9 cm, Ø 8 μm filter paper (Whatman No. 1) and 250 mL graduated cylinder in which a vacuum is applied a pressure of 460 mm Hg, until the cake is broken as result of a pressure drop (ΔP). The Ft is calculated by measuring the time required to reduce the volume to 50% of the initial sample (Cornell *et al.*, 1987). b) Compressibility: The specific resistance to filtration (r) at different pressures (P) was measured: 420–460–520 mm Hg, a 100 mL volume until dehydration by crushing the sludge cake. c) Distribution of moisture: The Ohaus oven was used for determining the different types of moisture fractions in sludge. It consists of a thermostatically controlled heating oven inside which was placed a digital balance. A sample of 20 g sample of $FeCl_3$ sludge was weighed and dried at a temperature of 103–105°C/100 min. The sludge mass was sampled at 10 min intervals and recording the weight, until there was no change in sludge mass.

Table 1. Residuals produced from arsenic removal in drinking water.

Technology	Removal As (%)	Volume of residuals produced (L/m³)	As concentration in residuals (mg/L)	Quantity of solids produced (kg/m³)
Conventional[3] coagulation	As(V) >80 As(III)>20–80	4.3	9.25[1]	21.59
Coagulation[3+] microfiltraion	As(V) >95	52.6	0.76	13.50
Softening	As(V) >90	9.6	4.2	239.89
Ion exchange	As(V) >85	4.0	10	0.623
Activated alumina	As(V) >95	4.2	9.52	2.8
Iron oxides coated sand	As(V) >99 As(III) >80	21	1.9	2.8
Nanofiltration/RO	As(V) >90/95 As(III) >70	200–300[2]	0.098	–

[1]It is assumed that 40 mg/L of As in the treatment is removed; [2]Estimated; [3]FeCl₃ as coagulant.

3 RESULTS AND DISCUSSION

3.1 Physical-chemical quality of the FeCl₃ sludge obtained from the optimization design

Table 2 shows the physical-chemical quality of the FeCl₃ sludge. The results expressed as removing As(V) and settling solids (response factors) the optimum values were found: A: 7.20; B: 34.33 mg/L; C: 0.89 mg/L for an initial concentration of As(V) 0.150 mg/L. The final concentration of As(V) in the water of all experiments was < 0.003 mg/L.

3.2 Specific Resistance to Filtration (SRF) and Filtration time (Ft)

The SRF was calculated by Poiseuille-law D'Arcy, Cornwell *et al.* (1987) (Fig. 1). The filtration time (Ft) was 107 s, for 100 mL of initial volume. Yongjun *et al.* (2015), explained that SRF will normally be proportionally to the applied pressure (P) due to the high compressibility. From theoretical view point it is impossible for the SRF increase more than proportional to P; SRF is considered excellent for values between 1.10^{+11}–4.10^{+11} cm/g in homogeneous thickened and conditioning sludge in water treatment plants. Without sludge conditioning the SRF obtained was $1.8.10^{+12}$–$2.7.10^{+12}$ cm/g, these values obtained were slightly higher than theorical values (Fig. 2a).

3.3 Moisture distribution

Figure 2b shows the aqueous phase in FeCl₃ sludge separated into four forms: a. Water of hydration that is chemically bound to the particle, 5%; b. Vicinal water that is associated with solids particles is held on particle surface by molecular structure of the water molecules, 15%; c) Interstitial water

Table 2. Quality physico-chemical quality of the FeCl₃ sludge of the optimum values.

Parameter	Value
Willcomb index	6–8*
Total solids	0.48 mg/L
Total suspended solids	0.31 mg/L
Setteable solids	7.59 mL/L
pH	6.70
Total solids in wet filter cake	2.904 mg/L
Total solids in dry filter cake	0.405 mg/L
Initial sludge moisture content (c_i)	99.95%
Final sludge moisture content (c_f)	86.07%
Arsenic	19.76 mg/L

*Sized flocs precipitate relatively quickly.

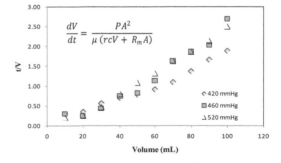

Figure 1. Specific resistance to filtration (SRF), at different pressure: 420, 460 and 520 mm Hg. Where: *r*: Specific sludge resistance (cm/g); *P*: pressure of filtration (g cm/s² cm²); *A*: filter area (cm²); *b*: slope(s/cm⁶); *V*: volume of filtrate (m³); *t*: filtration time (s); *μ*: viscosity of filtrate (g/cm s); *c*: weight of solids per unit volume of filtrate (g/cm³); R_m: total flow resistance to filtration.

491

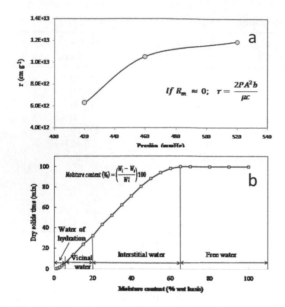

 shows graph b below graph a.

Figure 2.   a) Compressibility of FeCl$_3$ sludge at three pressures: A: 40.71 cm^2; μ (20°C): 0.01 Poise (g/cm s); c: 5.01^{-4} g/cm^3 and b) Drying characteristic curve, where W$_i$ is the initial weight and W$_f$ is the weight at any t after drying at 105°C.

that is trapped within the floc structure, 45%; d) Free water that is not attached to sludge solids, 35%. Verelli *et al.* (1990) found that the dewatering procedures can only remove part of the free and interstitial water; surface and bound water are not affected. Part of the free and interstitial moisture also remains with the sludge cake, which can be considered as the inherent inefficiency of the dewatering.

4   CONCLUSIONS

The optimum conditions for the coagulation-flocculation and sludge dewatering was evaluated in this study, in which a direct relationship between FeCl$_3$ dose, SRF and applied pressure was observed.

REFERENCES

Cornwell, D.A., Burmaster, J.W., Francis, J.L., Friedline, J.C. Jr., Houck, C., King, P.H., Knocke, W.R., Novak, J.T., Rolan, A.T., San Giacomo, R. 1987. Research needs for alum sludge discharge. *J. Am. Wat. Works Assoc.* 79(6): 99–104.

Garrido, S., Piña, M., López, I.; De La O, D., Rodríguez, R. 2013. Behavior of two filters media for to remove arsenic from drinking water. *Filtration* 13: 21–26.

Verelli, D.I., Dixon, D.R., Scales, P.J. 2010. Assessing dewatering performance of drinking water treatment sludge. *Water Res.* 44: 1542–1552.

Yongjun, S., Wei, F., Huaili, Z., Yuxin, Z., Fengting, L., Wei, C. 2015. Evaluation of dewatering performance and fractal characteristics of alum sludge. *PLosONE* DOI: 10.1371/journal.pone.0130683: 1–16.

Arsenic Research and Global Sustainability – Bhattacharya, Vahter, Jarsjö, Kumpiene, Ahmad, Sparrenbom, Jacks, Donselaar, Bundschuh & Naidu (Eds)
© 2016 Taylor & Francis Group, London, ISBN 978-1-138-02941-5

Effective and passive arsenic adsorption process for groundwater treatment

R.S. Dennis
De Nora Water Technologies, Inc., Rodenbach, Germany

ABSTRACT: After the World Health Organization recommended in 1993 that the drinking water arsenic standard of 50 µg/L be reduced to 10 µg/L, many countries have been implementing the new standard. Facing compliance in December, 2003 in the UK, Severn Trent Water (STW) developed a passive arsenic adsorption process for their plants needing treatment. They selected a Granulated Iron Oxide Media (GIM) as the most selective arsenic (As) adsorbent De Nora Water Technologies, formerly STW's sister company, advanced STW's arsenic adsorption technology for applications in other regions of the world. They partnered with Lanxess, a technical iron oxide producer, who developed and produces the Bayoxide® E33 media for De Nora's treatment systems. This paper will describe how De Nora approaches adsorption treatment development and designs for various water qualities to minimize the effect of interferents that may be present in the water in order to provide an As treatment system that yields low operating costs.

1 INTRODUCTION

When the WHO announced reduction of the arsenic standard to less than 10 µg/L, Severn Trent Water (STW) in Birmingham, UK realized that they had nearly 60 wells whose As levels exceeded the new standard. Having researched the available technologies that could treat drinking water to below this level, they determined that a more cost effective process would have to be developed. Over a four year period, STW decided on an iron oxide adsorbent as the most economical and simplest process to meet their objectives.

2 METHODS/EXPERIMENTAL

After conducting laboratory studies in 1995, STW selected an adsorption process over other technologies because it was more passive in operation and generated much lower residual volumes than any of the other processes. A testing facility was installed at one of their high As level well sites in Nottingham. A containerized unit containing eight 10 cm diameter columns were used to test various adsorbents under different conditions. The key variables that were measured during the hydraulic and chemistry (adsorption) testing included:

- Adsorption breakthrough
- Effect of hydraulic loading rate
- Empty bed contact time
- Pressure differential across bed
- Media As capacity

The types of media evaluated included activated alumina, iron impregnated alumina and several variants of iron oxide media. A Granular Ferric Oxide Media (GFO) was ultimately selected by STW. STW shared the results of its program after pilot testing was completed with its sister company, Severn Trent Water Purification (STWP), who took the technology to other markets where As standards were being lowered and treatment would be required including the United States. The water qualities here were much more diverse than those in STW's UK wells. Pilot test objectives were expanded in the US to include the following:

- The effect of pH adjustment on high pH waters
- The effect of interferents on media As capacity
- Adsorption of reduced As(III) by GFO
- The effect of iron & manganese on performance

3 RESULTS AND DISCUSSION

The extensive STW pilot program covered many test phases focusing on media evaluations, hydraulic testing and adsorption studies. As media selection was narrowed based upon physical properties and hydraulic tests, emphasis was placed on studying adsorption profiles on these media to determine which one had the highest capacity. Ultimately, three were tested and compared in dynamic column tests. The results of tests conducted on Nottingham water containing 20 µg/L As are illustrated graphically in Figure 1, a breakthrough curve.

As expected, the AA media performed poorly treating only a few thousand BV's before the column effluent broke through with over 10 µg/L As. It requires significant pH reduction to last an adequate amount of time. The other two medias are GFO's. The GFM performed well but not nearly as good as the Bayoxide. STW had worked with Lanxess on the development of this adsorbent, having specified the product requirements. This is the product STW decided to use in their commercial systems.

The US market for As treatment to <10 μg/L is much larger than that in the UK. DNWT first started pilot testing on a challenging high pH water in Rio Rancho, NM. The water also contained high levels of vanadium, an interferent that is co-adsorbed by the GFO. The Rio Rancho water contained an As level of 49 μg/L, a V level of 79 μg/L and a pH of 8.9. This pH is well above the 8.3 zero valence pH of Bayoxide meaning poor adsorption from the ambient pH water would occur. This was confirmed during initial column testing using the same type of equipment used by STW. pH reduction to 6.7 was then applied to a new column of media.

Some adsorption pilot test programs focused on other variables. For example, one test focused on removing reduced As(III) which in most well waters is un-ionized. As such, most As removal technologies won't remove As(III) including RO. A pilot test was conducted on RO permeate in Dare County, NC which contains 25 μg/L As, all of it reduced. The pilot treated over 150,000 BV's before being shut down, and breakthrough never occurred.

Complete performance analysis of adsorptive media when removing ppb level contaminants with high capacity media like Bayoxide can take several months in many cases when using conventional dynamic column testing. A test protocol was developed by Arizona State University to hasten the process to quantify media performance in terms of BV's. The Rapid Small Scale Column Test (RSSCT) is modeled after a similar protocol for evaluating activated carbon adsorption. Media is ground to a small particle size and packed in a small diameter laboratory column which is operated at low contact times to simulate the full scale pilot.

Third party testing is considered very important in evaluating and comparing adsorptive media. An example of this testing was conducted by a consulting engineer working for the city of Mesa, AZ. Their program included conducting an RSSCT on various media that were being considered. The results of the testing are graphically illustrated below in Figure 2.

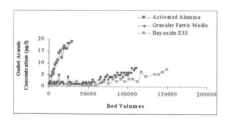

Figure 1. Adsorption breakthrough curves for three metal oxide media.

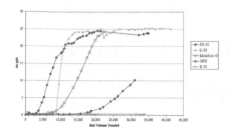

Figure 2. Mesa water RSSCT As breakthrough curve.

This test evaluated five different media including an iron impregnated activated alumina, a zeolite, titanium dioxide and two iron oxide medias. The GFO's exhibited the best performance for this water containing 40 μg/L As.

Since well water treatment for contaminants beyond hardness, iron and manganese was unusual and in order to evaluate the different available commercial technologies, the US EPA conducted a demonstration program for a number of small well systems throughout the US. DNWT with their Bayoxide GFO media was selected to provide systems for 25% of the first 12 EPA sites. Upon completion of the program and after 1 year of operation of the different systems, the EPA published a report on the technology's performances.

4 CONCLUSIONS

De Nora Water has successfully advanced Severn Trent Water's GIM adsorption technology for As treatment on a variety of water qualities not only in the United States but in Europe, Mexico, Chile, India and the Far East using Bayoxide E33 produced by Lanxess GmBh.

The key to effective, economical As treatment starts with a complete well water analysis from which the right solution can be applied. Sometimes, adsorption isn't the right technology because the water contains high levels of iron, manganese, phosphate or other interferents that make the process costly to operate. In many of these cases, coagulation filtration is used for As removal and may require the addition of coagulants. When fluoride is present and must be removed, a regenerative activated alumina adsorption process on pH reduced water is required.

ACKNOWLEDGEMENTS

The authors acknowledge the support of Severn Trent Water, Arizona State University, Mesa, Arizona and the US Environmental Protection Agency.

Arsenic Research and Global Sustainability – Bhattacharya, Vahter, Jarsjö, Kumpiene,
Ahmad, Sparrenbom, Jacks, Donselaar, Bundschuh & Naidu (Eds)
© *2016 Taylor & Francis Group, London, ISBN 978-1-138-02941-5*

IHE ADART: Field experiences from testing in Central and Eastern Europe and the Middle East

B. Petrusevski[1], Y.M. Slokar[1], F. Kruis[1] & J. Baker[2]
[1]*UNESCO-IHE Institute for Water Education, Delft, The Netherlands*
[2]*SELOR, Amsterdam, The Netherlands*

ABSTRACT: IHE ADART, arsenic removal technology utilizing adsorption on IOCS and an in-situ regeneration of exhausted adsorbent, had been extensively tested in Central-Eastern Europe and Middle East. Groundwater in these regions is typically anoxic and contains complex mixture of iron, manganese, phosphate, ammonia and occasionally methane. IHE ADART, supported by conventional pre-treatment (aeration-filtration) demonstrated potential to produce drinking water with low arsenic concentration from different groundwater quality matrixes. Approximately monthly regeneration cycles were sufficient for treatment of groundwater with moderate concentrations of arsenic and phosphate, and pH ≤8.2. Groundwater with high arsenic and phosphate concentrations, and high pH requires more frequent regeneration cycles. Successful prolonged operation of two full scale plants based on IHE ADART in Serbia and Jordan verified the suitability of the technology for drinking water production from arsenic contaminated groundwater.

1 INTRODUCTION

Presence of arsenic (As) in groundwater in Central and Eastern Europe and Middle East has been recently recognized as a serious threat to public health. Complex nature of Groundwater (GW) in this part of the world due to simultaneous presence of ammonia (NH_4^+), iron (Fe), Manganese (Mn), phosphate (PO_4^{3-}), Natural Organic Matter (NOM), and sometimes methane (CH_4), makes As removal difficult. Very often contaminated GW is used as 'drinking water' without dedicated As removal. If a treatment is applied, it commonly includes break-point chlorination, coagulation and (multi-media) filtration. In addition to intensive chemical usage, generation of hazardous by-products and excess toxic waste production, this treatment has difficulties to consistently produce drinking water with As <10 µg/L.

Adsorptive treatment can be an attractive option for drinking water production from As containing GW, given the process simplicity and high removal efficiency (Mohan & Pittman, 2007). Commercial As adsorbents are, however, expensive and in general cannot be regenerated once exhausted. Despite intensive research focused on development of new, low cost As adsorbents which demonstrated potential to effectively remove As under laboratory conditions, their prolonged testing under field conditions and full scale applications are not commonly reported. In addition effect of water quality matrix on treatment performance is often not investigated under field conditions.

IHE ADART is an As removal technology based on adsorption on Iron Oxide Coated Sand (IOCS), a by-product of treatment plants removing Fe from groundwater, in combination with an innovative *in-situ* regeneration procedure. Regeneration entails an in-situ exposure of (partially) As saturated IOCS to acidic Fe^{2+} solution, which results in creation of a new nano-layer of iron hydroxides over previously adsorbed As. Following extensive laboratory testing, performance of IHE ADART was further assessed in a mobile pilot plant on several locations in Hungary (Petrusevski *et al.*, 2007), Romania and Serbia. In addition, two full scale As removal plants based on IHE ADART have been operating in Jordan and Serbia for several years.

The objective of the paper is to introduce results and experiences from extensive pilot testing and full scale applications of As removal with IHE ADART. Specifically effect of groundwater water quality matrix on As removal will be discussed.

2 METHODS

Field testing of IHE ADART technology was conducted with mobile pilot plants (capacity 50 m³/day), with treatment scheme including aeration tower with co-current air-water flow or plate aerator, Rapid Sand Filtration (RSF, quartz sand 0.8–1.2 mm, 2.5 m bed depth) and two stage adsorptive filtration with IOCS as filter media (2.5 m of bed depth). The pilots operated continuously at filtration rate of 5 m/h. When head loss of sand filter reached 1.5 m (typically weekly) the filter was backwashed with water and air. IOCS filters were backwashed only after regeneration cycles (head loss in adsorptive filters was very low due to coarse IOCS). Regeneration of IOCS filters was carried out when an increase of As concentration in IOCS filters filtrate was observed, with typical duration of regeneration cycle of 30–60 min. Pilots operated (continuously) at more than 10 drinking water production sites in Hungary, Romania and Serbia, with operation time from a few months to more than a year. In addition, performance of two full scale plants in Serbia and Jordan, with capacities of 500 and 1000 m³/day, respectively, was monitored for over 20 months.

3 RESULTS AND DISCUSSION

In the pre-treatment (aeration followed by RSF), apart from removal of NH_4^+, Fe, Mn and occasionally CH_4, also As was partly removed (Fig. 1). The removal of As was limited to approximately 20%, due to low Fe/As ratio (0.6–3.1 with an average of 1.3 mg Fe/mg As). Arsenic removal in pre-treatment was achieved through coagulation with oxidized ferrous iron present in GW.

At all test sites complete As removal was achieved at the start-up of the plants. After prolonged operation, As concentration in treated water started to increase, due to saturation of easily accessible adsorption sites on IOCS. Consequently regeneration was required. Approximately monthly regeneration was sufficient at testing site characterized by relatively low As (≤ 50 µg/L), PO_4^{3-} (≤ 0.8 mg PO_4^{3-}/L), and pH ≤ 8.2 (Fig. 2).

Treatment of GW with high As in combination with very high pH (8.6–8.9) and high PO_4^{3-} (>1.4 mg PO_4^{3-}/L) resulted in rapid As breakthrough (Fig. 3). During the initial 42 days of the pilot operation, no regeneration was conducted. A single regeneration cycle conducted after day 42 resulted in only shortly lived recovery of IOCS As adsorption capacity. However, increasing the frequency of regeneration cycles to every 48 h resulted in consistently low As concentrations (≤ 10 µg/L) in treated water.

In-situ regeneration was found to be simple method to prolong useful life of IOCS, with no

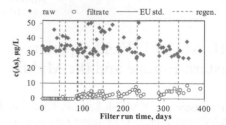

Figure 2.   Arsenic removal with mobile pilot plant at testing site Mako (Hungary).

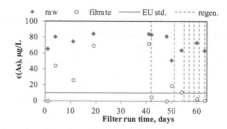

Figure 3.   Arsenic in ground- and treated water at testing site Ujkigyos (Hungary).

need for adsorbent replacement during at present over 20 months of continuous plants operation.

4 CONCLUSIONS

Results from prolonged operation of pilot and full scale plants have shown that IHE ADART technology can be successfully applied for treatment of complex anoxic groundwater common for Central and Easter Europe. Conventional pre-treatment (aeration-RSF) is required to eliminate in particular NH_4^+, Fe, Mn and CH_4, if present. Regeneration frequency of (partially) saturated IOCS required to keep As <10 µg/L was found to substantially vary (daily to monthly) and is a function of groundwater matrix, specifically concentrations of PO_4^{3-}, As and pH.

ACKNOWLEDGEMENTS

Results presented in this paper originate from projects financially supported by The Dutch Government, European Union and Water Authorities of Jordan.

REFERENCES

Mohan, D., and Pittman Jr., C.U. 2007. Arsenic removal from water/wastewater using adsorbents: A critical review *J. Hazard. Mater.* 142: 1–53.

Petrusevski, B., van der Meer, W.G.J., Baker, J., Kruis, F., Sharma, S.K., and Shippers, J.C. 2007. Innovative approach for treatment of arsenic contaminated groundwater in Central Europe. *Water Sci. Technol.: Water Supply* 7(3): 131–138.

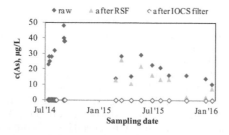

Figure 1.   Arsenic concentration in groundwater and after RASF and IOCS filters at full scale groundwater treatment plant Balad Jaber (Jordan).

4.2   Nanotechnology applications for treatment of arsenic

Arsenic Research and Global Sustainability – Bhattacharya, Vahter, Jarsjö, Kumpiene, Ahmad, Sparrenbom, Jacks, Donselaar, Bundschuh & Naidu (Eds)
© *2016 Taylor & Francis Group, London, ISBN 978-1-138-02941-5*

Transforming global arsenic crisis into an economic enterprise: Role of hybrid anion exchange nanotechnology (HAIX-Nano)

A.K. SenGupta[1,2,3], P.K. Chatterjee[3] & M. German[1]
[1]*Department of Civil and Environmental Engineering, Lehigh University, Pennsylvania, USA*
[2]*Department of Chemical Engineering, Lehigh University, Pennsylvania, USA*
[3]*Society for Technology with a Human Face (NGO), Kolkata, India*

ABSTRACT: Although unknown nearly twenty five years ago, natural arsenic (As) contamination of groundwater has emerged as a major global crisis affecting over twenty countries including the USA, Cambodia, Vietnam, Bangladesh, Nepal, India, Argentina, etc. Polymeric anion exchangers have fixed positive parent functional groups (e.g., quaternary ammonium groups) that create a Donnan Membrane Effect where like-charged cations are rejected, i.e., inorganic metal ions, making them seemingly incompatible. However, we have been able to successfully disperse nanoscale hydrated Fe(III) oxide and Zr(IV) nanoparticles within polymeric anion exchangers by controlling solution conditions. The resulting hybrid anion exchanger impregnated with metal oxide nanoparticles, or HAIX-Nano, is robust, mechanically strong and shows extraordinarily high sorption affinity for both As(V) and As(III) present in contaminated groundwater. HAIX-NanoZr also has high sorption capacity for fluoride. Over one million people in the developing and the developed world across over six countries routinely drink arsenic-safe water though use of this hybrid nanosorbent. In many arsenic-affected communities, the sustainable arsenic mitigation through use of HAIX-Nano has created employment and spurred economic growth while providing safe drinking water.

1 INTRODUCTION

The Donnan Membrane Effect was used to design a novel As and fluoride sorbent using a polymeric anion exchange resin. Where counterions present in the aqueous phase, e.g., As and fluoride, will be concentrated within an anion exchange resin phase. Inside the anion exchange resin, oxides of polyvalent metals, namely, Fe(III) and Zr(IV) were desired because they are known to exhibit ligand sorption properties through formation of inner-sphere complexes. Hydrated iron and zirconium oxides (HFOs and HZOs, respectively) are ideal sorbents from a chemical and economical perspective: stable over a wide pH range, innocuous, available globally at a moderate price and exhibit high sorption affinity for Lewis bases, e.g., As, F, and no affinity towards non-Lewis bases, e.g., SO_4^{2-}, Cl^-. The hybrid of a strong polymeric anion exchange scaffold that utilizes the Donnan Effect to attract trace ions of concern with nanoparticles of HZOs creates the HAIX-Nano resin for efficient As and fluoride removal. Equally important, HAIX-Nano stability makes it amenable to efficient regeneration and reuse (Cumbal & SenGupta, 2005, Padungthon et al., 2015).

HAIX-NanoFe has been commercially manufactured in the USA for almost a decade. Through funding from multiple private organizations and the US-India government, HAIX-NanoZr began small-scale production in Kolkata, India during the last 6 months.

2 METHODS/EXPERIMENTAL

2.1 HAIX-NanoZr synthesis

The preparation of HAIX-Nano was refined in two versions from 1998–2015 through multiple iterations and is carried out at ambient temperatures to promote formation of amorphous metal oxide nanoparticles within the anion exchange resin beads. The final synthesis protocol is included in detail in the patent for HAIX-Nano (SenGupta, 2007).

2.2 Field work

Arsenic treatment activity in rural regions of the Indian Subcontinent using HAIX-NanoFe resins has been on-going since 2004 and using HAIX-NanoZr resins since 2015. Tens of systems have been installed where local community members drank safe water and paid a nominal water tariff to the local water committee.

3 RESULTS AND DISCUSSION

3.1 Characterization of HAIX-NanoZr

Detailed material characterization has shown the metal oxide nanoparticles are equally dispersed at sub 50nm diameters throughout the resin bead diameter where intraparticle diffusion of trace ions, e.g., As(V) and As(III), is fastest (Figs. 1,2). Pre- and post- regeneration the metal oxide morphology remains amorphous and the weight content does not decrease.

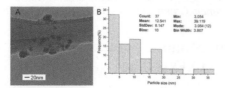

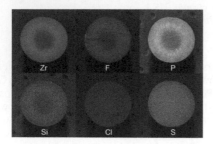

Figure 1.   TEM of HAIX-NanoZr and particle size distribution. In the scan, the mean diameter is 15±10 nm.

Figure 3.   A Ph.D. student from Lehigh University and a local engineer working to modify an HIX system (German, 2014).

Figure 2.   SEM-EDX of HAIX-NanoZr for zirconium, fluoride, phosphorus, silicon, chlorine, and sulfur.

3.2   Field performance

An HAIX-NanoZr system (Fig. 3) is not a magic solution for a community. Any piece of infrastructure requires investment of time and resources from the community in order for it to maintain effectiveness. A local water committee appoints a trusted individual to be the system caretaker to manage the daily system operation and maintenance and to collect the monthly water tariff. Monthly water testing of treated water samples is performed by an outside agency to confirm water meets WHO standards. Water delivery by a local rickshaw driver enables people to have water delivered to their doorstep. Training and process development has occurred through countless iterations and small steps in India (German *et al.*, 2014)

Material regeneration is required on an annual basis where the As or fluoride removed by HAIX-NanoZr is stripped from the material through treatment with alkali, followed by pH adjustment with acid. Regeneration, like all labor in the water ecosystem is reimbursed in an equitable manner.

HAIX-NanoFe systems have been in communities and providing safe water for up to 10 years of operation. HAIX-NanoZr systems have shown good performance over the last year of use. Today, HAIX-NanoZr production has reached up to several hundred liters per week, depending on demand; one system requires approx. 100L. A supply chain has been established through positive/negative experiences. Synthesis scale-up of HAIX-NanoZr in Kolkata and field-scale use throughout the Indian subcontinent has been increasing exponentially as part of several government contracts for municipality-scale As and fluoride treatment systems in collaboration with a commercialization partner, RiteWater (I), Pvt. Ltd.

4   CONCLUSIONS

HAIX-Nano has gone from idea to production in a short period of time to become a global solution to the As crisis. Key economic and groundwater conditions for self-sustainable microenterprise operations have been identified across multiple contexts and matched with qualitative results previously identified. Students, faculty, research labs, international partners, and a social business startup have all played important roles in transforming the global As crisis into an economic opportunity.

ACKNOWLEDGEMENTS

Numerous organizations have funded research and implementation including local, state, and national research grants, business competitions, private investments, and charitable donations. It is impossible to list them all in a short space without omitting several.

REFERENCES

Cumbal, L. and SenGupta, A.K. 2005. Arsenic removal using polymer-supported hydrated Fe(III) oxide nanoparticles: role of Donnan membrane effect. *Environ. Sci. Technol.* 39: 6508–6515.

German, M., Seingheng, H. and SenGupta, A.K. 2014. Mitigating arsenic crisis in the developing world: role of robust, reusable and selective hybrid anion exchanger (HAIX). *Sci. Total Environ.* 488–489: 547–553.

Padungthon, S., German, M., Wiriyathamcharoen, S. and SenGupta, A.K. 2015. Polymeric anion exchanger supported hydrated Zr(IV) oxide nanoparticles: a reusable hybrid sorbent for selective trace arsenic removal. *React. Funct. Polym.* 93: 84–94.

SenGupta, A.K. and Cumbal, L.H. 2007.Hybrid Anion Exchanger for Selective Removal of Contaminating Ligands. US Patent No. 7,291,578. 2007.

Arsenic Research and Global Sustainability – Bhattacharya, Vahter, Jarsjö, Kumpiene,
Ahmad, Sparrenbom, Jacks, Donselaar, Bundschuh & Naidu (Eds)
© *2016 Taylor & Francis Group, London, ISBN 978-1-138-02941-5*

Arsenic removal using "green" Nano Zero Valent Iron

A. Fiuza, A. Futuro, M.L. Dinis, M.C. Vila & R. Rios
Centro de Recursos Naturais e Ambiente (CERENA), Universidade do Porto,
Faculdade de Engenharia, Polo do Porto, Portugal

ABSTRACT: The research involved the production of different "green" nZVI using vegetable extracts, characterization of the texture of the nanoparticles using grain size distribution by laser diffraction, the study of the agglomeration and the role of surfactants, the characterization of the structure of nZVI by Scanning Electron Microscopy coupled with Energy Dispersive X-ray Spectroscopy and a study of the removal of As species (III and V) using the produced nZVI. The amorphous structure of the nZVI, coupled with a high content in carbon, probably of organic origin, creates a much lower capacity for As removal when compared to the one obtained by the usage of nZVI produced by the classical methods.

1 INTRODUCTION

Mining and processing plant operations generate liquid effluents that require treatment before recycling or partial discharge. In particular, liquid effluents resulting from gold mining and mineral processing plant may contain high levels of arsenic. Although there are several available technologies for arsenic removal from mining effluents, many of these technologies are expensive. Sorption by nanoparticles has been considered as a promising alternative due to the low cost, flexibility in design, simple operation and easy adaptation to in-situ remediation solutions.

The production of "green" nZVI (nano Zero Valent Iron) using vegetable extracts is known as a simple and low-cost method for producing iron nanoparticles. The application of these nanomaterials in the treatment of organic compounds, namely pharmaceuticals, has been deeply studied but its application in the removal of arsenic has not yet been assessed.

In our research we tested several methods for preparing nZVI. The texture of the produced nanoparticles was characterized by its grain size distribution. The time evolution of the texture, without and with surfactants was also assed. This data allowed for the selection of the most promising production methods. The nZVI was then characterized by Scanning Electron Microscopy coupled with Energy Dispersive X-ray spectroscopy (SEM/EDX). Finally, the removal of As species (III and V) was quantified by determining the "green" nZVI isotherms and comparing with nZI produced by classical chemical methods.

2 MATERIALS AND METHODS

2.1 Materials

Several vegetable extracts were considered for the production of the nZVI particles. The materials used in the experimental work were eucalyptus, oak and strawberry leafs, potatoes and carrot peels (Herlekar *et al.*, 2014, Machado *et al.*, 2013, Kharissova *et al.*, 2013). The peels were chosen by their antioxidant behavior resulting from their high content in phenolic compounds.

The following reagents were used without further purification, to obtain the reductive extract from the natural materials: iron (III) chloride hexahydrate (99%), Tween 20 and 80, Arsenic (III) oxide (99%), Arsenic (V) oxide and distilled water.

The mining effluent used in this study was collected in an abandoned mining area (Penedono) from former tailings and mine galleries.

2.2 Methods

The vegetables were dried at 50°C in an oven during 48 h and then milled using a cutting mill Retsch SM 2000. After, they were sieved and only the fraction with grain size below 4 mm was used. The extract was obtained by a simple procedure: the dried vegetables were weighed and transferred to Erlenmeyer flasks previously filled with 100 mL of distilled water warmed to a temperature of 80 °C. The contact time was 30 minutes with occasional stirring. The extract was then filtered and left to cool.

Chosen volumes of a solution 0.1 mol/L iron (III), made from ferric chloride hexahydrate, were added to the extracts and slightly mixed.

The reaction occurs immediately and the solution changes from an orange colour to a strong black, indicating the formation of iron particles. The same change in colour, has been reported by authors such as Pattanayak (Pattanayak *et al.*, 2013).

A previous characterization by visual inspection of the colour evidenced that some extracts were not suitable for the production of nZVI.

The remaining colored solutions were analyzed using a Laser Diffraction Malvern Mastersizer 2000 particle size analyser. The presence of nZVI was confirmed for extracts made with eucalyptus, oak and strawberry leafs. In all three cases the extract was made with 10 g of leafs in 100 mL of distilled water using a ratio of iron (III) solution/extract of 1:1.

The nZVIs produced from the eucalyptus leaves were selected for kinetic characterization studies. The role of two surfactants Tween 20 and Tween 80 in the stabilization of the nanoparticles, avoiding aggregation, was also assessed allowing to conclude that Tween 80 could partially avoid and delay the particle aggregation.

Two samples were submitted to Energy-Dispersive X-ray spectroscopy (EDX) and Scanning Electron Microscope (SEM) analysis using a JEOL JEM 6301F scanning electron microscope. The analysis evidenced the presence of iron and carbon and some minor elements such as P and Cl. The structure was not crystalline but evidenced an amorphous nature and carbon is probably of organic origin.

The second part of the experimental work was focused on the arsenic removal from the mining effluents. The solutions were prepared in different conditions: beakers with 100 mL of effluent each and different volumes of nZVIs, that were left to react for 8 hours with slight stirring. After, the solutions were filtered and analyzed by Hydride Generation Atomic Absorption Spectroscopy.

Industrial nZVIs were also tested for comparative purposes using an arsenic solution with a concentration of 500 µg/L.

3 RESULTS AND DISCUSSION

Isotherms were determined both for As(III) and As(V) using artificially prepared solutions with reagent grade reactants. The results showed that "green" ZVI was not effective in removing As(III) but it was relatively effective in the removal of As(V).

Another isotherm was determined using mining waste water with a total concentration in arsenic of 261 µg/L. The efficiencies of removal did not exceed 50% in any case, which was a disappointing result.

For comparative results a new set of tests using industrial nZVI (Toda Kogyo Corporation, Japan) was performed. The industrial nanoparticles were able to decrease the arsenic concentration continuously for 24 hours reaching a minimum value of 20 µg/L, corresponding to an efficiency of 92.3%

4 CONCLUSIONS

The obtained results showed that it is possible to produce nZVIs from natural materials. The best extractions were obtained with oak, strawberry and eucalyptus leafs. A short residence time of 5 minutes is sufficient for the reduction reaction to occur

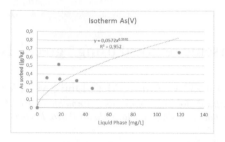

Figure 1. Isotherm As(V).

in full extent, producing the smaller nanoparticles. The conditions of production were optimized. The kinetics of aggregation was studied in detail. Over time nZVI particles have the tendency to aggregate and form clusters, reducing its surface area, decreasing its reactivity. The addition of Tween 80 stabilized the nanoparticles for longer periods of time. The removal of As(III) was not efficient while the removal of As(V) was slightly efficient. Calculation of the loading capacities are only possible if based on the assumption that all the iron that was added was transformed into nZVI. A final final conclusion is that "green" nZVI is not an efficient arsenic removal when compared to the crystalline nZVI produced by classical methods. The reason for such behavior probably lies in its amorphous structure with incorporation of organic carbon, as it was observed by EDX analysis and SEM images

ACKNOWLEDGEMENTS

This work is a part of a research project supported by the Portuguese National Funding Agency for Science, Research and Technology (FCT) within the project "Tools for sustainable gold mining in EU (SUSMIN)", a transnational cooperation project within the Network on the Industrial Handling of Raw Materials for European Industries.

REFERENCES

Herlekar, M. Siddhivinayak, B. & Rakesh, K. 2014. Plant-mediated green synthesis of iron nanoparticles. *J. Nanoparticles* Volume 2014 (2014), Article ID 140614, pp 1–9 pages.

Machado, S., Pinto, S.L., Grosso, J.P., Nouws, H.P., Albergaria, J.T., Delerue-Matos C. 2013. Green production of zero-valent iron nanoparticles using tree leaf extracts. *Sci. Total Environ.* 445–446 (2013):1–8.

Kharissova, O.V., Dias, H.V.R., Kharisov, B.I., Pérez, B.O. Pérez, V.M.J. 2013. The greener synthesis of nanoparticles. Trends In Biotechnol. 31: 240–248.

Pattanayak, M., Nayak, P.L. 2013. Green synthesis and characterization of zero valent iron nanoparticles from the leaf extract of *Azadirachta indica (Neem)*. *World Journal of Nano Science & Technology* 2(1): 06–09.

Arsenic Research and Global Sustainability – Bhattacharya, Vahter, Jarsjö, Kumpiene,
Ahmad, Sparrenbom, Jacks, Donselaar, Bundschuh & Naidu (Eds)
© 2016 Taylor & Francis Group, London, ISBN 978-1-138-02941-5

Synthesis of Fe-Cu bimetallic nanoparticles: Effect of percentage of Cu in the removal of As present in aqueous matrices

P. Sepúlveda[1], D. Muñoz Lira[1], J. Suazo[1], N. Arancibia-Miranda[1] & M.A. Rubio[2]

[1]*Department of Materials Chemistry, Faculty of Chemistry and Biology, University of Santiago, Chile*
[2]*Department of Environment, Faculty of Chemistry and Biology Sciences, University of Santiago, Chile*

ABSTRACT: In this work, three Fe-Cu bimetallic nanomaterials with different percentages of Cu were synthesised, to evaluate the effect generated by Cu in the removal of arsenate present in aqueous matrices. The sorption capacities were studied developing sorption kinetic, obtained that nanoscale zero valent iron showed the highest removal followed by bimetallic nanoparticles with a ratio of 0.9:0.1 Fe-Cu. Through of characterization technique, SEM, was possible determine that all materials studied have a spherical morphology with formation of chains.

1 INTRODUCTION

Arsenic (As) is present in surface and groundwater principally in organic and inorganic form, like arsenite (As III) and arsenate (As V) (Manning et al., 2002). Arsenic accumulates in the earth's crust by natural processes, such as volcanic erosions and, geochemical reactions and anthropogenic processes, like mining and agriculture activity, latter having caused an exponential increase of pollution in aqueous matrices (Mohan & Pittman, 2007)

Arsenic is considered one of the most toxic pollutants for human beings, due it is cytotoxic and genotoxic (Joseph et al., 2015). Arsenic causes different diseases when incorporated through water intake, such as diabetes, cancer and lung diseases (Luther et al., 2012).

Nanoscale zero valent iron (nZVI), has been widely used for the removal of this metalloid due to its high surface area and number of specific adsorption sites (Kanel et al., 2005). Nonetheless, the sorption capacities of this material are strongly diminished by agglomeration and oxidation processes (Luther et al., 2012). Different materials have been used to solve this problem, which serve as a support of nZVI. Impregnation techniques have had strong positive effects to date. Bimetallic nanoparticles composed of corrosive Fe and a precious metal (e.g., Pt, Pd, Ni or Cu), are combined to catalyze contaminant reduction and to retard the reoxidation process of nZVI. (Fu et al., 2014).

The aim of this work was to synthesize and characterize bimetallic nanomaterials Fe-Cu, evaluating the effect of the percentage of Cu in the removal of As(V) present in aqueous matrices.

2 METHODS/EXPERIMENTAL

2.1 Synthesis and characterization of Fe-Cu bimetallic nanoparticles

The synthesis of nanomaterials was performed following the experimental procedure proposed by Wang & Zhang (1997), which established a 1 M: 1.6 M, metal: reducing agent ratio. The change to the proposed synthesis was performed considering the simultaneous reduction of both metals, using $NaBH_4$ as a reducing agent.

The bimetallic nanoparticles were synthesized with: 0.9:0.1 (B1) 0.75:0.25 (B2) and 0.5:0.5 (B3), mass proportions of Fe-Cu. The three synthetic materials (B1, B2 and B3) were characterized by ZEISS EVO/MA10 Scanning Electron Microscopy (SEM).

2.2 Sorption kinetics

Sorption kinetics were carried out in batch systems working with 50 mg of nanomaterial and a solution of 200 mg/L As V (As_2O_5), 0.01 M NaCl (background electrolyte) and pH 7 (adjusted with NaOH and 0.1 M HCl). Each sample was stirred for 0, 5, 15, 40, 60, 120 and 180 minutes, subsequently centrifuged for 10 minutes at 6000 rpm for solid-liquid separation. Finally, the samples were filtered and stored before analysis by Atomic Absorbance Spectrometry (AAS).

3 RESULTS AND DISCUSSION

3.1 Characterization of pristine bimetallic nanoparticles

The images of nanoparticles a) nZVI, b) B1, c) B2 and d) B3 obtained by SEM reveal spherical particle morphology in the four materials (Fig. 1).

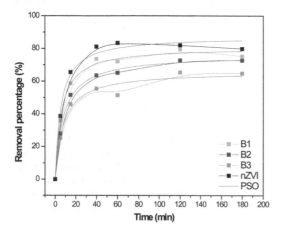

Figure 1. SEM images of nZVI (a), B1 (b), B2 (c) y B3 (d).

Figure 2. Sorption kinetics of As(V) using bimetallic (B1, B2, B3) and monometalic (nZVI) nanomaterials.

In the case of nZVI (Figure 1a), SEM images showed the formation of characteristic agglomerate of this magnetic material. For the cases of bimetallic nanoparticles, the formation of branches (chains) of greater length was observed. Similarly, it was possible to determine the approximate particle size, for nZVI (82 nm), B1 (64.0 nm), B2 (53.4 nm) and for B3 could not, because it presented two types of particles of different size (d).

3.2 Sorption kinetic

Figure 2 shows the sorption kinetics of As V using adsorbent materials B1, B2, B3 and nZVI, noting that the saturation time of the four nanomaterials was reached at 60 minutes, with a maximum removal capacity (qe) 65.5% (nZVI), 63.3% (B1), 57.9% (B2) and 52.1% (B3).

The data were mathematically fitted using kinetic models of Pseudo First (PFO) and Second Order (PSO), with the latter producing better fits. The PSO computed h parameter, which reports the sorption rate as time approaches zero, showed that bimetallic materials were approximately 2 times faster compared to nZVI.

Increasing Cu loading in the bimetallic nano-particles suppressed maximum As sorption capacities (Fig. 2). This behaviour was attributed to an interference effect of Cu on the removal processes of As by nZVI. A ratio of 0.9: 0.1 Fe-Cu (B1) is thus suitable for enhancing the nZVI sorption characteristics by preventing agglomeration and oxidative processes.

4 CONCLUSIONS

The study indicates that the percentage of Cu in the Fe-Cu bimetallic nano-particles is a parameter that positively affects the capacity of As removal, by nZVI.

At present, we are performing theoretical studies for determine bimetallic nanoparticles structure (core-shell heterostructure, intermetallic or multi-shell) for further applications for arsenic removal.

ACKNOWLEDGEMENTS

This research was supported by CORFO 12IDL2-16251 and Basal Project CEDENNA FB-0807.

REFERENCES

Fu, F., Dionysious, D.D. and Hong, L. 2014. The use of zero- valent iron for groundwater remediation and wastewater treatment: A review. J. Hazard. Mater. 267: 194–205.

Joseph, T., Dubey, B. and McBean, E.A. 2015. A critical review of arsenic exposures for Bangladeshi adults. Sci. Total Environ. 527–528: 540–551.

Kanel, S., Manning, B.A. and Charlet, L. 2005. Removal of arsenic(III) from groundwater by nanoscale zero-valent iron. Environ. Sci. Technol. 39: 1291–1298.

Luther, S., Borgfeld, N., Kim, J., Parsons, J.G. and Microche, J. 2012. Removal of arsenic from aqueous solution: A study of the effects of pH and interfering ions using iron oxide nanomaterials. Microchem. J. 101: 30–36.

Manning, B.A., Hunt, M.L., Amrhein, C. and Yarmoff, J.A. 2002. Arsenic (III) and arsenic (V) Reactions with zerovalent iron corrosion products. Environ. Sci. Technol. 36: 5456–5461.

Mohan, D. and Pittman Jr., C.U. 2007. Arsenic removal from water/wastewater using adsorbents—A critical review. J. Hazard. Mater. 142: 1–53.

Arsenic Research and Global Sustainability – Bhattacharya, Vahter, Jarsjö, Kumpiene,
Ahmad, Sparrenbom, Jacks, Donselaar, Bundschuh & Naidu (Eds)
© 2016 Taylor & Francis Group, London, ISBN 978-1-138-02941-5

A new meso-porous ceramics material for arsenic removal from drinking waters

L.J. Dong[1], H. Ishii[2], D. Wang[3], B.Y.G. Sun[3]
[1]*Hawaii Institute of Interdisciplinary Research, USA*
[2]*University of Hawaii, Hawaii, USA*
[3]*School of Public Health, China Medical University, Shenyang, P.R. China*

ABSTRACT: It is important to consider most problems and their priority in addressing the arsenic removal from field groundwater. A new material has been developed that integrates ceramics, pH buffer, nanotechnology and zero valent iron technology together to produce a meso-porous granular material, in which 40–60 nm sub-holes with high arsenic absorption capacity reside inside microscopic layers. In the column test, it can produce effluent water less than 10 µg/L when test synthesis' Arsenic (V) and Arsenic (III) water in laboratory with 50 µg/L and 300 µg/L influent without pH adjustments. The select adsorption capacity is also confirmed by field tests with influents of groundwater, spring water and industry mining-affected surface water, in which the water has arsenic concen-tration 960 µg/L, 750 µg/L and 2050 µg/L, respectively. The TCLP test shows that spent media will not leach arsenic back to the en-vironment. The material may provide industrialized solutions to arsenic pollution problems.

1 INTRODUCTION

Naturally occurring Arsenic concentration in groundwater varies from a few µg/L to as high as 10 mg/L (Li *et al.*, 2013) in different areas, but in most cases of arsenic pollution, like Bangladesh, West Bangle, India and Datong Basin, China, the arsenic concentration is on average 300 µg/L in the form of both arsenite and arsenate. Since the reverse osmosis technology has poor performance in arsenite removal (Walker *et al.*, 2008) and generates high concentration of wastewater, much of new technology development efforts has been focused on adsorptive media (Dong *et al.*, 2009), especially in the Point of Use water filtration systems. However, in order to solve arsenic pollution, there are many obstacles from material limitations, industrial scaleup, waste management, users education, operation and maintenance, cost effectiveness and even social habits. In 2005, the US National Academy of Engineering summarized five important criteria in the Grainger Challenge documents.

According to these criteria, other review papers and our experiences, we summarize these problems new technology must address in the following priorities: 1) Remove both arsenic (V) and arsenic (III) with efficiency rate of at least >97% to treat average 300 µg/L arsenic; 2) Spent waste must be safely handled and disposed of to avoid the risk of poisoning, especially to avoid the phenomenon of "accumulation and high concentration leach ate" in the operation process; 3) Address compete ions such as NO_3^-, $H2PO_4^-$, SO_4^{2-} and humic acid that are present in the groundwater; 4) Achieve a short contact time of less than 20 minutes that will avoid large containers and other engineering cost burdens and wear out the user's patience; 5) Filtration devices should achieve filtration via non-chemical solutions such as pH up and down adjustments in low-educated societies; 6) Media should avoid secondary problems such as high concentrations of Fe Al, Cu, and Zn etc that are within parameters of safe drinking water standard(s); 7) Filtration should achieve simple Operation and Maintenances because most arsenic pol-lution affected areas are poor societies with low-education residents; 8) New material production should be easily scalable to achieve cost effective-ness, and main raw material should sustainably procurable; 9) Filtration needs to balance the effluent water's safety, health and taste between long-term application deliverability and short-term spreads; 10) Simultaneously solving other accompanying problems such as high concentration of fluoride frequently present in natural groundwater, and heavy metals pollution in groundwater that is affected by mining industry activities.

Iron Coated Pottery Granular (ICPG) material can achieve both arsenite and arsenate removal efficiency of 98% in column tested (15 minutes contact time) by US NAE Grainger Challenge program (Dong *et al.*, 2009, Habuda-Stanić & Nujic, 2015). However, since the ICPG media functioned mostly through micro-scale porous holes and activated ad-sorptive points were mostly on the surface, the ad-sorptive capacity was low and contact time was still too long to address the field pollution waters. With the consideration of 10 challenges listed above, a new material has been redesigned with mesoporous (<50 nm) holes to increase the removal efficiency and capacity. This paper investigates the laboratory-synthesized water and field water performance and related engineering criteria.

2 METHODS/EXPERIMENTAL

2.1 Material preparation

Diatomaceous earth powders from bauxite mining site treated by desilicication was grinded into 1200 standard mesh, and mixed with 5% of starch as carbon sources. The mixture powder was granulated in size of 0.5mm~1.0mm raw pottery by adding 12%–15% pure water. The raw pottery granular was fired in 500° C for three hours with a temperature increase rate of 2 degrees/minute. The fired media was sub-merged in 2% $FeSO_4$ solution for 15 minutes and naturally leach ate the water, and then put into 2% of $NaBH_4$ for 30 minutes for zero valent iron crystallization to occur inside the porous. The ZVI solution treated media was fired again in the oven at 480–500°C for 3 hours with protection of 99.999% nitrogen during the entire firing process. When treated pottery was down to room temperature, it was stored and ready to use for Batch tests and column tests.

2.2 Batch test and colunm test

The adsorption capacity for arsenite and arsenate were tested by using the Freundlich equation under natural pH. While the equation is empirical, it is nonetheless widely used and has been found to describe adequately the adsorption process in dilute solution.

The media was packaged in two types of water filter cartridges; they were tested under gravity flow for both laboratory water and field water tests. Additional columns with 200 g media to determine the removal efficiency vs. empty bed contact time with 50 µg/L and 350 µg/L solutions.

3 RESULTS AND DISCUSSION

3.1 Adsorption capacity under different pH

The new material has similar adsorption capacity for arsenite and arsenate 8.5 mg/g- 9.2mg/g. But the at low pH, arsenate adsorption increase; at high pH, arsenite adsorption capacity increased.

3.2 Removal efficiency and the contact time

The arsenic removal efficiency is affected by initial influent concentration and other parameters; the

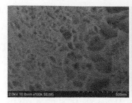

Figure 1. The SEM image comparison of before (left) and after (right) used, pores are not visible. Structure instead appears platy.

column test data shows their relation ships in treatment of 50 µg/L influent and 320 µg/L. It shows that 90 seconds and 75 seconds are required to achieve removal rate >97%. The field tests in different location showed the similar parameters.

3.3 TCLP, SEM and TEM analysis

The TCLP tresults of the exhausted media shows that arsenic leach rates are at non-detectable levels. SEM photos show the new material porous inside the media are in the range of 20–70 nm and filled with arsenic after adsorption shown in Figure 1.

CONCLUSIONS

The new media filtration performance is capable of addressing most challenges of arsenic pollution, especially its manufacturing process that can be easily adapted to industrial scale and compatible with industry standard products. More field tests are needed to prove its sustainability.

REFERENCES

Dong, L., Zinin, P.V., Cowen, J.P., Ming, L.C. 2009. Iron coated pottery granules for arsenic removal from drinking water. *J. Hazard. Mater*. 168: 626–632.

Habuda-Stanić, M., Nujic, M. 2015. Arsenic removal by nano-particles: a review. *Environ. Sci. Pollut. Res*. 22: 8094–8123.

Jadhav, S.V., Bringas, E., Yadav, G.D., Rathod, V.K., Ortiz, I., Marathe, K.V. 2015. Arsenic and fluoride contaminated groundwaters: A review of current technologies for contaminants removal. *J. Environ. Manag*.162: 306-325.

Li, S., Wang, M., Yang, Q., Wang, H., Zhu, J., Zheng, B., Zheng, Y. 2013. Enrichment of a rsenic in surface water, stream sediments and soils in Tibet, *J. Geochem. Explor*. 135: 104–116.

Walker, M., Seiler, J.R., Meinert, M. 2008. Effectiveness of household reverse osmosis systems in a Western U.S. region with high arsenic in groundwater. *Sci. Total Environ*. 389, 245-252.

Arsenic Research and Global Sustainability – Bhattacharya, Vahter, Jarsjö, Kumpiene, Ahmad, Sparrenbom, Jacks, Donselaar, Bundschuh & Naidu (Eds)
© *2016 Taylor & Francis Group, London, ISBN 978-1-138-02941-5*

Removal of As(V) from drinking water by adsorption on iron nanoparticles supported on bone char

D.E. Villela-Martinez, R. Leyva-Ramos & J. Mendoza-Barron
Centro de Investigación y Estudios de Posgrado, Facultad de Ciencias Químicas, Universidad Autónoma de San Luis Potosí, San Luis Potosí, México

ABSTRACT: The adsorption of arsenic (V) from drinking water on Modified Bone Char (MBC) was investigated in this work. The bone Char (CH) was modified by supporting zero-valent Iron Nanoparticles (INP) on its surface. The surface morphology of MBC was examined by SEM, and it was observed the formation of a coating on the surface of the BC particles. The TEM analysis showed clearly the presence of INPs in the MBC. The adsorption capacity of MBC towards As(V) was dependent on the synthesis conditions and the adsorption capacity of the optimal MBC (MBCOp) towards As(V) was enhanced more than 7 times than that of BC. The effects of pH and temperature on the adsorption capacity as well as desorption of As(V) adsorbed on MBCOp were also studied in detail. The removal of As(V) from polluted drinking water was performed and the removal of As(V) was carried out very efficiently.

1 INTRODUCTION

Arsenic is a major public health hazard when it is present in contaminated groundwater. The main sources of exposure to As(V) are water for human consumption, irrigation of crops and food preparation. Due to its toxicity, the concentration of As(V) in drinking water is being regulated according to official standards of each country and the maximum permissible limit of arsenic is currently set at 10 μg/L by the World Health Organization. Arsenic has been classified as a carcinogen because the presence of arsenic in drinking water increases the risk of skin, lung, kidney and bladder cancer.

Several technologies have been developed for the removal of arsenic from drinking water and reported that the As(V) can be efficiently removed by adsorption, ion exchange, precipitation, coagulation and filtration, reverse osmosis and electrodialysis (Mendoza-Barron et al., 2011). Adsorption is defined as the preferential accumulation of a substance or anion on the surface of a porous solid (adsorbent), and it is a surface phenomenon because the adsorbate accumulates on the adsorbent surface through interaction between the surface of the adsorbent and adsorbate. Adsorption is one of the most efficient separation processes for removing trace concentrations of As(V) from aqueous solution.

BC has been tested for removing As(V) from water solutions, but exhibits low adsorption capacity. The BC has to be modified to enhance its adsorption capacity towards As(V).

2 EXPERIMENTAL METHODS

2.1 Procedure for modification of BC

The BC was modified by supporting Iron Nanoparticles (INPs) on its surface, and the INPs were synthesized by mixing 0.045 M $FeCl_3$ and 0.25 M $NaBH_4$ solutions accordingly to the procedure reported by Li et al. (2006). The novel adsorbent was designated as MBC. The preparation conditions were optimized to maximize the adsorption capacity. The ratio of the mass of BC to the volume of $FeCl_3$ and $NaBH_4$ solution. This ratio was 1/5, 1/20 and 1/30 g/mL and the MBCs were designated as MBC1, MBC2 and MBCOp, respectively. The latter had the maximum capacity for adsorbing As(V).

2.2 Characterization of BC and MBC

The concentrations of active sites present on the surface of BC and MBCs were determined using the titration method proposed by Boehm (Mendoza-Barron et al., 2011), the Point of Zero Charge (PZC) and surface charge distribution of the adsorbent was evaluated using the method described elsewhere (Mendoza-Barron et al., 2011). The textural properties (surface area, pore volume and mean pore diameter) were determined using a surface area and porosimeter analyzer, Micromeritics, ASAP model 2020. The surface morphology of MBC was examined with a scanning electron microscope, Philips, model XL–30. The NPHs in MCB were observed with a transmission electron microscope, JEOL, model JEM 1230.

2.3 Determination of As(V) concentration

The concentration of As(V) in an aqueous solution was determined by an Atomic Fluorescence Spectrophotometry method. The emission intensity was determined in an atomic fluorescence spectrophotometer, Millennium Excalibur, PSA 10.055.

2.4 Adsorption equilibrium data

The experimental adsorption equilibrium data were obtained in a batch adsorber, and the procedure was very similar to that described by Mendoza-Barron et al. (2011). The adsorption experiments were carried using a solution of As(V) prepared using deionized water or water samples from wells polluted with As(V).

3 RESULTS AND DISCUSSION

3.1 Characterization of MCH

The textural characterization of MBCOp showed that its surface area and pore diameter were 91 m^2/g, and 10.3 nm. The quantification of the total acidic and basic sites revealed that the surface of MBCOp is basic. The BCM was examined by SEM, and a coating was formed on the surface of the particles of BC. The TEM analysis of MBCOp confirmed presence of INPs with diameters between 2 and 7 nm.

3.2 Effect of the modification on the adsorption capacity

The adsorption isotherms of As(V) on BC, MBC1, MBC2 and MBCOp are depicted in Figure 1, and the adsorption capacities towards As(V) decreased in the following order: MBCOp > MBC1 > MBC2 > BC. At a concentration of As(V) at equilibrium of 300 µg/L, the adsorption capacities of MBCOp, MBC1, MBC2 and BC were 412.5, 248.3, 93.5 and 38.6 µg/g, respectively. The modification of the BC enhanced the adsorption capacity up to seven times. The adsorption capacity of the MBC depends on the synthesis conditions, and the optimal conditions were those used for preparing MBCOp.

3.3 Effect of solution ph and temperature on adsorption capacity of mbcop

The adsorption capacity of MBCOp was dependent on the solution pH and temperature. The adsorption capacity of MBCOp towards As(V) was decreased by raising the solution pH, this behavior was attributed to the electrostatic interaction between the surface of the MBCOp and the arsenate anion present

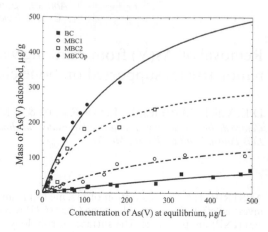

Figure 1. Adsorption isotherms of As(V) from aqueous solution on BC, MBC1, MBC2 and MBCOp at T = 25 °C.

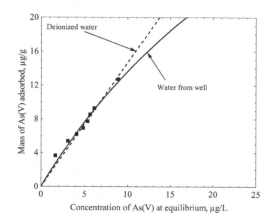

Figure 2. Adsorption isotherms of As(V) on MBCOp using As(V) solutions prepared using deionized water and water samples from a well naturally polluted with arsenic.

in the solution. The adsorption of As(V) on MBCOp was endothermic because its adsorption capacity augmented approximately 3.3 times when the temperature was increased from 15 °C to 35 °C.

3.4 Removal of As(V) from water for human consumption

The MBCOp was tested for removing As(V) from a sample extracted from a well and used for human consumption. The concentration of As(V) in this sample was 24.8 µg/L. The adsorption isotherm of As(V) on MBCOp was obtained using this water sample and is shown in Figure 2. The adsorption capacity of MBCOp towards As(V) was not affected by the presence of other anions in the water for human consumption, revealing that the adsorption of As(V) on MBCOp was very selective. The percentage removal of As(V) from several samples of drinking water ranged from 39.9 to 65.1%.

4 CONCLUSIONS

The modification of BC by supporting INPs enhanced considerably the adsorption capacity of MBCOp. The As(V) present in drinking water samples collected from wells was effectively removed by adsorption on MBCOp.

ACKNOWLEDGEMENTS

This work was funded by Consejo Nacional de Ciencia y Tecnologia, CONACyT, Mexico, through grant No.: PDCPN-2014-248395.

REFERENCES

Li, X.Q., Elliott, D.W. and Zhang, W.X. 2006. Zero-valent iron nanoparticles for abatement of environmental pollutants: materials and engineering aspects. *Crit. Rev. Solid State Mater. Sci.* 31: 111–122.

Mendoza-Barrón, J., Jacobo-Azuara, A., Leyva-Ramos, R., Berber-Mendoza, M.S., Guerrero-Coronado, R.M., Fuentes-Rubio, L. and Martínez-Rosales, J.M. 2011 Adsorption of arsenic(V) from a water solution onto a surfactant-modified zeolite. *Adsorption* 17(3): 489–496.

Arsenic Research and Global Sustainability – Bhattacharya, Vahter, Jarsjö, Kumpiene,
Ahmad, Sparrenbom, Jacks, Donselaar, Bundschuh & Naidu (Eds)
© 2016 Taylor & Francis Group, London, ISBN 978-1-138-02941-5

Montmorillonite-supported nZVI: A novel adsorbent for arsenic removal from aqueous solution

S. Bhowmick[1], D. Chatterjee[1], P. Mondal[2], Sudipta Chakraborty[3] & M. Iglesias[4]

[1]*Department of Chemistry, University of Kalyani, Kalyani, West Bengal, India*
[2]*Ceramic Membrane Division, CSIR-Central Glass and Ceramic Research Institute, Kolkata, India*
[3]*Department of Chemistry, Kanchrapara College, Kanchrapara, West Bengal, India*
[4]*Department of Chemistry, University of Girona, Campus de Montilivi, Girona, Spain*

ABSTRACT: A novel adsorbent consisting of montmorillonite supported nanoscale Zero-Valent Iron (Mt-nZVI) was synthesized to remove inorganic Arsenic (As) from aqueous solution. Characterization of the synthesized clay supported material demonstrated core shell Fe(0) structure with an outer oxide/hydroxide shell. Batch experiments showed that the adsorption kinetics followed pseudo-second order rate equation and the adsorption affinity of Mt-nZVI for both As(V) and As(III) decreased at pH above 9. Competing anion PO_4^{-3} showed strong inhibitory effect on adsorption while $SO4^{-2}$, HCO_3^{-1} and NO_3^{-1} did not show significant effect on As adsorption. XPS analyses of the spent Mt-nZVI indicated probable surface catalyzed oxidation of As(III) to As(V). Mt-ZVI was also potentially regenerated using NaOH solution. This study illustrates great potential of Mt-nZVI for reducing As from drinking water.

1 INTRODUCTION

Arsenic (As) is a toxic and carcinogenic element and is categorized as the first priority toxic substance by WHO. Chronic exposure to As via drinking water is the major cause of As poisoning and currently millions of people are at risk from consuming As contaminated water. Nanoscale Zero-Valent Iron (nZVI) has been effective to reduce the concentration of As in aqueous solution. However, tiny particle size along with the lack of durability and mechanical strength generally restricts the direct use of nZVI (Cumbal *et al.*, 2003). Hence, loading nZVI onto supporting material is generally the most frequently adopted methodology to decipher the problem. Clay-nZVI is known to remove heavy metals from aqueous solutions, but a very little is known for As removal by such materials. The present study aims to synthesize Montmorillonite-supported nanoscale Zero-Valent (Mt-nZVI) and to test the adsorption capability towards inorganic As in aqueous solution. A probable mechanism of As removal and method for regeneration of Mt-nZVI after repetitive use have also been proposed.

2 EXPERIMENTAL

2.1 *Materials and methods*

The natural Na-Montmorillonite (Mt) (Source Clay Mineral Repository, University of Missouri, Columbia, USA) was used without any pre-treatment with chemicals. Stock solutions of As(V) and As(III) were prepared by dissolving appropriate amounts of $Na_2HAsO_4.7H_2O$ (Merck) and $(NaAsO_2)$ (Fluka) respectively in Milli-Q water.

2.2 *Characterization and adsorption experiments*

The materials were characterized by using SEM, BET and XPS analysis. The adsorption experiments were performed in batch mode at room temperature ($22 \pm 1°C$) under aerobic conditions. Mt-nZVI dose of 1 g/L and an aqueous solution of either As(III) and As(V) of 5 mg/L was taken for the kinetic experiment with initial pH 7.00. The effects of oxy-anion (PO_4^{3-}, SO_4^{2-}, HCO_3^- and NO_3^-) on As adsorption and regeneration of Mt-nZVI was also investigated. All the experiments were performed in triplicates.

3 RESULTS AND DISCUSSION

3.1 *Characterization and adsorption kinetics*

Calculated BET value shows that the surface area was enhanced by incorporating nZVI on the montmorillonite (from 14.85 m²/g of Mt to 36.97 m²/g of Mt-nZVI), thereby increasing the reactivity. SEM images show the agglomeration of the nZVI decreases with increase of dispersion and mechanical strength when Mt was used as a supporting material (Fig. 1). An XPS spectrum indicates that Mt-nZVI consisted primarily of Fe and O, while a small amount of C, Cl, Si and Al also present (Fig. 2a). The Fe2p photoelectrons shows distinct peaks corresponding to the binding energy of oxidized iron [Fe(III)] (ca. 711 eV; 725 eV) while, a small shoulder at 707.2 eV indicate the peak of zero-valent iron (Fig. 2b).

Therefore, the above analyses conclusively demonstrate that the Mt loaded with nZVI that has an envelope of Fe oxide (Fe$_2$O$_3$, Fe$_3$O$_4$ and FeOOH).

For both As(III) and As(V), the initial fast uptake of As was followed by a much slower step while the removal of As(III) and As(V) also decreased with decreasing concentration of Mt-nZVI. The data obtained from adsorption kinetics were fitted well with pseudo-second order kinetic model:

$$t/q_t = 1/k_2 q_e^2 + t/q_e \qquad (1)$$

where, q_e and q_t are the amount of As adsorbed (mg g^{-1}) at equilibrium and at time t (h) respectively and k_2 (g/mg/h) represents the pseudo-second-order rate constant.

The effect of pH on As adsorption were performed in the pH range from 4 to 10 for both As(III) and As(V). Mt-nZVI showed higher affinity towards As(III) than As(V) and the adsorbed amount remained nearly constant from pH 4–8.5 for both As(III) and As(V) while decreasing sharply at higher pH.

3.2 Mechanism of arsenic removal

To delineate the removal mechanism, XPS analysis of the reacted Mt-nZVI was performed to charac-

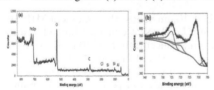

Figure 1. SEM images of (a) nZVI; (b) Mt-nZVI.

Figure 2. XPS spectra of (a) survey scan; (b) Fe2p region.

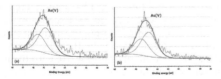

Figure 3. As3d XPS spectra of Mt-nZVI after reaction with 100 mg L^{-1} (a) As(V) and (b) As(III) at pH 7.0.

terize the As species after reaction. XPS analysis for both As(III) and As(V) reacted Mt-nZVI showed only the presence of doublet As(V) (Fig. 3). This suggest complete oxidation of As(III) by Mt-nZVI surface. The Fe^{2+} formed from oxidation of Fe(0) reacts with water and O$_2$ to produce oxidizing intermediate in solution such as H$_2$O$_2$, HO$^\circ$, O$_2^{\circ-}$. Such oxidizing intermediate are responsible for the oxidation of As(III) to As(V) following Fenton like reaction mechanism (Katsoyiannis et al., 2008). Thus our experimental finding shows that As(V) and As(III) in aqueous solution forms surface complexes with Mt-nZVI and the As(III) are consecutively oxidized to As(V) by oxidizing intermediate that are produced *in- situ* from the oxidation of Fe(0).

3.3 Effect of co-existing oxyanions and reusability

Our results suggest that NO$_3^-$, SO$_4^{2-}$, HCO$_3^-$ have insignificant effect on adsorption for both the species of As on Mt-nZVI. However, the adsorption of As decreased with the increase of PO$_4^{3-}$ concentration.

The spent Mt-nZVI was regenerated by shaking the As loaded adsorbent with 0.1 M NaOH solution for 4 h. Around 50% of the total adsorbed As was desorbed within 4 h. The material was stable even after 5 repeated adsorption-desorption cycles.

4 CONCLUSIONS

The synthesized adsorbent prepared by loading nZVI onto Mt were effective in removal of both As(III) and As(V) from aqueous solution with fast kinetics and the rate of adsorption followed pseudo-second order with slightly higher rate constant (K_{ads}) for As(III) than As(V). While reduction of As(V) was not observed in our experimental condition, complete oxidation of As(III) to As(V) took place in presence of Mt-nZVI, as evidenced from XPS analyses. Phosphate in solution had an inhibitory effect on the adsorption of As, suggesting competition for adsorption site. Thus we hypothesize that the Mt-nZVI reported in this work are relevant for preparing low cost adsorbent to remediate inorganic As from contaminated water.

REFERENCES

Cumbal, L., Greenleaf, J., Leun, D. and Sen Gupta, A.K. 2003. Polymer supported inorganic nanoparticles: characterization and environmental applications. *React. Funct. Polym.* 54 (1–3): 167–180.
Katsoyiannis, I.A., Ruettimann, T. and Hug, S.J. 2008. pH dependence of fenton reagent generation and As(III) oxidation and removal by corrosion of zero valent iron in aerated water. *Environ. Sci. Technol.* 42 (19): 7424–7430.

Arsenic Research and Global Sustainability – Bhattacharya, Vahter, Jarsjö, Kumpiene,
Ahmad, Sparrenbom, Jacks, Donselaar, Bundschuh & Naidu (Eds)
© 2016 Taylor & Francis Group, London, ISBN 978-1-138-02941-5

Encapsulation of Fe_3O_4 nanoparticles in porous materials for removal of arsenic from water

G. Du Laing[1], K. Folens[1], N.R. Nicomel[1], K. Leus[2] & P. Van Der Voort[2]
[1]*Laboratory of Analytical Chemistry and Applied Ecochemistry (ECOCHEM), Ghent University, Ghent, Belgium*
[2]*Center for Ordered Materials, Organometallics and Catalysis (COMOC), Ghent University, Ghent, Belgium*

ABSTRACT: We developed and tested novel composite materials based on the encapsulation of Fe_3O_4 nanoparticles (NPs) in Metal Organic Frameworks (MOFs) and Covalent Organic Frameworks (COFs) for removal of arsenic (As) species from water. The materials exhibit superior performance for removal of arsenic from arsenic-contaminated water. Aside from good removal efficiencies, the materials also have the advantages of not requiring pre-oxidation and pH adjustment prior to As removal.

1 INTRODUCTION

Arsenic (As) contamination in groundwater and surface water has emerged as a major environmental and human health problem worldwide, requiring an effective remediation technology. Among the available technologies for As removal from aqueous media, adsorption is the most widely studied because it is efficient, cost-effective, easy to operate, environment-friendly, and easy to disseminate. Furthermore, most adsorption processes, being reversible in nature, allow the regeneration of adsorbents through desorption processes for potential reuse. However, most adsorbents that have been studied for As adsorption have a relatively low capacity and cannot be successfully regenerated. This results in a need for frequent replacement of the adsorbents, which would be uneconomical. In this study, Fe_3O_4 nanoparticles were incorporated into the matrix of a chromium-based metal organic framework (MIL-101(Cr)) and a covalent organic framework (CTF-1) to synthesize $Fe_3O_4@MIL-101(Cr)$ and $Fe_3O_4@$CTF-1. These novel composite materials were subsequently tested as adsorbents for the removal of arsenite an arsenate from water, and their performance was compared to the performance of pure MIL-101(Cr), CTF-1 and the zirconium-based MOF UiO-66, which is also stable in water.

2 METHODS/EXPERIMENTAL

UiO-66 and CTF-1 were synthesized based on the *in situ* procedures reported by Shearer *et al.* (2014) and Kuhn *et al.* (2008) with some modifications. MIL-101(Cr) was synthesized using a hydrothermal method using $Cr(NO_3)_3·9H_2O$ and H_2BDC as precursors. Fe_2O_3 nanoparticles (NPs) were encapsulated in the MOFs and COF using the bottle-around-the-ship or *de novo* approach (Hu, 2015).

All of these materials, as well as pure Fe_2O_3 NPs were subjected to adsorption tests, performed in 12-mL centrifuge tubes at a shaking speed of 30 rpm at RT. The amount of UiO-66, MIL-101(Cr), and $Fe_3O_4@MIL-101(Cr)$ used was 38 mg. Compared to $Fe_3O_4@MIL-101(Cr)$, the relative Fe loading is lower for $Fe_3O_4@$CTF-1, thus, the amount of $Fe_3O_4@$CTF-1 to be used in the experiment (47 mg) was calculated relative to that of $Fe_3O_4@MIL-101-$(Cr) to obtain a similar Fe_3O_4 dose. Ten mL of solution containing arsenite (As(III)) or arsenate (As(V)) was added to the materials and the suspensions were shaken for different time intervals ranging from 10 minutes to 24 hours. After selected time periods, the tubes were centrifuged at 4500 rpm for 10 minutes to separate the adsorbent again from the liquid phase, and the supernatant was analysed for As content. These experiments were conducted at different pH levels, and also using real surface water and groundwater spiked with As. Moreover, Langmuir and Freundlich isotherms were constructed by varying the arsenic concentrations.

3 RESULTS AND DISCUSSION

At pH 9–10, most adsorbents, except MIL-101(Cr) and pure Fe_3O_4 NPs, considerably reduced the initial arsenic concentration of 10 mg/L by more than 60% for As(III) and 91% for As(V), within the first 10 minutes of contact (Fig. 1&2). Whereas the NPs alone showed a good performance for removal of As(III), but poor performance for As(V) (Fig. 1a) and the MIL-101(Cr) removed As(V) well but As(III) very poorly (Fig. 1c), the composite $Fe_3O_4@$MIL-101(Cr) and $Fe_3O_4@$CTF-1 material removed all species very well (Fig. 2a,b). Except for As(III) removal at pH 4 using UiO-66, the removal efficiencies of both arsenic species by the adsorbents were all above 92% in the pH range 4–10, with the maximum removal efficiency of 100% achieved at pH 7 (data not shown). The obtained results imply that As(III) and As(V) can be adsorbed by these materials in a wide range of pH. Moreover, since maximal

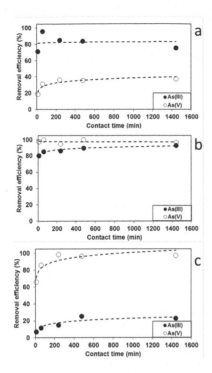

Figure 1.   Adsorption kinetics of arsenic in laboratory-prepared water samples for adsorption on: a) pure Fe_3O_4 nanoparticles, b) UiO-66, c) MIL-101(Cr) (initial As(III) and As(V) concentration = 10 mg/L, solution pH = 9–10; L:S ratio = 263).

adsorption was observed at pH 7, no pH adjustment would be necessary during the treatment of arsenic-contaminated water considering that most natural water bodies are at near-neutral pH.

At pH 7, UiO-66 revealed the highest As(III) and As(V) adsorption capacities of 253.9 mg/g and 255.3 mg/g, respectively. This was followed by Fe_3O_4@CTF-1 with As(III) adsorption capacity of 198 mg/g and As(V) adsorption capacity of 102.3. Fe_3O_4@MIL-101(Cr) was determined to have the lowest capacities of 121.5 mg/g and 80.0 mg/g for the adsorption of As(III) and As(V), respectively.

If the adsorption capacities of the two iron-based adsorbents would be normalized with respect to their iron contents, Fe_3O_4@CTF-1 would have values of 1065.1 mg/g Fe and 550.3 mg/g Fe for As(III) and As(V), respectively, while those of Fe_3O_4@MIL-101(Cr) would be 544.6 mg/g Fe and 358.6 mg/g Fe.

The adsorbents can also achieve high removal efficiencies in natural waters spiked with initial As(III) and As(V) concentrations of 10 mg/L^{-1}. In terms of As(V) adsorption, no considerable differences were observed among the removal efficiencies of the three adsorbents, which all exceeded 97%. However, for As(III) adsorption, the removal efficiencies were in the order of Fe_3O_4@CTF-1 > Fe_3O_4@MIL-101-(Cr) > UiO-66 with the corresponding efficiencies of 99.1%, 93.7%, and 91.6% for surface water and 99.3%, 98.1%, and 92.3% for groundwater.

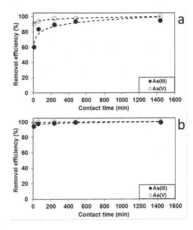

Figure 2.   Adsorption kinetics of arsenic in laboratory-prepared water samples for adsorption on: a) Fe_3O_4@ MIL-101(Cr) (L:S ratio = 263), and b) Fe_3O_4@CTF-1 (L:S ratio = 225) (initial As(III) and As(V) concentration = 10 mg/L, solution pH = 9–10).

4   CONCLUSIONS

The developed MOFs and COFs with and without encapsulated nanoparticles exhibit superior performance for removal of arsenic from arsenic-contaminated groundwater and surface water. Aside from good removal efficiencies, the materials also have the advantages of not requiring pre-oxidation and pH adjustment. Future studies will focus on the regeneration potential of these materials and arsenic adsorption at lower, environmentally relevant doses. Moreover, the performance of the organic frameworks shaped for practical applications, e.g. in the form of pellets (see Silva et al., 2015), should also be further investigated.

ACKNOWLEDGEMENTS

Nina Ricci Nicomel and Karen Leus acknowledge financial support from VLIR-UOS and Ghent University (BOF postdoctoral Grant 01P06813T and UGent GOA Grant 01G00710), respectively.

REFERENCES

Hu, P. 2015. Surfactant Directed Encapsulation of Metal Nanocrystals in Metal-Organic Frameworks. Boston College University Libraries.

Kuhn, P., Antonietti, M. and Thomas, A. 2008. Porous, covalent triazine-based frameworks prepared by ionothermal synthesis. Angew. Chem. Int. Edit. 47: 3450–3453.

Shearer, G.C., Chavan, S., Ethiraj, J., Vitillo, J.G., Svelle, S., Olsbye, U., Lamberti, C., Bordiga, S. and Lillerud, K.P. 2014. Tuned to perfection: ironing out the defects in metal–organic framework UiO-66. Chem. Mater. 26: 4068–4071.

Silva, P., Vilela, S.M.F., Tomé, J.P.C. and Almeida Paz, F.A. 2015. Multifunctional metal–organic frameworks: from academia to industrial applications. Chem. Soc. Rev. 44: 6774–6803.

Arsenic Research and Global Sustainability – Bhattacharya, Vahter, Jarsjö, Kumpiene,
Ahmad, Sparrenbom, Jacks, Donselaar, Bundschuh & Naidu (Eds)
© *2016 Taylor & Francis Group, London, ISBN 978-1-138-02941-5*

Arsenic removal using nano-TiO$_2$/chitosan/feldspar hybrid: Bio-bead development and arsenate adsorption

M. Yazdani[1], H. Haimi[1], R. Vahala[1], M. Janiszewski[2] & A. Bhatnagar[3]
[1]*Department of Built Environment, School of Engineering, Aalto University, Aalto, Finland*
[2]*Department of Civil Engineering, School of Engineering, Aalto University, Aalto, Finland*
[3]*Department of Environmental Science, University of Eastern Finland, Kuopio, Finland*

ABSTRACT: Nano-TiO2/feldspar supported chitosan bio-beads were developed and successfully tested for arsenic removal. Experiments were conducted to examine arsenic removal as a function of pH and bio-bead dose. The effect of initial arsenate concentration and operational time are also reported. Isotherm and kinetic data are presented. The results revealed that the bio-beads are able to remove arsenic in a wide range of pH. The adsorption equilibrium study demonstrated that equilibrium is achieved after 3 h. Kinetic data followed pseudo-second order kinetic model. The Freundlich isotherm model was best fitted to the equilibrium data.

1 INTRODUCTION

Arsenic in the environment originates mainly from the natural weathering of arsenic-containing rocks, but also from human activities including mining. Long-term exposures to arsenic pose acute and chronic health problems. In order to minimize the health risk, the World Health Organization (WHO) has set the allowed arsenic limits at 10 μg/L (WHO, 1993). The predominant arsenic species for natural exposure include the inorganic arsenite (As(III)) and arsenate (As(V)). Since As(III) is more toxic and more difficult to remove due to its uncharged form at most pHs, an effective removal requires to oxidize As(III) to As(V). In the context of developing efficient and low-cost technique for arsenic removal, Titanium Dioxide (TiO$_2$) has showed a high capacity for As(V), demobilizing it via the formation of oxygen bindings. Nano-sized TiO$_2$ (n.TiO$_2$) supplies a large surface area for enhanced adsorption and oxidation capacities, though the nanoparticles create a need for post-filtration to remove them from the treated water. To address this obstacle, combining TiO$_2$ with other adsorbents such as chitosan (CS) (Yamani et al., 2012) and geomaterials (Wan Ngah et al., 2011) has attracted a great interest. Embedding n.TiO$_2$ in an adsorbent would not only provide a larger surface area but would also eliminate the need for energy-intensive post-filtration, whereas a simple separation of the adsorbent from treated water is possible by gravity. A simple separation method can be utilized when considering the technology transfer from a lab phase to a field application. CS is a biopolymer derived from insect and shellfish exoskeletons. It has many advantages including biocompatibility and availability with a minimal environmental cost. However, CS as an adsorbent has the disadvantages of easy agglomeration and low mechanical stability. Supporting CS with geomaterials (Yazdani et al., 2014), like feldspars, the most abundant group of minerals distinguished by the presence of alumina in their chemistry, can increase its chemical and physical stability while reducing the required amount of CS without affecting the overall adsorption capacity. As the individual features of n.TiO$_2$, CS, and feldspars toward adsorption are not optimal and identical, the idea of combining them to make an effective bio-hybrid for arsenic removal is conceptualized. In this manner, inherited problems related to the individual materials can be minimized. This study aims to develop n.TiO$_2$/CS/feldspar bio-beads via a simple fabrication method and to evaluate them for As(V) removal.

2 EXPERIMENTAL

2.1 Preparation of n.TiO$_2$/CS/feldspar hybrid

CS was dissolved in 0.1 M CH$_3$COOH, creating a viscous solution into which n-TiO$_2$ was added and agitated overnight. Feldspar immersed in 0.2M NaCl was added to the n.TiO$_2$/CS solution and agitated for 24 h. The final solution was pumped via a syringe into 0.1 M NaOH, resulting in n.TiO$_2$/CS/feldspar beads (0.5–1.0 mm diameter), where both n.TiO$_2$ and feldspar are embedded in the CS matrix.

2.2 Adsorption experiments

Adsorption experiments were conducted to examine As(V) removal using the prepared bio-beads. The experiments were carried out with adding known amount of beads in 50 mL of 1 mg/L As(V) solutions. The pH of the solutions was adjusted using HCl or NaOH. Isotherm studies were conducted with the As(V) solutions of initial concentrations varying from 200 to 10000 µg/L. For the kinetics experiments, the As(V) solutions with initial concentrations of 1 mg/L were adjusted to pH 4. All samples are prepared and analyzed in duplicate and the average values with standard deviations are reported.

3 RESULTS AND DISCUSSION

3.1 Characterization of n.TiO$_2$/CS/feldspar hybrid

The FTIR spectrum of prepared bio-beads revealed a matching characteristic of the precursor materials. The typical peaks of CS were seen at 3400 cm^{-1} for –OH stretching vibrations and at 1600 cm^{-1} for –NH bending vibrations. The existence of TiO$_2$ nanoparticles was confirmed by the appearance of the band at 500 cm^{-1}. The presence of the TiO$_2$ compound along with the active hydroxyl and amine groups can accelerate an effective pollutant remediation via an adsorption-photodegradation mechanism. Further work is ongoing to explore the bio-beads for As(III) and As(V) removal in the presence of UV light.

3.2 Arsenic adsorption

It is well known that pH is a critical factor in arsenic adsorption onto TiO$_2$. Like other metal oxides, TiO$_2$ works ineffectively at most alkaline pHs, where electrostatic repulsion prevents the adsorption of negatively charged species on the negatively charged surface of TiO$_2$. Experimental results indicated bio-beads are able to effectively adsorb As(V) in a wide range of pH (pH range 4–12: removal 93–70%), while the capacity was 830 ± 5.0 µg/g in pH 7 and at its maximum in pH 4 (935 ± 0.2 µg/g). The effect of solid to liquid ratio was studied to optimize bead dose for As(V) removal. The ratio 1 g/L was found to be optimal considering the removal percentage and adsorption capacity (removal 78 ± 2.0%).

Kinetic data for arsenic adsorption by the bio-beads is presented in Figure 1, indicating that equilibrium is reached after 3 h (180 min) of contact time. Pseudo-first order and pseudo-second order kinetic models were selected to study adsorption kinetics. High linear regression coefficients ($R^2 \geq 0.999$) along

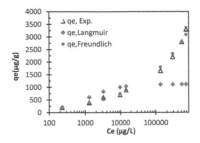

Figure 2. Adsorption isotherm of As(V) by n.TiO$_2$/CS/feldspar.

with reasonable agreement between the values of the calculated adsorption capacity ($q_{e,cal}$: 1000 µg/g) and experimental adsorption capacity ($q_{e,exp}$: 954 µg/g) indicate the adsorption of As(V) onto the bio-beads followed the pseudo-second order rate model. The equilibrium data were described using Freundlich and Langmuir isotherm models. Figure 2 indicates that the Freundlich model fits the adsorption equilibrium data better than the Langmuir model. This confirms the heterogeneous distribution of adsorptive functional groups on the surface of the bio-hybrid, as the Freundlich isotherm assumes a heterogeneous surface for the adsorbent.

4 CONCLUSIONS

Bio-beads, n.TiO$_2$/CS/feldspar hybrid, comprised of environmentally friendly precursor materials were synthesized and tested for As(V) removal. The bio-beads could adsorb 1111 µg As(V)/g ($q_{max, Langmuir}$) bead in batch experiments. Unlike other TiO$_2$-based adsorbents, the beads were able to remove As(V) in a wide range of pH. The Freundlich isotherm model was indicated to fit best the equilibrium data. The pseudo-second order kinetic model could be used to describe the adsorption behavior of the hybrid.

ACKNOWLEDGEMENTS

Finnish Maa- ja vesitekniikan tuki ry. is acknowledged for the financial support of the research.

REFERENCES

Wan Ngah, W.S., Teong, L. and Hanafiah, M. 2011. Adsorption of dyes and heavy metal ions by chitosan composites: A review. Carbohydrate Polymers 83: 1446–1456.

WHO (1993). Guidelines for Drinking Water Quality. 4th Edn. World Health Organization, Geneva, Switzerland.

Yamani, J.S., Miller, S.M., Spaulding, M.L. and Zimmerman, J.B. 2012. Enhanced arsenic removal using mixed metal oxide impregnated chitosan beads. Wat. Res. 46: 4427–4434.

Yazdani, M., Bahrami, H. and Arami, M. 2014. Feldspar/Titanium Dioxide/Chitosan as a biophotocatalyst hybrid for the removal of organic dyes from aquatic phases. Journal of Applied Polymer Science 131: 40247–40256.

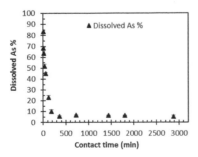

Figure 1. Adsorption kinetics of As(V) by n.TiO$_2$/CS/feldspar.

4.3 Innovative technologies

Arsenic Research and Global Sustainability – Bhattacharya, Vahter, Jarsjö, Kumpiene,
Ahmad, Sparrenbom, Jacks, Donselaar, Bundschuh & Naidu (Eds)
© 2016 Taylor & Francis Group, London, ISBN 978-1-138-02941-5

Removing arsenic to <1 µg/L in conventional groundwater treatment plants: Practical tips and tricks

M. Groenendijk[1], S. van de Wetering[2] & T. van Dijk[2]
[1]*Evides Waterbedrijf N.V., Rotterdam, The Netherlands*
[2]*Brabant Water N.V., 's-Hertogenbosch, The Netherlands*

ABSTRACT: Producing drinking water with arsenic (As) concentration of <1 µg/L is a new challenge embraced by the drinking water companies in the Netherlands. Advanced Oxidation-Coagulation-Filtration technology, abbreviated as AOCF, has been found effective to achieve 1 µg/L target. The technique has been extensively researched at lab-, pilot- and demonstration-scale at Brabant Water. AOCF can be easily integrated within the existing process configurations of conventional ground water treatment plants. In the beginning of 2016 the first fully operated plant in the Netherlands with the AOCF process will start-up and produce drinking water with <1 µg/L As. In this article the authors discuss the technicalities and practical "tips and tricks" of implementing AOCF to achieve ultra-low levels of As in drinking water.

1 INTRODUCTION

In the Netherlands, drinking water is of prime quality and arsenic (As) concentrations in the effluents of drinking water treatment plants are well below the Dutch maximum concentration limit, i.e. 10 µg/L. Arsenic is a highly toxic metal. There is a general consensus in the Netherlands that if possible As should be removed from drinking water supplies as far as possible. At present the drinking water companies in the Netherlands aim for a maximum As concentration of <1 µg/L in their product, i.e., the drinking water.

Over the past few years Brabant Water N.V. has been actively investigating different treatment approaches to remove As to lower than 1 µg/L from drinking water. The goal was to develop an innovative solution that lowers As concentrations in conventional groundwater treatment systems without encountering significant changes in the existing treatment configurations. Furthermore, simple to operate and cost-effective solution was desired. This article discusses the technicalities and application case studies of AOCF, our simple and affordable solution, that can be easily implemented at the conventional groundwater treatment plants to achieve <1 µg/L of As in the drinking water.

2 THE AOCF TECHNIQUE

AOCF stands for Advanced Oxidation-Coagulation Filtration. As the name indicates, it is a 3 step process, including (i) pre-oxidation of As(III) to As(V), (ii) adsorption of As(V) to $Fe(OH)_{3(s)}$ and finally (iii) filtering $Fe(OH)_3$-$As(V)_{(s)}$ matrix by granular media filtration. In deep groundwaters of Brabant Water, As(III) is the dominant form of As. We have observed that the oxidation of As(III) by conventional aeration (cascade, sprinkler etc.) is ineffective in oxidizing As(III) to As(V). Therefore, we use permanganate in the form of either $KMnO_4$ or $NaMnO_4$ as chemical oxidant to rapidly oxidize the As(III) concentration in raw water. Regarding the permanganate, the selection of this oxidizing agent was made after carrying out a series of lab-scale investigations which provided us with a clear evidence of its superior performance compared to other commonly used chemical oxidants in drinking water production industry. It must be noted that in comparison to $KMnO_4$, $NaMnO_4$ is more easy and safe to handle during operation. Therefore, if available, use of $NaMnO_4$ should get priority. Dosing of permanganate not only increases the rate of As(III) oxidation to As(V), but also Fe(II) to Fe(III). The iron hydroxide flocs [$Fe(OH)_{3(s)}$] sorb As(V). A significant amount of As(V) is also co-precipitated with $Fe(OH)_{3(s)}$. When As is present in relatively high concentration, addition of additional Fe becomes necessary and we have observed in our lab-scale experiments that adding Fe as Fe(III) is more effective compared to adding Fe(II).

AOCF increases flocculative iron removal in conventional treatment systems. This may have negative influence on the operation of rapid sand filters, for example, reduction of filter run time. In that case, optimizing the operation of filters becomes necessary and therefore we strongly recommend piloting the process at a scale of filter columns for complete scenario analysis. At the Drinking Water Treatment Plant (DWTP) Brabant Water, we have decided to optimize our groundwater treatment plants that produce drinking water with higher than 1 µg/L As. At any location, we start with a series of (preliminary) bench-scale experiments which are followed by a relatively long term pilot-scale validation/optimization phase. For the interest of readers, two of our case studies are described below.

3 CASE STUDIES

3.1 *DWTP Dorst*

DWTP Dorst produces 10.5 Mm^3 of drinking water per year from deep groundwater. The treatment scheme includes 10 parallel treatment trains, each consisting of a raw water intake from a common groundwater reservoir (13 µg/L As), a cascade aerator, a rapid sand filter and an effluent discharge to a common reservoir (6 µg/L As) from where the water is subsequently distributed to communities in the southern part of the Netherlands. The As concentration in the treated water of DWTP Dorst was one of the highest in the Netherlands. At lab-scale, number of trials with various treatment chemicals and contact times were studied. Arsenic concentrations below 1 µg/L could be achieved at various dosing combinations of $KMnO_4$ and $FeCl_3$, however, 0.45 µm (disc) membrane filters were used to separate the flocs from solution. The results from the lab-scale experiments were validated at pilot plant. Interestingly, lower dose of treatment chemicals was needed to remove As to <1 µg/L at pilot-scale compared to what we expected based on the results of our bench-scale tests. Moreover, during pilot testing we found that implementation of AOCF decreased average filter run time from 96 to 24 h. To resolve this issue, dual media/double layer filtration with anthracite (1–1.6 mm) and finer sand (0.5–0.8 mm) was used with the optimum chemical dosing combination. Average filter run time increased to more than 48 h. AOCF proved to be feasible for DWTP Dorst, therefore, we are now in the process of designing and up-gradation. In the beginning of 2016 the plant will be fully operational and it will be the first plant in the Netherlands (and perhaps the world) that is upgraded to produce drinking water with <1 µg/L As. Figure 1 is a graphical presentation of (part of) our data that we collected for several months during

demonstration of the AOCF technique using one of the full-scale treatment trains of DWTP Dorst.

3.2 *DWTP Prinsenbosch*

As we are upgrading the full-scale installation of DWTP Dorst, in parallel we have started investigating applicability of AOCF for another treatment location called DWTP of Prinsenbosch. DWTP Prinsenbosch is also conventional groundwater treatment plant and the process configuration is similar to Dorst. We have observed in our preliminary tests at bench-scale that As concentration of <1 µg/L can be realized at DWTP Prinsenbosch by only dosing permanganate in the raw water.

4 CONCLUSIONS

Producing drinking water with <1 µg/L As is technically possible and is demonstrated at DWTPs Dorst and Prinsenbosch. AOCF is a reliable, simple and affordable technology which can guarantee As concentrations below 1 µg/L. The technology can be easily implemented at the conventional groundwater treatment plants requiring only an addition of a chemical dosing setup and, in some cases, replacement of the filtration media. No evidence of disturbance has been noticed for the pre-existing removal processes. The total costs associated with the application of AOCF technology at the DWTP (10 Mm^3/year) have been estimated at approx. 0.02 €/m^3. In the beginning of 2016 the first fully operated plant in the Netherlands with the AOCF process will start-up and produce drinking water with <1 µg/L As.

ACKNOWLEDGEMENTS

We deeply appreciate Arslan Ahmad (Scientific Researcher at KWR Water Cycle Research Institute) and Jink Gude (PhD candidate at TU Delft) for their important contribution towards this project.

REFERENCES

Ahmad, A. 2014. Evaluation and optimization of advanced oxidation coagulation filtration (AOCF) to produce drinking water with less than 1 µg/L of arsenic. *TRITA-LWR Report* 2014:1, 101p.
Ahmad, A., van de Wetering, S., Groenendijk, M., Bhattacharya, P. 2014. Advanced Oxidation-Coagulation-Filtration (AOCF)—an innovative treatment technology for targeting drinking water with <1 µg/L of arsenic. In: M.I. Litter, H.B. Nicolli, M. Meichtry, N. Quici, J. Bundschuh, P. Bhattacharya & R. Naidu (eds.) *"One Century of the Discovery of Arsenicosis in Latin America (1914–2014) As 2014". Interdisciplinary Book Series: "Arsenic in the Environment—Proceedings".* Series Editors: J. Bundschuh & P. Bhattacharya, CRC Press/Taylor and Francis (ISBN 978-1-138-00141-1), pp. 817–819.

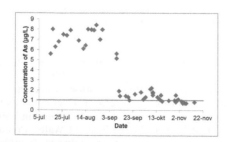

Figure 1. Total As concentration in the effluent of full-scale filter at DWTP Dorst. AOCF implemented at 3 Sep.

Arsenic Research and Global Sustainability – Bhattacharya, Vahter, Jarsjö, Kumpiene,
Ahmad, Sparrenbom, Jacks, Donselaar, Bundschuh & Naidu (Eds)
© *2016 Taylor & Francis Group, London, ISBN 978-1-138-02941-5*

Capacitive Deionization for removal of arsenic from water

K. Laxman[1], P. Bhattacharya[2], T. Ahmed[3], A.S. Uheida[3] & J. Dutta[3]
[1] *Water Research Center, Sultan Qaboos University, Muscat, Oman*
[2] *KTH-International Groundwater Arsenic Research Group, Department of Sustainable Development*
Environmental Science and Technology, KTH Royal Institute of Technology, Stockholm, Sweden
[3] *Functional Materials Division, Materials and Nano Physics Department, School of ICT, KTH Royal Institute*
of Technology, Kista, Stockholm, Sweden

ABSTRACT: Capacitive Deionization (CDI) is an electrochemical technique for the removal of charged species from aqueous media. It has been typically used for the desalination of low to medium salinity water, working on potential mediated electrosorption of ions. Herespecific discussion on the utilization of CDI constructed with nanoelectrodes for the removal of arsenic from water is introduced.

1 INTRODUCTION

Arsenic (As) is a ubiquitous trace element with typical soil concentrations of 5–10 mg/kg. In water, especially in ground and surface water, As content varies widely with reports ranging from less than 0.005 µg/L to as high as 21800 µg/L in river waters from Chile (Smedley & Kinniburgh, 2002, Bhattacharya *et al.*, 2002). Arsenic compounds arecarcinogenic to humans and classified in Group 1, by the International Agency for Research on Cancer (IARC, 1987). The WHO limit for As in drinking water is 10 µg/L. While most countries adhere to this limit, some like Canada have adopted a limit of 25 µg/L, while Bangladesh has adopted 50 µg/L as the maximum limit in drinking water.

Arsenic is unique compared to other heavy metals in that its oxidation state governs its toxicity, charge and solubility in aqueous media. The two dominant inorganic As oxidation states (oxyanions) are pentavalent As or arsenate [As(V)] and trivalent As or arsenite [As(III)]. As(V) is the charged form of As present as $H_2AsO_4^-$ in oxidizing conditions at pH<6.9, while $HAsO_4^{2-}$ is the dominant species at higher pH and$H_3AsO_4^0$ and AsO_4^{3-} only can be found under extremely acidic or alkaline conditions (Mehmood *et al.*, 2009). In reducing conditions, arsenite (trivalent As) takes the form of $H_3AsO_3^0$ (uncharged arsenite species which is more toxic) at pH values below approximately 9.2. Thus the removal of As species in water is a complex process which involves working with the pH and oxidation state of As. Most of the currently available technologies are less effective in removing As(III) than As(V) which is the reason why the oxidation of As(III) as a pretreatment is often applied to enhance the As removal efficiency (Geuckee*t al.*, 2009, Bissen & Frimmel, 2003, Leupin & Hug, 2005).

This is where Capacitive Deionization (CDI) has an advantage as it can combine both processes in a single device. While CDI cells typically work in the non-Faradaic domain of potential dependant electric field mediated ion adsorption as Electrical Double Layers (EDL) at the electrode surface, it can also be used in the Faradaic domain to promote redox reactions at the electrode surfaces. Literature shows that by using capacitive deionization technique, approximately two-fold higher As adsorption can be achieved (Fan *et al.*, 2016). A CDI device will electrosorb any anionic or cationic species in the aqueous medium, making it effective for removing not just arsenate but also other harmful heavy metal ions like mercury (Hg), lead (Pb), tin (Sn), nickel (Ni), chromium (Cr) etc. (Huang *et al.*, 2016). In fact higher the oxidation numbers of these cations, the better their adsorbing capability by the CDI device.

Here we focus on using commercial Activated Carbon Cloth (ACC) as the electrode material within a CDI cell for the removal of As(V) and other anions and cations from water.

2 METHODS/EXPERIMENTAL

2.1 CDI experiments

The desalination experiments were conducted using a flow through CDI cell. The CDI cell consists of symmetrical ACC electrodes of approximately 9 cm² area, a spacer of ca. 650 µm thickness, a reservoir made up of Poly-Methyl Methacrylate (PMMA) (Dow Corning SYLGARD® 184 Silicone elastomer kit), a current collector (graphite electrode) and an acrylic plate to organize the CDI cell (Laxman *et al.*, 2015a). The current collectors were used as electrical connection from power supply (DC) to the ACC electrodes. The experiments were initially carried out with1000 mg/L Sodium Chloride (NaCl) solution with a flow rate of 3 ml/min under an applied potential of 1.6 VDC to test and optimize the device performance.100 µg/L As (V) was prepared from sodium arsenate heptahydrate ($Na_2HAsO_4.H_2O$) and tested for its removal capacity.

Since no conductivity data could be observed at such low concentrations, the As(V) content was increased to 100 mg/L prior to running the experiment.

2.2 *Characterization*

Conductivity measurements were carried out using an online conductivity probe (eDAQ ET916) with a cell volume of 93 μL coupled to a conductivity iso-pod (EPU357) with single channel PodVu software for real time recording of the conductivity changes. Arsenic measurements were carried out in Varian 710-ES ICP Optical emission spectrometer.

3 RESULTS AND DISCUSSION

3.1 *Salt removal*

A CDI cell was fabricated with ACC as the electrodes (Fig. 1a). Upon passing ground water containing Na^+, Ca^{2+}, K^+ and Al^{3+} through the CDI cell, it was observed that the removal capacity was governed considerably by the ionic charges (Laxman *et al.*, 2015b). Al^{3+} was removed to the largest extent, followed by Ca^{2+} and then K^+ and Na^+. Also the size of the ion plays an important role, wherein ions with the same charge but smaller in size can be more efficiently removed. This gives us valuable insights into how a CDI cell might behave for As removal in a water matrix containing

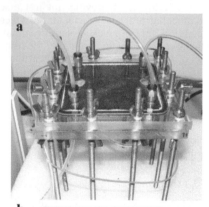

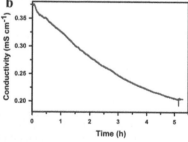

Figure 1. (a) Photo of the CDI cell comprising of Activated Carbon Cloth (ACC) as the electrodes and operated at an applied potential of 1.6 VDC and (b) As adsorption in with time in the CDI cell.

different ions of varying concentrations, sizes and charge. Also any other organic contaminants will have an adverse effect in the removal of the ions.

3.2 *Arsenic removal*

Upon introducing As(V) as the primary species in the 100 ml water flowing through the CDI cell, a substantial reduction in the conductivity of the solution was observed, indicating As removal by electrosorption (Fig. 1b). However the removal capacity is a function of the applied potential, flow rate and localized interfacial solution pH, effects of which will be discussed.

4 CONCLUSIONS

Capacitive deionization is demonstrated to be a viable technique for the removal of As(V) from water. When a potential is applied to the ACC electrodes, the CDI device removes As(V) via electrosorption of the As oxyanions at the positive electrode of the cell at low power consumptions.

REFERENCES

Bhattacharya, P., Frisbie, S.H., Smith, E., Naidu, R., Jacks, G., Sarkar B. 2002. Arsenic in the Environment: A Global Perspective. In: B.Sarkar (Ed.) *Handbook of Heavy Metals in the Environment*. Marcell Dekker Inc., New York, pp. 147–215.

Bissen, M., Frimmel, F.H. 2003. Arsenic—a review. Part II: oxidation of arsenic and itsremoval in water treatment. *Acta Hydrochim. Hydrobiol.* 31: 97–107.

Fan, C.S., Tseng, S.C., Li, K.C., Hou, C.H. 2016. Electr-removal of arsenic(III) and arsenic(V) from aqueous solutions by capacitive deionization. *J. Hazard. Mater.* 312: 208–215.

Geucke, T., Deowan, S.A., Hoinkis, J., Pätzold, C. 2009. Performance of a small-scale RO desalinator for arsenic removal. *Desalination* 239: 198–206.

Huang, Z., Lu, L., Cai, Z., Ren, Z.J., 2016. Individual and competitive removal of heavy metals using capacitive deionization.*J. Hazard. Mater.*302: 323–331.

IARC. 1987. Overall Evaluations of Carcinogenicity; an Updating of IARC Monographs. vol. 1–42, International Agency for Research on Cancer, Lyon, France.

Laxman, K., Al Gharibi, L. and Dutta, J., 2015a. Capacitive deionization with asymmetric electrodes: Electrode capacitance vs electrode surface area.*Electrochim.Acta* 176: 420–425.

Laxman, K., Myint, MTZ., Sathe, P., Al Abri, M., Dobretsov, S. and Dutta, J., 2015b. Desalination of inland brackish ground water in a capacitive desalination cell using nanoporous activated carbon cloth electrodes. *Desalination* 362: 126–132.

Leupin, O.X., Hug, S.J. 2005. Oxidation and removal of arsenic(III) from aerated groundwater by filtration through sand and zero-valent iron. *Water Res.* 39: 1729–1740.

Mehmood, A., Hayat, R., Wasim, M., Aktar, M.S. 2009. Mechanisms of arsenic adsorption in calcareous soils. *J. Agric. Biol. Sci.* 1(1): 59–65.

Smedley, P.L., Kinniburgh, D.G. 2002. A review of the source, behaviour and distribution of arsenic in natural waters.*Appl. Geochem.* 17(5): 517–568.

Arsenic Research and Global Sustainability – Bhattacharya, Vahter, Jarsjö, Kumpiene,
Ahmad, Sparrenbom, Jacks, Donselaar, Bundschuh & Naidu (Eds)
© 2016 Taylor & Francis Group, London, ISBN 978-1-138-02941-5

Forward osmosis—challenges and opportunities of a novel technology for arsenic removal from groundwater

J. Hoinkis[1], J. Jentner[1], S. Buccheri[1], S.A. Deowan[2], A. Figoli[3] & J. Bundschuh[4]
[1]*Institute of Applied Research, Karlsruhe University of Applied Sciences, Karlsruhe, Germany*
[2]*Mechatronics Engineering Department, Dhaka University, Dhaka, Bangladesh*
[3]*Institute on Membrane Technology ITM-CNR, Rende (CS), Italy*
[4]*Faculty of Health, Engineering and Sciences, University of Southern Queensland, Toowoomba, Australia*

ABSTRACT: Forward Osmosis (FO) represents a innovative membrane technology for liquid separation driven by the osmotic pressure of aqueous solutions. This work studied at laboratory scale commercial FO membranes regarding water flux, reverse salt diffusion as well as As(III) and As(V) rejection. The findings showed that As(V) is significantly better rejected than As(III) what is in agreement with the findings on Reverse Osmosis (RO) and Nanofiltration (NF) membranes. Since FO has to be operated as combined process with regeneration of the draw solution (e.g. RO) eventually rejection is higher due to the second barrier. Hence FO may offer an alternative solution with lower energy consumption for conventional two stage membrane processes (RO, NF).

1 INTRODUCTION

In the last decade, Forward Osmosis (FO) has gained increased attention as an emerging membrane technology. Therefore, many contributions have been made to improve the overall FO process efficiency from both academic researchers and industries (Chekli *et al.*, 2016). The principle of FO process relies on using the natural osmotic process to draw water across a semipermeable membrane from a saline feed water to a higher concentrated solution, namely the Draw Solution (DS). The driving force is therefore created naturally by the difference in osmotic pressure between the DS and the Feed Solution (FS). This offers several advantages over conventional hydraulic pressure-driven membrane processes (e.g. RO) such as lower energy requirements and reduced membrane fouling potential (McCutcheon & Elimelech, 2006). However, regarding drinking water production it should be pointed out that due to the fact that water is always drawn into a saline DS an additional separation step to recover the water from the DS is needed. Hence application of FO for drinking water production e.g. from groundwater always requires a hybrid process. FO generally has to be coupled with other process to separate the DS from the product water. This could be a pressure driven process such as Reverse Osmosis (RO) or a thermal processes such as distillation. Consequently, the economic suitability of such hybrid process needs to be studied in detail for each particular application. The most salient advantage of FO is its better resistance to fouling and scaling (Bamaga *et al.*, 2011) and due to hybrid process

a second barrier will provide higher water quality. However, the water fluxes of the commercially available FO membranes are still low. The objective of this work is to study arsenic removal efficiency of commercially available FO membranes.

2 METHODS/EXPERIMENTAL

2.1 Experimental rig

The set up consists of a stainless steel flat sheet module which was fitted with 86 cm² active membrane area (Fig. 1). The feed and draw solutions (1000 mL and 500 mL, respectively) are passed by peristaltic pumps through the module and recirculated into the feed glass beakers. The flow was controlled by two manual flowmeters. The feed and draw solution were recirculated into the respective beakers. The beaker with draw solution was set on a balance which was connected to PC. Additionally electrical conductivity with temperature sensors were immersed in the glass beakers. All data were acquired by LabView software (National Instruments).

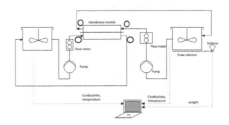

Figure 1. Experimental rig.

2.2 Test solutions and membranes

A liquid As(V) standard (1000 mg/L) purchased from the company Merck served as the parent solution to spike DI and local tap water and to prepare the calibration curves. Arsenic trioxide (As_2O_3) in powder form, also delivered by Merck, was used for the preparation of the As(III) parent solution (350 mg/L) in line with DIN 38405 D12. The arsenic trioxide was dried for 24 hours before preparing the standard solution. 0.462 g of As_2O_3 was dissolved with 12 mL of NaOH (2 mol/L); the solution was subsequently neutralized with sulfuric acid. Finally, the solution was filled up to 1000 mL with DI water. All samples were measured by AAS type contrAA 300 supplied by Analytic Jena. 1 M NaCl was used as draw solution. Three commercial FO membranes have been studied: 1) Porifera (Active layer: Polyamide), 2) HTI (Acitve layer: Cellulose acetate), and 3) Aquaporin (Active layer: Biomimetic aquaporin protein).

3 RESULTS AND DISCUSSION

3.1 Water flux

Figure 2 shows the water flux and reverse salt passage of the tested membranes. As can be seen, performance of the Porifera membrane is significantly better than for the HTI and Aquaporin. Moreover, Aquaporin membrane offer only low mechanical stability and was frequently ripped during experimental period of 2 hours. Therefore for the subsequent experiments with arsenic spiked solutions only the Porifera membrane was selected for the subsequent experiments with arsenic spiked solutions.

3.2 Arsenic removal efficiency

Figures 3 and 4 show arsenic concentration in the draw solution for the Porifera membrane. It is obvious that arsenic removal of As(III) is significantly lower than for As(V). This is in agreement with the findings of RO and NF membranes (Geucke et al., 2009). This can be explained as follows. The active layer of the membranes consists of a polymer with negatively charged groups. A charge exclusion effect enhances the rejection of negatively charged ions

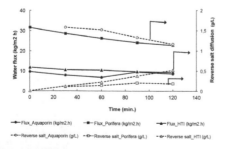

Figure 2. Water flux and reverse salt diffusion of the tested membranes (pH 7±0.5, 25±2°C).

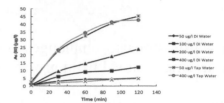

Figure 3. As (III) concentration in draw solution (pH 7±0.5, 20±2°C).

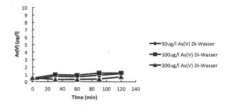

Figure 4. As(V) concentration in draw solution (pH 7±0.5, 20±2°C).

such as As(V). In the observed pH range, As(III) is only neutrally charged, hence it is less rejected. There was no significant difference between removal efficiency for DI and tap water (Fig. 3). For As(III) from 50 µg/L in feed solution arsenic level in draw solution exceeds the MCL of 10 µg/L. However, it should be noted that FO has to be run as combined process with e.g. RO which serves as second barrier and further reduces arsenic.

4 CONCLUSIONS

This experimental work which was conducted at laboratory scale with a commercial membrane and As(III) and As(V) model solutions showed that rejection of As(V) is significantly better than of As(III) what is in agreement with findings for pressure driven RO and NF membranes.

REFERENCES

Bamaga, O.A., Yokochi, A., Zabara, B., Babaqi, A.S. 2011. Hybrid FO/RO desalination system: preliminary assessment of osmotic energy recovery and designs of new FO membrane module configurations. *Desalination* 268: 163–169.

Chekli, L., Phuntsho, S., Kim, J.E., Kim, J., Choi, J.Y., Choi, J.-S., Kim, S., Kim, J.H. Hong, S., Sohn, J., Shon, H.K. 2016. A comprehensive review of hybrid forward osmosis systems: Performance, applications and future prospects. *J. Membr. Sci.* 497: 430–449.

McCutcheon, J.R., Elimelech, M. 2006. Influence of concentrative and dilutive internal concentration polarization on flux behavior in forward osmosis. *J. Membr. Sci.* 284: 237–247.

Geucke, T., Deowan, S.A., Hoinkis J., Pätzold, Ch. 2009. Performance of a small-scale RO desalinator for arsenic removal. Desalination 249: 198–206.

Arsenic Research and Global Sustainability – Bhattacharya, Vahter, Jarsjö, Kumpiene,
Ahmad, Sparrenbom, Jacks, Donselaar, Bundschuh & Naidu (Eds)
© *2016 Taylor & Francis Group, London, ISBN 978-1-138-02941-5*

In-situ formation of Fe(II)-sulfide coatings on aquifer sediments to remediate high arsenic groundwater

K. Pi, Y. Wang, X. Xie, T. Ma & C. Su
School of Environmental Studies and State Key Laboratory of Biogeology and Environmental Geology, China University of Geosciences, Wuhan, P.R. China

ABSTRACT: A field test was conducted to evaluate a novel technology of in-situ remediation of high arsenic aquifer based on microbially-stimulated Fe(II)-sulfide coating formation under reducing conditions. After multiple supply of $FeSO_4$ into the aquifer for 25 days, groundwater arsenic decreased from an initial concentration of 593 μg/L to 159 μg/L, and further lowered to 136 μg/L 30 days later after field implementation. Microbial sulfate reduction produced high aqueous sulfide that reacted with Fe(II) to firstly form mackinawite on sediment surfaces. Aqueous arsenic was sequestered via surface adsorption and co-precipitation with mackinawite that could be transformed to arsenic-bearing pyrite. Furthermore, As(V) can be reduced by sulfide into As(III) to facilitate its incorporation into the newly-formed Fe(II)-sulfide phases in the environment of enhanced reducing condition.

1 INTRODUCTION

Arsenic (As) contaminated groundwater has been a worldwide concern for decades (Sharma *et al.*, 2014). Recently, in-situ As removal based on interactions between As and different Fe-bearing minerals was proposed (Mondal *et al.*, 2013). Under reducing conditions, Fe(II) sulfides can be effective materials for As immobilization (Omoregie *et al.*, 2013). However, few work has been done on in-situ Fe(II)-sulfide coating onto aquifer matrix for high-As groundwater remediation. Therefore, this study aims to (1) investigate the feasibility of a novel technology for arsenic sequestration via forming Fe(II)-sulfide coating on sediments under reducing conditions and (2) understand the associated major (bio)geochemical processes.

2 METHODS/EXPERIMENTAL

2.1 *Field test*

An injection-pumping system comprising four injection wells equidistantly distributing at the upstream and one pumping well at the midpoint of downstream were constructed to cover the aimed aquifer region (75×50 m^2) located at central part of Datong Basin, China. After the build-up of a steady local groundwater flow regime, 23.2 L of 10 mmol/L $FeSO_4$ solution and 17 L of anoxic water were alternately introduced (11.6 L/h) for 25 d until Fe(II) concentration in monitoring wells showed no significant changes. During and after the field test, variations in hydrochemical parameters were monitored. Before (reference sample) and 30 d after the field test, sediment samples were retrieved from the aquifer.

2.2 *Analytical methods*

Total As and As species concentrations were measured using HG-AFS (AFS-820, Titan). Standards (Sigma-Aldrich) and sample replicates were routinely tested to produce reproducibility within ±5%. Sediment samples were air-dried, gold-coated and analyzed using SEM (Hitachi, SU8010) and EDS (Ametek, EDAX APPOLLO XP). Aquifer sediment samples were analyzed using XRD (Bruker AXS, D8-Focus) with a CuKα radiation.

3 RESULTS AND DISCUSSION

3.1 *Variations of groundwater arsenic*

Aqueous As concentration in the targeted aquifer decreased from an initial average concentration of 593 μg/L to 159 μg/L (73.18% removal rate) during the 25-d supply of $FeSO_4$, and further decreased to 136 μg/L (81.16% removal rate) on the 30th day after the field test. Significant decrease in As concentration firstly occurred in the groundwater near the injection wells. The As retention front gradually moved forward from the injection wells to the pumping well, and low As groundwater generally spread throughout the aquifer region by the end of the field test (Fig. 1).

Initially, aqueous As existed as As(V) and As(III) with a ratio of 52.74% and 47.26%, respectively. After onset of the field test, As(V) concentration quickly decreased with time, and remained a low level to the end of the test. Similarly, decrease in As(III) concentration with time was observed, though it is not equally obvious at the early stage.

Both groundwater pH and Eh decreased during the field test. Drop in pH can be ascribed to the

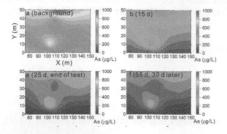

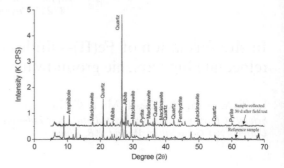

Figure 1. Time-dependent changes of aqueous As concentration in the remediation zone.

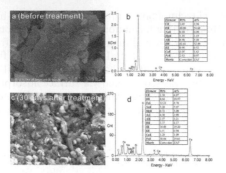

Figure 3. Refinement results of XRD analysis on sediments.

Figure 2. SEM images and EDS analysis results of sediment samples before and after treatment.

dissociation of dissolved hydrogen sulfide produced from microbial sulfate reduction and further proton release from the precipitation of Fe(II) sulfides (Omoregie *et al.*, 2013). Significant Eh decrease demonstrates the enhancement of reducing conditions as a result of induced sulfidogenesis.

3.2 *Sediment geochemistry*

SEM-EDS results indicate that the reference sample surfaces showed no discernable signal of Fe(II)-sulfide precipitates (Fig. 2), with relatively low Fe (2.5 wt.%) and undetectable S and As content. By contrast, in the sediment sample collected after treatment, coating precipitates were observed on the surfaces, with substantial increase in Fe, S and As contents up to 30.0, 34.5 and 4.3 wt.%, respectively.

The XRD results (Fig. 3) indicate that compared to the reference sample, sediment surfaces subject to Fe(II) sulfidogenesis showed enhanced signals of mackinawite and peaks of pyrite. This suggests that mackinawite was coated on sediment particles and represented a significant portion of Fe(II)-sulfide precipitates. However, the pyrite peaks in the sample imply that despite with a less amount, this crystalline mineral had been formed in-situ.

3.3 *Mechanisms of arsenic immobilization*

Our field test results confirm that aqueous As had been sequestered into the newly-formed Fe(II) sulfides. The produced sulfide reacted with Fe(II) to firstly generate mackinawite on sediments, and its

high surface area provided sufficient binding sites for As. Furthermore, the As(III) species produced from As(V) reduction, could be co-precipitated during mackinawite formation. Under continuous sulfidic conditions, mineralization may promote the transformation of the precipitates into As-bearing (arseno)pyrite (Bostick and Fendorf, 2003).

4 CONCLUSIONS

Groundwater As has been substantially sequestered (81.16% As removal rate) in the field trail of in-situ aquifer Fe(II)-sulfide coating via adsorption and co-precipitation with newly-formed mackinawite that can be further transformed into As-bearing pyrites. The proposed novel technology thus has broad application in remediation of high As groundwater under reducing conditions.

ACKNOWLEDGEMENTS

This research was financially supported by the Ministry of Science and Technology of China (2012 AA062602), the National Natural Science Foundation of China (41521001) and the China Scholarship Council (201506410019).

REFERENCES

Bostick, B.C. and Fendorf, S. 2003. Arsenite sorption on troilite (FeS) and pyrite (FeS₂). *Geochim. Cosmochim. Acta* 67: 909–921.
Mondal, P., Bhowmick, S., Chatterjee, D., Figoli, A. and Van der Bruggen, B., 2013. Remediation of inorganic arsenic in groundwater for safe water supply: A critical assessment of technological solutions. *Chemosphere* 92: 157–170.
Omoregie, E.O., Couture, R.M., Van Cappellen, P., Corkhill, C.L., Charnock, J.M., Polya, D.A., Vaughan, D., Vanbroekhoven, K. and Lloyd, J.R. 2013. Arsenic bioremediation by biogenic iron oxides and sulfides. *Appl. Environ. Microbiol.* 79: 4325–4335.
Sharma, A.K., Tjell, J.C., Sloth, J.J. and Holm, P.E. 2014. Review of arsenic contamination, exposure through water and food and low cost mitigation options for rural areas. *Appl. Geochem.* 41: 11–33.

Arsenic Research and Global Sustainability – Bhattacharya, Vahter, Jarsjö, Kumpiene,
Ahmad, Sparrenbom, Jacks, Donselaar, Bundschuh & Naidu (Eds)
© 2016 Taylor & Francis Group, London, ISBN 978-1-138-02941-5

TiO$_2$ facets determine arsenic adsorption and photo-oxidation

C. Jing & S. Liu

Research Center for Eco-Environmental Sciences, Chinese Academy of Sciences, Beijing, China

ABSTRACT: Anatase (TiO$_2$) nanomaterials have been widely used in arsenic (As) remediation, although reports on their adsorption and photocatalytic capacity have been controversial. The motivation for our study is to explore the As adsorption and photooxidation processes on different TiO$_2$ facets at the molecular-level. Our results from multiple complementary characterization techniques suggest that anatase {001} facets have stronger Lewis acid sites than those on {101} facets, resulting in a higher As adsorption affinity. Density Functional Theory (DFT) calculations confirmed that the As surface complex is more energetically favorable on {001} than on {101} facets. In addition, the strong interaction of {001} facets with molecular O$_2$ facilitates the transfer of photo-excited electrons to the adsorbed O$_2$ to generate superoxide radical, which is the primary As(III) oxidant as evidenced by our radical-trapping experiments.

1 INTRODUCTION

Adsorption and photooxidation on TiO$_2$ nanomaterials provide a promising technique for arsenic (As) removal (Guan *et al.*, 2012). Though the As adsorption mechanism is well known to involve the formation of a bidentate binuclear surface complex (Pena *et al.*, 2006), even TiO$_2$ of the same anatase phase from different sources exhibit distinct adsorption and photo-catalytic capacities (Dutta *et al.*, 2005, Jegadeesan *et al.*, 2010). The lack of conformity of many experimental observations with the general belief that particle size or surface area regulates TiO$_2$ adsorption and photo-catalysis motivates our study. The molecular-level insights obtained in this study should be of paramount importance in the design and implementation of commercially available TiO$_2$ in As remediation as well as in other environmental applications.

The objective of this study was to identify the primary factor influencing As adsorption and photo-catalysis on anatase TiO$_2$, as well as its mechanism. Multiple complementary characterization methods and Density Functional Theory (DFT) calculations were employed to explore the molecular-level processes on the {001} and {101} facets. The intrinsic facet-dependent mechanism obtained from this study would be fundamental in developing TiO$_2$-based environmental technologies with high adsorption capacity and photocatalytic activity.

2 METHODS/EXPERIMENTAL

2.1 *Materials*

Three different kinds of TiO$_2$ nanoparticles (NPs), including one homemade (HM) and two commercial samples, were used in this study. The HM anatase TiO$_2$ was prepared by hydrolysis of titanyl sulfate as detailed in our previous study (Luo *et al.*, 2010). Commercial JR05 and TG01 anatase were obtained from Xuancheng Jingrui New Material Co., LTD (Anhui, China).

2.2 *As adsorption and photooxidation*

Adsorption isotherm experiments were performed to determine the As(III) and As(V) adsorption capacity on the three types of TiO$_2$ in 0.04 M NaCl solution. As(III) photooxidation experiments were carried out in an Erlenmeyer flask with 14 mg As(III)/L in a suspension containing 0.3 g/L TiO$_2$ and 0.04 M NaCl at pH 7. The sample was stirred in the dark for 2 h to achieve equilibrium before being illuminated by a tubular mercury UV lamp (CEL-WLPM10-254, wavelength 254 nm) with an incident light intensity of about 4000 µW/cm^2. A control experiment in the dark was also performed. The As concentration and speciation were determined using High Performance Liquid Chromatography (HPLC) coupled with a hydride generation atomic fluorescence spectrometer (HG-AFS, Jitian, P.R. China) with a detection limit of 0.7 µg/L for As(III) and 1.7 µg/L for As(V). The electron spin resonance (ESR) was used to detect Reactive Oxygen Species (ROS) on a BRUKER EMX plus spectrometer after 30 s photo-irradiation with an 100 W mercury lamp (LOT-Oriel Gmbh & Co. KG).

2.3 *Characterization and computation*

The reaction mechanisms were investigated using XRD, BET, SEM, TEM, NH$_3$-TPD, FTIR, and Raman. DFT calculations were conducted to investigate the effect of TiO$_2$ facets on the As adsorption and photo-catalysis. The plane-wave based calculations were performed using the Castep package in Materials Studio (Accelrys, San Diego, CA).

3 RESULTS AND DISCUSSION

3.1 S_{BET} *effect on adsorption capacity*

The adsorption isotherms of As(III) and As(V) on the three TiO$_2$ NPs conformed to the Langmuir model. Unexpectedly, no linear dependence was observed for the maximum adsorption capacity (q_m) on S_{BET} or particle size. In contrast to the general belief that q_m depends on S_{BET} or particle size, our results suggest that S_{BET} or particle size may not be the key factor influencing the As adsorption capacity on TiO_2 NPs.

3.2 Effect of surface acid sites

Only the Lewis type of acid sites existed on the surfaces of the three TiO_2 NPs, and the acid site concentration was not proportional to the adsorption capacity. The distribution of acid sites is closely related to surface atomic coordination and arrangement, indicating that different crystal facets may endow the same Lewis acid sites with different strengths. We proposed that the high As adsorption capacity on JR05 should be related to its exposed facets and their relative proportions, and this hypothesis was justified by the following quantum chemistry calculations.

3.3 Facet dependence of As adsorption energy

TiO_2 {001} and {101} facets, having different surface atomic structures, exhibited distinct abilities in anchoring As. Notably, TiO_2 {001} facets provide 100% five coordinated (Ti_{5c}) sites as compared to 50% Ti_{5c} on the {101} facets (Fig. 1). The different density of these highly reactive unsaturated Ti_{5c} sites contributes to the variation in adsorption capacity for the facets. Thus, the higher proportion of exposed {001} facets with strong Lewis acid sites on JR05 gave rise to its high As adsorption capacity.

3.4 Identify ROS in As(III)-TiO₂/UV system

Our ESR and fluorescence experimental results show that both •OH and $O_2^{\bullet-}$ existed for all TiO_2 samples. To compare the contributions of •OH, $O_2^{\bullet-}$, and photogenerated holes (h+) to As(III) photo-oxidation, the photo-catalytic reaction was investigated using the radical trapping technique with three selective radical scavengers and N_2 purging. The results strongly indicate that $O_2^{\bullet-}$ dominated As(III) photo-oxidation on TiO_2.

3.5 Facet effect on As(III) photo-oxidation

Our molecular-level experimental results show that As(III) photo-oxidation on TiO_2 can be simultaneously tuned through the synergistic effects of surface atomic structure and surface electronic structure. The {001} facets with 100% Ti_{5c} atoms exhibited stronger interactions with molecular O_2 as evidenced by the shorter Ti-O bond distance (2.012 Å) than that on {101} facets (2.033 Å). Its unique surface atomic structure resulted in a more negative adsorption energy (−2.48 eV) compared with that on {101} facets (−1.79 eV), which could facilitate the charge transfer (CT = 0.39) from {001} facets to adsorbed O_2 to generate $O_2^{\bullet-}$. In addition, the electronic structure of {001} facets with abundant oxygen vacancies facilitated the mobility of the charge carriers, which should expedite the interfacial electron transfer and electron-hole separation, and subsequently promote $O_2^{\bullet-}$ generation and As(III) oxidation. This synergistic effect could explain the higher photo-reactivity of JR05 (0.0213 min⁻¹) with 17% {001} facets compared with HM (0.006 min⁻¹) with only 4% {001} facets.

4 CONCLUSIONS

TiO_2{001} facets, with stronger Lewis acid Ti^4 atoms, exhibit higher adsorption affinity than {101} facets. Furthermore, the strong interactions of {001} facets with oxygen molecules lead to more photoelectrons being transferred to the adsorbed O_2 to generate superoxide radical, which is the primary oxidant in the As(III) photo-oxidation.

ACKNOWLEDGEMENTS

We acknowledge the financial support of the National Basic Research Program of China (2015CB932003).

REFERENCES

Dutta, P.K., Pehkonen, S.O., Sharma, V.K. and Ray, A.K. 2005. Photocatalytic oxidation of arsenic(III): Evidence of hydroxyl radicals. *Environ. Sci. Technol.* 39(6): 1827–1834.
Guan, X., Du, J., Meng, X., Sun, Y., Sun, B. and Hu, Q. 2012. Application of titanium dioxide in arsenic removal from water: a review. *J. Hazard. Mater.* 215: 1–16.
Jegadeesan, G., Al-Abed, S.R., Sundaram, V., Choi, H., Scheckel, K.G. and Dionysiou, D.D. 2010. Arsenic sorption on TiO 2 nanoparticles: size and crystallinity effects. *Water Res.* 44(3): 965–973.
Luo, T., Cui, J., Hu, S., Huang, Y. and Jing, C. 2010. Arsenic removal and recovery from copper smelting wastewater using TiO_2. *Environ. Sci. Technol.* 44(23): 9094–9098.
Pena, M., Meng, X., Korfiatis, G.P. and Jing, C. 2006. Adsorption mechanism of arsenic on nanocrystalline titanium dioxide. *Environ. Sci. Technol.* 40(4): 1257–1262.

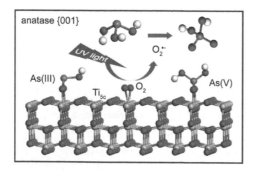

Figure 1. Mechanism of As adsorption and As(III) photooxidation on TiO_2{001}.

Arsenic Research and Global Sustainability – Bhattacharya, Vahter, Jarsjö, Kumpiene,
Ahmad, Sparrenbom, Jacks, Donselaar, Bundschuh & Naidu (Eds)
© *2016 Taylor & Francis Group, London, ISBN 978-1-138-02941-5*

Biological detoxification of As(III) and As(V) using tea waste/MnFe$_2$O$_4$ immobilized *Corynebacterium glutamicum* MTCC 2745

M.S. Podder & C. Balomajumder
Department of Chemical Engineering, Indian Institute of Technology, Roorkee. Roorkee, India

ABSTRACT: The present study has dealt with the design of Simultaneous Biosorption and Bioaccumulation (SBB) batch system for arsenic (As(III) or As(V)) ion removal from wastewater. Tea waste/MnFe$_2$O$_4$ composite was used as carrier to immobilize *Corynebacterium glutamicum* MTCC 2745. This approach was accepted for SBB of arsenic ion from wastewater. The surface texture of the biomass was investigated through Fourier Transform Infrared (FT-IR), Scanning Electron Microscopy (SEM) analysis. Various parameters which govern their optimal removal were optimized. The minimum contact time to reach equilibrium is about 240 min at pH 7.0 at 30 °C temperature using 1 g/L biosorbent for both ions. The use of 0.1% (v/v) formaldehyde as a disinfecting agent inhibited the growth of bacteria existing in the final wastewater discarded.

1 INTRODUCTION

Arsenic is extremely poisonous and has historically been utilized as a poison. Arsenic is categorized as a category 1 and group A human carcinogen by the International Association for Research on Cancer (IARC, 2004) and the US Environmental Protection Agency (US EPA, 1997), respectively. A novel technique for an efficient metal ion removal mediated by immobilized bacterial cells is designed. This metal ion removal system is called Simultaneous Biosorption and Bioaccumulation (SBB) system. In the present SBB system, both non-living biomass and living microbial cells are used simultaneously. The biosorption across liquid phase is a very cheap, robust, versatile, and eco-friendly technology of remediation of heavy metal pollution (Mishra et al., 2013). In the present work, a composite of Tea waste/MnFe$_2$O$_4$ (TW/MnFe$_2$O$_4$) has been synthesized in order to hybridize high adsorption capacity of MnFe$_2$O$_4$ with biosorptivity of tea waste.

2 METHODS/EXPERIMENTAL

2.1 *Materials*

Tea Waste (TW) was collected from the tea stall located in the campus of Indian Institute of Technology, Roorkee, India. NaAsO$_2$, Na$_2$HAsO$_4$, 7H$_2$O were purchased from Himedia Laboratories Pvt. Ltd. Mumbai, India. The synthesis of TW/MnFe$_2$O$_4$ using coprecipitation method follows according to the reactions:

$$Mn^{2+} + 2Fe^{3+} + 8OH^- \rightarrow Mn(OH)_2\downarrow + 2Fe(OH)_3\downarrow$$
$$\rightarrow MnFe_2O_4 + 4H_2O \qquad (1)$$

$$MnFe_2O_4 + TW \rightarrow TW-MnFe_2O_4 \qquad (2)$$

The microorganism used was the arsenic–resistant bacterium *B. arsenicus* MTCC 4380 (MTCC, Chandigarh, India).

2.2 *Batch experimental study*

A medium with 1.0 g/L of beef extract and 2.0 g/L of yeast extract, 5.0 g/L of peptone and 5.0 g/L of NaCl was utilized for the growth of the microorganism. The media was sterilized at 121 °C for 15 min, cooled to room temperature, inoculated with bacteria and kept at 30°C for 24 h with moderate shaking (120 rpm) in an incubator cum orbital shaker (REMI Laboratory instruments). Batch biosorption/bioaccumulation studies for optimizing process parameters were performed in round bottom flasks by taking 100 mL C. *glutamicum* MTCC 2745 bacterial suspension as a test solution in 250 mL round bottom flask closed with cotton plug tightly. After 24 h of immobilization of bacteria on TW/MnFe$_2$O$_4$ composite as described above the effect of difference process parameters (such as pH, biosorbent dose, contact time, temperature and initial adsorbate (As(III)/As(V)) concentration) were studied adding calculated amount of arsenic (As(III)/As(V)). To regulate the initial pH of the solution using a digital pH meter (HACH® India) 1.0 N NaOH and 1.0 N HCl solutions were used. The flasks were moderately agitated in an incubator cum orbital shaker working at 120 rpm. All the experiments were carried out in duplicates and average results were used. Initial arsenic (As(III)/As(V)) concentrations were selected as 50 mg/L. The optimum pH, TW/MnFe$_2$O$_4$ dose, contact time and temperature were selected from a range of pH 2.0–12.0, 0.1–10 g/L, 5–300 min and temperature 20–40°C, respectively. Initial arsenic concentration of 50–2000 mg/L were taken for SBB of arsenic (As(III)/As(V)) and uptake capacity was decided on the basis of maximum% removal of arsenic (As(III)/As(V)).

The samples were filtered through Whatman Filter paper (Cat No 1001 125) at predetermined time intervals and then centrifuged at 10,000 rpm for 10 min to investigate the influence of parameters and also for isotherm studies, a portion of filtrate was diluted with HNO_3 solution (10%, v/v). The filtrate was analysed for determination of arsenic concentration using ThermoFisher Scientific iCE 3000 Series AA Graphite Furnace Atomic Absorption (GFAA) spectrometer (detection limit 20 µg/L).

3 RESULTS AND DISCUSSION

3.1 Characterization of biosorbents

The Fourier Transform Infrared spectra (FT–IR) study of fresh $TW/MnFe_2O_4$ composite and immobilized bacterial cells at unloaded and arsenic (As(III)/As(V)) loaded stage at optimized batch experimental condition were studied for detecting the function groups responsible mainly for the SBB process. The SEM images of the prepared fresh TW/$MnFe_2O_4$ composite and immobilized bacterial cells at unloaded and arsenic (As(III) or As(V)) loaded stage were examined.

3.2 Optimization of process parameters

At the optimized conditions (pH 7.0 (Fig. 1a), biosorbent dose (Fig. 1b), contact time 280 min (Fig. 1c), and temperature 30°C (Fig. 1d), the uptake of As(III) and As(V) increased with increasing initial concentration of As(III) or As(V) (Fig. 2).

3.3 Disinfection of arsenic deficient wastewater

Considering the possible detachment of bacteria from the biofilm, disinfection of wastewater was conducted as a preventive measure for bacterial contamination of the final wastewater. The use of 0.1% (v/v) formaldehyde as a disinfecting agent inhibited the growth of bacteria existing in the final wastewater discarded.

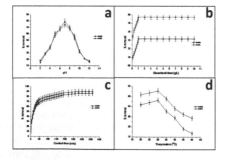

Figure 1.   Effect of: a) pH, b) biosorbent dose, c) contact time, and d) temperature on the removal of As(III) and As(V).

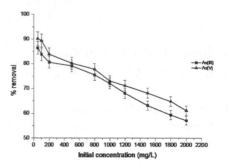

Figure 2.   Effect of initial concentration of As(III) and As(V) on removal.

4 CONCLUSIONS

The biosorption/bioaccumulation mechanism of As(III) and As(V) on the C. glutamicum MTCC 2745 immobilized on surface of $TW/MnFe_2O_4$ composite was complex. Electrostatic interaction occurred in biosorption/bioaccumulation of As(III) and As(V) and the hydroxyl groups bonded to $MnFe_2O_4$ and surface of bacteria played a vital role in the biosorption/bioaccumulation of As(III) and As(V) via electrostatic interaction or ligand exchange. The optimum biosorbent dose and contact time for biosorption/bioaccumulation of As(III) and As(V) were 1 g/L and 280 min, respectively. The rate of biosorption/bioaccumulation of both As(III) and As(V) by C. glutamicum MTCC 2745 immobilized on surface of $TW/MnFe_2O_4$ composite decreased with increasing concentration. C. glutamicum MTCC 2745 immobilized on surface of $TW/MnFe_2O_4$ composite, hybrid biosorbent has a prospective application for removal of arsenic in water treatment.

ACKNOWLEDGEMENTS

Our thanks to Indian Institute of Technology, Roorkee for providing necessary facilities and to Ministry of Human Resource Development, Government of India for financial support.

REFERENCES

IARC 2004. Evaluation of Carcinogenic Risks to Humans. Some Drinking-Water Disinfectants and Contaminants, including Arsenic, IARC Monograph Vol. 86. IARC Press, Lyon.

Mishra, V., Balomajumder, C., garwal, V.K. 2013. Design and optimization of simultaneous biosorption and bioaccumulation (SBB) system: a potential method for removal of Zn(II) ion from liquid phase, Desalin. Water Treat. 51: 3179–3188.

US EPA (US Environmental Protection Agency) 1997. IRIS (Integrated Risk Information System) Online Database Maintained in Toxicology Data Network (TOXNET) by the National Library of Medicine, Bethesda, Maryland.

Arsenic Research and Global Sustainability – Bhattacharya, Vahter, Jarsjö, Kumpiene, Ahmad, Sparrenbom, Jacks, Donselaar, Bundschuh & Naidu (Eds)
© 2016 Taylor & Francis Group, London, ISBN 978-1-138-02941-5

Preparation and evaluation of iron particles impregnated chitosan beads for arsenic removal from groundwater and surface water

G. Villa, C. Huamaní, M. Chavez & J. Huamaní

Laboratorio Químico Toxicológico, Centro Nacional de Salud Ocupacional y Protección del Ambiente para la salud (CENSOPAS), Instituto Nacional de Salud, Lima, Perú

ABSTRACT: Chitosan was extracted from squid pen (Dosidicus gigas) by cold homogeneous deacetylation process and characterized by FT-IR, SEM. Chitosan iron nanoparticles (Ch:nZVI:Ch) was prepared by reducing Fe^{+3} with $NaBH_4$ in the presence of chitosan as stabilizer. The images by SEM show that the area of chitosan beads are highly porous and zero-valent iron is loaded correctly and efficiently dispersed. To evaluate the efficiency of removal 100 mg of synthesized material was taken to remove As from 1000 µg/L to < 10 µg/L in 120 minutes. When the material is saturated, the regeneration of the adsorption was carried out using 0.1 M NaOH. Results obtained from batch studies of groundwater and surface water contaminated by arsenic obtained from Tacna—Perú, were reduced from 500 µg/L to < 10 µg/L. In conclusion, ions present in both groundwater and surface water do not interfere with the removal of As in drinking water sources.

1 INTRODUCTION

Millions of people in developing countries have ingested drinking water at a concentration several times higher than the recommended maximum of 10 µg/L of arsenic (As), such as Peru, Bolivia, Mexico and El Salvador (Bundschuh *et al.*, 2010; Mandal & Suzuki, 2002). In Peru, As is naturally through surface water and groundwater from the volcanic activity in the Andes Mountains, and this in turn used for domestic activities (Sancha, 2000).

Zero-valent iron stabilized in a matrix of chitosan as adsorbent has proven an efficient As removal from water. In this paper, we present the evaluation and application of Ch:nZVI:Ch towards the removal of As from contaminated water samples. (Gupta *et al.*, 2012; Geng *et al.*, 2009).

2 MATERIALS AND METHODS

2.1 *Materials*

Squid pen (*Dosidicus gigas*) were collected from the fish market of Villa Maria del Triunfo (Lima-Peru) and sieved through a 10 mesh screen. Reagents used as sodium hydroxide (NaOH), sodium borohydride (NaBH₄), ferric chloride ($FeCl_3.6H_2O$), ethanol (C_2H_5OH) were acquired from Merck Millipore and other reagents as acetic acid (CH_3COOH), Sodium arsenate ($HAsNa_2O_4.7H_2O$) from Sigma-Aldrich. All reagents were prepared with Millipore deionized water. Samples of surface waters and groundwater were collected from Tacna-Peru for testing under real conditions.

2.2 *Preparation of Ch:Nzvi:Ch*

Extraction of chitosan was by cold homogeneous deacetylation. It consists of two steps: Deproteinization (β-chitin) followed by deacetylation.

The preparation of nZVI (Zero-valent iron) on chitosan matrix was performed by the "sol-gel" process. Chitosan was dissolved in dilute acetic acid. To this solution added $FeCl_3.6H_2O$ and the solution was stirred overnight. Then, was added dropwise a fresh aqueous solution of NaBH4 and black precipitate was observed which is allowed to stir for 90 min. Finally, the resulted black precipitate was collected and washed by ethanol and filtered with filter paper Whatman® # 41. The whole process was carried out in a nitrogen atmosphere.

The preparation of Ch:nZVI:Ch was performed when chitosan was dissolved in dilute acetic acid. To this solution added the black precipitate and the solution was stirred for few hours in nitrogen atmosphere. Finally, the solution was added dropwise in alkaline medium.

3 RESULTS AND DISCUSSION

3.1 *Evaluation of chitosan*

The degree of deacetylation obtained by FT-IR (Thermo Scientific) and potentiometric titration (HANNA instruments) was >80% DA as shown in Figure 1. The surface analysis by scanning electron microscopy (SEM) shows parallel fibers and a high content of carbon and oxygen (Fig. 2).

Chitosan obtained has similar characteristic peaks as the control. The superficial analysis

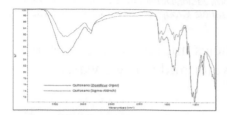

Figure 1. Comparison in FT-IR spectra between chitosan from squid pen and chitosan of sigma-Aldrich.

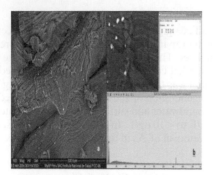

Figure 2. a) Superficial image of chitosan b) Surface composition of chitosan.

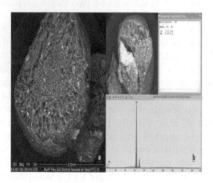

Figure 3. a) Chitosan beads highly porous. b) Surface composition of the bead.

showed a high content of carbon and oxygen, typical to its structure.

3.2 *Evaluation of Ch:nZVI:Ch*

Images by scanning electron microscopy showed a highly porous material. Also showed that zerovalent iron was loaded and correctly dispersed on the chitosan beads as shown in Figure 3.

3.3 *Evaluation of behavior*

The tests were performed in a range of pH 7–8, the kinetic assay performed with 100 mg of the material achieves remove of 1000 µg/L to < 10 µg/L in 120 minutes (Table 1). Also remove (same weight) from 10 mg/L to less than 50% in 30 minutes.

Table 1. Evaluation of contact time.

Time (min)	[C]i	[C]f	% Removal.
0	973.11	973.11	0.00
10	973.11	616.78	36.62
30	973.11	417.84	57.06
60	973.11	174.14	82.10
90	973.11	62.83	93.54
120	973.11	7.06	99.27

[C]i: initial concentration, [C]f: final concentration

3.4 *Desorption*

The material can be regenerated with 0.1M NaOH and can be used again showing high efficiency after desorption.

3.5 *Removal of arsenic from water of Tacna-Peru*

The material obtained was used as filter. The system achieved remove As from contaminate water (400–500 µg/L) to < 10 µg/L.

4 CONCLUSIONS

Chitosan obtained is a good support for the iron nanoparticles because prevents agglomeration. Chitosan beads provide wide surface area due to the large number of pores. Ions of groundwater and surface water do not prevent the removal of the pollutant.

ACKNOWLEDGEMENTS

Funding from Grand Challenges Canada (GCC) to carry out this work is gratefully acknowledged.

REFERENCES

Bundschuh, J., Litter, M., Ciminelli, V., Morgada, M, Cornejo, L., Garrido Hoyos, S., Hoinkis, J., Alarcón-Herrera, M.T., Armienta, M.A. & Bhattacharya, P. 2010. Emerging mitigation needs and sustainable options for solving the arsenic problems of rural and isolated urban areas in Latin America—A critical analysis. *Water Res.* 44(19): 5828–5845.

Gupta, A., Yunus, M., Sankararamakrishnan, N. 2012. Zerovalent iron encapsulated chitosan nanospheres—A novel adsorbent for the removal of total inorganic Arsenic from aqueous systems. *Chemosphere* 86: 150–155.

Geng, B., Jin, Z., Li, T., Qi, X. 2009. Preparation of chitosan-stabilized Fe0 nanoparticlesfor removal of hexavalent chromium in water. *Sci. Total Environ.* 407: 4994–5000.

Mandal, B., Suzuki, K. 2002. Arsenic around the world: a review. *Talanta* 58: 201–235.

Sancha, M., Esparza Castro de M. 2000. Arsenic status and handling in Latin America. Universidad de Chile, Grupo As de AIDIS/DIAGUA, CEPIS/OPS, Lima, Perú.

Arsenic Research and Global Sustainability – Bhattacharya, Vahter, Jarsjö, Kumpiene,
Ahmad, Sparrenbom, Jacks, Donselaar, Bundschuh & Naidu (Eds)
© 2016 Taylor & Francis Group, London, ISBN 978-1-138-02941-5

Investigating arsenic removal from water by the use of drinking water treatment residuals

A. Ahmad[1], V.A. Salgado[2], W. Siegers[1] & R. Hofman-Caris[1]
[1]*KWR Watercycle Research Institute. Nieuwegein, The Netherlands*
[2]*Sanitary Engineering Section, Delft University of Technology, Delft, The Netherlands*

ABSTRACT: Every year approximately 80,000 tonne of iron rich wet residuals (sludge) from drinking water treatment processes are generated by the Dutch drinking water companies. In current study the As adsorption potential of stabilized iron rich residuals has been investigated by a series of adsorption kinetics and isotherm experiments. The residuals based granules performed well in removing As(V) from aqueous solution at typical drinking water pH of 8.2. The initial rate of adsorption of As was rapid, followed by a slower rate that gradually approached a plateau. Adsorption rate data were best described by a pseudo second order kinetic model. According to estimation based on Freundlich equation, approximately 580 µg of As(V) can be adsorbed per g of stabilized residuals.

1 INTRODUCTION

A number of arsenic (As) removal techniques are available today, however, the performance of these methods in removing low As concentrations is not well-known. Among the many available As removal technologies, adsorption of As on packed beds of granular media can be regarded as the most straight-forward and easy method. Adsorbents that have been successfully demonstrated for the removal of As concentrations include a variety of metal oxides such as Activated Alumina (AA), iron oxides, Titanium Oxides (TiO_2) and Zirconium Oxides (ZrO_2). Presently, the adsorptive media based on iron oxides and hydroxides (e.g. GFH®, IOCS etc.) are most widely used in removing As because of their high adsorption capacity, robustness and relatively lower operational costs.

Every year approximately 80,000 ton of iron rich wet residuals (sludge) from drinking water treatment processes are generated by the Dutch drinking water companies. In the current study, As adsorption potential of stabilized iron rich residuals from 2 treatment plants in the Netherlands has been investigated and compared with an iron oxide based commercial adsorbent.

2 MATERIALS AND METHODS

2.1 Stabilization of iron rich residuals and preparation for adsorption tests

Residuals (8–12% dry matter) from two groundwater treatment plants (GWTP1 and GWTP2) in the Netherlands were collected, dried and pelletized by a special lab-scale process to form granules having 0.5–2 mm diameter. An iron oxide based commercial adsorbent was requested from the supplier to use in the study for comparison. All the 3 adsorbents, GWTP1, GWTP2 and commercial adsorbent, were pre-treated by washing with demineralized water and subsequent drying at 105 degree Celsius, before using the adsorbents in the batch tests for determining As(V) adsorption kinetics and isotherms.

2.2 Bach adsorption tests

To determine the As(V) adsorption potential of the residuals based granules, a series of adsorption kinetic and adsorption isotherm experiments were carried out. To determine As(V) adsorption kinetics, 0.15 g of each adsorbent was added to 300 mL of semi-synthetic (tap water spiked with 100 µg/L As(V)) solution, at pH 8.2. The batch reactors (300 mL bottles) were continuously mixed at 200 RPM by magnetic stirrers and placed in an incubator at 15 degree Celsius. Samples were collected at 1, 2, 5, 8, 24, 48, 72 and 144 h. Adsorption isotherms for all the 3 adsorbents were determined at 15 degree Celsius, initial As concentration of 100 µg/L and by varying the mass of adsorbents from 0.1 to 1.5 g/L. Samples were collected at 72 hours, i.e. at equilibrium pre-determined from the adsorption kinetic experiments. All the samples were analysed by ICP-MS at KWR Watercycle Research Institute.

Adsorption kinetic data were fitted to pseudo-first order and pseudo-second order kinetic equations and the data from adsorption isotherm experiments were fitted to Langmuir and Freundlich equations.

3 RESULTS AND DISCUSSION

Figure 1 illustrates the uptake of As(V) by the 3 granules as a function of time. It can be seen that the initial uptake rate was considerably fast for all

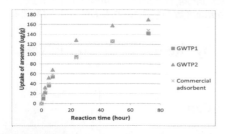

Figure 1. Uptake of As(V) as a function of time.

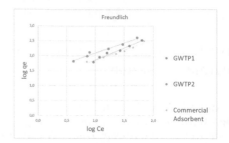

Figure 2. Linearized Freundlich adsorption isotherm.

Table 1. Estimated pseudo-second order model parameters for As(V) adsorption.

Adsorbent	K_2 g/(µg.min)	R^2	h µg/(g.min)	q_e µg/g
GWTP1	4.79E–06	0.997	0.1526	178.6
GWTP2	5.84E–06	0.999	0.2432	204.1
Com Ads	6.19E–06	0.987	0.1840	172.4

Table 2. Estimated Freundlich model parameters for As(V) adsorption.

Adsorbent	R^2	K_F (µg/g)	slope = 1/n	qm = $K_F C_0^{1/n}$ (µg/g)
GWTP1	0.972	11.3	0.806	463.1
GWTP2	0.975	26.7	0.669	579.7
Com Ads	0.966	15.1	0.674	337.4

the adsorbents. The curves started levelling-off after 24 h. The initial fast rate of As(V) adsorption is typical for porous adsorbents. It has been reported previously that the adsorbed As slowly diffuses towards the internal adsorption sites, and as a result the adsorption sites on the exterior become available for more As molecules (Fuller *et al.*, 1993; Banerjee *et al.*, 2008; Bhatnagar *et al.*, 2010). Moreover, it can be seen in Figure 1 that the As(V) uptake by the residuals from the GWTP2 is higher compared to the residuals from GWTP1 and the commercial adsorbent. This can be attributed to higher total iron oxide (86%) content of the residuals from the GWTP2 and comparable total iron oxide content of GWTP1 (73%) and the commercial adsorbent (72%). The adsorption kinetic data best fitted to a pseudo-second order model (Table 1). The estimated parameters K_2: rate constant of pseudo-second order adsorption, h: initial adsorption rate, q_e: adsorption capacity at equilibrium and the correlation coefficient R^2 are shown in Table 1.

The results of isotherm experiments are shown in Figure 2. It can be seen that the granules produced from the residuals of GWTP2 performed better compared to the other 2 adsorbents, although the isotherms were similar in shape. The Freundlich model fitted the equilibrium isotherm data better compared to the Langmuir model. Therefore, only the estimated Freundlich model parameters K_F: Freundlich dimensionless constant, q_{max}: maximum adsorption capacity, 1/n: a measure of adsorption intensity or surface heterogeneity and R^2 have been provided in Table 2.

4 CONCLUSIONS

Both the residuals based granules performed well in removing As(V) from aqueous solution at typical drinking water pH of 8.2. The initial rate of adsorption of As was fast, followed by a slower rate that gradually approached a plateau. Adsorption rate data were best described by a pseudo-second order kinetic model. According to an estimation based on Freundlich equation, approximately 580 µg of As(V) can be adsorbed per g of stabilized residuals, which is a higher adsorption capacity compared to the reference commercial adsorbent used in this study. The experimental data obtained in this study will be very useful in establishing a scientific framework for the use of stabilized residuals from the drinking water treatment processes as a cost effective adsorbent for As contaminated aqueous streams.

ACKNOWLEDGEMENTS

This study has been conducted within the Dutch Government's Top consortia Knowledge and Innovation (TKI) scheme with partners Brabant Water, Dunea Duin & Water, Evides Waterbedrijf, Reststoffenunie and Waternet.

REFERENCES

Banerjee, K., Amy, G., Prevost, M., Nour, S., Jekel, M., Gallagher, P., Blumenschein, C. 2008. Kinetic and thermodynamic aspects of adsorption of arsenic onto granular ferric hydroxide (GFH). *Water Res.* 42: 3371–3378.

Bhatnagar, A., Minocha, A., Sillanpää, M. 2010. Adsorptive removal of cobalt from aqueous solution by utilizing lemon peel as biosorbent. *Biochem. Eng. J.* 48: 181–186.

Fuller, C.C, Davis J.A, Waychunas, G.A. 1993. Surface chemistry of ferrihydrite: Part 2. Kinetics of arsenate adsorption and coprecipitation. *Geochim. Cosmochim. Acta* 57: 2217–2282.

Arsenic Research and Global Sustainability – Bhattacharya, Vahter, Jarsjö, Kumpiene,
Ahmad, Sparrenbom, Jacks, Donselaar, Bundschuh & Naidu (Eds)
© *2016 Taylor & Francis Group, London, ISBN 978-1-138-02941-5*

Improved drinking water treatment for arsenic removal by use of a combined biological-adsorptive iron removal step

J. Koen & H. Koen

PIDPA Department of Process Technology and Water Quality, Antwerp, Belgium

ABSTRACT: In this study the removal of arsenic during biological-adsorptive iron removal was investigated. Biological-adsorptive iron removal is an efficient way to remove arsenic without the use of chemical agents. Due to adaptation of the bacterial culture and/or formation of IOCS, the removal efficiency increased up until 70 days of operation and then remained constant afterwards (>95% removal). Filtration rates up to 12 m/h had no significant influence on the arsenic removal. After treatment arsenic concentrations of 1.0 µg/L are feasible (98% removal) when the Fe/As weight ratio in the raw water is above 250. If there is no iron rich groundwater source available (Fe/As ratio 40), a decrease in pH of the raw water can improve arsenic removal.

1 INTRODUCTION

The overall objective of this study was to investigate the removal of Arsenic (As) in combination with a biological-adsorptive iron removal step without the use of chemical dosage. Specific attention was paid to (1) the minimum As level that can be achieved after treatment by variation of the Fe/As ratios of the raw water, (2) the influence of the filtration rate and (3) the influence of unavailability of iron-rich water combined with correction of the pH on controlling As removal efficiency.

In groundwater treatment for drinking water, rapid sand filtration is a common step for iron removal. In the oxidation and precipitation of iron(II) into Hydrous Ferric Oxides (HFO), three processes may be distinguished: homogeneous (flocculent), adsorptive and biological oxidation (Sharma, 2001). Application of adsorptive iron oxidation forms an Iron Oxide-Coated Sand (IOCS) that can be used for adsorptive As removal (Petrusevski *et al.*, 2007). In biological oxidation, Iron-Oxidizing Bacteria (IOB) form biogenic HFO on which additional adsorption of As can occur (Lehimas *et al.*, 2001). Another advantage of the presence of IOB is that As(III) can be oxidized to As(V). As(V) is more easily removed by adsorption or co-precipitation than the As(III) form (Zouboulis & Katsoyiannis, 2005).

2 METHODS

All the experiments were performed with water from three different groundwater sources at the Water Treatment Plant (WTP) of Oud-Turnhout. In Table 1 the average water quality of the three sources is given. The pilot plant consists of a double rapid sand filtration system. Inoculation of the filter was carried out with rinsing water from a WTP that contains a large concentration of Gallionella species. Determination of total iron content was performed spectrofotometrically by the phenanthroline method (American Public Health Association, 1985). Total As content was measured by ICP-MS. Oxygen was measured with a WTW FDO 925 electrode and pH was determined with a WTW Sentix 940-3 electrode. Statistical analysis of all data was performed with Minitab.

3 RESULTS AND DISCUSSION

In the first part of the research the Fe/As weight ratio was varied from 75 to 300 and the filtration rate was varied from 6 to 12 m/h. The As level of the raw water was kept constant at around 65 µg/L. Interpretation of the experimental results was done by using a multiple linear regression model. During this period (150 days) the model only showed a significant influence on the iron concentration in the raw water. The model explained 95% of the variation in As concentration (Fig. 1). A minimum As level of 1.0 µg/L could be obtained after a second filtration when the Fe/As ratio was above 250.

Because of the large influence of the iron content in the raw water, other parameters seemed of no significance in the model. For this reason the data on the iron content in the raw water of 13 mg/L will be discussed separately. The model of this data showed a significant influence on the Fe/As ratio after the first filtration and the number of days the filter was in service. This model explained 65% of the variation in the As concentration. After 70 days of operation the As concentration always remained below 2.5 µg/L (Fig. 2). This improvement in As removal could be due to adaptation and growth of the inoculated bacteria (Lehimas *et al.*, 2001, Katsoyiannis & Zouboulis, 2004) or coating of the filter medium with

Table 1. Average water quality parameters for 3 water types.

Parameter	Oud-Turnhout	Ravels	Arendonk
Fe (mg/L)	23.0	4.0	3.8
As (µg/L)	39	75	46
pH	6.8	7.6	7.5

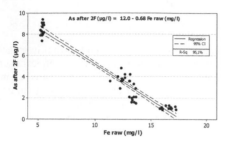

Figure 1. Linear regression of the arsenic concentration after second filtration (2F) versus iron content in raw water.

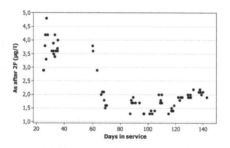

Figure 2. Arsenic concentration after second filtration (2F) versus number of days in service.

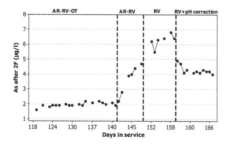

Figure 3. Effect of lowering the Fe/As ratio and subsequent pH correction.

iron oxides (IOCS) (Ahmad et al., 2014, Sharma, 2001). The filtration rate was varied between 7 and 12 m/h during this period which gave no significant change in As removal.

In addition the worst-case scenario, the absence of groundwater with the highest iron content from Oud-Turnhout, was investigated. This resulted in an iron content of the raw water of 3.5–4.0 mg/L, an As concentration of 75–90 µg/L (Fe/As ratio 40–50)

and a pH around 7.7. After 15 days of operation the As concentration after the second filtration was 6.6 µg/L (Fig. 3). When a pH correction is applied to the raw water from 7.7 to 7.0, the As concentration reduced from 6.6 to a constant value of 4.4 µg/L. The effect of pH reduction can be explained by a positive effect on the adsorption of As on IOCS (Edwards, 1994, Lenoble et al., 2002) and a negative effect on homogeneous iron oxidation, which gives the advantage to biological-adsorptive oxidation.

4 CONCLUSIONS

Changing the Fe/As ratio by increasing the amount of iron in the raw water was a very efficient way to lower the concentration of As after treatment with a biological-adsorptive iron removal method. A minimum As level of 1.0 µg/L could be obtained when the Fe/As ratio is >250 with filtration rates up to 12 m/h. The number of days a filter was in service had a significant influence on the final concentration of As. Creating a worst-case treatment scenario with low iron content in the raw water resulted in an increase of As up to 6.6 µg/L. Adjustment of the pH of the raw water from 7.7 to 7.0 reduced the As concentration after treatment to 4.4 µg/L.

REFERENCES

Ahmad, A., van de Wetering, S., Groenendijk, M., Bhattacharya, P. 2014. Advanced Oxidation-Coagulation-Filtration (AOCF)—an innovative treatment technology for targeting drinking water with <1 µg/L of arsenic. In: M.I. Litter, H.B. Nicolli, M. Meichtry, N. Quici, J. Bundschuh, P. Bhattacharya & R. Naidu (eds.) One Century of the Dis-covery of Arsenicosis in Latin America (1914–2014) As 2014. CRC Press/Taylor and Francis, pp. 817–819.
American Public Health Association. 1985. Standard Methods for the Examination of Water and Wastewater. 16th Ed., Washington, D.C., 215–220.
Edwards, M. 1994. Chemistry of arsenic: Removal during coagulation and Fe-Mn oxidation. Journal AWWA 86: 64–78.
Katsoyiannis I., Zouboulis A. (2004), Application of biological processes for the removal of arsenic from groundwaters. Water Res.38: 17–26.
Lehimas, G.F.D., Chapman, J.I., Bourgine, F.P. 2001. Arsenic removal from groundwater in conjunction with biological iron removal. Water Environ. J. 15: 190.
Lenoble, V. Bouras, O., Deluchat, V. Serpaud, B., Bollinger, J. C. 2002. Arsenic adsorption onto pillared clay and iron oxides. J. Colloid Interf. Sci. 255 (1): 52–58.
Petrusevski, B., van der Meer, W., Baker, J., Kruis, F., Sharma, S.K., Schippers, J.C. 2007. Innovative approach for treatment of arsenic contaminated groundwater in central Europe. Water Sci. Technol. 7: 131–138.
Sharma, S.K. 2001. Adsorptive iron removal from groundwater. PhD Thesis, IHE Delft, The Netherlands.
Zouboulis, A., Katsoyiannis, I. 2005. Recent advances in the bioremediation of arsenic-contaminated groundwaters. Environ. Int. 31: 213–219.

Arsenic Research and Global Sustainability – Bhattacharya, Vahter, Jarsjö, Kumpiene,
Ahmad, Sparrenbom, Jacks, Donselaar, Bundschuh & Naidu (Eds)
© 2016 Taylor & Francis Group, London, ISBN 978-1-138-02941-5

Fundamental study on arsenic removal from water/wastewater by using Hybrid adsorbent

J. Sareewan[1], M. Ohno[2], H. Suhara[2], Y. Mishima[2] & H. Araki[2]
[1]*Graduate School of Science and Engineering, Saga University, Japan*
[2]*JDC Corporation/Saga University, Japan*

ABSTRACT: A new adsorbent, Hybrid adsorbent (HB) has been synthesized for ion removal from water and wastewater. Nano-size Layered Double Hydrotalcite (NLDH) as anionic adsorbent and Zeolite (Ze) as cationic adsorbent have been combined to synthesize a more powerful adsorbent is called Hybrid adsorbent (HB) the NLDH:Ze weight ratio was selected as 1.14:1, which successfully adsorb both anion and cation at the same time. This study was investigated adsorption potency for Arsenic removal. Both *As*(III) and *As*(V) were examined by adsorption method using HB adsorbent and HB initial substances. From a result, HB showed high potential compared to NLDH alone or mixed materials. In addition, artificial leachate was used to confirm applicability of HB and another commercial adsorbents was also used to validate the efficiency of HB for toxic metals removal use.

1 INTRODUCTION

Arsenic (As) has been known as a toxic element. Toxicity of As is result in healthy effects when it up taken to the human body over a long period including skin problems, cancer, and diseases of the blood vessels of the legs and feet, and possibly also diabetes, high blood pressure and reproductive disorders (WHO, 1993). Recently, The most common As removal technologies can be grouped into the following four categories: (a) oxidation and sedimentation; (b) coagulation and filtration; (c) sorptive media filtration; and (d) membrane filtration (Cheng *et al.*, 1996, Hering *et al.*, 1997, Ahmed *et al.*, 2006). Oxidation of *As*(III) to *As*(V) is usually needed for effective removal of As from groundwater by most treatment methods. In the process of coagulation, *As* is removed from solution by three mechanisms: precipitation, co-precipitation, and adsorption. In this paper, we demonstrate a new adsorbent for the removal of As from water and wastewater as leachate by adsorption method using hybrid adsorbent.

2 METHODS/EXPERIMENTAL

2.1 *Artificial leachate*

All chemicals were prepared for immediate use using standard solution with deionized water the initial concentrations and chemicals as shown in Table 1. The pH of the artificial leachate was adjusted using 0.1 mol/L NaOH or 0.1 mol/L HCl solutions.

2.2 *Adsorbents*

This study used five types of adsorbent consisting of Powdered Activated Carbon (PAC), resin (cation and anion resin types and chelating type), Nano-size

Table 1. Initial concentrations of chemical compounds in artificial leachate. (Kashiwada et al. 2005)[a] [5].

Chemicals	Initial concentration (ppm)[a]	Standard solution and chemical reagent
F	1.74	F standard solution
NO_3^-	468	KNO_3
B	2.242	B standard solution
Na	1503	NaCl
Mg	106.9	$MgCl_2–H_2O$
Ca	216.4	$CaCl_2$
K	706.4	KI
Cd	0.00117	Cd standard solution
Cr	0.00121	Cr standard solution
Pb	0.00249	Pb standard solution
As	0.00065	As standard solution
Hg	0.0005	Hg standard solution
P	0.3633	P standard solution
Se	0.001	Se standard solution

Layered Double Hydroxide (NLDH), Zeolite (Ze), and Hybrid adsorbent (HB), which was synthesized by combining appropriate ratios of NLDH with Ze (1.14:1) then grinding the compound to powder.

2.3 *Adsorption experiments*

The efficiency of the adsorbent quantity on ions removal and *As* removal was investigated using 1 L of solutions. The mixture was stirred 24 hrs. in 2L beakers covered with a plastic film at 20°C in a temperature-controlled room. The mixture was filtrated and analyzed using chemical analysis equipment. Inductively Coupled Plasma-Atomic Emission Spectrometry (ICP-AES) and Ion spectrometer were used to determine the amount of ions in the mixture at difference wave ranges.

3 RESULTS AND DISCUSSION

3.1 Arsenic adsorption capacity test

Arsenite ion (AsO(OH)$_2$$^-$) and Arsenic acid (H$_2AsO_4$$^-$) solutions were prepared to use in adsorption test against NLDH, Hybrid adsorbent (HB), Zeolite (Ze) and 50% mixture of NLDH and Ze. To investigate the adsorption capacity for Arsenic the adsorption capacity, q$_e$ (mg/g), was determined from the difference between initial concentration (C$_0$, mg/L) and equilibrium concentration (C$_e$, mg/L) per gram of solid adsorbent;

$$q_e = \frac{(C_0 - C_e)V}{W} \quad (1)$$

Thus, W is the mass of adsorbent (g); V is the volume of solution (L)

Figures 1 and 2 show the results of equilibrium experiment against initial concentration of As(III) and As(V), respectively. In case of Arsenite ion adsorption (Figure 1), HB exhibit highest potential compared to pure NLDH or mixed compounds. Arsenic acid adsorption, as non-ionic it is excellent to adsorb by mixed compounds compared to pure NLDH. In addition, HB showed the same degree of adsorption capacity as compared with the mixed compounds as shown in Figure 2. Because of the surface of HB is rough, therefore the exchangeable area is expanded compared to the exchangeable surfaces of pure NLDH and pure Ze.

3.2 Comparison of arsenic removal efficiency between HB and commercial adsorbents by using artificial leachate

Artificial leachate was use in the study to approve the appreciable of HB and to compare with commercial adsorbent. The efficiency of the total As was determined using the following equation;

$$\%removal = \frac{C_0 - C}{C_0} \times 100 \quad (2)$$

where C$_0$ is initial concentration (mg/L) and C is residual concentration (mg/L).

Figure 3 shows the percentage of ion removal efficiency with each adsorbent. From this figure,

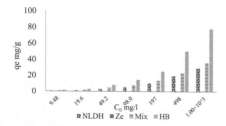

Figure 1. Arsenite ion adsorption.

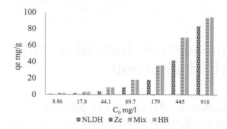

Figure 2. Arsenic acid adsorption.

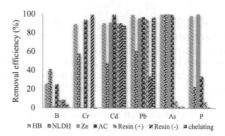

Figure 3. Comparison of ions removal in artificial leachate.

it is indicate HB's high potential for ions removal almost as equal as PAC. In addition, from this experiment, As removal by using HB, NLDH or PAC given a good result more than resin.

4 CONCLUSIONS

In this study investigation of Arsenic removal potential by HB was conducted. Both type of As(III) and As(V) were used and was found that HB exhibit high performance compared to pure NLDH. In addition, HB can be use to remove As in artificial leachate. However, the performance examination with actual leachate should be investigated for further study.

REFERENCES

Ahmed, F., Minnatullah, K., Talbi, A. 2006. Paper III. Arsenic Mitigation Technologies in South and East Asia in Toward a More Effective Operational Response, *Arsenic Contamination in Ground Water in South and East Asian Countries, vol. II, Technical Report.* Report No. 31303, The World Bank.

Cheng, R.C., Wang, H.C., Beuhler, M.D. 1996. Enhanced coagulation for arsenic removal, *J. Am. Water Works Assoc.* 86: 79–90.

Hering, J.G., Chen, P.Y., Wilkie, J.A., Elimelech, M. 1997. Arsenic removal from drinking water during coagulation, *J. Environ. Eng.* 123: 800–807.

Kashiwada, S., Osaki, K., Yasuhara, A., Ono, Y. 2005. Toxicity studies of landfill leachates using Japanese medaka (Oryzias Laptipes). *Australasian J. Ecotoxicol.* 11(5): 60–64.

WHO, 2003. Water Sanitation and Health, in Report on Intercountry Consultation, 9–12 December 2002. New Delhi: *WHO Regional Office for South-East Asia 2003*, Kolkata, India.

Arsenic Research and Global Sustainability – Bhattacharya, Vahter, Jarsjö, Kumpiene,
Ahmad, Sparrenbom, Jacks, Donselaar, Bundschuh & Naidu (Eds)
© 2016 Taylor & Francis Group, London, ISBN 978-1-138-02941-5

Development and application of Fe-biochar composite for arsenic removal from water

P. Singh & D. Mohan

School of Environmental Sciences, Jawaharlal Nehru University, New Delhi, India

ABSTRACT: Arsenic in groundwater is a serious problem. This paper investigates the role of biochar derived from agricultural byproduct to remediate arsenic from water. Biochar is developed using slow pyrolysis in a muffle furnace. The properties of biochar can be enhanced using physical and chemical treatment methods. Chemical modification is done by impregnating iron oxide in biomass. pH studies, sorption equilibrium and kinetic studies were conducted using various mathematical equations. Developed biochar composite was applied to remove arsenic from actual groundwater samples. Regeneration of exhausted biochar composite using suitable solvents were carried out. Biochar composites exhibit comparable sorption efficiency for arsenic removal. Therefore, it can be considered to replace commercial adsorbents.

1 INTRODUCTION

Arsenic in water is one of the major environmental concerns. WHO has reported that arsenic contamination in drinking water has caused greatest number of deaths as compared to the deaths reported by other heavy metals (Ahmed, 2000). According to the US Environmental Protection Agency, arsenic is amongst top most carcinogens. The arsenic permissible limit as prescribed by Bureau of Indian Standards (BIS) is 10 µg/L. Arsenic is prevalent in India and widely occur in the Gangetic flood plains in Uttar Pradesh, Bihar, Jharkhand and West Bengal, flood plains of Brahmaputra river in Assam and Manipur and in the regions of Chhattisgarh (Bhattacharya *et al.*, 2011).

2 METHODS/EXPERIMENTAL

2.1 *Material*

Several experiments and studies have been conducted to remove arsenic from water. Adsorption technique because of its simplicity and cost effectiveness has been used to remove arsenic. Several biological materials, mineral oxides, activated carbons, and polymer resins have been used to remove aqueous arsenic (Mohan et al., 2007). The biochar utilization for water decontamination is a relatively new practice. It is a fine grained porous substance developed using slow or fast/flash pyrolysis (Singh et al., 2015). In the present study, agricultural by-product was slow pyrolyzed. The resulting biochar was further modified and used for arsenic removal.

2.2 *Method*

Iron biochar composite is developed using slow pyrolysis method in a muffle furnace. Both physical and chemical activation of biomass has been done.

Bio-masses were soaked in solution prepared from a known iron compound. The soaked biomasses in iron compound solution were then subjected to slow pyrolysis in muffle furnace under suitable thermal and temporal conditions. Iron biochar composite so obtained were washed, dried and used for conducting pH studies on preliminary grounds. pH optimization studies have been carried out with the solution of given concentration containing As (III) in the pH range of 2–10. This will be followed by adsorption studies and isotherm studies. Characterization of material will also be carried be-fore after arsenic adsorption to find probable mechanism.

3 RESULTS AND DISCUSSION

3.1 *Sorption studies*

Sorption studies were performed in batch mode to obtain the equilibrium and kinetic constants and pH impact over arsenic adsorption from water on bio-char composites. Batch studies are considered be-cause of their simplicity.

3.2 *pH studies*

Preliminary results have shown good removal and efficiency so far. The maximum arsenic occurred in a pH range of 6-8 which is near to neutral pH. This is advantageous as this will not require the adjustment of pH of water after arsenic removal.

3.3 *Kinetic and isotherm studies*

Kinetic studies were performed to examine the effect of adsorbent dose, contact time and temperature on arsenic adsorption rate. To establish adsorption mechanism and to find the affinity between sorbate and sorbent, isotherm studies were conducted. Effect

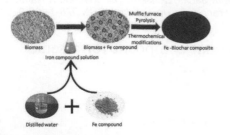

Figure 1. Scheme showing preparation of iron-biochar compo-site from biomass. This composite is used to conduct removal studies of As (III) from water.

Table 1. Comparative removal capacities and pH by various adsorbents used for arsenic removal.

Adsorbents	pH	Capacity removal (mg g-)	Models
Modified iron oxide coated sand	7.2	5.4	Freundlich
Iron hydroxide coated alumina	6.62	36.64	Langmuir
FeCl$_3$ treated tea fungal biomass	7.2	10.26	Freundlich
Iron oxide coated cement	7	0.67	Langmuir
Nano scale zerovalent Iron	7	2.47	Langmuir
Iron impregnated hickory biochar	5.8	2.16	Langmuir
Magnetic Iron oxide	5.5	9.74	-

of adsorbent dose (2 g/L, 1 g/L and 0.5 g/L) on arsenic removal was studied. Kinetic experiments have showed good arsenic removal at low adsorbent dose.

Adsorption data was modeled using Freundlich, Langmuir, Redlich-Peterson, Sips, Koble Corrigan, Radke-Prausnitz and Toth models to find various parameters necessary to design fixed bed reactors and find monolayer adsorption capacity. Information obtained from characterization will be applied to establish the probable mechanism for arsenic adsorption.

3.4 Thermodynamic studies and desorption—recovery

Thermodynamic studies were conducted to examine the adsorption behavior of arsenic onto the biochar surface. The parameters give the information about the feasibility and the spontaneity of the adsorption in terms whether adsorption process is endothermic or exothermic in nature.

Desorption studies recover the toxic heavy metals from adsorbent for safe disposal as well as in keeping the process cost down.

4 CONCLUSIONS

Preliminary results from pH studies conducted so far have shown maximum removal of As (III). Kinetic

Table 2. Instruments used for adsorbent characterization.

Characterization techniques	Purpose
Quantachrome surface area analyser	Specific surface area
SEM/EDX	Surface Morphology
TEM	Crystallinity, magnetic domains & stress
FT-IR	Functional group identification
FT-Raman	Compositional studies
XRD	Mineral composition
EDXRF	Elemental composition
Vibrating sample magnetometer	Magnetic moment

have shown good adsorption capacity at lower optimum dose. Further studies and characterization will provide information about probable mechanisms, conditions, properties etc. This will be utilized to con-duct experiments with real ground water samples to be collected from study area.

Here biochar is obtained from agricultural waste product that is very cost effective. These biochars can easily be used by developing modified biochar candles or columns to obtain arsenic free drinking water. Thus, composite biochars is considered to be a novel method for arsenic remediation. The biochar composites exhibited comparable sorption efficiency for arsenic removal. Therefore, this can easily replace commercial adsorbents.

ACKNOWLEDGEMENTS

One of the authors (PS) thanks to DST-INSPIRE for providing financial support to this work.

REFERENCES

Bhattacharya, P., Mukherjee, A., Mukherjee, A.B. 2011. Ar-senic in Groundwater of India. In: J.O. Nriagu (ed.) Encyclopedia of Environmental Health, vol. 1, pp. 150–164 Bur-lington: Elsevier.

Hu, X., Ding, Z., Zimmerman, A. R., Wang, S., Gao, B. 2015. Batch and column sorption of arsenic onto iron impreg-nated biochar synthesized through hydrolysis. Water Res. 68: 206–216.

Ming, Z., Gao, B., Varnoosfaderani, S., Hebard, A., Yao, Y., Inyang, M. 2013. Preparation and characterization of a novel magnetic biochar for arsenic removal. Bioresour. Technol. 130: 4457–462.

Mohan, D. 2007. Arsenic removal from water/waste using ad-sorbents—A critical review, Jour. Haz. Mater. 142: 1–53.

Singh, S.K. 2015. Groundwater Arsenic Con-tamination in the Middle-Gangetic plain, Bihar (India): The Danger Arrived. Int. Res. J. Environ. Sci. 4(2): 70–76.

Sun, L., Chen, D., Shungang W, Zebin, Y. 2015. Performance, kinetics and equilibrium of methylene blue adsorption on biochar derived from eucalyptus saw dust modified with citric, tartaric, and acetic acids. Bioresour. Technol. 198: 300–308.

Arsenic Research and Global Sustainability – Bhattacharya, Vahter, Jarsjö, Kumpiene,
Ahmad, Sparrenbom, Jacks, Donselaar, Bundschuh & Naidu (Eds)
© 2016 Taylor & Francis Group, London, ISBN 978-1-138-02941-5

Arsenic adsorption behavior on aluminum substituted cobalt ferrite adsorbents for drinking water application

Y. K. Penke[1], G. Anantharaman[2] & J. Ramkumar[3]
[1]*Materials Science Programme, IIT Kanpur, Kanpur, India*
[2]*Department of Chemistry, IIT Kanpur, Kanpur, India*
[3]*Department of Mechanical Engineering, IIT Kanpur, Kanpur, India*

ABSTRACT: Arsenic (As) adsorption on aluminum substituted cobalt ferrite (Co-Al-Fe) particles under different pH, time and arsenic concentrations were studied. The resultant adsorbed material was characterized using qualitative (using FTIR, Raman, and XPS) and quantitative (ICP) methods. The peak (around 840 cm^{-1}) in Raman spectra reveals the formation of inner sphere complex structures on top of the adsorbent. A multiplet peak in range of 42 to 46 eV in XPS spectrum confirms the occurrence of redox reactions in supporting the adsorption phenomenon. The pH based studies indicate the possible electrostatic interactions during the adsorption process. Adsorption isothermal studies were observed with Langmuir and Freundlich models for both As(III) and As(V) systems respectively. Adsorption kinetics was better fitted with Pseudo Second Order (PSO) model implying the chemisorption based adsorption phenomenon.

1 INTRODUCTION

Arsenic (As) is one of the naturally existing carcinogenic agents found in soil and aquifer systems. Generally arsenic transmits to atmosphere by natural processes like weathering, geogenic reactions and anthropogenic activities like glass and semiconductor based industries (Mohan & Pittman, 2007). Arsenic toxicity was well understood in causing several health disorders like hyperkeratosis (thickening of the skin), cardiovascular, neurological effects and even some times lead to different kind of cancers. Adsorbents, such as mono-metal oxide and hydroxide containing iron (Fe), aluminum (Al), copper (Cu), manganese (Mn) and magnesium (Mg) elements, were used for arsenic adsorption. Compared to this, binary and/or bimetallic oxide type adsorbents exhibit higher efficiency in arsenic adsorption. In the present case ternary metal oxide (Co-Al-Fe) adsorbent is been studied for their arsenic adsorption both in qualitative and quantitative methods.

2 EXPERIMENTAL

2.1 *Adsorbent synthesis*

All chemical reagents used here are of laboratory grade (A.R) and used without any further purification. Aluminum substituted cobalt ferrite (Co-Al-Fe) adsorbent was synthesized as per the reported procedure (Gul, 2008).

2.2 *Adsorption experiments*

In studying qualitative behavior the Co-Al-Fe adsorbent [1 g/L] and standard arsenic solutions [57 mg/L for As(III) and 24 mg/L for As(V)] were stirred for 8 h using an orbital rotary shaker. The supernatant solutions were vacuum filtered using membrane filters. The separated adsorbent powders were dried and further preceded for IR, Raman and XPS studies. The supernatant aqueous solutions were advanced to ICP-OES based quantitative analysis. In quantitative analysis adsorption isotherm (24 h), absorption kinetics (5 h) and pH variation studies were performed. In all adsorption systems pH adjustment was done using standard HCl (1 M) and NaOH (1M) solutions.

3 RESULTS AND DISCUSSION

3.1 *Vibrational spectroscopy results*

The adsorbent (Co-Al-Fe) and arsenic adsorbed material were characterized by qualitative and quantitative analysis. The IR active signals of As(III)/Ar(V) adsorbed materials at different pH conditions show very weak intensities (between 800–900 cm^{-1}) and the intense band for vs stretching vibrations of As-O signals around 810 cm^{-1} indicating the adsorption of these compounds (Müller, 2010). The variation in the intensity of signals corresponding to OH—ligand indicates the existence of ligand exchange mechanism during adsorption. The Raman signals for As(III) and As(V) systems were observed in the range of 800–900 cm^{-1} band corresponding to different vs(As-OH), vs(As-O), vas(As-OX)complexed [X = Fe, Co, Al] and vs of As-Ouncomplexed (Fig. 1). Raman signals around 810–830 cm-1 assigned to symmetric (vs) stretching vibrations of As-OH and As-OXcomplexed (X = Fe, Co, Al). Existence of multiple arsenic

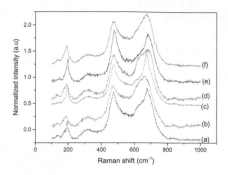

Figure 1. Raman spectra of pristine and As (III) adsorbed Co-Al-Fe particles at varying pH conditions. (a) Pristine sample (b) pH 2.0 (c) pH 5.0 (d) pH 7.0 (e) pH 9.0 (f) pH 12.0.

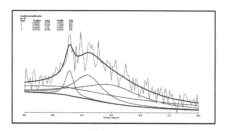

Figure 2. Individual As(3d) XPS spectrum of As (V) adsorbed Co-Al-Fe adsorbent in pH 7 system, (Adsorbent dose: 1.75 g/L, As(V) = 24 mg/L, t = 24 h, T = 27 ± 1 °C).

species and symmetry reduction behavior of As(III) and As(V) species resulted in peak splitting behavior in higher pH systems. The similarity in peak showing behavior in Raman spectra around 840 cm^{-1} observed in As(V) systems indicates the lack of pH effect on ν(As-O) can be attributed the formation of inner sphere surface complex structures onto the adsorbent (Goldberg, 2001).

3.2 X-ray Photo Spectroscopy (XPS) results

The adsorption of As(III)/As(V) on the adsorbent was further analyzed using the XPS. Different binding energy (B.E) values around 43.0 ± 0.5 eV and 44.0 ± 0.5 eV represents the active presence of arsenic species on top of adsorbent (Fig. 2). The existence of As(V) species in As(III) systems is due to the partial oxidation behavior of As(III) species during the adsorption process. In the same way the presence of As(III) species in As(V) systems indicates the reduction behavior in As(V) species.

The multiplet peak behavior in XPS spectra indicates the occurrence of redox reactions supporting the adsorption phenomenon. The weakness in the intensity of individual As(3d) XPS signal relative to the high intense Raman signals could be illustrated that apart from the surface complexation and redox reactions even intra-particle or inter-particle diffusion are also happening during the adsorption experiments.

Table 1. Freundlich adsorption isotherm parameters and maximum adsorption capacity (Q_{max}) of the adsorbent.

Arsenic Type	$K_{freundlich}$ ($mg^{l-n}\,L^n/g$)	n	R^2	$Q_{max-Langmuir}$ (mg/g)
As(III)	9.42	1.43	0.99	25
As(V)	5.67	1.51	0.98	16

3.3 Adsorption isotherm and mechanism of adsorption

The adsorption phenomena of As(III)/As(V) were evaluated using Inductively coupled plasma emission spectroscopy (ICP-OES). The outcome of the results suggests that Freundlich adsorption isotherm model is best suited for As(III) systems and As(V) adsorption systems is better suited with both Langmuir and Freundlich models. The Langmuir isotherm indicates the monolayer or homogenous layer adsorption phenomenon where as Freundlich isotherm suggests the multilayer adsorption or sorption on a heterogeneous surface i.e. surface with multiple sorption sites of different active energies.

4 CONCLUSIONS

Ternary metal oxide adsorbent particles (Co-Al-Fe) are verified for as probable arsenic remediation systems from aqueous systems. IR and Raman signals around 800–900 cm^{-1} band in various pH systems' indicated the presence of As-OH, As-OX (X = Fe, Al, Co) and As-O complex structures on top of the adsorbent. Redox, multilayer behavior and spontaneous nature of the adsorption was mostly observed on top of the adsorbent for both As(III) and As(V) systems through XPS and ICP-OES studies.

ACKNOWLEDGEMENTS

We would like to thank Indian Institute of Technology Kanpur, MHRD, Government of India.

REFERENCES

Goldberg, S. 2001. Mechanism of arsenic adsorption on amorphous oxides evaluated using macroscopic measurements, vibrational spectroscopy, and surface complexation modelling. J. Colloid Interf. Sci. 234: 204–216.
Gul, I.H. 2008. Structural, magnetic and electrical properties of cobalt ferrites prepared by the sol–gel route. J. Alloy Compd. 465: 227–231.
Müller, K. 2010. A comparative study of As(III) and As(V) in aqueous solutions and adsorbed on iron oxy-hydroxides by Raman spectroscopy. Water Res. 44: 5660–5672.
Mohan, D. & Pittman Jr., D. 2007. Arsenic removal from water/wastewater using adsorbents-A critical review. J. Hazard. Mater. 142: 1–53.

Arsenic Research and Global Sustainability – Bhattacharya, Vahter, Jarsjö, Kumpiene,
Ahmad, Sparrenbom, Jacks, Donselaar, Bundschuh & Naidu (Eds)
© 2016 Taylor & Francis Group, London, ISBN 978-1-138-02941-5

Synthesis, characterization and As(III) adsorption behaviour of β-cyclodextrin modified hydrous ferric oxide

I. Saha[1,2,3], K. Gupta[2], Sudipta Chakraborty[4], P. Bhattacharya[5], D. Chatterjee[3] & U.C. Ghosh[2]

[1]*Department of Chemistry, Sripat Singh College, Jiaganj, Murshidabad, West Bengal, India*
[2]*Department of Chemistry, Presidency University, Kolkata, India*
[3]*Department of Chemistry, University of Kalyani, Kalyani, West Bengal, India*
[4]*Department of Chemistry, Kanchrapara College, Kanchrapara, West Bengal, India*
[5]*KTH-International Groundwater Arsenic Research Group, Department of Sustainable Development,*
Environmental Science and Engineering, KTH Royal Institute of Technology, Stockholm, Sweden

ABSTRACT: This study investigates the adsorption of As(III) on β-cyclodextrin modified hydrous ferric oxide (HCC).The Langmuir monolayer adsorption capacity is 66.96 ± 9.16 (mg As/g of HCC) at 303 K. The high adsorption efficiency of the adsorbent is due to the modification of Hydrous Ferric Oxide (HFO) surface by β-cyclodextrin which provides ample—OH groups which in turn increase As(III) adsorption on HCC compared to HFO. The adsorbent can be used for a wide pH range(3–8)and spontaneous in nature which suggest wide applicability of the adsorbent. ThusHCC is found to be more efficient adsorbent than HFO.

1 INTRODUCTION

The metalloid Aarsenic (As) which has been inflicting havoc on mankind for the past few decades and has assumed global proportions transcending geographical boundaries and communities, has targeted water which is the support system of a civilization as the medium for its transportation into the environment and biological systems which explains the countless deaths it has left behind in its wake (Smedley & Kinniburgh, 2002, Vaughan, 2006). The cases of As poisoning through drinking water can be traced long back in the world history. Since then it has been conclusively established that unabated ingestion of As via drinking water causes a variety of carcinogenic effects of skin, liver, kidney and other organs which ultimately leads to death. All these incidents have led to World Health Organization (WHO) and other national agencies to fix the permissible limit to 10 µg/L of As in drinking water. The principle source of As contamination is groundwater and the majority of rural and semi-urban population in vast parts of the globe depend on this form of water for their drinking water and other household needs. The problem associated with As contaminationis severe in Bangladesh and major parts of West Bengal,India (Biswas *et al.*, 1998, Guha Mazumder *et al.*, 2010). In West Bengal, more than 6 million people from 12districts covering 111 blocks are worst affected are consumingdrinking water with

As concentration >50 µg/L.The present investigationsprovide promising results demonstrating that HFO modifiedby β-cyclodextrin (HCC) would be a better scavenger of As(III) from contaminated groundwater as against conventional pureHFO.

2 MATERIALS AND METHODS

A ferric chloride solution was prepared in 0.1 M HCl and mixedwith an aqueous solution of β-cyclodextrin in a 1:10 weight ratio.The mixture was warmed at 90°C accompanied by gradualaddition of dilute (1:1) NH_3 solution with constant stirring till thepH reached around 7. The dark brown precipitate formed was agedfor 48 h, filtered, and washed repeatedly with deionized water. Thefiltered mass obtained was dried at 100°C into an air oven. Thedried mass was ground in a mortar and sieved to obtain aparticle size between 60–100 mesh (250–150 microns) and subsequently used for adsorption studies.

3 RESULTS AND DISCUSSION

The FESEM image of HCC (Fig. 1A), indicates that the particles do not have any definite surface morphology. The EDAX spectrum (Fig. 1B) of the material shows thesurface composition of the material. Three main elements O, Fe and C are

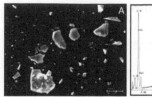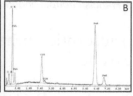

Figure1. A: Field Emission Scanning Electron Microscopy (FESEM) of HCC, B: EDX spectrum of HCC.

present on the surface. The spectrum indicates the absence of impurities on the composite surface.

The plot (adsorption vs pH) indicates that As(III) adsorptionis pH dependent and the remarkable decrease in concentration has been noticed in between the pH range 6.5–8.5. This is important because the pH range (6.5–8.5) has been commonly recommended for drinking water supply. At pH7.0, 65.8% As(III) uptake was observed for pure HFO whereas for HCC 81.5% of the total As(III) were removed from solution. The pH_{ZPC} value of HCC lied within the range 7.6–7.9 indicating that below this pH range the surface charge of HCC was positive and above it surface becomes negative. At low pH the most dominant form of As(III) species present in the solution is $As(OH)_3$ ($pK_1 = 9.2$). Due to the presence of large number of surface –OH groups the neutral $As(OH)_3$ species form strong H-bond with HCC surface at low pH region.

4 CONCLUSIONS

Hydrous Ferric Oxide (HFO) has proven to be a potent adsorbent for dissolved As(III) in contaminated water. However, surface modification of HFO with β-cyclodextrin by a simple and economical route shows large enhancement of As(III) scavenging power. The systematic As(III) adsorption by this material has showed that the optimum pH and equilibrium contact time are ~7.0±0.1 and 120 mins, respectively. The pseudo first-order equation describes the kinetic data (pH, 7.0±0.1; temperature, 303±1.6K) well. The equilibrium data

(pH, 7.0 ± 0.2; temperatures (± 1.6 K), 288, 303, 318 and 333) fit very well with the Langmuir isotherm model. The Langmuir monolayer adsorption capacity is 66.96 ± 9.16 (mg As/g. of HCC) at 303 K, and that increases with increasing temperature. The adsorption reaction is spontaneous ($\Delta G° $ = negative) and exothermic ($\Delta H°$ = negative), and that takes place with increasing entropy. The energy (kJ/mol) of As(III) adsorption and the FTIR analysis have suggested that the As(III) adsorption on HCC is of both physisorption as well as chemisorption in nature (Wang et al., 2007). The As(III) adsorption by HCC is negatively influenced by phosphate and sulphate ions. The regeneration of As(III) adsorbed material is possible maximum up to 75% with 1 M NaOH solution.

ACKNOWLEDGEMENTS

The authors are thankful to Sripat Singh College, Presidency University, Kolkata and University of Kalyani for providing laboratory facilities.

REFERENCES

Biswas, B.K., Dhar, R.K., Samanta, G., Mandal, B.K., Faruk, I., Islam, K.S., Chowdhury, M.M., Islam, A., Roy, S., Chakraborti, D. 1998. Detailed study report of Samata, one of the arsenic affected villages of Jessore district, Bangladesh.Curr. Sci. 74(2): 134–145.
Guha Mazumder, D.N., Ghosh, A., Majumdar, K.K., Ghosh, N.; Saha, C.; Guha Mazumder, R.N. 2010. Arsenic contamination of ground water and its health impact on population of district of Nadia, West Bengal, India. Ind. J. Community Med. 35(2): 331–338.
Smedley, P.L.; Kinniburgh, D.G. 2002. A review of the source, and distribution of arsenic in natural waters. Appl. Geochem. 17: 517–568.
Vaughan, D.J. 2006.Arsenic.Elements 2:71–75.
Wang, L., Chen, A.S.C., Tong, N., Coonfare, C.T. 2007. Arsenic removal from drinking water by ion exchange. U.S. EPA demonstration project at Fruitland, ID. Six month Evaluation Report. EPA/600/R-07/017. United States Environmental Protection Agency, Water Supply and Water Resources Division, National Risk Management Research Laboratory, Cincinnati, OH.

Arsenic Research and Global Sustainability – Bhattacharya, Vahter, Jarsjö, Kumpiene,
Ahmad, Sparrenbom, Jacks, Donselaar, Bundschuh & Naidu (Eds)
© 2016 Taylor & Francis Group, London, ISBN 978-1-138-02941-5

Ergonomic design of a system for removal of arsenic for household use

F. Yonni[1], H.J. Fasoli[1], A.E. Fernández[2], L.C. Martinez[3] & J.H. Alvarez[3]
[1]*Facultad de CFM e Ingeniería (UCA) y Facultad de Ingeniería del Ejercito, Escuela Superior Técnica,*
M.N. Savio (IUE), Argentina
[2]*SEGEMAR—Centro de Procesamiento de Minarales, Buenos Aires, Argentina*
[3]*INTI Textiles—UT Comercialización y Diseño—Laboratorio Químico Textil, Buenos Aires, Argentina*

ABSTRACT: A portable and for household use device has been designed, which allows to take down the concentration of arsenic in the water below the limits established by WHO. This device also allows to optimize three aspects of the system (human-process—environment). Samples of clays, zeolites and bentonites that are currently in Buenos Aires province has been tested with different grain size, to manage the maximun efficiency according to: speed of the water flow, time of residence of the solution and volume of water filtered. The results obtained indicate that the device is viable for the acquisition and the use of the middle social classes, where the design is a determining variable for the acceptance of the product.

1 INTRODUCTION

The issue of the human use of the underground aquifers that go through large areas of Argentina and which possess contamination with arsenic in amounts ranging from 0.01 to 1.6 mg/L, is an unresolved difficulty mainly in towns having less than 5,000 inhabitants settled in contaminated areas. Regarding their socio-economic characteristics, most of them do not have treatment plants that may decrease the concentration of arsenic below the limit established by World Health Organization (10 μg/L).

Although there are bibliographic records of different devices that can be used for the removal of arsenic in water destined for human consumption, to our knowledge there is no system that can accomplish the simultaneous conditions of being simple, cheap and not provoke significant flavor changes in the water. In addition, aesthetic and ergonomic designs are required to facilitate its implementation among middle and upper class people.

The main goals of this work focus on: 1) the study of naturally occurred zeolite and clays for the removal of arsenic from contaminated groundwater, 2) the design of a device that meets the conditions of being cheap, easy to operate and aesthetically pleasing.

2 METHODS/EXPERIMENTAL

The tested sediments include bentonite, zeolite and three different clays which have been identified as red clay 1, 2 and 3. These sediments were tested individually way or in the way of clay (1, 2, and 3)-zeolite mixture.

It were utilized different quantities of adsorbent material/ water simple for a time of permanence of 4 hours (which has set as the standard time).

It were utilized different quantities of adsorbent material/ wáter simple over a period of 4 hours (wich has set as the standard time). The filtration of the prepared As solution was carried out in the same place by using a wool's knitting synthetic material, which at the same time acts as a support of the adsorbent sediment that is going to be test. (Fig. 1).

The removable system is composed of a plastic container that contains in the inside, the sorbent material of zeolite-clay in separate multilayers with a filtrated material support.

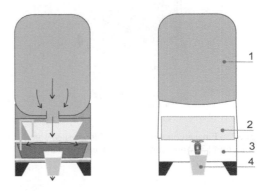

Figure 1. Compenents of filtered system. 1: plastic pan of 25 L of original non treated water; 2: Removable system with filtrate multilayer component (2L); 3: Plastic pan of filtrated water (2 L); 4: Outlet tap for purified water.

For the selection of the eventually adopted design, subjective proofs preferred by the user were made, they were done about 3 different models and over a total of 30 persons.

3 RESULTS AND DISCUSSION

An optimal relation was found 80:20 clay: zeolite, which was also respectively maintained in all cases. in a relation of 100 g of absorbent material/kg watery solution to be treated.

The absorbent that ended up being the most efficient in the retention of the arsenic for the established work's conditions, turned out to be the red clay 3 – zeolite, in a masses relation from 4 to 1.

As well, it was established that the relation of the masses between the absorbent and the water to be treated is 1:10.

The subjective study lets select the ergonomic design for the device that ended up being the most satisfactory for the public. Figure 2 shows the resultant option.

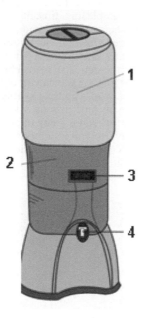

Figure 2.   Final ergonomic design. 1. Non treated water; 2. Purifier unit.; 3. Water consumption-meter; 4. Outlet tap.

The device made by plastic material presents appropriate size and weight. In the higher part, the original non treated water's pans are placed, and in the lower part the filtrate water's pan is put (here 2 pans are utilized, one for receipt of filtrate water, meanwihle the other is used for drinking).

4 CONCLUSIONS

The tests carried out show a strong absorption of Arsenic in the clay-zeolite samples, which coincide with previous studies of other authors.

The higher capacity of absorption can be reached with a type of clay called red clay 3.

The ergonomic design selected by the public, presents the typical limitations of a customer satisfaction survey.

The ergonomic design selected by the public, presents the typical limitations of the satisfaction's survey made for the population subject to enquiry.

REFERENCES

Bocanegra, O.C., Bocanegra, E.M., Alvarez, A.A. 2002. Arsenic in groundwater. In: O.C. Bocanegra, E.M. Martínez, D.Massone (Eds.) Groundwater and Human Dvelopment (ISBN 987-544-063-9).

Castro de Esparza, M.L., Wong de Medina, M. 1998. Abatement of arsenic in groundwater for rural area. 26th Inter-American Congress of Sanitary Engineering and Environment. (http://www.bvsde.paho.org/acrobat/percca02.pdf).

Castro de Esparza, M.L. 2006. Removal of arsenic in water to drink and bioremediation of soils. natural Arsenic in Groundwaters of Latin America International Congress, Mexico City. http://www.bvsde.opsoms.org/bvsacd/cd51/ remocion-agua.pdf.

Chavez jaw, M.L. Miglio Toledo, M. 2011. Removal of arsenic by solar oxidation in water destined to human consumption. *Rev. Soc. Quim. Peru* 77 n.4 Lima.

Gay, A, Samar, L. 2007. Industrial design in the historia. In. Ed. Tec. Association cooperative of the Faculty of Economics of the National University of Córdoba.

Simsek, E. B., Ercan Özdemir, E. and Beker, U. 2013. process optimization for arsenic adsorption onto natural zeolite incorporating metal oxides by response surface methodology. *Water Air Soil Pollut.* 224:1614.

Rodríguez, R., Echeverria, M. 2008. Reduction of arsenic in water by using a domestic method. ed. at the national technological University (dUTecNe) - Argentina. http://www.edutecne.utn.edu.ar.

Arsenic Research and Global Sustainability – Bhattacharya, Vahter, Jarsjö, Kumpiene,
Ahmad, Sparrenbom, Jacks, Donselaar, Bundschuh & Naidu (Eds)
© 2016 Taylor & Francis Group, London, ISBN 978-1-138-02941-5

Iron Oxide Coated Pumice: Promising low cost arsenic adsorbent

Y.M. Slokar & B. Petrusevski
UNESCO-IHE Institute for Water Education, Delft, The Netherlands

ABSTRACT: Adsorptive filtration with low cost adsorbents is one of attractive options for production of drinking water from arsenic containing groundwater. Such adsorbent can be produced by filtering Fe(II) solution through commonly used filter media. In this research, the tested media were pumice, quartz sand and garnet. Coating efficiency expressed as iron content of coated media was the highest for pumice. Iron content and arsenic adsorption capacity of produced Iron Oxide Coated Pumice (IOCP) strongly increased with prolonged coating time till about 10 days. Coating of pumice at pH of 6.5 resulted in ~25% higher Fe content and ~40% higher As adsorption capacity of IOCP, in comparison to IOCP produced at pH of 5.5 and 6.0. As adsorption capacity of IOCP is inferior to capacity of commercial iron oxide based adsorbents and IOCS, but the coating procedure for production of this low cost arsenic adsorbent shows potential for further improvements.

1 INTRODUCTION

Presence of arsenic (As) in groundwater used for drinking water production is a global problem. Different disadvantages of existing technologies for As removal makes them unattractive for application in developing countries. Adsorptive As removal is very effective and relatively simple method that is increasingly gaining the market. However, application of this technology could be limited in developing countries due to high costs of commercial As adsorbents.

As has strong affinity for iron oxide minerals, and consequently most of commercially available arsenic adsorbents are iron-oxide based composites. For the same reason, Iron Oxide Coated Sand (IOCS) has proven its potential to effectively remove As. Being it is a by-product of iron removal treatment plants, it is also very cheap. However, quantities of IOCS are limited. Given the global scale of As presence in groundwater, search for another low cost As adsorbent is a very contemporary research topic.

Objective of the study presented in this paper was to assess feasibility of producing low costs As adsorbent based on *in-situ* iron oxide coating of three common filter media: quartz sand, pumice, and garnet. In addition the study investigated the effect of coating pH and coating time on As adsorption capacity of produced media.

2 METHODS

For coating experiments, three common and commercially available filter media were used:

- quartz sand: 0.4–0.6 mm;
- pumice: 0.09–0.60 mm; and
- garnet: 0.4–0.8 mm.

Coating was conducted in laboratory filter columns with bed depth of 80–90 cm. The filters were operated in an up-flow mode, with filter bed expansion of approximately 40%. Acidified solution of Fe(II) was continuously pumped into feed (tap) water. The extent of coating was quantified by SEM-EDX, and by coating extraction to determine the amount of Fe present on the media. As adsorption capacity was assessed in short and long (isotherm) batch adsorption experiments.

Since pH plays an important role in Fe(II) oxidation kinetics and its adsorption efficiency, the best performing media was tested at three different coating pH values: 5.5 | 6.0 | 6.5. This specific pH range was selected having in mind that Fe(II) oxidation in feed water might be too fast at pH > 6.5 (and Fe(III) will precipitate and consequently not contribute to pumice coating), and that oxidation of adsorbed Fe(II) on the media surface will be too slow at pH < 5.5 (Stumm and Lee, 1961).

3 RESULTS AND DISCUSSION

Under identical conditions applied, iron coating of pumice was much more effective then it was the case with sand and garnet (Fig. 1).

This was likely due to large surface area and macro-porous structure of pumice. It is well known that formation of iron oxide layers on a surface is a catalytic process that strongly accelerates with creation of an initial iron-oxide layer. Apparently within the investigated coating time such catalytic layer was not sufficiently developed on sand and garnet. Prolonged coating time would likely increase the effectiveness of sand and garnet coating, but it would consequently also increase coat-

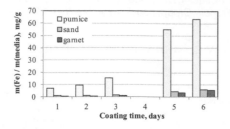

Figure 1. Iron content in coated filter media as a function of coating time and type of virgin filter media used.

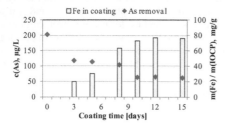

Figure 2. Effect of coating time on iron content and efficiency of arsenic adsorption of IOCP.

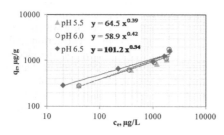

Figure 3. Freundlich isotherms for As(III) adsorption on IOCP produced at pH 5.5, 6.0 and 6.5. Model water pH 7; Contact time 28 days.

ing (adsorbent) costs. Due to the fast progress of iron coating, its wide availability and relatively low price, pumice was selected as the base filter media for further optimization of the coating procedure.

Given the affinity of arsenic for iron oxides it was stipulated that the amount of iron in the coating would directly influence arsenic adsorption potential of Iron Oxide Coated Pumice (IOCP). Results showed that prolonging coating time up to approximately 10 days resulted in rapid increase of Fe content on IOCP. Further extension of coating time, however, did not significantly further increase Fe content (Fig. 2). Such result was attributed to blocking of pumice pores with iron oxides, and related reduction of surface area available for further Fe adsorption. As removal in batch adsorption experiments with IOCP produced during different coating time showed similar trend as Fe adsorption: improved As adsorption with IOCP coated

up to about 10 days, after which no significant increase in As removal efficiency was observed.

With regard to different pH of coating solution, coating at pH of 6.5 resulted in approximately 25% higher Fe content on IOCP in comparison to coating at pH of 5.5 and 6.0. IOCP produced at pH of 6.5 also demonstrated to have the highest arsenic adsorption capacity based on Freundlich isotherm (Fig.3), confirming that it is Fe content of an adsorbent that controls As adsorption capacity.

Produced IOCP still has an inferior As adsorption capacity in comparison to commercial adsorbents and IOCS (e.g., some of the tested IOCS had a K value as high as 2500). Nevertheless, given promising first results, combined with room for further optimization of the coating process, and possibility to regenerate As saturated IOCP with IHE ADART technology (Petrusevski et al, 2007), IOCP can be consider as a promising low cost arsenic adsorbent that can be produced directly at the As removal sites.

4 CONCLUSIONS

Low cost arsenic adsorbent can be produced by filtration of Fe(II) solution through a filter with common filter media. Pumice demonstrated to be more suitable for rapid coating than quartz sand and garnet, due to its morphology. With prolonging coating time up to 10 days, substantial increase in Fe content and As adsorption capacity was observed. Further extension of coating time did not result in significant improvement of either. Coating at pH 6.5 resulted in approximately 25% higher Fe content and 40% higher As adsorption capacity in comparison to IOCP produced at pHs of 5.5 and 6.0.

Produced IOCP As adsorption capacity is still significantly lower than the capacity of most commercial iron oxide based adsorbents and IOCS. Nevertheless, given the wide availability of pumice, its very low cost, as well as the low cost of Fe(II) salts, which is the only chemical required in the process, IOCP shows great potential for developing an efficient low cost As removal adsorbent

ACKNOWLEDGEMENTS

The authors would like to acknowledge the efforts of UNESCO-IHE MSc participants who carried out the researched presented in this paper.

REFERENCES

Petrusevski, B., van der Meer, W.G.J., Baker, J., Kruis, F., Sharma, S.K. & Shippers, J.C. 2007. Innovative approach for treatment of arsenic contaminated groundwater in Central Europe. Water Science &Technology: Water Supply 7(3): 131–138.

Stumm, W. and Lee, G.F. (1961) Oxygenation of ferrous iron. Industrial Engineering and Chemistry, 53 (2), 143–146.

Arsenic Research and Global Sustainability – Bhattacharya, Vahter, Jarsjö, Kumpiene,
Ahmad, Sparrenbom, Jacks, Donselaar, Bundschuh & Naidu (Eds)
© 2016 Taylor & Francis Group, London, ISBN 978-1-138-02941-5

Removal of low levels of arsenic contamination from water by cysteine coated silica microspheres

F. Makavipour & R.M. Pashley
Department of Chemistry, School of Physical, Environmental and Mathematical Science, University of New South Wales, Canberra, Australia

ABSTRACT: The surface charging properties of silica micro-particles in aqueous solutions before and after surface chemical amination were measured and analyzed using a new approach to the theoretical modelling of ion adsorption. The surface of aminated silica microspheres was further modified by chemical adsorption of L-cysteine groups to determine their ability to adsorb low levels of arsenic ions from aqueous solution. The adsorption of arsenic ions was observed both from changes in particle surface charge and from direct solution analysis, using ICP-OES. The results obtained indicate adsorption levels of about 2–5% of the cysteine surface density, from low-level arsenic solutions (i.e. less than 1 mg/L). These initial results indicate that a suitable silica-based depth filter could be developed for the removal of low—level arsenic ions from contaminated water.

1 INTRODUCTION

Arsenic is a strong human carcinogen and metalloid, which is a common component of natural rocks producing low contamination levels in both ground and drinking water in many parts of the world (Choong *et al.*, 2007; Smedley & Kinniburgh, 2002). Contamination of ground water can also lead to agricultural contamination. (Mukherjee *et al.*, 2015) The World Health Organization guideline value for arsenic is 10 μg/L (WHO, 2016). Low level arsenic removal needs complicated and costly instrumental analysis, in addition there are other difficulties involved in arsenic extraction and detection processes such as: preparation steps (reduction or oxidation or pH control), slow adsorption process (iron oxides and hydrous iron oxides, which are the most common arsenic adsorbents used in previous studies, (Sigdel *et al.*, 2016) need a long equilibration time to be effective), high costs of chemicals used in filters or adsorbents, the possibility of chemicals entering the water supply and the low daily capacity of filters in common use (Hashmi & Pearce, 2011; Mohan & Pittman Jr, 2007).

In this study, the potential for removal of low concentrations of arsenic ions from aqueous solution using a novel adsorption process based on cysteine-coated silica microspheres was investigated. A novel surface charging model for silica and aminated silica particles has also been proposed.

2 METHODS/EXPERIMENTAL

2.1 *Surface coating*

Cysteine coated silica microparticles were prepared using glutaraldehyde as a crosslinking ligand between L-cysteine and aminated silica microspheres, according to the method described by Dakova et al. (Dakova *et al.*, 2011).

2.2 *Surface charge study*

25 ml samples of 0.4 ppm arsenic solution containing 10 ml NaCl 0.01M and cysteine-silica particles at a concentration of 0.16% were stirred for 4 hours, in different pH solutions: 3.63, 4.78 and 6.86. Then the zeta potentials (mV) of the cysteine-silica particles were measured using a Nano Zeta Sizer (Malvern Zeta Sizer).

2.3 *Adsorption study*

Surface charge variation on cysteine-coated micro silica particles in different solution conditions and in arsenic solutions was investigated. This was used as an indirect method of analysis of the removal of arsenic from water. Also, the change in supernatant arsenic concentration was measured using Inductively Coupled Plasma Optical Emission Spectrometry (ICP-OES) to determine adsorption directly.

3 RESULTS AND DISCUSSION

3.1 *Surface charge study*

The zeta potentials of silica particles in a range of electrolyte solutions and pH values were studied at each stage of the surface chemical modification, which led to the development of a new adsorption model (Fig. 1). The zeta potentials (Ψ_0) and the Debye lengths (K^{-1}) of each electrolyte solution were used to calculate the corresponding surface charge densities (σ_0) using the equation below,

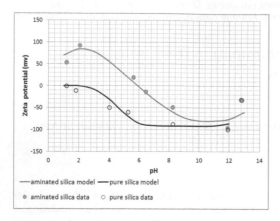

Figure 1. The new adsorption model.

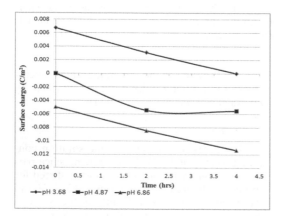

Figure 2. Changes in surface charge density on cysteine-coated silica particles on immersion in 0.4 ppm arsenic solution at different pH values in 0.01 M NaCl at 25°C.

assuming that the solution K^{-1} was significantly less than the radius of the silica particles.

$$\sigma_0(C/m^2) = \frac{3.571 \times 10^{-11}}{K_{(m)}^{-1}} \times \sinh[\Psi_0(mv) \times 0.01946]$$

(1)

The results, summarized in Figure 2, indicate that in the pH range 3.7 to 7 there was significant arsenic adsorption, as the surface of the adsorbent became more negatively charged with time.

3.2 Adsorption studies

The differences between the methods in Table 1 are related to several factors, such as chloride and arsenic ions competing for binding site on the

Table 1. A comparison of the adsorption of arsenic measured using two different analysis methods.

	From ICP-OES	From surface zeta potentials
Adsorbed amount of As on the surface	1.5×10^{-7} mol/m^2	7×10^{-8} mol/m^2
Equilibrium As concentration in solution	3.60×10^2 µg/L	3.80×10^2 µg/L
Adsorbed As species relative to the number density of surface cysteine groups	5%	2.5%

positively charged surface, which would reduce the change in surface charge with arsenic adsorption. Also, the surface roughness of the silica particles should be considered because the ICP-OES estimates used in this study were obtained from the geometric area of the particles, rather than the measured areas, while the charge densities were obtained from measured zeta potentials.

Arsenic binding studies in other systems (Shen et al., 2013) indicate that typically about four cysteine groups bind to each As (V) ion namely, $H_2AsO_4^-$ or $HAsO_4^{2-}$. Thus the number of potential binding sites should be less than about 25% of the total cysteine surface density. Therefore about 10–20% of potential sites appear to be active on these silica surfaces, when equilibrated with low level arsenic solutions.

4 CONCLUSIONS

The modified cysteine-coated silica microsphere particles could be used for low level arsenic removal from contaminated water. These observations are supported both by direct adsorption measurements using ICP-OES and through the analysis of surface zeta potential measurements. This method is considered more environmental friendly as it uses a natural amino acid; also, it has a short equilibrium time in pH close to natural water pH. In addition, the surface charging behavior observed was described using a new modified approach for the theoretical modelling of surface ion adsorption.

ACKNOWLEDGEMENTS

The authors would like to thank Dr M. Francis for his help to provide ICP-OES measurements in the CSIRO.

REFERENCES

Choong, T.S.Y., Chuah, T.G., Robiah, Y., Gregory Koay, F.L., Azni, I. 2007. Arsenic toxicity, health hazards and removal techniques from water: an overview. *Desalination* 217(1–3): 139–166.

Dakova, I., Vasileva, P., Karadjova, I. 2011. Cysteine modified silica submicrospheres as a new sorbent for preconcentration of Cd (II) and Pb (II). *Bulgarian Chemical Communications* 43: 210–216.

Hashmi, F., Pearce, J.M. 2011. Viability of small-scale arsenic-contaminated-water purification technologies for sustainable development in Pakistan. *Sustainable Development* 19(4): 223–234.

Mohan, D., Pittman Jr, C.U. (2007). Arsenic removal from water/wastewater using adsorbents—A critical review. *J. Hazard. Mater.* 142(1–2): 1–53.

Mukherjee, A., Saha, D., Harvey, C.F., Taylor, R.G., Ahmed, K.M., Bhanja, S.N. 2015. Groundwater systems of the Indian Sub-Continent. *J. Hydrol.: Regional Studies* 1(4): 1–14.

Shen, S., Li, X.-F., Cullen, W.R., Weinfeld, M., Le, X.C. 2013. Arsenic binding to proteins. *Chem. Rev.* 113(10): 7769–7792.

Sigdel, A., Park, J., Kwak, H., Park, P.K. 2016. Arsenic removal from aqueous solutions by adsorption onto hydrous iron oxide-impregnated alginate beads. *J. Ind. Eng. Chem.* 35: 277–286.

Smedley, P.L., Kinniburgh, D.G. 2002. A review of the source, behaviour and distribution of arsenic in natural waters. *Appl. Geochem.* 17(5): 517–568.

WHO. 2016. Arsenic. World Health Organization, Geneva, Switzerland. (Retrieved from http://www.who.int/mediacentre/factsheets/fs372/en/)

Arsenic Research and Global Sustainability – Bhattacharya, Vahter, Jarsjö, Kumpiene,
Ahmad, Sparrenbom, Jacks, Donselaar, Bundschuh & Naidu (Eds)
© 2016 Taylor & Francis Group, London, ISBN 978-1-138-02941-5

Effect of competing and coexisting solutes on As(V) removal by forward osmosis

P. Mondal[1], A.T.K. Tran[2] & B. Van der Bruggen[3]
[1]*CSIR-Central Glass and Ceramic Research Institute, Raja S.C. Mullick Road, Kolkata, India*
[2]*Faculty of Chemical and Food Technology, HCM University of Technical Education, Vietnam*
[3]*Department of Chemical Engineering, Process Engineering for Sustainable Systems (ProCESS),*
KU Leuven, Leuven, Belgium

ABSTRACT: Forward osmosis can effectively remove several inorganic and organic contaminants from aqueous solution. Therefore, the rejection behavior of As(V) in presence of several co-occurring solutes (nitrate, fluoride, sulfate, phosphate, silicate, bicarbonate and humic acid) are investigated in this study. Rejection of As(V) was decreases extensively in presence of phosphate and follows the sequence humic acid > bicarbonate > nitrate > fluoride > sulfate > phosphate, in presence of co-existing solutes. Donnan exclusion, size exclusion, and diffusion coefficients are factors that may explain the rejection. Moreover, the fouling layer formed by humic acid and higher pH increases the removal efficiency. Results showed that forward osmosis can successfully remove arsenic from contaminated water containing various coexisting and competing solutes and therefore, it can be considered as an alternative for other membrane based technologies in a hybrid treatment system.

1 INTRODUCTION

Arsenic (As) poisoning and its removal from drinking water is a critical challenge that needs to solve. Forward Osmosis (FO) is latest addition among the membrane based technologies that are applied for As removal due to its several advantages (Mondal *et al.*, 2013). Although FO can efficiently remove As from contaminated water, its removal in the presence of coexisting oxyanions, bicarbonate and hardness has not yet been investigated. These ions suppress As adsorption on iron based adsorbents due to competition, leading to a decrease of available adsorption sites (Holmes, 2002). Therefore, in this study we investigate the effect of competing and coexisting inorganic solutes (nitrate, fluoride, sulfate, bicarbonate, phosphate and silicate) on the efficiency of As removal. The consequence of fouling and hardness on As rejection using a FO membrane was also studied. Additionally, the effect of predominant co-occurring solutes on As rejection was evaluated in a range near the typical groundwater composition.

2 MATERIALS AND METHODS

2.1 Chemicals and FO membrane system

The draw solution was prepared using glucose. All the used chemicals during experiment were analytical grade and used without further treatment. The FO membrane was provided by Hydration Technology Innovation (HTI, Scottsdale, AZ). A laboratory scale plate and frame membrane cell was used with 64 cm² active surface area for experiment.

2.2 Rejection experiments and analytical methods

Required amounts of the As stock solution [$Na_2AsO_4 \cdot 7H_2O$ (RPL, Belgium)] were added into the feed solution for all the experiments. The concentrations of the solutes in synthetic water were selected according to the anticipated concentration variation in groundwater. Total As was quantified using ICP-MS (Thermo Electron Corporation X series).

3 RESULTS AND DISCUSSION

3.1 Effect of hardness

The positively charged Calcium and Magnesium ions shielded the negative charge of the membrane which lowers the electrostatic repulsion between membrane and As anions.

Additionally large molecular size hydrated solutes are strongly retained by membrane and thereby lowers As(V) rejection from 95.5 to 87.5% and 95.5 to 89.7% in presence of Ca^{2+} and Mg^{2+}, respectively (Fig. 1).

3.2 Effect of anionic solute

Effect of monovalent solute on As(V) removal was studied using NO_3^- and F^-. The water flux and As rejection were remain unaffected as the driving force was relatively constant.

Divalent ions have significant effect on As(V) rejection. Phosphate and sulfate have negative impact on rejection whereas bicarbonate and HA have positive impact. The rejection decreases from

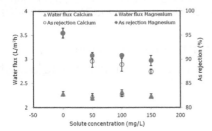

Figure 1. Effect of hardness on permeate flux and removal of As(V) (As(V) concentration = 30 ± 0.5 mg/L).

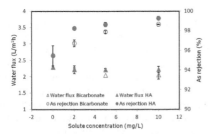

Figure 2. Effect of bicarbonate and humic acid (HA) on permeate flux and removal of As(V) (As(V) concentration = 30 ± 0.5 mg/L).

95.5 to 89.4% in the presence of 1600 ± 2.5 mg/L sulfate in the solution. The transport of phosphate ions is slower than that of arsenate ions due to difference in hydrated radius ($P = 4.90$; As = >2.00–2.20 Å). Therefore, rejection decreases significantly from 95.5 to 80.5% in presence of 0 to 1 ± 0.2 mg/L phosphate. Moreover, structural resemblance of arsenate and phosphate ions also has significant effect on removal due to their competition for rejection by the FO membrane. However, the effect of silica on As(V) removal was insignificant as it remains almost undissociated below pH 9.

Bicarbonate in solution generally acts as a buffering agent and emphasizes the removal of As(V) (Fig. 2). The solution pH increases due to elimination of excess CO_2 since at pH 7 ± 0.5 most of the HCO_3^- was transformed to H_2CO_3 at dissociation constant (6.35) which accelerate the rejection. Furthermore, As(V) rejection increases due to increase of electrostatic repulsion between the slightly negatively charged membrane and negatively charged As species ($H_2AsO_4^-$ and $HAsO_4^{2-}$) present in solution. Additionally, fouling helps to diffuse water molecules through the membrane as the hydrated layer of humic acids eases the diffusion process from 95.5 to 99.3%, due to fouling (Fig. 2). Cake like fouling layer increases the negative charge on membrane surface (Xie et al., 2013) that simultaneously increases the electrostatic interaction and improved As removal. Additionally, the presence of free hydroxyl and carboxyl groups in humic

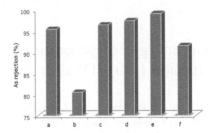

Figure 3. Rejection of As(V) in presence of influential co-existing solutes (As concentration = 30 ± 0.5 mg/L) [a = As(V), b = As(V) + P(V), c = As(V) + P(V) + bicarbonate, d = As(V) + P(V) + HA, e = As(V) + P(V) + bicarbonate + HA, f = As(V) + P(V) + bicarbonate + HA + sulfate + silicate + fluoride + nitrate + calcium + magnesium].

acid helps to improve the removal efficiency due to strong steric repulsion.

3.3 As rejection from simulated groundwater

Since groundwater is contaminated with As together with above mentioned ionic solutes, the combined effect of the solutes on As rejection is shown in Fig. 3. The observation repeats the findings as individual ions.

4 CONCLUSIONS

The removal efficiency of As(V) in presence of coexisting anionic solutes can be expressed as follows- humic acid > bicarbonate > nitrate > fluoride > sulfate > phosphate. Similar selectivity was also observed for synthetic groundwater. It was found that humic acid fouling helps to remove more As(V) than bicarbonate and concentration polarization is insignificant due to the lower water flux in FO during the use of glucose as a draw solution. Therefore, it is assumed that FO can be successfully used for As contaminated water treatment.

ACKNOWLEDGEMENTS

P.M would like to acknowledge SBO project, FunMem4 Affinity (IWT) for funding.

REFERENCES

Holm, T.R. 2002. Effect of CO_3^{2-}/bicarbonate, Si, and PO_4^{3-} on arsenic adsorption to HFO, J. Am. Water Works Assoc. 94(4): 174–181.

Mondal, P., Bhowmick, S., Chatterjee, D., Figoli, A., Van der Bruggen, B. 2013. Remediation of inorganic arsenic in groundwater for safe water supply: a critical assessment of technological solutions, Chemosphere 92 (2): 157–170.

Xie, M., Nghiem, L.D., Price, W.E., Elimelech, M. 2013. Impact of humic acid fouling on membrane performance and transport of pharmaceutically active compounds in forward osmosis, Water Res. 47(13) 4567–4575.

Arsenic Research and Global Sustainability – Bhattacharya, Vahter, Jarsjö, Kumpiene,
Ahmad, Sparrenbom, Jacks, Donselaar, Bundschuh & Naidu (Eds)
© 2016 Taylor & Francis Group, London, ISBN 978-1-138-02941-5

Biogas energy polygeneration integrated with Air-Gap Membrane Distillation (AGMD) as arsenic mitigation option in rural Bangladesh

E.U. Khan[1], A.R. Martin[1] & J. Bundschuh[2]
[1]*Department of Energy Technology, KTH Royal Institute of Technology, Stockholm, Sweden*
[2]*Deputy Vice-Chancellor's Office (Research and Innovation) and Faculty of Health, Engineering and Sciences*
University of Southern Queensland (USQ), Toowoomba, Queensland, Australia

ABSTRACT   Sustainable energy and drinking water access have been seen as major challenges for rural households in Bangladesh despite of governmental and non-governmental organizations have been made extensive efforts. This study contemplates a universal approach towards tackling both of these issues via biogas based polygeneration integrated with membrane distillation employed at the village level. The specific technologies chosen for the key energy conversion steps are as follows: plug-flow digester (co-digestion and mesophilic condition); internal combustion engine; and air-gap membrane distillation. The proposed techno-economic results show that daily electricity demand can be met with such a system while simultaneously providing 0.4 m³ cooking fuel and 2–3 L pure drinking water. Cost analysis illustrates that the approach is highly favorable compare to other available system. The payback time of such system is between 2 and 2.5 years.

1 INTRODUCTION

In Bangladesh, only about 25% of the rural population is connected to grid electricity and no grid gas supply for cooking, where about 75% of the total population (164 million in total) resides. Arsenic (As) contamination in shallow tube-well water is a serious health issue in rural Bangladesh (Khan & Martin, 2015, Ahmed *et al.*, 2004) and according to Tan et al. (2010), 20% of deaths can be attributed to arsenic poisoning due to various diseases including lung, skin, and bladder and kidney cancers. Uddin *et al.* (2007) report that the range of As concentrations in groundwater of Bangladesh is between 0.25 µg/L and 1600 µg/L and in around 60 districts (out of 64) groundwater is contaminated with As levels exceeding the WHO guideline value (10 µg/L) (Khan *et al.*, 2014). Several technologies have been tried for removal of As from tube-well groundwater; however, key challenges still remain towards overcoming As poisoning. Until now, biogas co-digestion with locally available animal-agro wastes feed, has not been considered in a polygeneration/multigeneration (renewable energy integrated with membrane distillation) context in order to solve the energy and water problems.

Locally available animals and agriculture wastes feedstock's are a suitable candidate for biogas co-digestion plant. Moreover, biogas plants to provide gas for cooking, lighting and water purification systems, and their combined applications are particularly limited. On the other hand, Membrane Distillation (MD) is a thermally driven water purification and separation technology, which has significant advantages over other processes such as it is operating at low temperature and pressure, has a low electricity consumption, a high separation efficiency, integrates production of useful waste heat, etc. (Khan & Martin, 2014, 2015, Khan *et al.*, 2014)

2 METHODOLOGY AND PROCESS DESCRIPTION

The proposed method of polygeneration and simultaneous freshwater production can be applied in the remote village of Sherpur (Bangladesh) which consists of 219 households (average: five person per household) to obtain the agro-waste residues potential and checking their techno-economic viability for energy production. A household survey had been conducted using a specific questionnaire to assess the biomass potential and energy demand. In the energy and mass analysis (energy and mass balances) the digester has been designed to fully accommodate all demands for cooking gas and/or electricity generation during the day. The generator supplies electricity according to estimated community demand, therefore no battery is considered for electricity storage. Experimental data have been used to evaluate the performance (product water, As removal efficiency and thermal energy consumption) of membrane distillation (Elixir500 unit, 550–600 L/day, 40–50 kWh/m³). Levelized Cost Of Energy (LCOE) and life cycle costs of water have been used as a tool for estimating production costs.

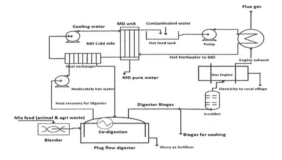

Figure 1. Biogas polygeneration integrated with MD water treatment process.

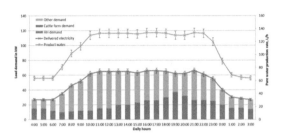

Figure 2. Electricity demand, supply and water production.

A sensitivity analysis has been performed considering the inherent uncertainty in assumptions and key parameters.

A plug flow digester, fed with animal and agriculture waste, is considered for delivering the required amount of biogas for cooking and fuel for a gas engine generating electricity. Exhaust flue gas from gas engine passes through a heat exchanger for heating up As-contaminated feed water, which is supplied to a MD unit. The techno-economic analysis of renewable energy and integrated MD system identifies that Sherpur village could be a potential candidate for the use of biogas and MD system for energy provision and pure water production.

3 RESULTS AND DISCUSSION

The energy and mass balance calculations were based on data from existing plug flow digesters, biogas engines, and available experimental data from a semi-commercial Elixir MD unit (HVR model, Scarab AB). Considering mass flows, there are two main inputs (cow manure as a feedstock for digester and contaminated feed water for MD) and four major outputs (biogas, electricity, safe water and fertilizer).

It was assumed that the total electricity demand of households in the village is 27 kWh/day (0.9 kWh/household/day, with an average of 6–8 h lighting, listening to radio and watching TV),

cooking gas demand of 0.4 m³/day/person (two meals per day) and drinking water demand of 3 L per person per day (Khan & Martin, 2015).

One of the key promising issues of the integrated system configuration is the levelized cost of the three major outputs: biogas for cooking (about 0.015 USD/kWh, depending upon the feedstocks handling charge), biogas generation of electricity (0.044–0.048 USD/kWh) and MD pure water (0.003 USD/liter). A sensitivity analysis has been performed for levelized costs of biogas production with feedstock handling costs from 1.25 USD/ton to 6.25 USD/ton and capital costs of digester (−20% to +50% change of digester capital costs) to demonstrate in what extent the levelized costs of biogas varies with these changes. The payback period has been estimated in two different cases (with and without slurry sales) and the minimum and maximum payback period in between 2 to 3 years (total investment costs about 140,000–150,000 USD, maintenance costs about 10% of Inv.) with lower and higher feedstock handling costs.

However, the real challenge of the integrate system is handling the As-rich liquid sludge of MD treatment. Arsenic removal from groundwater via MD results in liquid waste with high As concentration, caustic, and salinity. Hence, disposal of As-rich sludge is of environmental concern. Technologies currently available for treating the As-contaminated sludge are expensive, environmentally disruptive, and the waste handling during applying the technologies potentially hazardous. There are several remediation processes available: a) capping method; b) solidification and stabilization; c) phytoremediation/phytoextraction method; and d) sea as a sink.

4 CONCLUSIONS

The analysis of co-digestion and integrated MD system identifies that Sherpur village is a potential site for the use of biogas and MD system for energy generation and pure water production. From above economic analysis, it can be concluded that the cost benefit of biogas technology is largely increased if the slurry byproduct is used effectively on farms. Bioslurry may be considered as a good quality organic fertilizer for agriculture with a potential market in Bangladesh. However, the management of feedstock, slurry and arsenic rich sludge handling are challenging factors to the users and many of them are reluctant and are not willing to do the dung mixing job every day as they consider this as dirty job but such problem can be avoided by hiring full time operators for the plant. Social and non-pay attitude problems have been identified as barriers in rural areas, which in turn have made private investors reluctant to be active in biogas projects.

REFERENCES

Khan, E., Martin, A. 2015. Optimization of hybrid renewable energy polygeneration system with membrane distillation for rural households in Bangladesh. *Energy* 93: 1116–1127.

Tan, S.N., Yong, J,W., Ng, Y.F. 2010. Arsenic exposure from drinking water and mortality in Bangladesh. *Lancet* 376: 1641–1642.

Uddin, M.T., Mozumder, M.S.I., Figoli, A., Islam, M.A., Drioli, E. 2007. Arsenic removal by conventional and membrane technology: An overview. *Ind. J. Chem. Technol.* 14: 441–450.

Khan EU, Mainali B, Martin A, Silveira S. 2014. Techno-economic analysis of small scale biogas based polygeneration systems: Bangladesh case study. *Sust. Energy Technol. Assess.* 7: 68–78.

Ahmed, K.M., Bhattacharya, P., Hasan, M.A., Akhter, S.H., Alam, S.M.M., Bhuyian, M.A.H., Imam, M.B., Khan, A.A., Sracek, O. 2004. Arsenic enrichment in groundwater of the alluvial aquifers in Bangladesh: an overview. *Appl. Geochem.* 19(2): 181–200.

Khan, E.U., Martin, A.R. 2014. Water purification of arseniccontaminated drinking water via air gap membrane distillation (AGMD). *Period. Polytech. Mech. Eng.* 58: 47–53.

Arsenic Research and Global Sustainability – Bhattacharya, Vahter, Jarsjö, Kumpiene,
Ahmad, Sparrenbom, Jacks, Donselaar, Bundschuh & Naidu (Eds)
© *2016 Taylor & Francis Group, London, ISBN 978-1-138-02941-5*

Removal of arsenic from drinking water by *Epipremnumaureum* (L.) Engl. plant

H.B. Ahmad & A. Zaib

Institute of Chemical Sciences, Bahauddin Zakariya University Multan, Pakistan

ABSTRACT: *Epipremnumaureum* (L.) Engl. plant has been used for the removal of arsenic from water. Equally sized plants were grown in the solutions of arsenic with different concentrations. Removal of the contaminants has been checked by analyzing samples of water after fixed time intervals. Plots of concentration of arsenic versus time have been drawn to investigate the purification time up to safe limit. The sustainability of plants in different concentrations of arsenic has also been noted. Accumulation of arsenic in different parts of plant has been determined by digesting the parts separately and estimating the amounts in the extract of plant. In the light of experimental results, a prototype plant has been suggested for the purification of drinking water which can not only be used to remove the dissolved ionic contaminants but also the turbidity and microbes from water.

1 INTRODUCTION

Arsenic contamination in drinking water has been considered a serious issue for the last twenty years. Presence of arsenic in drinking water is very much alarming especially in Asian countries (Mukherjee *et al.*, 2006). Arsenic may cause many serious diseases in human bodies (Saha *et al.*, 1999). Presence of arsenic can be determined by test kits, atomic absorption spectroscopy, inductively coupled plasma, surface plasmon sensor and electrochemically. Recently we have suggested the determination of arsenic by silver nanoparticals (Ali, 2015).

Various methods have been adopted to remove arsenic from water. Our research group has also used different adsorbents for the removal of arsenic (Ahmad *et al.*, 2014, 2015). Another technique which is being used to remove arsenic from water is hydroponic technique. In present study, we have attempted to introduce *Epipremnumaureum* grown in many houses as indoor plant for the removal of arsenic from drinking water.

2 METHODS/EXPERIMENTAL

2.1 *The experimental setup*

Forty plants of equal size of *Epipremnumaureum* were selected for experimentation and grown in glass jars. The plants were inserted into the jars by piercing a hole in the lids of jars. All the plants were kept under the same conditions for fifteen days until the roots were established and plants started to grow normally in all the jars. Four treatment setups of Arsenic (0.01M, 0.006M, 0.003M and 0.001M) were used with 10 replicates of each treatment. Samples were taken from each jar after every six hours and the amount of arsenic in the solution was determined by titrating with potassium permanganate solution.

2.2 *Determination of arsenic after fixed time intervals*

Amount of arsenic was determined by taking 10 ml sample of water from each jar and titrated against $KMnO_4$ after adding 1–2 drops of potassium iodate solution (0.002 M). Concentration of arsenic in each jar was calculated at zero time and after every six hours up till 48 hours by taking 10 ml sample of water from fresh jar.

2.3 *Analysis of plant samples*

The plants from the experiments were removed from jars after 48 hours and washed the roots of the plants with deionized water. The roots, stems and leaves of every plant were separated and dried in oven at 80 °C for 48 hours. Samples were ground thoroughly and 100 mg of each were taken in a flask and digested according to literature procedure.

3 RESULTS AND DISCUSSION

The data obtained shows that the concentration of arsenic decreases with time (Fig 1a). Plot of percentage decrease of arsenic against time for different

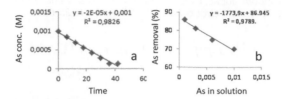

$y = -2E-05x + 0,001$
$R^2 = 0,9826$

$y = -1773,9x + 86.945$
$R^2 = 0,9789.$

Figure 1. a) Plot of removal of 0.001M arsenic solution with time; b) trend for removal of arsenic with initial concentration of arsenic.

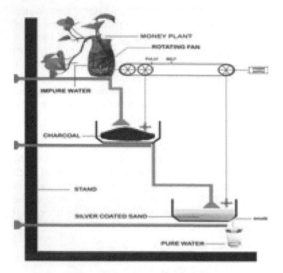

Figure 2. Four different classes of arsenolipids: Arsenic containing hydrocarbons (AsHC), arsenic containing fatty acids (AsFA), arsenic containing fatty alcohols (AsFAl) and arsenic containing phospholipids (AsPL).

concentrations (Fig. 1b), shows that the decrease in concentration of arsenic is more prominent in low concentrated solution. If the graph is extrapolated, it indicates that for very low concentrations, the decrease in arsenic will be very much significant (8.94% at zero concentration). This shows that the hydroponic technique can be adopted for the removal of arsenic from drinking water in which its quantity is very low. Analysis of different parts of the plants indicate that arsenic was accumulated more in stem cells than in roots and leaves.

3.1 *Design of prototype water purification plant*

In the light of arsenic removal results of Epipremnumaureum, a prototype treatment setup has been designed for the purification of drinking water (Fig. 3). The purification plant consists of three

chambers. In first chamber, *Epipremnumaureum* have been embedded in water tank. Water will remain 48 hours in this chamber. Mechanical stirrers are fixed at the bottom of this chamber. The water then will be transferred to next chamber in which granular charcoal is present which will decolorize water and remove particulate impurities. Mechanical stirrer is suspended in the chamber for smooth mixing of water with charcoal. In the third chamber, silver coated sand is added to remove microbes from water. Retention time in second and third chambers is one hour each.

4 CONCLUSIONS

The hydroponic technique in general and Epipremnumaureum in particular can be used be for the removal of arsenic from drinking water. Used plants can be reclaimed if they are kept in fresh water for a week. The proposed prototype of treatment setup can be a good solution for remediating drinking water from arsenic.

ACKNOWLEDGEMENTS

All the funding and facilities for this project were provided by the Institute of Chemical Sciences, Bahauddin Zakariya University Multan, Pakistan..

REFERENCES

Ahmad, H.B., Abbas, Y., Hussain M. & Arain, S.A. 2014. Synthesis and application of alumina supported nano zero valent zinc as adsorbent for the removal of arsenic and nitrate. *Korean J. Chem. Eng.* 31(2): 284–288.

Ahmad, H.B., Gul-E-Yasmin, Arain, S.A., Bhatti I.A. & Hussain, M. 2015. Synthesis of some novel adsorbents for antimicrobial activity and removal of arsenic from drinking water. *Korean J. Chem. Eng.* 32(4): 661–666.

Ali, W. 2015. Synthesis, characterization and applications of some novel silver nanoparticles. *M. Phil Thesis* in Physical Chemistry, Institute of Chemical Sciences, Bahauddin Zakariya University Multan, Pakistan.

Ma, L.Q., Komar, K.M., Tu, C., Zhang, W., Cai Y., Kennelley, E.D. 2001. A fern that hyperaccumulates arsenic. *Nature* 409: 579.

Mukherjee, A., Sengupta, M.K., Hossain M.A., Ahmed, S., Das, B., Nayak, B., Lodhi, D., Rahman, M.M., Chakraborti D. 2006. Arsenic contamination in groundwater: a global perspective with emphasis on the Asian scenario. *J. Health Popul. Nutr.* 24(2): 142–163.

Saha K.C., Dikshit A.K., Bandyopadhyay M.A. 1999. A review of arsenic poisoning and its effect onhuman health. *Crit. Rev. Environ. Sci. Technol.* 29: 281–313.

Arsenic Research and Global Sustainability – Bhattacharya, Vahter, Jarsjö, Kumpiene,
Ahmad, Sparrenbom, Jacks, Donselaar, Bundschuh & Naidu (Eds)
© 2016 Taylor & Francis Group, London, ISBN 978-1-138-02941-5

Developing innovations for adsorptive removal of arsenic from drinking water sources in North Mara gold mining area, Tanzania

R. Irunde[1], F.J. Lesafi[1], F. Mtalo[2], P. Bhattacharya[3,4], J. Dutta[5] & J. Bundschuh[6]

[1]*Department of Chemistry, College of Natural and Applied Sciences, University of Dar es Salaam, Dar es Salaam, Tanzania*
[2]*Department of Water Resources Engineering, College of Engineering and Technology, University of Dar es Salaam, Dar es Salaam, Tanzania,*
[3]*KTH-International Groundwater Arsenic Research Group, Department of Sustainable Development, Environmental Science and Engineering, KTH Royal Institute of Technology, Stockholm, Sweden*
[4]*International Center for Applied Climate Science, University of Southern Queensland, Toowoomba, Queensland, Australia*
[5]*Functional Materials Division, Materials and Nanophysics, School of ICT, KTH Royal Institute of Technology, Kista, Stockholm*
[6]*Faculty of Health, Engineering and Sciences, University of Southern Queensland, Toowoomba, Queensland, Australia*

ABSTRACT: Arsenic (As) contamination in drinking water have been reported to occur in areas where mining are practiced such as North Mara, lake Victoria basin in Mwanza, Geita and Rukwa. The removal of As requires methods such as ion-exchange, reverse osmosis, electrodialysis, contact precipitation, activated alumina, bone charcoal and activated clay. The use of ion-exchange, reverse osmosis and electrodialysis are expensive to be implemented in Tanzania. In order to safeguard the drinking water supplies, it is important to explore the low-cost and efficient locally available adsorbents such as activated alumina, bone char and clay for the removal of As for drinking water consumption.

1 INTRODUCTION

Arsenic (As) occurs as a trace element with variable concentrations in different environmental realms such as natural waters, soils and rocks as well as in the atmosphere (Bhattacharya et al., 2002, BGS, 2008). Gold mining areas in Tanzania has been reported to have high concentration levels of As in water sources (Mnali, 1999, Kassenga & Mato, 2008, IPM Report, 2013). The earliest studies on assessment for As levels from Lupa Gold Fields in south-west Tanzania, Mnali (1999) detected As concentrations in the range of 36 µg/L to 70 µg/L, which is associated with decomposition of gold ores that consist arsenopyrite (FeAsS), enargite ($Cu_3 AsS_4$) and tennantite ($Cu_{12} As_4S_{13}$). Nyanza et al. (2014) reported that Rwamagasa village in Geita (northwestern Tanzania) has high concentration levels of As above WHO and Tanzania Bureau of Standards. The study done on mining areas of Lake Victoria basin also revealed high concentration of As from 30 µg/L to 70 µg/L which is above Tanzania drinking water quality standard of 10 µg/L (Kassenga & Mato, 2008).

Exposure to As leads to an accumulation of As in tissues such as skin, hair and nails, resulting in various clinical symptoms such as hyperpigmentation and keratosis (Kapaj et al, 2006). There is also an increased risk of skin, internal organ, and lung cancers (Smith et al., 1998). Cardiovascular disease and neuropathy have also been linked to As consumption. Arsenic can also suppress hormone regulation and hormone mediated gene transcription. Increases in fetal loss and premature delivery, and decreased birth weights of infants, can occur even at low (<10 µg/L) exposure levels.

The arsenate (AsO_4^{3-}) species are strongly sorbed onto clays, Fe-Mn oxides/hydroxides and organic matter in which soil microbiota can accelerate the oxidation of soluble As(III), the toxic available form to a less soluble arsenate (Aswatharayana, 2001). The use of lime softening, ion-exchange resins, membrane methods and activated alumina in As removal have been done in developed countries but in Tanzania has not reported (Kassenga & Mato, 2008). The efficient developed method by Kassenga & Mato (2008) was reported to be a good adsorbent material. They used ferralsols (the red or yellow oxidized soil) which was prepared by mixing 1:1 locally available ferralsol and Portland cement in distilled water to make slurry that was air dried and cut into pieces to be used for adsorption of As at various pH conditions. The Portland cement aimed to improve structure due to presence of CaO and enhance the adsorption capacity for As.

The improved ferralsol had Fe-oxide concentration of 1560 mmol/kg which ensures high capacity of adsorbing As and hence could adsorb As up to 97% of the aqueous concentration independent of water pH. This study aims to prepare more efficient adsorbents such as activated alumina, activated born char and activated red clay for removal of As from drinking water sources in Tanzania.

2 DEVELOPING INNOVATIONS FOR ARSENIC REMOVAL

2.1 *Preparation of water sample*

The 50 mL of water sample will be digested by using 10 mL nitric acid under heat to remove organic material before analyzing As by Flamed Atomic Absorption Spectrophotometry or ICP-AES.

2.2 *Preparation of activated alumina, clay and bone char*

The red top soil (laterite) (about 10 g) will be dried and subsequently fired at 570°C for about three hours. The fired clay will be crushed and placed in plastic bottles. About 100 mL of the representative water samples (of known As concentration) brought from the site will be poured into the plastic bottle containing fired crushed clay materials. At given time intervals the decant solution will be extracted and analyzed for As concentration and compared to the concentration of the origin analyte.

Bauxite will be calcinated at 400 to 600°C to make activated alumina. The water sample will be passed through a packed column of activated alumina to remove As.

Bones also will be fired at 500 to 650°C. In contact with water the bone char will be able to adsorb As.

3 APPROACH AND OUTLOOK

3.1 *Data analysis*

The data obtained will be analyzed using statistical techniques such as Analysis of Variance (ANOVA). The equilibrium isotherms will be analyzed with Temkin, Langmuir and Freundlich equations. The goodness of data fittings will be inspected by linear regression analysis. In addition, adsorption data obtained by evaluating adsorbent dosage and contact time will be plotted as percent removal of adsorbate versus adsorbent dosage, percent removal of adsorbate versus contact time in minutes and amount of adsorbate adsorbed (mg/g) versus time in minutes

3.2 *Efficient of adsorbents*

The activated alumina, clay and bone char are expected to remove As at high percentage than method used by Kassenga & Mato (2008).

3.3 *Adsorption modeling*

The higher coefficient determination of Arsenic adsorption will be shown by Freundlich equations than Temkin and Langmuir.

4 CONCLUSIONS

Although various studies has been conducted to investigate the levels of As in gold mining areas such as in Mwanza, Geita, Mara and Rukwa but still less removal methods has been tested and found efficient on removing As especially stabilized ferralsols by Kassenga & Mato, 2008. More studies in other small gold mining areas should be conducted and various adsorbents such as activated alumina, bone char, red clay should be prepared to meet the requirement of the population living in areas contaminated with As.

ACKNOWLEDGEMENTS

The author would like to thank SIDA-UDSM program under DAFWAT-2235 project for financial support.

REFERENCES

Aswathanarayana, U. 2001. Water Resources Management and the Environment. CRC Press, 446p.
BGS. 2008. Water Quality Fact Sheet. Natural Environment Research Council. British Geology Survey.
Kapaj, S., Peterson, H., Liber, K., Bhattacharya, P. 2006. Human health effects from chronic arsenic poisoning. *J. Environ. Sci. Heal. A* 41(10): 2399–428.
Kassenga, G.R., Mato, R.R. 2008. Arsenic contamination levels in drinking water sources in mining areas in Lake Victoria Basin, Tanzania, and its removal using stabilized ferralsols. *Int. J. Biol. Chem. Sci.* 2(4): 389–400.
IPM Report. 2009. Investigation of Trace metal concentrations in soil, sedments and waters in the vicinity of Geita Gold Mine and North Mara Gold Mine in North West Tanzania.
Mnali, S.R. 1999. Palaeoproterozoic felsic magmatism and associated gold Quartz vein mineralization in the Western part of the Lupa Gold Field, South Western Tanzania. PhD. Thesis, University f Dar es Salaam, Tanzania (Unpublished).
Nyanza, C.E., Dewey, D., Thomas, D.S., Davey, M., Ngallaba, E.S. 2014. Spatial distribution of mercury and Arsenic levels in water, soil, and cassava plants in a community with long history of gold Mining in Tanzania. *Bull. Environ. Contam. Toxicol.* 93(6): 716–721.
Smith, A., Goycolea M, Haque R, Biggs ML, 1998. Marked increase in bladder and lung cancer mortality in region of Northern Chile due to Arsenic in drinking water. *Am. J. Epidemiol.* 147: 660–669.

Section 5: Societal and policy implications, mitigation and management

5.1   Societal and policy implications of long term exposure

Arsenic Research and Global Sustainability – Bhattacharya, Vahter, Jarsjö, Kumpiene,
Ahmad, Sparrenbom, Jacks, Donselaar, Bundschuh & Naidu (Eds)
© *2016 Taylor & Francis Group, London, ISBN 978-1-138-02941-5*

Arsenic at low concentrations in Dutch drinking water: Assessment of removal costs and health benefits

P. van der Wens[1,2], K. Baken[3] & M. Schriks[3]

[1]*Brabant Water, Den Bosch, The Netherlands*
[2]*Vewin, Den Haag, The Netherlands*
[3]*KWR Watercycle Research Institute, Nieuwegein, The Netherlands*

ABSTRACT: In 2015 the Association of Dutch Drinking water Companies (Vewin) concluded that a guideline value of arsenic <1.0 µg/L is both achievable and appropriate for the Dutch situation, thus following the WHO-guideline (WHO, 2011) to keep concentrations of arsenic as low as reasonably possible. The decision was made as a precautionary measure based on a two-tiered assessment of arsenic in Dutch drinking water. In the first tier a conservative estimation on human health effects of arsenic exposure via drinking water was made using data from an epidemiological study from Chili (Ferrecio et al., 2000). In the second tier a removal cost/health benefit analysis was carried out. The costs for arsenic removal in the Dutch treatment works were estimated based on pilot research and design by Brabant Water and Vitens. Health benefit analysis was based on the DALY-approach used by the Dutch National Health Council (NHC, 2007) for societal cost/benefit analysis.

1 INTRODUCTION

In 1993 WHO lowered the guideline value for arsenic in drinking water to 10 µg/L on the basis of growing concern with regard to carcinogenicity of arsenic in humans (Chappells et al., 2014). WHO denotes this guideline value as provisional and states that every effort should be undertaken to keep the concentrations as low as reasonably achievable. In many European countries—including the Netherlands—the WHO guideline value has been adopted in national drinking water legislation. Considering the continuous research providing novel insights into arsenic toxicity, the derivation of an adequate drinking water guideline should remain an ongoing process to strengthen water suppliers in their risk management.

2 APPROACH

2.1 *Dose-effect modeling of arsenic in low concentrations in drinking water*

In a first tier an elaborate literature survey was carried out into the human health effects of arsenic exposure via drinking water. In spite of a substantial database on the association between cancer and the consumption of arsenic in drinking water, there remains considerable uncertainty on the chronic risks at low concentrations (<50 µg/L) and the shape of the dose-response relationship. The excess lifetime lung cancer risk of arsenic in drinking water in the Netherlands was conservatively approximated by linear extrapolation of data from of an epidemiological study from Chili (Ferrecio et al., 2000). Lung cancer was the endpoint of choice since this effect resulted in the lowest $BMDL_1$ in an evaluation of the health risks of arsenic exposure by the European Food Safety Authority (EFSA, 2009). The National Research Council (NRC, 2001) previously concluded that based on the Chilean study, a concentration of 50 µg/L arsenic would result in an estimated 41% increase in population lung cancer risk.

Linear extrapolation to lower concentrations was applied as a conservative, non-threshold approach in absence of consensus on low dose effects. This approach resulted in a concentration of 0.004 µg/L that theoretically corresponds to an excess lifetime cancer risk of 10^{-6} in the Netherlands (Fig. 1).

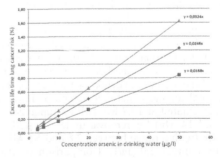

Figure 1. Estimated excess life time lung cancer risk (95% c.i.) in the Dutch population related to arsenic concentrations in drinking water.

Alternatively, when a safe Margin of Exposure would be applied between actual exposure levels and the $BMDL_{10}$, an arsenic concentration of 0.3 μg/L in Dutch drinking water would conservatively be considered to be of negligible health risk. These estimates indicate that the current guideline value of 10 μg/L would not be sufficiently protective of health effects of arsenic in drinking water.

2.2 Cost-benefit analysis of removal of arsenic

In the second tier a removal cost/health benefit analysis was carried out using target levels of <1,0 and <2,0 μg/L arsenic in drinking water. The costs were based on pilot scale research carried out by Brabant Water who used the Advanced Oxidation-Coagulation Filtration process added to existing treatment plants. Vitens applied a comparative approach using the Iron Oxide-Coated Sand filtration process. To deal with the uncertainty of adequate removal within some existing treatment plants an optional cost calculation was made based on an additional filtration step. The total annual costs for achieving the target level of <1,0 μg/L As in Dutch drinking water was calculated to amount to 3,8 to 7,2 M€; for a level <2,0 μg/L costs were estimated at 1,4 to 4,0 M€.

The health impact of arsenic concentrations in Dutch drinking water was estimated using the Disability Adjusted Life Years (DALY) approach of the Dutch National Health Council (NHC, 2007), based on the worst case linear dose response relation described above and assumptions for the years of life lost due to disability and death caused by lung cancer. It was conservatively modelled that every μg/L arsenic in drinking water would annually result in 51 (35–68) cases of lung cancer and 756 (516–996) DALYs (provided that the total population would lifelong be exposed to this concentration). In comparison, the total burden of disease in the Netherlands is estimated to be 3,272,640 DALYs per year (IHME, 2010) and the burden of disease that can be ascribed to environmental factors 175,734 DALYs (Hänninen et al. 2014).

Taking into account the costs of health care for cases of lung cancer theoretically caused by arsenic (approximately 1,0 M€ for each μg/L arsenic in Dutch drinking water) and the value of avoidance of a DALY (estimated to be € 60,000 on average in the Netherlands), a cost-benefit analysis of reduction of arsenic levels in drinking water can be made. The total annual benefit of arsenic removal was thus calculated at 10,7 M€ (with a range between 7,2 and 14 M€) for removal under 1,0 μg/L and 4,0 M€ (with a range between 2,8 and 5,5 M€) for removal under 2,0 μg/L. Comparison of (treatment) costs en modelled health benefits show that the benefits are approximately twice the costs.

3 DISCUSSION AND CONCLUSIONS

The presented cost-benefit approach supports—despite uncertainties in estimating the health effect of arsenic in drinking water—the initiative to produce drinking water with an arsenic concentration of <1.0 μg/L in The Netherlands. This concentration level is based on (i) technical removal efficiency < 1 μg/L, (ii) an adequate analytical Limit Of Detection (< 0,03 μg/L), (iii) a positive cost-health benefit ratio and (iv) the precautionary principal.

The main uncertainty in our analysis lies in the model selected to estimate the health effects of low doses of arsenic. By absence of a proven threshold below which effects will not occur, a linear non-threshold approach was used, resulting in a conservative quantification of the health benefit of arsenic removal from drinking water. It cannot be excluded that future research will point out that the health risk of arsenic was overestimated and more effort than necessary is made to safeguard the quality of the Dutch drinking water. Using the precautionary principle we conclude that effective measures will be taken to reduce arsenic in Dutch drinking water to levels ´as low as reasonably possible.

ACKNOWLEDGEMENTS

The authors would like to thank Vewin for their assignment and decisions based on the research.

REFERENCES

Chappells, H. et al., 2014. Arsenic in private drinking water wells: an assessment of jurisdictional regulations and guideline for risk remediation in North America. J. Water Health 12 (3), 372–392.

EFSA (European Food Safety Authority), 2009. Scientific opinion on Arsenic in food. EFSA Journal 7 (10), 1351.

Ferreccio, C., et al., 2000. Lung cancer and arsenic concentrations in drinking water in Chile. Epidemiology 11 (6), 673–679.

NHC (National Health Council), 2007. Maten voor milieug-ezondheidseffecten. Publication number: 2007/21

IHME (Institute for Health Metrics and Evaluation), 2010. Global Burden of Disease Study 2010 (2013). Results 1990–2010. Seattle, United States.

Hänninen, O.et al., 2014. Environmental burden of disease in Europe: Assessing nine risk factors in six countries. Environ Health Perspect 122 (5), 439–446.

NRC (National Research Council), 2001. Arsenic in Drinking Water: 2001 Update. The National Academies Press., Washington, DC.

Pomp, M., Schoenmaker, C.G., Polder, J.J., 2014. Op weg naar maatschappelijke kosten-batenanalyses voor preventie en zorg: Themarapport Volksgezondheid Toekomst Verkenning 2014. RIVM Rapport 010003003.

WHO, 2011. Guidelines for Drinking Water Quality, Fourth Edition. World Health Organization, Genève, Switzerland.

Arsenic Research and Global Sustainability – Bhattacharya, Vahter, Jarsjö, Kumpiene,
Ahmad, Sparrenbom, Jacks, Donselaar, Bundschuh & Naidu (Eds)
© 2016 Taylor & Francis Group, London, ISBN 978-1-138-02941-5

Regional variation of arsenic concentration in drinking water: A cross-sectional study in rural Bangladesh

T. Akter, F.T. Jhohura, F. Akter, T.R. Chowdhury, S.K. Mistry & M. Rahman
Research and Evaluation Division, BRAC, Mohakhali, Dhaka, Bangladesh

ABSTRACT: Many of the households in rural Bangladesh use shallow tubewells for drinking water which are mostly affected by high levels of arsenic. Women and children living in such households are at risk to high exposure to arsenic being associated with reproductive and developmental consequences. As part of the cross-sectional social survey across the country we visited 960 randomly selected households to check the permissible safe drinking water limits set by the WHO and Bangladesh government. We tested arsenic on the spot using Econo-Quick arsenic test kit and a part of the water samples were verified in the laboratory. We found that 68% and 77% of household members in Dhaka division exposed to higher levels of arsenic exceeded both WHO and Bangladesh standards, respectively. In some cases (e.g., Sonargaon), people used arsenic-affected drinking water sources even though they knew that the water was contaminated and damaging to health.

1 INTRODUCTION

It is reported that high exposure to arsenic is associated with cancer, reproductive and developmental effects and cardiovascular diseases. Millions of Bangladeshis whose health are at risk to high levels of arsenic in drinking water (Aziz *et al.*, 2014), while women and children living in such contaminated environment are particularly susceptible. Most of the rural people in Bangladesh do not have the choice, but to collect water from the available sources. In this study we assessed household's exposure to arsenic with respect to their demographic, socioeconomic and locational factors. We examined households' exposure to arsenic contamination according to the drinking water standard set by WHO and the Government of Bangladesh.

2 METHODS/EXPERIMENTAL

2.1 *Study design and area*

This study was part of our research on "The status of households WASH behaviors in rural Bangladesh" which was conducted in 960 households from 24 randomly selected *upazilas* (5% of the total). The current study on the assessment of drinking water quality embraced cross-sectional study design and was conducted in 12 out of 24 upazilas across the country including: Alfadanga, Kendua, Shibchar, Rupsha, Debhata, Patharghata, Rangabali, Anwara, Bijoynagar, Shajahanpur, Kamalganj and Kurigram Sadar.

2.2 *Testing and validation of water samples*

Out of 960 households, tubewells in 645 households (67%) were tested for arsenic in the field using "Econo Quick Arsenic Test Kit" made in USA by ITS (Innovators for water quality testing). Among the remaining 33% households, 29% of them had already tested arsenic of their tubewells in recent past and declared arsenic free. Besides, few of them (4%) used pond water for drinking, therefore, they were excluded for arsenic test on the spot.

Arsenic results measured in the field using test kit were checked in the laboratory for verification. Water samples with certain values such as 0.025, 0.050 mg/L were randomly picked up to crosscheck the values in the laboratory.

3 RESULTS AND DISCUSSION

Arsenic test-kit results on the spot indicates high arsenic concentration in Shibchar (Madaripur) (Mean: 0.07±0.09; Min: 0.01; Max: 0.50), Biswanath (Sylhet) (Mean: 0.09±0.08; Min: 0.00; Max: 0.20) and Dhaka (Mean: 0.06±0.11; Min: 0.01; Max: 0.30) which exceeded Bangladesh standard (0.05 mg/L) (Table not shown here). Validation results showed that water samples collected from Shibchar (Madaripur) (0.06±0.04) exceeded Bangladesh standard of arsenic concentration keeping consistency with the earlier test in the field (Table 1). The results showed that 68% and 77% of household members in Dhaka division exposed to

Table 1. Regional variation in the values of arsenic content of drinking water (results validated in the laboratory).

Sample site	Arsenic (mg/L)		
	Min	Max	Mean ± SD
Alfadanga (Faridpur)	0.022	0.085	0.047±0.034
Kendua (Netrokona)	0.003	0.126	0.031±0.053
Shibchar	0.020	0.141	0.057±0.037
Rupsha (Khulna)	0.003	0.017	0.009±0.006
Debhata (Satkhira)	0.003	0.021	0.006±0.006
Patharghata (Barguna)	0.004	0.004	0.004
Rangabali (Patuakhali)	0.003	0.005	0.004±0.001
Anwara (Chittagong)	0.003	0.025	0.009±0.008
Bangladesh Drinking water standard			0.05
WHO guideline			0.01

Table 2. Status of arsenic concentration by WHO and Bangladesh Drinking Water Standard (BDWS) (%).

Characteristics	Drinking water standard (mg/L)			
	WHO		BDWS	
	≤0.01	>0.01	≤0.05	>0.05
Gender				
Male	50.0	52.3	50.2	56.7
Female	50.0	47.7	49.8	43.3
Age				
≤4	7.1	6.5	7.0	6.7
5–20	35.6	39.3	35.1	53.3
21–40	31.8	23.4	30.1	20.0
41–60	17.2	24.3	19.3	20.0
≥61	8.4	6.5	8.5	90.0
Education				
No education	7.5	5.0	6.9	5.0
Primary	15.0	28.3	17.9	30.0
Secondary	21.1	20.0	20.8	20.0
Higher Secondary & above	56.4	46.7	54.3	45.0
Location by division				
Dhaka	9.1	67.9	22.7	76.7
Chittagong	16.5	9.2	15.6	0.0
Rajshahi	–	–	–	–
Khulna	20.2	16.5	20.9	0.0
Barisal	54.1	0.0	40.8	0.0
Sylhet	0.0	6.4	0.0	23.3
Wealth quantile				
Lowest	38.4	0.0	28.8	0.0
Second	12.2	9.2	11.1	13.3
Middle	28.3	20.2	26.9	13.3
Fourth	5.5	48.6	15.2	60.0
Highest	15.6	22.0	18.0	13.3
Water source by type				
Shallow tubewell	31.0	70.6	40.2	76.7
Deep tubewell	69.0	29.4	59.8	23.3
HH member (%)	68.9	31.1	91.5	8.5
HH (%)	69	31	91.5	8.5

higher levels of arsenic exceeded both WHO and Bangladesh standards, respectively (Table 2).

About 48% and 43% female participants are exposed to high arsenic concentrations which exceeded WHO and Bangladesh standard, respectively. In Bangladesh more than 50 million people including many pregnant women were exposed to drinking water exceeding WHO standard of arsenic concentration (McLellan, 2002). In our study, about 7% under five children and a major proportions of participants aged between 5 and 40 years were exposed to high arsenic levels exceeding WHO (63%) and Bangladesh (73%) standard. Rahman *et al.* (2015) found a significant association between prenatal arsenic exposure and drowning among children in Bangladesh. They also put further investigation if there is any relationship between children's cognitive function affected by arsenic exposure and drowning deaths.

Most households in Dohar, Shibchar, and Sonargaon used shallow tubewells for drinking, which were affected by high levels of arsenic. In Shibchar (West Kakor village), most tubewells were affected by arsenic, and the villagers were unaware of which tubewell was arsenic free; therefore, they collected drinking water from any tubewell. In some cases (e.g., Sonargaon), people used arsenic-affected drinking water sources even though they knew that the water was contaminated and damaging to health. Saint-Jacques *et al.* (2014) showed that bladder cancer risk increased 2.7 and 4.2 times by arsenic exposure of 10 and 50 µg/L in water, respectively. About 83% chance of developing bladder cancer and 74% probability of mortality at a 50 µg/L exposure level were reported in the same study. Moreover, mortality rates were 30% higher at 150 µg/L than 10 µg/L. According to a national survey conducted in 2009 by UNICEF/BBS, 53 and

22 million people were exposed to arsenic according to WHO and BDWS standards, respectively. Arsenic has been detected in the groundwater of 322 *upazilas* (sub-districts) and 61 districts in Bangladesh (Hossain *et al.*, 2015). The health effects of prolonged and excessive inorganic arsenic exposure include arsenicosis, skin diseases, skin cancers, internal cancers (bladder, kidney, and lung), diabetes, raised blood pressure, and reproductive disorders (WHO, 2011, Shankar *et al.*, 2014).

4 CONCLUSIONS

Most of the shallow tubewell users including about half of the female participants are exposed to high arsenic concentration. Consequently, a considerable health consequences are assumed to be on children's developmental and women's pregnancy related complications.

ACKNOWLEDGEMENTS

We acknowledge the Government of Royal Netherlands for the financial support.

REFERENCES

Aziz, S.N., Aziz, K.M.S., Boyle, K.J. 2014. Arsenic in drinking water in Bangladesh: factors affecting child health. *Front. Public Health* 2(57): 1–5.

Hossain, M., Rahman, S.N., Bhattacharya, P., Jacks, G., Saha, R., Rahman, M. 2015. Sustainability of arsenic mitigation interventions—an evaluation of different alternative safe drinking water options provided in Matlab, an arsenic hot spot in Bangladesh. *Front. Environ. Sci.* 3:30.

McLellan, F. 2002. Arsenic contamination affects millions in Bangladesh. *Lancet* 359: 1127

Rahman, M., Sohel, N., Hore, S.K., Yunus, M., Bhuiya, A. & Streatfield, P.K. 2015. Prenatal arsenic exposure and drowning among children in Bangladesh. *Global Health Action* 8:28702. (doi:10.3402/gha.v8.28702)

Saint-Jacques, N., Parker, L., Brown, P., Dummer, T. 2014. Arsenic in drinking water and urinary tract cancers: a systematic review of 30 years of epidemiological evidence. Environ. Health 13: 44.

Shankar S, Shanker U, Shikha 2014. Arsenic contamination of groundwater: a review of sources, prevalence, health risks and strategies for mitigation. *The Scientific World Journal* 2014: 18.

WHO 2011. *Guidelines for Drinking-Water Quality.* World Health Organization, Geneva, 541 p.

Arsenic Research and Global Sustainability – Bhattacharya, Vahter, Jarsjö, Kumpiene,
Ahmad, Sparrenbom, Jacks, Donselaar, Bundschuh & Naidu (Eds)
© 2016 Taylor & Francis Group, London, ISBN 978-1-138-02941-5

Recent advances and additional needs in science and policy fronts for arsenic mitigation in Bangladesh

K.M. Ahmed

Department of Geology, Faculty of Earth and Environmental Sciences, University of Dhaka, Dhaka, Bangladesh

ABSTRACT: Latest survey reveals that 19.2 million people in Bangladesh are still drinking water with arsenic concentrations exceeding 0.05 mg/L national drinking water limit. Mitigation efforts have stagnated over the recent past resulting into very slow progress in providing safe water to the most vulnerable communities. This stagnation has taken place at times when significant advancements have been made towards arsenic mitigation in science and policy fronts. Hydrostratigraphic mapping of safer intermediate aquifers, targeting safe aquifers based on sediment color, field trials of MAR and SAR in arsenic lowering are some areas of scientific developments. Revision of IPAM and progress towards formulation of regulations for enforcing Bangladesh Water Act 2013 highlights the major achievements in policy fronts. Renewed efforts powered by recent scientific and policy developments can ensure sustainable arsenic mitigation in Bangladesh. Combined actions by scientists and policy makers can speed up the pace of mitigation and break the stagnation.

1 INTRODUCTION

Efforts have been underway following detection of widespread arsenic contamination in Bangladesh occurring mostly in shallow Holocene alluvial aquifers under reducing conditions although this remains a major public groundwater management issue (Edmund *et al.*, 2015). The Pleistocene oxidized aquifers and deeper aquifers are mostly arsenic safe except in few localized areas. Various studies have been completed to understand the mechanisms of mobilization and demobilization of arsenic in certain hydro geological conditions. Policies have been formulated to provide As safe water to more than 22 million people exposed to arsenic above the national drinking water limit. However, still today more than 19 million people are exposed to same level as reported by the latest MICS survey (BBS/UNICEF, 2014). Nationally 14% of the tested water sources are found to have arsenic above

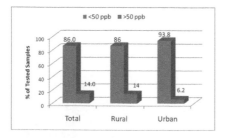

Figure 1. Percentages of sampled water sources exceeding Bangladesh drinking water limit as revealed by testing of 13,234 samples nationally under the MICS 2013.

0.05 mg/L; situations are better in urban areas compared to rural areas as shown in Figure 1.

2 RECENT ADVANCES TOWARDS FINDING ARSENIC SAFE WATER

2.1 *Oxidized aquifers*

Oxidized aquifers have long been quoted as source of safe water. Documentation on a national scale has been reported by Hoque *et al.* (2011). A color tool for use as guide in locating arsenic safe aquifer based on the links between arsenic concentrations and sediment color has been described by Hossain *et al.* (2014). Oxidized sediments have the potential for providing low arsenic water over many parts of the country.

2.2 *Intermediate deep aquifer*

Over most of Bangladesh the aquifer occurring between 100 and 150 m has the potentiality to provide low arsenic and manganese water as demonstrated by SASMIT studies in central Bangladesh (Hossain *et al.* 2014).

2.3 *Subsurface arsenic removal and managed aquifer recharge*

There have been recent investigations on the application of subsurface arsenic removal as described by Rahman *et al.* (2015). This has potential for use in areas where As safe deep aquifer or oxidized aquifer do not occur. Managed aquifer recharge also has the potentiality to reduce arsenic in the shallow aquifers. However, careful investigations are needed to assess the possibility of arsenic mobilization due to injection of high oxygen water in aquifers containing high total organic carbons as reported by Sultana et al. (2015).

3 RECENT DEVELOPMENTS IN POLICY FRONT

3.1 Bangladesh Water Act 2013

In Bangladesh there is no system for abstraction control and licensing for groundwater except only in Dhaka city. Indiscriminate abstractions for various sources resulting into adverse impacts like lowering of water table, drying up of wells and intrusion of saline and poor quality water etc. The Bangladesh Water Act 2013 has created the legal framework for better management of the vital natural resources. However, various regulations needed for effective implementation of the water act are yet to be prepared.

3.2 Revised Implementation Plan for Arsenic Mitigation (IPAM)

Government of Bangladesh adopted National Arsenic Policy and Implementation Plan for Arsenic Mitigation in 2004 with objectives to mitigate impacts of arsenic on drinking water supply, irrigation and health sectors. However, the IPAM 2004 was prepared based on limited knowledge on water supply options, impacts on health and agriculture. As our knowledge base has improved significantly over the last decade, it is urgent to revise the IPAM for effective arsenic mitigation. Suggested framework of the revised IPAM was provided by Ahmed & Ravencroft (2009). Very recently the Policy Support Unit of the Ministry of Local Government has prepared the draft of the revised IPAM with time bound actions for arsenic mitigation in water supply sector. Similar efforts are underway for producing separate IPAMs for Health, Agriculture and Water Resources. Implementation of the revised IPAMs needs close coordination and monitoring at national level. National Arsenic Mitigation Coordination body is needed to ensure arsenic safe environment for all sectors.

3.3 Need for groundwater governance

Although groundwater is a vital resource in Bangladesh in providing safe water access for protection of public health and also for ensuring food security, there is a crying need for introducing groundwater governance. Groundwater governance can ensure integrated water resources management to avoid conflicts among the various stakeholders. Also this is needed in the context of major challenges ahead such as population growth, urbanization, industrialization and possible impacts of groundwater. Proper institutional arrangement is very important in ensuring good governance. A specific organization with legal power, technical capability and adequate budget allocation need to be set up for steer heading groundwater governance including arsenic mitigation and protection of resource from degradation.

4 CONCLUSIONS

There is need for updating policies and strategies for arsenic mitigation in the light of recent scientific findings for safer sources of water and guides for targeting aquifers. Pragmatic policy based on scientific findings remains as the key element of successful arsenic mitigation in Bangladesh. Proper groundwater governance has to be initiated through a competent authority in order to mitigate arsenic as well as ensuring sustainable supply for meeting increased demands in relation to population growth, urbanization and industrialization.

REFERENCES

Ahmed, K.M. Ravenscroft, P. 2009. Recommendations for a Revised Implementation Plan for Arsenic Mitigation (IPAM 2009). Water and Sanitation Program, World Bank and Policy Support Unit, Local Government Division, Ministry of Local Government, Rural Development and Cooperatives, Government of Bangladesh.

BBS and UNICEF. 2014. Progotir Pathey Multiple Indicator Cluster Survey 2012–13, Key Findings. Bangladesh Bureau of Statistics and UNICEF Bangladesh, Dhaka.

Edmunds, W.M., Ahmed, K.M., Whitehead, P.G. 2015. A review of arsenic and its impacts in groundwater of the Ganges–Brahmaputra–Meghna delta, Bangladesh. Environmental Science Processes & Impacts DOI: 10.1039/c4em 00673a.

Hossain, M., Bhattacharya, P., Frape, S.K., Jacks, G., Islam, M.M., Rahman, M.M, von Brömssen, M., Hasan, M.A., Ahmed, K.M. 2014. Sediment color tool for targeting arsenic-safe aquifers for the installation of shallow drinking water tubewells. Sci. Total Environ. 493: 615–625.

Hossain, M., Haque, A., Alam, S., Rahman, R., Uddin, M.R., Sarwar, S.G., Kibria, M.G., Hasan, R., Ahmed, K.M., Hasan, M.A., Alam, J., Bhattacharya, P., Jacks, G., von Brömssen, M. 2012. Potentiality of intermediate depth aquifer as a source of arsenic and manganese safe tubewells in Bangladesh. In: J.C. Ng, B.N. Noller, R. Naidu, J. Bundschuh & P. Bhattacharya (Eds.) Understanding the Geological and Medical Interface of Arsenic, As2012. CRC Press/Taylor and Francis, pp. 71–73.

Hoque, M.A., Burgess, W.G., Shamsudduha, M., Ahmed, K.M. 2011. Delineating low-arsenic groundwater environments in the Bengal Aquifer System, Bangladesh. Appl. Geochem. 26: 614–623.

Rahman, M.M., Bakker, M., Borges Freitas, S.C., van Halem, D., van Breukelen, B.K., Ahmed, K.M., Badruzzaman, A.B.M. 2015. Exploratory experiments to determine the effect of alternative operations on the efficiency of subsurface arsenic removal in rural Bangladesh. Hydrogeol. J. 23: 19–34.

Sultana, S; Ahmed, K M; Mahtab-Ul-Alam, S M; Hasan M; Tuinhof, A; Ghosh, S K; Rahman, M S; Ravenscroft, P and Zheng, Y, 2015. Low-cost aquifer storage and recovery: implications for improving drinking water access for rural communities in coastal Bangladesh. J. Hydrol. Eng. B5014007(12).

Arsenic Research and Global Sustainability – Bhattacharya, Vahter, Jarsjö, Kumpiene,
Ahmad, Sparrenbom, Jacks, Donselaar, Bundschuh & Naidu (Eds)
© 2016 Taylor & Francis Group, London, ISBN 978-1-138-02941-5

Technological, social and policy aspects in Bangladesh arsenic mitigation and water supply: Connections and disconnections

B.A. Hoque[1,2], S. Khanam[1], M.A. Siddik[1], S. Huque[1], A. Rahman[1] & M.A. Zahid[1]
[1]*Environment and Population Research Centre (EPRC), Dhaka, Bangladesh*
[2]*Rollins School of Public Health, Emory University, Atlanta, GA, USA*

ABSTRACT: Millions of populations have been exposed to the risks of drinking arsenic contaminated water over a decade in Bangladesh. Here we have shared experiences gained in community-based interventions for safe water supply during applied research from 1999 to 2015 by us. Almost all of the alternative and arsenic treatment water technologies installed in the arsenic affected areas, except deep tube wells, were abandoned by the people. We have observed lack of: adequate knowledge about impacts of drinking arsenic contaminated water, appropriate technologies, access to safe water technologies, women access to decision making processes/leading roles in water management and interests in safe drinking water issues among the organizations involved in water issues. Arsenic mitigation water supply should be revisited.

1 INTRODUCTION

The risks for the exposure to drinking of arsenic contaminated water have been a continued challenge in rural Bangladesh and other countries (Singh, 2015, Hossain et al., 2015). Drinking of safe water is essential for both protection and cure of arsenic consumption related health problems. According to MICS UNICEF Bangladesh 2015 (more than a decade after the country was identified to be at the risk of exposure to the worst arsenic contamination of groundwater in the world UNICEF report) about 19.7 million people drank water with arsenic concentrations that exceeded Bangladesh standard (5 times higher than WHO GV). Here we have summarized experiences gained in technological, social and policy aspects of arsenic safe drinking water supply systems during applied researches from 1999 to 2015 by us in rural Bangladesh, so that the researchers, concerned professionals and stakeholders can discuss/find ways to address the water problem issues based on the realities.

2 METHODS

We have presented results from the following research: community-based action research on arsenic contamination assessment and safe water supply in *Srinagar*-Munshiganj in 1999–2002 (funded by Rotary International); community-capacity building arsenic mitigation water supply in Kalia in 2001–2005 (funded by Department of Public Health Engineering and UNICEF); acceptability of alternative and arsenic removal technology in Kalia in 2009–2012 (funded by ADB, BWDB, Dutch Embassy); and creating arsenic safe villages with user-women led sustainable improvement in water, sanitation and hygiene

in Narail in 2013–2015 (funded by UNICEF and the Dutch *Embassy*). We have installed and/or observed performances of various kinds of alternative and arsenic removal technology systems for arsenic safe water supply during the projects, in addition to other water, sanitation, and hygiene interventions

3 RESULTS AND DISCUSSION

More than 70% of the tube-wells were found contaminated in the project Srinagar-Munshiganj (1999–2002). We had promoted safe tube-wells, rain water harvesters (RWH) and few Arsenic Removal Technologies (ART). But the local women in a planning meeting demanded for pipe-water supply like 'the towns'. We undertook the pioneering rural pipe-water supply action research and achieved satisfactory results (Hoque et al., 2004). DTW was on the highest demand. None of the promoted/installed arsenic removal household or community technologies was found in use after a year. The National Policy for Arsenic Mitigation 2004 & Implementation Plan for Arsenic Mitigation in Bangladesh incorporated rural pipe-water as an option.

In Kalia (2001–2005) about 67% of >12000 tube-wells tested were found arsenic contaminated. Approximately a total 532 numbers of DTWs, PSFs, RWHs, Dug Wells (DW) and community pipe water systems were installed to provide water to about 46,000 population. The functional performance of most the systems were found satisfactory at the end of that project. Although women were the main caretakers of the technologies; about one-third women and rest men participated in the O&M trainings. The local leaders' responses were that national policy suggests at least about one-third women participation and those women participated after community decisions.

The safe water supply system acceptability research (2009–2012) in the same areas observed that additional few hundreds of DTWs and Bangladesh Government certified Arsenic Removal Technologies (ART) were promoted/installed in Kalia by another project by another group. We found that; only DTWs were in use and almost all of the earlier installed RWHs, PSFs and DWs (Dug Wells) were abandoned by the beneficiaries. The main reasons for the discontinuation in use of the other alternative technologies were: weak performance in competition with DWTs, sand boiling and difficulties in maintenance of DWs, uncertainty or scarcity in rain water and difficulties in maintenance of RWH, and poor availability of pond (source) water and difficulties in maintenance of PSFs. The pipe water system was still working, used and managed more or less satisfactorily by the community. The Arsenic Removal Technology (ART) was distributed in the complex areas where DTW technology was not feasible due to other water quality and/or technical problems (FAO, UNICEF, WHO and WSP, 2010). Our ART related results indicated that more than 90% of the households had discontinued its use due to smell in the treated water, problems in maintenance, and doubts about the safety of the water among the beneficiaries. In the ART area arsenic contaminated water was found in about 80% of the household stored drinking water.

The main objective of the arsenic safe village project was to improve health and livelihood scopes through gender mainstreamed community based safe drinking water supply and WASH education among arsenic affected poor populations in Narail (2013–2015). The project has installed about 506 DTWs, 63 RWHs and 2 solar pump based pipe water systems and supplied safe drinking water to about 56,000 populations after the UNICEF guideline in the complex areas of Kalia and Lohagora areas. We have found all the systems functioning satisfactorily in 2015. Manganese, iron and fecal coliform bacteriological contaminations were found in substantial number of tube well water samples. The local women showed promising results in decision making, leading and participating roles & responsibilities in planning, installation and O&M of the water systems. The local women were empowered as an institution of Women for Environment and Livelihood (WEL). But we cannot claim the results as evidence and draw guidance about WEL and/or women leadership in safe water supply system based on the observation over several months. We are searching fund for the important follow-up of the project which, can contribute towards sustainable development of safe water supply systems.

4 CONCLUSIONS

We have found that: people were exposed disproportionately to the risks of drinking arsenic contaminated water due to both lack of access to improved technology and lack of appropriate technology (design);

more than 70% of the people did not mention cancer in health impacts; of the installed technologies more than 80% of the kinds of alternative technologies (except DTWs) and more than 80% of all kinds of arsenic removal technologies showed unacceptable performances or were not used; levels of O&M of the installed arsenic mitigation water supply systems under the existing norms were questionable/poor; recognition as well as participation of the arsenic affected women in leading water management roles were negligible; and our pilot-study on women-led O&M of the installed water systems showed promising results. Most of the study areas falls under the national policy designated 'emergency response' areas. There remain questions, such as: do we have adequate appropriate technologies? How technologies-systems can be connected to proper O&M by users/communities/service providers to become sustainable? How to empower women, the managers of domestic water, to undertake leading roles and responsibilities in the decision making and implementation processes of water systems? How to improve the knowledge and interests about the health impacts for effective actions? How to connect the technological, social and policy-institutional aspects for sustainable safe water supply?

It is high time that the researchers, policy makers and program professionals from related multiple disciplines and countries revisit the opportunities, realties and challenges and, act to develop sound technological, social and policy-institutional issues to achieve universal and equitable access to *safe* and affordable *drinking water* for all by 2030.

ACKNOWLEDGEMENTS

The authors are grateful to: DPHE, UNICEF-Bangladesh, the Dutch Embassy in Bangladesh, ADB, EPRC and the communities for funding and/or participating in the projects. We thank Professor P. Bhattacharya for his inputs to the paper.

REFERENCES

FAO, UNICEF, WHO and WSP. 2010. Towards an arsenic safe environment in Bangladesh. A Joint Publication. http://www.unicef.org/bangladesh/Towards_an_arsenic_safe_environ_report_22Mar2010.pdf). (Accessed April 5, 2016).

Hoque, B.A., Hoque, M.M., Ahmed, T., Islam, S., Ali, N. et al. 2004. Demand based water options for arsenic mitigation: an experience from rural Bangladesh. *Public Health* 118: 70–77.

Hossain, M., Rahman, S.N., Bhattacharya, P., Jacks, G., Saha R., Rahman, M. 2015. Sustainability of arsenic mitigation interventions-an evaluation of different alternative safe drinking water options provided in Matlab, an arsenic hot spot in Bangladesh. *Front. Environ. Sci.* 3: 30. doi: 10.3389 /fenvs.2015.00030.

Singh, S.K. 2015. Groundwater arsenic contamination in the Middle-Gangetic Plain, Bihar (India): the danger arrived. *Int. Res. J. Environ. Sci.* 4(2): 70–76.

Arsenic Research and Global Sustainability – Bhattacharya, Vahter, Jarsjö, Kumpiene,
Ahmad, Sparrenbom, Jacks, Donselaar, Bundschuh & Naidu (Eds)
© *2016 Taylor & Francis Group, London, ISBN 978-1-138-02941-5*

District, division and regional distribution of arsenic affected house-holds and implications for policy development in Bangladesh

M.M. Rahman Sarker[1] & Y. Hidetoshi[2]

[1]*Department of Agricultural Statistics, Sher-e-Bangla Agricultural University, Dhaka, Bangladesh*
[2]*Graduate School of Economics, Hitotsubashi University, Kunitachi, Tokyo, Japan*

ABSTRACT: Bangladesh achieved a significant success providing 97% rural population and bacteriologically safe tubewell water. Unfortunately, arsenic pollution in the drinking water creates a great threat for public health. This study contributes to basic knowledge of the determinants of arsenic polluted households in Bangladesh by analyzing the expanded set of determinants at district, division and regional level. In this study, spatial statistical models were used to investigate the determinants and spatial dependence of arsenic pollution. This analysis extends the spatial model by allowing spatial dependence to vary across divisions and regions. A positive spatial correlation was found in arsenic pollution of households across neighboring districts at district, divisional and regional levels, but the strength of this spatial correlation varies considerably by spatial weight. These findings have policy implications among at district, divisional and regional levels to overcome the present crisis and achieve the post 2015 development goals.

1 INTRODUCTION

Bangladesh has maintained an impressive track record on growth and development. In the past decade, the economy has grown at nearly 6 percent per year and human development went hand-in-hand with economic growth. Similar progress has also been observed regarding the bacteriologically safe tube well water for rural population (97%) of Bangladesh. Unfortunately, the country has been hit by another environmental catastrophe in last decade. A large volume of its ground water, the major source of drinking water in the country, has been severely contaminated by arsenic. The high arsenic in the groundwater in South Asia has become a priority environmental and health issue. Econometric models that examine spatial interactions among economic units are used increasingly by economists and in many scientific fields, including social sciences, with important advances reported both in theoretical and empirical studies (Anselin, 2001, Anselin & Bera, 1998). Spatial dependence means a relationship in which observations both lack independence and have a spatial structure underlying these correlations (Anselin & Florax, 1995). When the spatial dependence is ignored, the estimates will be inefficient and the R^2 measure of fit will be misleading (LeSage & Pace, 2004). In other words, the statistical interpretation of the regression model will be wrong. However, the OLS estimates themselves remain unbiased, contrary to what is sometimes suggested in the literature. Since the attributes associated with the built environment and natural amenities are spatially located, it is reasonable to hypothesize that socioeconomic phenomena like arsenic pollution are spatially clustered based on neighboring socioeconomic, demographic and environmental attributes. While the application of explicit spatial econometric methods has recently shown a tremendous increase in the social sciences in general and economics in particular (Anselin, 2001), to date, there have been only a few studies that employed spatial regression analysis in the study of arsenic pollution-related data in South Asia. With this background, study focuses on detecting spatial dependence, and investigating the determinants of arsenic polluted households.

2 METHODS

2.1 Sources of data

Secondary data were used for the study from the five sources; such as (i) Bangladesh Multiple Indicator Cluster Survey (MICS) 2012–13 (BBS & UNICEF, 2014) for arsenic polluted and access to mass media and ICT of women, (ii) Statistical Pocketbook of Bangladesh for average household size of different districts, (iii) Statistical Yearbook of Bangladesh for literacy rate, (iv) Population Census 2011 for population density characteristics of different districts and (v) Directorate of Agricultural Extension personnel for daily average wage rate of agricultural labour.

2.2 Analytical models

The following models were used for detecting the determinants and spatial effect:

$$y = \rho Wy + X\beta + \varepsilon \quad \text{and} \quad \varepsilon = \lambda W \varepsilon + u \qquad (1)$$

where X is an $(n \times (k+1))$ matrix of observations on the explanatory variables, y and ε are $n \times 1$ vectors and β is a $(k+1)$ vector. If $E(\varepsilon\varepsilon) = \sigma^2 I$, where I is the nxn identity matrix then the arrangement properties of the attributes are irrelevant to the specification of the model. W_y and W_e are exogenously specified nxn weights matrices, lag and error, respectively. The scalar ρ and λ are spatial lag and spatial error coefficients, respectively. The second equation is necessary for estimating the spatial error coefficient lamda (λ). Moreover, the district, divisional and regional models (Sarker, 2012) with the binary join/contiguity matrix and the inverse distance spatial weights matrix specifications of W were used in the empirical models to highlight any differences in spatial patterns of arsenic polluted households.

2.3 Relation between spatial lag (ρ) and spatial error (λ) coefficients

The following four cases were consider for the modeling (i) $\rho = 0$, $\lambda = 0$, (ii) $\rho \neq 0$, $\lambda = 0$, (iii) $\rho = 0$, $\lambda \neq 0$ and (iv) $\rho \neq 0$, $\lambda \neq 0$

3 RESULTS AND DISCUSSION

3.1 Coefficients of determinates with spatial estimates using binary join weights

The results from the models that both the spatial lag and error models reveal that the spatial lag and error coefficients are significantly different from zero. However, the Akaike Information Criterion (AIC), Schwarz Criterion (SC) and the log-likelihood statistics all reveal that the spatial lag model is provided a better fit than the spatial error model. The results from the model that include both the spatial lag and error term reveals that only the spatial lag coefficient is significantly different from zero. The findings suggest that spatial dependence in arsenic polluted households of a district may be the best model using a spatial lag. This is supported by the AIC and SC, which are directly comparable across models and weigh the explanatory power of a model (based on the maximized value of the log-likelihood function).

3.2 Determinants with spatial estimates with inverse distance weights

The use of inverse-distance spatial weights is more appropriate to identify spatial interactions among the distance spatial weights of districts. Among the four specifications, log likelihood value is lowest for join (spatial lag + error) model and AIC and SC are lowest in spatial lag model. The findings suggest that the spatial lag model is the best model for expressing the spatial dependence of district arsenic polluted households.

3.3 Divisional and regional spatial estimates

Estimates of positive and significant spatial correlation in four divisions range from 0.6923 to 0.6236 when neighbors are defined by binary joins matrix and 0.3760 to 0.3166 when spatial weight are defined by inverse distance matrix. Estimates of spatial correlation in two regions are 0.6197 and 0.6526 when neighbors are defined by binary joins matrix and 0.3332 and 0.3328 when spatial weight are defined by inverse distance matrix. All spatial coefficients for regional are positive and statistically significant (P < 0.01). The AIC, BIC and log likelihood statistics suggest that the binary joins matrix contiguity spatial models show better fit than inverse distance spatial weights matrix models in both regional and divisional models

3.4 Equality of spatial coefficients of divisional and regional models

The spatial coefficients in the Dhaka, Chittagong, Khulna and Rajshahi divisions as well as Northern and Southern regions of Bangladesh are substantially larger when neighbors are defined by common border as opposed to inverse distance. Using the common-border (binary join matrix) neighbor definition, the test results are shown that the spatial correlation for districts in each division is statistically homogenous from the correlation in other divisions, with the exception of the Rajshahi and Khulna divisions.

4 CONCLUSIONS

Using either a binary or inverse distance weights matrix in the estimation of spatial effects, this results provide strong evidence that significant spatial correlation exists in district, divisional and regional level model. These unbiased efficient and consistent estimators will be useful for policy intervention to achieve two of the post 2015 development goals such health and well-being and clean water and sanitation for sustainable development through arsenic free water.

REFERENCES

Anselin, L. 2001. Rao's score test in spatial econometrics. *J. Stat. Plan. Infer.* 97: 113–139.
Anselin, L., Bera, A. 1998. Spatial dependence in linear regression models with an introduction to spatial econometrics. In A. Ullah, D.E.A. Giles (Eds.) *Handbook of Applied Economic Statistics.* Marcel Dekker, New York, pp. 237–289.
Anselin, L., Florax, R. 1995. Introduction. In: L. Anselin, R. Florax (Eds.) *New Directions in Spatial Econometrics.* Springer-Verlag, Berlin, pp. 3–18.
BBS & UNICEF. 2014. Bangladesh multiple indicator cluster survey. Progotir Pathey 2012–13. BBS-UNICEF, Bangladesh Bureau of Statistics, Dhaka.
LeSage, J.P., Kelley Pace, R. 2004. Introduction. *Advances in Econometrics. Volume 18: Spatial and Spatiotemporal Econometrics.* Oxford: Elsevier Ltd, UK, pp. 1–32.
Sarker, M.M.R. 2012. Spatial modeling of households' knowledge about pollution in Bangladesh. *Soc. Sci. Med.* 74: 1232–1239.

Arsenic Research and Global Sustainability – Bhattacharya, Vahter, Jarsjö, Kumpiene,
Ahmad, Sparrenbom, Jacks, Donselaar, Bundschuh & Naidu (Eds)
© 2016 Taylor & Francis Group, London, ISBN 978-1-138-02941-5

Do socio-economic characteristics influence households' willingness to pay for arsenic free drinking water in Bihar?

B.K. Thakur & V. Gupta
National Institute of Industrial Engineering (NITIE), Mumbai, India

ABSTRACT: The present study examines the influence of socio-economic variables on households' Willingness To Pay (WTP) for arsenic safe drinking water in Bihar from two districts by applying Contingent Valuation Method (CVM). Majority of the households (84%) were willing to pay for improved water quality. Socioeconomic variables which influence overall WTP are incomes of the household, age, education, awareness, contamination level and doctor visit. The mean annual WTP is obtained at INR 240 and the annual WTP for the entire study area is estimated at INR 17.1 million. The results of the study may be used by the policy maker to provide arsenic free drinking water to affected areas.

1 INTRODUCTION

There is an evidence of arsenic (As) contamination in many states of India, among which Bihar is severally affected. Existence of excess As in drinking water and its associated human health problems have been reported across the globe (Kapaj *et al.*, 2006, Ravenscroft *et al.*, 2009), from India (Bhattacharya *et al.*, 2011, Chakraborty *et al.*, 2003, 2010, 2015), and in Bihar (Ghosh *et al.*, 2007, 2012, Saha, 2009, Thakur *et al.*, 2013). High arsenic levels have been associated with various diseases, including skin lesions, and various forms of cancer. Previous research indicated that, over 10 million people in Bihar are drinking water containing more than 10 μg/L, and are living in the northern Bihar Gangetic plain. Several researchers used Contingent Valuation Method (CVM) to find the Willingness To Pay (WTP) for safe water (Chowdhury, 1999, Maddison *et al.*, 2005, Ahmed *et al.*, 2002, Haque *et al.*, 2007, Khan, 2006, Khan *et al.*, 2014). Recent study by Roy (2008) and Chowdhury *et al.* (2015) estimated annual health cost due to As affected population in West Bengal and Assam, India are INR 0.76 million and 229 million. Given the background, the objectives of the paper were: a) to examine the causal relationship between socio-economic variables and their effects on WTP; and b) to estimate the gross WTP for the study area.

2 METHODOLOGY AND DATA SOURCES

The study was conducted using a questionnaire through primary survey of household to examine the key socio-economic variables which influenced households WTP in Bihar. The CVM was used to determine WTP in 420 households, from two districts (Patna and Bhojpur) of Bihar. The study used multistage sampling framework. Both stratified random sampling and systematic random sampling were used for the selection of sample. Actual household selection was based on systematic random sampling. To establish the causal influence of socio-economic characteristics on households WTP, study used OLS linear regression. Study also used a field test kit to test As concentration in the household hand tube well. 420 water samples were collected and tested immediately through using field test kit. As concentration were identified on the basis of the field test kit ranges from 0 to 500 μg/L.

3 RESULTS AND DISCUSSION

The average household size is of eight family members due to more number of joint families with an average per capita income of INR 2000 and per capita expenditure of INR 1470. More than 40% of households reside below poverty line. Almost half of the household primary occupation is agriculture and allied activities. Majority of the household expenditure on food items (46%) were higher than the expenditure on non-food items. Literacy rate of the households is around 68% and there is a positive association between education and WTP. Household do not want to expend more on water but certainly the value if provided with As free water which they considered good quality water.

Majority of the households (98.71%) have access to drinking water source through hand tube wells. Only 12% households used deep hand tube wells and most of the households used shallow tube well as their drinking water. Around 9% households use water filter or As removal filters for drinking water which means it is not popular among households. The As concentration in the hand tube well water are ranged between o to 500 µg/L. From the analysis of water samples, 21.90% of water samples contain As less than WHO guideline of 10 µg L⁻¹ and 14.76% samples was below the BIS guidelines of 50 µg/L. The rest of the water samples contain As more than these limits. Around 17.86% of water samples contain As between 300 µg/L and 500 µg/L.

Deep tube well (46%) as mitigation measures are most preferable mitigation measures in the study area followed by piped water system (18%), rain water harvesting (15%), dug well (13%), and As removal filter (9%) among other. This findings are similar than the previous studies by Khan (2014), Khan & Hong (2012) and Ravenscroft et al. (2009) which also found that deep tube wells are most preferred mitigation options amongst the alternative availability.

From the survey, it was found that most of the households (84%) willing to pay for As safe drinking water. This indicates that households give more priority for safe drinking water. Income of the households is positively associated with its WTP. In order to see the importance of each variable in determining the maximum WTP the standardized beta coefficient is estimated which gives relative importance of each variable. We checked heteroscedasticity for all the explanatory variables individually and together using Breusch-Pagan/Cook-Weisberg test and found the chi square value is accepted and therefore, the null hypothesis is not rejected. The results show that the variables AGE, EDU, CNT, INC, MGR, and DVT are statistically significant and are with their expected signs of the WTP of households. Variables MGR and LND have expected negative sign whereas rests have positive causal influence on WTP. From the result it can be noticed that variables EDU and LND have expected signs but they are not statistically significant. Income of the household and doctor visit is a principal variable, which determines the changes in the decision of WTP of the household which beholds the existing CV literature on socioeconomic variables. The contamination level of the household is also a dominant variable to determine the changes in WTP of the household, but information on contamination level is difficult to be known to the household. Awareness and education are interlinked and it is expected that higher the education level in the family more awareness beholds. Surprisingly awareness of the household does not influence the changes in the WTP. The probable reason for this could be petite awareness about the contamination level of the household drinking water and lack of availability of alternative sources.

The gross WTP were estimated of all the stakeholders. The mean monthly WTP is estimated at INR 20 and annual INR 240. The gross WTP for entire study area is estimated at INR 17.1 million.

4 CONCLUSIONS

The results indicated that households value good quality water for ensuring good health and avoiding As exposure, and therefore, willing to pay, provided, for improved water quality. Socio-economic factors such as incomes of the household, age, education, awareness, contamination level and doctor visit influences households willingness to pay. The result also provided with an estimated value of gross willingness to pay to taking into account all the stakeholders of the surveyed household and the entire study area. The findings of the study help policy makers to adopt a better approach to provide improved drinking water to the affected areas.

ACKNOWLEDGEMENTS

The authors thank MHRD, Government of India, for providing financial support and doctoral scholarship. All the participants and villagers during field work for their active participation are acknowledged.

REFERENCES

Chakraborti, D., Mukherjee, S.C., Pati, S., Sengupta, M.K, Rahman, M.M., Chowdhury, U.K., Lodh, D., Chanda, R.C., Chakraborti, A.K., Basu, G.K. 2003. Arsenic groundwater contamination in middle Ganga plain, Bihar India: A future danger, *Environ. Health Persp.* 119 (9): 1194–1201.

Ghosh, A.K., Bose, N., Kumar, R., Bruining, H., Lourma, S., Donselaar, M.E.,Bhatt, A.G. 2012. Geological origin of arsenic groundwater contamination in Bihar, India. In.: J.C. Ng, B.N. Noller, R. Naidu, J. Bundschuh & P. Bhattacharya (Eds.) *Understanding the Geological and Medical Interface of Arsenic, As 2012*: 85–87. Leiden, CRC Press/ Taylor and Francis, pp. 522–525.

Ghosh, A.K., Singh, S.K., Bose, N., Chaudhary, S. 2007. Arsenic contaminated aquifers: a study of the Ganga levee zones in Bihar, India. *Symposium on Arsenic: The Geography of a global problem*, Royal Geographical Society, London.

Kapaj, S., Peterson, H., Liber, K., Bhattacharya, P. 2006. Human health effects from chronic arsenic poisoning: a re-view. *J. Environ. Sci. Heal.* A 41: 2399–2428.

Khan, N.I., Roy, B., Hong, Y. 2014. Households's willingness to pay for arsenic safe drinking water in Bangladesh. *J. Environ. Manage.* 143: 151–161.

Maddison, D., Catala-Luque, R., David, P., 2005. Valuing the arsenic contamination of groundwater in Bangladesh. *Environ. Resour. Econ.* 31: 459–476.

Roy, J. 2008. Economic benefits of arsenic removal from ground water—A case study from West Bengal, India. *Sci. Total Environ.* 397: 1–12.

Saha, D. 2009. Arsenic groundwater contamination in parts of middle Ganga plain, Bihar. *Curr. Sci. India* 96(6): 1–3.

Thakur, B.K., Gupta, V., Chattopadhyay, U. 2013. Arsenic Groundwater Contamination Related Socio—Economic Problems in India: Issues and Challenges, In: S. Nautiyal (Ed..) *Knowledge Systems of Societies for Adaptation and Mitigation of Impacts of Climate Change. Environmental Science and Engineering* Springer Berlin Heidelberg, pp. 163–182.

5.2   Risk assessment and remediation of contaminated land and water environments—case studies

Arsenic Research and Global Sustainability – Bhattacharya, Vahter, Jarsjö, Kumpiene,
Ahmad, Sparrenbom, Jacks, Donselaar, Bundschuh & Naidu (Eds)
© 2016 Taylor & Francis Group, London, ISBN 978-1-138-02941-5

Contaminated soil for landfill covers: Risk mitigation by arsenic immobilization

J. Kumpiene & L. Niero
Waste Science and Technology, Luleå University of Technology, Luleå, Sweden

ABSTRACT: The aim of the study was to evaluate the risks of arsenic (As) spreading from soil that was used in a landfill cover. The As-contaminated soil was treated with iron, and a combination of iron and peat and placed in a pilot-scale landfill cover. The main exposure pathways concerning risks to human health and the environment were studied by analysing dissolved As in soil pore water, As phytotoxicity and bioaccessibility. The results showed that the stabilization of As-contaminated soil with a combination of Fe^0 and peat significantly reduced the As concentration in soil pore water, uptake by plants and improved the main morphological parameters of plants. The soil treatment also reduced the bioaccessibility As indicating the reduced risks to human health. Using Fe^0 amendment alone, the positive impact on the measured indicators was considerably smaller or not significant.

1 INTRODUCTION

Closing landfills with a thick, nearly impermeable cover requires millions of tons of construction materials including clay, sand, gravel, and soil. At the same time, thousands of tons of contaminated soil are excavated during the remediation of contaminated sites and transported to landfills for disposal. The demand for soil as a landfill cover material may be partially fulfilled by replacing clean soil with contaminated masses that would otherwise be disposed of in a landfill. But the contaminated soil needs to be treated first in order to prevent contaminant spread to the environment. Stabilization or immobilization of Trace Elements (TE) can be applied in such cases using soil amendments to decrease TE mobility and bioavailability (Adriano *et al.*, 2004).

The aim of this study was to evaluate the risks of arsenic (As) spreading from soil that was used in a landfill cover. The As-contaminated soil was treated with iron, and a combination of iron and peat and placed in a pilot-scale landfill cover as a vegeta-tion/protection layer.

2 METHODS/EXPERIMENTAL

2.1 Soil treatment

The soil originated from a former industrial site in Northern Sweden where timber was treated with Chromated Copper Arsenate (CCA) chemical, which caused the soil contamination with As. About 800 t of soil was excavated from the site and transported to a landfill. The total As concentration was 137 ± 40 mg/kg, soil pH 7.8 and organic matter content 1.3%. The total amount of soil (800 t) was homogenized and divided into three parts, the first was kept untreated (ca. 267 t), while the remaining quantity was amended with 1 wt% spent Blasting Sand (BS) containing 98.3% of Fe^0. A half of this volume (one third of the overall amount) was further mixed with 5 wt% of peat. The soil was used to form a vegetation/protection layer on the top of the experimental landfill cover.

2.2 Analyses

Soil samples were collected directly after mixing the materials in field and transported to laboratory for phytotoxicity test with dwarf beans (*phaseolus vulgaris*) (Vangronsveld and Clijsters, 1992), soil pore water analysis and the bioaccessibility test (Juhasz et al., 2009). Soil pore water was collected the third day and the fourteenth day of the phytotoxicity test to assess the element availability in soil. The pH and EC were measured immediately after the sampling on a small aliquot of the sample. The remaining volume was stored at 4°C until element concentration analysis using ICP-OES.

Soil samples were taken one year later from the field and used for the repeated bioaccessibility tests. Plants that spontaneously vegetated the site were collected at the same occasion and analyzed for As content in roots and shoots.

3 RESULTS AND DISCUSSION

3.1 Dissolved arsenic in soil pore water

The As concentration in soil pore water decreased in both Fe^0 (by 39%) and Fe^0-peat treated soils (by 51%) compared to the untreated soil (Fig. 1). The Fe concentration did not show any significant differences

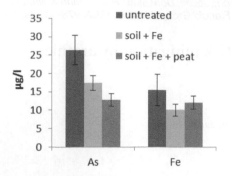

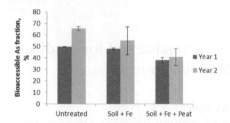

Figure 1. As and Fe concentrations in soil pore water collected at the end of the phytotoxicity test. The error bars represent standard deviations of means, n = 3.

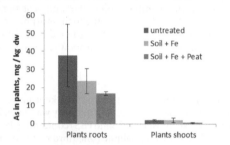

Figure 2. Total arsenic concentrations measured in plant roots and shoots. The error bars represent standard deviations of means, n = 4.

among the treatments. Concentration of dissolved Cu and Cr were low and likely had no effect on plant toxicity; Cu: 19 μg/L in untreated soil and 5.4–7.5 μg/L in treated soils; Cr: 5.4 μg/L in untreated soil and 1.3–2.8 μg/L in treated soils.

3.2 Soil phytotoxicity and As accumulation in plants

Generally, soil treatment with Fe^0-peat improved morphological parameters of plants and their growth conditions, while Fe^0 treatment impaired the plant growth compared to the untreated soil. The As concentration measured in plant roots was by one order of magnitude higher than in the shoots (Fig. 2). The largest decrease in accumulated As, both in roots (by 55%) and shoots (by 70%) of beans, was in soil treated with Fe^0 + peat.

Two spontaneously established plant species dominated the field plots: white melilot (*Melilotus albus*) and red clover (*Trifolium pretense*). None of the plants have taken up As from soil and As concentrations in both roots and shoots of plants from all three field plots were below instrument detection limits (i.e. <2 mg/kg dw).

3.3 Arsenic bioaccessibility

The As bioaccessibility decreased in both treated soils: by 3% in the Fe^0-treated soil and by 23% in the

Figure 3. Bioaccessible As fraction measured in untreated and treated soils as percent of the total As concentration. The error bars represent standard deviations of means, n = 3.

Fe^0-peat treated soil compared to untreated soil year 1 (Fig. 3). The repeated bioaccessibility measurement one year later showed somewhat larger bioaccessible As fraction in the untreated soil, while soil treatments showed similar results to those of the first year.

4 CONCLUSIONS

The stabilization of As-contaminated soil using a combination of zerovalent iron and peat effectively reduced As phytotoxicity, bioaccessibility and accumulation in plant shoots and roots. Hence, the risks connected to the As spreading to the environment from a landfill cover through uptake by plants and the direct ingestion of contaminated soil have substantially decreased due to this soil treatment. Using Fe^0 amendment alone, the positive impact on the measured indicators was considerably smaller or not significant.

ACKNOWLEDGEMENTS

This study was financially supported by the EU Regional Development Fund Objective 2 project *North Waste Infrastructure*, the Swedish Foundation for Strategic Research and the Swedish Research Council for Environment, Agricultural Sciences and Spatial Planning (FORMAS).

REFERENCES

Adriano, D.C., Wenzel, W.W., Vangronsveld, J., Bolan, N.S. 2004. Role of assisted natural remediation in environmental cleanup. *Geoderma.* 122(2): 121–142.

Juhasz, A.L., Weber, J., Smith, E., Naidu, R., Rees, M., Rofe, A., Kuchel, T., Sansom, L. 2009. Assessment of four commonly employed in vitro arsenic bioaccessibility assays for predicting in vivo relative arsenic bioavailability in contaminated soils. *Environ. Sci. Technol.* 43(24): 9487–9494.

Vangronsveld, J., Clijsters, H. 1994. Toxic effects of metals. In: M. Farago (Ed.). *Plants and the Chemical Elements.* VCH Verlagsgesellshaft, Weinheim, Germany. pp. 149–177.

Arsenic Research and Global Sustainability – Bhattacharya, Vahter, Jarsjö, Kumpiene,
Ahmad, Sparrenbom, Jacks, Donselaar, Bundschuh & Naidu (Eds)
© 2016 Taylor & Francis Group, London, ISBN 978-1-138-02941-5

Arsenic in drinking water and water safety plan in lowland of Nepal

S.K. Shakya
School of Environmental Science and Management, Kathmandu, Nepal

ABSTRACT: Water source for all purposes in lowland Terai of Nepal with more than half of the total population of the area is primarily groundwater extracted from shallow or deep aquifers. According to blanket arsenic testing results of DWSS, 7.5% of 11,01,536 samples exceeded WHO guideline value (10 µg/L) with 1.8% samples above the national standard (50 µg/L). Studies in some of the arsenic affected communities have identified several hundred arsenicosis cases. Government, national and international organizations have implemented arsenic mitigation programs aiming to prevent health damage and have provided different arsenic safe water alternatives to high arsenic exposed households. Thus, the integrated approach with water safety plan for wider awareness generation on arsenic mitigation, resources mobilization and proper management with active community involvement is necessary for addressing the arsenic problem in a sustainable way in lowland of Nepal.

1 INTRODUCTION

The acute toxicity of arsenic at high concentrations has been known about for centuries. It was only relatively and recently that a strong adverse effect on health was discovered to be associated with long-term exposure to even very low arsenic concentrations (Abernathy *et al.*, 2001).

The first major investigation of arsenic in ground water was completed in 2001 by DWSS/UNICEF, testing 4,000 tube wells in all 20 Terai districts. The study found over 3% tested tube wells above 50 µg/L and over 10% exceeded WHO guideline of 10 µg/L. The State of Arsenic in Nepal—2003 study by NASC/ENPHO in 2003 tested 18,635 tube wells in 20 Terai districts and reported 23.7% tube wells above WHO guideline of 10 µg/L and 7.4% exceeding Nepal Interim Standard guideline of 50 µg/L. Significant investigations have also conducted by Tandukar (2000), Tandukar *et al.* (2001), Shrestha *et al.* (2003), Bhattacharya *et al.* (2003), Gurung *et al.* (2005) and Shakya *et al.* (2012/13).

DWSS is also implementing water safety plan. WSP is a plan to ensure the safety of drinking water through the use of a comprehensive risk assessment and risk management approach, which covers all steps in drinking water supply from source to mouth (WHO, 2008). Water safety can be assured through a variety of interventions at the level of households, community, water supplier and regulator, often with an excellent cost-benefit ratio. The most effective means of consistently ensuring the safety of a drinking water supply.

2 METHODS/EXPERIMENTAL

2.1 The study area

The study was carried out in six districts namely Nawalparasi, Bara, Parsa, Rautahat, Rupandehi and Kapilbastu of terai region of Nepal (Fig. 1).

2.2 Social survey and health surveillance

A study on health impact survey in arsenic affected area was conducted to identify arsenicosis cases, know the extent of manifestations and status of arsenic exposure among risk population in study area. Identification of arsenicosis cases was done according to "A Field Guide for Detection, Management and Surveillance of Arsenicosis Cases".

2.3 Water quality surveillance

The water quality parameters were tested by using Merck & HACH/HANA field test kit and suspected high arsenic concentration water samples were reconfirmed by analysis from Atomic Absorption Spectrophotometer (AAS) in laboratory.

3 RESULTS AND DISCUSSION

3.1 Health impact

The prevalence of arsenicosis in six districts, on average was 2.2% but ranged from 0.7% to 3.6%. The highest prevalence of arsenicosis was found in Nawal-

Figure 1. Map of Nepal showing the location of study areas.

Table 1. Statistical summary of basic water chemistry data including pH, Eh and As, NH$_4$-N, NO$_3$-N concentrations of tubewell water.

	As conc. (µg/L)	pH	Eh mV	NH$_4$-N conc. ppm	NO$_3$-N conc.ppm
Max	1150	7.65	261	4.0	0.05
Min	4	6.74	60	0.0	0.00
Mean	542	7.37	97	1.6	0.003

parasi district (3.6%), which was also reported to be a highly arsenic contaminated district. The prevalence of arsenicosis was lower in Rupandehi (0.9%) and Kapilbastu (0.7%) among people who were aged 50 years and above. In the age group of 5–14 yrs, five (0.1%) cases of arsenicosis were reported and in the age group of below five years, only one (0.3%) person had arsenicosis. Of the total arsenicosis patients identified in the six districts, 268 (67%) were male and 132 (33%) were female.

3.2 Water quality surveillance results

The water quality result reveals the tested water quality parameters were within guideline value except few samples. The maximum, minimum and mean values of As, NH$_4$-N, NO$_3$-N concentrations pH, Eh of tube wells water of the study area is presented in Table 1.

The table shows the outline of values at the measurement. According to their mean values, As concentration of 542 µg/L was nearly 10 times of the Nepalese drinking water standard for As 50 µg/L. The pH of 7.37 showed a little alkalinity, Eh of 97 showed that the water was reduced condition, because Eh of <200 mV was categorized as the reduced condition. The NH$_4$-N concentration was a high of 1.6 ppm, while NO$_3$-N concentration was below level of detection.

The prevalence of arsenicosis was found highest (18.6%) in Patkhouli village where 95.8% of tube wells were arsenic contaminated i.e. >50 µg/L while none of the tube wells in Goini village was arsenic safe. It is also reported that among the cases, the most common manifestation was melanosis (95.7%), followed by leukomelanosis (57.7%) and keratosis (55.9%). Arsenic contamination was high in Nawalparasi, the prevalence of arsenicosis was also highest. Next to Nawalparasi, the prevalence of arsenicosis was found to be high in Rautahat.

4 CONCLUSIONS

Groundwater in lowland Terai region of Nepal is arsenic contaminated and visible symptoms of arsenic poisoning are appearing in many members of exposed population. Both government and non-governmental organizations have made continuous efforts for arsenic mitigation. In context of mitigation on arsenic, practical implementation in the actual field was found conducted by different Water Supply, Sanitation and Hygiene (WASH) stakeholders in order to reduce the concentration of arsenic in the ground water and vulnerability of population exposed to the arsenic. More than 7000 of improved tube wells, Biosand filters, Kanchan Arsenic filters have installed in the arsenic affected areas. Lack of monitoring and awareness on arsenic related problems have recently raised the concerns on sustainable use of the provided arsenic safe water alternatives. The experience has suggested the need of the integrated approach with appropriate water safety plan for wider awareness generation, local capacity building, resource mobilization and proper management with active community involvement for sustainable arsenic mitigation in lowland of Nepal.

ACKNOWLEDGEMENTS

Author is thankful to Department of Water Supply and Sewerage (DWSS), ENPHO and Nepal Red Cross Society for providing the technical support and information on various initiatives on arsenic during this case study.

REFERENCES

Abernathy, C., Morgan, A. 2001. Exposure and health effect. In: UN (2001). UN Synthesis Report on Arsenic in Drinking Water. Geneva, Switzerland: WHO, Chapter 3.

Bhattacharya, P., Tandukar, N., Neku, A., Valero, A.A., Mukherjee, A.B., Jacks, G. 2003. Geogenic arsenic in groundwaters from Terai alluvial plain of Nepal. J. Phys. IV 107: 173–176.

Gurung, J.K., Ishiga, H., Khadka, M.S. 2005. Geological and geochemical examination of arsenic contamination in groundwater in the Holocene Terai Basin, Nepal. Environ. Geol. 49: 98–113.

Kurosawa, K., Uddin M.S., Yiping, X., Tani, M., Shakya, S.K., 2013. Arsenic concentration and related water quality parameters of well water in the Terai Plain of Nepal. Japan Arsenic Symposium Miyazaki, Japan November, 2013.

NASC/ENPHO 2004. State of Arsenic in Nepal–2003.

Shakya, S.K., Maharjan, M. 2012. Local capacity building for sustainable arsenic mitigation in lowland, Nepal. Proceedings, "18th Japan Arsenic Symposium", Miyazaki, Japan.

Shrestha, R..R., Shrestha, M.P., Upadhyaya, N.P., Pradhan, R., Khadka, R., Maskey, A., Maharjan, M., Tuladhar, S., Dahal, B.M., Shrestha, K. 2003. Groundwater Arsenic Contamination, its Health Impact and Mitigation Program in Nepal.

Tandukar, N. 2000. Arsenic Contamination in Groundwater in Rautahat District of Nepal: An Assessment and Treatment. M.Sc. Thesis, Institute of Engineering, Lalitpur, Nepal.

Tandukar, N., Bhattacharya, P., Mukherjee, A.B. 2001. Preliminary assessment of arsenic contamination in groundwater in Nepal. In Arsenic in the Asia-Pacific Region: Managing Arsenic for our Future (Book of Abstracts), Adelaide, South Australia: 103–105.

WHO. Arsenic in drinking water (Fact sheet No. 210, Rev. ed) 2001.

WHO. 2008. Guidelines for Drinking Water Quality. 3rd Edition, World Health Organization, Geneva.

Arsenic Research and Global Sustainability – Bhattacharya, Vahter, Jarsjö, Kumpiene,
Ahmad, Sparrenbom, Jacks, Donselaar, Bundschuh & Naidu (Eds)
© 2016 Taylor & Francis Group, London, ISBN 978-1-138-02941-5

Impact of arsenic-contaminated irrigation water on food chain in GMB plain and possible mitigation options

T. Roychowdhury, S. Bhattacharya, N. Roychowdhury, S. Bhowmick, P. Dhali & K. Goswami
School of Environmental Studies, Jadavpur University, Kolkata, West Bengal, India

ABSTRACT: It is becoming apparent that ingestion of drinking water is not the only elevated source of arsenic (As) to the diet in Bengal delta. Irrigation of agricultural fields with As-contaminated groundwater has led to As build-up in soil, with subsequent elevation of As in crops grown on these soils. Increase in soil As concentration can cause an enhancement of As accumulation in rice grain and in turn threaten human health through food chain. Inorganic As, mainly arsenite and arsenate contribute more than 90% of total content of As in crops, grown on As-contaminated soil. Cooked rice and discarded water comes out during cooking rice preparation contribute a considerable amount of As. A pilot and field level approach has been taken for As reduction from irrigated water-soil system by using phytoremediation and indigenous microbial remediation in reducing As contents in food chain.

1 INTRODUCTION

Natural groundwater arsenic (As)-contamination and the illnesses of people as a result, has become one of the staggering challenges to human health in modern times, as large parts of the Ganga-Meghna-Brahmaputra plain with an area 569,749 km² and population over 500 million might be at risk (Chakraborti *et al.*, 2013). The situation is at its worst in the Bengal delta, with over 100 million people living in zones with As above 50 µg/L. Groundwater in appx. 3417 villages from 111 blocks in 9 districts of West Bengal has been reported as As-contaminated. The calamity in these places has assumed gargantuan proportion as As levels in drinking water has far overshot the guidelines set by World Health Organization (WHO) at 10 µg/L. What is further worrisome is that As-contaminated groundwater is increasingly being used to irrigate the fields. This means that As has begun to seep into the crops and food chain and could trigger a large scale environmental tragedy, fatally affecting future generations (Roychowdhury, 2008, 2010). Detection of As in rice, the main staple crop, an element known to be carcinogenic to humans, has been the topic of high public interest. Thus, other than drinking water, As poisoning in food chain is a potential health risk for the human continuum and other animals in Bengal delta.

Several mitigation options have been implemented to reduce As concentration in food chain. A pilot and field level approach has been taken for As reduction from irrigated water-soil system by using phytoremediation and indigenous microbial remediation in reducing As contents in food chain.

2 METHODS/EXPERIMENTAL

2.1 *Sample collection and preparation*

Two As-affected blocks, namely, Deganga and Gaighata located in North 24-Parganas district of West Bengal were selected as study areas. Water samples were collected from the household hand tubewells, used for mainly drinking and cooking and shallow large-diameter tubewells, used for agricultural irrigation. Food samples were collected from the agricultural fields and families living in As-affected zones. The collection and preservation of water and food samples were described in earlier publications (Roychowdhury *et al.*, 2003, Roychowdhury, 2008, 2010). Agricultural land soils were collected from upper surface of the lands. Several aquatic plant species available in study areas were treated for dearsenification. The experimental methodology involves both batch and column study. Both living and non-living dried/biomass plant materials were studied for As reduction from contaminated water-soil system.

2.2 *Chemicals and reagents*

All the reagents were of analytical grade. Distilled de-ionized water was used throughout. A solution of 1.25% NaBH₄ (Merck, Schuchardt, Germany) in 0.5% NaOH and 5.0M HCl were used for Flow Injection-Hydride Generation-Atomic Absorption

Spectrometry method. Standard stock solutions of arsenite and arsenate (1000 mg/L as As) were prepared by dissolving appropriate amounts of As_2O_3 and $Na_2HAsO_4·7H_2O$ (Merck, St. Louis, USA).

For HPLC, mixed standard solutions containing 30 µg/L and 150 µg/L of each As species (Arsenite, arsenate, DMA and MMA) were prepared daily from stock standard solutions, by appropriate dilution. San Joaquin soil (SRM 2709) from the National Institute of Standards and Technology (Gaithersburg, MD, USA); Rice flour (SRM 1568a) and wheat flour (SRM 1567a) from the National Bureau of Standards (Gaithersburg, MD, USA) were used as Standard Reference Materials (SRMs). Details of the reagents and standard reference materials were described elsewhere (Roychowdhury, 2008, 2010).

2.3 Digestion

Total As was determined after sample digestion in a Teflon bomb at about 100 °C (Roychowdhury, 2010). A mixture of nitric acid and hydrogen peroxide, 30% (v/v) was used for sample digestion in 2:1 ratio. Analytical procedure for As species extraction was described in elsewhere (Roychowdhury, 2008).

2.4 Analysis

Flow Injection-Hydride Generation-Atomic Absorption Spectrometer (Model: Varian AA140) was used for total As quantification. HPLC coupled with ICP-MS system (Agilent 7500, DE, USA) was used for quantification of As species.

3 RESULTS AND DISCUSSION

3.1 Food chain arsenic contamination

Considerable higher amounts of As are accumulated in paddy field soil and paddy plants. Mean As concentrations in vegetables, cereals and spices are 59 µg/kg (range: 27–608 µg/kg, n = 100), 125 µg/kg (range: 16–505 µg/kg, n = 88), and 98 µg/kg, (range: 25–378 µg/kg, n = 82), respectively in the study areas. Inorganic As (arsenite and arsenate) and DMA contribute 90% and 10% of total content of As in cultivated crops and vegetables, respectively.

3.2 Additional burden from raw and cooked rice

Rice grain plays a major role for As exposure. Accumulation of As is quite high in cultivated rice grain collected from agricultural fields (mean: 150 µg/kg, range: 29–365 µg/kg, n = 12). The daily dietary intake of inorganic As (µg/kg body wt./day) by an adult from rice grain itself (2.35) is higher than the

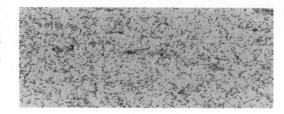

Figure 1. As in different stages during cooked rice preparation.

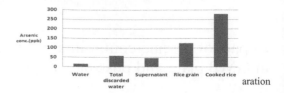

Figure 2. As accumulating indigenous microbe species.

WHO recommended PTDI value of inorganic As (2.1 µg/kg body wt./day).

Mean As concentration in cooked rice collected from the families in these two blocks is 300 µg/kg (range: 170–575 µg/kg, n = 20) and it is 2.1 times higher than that of raw rice grain (mean: 142 µg/kg, range: 53–348 µg/kg, n = 25). Mean As concentration in cooking water was 40 µg/L (range: 6–90 µg/L, n = 12). Thus, As concentration in cooked rice increases with water As concentration. Discarded water itself contributes a considerable amount of As during cooked rice preparation (Fig. 1).

3.3 As reduction from irrigated water-soil system

As removal efficiency of living and dried root of locally grown aquatic plants has been observed about 80%, and 95%, respectively. As adsorption capacity of several terrestrial plants has been found >60 µg/g. Organic (like vermin compost, farm yard manure) and inorganic amendments (silicate, Fe-sulphate) reduce the grain As content of rice over non amended soil. Pile up of several plant biomass (like rice husk) with inorganic amendments increases the reduction efficiency.

A further approach has been taken to reduce As concentration from contaminated irrigated soil by using indigenous microorganisms to resist the entry of As in food chain. Six bacterial strains have been isolated in soil, collected from paddy fields. They have been grown in As (III) spiked media (20 ppm) in laboratory condition. Three, out of six different microorganism species showed greater affinity for As accumulation. In a particular organism, appx. 42% As was reduced from aqueous solution with an adsorption rate of 78 µg As/g in biomass (Fig. 2).

4 CONCLUSIONS

Plants might be considered as useful weapons in the fight against this global epidemic of As poisoning. Application of bacterial consortia can be experimented in As contaminated soil in pot study to reduce its entry in soil, plants and consecutive food chain, followed by its application in field level.

ACKNOWLEDGEMENTS

Authors acknowledge the supporting members.

REFERENCES

Chakraborti, D., Rahman, M.M., et al. 2013. Groundwater arsenic contamination in Ganga-Meghna-Brahmaputra plain, its health effects and an approach for mitigation. *Environ. Earth Sci.* 70: 1993–2008.

Roychowdhury, T., Tokunaga, H., et al. 2003. Survey of arsenic and other heavy metals in food composites and drinking water and estimation of dietary intake by the villagers from an arsenic-affected area of West Bengal, India. *Sci. Total Environ.* 308: 15–35.

Roychowdhury, T. 2008. Impact of sedimentary arsenic through irrigated groundwater on soil, plant, crops, and human continuum from Bengal delta: special reference to raw and cooked rice. *Food Chem. Toxicol.* 46: 2856–2864.

Roychowdhury, T. 2010. Groundwater arsenic contamination in one of the 107 As-affected blocks in West Bengal, India: Status, distribution, health effects and factors responsible for As-poisoning. *Int. J. Hyg. Env. Health* 213: 414–427.

Arsenic Research and Global Sustainability – Bhattacharya, Vahter, Jarsjö, Kumpiene,
Ahmad, Sparrenbom, Jacks, Donselaar, Bundschuh & Naidu (Eds)
© *2016 Taylor & Francis Group, London, ISBN 978-1-138-02941-5*

Arsenic contamination in groundwater in rice farming region of Ganga basin—a study for resource sustainability

D. Dutta[1] & N. Mandal[2]

[1]*All India Coordinated Research Project on Water Management, Bidhan Chandra Krishi Viswavidyalaya, Gayeshpur, Nadia, West Bengal, India*
[2]*Department of Environmental Science, Bidhan Chandra Krishi Viswavidyalaya, Mohanpur, Nadia, West Bengal, India*

ABSTRACT: Investigations reveal a vertical boom in groundwater irrigation propels the unprecedented growth of summer rice cultivation in lower Ganga basin of India. The excessive withdrawal of ground-water for water-intensive summer rice causes substantial lowering of water-table and severe arsenic con-tamination of tube-well drinking water in this region. The study also explores the long-term negative externalities and arsenic health hazards that may result unless effective regulation on water extraction is ensured. Finally, institutional weaknesses that need to be strengthened are identified and the region-specific supply and demand-side management interventions are suggested to reduce arsenic pollution and ensure protection of the critical resources in this part of the world.

1 INTRODUCTION

The lower basin of the river Ganga is character-ized by high potential of groundwater irrigation and pre-dominance of rice farming. Rice is the main staple food to the population of high density. The unregulated withdrawal of groundwater for extensive summer rice cultivation (January-May) can wreak havoc on the environment (Singh, 2006). Aquifer pollution with leached-out arsenic is a direct off-shoot of groundwater over-abstraction (Chakraborti, 2003). Some studies have focused on the problems, but lacking the information of inte-grated approaches for sustainability of resources. Therefore, the study attempts to fill up the gap through a systematic analysis of the present trends of groundwater irrigation, arsenic contamination and also to suggest the reforms for sustainable management of water resources in this region.

2 METHODS

Field studies were carried out in 5 districts namely Murshidabad, Nadia, Burdhaman, Hooghli and North 24 Parganas of West Bengal situated in lower Gangetic plan of India. Survey sample consists of 3 blocks (sub-district level) from each district, 5 villages from each block and 20 farmer house-holds from each village (total 1500 respondents) selected at random for in-depth interviews (ques-tionnaire method) and primary data generation.

Observations also include monitoring of water-table during post and pre-monsoon periods (in November and May months respectively) at 45 hydrological points using 'Steel tape and chalk' method(lowering the graduated tape slowly into the pre-fixed wells and the water-level is measured by subtracting the submerged distance, as indicated by the absence of change in chalk colour, from the reference point at the top of the well)and arsenic analysis of 3,000 tube-wells drinking water sam-ples by standard methods (AAS-FIAS 400-HG). 15 farmers practicing SRI (system of rice intensifi-cation) and conventional methods of summer rice cultivation in the same village of Nadia are studied further. In SRI innovation, the rice crop is watered by applying intermittent irrigation with alternate wet and dry cycle (no stagnation of water on fields), but conventional practices always maintain 5–10 cm of standing water on fields by continu-ous flood irrigation. Secondary data are compiled from the government literatures.

3 RESULTS AND DISCUSSION

Study reveals that a considerable development (80% growth) in tube-well irrigation system accelerates summer rice cultivation (300% area expansion) during the last three decades in this region. About 86% cultivating area of high yielding summer rice is largely dependent on groundwater resources and this rice requires 35000–40000 m3 water per

Table 1. Linear combination fit results for arsenic XANES.

Effects in pre-moon period	Total observations	Number showing the effects	% showing the effects
Difficult to pump hand tube-wells for drinking water	900	576	64
Non-functioning of tube-wells	900	153	17
Shallow tube-wells replaced by submersible pumps	350	245	70
Drying up of dug-wells	80	25	31
People awareness of arsenic contamination*	1500	270	18

*Number of people surveyed

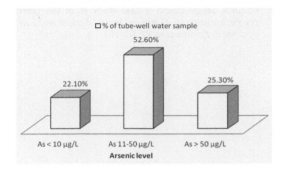

Figure 1. Arsenic concentration in tube-well drinking water.

hectare cultivation. Hence, the water-table is more or less declining due to exhaustive pumping during the lean period (November to May) without any efficient recharging system, except normal rainfall in monsoon. Increasing number of white, grey and dark blocks has witnessed the aggravating imbalance of groundwater withdrawal to recharge rate. The substantial lowering of water-table (5 ± 0.8 m) is also evident from drying up of dug-wells (31%), difficulty in operating hand pump tube-wells (64%), non-functioning of tube-wells (17%) and replacement of shallow tube-wells by deep submersible pumps (70%) in the surveyed wells (Table 1). Eleven blocks out of 15 surveyed in this region are exposed to arsenic contamination beyond WHO permissible limit for drinking water (10μg/L). Fifty three percent of tube-well drinking water contains arsenic above 10μg/L and 25% above 50μg/L (Fig. 1). However, the villagers are mostly unaware of the adverse impacts of unscientific extraction (Table 1) even suffering from health hazards and 'arsenicosis'. The state government has enacted 'Groundwater Regulation Laws 2004' to control unrestrained withdrawal of water by putting a check on excessive installation of private tubewells. But most of the village panchayats (the local government institutions) remain reluctant to implement the laws for socio-political interests.

Studies on supply-side management ensuring sus-tainability of the resources emphasize the need of strong institutional mechanisms in enforcing the regulatory laws, and collaborative involvement of all stakeholders and public agencies in decision making process. With such an interactive process the user community can develop asense of ownership over decision-making. Instead of panchayats, the block development officer (administrative unit just above village level) would have to be entrusted of issuing new permits to the users for power-driven withdraw-al of groundwater backed by data on block level hy-drology. Compliance with rules can be achieved with consensus building by raising local people awareness about regional hydrology, mechanisms of water depletion and consequences of arsenic contamination (Burchi & Nanni, 2001). In-situ rainwater harvesting by maintenance and re-digging of age-old ponds and tanks can contribute significantly to recharge aquifers. Scope for development of river based surface irrigation may also reduce pressure on groundwater. Further, demand-side management de-pends largely on the alternative avenues such as switching over to low water-requiring remunerative crops like wheat, maize, pulses, oilseeds by replacing summer rice and cultivating rice with SRI technique to save 40% water without affecting productivity.

4 CONCLUSIONS

From evidences of the deleterious effects of intensive rice irrigation on arsenic contamination of groundwater in the river basin, the study calls for an immediate paradigm shift in policy from further groundwater development to sustainable water resources management. Evolution of best strategy integrating supply and demand side management will create a regime which will foster the building-up of people knowledge base about local hydrology, arsenic pollution and change in societal mindset about water usage, enforce the water-related structural (institutional) reforms and public regulation involving all stakeholders and policy makers, promote large-scale adoption of water-saving cropping pattern and SRI irrigation technology, and revive the traditional rainwater harvesting systems for surface water development. The mitigation of arsenic hazards and the sustainability of water resources of the region will largely remain a function of the working of this regime.

ACKNOWLEDGEMENTS

Authors wish to acknowledge the assistance from the All India Coordinated Research Project on Water Management (Project code: 912 at BCKV Centre) to carry out the research work in the arsenic affected regions of West Bengal during 2007–2012.

REFERENCES

Burchi, S., Nanni, M. 2003. How ground water ownership and rights influence ground water intensive use management. In: R. Llamas, E. Custodio (Eds.) *Intensive Use of Ground Water: Challenges and Opportunities.* A.A Balkema Publishers Leiden, The Netherlands. pp. 227–240.

Chakraborti, D., Mukherjee, S.C., Pati, S., Sengupta, M.K., Rahman, M.M., Chowdhury, U.K. et al. 2003. Arsenic groundwater contamination in Middle Ganga Plain, Bihar, India: a future danger? *Environ. Health Perspect.* 111: 1194–1201.

Singh, R.B. 2006. Environmental consequences of agricultural development: a case study from the green revolution state of Haryana, India. *Agriculture, Ecosystems & Environment* 82: 97–103.

Arsenic Research and Global Sustainability – Bhattacharya, Vahter, Jarsjö, Kumpiene,
Ahmad, Sparrenbom, Jacks, Donselaar, Bundschuh & Naidu (Eds)
© *2016 Taylor & Francis Group, London, ISBN 978-1-138-02941-5*

Arsenic contaminated aquifers and status of mitigation in Bihar

S.K. Singh, A.K. Ghosh & N. Bose
Department of Environment and Water Management, A.N. College, Patna, India

ABSTRACT: Arsenic contaminated aquifers in Bihar, provide water for both drinking and irrigation purpose. This research group till date has tested more than 40 thousand drinking water and irrigation water sources. The total vulnerable population is over 725,000 (Ghosh *et al.*, 2009). Surveys conducted show inherent flaws in the organisation of mitigation work, the contributory effect of the state's population burden, socio-economic backwardness, lack of accountability in implementing different mitigation projects, as well as the failure of the decision makers to seek community awareness and cooperation for faster mitigation delivery techniques. Bihar is in the danger of repeating Bengal's mistakes in tackling the arsenic problem.

1 INTRODUCTION

The threat of microbial contamination in drinking water led to shift from surface water to ground water in developing countries including Bihar. Although Bihar is drained by numerous perennial Himalayan streams, ease of accessing its abundant ground water resources has promoted proliferation of irrigation bore wells. This overexploitation of ground water for both drinking and irrigation led to release of arsenic in the aquifers of Bihar. More than 40 thousand drinking water and irrigation water were tested till date starting from 2005. About 32% of water samples tested positive for As level of more than the WHO drinking water guideline value 10 μg/L. The estimated total vulnerable population is over 725,000 (Ghosh *et al.*, 2009). Many mitigation structures were put in arsenic contaminated districts of Bihar by government and NGOs. The status of mitigation status was studied in arsenic contaminated districts. Surveys conducted show inherent flaws in the organisation of mitigation work, the contributory effect of the state's population burden, socio-economic backwardness, lack of accountability in implementing different mitigation projects, as well as the failure of the decision makers to seek community awareness and cooperation for faster mitigation delivery techniques.

2 METHODS/EXPERIMENTAL

Each water sample was provided with an identification number (ID) which conveyed the type source of water, the date of testing, name initials of the tester and the serial number. This information was used to generate a unique identity. Recording of the location of arsenic-affected tube wells using Global Positioning System and Geographical Information Systems (GPS/GIS) was done as part of a longer term water quality monitoring system. The water samples were preserved and analyzed as per Standard Method (APHA, 2012).

All samples having arsenic level of 10 μg/L and above were analyzed by UV-VIS. 10% of the samples with more than 40 μg/L were also subject to confirmatory tests by AAS. Choropleth mapping of data obtained aided in determining the spatiality and trends of intensity of the contaminations. Status of mitigation initiatives were studied through questionnaire based survey, and random testing of treated water.

3 RESULTS AND DISCUSSION

3.1 *Status of arsenic contaminated drinking water sources*

Since 2004, random drinking water samples in 18 out of 38 districts have been tested positive for arsenic contamination. Public drinking water hand pumps located within 5 km. on either bank of river Ganga in the floodplains of Bhojpur, Patna, Vaishali, and Bhagalpur yielded arsenic levels in the range of 10μg/L to 1861μg/L in 8671 out of 28,500 hand pump waters tested (Table 1). However the frequency and intensity have been found to be the highest in the flood plains of the Ganga river and in segments of the Himalayan feeder streams in north Bihar. In the western floodplain of Bhojpur district, 31% of 6000 hand pumps had more than 10 μg/L, with the contamination intensity reaching a maximum of 1861 μg/L in Barhara. Lower level, but more widespread, arsenic contaminations exist in Vaishali and Bhagalpur districts. Characterized by predominance of urban settlement dependent upon deep bore wells for its drinking water, Patna district on the other hand, has confined areas of high arsenic contamination in its floodplains, namely at Maner, Barh and Bakhtiarpur.

Health impacts of these contaminations, which were confined to a few cancer patients, and dispersed cases

Table 1. District-wise severity of arsenic contaminated drinking water sources in the Ganga flood plains.

Districts	Arsenic content within 10 μg/L	As content > 10 μg/L	Highest As. Content (μg/L)
Bhojpur	310	2496	1861
Patna	389	671	724
Vaishali	1749	1513	325
Bhagalpur	268	1275	608

of skin diseases mostly among children and women in 2004–2005 (Ghosh *et al.*, 2009), have morphed to an ever-increasing incidences of skin and gall bladder cancers; keratosis; melanosis; infertility; and even psychosomatic disorders in persons with high arsenic levels in their blood samples, but otherwise having healthy appearances (Kumar *et al.*, 2015).

3.2 *Status of arsenic contamination in the food chain*

Indirect ingestion of arsenic through the water-soil-crop route is common as the state relies heavily on shallow ground water irrigation schemes.

Collection of water samples of 406 irrigation bore wells were undertaken from November 2009 till April 2012. The presence of arsenic in shallow aquifers were confirmed from the Indo-Nepal border in the north to the Ganga banks, along the Raxaul—Patna line in western Bihar, and the Forbesganj—Bhagalpur line in eastern Bihar. Maximum depth of contaminated bore wells were 80 feet in North Bihar Plains, and upto 300 feet in the South Bihar Plains. In all, 36 of the 38 districts of Bihar had arsenic contaminated irrigation bore well water. Arsenic contamination levels in the floodplain tracts averaged to around 50 μg/L, maximum being 857 μg/L in Khagaria district (Fig. 1).

3.3 *Population affected*

Population vulnerable to geogenic arsenic poisoning has increased from around 0.7 million persons in 2007 (Bose *et al.*, 2007) to over 1.5 million persons, based upon the contaminated rural habitation sites in the flood plains of the 4 districts (Table 2).

4 CONCLUSION

Mitigation strategies, including use of renovated open wells and infiltration units based on RO and adsorption-based technologies, have been implemented. However, most of these initiatives have failed and a few have limited and declining functioning capacities due to lack of surveillance, maintenance and other management issues. These aspects are increasing pointers to earliest adoption of mitigation models that are decentralized and community-driven and adopt green technologies which incorporates ICT methodologies.

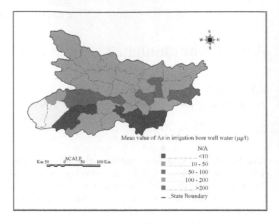

Figure 1. Arsenic contaminated irrigation bore well water: trends in the districts of Bihar.

Table 2. Arsenic contamination in the Ganga floodplains of Bihar and estimates of the affected population.

District	Population in the arsenic hotspots*
Bhojpur	366,220
Patna	101,176
Vaishali	479,138
Bhagalpur	346,633
Others	300,500 (approx.)
Total	1,593,667

* based on Census of India figures, and field work

ACKNOWLEDGEMENTS

This study was funded by UNICEF, Bihar and UGC, New Delhi.

REFERENCES

Ghosh, A.K., Singh, S.K., Bose, N., Singh, K. 2009. Arsenic hot spots detected in the state of Bihar (India) -a serious health hazard for estimated human population of 5.5 Lakh. In: AL. Ramanathan, P. Bhattacharya, A.K. Keshari, J. Bundschuh, D. Chandrashekharam, S.K. Singh (Eds.) *Assessment of Ground Water Resources and Management.* I.K. International Publishing House Pvt. Ltd., New Delhi, pp. 62–70.

Kumar, A., Ali, M., Rahman, S.M., Iqubal, A.M., Anand, G., Niraj, P.K. et al. 2015: Ground water arsenic poisoning in "Tilak Rai Ka Hatta" village of Buxar District, Bihar, India causing severe health hazards and hormonal imbalance. *J. Environ. Anal. Toxicol.* 5: 290 (doi: 10.4172/2161–0525.1000290).

Standard Methods for the Examination of Water and Wastewater. 22nd Edition. 2012. Edited by L. Bridgewater, E.W. Rice, R. B, Baird, A.D. Eaton, L.S. Clesceri. American Public Health Association, Washington.

Arsenic Research and Global Sustainability – Bhattacharya, Vahter, Jarsjö, Kumpiene,
Ahmad, Sparrenbom, Jacks, Donselaar, Bundschuh & Naidu (Eds)
© 2016 Taylor & Francis Group, London, ISBN 978-1-138-02941-5

Groundwater arsenic in alluvial aquifers: Occurrence, exposure, vulnerability and human health implications

A. Singh[1], N. Bose[2], S.K. Singh[2] & A.K. Ghosh[2]

[1]*Department of Research and Planning, Xavier Institute of Social Service, Ranchi, Jharkhand, India*
[2]*Department of Environment and Water Management, A.N. College, Patna, Magadh University, Bodh-Gaya, India*

ABSTRACT: Groundwater arsenic (As) contamination is now recognized as worldwide problem, which has several implications like serious health issues, contaminating food chain, social and behavioral complexities etc. In the present investigation, Barhara block of Bhojpur district of Bihar (India) has selected for assessment of groundwater As contamination, estimation of agricultural area affected by As due to groundwater irrigation, average daily As intake and its vulnerability on human health. Barhara block is composed of 58 revenue villages, out of them, 28 villages were selected for the study. Study findings reveal that out of 28 revenue villages, 26 villages were severely affected by groundwater As contamination. About 2,026.55 hectares irrigated agricultural land (23.19% of blocks irrigated land) of 26 villages is irrigated directly or indirectly by As contaminated groundwater. Agricultural crops cultivated in As affected area are vulnerable to assimilation of groundwater As through irrigation. Study findings show that per capita consumption of As by children, young, adults and old persons are above the permissible limit.

1 INTRODUCTION

Arsenic (As) is one of the most toxic and carcinogenic of all the natural groundwater contaminants, available for ingestion directly in drinking water. Chronic As pollution is now recognised as a worldwide problem, with 21 countries experiencing As groundwater contamination. The largest population currently at risk is in Bangladesh, followed by West Bengal in India where groundwater concentrations frequently exceed the WHO guideline (10 µg/L) more than 10 fold (Rahman, 2002, Sarkar, 2008).

In recent study, fifteen districts of Bihar namely Begusarai, Bhagalpur, Bhojpur, Buxar, Darbhanga, Katihar, Khagaria, Kishanganj, Lakhisarai, Munger, Patna, Purnea, Samastipur, Saran and Vaishali are identified as As affected districts (CGWB, 2010).

In the present study, Barhara block of Bhojpur district of Bihar (India) has been selected for assessment of groundwater As contamination, estimation of agricultural land affected by As due to groundwater irrigation, average daily As consumption and its vulnerability on human health.

Barhara is one of the 14 blocks of Bhojpur district of Bihar, situated on the bank of river Ganga and located at 25⁰40.424' N latitude and 84⁰43.465' E longitude. The block is composed of 20 panchayats, 58 villages and 145 habitations. The study area in the block is spread over a 10 km belt running roughly parallel to the river Ganga (southern boundary).

2 METHODS

The initial assessment of groundwater As contamination has been done by Field Test Kit (FTK) supplied by National Chemical Laboratory (NCL), Pune. Because the study area is huge, FTKs are essential for quick preliminary survey. Later, the most contaminated samples (concentration > 50 µg/L) identified by FTKs were retested by Atomic Absorption Spectrophotometry (AAS) method. Focus Group Discussion (FGD) has made in selected villages for collection of qualitative data. For secondary data, Census of India (2001 and 2011) has been used.

3 RESULTS AND DISCUSSION

3.1 *Arsenic affected villages in Barhara block*

Out of 58 revenue villages of Barhara block, 28 villages were selected for the study. Among 28 villages, 02 villages are free from groundwater As. Remaining 26 villages were identified as As contaminated villages. Based on the study, 1,14,369 people and 19,244 (16.82%) children of 0–6 years age group in 26 villages are vulnerable by consuming As contaminated groundwater. The As contamination in 1,086 (88.07%) hand pumps water is higher than the World Health Organisation (WHO) limit of 10 µg/L (WHO, 1993).

During the study 1,233 groundwater samples of public hand pumps were tested through FTK and 147 samples shown positive results of As concentration between 01 and 10 µg/L, 576 samples shown between 11 and 50 µg/L and 510 samples shown between 51 and 110 µg/L respectively.

Out of total 1,233 groundwater samples, 348 samples (As content >50 µg/L) were retested through AAS-HG. Maximum concentration of As was 1861 µg/L followed by 1064 µg/L in hand pump of Pandey Tola in Sinha village (Ghosh, 2008). Out of total 348 samples, 23 (6.60%) samples shown As below detective level (BDL), 19 (5.45%), 78 (22.41%) and 67 (19.25%) samples shown between 1 and 9 µg/L, 10 and 49 µg/L and 50 and 99 µg/L respectively. Similarly, 108 (31.03%), 31 (8.90%) and 07 (2.01%) samples shown between 100 and 199 µg/L, 200 and 299 µg/L and 300 and 399 µg/L respectively and 08 (2.29%) samples shown between 400 and 499 µg/L. Further, 03 (0.86%), 02 (0.57%) and 01 (0.28%) sample shown between 500 and 599 µg/L, 800 and 899 µg/L and 1064 µg/L respectively and another 01 (0.28%) sample shown 1861 µg/L. No any sample shown As content between 600 and 699 µg/L and 700 and 799 µg/L respectively.

3.2 Vulnerability of agricultural land and crops

Based on Census of India 2001, 2,026.55 hectares (23.19% of blocks irrigated land) of agricultural land in Barhara block is irrigated directly and indirectly by As contaminated groundwater. The cultivable crops in these lands may be vulnerable due to assimilation of As through plant roots by irrigation of As contaminated groundwater.

3.3 Vulnerability of human health

Study findings reveals that population of 26 villages of Barhara block is critically vulnerable to As exposure. Out of total population (1,14,369), 19,244 (16.82%) children below 06 years, 13,814 (12.07%) SC, 1520 (1.32%) ST and 99,035 (86.59%) other category population is vulnerable to As.

3.4 Daily intake of arsenic

Based on the FGD and average groundwater As concentration, average per capita intake of As from various sources by male children, young, adult and old population were comparatively more than female children, young, adult and old population

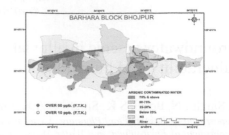

Figure 1. Arsenic contaminated village of Barhara block, based on FTK data and GPS fixes.

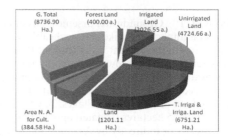

Figure 2. Land use pattern in arsenic contaminated villages of Barhara block.

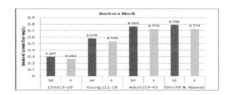

Figure 3. Arsenic consumption by different age groups.

(Fig. 3). Average per capita As consumption from various sources by children, young, adults and old persons are 0.280 mg/L, 0.558 mg/L, 0.746 mg/L and 0.757 mg/L respectively. The above findings have exceeded the maximum allowable limit of As consumption through water and food (0.2 mg/kg/day).

4 CONCLUSIONS

Out of 28 selected revenue villages of Barhara block, 26 villages are severely affected by groundwater As contamination. The average per capita As intake from all sources (drinking water, rice, pulse, chapatti and vegetable) by children, young, adults and old persons are greater than its permissible limit (0.2 mg/kg/day by water and food). In As contaminated villages, 16.82% children (0–6 years), 12.07% SC, 1.32% ST and 86.59% other category population is vulnerable to As.

The study finding shows that 2,026.55 hectares agricultural land of 26 villages is irrigated by As contaminated groundwater directly or indirectly. The cultivated food grain bearing crops and vegetables in these lands may be vulnerable to containing substantial amount of As due to its assimilation.

ACKNOWLEDGEMENTS

We acknowledge the support of UNICEF for this study through the project "Assessment of Arsenic Contamination in the Ground Water" (reference number: BH/WASH /2005/304).

REFERENCES

CGWB. 2010. *Ground Water Quality in Shallow Aquifers of India.* Central Ground Water Board, Ministry of Water Resources, Government of India, Faridabad, India, 107p.

Ghosh, A.K., Singh, S.K., Singh, S.K., Singh, A., Roy, N.P., Upadhyaya, A., Bose, N., Chaudhary, S., Mishra, R. 2008. Study of arsenic contamination in ground water of Bihar, (India) along the river Ganges. In: V.K. Kanjlia, A.C. Gupta (Eds.) *3rd Int. Conference on Water Quality Management.* pp. 293–303.

Rahman, M.M., Mukherjee, D., Sengupta, M.K., Chowdhury, U.K., Lodh, D., Chanda, C.R., Roy, S., Selim, M., et al. 2002. Effectiveness and reliability of arsenic field testing kits: Are the million dollar screening projects effective or not? *Environ. Sci. Technol.* 36(24): 5385–5394.

Sarkar, S., Blaney, L., Gupta, A., Ghosh, D., Sengupta, A. 2008. Arsenic removal from groundwater & its safe containment in a rural environment: Validation of a sustainable approach. *Environ. Sci. Technol.* 42: 4268–4273.

WHO 1993. *Guidelines for Drinking Water Quality. Recommendations.* 2nd Ed., World Health Organization, Geneva.

Arsenic Research and Global Sustainability – Bhattacharya, Vahter, Jarsjö, Kumpiene,
Ahmad, Sparrenbom, Jacks, Donselaar, Bundschuh & Naidu (Eds)
© 2016 Taylor & Francis Group, London, ISBN 978-1-138-02941-5

Removal of arsenic from acid mine drainage by indigenous limestones

M.A. Armienta[1], I. Labastida[2], R.H. Lara[3], O. Cruz[1], A. Aguayo[1] & N. Ceniceros[1]

[1]*Instituto de Geofísica, Universidad Nacional Autónoma de México, Ciudad de México, D.F., Mexico*
[2]*Universidad Autónoma Metropolitana Unidad Azcapotzalco, Reynosa Tamaulipas, Ciudad de México, D.F., Mexico*
[3]*Universidad Juárez del Estado de Durango, Durango, Mexico*

ABSTRACT: This study was carried out with the aim of evaluating the use of indigenous limestones in an open system channel to retain As from acid mine drainage. To achieve this goal, limestones were collected at the historical mining zone of Zimapán, México. Batch and column experiments were performed with synthetic leachates obtained from two tailing ponds with different oxidation degrees. Results of mineralogical (XRD, FTIR-ATR and SEM-EDS), physico-chemical parameters, arsenic and heavy metals concentrations were interpreted to determine removal mechanisms. Formation and sorption on iron oxyhydroxides and on calcite, and probably arsenates precipitation were identified as the main processes influencing As retention. Relevance of oxidation conditions is highlighted for an adequate operation of the system.

1 INTRODUCTION

Acid mine drainage is one of the main anthropogenic arsenic contamination sources. Once released from mining wastes, arsenic may pollute superficial and ground waters, and soils. Several treatment procedures may be applied to control arsenic and heavy metals environmental impact. However, in locations with scarce economic and technological resources, passive treatment systems constitute a potential option. Indigenous limestones from the historical mining area of Zimapán, México, have proven effective to retain arsenic (Romero et al., 2004). Besides, batch experiments with synthetic acid leachates obtained from tailings impoundments showed also the indigenous rocks capability to remove arsenic (Labastida et al., 2013). Many mining zones are located in the high-temperature carbonate-hosted Ag–Zn–Pb(Cu) deposits zone of northern Mexico. Use of limestones in an open treatment channel system may thus constitute a feasible option to treat acid mine drainage in those areas and others with similar geological environment. This study was carried out with the aim of evaluating the use of indigenous limestones from Zimapán to remove As from acid mine drainage and determine the retention mechanisms. To achieve this goal batch and column experiments were performed with rocks collected from outcrops in the Zimapán area.

2 METHODS

2.1 Samples collection and leachates preparation

Limestone rock samples from the Tamaulipas formation were collected at previously identified sites (Romero et al., 2004). Tailings samples were obtained from two deposits showing a different oxidation degree located at Zimapán outskirts CMZ (Compañía Minera Zimapán, highly oxidized) and SMN (San Miguel Nuevo, less oxidized). Rock samples were crushed, milled, quartered and sieved to obtain particles between 0.84 and 1.41 mm. Tailings were treated with water following the extraction protocol ASTM D3987–85 to simulate acid mine drainage.

2.2 Batch and column experiments

Batch batch and column experiments were done at room temperature with continuous agitation putting in contact 6 g rock with 200 mL acid leachate in a 250 mL Erlenmeyer flask. Column experiments were carried out in acrylic cylinders 50 cm length, 10 cm width with five sampling ports at 10 cm distance. Leachates were delivered by a peristaltic pump (0.5 L/d flow).

2.3 Chemical and mineralogical determinations

Arsenic and metals were determined in the solutions by flame atomic absorption spectrometry and hydride generation (As) AAnalyst 100, MHS-15. Major ions were measured following standard methods. Mineralogical determinations included XRD, FTIR-ATR, and SEM-EDS.

3 RESULTS AND DISCUSSION

3.1 Batch experiments

Concentrations of leachates and treated solutions are shown in Table 1. The pH increased at the end of

Table 1. Concentration (mg/L) in leachates and treated solutions in batch experiments CMZ = Compañía Minera Zimapán. SMN = San Miguel Nuevo.

Species	Cleachate CMZ	Ctreatment CMZ	Cleachate SMN	Ctreatment SMN
As	34	0.040	6.45	2.79
Fetot	705	<0.05	7.4	<0.05
Al	22	<1.0	11.4	<1.0
SO_4^{2-}	3975	1777	1604	1576
Ca^{2+}	383	695	604	674

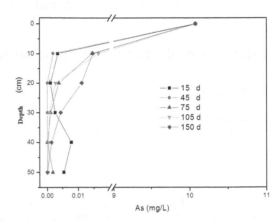

Figure 1. Arsenic (mg/L) concentration profile along the column treating SMN leachates at the end of 15, 45, 75, 105 and 150 days.

the experiment, from 2.18 (initial) to 6.10 (final) for CMZ, and from 4.0 (initial) to 6.24 (final) for SMN.

Although As was removed by the rocks from both leachates, a higher retention was achieved upon CMZ treatment. XRD and SEM-EDS analyses evidenced the formation of iron precipitates (mainly ferrihydrite) containing As on the rocks after treatment of CMZ leachates. Geochemical modeling, with visual MINTEQ, also indicated oversaturation of Fe and Al minerals. Arsenates (scorodite (FeAsS), simplesite $(Fe_3(AsO_4)_2 \cdot 8H_2O)$ and köttigite $(Zn_3(AsO_4)_2 \cdot 8H_2O)$ were also predicted to precipitate by the modeling. Absorption bands corresponding to As sorption on Fe-oxyhydroxides were identified in FTIR-ATR spectra. These results indicate that As may have been removed by sorption on Fe minerals or by arsenates precipitation. Specific minerals besides those identified in the raw rock could not be observed after treatment of SMN leachates. However, geochemical modeling suggested the formation of Al minerals that may be responsible of the removal of Al from the solution. The decrease of As concentration may have resulted from sorption on calcite as reported by Romero et al. (2004) or on Fe-oxyhydroxides that could not be observed with the mineralogical methods used.

3.2 Column experiments

Column experiments under oxic conditions (2 columns for each tailings' leachate) carried out over 8–20 weeks showed also As retention on limestones. A decrease of almost 100% of Fe and As was obtained in the lowest port of the columns treating CMZ leachates on the first and 5th weeks respectively. Iron also decreased to concentrations below detection level after de first week in the columns treating SMN leachates. Arsenic decreased from 10 mg/L to concentration below detection level (0.0005 mg/L) on the fourth week. However, it showed peaks (up to 0.018 mg/L) on the 10th, 17th, 19th and 20th weeks.

Arsenic concentration strongly decreased after leachate flow through the first 10 cm of the column (Fig. 1) and maintained low values downward.

This shows the efficiency of limestone to retain As. Diffractograms of the rocks after the experiment showed ferrihydrite formation that may have been formed at the pH measured on that section (about 6.5) and retain As. In addition, pH decreased with time to 5.3 and 4.5 after 105 and 150 days when As increased again, possibly due to a partial ferryhidrite dissolution. SEM-EDS analyses also showed the presence of Fe rich particles with As traces.

4 CONCLUSIONS

Batch and column experiments showed that indigenous limestones from a Mexican historical mining zone constitute a promising option to remove As from acid mine drainage. The rocks may be used in an open channel system that has the advantage of requiring limited maintenance, besides its low cost. Formation of Fe-oxyhydroxides and arsenates contribute to As retention, besides sorption on limestone may also participate in the removal process.

ACKNOWLEDGEMENTS

The authors acknowledge UNAM, DGAPA PAPIIT (Project IN 103114) for funding.

REFERENCES

Labastida, I., Armienta M.A., Lara-Castro R.H., et al., 2013. Treatment of mining acidic leachates with indigenous limestone, Zimapan Mexico. *J. Hazard. Mater.* 262: 1187–1195.
Romero, F.M., Armienta M.A., Carrillo-Chavez A. 2004. Arsenic Sorption by Carbonate-Rich Aquifer Material, a Control on Arsenic Mobility at Zimapán, Mexico. *Arch. Environ. Con.Tox.* 47: 1–13.

Arsenic Research and Global Sustainability – Bhattacharya, Vahter, Jarsjö, Kumpiene,
Ahmad, Sparrenbom, Jacks, Donselaar, Bundschuh & Naidu (Eds)
© *2016 Taylor & Francis Group, London, ISBN 978-1-138-02941-5*

Interaction of arsenic with biochar in soil

M. Vithanage[1,2]

[1]*Chemical and Environmental Systems Modeling Research Group, National Institute of Fundamental Studies, Kandy, Sri Lanka*
[2]*International Center for Applied Climate Science, University of Southern Queensland, Toowoomba, Queensland, Australia*

ABSTRACT: This review provides an overview of recent advances in the removal of As in soil and microbial properties of soil due to application of biochar. Biochar has been widely used as an amendment for many different soils including As. Changes in soil physio-chemical properties may influence the As bioavailability and phytotoxicity. Contrasting results have been reported in terms of As mobility in soil due to changes in ions in soil solution, DOC, pH, CEC, P and etc. Despite the use of biochar as adsorbent is increasing, a number of research gaps and uncertainties still exist as identified in this review in terms of As. Thus more relevant investigations are needed in further research especially on ageing, mechanisms and enhancing adsorption capacities.

1 INTRODUCTION

Biochar created a center of attention by being a renewable resource and due to its economic and environmental benefits as a promising resource for environmental technology used for water contaminants treatment (Inyang & Dickenson, 2015). Many studies have reported an excellent ability of biochars to remove heavy metals, organic pollutants and other pollutants from aqueous solutions (Ahmad et al., 2014, Inyang & Dickenson, 2015). Due to the above factors, an increasing global interest in biochar research is observed in past few years.

"Biochar + Soil" publications were started from 2006 and similarly in 2009 quite significant increase in publications were seen (Fig. 1). "Biochar + Arsenic" publications have a recent history of half a decade from 2009 onwards with a noteworthy increase is seen from 2013.

With the increasing interest of scientific research on biochar and its surface modifications, an integrated understanding of biochar's function is urgently needed for future engineering applications of biochar for As remediation in soil. Hence, the aim of the review is to analyze the literature related to the application of biochar for the removal of As in soil environment. This review summarizes the recent developments in the preparation and properties of raw and engineered biochar, discusses the adsorption mechanisms involved and regarding the soil related aspect in terms of As remediation.

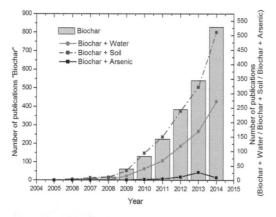

Figure 1. Science citation indexed publications on arsenic (a) and biochar (b) in SCOPUS database.

2 ARSENIC REMOVAL BY BIOCHAR

2.1 Influencing factors

The capacities of As removal appear to be inter-dependant with the feedstocks, production technology, temperature etc. A range of feedstocks have been used to test their effectiveness on As remediation in water. Those include, pine wood and bark, oak wood and bark, solid waste, rice husk, sewage sludge etc. However, there are still many other feedstocks needed to assess. Additionally, it is established that both, As(III) and As(V) have to be been studied in balance (Samsuri et al., 2013). However, many researchers have consider them in single ion studies so that up to date only one comparison study have been conducted to understand the effect of the same biochar on both inorganic speciation (Samsuri *et al.*, 2013). Moreover, to our best knowledge, there is no study conducted until now, which has investigated the organic forms of arsenic and the interactions with BC. This is a clear challenge for the near future.

2.1.1 *Effect of pH*

A study conducted on pyrolysed sewage sludge on As(III) removal showed that the two acidic pHs they studies, pH 3–3.5 and pH 6–6.5 do not show a distinct difference (Tavares *et al.*, 2012). However, pH essentially affects the equilibrium time of adsorption into sludge biochar demonstrating slightly higher and quicker removal in pH 6–6.5 than the low pH (Tavares *et al.*, 2012).

2.1.2 *Redox condition*

Redox conditions are also considered to affect biochar and the interaction with As (Rinklebe *et al.*, 2015). In opposite—several elements such as As are known to be mobile under anoxic conditions Recently, Rinklebe *et al.* (2015) reported that the pattern of Eh/pH and the release mechanisms of As in a highly contaminated floodplain soil and in the same soil treated with biochar based material are basically similar which is an important scientific finding also with view to a sustainable management of these ecosystems. However, to elucidate underlying mechanisms of the effect of various BC's on the interactions with As on various soils remains a challenge to be explore in future.

3 ROLE OF BIOCHAR FOR ARSENIC IN SOIL

Many different pot experiments have been conducted with As contaminated soils with various biochars produced in different temperatures with different plants such as ryegrass (*Lolium perenne* L.), tomato (*Solanum lycopersicum* L.), *Miscanthus giganteus* and *Zea mays* L. Soils that have been used are naturally contaminated soil such as mine soil (Beesley et al., 2013, Zhang *et al.*, 2013), industrially contaminated soil (Hartley *et al.*, 2009), *Lolium perenne* (Beesley et al., 2014) and As based pesticide applied soils (Gregory *et al.*, 2015, Gregory *et al.*, 2014). These pot experiments were conducted in different application rates and in different time periods.

Conflicting results have been reported in literature regarding As mobility due to the application of biochar (Beesley *et al.*, 2014, Gregory *et al.*, 2015, Namgay *et al.*, 2010). In spite of similar As concentrations in the soils, inherent differences of As mobility and bioavailability existed between the soils which was probably a due to particular differences in P, pH, Fe, OM and DOC concentrations before and after BC additions (Hartley *et al.*, 2009). High CEC in soil possibly lead to the formation of stable biochar–trace element complexes in soil. It has been reported by the biochar ageing studies.

3.1 *Response of soil physio-chemical properties of As contaminated soils to BC*

Competition between DOC, P and As for retention sites on soil surfaces can increase in soluble As with increasing concentrations of DOC and P (Beesley *et al.*, 2014, Hartley *et al.*, 2009). And in some cases BC application may induced considerable solubilisation of arsenic to pore water (>2500 mgl) related to pH and soluble phosphate but combining amendments with compost most effectively reduced toxicity due to simultaneous reductions in extractable metals and increases in soluble nutrients (P) (Beesley *et al.*, 2014, Hartley *et al.*, 2009).

3.2 *Response of microbiological properties in biochar amended As contaminated soils*

Only a very few research work have been carried on the response of microbes on biochar amended As contaminated soils, however, the addition of both low and high temperature biochar caused a significant increase in DHA relative to the control however no significant difference was observed with the application rate (Gregory *et al.*, 2014).

4 CONCLUSIONS

There is a consensus that biochar amendment changes the chemistry of As in soil. More research is needed to develop greater mechanistic understanding of the soil–plant processes that are associated with the amendment of soil with biochar in As contaminated soils.

REFERENCES

Ahmad, M., Rajapaksha, A.U., Lim, J.E., Zhang, M., Bolan, N., Mohan, D., Vithanage, M., Lee, S.S., Ok, Y.S., 2014. Biochar as a sorbent for contaminant management in soil and water: A review. *Chemosphere* 99: 19–33.

Beesley, L., Inneh, O.S., Norton, G.J., Moreno-Jimenez, E., Pardo, T., Clemente, R., Dawson, J.J.C., 2014. Assessing the influence of compost and biochar amendments on the mobility and toxicity of metals and arsenic in a naturally contaminated mine soil. *Environ. Pollut.* 186: 195–202.

Gregory, S.J., Anderson, C.W.N., Camps-Arbestain, M., Biggs, P.J., Ganley, A.R.D., O'Sullivan, J.M.,

McManus, M.T., 2015. biochar in co-contaminated soil manipulates arsenic solubility and microbiological community structure, and promotes organochlorine degradation. *PLoS ONE* 10: e0125393.

Hartley, W., Dickinson, N.M., Riby, P., Lepp, N.W., 2009. Arsenic mobility in brownfield soils amended with green waste compost or biochar and planted with Miscanthus. *Environ. Pollut.* 157: 2654–2662.

Inyang, M., Dickenson, E., 2015. The potential role of biochar in the removal of organic and microbial contaminants from potable and reuse water: A review. *Chemosphere* 134: 232–240.

Namgay, T., Singh, B., Singh, B.P., 2010. Influence of biochar application to soil on the availability of As, Cd, Cu, Pb, and Zn to maize (Zea mays L.). *Soil Res.* 48, 638–647.

Rinklebe, J., Shaheen, S.M., Frohne, T., 2016. Amendment of biochar reduces the release of toxic elements under dynamic redox conditions in a contaminated floodplain soil. *Chemosphere* 142: 41–47.

Samsuri, A.W., Sadegh-Zadeh, F., Seh-Bardan, B.J., 2013. Adsorption of As(III) and As(V) by Fe coated biochars and biochars produced from empty fruit bunch and rice husk. *J. Environ. Chem. Eng.* 1: 981–988.

Tavares, D., Lopes, C., Coelho, J., Sánchez, M., Garcia, A., Duarte, A., Otero, M., Pereira, E., 2012. Removal of Arsenic from Aqueous Solutions by Sorption onto Sewage Sludge-Based Sorbent. *Water Air Soil Poll.* 223: 2311–2321.

Zhang, X., Wang, H., He, L., Lu, K., Sarmah, A., Li, J., Bolan, N.S., Pei, J., Huang, H. 2013. Using biochar for remediation of soils contaminated with heavy metals and organic pollutants. *Environ. Sci. Pollut. Res.* 20(12): 8472–8483.

Arsenic Research and Global Sustainability – Bhattacharya, Vahter, Jarsjö, Kumpiene,
Ahmad, Sparrenbom, Jacks, Donselaar, Bundschuh & Naidu (Eds)
© 2016 Taylor & Francis Group, London, ISBN 978-1-138-02941-5

Citric acid assisted phytoremediation of arsenic through *Brassica napus* L.

M. Farid[1,2], S. Ali[1] & M. Rizwan[1]

[1]*Department of Environmental Sciences and Engineering, Government College University, Faisalabad, Pakistan*
[2]*Department of Environmental Sciences, University of Gujrat, Hafiz Hayat Campus, Gujrat, Pakistan*

ABSTRACT: Arsenic (As) contamination of waters and soils is an environmental and health issue worldwide due to the carcinogenic nature and toxic effects of As. As-contaminated water used for irrigation purposes enhancing As in food chain causing serious health issues to consumers. The present study was planned to remediate the As from soil by using the hyper-accumulator plant, such as *Brassica napus* L, in the presence of organic chelator (citric acid) at different levels. The results revealed that the citric acid significantly enhances the As concentration in different parts of plant such as root, stem and leaves respectively and also regulates plant growth suggesting the suitability of this technique for As remediation.

1 INTRODUCTION

Arsenic (As) is a highly toxic metalloid that forms approx. 0.00005% of the Earth's crust and also 20th naturally abundant element in earth crust (Jain & Ali, 2000, Smedley & Kinniburgh, 2002). The most abundant inorganic forms of As are arsenite [As(III)] and arsenate [As(V)]. Naturally, As released by volcanic eruptions and weathering of parent material (Mohan & Pittman, 2007, Smedley & Kinniburgh, 2002). In groundwater, As concentration has been found up to 4000 µg/L while the WHO safe limit for As is 10 µg/L (Rahman *et al.*, 2010). Arsenic accumulates in the human body via consuming As-contaminated food, vegetables and drinking water (Rahman *et al.*, 2010).

Various remediation methods have been used to reduce or eliminate As from water and soil (Banerjee *et al.*, 2012). Phytoremediation is one of the environment friendly and solar energy driven method to remediate the As from As-contaminated waters and soils (Ehsan *et al.*, 2014). *Brassica napus* L. is hyper-accumulatore plant and potential candidate for metal phytoextraction because having capacity to accumulate larger quantity of metals in its roots and above ground biomass (Farid *et al.*, 2015). Chelators have the ability to enhance the bioavailability of metals in water and soil by making complexes with metals and also provide support to plants in terms of growth (Afshan *et al.*, 2015). Keeping in view the above scenario, the present study was planned to remediate the As-contaminated soils by using *Brassica napus* L in the presence of citric acid.

2 METHODS/EXPERIMENTAL

A clay loam (25% sand, 12% silt, 51% clay) soil was used in this study. Mature seeds of *Brassica napus* L.

genotype (Faisal Canola (RBN-03060) were washed with distilled water thoroughly and sown in earthen pots each containing 5 kg of soil (dry weight) under wire house conditions. After four weeks of germination, uniform plants were treated with increasing concentrations of As (0, 100 and 500 µM) along with CA (0, 2.5 and 5.0 mM) in different combinations. Each treatment was triplicated.

After the 8 weeks of treatment application, different parameter were measured to check the effects of As and citric acid on *Brassica napus* L. such as, plant growth, biomass, chlorophylls, carotenoids, gas exchange attributes, activity of antioxidant enzymes (SOD, POD, CAT, APX) and ROS along with EC and SPAD value. The concentration of As and its accumulation in different plant organs, i.e., leaf, stem and root was also measured.

3 RESULTS AND DISCUSSION

Increasing As concentration caused toxic effects to plants and reduced plant growth by limiting its enzymatic activities. The As significantly reduced the plant height, root length, number of leaves per plant, fresh and dry biomass of root, stem and leaf, chlorophylls, SPAD value, gas exchange attributes and activities of antioxidant enzymes (Fig. 1 a,b). On other hand the addition of citric acid at different levels significantly reduced the toxic effects of As on *Brassica napus* L (Fig. 1 a,b).

Increasing concentration of As also enhanced the As in different parts of plants while the addition of CA boosted the concentration of As and also increase the As uptake in roots, leaves and stems which can be seen more clearly in Table 1.

3.1 *Discussion*

Phytoremediation has previously been used with some success to remove As from contaminated soils and water. For example, Feller (2000) investigated the removal of As from a site in Germany by using plant, *Reynoutria sachalinensis* which accumulated a maximum of 1900 μg As g^{-1} dry mass in shoots and up to 430 μg As g^{-1} in root tissue. Many other species such as *Pityrogramma calomelanos* and *Pteris vittata*, are also potential candidate for the phytoremediation of As (Visoottiviseth et al., 2002).

Two major classes of heavy metal chelating peptides are known to exist in plants, namely metallothioneins and phytochelatins. (Robinson et al., 1993). Addition of organic chelating acid (citric acid) dramatically enhances the bioavailability

and uptake of As and maintain its accumulation in *Brassica* plants which results in higher As accumulation.

4 CONCLUSIONS

Addition of citric acid ameliorates the negative effects of As on plants and enhanced As uptake. However, the ability of *Brassica napus* L. to grow healthily on As contaminated sites and to hyperaccumulate As indicates their unprecedented potential for phytoremediation of As.

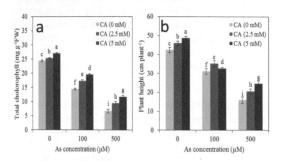

Figure 1. The effects of As and CA application on height and total chlorophylls of *Brassica napus* L.

Table 1. Effect of different concentration of chromium on As concentrations and accumulation/uptake by *B. napus* plants.

Treatments	As concentrations (mg/kg^1 DW)		
	Roots	Stem	Leaves
Control	0.00 ± 0.00^a	0.00 ± 0.0^a	0.01 ± 0.0^a
CA (2.5 mM)	0.41 ± 0.36^a	0.55 ± 0.0^a	0.01 ± 0.0^a
CA (5.0 mM)	0.07 ± 0.0^a	0.67 ± 0.0^a	0.1 ± 0.0^a
As (100 μM)	322.8 ± 10.2^b	125.6 ± 13.1^b	71.2 ± 5.2^b
As 100 + 2.5 CA	267.0 ± 11.1^c	102.6 ± 8.0^c	63.6 ± 1.3^c
As 100 + 5.0 CA	234.7 ± 9.8^d	83.6 ± 4.8^d	53.4 ± 3.2^d
As (500 μM)	585.4 ± 21.8^e	328.4 ± 24.7^e	180.7 ± 4.2^e
As 500 + 2.5 CA	487.2 ± 28.0^f	215.3 ± 13.1^f	134.5 ± 4.54^f
As 500 + 5.0 CA	423.6 ± 14.5^g	176.7 ± 11.8^g	111.2 ± 4.0^g

Treatments	As accumulation (mg/plant)		
	Roots	Stem	Leaves
Control	0.0 ± 0.0^a	0.00 ± 0.00^a	0.00 ± 0.00^a
CA (2.5 mM)	0.0 ± 0.0^a	0.001 ± 0.00^a	0.00 ± 0.00^a
CA (5.0 mM)	0.0 ± 0.0^a	0.002 ± 0.00^a	0.00 ± 0.00^a
As (100 μM)	0.153 ± 0.002^b	0.208 ± 0.02^b	0.678±0.012^b
As 100 + 2.5 CA	0.156 ± 0.003^b	0.210 ± 0.01^b	0.710±0.018^c
As 100 + 5.0 CA	0.167 ± 0.004^c	0.218 ± 0.00^b	0.720±0.006^c
As (500 μM)	0.117 ± 0.004^d	0.214 ± 0.01^b	0.722±0.020^c
As 500 + 2.5 CA	0.139 ± 0.006^e	0.20 ± 0.006^b	0.758±0.010^d
As 500 + 5.0 CA	0.163± 0.009^c	0.216 ± 0.01^b	0.816±0.024^e

ACKNOWLEDGEMENTS

Special thanks to Higher Education Commission (HEC) of Pakistan for the financial and experimental support in this research.

REFERENCES

Afshan, S. et al., 2015. Citric acid enhances the phytoextraction of chromium, plant growth, and photosynthesis by alleviating the oxidative damages in *Brassica napus* L. *Environ. Sci. Pollut. Res.* 1–11.

Banerjee, K. et al., (2012). A Novel Agricultural Waste Adsorbent, Watermelon Shell for the Removal of Copper from Aqueous Solutions. *Iranica. J. Energy Environ.* 3: 143–156.

Ehsan, S., et al., 2014. Citric acid assisted phytoremediation of cadmium by *Brassica napus* L. *Ecotoxicol. Environ. Safe* 106, 164–172.

Farid, M., Ali, S., et al. 2015. Exogenous application of EDTA enhanced phytoremediation of cadmium by *Brassica napus* L. *Int. J. Environ. Sci. Technol.* 12(12):3981–3992

Feller, K.A. 2000. Phytoremedaition of soils and waters contaminated with arsenicals from former chemical warfare installations. In: D.L. Wise, D.J. Trantolo, E.J. Cichon, H.I. Inyang, U. Stottmeister (Eds.) *Bioremediation of Contaminated Soils.* Marcel Dekker, New York.

Jain, C., Ali, I. 2000. Arsenic: occurrence, toxicity and speciation techniques. *Water Res.* 34: 4304–4312.

Mohan, D., Pittman, C.U. 2007. Arsenic removal from water/wastewater using adsorbents, a critical review. *J. Hazard. Mater.* 142(1): 1–53.

Rahman, M.M. et al., (2010). Arsenic exposure from rice and water sources in the Noakhali District of Bangladesh. *Water Qual., Expos Health* 3, 1–10.

Robinson, N.J., et al. 1993. Plant metallothioneins. *Biochem. J.* 295: 1–10.

Smedley, P.L., Kinniburgh, D. 2002. A review of the source, behaviour and distribution of arsenic in natural waters. *Appl. Geochem.* 17: 517–568.

Visoottiviseth, P. et al., 2002. The potential of Thai indigenous plant species for the phytoremediation of arsenic contaminated land. *Environ. Pollut.* 118(3): 453–461.

5.3   Mitigation and management

Arsenic Research and Global Sustainability – Bhattacharya, Vahter, Jarsjö, Kumpiene,
Ahmad, Sparrenbom, Jacks, Donselaar, Bundschuh & Naidu (Eds)
© 2016 Taylor & Francis Group, London, ISBN 978-1-138-02941-5

The role of hydrogeologists in the management and remediation of aquifers with arsenic

R. Rodriguez[†], M.A. Armienta & I. Morales
Instituto de Geofísica, Universidad Nacional Autónoma de México, Ciudad de México, D.F., Mexico

ABSTRACT: The main focus of hydrogeological research related to groundwater arsenic in developing countries, is to determine the concentrations, affected area or/and, origin of the element. Health effects, if detected, are reported to Water authorities. These studies are published in journals or technical Reports. The actual application of the research results in the solution of water problems is usually unknown by hydrogeologists. Water management authorities must look for integral solutions, not only to identify sites to drill new wells. Occasionally, hydrogeologists participate in epidemiological studies to evaluate health affectations and recognize communities at risk. Some Mexican examples are presented in this work.

1 INTRODUCTION

Arsenic may occur in diverse geological environments in urban and rural areas. Many studies have been carried out in populated sites. When contaminated groundwater is the only source of water supply, the relevance of these studies increases. Groundwater monitoring campaigns are carried out to determine concentrations, temporal and spatial evolution and the extension of affected aquifers. It is not easy to recognize the health affectations related to arsenic exposure. One of the first manifestations is keratosis. In low-income areas hospitals lack of epidemiologists. It is common that medical doctors cannot identify affectations due to contaminated water consumption. Keratosis is often confused whit skin irritation. Organic fluids monitorig is not a common practice. Water treatment systems to remove As are usually view as a chemistry research line. In few sites of Mexico multidisciplinary and integral studies have been carried out.

2 METHODOLOGY

2.1 Bibliographic review

A review of journal articles and technical reports related with As in Mexican groundwater was performed. The main objetive in most cases was to define the extension of the aquifer with arsenic concentrations over the National Standars for drinking water. The location of places for new wells is other of the objetives. Multidisciplinary

groups look for health affectations and alternatives of water treatment. Participation of hydrogeologists in this research area is scarce. We estimated the number of hydrogeologists working in the Water Commissions. Some medical studies were reviewed.

3 RESULTS AND DISCUSSION

3.1 Discussion

The quantity of hydrogeologists, HG's, dedicated to groundwater arsenic research is limited. In Mexico formal HG's graduated from academic institutions are scarce. Most of the professionals dealing with groundwater issues do not have a hydrogeological formation. A low number of postgraduate Programs has been established, no more than six in all the country.

There is a lack of strategies to prepare HG's, and not well-defined policies for management of water supply in areas with aquifers containing arsenic. Water monitoring is carried out with regularity in big cities. In small communities and rural setlements the main concern is the bacteriological contamination and in consequence only the chloride content, mainly residual chlorine. The medical sector has not enough information to diagnose health affectations related to As ingestion, arsenicosis. However, it is relatively easy to identify skin anomalies like keratosis, mainly hiperkeratosis. Cancer and cardiovascular diseases require clinical studies. Not all HG focused on As research look for healh affectations. One of the problems that have most of hydrogeologists is the scarce knowledge of arsenic hydrogeochemistry.

†Deceased

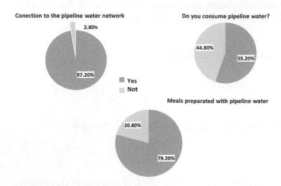

Figure 1. Pattern of water consumption obtained for Juventino Rosas, Central Mexico.

The level of health affectations depends mainly on the quality of water and the pattern of water consumption (volume, time) (Fig 1). For most HG the definition of water consumption is not one of their main concerns.

In remediation and management strategies, the participation of HG is limited to reports of As concentrations and water availability. Mathematical modelling of flow and transport of contaminants should incorporate geological models and hydrogeological parameters. The participation of HG in the process of modelling generally seems to be out of the field of groundwater professionals, but the inclusion of them is fundamental to obtain results that are more realistic.

In low-income areas, some methodologies to treat contaminated water are being developed, using local rocks. The interaction of HG with the remediation researchers can help to define the outcrops of suitable geological material. More sophisticated remediation schemes, like hydraulic barriers with arsenic adsorbing materials, may also be applied. In these cases the preferential flow direction and the watertable depletion must be defined by HG.

3.2 Improving the role of hydrogeologists in studies concerning groundwater

The participation of HG in integral groundwater As studies depends on personal interest, the academic profile of the institutions, and their social compromise.

Environmental groups looking for solutions must incorporate HG´s and hydrogeochemical researchers. Publication of papers in indexed journals must not be their main objective. The contribution to provide integral solutions is more than

a technical problem; it is a social compromise. Some affected people can develop serious diseases. Hyperkeratosis can lead to the amputation of extremities. School and profesors should inculcate moral and ethical principles. The vision of HG must not be only geology or hydrogeology.

HG's must communicate their results in seminars or conferences with the participation of local authorities. People must know the risks of water consumption.

4 CONCLUSIONS

The HG commitment goes beyond papers or academic recognition. Our participation in integral solutions facilitates the management, policy and water availability. In many cases, we must perform epidemiological studies. Results must reach the local, regional and federal Water and Health authorities. In many cases, we have been forced to confront authorities to carry out our proposals.

The hydrogeologist's vision should be expanded. HG's have the compromise to participate in solutions according to local infrastructure.

ACKNOWLEDGEMENTS

Finantial support was obtained from DGAPA-UNAM grants IN101715 and IN103114.

REFERENCES

Armienta, A., Rodriguez, R., Cruz, O. 1997. Arsenic content in hair and people exposed to natural arsenic polluted groundwater at Zimapán, Mexico. *Bull. Environ. Contam. Toxicol.* 59(4): 538–589.

Charles, E., Thomas, D.S.K., Dewey, D., Davey, M., Ngallaba, S.E., Konje, E. 2013. A cross-sectional survey on knowledge and perceptions of health risks associated with arsenic and mercury contamination from artisanal gold mining in Tanzania. *BMC Public Health* 13: 74 (DOI: 10.1186/1471-2458-13-74).

Grima, J., Luque-Espinar, J.A., Mejia, J.A., Rodriguez, R. 2015. Methodological approach for the analysis of groundwater quality in the framework of the Groundwater Directive. *Environ. Earth Sci.* 74(5): 4039–4051.

Hassan, M.M., Atkins, P.J., Dunn, C.E. 2005. Social implications of arsenic poisoning in Bangladesh. *Soc. Sci. Med.* 61(10): 2201–2211

Rodriguez, R., Berlín, J., Mejia, J.A. 2002. Percepción social de la contaminación del agua. Salamanca, Un caso Anómalo? In: B. Boehm, J.L.M. Duran, R. Sánchez, R.A Torres (Eds.) *Los Estudios del Agua en la Cuenca Lerma Chapala Santiago.* Col. Michoacán, Universidad de Guadalajara. 317–330 pp.

Arsenic Research and Global Sustainability – Bhattacharya, Vahter, Jarsjö, Kumpiene,
Ahmad, Sparrenbom, Jacks, Donselaar, Bundschuh & Naidu (Eds)
© *2016 Taylor & Francis Group, London, ISBN 978-1-138-02941-5*

Arsenic poisoning of drinking water and associated challenges of a developing nation for its mitigation and management—a case study from Central India

Arunangshu Mukherjee

Central Ground Water Board, Bhujal Bhawan, Faridabad, India

ABSTRACT: The hydrothermal phase following the batholithic intrusion is responsible for sulfide mineralization and related enrichment of Arsenic (As) concentration in the granite-meta rhyolite rocks along the Kotri lineament of central Indian craton. Isolated clusters of high As ground water were found in an area of 10 km radius, where ground water is the only source of available water. Remediation through biochemical and adsorption technology and options of supplying surface water were exhausted in the area. In-situ relief through micro level management applying prevailing geohydrological controls over distribution of As rich ground water have proved to be better option. The relatively younger intrusive basic rocks, which occupy nearly 28% of the geographical area, have invariably low As concentrations in ground water which were targeted with specially designed wells for providing safe and sustainable drinking water.

1   INTRODUCTION

Rocks of the "Kotri lineament" stretching over 90 km have been marked as Arsenic Advisory Zone in the south-central part of Chhattisgarh State in central Indian craton (Mukherjee *et al.*, 2009). The area is occupied regionally by granite-rhyolite batholithic rocks (2485 Ma, U-Pb Sensitive High Resolution Ion Microprobe age, Bickford et al., 2014) and is intruded by multi phased basic intrusions. Known for its economic mineral repository, the area is having copper, manganese and uranium deposits and proven reserves of Pb-Zn and Au-Ag.

Geogenic Arsenic (As) contamination has been reported in groundwater within a radius of 10 km area in and around 30 villages. This densely forested, remotely located and under-developed area is facing social agitations of "naxalites". In such scenario in India, groundwater is the only safe and reliable source of water supply. However, as As contamination in ground water has been reported, the ground water supply as safe drinking water needs scientific intervention.

2   METHODOLOGY

Groundwater condition of the area was investigated in detail including in-situ measurements of pH, Electrical Conductivity (EC) and As using As analysis kit. 250 mL water samples were collected in polyethylene bottle monthly for arsenic analysis in after filtering Sartorius cellulose nitrate (through 0.45 µm membrane filter of made of make) using hand operated vacuum pump (Tarson make). The samples were then acidified. Coordinates of all the sampling points were taken through GPS. Reduced levels of all the sampling points were determined through surveying from nearest bench mark. Arsenic analysis was carried out on a AAS with hydride generation technique. The detection limit of the instrument is 1 µg/L. Data were plotted on GIS platform and also superimpose on other thematic maps.

3   RESULTS AND DISCUSSIONS

The source of As in rocks of the area is arsenic bearing pyrite minerals. Enrichment of arsenic is also observed preferentially with Fe-hydroxide, coating on clay minerals, biotite and chloride on weathered granite and rhyolite (Ashiya & Patel, 1998, Pandey et al., 2002). Released from these sources, the concentration of As in ground water is recorded as high as 1025 µg/L against the prescribed permissible limit of 10 µg/l. The high temporal and spatial variation in As concentration complicated the situation further and interrupted supply of safe ground water. The hand pumps of the As effected wells were sealed. Even supply from bore wells to communities was suspended due to rise in As level in ground water subsequently after installation of pump. The problem also aggravated due to the paucity of good laboratory facilities at local level to detect As concentrations reliably and repeatedly up to 10 µg/l. This has been considered as one of the major constraints to combat the As poisoning through drinking water.

Biochemical remediation plants were installed in the worst effected villages- Kaudikasa and Mureti-tola, but low water yield at the peak hours and over all high maintenance became main cause of dislikeand unacceptability of plant among villagers. The As adsorption kits installed with few hand pumps were all defunct soon due to low sustainability. State Government also looked for supply through surface water source by developing canal network connected to Mogra Dam on Seonath River. However, supply through surface water required establishment of filter plant and network of distribution system through highly undulating terrain. This required huge investment and maintenance with recurring budget. Thus, in-situ relief Appears to be the most viable option for local supply of As free water.

A number of alternative safe ground water source have been identified through initiatives of Central Ground Water Board in the area. The structural, lithological and geochemical controls over the occurrence of As distribution were identified for providing required scientific inputs. Investigations show As contamination occurs in isolated clusters along the N-S trending shear zone. Inventory of all drinking water well by -NEERI (a premier institute) in the area reveals not all the wells, even in the worst effected village, are contaminated with high arsenic. Investigation has pointed out that wells situated over basic rocks are invariably low in As concentration and come under safe category. Basic rocks are found covering 28% of the geographical area. Delineation of arsenic free aquifers within the arsenic advisory zone has provided cheaper and sustainable alternative. The specially designed bore wells of Central Ground Water Board were used for community water supply. Such a micro level management was found suitable for the area.

4 CONCLUSIONS

The structural, lithological and geochemical control on occurrence of high As ground water along the Kotri lineament of central Indian craton has been identified. These controls were applied for delineation of As free aquifers and micro level management for providing safe drinking water which is sustainable. The method is found cheaper and suitable for remotely located, undulating rural area over other remediation methods.

ACKNOWLEDGMENT

Author is highly obliged to the Chairman, CGWB who kindly accorded permission to use and publish the data. The opinion expressed in the paper is authors own and may not be the same of his organization.

REFERENCES

Ashiya, I.D., Patel, M.K. 1998. Preliminary prospecting for gold mineralization in pyritiferous gabbro and tuffite in Dongargarh Chowki area, Bastar and Rajnandgaon districts, Madhya Pradesh. *Rec. Geol. Surv. India* 131 pt. 6: 102–104.

Bickford, M.E, Basu, Abhijit, Kamenov, George, D., Mueller, P.A., Patranabis-Deb, S., Mukherjee, A. 2014. Petrogenesis of 1000 Ma felsic tuffs, Chhattisgarh and Indravati Basins, Bastar craton, India: Geochemical and Hf isotope constraints. *Jour. Geol.* 122: 43–54.

Mukherjee, A., Tewari, D., Verma, J.R., Subramanian S., Roy, R.K., Devangan, R. 2009. Geogenic arsenic contamination to ground water in parts of Ambagarh Chowki block, Rajnandgaon district, Chhattisgarh. *Bhu Jal News* 24(2–3): 40–59.

Pandey, Y., Mishra, H., Mishra, V.P., Ashiya, I.D. 2002. Arsenic contamination in water in Rajnandgaon district, Chhattisgarh. *IGCP-454 Proc. Workshop on Medical Geology*, pp. 211–220.

Arsenic Research and Global Sustainability – Bhattacharya, Vahter, Jarsjö, Kumpiene,
Ahmad, Sparrenbom, Jacks, Donselaar, Bundschuh & Naidu (Eds)
© 2016 Taylor & Francis Group, London, ISBN 978-1-138-02941-5

Groundwater management options for augmentation of agricultural growth in the arsenic infested Gangetic alluvium of Nadia district, West Bengal, India

A. Gayen[1] & A. Zaman[2]
[1]*Rajiv Gandhi National Ground Water Training and Research Institute (RGNGWTRI), Raipur, Chhattisgarh, India*
[2]*Department of Agronomy, Bidhan Chandra Krishi Viswavidyalaya (BCKV), Mohanpur, West Bengal, India*

ABSTRACT: Nadia district of West Bengal State forms lower part of Indo-Gangetic alluvium of Quaternary age. Population explosion, rapid agricultural growth and inadequate surface water resources has put tremendous thrust on groundwater development. Irrigation potential of the district is 3356 MCM of which 3100 MCM belongs to groundwater. Summer (*boro*) paddy cultivation demands huge groundwater. Shallow tube wells within 80 meter are playing an important role in groundwater irrigation. Net groundwater available is 1990 MCM, groundwater draft for irrigation purposes is 1676 MCM. Net groundwater available for future irrigation development is 350 MCM. Maximum arsenic in groundwater is 1.18 mg/l. Crop planning in irrigated area needs to be paid proper attention to implement location specific proven water saving technologies with conjunctive application of groundwater and surface water. Low arsenic accumulating crops may help to reduce arsenic entry into the food chain. River lift irrigation may be adopted as alternative of arsenic contaminated groundwater.

1 INTRODUCTION

Nadia district forms lower part of the Ganga basin and is bounded by the latitude 22° 25′ 30″ N to 42° 05′ 40″ N and Longitudes 88° 07′ 30″ E to 88° 47′ 45″ E. Entire district is arsenic affected. Arsenic contamination in agro-ecosystem acts as a conduit for the passage of the toxicant to human population *via* food web came under serious consideration since 2002 (Sanyal and Nasar, 2002, Zaman *et al.*, 2013). The main agricultural crops grow have arsenic content within the range of 0.10 to 5.60 mg/ kg (Reported from SAU (BCKV), West Bengal). Major constraints for agricultural growth are inexplicable change in monsoon and occurrence of dry spells. The arsenic toxicity occurs in groundwater in shallow aquifer within depth of 80 m bgl (CGWB).

Reportedly in the district, shallow tube wells are major in numbers that play an important role in irrigation. Consequently, the groundwater resource in shallow aquifer is being exhausted in alarming extent. Hence, in future for the required agricultural production in the district, people have to depend on the deeper aquifers, which is a costly affair. At present the district has a groundwater resource of 2172 MCM, of which gross groundwater draft is 1653 MCM (CGWB, MoWR, 2013). Out of 17 blocks, six blocks of the district are semi-critical as groundwater development is beyond 87.43% and long term decadal trend indicates the falling of groundwater level during both pre-monsoon and post-monsoon periods.

2 METHODS/EXPERIMENTAL

2.1 Methods

Water samples from 83 Network hydrograph monitoring stations (dug wells, piezometers and tube wells) were collected in 250 ml polythene bottles during pre-monsoon (April). The bottles were tightly capped without air gap and immediately shifted to the chemical laboratory. Physical characteristics of water samples (temperature, specific conductance, pH) were measured at the well-site by a field kit. All types of abstraction structures were sampled for analysis of arsenic. Dynamic ground water resources of the district were estimated as per the guidelines of Ground Estimation Committee (GEC)-1997 methodology. Detailed survey had been conducted to understand impact of arsenic contaminated groundwater in agro-ecosystem.

2.2 Hydrogeology

Thick Gangetic alluvium of quaternary age is composed of sand of various grades, silt, clay, gravel and kankar and their various admixtures deposited by the Ganga River and its tributaries. In general three aquifers have been identified. The shallow aquifer exists down to a depth of 80 m bgl. The next aquifer occurs within the depth range of 100 to 162 m bgl and the deepest one exists within the depth range of 200 to 335 m bgl. But at places, the demarcating horizon in between first and second aquifer could not clearly be detectable owing to pinching and swelling nature of Indo-Gangetic riverine deposits. The flow of ground water is towards south and south-east. Depth to water level ranges from 5 to 10 m bgl in during pre-monsoon (April) period, whereas during post-monsoon (November) water level ranges from 2 to 5 m bgl. Shallow tube wells yield is ranging from 3.21 to 62.11 lps. Transmissivity (T) ranges from 1487 to 8607 m²/day. Net ground water availability is 1990 MCM, wherein existing groundwater draft for Irrigation is 1676 MCM. Stage of ground water development is 87%. Net ground water availability for future irrigation development has been calculated as 350 MCM (CGWB).

3 RESULTS AND DISCUSSION

3.1 *Ground water for agricultural use*

Summer rice crop (*boro*) is essentially dry season rice crop (spring-summer) under irrigated ecology contributes significantly to food basket of the country as well as making sustained self-sufficient in meeting demand of grain production for ever-increasing population. The extent of summer rice crop indicated farm level groundwater exploitation. Ground water development by shallow tube wells (depth 18 to 30 m bgl) is more as compared to deep tube wells.

3.2 *Results*

The arsenic in shallow tube well water samples ranges from 0.02 to as high as 1.18 mg/l. Source of arsenic contamination in groundwater is geogenic.

The application of Farm Yard Manure (FYM) and phosphate was found to have opposing effect on release of native and applied arsenic in the contaminated soils. The crop with less water requirement also might be helpful in lowering down the effect of arsenic toxicity. The crop which require less water are considered as substitute for the dominant rice-rice-fallow sequence in the affected areas to minimize the arsenic uptake by food crops. It demonstrates the significance of crop diversification during summer season. Inclusion of pulses/other legumes/green manure crops in the cropping sequence, coupled with organic manure addition, was found helpful in moderating arsenic build-up in soil and plant parts. Blue Green Algae (*Anabaena sp. and Nostoc sp.*) and some bacteria were shown the effect having decontamination ability. Considering the severity of arsenic contamination in groundwater, the tube wells fitted with arsenic removal plants were recommended. Growing crops with surface water was advisable to minimize the arsenic problems.

3.3 *Arsenic in soil and crops*

The input of arsenic to soil from various sources may prove detrimental to plant as arsenic uptake would reach to the toxic limit, thereby facilitating its entry into the food chain and human body. The top soil contains arsenic within the range of 2.02 to 21.08 mg/kg. These crops (such as elephant-foot-yam, green gram, cowpea, sesame, groundnut, etc.) tended to show a build-up of arsenic in substantial quantities in different plant parts. A number of other vegetables, namely cauliflower, tomato, bitter gourd were also noted to accumulate arsenic in their economic produce. Zaman *et al.* (2013) stated that the distribution of arsenic content in plant parts generally followed the order: root > stem > leaf >economic produce.

4 CONCLUSIONS

Occurrence of arsenic in groundwater is sporadic and geogenic in nature throughout the district. A total population of 17,43,889 is residing in the risk zone of arsenic poisoning. At present, groundwater is used for irrigation through shallow tube wells and partly by deep tube wells. River lift irrigation (RLI), which is already in use (370 nos. RLI) may be enhanced to avoid arsenic rich groundwater from shallow aquifer for irrigation. In places, deep tube well groundwater may be used for irrigation in judicious manner. There are six semi-critical blocks in the district, which need special attention for future groundwater development and management. Crop planning in irrigated area is to be paid proper attention to implement location specific proven water saving technologies with conjunctive application of ground water and surface water. Summer season (*boro*) paddy cultivation demands huge groundwater. Low arsenic accumulating crops can safely be included in cropping sequence to reduce the arsenic entry into the food-chain. The reduction of poverty may also help to mitigate the arsenic problem by increasing the living standard (D. Halder *et al*, 2012).

ACKNOWLEDGEMENTS

The main author Dr. Anadi Gayen is thankful to the Regional Directors, RGNGWTRI, Raipur and the Chairman of Central Ground Water Board for according kind permission to participate and present this paper in the 6th International Congress: As2016.

REFERENCES

Biswas, A., Samal, A.C., Santra, S.C. 2012. Arsenic in relation to protein content of rice and vegetables. *Res. J. Agric. Sci.* 3(1): 80–83.
Chatterjee, D., Halder, D., Majumder, S., Biswas, A., Nath, B., Bhattacharya, P. et al. 2010. Assessment of arsenic exposure from groundwater and rice in Bengal Delta Region, West Bengal, India. *Water Res.* 44: 5803–5812.
Jean, J.-S. Bundschuh, J., Bhattacharya, P. (Eds.) 2010. CRC Press; 2010, p.197–199.
Zaman, A., Patra, S.K., Hedayetullah, M. 2013. Arsenic contamination in groundwater and suggestive measures for agricultural sustainability. *Proceedings of the Workshop of CGWB*, February 19, 2013, pp 1–2.

Arsenic Research and Global Sustainability – Bhattacharya, Vahter, Jarsjö, Kumpiene,
Ahmad, Sparrenbom, Jacks, Donselaar, Bundschuh & Naidu (Eds)
© *2016 Taylor & Francis Group, London, ISBN 978-1-138-02941-5*

Arsenic mitigation through Behavior Change Communication

C. Das[1], S.N. Dave[2], A. Bhattacharya[3] & V. John[4]
[1]*Hijli Inspiration, Kolkata, India*
[2]*Unicef, Kolkata, India*
[3]*Public Health Engineering Department, Government of West Bengal, Kolkata, India*
[4]*WSSO, Public Health Engineering Department, Government of West Bengal, Kolkata, India*

ABSTRACT: A "pilot for replication" approach was adopted in Sayestanagar II GP (conglomeration of villages), District North 24 Paraganas, West Bengal, India covering a population of 11000 served by public handpump tube-wells of which 58.8% (233) are arsenic contaminated (conc >50 mg/l). The methodological approach comprised a participatory investigation followed by a design of a basket of Behaviour Change Communication activities ranging from print, electronic and cultural media to interpersonal communication (BCC) and stakeholder convergence, targeting the mass and vulnerable groups. The uniqueness of the approach lay in rightly timing and spacing the activities with Administrative Initiatives related to red marking of arsenic contaminated unsafe public handpump tube-wells. The communication strategy had a dramatic impact at the community and household level in terms of (near 100%) shift to safe drinking sources, people's intent in getting private sources tested against payment of moderate amount, student involvement through Action Groups and mitigating arsenic through neighbourhood cooperation.

1 INTRODUCTION

In West Bengal, 83 out of 341 administrative blocks are arsenic (As) affected and about 20.5 million rural population is at risk of consuming high levels of As through drinking water. Baduria is one of the worst As affected blocks in North 24 Paraganas district in West Bengal (PHED, 2006).

The JPOA report (2006–07) and subsequent water testing in the area (PHED, 2006), shows high levels of As (0.01–3.7 mg/L) in spot sources in Sayestanagar II GP, Baduria block. Therefore, three severely As affected villages (Bakra Chandirboar, Gandharbapur and Piyara) in Sayestanagar II GP (Gram Panchayat), Baduria block were considered for the intervention for the "pilot for replication" project through a Behaviour Change Communication (BCC) based mitigation approach. The target area covers a population of 10,156 (Census, 2011) served by 396 public handpumped tubewells out of which 233 are As contaminated (concentration > 50 μg/L).

2 METHODS/EXPERIMENTAL

The current BCC approach is based on the Drinking Water Advocacy Communication Strategy (DWACS) and Framework 2013–2022 under the combined efforts of Ministry of Drinking Water and Sanitation, Government of India and the UNICEF (MDWS-GoI/UNICEF. 2013).

2.1 *Challenges*

Promoting behavioural change is always a challenging process, especially in rural areas, as there is little or no awareness about the chronic toxicity due to consumption of As through drinking water. Further immediate mitigation options are limited and constrained by the fact that it may mean giving up private and/or nearby As contaminated water sources for more distant but safe public sources.

2.2 *Work process*

WSSO–UNICEF–Hijli INSPIRATION (a partner NGO) has adopted systematic, yet conscious steps to deal with the As mitigation process. The steps are as follows:

- Reconnaissance and piloting
- Baseline survey
- Analysis–community needs assessment
- Facilitating administrative initiative
- Communication strategy design
- Communication strategy implementation
- Concurrent monitoring and assessment

The uniqueness of the approach lay in spacing and timing of communication activities with survey and the administrative initiative. The PHED started the awareness initiative by red marking with suitable message inscribed on unsafe tubewells to prevent people from collecting drinking (and cooking) water from such sources.

3 RESULTS AND DISCUSSION

3.1 *Community interventions*

Community meetings, Focused Group Discussions, Participatory Rural Appraisal (PRA) mapping, red marking of unsafe handpump tube-wells on a pilot scale (Fig. 1), health awareness checkup camp, stakeholder meetings were conducted before the baseline survey for reconnaissance and pilot testing.

3.2 *Baseline survey*

The baseline survey was conducted in 100% (n = 2300) households in 26 habitations in Baduria block. Respondents were asked questions on Knowledge, Attitude and Practice (KAP). The baseline survey provided an opportunity for interpersonal communication along with handing over of awareness flyer. PRA mapping led to a community consensus on neighbourhood specific mitigation. Hence the survey in itself was an awareness approach.

Baseline results reveal that respondents are aware of As (65%) but knowledge about the presence of As in their water source is limited (<20%). Respondents are aware of the health implications (especially skin disease) due to consumption of As contaminated water (>50%).

About 82% know that water quality can be tested to ensure the presence of As in water. It was also observed that women use either pond water or water from (untested) private source for cooking whereas drinking water was collected from public sources. The general belief that boiling removes all impurities including As is leading to this practice.

Water quality testing of private sources is not very popular (10%). But, when asked 80% said that they are willing to test their water sources and 76% even claimed to pay for the test. About 77% support the red painting initiative and said that it is useful for them to identify the unsafe source.

Figure 1. Red marking of arsenic contaminated handpump tubewell.

The baseline analysis led to the identification of education needs, design and implementation of communication strategy. Post baseline survey a series of BCC activities took place. 'Jal Bandhu' mascot with a tagline was used for a prompt recall. School action groups were formed and children were oriented on use of field testing kits. Ventriloquism was adopted as a means for educating primary school children linking it with WASH initiatives. Laminated flash cards on As awareness were distributed to grassroots health workers, for a close interpersonal interaction with the vulnerable population. Street plays as cultural medium that served as a platform for addressing social vulnerabilities. Wall paintings were done in strategic locations with As awareness message. Public transport, local mass media were also used for disseminating awareness messages. Health Awareness Camps conducted by eminent doctors were organized for check-up as well as raising awareness among the communities.

3.3 *Assessment of change in attitude and behaviour over the baseline situation*

After the awareness and orientation of villagers on As, within a three month time-span, change in behaviour perspectives were evident. Some of the positive changes are listed below:

- Shift to safe drinking water source: 97%
- Motivation to get private water source tested: 80% – 99%
- Willingness to pay for testing water source: 76% – 96%
- Student/Youth action groups formed for community monitoring: 46 teams for 46 sub neighbourhoods
- Willingness to help neighbor in mitigating As: 86% – 100%

4 CONCLUSIONS

Well designed and rightly spaced, BCC has an immense impact in As mitigation. It needs to be complemented by conditions of access to safe drinking water sources through infrastructural improvements. An aware community with an access to an alternative safe water source within an optimal distance will lead to a permanent behavior change towards a safe water practice in As contaminated areas. However the key caveat is the likelihood to use of water (primarily for cooking) from untested private individual sources which is weighed by convenience. Awareness and attitude is expected to lead to a practice related to testing of private sources against charges through individual initiative.

ACKNOWLEDGEMENTS

The Authors wish to thank all institutional stake-holders, Public Health Engineering Department, WSSO, District Administration, North 24 Para-ganas, UNICEF and Hijli Inspiration, the NGO staff involved in the research and the community, in particular for their valuable feedback, coopera-tion and support.

REFERENCES

Census. 2011. http://winweb2.wbphed.gov.in/phed_v2view/CVF00001/view.html?index=1&mN=2.

MDWS-GoI/UNICEF. 2013. Drinking Water Advocacy Communication Strategy (DWACS) and Framework 2013–2022. Ministry of Drinking Water and Sani-tation (MDWS), Government of India (GoI) and UNICEF.

PHED. 2006. *Results of tubewells tested for arsenic under UNICEF supported JPOA for arsenic mitigation.* JPOA Survey Report, Public Health Engineering Depart-ment, Government of West Bengal, Kolkata, India.

Arsenic Research and Global Sustainability – Bhattacharya, Vahter, Jarsjö, Kumpiene,
Ahmad, Sparrenbom, Jacks, Donselaar, Bundschuh & Naidu (Eds)
© 2016 Taylor & Francis Group, London, ISBN 978-1-138-02941-5

Arsenic contamination in groundwater of Indo-Gangetic plains and mitigation measures adopted in Bihar, India—lessons learnt

N.T. Santdasani[1] & S. Mojumdar[2]
[1]*UNICEF, Patna, India*
[2]*UNICEF, ICO, New Delhi, India*

ABSTRACT: Arsenic (As) contamination in ground water above permissible limits in Ganga-Brahmaputra and Meghna delta is now widely recognized as a threat to public health in India and Bangladesh. A significant proportion of 500 million people, constituting 30% of people living in the delta region, live in danger of drinking As contaminated water. The population vulnerable to arsenic poisoning are the economically challenged, impoverished villagers with low literacy levels, with its susceptibility increasing among children and infants. With every new survey, new areas and population suffering from Arsenicosis are being reported. Despite number of corrective and precautionary measures, the spread of As contamination in groundwater continues to grow and is a matter of concern in the region. The measures taken have appeared to be partial and inadequate. UNICEF is using the lessons learned through different studies for furthering the achievement of a sustainable and holistic solution to this problem.

1 INTRODUCTION

Arsenic (As) in groundwater has been reported, mostly in areas formed by recent alluvial sediments located in Holocene aquifers of the Ganga-Meghna-Brahmaputra plains, where concentration is exceeding the WHO drinking water guideline value of 10 µg/L (Bhattacharya *et al.*, 2011). A total of seven states -West Bengal, Jharkhand, Bihar, Uttar Pradesh in the flood plain of the Ganga River; Assam and Manipur in the flood plain of the Brahamaputra and Imphal rivers and Rajnandgaon village in Chhattisgarh state have been reported for As contamination in groundwater above the 10 µg/L (Saha & Sahu, 2015). Ironically, all the As affected river plains have the river route originating from the Himalayan region. However, over the years, the problem of As contamination in groundwater has been complicated, due to a large variability at both the local and regional scale, by a number of unknown factors (Mishra, 2008). Testing of groundwater used for drinking for As has been undertaken more widely by the state governments of seven affected states with support from UNICEF (Nickson *et al.*, 2007). Since 1983, with every new survey, new areas and people suffering from Arsenicosis are being reported.

2 METHODS

With the background of severity of As in West Bengal and the adjoining states and also keeping in view the 432 km stretch of river Ganga in the state, School of Environmental Studies, Jadavpur University in the year 2003, conducted a random testing of 206 groundwater samples in Bhojpur district of Bihar and found 57% samples with As concentration above 50 µg/L. Thereafter, Shriram Institute for Industrial Research, New Delhi, with the support of UNICEF, conducted random testing of groundwater samples in 11 districts including Bhojpur. But, the results were not so alarming, as envisaged. Later, the Public Health Engineering Department (PHED), Bihar, in collaboration with A.N. College, Patna and the support of UNICEF, conducted rapid testing in 11 districts of the State during the year 2005–06. The testing was done using field test kits (FTKs), followed by laboratory analysis of the samples exceeding 40 µg/L by UV spectrophotometer and Atomic Absorption Spectrophotometry (HG-AAS) (PHED, 2009).

3 RESULTS AND DISCUSSION

3.1 *Overall status of arsenic contamination in Bihar*

Based on the last survey, as on 1st April 2010, 13 districts covering 1590 habitations were declared As affected by PHED.

However, other agencies have claimed that the affected area is higher than what was reported by the PHED. UNICEF has advocated for the blanket testing of all sources in the affected areas so as to strategize mitigation measures, which is under progress and will be completed in January 2016.

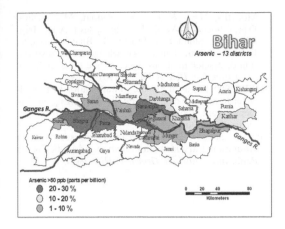

Figure 1. Map of the state of Bihar with the status of arsenic contamination in groundwater (PHED, 2009).

Table 1. Summary of the affected districts with As contaminated habitations in Bihar (as on 1st April, 2015).

S. No.	District e	As >50 µg/L Habitations/Pop. affected		As 10–50 µg/L Habitations
1	Begusarai	7	16193	49
2	Bhagalpur	2	271	149
3	Bhojpur	0	0	19
4	Buxar	15	6990	252
5	Darbhanga	4	36753	4
6	Katihar	0	0	23
7	Khagaria	0	0	246
8	Lakhisarai	8	2487	203
9	Munger	25	19666	118
10	Patna	3	4219	44
11	Samastipur	27	6851	129
12	Saran	4	6233	31
13	Vaishali	0	0	74
	Total:	95	99663	1341

3.2 Mitigation strategy adopted

In order to provide safe water in the affected habitations, PHED has adopted mitigation measures as below:

a) Short term/Midterm mitigation measures:
  • Dug wells converted into sanitary wells
  • Replacement of shallow TWs with IM-II HP
  • Construction of new sanitary wells
  • Provision of HP with As removal units
b) Long term mitigation measures:

  • Deep Hand Pumps from deeper aquifers
  • Mini WSS with treatment plant, solar based
  • Surface water based multi village PWS
  • Rain water harvesting

After undertaking the above mitigation measures, out of the affected 1590 habitations, only 95 habitations (> 50 µg/L) remains to be covered as on 01.04.2015.

However sustainability of these measures remain a cause for concern. Additionally, there are 1341 more habitations in 13 arsenic affected districts in the State, where the As content in groundwater has been detected to be between 10 to 50 ppb. However, an action plan has been formulated to mitigate arsenic contamination issues in both the 95 as well as 1341 habitations with lesser contamination.

3.3 Lessons learnt

In spite of the execution of short-term, medium term and long term mitigation measures at crores of rupees, people of the affected area are often still exposed to risk, due to non-adherence to the stipulated performance standards and lack of involvement of the community in such measures.

UNICEF conducted a randomized study on performance of water treatment plants in Bihar in 2010, which painted a grim picture of the intended outcome of access to safe water for the community. It advocated for reconstitution of Water Quality Task Force in Bihar to strategize mitigation measures & monitor the intended outcomes. UNICEF also supports PHED in finalizing the State Water Quality Mitigation Strategy and in formulating guidelines for ensuring community involvement in O&M of Rural Water Supply Schemes (RWSS) in the state. It has also supported a research proposal for resolving the issue, through smart small scale Piped Water Supply (PWS) systems with consortium of academic institutions from Netherlands, India and Bangladesh.

4 CONCLUSIONS

The present state of affairs of Arsenic menace in the region demands a systematic translation of success stories of one place to another, and overcoming shortfalls by conceiving R&D studies in the area deemed fit.

ACKNOWLEDGEMENTS

Authors are thankful to Dr. Ashok Ghosh, AN College, Patna & Francis Odhiambo, WASH Specialist, UNICEF, ICO, Delhi for their guidance, Mr. S.N Mishra, WQ Director & Mr. D. S Mishra, Former WQ Director, PHED for their inputs and Ms Suzanne Joan Coates, Chief, WASH, ICO, UNICEF & Dr. Yameen Mazumder, CFO,BFO,UNICEF for all the support.

REFERENCES

Bhattacharya, P., Mukherjee, A., Mukherjee, A.B. 2011. Ar-senic in Groundwater of India. In: J.O. Nriagu (ed.) *Encyc-lopedia of Environmental Health*, Vol. 1, Burlington: Elsevier, pp. 150–164.

Mishra, D.S. 2008. As contamination of groundwater in Mid Ganga Basin in the state of Bihar, India. Arsenic: From Nature to Humans. *2nd International Congress on Arsenic in the Environment*. May 21–23, 2008. Valencia, Spain.

Nickson, R., Sengupta, C., Mitra, P., Dave, S.N., Banerjee, A.K., Bhattacharya, A., Basu, S., Kakoti, N.,

Moorthy, N.S., Wasuja, M., Kumar, M., Mishra, D.S., Ghosh, A., Vaish, D.P., Srivastava, A.K., Tripathi, R.M., Singh, S.N., Prasad, R., Bhattacharya, S., Deverill, P. 2007. Current knowledge on the distribution of arsenic in groundwater in five states of India. *J. Environ. Sci. Heal A* 42(12): 1707–1718.

PHED. 2009. Status of arsenic contamination in groundwater in he state of Bihar. Public Health Engineering Department, Govt of Bihar, Patna. http://phed.bih. nic.in/ WaterQuality.htm

Saha, D., Sahu, S. 2015. A decade of investigations on groundwater arsenic contamination in Middle Ganga Plain, India. *Environ. Geochem. Hlth* 38(2): 315–337.

Arsenic Research and Global Sustainability – Bhattacharya, Vahter, Jarsjö, Kumpiene,
Ahmad, Sparrenbom, Jacks, Donselaar, Bundschuh & Naidu (Eds)
© 2016 Taylor & Francis Group, London, ISBN 978-1-138-02941-5

Geogenic arsenic contamination in the groundwater of Bhagalpur district in Bihar, India: A challenge for sustainable development

N. Bose[1], S. Chaudhary[2] & A.K. Ghosh[1]

[1]*Department of Environment and Water Management, A.N. College, Patna, India*
[2]*Tilka Manjhi Bhagalpur University, Bhagalpur, India*

ABSTRACT: Ground water quality in Ganga floodplains of Bhagalpur has registered high levels of arsenic. Sources of arsenic ingestion by humans include contaminated drinking and irrigation ground waters. As health issues emerge, arsenic mitigation strategies have not obtained the desired result in the area. This study undertakes a cohesive view of the arsenic problem in the Ganga floodplains of Bhagalpur as a precursor to preparation of a sustainable arsenic mitigation plan based upon participatory approach against the backdrop of the socio-economic milieu of the study area. Methods of study included determination of arsenic levels in pumped ground water, assessment of mitigation techniques, and interviews of affected population to derive a suitable model of implementation of mitigation. Mapping of contamination sources has been done to facilitate monitoring, surveillance and further research.

1 INTRODUCTION

Bhagalpur is the eastern-most district of Bihar state in India having an area of 2570 km², and forming the eastern boundary of the Middle Ganga Plains. It is crossed by the west-to-east flowing Ganga River, whose braided pattern forms a wide flood plain of recent alluvium.

Preliminary studies indicate that the geological and geomorphological environments promoting the leaching of Arsenic (As) into its groundwater is distinct from those of the adjacent As contaminated Bengal Delta Plains. An assessment of a geological transect in this region revealed that this geogenic As is present in the young alluvial deposits of the flood plain, 80% of shallow hand pump water samples testing positive for As(III) (Mukherjee *et al.*, 2012). These findings, therefore, necessitated a cohesive assessment of the groundwater As contamination scenario within a 5 km belt along the southern bank of the Ganga River in Bhagalpur flood plains. This belt has a predominantly rural agricultural population, already facing several economic barriers in the form of limited infrastructural development and employment opportunities. The purpose of this study was to assess the extent of groundwater As contamination in the Bhagalpur flood plains, along the southern bank of the river Ganga, to aid in the implementation of appropriate mitigation mechanisms under the existing socio-economic systems.

2 METHODS

The methodology thus involved collection of 8128 hand pump water samples and 36 shallow irrigation bore well water samples, and collection of seven plant samples from standing edible crops. Of the to-tal 8128 hand pump water samples, 914 samples were located in Bhagalpur Municipal Corporation (urban) zone. Spectrophotometric analyses of all the samples as per APHA protocol were done for determining the presence and quantum of As (Bose *et al.*, 2015). Field interviews were undertaken to assess the socio-economic status and responses to the problems arising from As ingestion. Finally, an assessment of the functioning mitigation techniques was undertaken.

3 RESULTS AND DISCUSSION

3.1 *Arsenic in groundwater*

Out of 184 villages in the study belt, 93 were identified with As contaminated drinking water sources. 429 water samples had As in the range of 10 μg/L – 50 μg/L, while 846 samples had > 50 μg/L of As (Table 1). pH values of water samples at source were between 6.8 to 7.5. All the samples had high iron content. Highest concentration of As was found in the wells of Nathnagar (608 μg/L), followed by Kahalgaon (608 μg/L), Pirpainti (479 μg/L), Sabour (311 μg/L), and Sultanganj (212 μg/L). The number of affected hand pumps varied considerably between from one village to another (Table 1).

Despite its close proximity to the river Ganga, this area is heavily dependent on groundwater for irrigation. Analysis of groundwater samples from 36 random irrigation bore wells (except from Sabour) revealed moderate to very high As levels (Table 2), the trend being similar to that of the

Table 1. Distribution of arsenic in drinking water hand pumps, in Ganga flood plains, Bhagalpur.

Adminstrative Blocks	Villages with As. contaminated hand pumps	As concentration (µg/L)		Highest As level (µg/L)
		<50 µg/L	>50 µg/L	
Sultanganj	11	49	8	212
Nathnagar	25	156	477	608
Sabour	10	71	45	311
Kahalgaon	25	81	165	608
Pirpainti	22	72	151	479
Total (n)	93	429	846	–

Table 2. Arsenic contaminated irrigation bore well sources, Ganga flood plains, Bhagalpur.

Adminstrative Blocks	As in irrigation bore well sources	As concentration in irrigation bore well water		Highest As level (µg/L)
		<50 µg/L	>50 µg/L	
Sultanganj	06	–	06	117
Nathnagar	10	–	10	431
Sabour	–	–	45	311
Kahalgaon	10	81	165	608
Pirpainti	10	72	151	479
Total (n)	36	153	377	–

drinking water samples where Pirpainti, Kahalgaon and Nathnagar had the highest frequencies and levels of As concentration. These have been instrumental in transfer of As into the food chain, particularly in standing maize, rice, sugarcane and leafy vegetables grown in the area (Bose et al., 2015). Typical depths of both the hand pumps and bore well pumps are from 12–15 m.

Areas of arsenic ingestion through the foodchain was difficult to assess as agriculture produce were being marketed locally and across the states. It was found that sugarcane juice from which jaggery was being made had arsenic levels of 52 µg/L.

For the first time, arsenic contaminated surface water source was detected in a north bank inlet of the river Ganga in Bhagalpur, known as the Jamunia Chharan/Dhar. This water is used for irrigating the adjacent fields in the dry season. Repetitive tests of the Jamunia water samples from different spots revealed the contamination levels up to 66 µg/L. Further investigation warrants this finding as it contradicts the general premise that surface water sources are free from arsenic in the Bihar plains.

3.2 Affected population

The 93 impacted villages have a population of 346,633 persons. As annual flooding forces local migratory trends, this population redistributes itself within the fertile floodplains. However, vulnerability to mass arsenic poisoning persists in this area as levels of arsenic spread to newer hand pumps and ground water irrigation sources.

3.3 Status of arsenic mitigation in Bhagalpur

Consumption of arsenic contaminated ground water were already impacting upon community health. Villagers accessing contaminated pump water had skin diseases, rising cases of cancer and other physical ailments that were limiting their employment options, and increasing poverty levels.

Responses to gradual mass poisoning from geogenic arsenic contamination in Bihar is greatly dependent on community awareness and demand for clean water, along with the existence, adequacies, and implementation of water-related policies. Mitigation strategies implemented in the area includes sealing of identified high arsenic contaminated hand pumps, occasional installation of deeper hand pumps for clean water, open wells and filtration techniques based on adsorption (Brouns et al., 2013). However, operational and maintenance issues, social segregation along caste lines, lack of awareness of the arsenic problem, and financial viability have compromised the sustainability of mitigation initiatives undertaken so far. A severe environmental problem has emanated from lack of proper sludge disposal from the filters, the sludge being either drained into agricultural fields or into unlined subsurface tanks.

4 CONCLUSIONS

This study has identified lack of ownership and community participation as the most important contributory factors in accessing clean water by the affected population. Studies are underway to develop a participatory arsenic mitigation model that, once set up, have prospects of sustained clean water supply to the arsenic affected community in the study area.

REFERENCES

Bose, N., Ghosh, A.K., Kumar, R., Singh, A. 2015. Impact of arsenic contaminated irrigation water on some edible crops in the fluvial plains of Bihar. In A. Ramanathan, S. Johnston, A. Mukherjee, B. Nath (eds.), *Safe and Sustainable Use of Arsenic-Contaminated Aquifers in the Gangetic Plain: A Multidisciplinary Approach*. Springer. ISBN: 978-3-319-16124-2

Brouns, M., Janssen, M. Wong, A. 2013. Dealing with arsenic in rural Bihar, India: evaluating the successes and failures and providing a long-term mitigation strategy. Dept. of TPM, Delft Institute of Technology, The Netherlands. (Access: http://www.indiawaterportal.org/articles/dealing-arsenic-rural-bihar-evaluating-successes-and-failures-mitigation-projects)

Mukherjee, A., Scanlon, B.R., Fryar, A.E. Saha, D., Ghosh, A., Choudhuri, S., Mishra, R. 2012. Solute chemistry and arsenic fate in aquifers between the Himalayan foothills and Indian craton (including central Gangetic plain): Influence of geology and geomorphology. *Geochim. Cosmochim. Acta* 90: 283–302.

Arsenic Research and Global Sustainability – Bhattacharya, Vahter, Jarsjö, Kumpiene,
Ahmad, Sparrenbom, Jacks, Donselaar, Bundschuh & Naidu (Eds)
© 2016 Taylor & Francis Group, London, ISBN 978-1-138-02941-5

Mitigating arsenic health impacts through Community Arsenic Mitigation Project (CAMP)

S. Bhatia[1], K. Sheshan[2], E. Prasad[3] & V. Kumar[4]
[1]Disaster Management, Tata Institute of Social Sciences, Mumbai, India
[2]Arghyam, Bengaluru, India
[3]Megh Pyne Abhiyan, Patna, India
[4]Water Action, West Champaran, India

ABSTRACT: The area Khap Tola is a remote, backward and highly flood-prone area in North Bihar. Due to lack of awareness, scientific know-how, public health facilities and district administrations negligence, people are forced to drink contaminated water in the absence of any other clean drinking water source. The area has high arsenic contamination, maximum value being 397 ppb. A pilot was done to provide clean drinking water in the village. A systems approach was followed looking at the processes, starting from sample testing, making the community aware of the health impacts, community health camp in the village and finally exploring the possibility of dug wells as potential source to get arsenic-free drinking water. The interventions done will be discussed in the paper using a social exclusion model and the dimensions for arsenic mitigation. Further, the need for developing a sustainable model for mitigation by community involvement and their capacity building both in terms of local knowledge as well as health issues will be discussed.

1 INTRODUCTION

High Arsenic (As) contamination was reported in Khap Tola, West Champaran, Bihar, India (Bhatia et al., 2014). The maximum value reported was 397 ppb and it was observed that 57% of the samples from private hand-pumps in the shallow aquifer zone of 15–35 m have As greater than 200 ppb. (Bhatia et al., 2014). Further using USEPA guidelines, it was calculated that children age group 5–10 years are under high risk of getting cancer. The Hazard Quotient calculated for 21 children taken for study, indicated that children may have adverse non-carcinogenic health impacts, in the future, with continued exposure. Based on the results of sample testing and analysis, village meetings and awareness was done. With the danger that As poses to the health of communities living in these areas, it was essential to conduct a health impact assessment for symptoms of As poisoning. A total of 18 suspected cases of As poisoning were identified in Khap Tola. Finally, with the water quality data, two dug wells were revived for As free water by the villagers. These are being managed by the people.

2 METHODS

Water samples from all 20 hand-pumps in the Khap Tola were tested for As. Standard water testing methodology was followed and the samples

Figure 1. Systems approach explaining the interventions.

were acidified with two drops of HCl to maintain a pH < 2. The samples were tested in the laboratories of the Department of Environment and Water Management, A.N. College, Patna using Atomic Absorption Spectrophotometer (AAS).

The approach for sustainable mitigation measures was based on the concept of systems approach for community based sustainable mitigation measures. The approach is shown below in the Figure 1.

3 RESULTS AND DISCUSSION

3.1 Detection of arsenic in ground water

High As contamination with maximum value of 397 µg/L (Fig. 2) was reported in Khap Tola Village (Bhatia et al., 2015).

3.2 Community awareness

Based on the results of sample testing and analysis, village meetings and awareness was done. A movie was screened to showcase the health impacts of As in drinking water.

3.3 Health assessment camp

With the danger that As poses to the health of communities living in these areas, it was essential to conduct a health impact assessment for symptoms of As poisoning. Prof. (Dr.) Kunal Kanti Mazumder, a renowned arsenicosis expert from KPC Medical College and Hospital, Kolkata. A total of 18 suspected cases of As poisoning were identified in Khap Tola (Fig. 3).

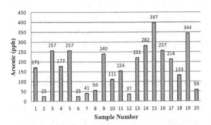

Figure 2. High arsenic contamination in Khap Tola, Bihar.

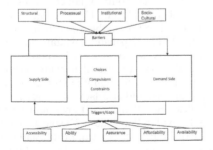

Figure 3. Suspected case of arsenicosis and mild keratosis, drinking water from hand-pump with As concentration of 200 μg/L.

Figure 4. Social exclusion framework explaining the interventions.

3.4 Exploring well revival for clean drinking water

Based on the sample testing and analysis, the dug wells were found to be a potential source of As-free drinking water. This included involving the community and working through the available technical and local community knowledge.

The interventions were done using a social exclusion framework (see Fig. 4) to see the demand and supply side and dimensions for sustainable community based mitigation measures. This will include review of the existing mitigation concepts like SASMIT, Project Well, etc.

4 CONCLUSIONS

Based on the interventions, it was observed that it is very important to involve people in the process of As mitigation to have sustainable solutions. Also, we noted that a systems approach for understanding the As impact to a community is important. Further, for proposing mitigation measures, a social exclusion model needs to be developed, which includes technical as well as locally based solutions for As mitigation. Further action research needs to be done for making As mitigation measures sustainable.

ACKNOWLEDGEMENTS

We would like to thank all the staff of Megh Pyne Abhiyan and Water Action. We would like to thank Dr Ashok Ghosh for providing us the platform to do our project. We thank Pravah and PACS for supporting us financially through changelooms with.in fellowship.

REFERENCES

Bhatia, S., Balamurugan, G., Baranwal, A. 2014. High arsenic contamination in drinking water hand-pumps in Khap Tola, West Champaran, Bihar, India. *Front. Environ. Sci.* 2: 49.

Bhattacharya, P., Jacks, G., Ahmed, K.M., Hasan, M.A., von Brömssen, M. 2014. Sustainable Arsenic Mitigation (SASMIT) – Community driven initiatives to target arsenic safe groundwater as sustainable mitigation strategy DOI: 10.13140/2.1.1290.3200.

Guha Mazumder, D.N.G. (2008). Chronic arsenic toxicity & human health. *Ind. J. Med. Res.* 128: 436–447.

Singh, S. K., Ghosh, A.K. 2012. Health risk assessment due to groundwater arsenic contamination: children are at high risk. *Hum. Ecol. Risk Assess. Int. J.* 18: 751–766.

Smith, A. H., Lingas, E. O., Rahman, M. (2000). Contamination of drinking-water by arsenic in Bangladesh: a public health emergency. *Bull. World Health Organ.* 78: 1093–1103.

U.S. EPA. (2007). Inorganic Arsenic; Chemical Summary. Vol. 23. Chicago: The U.S. EPA Region 5 Toxicity and Exposure Assessment for Children's Health.

Arsenic Research and Global Sustainability – Bhattacharya, Vahter, Jarsjö, Kumpiene,
Ahmad, Sparrenbom, Jacks, Donselaar, Bundschuh & Naidu (Eds)
© 2016 Taylor & Francis Group, London, ISBN 978-1-138-02941-5

Network approach to the wicked problems of arsenic—Arsenic Knowledge and Action Network

S. Fanaian[1] & A. Biswas[2]

[1]*SaciWATERs, Hyderabad, India*
[2]*AKVO, New Delhi, India*

ABSTRACT: Water quality problems can rightly be called as wicked problems. Ground water quality is a rising concern and there seems to be no one solution. Furthermore, within the spectrum of water quality, arsenic becomes additionally challenging, as it is not visible to the naked eye and has no taste or smell to mention off in drinking water. Compounding this is the fact that it is poisonous in such small quantities that it is difficult to measure on a large scale. Resolving this wicked problem requires multi-pronged effort from different disciplines. This paper strives to understand water quality problems from a wicked problem perspective. With the coping mechanism of collaborative strategy we analyze the efforts of Arsenic Knowledge and Action Network.

1 INTRODUCTION

Wicked problems have been a tag line for many issues and mostly linked to those tricky situations/issues that have confounded solutions. They appear in various fields ranging from environmental problems to urban planning to business management. Within that spectrum and from the criteria's defined by Rittel & Webber (1973), water quality problems fall into the purview of wicked problems.

Looking at its explanation and criteria, one of its inherent characteristic is that wicked problems have no defined solutions and their circular nature usually brings new problems the moment we think we have a solution. For example, to avoid consumption of water that had microbial contamination, there were massive concerted and well-meaning efforts to shift usage pattern from surface water to tube-wells so as to tap "pure" groundwater. In many parts of Bangladesh and India, this large-scale effort resulted in exclusive usage of groundwater for drinking water.

Resolving one problem, what was later found out was that more than 30% of tubewells in Bangladesh and many million in India contained arsenic levels exceeding the Bangladeshi drinking water standard (50 µg/L) as well as the recommended guideline value of WHO (10 µg/L) leading to several other health complications. As seen here, solving one problem led to many new ones (Khan *et al.*, 2009, Chakraborti *et al.*, 2009, Saha & Sahu, 2015).

2 APPORACHES

2.1 *Wicked problem—solution options*

One easy approach is to designate these wicked problems as insurmountable and give up on them altogether. However, problems linked to water quality are serious and their implications affect millions of lives daily. Resolving this wicked problem is convoluted and tricky, nevertheless essential. There are three strategies used to tackle wicked problems (Roberts, 2000).

1. Authoritative strategies
2. Competitive strategies
3. Collaborative strategies

2.2 *Arsenic problem—network solution*

All the strategies designed and used to resolve wicked problems have their share of flaws and difficulties. One strategy that has been known to be 'least bad' way of 'making wicked problems governable" is that of collaboration or through open networks (Roberts, 2000, Bandelli *et al.*, 2009, Ferlie & Fitzgerald, 2011).

Bangladesh was the one of the earliest country to acknowledge the problem of arsenic in water and its severity. The spillover from Bangladesh shed light for testing and mapping arsenic contamination in other South Asian countries such as India, Nepal and Pakistan. However, arsenic as an element is fairly recent in the water quality vocabulary within India, Nepal and Pakistan, dating back to over 25 years (Mukherjee *et al.*, 2006).

Due to the complexity in identifying arsenic and conveying its threats to communities there is general gap in awareness and also lack of policy intervention. Policies even if they exist are country specific such as that within Bangladesh. Despite much stellar work carried out there is a gap in translating these to large-scale action and sustained mitigation.

To move in the direction of mitigating this problem and create a larger knowledge system for effective development and implementation of solutions, the

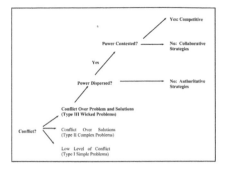

Figure 1. Coping strategies to deal with wicked problems (Roberts, 2000).

idea emerged for collaborative network approach namely: Arsenic Knowledge and Action Network.

3 RESULTS AND DISCUSSION

3.1 Arsenic network—initiation

The arsenic network from the beginning had been provided with a flexible structure within the framework to enable collaborative approach for resolving arsenic problems. The Arsenic Network has a secretariat that coordinates efforts with the guidance of a broad spectrum of experts depending on the need that emerges from regions.

The purpose of Arsenic Network at the foremost was to bring attention and knowledge collation on arsenic issues. It is also meant to assist in sharing knowledge; build capacity, further advocacy, and creating spaces for interdisciplinary and multisectoral joint consultation and action.

3.2 Experiences of networks towards wicked problem

Knowledge collation and sharing: Arsenic Network has succeeded in bringing together a comprehensive website and regular newsletters that share ongoing efforts from the ground and from the lab. These include interactive maps that showcase the data that is shared and overlaps it with governmental data.

Joint consultation: Spaces have been created through consultation meeting and workshops for joint deliberation among various sectors to recognize existing gaps, and define actions to be taken along with mechanism to share responsibilities. This effort has successfully occurred within three of the six affected states in India.

Building capacity: Recognizing the multi sectorial nature, there were also gaps identified in capacities of different sectors. Taking this ahead there have been three trainings held for around 120 doctors on the health aspect of arsenic. Further there

has been awareness program held for Civil Society Organization on mitigation options.

Advocacy: Taking the concerns of the grassroots and issues to the policy level has been another mandate of the network. The Arsenic Network through its advocacy effort has been able to interact and inform Members of Parliament within India, the issues of Arsenic as a serious health risk within India was also raised within the parliament discussion as a result. At the state level there have also been interaction with multiple level of government bureaucracies.

These are some of the interventions that are being taken up by the Network at this stage to address arsenic mitigation

4 CONCLUSIONS

The Arsenic Knowledge and Action Network at its inception in 2014started with 30 odd members and today has a growing number of around 400 individual, governmental, non-governmental and private institutions. The wicked problem of arsenic requires everyone's attention, it requires coming together of multiple sectors to address its multi-pronged challenge. These efforts have now resulted in new pilots that take into consideration previous learning and include more robust framework for mitigation.

ACKNOWLEDGEMENTS

We would like to thank Arghyam in providing the seed fund to support the secretariat and all the multiple partners that support its various activities.

REFERENCES

Bandelli, A., Konijn, E., Willems, J. 2009. The need for public participation in the governance of science centers. *Museum Management and Curatorship* 24(2): 89–104.
Conklin, J. 2006. *Dialogue Mapping. Building Shared Understanding of Wicked Problems.* West Sussex, England: John Wiley & Sons.
Ferlie, E., Fitzgerald, L., McGivern, G., Dopson, S., Bennett, C. 2011. Public policy networks and 'wicked problems': a nascent solution? *Public Administration* 89(2): 307–324.
Khan, N.I., Owens, G., Bruce, D., Naidu, R. 2009. An effective dietary survey framework for the assessment of total dietary arsenic intake in Bangladesh: Part-A—FFQ design. *Environ. Geochem. Heal.* 31(1): 207–220.
Mukherjee, A., Sengupta, M.K., Hossain, M.A., Ahamed, S., Das, B., Nayak, B. Lodh, D., Rahman, M.M., Chakraborti, D. 2006. Arsenic contamination in groundwater: a global perspective with emphasis on the Asian scenario. *J. Health Popul. Nutr.* 24(2): 142–163.
Rittel, H.W., Webber, M.M. 1973. Dilemmas in a general theory of planning. *Policy Sci.* 4(2): 155–169.
Roberts, N. 2000. Wicked problems and network approaches to resolution. *International Public Management Review* 1(1): Online Journal at http://www.ipmr.net.

5.4   Arsenic in drinking water and the Agenda 2030
Sustainable Development Goals

Arsenic Research and Global Sustainability – Bhattacharya, Vahter, Jarsjö, Kumpiene,
Ahmad, Sparrenbom, Jacks, Donselaar, Bundschuh & Naidu (Eds)
© 2016 Taylor & Francis Group, London, ISBN 978-1-138-02941-5

Mitigation of the impact of groundwater arsenic on human health and rice yield in Bangladesh: Solutions for overcoming the current stagnation

A. van Geen[1] & K.M. Ahmed[2]
[1]Lamont-Doherty Earth Observatory of Columbia University, Palisades, New York, USA
[2]Department of Geology, University of Dhaka, Dhaka, Bangladesh

ABSTRACT: Widespread groundwater contamination with arsenic in Bangladesh was recognized 15 years ago but over 40 million villagers remain exposed today. Over the same period, irrefutable evidence of the health impacts and economic costs of exposure has accumulated. A reinvigorated mitigation effort requires recognition at the national level of the need for (1) a permanent network of reliable arsenic testers responding to any household owning a well, and (2) the installation of many more low-arsenic public wells, with their siting more strongly linked to need and public access. Field experiments comparing different ways of meeting these objectives at scale are urgently needed. The reduction in rice yield caused by the accumulation in paddy soil of arsenic contained in irrigation water is an emerging issue. The benefit to farmers of mapping affected portions of their fields for soil removal using an arsenic kit could help sustain a network of well testers.

1 TIMELINE

It is disturbing to enter almost any village of the Bengal basin today and find that groundwater drawn from untested shallow wells continues to be used routinely for drinking and cooking, given that the arsenic problem was already recognized in the mid-1980s in West Bengal and the mid-1990s in Bangladesh. Naturally elevated arsenic levels have since been documented in groundwater across the world, but Bangladesh remains the most affected single country (Ravenscroft et al., 2009). A first national survey completed in 2000 showed that a population close to 57 million was exposed to arsenic levels exceeding the WHO guideline for drinking water of 10 µg/L by up to two orders of magnitude. Two representative drinking-water surveys conducted since have indicated a decline in the exposed population to 52 million in 2009 and 40 million in 2013, which is still an enormous number (BBS/UNICEF, 2015). Relative to the Bangladesh standard of 50 µg/L, the corresponding decline in the exposed population over the same period has been from an initial 35 to 22 and 20 million, respectively. After some initial successes, arsenic mitigation has clearly stagnated.

2 HEALTH AND ECONOMIC IMPACTS

While the decline in arsenic exposure in Bangladesh over the past 15 years has been disappointing, indications of the significant health implications have only become stronger and now include mortality due to cardio-vascular disease and cancers of the lung, liver, and bladder in adults, as well as diminished intellectual and motor function in children. Flanagan et al. (2012) associate an arsenic-related mortality rate of 1 in every 18 adults over the next 20 years with US$13 billion in economic losses, to which Pitt et al. (2015) add an estimated 9% reduction in household income associated with each exposed income earner across his or her lifetime. Whereas considerable attention has been paid to the transfer of arsenic contained in irrigation water to rice, the considerable loss in yield that results from the toxicity of arsenic to rice plants may be more significant, at least from an economic perspective (Panaullah et al. 2009). Rice production in Bangladesh today, which still accounts for about 5% of GDP, may be as much as 20% lower than it would have been without the accumulation of arsenic in paddy soil over the past several decades.

3 REASONS FOR CONTINUED EXPOSURE

Perhaps the leading reason for continued exposure is that shallow wells continue to be installed by individual households throughout Bangladesh, either as additional wells or to replace failed wells, and that these wells remain largely untested (Fig. 1). Most rural households in Bangladesh have nowhere to turn if they install a new well and want to have it tested. The growing pool of untested wells may already have reversed the reduction in exposure gained from wells-switching following the testing of almost 5 million wells throughout the country over a decade ago (Ahmed et al., 2006).

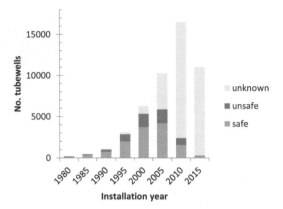

Figure 1. Histogram of when 48,790 wells in Araihazar upazilla tested with a field kit in 2012–13 were installed (adapted from van Geen et al., 2014). Note that the last bin doesn't span an entire 5 years. Shown in color is the corresponding status of wells reported by well owners or users. BAMWSP workers last swept through the area to test all wells in 2003.

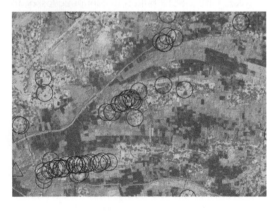

Figure 2. Close-up of surveyed portion of Araihazar upazila. The satellite image converted to grey-scale shows a mosaic of rectangular paddy fields separating inhabited areas where the wells are concentrated. Light blue dots indicate wells that meet the WHO guideline of 10 µg/L and red dots wells that do not meet the Bangladesh standard of 50 µg/L, respectively (wells in the intermediate category are shown in green). The larger circles 200 m in diameter are centered on the subset of wells >90 m deep, most of which are government-installed community wells that are low in arsenic. From van Geen et al. (2015).

The second leading reason for continued exposure is that a particularly effective form of mitigation, the installation of several hundred thousand deep community well, has not been optimized. At the national level, many of the community wells were allocated to areas where the need was not necessarily the greatest (JICA/DPHE, 2009). At the local levels too, a recent survey conducted in Araihazar has revealed that community well installations could have had a much bigger impact if their locations had been selected more equitably. Possibly through political connections, some villages have received an abundance of community wells whereas other equally affected neighboring villages still don't have any (Fig. 2). There is no reason to believe that the allocations took place differently in other parts of the country.

4 SOLUTIONS

If lack of testing and the inequitable allocation of community wells are indeed the two main obstacles to further reductions in exposure in rural Bangladesh, this attribution also point to solutions. A network of arsenic testers relying on field kits that reaches into the villages urgently needs to be established at the national scale. By requiring certified testers to enter each geo-referenced data point on a smartphone, quality control could be maintained through random checks. Data entry on smartphone phones would also facilitate centralized payment according to the number of wells tested, as a financial incentive for seeking out untested wells. Farmers might also be willing to pay considerably more for use of the same kit to identify the most contaminated portions of their paddy field. This would allow them to remove the most contaminated soil from their field and increase their rice yield. With respect to the many more community wells that are needed throughout the country, a requirement to post publicly post past and future allocations at the local government (union) levels in the form of legible maps might generate the grassroots pressure needed to allocate wells more equitably.

ACKNOWLEDGMENTS

Our work was supported by NIH grant P42 ES010349, NSF grant ICER 1414131, IGC grants 13854 and 14347, and several internal grants from the Earth Institute, Columbia University. We thank the many colleagues and students who contributed to this effort over the years.

REFERENCES

Ahmed, M.F., Ahuja, S., Alauddin, M., Hug, S.J., Lloyd, J.R., Pfaff, A., Pichler, T., Saltikov, C., Stute, M., van Geen, A. 2006. Ensuring safe drinking water in Bangladesh. *Science* 314, 1687–1688.

BBS/UNICEF. 2015. Bangladesh Multiple Indicator Cluster Survey 2012–2013, Progotir Pathey: Final Report. Bangladesh Bureau of Statistics (BBS) and UNICEF.

DPHE/JICA. 2009 Situation analysis of arsenic mitigation. Dhaka, Bangladesh: Department of Public Health Engineering, Government of Bangladesh, and Japan International Cooperation Agency.

Flanagan, S.V., Johnston, R.B., Zheng, Y. 2012. Arsenic in tube well water in Bangladesh: health and economic impacts and implications for arsenic mitigation. *Bull. World Health Org.* 90: 839–846.

Panaullah, G.M., Alam, T., Hossain, M.B., Loeppert, R.H., Lauren, J.G., Meisner, C.A., Ahmed, Z.A., Duxbury, J.M. 2008. Arsenic toxicity to rice (*Oryza sativa L.*) in Bangladesh. *Plant Soil* 317: 31–39.

Pitt, M.M., Rosenzweig, M.R., Hassan, N. 2015. Identifying the cost of a public health success: Arsenic well water contamination and productivity in Bangladesh. *NBER Working Paper No. 21741.*

Ravenscroft, P., Brammer, H., Richards, K. 2009 *Arsenic Pollution: A Global Synthesis.* RGS-IBG Book Series, Wiley-Blackwell, Chichester, UK.

van Geen, A., Ahmed, E.B., Pitcher, L., Mey, J.L., Ahsan, H., Graziano, J.H., Ahmed, K.M. 2014. Comparison of two blanket surveys of arsenic in tubewells conducted 12 years apart in a 25 km² area of Bangladesh. *Sci. Total Environ.* 488–489.

van Geen, A., Ahmed, K.M., Ahmed, E.B., Choudhury, I., Mozumder, M.R., Bostick, B.C., Mailloux, B.J. 2015. Inequitable allocation of deep community wells for reducing arsenic exposure in Bangladesh. *J. Water, Sanitation and Hygiene for Development,* (in press).

*Arsenic Research and Global Sustainability – Bhattacharya, Vahter, Jarsjö, Kumpiene,
Ahmad, Sparrenbom, Jacks, Donselaar, Bundschuh & Naidu (Eds)*
© 2016 Taylor & Francis Group, London, ISBN 978-1-138-02941-5

Delineating sustainable low-arsenic drinking water sources in South Asia

A. Mukherjee[1], P. Bhattacharya[2,3], M. von Brömssen[4] & G. Jacks[2]

[1]*Department of Geology and Geophysics, Indian Institute of Technology (IIT), Kharagpur, Kharagpur, India*
[2]*KTH-International Groundwater Arsenic Research Group, Department of Sustainable Development,
Environmental Science and Engineering, KTH Royal Institute of Technology, Stockholm, Sweden*
[3]*International Center for Applied Climate Science, University of Southern Queensland, Toowoomba,
Queensland, Australia*
[4]*Department of Water Resources, Ramböll Water, Ramböll Sweden AB, Stockholm, Sweden*

ABSTRACT: Access to safe and sustainable drinking water is a basic human. In present times, when much of the world is reeling through severe groundwater availability stress, large parts of South Asia that hosts the aquifers of the three of the largest global river systems are bountiful with groundwater. However, much of this groundwater is enriched with carcinogenic arsenic. The present study aims to rapidly delineating sustainable safe drinking water sources in regionally arsenic-unsafe groundwater areas of Bangladesh and West Bengal. The study validates the correlation between aquifer sediment colours obtained from local drillers and quantified groundwater chemical composition, characterize aqueous and solid phase geochemistry and dynamics of As mobility and to assess the risk for cross-contamination of As between aquifers in study areas. The optimistic outcome of the study provides an unique opportunity for the local drillers in rural communities to target As-safe aquifers for well installations in Bengal basin.

1 INTRODUCTION

Meeting growing demands for water resources in a sustainable manner is a critical global challenge (Postel, 2003). The availability of sufficient quantities of clean water is a major issue in water-poor, is a consequence of uneven rainfall patterns and increasing contamination of surface and ground waters, (Mukherjee *et al.*, 2015). Lack of awareness of the naturally occurring, elevated concentrations of arsenic (As) in groundwater in South Asia has caused the death of thousands of people and exposed millions to serious health hazards (Bhattacharya *et al.*, 2011). Today a score of developing countries are affected by As in drinking water exceeding WHO drinking water guideline of 10 mg/L and the global impact now makes it a top priority water quality issue, second in order to microbiological contamination (Mukherjee *et al.*, 2011a). Early studies were mostly aimed at understanding the scale and the primary drivers of this enormous problem. However, finding out sustainable, safe drinking water sources in these As-unsafe areas are of more important for human society (von Bromssen *et al.*, 2007).

The extent of the human tragedy depends on the pace at which As-mitigation programs can be implemented. The sensitivity to As poisoning is clearly related to the economic situation of the individual. The water handling in developing countries is generally the task of women (Singh *et al.*, 2008). Women are more aware than men about the As problem and would like switching to safe wells if properly informed and motivated. However, switching to safe wells generally implies longer distances to the water source and more work for them. The As poisoning case is thus structured in several ways by class and gender. Only rich people can afford the higher cost involvement in installing deeper wells, that are As safe, leaving behind the poor communities more vulnerable. The worse nutritional status of poor households, and particularly the women of those households, may mean that As contamination has more severe social consequences for them. The identification of contaminated wells may lead to greater conflict over uncontaminated water and greater hardship for women fetching water. The gender impacts of As contamination of water are also becoming evident in other areas of women's lives: health and social status.

2 METHODS

In West Bengal, generally, As enrichment is largely believed restricted to <100 m depth and deeper aquifers are safe. However, dissolve As in drinking water, have been observed up to 250 m depth in ~80% of deep wells, lying north of the latitude 22.75 °N in a regional-scale. The reason is probably a combination of geologic structure of subsurface, sediment and water chemistry and land-use such as irrigational pumping; though south of it are more suitable and sustainable (Mukherjee *et al.*, 2011b). However, in a local-scale, the distribution of As in groundwater is highly heterogeneous in a local-scale. The concentration may vary widely both spatially and vertically within a scale of few meters (Chakraborty *et al.*, 2015). Thus currently mapping of this aquifer requires testing of groundwater samples from millions of TWs and drilling of several hundreds of exploratory boreholes in a region. Testing of groundwater from millions of wells is a challenging task from the point of view of technology, manpower, time, social acceptance, and economy for the developing countries like India. Moreover the concentration of As in groundwater shows temporal variability (Mukherjee *et al.*, 2011b). Our collaborative study was focused on assessing the potential for local drillers to target As safe groundwater in South Asia, by validating the correlation between aquifer sediment colours and groundwater chemical composition, characterize aqueous and solid phase geochemistry and dynamics of As mobility and to assess the risk for cross-contamination of As between aquifers in areas of southeastern Bangladesh and southern West Bengal.

3 RESULTS AND DISCUSSION

Drillings in Bangladesh to a depth of 60 m revealed two distinct hydrostratigraphic units, a strongly reducing aquifer unit with black to grey sediments overlying an isolated, patchy sequences of weathered and oxidized white, yellowish-grey to reddish-brown sediment. The aquifers are separated by an impervious clay unit. The reducing aquifer is characterized by high concentrations of dissolved As, DOC, Fe and PO_4^{3-} tot. On the other hand, the off-white and red sediments contain relatively higher concentrations of Mn and SO_4^{2-} and low As. Groundwater chemistry correlates well with the colors of the aquifer sediments. Geochemical investigations indicate that secondary mineral phases control dissolved concentrations of Mn, Fe and PO_4^{3-}-tot. Dissolved As is influenced by the amount of Hfo, pH and PO_4^{3-} tot as a competing ion. Laboratory studies suggest that oxidised sediments have a higher capacity to absorb As (von Brömssen *et al.*, 2007). In order to replicate the salient outcomes of the Bangladesh study results, an investigation was carried out in West Bengal to investigate the potentiality of similarly spaced, isolated light-colored sand aquifers as a safe drinking water source, in contrast to the generally available grey-dark sand aquifers (Biswas *et al.*, 2012). Although the major ion compositions indicated close similarity, the redox conditions were markedly different in groundwater abstracted from the two groups of aquifers (Biswas *et al.*, 2012). These two sand aquifer groups were analogous to the reducing and the oxidized sediment sequences in Bangladesh (Fig. 1).

4 CONCLUSIONS

The outcomes of the study has thus established a scientific knowledge linking relationship between the color of aquifer sediments, redox conditions and hydrogeochemical parameters that provides unique opportunity for the local drillers in rural

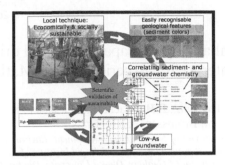

Figure 1. Conceptual framework for targeting safe aquifers by local drillers for tubewell installation.

communities to target As-safe aquifers for well installations in Bengal basin.

ACKNOWLEDGEMENTS

The authors acknowledges the financial support provided by MISTRA and Sida to PB and long, persistence effort of their respective research groups, institutions and funding agencies.

REFERENCES

Biswas, A., Nath, B., Bhattacharya, P., Halder, D., Kundu, A.K., Mandal, U., Mukherjee, A., Chatterjee, D., Mörth, C.M., Jacks, G., 2012. Hydrogeochemical contrast between brown and grey sand aquifers in shallow depth of Bengal Basin: consequences for sustainable drinking water supply. *Sci. Total Environ.* 431: 402–412.

Chakraborty, M., Mukherjee, A. and Ahmed, K.M. 2015. A review of groundwater arsenic in the bengal basin, bangladesh and india: from source to sink. *Curr. Pollut. Rep.* 1(4): 220–247.

Mukherjee, A., Bhattacharya, P., Fryar, A.E. 2011a. Arsenic and other toxic elements in surface and groundwater systems. *Appl. Geochem.* 26(4): 415–420.

Mukherjee, A., Fryar, A.E., Scanlon, B.R., Bhattacharya, P. and Bhattacharya, A., 2011b. Elevated arsenic in deeper groundwater of the western Bengal basin, India: Extent and controls from regional to local scale. *Appl. Geochem.* 26(4): 600–613.

Mukherjee, A., Saha, D., Harvey, C.F., Taylor, R.G., Ahmed, K.M. 2015. Groundwater systems of Indian Sub-Continent. *J. Hydrol.- Regional Studies* 4: 1–14.

Postel, S.L., 2003, May. Securing water for people, crops, and ecosystems: new mindset and new priorities. *Natural Resources Forum* 27(2): 89–98.

Singh, N., Jacks, G., Bhattacharya, P. 2008. Ensuring arsenic-safe water supply in local communities: Emergent concerns in West Bengal, India. In: P. Bhattacharya, AL. Ramanathan, A.B. Mukherjee, J. Bundschuh, D. Chandrasekharam, A.K. Keshari (Eds.) *Groundwater for Sustainable Development: Problems, Perspectives and Challenges.* Taylor and Francis/A.A. Balkema, The Netherlands, pp. 357–364.

von Brömssen, M., Jakariya, M., Bhattacharya, P., Ahmed, K.M., Hasan, M.A., Sracek, O., Jonsson, L., Lundell, L, Jacks, G., 2007. Targeting low-arsenic aquifers in Matlab Upazila, southeastern Bangladesh. *Sci. Total Environ.* 379(2): 121–132.

Arsenic Research and Global Sustainability – Bhattacharya, Vahter, Jarsjö, Kumpiene,
Ahmad, Sparrenbom, Jacks, Donselaar, Bundschuh & Naidu (Eds)
© 2016 Taylor & Francis Group, London, ISBN 978-1-138-02941-5

Enhancing the capacity of local drillers for installing arsenic-safe drinking water wells—experience from Matlab, Bangladesh

M. Hossain[1], P. Bhattacharya[1], G. Jacks[1], K. Matin Ahmed[2], M.A. Hasan[2],
M. von Brömssen[3] & S.K. Frape[4]

[1]*Department of Sustainable Development, Environmental Science and Technology, KTH-International Groundwater Arsenic Research Group, KTH Royal Institute of Technology, Stockholm, Sweden*
[2]*Department of Geology, Dhaka University, Dhaka, Bangladesh*
[3]*Department of Water Resources, Ramböll Water, Ramböll Sweden AB, Stockholm, Sweden*
[4]*Department of Earth and Environmental Sciences, University of Waterloo, Waterloo, Canada*

ABSTRACT: Nearly 90% of the estimated 10 million tubewells in Bangladesh are installed privately by local drillers for rural drinking water supplies in Bangladesh. The awareness of local drillers on elevated Arsenic (As) concentrations in tubewell water at shallow depths have made them change their practice of installation of tubewells. Using the visual color attributes of the shallow sediments (<100 m) and content of dissolved iron, generally associated with high As concentrations, the local drillers presently install community tubewells at depths targeting red/brownish or off-white sediments. This study recognizes the local tubewell drillers as important stakeholder in the business of tubewell installation. A Sediment Color Tool has been developed to enhance the local driller's capacity to identify the As-safe aquifers that would bring significant change to reduce As exposure and scale-up safe water access in rural Bangladesh.

1 INTRODUCTION

Access to safe drinking water is a basic human right and an important component for effective public health protection. The widespread occurrence of natural arsenic (As) in groundwater in Bangladesh has drastically reduced the safe water access across the country. Since the discovery of As in the country in 1993; tens of millions of people are still exposed to As at levels above the WHO guideline (10 µg/L) and the Bangladesh Drinking Water Standard (BDWS; 50 µg/L). Keeping in view the magnitude of the human health impacts and the outcomes of the mitigation programs, the main challenge is to develop a sustainable mitigation option for scaling up safe water access. Tubewells are most widely accepted safe drinking water option (Jakariya et al., 2007, Hossain et al., 2015), and ~90% of these tubewells are installed by local tubewell drillers, which emphasizes the important role of the tubewell drillers.

The mechanism behind the naturally occurring excess As concentrations in groundwater in the Bengal delta is the reducing condition in some sediment sections where Fe-oxyhydroxides are reduced releasing As to the groundwater (Bhattacharya et al., 1997). This implies that there is a covariation of As and ferrous iron in the groundwater. The local drillers have learnt to search for low iron groundwater and this coincides with low As. This was discovered by Jonsson & Lundell (2004). The low iron and As groundwater occurs in oxidized sediments which can be characterized by the color of the sediment (Hossain et al., 2014). The cost of drilling depends on the depth. The depth accessible for the local drillers with their hand percussion drilling normally ranges to a depth of 100 m. The cost for drilling to that depth with hand-percussion

technology is affordable for most households. Thus in the process of giving a large fraction of the population As-safe water the local well drillers and their ability to identify safe aquifers could play an extremely important role. A simple sediment color tool is needed that can be handled by the drillers.

2 METHODS

2.1 Testing of the color code

A four color code based on the local driller's perception of sediment color (Fig. 1a) was proposed by von Brömssen et al. (2007). This code has been further validated through an extensive hydrogeological investigation carried out in Matlab by comparing a total of 2240 sediment samples using Munsell Color Chart and the chemical data of 521 groundwater samples from 144 wells (Hossain et al., 2014).

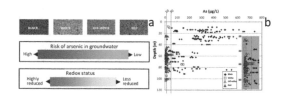

Figure 1. a) Four color sands (driller's perception) with corresponding risks of As concentration in water under varying redox status; b) As concentration in groundwater derived from the sediments of the four color groups with depth. Inset: enlarged view of data points with As below 10 µg/L.

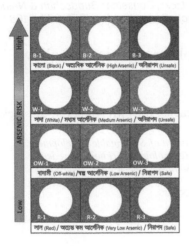

Figure 2. Simplified sediment color tool developed for the local drillers to identify As-safe aquifers.

2.2 Participation of the local drillers

Considering the role of local drillers during tubewell installation, their perception of sediment color and experience in the local variation of targeting sediments, they were largely involved in the study. Information collected on the distribution of sediments of different colors from the local drillers was taken into consideration for the installation of test wells. In the narrow down process for assigning the four colors (black, white, off-white and red) to all 2240 sediment samples, local drillers opinions were sought as an integral part of the method. At the end of the study, the outcomes of the research were shared with the drillers in the context of the enhancement of their knowledge and capacity.

3 LINKING SCIENCE WITH LOCAL CAPACITY

3.1 Local drillers understanding of the aquifers

A first overview of the hydrogeology of the test area was obtained from the local drillers. They describe the aquifers in the region comprising two separate lithological units, the shallow grey to black sediments overlying a sequence of sediments with yellowish-grey to reddish-brown color at depth separated by a thick clay layer. Off-white and white colored sediments recognized by the drillers occur sporadically. The local drillers felt comfortable with a four color code comprising black, white, off-white and red.

3.2 Redox chemistry and sediment color code

Groundwater chemistry and sediment color showed a good correlation. Black (grey) sediments showed elevated Fe, As, total PO_4^{3-}, NH_4^+, DOC and low SO_4^{2-} while the red and off-white (oxidized) sediments are characterized by higher Mn, low NH_4^+, DOC, total PO_4^{3-}, Fe and As. Thus, it is apparent, that the redox

chemistry is well mirrored by the sediment color. Black sediments showed a wide range of variability in terms of depth and As concentration. Sixty six wells were identified as being screened in black sediments, with a median As concentration of 240 µg/L. Almost 50% of these wells were within a very shallow depth range of 10–40 m and assigned as 'high-risk' for the installation of shallow drinking water wells. Average and median values of As in groundwater monitored from 39 wells screened in red and off-white sediments were found to be less than the WHO guideline value of 10 µg/L. Depth of these wells installed in red and off-white sediments indicate that these oxidized sediments mostly occur between the depth of 50 and 80 m (Fig. 1b).

3.3 Sediment color tool for field deployment

Sixty color shades devised from 2240 sediment samples, each of which were also assigned with four (black, white, off-white and red) were further grouped into light, moderate and deep shades so that the end users can have the convenience of using this tool to consider the whole range of the shades within each of four colors (Fig. 2).

4 CONCLUSIONS

Tubewell as the most preferred drinking water option could be installed by the local tubewell drillers through identification of the aquifers that produce As-safe drinking water. This study recognized the local tubewell drillers as important stakeholders in the business of tubewell installation. If the local driller's capacity can be enhanced to identify the As-safe aquifers in the field, it would bring a significant change to reduce the As exposure and thereby in the scaling-up safe water access in rural Bangladesh.

ACKNOWLEDGEMENTS

We acknowledge the Swedish International Development Cooperation Agency (Sida) Global Program (Contribution 75000854). We also thank the teams of local drillers for their cooperation and active help.

REFERENCES

Bhattacharya, P., Chatterjee, D., Jacks, G. 1997. Occurrence of arsenic contaminated in groundwater alluvial aquifers from Delta Plains, Eastern India: options for safe drinking water supply. *Int. J. Water Res. Development.* 13(1): 79–92.
Jakariya, M., von Brömssen, M., Jacks, G., Chowdhury, A.M.K., Ahmed, K.M., Bhattacharya, P. 2007. Searching for a sustainable arsenic mitigation strategy in Bangladesh: experience from two upazilas. *Int. J. Environ. Poll.* 31(3/4): 415–39.
Jonsson, L., Lundell, L. 2004. Targeting safe aquifers in regions with arsenic rich groundwater in Bangladesh. MinorField Studies 277. Swedish University of Agricultural Sciences, SLU External Relations, Uppsala. ISSN 1402-3237.
Hossain, M., Bhattacharya, P., Frape, S.K., Jacks, G., Islam, M.M., Rahman, M.M., et al. 2014. Sediment color tool for targeting arsenic-safe aquifers for the installation of shallow drinking water tubewells. *Sci. Total Environ.* 493: 615–625.

Arsenic Research and Global Sustainability – Bhattacharya, Vahter, Jarsjö, Kumpiene,
Ahmad, Sparrenbom, Jacks, Donselaar, Bundschuh & Naidu (Eds)
© 2016 Taylor & Francis Group, London, ISBN 978-1-138-02941-5

Tubewell platform color as a screening tool for arsenic in shallow drinking water wells in Bangladesh

M. Annaduzzaman[1], A. Biswas[1], M. Hossain[1], P. Bhattacharya[1,2], M. Alauddin[3], R. Cekovic[3], S. Alauddin[3] & S. Shaha[4]

[1]*KTH-International Groundwater Arsenic Research Group, Department of Sustainable Development, Environmental Science and Engineering, KTH Royal Institute of Technology, Stockholm, Sweden*
[2]*International Center for Applied Climate Science, University of Southern Queensland, Toowoomba, Queensland, Australia*
[3]*Department of Chemistry, Wagner College, Staten Island, NY, USA*
[4]*Exonics Technology Center, Dhaka, Bangladesh*

ABSTRACT: The development of a simple and low cost technique for determination of arsenic (As) in drinking water wells is an urgent need to accelerate As mitigation policy. The aim of this study was to evaluate the potentiality of tubewell platform color as low-cost, quick and convenient screening tool for As. The result shows strong correlation between the development of red color stain on tubewells platform and As enrichment in the corresponding tubewells water compared to WHO (10 µg/L) and BDWS (50 µg/L), with 99% certainty. The red color stain in the platform indicates 98% sensitivity with WHO (10 µg/L) and BDWS (50 µg/L). With regard to WHO and BDWS, the corresponding efficiency of the platform color as screening tool for As are 97.3% and 97%. This study suggests that platform color can be potentially used for screening tubewells, help users switch to tube wells with low As and facilitate sustainable As mitigation efforts in developing countries.

1 INTRODUCTION

The geogenic arsenic (As) in shallow depth groundwater well is a severe drinking water quality problem in many countries around the world (Bhattacharya *et al.*, 2015; Herath *et al.*, 2016), especially in Bengal Delta Plan (BDP) consisting of West Bengal, India and Bangladesh, where millions of people depend on these tubewells (TW) as their main drinking water source (Hossain *et al.*, 2015). These shallow depth wells exceed the WHO drinking water guideline value of 10 µg/L, as well as the Bangladesh Drinking Water Standard (BDWS) which is maintained at 50 µg/L. Arsenic concentration vary widely within a short distance between TWs and it is not possible to predict As level by testing a small number of wells. Regular monitoring of As level is a difficult task due to lack of skilled manpower and technology in a country like Bangladesh. Various field test kits have been proposed by researchers; however, they are time consuming and generate harmful toxic arsine gas (Hussam *et al.*, 1999).

The applicability of TW platform color stain as a screening tool for As and Mn in shallow TWs (depth <70 m) has been demonstrated in a recent study by Biswas *et al.* (2012). In West Bengal, India, they observed that red stain that develops in a concrete platform at the base of a TW due to precipitation of Fe-oxides are indicative of As enriched (>10 µg/L) groundwater. Our study was carried out in Bangladesh to investigate the feasibility of deployment of platform color as a screening tool for identifying the wells with elevated As in groundwater.

2 METHODS AND MATERIALS

The study area is located in Uttar Matlab and Shahrasti Upazila in Chandpur District, a well-known As affected area in Bangladesh. The groundwater sampling was carried out from 372 shallow (<70 m) aquifer wells (272 in Matlab and 100 from Shahrasti). The water samples were filtered through 0.45 µm Sartorius membrane filters and stored in 20 ml prewashed polyethylene vials and the sample water was acidified by ultrapure nitric acid. The geographical coordinates of each TW was collected by Global Positioning System (GPS, Garmin-GPS60) and the picture of wells platform were taken by a digital camera (Sony Cyber Shot-W220, 12MP, 4x optical zoom) to validate the color. The concentration of As, Fe and Mn in samples of water were determined by the ICP-OES. The statistical analysis was carried out by Baysian method for the evaluation of effectivity of the tool.

3 RESULTS AND DISCUSSIONS

3.1 Tubewells classification according to platform color

Tubewell platforms were classified into three different color group namely red, black and Not Identified (NI) (Figure 1). Of the 272 wells, 233 (86%) wells platform were classified as red, 4 (1%) as black, and rest 35 (13%) as NI due to undeveloped platform color. Undeveloped color may be the result of regular cleaning of platform, biofilm formation or algal growth on platform. The age of the platform has also an effect on color formation on platform, because newly built platforms are less exposed to build coloration on platform regardless of excessive presence of Mn or Fe in well water. Some platform has complex color (not black or not red), which may occur due to overlapping redox transition of Mn and Fe in aquifer.

3.2 Evaluating platform color as a screening tool for arsenic

Measurement of As in the surveyed (n = 372) wells without considering platform color indicated that 99% (n = 369) water samples exceeded the WHO guideline value (10 µg/L) and 98% (n = 367) samples exceeded BDWS (50 µg/L). However, according to platform color, in 330 red colored platform wells, 99.1% (n = 327) exceeded WHO guideline value and 98.8% (n = 326) exceeded BDWS. Six TW with black color platform are found to be As rich and exceeded BDWS value. However, in NI labelled TW 97% (n = 34) exceeded BDWS and all (n = 36) exceeded WHO guideline value. Without considering platform color, it is observed that only 1% (n = 5) in 372 TWs are safe with respect to BDWS (50 µg/L).

3.3 Effectiveness as arsenic screening tool

Effectiveness of platform color as As screening tool in TWs depends on high probability of true-positive and negative values and correspondingly on low probability of false-positive and negative values relating to specific drinking water standard. The sensitivity, specificity, efficiency, Positive Predictive Value (PPV) and Negative Predictive Value (NPV) indicate the effectiveness of platform color as a screening tool. The results (Table 1) indicate

Figure 1. Development of typical colors on tubewell platforms. a) red colored platforms, b) Non Identifiable (NI) platforms, and c) black colored platforms.

Table 1. Assessment of the effectiveness of TW platform color as As screening tool.

Indices for Validation of Platform Color Tool	Basic Calculation	WHO Guideline Value (>10 µg/L)	BDWS (>50 µg/L)
True Positive	A	327	326
False Positive	B	3	4
False Negative	C	6	6
True Negative	D	0	0
Sensitivity	A/(A+C)	0.98	0.98
Specificity	D/(B+D)		
Efficiency	(A+D)/ (A+B+C+D)	0.973	0.97
PPV	A/(AB)	0.99	0.99
NPV	D/(A+B)	0	0

that at WHO guideline and BDWS, red color platform can be used as screening tool for As unsafe TW.

3.4 Predictive value and prevalence of As screening tool

Bayesian model is used at different cutoff level to evaluate the effect of prevalence and predictive values. The results show that the PPV varies linearly with prevalence for both As standard (50 µg/L and 10 µg/L) due to very few black colored platform tubewells (n = 6). That means, in a particular area, the performance of the tool to identify as As unsafe tubewells increases with As concentration level.

4 CONCLUSIONS

The present study indicates that TW platform color can be used as rapid screening tool for As. The platform stain screening offers a simple method for villagers to identify TWs with high As and help them switch to a safe TW for their drinking water supply. The method has no associated cost and it is a simple and time saving tool and it may help villagers to screen TW quickly and find safe drinking water. This is particularly important for villagers who are ill-educated and have no access to kits to screen the existing TW. We believe, the platform color screening tool is an effective mitigation effort along with others to ensure reduced exposure to high level of As for many villagers in Bangladesh.

ACKNOWLEDGEMENTS

This study was supported by the Swedish International Development Cooperation Agency (SIDA) and Linnaeus Palme (LP) Academic Exchange programme. Partial support to MA was provided by a grant from Wagner College, New York, USA.

REFERENCES

Bhattacharya, P., Thunvik, R., Jacks, G., von Brömssen, M. 2015. Targeting Arsenic-Safe Aquifers in Regions with High Arsenic Groundwater and its Worldwide Implications (TASA). Project Report, MISTRA Idea Support Grant. TRITA LWR Report: 2015:01, 100p.

Biswas, A., Nath, B., Bhattacharya, P., Halder, D., Kundu, A.K., Mandal, U., Mukherjee, A., Chatterjee, D., Jacks, G. 2012. Testing tubewell platform color as a rapid screening tool for arsenic and manganese in drinking water wells. Environ. Sci. Technol. 46: 435–440.

Herath, I., Vithanage, M., Bundschuh, J., Maity, J.P., Bhattacharya, P. 2016. Natural arsenic in global groundwaters: distribution and geochemical triggers for mobilization. Current Pollution Reports doi:10.1007/s40726-016-0028-2.

Hossain, M., Rahman, S.N., Bhattacharya, P., Jacks, G., Saha, R., Rahman, M. 2015. Sustainability of arsenic mitigation interventions—an evaluation of different alternative safe drinking water options provided in Matlab, an arsenic hot spot in Bangladesh. Fronti. Environ. Sci. 3:30.

Hussam, A., Alauddin, M., Khan, A.H., Rasul, S.B. and Munir, A.K.M. 1999. Evaluation of arsine generation in arsenic field kit. Environ Sci. Technol. 33: 3686–3688.

Arsenic Research and Global Sustainability – Bhattacharya, Vahter, Jarsjö, Kumpiene,
Ahmad, Sparrenbom, Jacks, Donselaar, Bundschuh & Naidu (Eds)
© *2016 Taylor & Francis Group, London, ISBN 978-1-138-02941-5*

Subsurface arsenic removal—experiences from Bangladesh

D. van Halem

Water Management Department, Faculty of Civil Engineering and Geosciences, Delft University of Technology,
Delft, The Netherlands

ABSTRACT: Subsurface arsenic (As) removal is a technology based on injection of aerated water into an anoxic or anaerobic aquifer that is rich in both iron and As. This injection procedure will onset the oxidation of adsorbed Fe(II), resulting in the formation of Fe(III) oxides for subsequent As adsorption during abstraction. The series of studies executed between 2008–2014 have shown that achieving high V/Vi ratios, i.e., ratio between abstracted and injected volume, with low As concentrations is challenging. Experimenting with operational conditions can enhance performance, as well as choosing geochemically favorable conditions, however, in this study water revenues remained below V/Vi = 2.

1 INTRODUCTION

Arsenic (As) contamination of shallow tube well drinking water in Bangladesh is an urgent developmental and health problem, disproportionately affecting the rural poor, i.e. those most reliant on this source of drinking water.

Subsurface Arsenic Removal (SAR, Fig. 1) relies on existing infrastructure of a shallow hand-pump based tube well and retains As in the subsurface. As such, it has crucial advantages over other household/community As removal filters (no costly filter media), tube well is the first preferred drinking water option, (minimum) additional hardware widely available, additional removal of iron, and potential for post-deployment monitoring of water quality).

During subsurface As removal aerated water is injected into an anoxic or anaerobic aquifer that is rich in both iron and As. This injection procedure will onset the oxidation of adsorbed Fe(II), resulting in the formation of Fe(III) oxides for subsequent As adsorption during abstraction. This abstract presents the findings of a series of studies conducted between 2008–2014 to assess the potential of SAR in the diverse geochemical settings in Bangladesh.

2 METHODS/EXPERIMENTAL

Subsurface As removal was investigated in Bangladesh with test sites in Manikganj and Muradnagar in the period between 2007–2014. Different sample campaigns were executed to determine the water quality improvements that could be achieved with this injection-abstraction technology under different operational conditions: both on short and longer term. The experimental set-up in Muradnagar consisted of an

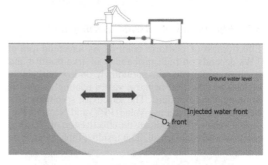

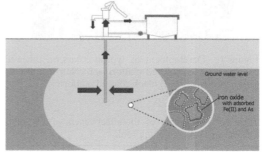

Figure 1. Principle of a subsurface arsenic removal cycle: injection of aerated water (top) and abstraction phase (bottom).

elevated storage tank for the injection water (Fig. 2), connected to a newly constructed shallow tube well. Injection water was aerated with a shower head and a bubble aerator in the bottom of the tank.

Water quality samples were collected for analyses in the laboratory (ICP-MS) and inline measurements were done for pH, Eh, EC and DO (WTW electrodes).

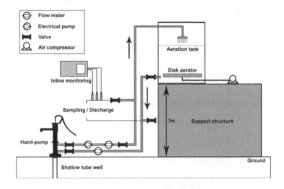

Figure 2. Experimental set-up of subsurface arsenic removal in Muradnagar, Bangladesh (Freitas *et al.*, 2013).

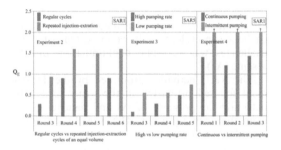

Figure 3. Overview of SAR efficiency (Q_E) for different operational modes (Rahman *et al.*, 2015).

3 RESULTS AND DISCUSSION

Initial results in Muradnagar (showed that with a regular push-pull configuration, also after multiple injection-abstractions cycles or rounds, the As concentration reached C/C0 = 0.5 after V/Vi = 1–1.5; V/Vi corresponds to the ratio between abstracted and injected volume (Fig. 2). These results are similar to the findings during earlier experiments in Manikganj (van Halem et al, 2010), and would in both cases provide <0.5Vi of safe water. However, by changing the operational modes the SAR performance was increased, with particularly good results for intermittent operation versus continuous pumping (Fig. 3).

The extraction efficiency (Q_E), which corresponds to the volume that was abstracted below the Bangladesh guideline of 50 ug/L minus Vi, exceeded 2 times Vi.

Although different operational modes have shown to increase SAR efficiency, it may also be hypothesized that the co-occurrence of competing anions (phosphate, bicarbonate, silicate) at both research sites had limited SAR performance. Therefore a PhreeqC model was developed to predict SAR performance in the different geochemical settings in Bangladesh. The model was on beforehand calibrated with the data from one site and verified with the data from two other sites Based on the model no specific geographical areas were identified that are suitable for SAR, however a higher Q_E (3–4Vi) is expected for SAR particularly at locations with lower As concentrations (10–100ug/L).

4 CONCLUSIONS

SAR is a low-cost and easy-to-operate technology, however, the series of studies executed between 2008–2014 have shown that achieving high V/Vi ratios with low As concentrations is challenging.. Experimenting with operational conditions can enhance performance, as well as choosing geochemically favorable conditions, however, in this study water revenues remained below V/Vi = 2.

ACKNOWLEDGEMENTS

The authors would like to thank the key partners in Bangladesh, Department of Public Health Engineering, Dhaka University and Bangladesh University of Engineering and Technology, for a fruitful collaboration. The authors are grateful for the financial assistance by the Netherlands Organization for Scientific Research (NWO) in the integrated programme of WOTRO Science for Global Development.

REFERENCES

Freitas, S.C.B., van Halem, D., Rahman, M.M., Verberk, J.Q.J.C., Badruzzaman, A.B.M., van der Meer, W.G.J. 2014. Hand-pump subsurface arsenic removal: the effect of groundwater conditions and intermittent operation. *Water Sci. Technol. Water Supply* 14(1): 119–126.

Rahman, M.M., Bakker, M., Borges Freitas, S.C., van Halem, D, van Breukelen, B.M., Ahmed, K.M., Badruzzaman, A.B.M. 2015. Exploratory experiments to determine the effect of alternative operations on the efficiency of subsurface arsenic removal in rural Bangladesh. *Hydrogeol. J.* 23: 19–34.

van Halem, D., Olivero, S., de Vet, W.W.J.M., Verberk, J.Q.J.C., Amy, G.L., van Dijk, J.C. 2010. Subsurface iron and arsenic removal for shallow tube well drinking water supply in rural Bangladesh. *Water Res.* 44(19): 5761–5769.

Arsenic Research and Global Sustainability – Bhattacharya, Vahter, Jarsjö, Kumpiene,
Ahmad, Sparrenbom, Jacks, Donselaar, Bundschuh & Naidu (Eds)
© *2016 Taylor & Francis Group, London, ISBN 978-1-138-02941-5*

Groundwater arsenic problem: An overview of Indian R&D initiatives

K. Paknikar[1], N. Alam[2] & S. Bajpai[2]
[1]*Agharkar Research Institute (An Autonomous Body under the Department of Science and Technology, Government of India), Pune, India*
[2]*Department of Science and Technology, Government of India, New Delhi, India*

ABSTRACT: This presentation attempts to take an overview of the R & D initiatives spearheaded by the Department of Science and Technology, Government of India in tackling the groundwater arsenic problem in the country. It outlines different technologies based on various scientific principles, discusses their key features, operation and maintenance issues as well as monitoring and evaluating aspects.

1 INTRODUCTION

Groundwater has been thought of as one of the safest sources of drinking water, especially in rural areas of India where piped water is not available. However, it is now recognized that in West Bengal and several other states of India millions of people are drinking arsenic (As) contaminated water with levels of above 100 µg/L and are or are at risk of suffering from As toxicity. It is reported by the US Environmental Protection Agency (US EPA, 2003) that As is a persistent, bioaccumulative carcinogen and is among the top 20 toxic substances identified by the US Agency for Toxic Substances and Disease Registry. So far, no treatment is available for chronic As toxicity. Therefore, provision of As-safe water to people is essential.

2 INITIATIVES OF THE GOVERNMENT

Realising the need to address various Research and Development (R & D) issues in providing safe water, the Department of Science and Technology (DST), Government of India launched the Water Technology Initiative (WTI) in the year 2007–08, with an aim to design and develop low cost solutions for domestic use of safe drinking water, referencing of technologies to social context, capacity building of water managers and encouraging new research ideas.

In order to develop holistic and viable research and technology-based solutions for tackling problems of water quality and water scarcity, the Department promoted activities so as to address issues related to drinking water in terms of purification, availability, reuse and recycling under the aegis of Technology Mission "Winning, Augmentation and Renovation (WAR) for Water".

A report submitted in August 2009 by the Technical Expert Committee (TEC) constituted by the Government of India, identified 26 major water challenges prevalent in the country. One of these challenges was geogenic contamination due to arsenic (As).

The R&D activities promoted by the department focused primarily on the issues related to detecting the presence of As and removal of As from the drinking water. The R & D challenges identified included.

i. Development of cost-effective detection techniques with technical performance better or comparable to currently available alternatives
ii. Development of cost-effective and efficient materials for As removal based on locally available resources.
iii. Development of household and community As removal systems based on indigenously developed materials
iv. Field demonstration of developed systems to assess their suitability in specific social context.
v. Sludge management.

The objective of the R & D activities was to develop adsorbents, which were cheaper yet efficient for removal of As from drinking water. Various types of adsorbents were prepared, and their loading capacity was compared with other alternatives. Attempts were made to develop adsorbents superior to commercially available adsorbents for a broad range of operating condition i.e. pH & temperature, having larger surface area and higher adsorption capacity. Further, suitable encapsulation and stabilization of the developed material through appropriate techniques were undertaken to enable their prolonged use with least adverse effect on the efficiency.

Because As (III) removal is quite difficult, efforts were also made to develop improved oxidation methods besides chemical oxidant for the faster conversion of As (III) to As (V). Recognising the importance of developing low-cost detection techniques for estimation of As in ground water, efforts were made to develop cost effective field test kits having a shelf life, detector strip sensitivity, incubation time and interference due to the presence of other contaminants, better than or, at least, comparable to commercially available imported test kits.

The promising leads obtained from the investigations of various materials were taken to the next logical steps which focused on lab scale studies of these materials for spiked as well as naturally contaminated As laden water.

The successful lab scale systems were upscaled, and prototypes were tested in real field conditions (Table 1). The systems, which conformed to technical performance parameters and found large community acceptance were replicated in larger numbers to generate enough scientific data for validation.

The findings of these R & D projects have resulted in several scientific publications in

Table 1. Selected technologies supported under WTI Programme of DST, Government of India.

Laterite based Arsenic Filter by Indian Institute of
Technology—Kharagpur
Arsenic Filter by Indian Institute of
Technology—Bombay
AMRIT—Arsenic and Metal Removal by Indian Technology—Indian Institute of Technology—Madras
Arsiron Nilogon Arsenic Filter by Tezpur University
DRDO Arsenic Removal Filter by Defense Research
and Development Organization
ARI Groundwater Arsenic Treatment Plant by
Agharkar Research Institute, Pune
Hand Pump Attached Arsenic Removal Unit by
Jadavpur University

Figure 1. Groundwater arsenic treatment plant based on the 'Integrated microbial oxidation alumina adsorption process' developed by Agharkar Research Institute, Pune.

international journals of repute and patents on materials, techniques and processes. Besides the institutional mechanisms of individual institutions to take these projects to the next level, the Department is consciously making attempts to encourage individual researchers to further their research so that the research efforts could culminate into a socially useful output in the field. The outcomes of these research efforts are also shared at various inter-ministerial forums including core committee on As mitigation.

However, the limited experience of the department has revealed the need to have last mile connectivity to translate the research outputs to field. While these R & D projects have proven their potential at lab scale, demonstration of capabilities of these technologies to provide convergent solutions with possible up-scaling needs sustained efforts.

Finally, based on the long-term association of the authors with the Government of India's programmes, viz. Water Technology Initiative (WTI) and "Winning, Augmentation and Renovation (WAR) for Water", an attempt will be made to prescribe Short Term/Medium Term/Long Term measures for addressing the problem of As contamination in drinking water.

ACKNOWLEDGEMENTS

The authors are grateful to the Secretary, Department of Science and Technology, Government of India for his unreserved support and encouragement. They also thank their colleagues in the Project Advisory Committee of WTI for lively discussions. Thanks are also due to several Project Investigators who by the dint of their hard work generated meaningful data making this presentation possible.

REFERENCES

Anon. 2009. Technology Mission: "Winning, Augmentation and Renovation" Technology Mission: WAR for Water, Plan document prepared by the Union Ministry of Science and Technology, Government of India, New Delhi.
Anon. 2015. R & D Activities related to Arsenic Contamination in Drinking Water—Salient Efforts of Department of Science and Technology (DST), New Delhi.
Bhattacharya, P., Mukherjee, A., Mukherjee, A.B. 2011. arsenic in groundwater of India. In: J.O. Nriagu (Ed.) Encyclopedia of Environmental Health, vol. 1, Elsevier, Burlington. pp. 150–164.
Mokashi S.A., Paknikar, K.M. 2002. Arsenic (III) oxidizing Microbacterium lacticum and its use in the treatment of arsenic contaminated groundwater. Lett. Appl. Microbiol. 34(4): 258–262.
US EPA. 2003. National Primary and Secondary Standards. Office of Water (4606M) EPA 816-F-03-016, June 2003, 6p. (Accessed on www.epa.gov/safewater on 28 November 2015).

Arsenic Research and Global Sustainability – Bhattacharya, Vahter, Jarsjö, Kumpiene,
Ahmad, Sparrenbom, Jacks, Donselaar, Bundschuh & Naidu (Eds)
© 2016 Taylor & Francis Group, London, ISBN 978-1-138-02941-5

Removing arsenic to <1 µg/L during drinking water treatment: Application of AOCF at pellet softening plant of Oosterhout, The Netherlands

A. Ahmad[1], S. van de Wetering[2] & T. van Dijk[2]
[1]*KWR Watercycle Research Institute, Nieuwegein, The Netherlands*
[2]*Brabant Water, 's-Hertogenbosch, The Netherlands*

ABSTRACT: The current study is aimed at investigating the potential of AOCF to achieve <1 µg/L effluent As concentration at drinking water treatment plan of Oosterhout which is conventional pellet softening plant. Firstly, a sampling campaign was carried out at the full-scale treatment process of Oosterhout. Subsequently, a series of specialized jar tests was performed on-site with actual segments of the process stream to determine if desired As concentrations (<1 µg/L) could be achieved by AOCF. Dosing of $KMnO_4$ in anaerobic raw water could effectively reduce As concentrations to the desired level. It is highly likely that the removal of As will be higher in the full-scale treatment plant because of the adsorptive capacity of the filtration media.

1 INTRODUCTION

The current study is aimed at investigating the potential of AOCF (Advanced Oxidation Coagulation Filtration) to achieve <1 µg/L effluent As concentration at Drinking Water Treatment Plan (DWTP) of Oosterhout. AOCF has already been proven effective for removing As to <1 µg/L at two of the conventional aeration-filtration plants of Brabant Water (Ahmad et al., 2014; Ahmad, 2015). DWTP of Oosterhout is a different sort of conventional groundwater treatment plant. It consists of pellet softening process, for $CaCO_3$-hardness removal, between the aeration and filtration steps. The uncertainty regarding the performance of the AOCF process at the higher pH conditions (> pH 9) present during softening rationalized the investigations described in this paper.

2 MATERIALS AND METHODS

2.1 Sampling campaign at the full-scale plant

Process water samples were collected at predetermined points within the full-scale treatment process of Oosterhout in duplicate. The objectives of the sampling campaign were to study (1) the overall As removal efficiency of the existing treatment system, (2) the variation/interchanging of the As species during the treatment and (3) the interaction of As with Fe, Mn and Ca based solids that precipitated during the treatment process. Figure 1 shows the simplified process scheme of DWTP Oosterhout and the sampling points. At both the instances, samples were collected from the same treatment train.

2.2 Jar tests

In order to determine the applicability of AOCF approach at Oosterhout, jar tests were performed on-site with actual segments of the process stream. Firstly, a set of preliminary jar tests was performed to study the influence on As uptake when As(III) was oxidized to As(V) at different locations within the treatment process by dosing $KMnO_4$. Secondly, a protocol was developed and tested to effectively simulate the complete full-scale treatment process of DWTP Oosterhout by the help of a jar test apparatus. This study is probably the first attempt to simulate the conventional pellet softening process at jar scale including the by-pass configuration. One jar in the jar test apparatus was designated to perform aeration + softening, a second jar was designated to perform aeration + by-pass, and a third jar was designated to receive the softened stream and the by-pass water from the first two jars. The mixed stream in the third jar contained 56% of the softened water and 44% of the by-pass water. This mixing ratio was adopted from the full-scale treatment process. Finally, detailed investigations by additional jar tests were carried out to determine the extent of treatment chemical(s) required to realize effluent As concentration of <1 µg/L.

3 RESULTS AND DISCUSSION

3.1 Sampling campaign

The concentrations of As, Ca, Fe and Mn at different sampling points slightly varied between the two sampling occasions. This variation could be attributed to differences in pumping wells, supernatant depths and filtration duration after the last backwash at the two occasions. The raw water contained 7 µg/L As, on average, which was entirely dissolved. The principle form of As was As(III). The presence of As(III) as the dominant As specie is consistent with the observations at other groundwater treatment plants of Brabant Water. The aeration and softening processes

Figure 1. The process scheme of DWTP Oosterhout with sampling points.

could not oxidize As(III) to As(V) significantly. However, the effluent of the rapid sand filters contained 4.7 µg/L As which was entirely present as As(V). Generally, when atmospheric oxygen is the only oxidizing agent, as the case at Oosterhout, transformation of As(III) to As(V) is expected to be very slow (Bissen and Frimmel, 2003). However, certain solid-liquid interfaces in the presence of dissolved oxygen may catalyze the oxidation process of As(III). It has also been reported that Mn-oxides produced during auto-catalytic Mn removal process can lead to a faster oxidation of As(III) in a time scale of only few minutes. Besides physicochemical processes, biological mechanisms in the filter bed may also have played a role in the oxidation of As(III).

3.2 Jar tests

In preliminary jar tests when $KMnO_4$ was dosed in the anaerobic raw water, dissolved As concentrations significantly decreased in the effluent of the simulated softening process. It was not due to an increased As(V) uptake in the softening reactor for As(V). Rather, it referred to the fact that the oxidation rate of As(III) to As(V) was so rapid that the oxidation and subsequent sorption reaction got completed before the $Ca(OH)_2$ was dosed. The following 10–45 min exposure of As containing ironhydroxide flocs to high softening pH (9.5–10) in one of the jar tests could desorb only an insignificant concentration of As (around 0.3 µg/l). Dosing $KMnO_4$ in the by-pass and the mixed stream could not increase As removal significantly. Therefore, application of $KMnO_4$ to the anaerobic raw water of DWTP Oosterhout was selected for further investigations to remove As to <1 µg/L (desired removal ≈90%).

Figure 2 shows the results of additional jar tests performed with the special jar test procedure developed for DWTP Oosterhout. It can be observed that As removal drastically increased by the addition of 0.5 mg/L of $KMnO_4$. It was due to the conversion of As(III) to As(V). When $KMnO_4$ dose was increased from 0.5 mg/L to 1 mg/L, the As concentration in the filtered sample of the mixed stream decreased from 2 µg/L to 1.7 µg/L. Only a slight increase in As removal with a 0.5 mg/L increase in $KMnO_4$ dosing suggests that the adsorptive solids (e.g. Fe-hydroxides) were limiting the further removal of As. However, if the adsorptive capacity of the full-scale filtration media is considered in the calculations, only dosing 0.5–1.0 mg/L $KMnO_4$ should be adequate to lower the As concentrations to lower than 1 µg/L. Figure 2 also shows the effect of doing Fe(III) together with $KMnO_4$. Arsenic removal increased, as expected.

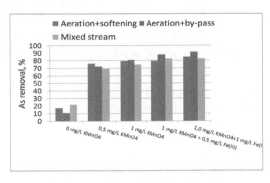

Figure 2. Arsenic removal by kmno4 and Fe(III) dose.

4 CONCLUSIONS

Based on the sampling campaign and series of jar tests performed in this study it can be concluded that AOCF is an effective treatment process to remove As at DWTP Oosterhout, and most probably at other drinking water treatment plants which apply pellet softening for removing $CaCO_3$-hardness. Dosing of $KMnO_4$ in anaerobic raw water could effectively reduce As concentrations to the desired level. It is highly likely that the removal of As will be higher in the full-scale treatment plant because of the adsorptive capacity of the filtration media and lower mixed stream pH compared to what could be achieved in the jar tests.

ACKNOWLEDGEMENTS

Jack Hopman from Aqualab Zuid is deeply appreciated for support during the experiments.

REFERENCES

Ahmad, A., van de Wetering, S., Groenendijk, M. and Bhattacharya, P. 2014. Advanced Oxidation-Coagulation-Filtration (AOCF)—an innovative treatment technology for targeting drinking water with <1 µg/L of arsenic. In: M.I. Litter, H:B. Nicolli, M. Meichtry, N. Quici, J. Bundschuh, P. Bhattacharya, R. Naidu (Eds.) *Proceedings of the 5th International Congress on Arsenic in the Environment, May 11–16, 2014, Buenos Aires, Argentina.* CRC Press 2014 pp. 817–819.

Ahmad, A. 2015. Arsenic removal to <1ug/L at drinking water treatment plant of Prinsenbosch by AOCF. *BTO report 2015.072.* KWR Watercycle Reseach Institute. The Netherlands, 30 p.

Bissen M., Frimmel F.H. 2003. Arsenic- a review. Oxidation of arsenic and its removal in water treatment. *Acta Hydrochim. Hydrobiol.* 31: 97–107.

Buamah, R. Petrusevski, B. and Schippers, J.C. 2008. Adsorptive removal of manganese (II) from the aqueous phase using iron oxide coated sand. *J. Water Supply: Research and Technology—AQUA* 57:1–12.

Arsenic Research and Global Sustainability – Bhattacharya, Vahter, Jarsjö, Kumpiene,
Ahmad, Sparrenbom, Jacks, Donselaar, Bundschuh & Naidu (Eds)
© 2016 Taylor & Francis Group, London, ISBN 978-1-138-02941-5

Arsenic removal with Zero-Valent Iron amended Ceramic Pot Filters

W.A.C. van Hoorn, L.C. Rietveld & D. van Halem
*Sanitary Engineering Section, Department of Water Management, Delft University of Technology, Delft,
The Netherlands*

ABSTRACT: The objective of this study was to extend the properties of Ceramic Pot Filters (CPF) with
arsenic removal capabilities by incorporating Nano Zero Valent Iron (nZVI) and Composite Iron Matrix
(CIM) in the clay mixture before firing. In batch experiments it was found that with 50g/L crushed material
of a fired blank filter and of fired filters amended with ZVI the same small amount of arsenic was removed,
while ZVI itself is a good arsenic adsorbent. Crushed material of unfired filters containing ZVI showed much
faster arsenic removal compared to fired filters. The loose of effectiveness of the fired-in ZVI is probably
attributed to the intensive oxidation of the ZVI that took place during firing and to the less available surface
sites due to incorporation in the clay. It is therefore not recommended to incorporate ZVI in CPF before
firing.

1 INTRODUCTION

Since it will take years before households in developing
countries have access to reliable piped water systems,
there is a need for simple, socially accepted and low-
cost Household Water Treatment Systems (HWTS).
One of the most socially accepted and effective HWTS
in terms of bacteria removal is the CPF. CPFs are
locally manufactured from a mixture of locally available clay, burnout material and water, which are pressed
into a pot-shape and are subsequently fired. The CPF
is chosen as a host filter for the incorporation of ZVI,
which is a well-known adsorbent for arsenic (Kanel
et al., 2005, Neumann *et al.*, 2013, Sun *et al.*, 2006)
both as nZVI particles and low-cost CIM powder. The
incorporation of ZVI in CPF would result in a multifunctional HWTS removing both pathogenic-microor-
ganisms and arsenic.

2 MATERIALS AND METHODS

2.1 *Ceramic filter disks*

CDF were manufactured at the Filterpure CPF factory
(Dominican Republic) according to the standard pot
filter protocol: a mixture of 333 g clay soil (35-mesh),
67 g sawdust (35-mesh) and 110 g demineralized water
per filter. Additionally, a nZVI dispersion (NanoIron
S.R.O., Rajhrad, Czech Republic) or CIM powder was
added to the clay mixture to obtain an iron content of
0.05% or 5% based on the weight of a dry disk. The
mixture was mechanically mixed for 20 minutes and
subsequently, pressed with 220 psi into disks with a
diameter of 13 cm and a thickness of 1.9 cm by means
of a mold. The disks were dried for 2–3 days and fired
in a kiln: ramp rate was 2.4°C/minute, peak temperature was 860°C and dwell time was 40 minutes. Afterwards, the disks were allowed to cool overnight.

2.2 *Ceramic filter characterization*

XRD measurements were performed with a BrukerD8
Advance diffractometer with Bragg-Brentano geometry with K_α energy and a coupled θ-2θ scan over a
range of 10° to 110°. In addition, transmission ^{57}Fe
Mössbauer absorption spectra were collected at 300 K
with a conventional constant-acceleration spectrometer using a ^{57}Co(Rh) source. For both type of analyses both fired as unfired powdered filter material were
inspected to gain insight of the effect of incorporating
iron particles in the clay mixture before firing.

Furthermore, the flow rates of the filters were
determined. For this purpose an experimental set-up
was used that consists of a couple of PVC pipes of 1 m
length and a diameter of 125 mm with a pipe connec-
tor attached to each pipe, which together functions as
water column. The ceramic disks were placed on the
bottom of the pipe connectors with a rubber O-ring
and secured with a bead of silicone. A funnel was connected to the filter holders, so that to the water could
flow into the collection receptacles.

2.3 *Kinetic batch experiments*

Arsenite and arsenate stock solutions were prepared by mixing sodium arsenite (NaAsO$_2$) and
(Na$_2$HAsO$_4$·7H$_2$O), respectively with degassed
ultrapure water. The As adsorption kinetics was investigated in 250 mL closed bottles containing 250 mL
synthetic water (1 mM NaCl, 4 mM NaHCO$_3^-$,
pH 7) with an initial As concentration of 200 μg/L.
To this solution ground ceramic material, nZVI aqueous dispersion or CIM powder was added. The bottles were stirred at 150 rpm with a temperature of
28–29°C. Periodically, 10 mL samples were taken and
filtered through a 0.45 μm membrane filter (Whatman). Each experiment was run in duplicate. The total
arsenic (As$_T$) concentrations were quantified by means

of Inductively Coupled Plasma Mass Spectrometry (NEN-EN-ISO 17294-2). All chemicals were reagent grade obtained from Sigma Aldrich.

3 RESULTS AND DISCUSSION

3.1 Material characterization

The XRD and 57Fe Mössbauer spectroscopy results revealed that the ZVI was oxidized to iron oxide hydroxide and hematite during firing. But, also when the filter was not fired a large part of the added nZVI was oxidized to hematite, according to the 57Fe Mössbauer spectra. It was expected that the addition of metals could cause a decrease in flow rate. The results of this study do not confirm this. The flow rates of the filter with 5% nZVI even suggest the opposite: the addition of nZVI increases the flow rate significantly. This might be due to the successive expansion and shrinking of nZVI.

3.2 Arsenic experiments

Batch experiments were conducted with 50 g/L crushed fired ceramic filter material. The removal of As_T in time with addition of 50g/L ground material of a fired blank filter, a fired and unfired filter with 5% nZVI, a fired and unfired filter with 5% CIM and the addition of 2.5g/L of solely nZVI are presented in Figure 1.

There is no considerable difference between the adsorption capacities of the fired blank filter and the fired filters with iron incorporated. This is remarkable, because the filters with nZVI contain 5% nZVI, which corresponds 2.5g/L nZVI, and when solely 2.5g nZVI/L was added to the arsenic solution no arsenic was measured after only 5 minutes of contact time. This large difference in arsenic adsorption by the same amounts of nZVI suggests that due to the incorporation into the clay before firing the nZVI gets less effective against arsenic.

To gain inside on the effect of the firing process the effectiveness of unfired ceramic filters was also studied. It can be seen in Figure 1 that crushed material of the unfired filters containing 0.05% nZVI, 5% nZVI and 5% CIM remove As(III) much faster than the than the fired filters. The difference in As_T adsorption between the different unfired filters proves that not only the clay, but also the iron, which was incorporated in the clay mixture, is responsible for the faster As sorption of unfired filters compared to fired filters. This implies that the firing process causes the ZVI to loose its arsenic adsorption capacities. There are several reactions that take place during firing that might influence the reactivity of the zero valent iron. First of all, oxidation takes place. According to the 57Fe Mössbauer spectra all nZVI corroded into hematite. While in water, the main corrosion products of ZVI are oxyhydroxides. Different ZVI corrosion products have also a different ability to adsorb arsenic. Manning et al. (2002), for instance, reported that goethite, lepidocrocite, and maghemite adsorbed As(III) within 24 h of contact time, while magnetite and hematite did not. Secondly, it is likely that during the firing process a link between

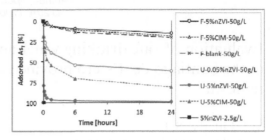

Figure 1. The adsorbed amount of As_T in time with addition of 50g/L ground material of a blank fired filter, of a fired and unfired filter containing 5% nZVI, of a fired and unfired filter containing 5% CIM and 2.5g/L of solely nZVI. The F in the legend stands for crushed material of a fired filter and the U for crushed material of an unfired filter.

the ZVI and clay develops, which leads to less available surface area for As adsorption. Hence, in an actual filter part of the iron oxide particles are probably not located on the outside surface of the pores of the filter, but are hidden in the clay structure, the As is in that case not able to adsorb onto the iron corrosion products.

4 CONCLUSIONS

It can be concluded it is not recommended to incorporated ZVI in CPFs before firing. However, the results showed that ZVI on it self is a good As adsorbents and that the unfired filters were much better able to remove As compared to the fired filters. It is recommended to search for alternative solutions such as an iron coating or a separate As removal element with ZVI, which can be used in combination with the CPF.

ACKNOWLEDGEMENTS

R.W.A. Hendrikx and A.I. Dugulan of the Delft University of Technology are acknowledged for the X-ray analysis and 57Fe Mössbauer spectroscopy analysis, respectively. Furthermore, the authors thank the laboratory staff of the Sanitary Engineering Section of the Delft University of Technology.

REFERENCES

Kanel, S.R., Manning, B., Charlet, L., Choi, H. 2005. Removal of arsenic (III) from groundwater by nanoscale zero-valent iron. Environ. Sci. Technol. 39(5): 1291–1298.

Manning, B.A., Hunt, M.L., Amrhein, C., Yarmoff, J.A., 2002. Arsenic(III) and arsenic(V) reactions with zerovalent iron corrosion products. Environ. Sci. Technol 36: 5455–5461.

Neumann, A. Kaegi, R., Voegelin, A., Hussam, A., Munir, A.K.M., Hug, S.J. 2013. Arsenic removal with composite iron matrix filters in Bangladesh: A field and laboratory study. Environ. Sci. Technol. 47: 4544–4554.

Sun, H., Wang, L., Zhang, R., Sui, J., Xu, G., 2006. Treatment of groundwater polluted by arsenic compounds by zero valent iron. J. Hazard. Mater. 129: 297–303.

Arsenic Research and Global Sustainability – Bhattacharya, Vahter, Jarsjö, Kumpiene,
Ahmad, Sparrenbom, Jacks, Donselaar, Bundschuh & Naidu (Eds)
© *2016 Taylor & Francis Group, London, ISBN 978-1-138-02941-5*

Evaluation of safe drinking water access for dispersed rural populations in the Santiago del Estero province, Argentina—a challenge for arsenic removal

M.I. Litter[1] & S. Pereyra[2]

[1]*Comisión Nacional de Energía Atómica, San Martín, Provincia de Buenos Aires, Argentina*
[2]*Instituto de Altos Estudios Sociales, Universidad Nacional de San Martín, San Martín, Provincia de Buenos Aires, Argentina*

ABSTRACT: A joint work was performed by a transdisciplinary group involving chemical and social scientists about arsenic removal and provision of safe water in two localities of the Santiago del Estero province, Argentina, one close to the capital, in the Banda Department and another one in the north, in the Copo Department. A diagnosis about access, uses and water quality indicated that the northern zone presented serious problems in the quality of water for human consumption, deepened by difficult socioeconomic conditions, whereas, in the Banda zone, the waters presented neither microbiological contamination nor high arsenic levels. Samples of water were taken in the different localities, their properties were measured and, in the samples containing the highest arsenic levels, the capability of some low-cost materials for arsenic removal were evaluated envisaging the design of useful devices for the affected localities. Further replicates of this project can be performed in similar regions.

1 INTRODUCTION

In many isolated rural and periurban places of Latin America and particularly in Argentina, access, use and quality of water is rather poor, in agreement with the difficult socioeconomic life conditions and low income of the population. Waters are generally affected by the presence of biological and chemical pollution including organics and toxic metals or metalloids such as arsenic (As). Fatal endemic diseases (hepatitis, typhus or cholera) are current and represent a dangerous health risk. As known, the presence of As in groundwaters causes the incidence of arsenicosis (HACRE, chronic regional endemic hydroarsenicism). Thus, ways to facilitate the access to safe water to the populations, economically and socially accepted by the people, are imperative (Bundschuh *et al.*, 2010, Litter *et al.*, 2010).

2 OBJECTIVES AND STUDY AREA

The project was developed by a research team formed by chemists, engineers, sociologists and anthropologists, and linked advances and development of remediation technologies for As polluted water with social studies on life conditions and conflicts related to the access to water (Neiburg & Nicaise, 2009) in Argentina. The focus was on populations of peripheral localization and low socioeconomic resources where HACRE cases have been detected, especially in Santiago del Estero. According to the 2010 Census (Censo 2010, INDEC), more than 41% of households of the province have no access to water network inside the house.

Two main areas were selected: a) a rural periphery zone close to the locality of Clodomira, Banda Department (20 km north of the capital of the province), and b) another rural periphery zone close to San José del Boquerón, Copo Department, northern part of the province. In Banda, around 19,000 disperse inhabitants live in rural populations, hamlets, villages or isolated settlements of few houses, with an economy of subsistence and important difficulties of access to water. The area is located in the Río Dulce basin, where 95% of the inhabitants could be affected by consuming well waters containing As (0.05–2 mg/L, Bundschuh *et al.*, 2004, 2010, Bhattacharya *et al.*, 2006, Litter *et al.*, 2010). In Copo, the information is more scarce, but around 31,000 inhabitants and 28% of the houses (ca. 9,000 inhabitants) are in rural disperse zones, and 36% of the households have no access to the public drinking water network.

The project comprised two stages. In the first stage, of diagnosis, the daily practices of the population related to the access, uses and habits of water have been collected. The social mediations involved, such as present and historical conflicts related to water, map of collective actors, related cultural traditions and water circuits in the domestic economy and in the productive activities have been analyzed. The objective was the elaboration and analysis of indicators of economic, social and human development that allowed to define the most relevant characteristics of the zone (employment and working market, education, health, poverty, etc.). However, as the census data have a strong bias towards the urban population and the project is referred specifically to the peripheral rural population, a qualitative field work has been designed, combining ethnographic observations in environments and situations significant for the research. Between 2013 and 2015, 57 interviews to people of both zones were performed (30 in Banda and 27 in Copo). In parallel, water samples have been taken to evaluate the conditions for the implementation of remediation strategies. A first screening campaign was performed in August 2013 with visits to four schools of the rural peripheral zone of Clodomira, and interviews with key informants. This exploratory work indicated that most of the population has now access to safe water, free of As, coming from a network recently provided by the new government of the province (Ragno, 2010). Four other campaigns were carried out later in both regions, two in Banda and two in Copo, where 18 samples of well waters were collected and submitted to a complete physicochemical characterization. The results indicated that the Banda waters were of better quality and did not contained important As levels.

As conclusions of the five campaigns, differences in the economic activity of the families, the general infrastructure, the land tenure and conflicts relative to HACRE and access to water were registered in both regions, analyzed and compared. The diagnosis pointed out that the northern zone presented dramatic problems in the quality of water for human consumption, aggravated by difficult socioeconomic conditions, while the Banda waters presented neither microbiological pollution nor important As levels.

3 REMEDIATION STRATEGIES

Innovative procedures to provide safe water should be accessible, environmentally friendly and adapted to the cultural practices and rules of the population.

In this case, the capability of some low-cost materials for arsenic removal was evaluated envisaging the design of useful devices for the affected localities.

Six commercial and two laboratory prepared materials based on iron and titanium were tested. The materials were divided in two groups regarding their morphology: granular and nanoestructurated particles. First, all the materials were tested in synthetic waters containing 0.1–0.2 mg/L of As (V) in a batch system using different concentrations of the treating material. The best six materials were later tested on natural waters of Copo, which contained the highest As levels. It was concluded that the nanoparticulated materials presented a higher capacity for As removal. However, as these materials are still in a stage of research, especially concerning their immobilization to adequate supports to be used in filter devices (Tesh & Scott, 2014), the granular materials can be preferred for an immediate use and will be selected for further field tests.

4 CONCLUSIONS

The scope and first advances of a transdisciplinary project about access, uses and quality of water in two different rural zones of the Santiago del Estero Province are presented. The daily practices of the population related to the use and habits of water were considered. Samples of water were collected and their analysis included in the diagnosis, while low-cost technologies with costless materials have been tested.

An intervention in the zone will be performed in the future in few selected places with the implementation of devices for disinfection (when necessary) and arsenic removal. At the same time, the design should accomplish certain requirements related to the ways of life of the people object of this work. The goal is that the technologies could be used and appropriated by the people, with the less possible alteration of uses and habits related to water. This methodology can be replicated in other similar zones of the country and other isolated regions of the world.

ACKNOWLEDGEMENTS

This work was funded by UNSAM *Diálogo entre las Ciencias* Project, and Alimentaris Foundation (Argentina). The authors are members of CONICET. To C.E. López Pasquali, A. Iriel, A.M. Senn, F.E. García, M.F. Blanco Esmoris, K. Rondano, D.C. Pabón, L.E. Dicelio, M.G. Lagorio, G.D. Noel, J.J. Ochoa and M. Benhamou for collaboration in different parts of this work. Special thanks to Dr. C. Padial for bringing information and contacts with Santiago del Estero schools.

REFERENCES

Bhattacharya, P., Claesson, M., Bundschuh, J., Sracek, O., Fagerberg, J., Jacks, G., Martin, R.A., Storniolo, A.R., Thir, J.M. 2006. Distribution and mobility of arsenic in the Río Dulce Alluvial aquifers in Santiago del Estero Province, Argentina. *Sci. Total Environ.* 358(1–3): 97–120.

Bundschuh, J., Farias, B., Martin, R., Storniolo, A., Bhattacharya, P., Cortes, J., Bonorino, G., Alboury, R. 2004. Grounwater arsenic in the Chaco-Pampean Plain, Argentina: Case study from Robles County, Santiago del Estero Province. *Appl. Geochem.* 19(2): 231–243.

Bundschuh, J., Litter, M., Ciminelli, V., Morgada, M.E., Cornejo, L., Garrido Hoyos, S., Hoinkis, J., Alarcón-Herrera, M.T., Armienta, M.A., Bhattacharya, P. 2010. Emerging mitigation needs and sustainable options for solving the arsenic problems of rural and isolated urban areas in Iberoamerica—A critical analysis. *Water Res.* 44: 5828–5845.

Censo INDEC 2010. http://www.censo2010.indec.gov.ar/ CuadrosDefinitivos/H2-P_Santiago_del_estero.pdf. (accessed 8–10-2013).

Litter, M.I., Morgada, M.E., Bundschuh, J. 2010. Possible treatments for arsenic removal in Latin American waters for human consumption. *Environ. Pollut.* 158: 1105–1118.

Neiburg, F., Nicaise, N. 2009. *The social life of water*, Río de Janeiro, Viva Rio.

Ragno, J.D. 2010. *El hidroarsenicismo en la zona de riego está siendo vencido*, personal communication.

Tesh, S.J., Scott, T.B. 2014. Nano-composites for water remediation: a review. *Adv. Mater.* 26: 6056–6068.

Arsenic Research and Global Sustainability – Bhattacharya, Vahter, Jarsjö, Kumpiene,
Ahmad, Sparrenbom, Jacks, Donselaar, Bundschuh & Naidu (Eds)
© 2016 Taylor & Francis Group, London, ISBN 978-1-138-02941-5

Arsenic accumulation in drinking water distribution networks

E.J.M. Blokker[1], J. van Vossen[1], M. Schriks[1], J. Bosch[2] & M. van der Haar[2]

[1]*KWR Watercycle Research Institute, Nieuwegein, The Netherlands*
[2]*Brabant Water,'s-Hertogenbosch, The Netherlands*

ABSTRACT: As is contained in particulate matter and biofilm. It is unclear if and how the network
A measurement campaign showed that arsenic (As) can accumulate in drinking water distribution networks, potentially leading to 40 times higher As concentrations during hydraulic disturbances. Most of
the accumulated configuration or pipe material affects As concentrations. There is no correlation between
As and total amount of sediment. However the increased health risk according to a margin-of-exposure
approach from the arsenic accumulation is limited because hydraulic disturbances seldom occur. The sediment accumulation is controlled using a systematic flushing programme of the Dutch water companies.

1 INTRODUCTION

In The Netherlands, raw water is treated to meet
the arsenic (As) standard of 10 µg/L before entering the Drinking Water Distribution Network
(DWDN). It is assumed that between Water Treatment Works (WTW) and the consumers tap no As
is added to the drinking water and the concentrations measured at the WTW are indicative of the
As concentrations at the consumers taps.

However, Lytle *et al.* (2004, 2010) showed that
As has the potential to accumulate in the DWDN.
This was supported by some preliminary measurements at a water utility, showing that this could
lead to temporary exceedances of the 10 µg/l
standard. This raises the question if there is health
risk related to temporary exceedances of the standard and if this risk can be reduced.

2 METHODS

2.1 *Measurements*

A measurement campaign was conducted in November and December 2014 in a city which is fed by three
WTWs. Of these WTWs "X" is known to have a relatively high total As concentration of around 5.8 µg/l
after treatment. WTW "Y" has a lower concentration
of ca. 3 µg/l and WTW "Z" of ca. 2 µg/L. The measurement campaign concentrated on the areas fed by
WTW X, but included a few locations from WTW Y.

To gain insight into the effect of different parts
of the DWDN on As concentrations, measurements were taken in three parts of the DWDN:
WTW, transport mains and distribution mains
(both in self-cleaning and conventional networks).
In addition several pipe materials were considered
(cast iron, PVC and AC).

Two types of measurements were carried out. Type
A samples were taken during normal operating con-

ditions at the consumers tap at 36 locations. Type B
samples were taken during intentional hydraulic disturbances (flow velocity increase with 0.3 and 1.5 m/s)
of the system at 12 hydrant locations. The samples
were analysed for a.o. total As, turbidity and TOC.

2.2 *Health risk assessment*

Lung cancer is the most relevant and sensitive
endpoint for chronic As exposure at relative low
concentrations (<10 µg/L). For an indicative risk
assessment we followed the method used by the
EFSA (2008), which is based on the Margin-Of-
Exposure (MOE) approach recommended for
genotoxic carcinogens in food. The MOE is the
ratio between a predefined reference dose and the
estimated human intake or exposure. A $BMCL_{01}$
(Benchmark Concentration Level which provides
1% increase in lung cancer) of 14 µg As/L is used
as a reference point. Here the $BMCL_{01}$ is based on a
human epidemiological study and a MOE of ≥ 1000
is considered as of low concern for public health
and low priority for risk management action.

The average As concentration was multiplied by
a worst-case estimate of the duration of exposure
in order to derive an estimate of the average exposure per year. In accordance with Blokker *et al.*
(2013) the exposure for hydraulic disturbance conditions was estimated at 4 days a year. We assumed
normal operating conditions during the remaining
361 days of the year.

3 RESULTS AND DISCUSSION

3.1 *Measurement results—tap samples*

Table 1 shows the results of type A samples. The
average for all measurements is 6.0 µg/L with a
variance of 1.5%. There is no significant difference
between self-cleaning and conventional systems or

Table 1. Total As (µg/L) in type A samples.

Network	Average	Variance	Min	Max
Total set (31)	6.0	0.09	5.2	6.6
Divided over network type				
Self-cleaning (6)	6.1	0.06	5.8	6.5
Conventional (25)	6.0	0.10	5.2	6.6
Divided over pipe material				
Cast Iron (6)	6.1	0.05	5.9	6.6
PVC (13)	6.0	0.06	5.6	6.3
AC (10)	5.8	0.08	5.2	6.4
Unknown (2)	6.6	-	6.5	6.6

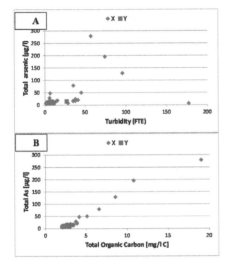

Figure 1. Scatter plots of (A) total As [µg/l] versus turbidity [FTU] and (B) total organic carbon [mg/l C], type B samples.

the different pipe materials. There was no correlation between total As and any of the analysed substances or residence time. As concentrations further from the WTW were not lower than close to the WTW.

3.2 Measurement results—disturbance samples

Figure 1 shows the results of type B samples. Figure 1A illustrates that values of As are much higher during the disturbed conditions than during normal operations with a maximum of 279 µg/L. There is a significant difference between total As and dissolved As. This indicates that a large proportion of the As is contained in sediment and biofilm in the DWDN. There is a clear correlation between total As and iron, manganese and TOC and less so for turbidity (Figure 1B). The dissolved As(V) has no correlation to iron, TOC, colour nor to total As. There is little dissolved As(III) found above the detection limit of 0.1 µg/L.

The measurements show that during disturbances the concentrations of As are much higher than the source concentrations. Therefore, As must have accumulated over time in the DWDN.

The measurements for locations supplied mainly by WTW X show lower total As concentrations than the locations supplied mainly by WTW Y.

3.3 Health risk

During normal operating conditions the average As concentration was 6 µg/l. The highest risk from As occurs when the As is dissolved. The highest measured value for dissolved As was in the order of 120 µg/l. We assume this value represents a worst-case concentration that might be consumed during disturbances. This leads to an estimated average annual As concentration of 7.3 µg/l and a corresponding MOE of ~2. 6.0 µg/l during 365 days per year corresponds to a MOE of ~2.3. This is considerably lower than a MOE of ≥ 1000, which is a value considered to be of low concern.

4 CONCLUSIONS

The present study illustrates that As accumulates in the DWDN. The tap samples reflect the concentrations of the source with some local variations. The hydrant samples taken during hydraulic disturbances clearly show As concentrations higher than the source concentrations with values of total As reaching 279 µg/L. The accumulated As is presumably contained in the particulate matter and biofilm. There is no correlation between As and sediment mass.

The study illustrates that temporary exceedances of the As drinking water standard can occur after disturbances in the drinking water distribution network. The increased health risk from these disturbances is limited. The systematic flushing programme of the Dutch water companies ensures removal of sediment and therefore limits As accumulation in the DWDN

REFERENCES

Blokker, E.J.M., van de Ven, B.M., de Jongh, C.M., Slaats, P. G.G. 2013. Health implications of PAH release from coated cast iron drinking water distribution systems in the Netherlands. *Environ. Health Perspect.* 121: 600–606.

FSA, 2008. Polycyclic aromatic hydrocarbons in food. Scientific opinion of the panel on contaminants in the food chain. Adopted on 9 June 2008; Question No. EFSA-Q-2007-136. *The EFSA Journal* 742: 1–114.

Lytle, D.A., Sorg, T.J., Frietch, C. 2004. Accumulation of arsenic in drinking water distribution systems, *Environ. Sci. Technol.* 38(20): 5365–5372.

Lytle, D.A., Sorg, T.J., Muhlen, C., Wang, L. 2010. Particulate arsenic release in a drinking water distribution system, *American Water Works Association. Journal* 102(3): 87.

Arsenic Research and Global Sustainability – Bhattacharya, Vahter, Jarsjö, Kumpiene,
Ahmad, Sparrenbom, Jacks, Donselaar, Bundschuh & Naidu (Eds)
© 2016 Taylor & Francis Group, London, ISBN 978-1-138-02941-5

Inorganic arsenic in rice and rice products on the Swedish market 2015 Part 3: Risk management

E. Halldin Ankarberg, R. Bjerselius & K. Gustafsson
National Food Agency, Uppsala, Sweden

ABSTRACT: It has been documented globally that inorganic arsenic exposure from rice and rice products can be significant. Codex Alimentarius and the EU have therefore been working on defining maximum levels for how much inorganic arsenic can be present in rice and certain rice products. In a project at the Swedish National Food Agency (NFA), the inorganic arsenic levels in 102 different rice and rice products have been investigated. The results show that the intake of inorganic arsenic from certain products is high, and in some cases exceeded health based reference points, even though the coming maximum levels were not exceeded. These results obliged the Swedish NFA to take management actions in form of dietary advice for rice and rice products as well as pleading for lower maximum levels for inorganic arsenic in rice and rice products at an international level.

1 INTRODUCTION

In 2015, the Swedish National Food Agency (NFA) investigated the inorganic arsenic content of 102 different rice and rice products (Kollander & Sundström, 2015). The results show that rice contains inorganic arsenic in various concentrations and that the levels of inorganic arsenic are high, especially in rice cakes and whole grain rice. It has been documented globally that inorganic arsenic exposure from rice and rice products can be significant. Codex Alimentarius and the EU have therefore been working on defining maximum levels for how much inorganic arsenic can be present in rice and certain rice products. However, scenario analyses of intakes in relation to health based reference points show that the exposure of inorganic arsenic from rice and rice products is high. These results also reveal that the new maximum levels in the EU for inorganic arsenic in rice and rice products will not have the desired effect of reducing arsenic exposure from rice and rice products in Sweden, in a satisfactory way.

Because of this, the Swedish NFA decided to take management actions to reduce the exposure of inorganic arsenic for Swedish consumers. The Swedish NFA have issued consumer advice, informed producing companies and urged the EU-Commission and Codex to try to lower the maximum limits for inorganic arsenic in rice and rice products.

2 METHODS

2.1 Risk management

All work at the Swedish NFA follows the Principles of Risk analysis by Codex.

The results from the risk assessment show that the intake of inorganic arsenic from certain products is high, and in some cases it exceeds health based reference points, even though the coming maximum levels are not exceeded. About a third of the inorganic arsenic we are exposed to from food in Sweden comes from rice and rice products, which is the single largest contribution from a food commodity. In addition, rice and rice products have ten times higher inorganic arsenic levels relative to comparable foods such as potato, pasta etc. Altogether, this led NFA to decide on actions to reduce the exposure of inorganic arsenic in rice and rice products for Swedish consumers.

The occurrence of certain contaminants in food is regulated by regulation (EC) No. 1881/2006. Until now, no maximum level for arsenic in food has applied, but maximum levels for inorganic arsenic in rice and certain rice products will come into force from the 1st of January 2016 (Regulation No 2015/1006). In the preceding work, Sweden has contributed to the discussions and has presented all previous analytical results. The results emphasize the need for lower maximum levels than those now decided. Maximum levels for contaminants are set according to the ALARA principle, which means that maximum levels are to be set as low as practically possible without excluding the market. Thus, the ALARA principle does not take into consideration the actual exposure to arsenic from rice and rice products, but rather seeks to force levels down from the current levels.

Other legitimate factors that have been considered:

- Environmental aspects
  Comparing rice with potatoes and pasta, rice has a higher negative impact on the environment. Reducing rice consumption would thus not have negative effects on the environment.
- International Management
  - In 2015 the German risk assessment institute BfR concluded that the inorganic arsenic exposure of the German population is close to or somewhat over the health-based guidelines that have been determined. Its conclusion and advice is that rice can be included in a varied and balanced diet, but that certain rice products such should be eaten in moderation.
  - In 2013, the Danish Veterinary and Food Administration in Denmark issued advice that rice can be included in a varied healthy diet, but that rice cakes and rice drinks should not be given to children.
  - Great Britain advises children aged between 1 and 4.5 years not to drink rice drinks as a substitute for breast milk, formula milk or cow milk.

3 RESULTS AND DISCUSSION

This study shows that parts of the Swedish population have an exposure to inorganic arsenic that in certain cases is at or over the limit of what can be considered acceptable from a long-term perspective. For those who eat rice and rice products, this consumption contributes to a large part of the arsenic exposure.

Therefore, the Swedish NFA considers that it is justified to inform consumers about levels of inorganic arsenic in rice and rice products and giving advice about appropriate rice consumption on the basis of the levels found and the intake calculations made. This is to reduce exposure to arsenic from foods in Sweden.

NFA advice:

- Do not give rice drinks to children aged under six
- Do not give rice cakes to children under 6. Children aged over seven and adults may eat rice cakes occasionally, depending on how much other rice products are being eaten.
- Limit consumption of rice and rice products (such as rice porridge, rice noodles and rice porridge snacks) to four times per week (children) or six times per week (adults).
- When eating rice, you should not always choose whole grain rice.

- Cooking rice with a substantial surplus of water that is discarded reduces the arsenic content in rice by up to 70 per cent.

4 CONCLUSIONS

The Swedish NFA considers that it is justified to inform consumers about levels of inorganic arsenic in rice and rice products and in giving advice about appropriate rice consumption on the basis of found levels.

The Swedish NFA is also informing companies about the results and project conclusions. The NFA is encouraging companies to seek rice raw material with as low inorganic arsenic content as possible, or to develop methods to reduce the level of arsenic in products. In this way, companies are themselves given the opportunity to influence consumer exposure to inorganic arsenic.

The Swedish NFA is also informing the European Commission and Codex of the results and project conclusions. The results clearly show that the maximum levels shortly to be introduced do not satisfactorily reduce exposure to inorganic arsenic from food in general and from rice as a raw material in particular. The purpose of the Swedish NFA action to the European Commission and Codex is in a long term perspective to be able to decrease consumer exposure to inorganic arsenic nationally and internationally.

The Swedish NFA is also informing Efsa about the results and project conclusions. The results clearly show that parts of the Swedish population have an exposure to inorganic arsenic from rice that in certain cases is on the limit of what can be considered acceptable from a long-term health perspective.

REFERENCES

Halldin Ankarberg, E., Foghelberg, P., Gustafsson, K., Nordenfors, H., Bjerselius, R., 2015. Inorganic arsenic in rice and rice products on the Swedish Market—Part 1. A Survey of inorganic arsenic in rice and rice products. *Swedish National Food Agency Report* 16/2015, NFA, Uppsala, Sweden.

Kollander, B., Sundstrom, B. 2015. Inorganic arsenic in rice and rice products on the Swedish Market—Part 1. A Survey of inorganic arsenic in rice and rice products. *Swedish National Food Agency Report* 16/2015, NFA, Uppsala, Sweden.

Sand, S., Concha, G., Öhrvik, V., Abramsson, L., 2015. Inorganic arsenic in rice and rice products on the Swedish Market—Part 2. Risk Assessment. *Swedish National Food Agency Report* 16/2015, NFA, Uppsala, Sweden.

Author index

650